GEOTECHNICAL SPECIAL PUBLICATION NO. 177

GeoCongress 2008

Geotechnics of Waste Management and Remediation

PROCEEDINGS OF SESSIONS OF GEOCONGRESS 2008

March 9–12, 2008
New Orleans, Louisiana

SPONSORED BY
The Geo-Institute of the American Society of Civil Engineers

EDITED BY
Milind V. Khire
Akram N. Alshawabkeh
Krishna R. Reddy

Published by the American Society of Civil Engineers

Library of Congress Cataloging-in-Publication Data

Geocongress-2008 (2008 : New Orleans, Louisiana)
Geotechnics of waste management and remediation : proceedings of sessions of Geocongress 2008 sponsored by the Geo-Institute of the American Society of Civil Engineers March 9-12, 2008 New Orleans, Louisiana / edited by Milind V. Khire, Akram N. Alshawabkeh, Krishna R. Reddy.
p. cm. -- (Geotechnical special publication ; no. 177)
Includes bibliographical references and index.
ISBN 978-0-7844-0970-1
I. Khire, Milind V. II. Alshawabkeh, Akram N. III. Reddy, Krishna R. IV. Title.

TD171.9.G43 2008
628.4--dc22 2008001081

American Society of Civil Engineers
1801 Alexander Bell Drive
Reston, Virginia, 20191-4400

www.pubs.asce.org

ISBN 978-0-7844-0970-1
Manufactured in the United States of America.

Geotechnical Special Publications

1 *Terzaghi Lectures*
2 *Geotechnical Aspects of Stiff and Hard Clays*
3 *Landslide Dams: Processes, Risk, and Mitigation*
7 *Timber Bulkheads*
9 *Foundations & Excavations in Decomposed Rock of the Piedmont Province*
11 *Dynamic Response of Pile Foundations - Experiment, Analysis and Observation*
14 *Geotechnical Aspects of Karst Terrains*
15 *Measured Performance Shallow Foundations*
16 *Special Topics in Foundations*
17 *Soil Properties Evaluation from Centrifugal Models*
18 *Geosynthetics for Soil Improvement*
19 *Mine Induced Subsidence: Effects on Engineered Structures*
21 *Hydraulic Fill Structures*
22 *Foundation Engineering*
23 *Predicted and Observed Axial Behavior of Piles*
24 *Resilient Moduli of Soils: Laboratory Conditions*
25 *Design and Performance of Earth Retaining Structures*
27 *Geotechnical Engineering Congress*
28 *Detection of and Construction at the Soil/Rock Interface*
29 *Recent Advances in Instrumentation, Data Acquisition and Testing in Soil Dynamics*
32 *Embankment of Dams-James L. Sherard Contributions*
33 *Excavation and Support for the Urban Infrastructure*
34 *Piles Under Dynamic Loads*
35 *Geotechnical Practice in Dam Rehabilitation*
37 *Advances in Site Characterization: Data Acquisition, Data Management and Data Interpretation*
39 *Unsaturated Soils*
40 *Vertical and Horizontal Deformations of Foundations and Embankments*
41 *Predicted and Measured Behavior of Five Spread Footings on Sand*
42 *Serviceability of Earth Retaining Structures*
43 *Fracture Mechanics Applied to Geotechnical Engineering*
44 *Ground Failures Under Seismic Conditions*
45 *In Situ Deep Soil Improvement*
46 *Geoenvironment 2000*
47 *Geo-Environmental Issues Facing the Americas*
48 *Soil Suction Applications in Geotechnical Engineering*
49 *Soil Improvement for Earthquake Hazard Mitigation*
50 *Foundation Upgrading and Repair for Infrastructure Improvement*
51 *Performance of Deep Foundations Under Seismic Loading*
52 *Landslides Under Static and Dynamic Conditions-Analysis, Monitoring, and Mitigation*
53 *Landfill Closures-Environmental Protection and Land Recovery*
54 *Earthquake Design and Performance of Solid Waste Landfills*
55 *Earthquake-Induced Movements and Seismic Remediation of Existing Foundations and Abutments*
56 *Static and Dynamic Properties of Gravelly Soils*
57 *Verification of Geotechnical Grouting*
58 *Uncertainty in the Geologic Environment*
59 *Engineered Contaminated Soils and Interaction of Soil Geomembranes*
60 *Analysis and Design of Retaining Structures Against Earthquakes*
61 *Measuring and Modeling Time Dependent Soil Behavior*
62 *Case Histories of Geophysics Applied to Civil Engineering and Public Policy*
63 *Design with Residual Materials: Geotechnical and Construction Considerations*
64 *Observation and Modeling in Numerical Analysis and Model Tests in Dynamic Soil-Structure Interaction Problems*
65 *Dredging and Management of Dredged Material*
66 *Grouting: Compaction, Remediation and Testing*
67 *Spatial Analysis in Soil Dynamics and Earthquake Engineering*
68 *Unsaturated Soil Engineering Practice*
69 *Ground Improvement, Ground Reinforcement, Ground Treatment: Developments 1987-1997*
70 *Seismic Analysis and Design for Soil-Pile-Structure Interactions*
71 *In Situ Remediation of the Geoenvironment*
72 *Degradation of Natural Building Stone*
73 *Innovative Design and Construction for Foundations and Substructures Subject to Freezing and Frost*

137 *Legal and Liability Issues in Geotechnical Engineering*
138 *Site Characterization and Modeling*
139 *Calibration of Constitutive Models*
140 *Slopes and Retaining Structures under Seismic and Static Conditions*
141 *International Perspectives on Soil Reinforcement Applications*
142 *Waste Containment and Remediation*
143 *Geomechanics: Testing, Modeling, and Simulation*
144 *Sinkholes and the Engineering and Environmental Impacts of Karst*
145 *Seismic Performance and Simulation of Pile Foundations in Liquefied and Laterally Spreading Ground*
146 *Asphalt Concrete: Simulation, Modeling and Experimental Characterization*
147 *Unsaturated Soils 2006*
148 *Advances in Unsaturated Soil, Seepage, and Environmental Geotechnics*
149 *Site and Geomaterial Characterization*
150 *Soil and Rock Behavior and Modeling*
151 *Advances in Earth Structures: Research to Practice*
152 *Ground Modification and Seismic Mitigation*
153 *Foundation Analysis and Design: Innovative Methods*
154 *Pavement Mechanics and Performance*
155 *Underground Construction and Ground Movement*
156 *Geomechanics II: Testing, Modeling, and Simulation*
157 *Computer Applications in Geotechnical Engineering*
158 *Contemporary Issues in Deep Foundations*
159 *Case Studies in Earth Retaining Structures*
160 *Dynamic Response and Soil Properties*
161 *Embankments, Dams, and Slopes: Lessons from the New Orleans Levee Failures and Other Issues*
162 *Problematic Soils and Rocks and In Situ Characterization*
163 *Geoenvironmental Engineering*
164 *Innovative Applications of Geophysics in Civil Engineering*
165 *Geosynthetics in Reinforcement and Hydraulic Applications*
166 *Educational Activities in Geotechnical Engineering*
167 *Geotechnics of Soil Erosion*
168 *Grouting for Ground Improvement: Innovative Concepts and Applications*
169 *Soil and Material Inputs for Mechanistic-Empirical Pavement Design*
170 *Probabilistic Applications in Geotechnical Engineering*
171 *Advances in Shallow Foundations*
172 *Soil Improvement*
173 *Advances in Measurement and Modeling of Soil Behavior*
174 *Designing Our Underground Space*
175 *Field Measurements in Geomechanics 2007*
176 *Analysis of Asphalt Pavement Materials and Systems: Emerging Methods*

Preface

Geo-engineers and geo-scientists are increasingly confronted by new challenges in protecting and preserving our environment and infrastructure. Innovative and emerging scientific concepts and technologies are always needed to address a wide range of complex geoenvironmental issues such as groundwater protection and remediation, management of waste disposal, sustainable development and mitigation against natural and man-made geohazards.

In 1995, ASCE organized a conference on "Geoenvironment 2000: Characterization, Containment, Remediation and Performance of Environmental Geotechnics" in New Orleans, Louisiana. This conference highlighted the role of geo-professionals and the wide range of geoenvironmental engineering challenges, such as the design, construction, and monitoring of containment facilities as well as engineering mitigation of environmental geo-hazards. In 1997, ASCE organized another specialty conference on "In Situ Remediation of the Geoenvironment," in Minneapolis, Minnesota.

Due to rapid industrialization, increased environmental awareness, and the complexity of tackling past unsafe waste disposal practices; geoenvironmental engineering has continued to evolve as a discipline that bridges between engineering and basic sciences to address geo-specific environmental problems. It was imperative to organize Geocongress2008, March 9 – 12, 2008 in New Orleans, Louisiana to highlight recent advances, new directions and opportunities for sustainable engineering to protect the environment and infrastructure.

The Congress attracted a significant number of papers and more than 400 were accepted. The papers were divided into three Geotechnical Special Publications (GSPs) that capture the multidisciplinary aspects and the challenges of the sustainability of the geoenvironment.

The first GSP, *Geotechnics of Waste Management and Remediation*, tackles the challenges of sustainability in remediation and waste management, and covers topics on new and conventional remediation technologies, design and operational aspects of bioreactor landfills, innovations in design and assessment of covers and liners, management of mining wastes, and recycle and reuse of waste materials.

The second GSP, *Geosustainability and Geohazard Mitigation*, tackles the challenges of sustainability in geotechnics, and covers topics on education, sustainable materials and infrastructure, risk-based analysis and design, and impacts and mitigation of geohazards.

The third GSP, *Characterization, Monitoring and Modeling of GeoSystems*, covers mechanical and chemical soil behavior, testing, and modeling. The GSP presents innovations on subsurface characterization and monitoring, characterization of rocks, problematic Soils and waste materials; and sensor technologies. Recent developments in numerical and computational geotechnics, emerging technologies, fate and transport modeling, uncertainty modeling, and micro- and environmental geo-mechanics are also covered in this GSP.

The paper review process was managed by the editors and Paper Review Board. The review board had a very active and essential role in reviewing papers, organizing the conference sessions, and making the GSPs possible. The editors sincerely appreciate the help and patience of the Review Board. The editors also appreciate the help of Ms. Sheana Singletary of ASCE for her help in managing paper submissions and dealing with the glitches of the database.

We hope that these GSPs will serve as valuable references to all working in geoengineering.

Editors
Milind V. Khire
Akram N. Alshawabkeh
Krishna R. Reddy

Review Board

Stephan Jefferis, Environmental Geotechnics, Ltd.
Hsein Juang, Clemson University
Edward Kavazanjian, Arizona State University
Amid Khodadoust, University of Illinois Chicago
Debra Laefer, University College Dublin
Mengjia Li, Gundle/SLT Environmental, Inc. (GSE)
Shu-Guang Li, Michigan State University
William Likos, University of Missouri
Ning Lu, Colorado School of Mines
Ronaldo Luna, University of Missouri-Rolla
Laloui Lyesse, Swiss Federal Institute of Technology Lausanne
Krishnanand Maillacheruvu, Bradley University
Jay Meegoda, New Jersey Institute of Technology
Mysore Nataraj, University of New Orleans
Vincent Ogunro, University of North Carolina at Charlotte
Jun Otani, Kumamoto University
Sibel Pamukcu, Lehigh University
Junboum Park, Seoul National University
Kok-Kwang Phoon, National University of Singapore
Ali Porbaha, California State University, Sacramento
Anand Puppala, University of Texas-Arlington
Ellen Rathje, University of Texas-Austin
Chris Ryan, Geo Solutions, Inc.
Shadi Saadeh, California State University, Long Beach
Rodrigo Salgado, Purdue University
Charles Shackelford, Colorado State University
Joseph Sinfield, Purdue University
Chung Song, The University of Mississippi
Burak Tanyu, GeoSyntec Consultants, Inc.
Hani Titi, University of Wisconsin Milwaukee
Binod Tiwari, California State University, Fullerton
David Toll, Durham University
Erol Tutumluer, University of Illinois Urbana-Champaign
Cumaraswamy Vipulanandan, University of Houston
Andrea Welker, Villanova University
Nazli Yesiller, Consultant, San Luis Obispo
Zhanping You, Michigan Technological University
Lei Yuan, GeoSyntec Consultants, Inc.
Mourad Zeghal, Rensselaer Polytechnic Institute
Dimitrios Zekkos, Geosyntec Consultants, Inc.
Limin Zhang, The Hong Kong University of Science and Technology
Xiong Zhang, University of Alaska Fairbanks
Thomas Zimmie, Rensselaer Polytechnic Institute

Contents

Waste Management

Physical, Chemical, and Biological Aspects of Liners, Covers, and Waste

Recycled and Waste Materials

Indexes

Performance of Engineered Waste Containment Barriers

James K. Mitchell[1], Hon. M. ASCE

[1]University Distinguished Prof. Emeritus, Virginia Tech; Patton Hall, Blacksburg, VA 24061-0105, jkm@vt.edu

ABSTRACT: Up to 20 years of field observations suggest that engineered waste containment barrier systems that are designed, constructed, operated, and maintained in accordance with current statutory regulations and requirements have so far provided environmental protection at or above specified levels. However, these barriers have not been in existence long enough to assess long-term (post closure) performance for periods that may need to extend for hundreds of years. Much of the available data are from monitoring the environment downstream of the barrier, and there is little direct monitoring of the barriers themselves. New monitoring techniques, expanded collection and reporting of data and periodic assessments of barrier system effectiveness are needed to further increase confidence in barrier system performance.

This paper summarizes the findings and recommendations of a recently completed study by the Committee to Assess the Performance of Engineered Barriers appointed by the U.S. National Research Council and sponsored by four federal agencies (NRC, 2007).

INTRODUCTION

As of 2002, there were approximately 4,000 landfills used for containment of hazardous and non-hazardous waste and about 7,800 contaminated sites still awaiting corrective action and cleanup in the U.S (Bonaparte, et al., 2002). Modern landfills are required to have engineered barrier systems to contain the waste, prevent offsite movement of contaminants, and minimize infiltration of outside water. Engineered barriers are also used to isolate and contain waste at abandoned hazardous waste sites. The designs, initial performance criteria, and monitoring requirements for these containment systems are governed by federal and state regulations, the two most important being RCRA, the Resource Conservation and Recovery Act, enacted about thirty years ago, and CERCLA, the Comprehensive Environmental Restoration, Compensation, and Liability Act (Superfund), enacted in 1980. The major federal statutes governing waste classification and containment are listed in Table 1.

Table 1. Major U.S. Federal Statutes Governing Waste Classification and Containment

Waste Classification	Description	Federal Regulation
Municipal solid waste	Residential and commercial refuse such as food, paper, glass, plastic, textile, grass, wood, metal	Resource Conservation and Recovery Act (RCRA) as amended (42 USC 6901 et seq.) Subtitle D
Hazardous waste	Industrial waste such as refining and manufacturing by-products, paint, solvents, pesticides, and ashes; hazardous wastes must be stabilized before disposal in a containment system	RCRA as amended (42 USC 6901 et seq.) Subtitle C
Wastes associated with cleanup of abandoned hazardous waste sites	Applies to hazardous contaminated soils and liquids removed from the ground or generated from treatment	Comprehensive Environmental Restoration, Compensation, and Liability Act (CERCLA) (1986) as amended
Low-level radioactive and mixed waste from production of nuclear power (USNRC) and weapons (DOE)	Requires each state to provide disposal capacity for commercial low level radioactive waste, such as protective clothing, filters, containers, tools, laboratory equipment, and piping	Low Level Waste Policy Act as amended in 1985 (42 USC 2021b et seq.)
Uranium tailings and other contaminated materials at uranium mill processing sites and adjacent properties	Waste rock, tailings from ore processing, industrial waste, waste water	Uranium Mill Tailings Radiation Control Act of 1978 (42 USC 2022 et seq.)

At the request of the United States Environmental Protection Agency (EPA), the Department of Energy (DOE), the Nuclear Regulatory Commission (USNRC), and the National Science Foundation (NSF) a study was undertaken by a committee appointed by the U.S. National Academies to assess the performance of these barriers over time. The following two questions provided the primary focus for the study:

1. How well are engineered barrier systems working?

2. How long are they likely to work effectively?

The following tasks were undertaken to develop the information needed to address these questions:

1. Identify engineered barrier systems used for surface and subsurface waste containment.
2. Describe how performance is defined, predicted, measured, and monitored.
3. Present information on field performance of engineered barrier systems.
4. Evaluate available information on field performance.
5. Assess methodologies and capabilities for predicting and monitoring performance and for assessing risk.
6. Identify information needed to fill the knowledge gaps.

The information and data needed to accomplish these tasks were obtained from review of field performance observations, monitoring data, and models. To the extent possible, both barrier components and the total system at a site were considered. Data sources included publications, presentation to the committee, several field trips, data summaries provided by regulators, and committee member experiences. The results of this study are summarized in this paper.[1]

TIME FRAMES

Because engineered barrier systems constructed in compliance with current regulations have been in place for only a few decades at most, definitive observations and data are available only for the short and, in a few cases, the medium term, according to the following definitions:

- *Short term:* the period until completion of construction of the barrier component.
- *Medium term:* the operating period of the waste unit.
- *Long term:* the post-closure period.

Based on these definitions, predictions of the long term performance must be based on models, data extrapolation and observations made over much shorter time periods.

WASTE TYPES, BARRIER SYSTEMS, BARRIER MATERIALS, AND FUNCTIONAL MECHANISMS

The waste types for which barrier effectiveness was evaluated included municipal solid waste, other non-hazardous solid waste, hazardous and toxic waste, and low-level radioactive waste. High-level radioactive waste, mining and agricultural wastes, and liquid impoundments were not considered.

A modern waste containment system may contain bottom and side slope liners, covers, vertical barriers, hydraulic and pneumatic extraction and injection systems, and natural barriers, as shown schematically in Figs. 1 and 2. Reactive barriers; e.g., permeable reactive walls that contain materials to neutralize the hazardous and toxic compounds in migrating contaminated groundwater, chemical spills, etc., were not considered in this study.

---

[1] Space limitations preclude inclusion of the extensive list of references used in development and support of the findings presented in this paper. It is available in NRC (2007).

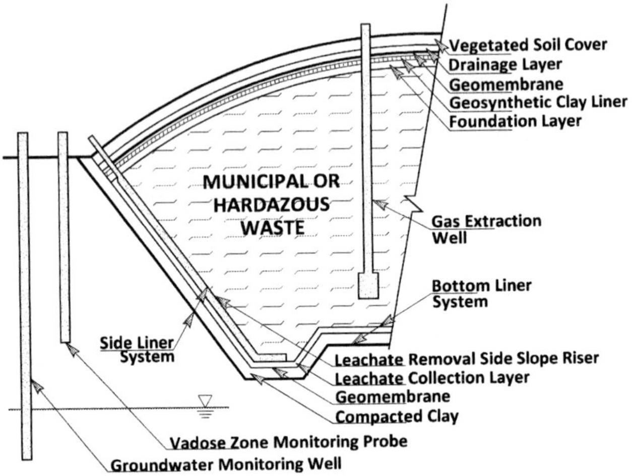

Fig. 1. Schematic diagram of the components of a new engineered landfill for containment of municipal solid waste

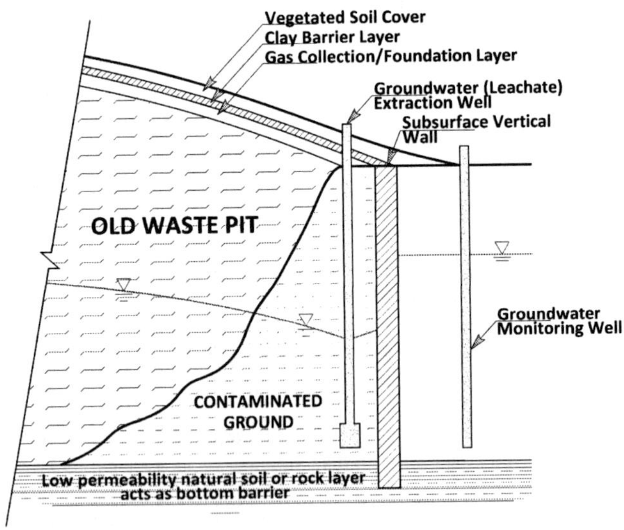

Fig. 2. Schematic diagram of the components of an engineered containment system for a preexisting waste site.

The materials that are most extensively used for different components of a waste containment system are soil, geosynthetics, and cementitious materials. Low permeability soil, usually compacted clay, is a common component of liners and covers. Granular soils; i.e., sands and gravels, are used in drainage layers. Geomembranes (flexible membrane layers) are used to contain leachate, geotextiles are used for filtration, and geonets are used for drainage. Soil-bentonite is used in vertical cut-off walls. Geosynthetic clay liners (GCL) containing low permeability clay, usually a bentonite, in combination with a geomembrane or geotextile barrier are now widely used in landfill liners and covers. Portland cement concrete, asphalt cement concrete, cement-bentonite and soil-cement-bentonite are used for vertical barrier walls, as well as horizontal floors, roofs and caps.. Cement and chemical grouts are sometimes used for sealing and for keying walls and bottom barriers.

Engineered barriers can contain waste by several mechanisms, as illustrated in Fig. 3. *Resistive barriers* are the most used. They contain waste by their inherent resistance to advective and/or diffusive flow of contaminants through them. *Capacitive barriers* retain contaminants or contaminant transport media by adsorptive processes or retention in pore spaces. As the retention capacity of this type of barrier is limited, a supplemental mechanism is required if it is to be effective over the long term. In an *advective barrier* gradients are imposed to generated a flow counter to the indigenous direction of contaminant transport. *Extractive barriers* are used in combination with advective transport to remove contaminated liquid or gas for treatment and/or disposal. Typical components, their orientations and functional mechanisms for waste containment barriers are listed in Table 2.

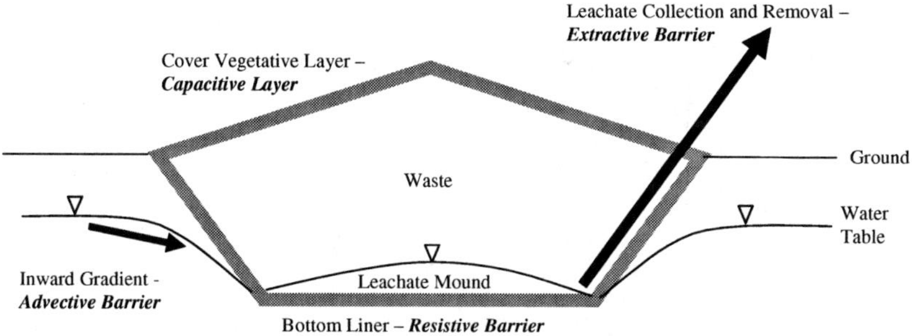

Fig. 3. Schematic illustration of functional mechanisms for engineered barriers

MONITORING BARRIER PERFORMANCE

Most waste containment systems are buried, resulting in indirect monitoring of their components during their service life. Direct assessment is usually limited to QA/QC measurements during and immediately following construction. Observations have shown that modern construction and QA procedures are effective in ensuring the short-term integrity of barrier components. Electrical leak detection systems have proven useful for determining the integrity of geomembrane liners after installation. Total system performance evaluation is usually indirect and limited to down-gradient piezometers, wells, and sampling points.

The primary (top) liner in a double liner system is usually the only type of engineered barrier in which the post-construction behavior is monitored directly. Liquids collected in the leak detection layer (see Fig. 1) between the primary and secondary bottom liners provide a direct measure of the primary liner effectiveness. The post-construction condition of caps (top covers) can be monitored by exhumation and testing of cap material. In-situ water content monitoring of layers within and beneath covers can provide an indirect measure of their performance. Lysimeters have been used beneath the barrier layer in covers to assess their performance; however, there may be uncertainties in the interpretation of such measurements owing to the influence of the lysimeter on the fluid and vapor profiles.

Geophysical techniques offer considerable potential for cost-effective, long-term indirect monitoring of barrier systems; however, they have been little used to date to provide quantitative and reliable data for performance evaluation.

Table 2. Orientation, Components, and Functional Mechanisms of Engineered Barrier Systems

Barrier Orientation	Typical Components	Functional Mechanism
Final covers	Low permeability soil layers, geosynthetic clay liners, geomembranes, ACC and PCC layers	Resistance
	Vegetated evapotranspirative soil layers,	Capacitance
	Blanket drainage layers	Extraction
Bottom barriers	Low permeability soil liners, geomembranes, geosynthetic clay liners, ACC liners, PCC floors	Resistance
	Compacted soil attenuation layers	Capacitance
	Blanket leachate collection layers, hydraulic control layers	Extraction and/or injection
	Existing in-situ low permeability soil	Resistance and capacitance
Lateral barriers	Soil-bentonite and cement-bentonite walls, PCC walls, soil-cement and jet-grouted walls, sheet pile walls, vertical geomembranes	Resistance
	Low permeability "treatment" walls	Capacitance
	Vertical wells and trenches	Extraction and/or Injection

NOTE: ACC = asphalt concrete cement; PCC = Portland cement concrete

Electrical resistivity and electromagnetic surveys may detect gross defects leading to concentrated flow through vertical barriers. Tomographic imaging and seismic velocity surveys may detect changes in properties caused by vertical barrier degradation. Multi-spectral imaging may be useful for detecting changes in vegetation, water content and temperature in near surface soils caused by defects in caps and seepage through inclined and vertical barriers. Interferometric synthetic aperture radar, light detection and ranging (LIDAR), and other airborne or satellite techniques can detect centimeter-scale deformations that could have been caused by local or global slope instability or other problems. Further development of these tools for long-term performance monitoring of waste containment systems is needed.

BARRIER SYSTEM COMPONENT PERFORMANCE

The most commonly used barrier components are earthen barriers (usually compacted clay liners), geomembranes, geosynthetic clay liners, granular and geosynthetics drainage layers, evapotranspirative covers, vertical barriers, and Portland cement or asphalt concrete barriers. Most performance information to date is on covers and liners. Schematic cross sections of prescriptive minimum bottom barrier (liner) systems and final cover systems for municipal solid waste landfills and hazardous waste landfills in the U.S.A. are shown in Figs. 4. and 5.

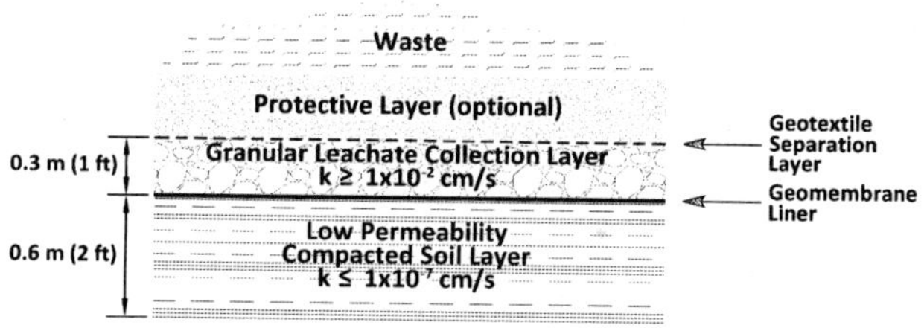

(a)

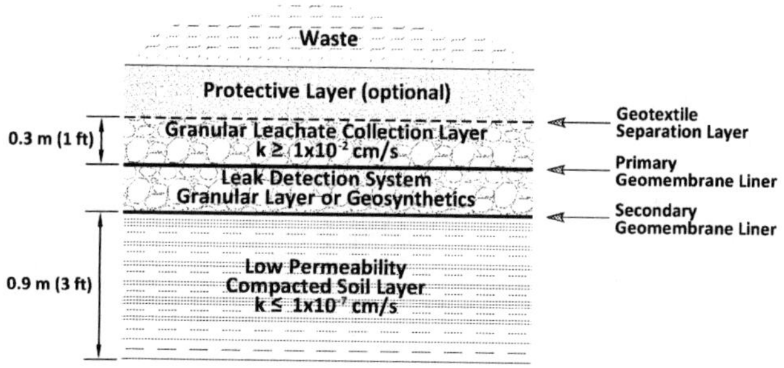

(b)

Fig. 4. Prescriptive minimum bottom barrier system for (a) municipal solid waste landfills and (b) hazardous waste landfills under RCRA regulations in 40 CFR §264 and §258. The protective layer between the waste and the geotextile separation layer is typical but not part of the prescriptive standard.

Little quantitative performance data is available for vertical barrier walls constructed in-situ for containment of lateral flow out of or into contaminated subsurface zones.

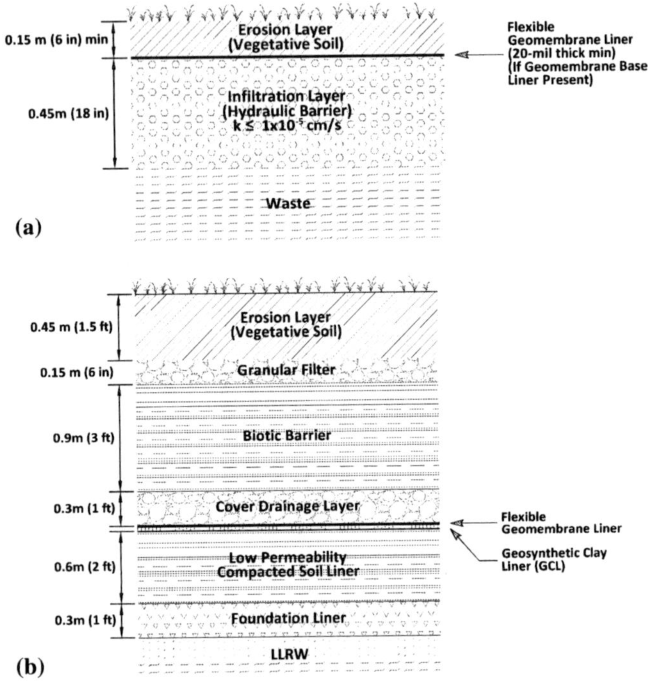

Fig. 5. Prescriptive minimum final cover system for (a) municipal solid waste landfills and (b) hazardous waste and low-level radioactive waste landfills under RCRA 40 CFR §258.

Compacted Clay Liners

Compacted clay liners have been found to work well as components within barrier systems provided good construction and operational practices are followed and they are protected against desiccation (Rowe, 2005). Secondary permeability can develop subsequent to construction as a result of wetting and drying, freezing and thawing, and deformation. Root penetration can be a problem in covers. Diffusion can contribute significantly to the total transport of chemical contaminants through well-constructed low permeability clay barriers. High temperatures near the barrier and reactions between migrating chemicals and clay, especially bentonite, may increase the hydraulic conductivity above the usual specified maximum of $<1x10^{-9}$ m/s over the medium and long terms. Whether compacted clay and composite barriers effectively prevent the migration of volatile organics over the long term is not yet known..

Geomembranes

Available data show that geomembranes installed in accordance with strict construction quality control requirements exhibit significantly fewer leaks than those without such requirements. Defective materials and seams, as well as physical damage during construction, all degrade short-term performance. Punctures caused by increased overburden pressure and material degradation, accelerated by high temperatures adversely impact the effectiveness of geomembranes over the medium and long terms. Estimated service lives decrease from 1,000 years at 10°C to only about 15 years at 60°C (Rowe, 2005). Geomembranes have little resistance to diffusive migration of several types of volatile organic compounds (VOC). This can be a short-term problem if a geomembrane is used as the sole barrier, or a medium to long-term problem in a composite barrier of different materials.

Geosynthetic Clay Liners (GCL)

Installation of defective materials or separation of overlapped panels decreases the short-term effectiveness of GCLs. Exposure of GCLs to liquid chemicals of high ionic strength and solvents is a performance concern over all timescales. Desiccation and local and global slope instability are potential medium- and long-term problems. Chemical transport through GCLs can be a problem when holes are too large to permit self-healing by swelling of the clay or when the GCL is the sole barrier and it is susceptible to diffusion.

Granular Drainage Layers

Granular drainage layers enhance stability, cut off advective and diffusive flow, and reduce leachate heads in liners and covers. Their short-term performance can be degraded by inadequate discharge capacity and by clogging. Clogging can develop over the medium- and long-term as a result of soil infiltration into the drainage layer, biological activity, and chemical precipitation. Both granular soil and geosynthetic drainage layers are susceptible to similar problems. Installation damage will adversely impact short-term performance.

Evapotranspirative Barriers

Evapotranspirative barriers are now being more used as capacitive cover systems. Although performance data have only become available in the past few years, it suggests that an evapotranspirative barrier can be an effective alternative to a compacted clay or composite cover in arid and semi-arid areas. Considerably more data accumulated over much longer time frames and/or studies of natural analogs that have functioned well for hundreds or thousands of years are needed to enable reliable predictions of the long-term performance of these barriers.

Vertical Cutoff Walls

Soil-bentonite, cement-bentonite, soil-cement-bentonite, high density polyethylene (HPDE), and steel sheet pile vertical barriers appear to be performing well, but little quantitative data is available. There has been almost no direct monitoring of vertical barriers after construction. It seems evident, however, that the short-term performance is affected primarily by the quality of construction. Construction defects that can adversely affect the wall performance include gaps in the wall as a result of poor material mixing and defective material and high-permeability zones caused by caving of the adjacent trench wall or trapping of low quality material at joints between panels. Chemical incompatibility, desiccation above the water table, and cracking all adversely affect the medium- and long-term performance.

Portland Cement and Asphalt Cement Floors, Walls and Roofs

Cracking due to shrinkage, expansion, or differential settlement is the major concern with Portland cement and asphalt concrete barriers. Additives may improve resistance to cracking and durability; however, there is little medium- and long-term data.

CONTAINMENT SYSTEM PERFORMANCE

The results of this study led to several conclusions concerning the performance of engineered waste containment systems.

1. Existing data indicate that modern waste containment systems are performing well and as designed so far. However, they have not been in service long enough to allow a direct assessment of long-term performance.
2. Predictive models may successfully predict long-term behavior; however, few data are available to verify them. Predictive models for long-term performance of a containment system depend on predictions of the long-term integrity of the system elements. Thus maintaining the integrity of the containment system elements over the active life of the wastes contained appears essential.
3. Liners that are constructed following rigorous quality assurance guidelines provide protection against offsite contaminant leakage.
4. Redundant design appears to enable the containment system to serve as an effective barrier to contaminant transport even if an individual component degrades with time.
5. Composite liners composed of both compacted clay and geomembranes or geosynthetic clay liners and geomembranes provide better protection than any single component acting alone.
6. Reliable predictions of leakage rates through composite liners should take into account holes and wrinkles in geomembranes and elevated leachate heads.
7. Cover systems can be effective in maintaining waste isolation as long as periodic maintenance is performed.
8. Vertical waste containment barriers have not been monitored sufficiently to enable definitive conclusions about their field performance or the accuracy of predictions of the transport of contaminants through them.

RECOMMENDATIONS

Several specific recommendations for better assessment of the condition and effectiveness of engineered waste containment barrier systems and their probable performance over their lifetimes were developed based on the findings in this study. Paramount among them is the need for more and better performance data. In addition to more direct monitoring data on the integrity of barrier system components in the medium and long term, key parameters that should be monitored include groundwater quality in the saturated zone at the down-gradient edge of the containment facility, gas emissions in the vadose zone around the site if the waste has potential for generating harmful gases such as methane, leachate head acting on liners, temperature on geomembrane liners, and the quality and quantity of leachate being generated. To facilitate the gathering and evaluation of this information and data, several specific actions were called for, including the following:

1. Monitoring programs for new facilities should include provisions for collecting data for long-term performance assessment and operators of existing facilities should collect these data using in-place monitoring systems to the extent practical. Key types of data that should be collected are listed in Table 3.
2. Regulatory agencies should develop guidelines for increased direct monitoring of barrier systems and their components, and the National Science Foundation should sponsor research for development of new cost-effective monitoring methods, especially for assessing the effectiveness of vertical barriers.
3. Federal agencies responsible for engineered barrier systems should commission and fund assessments on a regular basis, probably once every 5 to 10 years. Assessment results should be placed in the public domain in a readily accessible form.
4. The sponsoring and responsible federal agencies (EPA, DOE, USNRC, and NSF) should establish a set of observatories at operational containment facilities to assess long-term performance of waste containment facilities at field scale.

Additional recommendations made by the committee relate to models, monitoring periods and performance criteria:

1. Regulatory agencies and research sponsors should support the validation, calibration, and improvement of models to predict the behavior of containment system components and the composite system over long periods of time. Field observations and measurements should be used for validation and calibration of these models.
2. Financial assurance mechanisms to ensure funding for monitoring and care for as long as the waste poses a threat to human health and the environment should be developed.
3. Guidance should be developed for the practical implementation of performance-based criteria for assessment of containment system performance as an alternative to prescriptive designs.

Table 3. Recommended Data and Information Collection for Long-Term Assessment of Engineered Barrier Performance

Parameter[a]	Measurement technique	Purpose	Frequency	Location
Existing, but should be more accessible				
Leachate flow rate	Lysimeters, LCRS flow rate measurements	Cover and LCRS effectiveness; demand on liner or barrier	Collect continuously, report monthly averages and peak flows annually	At collection and discharge points
Leachate composition	Chemical analyses	Demand on liner or barrier	Collect indicators semi-annually	At collection points
Leak detection system flow rate	Fluid levels, piezometers	Effectiveness of primary liner	Collect monthly peaks and averages	LDS collection or discharge points
Composition of LDS liquid	Chemical analyses	Source of leakage	Collect indicators semi-annually	LDS sumps
Geomembrane defect frequency	Electrical leak detection	Short term integrity of geomembrane	After placement of LCRS	Sensors on membrane
Leachate head in sumps	Observation wells, piezometers	Head on sump liner	Monthly average	At the sumps
Hydraulic head and concentration differences across vertical barriers	Piezometers, groundwater wells	Hydraulic gradient, concentration gradient, mass flux across barrier	Semi-annually	Opposite sides of the barrier
Physical condition of cap (cracking, settlement, erosion, stability)	Visual observations, surveys, photographs, LIDAR surveys	Vegetative health, erosion, demand on barrier layers	Quarterly and after extreme events	Full site

Parameter[a]	**Measurement technique**	**Purpose**	**Frequency**	**Location**
Physical condition of vertical barrier (at the surface)	Visual observations	Cracks and settlement along the wall	Quarterly	Along the entire alignment
Gas emissions through cap	Hand-held probes, flux box	Effectiveness of cap at gas containment	Monthly	Specified distributed measurement points
Subsurface gas migration	Perimeter gas probe measurements	Effectiveness of containment system for gas migration control	Monthly	Multi-depth probes at specified spacing around perimeter
Groundwater monitoring of sample compositions	Chemical analyses	Containment system effectiveness—assurance that MCL are not exceeded	Semi-annually	Groundwater monitoring points based on site hydro-geology
Proposed				
Liner temperature	Temperature sensors	Thermal environment on liner (for degradation prediction)	Daily (average and/or max. and min. reported monthly)	Multiple points from the edge to center of a cell at selected MSW and ash landfills
Head on liner	Leachate level measurement in wells, piezometers	Demand on liner, effectiveness of LCRS	Monthly peak and average values	Representa-tive points on MSW, ash, and hazardous waste landfills
Leakage (lysimeter) beneath sump	Leachate collection in lysimeter	Fluid flow and mass flux through area over lysimeter	Monthly totals	All single-lined landfills

Parameter[a]	Measurement technique	Purpose	Frequency	Location
Defects in vertical barriers	Geophysical techniques (e.g., electrical imaging of gas tracers or injected brine), fluid head and chemical concentration on opposite sides of the barrier	Barrier integrity	End of construction and periodically thereafter	All vertical barriers
Change in hydraulic conductivity of vertical barriers	Coring and sampling; in situ testing	Barrier effectiveness	Every 5 years	Soil-bentonite and cement-bentonite barriers
Geomembrane oxidation induction time	Testing of sacrificial coupons[b]	Geomembrane aging	Every 5 years	Coupons placed in sumps and in the cover
GCL hydraulic conductivity	Testing of sacrificial coupons	GCL degradation	Every 3 years	Panels buried in landfills with single GCL covers

NOTES: LCR = leachate collection and removal system; LDS = leak detection system; LIDAR = Light Detection and Ranging; MSW = municipal solid waste
[a]Accompanying metadata are essential (e.g., rainfall, temperature, location of facility, site activity reports).
[b]Sacrificial coupons are loose pieces of geomembranes that can be retrieved periodically for testing.

CONCLUSION

Up to 20 years of field observations suggest that engineered waste containment barrier systems designed, constructed, operated, and maintained in accordance with current statutory regulations and requirements have so far provided environmental protection at or above specified levels. Extrapolations of long-term performance made using existing data and both empirical and theoretical physical and chemical models will have high uncertainty until field data are available for longer periods, on the order of 100 years or more for some systems.

Successful prediction of the long-term performance of waste containment barrier systems depends strongly on the accurate prediction of the service lives of the individual components, including compacted clay layers, geomembranes, geosynthetic clay liners, vertical walls of various types, and leachate and gas collection and removal systems. Specific conclusions about the performance to date of these components and overall containment systems and recommendations for needed actions for better assessment of conditions and prediction of future behavior are given in the paper and the report on which it is based.

ACKNOWLEDGMENTS

This paper is derived from the report, *Assessment of the Performance of Engineered Waste Containment Barriers,* National Research Council National Academy Press, 2007, 119 pp. Members of the committee, chaired by the author, that prepared this report are Lisa Alvarez-Cohen, University of California, Berkeley; Estella A. Atekwana, Oklahoma State University, Stillwater; Susan E. Burns, Georgia Institute of Technology, Atlanta; Robert B. Gilbert, University of Texas, Austin; Edward Kavazanjian, Jr., Arizona State University, Tempe; W. Hugh Riordan, Givens, Pursley LLP, Boise, Idaho; R. Kerry Rowe, Queens University, Kingston, Ont.; Charles D. Shackelford, Colorado State University, Fort Collins; Hari D. Sharma, GeoSyntec Consultants, Oakland, California; Nazli Yesiller, San Luis Obispo, California; Anne Linn (*Study Director),* the National Research Council, Washington, DC.

Nestor Suarez provided assistance in the preparation of figures.

REFERENCES

Bonaparte, R., Daniel, D.E., and Koerner, R. M. (2002). *Assessment and recommendations for improving the performance of waste containment systems,* CR-821448-01-0, Environmental Protection Agency, Washington, DC, 1,039 pp.

NRC (National Research Council) (2007). *Assessment of the performance of engineered waste containment barriers,* National Academy Press, Washington, DC, 119 pp.

Rowe, R.K. (2005). "Long-term performance of contaminant barrier systems," 45th Rankine Lecture, *Geotechnique,* Vol. 55 (9), 631-678.

An Innovative Ashfill Expansion at the Town of Babylon Landfill Site

Basak Gulec[1], A.M. ASCE, Kwasi Badu-Tweneboah[2], M. ASCE, P.E., Tom Vetri[3], and Alexander Maestre[4]

[1]Engineer, Geosyntec Consultants, 1255 Roberts Boulevard, NW, Suite 200, Kennesaw, GA, 30144; bgulec@geosyntec.com
[2]Associate, Geosyntec Consultants, 2258 Riverside Avenue, Jacksonville, FL, 32204; kbadu-tweneboah@geosyntec.com
[3]Deputy Commissioner, Town of Babylon, Department of Environmental Control, 57 Field Street, West Babylon, NY, 11704; tvetri@optonline.net
[4]Senior Staff Engineer, Geosyntec Consultants, 1255 Roberts Boulevard, NW, Suite 200, Kennesaw, GA, 30144; amaestre@geosyntec.com

ABSTRACT: A lateral and vertical expansion of an existing ashfill was designed for the Town of Babylon Landfill (TBL) site in New York. The proposed expansion consists of the construction of a new 2-hectare (5-acre) lateral expansion cell in the area of the existing on-site recharge basin, followed by vertical expansion over an existing ashfill using reinforced soil wall (RSW) systems. This proposed expansion, with the use of innovative surface water management (SWM) and RSW systems, will provide additional ashfill airspace of approximately 612,000 cubic meters (800,000 cubic yards) and extend the active life of the ashfill by approximately nine years. The design and permitting of the RSW and SWM systems for the ashfill expansion at the TBL site are presented in this paper. Global stability analysis and liquefaction potential evaluation performed for the proposed expansion are also briefly discussed in the paper.

INTRODUCTION

The TBL site in Suffolk County, New York has been used for waste disposal since 1947. The majority of the landfill site was originally used for sand and gravel pit mining. The pits initially provided convenient below ground disposal areas, but were eventually filled. The landfill subsequently developed into a mound with waste deposited as deep as 21 m (70 ft) below and up to 42 m (140 ft) above the surrounding ground surface with waste slopes as steep as 1 horizontal to 2 vertical (1H:2V).

Since 1986, innovative design, permitting, construction support, and consulting services have been provided for the Town on the TBL site. For example, in 1995, the landfill was closed using an innovative closure system including reinforced soil slopes (Cargill and Olen, 1997 and 1998). Since the closure of the landfill, the site has been used for the construction and operation of ashfills. The Town constructed a Resource Recovery Facility (RRF) adjacent to the landfill for municipal solid waste (MSW) incineration as part of the resource recovery program. In order to provide efficient and economical disposal of ash produced by the RRF, an ash monofill (now called Southern Ashfill) consisting of six cells, located to the south and partially on the side slopes of the existing landfill, was designed and constructed in phases between 1989 and 1995. Operation of the Southern Ashfill began in 1990.

The design and construction of the Northern U Landfill took place in 1993. The Northern U Landfill was constructed with a double-composite liner system, and was designed and intended for disposal of waste which must by-pass the RRF during periods of maintenance or other shutdowns. The Town subsequently arranged for the disposal of by-pass waste at RRFs and transfer stations in neighboring communities, and the Northern U Landfill was redesignated for ash disposal, and thereafter called Northern U Ashfill, in accordance with a new disposal plan (Geosyntec Consultants, 1993).

In 1998, the design and construction plans for the expansion and closure of the Southern Ashfill using a two-tiered RSW system along the western and southern perimeters of the ashfill were completed (Gopalakrishnan and Cargill, 2000). The proposed system increased the capacity of the ashfill by approximately 298,000 cubic meters (390,000 cubic yards) for an additional three years of airspace for disposal operations. Also in 2000, the engineering design and construction for the vertical and lateral expansion of the Northern U Ashfill using a two-tiered RSW system was completed (Cargill et al., 2002). This expansion increased the active life of the Northern U Ashfill by seven years by adding approximately 642,000 cubic meters (840,000 cubic yards) of airspace for ash disposal until 2007.

In 2006, additional ashfill capacity was needed for the Town to meet Town's contractual obligations with the RRF operator. A feasibility study was initially conducted to evaluate alternatives for the expansion of the TBL site (Geosyntec Consultants, 2006). The results of the feasibility study indicated that the most viable option was for a 2-hectare (5-acre) lateral expansion (Cell 7) over the current on-site recharge basin followed by a vertical expansion over the existing Southern Ashfill using RSWs. The on-site recharge basin was used as a sedimentation pond to manage surface water runoff from the existing landfill and ashfills, as well as the Town roads and immediate areas surrounding the site. The design required backfilling the recharge basin with recycled concrete aggregate (RCA) material prior to cell construction. Figure 1 shows the layout of the TBL site and the final cover grading plan for the proposed expansion. A cross section (A-A') of the proposed ashfill expansion is illustrated in Figure 2.

Reinforced Soil Wall System Design

A two-tiered RSW system along the perimeter of Cell 7 and a third upper wall over the proposed Cell 7, and existing Southern Ashfill are proposed. The proposed RSWs consist of a RSW system with masonry facing units and geogrid reinforcement.

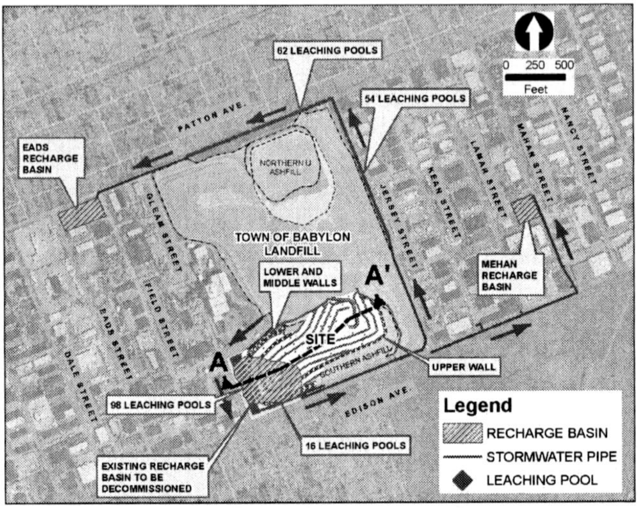

FIG. 1. Site Layout of the Town of Babylon Landfill.

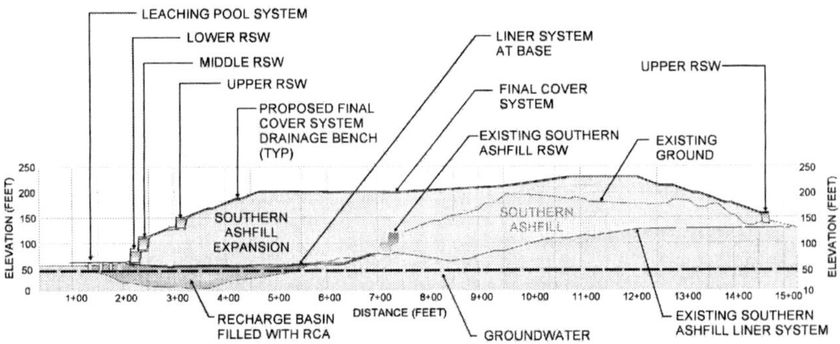

FIG. 2. Cross Section A-A'.

As shown in Figure 1, the lower and middle walls are terminated on the south side leaving the continuation of the upper wall as a single wall extending towards the existing Southern Ashfill. The two-tiered RSW system (i.e., lower and middle walls) is approximately 396 m (1,300 ft) long. The lower and middle walls are designed for heights ranging from 2.1 to 9.1 m (7 to 30 ft) and 1.5 to 8.2 m (5 to 27 ft), respectively. The upper wall, which is parallel to the two lower walls in the proposed expansion area is approximately 640 m (2,100 ft) long and is designed for heights ranging from 1.2 to 9.4 m (4 to 31 ft). The upper and middle walls support a 2.5H:1V slope with a 4.7 m (15.5 ft) setback from the face of the wall. Typical lower and middle, and upper wall cross sections are shown in Figure 3.

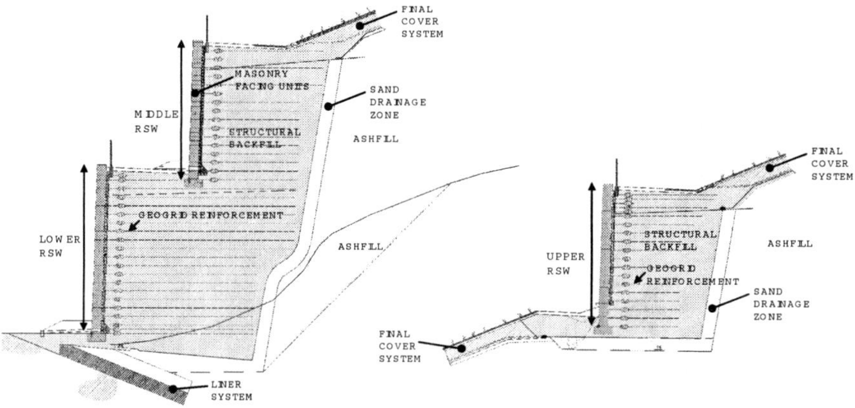

FIG. 3. Lower and Middle, and Upper Reinforced Soil Wall Cross Sections.

The design of the RSWs is based on the National Concrete Masonry Association (NCMA) design procedure and SRWall software, in which the NCMA procedure is implemented (NCMA, 2002). Using SRWall, the RSW system was designed to meet the required minimum factors of safety (FSs) against failure under different stability scenarios (i.e., external, internal, and facing) and failure modes (such as base sliding, overturning, bearing capacity, pullout, connection failure, etc.).

Based on this design, the geogrid reinforcement length and spacing vary between 1.5 to 9.8 m (5 to 32 ft); and 0.2 to 0.8 m (0.67 to 2.67 ft, which is one to four facing units), respectively. The long term design strength (LTDS) used for the analysis is 30.65 kN/m (2,100 lb/ft), which corresponds to an allowable strength (T_a) of 20.43 kN/m (1,400 lb/ft). The allowable tensile strength of the geogrid reinforcement incorporates a FS of 1.5 with respect to LTDS.

The wall design includes provisions for internal wall drainage, as well as leachate interception and drainage. A gravel drainage zone is included behind the masonry facing blocks to convey any potential infiltration through the wall backfill into a perforated drainage pipe near the base of the wall face. With regard to leachate seeps, it is noted that the new ash placed as part of the Southern Ashfill expansion will be contained by the double liner system, and thus any leachate will be collected by its leachate collection system. For the existing ash in the Southern Ashfill cells, any leachate seeping horizontally in the existing ash behind the walls will be intercepted by a sand drainage zone between the reinforced soil zone and the ash, and will be conveyed to the leachate collection layer at the existing liner system. Leachate management and control were an important design requirement during the initial permitting process with the New York State Department of Environment and

Conservation (NYSDEC) regulatory staff.

In addition to the design of the RSWs, as presented above, the global stability of the RSWs, foundation soils, and adjacent slopes is evaluated using the computer program SLIDE 5.00 developed by Rocscience Inc. (2006). Potential failure surfaces passing through and underneath the reinforced soil zone are considered in both static and seismic analyses.

Based on the results of the RSW design and global stability analyses, the calculated FSs were equal to or greater than the minimum required FSs.

Material Properties

Table 1 provides a summary of the material properties used for the RSW design and the global stability analyses.

Table 1. Summary of Material Properties used in Stability Analyses

Material	Normal Stress (kN/m^2)	γ (kN/m^3)	c (kN/m^2)	ϕ (°)
Ash [(1)]	$\sigma_v \leq 287$	18.86	45.49	37
	$287 < \sigma_v \leq 347$	18.86	0.26	0
	$\sigma_v > 347$	18.86	0	37
MSW Waste [(2)]	$\sigma_v \leq 37$	11.00	23.94	0
	$\sigma_v > 37$	11.00	0	33
Reinforced Soil [(1)]	N/A	18.08	0	35
Foundation Soils [(1)]	N/A	18.86	0	31
Recharge Basin Fill	N/A	18.86	0	35

(1) Material properties are based on the previous designs and testing programs at the TBL site and referenced in this paper.

(2) The total unit weight for MSW was assumed based on Fasset et al. (1994). The bilinear Mohr-Coulomb envelope in Kavazanjian et al. (1995) is used for the strength properties.

Liquefaction Analysis

The proposed expansion includes backfilling the on-site recharge basin with RCA material that consists essentially of silty sands and gravels. A review of the particle-size distribution curve of the RCA indicated the material falls within the range of soils that liquefied in the past from the 1983 Borah Peak, Idaho Moment Magnitude Mw 7.3 Earthquake, as cited in Ishihara et al. (1989). Also, a portion of the RCA will be fully saturated from placement below the groundwater table. Therefore, an analysis was required to evaluate the potential for liquefaction of foundation soils during the design earthquake and its effect on the integrity of the ashfill liner system.

Charts developed for evaluation of soil liquefaction from shear wave velocity (Stokoe et al., 1989) were used to evaluate the liquefaction potential for this site. The results indicate that soil liquefaction potential within the footprint of the proposed ashfill expansion is low. However, prior to construction, additional testing and evaluation will be performed to confirm the suitability of the fill area with respect to liquefaction to support the proposed ashfill expansion.

Surface Water Management System Design

The on-site recharge basin, with approximately 5-ha-m (40-acre-ft) capacity, will be replaced by an interconnected underground leaching pool system that will overflow to Eads Recharge Basin located west of TBL (Figure 1). The leaching pools are cylindrical open-bottom structures 3.1 m (10 ft) in diameter and 3.1 m (10 ft) in height. A typical leaching pool detail is shown in Figure 4. Additional flow from the Town roads that currently discharge into the basin will be discharged into the existing Mahan Recharge Basin east of TBL (Figure 1), using corrugated metal pipes. Over 200 leaching pools will be interconnected in the proposed design. The leaching pools and the two off-site recharge basins will provide detention and management of runoff from a 25-yr, 24-hr design storm event.

The interconnected leaching pool system was designed considering two specific objectives. The first objective was to create a system able to infiltrate a fraction of the total runoff generated at TBL. The remaining fraction of runoff that could not be handled by the leaching pool system should be diverted to Eads Recharge Basin. The second objective was to ensure that the water levels inside the leaching pools should not exceed 18 m (60 ft) above the mean sea level (MSL) to prevent flooding in the lowest area of the site, where the on-site recharge basin is currently located.

Infiltrating a fraction of the total runoff was possible because the soils located in the vicinity of the landfill are highly permeable (primarily sands and gravels). Infiltration rates close to 25 cm/hr (10 in./hr) were observed in several soil test borings performed by the Town inside and outside the TBL. A FS of two for infiltration was selected due to the variability of site conditions with respect to infiltration rates, potential siltation in the leaching pools, and uncertainty associated with the available capacity of the overall system.

The U.S. Environmental Protection Agency (USEPA) Storm Water Management Model (SWMM) was used to analyze the system especially because it allows the modeling of backwater effects (USEPA, 2006). The number of leaching pools, and the size and slope of the connecting pipes were modified until the system was able to handle the design storm.

The model demonstrated that the existing two off-site recharge basins and the interconnected underground leaching pool system would provide adequate capacity to convey and/or infiltrate the runoff from on-site and off-site areas.

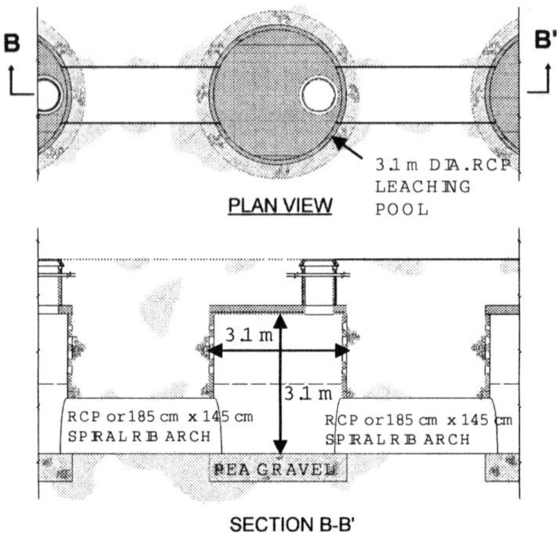

FIG. 4. Typical Leaching Pool Detail.

SUMMARY AND CONCLUSIONS

This paper discusses two unique design components of the proposed expansion, the proposed RSW and the SWM systems, for the TBL Southern Ashfill. In conjunction with the RSW design, global stability analysis and liquefaction potential evaluation are briefly discussed in this paper.

To provide additional ashfill capacity, the existing on-site recharge basin was backfilled and lined, and replaced with a series of leaching pools using a technical solution that takes advantage of site conditions such as the high infiltration rate and the additional capacity provided by two existing off-site recharge basins owned by the Town. This technical solution allowed the area to be used for the lateral expansion. In order to gain more airspace, the lateral expansion was coupled with a vertical expansion using a RSW system consisting of three walls (i.e., a two-tiered wall and an additional single wall), one of which extends over the existing Southern Ashfill. Approximately 612,000-cubic meter (800,000-cubic yard) airspace was achieved with the use of innovative RSW system design and SWM system design and modeling. The proposed expansion will extend the life of the ashfill by approximately nine years.

REFERENCES

Cargill, K. W. and Olen, K. L. (1997). "An Innovative Final Closure System," *Proceedings, 2nd Annual Landfill Symposium*, Solid Waste Association of North America, Sacramento, CA, pp. 79-85.

Cargill, K.W. and Olen, K. L. (1998). "Landfill Closure Using Reinforced Soil Slopes," *Sixth International Conference on Geosynthetics*, Atlanta, GA, pp. 481-486.

Cargill, K.W., Iravani, S., and Vetri, T. (2002). "The Walls of Babylon – The Vertical Expansion Answer," *Proceedings, 7th Annual Landfill Symposium*, Solid Waste Association of North America, Louisville, KY.

Fasset, J., Leonards, G., and Repetto, P. (1994). "Geotechnical Properties of Municipal Solid Wastes and Their Use in Landfill Design," *Proceedings of Waste Tech 94*, Charleston, SC, pp. 1-31.

Geosyntec Consultants (1993). "Evaluation of Ash Disposal in Northern "U" Landfill", *Technical Report,* prepared for the Town of Babylon, NY.

Geosyntec Consultants (2006). "Final Report, Feasibility Study for the Expansion of Ash Capacity, Town of Babylon Ashfill", *Technical Report,* prepared for the Town of Babylon, NY.

Gopalakrishnan, G. and Cargill, K.W. (2000). "Innovative Use of a Reinforced Soil Wall System for Ashfill Vertical Expansion," *Conference Proceedings, Waste Tech 2000,* Orlando, FL.

Ishara, K. , Kokusho, T., and Silver, M.L. (1989). "Recent Developments in Evaluating Liquefaction Characteristics of Local Soils", State-of-the-Art Report, *Proceedings of 12th International Conference on Soil Mechanics and Foundation Engineering*, Rio de Janeiro, Brazil, Vol. 4, pp. 2719-2734.

Kavazanjian Jr, E., Matasovic, N., Bonaparte, R., and Schmertmann, G. (1995). "Evaluation of MSW Properties for Seismic Analysis," *Proceedings of Geoenvironmental 2000*, Vol. II, New Orleans, LA, pp. 1126-1141.

NCMA (2002). "*SRWall – Design Software for Segmental Retaining Walls",* Version 3.0, National Concrete Masonry Association, Herndon, VA.

Rocscience, Inc. (2006). *"SLIDE 5.023: 2D Limit Equilibrium Slope Stability Software,"* Ontario, Canada.

Stokoe, K. H., Roesset, J. M., Bierschwale, J.G., and Aouad, M. (1989). "Liquefaction Potential of Sands from Shear Wave Velocity", *Proceedings of 9th World Conference on Earthquake Engineering*, Tokyo, Japan, Vol. 3, pp. 213-218.

USEPA (2006). *"Storm Water Management Model User's Manual Version 5.0,"* United States Environmental Protection Agency, Cincinnati, OH.

Analytical and Numerical Methodology for Modeling Temperatures in Landfills

James L. Hanson[1], M. ASCE, PE, Wei-Lien Liu[2], A.M. ASCE, and Nazli Yesiller[3], A.M. ASCE

[1]Associate Professor, Civil and Environmental Engineering Department, California Polytechnic State University, San Luis Obispo, CA 93407, jahanson@calpoly.edu
[2]Senior Staff Engineer, CTI and Associates, Inc., 12482 Emerson Dr., Brighton, MI 48116, lwl919@hotmail.com
[3]Independent Consultant, 590 Bluerock Dr., San Luis Obispo, CA 93401, nyesiller@gmail.com

ABSTRACT: Analytical and numerical approaches have been developed for modeling temperatures in municipal solid waste landfills. Steps for model formulation and details of boundary conditions are described. The formulation was based on a transient conductive heat transfer analysis. Conventional earth temperature theories were modified for landfill systems by incorporating heat generation functions representing biological decomposition of wastes. Finite element analysis was used for general modeling and parametric evaluations. Thermal properties of materials were determined using field observations and data reported in literature. The boundary conditions consisted of seasonal temperature cycles at the ground surface (established using near-surface field measurements) and constant temperatures at the far-field boundary (established using field measurements and maps of regional groundwater temperatures). For heat generation, first a step-function was developed to provide initial (aerobic) and residual (anaerobic) conditions. Second, an exponential growth-decay function was established; and third, the function was scaled for climatic conditions. The formulations developed can be used for prediction of temperatures within various components of landfill systems (liner, waste mass, cover, and surrounding subgrade), determination of frost depths, and determination of heat gain due to decomposition of wastes.

INTRODUCTION

Temperatures influence the engineering properties of geomaterials. Temperature extremes and thermal cycles affect the integrity and durability of earthen and geosynthetic components of waste containment barrier systems (Rowe 2005). Determination and prediction of temperature conditions within landfill systems are needed to properly evaluate the coupled geotechnical performance of these facilities.

Fully coupled geotechnical and thermal numerical models represent the state-of-the-art in landfill analysis.

Earth temperature theory is generally well established including coverage of temperature cycles, amplitude decrement and phase lag with depth, frost depths, ground surface temperatures, and layered systems (Carslaw and Jaeger 1959, ORNL 1981). Analytical solutions are available for thermal analysis under seasonally cyclic ground surface temperature conditions. Earth temperature theory can be applied to landfills with the addition of heat generation due to waste decomposition. Modeling the heat generation requires functions that can be incorporated using advanced numerical techniques such as finite element analysis (FEA) or finite difference approach. Numerous investigators have reported methods for modeling heat transfer in landfill systems (El Fadel et al. 1996, Doll 1997, Yoshida and Rowe 2003, Southen and Rowe 2005, Yesiller et al. 2005). Doll (1997) and Southen and Rowe (2005) used simplified constant elevated temperature conditions for modeling performance of liner systems and neglected to incorporate temporal trends of heat generation due to decomposition. El Fadel et al. (1996) developed a model for predicting heat transfer in landfills that included heat generation, but a simplified function was used (that was correlated only to estimated values of acetic acid generation rate) and losses were assumed to be linearly related to heat generation. Yoshida and Rowe (2003) developed a model that included heat generation as a function of gas production rate. Yesiller et al. (2005) quantified heat generation as compared to unheated ambient conditions, but included limited heat transfer analysis.

Limited information has been reported related to thermal analysis of municipal solid waste including heat transfer, heat generation, ground surface temperature conditions overlying wastes, and thermal properties. The objective of this paper is to outline the methodology for development of thermal modeling for landfill systems with heat generation in wastes due to decomposition, underlying native soil conditions, and overlying seasonal air temperature fluctuations. This investigation was conducted to provide material properties, model geometry and boundary conditions, heat generation functions, and overall methodology for model formulation. The analysis was conducted for landfills located in four different climatic regions in North America (Michigan, New Mexico, Alaska, and British Columbia). Transient analyses were conducted using finite element analysis.

MATERIAL PROPERTIES

For heat transfer analysis of landfills, material properties were required for the cover, waste, bottom liner, and subgrade. In particular, unit weight, thermal conductivity, and heat capacity were required. Unit weight (γ) was determined using operational site records for earthen barrier materials, wastes, and native subgrade soils. Thermal conductivity (k_t) was determined using laboratory and field thermal conductivity probe experiments (Hanson et al. 2000) and using data from literature (e.g., Andersland and Ladanyi 2003). Volumetric heat capacity (C) was calculated by summing heat capacity of individual components of the materials on a volumetric basis. Weight-volume relations of the soils were determined using site records and fractions of constituents of the wastes were obtained from U.S. EPA (2005). Thermal diffusivity

(α) was calculated as the quotient of thermal conductivity and volumetric heat capacity. A summary of material properties used for the study sites is presented in Table 1. The thermal properties varied between the sites due to soil type, construction conditions, precipitation, and waste placement conditions. In general, both k_t and C (for wastes and soils) increase with increasing γ and moisture content.

Table 1. Material Properties for Analyses

Property	Michigan	New Mexico	Alaska	British Columbia
γ_{waste} (kN/m^3)	9.8	7.4	5.2	9.8
$k_{t\text{-}waste}$ (W/mK)	1.0	0.6	0.3	1.5
C_{waste} (kJ/m^{3}K)	2000	1200	1000	2200
α_{waste} (m^2/s)	5.0 x 10^{-7}	5.0 x 10^{-7}	3.0 x 10^{-7}	7.0 x 10^{-7}
γ_{soil} (kN/m^3)	20.5	16.7	21.0	17.7
$k_{t\text{-}soil}$ (W/mK)	2.5	0.7	2.4	1.0
C_{soil} (kJ/m^{3}K)	2800	1300	1800	3100
α_{soil} (m^2/s)	9.0 x 10^{-7}	4.9 x 10^{-7}	1.3 x 10^{-6}	3.3 x 10^{-7}

MODEL GEOMETRY AND BOUNDARY CONDITIONS

The analysis has been formulated in 1-D using FEA (ABAQUS version 6.5). A series of simulations was conducted with varying degrees of mesh refinement using element sizes ranging from 0.1 m to 6 m. An element size of 0.5 m was selected based on a relative error analysis conducted using these simulations (Cook et al. 1989). The time step for transient analysis is in part controlled by element size, and the time step was established as 1 day for most analyses.

Boundary conditions required in modeling landfill heat transfer consisted of far field boundary at depth and the ground surface boundary. The bottom boundary was established 75 m below the bottom liner. This depth was determined by analyzing the temperature response of different model geometries using increasingly greater depths. Simulations were conducted using variable depths below the liner system ranging from 10 to 200 m for 30-year simulation periods. For this parametric evaluation, the liner temperatures were assumed to be constant at 30°C (consistent with long-term stable field measurements in a liner system). The distance from the liner to the bottom boundary was determined to be 75 m, the first depth at which the differential in response between the simulation for 200 m boundary and a nearer boundary was less than 0.1°C. The far field boundary was fixed at the mean annual earth temperature. Mean annual earth temperatures were obtained on a site-specific basis using field measurements and data from literature for groundwater temperatures (e.g., ORNL 1981).

The ground surface boundary for landfills is highly complicated to define due to the coupled influence of wind, precipitation, snow cover, moisture evaporation, solar radiation, and heat gain from decomposition of underlying wastes. A common approach to obtain ground surface temperatures is to modify air temperatures to account for radiation absorption and emission effects using ground surface freezing and thawing n-factors (Andersland and Ladanyi 2003). Freezing or thawing indices represent the area bound by the temperature-time curve and the 0°C baseline (area below 0°C for freezing, area above 0°C for thawing). The n-factors are determined as the quotient of paired surface and air indices for freeze or thaw. These factors are typically applied on a seasonal basis to provide idealized sinusoidal annual ground surface temperature fluctuations. Surface n-factors are affected by climatic and ground surface conditions.

Well established surface n-factors have not been reported for landfill conditions. For model formulation, ground surface temperatures were determined based on an analysis of measured ground surface and near-surface temperatures. Measured ground surface temperatures were highly variable due to surface weather effects and measurement frequency. Therefore, measured near surface temperature extremes (representing maximum and minimum temperatures at each depth) were extrapolated upward to obtain two idealized surface temperature functions at each site: one for native soils and one for wastes. The resulting idealized sinusoidal ground surface temperatures were used to back-calculate n-factors for the sites using conventional 30-year definition (Andersland and Ladanyi 2003). These n-factors are provided for use at other sites to determine ground surface temperatures using air temperature data (without the need for field ground surface temperature measurements). For all ground surface temperature waves, a phase lag of $\pi/8$ was applied (Carslaw and Jaeger 1959), such that the coldest surface temperature occurred 45.6 days after the winter solstice.

A sequential model development was used that accounted for model geometry modifications with time (an excavated landfill cell below grade, waste filling, and installation of landfill cover). Chronologically, first the excavated landfill cell was modeled. The initial temperature of native subgrade soil (from bottom liner system to the far-field boundary) was uniformly set to the mean annual earth temperature. The model was then allowed to run under application of the idealized ground surface temperature function for soil. The surface temperatures were applied until long-term annual temperature cycle stasis of the subgrade was obtained. Approximately 7-10 years of simulation was required to reach this condition. Second, waste filling was modeled. The waste placement sequences were obtained from site records including annual aerial fly-over surveys, land-based topographic measurements, and detailed waste placement records. Waste placement was modeled using individual compacted lifts of approximately 3 to 5 m. Heat generation functions are applied sequentially with waste placement to the entire column of waste. Based on field measurements of as-delivered waste temperatures, it was estimated that the waste was placed at a temperature equal to average daily air temperature. The idealized seasonal surface temperature wave was applied to the waste surface (i.e., daily cover). Third, the installation of cover system (either interim or final cover) was modeled. The initial temperature of the cover materials was assumed to be equal to the ground surface temperature on the day of installation. The idealized seasonal surface temperature wave was then applied to the cover. A summary of the parameters used for boundary conditions including back-calculated n-factors is presented in Table 2. Ground surface temperatures overlying wastes were generally warmer than ground temperatures of native soils, as indicated by the higher thawing n-factors and lower freezing n-factors.

Table 2. Summary of Boundary Parameters Used in the Model

Parameter	Michigan	New Mexico	Alaska	British Columbia
Mean Soil Temperature, T_m (°C)	12.3	19.0	5.4	12.1
Amplitude for Soil Temperatures, A_s (°C)	17.3	12.0	13.4	11.0
Thawing n-factor (soil)	1.23	1.24	1.15	1.07
Freezing n-factor (soil)	0.91	NA	0.53	NA
Mean Cover Temperature, T_m (°C)	13.0	20.0	6.6	17.3
Amplitude for Cover Temperatures, A_s (°C)	16.6	12.0	14.3	12.0
Thawing n-factor (waste)	1.27	1.31	1.31	1.53
Freezing n-factor (waste)	0.56	NA	0.48	NA

NA – not applicable due to lack of significant freezing at these sites

MODELING OF HEAT GENERATION

Modeling with FEA was conducted using the material properties, geometries, and boundary conditions described above. Model results were compared to measured field temperatures at each of the four extensively instrumented sites to develop heat generation functions (Liu 2007). Methodology for field measurements and analysis of results is presented elsewhere (Yesiller et al. 2005, Hanson et al. 2005). Nonlinear regression analysis was performed to establish the functions that provided the best agreement between modeled and measured temperatures (Liu 2007). The data set used for comparisons represented 4 to 8 years of temperature measurement. However, because the temperature sensors were placed in wastes of varying ages, the range of waste age documented extends from 0 to 40 years. This allowed for greater predictive significance. In general, the trends in modeled and measured temperature vs. time were similar throughout various components of the landfill systems and through the entire range of waste age available for comparison. Damping of seasonal temperature variations with depth, phase lag with depth, depth of frost penetration, and onset and presence of heat gain due to waste decomposition were captured by the modeling (Liu 2007). The modeled temperatures were generally within ±2°C of measured temperatures throughout the entire depth of the landfill.

Heat generation functions that varied with waste age were developed. For all heat generation functions, zero heat generation rate was specified when modeled waste temperatures were either less than 0°C or greater than 80°C. Heat generation was typically high for young wastes and low for old wastes. The heat generation functions were formulated to account for net heat gain due to decomposition under normal landfill operations. The functions therefore accounted for thermal losses such as convective heat flow due to leachate migration and removal. Initially, a step-function was used to model heat generation rate in wastes due to aerobic and anaerobic decomposition for the Michigan and Alaska sites. These sites were selected for the step function formulations as maximum and minimum heat gain (in relation to ambient temperature conditions) occurred at these sites (Yesiller et al. 2005). In this formulation it was assumed that the initial aerobic phase occurred during the first 4 months subsequent to placement of waste. Then, the waste entered the anaerobic phase and remained there for the balance of the analysis. The 4-month period was selected for the aerobic phase based on the analysis of gas concentration data from the sites (Hanson et al. 2005). The resulting limiting values (maximum and minimum) for aerobic heat generation were 11.3 and 2.5 W/m^3 and for anaerobic heat generation were 0.38 and 0.08 W/m^3, for Michigan and Alaska, respectively.

Next, a more sophisticated exponential growth and decay function (Equation 1) was used for modeling heat generation for all the sites. The format of Equation 1 was selected to provide more realistic response than a step function. The function resulted in growth of heat generation rate to a peak value followed by an exponential decay. The constants (A, B, and D) control the peak value, the shape of the peak, and the rate of decay. Parameters were determined using best fit nonlinear regressions with field data.

$$H = A\left[\frac{Bt}{B^2 + 2Bt + t^2}\right]e^{-\sqrt{\frac{t}{D}}} \tag{1}$$

where,

H = heat generation rate (W/m^3)
t = time (day)
A = peak heat generation rate factor (W/m^3)
B = shape factor (day)
D = decay rate factor (day)

The exponential growth and decay heat generation rate functions obtained for the four study sites are presented in Figure 1 and summarized in Table 3. The peak heat generation rate values varied significantly with the highest values for British Columbia followed by Michigan, New Mexico, and Alaska. The high rates are attributed to relatively wet and warm climates. Even though the maximum heat generation rate was obtained for British Columbia, the maximum heat gain occurred in Michigan due to thermal properties of the wastes (Yesiller et al. 2005). The variability in durations required to reach the modeled peak heat generation rate was less than the variability in peak values. Total energy expended represented the area under a given heat generation rate vs. time curve and was calculated on a per unit volume basis. The peak heat generation rate values ranged from 0.19 to 2.21 W/m^3, the times for peak heat generation rate ranged from 174 to 360 days, and the total energy expended ranged from 15 to 191 MJ/m^3.

Effects of climatic and operational conditions on the peak heat generation rate values were investigated to extend the applicability of the heat generation functions to other sites. Correlations were developed for the heat generation function parameters A and B to a composite climatic-operational condition factor, λ, that was calculated as the product of average daily air temperature and the average annual precipitation divided by the average compacted unit weight of the wastes. A correlation was also developed between parameter D and an operational condition factor, F, defined as the average vertical waste filling rate (m/year). The correlations are described by the following equations:

$$A = -7.92 + 0.12\lambda \tag{2}$$

$$B = -2027 + 20.47\lambda - 0.015\lambda^2 \tag{3}$$

$$D = 55.5 + 2.79F \tag{4}$$

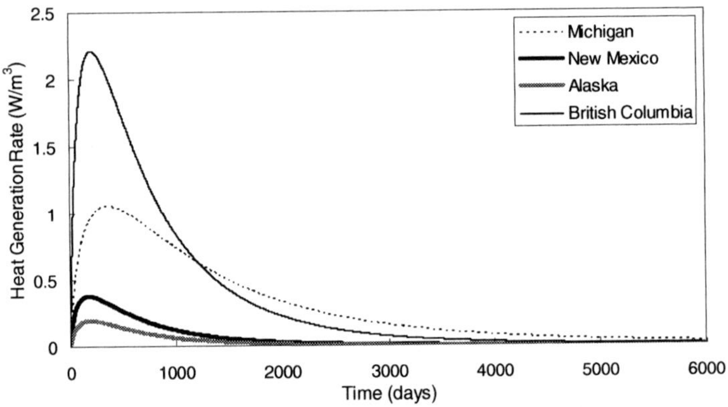

Figure 1. Exponential Growth and Decay Heat Generation Rate Functions

Table 3. Summary of Heat Generation Functions

Site	Peak Heat Generation Rate Value (W/m³)	Duration for Peak Value (days)	Total Energy Expended (MJ/m³)	*A*	*B*	*D*
Michigan	1.05	360	158.0	95	5000	120
New Mexico	0.38	174	28.8	75	5000	50
Alaska	0.19	190	15.0	7	1200	90
British Columbia	2.21	210	191.0	130	2000	80

CONCLUSIONS

Numerical modeling of temperatures in landfills requires transient, nonlinear analysis. Ground surface temperature functions overlying native soils and wastes were developed. Heat generation functions were developed that provide temporal variation in heat generation of wastes due to biological decomposition. The parameters used to define these functions have been correlated to climatic and operational conditions. The formulations described herein can be used for prediction of temperatures within various components of landfill systems (liner, waste mass, cover, and surrounding subgrade), determination of frost depths, and determination of net heat gain due to decomposition of wastes. The methodology is developed to be broadly applicable to municipal solid waste landfills in various climatic regions.

ACKNOWLEDGEMENT

This study was supported by the National Science Foundation (GOALI Grant: CMS-9813248, SGER Grant: CMS-0301032, and a 2004 AAAS/NSF WISC Grant). Assistance of the partner landfills (Sauk Trail Hills Development, Corralitos Regional Landfill, Anchorage Regional Landfill, and Vancouver Landfill) is greatly appreciated. The assistance of Mr. Nicolas Oettle with analysis is greatly appreciated.

REFERENCES

Andersland, O. B. and Ladanyi, B. (2003). *Frozen Ground Engineering*, 2nd Edition, Wiley, NY.

Carslaw, H. S. and Jaeger, J. C. (1959). "Conduction of Heat in Solids," second ed., *Oxford University Press*, Oxford.

Cook, D. R., Malkus, S. D., and Plesha, E. M. (1989). *Concepts and Applications of Finite Element Analysis* (3rd edition), Wiley, NY.

Doll, P. (1997). "Desiccation below mineral liners with heat production," *J. Geotech. Geoenviron. Eng.*, ASCE, Vol. 123, No. 11, p. 1001-1009

El Fadel, M., Findikakis, A. N., and Leckie, J. O. (1996a). "Numerical modelling of generation and transport of gas and heat in sanitary landfills: I. Model formulation," *Waste Management & Research*, ISWA, 14, 483-504.

Hanson, J. L., Yesiller, N., and Kendall, L. A. (2005). "Integrated Temperature and Gas Analysis at a Municipal Solid Waste Landfill," *Proceeding of the 16th International Conference on Soil Mechanics and Geotechnical Engineering,* Millpress Science Publishers, Rotterdam, the Netherlands, Vol. 4, pp. 2265-2268.

Hanson, J. L., Edil, T. B., and Yesiller, N. (2000). "Thermal properties of high water content materials," *ASTM Special Technical Publication 1374 Geotechnics of High Water Content Materials*, Edil, T. B. and Fox, P. J., eds., ASTM, West Conshohocken, PA, 137-151.

Liu, W.–L. (2007). *Thermal Analysis of Landfills,* Ph.D. Thesis, Wayne State University, Civil and Environmental Engineering Department, Detroit, Michigan.

ORNL (1981). *Regional analysis of ground and above-ground climate. Oak Ridge National Laboratory Report* No. ORNL/Sub-81/40451/1, U.S. Department of Energy, Office of Buildings Energy R&D.

Rowe, R. K. (2005). "Long-Term Performance of Contaminant Barrier Systems," *45th Rankine Lecture, Geotechnique*, 55 (9), 631-678.

Southen, J. M. and Rowe, R. K. (2005). "Modelling of thermally induced desiccation of geosynthetic clay liners," *Geotext. Geomembr.*, Elsevier, 23 (5), 425-442.

U.S. EPA (2005). *Municipal Solid Waste Generation, Recycling, and Disposal in the United States: Facts and Figures for 2003*, EPA530-F-05-003, Washington, D.C.

Yesiller, N., Hanson, J., and Liu, W. (2005). "Heat generation in municipal solid waste landfills," *J. Geotech. Geoenviron. Eng.*, ASCE, Vol. 131, Issue 11, 1330-1344

Yoshida, H. and Rowe, R. K. (2003). "Consideration of landfill liner temperature," *Proceedings, Sardinia 2003, Ninth International Waste Management and Landfill Symposium*, Christensen, T. H. et al., eds., CISA, Italy.

Capping a Toothpaste Landfill and Constructing a New Landfill on Top

Gerald L. DeMers[1], M. ASCE, PE

[1]Senior Engineer, Symbiont Science and Engineering, 6737 W. Washington Street, Suite 3440, West Allis, Wisconsin 53214, Jerry.demers@symbiontonline.com

ABSTRACT: A manufacturer of silica gel was required by state regulators to close its existing filter cake landfill. The closure design needed to be flexible so that a new landfill could be constructed on top of the old one. A geotechnical evaluation of the existing filter cake was performed to determine the feasibility of constructing a landfill cap or a new landfill on top of the existing facility and to determine the amount of consolidation expected. The geotechnical evaluation determined that either closure or expansion was feasible if the site was graded to accommodate the consolidation of waste materials. About 0.7 m of consolidation would occur due to construction of the cap, and about 1.5 m if the tallest possible new landfill were constructed on top. Construction of the multiple layer cap and liner was completed in spring 2007.

INTRODUCTION

An inorganic chemical manufacturer in the eastern United States produces amorphous silica dioxide (silica gel). For the past 20 years, it has used an on-site landfill for the disposal of filter cake generated from the wastewater treatment plant that treats effluent from this chemical process. Amorphous silica dioxide is a major component of toothpaste, and it is also used in a variety of applications as a desiccant because of its hygroscopic properties. The filter cake generated by the wastewater treatment system typically had a solids content of about 15 % when placed in the landfill. Dredged solids from an effluent settling pond were also occasionally placed into the landfill and had lower solids content. Filter cake that was this wet was too soft to be spread with a low ground pressure dozer; the moisture in the filter cake was allowed to evaporate before it was graded with a dozer.

The landfill was constructed in the mid-1980s without an engineered liner or leachate collection system. The landfill is a single cell, approximately 4.2 ha (10.3 ac) in size. It was excavated in the side of a hill, with a berm constructed opposite the hill to create a 6 m (20 ft) deep unlined basin. Soil borings indicated that the site is underlain mainly by silty sand (SM) with some gravel or clay. The design elevation of the base of the existing landfill is 2 to 3 m above the water table. Groundwater monitoring and modeling indicated that the impacts to groundwater and potential impacts to nearby surface water after 20 years of operation have been negligible.

Because the landfill was unlined, state regulators required its closure by capping with a low-permeability cover. As filter cake would continue to be generated and off-site disposal was substantially more costly, continued on-site disposal of the filter cake was preferred. However, the only available space was on top of the former filter cake landfill. As a result, the closure design for the landfill needed to provide flexibility so that a new landfill could be constructed on top of it if approved by state regulators. However, the decision by regulators to approve or deny the landfill expansion would not be made until construction of the cap/liner was well underway.

The challenges related to the design for this facility were:

- Flexibility must be maintained to be able to close or expand the site;
- The high moisture content of the filter cake (initially 15% solids) resulted in very soft material, unable to support the weight of construction equipment, and the filter cake was thixotropic, becoming unstable under vibration;
- Drying the filter cake resulted in considerable shrinkage and consolidation;
- The filter cake would consolidate under the weight of a landfill cap or a new landfill above the old;
- The landfill was not "full' in that the berm surrounding it was higher than the waste materials within it, thereby preventing stormwater from draining from its surface; and
- The closed landfill surface required a minimum slope of 4% in order to drain water and prevent ponding.

GEOTECHNICAL EVALUATION

Because of the characteristics of the filter cake, an investigation of the materials in the existing landfill was performed to determine how its properties would affect the closure and expansion designs. The following were performed to evaluate the geotechnical characteristics of the materials:

- Six boreholes through the filter cake and into native materials were drilled, logged and sampled;
- Laboratory samples were analyzed for moisture content, liquid limit, specific gravity and consolidation;

Additional boreholes and testing were conducted of the soil berm to determine its adequacy to withstand the weight of additional materials in an expanded landfill.

Results of Standard Penetration Tests in waste gel were very low, ranging from only the weight of the hammer pushing the sampler the full 0.6 m (24 in), to 2 or 3 blows. Moisture content of the waste gel ranged from 67% to 296% on a dry weight basis (see Table 1). Samples were found to be non-plastic, and plastic limit tests were not performed. Liquid limits test results ranged from 115% to 190%. Specific gravity was determined to range from 2.40 to 2.61. Ten consolidation tests were performed on Shelby Tube samples, with Compression Index (C_c) results ranging from 0.52 to 3.05. A relationship was established between the C_c and the moisture content of the samples. C_c generally increased with depth within the landfill, with a stiff layer present between 1.2 and 2 m (4-6 ft) at several locations. Coefficients of Consolidation could not reliably be calculated because the time-settlement curve exhibits a concave downward shape in the early portion of the curve.

Two scenarios were evaluated in the geotechnical analysis based upon site closure with a final surface slope of 2%, and an expanded landfill cell with side slopes of 33% (3:1 horizontal to vertical). These two slopes represent the extreme ranges for the final surface of the landfill, and were used in order to determine the range of consolidation that could be expected over this range of loading. The load on the waste materials from the landfill cap would consist of about 1.5 m (5 ft) of soils, while the expanded landfill could create a loading on the existing wastes equal to that of the cap plus up to 30 m (100 ft) of additional waste materials.

The assigned parameters in Table 1 were used to calculate new loading conditions for the waste profile at each borehole, and with the assigned C_c values and the initial void ratio values from the laboratory consolidation tests, the magnitude of the settlement of the filter cake was calculated. An example calculation for the maximum expected settlement is provided in Table 2. For the closure scenario, a maximum settlement of approximately 0.7 m (2.3 ft) was calculated. For an expanded landfill, a maximum of 1.5 m (4.5 ft) of consolidation was estimated. Additional filter cake material is necessary at the center of the site to accommodate the expected consolidation.

In order to construct either a cap or liner on top of the filter cake, a foundation fill layer would be needed to operate heavy equipment on the waste materials. Once the filter cake materials have been graded to their proper topography, placement of foundation fill soils could commence at one end of the site, working toward the other end (or to the center) by traveling over the newly placed compacted fill. Mudwaves of soft waste gel could be expected to well up in front of the advancing fill and these could be relieved from time to time by excavation and deposition in a designated area of the landfill which would be closed last. The use of a geofabric over the prepared landfill surface prior to placing foundation fill would minimize the generation of mudwaves and the amount of fill required.

LANDFILL CAP/LINER TOPOGRAPHY

The desired configuration for a final landfill surface is a mound whose surface is above the berm around the landfill. Based on the Geotechnical Evaluation, the initial slope for the mound would be 6%, resulting in a 4% slope after the filter cake has consolidated. However, the amount of filter cake was insufficient to create a mound

Table1. Assignment of Geotechnical Parameters

Boring	Depth (m)	Moisture Content (%)	Liquid Limit LL	Moist Density (kg/m³)	Compression Index Cc
B-2	0.6- 1.2	84, 115		1320	0.2 [b], 0.52 [a]
	1.2- 1.8	67			0.2 [c]
	1.8- 2.4	108	171	1250	1.45 [b]
	3.6- 4.2	201		1050	2.50 [a]
B-4	0.6- 1.2	158, 140	170	1090	1.81[a], 1.44[b], 1.7[c]
	1.2- 1.8	125		1200	0.2 [c]
	3.0- 3.6	221	148	960	1.24 [b], 2.4 [c]
	3.6- 4.2	202		1000	2.22 [a], 2.0 [c]
B-5	0.6- 1.2	237	152	930	1.27 [b], 2.75 [c]
	1.2- 1.8	178, 144		1120	1.48 [a], 1.4 [c]
	1.8- 2.4	124		1200	0.2 [c]
	2.4- 3.0	296	190	1090	1.62 [b]
	3.0- 3.6	253		1070	3.1 [c]
	3.6- 4.2	262		1080	3.05 [a]

[a] Laboratory Test
[b] Skempton Liquid Limit vs. C_c (Peck, 1974)
[c] Laboratory test vs. moisture content

shape; the actual quantity of material would result in a depression. If the landfill was to be closed, without the possibility of expansion, then site drainage could be attained by lowering the berm adjacent to the north side of the site. However, lowering the berm would significantly reduce the disposal capacity of a new landfill cell. Therefore, the final waste surface, and the resulting layers of the cap were designed with a shape resembling a "lemon juicer." The landfill surface has a central mound, falls to a low point in drainage channels, and then rises to the existing berm at the perimeter of the site.

The grading plan for the top of wastes has the following overall design features:

- The maximum slopes are 5:1 (20%). These maximum slopes are at the perimeter of the landfill, and the slopes are downward toward drainage swales that run parallel to the perimeter of the site;
- The minimum design slope is 6%. For a landfill cap the minimum acceptable slope after consolidation is 4%, and the maximum expected consolidation of 1.5 m (5 ft) requires that the slope is initially constructed with a 6% slope;
- The swales are underlain by an internal drainage pipe beneath the cap to collect liquids that may be squeezed from the existing filter cake landfill; and
- The swales are also underlain by a drainage pipe within the drainage layer to collect infiltrated precipitation (for the closed landfill) or leachate (from the expanded landfill).

Table 2. Settlement Analysis, Landfill Closure and Expansion

Boring No.	Layer No.	C_c, compression index	Unit Weight, γ, kN/m^3	Layer Thickness, m	Initial Void Ratio, e_0	P_0, initial pressure at layer midpoint, kN/m^2	H_c, thickness of cap, m	ΔP_c, pressure change due to cap, kN/m^2	H_e, thickness of new waste, m	ΔP_e, pressure change due to new waste, kN/m^2	S_c, settlement due to cap, m	S_e, settlement due to new waste, m
B-2	1	0.52	12.9	1.8	3.05	11.9	2.0	27.7	9.7	123	0.12	0.25
	2	1.45	12.3	0.6	3.05	27.5	2.0	27.7	9.7	123	0.07	0.16
	3	2.5	10.3	2.1	6.14	42.3	2.0	27.7	9.7	123	0.16	0.44
										TOTAL	0.35	0.85
B-4	1	1.8	10.7	1.2	4.73	6.6	2.1	30.2	11.0	138	0.29	0.52
	2	0.2	11.8	1.2	4.73	20.3	2.1	30.2	11.0	138	0.02	0.04
	3	2.22	9.8	2.9	6.64	41.7	2.1	30.2	11.0	138	0.20	0.54
										TOTAL	0.51	1.10
B-5	1	1.48	11.0	1.8	4.44	10.0	2.7	38.5	16.2	203	0.34	0.66
	2	0.2	11.8	0.6	4.44	23.6	2.7	38.5	16.2	203	0.01	0.02
	3	3.05	10.6	2.9	7.39	42.6	2.7	38.5	16.2	203	0.29	0.80
										TOTAL	0.64	1.48
Max[1]	1	1.48	11.0	0.6	4.44	3.3	4.6	64.7	30.8	388	0.22	0.34
	2	0.02	11.8	0.6	4.44	10.3	4.6	64.7	30.8	388	0.02	0.04
	3	3.05	10.6	2.3	7.39	26.0	4.6	64.7	30.8	388	0.45	1.00
										TOTAL	0.69	1.38

γ_c- Unit weight of new cap is 14.1 kN/m^3
[1]Max is maximum settlement at point of thickest new fill

γ_e - Unit weight of new waste is 12.6 kN/m^3

INTERNAL DRAINGE SYSTEM

Much of the anticipated consolidation will be the result of water being squeezed from the pore space between the filter cake particles. In order to prevent the buildup of water pressure within the foundation fill layer, it was designed to remove water more quickly than the rate it is generated from the filter cake below. This was accomplished by constructing the foundation fill layer with a hydraulic conductivity that is greater than the filter cake. The filter cake has a measured hydraulic conductivity of about 5 x 10^{-5} cm/sec; the foundation fill soils are specified to have a hydraulic conductivity of at least 1 x 10^{-3} cm/sec.

The water generated from consolidation of the wastes will be transported to near the perimeter of the landfill for removal by perforated drainage piping. The piping is directed to two sumps at low points in the landfill. Any liquid collected in the sumps will be pumped to a settling pond for treatment.

TEST PLOT CONSTRUCTION

In order to determine the response of the existing waste material to the placement of the foundation fill and to determine the accessibility of earthmoving equipment over the regraded waste material, a test plot was constructed on a typical portion of the landfill. The purpose of this test plot was to determine the thickness of the foundation fill layer needed for liner installation on the closed landfill cell.

The test plot area was approximately 23 m by 23 m (75 ft by 75 ft), and was divided into three pads, each about 8 m (25 ft) wide. A single layer of woven geotextile was placed over the test plot area, with 0.3 m (1 ft) wide overlaps, as recommended by the manufacturer. Three different materials were used: recycled crushed concrete, screened sand from a commercial gravel pit and a sandy, common fill from a borrow site. The three materials were selected to provide a reasonable range of available materials for a granular foundation fill layer.

Dump trucks initially emptied their respective loads immediately adjacent to the landfill perimeter, and the materials were spread with a front end loader. The materials were placed in 0.2 to 0.3 m (8 to 12 in) lifts, and each lift was compacted with a hydrostatic roller. Based on observations during material placement and compaction, each plot was tested by driving a loaded dump truck onto it to determine if the test plot was of sufficient thickness to support the truck. Moisture content and density tests were performed on the screened sand and common fill. In addition, two PVC standpipes were installed to the depth of the filter cake to detect any free moisture.

Based upon field observations, the waste solids were stable under the loads of the test plot fill, the construction equipment and the fully loaded ten-wheel dump trucks. No displacement of the existing waste material was observed during or after installation. Each test pad was capable of supporting construction traffic with no deflection or rutting of recycled concrete, slight wheel rutting of the sand, and slight wheel rutting and minor localized pumping in the common borrow. No moisture was detected at the completion of the test plots, or when checked two weeks later.

Based on the results of the test plot, earth moving equipment and truck access on top of the waste solids in the landfill is possible using crushed recycled concrete, compacted sand or a non-plastic sand and silt. The test pad suggested that 0.6 m (2 ft) of foundation fill would be adequate to support equipment and trucks, but intensely traveled areas such as haul roads may require greater thickness and routine maintenance. Because of the soft nature of the waste materials, a performance specification of at least three passes by non-vibratory compaction equipment was used for the initial lifts of foundation fill, rather than a minimum percent compaction. After a firm surface is created by these lifts of the foundation fill, then a percent compaction criterion was used with additional fill materials.

FINAL DESIGN

The cap for the closed landfill cell must also function as the liner of the expansion landfill cell. Portions of the cap system, such as the topsoil and part of the protective layer soils beneath it would be removed prior to the placement of wastes into the new landfill cell. The final design consisted of the following layers, from bottom to top:

- Woven geotextile to provide separation between the filter cake and the foundation layer soils;
- Drainage piping to collect liquids that may be squeezed from the filter cake;
- 0.6 m (24 in) granular foundation soils, with hydraulic conductivity at least ten times greater than the filter cake;
- Non-woven geotextile cushion to protect the lamination on the Geosynthetic Clay Liner (GCL) above it;
- Reinforced GCL with a composite lamination facing down;
- 1.50 mm (60-mil) Linear Low Density Polyethylene (LLDPE) membrane;
- 0.3 m (12 in) of drainage layer sand, which has a hydraulic conductivity of at least 1 x 10^{-3} cm/sec, and drainage piping;
- Non-woven geotextile filter to prevent fines from the protective layer from clogging the drainage layer;
- 0.5 m (18 in) protective layer soils, placed in a single lift and only nominally compacted
- 0.2 m (6 in) topsoil
- Seeding, fertilizing and mulch.

The filter cake contains a high concentration of calcium. GCLs contain sodium bentonite. If calcium is available to displace the sodium in the bentonite, the change in the chemical structure of the bentonite results in an increase in its hydraulic conductivity. The selected GCL included a laminated plastic layer on its underside to prevent calcium from the filter cake from reducing the conductivity of the GCL. The geomembrane above the GCL will prevent surface water (or leachate from the new landfill cell) from coming in contact with the GCL.

CONSTRUCTION

The filter cake surface was regraded over a period of several weeks in order to dry the filter cake and create the desired topography. Prior to construction, the waste surface was resurveyed. While additional wastes had been added to the site since the previous survey, the actual quantity of materials was significantly less than had been estimated. The regrading efforts to provide the proper shape to the waste surface had resulted in the filter cake drying out to a greater degree, with a resulting loss in volume. The dryer filter cake provided a firmer surface for cap construction and indicated that some of its anticipated consolidation had already occurred. Because of the reduced quantity of waste materials, the surface slope of the regraded waste was reduced to 5%. In order to provide sufficient slope for the landfill surface, additional foundation fill was placed at the center of the site. Construction began in mid-2006 and was completed in spring 2007.

LANDFILL EXPANSION APPROVAL

The landfill expansion was approved by state regulators during construction of the cap for the existing landfill. The expanded landfill will have 25% side slopes, and will be developed over four phases of operation. The first phase of operation consists of the northwest quadrant of the site. The topsoil and 0.2 m (6 in) of the protective layers soils were not placed on this portion of the site. The expanded landfill began operation in the spring of 2007.

SUMMARY AND CONCLUSIONS

A landfill cap was designed with the flexibility that a new landfill could be constructed on top of the old one. A geotechnical evaluation of the existing filter cake was performed to determine the amount of consolidation expected for either capping or expanding the landfill. The anticipated consolidation for both closure and expansion conditions were calculated using compression indices, with about 0.7 m of consolidation due to construction of the cap, and about 1.5 m if the tallest possible new landfill were constructed on top. The landfill was designed with an initial slope of 6%, which would be reduced to 4% as materials consolidated. A multilayer cap/liner was constructed, including an internal drainage system and foundation layer.

REFERENCES

Peck, R. B., Hanson, W. E., Thornburn, T. H., (1974), *Foundation Engineering*, Second Edition, John Wiley and Sons, Inc., New York, New York.

Dry disposal of bauxite residues in abandoned mine open pits

Anthimos Xenidis[1] and Dimitrios Boufounos[2]

[1]Ass. Professor, Lab. of Metallurgy, School of Mining and Metallurgical Engineering, National Technical University of Athens, GR 157 80 Zografos, Athens, Greece; axen@central.ntua.gr
[2]Director, Health, Safety and Environment, Aluminium of Greece SA, Agios Nikolaos, Paralia Distomou, GR 32003 Viotia, Greece; dimitri.boufounos@alhellas.gr

ABSTRACT. Bauxite residues exit the process stream as a highly alkaline slurry with low solids content rendering their land disposal a difficult issue. The objective of this study was to develop a methodology for environmentally safe dry disposal of bauxite residues in abandoned mine open pits. The methodology includes dewatering and controlled disposal of bauxite residues, capping with waste rock or treated bauxite residues and, finally, development of a vegetation cover. This methodology is mainly based on bauxite residue properties, which possess low hydraulic conductivity if compacted at optimum moisture content. Several parts of this scheme were investigated at laboratory and pilot field scale. Simulation of an abandoned mine that was planned to be rehabilitated was also performed. Based on the results obtained, the main risk due to bauxite residues disposal is associated with the alkalinity of the material and initially high Na and Al concentrations in the leachate that progressively decrease due to the depletion of these elements. Both the field tests and simulation indicate that the amount of drainage water will be minimal, i.e. approximately 3% of the annual precipitation.

INTRODUCTION

One of the most challenging problems the alumina industry faces today is the disposal of bauxite residues (BR), which are the main by-products of bauxite ore processing by the Bayer alumina extraction process. Bauxite residue, also called alumina red mud, is a red-colored, highly alkaline material composed of fine particles of silica, aluminium, iron, calcium and titanium oxides and hydroxides. The amount of bauxite residues is significant and depends on the quality of the bauxite ore used. Bauxite residues exit the process stream as a highly alkaline slurry (pH 10-12.5) with low solids content (15-30%) making surface disposal difficult.

The Aluminium of Greece refinery is the only aluminium refinery in Greece. This refinery produces almost 900 kg of bauxite residues (dry wt.) per ton of produced alumina, which corresponds to 680,000 tonnes of residue annually. The residue that has to be annually disposed of in slurry form with 300 - 400 g /L solids concentration is pumped through pipes from the refinery to the sea bed of the adjacent gulf of

Corinth. In order to avoid any kind of environmental impacts related to the disposal of bauxite residues in the sea, finding alternative, environmentally friendly, technologically feasible and cost effective land disposal methods is necessary. This study focuses on the development, pilot testing and field application of a method for disposing bauxite residues in abandoned surface mines. Although bauxite residue is an alkaline waste, the environmental risk associated with the containing alkalinity from such a disposal option is low. This is due to the very low hydraulic conductivity of bauxite residues, which is slightly higher than the limit posed for low permeability layers (i.e. 1×10^{-7} cm/s). However, densification of the bauxite residues is required in order to act as a low permeability layer. By acting as a low permeability layer and taking into account the low precipitation and high evapotranspiration rates at the sites, water percolation through bauxite residue layers is estimated to be minimal. In order to test this hypothesis, pilot field tests were conducted and numerical simulation of both the pilot tests and the abandoned Kleisoura mine proposed restoration were performed.

BAUXITE RESIDUES PROPERTIES

The main geotechnical parameters and geochemical properties of bauxite residues determined are summarized in Tables 1 and 2 respectively. Based on the Unified Soil Classification System (USCS, ASTM D2487),the bauxite residues are characterized as a fine-textured material given that 86% dry wt. of the residues are comprised of particle sizes finer than 75μm. According to the USCS, the material is classified as a ML category, i.e. silt with low plasticity.

The optimum moisture content (OMC) using the standard compaction (ASTM D698) was 28.4% wt dry basis. The corresponding maximum dry density was $\gamma_d =$ 1.506 t/m^3. The OMC is approximately 2% less than the plasticity index. Falling head permeability measurements were conducted according to ASTM 5084-90, with a flexible wall permeameter (Tri-Flex 2 ELE) using water as the permeant liquid and applying hydraulic gradient equal to 10 and 20 cm/cm and effective stress equal to 45 kPa. Permeability measurements indicate that bauxite residues possess low hydraulic conductivity (k) values ranging from 3×10^{-5} to 4.6×10^{-7} cm/s depending on the compaction conditions. The lower k value of 4.6×10^{-7} cm/s was obtained when bauxite residues were compacted at OMC. This k value is slightly higher than the limit posed in the European landfill directive for low permeability layers, i.e. 1×10^{-7}cm/s (1999/31/EC, 26-04-1999). By acting as a low permeability layer and taking into account the low precipitation rates at the sites into consideration, water percolation through bauxite residue layers is estimated to be minimal. When bauxite residues are compacted to 95% of the maximum dry density, the hydraulic conductivity was found to be 1.3×10^{-6} cm/s, whereas when the material was compacted at a moisture content of 30.4 % (2% above OMC), the permeability coefficient was 9.1×10^{-7} cm/s, i.e. approximately twice the value obtained when compaction was performed at optimum conditions. The variation in hydraulic conductivity with time, compaction and moisture content is given in Fig. 1. Permeability measurements conducted under various vertical stresses indicated that permeability coefficient as low as 1×10^{-7}cm/s can be obtained when vertical stress equal to 400 kPa is applied.

Table 1. Geotechnical properties of bauxite residues

Parameter / Test	Results
Composition (% wt.)	
Clay (<2 μm)	40
Silt (2-60 μm)	45
Sand (60-600 μm)	15
Atterberg Limits (ASTM D 4318)	
Liquid Limit, LL (%)	35
Plasticity Index, PI (%)	5
Classification (USCS)	ML
Standard Proctor method	
Optimum gravimetric moisture (% wt. db)	28.4
Maximum dry density (t/m^3)	1.506
Specific gravity (t/m^3)	2.92
Hydraulic conductivity, k (cm/s) *(standard compaction at optimum moisture)*	4.6×10^{-7}

Table 2. Leaching characteristics of bauxite residues

Test / Conditions	Results
US EPA TCLP: Single extraction with acetic acid pH:2.88, particle size: -9.5 mm, Liquid / Solid (L/S)=20 L/kg, 18h	Concentration (μg/L) of Cd (1.6), Cr (205), Pb (<5), Ag (4.5) < TCLP limits, < Greek effluent discharge limits Cd, Pb, Ag < Drinking water standards
DIN 38414 S4 (German): Single extraction with de-ionized water, particle size: -10 mm, L/S=10 L/kg, 24h	Concentration (μg/L) of Al (89500): 3-18 × Greek effluent discharge limit Concentration (μg/L) of Cd (<0.2), Cr (265), Cu (120), Ni (11), Pb (<5), Zn (<40) < effluent discharge limits Concentration (μg/L) of Cd, Cu, Ni, Pb, Zn < Drinking water standards
NEN 7341 (Dutch): Two extractions with HNO_3 solution: 1st stage pH:7, 2nd stage pH: 4, particle size: -125μm, L/S = 50 L/kg, 2×3 h	Availability (mg/kg) of Al (1988) > Cu (30) > Zn (8.5) > Cr (3.4) > Ni (1.6) > Pb (<0.5) > Cd (0.05)
EN 12457-02 (European): Single extraction with deionized water, particle size: -4 mm, Liquid / Solid (L/S)=10 L/kg, 24h)	Average leachability (mg/kg): As (0.37), < Cd (<0.004), Cr (<0.01), Cu (0.028), Mo (0.028), Ni (<0.05), Pb (<0.05), Zn (<0.4) < EC criteria for acceptance in inert waste landfills

Bauxite residues mainly consist of iron (48%), aluminium (15.85%), calcium (14.84%), silicon (6.96%) and titanium (7.06%) oxides. They also contain lower percentages of sodium (3.26% Na_2O), potassium (0.07% K_2O) and magnesium (0.24% MgO) and numerous trace elements (e.g. Cr, Ni, Mn, As, Zn) with Cr and Ni presenting the highest contents (1927 and 1145 mg/kg respectively). pH and electrical conductivity (EC) measured on a saturation extract of bauxite residues were found to be 12.81 and 7.66 mS/cm respectively. Based on the NEN 7341 test results, 0.7 m-greq HCl/g red mud are required to reduce bauxite residues pH to 7. The cation exchange capacity (CEC) was found equal to 38 meq/100 g. The leaching behavior of elements in the bauxite residues was examined by undertaking batch and kinetic tests. Standard batch tests performed include a) the US EPA TCLP b) the Dutch NEN 7341 for the determination of the maximum availability (long-term leachability) of elements, c) the German DIN 38414 S4 leaching test developed for characterisation of wastes, sludges and sediments, and d) the European EN 12457-02 test for the determination of leaching properties of granular waste materials. The leaching tests results were compared with drinking water limits, Greek effluents discharge limits, or EC criteria for acceptance of wastes at landfills (Table 2). Although there are differences on the results obtained from the various leaching tests due to the different conditions applied, the main elements dissolved from bauxite residues are sodium (Na) and aluminium (Al). For all of the other elements, leachability is very low compared to Al and Na.

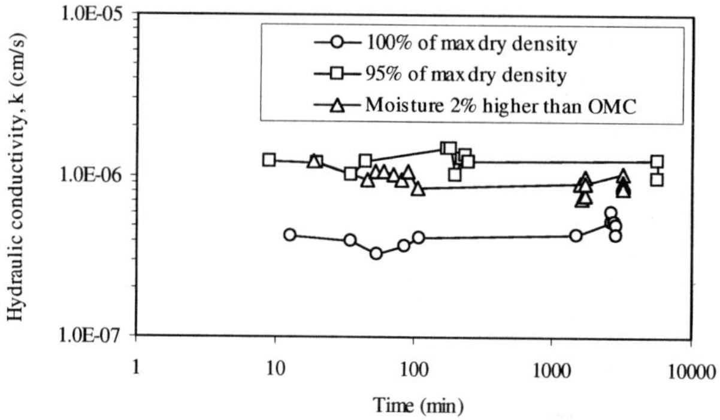

FIG. 1. Variation of hydraulic conductivity with time at various compaction rates and moisture contents (hydraulic gradient 10 cm/cm)

Kinetic tests were performed to estimate the behavior of bauxite residues under conditions simulating their disposal in open pits. The tests were undertaken in columns of diameter 20 cm and height 80 cm and cells of diameter 10 cm and height 15 cm. Cell tests were conducted primarily to evaluate the long-term behavior of bauxite residues, whereas column tests were performed to simulate bauxite residue disposal in mine open pits. For this reason, column testing involved consecutive wet-dry cycles and monitoring of the volume and quality of leachates produced.

The pH of leachates from the cell tests was initially very alkaline and progressively decreased to values close to 9.5. The Al and Na concentrations were high (800 and 4000 mg/L respectively) at a low liquid to solid ratio, L/S, of 1 L/kg and dropped dramatically at higher L/S ratios. These values exceed Greek effluents discharge limit and drinking water standards. Any soluble compounds in the bauxite residues were easily depleted. Therefore, drainage from bauxite residues at a disposal site is expected to be initially alkaline with high concentrations of Al and Na that will progressively decrease over time to lower values.

The column test results were similar to those obtained from the cell tests. However, the L/S ratios obtained (0.2-1.6 L/kg) were lower than those of the cell tests (2-12 L/kg). Due to the low permeability of bauxite residues, the infiltration rate was low and consequently the amount of leachates collected from column tests was lower than that obtained from the cell tests where additional pressure of 10 kPa was applied. Although there was a progressive decrease of pH, EC and Al and Na concentration, values for these parameters were lower than those obtained in the cell tests, due to the lower quantity of infiltration.

FIELD PILOT TESTS

Pilot tests were implemented in the abandoned Prosorema mine, situated near Desfina, Fokida, Central Greece. The test covers an area of 580 m^2 which was divided with a geomembrane into two equal sections or test pads (TP) (Fig. 2 and 3). Lysimeters were installed in the central part of each test pad to avoid edge effects and to collect and monitor the drainage water. PVC pipes were used to transfer drainage water out of the lysimeters for separate collection. Water-collection pipes and vessels were also installed at the sides of each test pad to collect and monitor the run-off water quantity and quality. In the first test pad, a 1.6 m thick bauxite residue layer was applied, whereas in the second a 0.6 m thick layer of waste rock was placed above a 1 m thick bauxite residues layer. The first test pad (TP 1) simulated conditions of the disposal site during operation, whereas the second pad simulated conditions at the disposal site after capping and the establishment of a vegetation cover. The field tests have a 10 × 10m base, 1.6m height and 3:1 side slopes to improve the physical stability of the material and decrease wind and water erosion rates (Boufounos et al., 2005, Fig.2 and 3).

Bauxite residues were disposed in layers of 20-30 cm with each layer compacted by a drum roller. Samples were taken during placement for moisture content determination. In-situ measurements of bulk density were carried out to ensure that dry density of the compacted material was at least 95% of the standard maximum ratio. In the second test pad, waste rock mixed with sewage sludge at a ratio of 6 kg of dry matter per square meter was placed on top of the 1 m thick bauxite residues layer up to a total height of 1.6 m. Finally, annual and perennial grass varieties were seeded to develop a vegetation cover. Field tests were monitored for two years (1-6-2004 to 31-5-2006). The main parameters monitored were drainage and run-off water quantity and quality, moisture content profiles by TRIME-FR TDR and meteorological data.

The results of drainage quantity collected from TP 1 and TP 2 as compared to precipitation are presented in Fig. 4. Although total precipitation in the area during the first year was 570 mm, only 15 and 17 mm of water percolated through the material in TP 1 and TP 2 respectively and was received in the drainage collection

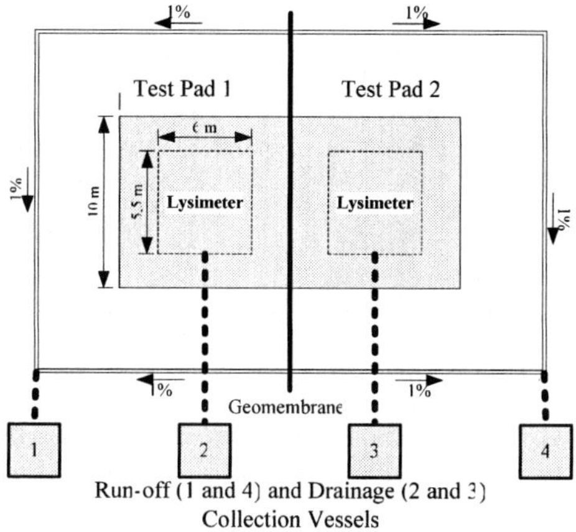

FIG. 2: Design of pilot tests

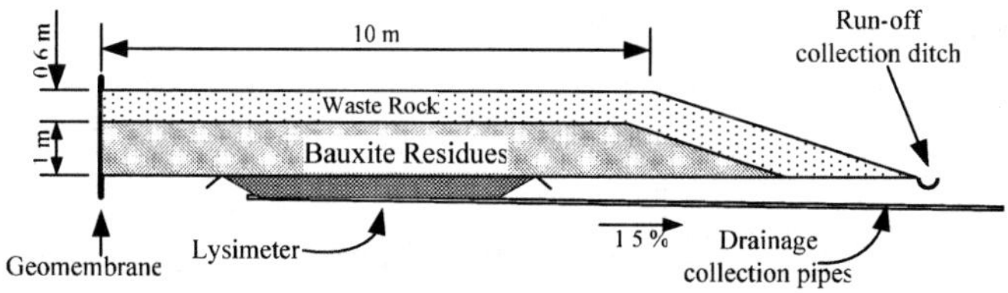

FIG. 3. Cross section of Test Pad 2 (waste rock layer on top of BR)

vessels in early April 2005. This amount of drainage water corresponds to less than 3% (2.7 and 2.9 % for TP 1 and TP 2 respectively) of the annual precipitation. The quantity of drainage water collected during the second year of monitoring was higher than this value. However, taking into account the data from two years of monitoring, the average annual percolation rates for the TP 1 and TP 2 were 23 and 30 mm, respectively, corresponding to only 4 and 5.3 % of the total precipitation. In accordance with kinetic test results, elevated pH (9-12.5), Na (2500±230 mg/L) and Al (250±17 mg/L) concentration in the leachates were monitored which were higher than the respective effluents discharge limits. Na and Al concentrations in the run-off water was above effluents discharge limits only in the case of TP 1

DESIGN OF KLEISOURA MINE RESTORATION

Full-scale restoration of an abandoned bauxite surface mine is planned in order to demonstrate the effectiveness of the proposed remediation scheme. This mine is located at Kleisoura, 4 km from the alumina refinery at Agios Nikolaos, Viotia. According to the rehabilitation design, an area of 14150 m^2 will be filled with dried bauxite residues up to a depth of 28 m (Fig. 5).

The final rehabilitation design of the Kleisoura mine involves near flat areas (2% slope, light grey areas in Fig. 5) and areas with slopes of 1:3 (vertical to horizontal, dark grey areas in Fig. 5). Geotechnical analyses indicated that a slope of 1:3 or less is required to ensure the stability of the impoundment. As in the case of almost all surface mines of this type, there is a significant altitude difference between the North-West and Southern parts of the pit. The design included a methodology for disposal starting from the lower part (South part) of the open pit taking advantage of existing roads to transfer and dispose the residues. Furthermore, water deviation works have been included in the design, to limit surface water access to the disposal site. Trenches are also proposed for the periphery of the disposal site and at the lower part of the slopes, to collect and remove run-off water from the disposal area. The scheme investigated for the final capping includes a drainage layer which will be constructed with the coarse fraction of waste rock and protective and vegetation layer that will be constructed with properly modified waste rock or bauxite residues.

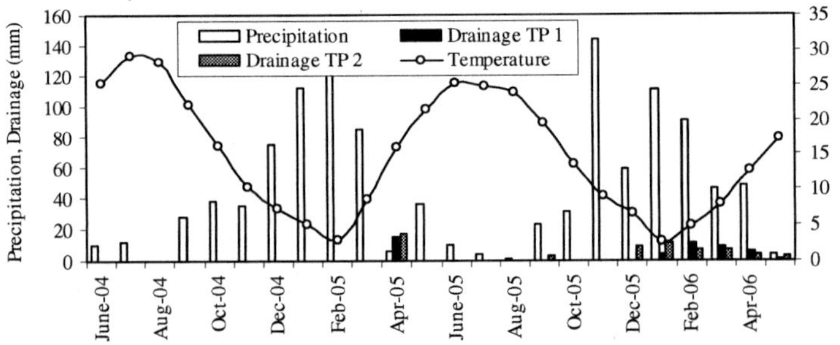

FIG. 4. Variation of temperature, precipitation and drainage from test pads

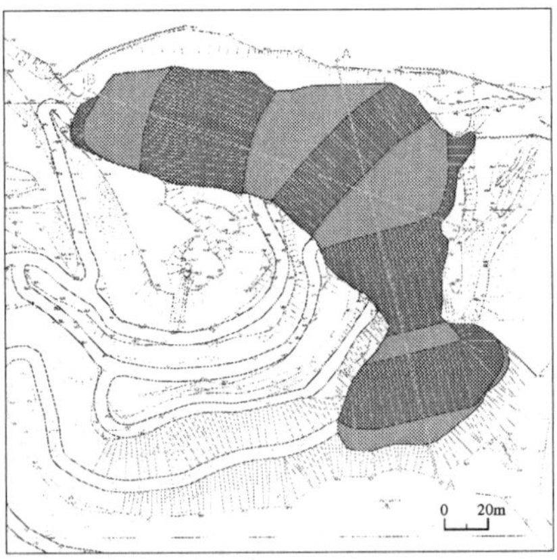

FIG. 5. Design of Kleisoura mine restoration

SIMULATION

Two models were used to determine a detailed hydrological balance under a variety of input parameters for the surface mines to be rehabilitated using bauxite residues as a filling material. These models were the finite difference code "Visual ModFlow" and the "Visual HELP" model, both developed by Waterloo Hydrogeologic Inc.

For the Kleisoura mine, a reference scenario was developed using ModFlow based on the most probable parameter values. Five additional scenarios were developed by varying one parameter at a time (Vavadakis et al., 2005). The parameters investigated include the hydraulic conductivity of all the layers (bauxite residues, gravel, vegetation layer), the specific porosity and water retention capacity, the annual precipitation, the percentage of vegetation coverage and the layers thickness. In all the cases, at least 3 values of each parameter were examined. The reference scenario resulted in total annual percolation rate corresponding to 3.1% of the annual precipitation. The respective value for all the other scenarios was between 2.3-4.6 %.

Simulation of the water infiltration for the pilot tests using ModFlow gave a prediction of 4.4% which is close to the infiltration values (less than 3% of the total annual precipitation) measured in corresponding field tests. Results obtained using Visual HELP simulation package were similar to ModFlow. The predicted by Visual Help total amount of drainage water from TP1 with a hydraulic conductivity of 4.6×10^{-7} cm/s, was estimated to 2.39% of the annual precipitation. This is close to the corresponing value obtained using ModFlow (4.4%).

CONCLUSIONS

A cost-effective method for restoring abandoned surface mines mainly using dried bauxite residues as a filling material was investigated. The results of field-test monitoring and simulation of the Kleisoura mine planned to be rehabilitated and pilot-scale tests using Visual ModFlow and Visual HELP confirmed the hypothesis that the amount of water infiltrating through bauxite residues and released to the environment should be minimal, i.e. approximately 3% of the annual precipitation. Leaching tests indicated that drainage from bauxite residues at a disposal site is expected to be initially highly alkaline with high concentrations of Al and Na that will progressively decrease over time to lower values.

ACKNOWLEDGEMENTS

The authors would like to acknowledge the financial support of the European Commission under the LIFE-Environment, Contract No. LIFE 03 ENV/GR/000213

REFERENCES

Boufounos, D., Pernientaki, I., Fafoutis, D., et al (2005). "Use of bauxite residues for the rehabilitation of abandoned surface mines" *Int. Conf. Geoenvironment and Geotechnics* (Z. Agioutantis & K. Komnitsas, eds.), Milos, Greece, pp. 183-188.

Vavadakis, D., Agioutantis, Z., A. Xenidis et al. (2006). "Simulation of bauxite residues disposal in abandoned mine open pits" *AMIREG 2006*, (Z. Agioutantis and K. Komnitsas eds.), Hania, Greece, 2006, pp. 219-226.

Studies on the Integrity of Clay - Rich Soil Liners for Cover Systems of Landfills

B. V. S Viswanadham[1], S. Rajesh[2] and S. S. Sengupta[3]

[1]Associate Professor, Department of Civil Engineering, Indian Institute of Technology Bombay, Mumbai – 400 076, India; viswam@civil.iitb.ac.in
[2]Research Scholar, Department of Civil Engineering, Indian Institute of Technology Bombay, Mumbai - 400 076, India; rajesh.s@iitb.ac.in
[3]Formerly Post-Graduate Student, Department of Civil Engineering, Indian Institute of Technology Bombay, Mumbai-400 076, India

ABSTRACT: An impervious barrier formed with clay-rich soil is being used widely in cover systems to contain municipal and hazardous wastes safely. The barrier layer is intended to control infiltration into the waste and hence leachate from the waste to the ground water. In this paper, experimental investigations on the study of failure of clay based landfill cover system due to non-uniform settlements is addressed. A short series of centrifuge model tests at 40g were carried-out on 30 mm thick clay-rich soil liners (without any overburden) by varying molding water content and compaction energies. With an increase in compaction energy, a slight delay in the initiation of cracking was observed. For clay-rich soil liner moist-compacted just above its optimum moisture content, a single crack was observed to occur up to full depth and widen gradually. Contrary to this, for a clay-rich soil liner compacted towards wet side of optimum moisture content, a slight delay in crack initiation with multiple cracking patterns extending up to full-depth was observed. Finally, measures for retaining the integrity of clay-rich soil liners for cover systems of landfills are highlighted.

INTRODUCTION

Landfill remains an essential part of any waste management system and in many countries the only economic form of Municipal Solid Waste (MSW) disposal. Once the filling of the MSW reached its limiting design value it has to be closed by providing recommended cover system. Nearly all currently used landfill cover systems (both in active and passive systems) incorporate an impermeable barrier (Benson et al. 2007). Various materials used for constructing barriers are clay-rich soils, synthetic membranes or other artificially manufactured materials, and amended soil or other admixtures. Clay rich soils can be considered as an effective and economical barrier material provided it has a coefficient of permeability of 1 x 10^{-9} m/s or less (due to its particle sizes, and plasticity characteristics). Long-term performance of Compacted Clay Liner (CCL) is a function of properties, such as low permeability, low diffusivity, relatively high ductility, high chemical compatibility and retardation, less degree of preferential flow paths, and good constructability. However, such CCLs may fail to perform satisfactorily for several reasons. For example, clay based cover systems are susceptible to climatic and frost action as well as deformations of the barrier itself caused by differential settlements of the contained wastes. Excessive differential settlements lead to cracking of clay barrier and impair the main function of a clay barrier. Other than this, cracking in clay can also be caused by seasonal wetting and drying cycles leading to desiccation cracks. Even though desiccation cracking due to seasonal variation severely affects the integrity of cover system, damage caused by settlements especially differential settlements is also more pronounced in cover systems (Sowers, 1972 and Ling et al. 1998). Hence, in this paper an attempt has been made to investigate the performance and integrity of clay based landfill cover system subjected to differential settlements in a centrifuge.

PREVIOUS STUDIES ON DIFFERENTIAL SETTLEMENTS

Sowers (1972) estimated that waste settles 10 – 30 % of its original height under its own weight within one to two years of its burial. Jessberger (1994) estimated that settlement can be as much as 40-50 % of original height. In Bioreactor landfill due to non-uniformity of recirculation of leachate by horizontal and vertical trenches, the occurrence of the differential settlements is quite natural and more pronounced. The problem due to differential settlements is more pronounced in case of a system than the bottom lining system because the former is subjected to low overburden pressure as well as settlement due to readjustment and ongoing bio-degradation of the wastes that lies underneath. Excessive differential settlements result in tension cracks in the zone of sharp curvatures thereby resulting in loss of integrity of the whole cover system. The cracking of CCL lead to a drastic increase of hydraulic conductivity of the clay barriers thereby increases the rainfall infiltration through the cover system and resulting in alarmingly large quantities of leachate as well as release of harmful gases into the environment. Hence, the whole lining system fails to perform satisfactorily.

Very few investigators have attempted to evaluate behavior of clay liners at the onset of differential settlements. Jessberger and Stone (1991) investigated the effect of overburden and the choice of barrier material on the response of model clay liners to the imposed deformations in centrifuge. Viswanadham and Mahesh (2002) have evaluated the deformation behavior of clay liners considering the effect of thickness, consistency and overburden in a small geotechnical centrifuge by simulating non-uniform settlements by dissolving a compacted salt placed in the center beneath the sacrificing layer at 150*g*. Viswanadham and Jessberger (2005) have brought out the significance of reinforcement effect on the performance of CCLs subjected to non-uniform settlements in a large beam centrifuge at 50*g*. To summarize from the literature review, it was reported that the depth of the cracks were found to decrease with an increase in the thickness of the clay liner. The suppression of cracks was reported for a clay-rich soil liner with nominal overburden but devoid of cracks was not observed. However, the effect of molding water content and compactive effort on the deformation behavior of clay-rich soil liners was not attempted. Hence, the scope of this present study is to investigate the effect of artificially induced differential settlements on the integrity and deformation behavior of a clay-rich soil liner in a large beam centrifuge.

CENTRIFUGE MODELING

Application of centrifuge modeling technique to the present study is relevant because of the loss of integrity of a clay-rich soil liner is highly influenced by the presence of prototype stress conditions. Centrifuge scaling factors relevant to modeling of clay liners and errors due to high acceleration field have been described elsewhere extensively by Viswanadham and Mahesh (2002) and Viswanadham and Jessberger (2005). The centrifuge tests reported here were performed at an acceleration field of 40*g*. The 4.5 m radius large beam centrifuge at Indian Institute of Technology Bombay (IIT Bombay) is being used for studying number of problems of importance in geotechnical and geo-environmental engineering. The centrifuge capacity is 250*g*-ton with a maximum payload of 2.5 t at 100*g* and at the higher acceleration of 200*g* the allowable payload is 0.625 t.

EXPERIMENTAL TEST SETUP

Model Material

A mixture of commercially available kaolin clay and locally available poorly graded sand in proportion of 80:20 (by dry weight) was used to serve as a model clay liner material. In the present study, model clay liner material was chosen such a way that it represents the properties of CCL material used in construction of lining systems for landfills (Benson et al. 1999). The properties of the material are in the same range as those of CCL material characteristics used in actual practice in many landfills in USA. The model clay liner was found to have a liquid limit of 38%, plasticity index of 16%, maximum dry unit weight and OMC with standard and modified Proctor compaction energies are 15.9 kN/m^3 and 22%; 17.6 kN/m^3 and 16% respectively.

Model Test-Package and Test Procedure

A container having internal dimensions of 720 mm x 450 mm x 410 mm was used in the present study. Figure 1 show a photograph of the test set-up used in the present study. The model test set-up includes two rigid supports of 180 mm length connected to two aluminum plates with a hinge connection. These hinged aluminum plates were made to rest on a trapdoor settlement plate connected to the piston of hydraulic cylinder symmetrically. The horizontal level surface was maintained between these two platforms and was verified using a spirit level. By using this arrangement it is possible to model a landfill area as large as 520m^2 in the prototype scale and simulates the ongoing differential settlements undergoing over large areas of landfills. A hydraulic cylinder was custom designed to induce differential settlements to the clay-rich model soil liner. The model was prepared at 1*g* with the cylinder at its full stoke under an operating pressure of 500 kN/m^2 and the flow control valves were kept closed and oil in the trapdoor cylinder was not allowed to move out of cylinder to cause premature settlements to the model soil liner. While increasing the gravity level from 1*g* to 40*g*, air pressure was increased from 500 – 550 kN/m^2 gradually to prevent any premature movement of the trapdoor plate. Desired settlement control rate was achieved by withdrawing the air pressure in the tank in steps and allowed the oil to flow out of the cylinder through a needle flow control valve. On an average, a settlement rate of 0.85 mm/min (in model dimensions) was achieved for all the tests and non-uniform settlements were induced continuously. This settlement rate may not be realistic when projected to the prototype dimensions but to some extent these settlement rates represent localized depressions and / or early period of settlements in bioreactor and/or sudden collapse of waste material or ground subsidence in the waste disposal sites. A miniature LVDT was attached to the trapdoor plate to monitor the movement of the trapdoor plate, which would provide the central settlement values starting from zero to a maximum of 25mm in model dimensions during the centrifuge test.

The model preparation involves placement of a thick non-woven geotextile layer cut into four pieces along the length of the container. A 30 mm thick coarse sand layer followed by 30 mm thick fine sand layer was placed in dry state by raining technique at a relative density of 55 %. These layers were introduced to eliminate any stress concentration, which may arise due to abrupt discontinuity at hinge locations and are referred herein as sacrificial layers. The kaolin-sand mixture was moist-compacted to achieve desired dry unit weight and molding water content (as given in Table 1) with proper care to make a model clay liner of uniform thickness of 30 mm (Models SSL-4, SSL-7 and SSL -10). Before placing the model clay liner, the underlying sand bed was pre-saturated and drained for about 9 – 10 hours duration at 1g. This pre-saturated drained sand provides a firm platform so that model clay liner can be prepared above this layer without any undulations.

Six numbers of Linearly Variable Differential Transformers (LVDTs) were used in all models except SSL-10 placed on the right half cross section of the clay liner at every 50 mm interval starting from its the center. Discrete markers were prepared using transparency strips having black circular portion (5mm diameter) with a white central dot on it. Such strips were folded in the form of L-shape and placed on 5mm below the top surface of the model clay liner (at 20mm center to center) such that black circular portion facing acrylic sheet (Fig. 1). Discrete markers are placed 5mm below the top surface of the liner to provide some confinement effect to the markers such that displacement of markers are purely due to the material behavior of the clay liner. In addition to front markers, a 20 mm square grid of top markers (made up of graphite pieces) of 2mm diameter were inserted on the liner surface along the zone of maximum curvature.

Two Charge Coupled Device (CCD) video cameras, one on the front side of the model and the other on the top over the model surface were installed to monitor the performance of the clay barrier during the centrifuge test. Through an image acquiring software, photographs were taken by these cameras and were later used for image analysis at different settlement levels. At the various stages of central settlements, photographs were taken through an image acquiring software and were later used for image analysis to compute deformation profiles and for arriving at strain distribution along the top fiber of the clay barrier. Figure 2 shows the front view of the deformed clay liner at the end of the centrifuge test.

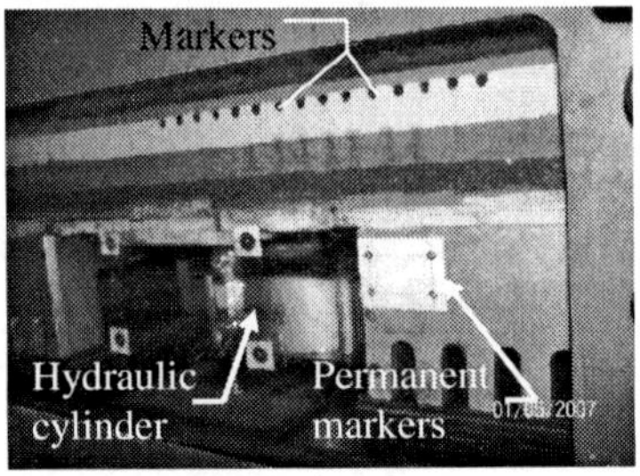

FIG. 1 View of the model before inducing differential settlements

FIG. 2 View of the model after inducing differential settlements

The images captured with the help of an image capturing software at various settlement stages were digitized. Position of permanent markers (Fig. 1) relative to each other was pre-measured and was used to scale and control the images captured. It was followed by tracking the location and analyzing the displacement of markers at each settlement stage was carried out. All the markers were digitized with an accuracy of ± 0.05 mm. The measured co-ordinates of all discrete markers at various settlement stages are approximated with an exponential equation of the normal distribution to get the settlement contours at various stages (Sengupta, 2005). From settlement contours, outer fiber strains were computed using combined bending and elongation method (Lee and Shen, 1968; Tognon et al. 2000). In addition to the strain distribution, crack initiation and propagation of cracks at the zone of maximum curvature were monitored optically.

RESULTS AND DISCUSSION

Table 1 summarizes details of centrifuge model tests. To investigate the influence of molding water content on the deformation behavior of clay-rich soil liners, two model tests namely, SSL-4 and SSL-10 were carried out on 30 mm thick liners moist-compacted at OMC and OMC+5% and corresponding dry unit weights (as per Standard Proctor compaction). It was observed that both the liners are subjected to cracking up to a full depth. However the modes of failure of the two liners are different. The liner with low molding water content, i.e., at OMC (SSL-4) failed by tension cracking, featured by a single prominent tension crack which originated symmetrically about the centerline of the liner at the zones of sharp curvatures. The liner constructed with high molding water content, i.e., OMC+5% (SSL-10) experienced multiple cracking pattern which spread at the zone of sharp curvatures over a wide area (Fig. 3a and 3b oriented along the breadth of the container). The width of cracks was found to be less than the clay liner in model test SSL-4. Ruptures were found to spread over the liner surface at more than four to five locations at the zone of maximum curvature. In model SSL-10, on the surface of a clay liner 30 mm high water was provided and this will result as an overburden of about 12 kN/m^2.

Both the model clay liners experienced initiation of cracking at identical strains ranging from 0.65 – 0.69 %, which is an outer fiber strain at the zone of maximum curvature (as shown in Fig. 2) and is referred herein as strain at crack initiation ε_c. The reason for the difference in observed cracking pattern is attributed to molding water content of the clay liner and development of pronounced peak surface strain for model SSL-4 at the zone of maximum curvatures. Whereas, for a clay liner with high molding water content, distribution of strain has occurred over a wide area at the zone of maximum curvatures. At higher molding water content, the strength of the material is low and allows formation of cracks at number of locations at the zone of maximum curvatures. These types of observations are seldom possible to make in landfill sites and results of centrifuge model tests are very significant.

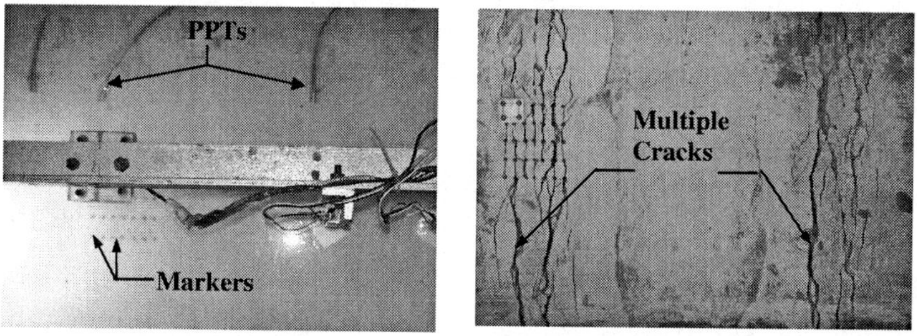

a) View of clay liner surface at 40*g* at *a* = 0 mm

b) View of clay liner surface at the end of the centrifuge test ($a_{max.}$= 25 mm)

FIG. 3. Status of model clay liner surface during centrifuge test [model: SSL-10]

Table. 1. Details of the centrifuge model test results

Parameters	SSL-4		SSL-7		SSL-10	
	Model	Prototype	Model	Prototype	Model	Prototype
d	30 mm	1.2# m	30 mm	1.2 m	30 mm	1.2 m
OMC (%)/ $(\gamma_d)_{max.}$ (kN/m^3)	22 / 15.9 (Std. Proctor)		16 / 17.6 (Modified Proctor)		27 / 14.2 (Std. Proctor, OMC+5)	
$a_{max.}$	25 mm	1 m	25 mm	1 m	25 mm	1 m
w_c (*)	4.5 mm	0.18 m	4.4 mm	0.176m	2.4 mm	0.096m
d_c (*)	30 mm	1.2 m	30 mm	1.2 m	30 mm	1.2 m
ε_c	0.65 %		1.04 %		0.69 %	
Max. ε_{of} (**)	3.53 %		4.03 %		2.76 %	

d -Thickness of the model clay liner; # prototype $d_p = Nd_m$; N = 40. suffix m = model; suffix p = prototype; Maximum central settlement = a_{max}.; Average crack width= w_c; Average crack depth = d_c; *Measured at the end of centrifuge test; **computed for $a_{max.}$ = 25 mm (1.0 m); Maximum outer fiber strain = Max. ε_{of} ; Strain at crack initiation = ε_c

To investigate the influence of compactive effort on the deformation behavior of clay-rich soil liners, two centrifuge model tests namely, SSL-4 and SSL-7 were carried out on 30 mm thick liners moist-compacted by varying compactive efforts (Table 1). Figure 4 presents variation of computed maximum outer fiber strain with the radius at the zone of maximum curvature for models SSL-4 and SSL-7. As can be noted, with an increase in the compactive effort, a slight delay in the crack initiation was observed. However, the prognosis of the cracking for model clay liner in SSL-7 was found to be rapid and was observed to have single and wide crack extending up to full thickness of the clay liner (Table 1). The observed results are justified since more compactive efforts lead to a brittle material (with more unit weight and low water content) with a higher strength in both tension and compression. Hence the initiation of cracking occurred at higher stress and strain as well as sharp curvatures and then material failed quickly without showing any resilience. Hence, this implies adoption of enhanced compactive effort for improving tensile strength-strain characteristics of clay liner material can aggravate the development of cracking in clay-rich soil liners. This implies that enhancement of ultimate tensile strain of the clay liner material may not be a viable option. This requirement is not only difficult to control in the field, but also seldom possible to implement. In such situations, this necessitates the requirement of strengthening of clay liner. Details of viable alternatives for strengthening clay liner with nominal overburden (like reinforcing clay-rich soil liner with a geogrid layer, reinforcing with discrete and randomly distributed fibers, etc.,) are discussed elsewhere by Viswanadham and Jessberger (2005) and Viswanadham et al. (2006).

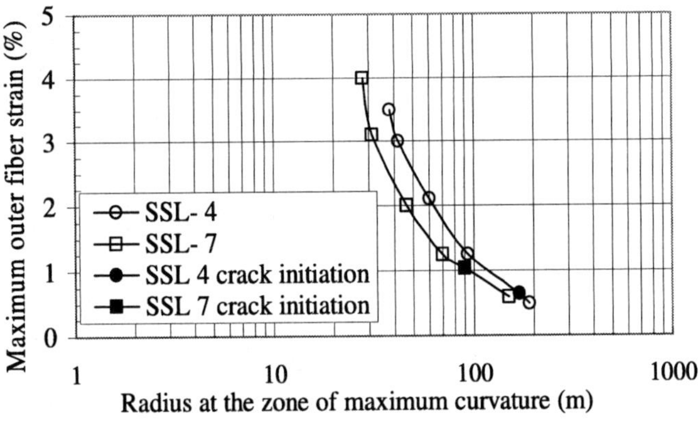

FIG. 4. Variation of maximum ε_{of} with radius at the zone of maximum curvature

CONCLUSIONS

Based on the analysis and interpretation of centrifuge model test results, the following conclusions can be drawn:

i) For a clay-rich soil liner moist-compacted at molding water content close to its OMC, it was observed to experience a single, wide crack extending up to full thickness of the clay liner. Contrary to this, for a clay-rich soil liner moist-compacted at molding water content towards wet side of OMC (OMC + 5 %), a multiple cracking pattern extending over a wide area at the zone of maximum curvatures was

noticed. Except a slight delay in the initiation of cracking in model SSL- 10, strain at crack initiation (for both SSL-4 and SSL–10) was found to be in the range of 0.65 – 0.69 %.
ii) With an increase in the compactive effort, the strain at crack initiation was found to be at 1.04 % and thereafter, the development of cracking was found to be very rapid. This implies that enhancement of ultimate tensile strain of the clay liner material may not be a viable option. This requirement is not only difficult to control in the field, but also seldom possible to implement. In such situations, this necessitates the requirement of strengthening of clay liner. Further studies are being carried out to address strengthening options for clay-rich soil liners with nominal overburden.

REFERENCES

Benson, C.H, Daniel, D.E, Boutwell, G.P. (1999). "Field performance of compacted clay liners." *Journal of Geotechnical and Geoenvironmental Engineering*, ASCE. 125 (5), pp. 390-403.

Benson, C.H., Barlaz, M.A., Lane, D.T., and Rawe, J.M. (2007). "Practice review of five bioreactor/recirculation landfills." *Elsevier Journal of Waste Management*, 27, pp.13-29.

Jessberger, H.L. (1994). "Geotechnical Aspects of Landfill Design and Construction. Part 1, Part 2 and Part 3". Proc., Institution of Civil Engineers, *Geotechnical Engineering*, 107, pp. 99-122.

Jessberger, H. L, and Stone, K. J. L. (1991) "Subsidence effect on clay barriers". *Geotechnique,* 41 (2) pp. 185-194.

Lee, K.L., Shen, C.K. (1969). "Horizontal movements related to subsidence." *Journal of Soil Mechanics and Foundation division*, ASCE, 94 (6), pp. 139-166.

Ling, H.I., Leshchinsky, D., Mohri, Y. and Kawabata, T. (1998). "Estimation of municipal solid waste landfill settlement." *Journal of Geotechnical and Geoenvironmental Engineering*, ASCE, 124 (1), pp. 21-28.

Sengupta, S.S. (2005). "Modeling deformation behaviour of compacted soil liners in a geotechnical centrifuge,'' Masters of Technology dissertation, Indian Institute of Technology Bombay, India.

Sowers, G.F (1972). "Settlement of waste disposal fills". Proc., *8th Int. Conf. On Soil Mechanics and Foundation Engineering*, Moscow, pp. 207-210.

Tognon, A.R., Rowe, R.K., Moore, I.D. (2000). "Geomembrane strain observed in large-scale testing of protection layers." *Journal of Geotechnical and Geoenvironmental Engineering*, ASCE, 126 (12), pp. 1194-1208.

Viswanadham, B.V.S and Mahesh, K.V. (2002). "Modeling deformation of clay liners in a small centrifuge." *Canadian Geotechnical Journal*, 39, pp. 1406 -1418.

Viswanadham, B.V.S and Jessberger, H.L. (2005). "Centrifuge Modelling of geosynthetic reinforced clay liners of landfills." *Journal of Geotechnical and Geoenvironmental Engineering*, ASCE. 131 (5), pp. 564-574.

Viswanadham, B.V.S., Sengupta, S.S., and Muthukumaran, A. E. (2006). "Studies on the Deformation Behaviour of Randomly Reinforced Soil Liners in a Geocentrifuge," *Proc. 8th Int. Conf. On Geosynthetics*, 18-22, September 2006, Yokohama, Japan, Vol. 4, pp. 1671-1674.

Geofiber Reinforced Soil Liner for Waste Containment Systems

B.V.S. Viswanadham[1], B.K. Jha[2], S. Pawar[3], and C. Kashyap[4]

[1]Associate Professor, Department of Civil Engineering, Indian Institute of Technology Bombay, Mumbai – 400 076, India; viswam@civil.iitb.ac.in (**Corresponding Author**)
[2 & 3]Formerly Post-Graduate students, Department of Civil Engineering, Indian Institute of Technology Bombay, Mumbai - 400 076, India
[4]Formerly Summer Intern, Department of Civil Engineering, Indian Institute of Technology Bombay, Mumbai-400 076, India

ABSTRACT: The objective of this paper is to evaluate the suitability of fiber-reinforced soil as a liner material for waste containment systems. This paper addresses the affect of aspect ratio and dosage of fibers on flexural and permeability characteristics of soil with and without discrete and randomly distributed fibers (DRDF). In addition, in order to verify the Velcro® effect like phenomenon along the Geofiber Reinforced Soil Liner (GRSL) and Geomembrane (GM) interface, a series of modified direct shear tests were conducted. For fibers with aspect ratio up to 45 and fiber content equal to 0.5 %, the coefficient of permeability was found to be about 1 x 10^{-9} m/s. Beyond 0.5 %, with an increase in fiber content, a gradual increase in the permeability was observed. Analysis and interpretation of modified direct shear tests conducted on soil and geomembrane with two textures (namely, smooth and rough) indicate that interface strength increases with an increase in fiber content and aspect ratio. For fibers with l/b = 45 and fiber content equal to 0.5 %, the strain at crack initiation was observed to be 2.75 %. Contrary to this, for an un-reinforced soil beam it is only 1.54 %. With the help of this study, fiber reinforced soil can be considered as a viable alternative to enhance the soil liner characteristics of waste-containment systems.

INTRODUCTION

Compacted clayey soils are commonly used as hydraulic barriers in waste containment systems such as in liners and covers for landfills, impoundments, hazardous liquid wastes, reservoirs, and ponds. Compacted clays can minimize infiltration of liquid into waste or control the release of contaminated liquids to the surrounding soils and groundwater. Conditions for crack formation are most favorable during the construction of the compacted soil liners (CSLs). Factors promoting crack formation during construction include exposed lift interfaces, existing clods, movement of heavy construction equipment, and non-uniform settlements. Cracking may also occur as a result of consolidation effects. The addition of waste above the CSL causes it to consolidate, which decreases the ductility, increases the stiffness of the clay liner and, thus, produces cracks. According to Daniel (1983), two predominant causes of failures of CSLs are: (i) Desiccation cracking due to moisture fluctuations, and (ii) Cracking due to excessive non-uniform settlements. Imperviousness of CSLs is essential, especially in landfills in order to safeguard the environment from contamination. A variety of research efforts have been attempted to address the problem of cracking of CSLs of lining systems due to non-uniform settlements and desiccation. The problems associated with desiccation cracks in soils include expansion of the soils upon wetting and softening of the soils as a result of water entering the soil structure. Miller et al. 1998 have reported that the hydraulic conductivity of CSLs having desiccation cracks has increased considerably. In addition, especially in compacted clayey soils at low overburden, thermally induced moisture changes in the CSL material can cause a reduction in plastic deformability and desiccation cracks. Desiccation of soil liners of landfills is continued to be major factor affecting landfill performance.

In addition to the above considerations, there could be sliding failure along interfaces within the composite, multilayered geosynthetic-compacted soil liner beneath the waste fill. Especially, interface between saturated compacted clay and a geomembrane layer was found to be one of the most critical surfaces. For a part of a landfill unit B-19 of a Class-I hazardous waste treatment and storage facility at Kettleman city, California, USA, interface friction angle between a clay liner and a geomembrane was reported to be as low as 8° (Mitchell et al. 1990). Although CSLs are most widely used all over the world, very few attempts have been made to develop a viable solution to retain their integrity at the onset of non-uniform settlements. It was reported that the cracking of CSL in a top lining system is more prominent than the bottom lining system (Viswanadham and Jessberger, 2005). The cracked CSL and other components of the top lining system are shown in Fig. 1 along with cracks in the clay liner due to non-uniform settlements of distinct variations, say Δs_1, Δs_2, with different ranges of curvature radii distributed over a large area of a landfill. The phenomenon of cracking leads to a drastic increase in permeability of the CSL, which affects the integrity and efficiency of a lining system.

In the recent past, the interest of using fibers has arisen to improve compacted clay performance as hydraulic barriers without changing physical properties of soil. Use of Discrete and Randomly Distributed Fiber (DRDF) reinforcement technique to decrease a soil's crack potential can be explored as one of the viable techniques to address the problem of cracking of CSL at the onset of non-uniform settlements. Figure 2 shows a schematic cross-section of top lining system along with a deformed fiber reinforced soil liner at the onset of non-uniform settlements. This technique was investigated earlier by Rodatz and Oltmanns (1997) to develop as a new liner construction material for landfill liners, especially when high mechanical and thermal loadings are anticipated. Recently, Miller and Rifai (2004) have attempted to evaluate the affect of discrete and randomly distributed fibers on the tensile strength-strain characteristics of clayey soil and to restrain desiccation cracking in CSLs of waste containment systems. This technique was also employed in road construction, and in strengthening backfill material (Miller and Rifai, 2004). Past research has demonstrated that the inclusion of randomly distributed discrete fibers significantly improve the engineering response of soils (Ziegler et al. 1998; and Miller and Rifai 2004). It was observed that the inclusion of fibers increased the tensile strength and hydraulic conductivity of clay and provides ductile behavior to material, but increase was more prominent at higher fiber contents (Ziegler et al. 1998). However, studies in evaluating the influence of discrete and randomly distributed fibers on the flexural tensile characteristics were limited.

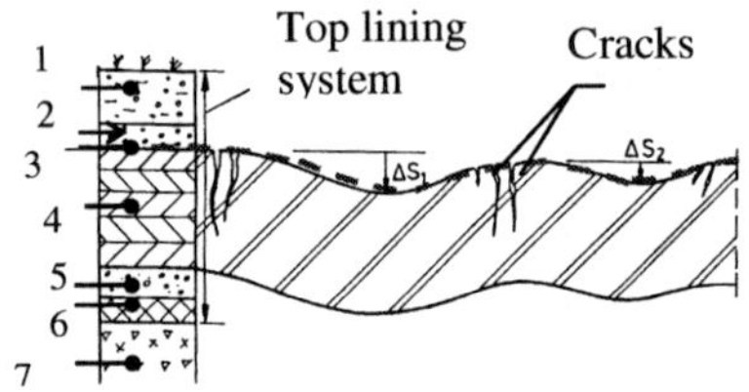

1: Cover soil; 2: Water drainage layer; 3: Geomembrane; 4: Compacted soil liner; 5: Gas drainage layer; 6: Restoration profile; and 7: Municipal/Hazardous waste.

FIG.1 Flexural distress in a top lining system

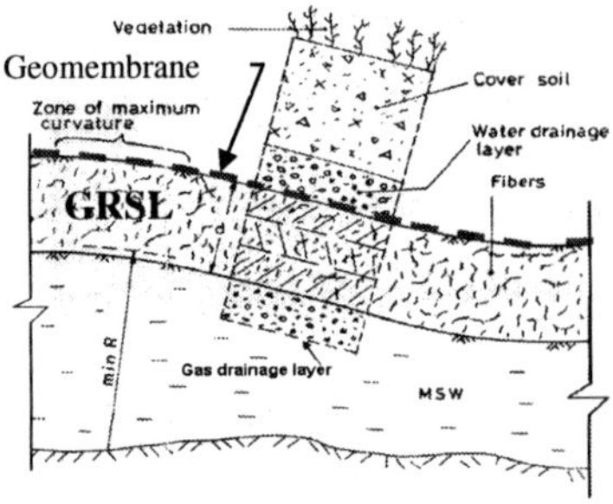

FIG. 2 Deformed GRSL in a top lining system

Ling et al. (2001) conducted some interface testing on PVC geomembranes using clays with different plasticity indices ranging from 35 to 100 %. They found that with an increase in plasticity index up to 70 %, the interface friction angle increases and beyond the plasticity index of 70% the interface friction angle decreases. The soil surface may have a good interaction with a smooth geomembrane depending upon the plasticity of soil and this may influence the shear behavior of smooth geomembrane-soil interfaces. However, based on the field data compiled and presented by Benson et al. (1999), the plasticity index of the clay liners range only from 10 % to 20 %. To some extent, attempts have been made to introduce rough texture to geomembrane surface to increase interface shear resistance characteristics. Recently, Zornberg (2005) reported application of randomly reinforced fiber soil technique for the base liner at Redwood landfill, California, USA. In this case study, use of fibrillated polypropylene fibers at a dosage of 0.1 % was reported. The main objective of using fibers in base liner construction was to increase friction angle between clay liner and textured geomembrane. Effect of increase in shear strength at the interface of clay liner and textured geomembrane is attributed to a Velcro® like effect. This is shown schematically in Fig. 2 depicting interaction between GRSL and geomembrane layers. However, limited attempts have been made to quantify this Velcro® like effect due to the presence of discrete and randomly distributed fibers in soil. In this regard, an attempt has been made by Welker and Josten (2005). A series of small-box direct shear tests were conducted to study the interface friction of a geomembrane with a fiber reinforced clay soil. The results indicate that to maximize shear strength, an optimum dosage of 0.2 % fiber content (by dry weight). Welker and Josten (2005) varied fiber content up to 0.3 % only. This present paper addresses the affect of aspect ratio and dosage of fibers on flexural and permeability characteristics of soil with and without discrete and randomly distributed fibers (DRDF). An attempt has also been made to bring-out the influence of randomly mixed fibers on the interface behaviour along smooth and textured geomembranes.

FLEXURAL AND PERMEABILITY CHARACTERISTICS OF SOIL WITH AND WITHOUT FIBERS

Most regulatory agencies require that CSLs to be designed to have a hydraulic conductivity of less than or equal to 1.0 x 10^{-9} m/s. It was observed that, the most landfill sites have CSL material with liquid limit ranging from 30 % - 40 % and plasticity index ranging from 10 % - 20 % (Benson et al. 1999). Based upon the above considerations, a blend of 80 % kaolin and 20 % fine sand (by dry weight) was used as a soil. The soil was found to have liquid limit of 36 % and plasticity index of 15%. The maximum dry unit weight and optimum moisture content (OMC) of the soil (as per standard Proctor compaction) are 16 kN/m^3 and 22 % respectively. In the present study, polypropylene tape fibers (having 1 - 2.2 mm width) having 890 denier (a measure of linear density of textile fiber mass) and with a breaking load equals to 48.54 N at 22 % breaking elongation were used. For the range of aspect ratios varied in this study, fiber lengths equal to 15 mm, 30 mm and 90 mm were considered and compaction characteristics of soil-fiber mixtures were found to have negligible variation when compared with soil alone. Aspect ratio (l/b) is defined herein as ratio of length of fiber to its breadth. Hence, it was decided to adopt compaction characteristics of soil liner material even for blends of soil-fiber mixtures. Variation of coefficient of permeability with fiber content was measured by conducting falling-head permeability tests in an oedometer over a test interval of 3 weeks. It was observed for the varied aspect ratios, once fiber content $f > 0.5$ %, an increase in the permeability by about one-and-half orders and is found to be more than 1 x 10^{-9} m/s. These results were also found to be in agreement with Miller and Rifai (2004).

Flexural behavior of soil liner with and without fibers was evaluated by executing a short series of flexural tensile tests on soil beams of 250 mm x 50 mm x 50 mm. Soil beams are prepared at moist-compacted conditions by compacting soil at a molding water content just above optimum moisture content (OMC+2 %) and its maximum dry unit weight as per standard-Proctor compaction. Figure 3 is a photograph of the fiber reinforced soil beam before and during the flexure tensile test. Moist-compacted soil beams were coated with a thin layer of silicone oil to prevent evaporation losses and pure bending was induced at a constant strain rate of 1.25 mm/min. As shown in Fig. 3a, three potentiometers were used to measure deflection of beam during test. In addition, 2.8 mm diameter and 3 mm graphite markers were embedded to evaluate strain in the soil beam during flexural tensile test. For this purpose, digital image analysis technique was adopted. Central deflection is indicated by Δ and distortion level is defined as Δ/L. Based on the radius of curvature R at centre of the beam, was computed using $[\Delta/2+L/(2\Delta/L)]$. For each test, moment-curvature relationships were established. Values of strain along the top row of markers (Fig. 3a) were determined using κd. Where κ is the curvature at center of the beam (= 1/R) and d = thickness of beam). It was assumed that neutral layer passes through the bottom of the beam. This was validated by computing strains along the depth of the beam by analyzing integral displacement of markers. Figure 4a and 4b present variation of computed strain at the centre of the beam. Results of the four flexural tensile tests are reported for soil beam: (i) without any fibers (CU), (ii) with fiber content = 0.25 % and l/b = 45 (CR1), (iii) with fiber content = 0.50 % and l/b = 30 (CR2), and (iv) with fiber content = 0.50 % and l/b = 45 (CR3). A maximum fiber content of 0.5 % was fixed based on the measured variation of permeability with fiber content. For an un-reinforced soil beam, strain at crack initiation ε_c was calculated as 1.54 % at a distortion level equal to 0.019. Contrary to this, for a soil beam reinforced with fibers having l/b = 45 and dosage of 0.5 %, ε_c was calculated as 2.75 % at a distortion level equal to 0.035. This implies that it requires higher aspect ratios to restrain cracking of soil at the onset of flexure. With an increase in aspect ratio, a considerable delay in the initiation of cracking and full depth penetration of crack was observed. Aspect ratio is defined as a ratio of length of fiber to its width (For example l/b = 45 implies length of fiber equals to 90 mm and with a width of 2 mm). Failure of fiber reinforced soil beams was found to be much more ductile when compared to an un-reinforced soil beam. These results are found to be in good agreement with those reported by Ziegler et al. (1998). Table 1 summarizes results of flexural tensile tests.

EFFECT OF FIBERS ON GEOMEMBRANE INTERFACE STRENGTH

Several researchers have attempted to measure interface shear strength characteristics (Mitchell et al. 1990, Ling et al. 2001, Ramirez and Gourc, 2003, and Fleming et al. 2006). Determination of interface geomembrane-soil interface characteristics involves (i) the use of a direct shear apparatus modified to facilitate shearing under a known normal stress (ASTM D5321) and (ii) the use of an inclined plane test (Ramirez and Gourc, 2003). The specimen size 100 mm x 100 mm has also been used (Ling et al. 2001, and Fleming et al. 2006), while many other studies were conducted using a shear box of 50 mm to 70 mm. However, effect of discrete and randomly distributed fibers within the soil on the interface characteristics was not studied. Recently, Viswanadham et al. 2006 reported the compacted soil liner (moist compacted at just above its optimum moisture content) was found to sustain large non-uniform settlements, when mixed with 0.5 % (by dry weight) polypropylene fibers having 90 mm length and 2 mm wide discretely. This was demonstrated through large beam centrifuge model tests in the laboratory. A 1.5 mm thick HDPE smooth geomembrane having average surface roughness equal to 0.420μ was used in the present study. Surface roughness of smooth geomembrane was measured using pertometer M2 as per the procedure outlined in B 46.1 of ASME (1995). In order to understand the affect of surface roughness on the soil-geomembrane interface shear strength, texture was created artificially by gluing a quartz sand having sizes greater than 1 mm and less than 2 mm and referred herein as rough geomembrane. Due to non-availability of a textured geomembrane with the authors, sand smeared geomembrane was used in the present study. An adequate care was taken to glue sand particles permanently to the smooth geomembrane surface. In the present study, a series of small-size modified direct shear tests (60 mm x 60 mm size) were carried-out on moist-compacted soil with and without fibers. From the flexural tensile strain point of view, it was noticed about the requirement of long fibers (i.e. l/b = 45). Hence majority of modified direct shear tests were carried-out with l/b = 45 and by varying fiber content (except with rough geomembrane l/b = 30 and f = 0.5 %).

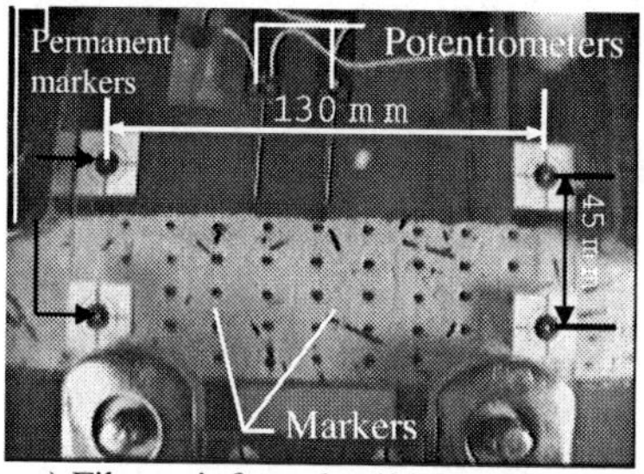

a) Fiber reinforced soil beam before commencement of test

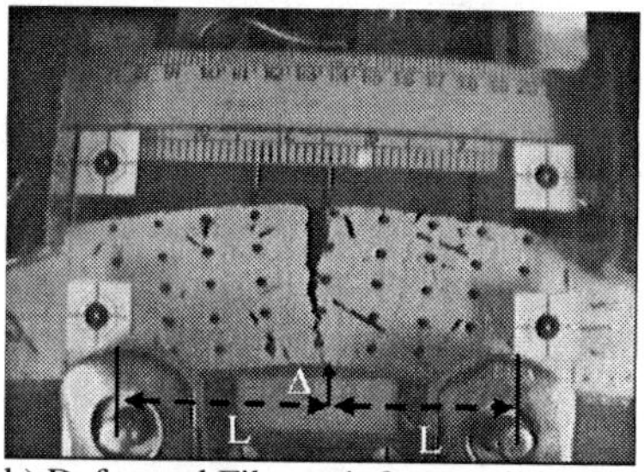

b) Deformed Fiber reinforce soil beam

FIG. 3 View of Fiber soil beam during flexural testing

Table 1. Summary of flexural beam tests results

Legend	*M_c (N.m)	*M_d (N.m)	*$(\Delta/L)_c$ (-)	Moisture Content (%) (Before/After test)	*ε_c (%)	*ε_d (%)
CU	0.31	0.42	0.019	24.2/23.8	1.54	7.4
CR1	0.478	1.28	0.027	23.9/23.6	2.3	10.7
CR2	0.5	1.33	0.030	24.1/23.4	2.45	12.7
CR3	0.625	2.0	0.035	24.35/23.75	2.75	13.2

ε_c and ε_d are strains at initiation and full depth crack, M_c and M_d are moments at the crack initiation and full depth crack and $(\Delta/L)_c$ is distortion level at the point of crack initiation.

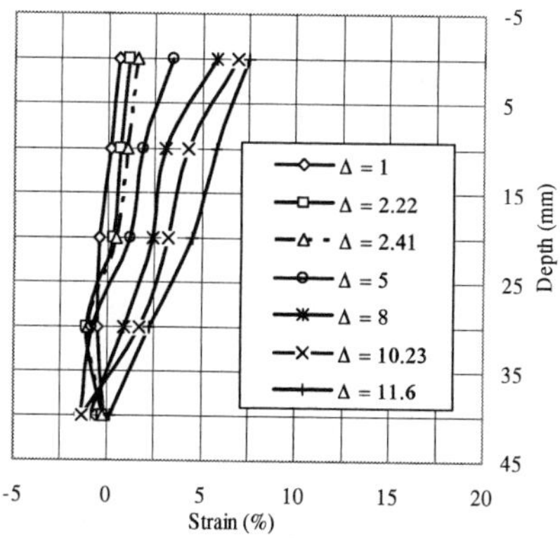

a) Test : CU (f = 0 %: l/b = 0: OMC+2)

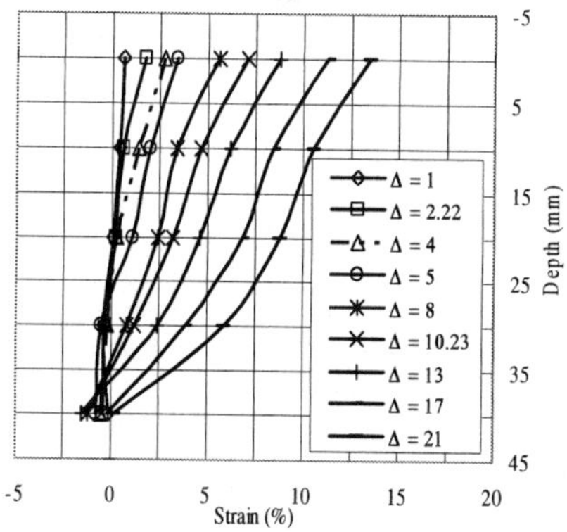

b) Test : CR3 (f = 0.5 %: l/b = 45: OMC+2)

FIG. 4 Strain distributions in un-reinforced and reinforced soil beams

Figure 5 presents variation of interface friction angles with fiber content for fibers having aspect ratios l/b = 30 and 45 respectively. Low values of interface friction angles could be attributed to small shear area in a modified direct shear test. With an increase in fiber content, an increase in interface friction angle and a decrease in the adhesion were observed. For fiber content up to 1 %, a 50 % increase in the interface friction angle was noted even with a smooth geoemembrane. For a rough geomembrane, an increase in interface friction angle by 2.6 times can be noted. Considering difficulties in manufacturing textured geomembranes (Fleming et al. 2006), adoption of geofiber reinforced soil liner (GRSL) is a viable alternative for enhancing interface frictional characteristics along soil-geomembrane interface. To some extent, this could be attributed to Velcro® effect, as described by Zornberg (2005).

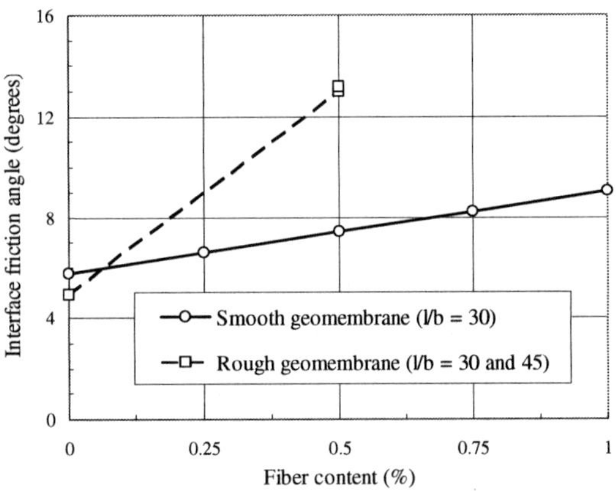

FIG. 5 Variation of interface friction angle with fiber content

CONCLUSIONS

In this paper, an attempt has been made to evaluate the suitability of fiber-reinforced soil as a liner material for waste containment systems. In order to achieve this, flexural, permeability of soil with and without discrete and randomly distributed fibers (DRDF), and geomembrane interface characteristics were evaluated in the laboratory. Based on the analysis and interpretation of test results, the following conclusions can be drawn:

i) For fibers with aspect ratio up to 45 and fiber content equal to 0.5 %, the coefficient of permeability was found to be about 1 x 10^{-9} m/s. Beyond 0.5 %, with an increase in fiber content, a gradual increase in the permeability was registered.

ii) For fibers with l/b = 45 and fiber content equal to 0.5 %, the strain at crack initiation was observed to be 2.75 %. In comparison, the strain at cracking for an un-reinforced soil beam it is only 1.54 %.

iii) With an increase in fiber content, an increase in interface friction angle and a decrease in the adhesion were observed. Interface friction angle for rough geomembrane and with f = 0.5 % and l/b = 45 was found to be 2.6 times that of a smooth geomebrane.

However, further studies are required in this direction to evolve at rational methodology and acceptancy. Nevertheless, this study demonstrates that fiber reinforced soil can be considered as a viable alternative to enhance the soil liner characteristics of waste-containment systems.

REFERENCES

ASTM (1992). "Standard test method for determining the coefficient of soil and geosynthetic or geosynthetic and geosynthetic friction by the direct shear method, " ASTM D5321-92, Philadelphia.

ASME (1995). "Surface Texture." American Society of Mechanical Engineers (ASME). B 46.1, New York.

Benson, C.H., Daniel, D.E., Boutwell, G.P. (1999). "Field performance of compacted clay liners," *Journal of Geotechnical and Geoenvironmental Engineering*, ASCE, 125(5), 390-403.

Daniel, D. E. (1983). "Shallow land burial of low-level radioactive waste", *J. of Geotechnical Engineering*, ASCE, 109 (1), 40-55.

Fleming, I.R., Sharma, J.S., and Jogi, M.B. (2006). "Shear strength of geomembrane-soil interface under unsaturated conditions." Geotextiles and Geomembranes, 24, 274-284.

Ling, H.I., Pamuk, A., Dechasakulsom, M., Mohri, Y., and Burke, C. (2001). "Interactions between PVC Geomembranes and Compacted clays" *Journal of Geotechnical and Geoenvironmental Engineering*," ASCE, 127(11), 950 - 954.

Miller, C.J., Hong, M., and Yesiller, N., (1998). "Experimental analysis of desiccation crack propagation in clay liners." J. American Water Resources Association, 34(3), 677-684.

Miller, C.J., and Rifai, S. (2004). "Fiber reinforcement for waste containment soil liners," *J. of Envirornmental Engineering*, ASCE, 130(8), 891-895.

Mitchell, J.K., Seed, R.B., and Seed, H.B. (1990). "Kettleman hills waste landfill slope failure I: Liner-system properties." J. Geotechnical Engineering, ASCE, 116 (4), 647-668.

Ramirez, R.R. and Gourc, J.P. (2003). "Use of the inclined plate test in measuring geosynthetic interface friction relationship." *Geosynthetics International Journal*, 10(5), 165-175.

Rodatz, W. and Oltmanns, W. (1997). "Permeability and stress-strain behaviour of fiber-reinforced soils for landfill liner systems," Advanced landfill liner systems, H. August, U. Holzlohner, and T. Meggyes (Eds.), Thomas Telford (Pubs.), 321-332.

Viswanadham, B.V.S. and Jessberger, H.L. (2005). "Centrifuge Modeling of geosynthetic reinforced clay liners of landfills," *Journal of Geotechnical and Geoenvironmental Engineering*, ASCE. 131(5), 564-574.

Viswanadham, B.V.S., Sengupta, S.S., and Muthukumaran, A. E. 2006. "Studies on the Deformation Behaviour of Randomly Reinforced Soil Liners in a Geocentrifuge," Proc. 8th International Conference on Geosynthetics, 18-22, September 2006, Yokohama, Japan, Volume 4, pp. 1671-1674.

Welker, A.L., and Josten, N. (2005). "Interface friction of a geomembrane with a fiber reinforced soil." Proc. GeoFrontiers 2005 Congress, Geotech Sp. Pub. 140 Slopes and Retaining structures under seismic and static conditions, ASCE, M.A. Gabr, D. Elton, J.J. Bowders, and J.G. Zornberg (Eds.), 1- 7.

Ziegler, S., Leshchinsky, D., Ling, H.I., and Perry, E.B. (1998). "Effect of short polymeric fibers on crack development in clays," *Soils and Foundations*, 38(1), 247-253.

Zornberg, J.G. (2005). "Geosynthetic reinforcement in landfill design: US perspectives," Proc. Geo-Frontiers 2005 Congress, ASCE, Geotechnical Special Publication No. 141, J.J. Zornberg, M. Gabr, and J.J. Bowders (Eds.), Austin, Texas, USA.

Development of Predictive Soil-Atmosphere Models for Test Plots at the Questa Mine, New Mexico

Matt Wickham[1], Kelly Greaser[1], Sarah Doyle[1], Gordan Gjerapic[1], and Jay Gear[2]

[1]Golder Associates Inc., Lakewood, Colorado 80228; PH (303) 980-0540; FAX (303) 985-2081; email: mwickham@golder.com, kgreaser@golder.com, sedoyle@golder.com, ggjerapic@golder.com
[2]Chevron Mining Inc., Questa, NM 87556; PH (575) 586-7638; FAX (575) 586-0811; jgear@chevron.com

ABSTRACT: This paper describes the development of predictive soil-atmosphere models for test plots at the Questa Mine located near Questa, New Mexico. The models were developed using data from instrumented test plots installed in 2000 in exposed mine rock materials. The models will be used to support decisions about mine water management and to guide final reclamation of mine rock piles. This paper describes site-specific considerations for model development including selection of the model code, model calibration to measured lysimeter outflows, and calibration targets. Results of the model predictions are summarized for one of the test plots.

INTRODUCTION

Methods to quantify or estimate infiltration into mine rock piles are often required to support evaluations of Acid Rock Drainage (ARD) and mine water management, as well as final cover design and reclamation. Predictive modeling guided by information from laboratory and field testing and instrumented test plots is commonly used for this purpose. The predictive model can be used to evaluate variability and uncertainty in material properties, climate conditions, and other site conditions affecting the water balance for uncovered and covered mine rock where direct measurements with test plots or other means are not practical or cost effective.

This paper describes the development of predictive soil-atmosphere models for test plots at the Questa Mine located near Questa, New Mexico. Instrumented test plots were installed in 2000 (Wels et al. 2001, 2002) in exposed mine rock materials to generate information required to develop predictive models. The models will be used to support decisions about mine water management and to guide final reclamation of mine rock piles. This paper describes site-specific considerations for model development including selection of model code, temporal discretization of climate inputs, calibration to measured lysimeter outflows, and calibration targets.

SITE DESCRIPTION AND TEST PLOT PROGRAM

The Questa Mine is located 5.8 km east of the town of Questa in Taos County, New Mexico. The mine rock piles cover a surface area of about 275 ha ranging in elevation from approximately 2470 m to 2990 m (Wels et al. 2001). The climate is semi-arid with mild summers and cold winters. Precipitation and potential evaporation data, based on the data from on-site weather stations collocated with lysimeter test plots, are summarized in Table 1. Daily potential evaporation values were calculated from the recorded temperature, humidity, wind speed and net radiation measurements using the FAO Penman-Monteith methodology (Allen et al. 2000).

Table 1. Average annual precipitation and evaporation from 2001 to 2006

Lysimeter ID	Precipitation (mm)	Potential Evaporation (mm)
TP-4	380	1280
TP-5	390	1050
TP-6 & 7	430	1210

Material properties of the mine rock at the site can be divided into four broad categories (Wels et al. 2002): well graded (finer) mixed volcanics, poorly-graded (coarser) aplite/black andesite, fine graded erosion material and very coarse rubble material. Initial mine rock characterization was described by Wels et al. (2001). A summary of the grain-size distributions for materials with laboratory determined soil water characteristic curves (SWCCs) is given in Table 2.

Table 2. Material fractions for materials with laboratory determined SWCCs

Fraction	Sample Designation					
	CCU-1	SG-2	SSL-1	SSM-3	SSU-3	SSW-2
%gravel (>2mm)	53	68	48	84	57	57
%sand	26	24	23	8	28	26
%silt	17	6	26	6	9	11
%clay	4	2	3	2	6	6

The test plot program at Questa Mine was initiated in 2000 to assess current uncovered conditions for the mine rock piles to provide information useful in planning for closure and final reclamation. The program includes two phases. Phase 1 includes construction and instrumentation of lysimeters, climate monitoring, and development and calibration of soil atmosphere models. Phase 2 includes completing a sensitivity analysis for net infiltration for site-wide conditions, and developing water balance calculations for all mine rock piles at the site. Wels et al. (2001) describe the installation of four barrel lysimeters constructed at the Questa Mine. The installation included excavating mine rock to a total depth of about 2.4 m and placing a 20 mm thick-walled HDPE tank with a diameter of 2.4 m and a height of 2.3 m. The lysimeter base provides lateral drainage towards a 50 mm (2 inch) PVC drainage collection pipe located in the center of the tank. The pipe drains to a tipping bucket in a concrete manhole with continuous monitoring of lysimeter outflow. The lysimeter was backfilled with the excavated mine rock in 30 to 60 cm lifts. Precipitation, temperature, humidity, net radiation and wind speed are monitored and recorded at each lysimeter. Two lysimeters (TP-4 and TP-5) were constructed using finer grained mine rock materials. In addition, one (TP-6) used a coarse mine rock and one (TP-7) was constructed using the coarse grained material with approximately 30 cm of the finer grained mine rock at the top as illustrated in Table 3.

Table 3. Lysimeter test plots

Lysimeter ID	Elevation (m)	Material
TP-4	2,820	well-graded (finer) mine rock
TP-5	2,990	well-graded (finer) mine rock
TP-6	2,660	poorly-graded (coarse) mine rock
TP-7	2,660	poorly-graded (coarse) mine rock capped with 30 cm of finer mine rock material

MODEL DEVELOPMENT

Soil-atmosphere models have been developed and calibrated to field data from each of the test plots. Important considerations for model development include model code selection and model calibration.

Model code

An inspection of the lysimeter outflow data for the test plots reveals a component of preferential and/or non-equilibrium flow characteristic of coarse grained materials. As such, a consideration for model selection is the relative importance of preferential flow to the overall water balance and the degree to which it can be represented using porous-type flow models based on some form of Richard's equation using area-averaged moisture contents and pressure heads. While a number of popularly used approaches to modeling preferential and non-equilibrium flow in the vadose zone exist (Šimunek et al. 2003), few are available in validated numerical models for commercial application and rarely do data exist outside academia with which to uniquely parameterize such models. The approach adopted for the test plot model development was to first attempt calibration using an available single porosity, porous-type flow model, then evaluate the benefit and cost of explicitly representing preferential and non-equilibrium processes.

Two model codes were evaluated for this application: UNSAT-H (Fayer 2000), a one-dimensional finite-difference code for simulating water and heat flow through soil (Fayer 2000); and HYDRUS-1D (Šimunek et al. 1998), which also solves a modified Richard's equation for liquid flow and a one-dimensional convection-dispersion equation for heat flow. The governing equations are solved numerically using Galerkin-type linear finite element schemes. In addition, UCODE was used to aid in model calibrations with both UNSAT-H and HYDRUS-1D. UCODE (Poeter and Hill 1998) and UCODE_2005 (Poeter et al. 2005) perform inverse modeling, posed as a parameter-estimation problem, using nonlinear regression. The nonlinear regression problem is solved by minimizing a weighted least squares objective function with respect to the parameter values using a modified Gauss-Newton method also referred to as a Levenberg-Marquardt method. UCODE was coupled with each model code for the evaluation which allowed for objective refinement of material property estimates based on comparisons between measured and predicted lysimeter outflows. The literature includes comparisons of various model codes for a variety of conditions (Bonaparte et al. 2002; Roesler et al, 2002; Benson et al. 2004).

HYDRUS-1D (Simunek et al. 2005) was ultimately chosen for the soil-atmosphere modeling. HYDRUS-1D could be used to simulate long-term time periods with hourly climate data; calculation of evaporation during rain events (which can be critical in the dry climate of the southwest where evaporation can be significant even during storm events); simulation of a seepage face at the model lower boundary in which the boundary condition varies based on the saturation condition (this option is ideal for simulating lysimeter outflow [Scanlon et al. 2002]); and provides numerical stability with acceptable computation times. In addition, a 2D version of HYDRUS is available that may be used for future modeling (e.g. the mine rock water balance) and dual-permeability models for HYDRUS are in development, which may provide the capability to model the dual-flow system occurring in the lysimeters (i.e. porous and preferential flow types). UNSAT-H provides for specifying a separate relationship between moisture content and suction, and moisture content and hydraulic conductivity, whereas HYDRUS-1D does not.

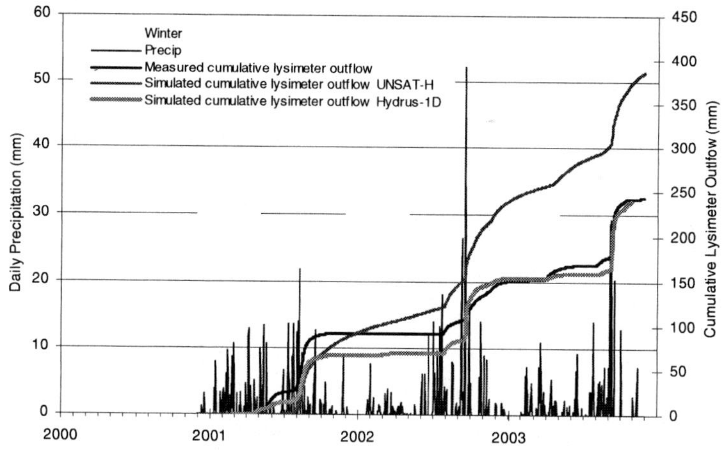

Figure 1. Comparison of calibrated lysimeter outflow, TP-6

The performance of the two model codes was compared using objective parameter estimation calibrated to lysimeter outflow for TP-6. A significant difference between the two model configurations was the temporal discretization of precipitation. Daily precipitation was applied by UNSAT-H beginning at midnight at a specified rate. Hourly precipitation was specified in HYDRUS-1D. HYDRUS-1D provided a better match to cumulative lysimeter outflow, as shown on Figure 1. This demonstrates the sensitivity to accurate representation of precipitation. This result was not unexpected, as runoff can be the dominant component of the water balance for fine grained materials, but a minor component for coarse mine rock materials. As a result, storm intensity can be a major factor controlling infiltration for coarse grained materials that exhibit a high saturated hydraulic conductivity (K_{sat}). Further differences can be attributed to representing evaporation during rainfall periods and realistically representing a seepage face at the base of the lysimeter.

Model calibration

Irrigation step tests, careful selection of calibration targets, and automated parameter estimation were critical to successful calibration of the soil-atmosphere models. Well controlled, stepped irrigation and draindown tests were conducted on the test plots to provide a direct measurement of effective material properties of the lysimeter materials (Gjerapic and Wickham 2006). The irrigation test provided, in part, information regarding the change in the volumetric moisture content (ΔY_v) and unsaturated hydraulic conductivity, $K(Y_v)$. The relationship $K(\Delta Y_v)$ was directly measured during the test assuming unit gradient conditions. This information was used to calibrate a numerical model, specifically by constraining the relationship $K(Y_v)$ where standard models (e.g. Burdine [1953], Mualem [1976], van Genuchten [1978, 1980]) based on the proposed soil water characteristic curves were not valid.

Calibration targets were selected to reflect the desired accuracy and intended use of the calibrated model, to effectively reduce and represent the copious field data, and to properly target the processes of interest. For example, cumulative precipitation across a season provided a more reasonable calibration target than each wetting front, particularly when preferential flow was significant. Calibration to test plot lysimeter outflow targeted, in order of decreasing priority: 1) timing and magnitude of the initial preferential flow response to storm events, 2) cumulative flow over the period of record, and 3) draindown of porous type flow following storm events. Parameter estimation was not constrained to "typical" ranges in an attempt to capture the component of preferential flow indicated by the measured lysimeter outflow.

An inversion program external to the code (e.g. UCODE [Poeter and Hill 1998] and UCODE_2005 [Poeter et al. 2005]) proved essential for effective calibration to field measurements, particularly given the large volume of field measurements and multiple calibration targets. The inversion programs provided objective parameter optimization compared to subjective, manual and often tedious calibration. The objective function was the sum of squared weighted daily flow residuals. Residuals were weighted by season, with lower weighting given to winter months when frozen soil conditions – which were not explicitly simulated – and snowpack processes affected moisture availability and infiltration.

RESULTS

Simulations were conducted for the period of record with the calibrated models. For TP-6, the final calibrated values were: K_{sat}= 1.1 cm/s, saturated moisture content

(θ_{sat})= 0.28, residual moisture content (θ_r) = 0.06, van Genuchten α= 1.22 1/cm, and van Genuchten n= 1.09, where the values for K_{sat}, θ_r, and α reflect the coarse nature of the materials and an attempt to capture the component of preferential flow. Figure 2 shows the daily measured and simulated lysimeter flows for TP-6. Timing of the majority of the storm responses is matched well, especially later in the period of record following a lysimeter retrofit in April 2004 to address lysimeter edge effects. Also, the model simulations match the overall magnitude of the storm responses.

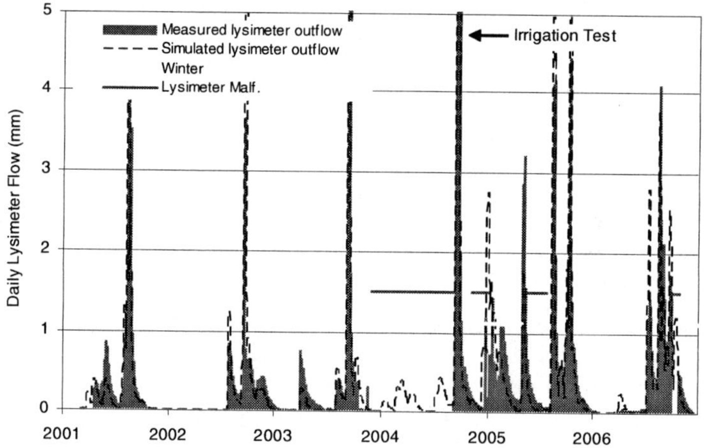

Figure 2. Calibrated Daily Flow for TP-6

Figure 3 shows the measured and simulated cumulative flow for the period of record. The model results are normalized after periods of missing lysimeter data due

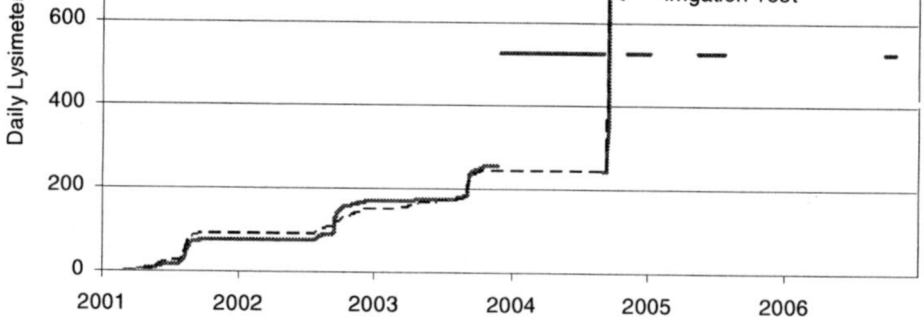

Figure 3. Calibrated Cumulative Flow for TP-6

to equipment malfunction. The figure shows that the model was capable of accurately simulating cumulative flow, especially after April 2004. The model simulates less accurately the third criterion, the porous type flow draindown – which represents a conscious tradeoff between matching the timing and magnitude of the initial preferential flow response and the draindown response.

CONCLUSIONS AND ADDITIONAL EFFORTS

Predictive soil-atmosphere modeling was developed for the test plots based on information from laboratory and field testing and instrumented test plots. A number of important considerations were evaluated during model development. A conscious decision was made to initially use single-porosity, porous-type flow models, despite evidence for a component of preferential and/or non equilibrium type flow in the lysimeter response. The commonly-used models UNSAT-H and HYDRUS-1D were evaluated. HYDRUS-1D was found to provide a better calibration to measured lysimeter outflows, by a) providing the means to represent hourly climate inputs which more accurately accounts for runoff for coarse grained materials in a more efficient manner, b) representing evaporation during rainfall periods, and c) realistically representing a seepage face at the base of the lysimeter.

Irrigation step tests, careful selection of calibration targets, and automated parameter estimation were critical to successful calibration of the soil-atmosphere models.

The calibrated model results for TP-6 are provided and summarized. The model was able to simulate accurately the timing of the majority of the storm responses, especially later in the period of record following a lysimeter retrofit in April 2004. The model simulations match the overall magnitude of the storm responses.

Additional efforts underway for the test plot program include construction and revegetation of final covers on each of the lysimeter test plots, continued model refinement (including a simple dual-porosity flow model), and snowpack monitoring to improve the conceptual and numerical models of winter conditions and spring snowmelt processes.

ACKNOWLEDGMENTS

The authors would like to thank the staff of the Chevron Mining Inc. (formerly Molycorp Inc.) Questa Mine for their help and support during ongoing projects.

REFERENCES

Allen, R.G., Pereira, L.S., Raes, D. and Smith M. (2000). "FAO Irrigation and Drainage Paper No. 56 – Crop Evapotranspiration", Food and Agriculture Organization of the United Nations.

Benson, C. H., Bohnhoff, G. L., Apiwantragoon, P., Ogorzalek, A. S., Shackelford, C. D., and Albright, W. H. (2004). Comparison of model predictions and field data for an ET cover. Tailings and Mine Waste '04, Balkema, Leiden, The Netherlands, 137-142.

Bonaparte, R., Gross, B. A., Daniel, D. E., Koerner, R. M., and Dwyer, S. F. (2002), "Technical Guidance for RCRA/CERCLA Final Covers," U.S. EPA Office of Solid Waste and Emergency Response, Washington, D.C. (in final review)

Burdine, N.T. (1953). "*Relative Permeability Calculations from Pore-Size Distribution Data*", Transactions AIME, Vol. 198, pp. 71-78.

Gjerapic, G. and Wickham, M. (2006). "Parameter Estimation from Stepped-Irrigation Tests on Instrumented Lysimeter Test Plots", Unsaturated Soils 2006 Conference Proceeding, Carefree, Arizona, April 2-6, pp. 135-146.

Mualem Y. (1976). "*A new model for predicting the hydraulic conductivity of unsaturated porous media.*" Water Resour. Res. 12(3):513-522.

Poeter, E. P., Hill, M. C., Banta, E. R., Mehl, S., Christensen, S., 2005, UCODE_2005 and Six Other Computer Codes for Universal Sensitivity Analysis, Calibration, and Uncertainty Evaluation, U.S.G.S Techniques and Methods 6-A11, 283 p.

Roesler, A. C., Benson, C. H., and Albright, W. H. (2002). Field Hydrology and Model Predictions for Final Covers in the Alternative Assessment Program. Geo Engineering Report No. 02-08, University of Wisconsin-Madison.

Scanlon, B. R., Christman, M., Reedy, R. C., and Gross, Beth, 2002, Intercode comparisons for simulating water balance in an engineered cover, *in* 2001 International Containment and Remediation Technology Conference, Orlando, Florida, Institute for International Cooperative Environmental Research, Florida State University, Paper ID. No. 148, http://www.iicer.fsu.edu, 3 p.

Šimunek, J., van Genuchten M. Th. and Sejna M. (2005). "The HYDRUS-1D Software Package for Simulating the Movement of Water, Heat, and Multiple Solutes in Variably Saturated Media, Version 3.0", HYDRUS Software Series 1, Dept. of Env. Sciences, University of California, Riverside, California, 270 pp.

Šimunek, J., N. J. Jarvis, M. Th. van Genuchten, and A. Gärdenäs, Review and comparison of models for describing non-equilibrium and preferential flow and transport in the vadose zone, *Journal of Hydrology*, 272, 14-35, 2003.

van Genuchten R. (1978). "*Calculating the unsaturated hydraulic conductivity with a new closed-form analytical model.*" Water Resources Program, Department of Civil Engineering, Princeton University, Princeton, New Jersey.

van Genuchten, M.T. (1980). "*A closed-form equation for predicting the hydraulic conductivity of unsaturated soils*", Soil Sci. Soc. Am. J., 44: 892-898.

Wels, C., Loudon, S. & Fortin, S. (2002). "Factors Influencing Net Infiltration into Mine Rock Piles at Questa Mine, New Mexico", Proc. 9th Int. Conf. on Tailings and Mine Waste, 27-30 January , Fort Collins, Colorado, USA, pp. 469-477.

Wels, C., O'Kane, M., Fortin, S. and Christensen, D. (2001). "Infiltration Test Plot Study for Mine Rock at Questa Mine, New Mexico", Proc. American Society for Surface Mining and Reclamation Conference, Volume 1, pp. 195-209, 27-30 January, Albuquerque, New Mexico.

ET Covers: Construction & Tree Development Inside & Outside of Lysimeters

Tarek Abichou[1], P.E and Jubily Musagasa[2], E.I

[1]Associate Professor, Florida Agricultural & Mechanical University/Florida State University, Civil and Environmental Engineering Dept., 2525 Pottsdamer Street , Tallahassee, FL 32310; abichou@eng.fsu.edu

[2]Research Assistant, Florida Agricultural & Mechanical University/Florida State University, Civil and Environmental Engineering Dept., 2525 Pottsdamer Street , Tallahassee, FL 32310; musagasa@eng.fsu.edu

ABSTRACT: Two evapotranspiration covers pan lysimeters were constructed side by side to two unlined cover test sections in 2004 at the Leon County Solid Waste Management Facility in Tallahassee, Florida. The unlined test sections were constructed in the same manner as the pan lysimeters, but without at a pan, i.e., the soil profile in the unlined test sections was in contact with landfill gas. One objective of this study was to contrast tree growth, soil temperature, and soil water storage inside and outside the pan lysimeter. This paper summarizes the construction of the lysimeters and unlined test sections compares tree development, soil water storage, and temperature inside and outside the lysimeters.

Analysis of variance showed no significant difference between eucalyptus trees inside the lysimeter to those outside. Cottonwood heights were significantly different during the first measurement but not significantly different during the next two measurements. Soil temperature inside the lysimeter was significantly lower than outside the lysimeter. Soil water storage inside the lysimeter was significantly higher than outside. The difference in water storage and temperature did not lead to difference in vegetation growth.

INTRODUCTION

Most field studies use lysimeters to assess performance of alternative landfill covers. Lysimeters provide an opportunity to directly measure percolation rate from alternative covers. Percolation rates can be measured with a precision of 0.5 mm/yr or better using lysimeters (Gee & Hillel, 1988; Benson et al.; 1994; Albright et al., 2004; Ward & Gee, 1997; Benson et al.; 2001). One of the disadvantages of lysimeters is the presence of an artificial no flow boundary induced by the lysimeter at the base of the profile which prevents upward and downward flow of vapor and liquid across the base of the lysimeter. All water that migrates downward to the base of the profile is collected and routed out of the system. Consequently, the collected water as well as the moisture under the lysimeter can never move upward as a result of natural upward gradients induced by evapotranspiration (Benson et al., 2001). Lysimeter and associated monitoring equipment installation disturb the soils inside and outside lysimeters. This disturbance is thought to affect the percolation rate through the soil cover and, consequently, affect the precision with which percolation rate is measured (Liu, 2004). Despite this, few studies have been conducted to assess the effect of the bottom boundary on water flow and solute transport in the lysimeters (Flury et al., 1999; Abdou & Flury, 2004). The effects of the lysimeter lower boundary on the temperature, water content and vegetation is not well known. This paper addresses the effect of the lysimeter lower boundary on the growth of trees, soil temperature and soil water storage. Water balance results are presented in other publications.

LYSIMETERS

Lysimeters are devices used to measure percolation of water infiltrating through soils. Lysimeters are widely used to study water flow and solute transport in soils (e.g. Bergstrom & Jarvis, 1993; Putz et al.., 1998). Lysimeter studies are considered to be an intermediate approach between field studies and small scale laboratory experiments. Because lysimeters are exposed to the same environmental conditions, they are more likely to mimic natural field conditions than laboratory studies (Abdou & Flury, 2004). Lysimeters consists of a base pan with or without sidewalls. The pan acts as a collector of percolation. The pans of most lysimeters used to monitor landfill cover sections are rectangular. The depth of a lysimeter is a critical design parameter and varies with the intended purposes of the tests. Because of the critical role that plants play in removing water from landfill cover, the depth, distribution and density of plant roots have to be taken into account in determining depth of a lysimeter in a landfill cover study. The depth should permit the development of normal rooting density and rooting depth and provide similar available water profiles to the field profile (Van Bavel et al., 1961). The areal extent of lysimeters depends on the spatial variability in the properties of cover soils and vegetation. The length and width are usually 5 times larger than depth to ensure that preferential flow processes (e.g., rapid flow in such features as cracks, fissures, root channels, and worm holes) are captured in the test and that the construction process would mimic full scale conditions (Bews et al., 1999; Benson et al., 2001).

MATERIALS AND METHODS

Construction of the Lysimeters and Unlined Test Sections

The schematic diagram of the ET cap lysimeters is shown on Fig. 1. The 7 m x 14 m lysimeter consists of a "bath tub" lined with geomembrane and geocomposite. At the down slope, a drainage sump was installed and connected to a drainage monitoring system. A geocomposite drainage layer followed by a thin gravel layer was placed on top of geomembrane in order to collect the percolating water from the cap soil and rapidly transmit to the sump area. This layer also acts as a cushion layer to protect the bottom geomembrane during soil placement. Root barrier (Biobarrier II, Reemay, Inc.) was placed on top of the gravel layer. The root barrier is intended to prevent the roots from reaching the deep drainage collection system. The alternative cap was then placed in three lifts of soils. A thin layer of compost was also added at each lift of soil to improve fertility to aid growth of trees and grasses and to improve the water storage capacity of the soil.. The unlined test sections were constructed to the same dimensions as the lysimeters, except that in the unlined test sections, the soil profile is in direct contact with landfill gas. The unlined test section consisted of a thin gravel layer placed on top of the existing interim cover. Several 8 inches wide holes filled with gravel were drilled through the interim cover (into the waste) to allow direct movement of landfill gas to the gravel layer below the cover.

Vegetation

Two species of trees were planted in the ET cover lysimeters. Eucalyptus trees were planted on one of the lysimeters and cottonwood trees were planted on the other. These species were used because they are native to the study area and their ability to remove water from the soil. Trees were also planted on the buffer area between the test pads to reduce the edge effects. Trees were planted in May 2004 as small containerized plants (300 mm on average). Tree survival during planting was kept to 100% by removing all dead cuttings and replacing with fresh tree cuttings. Grasses were seeded on both ET covers. Tree heights were measured in August 2004, December 2005, and January 2007. Tree height and diameter at breast height (DBH) were measured inside the lysimeter and inside the unlined test sections. The sample size used in this study included 35 and 30 cottonwood trees in unlined test section and lysimeter respectively. The number of eucalyptus trees measured was 36 for unlined test section and 38 for lysimeter.

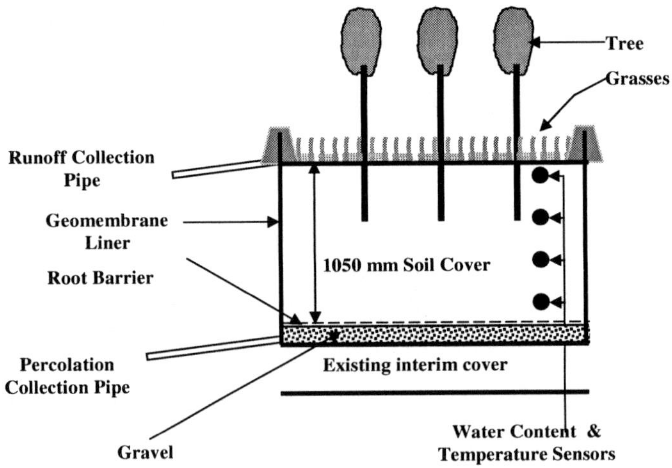

Fig. 1. Schematic of ET cap lysimeter

Water Content and Temperature Sensors

CS616 water content reflectometers (WCR) and CR 23X datalogger were used to record soil moisture at various depths for both the ET cover inside and outside the lysimeters. The CS616 probes are used to measure the probe output periods which were converted to volumetric water content using calibrations equations developed in the laboratory. The probes were installed in two nests at three locations along the center line of the lysimeter and unlined test sections. Each nest contains four probes alligned horizontally as shown on figure 1. Soil water storage was then computed by integrating the volumetric soil water content measured with the probes over the representative volumes of the cover profiles. The volumetric water content recorded by each sensor was multiplied to the thickness of the soil where the sensor is placed. The water storages in each soil layer were then added to obtain the water storage of the entire cover profile.

Thermocouples were used to measure the soil temperature at various depths of the covers. Temperature sensors were co-located with water content sensors at each nest. These sensors were also connected to the CR 23X Campbell Scientific datalogger. Water content and soil temperature data were recorded in the datalogger every two hours for the duration of three years. Soil water storage and soil temperature in the lysimeters were then compared with that in the unlined test sections. Analysis of variance was also carried out to assess the significance of the difference in water storage and soil temperature in the lysimeters as compared to that in the unlined test section.

RESULTS

Tree Growth

Eucalyptus trees grew to an average height of 830 mm (lysimeter) and 940 mm (unlined section) in August 2004 and to 5230 mm (lysimeter) and 5490 mm (unlined section) in December 2005 (Table 1). The eucalyptus trees further grew to an average height of 8040 mm (lysimeter) and 8820 (unlined section) in January 2007. ANOVA ($\alpha = 0.01$) of the tree height data and DBH (Table 3) data showed no significant difference between tree height and DBH inside and outside the lysimeter, suggesting that neither landfill gases nor the presence of the bottom pan impacted the eucalyptus tree growth rates.

The cottonwood trees grew to an average height of 1270 mm (lysimeter) and 1710 mm (unlined section) in August 2004 (Table 2). The difference is tree height was not significantly different outside and inside the lysimeter during this period. In December 2005, the cottonwood trees grew to an average height of 6350 mm (lysimeter) and 6310 mm (unlined section). In January 2007, the cottonwood trees reached an average height of 7110 mm (lysimeter) and was not significantly different than outside the lysimeter (8010 mm). The early growth of cottonwood trees was higher than that of the eucalyptus trees. Later on however, the eucalyptus trees grew to an average final height of 8420 mm whereas the cottonwood had a final average height of 7560 mm.

ANOVA of eucalyptus trees and cottonwood trees revealed no significant difference between DBH inside the lysimeter to that outside the lysimeter (December 2005 data). However, ANOVA of DBH measured on January 2007 showed a significant difference between trees inside the lysimeter to those outside the lysimeter for cottonwood trees. Additional monitoring is being performed to monitor the long term growth and tree development in all test sections.

Table 1. Eucalyptus Test Section Mean Tree Height Summary

Date	Note	Lysimeter		Unlined Section		P-Value
		Height (mm)	St Dev (mm)	Height (mm)	St Dev (mm)	
May-04	Planting	300	-	300	-	-
Aug-04	-	830	240	940	220	0.061
Dec-05	-	5230	1130	5490	1100	0.311
Jan-07	-	8040	1780	8820	790	0.019

Table 2. Cottonwood Test Section Mean Tree Height Summary

Date	Note	Lysimeter		Unlined Section		P-Value
		Height (mm)	St Dev (mm)	Height (mm)	St Dev (mm)	
May-2004	Planting	300	-	300	-	-
Aug-2004	-	1270	480	1710	360	-
Dec-2005	-	6350	630	6310	1450	0.877
Jan-2007	-	7110	1090	8010	1590	0.014

Table 3. Mean Diameter at Breast Height (DBH) of Trees

Date	Trees Type	Lysimeter		Unlined Section		P-Value
		DBH (mm)	Std Dev (mm)	DBH (mm)	Std Dev (mm)	
Dec-2005	Eucalyptus	14	4.6	16	5.1	0.082
	Cottonwood	14	3.4	17	5.6	0.061
Jan-2007	Eucalyptus	67	22.2	78	18.6	0.019
	Cottonwood	53	14.8	68	23.3	0.004

Soil Water Storage and Soil Temperature

Fig. 2 shows soil water storage inside and outside the lysimeter. Generally, the soil water storage in the lysimeter was slightly higher than the storage in the unlined test section. The average soil water storage in the lysimeter was 429 mm while that of the unlined test section was 408 mm and these values were significantly different ($P<0.001$). The higher soil water storage in the lysimeter may be due to the presence of the geocomposite at the bottom boundary. The presence of geocomposite may induce capillary effects and increase the water content right above it. The soil temperature inside the lysimeter was slightly lower than soil temperature in the unlined section (Fig. 3). The decreased temperature inside the lysimeter might be caused by the fact that the lysimeter act as a container for the soil with the liner acting as a thermal separation layer between the outside and the inside of the lysimeter. Analysis of variance (Table 4) indicates that there is a significant difference between the soil temperature in the lysimeters and in the unlined test sections.

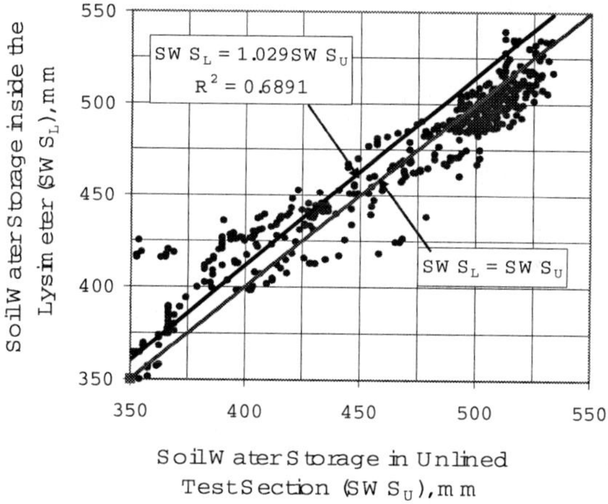

Fig. 2. Comparison of soil water storage inside and outside the lysimeter

Table 4. Soil Temperature in the Lysimeter and Unlined Section

Trees	Depth (mm)	Lysimeter		Unlined Test Section		P-Value
		Soil Temp. (^{o}C)	St Dev (^{o}C)	Soil Temp. (^{o}C)	St Dev (^{o}C)	
Eucalyptus	75	22.14	5.92	23.21	6.30	<0.01
	950	24.91	4.36	25.48	4.55	<0.01
Cottonwood	150	22.93	6.14	26.17	5.61	<0.01
	750	26.19	4.89	27.20	5.08	<0.01

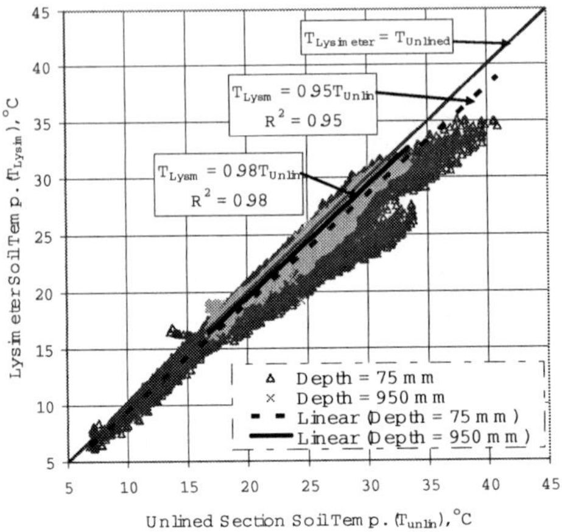

Fig. 3. Soil temperature inside and outside the eucalyptus trees lysimeter

SUMMARY AND CONCLUSIONS

Two ET cover lysimeters and two ET unlined test section were constructed in Tallahassee, Florida to assess the effect of bottom boundary on the performance of lysimeters. Assessment of the effect of the bottom boundary on the performance is achieved by comparing water storage, soil temperature, and trees growth in lysimeters to those measured in unlined test sections. Generally, the average heights of trees inside the lysimeters were slightly less that those outside the lysimeters. The diameters at breast height (DBH) inside the lysimeters were also smaller than those of the trees outside the lysimeters. Analysis of variance of tree growth inside and outside lysimeters, however, indicated that there is no significant difference in tree height for eucalyptus trees for all three measurement events. The growth rate of eucalyptus trees inside the lysimeter also showed no significant difference to that of trees outside the lysimeter. ANOVA of both eucalyptus trees and cottonwood trees revealed no significant difference between DBH inside the lysimeter to that outside the lysimeter (December 2005 data). However, ANOVA of DBH measured on January 2007 showed a significant difference between trees inside the lysimeter to those outside the lysimeter for cottonwood trees. Additional monitoring is being performed to monitor the long term growth and tree development in all test sections.

The soil water storage in the lysimeter is slightly higher than the storage in the unlined test section. The higher soil water storage in the lysimeter may be due to the presence of the geocomposite at the bottom boundary. However, ANOVA comparison of the soil water storage inside and outside the lysimeter showed no significant difference between them. The soil temperature inside the lysimeter seems to be slightly lower than the soil temperature in the unlined test section. ANOVA comparison of the soil temperature inside and outside the lysimeter also showed that there is a significant difference temperature inside and outside the lysimeter.

ACKNOWLEDGMENTS

Financial support was provided by the Florida Center of Solid and Hazardous Waste Management (FCSHWM), and the U.S. EPA-Science to Achieve Results (STAR) Program,

REFERENCES

Abdou, H. M., and Flury, M. (2004), Simulation of Water Flow and Solute Transport in Free-Drainage Lysimeters and Field Soils with Heterogeneous Structures. *European Journal of Soil Science,* Vol.55: 229-241

Albright, W., Benson, C., Gee, G., Roesler, A., and Abichou, T. (2001), Field Water Balance of Landfill Final Covers. *Journal of Environmental Quality,* Vol.33: 2317-2332

Benson, C., Abichou, T., Albright, W., Gee, G., and Roesler, A. (2001), Field Evaluation of Alternative Earthen Final Covers. *International Journal of Phytoremediation,* Vol.3: 105-127.

Benson, C., Bosscher, P., Lane, D., and Pliska, R. (1994), Monitoring System for Hydrologic Evaluation of Landfill Covers. *Geotechnical Testing Journal,* Vol.17: 138-149

Bergstrom, L., and Jarvis, N. (1993), Leaching of Dichlorprop, Bentazon, and 36Cl in Undisturbed Field Lysimeters of Different Agricultural Soils. *Weed Science,* Vol. 41: 251-261.

Bews, B., Wickland, B., and Barbour, S. (1999). Lysimeter Design in Theory and Practice. *Tailing and Mine Waste 1999*, Balkema, Rotterdam, 13-21

Liu, X. (2004), Design of Cost – Effective Lysimeter for Field Evaluation of Alternative Landfill Cover Projects Using Hydrus 2D Simulation. *MSc. Thesis*, Florida State University, Tallahassee

Putz, T., Brumhard, B., Dressel, J., Kaiser, R., Wu¨Stemeyer, A., and Scholz, K. (1998), FELS: A Comprehensive Approach to Studying the Fate of Pesticides in Soil at the Laboratory, Lysimeter and Field Scales. *ACS Symposium Series,* Vol.699: 152-162

Van Bavel, C. (1961), Lysimetric Measurements of Evapotranspiration in the Eastern United States. *Soil Science Society of America*, Vol. 25: 138 – 141

Wing, R. and Gee, G. (1994), Quest for the Perfect Cap. *Civil Engineering*, October, 1994, pp. 38-41

ESTIMATING METHANE EMISSION AND OXIDATION FROM EARTHEN LANDFILL COVERS

T. ABICHOU[1], L. YUAN[2], J. CHANTON[3]

[1]Associate Professor, Florida State University, Civil and Environmental Engineering Dept., 2525 Pottsdamer Street , Tallahassee, FL 32310; abichou@eng.fsu.edu

[2] Senior Staff Engineer, Geosyntec Consultants, Columbia, MD.

[3] Professor, Dept of Oceanography, Florida State University, Tallahassee, FL 32303.

ABSTRACT: Measuring methane emission and oxidation through landfill covers has been well studied. However, numerical methods to estimate methane emission and oxidation are very limited. A simulation model was developed that combined water and heat flow model and a gas transport and oxidation model. The gas transport and oxidation model is able to use dynamic parameters associated with water content and temperature and incorporate into dynamic methanotrophic activity. Four sites were selected to showcase how emissions and oxidation can be estimated knowing cover design, management practices, and climatic conditions. Simulations were performed for scenarios with and without an active gas collection system. Different simulations were performed with and without organic amendments to the soil cover. Thirty-two simulations were conducted under different locations, climate conditions, bottom pressure boundaries, and soil oxidation capacities.

Simulations showed that soil covers in subhumid areas can prevent high methane emission with blockage and decent oxidation capacity. In semiarid sites, higher emission was obtained due to the higher air filled void space of the soil. Oxidation capacities in semiarid sites are higher than those in subhumid sites since influxes of methane are higher in semiarid sites. High pressure underneath the cover caused higher emission in all sites. Even with active gas collection system (vacuum pressure), emissions were significant in semiarid climates. Soil oxidation is not only dependent on the potential methane oxidation capacity (Vmaxmax), but also depends on methane availability.

INTRODUCTION

The global annual input of methane to the atmosphere is estimated to be 535±125 Tg (IPCC, 1995), of which about half is considered to be both anthropogenic and originating from biospheric processes, particularly anaerobic bacterial fermentation. Decomposition of refuse in municipal landfills is believed to be one of the major components of this biogenic methane, but past estimates of the emissions from this source have varied greatly, from 9 to 70 Tg per year. More reliable estimates are clearly needed, but it appears that landfills are the largest anthropogenic source of atmospheric methane in the United States and European countries. Rates for both methane production and oxidation can exceed observed rates for other terrestrial ecosystems by large factors. Field flux measurements (net emissions) vary over 7 orders of magnitude, from less than 0.0004 to about 4,000 grams per square meter per day (Bogner et al., 1997). These net emissions, of course, are the result of methane production, oxidation, and gaseous transport processes in the cover soil. Hence, it is difficult to predict emission rates at sites with various cover types, climatic regimes, and management practices. Oxidation rates in these soils range up to over 100 g m^{-2} d^{-1}, and in some cases, the landfill can be a net sink for atmospheric methane oxidation. Emission data can be obtained by chamber, inert tracer, and micrometeorological methods.

Water content is a very important factor affecting CH_4 transport and oxidation in landfill cover soils. Boeckx et al. (1996) used a multiple linear regression analysis under different incubation conditions and concluded that water content has more influence on CH_4 oxidation than temperature. Christophersen et al. (2000) used statistical methods to analyze the effect of soil water content on CH_4 oxidation. They also concluded that water content can explain most of the variation observed in CH_4 emission data. Water plays three important roles. First, the optimum environment for CH_4 oxidizing bacteria (methanotrophy) is obtained at certain water content. Second, water content affects the penetration of oxygen (O_2) into the soils, which is the main reactor for CH_4 oxidation. As the water content increases in the soil, the O_2 diffusion into the soil is hindered. Thirdly, water content affects the air filled porosity of the soil and influences gas transport through the soil. As water fills the pores in the soil, it blocks the flow of gas upward. At the same time, the blocking of flow might lead to CH_4 emission due to the excess pressure built-up in the landfill (Boeckx et al., 1996).

Generally, CH_4 oxidation rate increases with increasing temperature (De Visscher and Van Cleemput, 2003). Low temperatures inhibit CH_4 oxidation (Whalen et al. 1990; Nozhevnikova et al., 1993; Borjesson and Svensson, 1997; Visvanathan et al., 1999). Borjesson and Svensson (1997) even reported that soil temperature is the controlling factor of CH_4 oxidation, and can explain 85% of the variation in measured CH_4 oxidation. Methanotrophic bacteria favor a certain range of temperatures. Czepiel et al. (1996) reported that oxidation rate increased as temperature increased to 36°C. They also reported that CH_4 oxidation essentially stopped when temperature reached 45°C. Humer and Lechner (2001) reported that CH_4 oxidation rate was 70-80% at 18°C. At a lower temperature of 4°C, little oxidation was observed. Borjesson and Svensson (1997) reported that the optimum temperatures for CH_4 oxidation were ~25 to 35°C.

This paper present a numerical model which combines water balance, heat transport models with a gas transport and oxidation model to estimate methane emission and oxidation from landfills with different climates. Four sites were selected to showcase how emissions and oxidation can be estimated knowing cover design, management practices, and climatic conditions. Simulations were performed for scenarios with and without an active gas collection system. Different simulations was performed with and without organic amendments to the soil cover.

MATERIALS AND METHODS

Four sites were selected to run the model for typical subhumid-warm weather at Florida (FL), subhumid-cold weather at Iowa (IA), semiarid-warm weather at California (CA), and semiarid-cold weather at Montana (MT). In the wet areas (FL and IA), the covers were built thicker, 135 cm and 165 cm, respectively, to limit percolation , while in the dry areas, the covers were build thinner of 120cm thick each. All the covers are well vegetated.

HYDRUS1D Modeling Description

The investigation of the volumetric water content and temperature profile of landfill cover on their performance involved simulations of water and heat flow in variably saturated soils using the computer program HYDRUS1D v3.0 (Simunek et al. 2005). The program numerically solves the Richards' equation for saturated–unsaturated water flow as follows:

$$\frac{\partial \theta}{\partial t} = \frac{\partial}{\partial x}[K(\frac{\partial h}{\partial x} + \cos\alpha)] - S \quad (1)$$

where θ is volumetric water content, h is pressure head [L], x are the spatial coordinates [L], t is time [T], S is the sink term [$L^3L^{-3}T^{-1}$], α is the angle between the flow direction and the vertical axis (i.e., α = 0 for vertical flow, 90 for horizontal flow, and 0 < α < 90 for inclined flow), The hydraulic properties K [m s^{-1}] were represented by the Mualem-van Genuchten function (Van Genuchten, 1980).

$$S_e = [1+(ah)^n]^{(1/n-1)} \tag{2}$$

$$K_r = K_s S_e^l [1-(1-S_e^{1/(1-1/n)})^{(1-1/n)}]^2 \tag{3}$$

where Ks is saturated hydraulic conductivity, Kr is relative hydraulic conductivity. l is a hydraulic conductivity parameter. a is related to the air-entry value [m^{-1}], and n is a pore size distribution parameter. *Se* in Eq. 2 is expressed by Eq. 4

$$S_e = \frac{\theta - \theta_r}{\theta_s - \theta_r} \tag{4}$$

where θ_r is the residual water content [m^3 m^{-3}]. θ_s is the saturated water content [m^3 m^{-3}]

S in Eq. 1 is a sink term to account for water uptake by plant roots and is defined as:

$$S(h) = \alpha(h) T_p / L_R \tag{5}$$

where, $\alpha(h)$ is the plant water stress function, T_p is the potential transpiration rate[LT^{-1}], L_R the depth [L] of the root zone.

Heat transport through the cover was coupled to water transport and is described with a convection-dispersion equation of the form:

$$\frac{\partial C_p(\theta)T}{\partial t} = \frac{\partial}{\partial x}[\lambda(\theta)\frac{\partial T}{\partial x}] - C_w \frac{\partial qT}{\partial x} - C_w ST \tag{6}$$

where $\lambda(\theta)$ is the coefficient of the apparent thermal conductivity of the soil [M m s^{-3} K^{-1}]. $C_p(\theta)$ and *Cw* is the volumetric heat capacities [M m^{-1}s^{-2} K^{-1}] of the porous medium and the liquid phase, respectively. q is Darcian fluid flux density [m s^{-1}]

Gas Transport Modeling

Volumetric water content and temperature were generated at each node each day by HYDRUS1D. The gas transport model used these dynamic results to simulate methane emission and oxidation at each day.

A continuity equation and a mass balance equation were used to describe the gas flow and reaction within the porous media.

$$\varepsilon \frac{dC_i}{dt} = -\frac{dJ_i}{dx} + r_i \tag{7}$$

where ε is the air filled porosity [m$^{-3}_{gas}$ m$^{-3}_{soil}$], C_i, the molar gas concentration [mol m^{-3}], J_i is the flux of gas component i including the diffusive and advective flux (mol m^{-2} s^{-1}), r_i is the reaction rate of gas component i [mol kg$^{-1}_{dry\ soil}$ s^{-1}], and dt [s] and dx [m], are time and vertical distance coordinates.

The flux of the gas, J_i, has two components: diffusion and advection. Gas diffusion in porous media is governed by Fick's law and gas advection is governed by Darcy's law, the total flux can then be expressed as follows:

$$J_i = -D_{soil,i}\frac{dC_i}{dx} - \frac{k}{\mu}\frac{dP}{dx}C_i \tag{8}$$

where the diffusion coefficient of gas component $i, D_{soil,i}$, in soil [m^2 s^{-1}]. k is the intrinsic permeability of the soil (m^2), μ is the gas mixture viscosity [N s m^{-2}] and P is the pressure [Pa].The pressure P is obtained by the ideal gas law:

$$P = \sum_{i=1}^{4} C_i RT \tag{9}$$

where T is the absolute temperature (K) and R is universal gas constant [8.314 J K^{-1} mol^{-1}].

Methanotrophic Reaction

The reaction component of the gas transport equation was assumed to be the following (De Visscher and Van Cleemput 2003):

$$CH_4 + aO_2 \rightarrow bCO_2 \tag{10}$$

where a, b are the stoichiometric factors for oxygen and carbon dioxide, and were assumed to be 1.5 and 0.5. This leads to the following reaction equation:

$$CH_4 + 1.5O_2 \rightarrow 0.5CO_2 + 0.5CH_2O + 1.5H_2O \tag{11}$$

where CH_2O represents biomass. r_i in Eq. 7 is the reaction rate of methanotrophic bacteria, which is calculated by Michaelis-Menton kinetics from the incubation experiment.

$$r_{CH_4} = V_{max} \frac{C_{CH_4}}{K_{m[CH_4]} + C_{CH_4}} \frac{C_{O_2}}{K_{m[O_2]} + C_{O_2}} \tag{12}$$

where V_{max} is the maximum methane consumption rate [nmol s^{-1} $kg^{-1}_{dry\ soil}$] and K_m is the half saturation constants [mol m^{-3}]. The Combined model simulation scheme is shown a flow chart (Fig. 1).

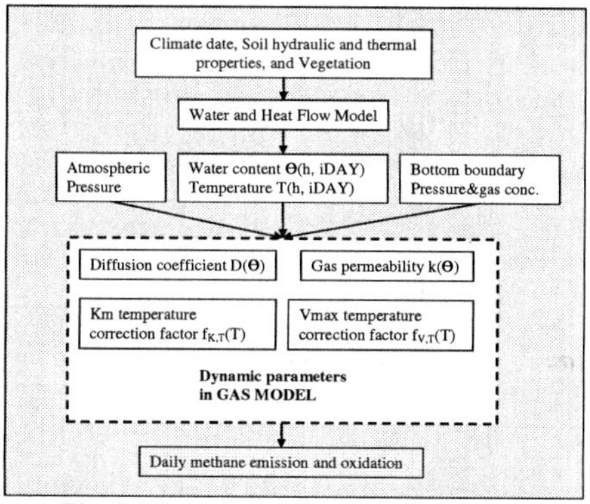

Fig. 1. Flow chart of model algorithm.

Boundary, Initial Conditions, and Input Assumptions

The gas composition at the surface node of the model was assumed to be the atmospheric gas compositions, which are 21.21% oxygen, 1.8 ppmv methane, 370 ppmv carbon dioxide and 78.75% nitrogen. For the bottom boundary, different constant pressures were used to describe the different situation of landfill pressure (as described below). Constant gas concentration boundary condition (60% methane and 40% carbon dioxide) was selected.

Climatic input data for FL site was obtained from a weatherstation build nearby the site. Other sites' historical data were obtained from regional climate centers. The data included daily precipitation, potential evapotranspiration (PET), Maximum temperature, minimum temperature, and average temperature. These climate data were used as a time-variable atmospheric boundary condition. The soil properties for HYDRUS1D were measured for IA, MT, and CA sites. The soil properties for the FL site were estimated based on soil classification. In order to minimize the influence of initial condition, a period of ten consecutive cycles of fourteen months period was modeled by HYDRUS1D code. At the end of this period, daily water content and temperature in each node were obtained for the input of gas transport and oxidation model. Thirty-two gas transport and oxidation simulations were conducted for different bottom pressure boundary condition and different soil oxidation capacity. Four kinds of pressure boundaries were used to represent different landfill conditions. (1) *High* pressure underneath the cover equal to constant 1.1 atm; (2) *Medium* pressure underneath the cover equal to constant 1.04 atm; (3) *Zero* pressure at bottom boundary equal to atmospheric pressure; (4) *Vacuum* pressure at the bottom boundary equal to -10 inches of water (2.49 kPa) as a typical vacuum pressure underneath cover when active landfill gas collection system is on. Two types of soil oxidation capacity were selected as: (1) High oxidation capacities as Vmaxmax=2000 nmol kg^{-1} s^{-1}; (2) low oxidation capacity as Vmaxmax=500 nmol kg^{-1} s^{-1}, where Vmaxmax is the maximum value that Vmax can be reached (De Visscher and Van Cleemput, 2003)

RESULTS AND DISCUSSION

Water Balance

Water balance results are shown in Table 1. The total water applied to the cover (precipitation) was divided into four parts: runoff, evaporation, transpiration, and percolation. For semiarid areas (CA and MT), the cumulative precipitation for the entire period was 503 and 357.9 cm, respectively. In these semiarid areas, the runoffs were zeros since the precipitation was low and no heavy density of rainfall event exist. The transpiration of MT site was only 140.5 cm compared to 217.9 of CA site since the vegetation grow season of MT is shorter than CA. Overall percolation ratio over precipitation for CA and MT were 5.78% and 4.36%. For the subhumid areas (FL and IA), the cumulative precipitation for the entire period was 1717.9cm and 977.1cm, respectively. The runoff of FL site was much higher than IA site since the rainfall was basically very dense in several events in FL. Transpiration and evaporation in FL was also higher than IA site, therefore the percentage of percolation over total precipitation for FL was 8.67%, which was lower than the IA site for 16.13%.

Table 1. HYDRUS1D modeling results.

Water balance (cm)	CA	FL	IA	MT
Precipitation	503	1717.9	977.1	357.9
Runoff	0	445.9	62.4	0
Evaporation	255.9	462.9	226.3	211.6
Transpiration	217.9	659.8	509.6	140.5
Percolation	29.1	149.3	157.7	15.6
Percentage of Percolation	5.78%	8.67%	16.13%	4.36%

Methane Emission and Oxidation Modeling

Fig. 2 shows 32 simulations with different pressure boundary condition and different Vmaxmax. Complete values are listed in Table 2. For semiarid sites (CA and MT), the overall emissions are higher than the wet sites (FL and IA), since there was more air filled void space for gas to migrate through the cover. Methane emission in MT site ranged from 2877.1~44904.1 g m-2 yr-1 and from 466~24152.6 g m-2 yr-1 in CA. However for the wet sites, the emission in FL was only 160.6~857.4 g m-2 yr-1 and 48.2~118 g m-2 yr-1 in IA. Lowest emissions were obtained when the bottom boundary is vacuum and with higher

Vmaxmax for all sites. Highest emissions were obtained when the bottom boundary is high pressure (1.1 atm) and with low Vmaxmax.For each site, high Vmaxmax leads to low emission and low Vmaxmax leads to high emission. The differences of methane emission due to Vmaxmax of Dry sites are larger than wet sites due to the higher influx of dry site. At medium pressure, the differences of emissions between high Vmaxmax and low Vmaxmax were 4.3, 163.5, 3520.1, and 5617.6 corresponding to influx values of 199.9, 3806.3, 21209, and 32744 for IA, FL, CA, and MT, respectively. The reason for phenomena is that low influx in wet sites had no enough methane for the methanotrophic bacterial to reach its oxidation capacity. So choosing a high oxidation capacity soil can be an efficient way to mitigate methane emissions, however for low emission site such IA, high oxidation capacity soil didn't significant reduced the emission.

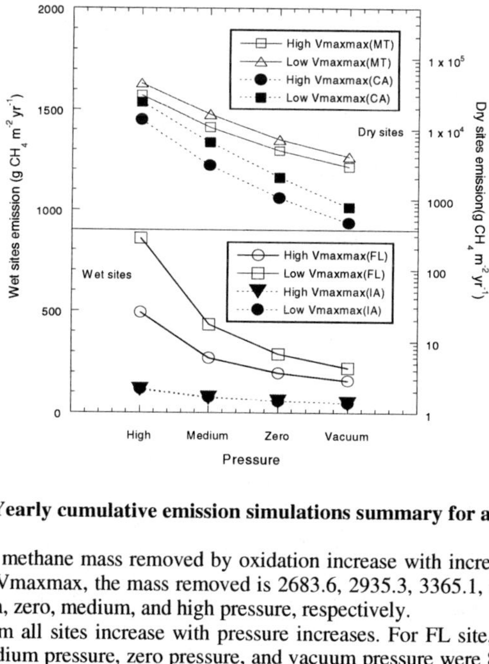

Fig. 2 Yearly cumulative emission simulations summary for all sites.

For each site, methane mass removed by oxidation increase with increasing of pressure. In FL site at low Vmaxmax, the mass removed is 2683.6, 2935.3, 3365.1, and 4098.1 g CH_4 m^{-2} yr^{-1} for vacuum, zero, medium, and high pressure, respectively.

Emissions from all sites increase with pressure increases. For FL site, the emissions of high pressure, medium pressure, zero pressure, and vacuum pressure were 857, 434.4, 290.9, and 221.9 g m^{-2} yr^{-1}, respectively (low Vmaxmax). This show even with active landfill gas collection system, there was still emission of methane into the atmosphere due to the diffusion of methane. In dry sites, this emissions can be as high as 785.5 and 4039 g m^{-2} yr^{-1} for CA and MT site, respectively (low Vmaxmax).

Table 2 summary of the yearly cumulative emission of all ET sites (g m^{-2})

	Pressure		IA	FL	CA	MT
High Vmaxmax	High	Influx	264.5	5173.4	42795.6	62726.3
		outflux	114.6	493.2	13724	30339
		mass removed	149.9	4680.2	29071.6	32387.3
		oxidation	56.67%	90.47%	67.93%	51.63%
	Medium	Influx	199.9	3806.3	21208.6	32744.4
		outflux	73.6	270.9	3014.5	10657
		mass removed	126.3	3535.4	18194.1	22087.4
		oxidation	63.18%	92.88%	85.79%	67.45%
	Zero	Influx	168.2	3230.6	15254.6	20954
		outflux	58.1	198	1056.8	4943.4
		mass removed	110.1	3032.6	14197.8	16010.6
		oxidation	65.46%	93.87%	93.07%	76.41%
	Vacuum	Influx	151.2	2909.4	10307.3	16412.2
		outflux	48.2	160.6	466	2877.1
		mass removed	103	2748.8	9841.3	13535.1
		oxidation	68.12%	94.48%	95.48%	82.47%
Low Vmaxmax	High	Influx	264	4955.5	42359	63566.5
		outflux	118	857.4	24152.6	44904.1
		mass removed	146	4098.1	18206.4	18662.4
		oxidation	55.30%	82.70%	42.98%	29.36%
	Medium	Influx	199.6	3799.5	21049	32277.8
		outflux	77.9	434.4	6534.6	16274.6
		mass removed	121.7	3365.1	14514.4	16003.2
		oxidation	60.97%	88.57%	68.96%	49.58%
	Zero	Influx	168	3226.2	15199.2	20759.5
		outflux	59.1	290.9	2030.8	7195
		mass removed	108.9	2935.3	13168.4	13564.5
		oxidation	64.82%	90.98%	86.64%	65.34%
	Vacuum	Influx	151	2905.5	10286	16300.4
		outflux	48.9	221.9	785.5	4039
		mass removed	102.1	2683.6	9500.5	12261.4
		oxidation	67.62%	92.36%	92.36%	75.22%

CONCLUSIONS

A model was developed to study how landfill final earthen cover's construction and climate conditions affect methane emission and oxidation from landfills. This numerical model combines the water and heat flow model (HYDRUS1D) and a gas transport and oxidation scheme. The gas transport and oxidation model is able to incorporate dynamic parameters associated with water content and temperature, such as, air filled porosity, diffusion coefficient, and gas permeability. A dynamic oxidation capacity parameter (Vmax) is also embedded in the gas transport scheme. Four sites were selected to showcase how emissions and oxidation can be estimated knowing cover design, management practices, and climatic conditions. Simulations were performed for scenarios with and without an active gas collection system. Different simulations was performed with and without organic amendments to the soil cover

Simulations showed that soil covers in subhumid areas can prevent high methane emission with blockage and decent oxidation capacity. In semiarid sites, higher emission was obtained due to the higher air filled void space of the soil. Oxidation capacities in semiarid sites are higher than those in subhumid sites since influxes of methane are higher in semiarid sites. High pressure underneath the cover caused higher emission in all sites. Even with active gas collection system (vacuum pressure), emissions were significant in semiarid climates. Soil oxidation is not only dependent on the potential methane oxidation capacity (Vmaxmax), but depends on methane availability. The simulations are however the results of a numerical analysis. A verification of the results of the predictions of the developed model should be performed.

REFERENCES

Boeckx, P., Cleemput, O., and Villaralvo, I. (1996), "Methane emission from a landfill and the methane oxidizing capacity of its covering soil." Soil Biol. Bioche., 28, 1397-1405.

Bogner, J.,, Meadows, M., and Czepiel, P. (1997), "Fluxes of CH4 between landfills and the atmosphere: Natural Engineered Control." Soil Use and Management, 3, 268-277.

Borjesson, G., and Svensson, B. (1997), "Seasonal and Diurnal Methane Emissions from a Landfill and their Regulation by Methane Oxidation." Waste Management Research, 15, 33-54.

Christophersen, M., Linderod, L., Jensen, P., and Kjeldsen, P. (2000), "Methane oxidation at low temperatures in soil exposed to landfill gas." J. Environ. Qual., 29, 1989-1997.

Czepiel, P , Mosher, B., Crill, P., and Harriss, R. (1996), "Quantifying the Effects of Oxidation on Landfill Methane Emissions." Journal of Geophysical Research, 101(D11), 16721-16729.

De Visscher, A., and Cleemput, O. (2003), "Simulation model for gas diffusion and methane oxidation in landfill cover soils." Waste Management 23, 581-591.

Humer, M., and Lechner, P. (2001), "Microbial Methane Oxidation for the Reduction of Landfill Gas Emissions." Journal of Solid Waste Technology and Management, 27, 3-4.

IPCC. (1995). "Climate change 1995: the science of climate change." Intergovernmental Panel on Climate Change, Cambridge, U.K.

Nozhevnikova, A., Nekrasova, V.., Lebedev, V., and Lifshit, A. (1993), "Microbiological Process in Landfills." Water Science and Technology, 27, 243-252.

Šim nek, J., van Genuchten, M. and Šejna, M. (2005), The HYDRUS-1D Software Package for Simulating the Movement of Water, Heat, and Multiple Solutes in Variably Saturated Media, Version 3.0, HYDRUS Software Series 1, Department of Environmental Sciences, University of California Riverside, Riverside, California, USA, 2005.

Van Genuchten, M. (1980), A closed-form equation for predicting the hydraulic conductivity of unsaturated soils. Soil Sci. Soc. Am. J., 44, 892-898

Visvanathan, C., Pokhrel, D., Cheimchaisri, W., Hettiaratchi, J. P. A., and Wu, J. S. (1999). "Mathanothropic activities in tropical landfill cover soils: Effects of temperature, moisture content and methane concentration." Waste Manage. Res., 17, 313-323.

Whalen, S. C., Reeburgh, W. S., and A.Sandbech, K. (1990). "Rapid methane oxidation in landfill cover soil." Applied Environ. Microbiol., 62, 3203-3209.

Influence of the Waste Layer on Percolation Estimates for Earthen Caps Located in a Sub-humid Climate

Milind V. Khire[1], Ph.D., P.E. and Ramil G. Mijares[2]

[1]Assistant Professor, Department of Civil & Environmental Engineering, Michigan State University, A135 Research Complex Engineering, E. Lansing, MI 48824; khire@egr.msu.edu
[2]Graduate Research Assistant, Department of Civil & Environmental Engineering, Michigan State University, A30 Research Complex Engineering, E. Lansing, MI 48824.

ABSTRACT: In this numerical study, an alternative cap consisting of a vegetative layer and a moisture storage layer was simulated with a bottom boundary consisting of either a lysimeter or an underlying waste layer for a sub-humid northern climate (Detroit, Michigan) using the water balance model Vadose/W. Various combinations of saturated and unsaturated hydraulic properties of the cap layers and the waste layer were simulated using the numerical model. In this study, simulated percolation rates were more for lysimeters compared to caps underlain by the waste.

INTRODUCTION

When landfill reaches its capacity and is scheduled for closure, a final capping system is installed. In the U.S., the minimum required design is a conventional cap which may consists of, among other components, a geomembrane liner and/or a compacted clay barrier layer as [Resource Conservation and Recovery Act (RCRA), Subtitle D, USEPA]. Capital construction cost of the conventional cap is usually greater than alternative earthen caps. RCRA includes a provision that allows alternative final cap designs as long as they are capable of achieving an equivalent reduction in infiltration and protection from erosion that prescriptive caps may provide.

Several types of alternative caps have been considered for landfills primarily located in areas with arid and semi-arid climates (Khire et. al. 1999; Albright et. al. 2004; Scanlon et. al. 2005). The best known method for assessing the performance of these alternative caps entails the installation of lysimeters at the proposed site to directly measure the percolation through the proposed cap. Numerical studies on lysimeters generally employ simulations neglecting the effect of the underlying waste (Khire et. al. 2000; Benson et. al. 2005) because percolation is collected through a drainage layer constructed above a geomembrane. While the geomembrane allows collection of infiltration through the cap, it prevents upward migration of moisture from the waste under evapotranspirative gradients. When an actual earthen cap is constructed, the geomembrane layer used for a lysimeter is not placed. Hence the key objective of this numerical study is to compare long-term predicted percolation rates through lysimeters and actual caps (without lysimeters). A two-layer earthen cap located in a sub-humid climate (Detroit, Michigan) was simulated.

WATER BALANCE MODEL

Commercially available finite element water balance model Vadose/W (Geo-Slope 2007) was used in this study. Vadose/W numerically solves a modified form of Richards' equation to estimate the flow of water through both saturated and unsaturated soil media in one or two-dimensions. While a few two-dimensional simulations were carried out, a majority of simulations were one-dimensional. The differences between the 2-d and 1-d simulated percolations were not significant.

Vadose/W has been selected for this numerical study because several studies have used Vadose/W to evaluate the effectiveness of earthen caps and have demonstrated that it predicts the overall water balance of earthen caps reasonably well (Benson et. al. 2005, among others). For this numerical study, the Vadose/W simulations were carried out using Vadose/W v.07.

NUMERICAL MODELING

An landfill final cap consisting of a vegetative layer (VL) and a storage layer (SL) was considered in this study. This cross section is a relatively basic form for an alternative cap. A schematic of the conceptual model used in this numerical study is presented in Fig. 1. To evaluate the impact of the underlying waste on the hydrology of the earthen cap, a waste layer (WL) was also included underneath the storage layer for all simulations except when a lysimeter was simulated.

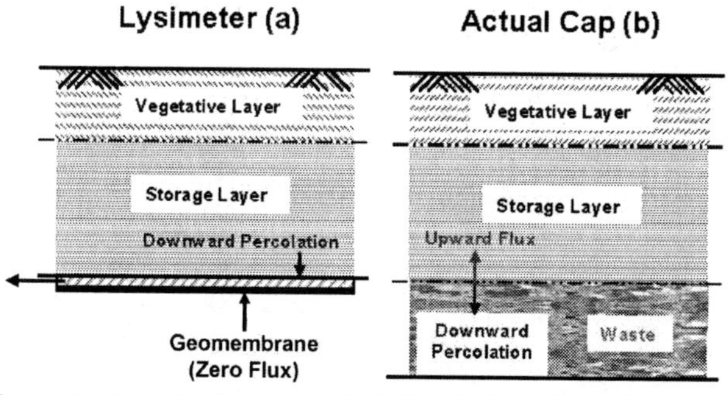

FIG. 1. Conceptual model for numerical simulation of an alternative capping system consisting of: (a) lysimeter only; and (b) cap with underlying waste.

In this numerical study, the thicknesses of the vegetative layer (T_{VL}) and the thickness of the storage layer (T_{SL}) were fixed at 0.3 m (1 ft) and 1.0 m (3.5 ft), respectively. Thickness of the waste layer (T_{WL}) was varied from 1.6 to 6.6 m.

Material Properties

Table 1 summarizes the saturated and unsaturated hydraulic properties of the soils used to represent the three layers in the model domain: VL, SL, and WL. Loose and compacted clays simulate the cap layers VL and SL, respectively. Hydraulic properties of municipal solid waste reported by Benson and Wang (1998) were assumed to represent WL.

Table 1. Saturated and unsaturated properties of the cap components

Layer	Material	Thickness (m)	θ_s (cm³/cm³)	a^1 (kPa)	n^1	m^1	K_{sat} (cm/s)
VL	Loose clay (ML-CL)	0.3	0.41	26.79	10.31	0.37	$5.0 \times 10^{-5\text{ to }-3}$
SL	Compacted clay (CL)	1.0	0.35	156.05	0.75	1.17	2.0×10^{-7}
	Silty Clay (ML)		0.52	3.08	1.31	0.44	2.0×10^{-6}
	Sandy Silt (SM-ML)		0.35	7.26	1.45	0.21	2.0×10^{-5}
WL	MSW (Benson & Wang 98)	1.6 to 6.6	0.53	0.29	3.28	0.60	$2.0 \times 10^{-6\text{ to }-2}$

Note: [1] Fredlund-Xing (1994) function was used for θ-ψ fitting.

Data points for the soil water characteristic curves (SWCC) for the cap layers were obtained from Hines (2006) for the ML-CL and CL soils used for simulating VL and SL, respectively and for the ML and SM-ML soils reported by Khire et al. (2000) for the SL. The Fredlund and Xing (1994) function fits describing the SWCC are presented in Table 1. The fitting parameters presented in Table 1 are: θ_s is the saturated volumetric water content, ψ_r is the matric suction corresponding to the residual water content θ_r with typical values in the range of 1500 – 3,000 kPa, e is the natural number equal to 2.719, a, n, and m are Fredlund and Xing (1994) fitting parameters.

The unsaturated hydraulic conductivity functions were predicted from the estimated SWCCs using the Fredlund et al. (1994) function option in Vadose/W with measured saturated hydraulic conductivities for the soils (Table 1).

In this study, the influence of hydraulic properties of the cap layers as well as of the underlying waste layer was evaluated by simulating alternative caps with and without waste underneath the cap for various combinations of the saturated hydraulic conductivities. This was accomplished by simulating increase in the saturated hydraulic conductivity (K_S) values of VL, SL, and WL by one to 1 to 3 orders of magnitude as presented in Table 1. The SWCCs were changed to reflect the change in the K_S value. Because Vadose/W couples heat and mass flows to predict water balance parameters, two thermal functions are required by the program for each material type, i.e. volumetric specific heat and thermal conductivity functions. In particular, these functions are required for modeling thermal processes such as those encountered on evaporative flux boundary conditions typically imposed on analyzing soil covers.

The volumetric specific heat functions were estimated using De Vries (1963) estimation method. A built-in Vadose/W tool estimates the said function from the specified SWCCs. A soil mineral specific heat value of 0.71 was assumed for the waste material, while mass specific heat values for the compacted and loose clay soils were assumed equal to 0.8 and 1.674, respectively.

The thermal conductivity functions for each material were estimated using the built-in Vadose/W estimation tool. The thermal conductivity – volumetric water content relations were predicted using the Johansen (1975) estimation method with assumed soil mineral thermal conductivity of 216, 130, and 280 kJ/day/m/°C for the WL, SL, and VL soils, respectively.

Climate Data

Hourly meteorological data were recorded for approximately two consecutive years, from February 2005 to December 2006, at the landfill site in Detroit, Michigan where the alternative cover has been constructed. Field measurements were obtained for maximum and minimum temperatures, wind speed, as well as for hourly precipitation. Meanwhile, corresponding maximum and minimum values of relative humidity were obtained from a nearby local weather station maintained by the National Oceanic and Atmospheric Administration (NOAA). These meteorological data were input to Vadose/W. Net solar radiation for the site was estimated by Vadose/W based on the latitude (42.38° North) input. The ground surface was assumed bare to ignore transpiration to simulate conservative and consistent evaporation rates.

A sinusoidal distribution pattern was selected to distribute daily values of air temperature, relative humidity and net radiation between the calculated sunrise and sunset times. Precipitation was applied at an hourly frequency specifically during the time intervals it occurred during a day. Wind speed was applied as a constant value (5 km/hr) throughout the day.

For all numerical analyses, four-year simulations were performed by applying the two-year meteorological data successively. This was done to minimize the effect of the initial conditions on the simulated annual average quantities of percolation. Reported percolation values are annual percolation rates averaged for all four simulation years.

Average annual precipitation (P) at the landfill site is about 730 mm. Average annual potential evapotranspiration (PET) estimated by Vadose/W using a modified Penman method is 1,080 mm (P/PET ~ 0.68).

Initial and Boundary Conditions and Numerical Control Parameters

A variable flux boundary was applied at the ground surface (Fig. 1). A unit gradient boundary condition was imposed on the lower boundary of the model domain for lysimeter as well as the actual cap. Fayer et. al. (1992), Khire et. al. (1997, 1999), and Benson et. al. (2005) have used this boundary condition for lysimeters. Benson et. al. (2005) has shown that there is no significant difference between seepage face or unit gradient boundary conditions when used for simulating percolation using Vadose/W for lysimeters. The nodal spacing between the nodes located near the lower boundary was relatively small (1 mm) to ensure as little numerical error as possible due to the assigned boundary condition. Initial conditions for Vadose/W were specified using matric suctions corresponding to in-situ water contents corresponding to average degree of saturation equal to 70%.

Spatial discretization of the model domain was optimized by conducting sensitivity analysis. This was done by repeatedly refining the mesh density until insignificant changes on calculated water balance parameters were achieved. An adaptive time stepping was used to discretize the time domain into a series of incremental time steps. The minimum and maximum time steps were set at 0.24 and 2 hours, respectively. Time steps were controlled by the vector norm of all nodal heads. A maximum of 1% change in head per step was set in most simulations. This was necessary to accurately evaluate the extreme drying and wetting conditions typically encountered on the ground surface of the model. For all numerical analyses, mass balance errors of less than 1 mm/yr were achieved.

RESULTS

In this study, about 25 simulations were carried out using Vadose/W to quantitatively evaluate the effect of the presence of the underlying waste layer on the percolation through the cap.

Waste Layer Thickness

T_{WL} values of 1.6 m, 3.3 m, and 6.6 m were simulated while keeping T_{VL} and T_{SL} fixed at 0.3 and 1 m, respectively. Various combinations of magnitudes of K_{VL} and K_{SL} were simulated when K_{WL} was varied from 2 x 10^{-6} cm/s to 2 x 10^{-2} cm/s. Saturated hydraulic conductivity of municipal solid waste (MSW) has been reported to lie in the range of 10^{-4} ~ 10^{-3} cm/s by Oweis et. al. (1990) and Jang et. al. (2002).

Initial simulations showed that annual percolation and other water balance parameters differed by less than 0.1% when T_{WL} is increased from 3.3 to 6.6 m. Hence, T_{WL} was fixed at 1.6 m for all succeeding simulations.

Waste Layer Hydraulic Conductivity

Fig. 2 presents the ratio of average annual percolation for a cap without lysimeter (Pr_{CAP}) to that with a lysimeter (Pr_{LYS}) versus the hydraulic conductivity of the waste layer. Pr_{CAP} represents the percolation into the waste layer from the storage layer (Fig. 1). K_{WL} was varied from 2 x 10^{-6} cm/s to 2 x 10^{-2} cm/s. K_{VL} and K_{SL} were varied from 5 x 10^{-5} cm/s to 5 x 10^{-3} and from 2 x 10^{-7} cm/s (denoted by X 1) to 2 x 10^{-5} cm/s, respectively. The SWCC was also varied to reflect the increase in K_s. Fig. 2 shows that the percolation ratio decreases as the hydraulic conductivity of the waste layer increases. When the ratio is less than one, lysimeter overestimates percolation and vice versa. While cracks were not simulated in the cap soils, these results indicate that lysimeter will likely further overestimate percolation if cracks form and the bulk hydraulic conductivity of the cap increases.

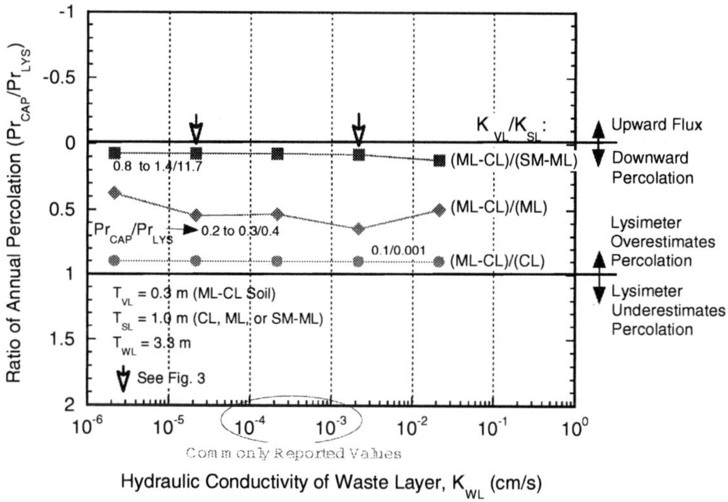

FIG. 2. Percolation ratio for various values of K_{WL}.

Fig. 3 presents cumulative percolation and surface runoff for K_{VL} and K_{SL} equal to 5 x 10^{-3} cm/s and 2 X 10^{-5} cm/s, respectively, and K_{WL} equal to 2 x 10^{-5} and 2 x 10^{-3} cm/s. Percolation ratios for these two simulations are represented by an arrow in Fig. 2. Fig. 3 shows that cumulative runoff for the lysimeter was about 40 mm more and cumulative evaporation was about 40 mm less compared to those for the caps underlain by the waste. Cumulative percolation for the caps underlain by the waste was less than that for lysimeter for all simulated K_{WL} values.

Partly the reason for the presence of a lysimeter overpredicting percolation is because the drainage collected from the lysimeter is immediately removed from the system and is not available for removal by evaporation (Fig. 3). When a geomembrane is not present, percolation into the waste may be temporarily stored in the underlying waste layer and can be potentially removed by evapotranspiration. Fig. 3 shows that between days 600 and 700 and 950 and 1000, all simulations except the lysimeter simulation indicated upward flux due to evaporation. While percolation ratio depends upon the hydraulic conductivities of all components of the cap, the simulations carried out in this study showed lysimeter conservatively estimates percolation for all values of K_{WL}. These results show that, because it is not possible to directly measure percolation without a lysimeter, the percolation values measured in the field need to be re-evaluated by carrying out numerical analysis that includes the underlying waste to simulate the long-term percolation scenario for the final cap.

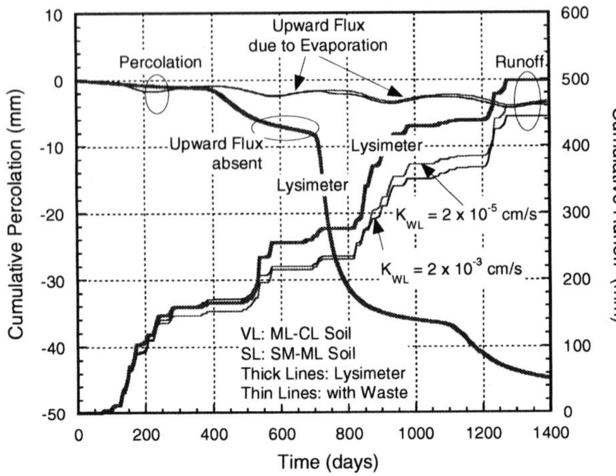

FIG. 3. Cumulative percolation and surface runoff with and without lysimeter

SUMMARY AND CONCLUSIONS

In this numerical study, an earthen final landfill cap consisting of a vegetative layer and a moisture storage layer was simulated with an underlying waste layer for a sub-humid climate (Detroit, Michigan) using the water balance numerical model Vadose/W. Flow through potential preferential paths was not simulated. Various combinations of saturated hydraulic conductivities of the cap layers and the waste layer and an equivalent lysimeter were simulated using the numerical model. The results indicate that the percolation magnitude is influenced not only by the hydraulic conductivities of the cap layers but also by the presence of the underlying waste layer.

For all combinations of K_{VL}, K_{SL}, and K_{WL}, the lysimeter percolation was more than that for the caps underlain by the waste. Because it is not possible to directly measure percolation without a lysimeter, the percolation values measured in the field should be re-evaluated by carrying out site-specific numerical analysis for representative field-scale hydraulic properties of the waste to simulate the long-term percolation scenario for the actual final cap.

ACKNOWLEDGEMENT

Financial support for this project has been provided by Waste Management, Inc (WMI). The authors gratefully acknowledge the technical assistance provided by Greg Newman of Geo-Slope with Vadose/W and Paul Mazanec and Terry Johnson of WMI for providing meteorological and soil data collected at the Woodland Meadows Landfill in Michigan. The findings and opinions presented in this manuscript are those of the authors.

REFERENCES

Albright, W., Benson, C., Gee, G., Roesler, A., Abichou, T., Apiwantragoon, P., Lyles, B., and Rock, S. (2004). "Field Water Balance of Landfill Final Covers." *J. Environmental Quality*, 33: 2317-2332.

Benson, C., Bohnhoff, G., Ogorzalek, A., Shackelford, C., Apinwantragoon, P., and Albright, W. (2005). "Field Data and Model Predictions for a Monolithic Alternative Cover." *Proc., Geo-Frontiers 2005* (CD-ROM), ASCE, Reston, VA.

Benson, C. and X. Wang (1998). "Soil Water Characteristic Curves for Solid Waste." Environmental Geotechnics Report, 98-3, UW-Madison.

De Vries, D. (1963). "Thermal Properties of Soils." *Physics of Plant Environment.* North-Holland Publishing Company, Amsterdam, The Netherlands: 210-235.

Fayer, M., Rockhold, M., and Campbell, M. (1992). "Hydrologic Modeling of Protective Barriers: Comparison of Field Data and Simulation Results." *Soil Science Society of America J.,* 56: 690-700.

Fredlund, D., and Xing, A. (1994). "Equations for the Soil-Water Characteristic Curve." *Canadian Geotechnical J.*, 31: 521-532.

Fredlund, D., Xing, A., and Huang, S. (1994). "Predicting the Permeability Function for Unsaturated Soils Using the Soil-Water Characteristic Curve." *Canadian Geotechnical J.*, 31: 533-546.

Geo-Slope (2007). "Vadose Zone Modeling with Vadose/W 2007: An Engineering Methodology." GEO-SLOPE International Ltd, Alberta, Canada.

Hines, J. (2006). "Laboratory Report for Waste Management, Inc.: Woodland Meadows Cover Soil." *December 2006 Report*, Daniel B. Stephens & Associates, Inc.

Jang, Y., Kim, Y., and Lee, S. (2002). "Hydraulic Properties and Leachate Level Analysis of Kimpo Metropolitan Landfill, Korea." *Waste Management*, 116 (4): 539-553.

Johansen, O. (1975). Thermal Conductivity of Soils. *Ph.D. Thesis*, University of Trondheim, Norway. CRREL Draft English Translation 637, US Army Corps of Engineers, Cold Regions Research and Engineering Laboratory, Hanover, N.H.

Khire, M., Benson, C., and Bosscher, P. (1997). "Water Balance Modeling of Earthen Final Covers." *J. Geotechnical and Geoenv. Engrg.*, 22 (3): 261-267.

Khire, M., Benson, C., and Bosscher, P. (1999). "Field Data from a Capillary Barrier and Model Predictions with UNSAT-H." *J. Geotechnical and Geoenv. Engrg.,* 125 (6): 518-527.

Khire, M., Benson, C., and Bosscher, P. (2000). "Capillary Barriers: Design Variables and Water Balance." *J. Geotechnical and Geoenv. Engrg.,* 126 (8): 695-708.

Oweis, I., Smith, D., Ellwood, R., and Greene, D. (1990). "Hydraulic Characteristic of Municipal Refuse." *J. Geotechnical Engrg.,* 116 (4): 539-553.

Scanlon, B., Reedy, R., Keese, K., and Dwyer, S. (2005). "Evaluation of Evapotranspirative Covers for Waste Containment in Arid and Semiarid Regions in the Southwestern USA." *Vadose Zone J.,* 4: 55-71.

Methane Oxidation on a Coverage Layer Study

Juliana L. Rose[1], Pedro Paulo F. Gouvêa[2] and Cláudio F. Mahler[3]

[1] Doctorating, Civil Engineering Department - COPPE/UFRJ, Rio de Janeiro, Brazil; julrose@gmail.com
[2] Civil Engeneering. Bentonit União Nordeste Ltda., Santo Amaro, São Paulo, Brazil; ppfgouveia@bentonit.com.br
[3] Prof. Dr., COPPE/UFRJ. Coordinator of the Group for Study of Residue Technology (GETRES) – Federal University of Rio de Janeiro (UFRJ), Rio de Janeiro, Brazil; mahler@coc.ufrj.br

ABSTRACT: Due to the major disasters recently occurred because of global warming, studies intended to minimize the release of polluting gases into the atmosphere are matter of global interest. This study is intended analyze a covering layer meant to oxidize methane (CH4), which is considered one of the strongest potentially triggering greenhouse effect gases. In this sense, covering layers have been studied both on site and in laboratories, using organic compost done by solid urban residues (RSU) and combinations of this compost with a soil-specific bentonit (BES), and/or a soil with plastic characteristics. This article provides a description of the first steps of the survey.

INTRODUCTION

Methane gas (CH_4) is generated by a variety of anthropogenic and natural sources, and it is considered one of the gases that most contribute to the global climate change. Its power to retain heat is 23 times higher than that of carbon dioxide (Prentice, 2001), and its time in the atmosphere extends to approximately 12 years (IPCC, 2006).

According to IPCC (2006), it is estimated that almost half of the global emissions of CH_4 are related to anthropogenic activities, and that 20% of all CH_4 released comes from the treatment and disposal of residues and leachate.

Among the anthropogenic activities that release significant amounts of CH_4 into the atmosphere, we can mention residue landfills, petrol and natural gas exploitation systems, agricultural activities, coal mining, livestock raising (enteric fermentation), and biomass burn.

Natural sources that release CH_4 include swamps, glaciers, and oceans.

The amount of CH_4 generated may vary from region to region or even from country to country, as the levels of emissions are related to several factors, such as weather, industrial development, and type of agriculture practiced in a given region.

The interference of the weather in the generation of $_{CH4}$ is observed during the anaerobic digestion process, as temperature and humidity are directly related to the biologic degradation rate. Other important factors, which cause CH_4 emission levels to differ from a site to another are the technology used to capture this gas at the generating source, and the way residues are handled.

This way, efforts to use this gas or prevent these gas emissions from escaping to the atmosphere can bring significant energetic, economic, and environmental benefits.
Focused on the environmental area, this work has the purpose of finding an alternative to minimize the atmospheric emissions of CH_4 through the use of a CH_4 oxidation layer on the surface of residue landfills.

The results of the first analysis of the mixture, the soil, the soil-specific BES (BES), and the soil-BES mixture are presented in this work.

METHODS

This work is divided in two stages: one of which to be carried out in the field (Metropolitan Landfill of Gramacho - RJ) and another to be carried out in the laboratory (Geotechniques and Environmental Geotechniques Laboratory - UFRJ).

Laboratory Essays

The laboratory essays are intended to serve as parameter to the field essays, as the laboratory conditions can be defined and controlled.
In order to simulate the methane oxidation layer, three sets of columns with approximately 60 cm in length and 20cm in diameter will be built (Figure 1) and filled with different matters, with the following possibilities: a - Organic compost generated by the composting of solid urban residues, produced at Usina de Triagem e Compostagem do Caju [Sorting and Composting Plant of Caju] (COMLURB - Rio de Janeiro - RJ); b - Mixture of BES and the organic compost; c - Mixture of clayish soil and the organic compost; d - Mixture of the three elements (organic compost, clayish soil, and BES).

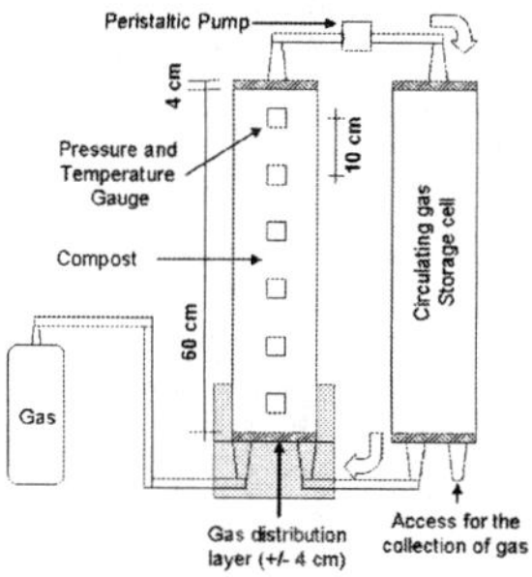

Figure 1 – Simplified chart of the experimental column.

In all cases, the elements will be introduced into the columns at an approximate humidity of 70 to 80% of the value of the material's field capacity value, and will be subject to a slight compaction.

The three columns will receive a fixed volume of a gaseous mixture ($CH_4 + O_2 + 5\%$ CO_2). Each column will receive a different volume of this mixture, and this volume will circulate inside the column for an equal Δ_t period for all columns. The gas circulation system will be closed, and for each Δ_{tx} small samples of the circulating gas will be collected. As the system is closed, in order to prevent pressure decrease, the same volume of the initial gaseous mixture collected for analysis will be reintroduced to the system.

In order to monitor the degradation process and the flow, pressure and temperature gauges will be installed at every 10.0 cm.

This essay is intended to measure the methane oxidation capacity of the mass contained in the columns.

PRELIMINARY RESULTS

Essays carried out with the compost

An essay was initially carried out to verify if the compost was microbiologically stable.

This essay used thermal bottles, in which a volume of the compost was introduced under humidity rates of 20, 35, and 50%, sufficient to fill up the entire bottle. These bottles remained open all the time and the internal temperatures were measured for 5 days. These temperatures did not rise at any time, indicating that microbiological activity no longer existed in the material.

After that, the compost was subject to a granulometric analysis. This analysis presented a granulometric curve, as demonstrated in Figure 2 below.

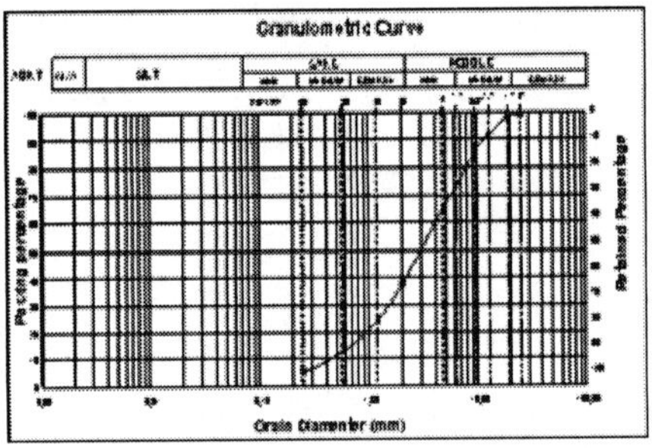

Figure 2 – Granulometric curve of the compost.

The methodology employed to perform the granulometric essay of the compost with sieves did not follow a Brazilian regulation, as such does not exist, but the methodology used in Germany to characterize residues.

According to essays on permeability to variable loads, this compost presents a permeability of 8.08 x 10^{-3} cm/s.

Essays carried out with clayish soil

Clayish soil obtained at the landfill of Gericinó will also be used in the methane oxidation study (COMLURB - RJ).

Essays intended to find some granulometric and geotechnical characteristics of this soil were carried out in laboratory.

Both for the clayish soil and the BES, the methodology employed in the essays were based in the respective ABNT regulations (NBRs 7180, 6459 e 7182).

Table 1 below shows the granulometric and geotechnical characteristics of this clayish soil. Analyzing the chart above, note that this soil is classified as sandy silt clay

Table 1 – Granulometric and Geotechnical caracteristic of the clayish soil

Granulometric characteristic (%)				Geotechnical characteristic (%)		
Clay	Silt	Sand	Gravel	Liquidity Limit	Plasticity Index	Plasticity Limit
39.0	29.0	30.0	2.0	63.0	38.2	24.8

Essays demonstrated that the permeability coefficient of this sandy silt clay is 1.3 x 10^{-7} cm/s, and that the optimal humidity of this material is 24.5% at $\gamma = 1.571$ g/cm^3.

Essays carried out with BES

Just like the sandy silt clay, some geotechnical and granulometry essays were also carried out with the BES, as this material will be in the combination of one of the mixtures to be used during the laboratory essays on methane oxidation.

Table 2 below shows the granulometric and Geotechnical characteristics of the BES.

Table 2 – Granulometric and Geotechnical caracteristic of the clayish soil

Granulometric characteristic (%))				Geotechnical characteristic (%)		
Clay	Silt	Sand	Gravel	Liquidity Limit	Plasticity Index	Plasticity Limit
86.6	13.0	0.4	0.0	505.6	46.3	459.3

According to Ferrari (2005), this BES has a permeability of 1.0 x 10^{-14} m/s.

Sandy silt clay and BES

Some essays were carried out with the mixture of these two materials: sandy silt clay and BES. The results of these essays are presented below.

The decision to work with a mixture of 95% of sandy silt clay and 5% of BES was due to studies reporting that BES amounts beyond 10% turn the project financially unattractive.

Table 4 below shows the results obtained in essays to determine the Atterberg limits of the mixture.

Table 4 – Geotechnical characteristics of the sandy silt clay added with 5% of BES

Characteristic	Value (%)
Liquidity Limit	98,0
Plasticity Index	24.3
Plasticity Limit	73.7

The essay to determine the optimal humidity of this mixture was also carried out. According this assay, optimal humidity of this mixture is 25.6% at γ = 1,501 g/cm^3. The permeability coefficient of this material is in the neighborhood of 4.8 x 10-8cm/s.

Metropolitan Landfill of Gramacho

In order to make an on site verification of the efficacy of the oxidation layer studied in laboratory, two areas of 10 x 10 meters will be analyzed, both located in the Metropolitan Landfill of Gramacho (RJ).

The Metropolitan Landfill of Gramacho, located in the neighborhood of Jardim Gramacho (Municipality of Duque de Caxias, Rio de Janeiro, Brazil) has an area of 1.3 million square meters, and has been in operation since 1976 (Figure 3). The landfill is the main destination of the residues generated in the Metropolitan Region of Rio de Janeiro, serving the municipalities of Rio de Janeiro, Duque de Caxias, Nilópolis, São João do Meriti, and Queimados (COMLURB, 2006).

In 1995, the landfill of Gramacho was recovered and started to be operated as a controlled landfill, receiving on average 6,300.00 t/day of urban and hospital residues (COMLURB, 2006).

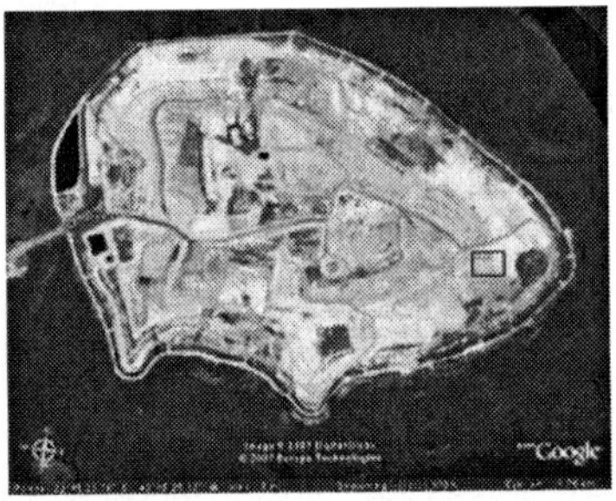

Figure 3 – View of the Metropolitan Landfill of Gramacho. The highlighted square shows the approximate location of the experimental area (not in scale) (Google Earth, 2007).

The experimental area is located in the north region of the landfill (highlighted square) and has not received residues for at least four years.

To test the presence of gas in the experimental area, four holes were made to collect gas samples.

These samples were collected with 20 ml syringes and taken for analysis by gas chromatography in laboratory.

As demonstrated on Table 5 below, out of the four sampling points analyzed, only two showed to be feasible for the construction of an oxidation layer. This way, our essays will be carried out in the areas of sampling points P1 and P4.

Table 5 – Table with percentage amounts (%) of gases analyzed.

Components	P1	P2	P3	P4
Ar	16.38	96.84	84.95	56.21
CH4	49.40	0.87	0.56	20.21
CO2	33.20	1.37	13.13	22.28
C2H6			0.00	0.00
H2S			0.00	0.00
H2O	1.02	0.92	1.35	1.30

Due to the time difference (15 days) between the collection of the gas and the chromatographic analysis, we are unable to say if the other two points are generating little gas, as loss or degradation of the gas may have occurred during the period in which the syringes were stored before the analysis.

For the same reason, we cannot make any other type of statement regarding the gas generated by this landfill, unless gas is being generated in the points P1 and P4 areas.

For the construction of the oxidation layer, the dug area will form a trench to the depth of approximately 1.2m, or until the residue mass is found. The sides of these trenches will receive a PVC geomembrane of approximately 1mm in thickness to prevent the gas from dissipating. The bottom will receive a gas-diffusing layer of 0.2m in height, built with rocks number 1 (dimension between 16 and 32 cm). The top of the gas distribution layer will receive the methane oxidation layer, with thickness of 1.0m, comprised only by the compound (the same used in the laboratory essays).

Figures 4 and 5 show a schematic model of how the methane oxidation layer will be disposed.

Figure 4 (below) shows the initial scheme. Most of the time, the area will have this configuration, allowing the entrance of water, insulation, and oxygen into the system, as they are essential factors for the existence of microbiologic activity.

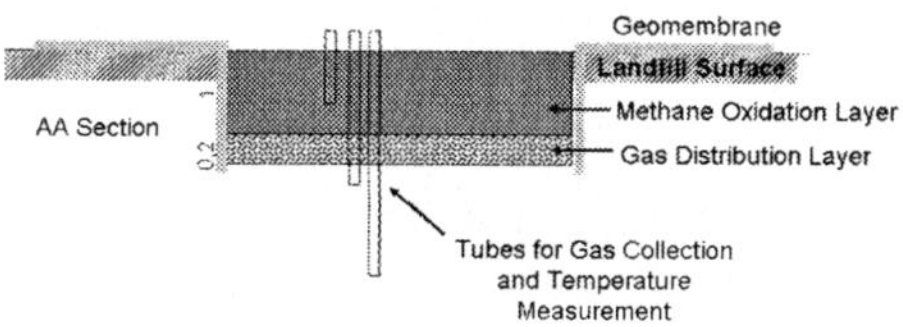

Figure 4 – Schematic drawing of the configuration of the methane oxidation layer installed on the landfill.

Five sets of tubes will be installed in the area. Each set will have at least eight tubes of different lengths. The objective of this set of tubes is to collect gas and measure the temperature at different depths. One of the tubes must necessarily be at half the height of the oxidation layer and the other must end a bit before the beginning of the gas distribution layer, that is, inside the residue mass. All other tubes end in different depths inside the residue mass.

Two days before the superficial gas collection the experimental area will be covered with the PVC geomembrane and a chimney will be installed to provide an access to the superficial gas collection (Figure 5). The gas collection will be carried out with 20.0ml syringes (the same used to collect the gas to define the experimental areas). After the gas collection, the gas will be analyzed through gas chromatography at the collection site itself. When the on-site analysis is not possible, these syringes will be stored in a dark and refrigerated location in order to prevent the loss of gas, and will then be taken to analysis in laboratory, also through gas chromatography.

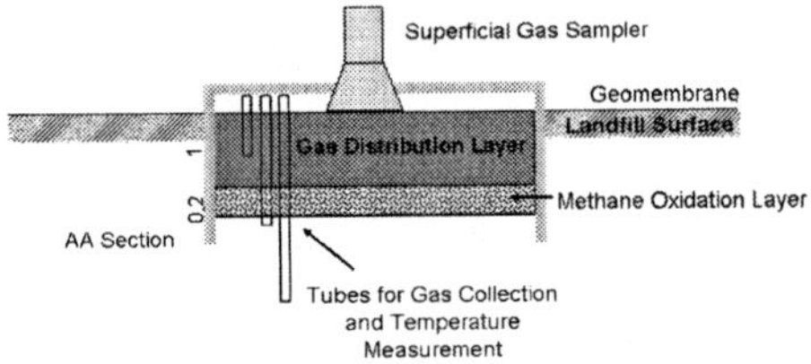

Figure 5 – Schematic drawing of the methane oxidation layer installed with the covering and the chimney for gas collection. Final comments

FINAL COMMENTS

The weather and the climate have deep influence in the life on earth. They are part of the daily experience of human beings, and essential for the human health, the production of food, and well being.

Any interference in this arrangement has an effect that can come to be hazardous to the environment in the long run and consequently to the human being.

Before the industrial era (XVIII century), the concentrations of greenhouse effect gases (CO2, CH4, and N2O) were relatively stable, causing the temperature on Earth to be constant.

The advent of industries brought an exaggerated increase of these gases, causing the Earth's capacity to retain heat to increase. This temperature increase, though small, generates several problems, such as the melting of the polar caps and the consequent rise of the sea level.

Actions intended to minimize the emission of these gases to the atmosphere would, in the long run, help reduce environmental problems resulting from the excess of greenhouse effects released in the past.

In spite of all that, many people could wonder why to strive efforts to create a methane oxidation layer on residue landfills if they could be used to generate energy, and consequently generate income to the group administrating the site.

Many are the reasons, but we can mention two:

- In Brazil, in 2000, 22.3 % of the solid residues produced were deposited in controlled landfills, and 30.5 % in sanitary landfills (IBGE, 2000). In these two cases, the construction of an efficient gas collector is very expensive, which renders such action unfeasible. The use of an oxidation layer would have the best cost-benefit relation in both cases.

- The sanitary landfills construction projects contemplate including gas drainage systems. In these cases, an oxidation layer would improve gas drainage, as the system is not 100% efficient and outflow of the generated gas ranges from 5 to 10%. This percentage that "escapes" from the collection would then go through the oxidation layer, contributing to the reduction of the hazardous environmental effects generated by the release of CH_4.

Thus, the presence of an oxidation layer on the surface of the landfill shows to be necessary, regardless of the way the landfill is managed.

ACKNOWLEDGEMENTS

The authors would like to register their gratitude to the following people and institutions: CNPq; COMLURB; Bentonit União Nordeste Ltda.; Dr. Kai Munnich, and Tobias Bahr from Technische Universität Braunschweig.

REFERENCES

COMLURB – Companhia Municipal de Limpeza Urbana do Município do Rio de Janeiro (2006). Available at <http://www.rio.rj.gov.br/comlurb/>. Accessed on January 23 2007.

FERRARI, A.A.P. (2005). Viabilidade da Utilização de Argila silte-arenosa Compactado como Material de Impermeabilização em Aterros de Resíduos. Strictu Sensu Specialization Work, UPS, São Paulo – SP, Brazil.

IBGE – Instituto Brasileiro de Geografia e Estatística (2002). Pesquisa Nacional de Saneamentos Básico 2000. Rio de Janeiro, 397p.

IPCC - Intergovernmental Panel on Climate Change (2006). IPCC Guidelines for National Greenhouse Gas Inventories. Available at <http://www.ipcc-nggip.iges.or.jp/public/2006gl/index.htm>. Accessed on January 23, 2007.

Prentice, I.C. (2001). The Carbon Cycle and Atmospheric Carbon Dioxide. Ed. Houghton, JT et al. In: Climate Change 2001: The Scientific Basis. Cambridge University Press, Cap. 3, p 786. Available at: <http://www.grida.no/climate/ipcc_tar/wg1/index.htm>. Accessed on January 23, 2007.

In-Situ Measurements of Pore Water Pressures in Landfilled Waste in Response to Liquids Addition

Timothy G. Townsend[1], PhD, P.E., A.M. ASCE, Ravi Kadambala[2], Sendhil Kumar[3] and Pradeep Jain[4], PhD, A.M. ASCE

[1]Associate Professor, Dept. of Environmental Engineering Sciences, Univ. of Florida, P.O. Box 116450, Gainesville, FL 32611-6450. E-mail: ttown@ufl.edu.
[2]Graduate Research Assistant, Dept. of Environmental Engineering Sciences, Univ. of Florida, P.O. Box 116450, Gainesville, FL 32611-6450. E-mail: ravi22k@ufl.edu.
[3]Graduate Research Assistant, Dept. of Environmental Engineering Sciences, Univ. of Florida, P.O. Box 116450, Gainesville, FL 32611-6450. E-mail:ssk82@ufl.edu.
[4]Environmental Engineer, Innovative Waste Consulting Services, LLC, 6628 NW 9th Blvd., Suite 3, Gainesville, FL 32606. E-mail: pjain@iwcs.biz

ABSTRACT: Bioreactor landfills differ from traditional landfills in that large volumes of liquids are added to the waste to promote waste stabilization. To achieve desired liquid addition rates, operators must often add liquids under pressure. One concern at any landfill where liquids are present under pressure is slope stability. An increase in pore water pressure can result in a decrease in shear strength and subsequently geotechnical stability of side slopes. Modeling techniques allow the engineers to predict possible pressure buildup and factors-of-safety for slope failure, but very little is known about what pressures actually develop within landfills as a result of added liquids. Two experiments are underway in Florida where pore water pressure transducers have been buried in the waste surrounding liquid addition devices. At one site, transducers were installed at different radial locations around a vertical leachate recirculation well. At another site, transducers were buried in a lift of waste amongst a series of horizontal liquids addition trenches; some transducers were placed in the trenches and others were placed in the waste at different distances away. The objective of the experiments is to measure the pore water pressure in the waste surrounding the devices as compared the pressure being delivered to the devices themselves. The results will provide insight to the pressure distribution within bioreactor landfills, nature of fluid flow and will assist in developing design and operational constraints. The paper describes the experiments and presents preliminary results.

INTRODUCTION

The stability of a side slope is one of the primary concerns at any landfills where liquids are present at elevated pressure. An increase in the liquid pressure in the interstitial spaces of the waste and cover soil (referred to as *pore water pressure*) decreases the effective stress between waste (and or soil) particles. The factors that can result in increase in pore water pressure and subsequent decrease in shear strength are: injection of liquids under pressure to operate a landfill as a bioreactor and (or) accumulation of water on top of the cover soil layer in case a low permeability soil is used as cover material. The decreases in effective stress between waste (and or soil) particles lead to reduction in the geotechnical stability of the landfill.

Liquids (leachate, groundwater and other sources of liquids) addition at municipal solid waste landfill has gained momentum because of its well-documented environmental and economic benefits (Pohland, 1975; Leckie et al., 1979; Watson, 1987; Townsend et al., 1996; and Jain, 2005). Liquids addition systems can be broadly grouped into two categories depending on the way liquids are introduced into the landfill: Surface systems and sub-surface systems. Sub-surface systems such as horizontal trenches (Townsend, 1995) and vertical wells (Jain, 2005; Jain et al., 2005a; Jain et al., 2007) are more popular as liquids can be added at higher flow rate than via surface systems as these systems can be operated at higher pressure. One of the primary inputs needed to evaluate the geotechnical stability of the slopes of a landfill is pore water pressure. What remains unknown is the change in pore water pressure of the surrounding waste due to liquids injection into a landfill. A focus of research at several landfills in Florida is the development and evaluation of methodologies for the safe design and operation of bioreactor landfills. As a part of this effort Jain (2005) and Jain et al. (2007) mathematically modeled the impact of liquids addition via vertical wells on the pore water pressure of the surrounding waste in the landfill. However, no field scale date pertaining to pore water pressure in the waste that is subjected to liquids addition is published to date. An examination of the temporal and spatial variation in pore water pressure of the waste subjected to liquids addition via vertical wells and horizontal trenches is the primary objective of the field-scale experiments discussed in this paper. These experiments consist of pore water pressure transducers buried in landfilled waste surrounding the leachate injection devices, adding leachate, and measuring the resulting pore water pressures in the surrounding waste. Two experiments are underway in Florida. At one site, transducers were installed at different radial locations around a vertical leachate recirculation well. At another site, transducers were buried in a lift of waste amongst a series of horizontal liquids addition trenches; some transducers were placed in the trenches and others were placed in the waste at different distances away. Leachate recirculation has begun at one site; leachate recirculation will commence in the fall of 2007 at the other site. This paper describes the experimental setup and provides some preliminary data to illustrate the output of the sensors. The ultimate results of the research will provide insight to the pressure distribution within bioreactor landfills, the nature of fluid flow and will assist in developing design and operational constraints.

MATERIALS AND METHODS

Site 1 Description

The Polk County North Central Landfill (NCLF) is located in Winter Haven, Florida. Phase 2, the active landfill site at Polk receives approximately 2,500 metric tons of waste per day which consists of mixed residential, commercial, and industrial waste. The site also contains several closed landfill units. A large number of horizontal injection trenches (or lines) were installed in Phase 2 cell. The pipes were, nominally, 4-inch-diameter high-density polyethylene (HDPE). The perforated pipe consisted of two 9.5 mm diameter perforations with a frequency of every 0.6 m linear meter of pipe. They were each oriented at 45° from either side of the vertical and were always placed in the downward direction. Each injection line was constructed by excavating approximately 1 m wide and 1 m deep trenches and backfilling it with a layer of a porous bedding medium, laying a perforated pipe and backfilling the rest of the trench with the bedding media. The experimental area is approximately 0.5 ha on the Phase 2 cell and consists of three horizontal trenches each with a different bedding media (tire chips, glass and excavated waste). These three injection trenches were surrounded by 81 pressure transducers. Three transducers were installed within each trench and the rest were installed at different distances from the lines, the piezometers are installed in a V formation around the lines as shown in the Figure 1 to minimize short-circuiting and have an undisturbed waste in the fluid path from the Horizontal Leachate Injection Lines (HLIL) to the piezometers. The injection lines are 110-m long. Figures 1 and 2 show the plan view and the cross section of the piezometer experiment layout. The density of the land-filled waste is approximated to be 880 kg/m^3. Vibrating wire (VW) piezometers were used for the experimental set-up (model 4500B and manufactured by Geokon, NH). These piezometers measure both pressure and temperature. The piezometers cables were routed through the schedule-80 PVC conduit and the cables were connected to a datalogger on the south slope of Phase 2 cell.

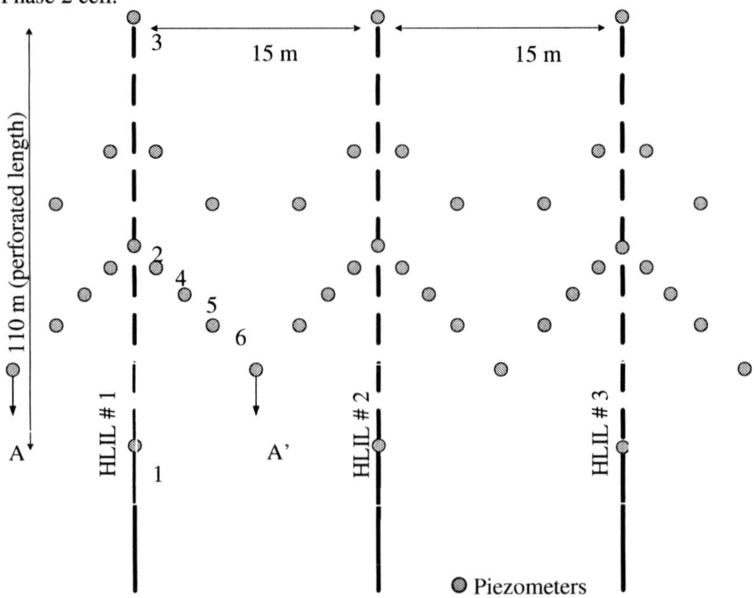

FIG. 1. Plan view of the installed piezometers

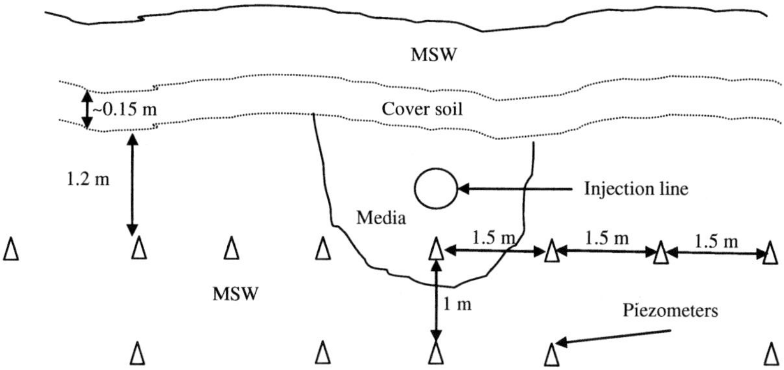

FIG. 2. Cross section of piezometers A-A' from FIG. 1.

Site 2 Description

The New River Regional Landfill (NRRL) is located in Union County, Florida. The NRRL receives approximately 730 metric tons per day of waste consisting of mixed residential, commercial, and industrial waste. The NRRL site is approximately 200 ha in total area. The landfill currently consists of five contiguous lined landfill cells, totaling approximately 32 ha. The experimental set-up was built on cell 4. The in-situ density of the waste is approximated to be 710 kg/m^3. The experimental set-up consists of two vertical injection wells, situated 7.6 m away from each other as shown in Figs 3 and 4. A 20-cm hollow stem open flight auger was used for drilling the wells. After all the piezometers were placed in a monitoring well, the bore hole was then grouted using a grout machine (ChemGrout, Illinois). The vertical injection wells are connected to their separate lateral leachate recirculation line as shown in Fig. 4. Vertical Well 2 is the main injection well. Surrounding this vertical well are several multi-level piezometer (model 52611199 and manufactured by Slope Indicator, WA) wells, located approximately at 3, 6, 9, 12, 15 m away from Well 2. Each multi-level piezometer well has 5 VW piezometers. Each of these piezometers is 3 meters away from each other in a multi-level piezometer wells. The piezometers measure both pressure and temperature. All the piezometers were connected to a datalogger through the multiplexers to collect data at regular intervals.

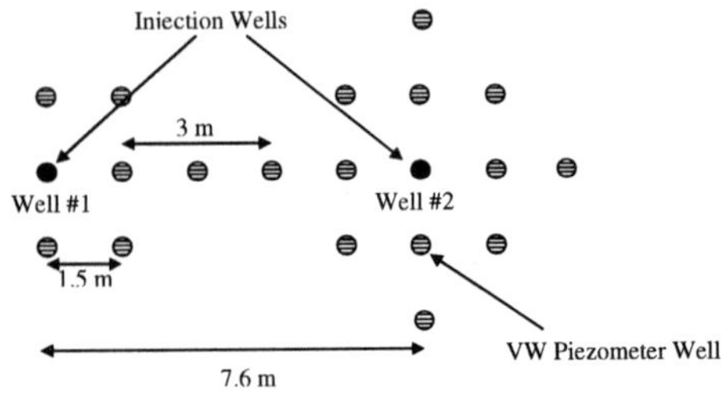

FIG. 3. Plan view of the VW piezometer experimental set-up

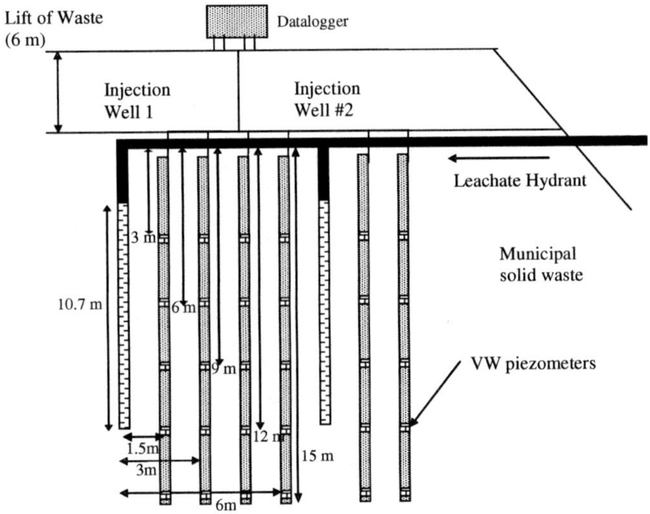

FIG. 4. Cross-sectional view of the VW piezometer experimental set-up

LEACHATE INJECTION TEST

The injection tests were only recently started at the NCLF and are yet to commence at the NRRL. The Horizontal Leachate Injection Lines at PNCL were operated individually for determining the pressure change due to injection in that particular line. HLIL # 1 was chosen first to operate at a flowrate of 9.4×10^{-5} m^3/s (15 gpm). Approximately 29.10 m^3 (76,900 gallons) of leachate were injected through this line over a period of 15 days. HLIL # 2 and 3 (Fig. 2) were operated for one day to test the lines. Table 1 summarizes volume of leachate added along with the duration of operation of these lines.

Table 1. Injection Tests Data

HLIL#	Bedding Media	Operation duration	Volume of leachate added (m^3)
1	Tire chips	15 days	29.10
2	Excavated waste	5 hrs	2.0
3	Crushed Glass	7 hrs	3.0

As presented in Table 1, only a small leachate volume is added to date in HLILs except for HLIL # 1 which was run until it reached the maximum permitted injection pressure. All the transducers that were installed within the trenches were impacted by leachate addition. Figure 5 presents the response of transducer 1 which is typical response of pressure transducers and temperature sensor installed within the trench. The figure shows a continuous increase in pressure from the time the line was started in the morning until the end of the day when it was shut down. The temperature decreased at the same time indicating the flow of leachate to this location. The temperature increased immediately after leachate addition to the line was stopped.

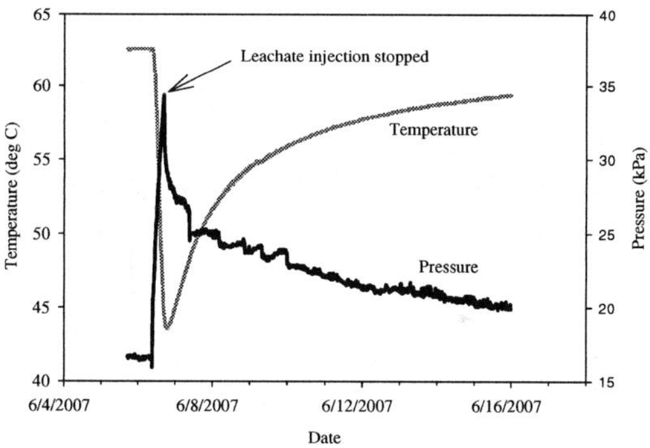

FIG. 5. Pressure and temperature change of a piezometer in trench

Figure 6 shows the pressure change of piezometers located in the HLIL# 1. As shown in Figure 2, transducer 1 is nearest and the transducer 3 is farthest in trench to the point leachate addition begins. As anticipated the pressure in the trench declined from the start to the end of the trench. This decline in pressure is associated with the pressure loss encountered for flow though the bedding media and perforated pipe. It can be seen that the pressure loss encountered across the trench length of 110 m is approximately 10 kPa. Figure 7 presents temporal variations of pressure transducers located at varying distances from HLIL#1 (transducers 4, 5, 6 in Figure 2) along with a transducer in the trench (transducer 2). Each peak in pressure represents an operational day. As anticipated, the data suggests that pressure dissipates within a small distance away from the injection trench. It can be seen from Figures 7 that the pressure in the trench, which is representative of injection pressure, increased up to 60 kPa, whereas the maximum pressure that the transducer just 1.5 m away from the trench registered was approximately 50 kPa indicating reduction in pressure by approximately 10 kPa within 1.5 m from the injection trench. The magnitude of pressure loss within the waste was found to be much higher than the pressure loss within the trench; a pressure loss of 10 kPa was registered across the 110 ft length of trench whereas similar pressure loss occurred for flow through only 1.5 m of waste. The pressure declined over 25 kPa (~3.5 psi) within 4.5 m of the injection trench.

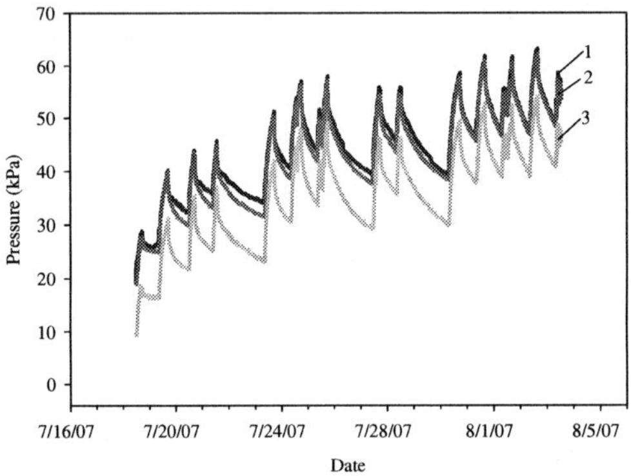

FIG. 6. Pressure variations recorded by Piezometers in the trench

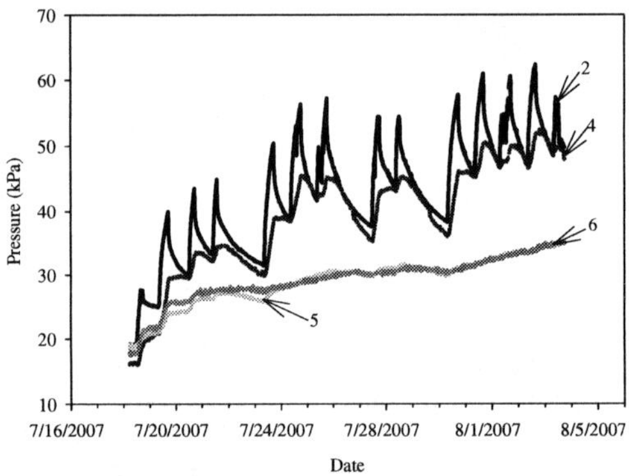

FIG. 7. Piezometers on the right side of the HLIL # 1

SUMMARY AND CONCLUSIONS

This paper describes two ongoing field-scale experiments in Florida to evaluate the distribution of pore water pressure in landfilled waste surrounding devices (wells and trenches) where liquids are added under pressure. At the NRRL, instruments were installed surrounding vertical wells, and at the NCLF, instruments were installed around horizontal trenches. The results on the pressures measured from selected instruments surrounding one of the horizontal injection trenches at the NCLF were presented. While only for one HLIL, the data gathered to date suggest that most of the instruments continue to provide readings and that the dissipation of pressure within a small distance around the horizontal injection trench.

ACKNOWLEDGMENTS
The authors appreciate the support of John Schert, director of Hinkley Center for Solid and Hazardous Waste Management, Fletcher Herrald of Florida Department of Environmental Protection, Darrell O'Neal, Director of New River Regional Landfill and Brooks Stayer of Polk County North Central Landfill.

REFERENCES

Jain, P. (2005). Moisture addition at bioreactor landfills using vertical wells: Mathematical modeling and field application. Ph.D. Dissertation. University of Florida, Gainesville, FL, USA.

Jain, P., Farfour, W. M., Jonnalagadda, S., Townsend, T. G., and Reinhart, D. R. (2005a). "Performance evaluation of vertical wells for landfill leachate recirculation." Proceedings of Geo Frontier 2005, ASCE conference, January 24-26, Austin, TX, USA.

Jain, P., Larson, J., Townsend, T.G. and Tolaymat, T. (2007). "Design of vertical wells for introduction at bioreactor landfill: Guidelines based on mathematical modeling" Submitted to Journal of Environmental Engineering, ASCE.

Koerner, R. M., and Soong, T.-Y. (2000). "Leachate in landfills: The stability issues." Geotext. Geomembrane, 18_5_, 293–309.

Leckie, J.D., Pacey, J.G. and Halvadakis C.P. (1979). Landfill management with moisture control. Journal of Environmental Engineering Division, ASCE EE2, 105, 337-355.

Pohland, F.G. (1975). Sanitary landfill stablization with leachate recycle and residual treatment. EPA-600/2-75-043. USEPA, Washington DC, USA.

Reinhart, D. R., and Townsend, T. G. (1997). Landfill bioreactor design and operation. Lewis Publishers, Boca Raton, FL, USA.

Townsend, T. G. (1995). Leachate recycle at solid waste landfills using horizontal injection. Ph.D. Dissertation, University of Florida, Gainesville, FL, USA.

Townsend, T. G., Miller, W. L, Lee, H. J., and Earle, J. F. K (1996). "Acceleration of landfill stabilization using leachate recycle." Journal of Environmental Engineering, ASCE, 122(4), 263-268.

Watson, R.P. (1987). "A case study of leachate generation and recycling at two sanitary landfills" in Proceedings from the Technical Sessions of the GRCDA 25th Annual International Seminar, Equipment, Services, and Systems Show. Vol. 1, August 11-13, Saint Paul, MN.

R.W. Beck. (2004). Final Report: Composition of Municipal Solid Waste Disposal in Polk County, Florida. Winter Haven, FL.

Lab-scale Liquid Injection Model of Bioreactor Landfill

Moumita Mukherjee[1], Milind V. Khire[2], Ph.D., P.E., and Xuede Qian[3], Ph.D.

[1]Graduate Research Assistant, Dept. of Civil and Environmental Engineering, Michigan State University, A30 Research Complex Engineering, E. Lansing, MI 48824, U.S.A.
[2]Assistant Professor, Dept. of Civil and Environmental Engineering, Michigan State University, A135 Research Complex Engineering, E. Lansing, MI 48824, U.S.A. Email: khire@egr.msu.edu.
[3]Geotechnical Engineering Specialist, Michigan Department of Environmental Quality, 525 W. Allegan Street, Lansing, MI 48913, U.S.A.

ABSTRACT: A lab-scale physical model of a bioreactor landfill with liquid recirculation is developed. The landfill model is 86 cm long x 30 cm wide x 56 cm high. A 50 cm long x 30 cm wide x 2 cm thick horizontal permeable blanket made up of pea gavel is built in the model for subsurface liquid injection. Deionized water injection in the blanket was monitored by pressure, flow, and water content sensors for injection rates ranging from 16.5 to 23.5 cm^3/s. Fine sand was used to simulate "waste" material. When water was injected in the blanket, all pressure transducers in the blanket registered increase in the pressure head and the pressure heads increased or decreased as the injection flow rate was varied. The increase in the pressure head was earliest and greatest for the sensor located closest to the liquid injection pipe. The sensor responses for the blanket in the lab-scale landfill model were similar to those in an instrumented field-scale (55 m long x 9 m wide x 0.15 m thick) crushed recycled glass blanket located in Michigan.

INTRODUCTION

Bioreactor landfills are designed and operated to accelerate the decomposition of organic constituents of municipal solid waste (MSW) and to decrease the time to achieve waste stabilization in the process by purposeful control of biological processes. The most common form of bioreactor landfill operation is moisture addition by recirculation of leachate (or injection of other liquids) as a means to enhance moisture levels within the landfill and creating an environment conducive to rapid degradation of waste.

Leachate recirculation (LR) or liquid injection (LI) can be performed using multiple techniques, both surface and subsurface. Subsurface application techniques are: (1) vertical wells (VWs); (2) horizontal trenches (HTs); and (3) permeable blankets (PBs). Haydar and Khire (2005), Haydar and Khire (2007) and Khire and Mukherjee (2007) have presented design guidelines for HTs, PBs and VWs, respectively, through numerical studies.

Haydar and Khire (2006) tested the concept of leachate recirculation using a field-scale instrumented permeable blanket located at McGill Landfill located in Jackson. The authors have reported the response of thermistors and pressure transducers embedded in the field-scale glass blanket which was 55 m long by 9 m wide by 0.15 m thick. The response of pressure transducers placed at 0.5, 4.5, and 14 m distances from the leachate injection pipe for two consecutive liquid injection events is presented in Fig. 1. All pressure transducers detected the migration of leachate by measuring a gradual increase in the hydraulic pressure head (henceforth referred to as pressure heads) in response to the leachate injection event. The pressure transducers indicated an immediate but gradual decrease in the leachate pressure head when the leachate injection stopped after about 140 minutes.

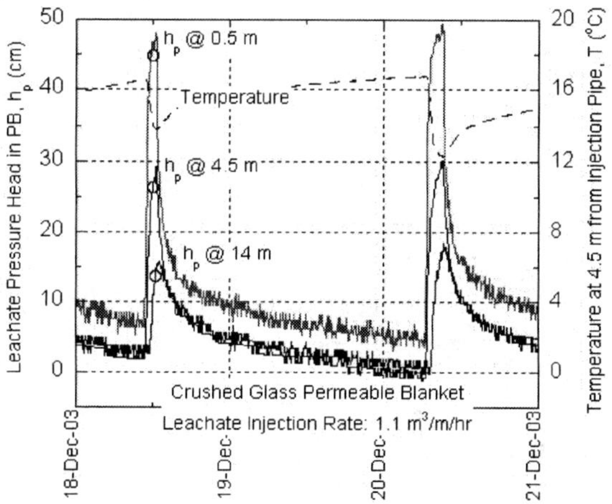

FIG. 1. Field response of pressure transducers and temperature sensors during leachate injection events at McGill Landfill, Jackson, Michigan.

The estimation of real-time field-scale hydraulic conductivity of the underlying waste is a challenging task. Hydraulic properties are critical for simulating and understanding landfill hydraulics. The long-term objective of the current research project is to estimate the hydraulic conductivity of the underlying waste using the pressure and flow data from the sensors embedded in the blanket. A laboratory scale physical model of landfill was developed to conduct controlled lab tests in order to independently verify the method before applying it to the field data. Unlike in the field, the lab model has sensors embedded in the sand underlying the blanket which will help understand the hydraulics of liquid flow due to subsurface injection. The data from the lab model would allow us to validate commonly used numerical models (e.g., HUDRUS, Vadose/W). Such models can be used to simulate long-term behavior of bioreactor landfills. The objective of the study presented in this paper is to replicate the field-scale blanket in a laboratory scale landfill model.

MATERIALS AND METHODS

Fig. 2 presents a schematic of the landfill model fabricated to simulate a horizontal permeable blanket. The plexi-glass box for housing the landfill model was designed partly by simulating the landfill model in saturated/unsaturated numerical model HYDRUS-2D such that liquid injection fluxes ranging from 8 cm^3/s to 250 cm^3/s could be tested for soils having conductivities ranging from 10^{-2} to 10^{-3} cm/s. The dimensions of the landfill model are presented in Fig. 2. The acrylic panels of the model were screwed together with rubber seals in-between the panels. The blanket was placed at about 17 cm from the top to carry out liquid injection without creating artesian conditions for the selected flow rates. A separate acrylic panel was used to make the bottom of the leachate collection system (LCS) raised to a slope of 3%. The perforated seepage pipes for LCS had at least 10 times higher flow capacity than the flows injected in the model to maintain the pressure head in the LCS within its thickness of 4 cm.

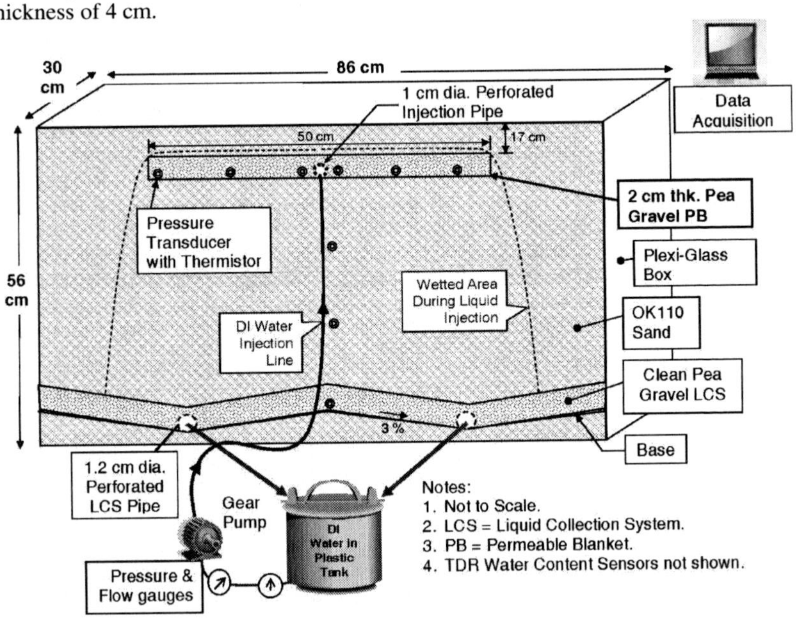

FIG. 2. Schematic of fabricated lab-scale landfill model.

Sensing System

The sensing system used in the landfill model consisted of these sensors: (1) pressure transducers with built-in thermistors; (2) time domain reflectometry (TDR) water content sensors; and (3) flow sensors. All sensors were connected to a datalogger through multiplexers to continuously monitor and log the data at frequencies ranging from 5 s to 30 min.

Pressure Transducers

The pressure transducers used in this study are vented and have a 0-5 VDC analog as well as a digital modbus-capable interface to read temperature and pressure. A thermistor is embedded within the pressure transducer to record temperature. The signal drift in the sensor performance resulted in pressure readings with errors ranging from 0.3 to 0.6 cm during the 6 month testing period. The galvanic corrosion was minimized by not grounding the shield wire connected to the housing of the sensor. The sensors have hermetically sealed cavity around the vibrating wire to minimize corrosion problems.

TDR Water Content Sensors

The mini-TDR water content sensors selected for this study consisted of three pronged 0.15 cm diameter stainless steel rods mounted into an encapsulated plastic head. The probe rod length is 6 cm and spacing between the probe rods is 0.6 cm. The TDR sensors are connected to the datalogger via an electro-magnetic pulse generator and multiplexer.

Flow Sensor

The flow sensor is capable of measuring relatively low flow rates ranging from 8 to 165 cm^3/s. The flow sensor incorporates a pelton-type turbine wheel to measure the flow rate of water. Electrical pulses are generated as the turbine wheel rotates in response to the rate of flow. The sensors provide analog DC voltage output proportional to the flow rate. All sensors were calibrated before installation in the model.

Materials

Due to the heterogeneity and anisotropy in the hydraulic properties of municipal solid waste, the measurement of representative saturated and unsaturated hydraulic properties of waste is challenging. Hence, in order to allow thorough characterization for the experiments and allow model verification, waste was simulated by homogeneous and isotropic sand. Table 1 shows the grain size and hydraulic characteristics of the soils used in the landfill model. The blanket and the LCS were made up of pea gravel. The waste was represented by uniform fine silica sand (OK110 sand). The saturated hydraulic conductivity (K) of the fine sand ranged from 1.8×10^{-3} to 4.5×10^{-3} cm/s. This range is consistent with the typical values of hydraulic conductivities of municipal solid waste published by Hughes et al. (1971), Fungaroli and Steiner (1979), Korfiatis et al. (1984), Oweis et al. (1990), Chen and Chynoweth (1995), and Jang et al. (2002) . The soil water retention characteristics were measured under static equilibrium by hanging column method (ASTM D 6836).

Landfill Model Components

Fig. 2 presents a schematic of the model. A 4-cm thick LCS drainage layer made up of washed pea gravel was constructed at the bottom of the plexi-glass model. Two 1.5-cm diameter perforated pipes discharging freely into the atmosphere were placed in the LCS pea gravel layer.

Table 1. Properties of soils used in the landfill model.

Soil Type	Grain Size Distribution				Saturated and Unsaturated Hydraulic Properties				
	D_{50} (mm)	C_u	C_c	ρ_d (g/cm^3)	K (cm/s)	θ_s	θ_r	α (1/cm)	n
OK110 Sand	0.11	1.34	1.06	1.55	1.8 to 4.5×10^{-3}	0.42	0.03	0.01 to 0.02	6.5
Pea Gravel	2.84	1.68	0.96	1.41	1.5	0.43	0.01	0.35	3.3

Notes: θ_s = saturated volumetric water content [dimensionless];
θ_r = residual volumetric water content [dimensionless]; and
α [1/L] and n are van Genuchten's fitting parameters (van Genuchten 1980)

The LCS and the "waste" sand were separated by a non-woven geotextile. The geotextile had a thickness of about 2 mm (ASTM D 5199), mass per unit area of 270 g/m^2, and a hydraulic conductivity of 0.3 cm/s (ASTM D 4491). About 38-cm thick dry OK110 sand having dry density equal to 1.6 g/cm^3 and porosity equal to 0.42 was placed below the permeable blanket. In the sand layer, two pressure transducers were embedded in vertically upright position at 10-cm intervals (Fig. 2). Each pressure sensor's tip was wrapped in the same geotextile as described before to prevent fine sand particles falling on the diaphragm without restricting the flow. The space above the diaphragm within the sensor was filled with DI water to remove air to reduce errors in pressure measurement. A TDR water content sensor was placed at the same location of the sensing tip of the pressure sensor at each level.

The blanket for the recirculation system was made up of the same pea gravel used in LCS. The blanket was about 50 cm long and 30 cm wide. The thickness of the blanket was 2.0 cm. The blanket was wrapped in a non-woven geotextile to prevent clogging. The pressure sensors were vertically placed in the sand below the blanket such that the tips of the sensors were in the blanket. TDR sensors were placed horizontally in the blanket at two locations (not shown in Fig. 2). Total six pressure transducers and two TDR water content sensors were embedded in the blanket. A 40-cm long perforated PVC pipe (two rows of holes having diameter equal to 1.6 mm, 10 holes per row) of 1 cm diameter was placed at the center of the blanket in the direction parallel to the width of the blanket where water was injected under pressure.

One end of the injection pipe inside the blanket was capped and the other end was connected to a pressure transducer and a flow sensor to measure the injection pressure and flow rate, respectively. The flow sensor was connected to a miniature gear pump to pump water from a storage tank into the blanket. A pressure transducer was also placed in the storage tank to monitor the change in head of water in the tank to monitor steady-state. The miniature gear pump was operated with a variable power supply to obtain variable injection flow rates. The gear pump was chosen because of its ability to deliver a pulseless flow.

RESULTS

Deionized (DI) water was pumped into the blanket through the injection pipe at various injection rates ranging from 16.5 to 23.5 cm^3/s. The rate flow of injected water was increased in steps until a steady-state for each flow rate was achieved. The steady-state was defined as when the injected flow in the blanket equated the outflow from the LCS and the pressure heads in the blanket did not show upward or downward trend during few hours after the flows became equal.

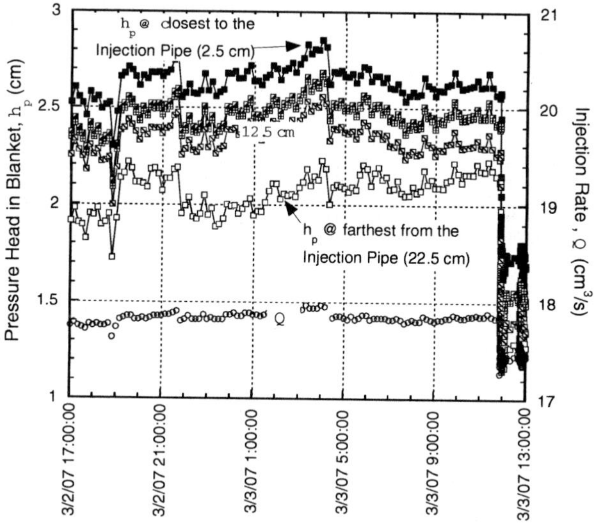

FIG. 3. Response of pressure transducers in the lab-scale blanket.

All pressure transducers detected the migration of water during the injection events by measuring increases or decreases in the pressure heads as the injected flux was varied. The pressure heads were maximum in the blanket and decreased downwards until they were almost zero in the LCS. Initially, the pressure heads measured by the sensors in the sand were relatively high and were found to be equivalent to the overburden stress in addition to the hydrostatic pore water pressure at that point. As the pore water pressures due to overburden stress slowly dissipated, the pressure readings gradually reduced and equated with the hydrostatic pressure due to the injection.

For flow rates less than about 18 cm^3/s, the pressure heads in the blanket were less than the thickness of the blanket (~2 cm) and hence water did not pool above the blanket. Fig. 3 shows the pressure heads in the blanket at steady-state when the rate of injection was maintained at about 18 cm^3/s. The increase in the pressure head was earliest and greatest for the sensors located closest to the injection pipe and decreased subsequently with distance away from the injection pipe.

This finding is similar to the data from the field-scale blanket (Fig. 1). When the injection was stopped in the lab model, there was a sharp drop in the pressure heads.

The pressure head momentarily became negative and then bounced back slightly above zero (Fig. 4). In the field-scale blanket, similar drop in the pressure was observed (Fig. 1). However, the drop was more gradual and the pressure heads did not become negative. Possible reasons for this difference in the lab data versus field data could be that the hydraulic conductivity of the waste underlying the field blanket was lower than OK110 sand used in the lab model and hence it would have taken significantly longer for the heads to drop. Field hydraulic conductivity of the waste was not measured because measurement of representative conductivity of waste is still a challenge. Numerical modeling needs to be carried out to identify the key reasons which are beyond the scope for this paper.

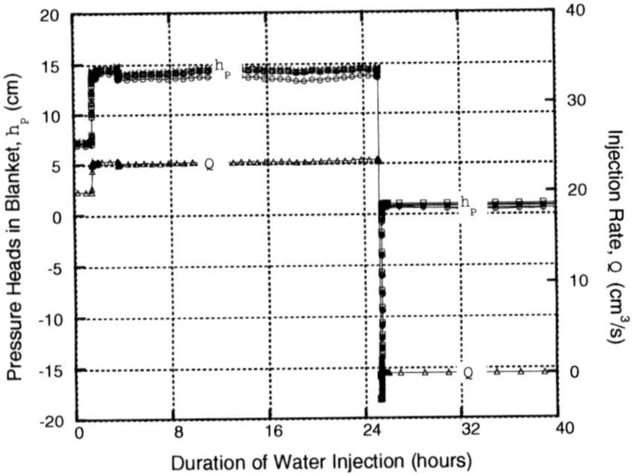

FIG. 4. Response of pressure transducers showing momentary suction when pumping was stopped.

CONLUSIONS

Pressure heads in a lab-scale blanket due to liquid injection were measured using an automated sensing system consisting of pressure sensors. The field responses of the pressure sensors in the blanket could be mimicked in a laboratory-scale landfill model. The pressure heads in the blanket at steady-state are dependent upon, among other factors, the rate of injection the blanket. Experiments using other soils and surrogate waste to explore other parameters associated with liquid injection systems are planned in the near future.

ACKNOWLEDGEMENT

Financial support for this project has been provided by the National Science Foundation (Grant No. CMS-0510091). Waste Management and Environmental Research and Education Foundation (EREF) provided site support and funding for the field-scale component of the previous but related project. We also express sincere appreciation to Jason Ritter of Campbell Scientific for his programming help for the data acquisition system and James Brenton for his help with the fabrication of the landfill model. The findings and opinions presented in this manuscript are those of the authors.

REFERENCES

Chen, T., and Chynoweth, D. P. (1995). "Hydraulic conductivity of compacted municipal solid waste." *Bioresour. Technol.*, Vol. 51(2–3), 205–212.

Fungaroli, A. and Steiner, R. (1979). "Investigation of sanitary landfill behavior." Vol. 1, Final Report, U.S. EPA 600/2-79-053a.

Haydar, M. and Khire, M. (2005). "Leachate Recirculation Using Horizontal Trenches in Bioreactor Landfills." *Journal of Geotechnical and Geoenvironmental Engineering*, Vol. 131(7): 837-847.

Haydar, M. and Khire, M. (2006). "Geotechnical Sensor System to Monitor Injected Liquids in Landfills." *Geotechnical Testing Journal*, ASTM, Vol. 29(1):37-44

Haydar, M. and Khire, M. (2007). "Leachate Recirculation Using Permeable Blankets in Engineered Landfills." *Journal of Geotechnical & Geoenvironmental Engineering*, Vol. 133(4): 360-371.

Hughes, G., Landon, R., and Farrolden, R. (1971). "Hydrogeology of Waste Disposal Sites in Northeastern Illinois." *EPA Solid Waste Management Series*, SW-12d.

Jang, Y. S., Kim, Y. W., and Lee, S. I. (2002). "Hydraulic properties and leachate level analysis of Kimpo Metropolitan Landfill, Korea." *Waste Management.*, Vol. 22(3), 261–267.

Khire, M. and Mukherjee, M. (2007). "Leachate Injection Using Vertical Wells in Bioreactor Landfills." *Waste Management*, Vol. 27(9): 1233-1247

Korfiatis, G., Demetracopoulos, A., Bourodimos, E., and Nawy, E. (1984). "Moisture transport in a Solid Waste Column." *Journal of Environmental Engineering*, Vol. 110(4): 789-796.

Oweis, I., Smith, D., Ellwood, R., and Greene, D. (1990). "Hydraulic Characteristic of Municipal Refuse." *Journal of Geotechnical Engineering,* Vol. 116(4): 539-553.

Simunek, J., Sejna M., and Van Genuchten, M. Th. (1999). HYDRUS 2D, Simulating water flow, heat, and solute transport in two-dimensional variably saturated media, Version 2.0, US Salinity Laboratory, ARS/USDA, Riverside, California and International Ground Water Modeling Center, IGWMC- TPS 53, Colorado School of Mines, Golden, Colorado.

Topp, G.C., Davis, J.L., and Chinnick, J.H., (1980). "Electromagnetic Determination of Soil Water Content: Measurements in Co-axial Transmission Lines." *Water Resources Research*, Vol. 16(3): 574-582

van Genuchten, M.Th. (1980). "A Closed-form Equation for Predicting the Hydraulic Conductivity of Unsaturated Soils." *Soil Science Society of America Journal*, Vol. 44: 892-898.

Permeability of Municipal Solid Waste (MSW) in Bioreactor Landfill with Degradation

MD Sahadat Hossain[1], Kiran Kumar Penmethsa[2], and Laureano Hoyos[3]

[1]Assistant Professor, Department of Civil and Environmental Engineering, University of Texas at Arlington, TX 76019, hossain@uta.edu
[2]Graduate Student, Department of Civil and Environmental Engineering, University of Texas at Arlington, TX 76019
[3]Associate Professor, Department of Civil and Environmental Engineering, University of Texas at Arlington, TX 76019, email: hoyos@uta.edu

ABSTRACT: Bioreactor landfills are operated for rapid stabilization of waste, increased landfill gas generation for cost-effective energy recovery, gain in landfill space, enhanced leachate treatment, and reduced post closure maintenance period. The fundamental process of waste stabilization in bioreactor landfill is recirculation of generated leachate back into the landfills, which creates a favorable environment for rapid microbial decomposition of the biodegradable solid waste. In order to better estimate the generated leachate and design of leachate recirculation system, clear understanding of permeability of MSW and the factors influencing the permeability is necessary. The objective of the paper is to determine the changes in hydraulic properties of MSW in bioreactor landfill with time and decomposition. Four small scale bioreactor landfills were built in laboratory to prepare samples representative of each phase of decomposition. A series of constant head permeability tests were performed on the samples generated in laboratory scale bioreactors to determine the variation of permeability with degradation. The test results indicated that permeability of MSW decreased from 0.0088 cm/sec, at initial phase, to 0.0013 cm/sec at advance stage of decomposition.

INTRODUCTION AND BACKGROUND

In a conventional landfill that is designed and operated in accordance with Subtitle D of the Resource Conservation and Recovery Act (RCRA), efforts are typically made to minimize moisture infiltration. Due to the absence of moistures, refuse decomposition proceeds at suboptimal rates for decades or even centuries. Recently, a new type of landfilling technology known as Bioreactors has emerged. The fundamental process of waste stabilization in bioreactor landfill is recirculation of generated leachate into the landfills. Leachate recirculation plays an important role in successful operation of the bioreactor landfill (Barlaz et al. 1990; Reinhart and Townsend 1998, Pohland, 1996, Hossain et al. 2003). Permeability of MSW in a bioreactor landfill is an important parameter to determine the recirculation rate and overall performance of landfill. In order to better estimate the leachate generated and design of leachate recirculation system, clear understanding of the permeability of the MSW and the factors influencing the permeability is necessary

Chen et al., (1995) studied the variation in permeability with density of the samples. Constant head tests were conducted on test columns at densities 160 kg/m^3, 320 kg/m^3 and 480 kg/m^3. The permeability reduced from 9.6 x 10^{-2} cm/s to 4.7x 10^{-5} cm/s when compacted to densities 160 to 480 kg/m^3, respectively. Powrie and Beaven (2005) indicated a decrease in permeability of MSW with increase in effective stress, density and decrease in porosity. Korman et al. (1987) conducted tests on fresh and older waste and indicated higher values of permeability of fresh waste compared to old waste. Chen (1995) from his study on the temporal variation of permeability of MSW indicated a decrease in permeability with time.

The physical properties that influence the permeability of MSW are density, particle size, porosity, material type, degree of saturation, stage of decomposition, and depth within the landfill. During the regular land filling the refuse is compacted in thin lifts leading to horizontal stratification within the landfill, which might result in greater horizontal permeability of the waste than the vertical permeability. In addition to the particle size, the application of daily and intermediate soil covers lead to anisotropy and heterogeneity within a landfill (Powrie and Beaven 1999; Hyder and Khire 2004). Oweis and Khera (1990) found that channeling results in the downward movement of leachate through interconnected pores at rates faster than the uniform flow under gravity. This variation in flow results in non uniform leachate recirculation and non uniform biodegradation leading to differential settlements (Hyder and Khire 2004).

However, there is a limited availability of literature on variation of permeability with degradation. With degradation the particle size of the degradable constituent in MSW was expected to reduce which might significantly influence the permeability of MSW. The objective of the paper is to prepare MSW at different stages of degradation and to determine the permeability of MSW as a function of degree of decomposition. MSW samples at different stages of degradation were generated in four laboratory scale bioreactors with leachate recirculation to simulate bioreactor landfills. The stage of decomposition was determined based on gas volume, composition, and by the volatile solids composition. Initially, particle size distribution was completed for MSW samples. Constant head permeability tests were performed on the remolded samples generated in simulated bioreactor landfills to determine the variation of permeability with degradation.

SAMPLE PREPARATION AND EXPERIMENTAL DESIGN

Preparation of MSW Samples at Different Stages of Decomposition

Municipal solid waste (MSW) samples were collected from a transfer station in Burlington, Texas. The standard collection procedure (Hossain et al, 2003) was followed for obtaining a representative well mixed sample. Initially, physical characterization of the waste, including visual inspection of refuse composition was completed. Based on the physical characteristics, the average paper, plastic and food waste was determined as 56, 13 and 16%, respectively.

Four 16-gallon bioreactor cells were built in the laboratory to generate samples at different stages of decomposition. The reactors were dismantled and destructively sampled at each phase of decomposition. The stage of decomposition was determined from the gas volume, composition, and by the volatile solids composition. The detail description of the sample preparation is presented in Hossain et al. (2007).

Geotechnical Testing Program

Initially the grain size distribution test was conducted. Considering the wide range of particle size and the nature of the waste, wet sieve analysis was conducted. For MSW particles passing US sieve No. 200, hydrometer tests were conducted. The wet sieve analyses on MSW samples at different stages of decomposition were conducted in accordance with standard test method of particle size analysis of soils (ASTM D422-63).

The coefficient of permeability of the MSW samples was measured at different phases of decomposition. The MSW particles are much thinner and flatter than the soil particles and the flow in waste might be through inter particle and intra particle voids. Therefore, the general principles applicable for soils may not be applicable to MSW. Also, due to the lack of standard test procedure for determination of permeability of MSW, the standard test procedure for determination of permeability of the granular soils (ASTM D 2434-68) is adopted for the current work. Due to presence of larger size MSW particles, the test procedure has not followed exact ASTM Standard guideline (minimum 6:1 ratio - sample diameter to max particle size). However, the test results provide a basic understanding of changes in permeability of MSW particles with degradation. Based on the literature review, the unit weight of MSW ranges between 20pcf to 80pcf depending on the compaction effort. For the current study, the permeability tests were conducted at three different unit weights: 40pcf (640.73 kg/m^3), 50 pcf (800.92 kg/m^3) and 60 pcf (961.10 kg/m^3). The test setup consists of an acrylic cylinder with inside diameter of 15.24 cm (6 inches) and a height of 24 cm (10 inches).

Figure 1 step by step procedure for determination of coefficient of permeability. (a) and (b) apparatus and experimental setup; (c) representative sample; (d), (e) and (f) sample placement and compaction; (g) sample mounted on the stand with porous stone on the top; (h) set up connected to water line (i) maintaining constant head; (j) allowing the flow in sample to stabilize (k) collecting the water drained at a given time; (l) measuring the water collected.

TEST RESULTS

Refuse Decomposition Results

Methane production rates and the pH of the produced leachate are presented in Figures 1 and 2 respectively. The anaerobic digester sludge and leachate neutralization with recirculation enhanced the refuse decomposition. Reactors were destructively sampled at days 25, 106, 225, and 253. Based on the test results, sample at day 25 was in anaerobic acid phase, at day 106 sample was at accelerated methane production phase. Samples at day 225 and 253 were in decelerated methane production phases and completely decomposed phase. The test results are presented in Table 1.

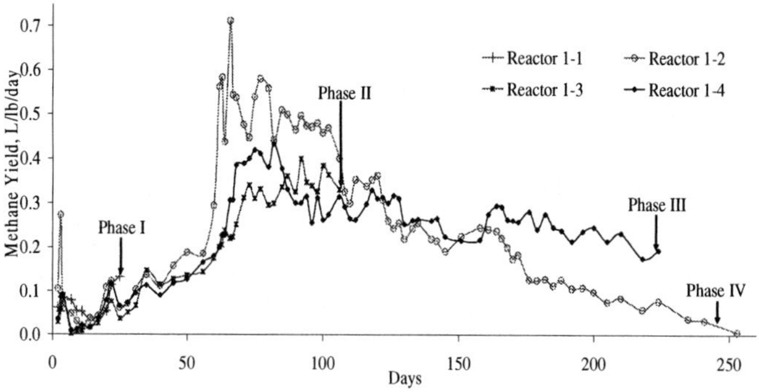

Figure 1 Rate of gas production from reactors at each phase of decomposition (Hossain et al, 2007)

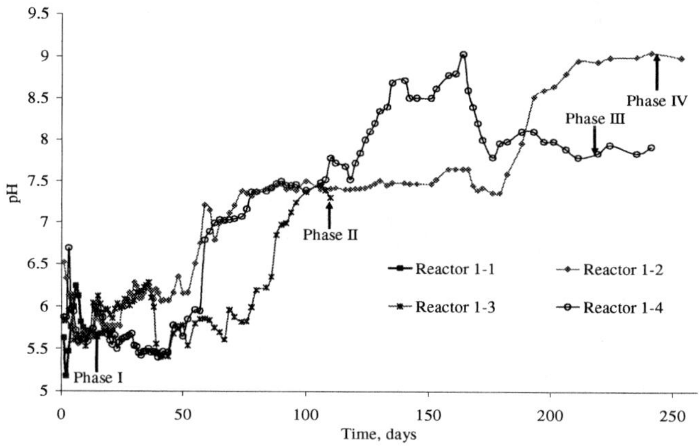

Figure 2 pH of generated leachate from reactors at each phase of decomposition (Hossain et al, 2007)

Table 1 Methane Production and Sample Composition in Bioreactors (Hossain et al, 2007)

Phase	Time of Reactor Operation	Weight of Solid Waste – wet basis (lb)	Cumulative Methane Production (L)	% Change in Volatile Solids
I	25	18.5	23.49	0.91
II	106	19.1	195.03	1.40
III	225	18.7	487.73	12.88
IV	253	18.6	515.41	39.24

Geotechnical Testing Results

Particle Size Analysis

The particle size distribution curves for municipal solid waste at different stages of decomposition are presented in Figure 3. Test results indicated an increase in percent of fraction passing US sieve No. 200 with degradation. The MSW particles were relatively larger during the initial stages of degradation and with degradation, the matrix structure of paper, textile and other degradable constituents were broken down into smaller particles. This resulted in an overall increase in percentage of finer fraction passing US sieve 200. The percentage of finer particles (US sieve No 200) for phase I is only 10% when compared to 39% in Phase IV. The increase in finer fraction with degradation is about 2-3 times, which is expected to change the hydraulic behavior of MSW in a landfill.

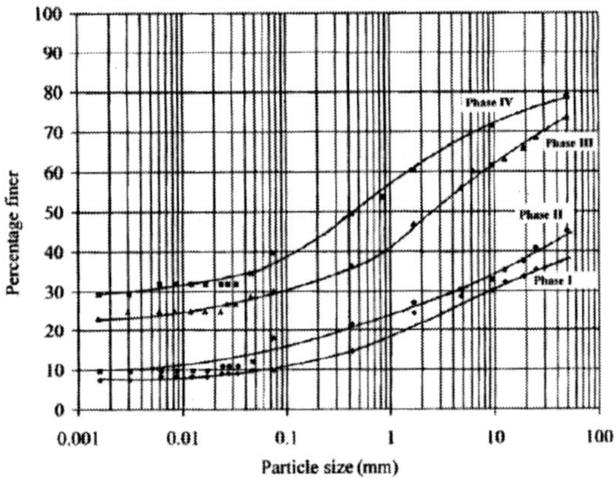

Figure 3 Particle size distribution of MSW with degradation

Permeability with Degradation

The change in permeability of MSW at each phase of degradation is presented Figure 4. The test results indicated that permeability of MSW decreased from 0.0088 cm/sec, at initial phase, to 0.0013 cm/sec at advance stage of decomposition. The tests were conducted at 40pcf (640.73 kg/m^3), 50 pcf (800.92 kg/m^3) and 60 pcf (961.10 kg/m^3). However, for comparison purpose, the permeability for 700 kg/m^3 (44 pcf) was determined using interpolation. The test results clearly demonstrated that the permeability of MSW decreases with decomposition. The test results presented here are average of two tests conducted at same density. The variations of test results within the same density were not significant. The decrease in permeability with decomposition was expected due to: (1) with degradation, the matrix structure of MSW particles were broken down into smaller particles and resulted in an overall increase in percentage of finer fraction passing US sieve 200 (as presented in Figure 3). Accordingly there might be a significant change in pore size, geometry and continuity resulting in the decrease of fluid flow. (2) Increase in density with degradation.

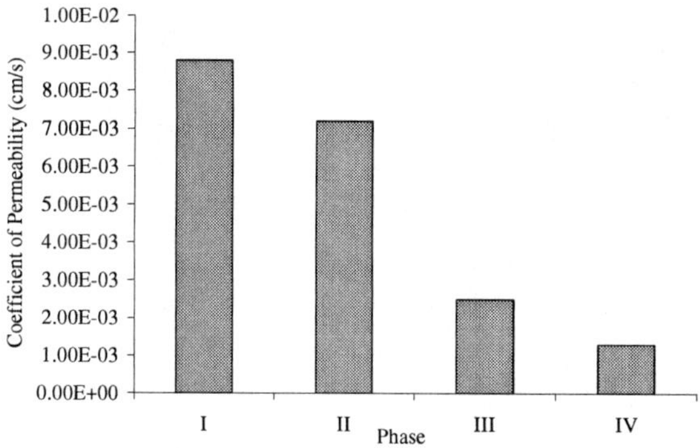

Figure 4 Permeability with degradation of the samples

Figure 5 compares the results obtained from current research with the results obtained by Powrie *et al.* (2005). The permeability results when plotted against corresponding densities determined by Powrie *et al.* (2005) gave some comparison of the upper and lower boundaries of the test results. There is wide variation within the results, however, for the current study, MSW permeabilities due to variation of degree of decompositions varied at same density.

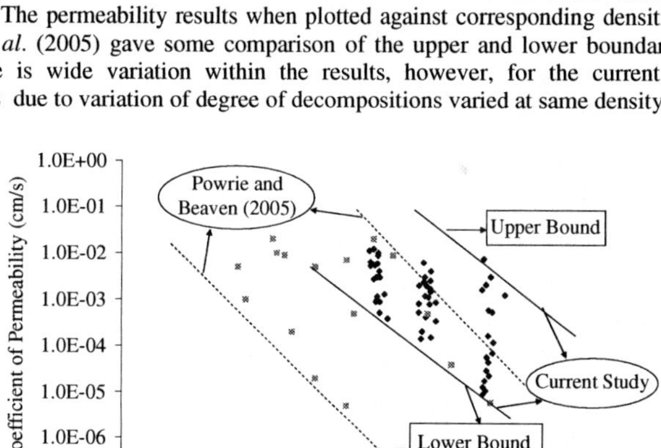

Figure 5 Comparison of the results between current study and results obtained by Powrie and Beaven

Summary and Conclusions

In the current study an experimental program was designed to determine the variation of permeability with degradation. The experimental results are summarized as follows:

1 The MSW particles were relatively larger at the initial stages of degradation and with degradation, the matrix structure of MSW particles were broken into smaller particles. This resulted in an overall increase in percentage of finer fraction passing US sieve 200. The percentage of finer particles (smaller than US sieve No 200) is only 10% in Phase I compared to 39% in Phase IV.

2 The permeability of MSW decreased from 0.0088 cm/s,at initial degradation phase, to 0.0013 cm/s, at advance degradation phase, and both the tests were conducted at 700 kg/cm^3. The decrease in permeability might be due to the increase in percentage of finer fraction passing US sieve no. 200. The increase in percentage of finer content is expected to increase chances of occupying the pore spaces and thus decreasing the size of the flow path, and finally causing a decrease in permeability.

Therefore, the variation in permeability with time and landfill depth should be considered during design of landfill rather than considering an average value for the entire landfill height and operation period.

References

Barlaz, M.A., Ham, R.K., and Schaefer, D.M. (1990). "Methane production from municipal refuse: a review of enhancement techniques and microbial dynamics." CRC Critical Reviews in Environmental Control, 19(3), 557-584.

Chen, H. T. (1995). "Hydraulic Conductivity of Compacted Municipal Solid Waste" *Bioresource Technology*, 51, 205-212.

Hossain, M.S., Haque, M.A., Qasim, S. R., and Hoyos, L. (2007). "Dynamic Characteristic of Municipal Solid Waste with Degradation in Bioreactor Landfills," 4th International Conference on Earthquake Geotechnical Engineering, June 25 - 28, 2007, Thessaloniki, Greece.

Hossain, M.S., Gabr, M.A., and Barlaz, M.A. (2003), "Relationship of Compressibility Parameters to Municipal Solid Waste Decomposition", ASCE Journal of Geotechnical and Geoenvironmental Engineering, Vol. 129, No. 12, December 1, 2003, pg. 1151 –1158

Hyder, M. M., and Khire M. V (2004) "Evaluation of heterogeneity and anisotropy if waste properties on leachate recirculation in bioreactor landfills" Journal of Solid Waste Technology and Management, 30(4) 233-242.

Korman, E. M, Horvitz, E.G., Kleppe H. J., and Dalton, G. M. (1987). "Material properties of landfilled primary waste" Environmental Conference : Portland Marriott, Portland, OR, April 27-29. 179-186.

Oweis, I. S., and Khera, R. P. (1990). "Geotechnology of waste management" Butterworths, Kent, England.

Pohland, F. (1996). "Landfill Bioreactors: Fundamentals and Practice", Water quality International, 18-22.

Powrie, W and Beaven R.P. (1999). "Hydraulic Properties of Household Waste and Applications for Landfills" Proceedings Institution of civil Engineers, Geotechnical Engineering, 137, 235-247.

Powrie, W., Beaven R.P., and Hudson, A. P., (2005). "Factors affecting the hydraulic conductivity of waste" International work shop, Hydro-Physico-Mechanics of Landfills, LIRIGM Grenoble University of France.

Reinhart D R., and Townsend, T. G. (1998). "Landfill Bioreactor Design and Operation" Lewis Publishers, New York.

The Influence of Leachate Recirculation and Air Flow on Aerobic Bioreactor Performance

M. Erfan Mostafid[1], Paul. T. Imhoff[2], Ramin Yazdani[3], Pei Chi[4], Don Augenstein[5], Harold W. Bentley[6], Stewart Smith[7], Bryan J.Travis[8]

[1]PhD Student, Department of Civil and Environmental Engineering, University of Delaware, Newark, DE 19716-3199; mostafid@udel.edu
[2]Associate Professor, Department of Civil and Environmental Engineering, University of Delaware, Newark, DE 19716-3199; imhoff@udel.edu
[3]Senior Engineer, Planning, Resources, and Public Works Department, Division of Integrated Waste Management, County of Yolo, 292 West Beamer Street, Woodland, CA 95695-2511, ramin.yazdani@yolocounty.org
[4]Associate Professor, Department of Civil and Environmental Engineering, University of Delaware, Newark, DE 19716-3199; pei@udel.edu
[5]Director, Institute for Environmental Management, Inc., 4277 Pomona Ave, Palo Alto, CA 94306-4313; iemdon@aol.com
[6]Principal Scientist, Hydro Geo Chem, Inc. 51 West Wetmore Road, Suite 101, Tucson, AZ 85705-1776 haroldb@hgcinc.com
[7]Hydrologist, Hydro Geo Chem, Inc. 51 West Wetmore Road, Suite 101, Tucson, AZ 85705-1776 stewarts@hgcinc.com
[8]Senior Modeler, Hydro Geo Chem, Inc. 51 West Wetmore Road, Suite 101, Tucson, AZ 85705-1776 bryant@hgcinc.com

ABSTRACT: The primary objective of aerobic landfill operations is to control air and moisture conditions such that aerobic waste degradation occurs throughout the landfill. While in a number of cases air has been injected into waste to enhance aerobic processes, the success of achieving aerobic waste degradation has been difficult to quantify. Complicating operations is the need to supply additional liquid to waste. Injection of water displaces gas from refuse pores, redistributes gas flow, and may create pockets of refuse that are not fully aerated. In this work we operated a bioreactor test cell located at Yolo County, California (USA) aerobically for almost six months in 2006 to evaluate the influence of air flow operations and liquid addition on aerobic refuse degradation. By measuring the gas composition data from the effluent gas and examining possible biodegradation reactions, the portion of the refuse degraded anaerobically was estimated and tracked through time. Over a six month period of operation, the percent of the waste degraded anaerobically ranged between 16 to 68%, suggesting that a significant fraction of the waste degraded anaerobically despite attempts to achieve uniform air flow. Anaerobic activity increased during periods of liquid addition. Local measurements of gas composition indicated that even when local oxygen concentrations were greater than ~15%

significant anaerobic activity persisted in some locations.

INTRODUCTION

Landfills are the largest contributor to anthropogenic methane emissions in the United States (USEPA 2001). Aerobic bioreactor landfilling is a relatively recent method that has been proposed to reduce methane emissions and accelerate waste biodegradation. The principle idea behind this technology is to inject air and add water into the landfill in a manner that aerobic biodegradation of waste materials is promoted. Although lab-scale studies have shown successful application of the idea of aerobic landfilling (Reinhart and Townsend 1998), several field–scale operational issues remain to be fully resolved. Dissipation of heat is a major challenge: unless heat is dissipated adequately, temperatures may increase sufficiently to cause fires. A recent study (Powell, Jain et al. 2006) showed that injection of air to landfills can result in high CO readings in some regions. High CO readings are usually considered an indicator subsurface fires (USEPA 1998), although they have also been attributed to biological activity (Powell, Jain et al. 2006).

One issue not previously explored is the degree to which aerobic biodegradation can be maintained in a landfill cell specifically designed for such operation. Can oxygen levels be maintained throughout the waste resulting in minimal anaerobic activity? How does the incremental addition of liquid affect gas flow and aerobic conditions? These questions were addressed by operating an aerobic bioreactor test cell at the Yolo County Central Landfill (Yolo County, California) for a six month period in 2006. Landfill gas (LFG) composition was used to evaluate the degree of anaerobic activity in this landfill cell.

METHODS

A heavily instrumented aerobic bioreactor test cell in Yolo County Central Landfill, California (USA) was studied. This 2.5 acre landfill cell was filled with 12 tonnes of municipal solid waste and instrumented with 59 temperature sensors, 52 resistivity moisture sensors, and 59 small-diameter tubes for collecting gas samples and measuring fluid pressures throughout the landfill. An array of horizontal wells located in three layers was used to add liquid (fresh water or recirculated leachate) or to pull air into the landfill. Because air entered the landfill through the landfill surface and was collected in a central vacuum line, the "effluent" gas from the landfill could be easily monitored during aerobic operations. A vertical cross-section through the landfill cell is shown in Figure 1. A horizontal cross-section showing the locations of gas collection lines in Layers 1 and 2 is shown in Figure 2. Landfill gas can be collected from the north and south sides of the gas collection lines. Waste placement in this aerobic cell occurred between August 8, 2001 and September 26, 2001.

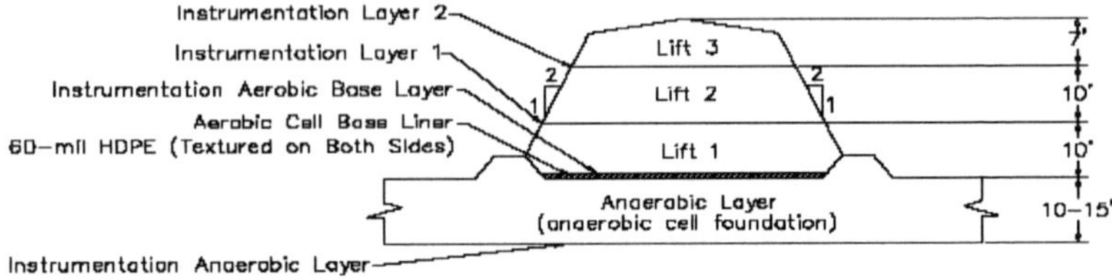

Figure 1. Cross-section of aerobic bioreactor at Yolo County. Instruments, gas collection wells, and liquid addition lines located in instrument layers.

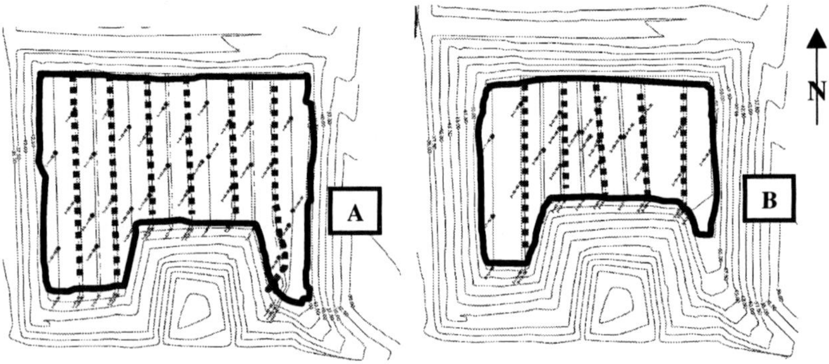

Figure 2. Horizontal cross sections through Layer 1 (A) and Layer 2 (B) of aerobic bioreactor landfill. Thick dashed lines are horizontal gas collection wells; thin solid lines between gas wells are leachate recirculation lines.

The main vacuum line connected to all gas collection wells was turned on from 03/28/06 through 09/12/06. Gas flow through individual wells varied depending on the local gas permeability, which fluctuated through time because of liquid addition. However, the overall gas flow from the landfill cell was relatively constant, as shown in Figure 3. The average gas flow was 1,000 SCFM.

The composition of the LFG in the main header line was measured frequently using a gas monitor (Landtech Inc., GEM 2000). Temperature and moisture sensor readings (resistivity sensors) were recorded continuously. Using plastic tubes located throughout Layers 1 and 2, LFG composition was analyzed locally within the waste on 08/23/06 to quantify the spatial variability in anaerobic activity on this date.

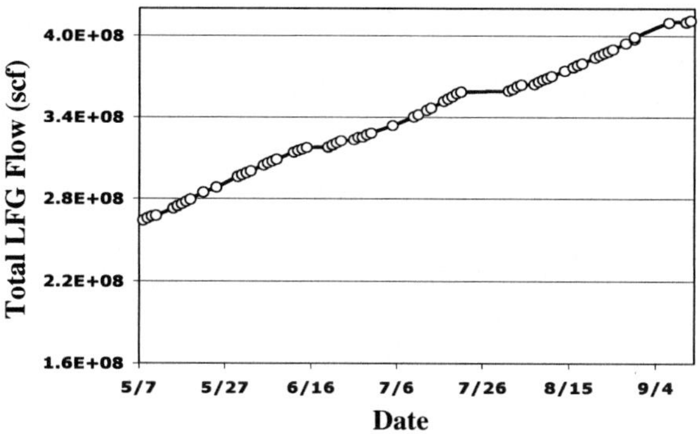

Figure 3. Cumulative LFG flow during the operational period.

Liquid was added to the cell during the operational period either as recirculated leachate or supplemental water through horizontal leachate recirculation lines. These lines were located in instrument Layers 1 and 2. During the eighteen-day period from 04/07/06 to 04/25/06, 13,250 gallons of leachate was recirculated in the cell. The recirculation was then stopped and started again on 05/15/06 and continued until 06/29/06, when 331,150 gallons of leachate were recirculated in the cell. Supplemental water (190,950 gallons) was added to the cell between 05/30/06 to 06/15/06. The cumulative volumes of recirculated leachate and supplemental water are shown in Figure 4.

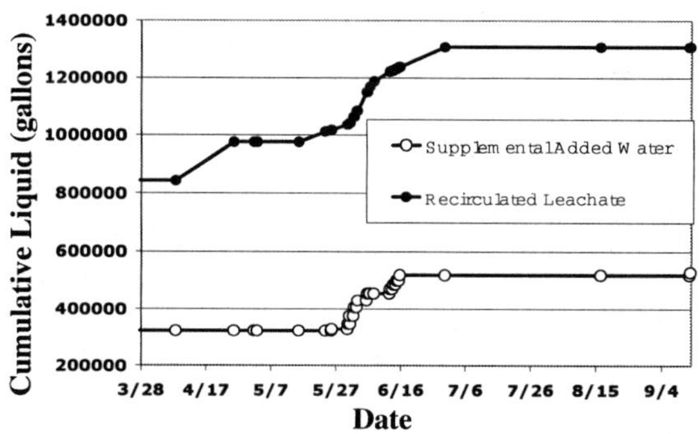

Figure 4. Cumulative liquid added during the operational period.

RESULTS AND DISCUSSION

The chemical composition of LFG was used to estimate the waste fraction that was degraded anaerobically. Landfill gas extracted from traditional anaerobic landfills is usually composed of 45-60% CH_4 and 40-60% CO_2 (Tchobanoglous, Theisen et al. 1993). Considering the redox state of carbon in CH_4 and CO_2, which are - 4 and + 4 respectively and approximately 1:1 mole ratio of these gases in LFG, we assume that the fraction of biodegradable waste within the landfill should have carbon atoms at an average redox state close to zero. Aerobic and anaerobic degradation equations for carbon-containing compounds at a redox state of zero are

$$C + H_2O \Rightarrow (1/2)CH_4 + (1/2)CO_2 \text{ (Anaerobic Degradation)} \quad (1)$$

$$C + O_2 \Rightarrow CO_2 + C_{biomass} \text{ (Aerobic Degradation)} \quad (2)$$

The amount of biomass produced because of anaerobic waste degradation is usually small and was neglected in equation (1). However this is not the case for aerobic reactions. In our analysis we assumed that the system reached a quasi-steady state, with biomass produced and consumed (producing CO_2) at roughly equal rates. The impact of this assumption on our analysis will be explored more fully in future work.

Based on the stoichiometry of these two reactions and the assumption of steady production and degradation of biomass, the percentage of waste degraded anaerobically, P, can be estimated from

$$P = \frac{2C_{CH_4}}{2C_{CH_4} + (C_{CO_2} - C_{CH_4})} \quad (3)$$

where C_{CH_4} and C_{CO_2} are the measured concentrations of CH_4 and CO_2. Equal production of CH_4 and CO_2 is assumed in equation (1). However, the composition of LFG in anaerobic landfills is variable. The two extreme cases of 60% CH_4 - 40% CO_2, and 40% CH_4 - 60% CO_2 were also examined and P calculated for these extremes using equations similar to equation (3)

The results from these methods of computing P are shown in Figure 5 for gas collected from the main header line over the six month operational period. The two alternative methods of estimating P yield results that are generally within +/- 10% of equation (3) estimates. Air began to be pulled into the cell from the landfill surface on 03/28/06. It took almost a week to achieve stabilized conditions in the cell, where P decreased from an initial value of 66% (equation 3) to 20% on 04/06/06. P remained relatively constant for the next month. The spikes in P that appear throughout Figure 5 are associated with short rest periods when the blower was turned off. During these periods, generally 24 hours or less, CH_4 concentrations rose and the percent anaerobic activity increased. Once the blower was restarted, pore gas was flushed out of the cell and values of P returned to levels observed before the blower was turned off.

Liquid began to be added to the cell on 05/30/06: 277,050 gallons of supplemental water and 289,820 gallons of leachate were added from 05/30/06 to 06/29/06. Soon after the onset of liquid addition, P increased and reached a maximum of ~ 65% on 06/26/06 (equation 3, ignoring peaks associated with blower shutoff). Water addition and leachate recirculation were terminated on 06/30/06 and P gradually decreased to 31% on 08/04/06. P fluctuated between 25% and ~ 40% through the remainder of the operational
period

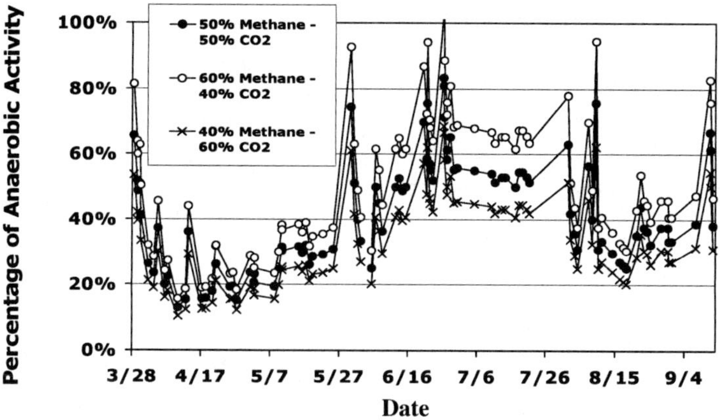

Figure 5. Percentage of anaerobic activity, *P*, determined from the main header line.

On 08/23/06 landfill gas samples were collected from tubes inserted throughout the landfill. Contours of anaerobic activity, oxygen concentration, and temperature from measurement locations in Layer 1 are shown in Figure 6. Regions with anaerobic activity higher than 50% are circled with dashed lines in Figure 6A and are correlated with locations where oxygen concentrations were less than 10%, circled in dashed lines in Figure 6B. Regions with temperatures greater than 70°C also corresponded to regions of small anaerobic activity (less than 50%), highlighted with dotted squares in Figure 6A and dashed lines in Figure 6C.

One puzzling observation is the large number of measurement points with oxygen concentrations close to 20%, many of them in the center of Layer 1 (see Figure 6B). These high concentrations are consistent with oxygen levels measured from horizontal gas extraction lines (located in the same vicinity of the gas sampling tubes, see Figure 2) at the point where the lines emerge from the side of the landfill. Since gas extraction lines and gas sampling tubes are located in the same horizontal trenches filled with high permeability tire shreds, we hypothesize that for some trenches gas preferentially entered the landfill from the south side slope and flowed through the trenches, bypassing much of the waste. While the ends of the trenches were covered with several feet of dirt to inhibit air entry, for some trenches, most notably those in the center of the landfill, the low permeability soil cover material might have been insufficient to inhibit preferential air entry from the south side slope.

Another interesting observation from the data shown in Figure 6 is that the landfill appears to be more anaerobic on the north side, where oxygen concentrations are lower and the percent anaerobic activity higher. This result is consistent with the gas flow measured from the horizontal gas collection lines: while a similar vacuum was imposed on both the north and south sides of each gas collection line, approximately 70% of the landfill gas was extracted from the southern end of the gas collection wells. The cause of the preferential gas flow through the southern side is unknown, although we hypothesize it is associated with (1) lower refuse permeability on the north side and (2) better sealing of the horizontal shredded tire trenches on the north side slopes.

Finally, it is interesting to note that even in aerobic regions where oxygen concentrations are high (greater than ~15%), the percent anaerobic activity is seldom below $P = 0.20$. This suggests that even when sufficient oxygen is supplied locally within the landfill, anaerobic pockets persist where a portion of the refuse degrades anaerobically. This is consistent with the overall performance of the landfill, where P never decreased below 16% (see Figure 5). Similar data trends shown for Layer 1 in Figure 6 were also observed for Layer 2.

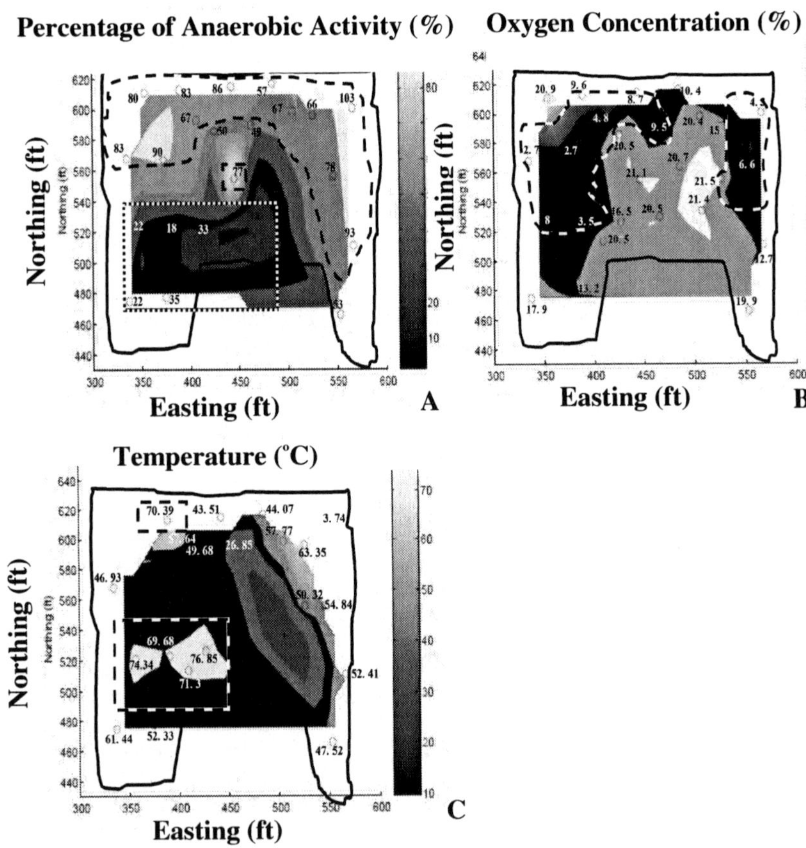

Figure 6. Contours of anaerobic activity, P (A); oxygen concentration (B), and temperature (C) within Layer 1.

SUMMARY

A primary objective of aerobic bioreactor operations is to maximize aerobic activity through air injection. Because of the design of the aerobic bioreactor test cell at Yolo County, essentially all air entering this landfill cell was collected and analyzed for CH_4 and CO_2. These measurements indicated that despite attempts to achieve uniform air flow and high oxygen concentration in the landfill cell, a significant portion of the waste was degraded anaerobically over a six month period of operations. Anaerobic activity increased during periods of liquid addition and persisted locally, even when oxygen concentrations were maintained at ~15%. These data illustrate the difficulty of maintaining a high degree of aerobic biodegradation during bioreactor landfill operations.

ACKNOWLEDGMENTS

This work was supported by the US Department of Energy under Cooperative Agreement DE-FC26-05NT42432.

REFERENCES

Augenstein, D. C. (2003). Landfill biofiltration system and methods for removal of gas-phase pollutants and contaminants. U. S. Patent. 20030008381

Powell, J., P. Jain, et al. (2006). "Changes in landfill gas quality as a result of controlled air injection." Environmental Science & Technology 40(3): 1029-1034.

Reinhart, D. R. and T. G. Townsend (1998). Landfill Bioreactor Design and Operation. Boca Raton, Lewis Publishers.

Tchobanoglous, G., H. Theisen, et al. (1993). Integrated Solid Waste Management, Engineering Principles and Management Issues. New York, McGraw-Hill, Inc.

USEPA (1998). Compilation of Air Pollutant Emission Factors, U.S. Environmental Protection Agency, Office of Air Quality Planning and Standards.

USEPA (2001). "Non-CO2 Greenhouse Gas Emissions from Developed Countries, EPA-430-R-01-007." U.S. EPA: Washington, DC.

Characterization of landfills in Central Asia by means of site investigations and landfill simulation in laboratory bioreactors

Marco Ritzkowski[1], Helmut Adwiraah[1], Wei Wu[2] and Rainer Stegmann[1]

[1]Hamburg University of Technology, Institute of WasteResourcesManagement. Harburger Schloßstr. 36, 21079 Hamburg, Germany; m.ritzkowski@tu-harburg.de
[2]University of Natural Resources and Applied Life Sciences, Institute of Geotechnical Engineering. Feistmantelstr. 4, 1180 Vienna, Austria; wei.wu@boku.ac.at

ABSTRACT: Landfill practices are often insufficient in the countries of Central Asia (CA). Municipal Solid Waste (MSW) is often disposed of together with medical waste, commercial, and industrial waste. Most of the existing MSW landfills can be considered as open dumps, causing severe problems with respect to environmental pollution and human health. In the framework of the international research project NISMIST these landfills have been subjected to closer investigations. Some of the main results are summarized in this paper. It became clear that many landfills possess both severe ongoing emissions in terms of polluted leachate and landfill gas as well as a significant emission potential. The situation deteriorates due to uncontrolled surface and subsurface fires on many sites, releasing high concentrations of trace components, e.g. carbon monoxide, and potentially toxic organic compounds. Only very few exceptions can be found, demonstrating that relatively simple and cost effective measures such as waste compaction and intermediate covering can contribute significantly to an improvement of the situation.

1. INTRODUCTION

Waste management in Central Asia (i.e. Kazakhstan, Kyrgyzstan, Uzbekistan, Tajikistan and Turkmenistan) is predominantly characterized by poor landfill practices, resource-intensive production, and very limited resources of arable soil and water. In addition, most of the industrial and mining activities in these regions are located where one could also find the main groundwater resources and the majority of landfill sites.

During the soviet time many MSW landfills were constructed outside the big cities, following an advised siting procedure. Accordingly landfills were constructed in large gullies, offering favorable geological and hydro-geological conditions and the possibility of collecting potential leachate flowing out at the lowest point by gravitational force.

In addition to the existing number of old landfills, the generation of MSW is bound to increase in the future due to introduction of a market-structured economy, which has the effect of promoting consumerism. This increase in MSW will lead to the installation of new and probably larger landfills.

In order to maintain and improve living standards in Central Asian countries it is essential to implement waste management legislation, infrastructure, and procedures. The steps include choosing landfill sites equipped with naturally occurring barriers in areas where groundwater pollution is not expected and the seismic hazard is relatively low. It is also necessary to reduce the risks and hazards associated with the existing landfills, based on a detailed characterization of the sites. This characterization work is part of the international project NISMIST, Management of environmental risks associated with landfills in seismically active regions in the New Independent States of Central Asia. (www.nismist.info)

2. MUNICIPAL SOLID WASTE LANDFILLS IN CENTRAL ASIA

2.1 Landfill classification catalogue

Statistical data on MSW landfills are often incomplete or inadequate in CA countries. Furthermore the existing information is generally not available at a national regulator or public authority, thus significantly aggravating the information search.

In order to improve the situation a landfill classification catalogue (LCC) has been developed within the framework of the NISMIST project. The LCC is subdivided into three levels, differing in information content and number of landfills included. The first level contains basic information like landfill name, location and address (including coordinates), operational period, landfill area, capacity and type as well as possibly protection barriers. At present more than 200 data sets on CA landfills are included in the LCC.

The second and third levels contain more detailed information on selected sites, i.e. five landfills per CA country in level 2 and one landfill per country in level 3. The special feature of level 3 is the fact that the presented information is not only based on data provided by the landfill operator or the competent authority, but also results from site investigations carried out in the course of the project.

2.2 Landfill practices

Landfill operation in CA countries can be generally characterized as insufficient. The majority of landfills have to be considered as open dumps, missing any kind of protection barriers, safety installations or an organized planning for waste deposition. Scavengers (often children) sorting the incoming wastes and collecting the recyclables are present on almost every landfill. Fences for the prevention of inadmissible entrance are missing just like any kind of litter control. After its delivery the wastes are distributed by bulldozers without sufficient compaction. Intermediate covers are very seldom installed. These practices lead to subsurface fires, especially near the slopes of the tipping area. In addition the low compaction of the wastes reduces the amount of surface runoff and contributes to the acceleration of wash out processes.

Exceptions from this landfill practices are very rare, but can be found, for instance, in Tashkent, Uzbekistan. Financed by World Bank credits, modern equipment has been acquired, including waste trucks, compactors and excavators. Landfill operation is well organized including waste registration, emplacement records and the adequate control of the landfill territory. However, as for all the other sites in CA a sufficient monitoring program for determination of the landfill impact is missing.

2.3 Site investigations

Risk analysis for landfill sites basically needs to consider both the current emissions and the specific emission potential of the chosen landfill. To obtain the necessary data several possibilities exist. Theoretical approaches help to estimate landfill gas production rates based on parameters like the amount of organic carbon, precipitation and evaporation, temperatures and time. The theoretical models can be evaluated and calibrated by means of control measurements on site. During a site investigation, values of gaseous and liquid emissions can be obtained through online measurements, e.g. gas quality and quantity, leachate, surface water and groundwater quality, and waste composition. Additional information can be gained through the analyses of solid waste samples in the laboratory by means of:

- Chemical analysis
- Biological tests
- Investigations in Landfill Simulation Reactors (LSR)

The following case study provides an exemplary summary of site investigations carried out at a landfill in CA and their results:

2.3.1 Case study: Karasai Landfill, Almaty, Kazakhstan

The Karasai landfill started its operation in 1989 serving as the main disposal site for the former capital of Kazakhstan, Almaty. The landfill covers an area of approx. 23 ha and has reached a maximum height of 105 m as it has been constructed in a natural Y-shaped broad gully extending from the north to the south. The sides of gully are steep and turf-covered. The width and depth decrease towards the northern part, from 350- 340 m to 150-140 m and from 95-90 m to 40-35 m correspondingly. At present a total volume of approx. 24 M m^3 of MSW has been disposed of at Karasai landfill. The remaining operation time until the site reaches its projected maximum capacity is estimated to be 5 years.

Geological and hydro-geological conditions

The landfill site is placed on lower-quaternary alluvial-proluvial deposits, represented by loess-like loams. Their specific feature is alternation of water permeable and poorly permeable layers in the vertical section with poorly permeable at evident predominance. This material is generally characterized by a vertical hydraulic conductivity of approx. $5 * 10^{-8}$ m/s. The ground water table is relatively deep for the respective territory and can be estimated to a depth of 120 to 200 m below soil surface. Due to the low vertical hydraulic conductivity, infiltrating leachate might reach the ground water only after a time period of 75 to 125 years in the worst case. However, perched ground water can be found at a depth of 1.5 to 4.5 m beneath surface at the northern end of the gully. Fluctuations amplitude of these ground waters is 0.8-1.2 m.

Meteorological conditions
The climate in the considered region is moderately continental with dry air and great number of sunny days. The precipitation rates average 509 mm per year at an annual evaporation rate of 452 mm. Average air temperatures vary significant over the year from – 9.9°C during January until + 29.5°C during July.

Leachate and ground water quality
On the Karasai site there are 3 ground water monitoring wells and one leachate collection pond situated in the northern part of the landfill. In the course of site investigations samples have been taken and instantly analysed for temperature, pH, redox potential, electrical conductivity, dissolved oxygen and total suspended solids. Table 1 shows analytical results from own measurements for the leachate as well as results extracted from existing records of the landfill operator for the perched ground water in the landfill proximity.

Table 1. Results of chemical analyses for leachate and ground water samples taken at the Karasai landfill near Almaty, Kazakhstan.

	pH [-]	Eh [mV]	Cond.el. [mS/cm]	BOD [mg/l]	Ammonium [mg/l]	Sulfate [mg/l]	Chloride [mg/l]
Leachate	7.66	- 383	42,5	4,920	970	6	10,780
Ground water	8.03	133	8.3	4	9.5	2,002	1,992

The leachate pollution pattern can be regarded as typical for MSW landfills filled with untreated wastes. The chloride concentration is quiet remarkable and leads to a electrical conductivity of more than 40 mS/cm. The concentration of organic pollutants (expressed as Biochemical Oxygen Demand, BOD) is high, and inorganic nitrogen (ammonium) appears to be at a moderate level. Under the prevailing reducing conditions (Eh = -383 mV) sulphate concentrations are low.

The results of ground water analyses, however, clearly show the pollution caused by infiltrating leachate. This is particularly evident for the parameters chloride and sulphate, but also the ammonium concentration is rather elevated.

Landfill gas production and composition
On site measurements for the determination of a potential LFG production have been carried out during the site investigation in autumn 2006, where a portable landfill gas analyser (GA2000, Geotechnical Instruments) and a portable flame ionisation detector (PORTAFID, Sewerin) have been used. These instruments allow the measurement of both, gas composition inside the landfill body (in this case in depth between 0.2 and 1.0 meter below landfill surface, components: CH_4, CO_2, O_2, CO and H_2S) and diffuse methane emission through the landfill surface.

As LFG measurements have never been applied to the Karasai landfill before, the data obtained could not be compared with existing values. Furthermore, the morphology of the wastes did now allow a deep penetration of the gas lance used for gas extraction. However, at different positions of the landfill methane concentrations averaging 43.9 % v/v (range: 34.6 to 55.2 % v/v) could be measured near the surface. The LFG found was free of oxygen and showed maximal CO_2 value up to 34.8 % v/v. Carbon monoxide concentrations were generally elevated, showing an average concentration of 245 ppm, measured at a total of 23 reading points. Maximum CO concentrations rose up to more than 2000 ppm, thus presenting a severe risk for the workers and scavengers on the site.

FID measurements have been carried out at different landfill areas. The results obtained are difficult to interpret due to the fact that only some areas had been covered (with soil) at that time. For a total of 53 reading points an average of 1,311 ppm CH_4 was detected. Peak concentrations rose up to more than 1.8 % v/v, indicating a significant production of LFG at certain parts of the landfill.

In parallel to the above mentioned calculation a LFG prognosis has been conducted based on the model introduced by Tabasaran and Rettenberger back in 1987 (Tabasaran and Rettenberger, 1987). For running the model, basic data gained during the site investigation as well as data from the landfill operators records have been used. Once the geometry of the site is known the actual waste mass can be calculated. From the analyses of solid waste samples average water content as well as the amount of organic carbon can be determined. The latter is the main parameter for a sufficient LFG prognosis, provided that its biodegradable fraction is adequately estimated. For a first approach the TOC concentration of recently dumped MSW had been used where the biodegradable portion was estimated to 60%.

According to results of the LFG production model production rates, between 10,000 to 12,500 m^3/h can be expected for the complete landfill in 2006. These figures clearly demonstrate the significant energy potential contained in the deposited wastes, but at the same time – as no LFG extraction and utilization system has been installed – give an impression of the significant release of greenhouse gases into the atmosphere.

Solid waste properties

In order to characterize the landfilled waste according to its composition, water content and temperature, two excavations were done down to a depth of 5 m below landfill surface by means of a backhoe. The selected excavation areas represented both rather fresh MSW (uncovered) and older wastes which had been landfilled several years ago (covered with soil). Solid waste samples were taken from different waste horizons and subjected to a sorting analysis. Temperature profiles were determined and optical and olfactory impressions noted. Figure 1 presents the results of sorting analyses for both excavations with the associated data from chemical/physical analyses.

Although no waste separation takes place in the city of Almaty and the MSW is landfilled without any pre-treatment, the TOC values analysed for the fresh waste are relatively low. At the same time fine, soil-like materials as well as inert substances like metals, stones and glass dominate the waste composition. An explanation for this can be found from the specific conditions at the Karasai landfill site. Recyclables are collected on site through a large number of scavengers, thus mainly reducing the amount of plastics, but paper, cardboard, glass and metals also are collected to a certain extent. In addition the landfill exhibits a number of severe open and underground fires, particularly near the steep slopes of the tipping areas. By means of these fires the waste composition is subject to changes and the amount of organic material decreases.

The wastes sampled on an older part of the landfill showed significantly decreased TOC contents and temperatures in the range of the ambient conditions. Taking the specific waste composition into consideration (relatively smaller amount of fines, relatively higher amount of plastics) it can be assumed that the fine fraction is mainly assigned to mineral (inorganic) material.

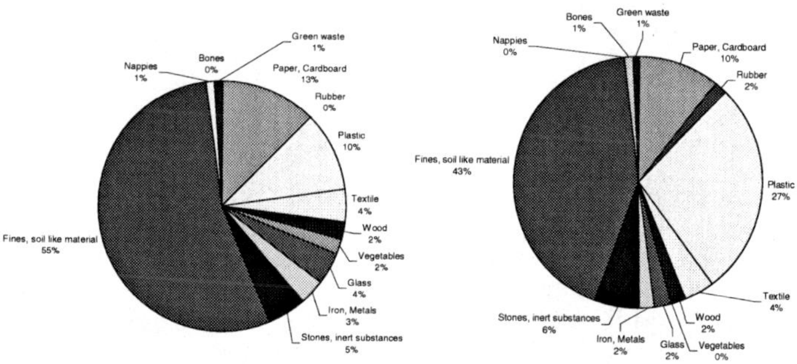

Characteristic:	fresh waste	Characteristic:	old waste
Water content:	26.2 % v/v	Water content:	33.2 % v/v
TOC:	112 kg/ton DM	TOC:	60 kg/ton DM
Temperature:	64 °C	Temperature:	25 °C

FIG. 1. Composition and chemical/physical characteristic of two waste samples taken by means of excavation at the Karasai landfill in autumn 2006.

2.4 Landfill simulation by means of lysimeter tests

2.4.1 Background

The basic conditions and processes that occur in a landfill can be simulated under laboratory conditions by means of landfill simulation reactors. These processes include anaerobic biodegradation processes leading to the production of landfill gas as well as to the formation of liquid effluent (i.e., leachate). The different phases of anaerobic fermentation, i.e., hydrolysis, acidification, acetogenic phase, and methane formation, can be reached allowing the determination of leachate pollution under changing environmental conditions (e.g., pH values, redox conditions). The laboratory simulation allows the simulation of these processes within a relatively short timeframe. The rapidity of the rates of reaction is achieved by the use of a high rate of water percolation through the mass in the reactor (a pump enables recirculation of leachate, thus enhancing the accelerated biological degradation of the waste in the reactor), following the assumption that under anaerobic conditions extraction processes dominate over those of biological (i.e., microbial) decomposition. Depending on the input material and the adjusted water exchange rate (ratio of liquid to solids or liquid/solid), processes, that under normal conditions in a landfill might last years or decades, can be modeled within months, allowing estimation of the emission behavior of the materials in the reactors over long periods of time. The investigation of solid waste samples, taken from existing landfills, in landfill simulation reactors enables the determination of both current landfill emissions as well as the emission potential. Moreover, synthetic waste mixtures can be used in order to simulate landfill processes, provided that basic information regarding the specific waste composition at a certain landfill is known.

In the framework of the NISMIST project LSR tests have been carried out using the morphological composition of MSW from Ashgabad, capital of Turkmenistan. Under simulated landfill conditions in the LSR specific emission pattern regarding the production of landfill gas and polluted leachate can be obtained. These data allow the prediction of maximal emissions to be expected from the landfilled waste under the assumption that the tests are performed under optimal ambience conditions.

2.4.2 Material and methods

Investigation were carried out in three LSR made of stainless steel, showing a volume of 8.8 l each (Figure 2). All of the reactors were kept in an environmentally-controlled room. The room was kept at a constant temperature of 35°C to simulate typical conditions in the interior of the landfill. Liquid and gas samples have been taken at regular intervals and analysed according to a specific sampling and analysis program.

FIG. 2. Landfill simulation reactors (LSR); V = 8.8 liter

2.4.2 Results

The landfill gas potential was found to be low, showing an average of approx. 37 liter per kilogram dry waste. This is due to the relatively small content of organic substances in the synthetic waste mixture, but in correspondence to the waste composition in Ashgabad. Taking the local climate of Turkmenistan into consideration (high temperatures and low precipitation rates), the gas production rate to be expected at the Ashgabad landfill tends to be very moderate.

The analyses of the leachate composition show moderate ammonium concentrations in the range of 200 to 300 mg/l and COD (Chemical Oxygen Demand) concentrations around 700 mg/l. Provided that the actual rates of precipitation and evaporation are known the results of the LSR can be used for an estimation of those time periods where a severe leachate pollution is to be expected.

3. RECOMMENDATIONS AND CONCLUSIONS

Based on the results of site investigations and existing records, CA landfills can be often characterized as open dumps, showing severe emissions and large emission potentials. The reasons for this situation is found in both insufficient finances in the waste management sector but often also results from a lack of experience and knowledge. Positive exceptions are rare, but can be found, e.g., in Tashkent, Uzbekistan. The international project NISMIST is dealing with the situation of MSW landfills in CA. This includes the exemplary elaboration of remediation measures for selected sites. At present the following recommendations can be made: For existing landfill sites control measures should be increased, including access to the site, inspection of incoming wastes, avoidance of co-disposal of hazardous and non-hazardous wastes, and organized waste placement. In terms of emission control the main objective should be the extinction of subsurface fires and their future avoidance. Therefore, intermediate surface covers (using soils from the surrounding areas) should be installed, and increased compaction work should be implemented. Leachate should be collected in controlled ponds and either treated or re-infiltrated into the waste mass. With respect to LFG the models applied often showed significant production rates. However, these values often overestimate the possible LFG collection rates by far. In order to verify the model results qualified site investigations including in situ LFG measurements, waste sampling and analyses (GP_{21}, TOC) as well as LFG extraction tests should be conducted. The results of these tests should subsequently act as the basis for final determination of the LFG generation rates. Currently, the energy contained in the LFG is not utilized at all. Furthermore, uncontrolled methane emissions significantly contribute to the global greenhouse effect. In this connection and against the background of the Kyoto mechanisms, possibilities for the implementation of future CDM projects can be seen. Through this mechanism the environmental impact of CA landfills could be significantly improved and technical measures could be, at least partly, refinanced.

REFERENCES

Krümpelbeck, I. (2000) Untersuchungen zum langfristigen Verhalten von Siedlungsabfalldeponien. Veröffentlichung des Lehrstuhls für Abfall- und Siedlungswasserwirtschaft der Bergischen Universität – Gesamthochschule Wuppertal, Bd. 3

Tabasaran, O., Rettenberger, G. (1987) Grundlage zur Planung von Entgasungs-anlagen. In: Müllhandbuch. Erich Schmidt Verlag, Berlin.

Geotechnical Properties of Municipal Solid Waste Subjected to Leachate Recirculation

Krishna R. Reddy[1], Janardhanan Gangathulasi[2], Hiroshan Hettiarachchi[3], and Jean Bogner[4]

[1]Professor, Department of Civil and Materials Engineering, University of Illinois at Chicago, 842 West Taylor Street, Chicago, Illinois 60607; kreddy@uic.edu
[2]Graduate Research Assistant, Department of Civil and Materials Engineering, University of Illinois at Chicago, 842 West Taylor Street, Chicago, Illinois 60607; kreddy@uic.edu
[3]Assistant Professor, Department of Civil Engineering, Lawrence Technological University, 21000, West Ten Mile Road, Southfield, Michigan 48075; hiroshan@ltu.edu
[4]President, Landfills +, Inc., 1144 N. Wheaton, Illinois 60187; jbogner@landfillsplus.com

ABSTRACT: This paper compares the geotechnical properties of fresh municipal solid waste (MSW) to landfilled MSW obtained from a bioreactor landfill. Fresh MSW samples were collected from the working face of Orchard Hills Landfill. Landfilled MSW (15-19 months old, subjected to intermittent very low flow leachate recirculation for a year) samples were recovered from a borehole located near a multi-level horizontal leachate recirculation system in the same landfill. Laboratory testing was performed on shredded samples to determine the variation in in-situ moisture content, organic content, biochemical methane potential (BMP), specific gravity, density, hydraulic conductivity, compressibility, and shear strength properties of fresh as well as landfilled MSW. In-situ dry gravimetric moisture content of fresh MSW was around 44% and landfilled MSW varied between 21–43%. Organic content of the fresh MSW was 76% whereas degradation and presence of other components in landfilled MSW resulted in organic content between 62-68%. Due to the presence of finer particles, reconstituted dry density of landfilled MSW was 601-712 kg/m^3 whereas fresh waste was 515 kg/m^3. For confining pressure greater than 69 kPa, both fresh and landfilled waste possessed similar hydraulic conductivity. Compressibility test results showed higher strains for fresh MSW than for the landfilled, with compression index of 0.28 for fresh and 0.22 for landfilled MSW. Based on direct shear testing, shear strength properties for fresh MSW were: c'=47 kPa and ϕ'=29^0 and for landfilled MSW were: c'=12 kPa and ϕ'=32^0. Overall, this study shows that the changes in geotechnical properties of waste subjected to low overall rates of leachate recirculation for a year are not significant; however, additional research is warranted to quantify changes in properties of waste under longer-term leachate recirculation.

INTRODUCTION

Bioreactor or leachate recirculation landfill technology is being explored to accelerate or enhance the anaerobic biodegradation of MSW. Injected additional moisture stimulates microbial activity by providing better contact between substrates, nutrients, and microorganisms. Thus, bioreactor landfills offer a way to achieve higher rates of MSW degradation, accelerates biogas production, limit the potential for settlement–induced damage to the final cover and increase landfill volumetric capacity. Although this technology has several advantages, the addition of moisture or leachate should be done in a controlled manner to avoid geotechnical instability. Several historic slope failures at landfills (e.g., Hendron et al. 1999; Eid et al. 2000) have emphasized the need for studying the change in moisture and solids contents of waste during decomposition because it has direct influence on the geotechnical stability of bioreactor landfills. Currently, there are no standard designs for bioreactor landfills, and there are a number of unresolved engineering issues relating to the design and operation of safe and effective bioreactor landfills.

Accurate determination of geotechnical properties of MSW is difficult due to high compressibility, microbial degradation and heterogeneous nature of MSW. Previous studies have discussed the geotechnical properties and mechanical behavior of MSW (Landva and Clark 1990; Gabr and Valero 1995; Kavazanjian et al. 2001; Hossain 2002; Zekkos 2005). General belief is that the geotechnical properties of MSW vary with age/decomposition. However, enhancement of this aging effect due to leachate recirculation has not been captured well in the literature. Although waste samples sometimes were obtained directly from landfills, it is extremely difficult to obtain and test large undisturbed samples; therefore, studies are often performed in the laboratory using shredded reconstituted field samples or synthetic waste samples. The objective of this study was to compare the geotechnical properties of fresh MSW and landfilled MSW obtained from a bioreactor landfill. Series of geotechnical tests were conducted to determine the in-situ moisture content, organic content, BMP, specific gravity, density, hydraulic conductivity, compressibility, and shear strength properties of fresh and 15-19 months old landfilled waste. The relative changes in the properties of the fresh and landfilled waste helped to assess the effects of enhanced degradation on geotechnical properties of waste due leachate recirculation.

WASTE SAMPLING

Waste samples were collected from Orchard Hills Landfill in Davis Junction, Illinois, USA, which is owned and operated by Veolia Environmental Services. The incoming waste consisted approximately 70% MSW, 17% construction and demolition debris (C&D), 11% soils, and 2% other types of waste. The waste components included 8% paper, 11% plastics, 12% wood, 4% glass, 4% metals and others (Grellier et al. 2007). Leachate recirculation was accomplished by spraying on the working face during filling operations and thereafter using a network of multi-level horizontal leachate recirculation lines (LRLs) installed within the waste. These consisted of 15 cm diameter perforated HDPE pipes in gravel-filled trenches spaced at 15-20 m between centers. Leachate was recirculated intermittently depending on the availability of leachate at the site.

Landfilled waste was collected while installing gas extraction well GEW16. The borehole was drilled using a bucket auger which was 0.9 m in diameter and 1.5 m long and samples were recovered at every 3 m depth intervals to a maximum depth of 29 m below the ground surface. Between 30 and 40 kg of waste were collected every 3 m, and about 100-150 kg of waste were collected at three selected depths (GEW16M1 at 14.6 m, GEW16M2 at 19.8 m, and GEW16M3 at 23.8 m). The two closest LRLs to the borehole location of GEW16 were LRL29 and LRL26 oriented in east-west direction. The LRL29 was located at a depth of 12 m and 7.5 m south of GEW16. The LRL26 was located at a depth of 22 m and 12 m north of GEW16. The last leachate recirculation occurred about 6 months before drilling the deep LRL26 and 4 months before drilling the shallower LRL29. Landfilled MSW at GEW16 was approximately 15 to 19 months old and represents landfilled waste subjected to leachate recirculation; 530 and 620 m^3 of leachate were recirculated at LRL26 and 29, respectively, for approximately one year.

For this study, MSW samples as collected from the field were shredded with a slow-speed, high-Torque shredder (Shred Pax Corp., AZ-7H, Wood Dale, IL) to suit for small-scale laboratory testing. The particle size distribution of waste before and after shredding is shown in Figure 1. It can be highlighted that fresh MSW had particles with larger diameters than the GEW16M2 sample before shredding. Higher finer proportion in the landfilled MSW sample could be a result of degradation and mixing up with daily cover material (Grellier et al. 2007). Figure 1 shows shredding resulted in a similar size distribution for both fresh and landfilled MSW samples.

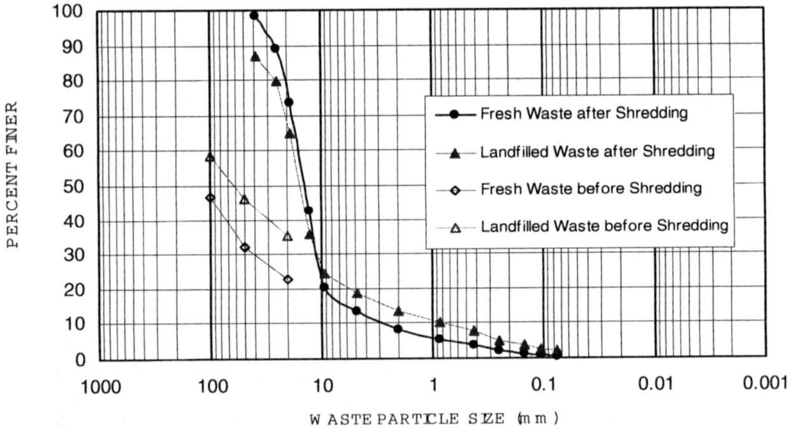

FIG. 1. Particle size distribution of field and landfilled waste (GEW16M2)

TESTING PROCEDURES

Testing on the waste samples was conducted per the American Society for Testing and Materials (ASTM) specification where applicable. Table 1 summarizes the testing procedures and deviations, if any. In-situ moisture content, organic content, BMP, specific gravity, and density tests were conducted on all waste samples collected from the field. Geotechnical properties (hydraulic conductivity, compressibility, and shear strength properties) were evaluated for fresh MSW and representative landfilled MSW sample GEW16M2.

Table 1: Standard procedures followed for testing of MSW

Test	Procedure	Deviation/modification
In-situ moisture content	ASTM D2216	Oven temperature maintained at 60°C
Organic content	ASTM D2974	Per standard
BMP	-	Per the procedure developed by Morton Barlaz at North Carolina State University
Specific gravity	ASTM D854	Entire shredded sample taken instead of sample passing over US sieve No.4
Bulk density	ASTM D698	Standard compaction testing procedure on sample at in-situ moisture content
Hydraulic conductivity – Rigid-wall permeameter	ASTM D2434	Per standard
Hydraulic conductivity – Triaxial permeameter	ASTM D5084	Per standard
Compressibility	ASTM D2435	Per standard, maximum pressure to simulate waste depths greater than 40 m
Direct shear	ASTM D3080	Per standard, maximum pressure to simulate waste depths greater than 40 m

RESULTS AND DISCUSSION

In-situ Moisture Content

Table 2 shows the in-situ moisture content of the fresh MSW and landfilled MSW obtained at different depths below the ground surface in borehole GEW16. Dry in-situ moisture content of the fresh MSW was 44%. For landfilled MSW, in-situ moisture content varied from 21% to 43%. Leachate recirculation occurred for four months prior to drilling near the GEW16, which resulted in higher moisture content of 43% at the depth level of LRLs. Less moisture content of 21% was also observed in the GEW16 at some depths which is attributed to the presence of C&D waste and/or daily soil cover. The heterogeneity of municipal waste resulted in the zone of varying moisture content and no particular trend was observed in the variation of moisture content with depth.

Table 2: Properties of fresh and landfilled MSW

Sample: Depth	In-situ moisture content (wet wt. basis %)	Organic content (%)	BMP (mL/g)	Specific Gravity	Dry density (kg/m^3)	Hydraulic conductivity (rigid wall) (m/s)
Fresh waste	43.7	76.2	135.4	0.85	515	1.18×10^{-4}
GEW16M1:14.6 m	21.1	68.5	62.5	1.24	627	0.43×10^{-4}
GEW16M2:19.8 m	43.3	63.1	108.8	0.97	712	3.04×10^{-4}
GEW16M3:23.8 m	34.8	62.4	98.4	1.06	601	0.17×10^{-4}

Organic Content (Volatile Solids)

Organic contents (also known as volatile solids or loss-on-ignition) of the fresh and landfilled waste samples are shown in Table 2. Fresh waste collected from the landfill had a higher organic content of 76%. Landfilled samples influenced by leachate recirculation at GEW16 had less organic content which varied from 62% to 68%. The decreasing organic content with depth suggests that waste degradation is occurring in the landfilled waste at this location.

Biochemical Methane Potential (BMP)

BMP of the fresh waste was 135.4 mL CH_4/gm, and it was higher than the landfilled waste (Table 2). Lower BMP values imply that degradation has been occurring in the landfilled waste. In general, BMP of waste should be directly proportional to organic content; however, for landfilled waste such correlation was not clearly evident perhaps due to heterogeneity of waste samples selected for testing.

Specific Gravity and Density

The specific gravity of fresh MSW was 0.85 and it varied from 0.97 to 1.24 for the landfilled MSW (Table 2). Decomposition of organic matter as explained by the reduction in organic content is the main reason for the increase in the average specific gravity of landfilled waste. The variability in specific gravity of landfilled MSW may be attributed to heterogeneous composition, including the presence of daily cover soil in the samples. It has been often observed that the decomposition of MSW eventually results in higher landfill densities. In the absence of a comparison of field density values, the waste samples, at their in-situ moisture content, were compacted into a standard Proctor mold (3 layers, 25 blows per layer) to measure their densities. Fresh MSW achieved a 515 kg/m^3 dry density and in landfilled MSW the dry density varied from 601 to 712 kg/m^3.

Hydraulic Conductivity

The rigid-wall permeameter estimated 1.18×10^{-4} m/s for the hydraulic conductivity of the fresh MSW (Table 2). For the landfilled MSW, the hydraulic conductivity varied in a wide range from $0.17\text{x}10^{-4}$ to 3.04×10^{-4} m/s and it did not follow any trend with respect to depth. Triaxial hydraulic conductivity tests at 35 kPa confining pressure resulted in hydraulic conductivity values on the same order of magnitude as rigid wall hydraulic conductivity tests. For the confining pressure greater than 69 kPa, hydraulic conductivity of both fresh and landfilled wastes was significantly reduced to the same order of approximately 10^{-7} m/s. Such hydraulic conductivity reduction is attributed to increased sample density to 880 kg/m^3 for fresh waste and 790 kg/m^3 for landfilled waste due to applied confining pressure. This reduction in hydraulic conductivity should be taken into account in the design of leachate recirculation systems.

Compressibility Characteristics

Figure 2 shows the compressibility behavior of fresh and landfilled MSW at their respective field moisture content under different normal pressures. Both fresh and landfilled MSW exhibited similar trends in compressibility, with a total strain of 55%

for fresh and 42% for landfilled MSW subjected to recirculation (GEW16M2). Relatively high immediate compressive response followed by a delayed compression response was observed for both types of waste. The strain versus time data provided in Figure 2 was used to plot the compression curve given in Figure 3. Compression ratio for landfilled MSW was found to be 0.22 which was lower than 0.28 obtained for fresh MSW. Partial degradation and presence of daily soil cover material in the landfilled MSW sample could be the reason for the reduction in compressibility.

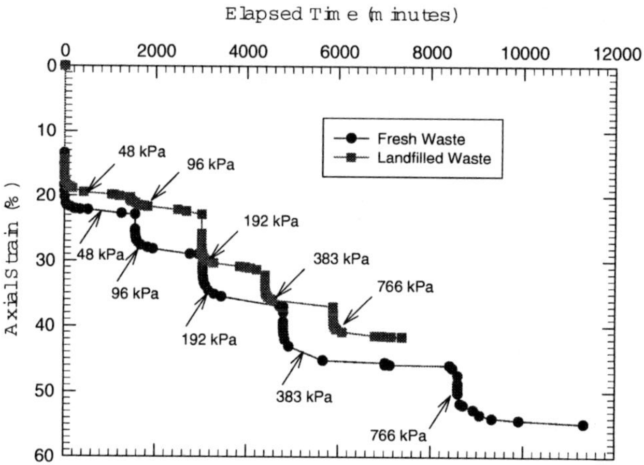

FIG. 2 . Strain versus time for fresh and landfilled waste (GEW16M2)

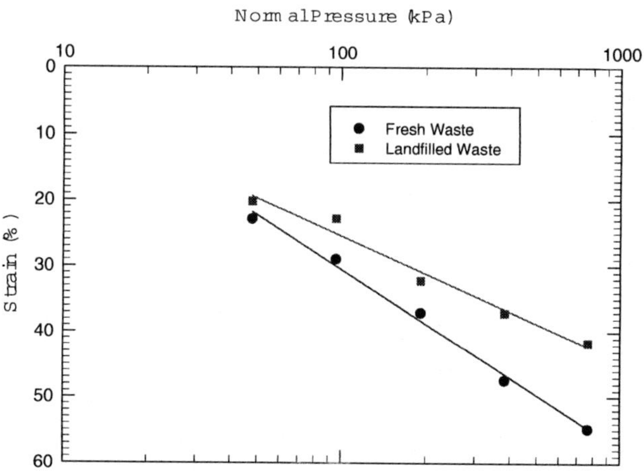

FIG. 3 . Compression curve for fresh and landfilled waste (GEW16M2)

Shear Strength Parameters by Direct Shear Testing

Shear strength properties of fresh and landfilled MSW samples were determined by direct shear tests. None of the samples reached peak strengths, and both fresh and landfilled MSW exhibited strength hardening with increased horizontal displacement well in excess of 15% of sample diameter. In the absence of peak strength, shear strengths reported at 15% displacement were considered to construct the shear strength envelopes for fresh and landfilled MSW shown in Figure 4. The shear strength parameters for fresh waste were: c'=47 kPa and $\phi'=29^0$ and for landfilled waste were: c'=12 kPa and $\phi'=32^0$. These strength parameters are within the values range of values reported by Eid et al. (2000) and Kavazanjian (2001). The results show that fresh and landfilled wastes have essentially the same strengths at normal stresses greater than 300 kPa. Landfilled waste is slightly weaker than fresh waste at low normal stresses. The landfilled waste degraded only slightly in 15-19 months; therefore, the differences in shear strengths between fresh waste and landfilled waste appear to be not significant. Additional research is being performed to determine shear strength of MSW at different levels of degradation.

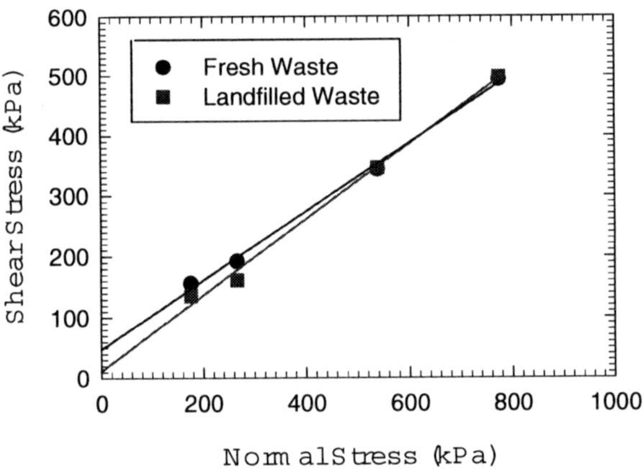

FIG. 4 . Shear strength of fresh and landfilled waste (GEW16M2)

CONCLUSIONS

Laboratory geotechnical testing was conducted on shredded fresh and landfilled MSW samples from a bioreactor landfill to assess the effects of enhanced degradation on geotechnical properties. Major findings of this study are: (1) In-situ moisture content of the fresh waste was higher than the landfilled waste; (2) Organic content of fresh MSW was 76% and it varied from 62% to 68% for landfilled MSW; (3) Average specific gravity increased from 0.85 for fresh to 0.97-1.24 for the landfilled MSW; (4) Fresh MSW yielded a dry density of 515 kg/m^3 when compacted into a standard Proctor mold (3 layers, 25 blows per layer) at it' field moisture content. In landfilled MSW it varied from 601 to 712 kg/m^3; (5) For confining pressure greater than 69 kPa, both fresh and landfilled wastes had similar hydraulic conductivity; (6) Compression behavior of fresh and landfilled wastes was similar with a maximum axial strain of 55-58% for fresh waste and 42% for landfilled waste under maximum normal pressure of 766 kPa. Compression ratios for fresh and landfilled MSW were 0.28 and 0.21, respectively; (7) Based on direct shear tests, the shear strength parameters for fresh waste were: c'=47 kPa and ϕ'= 29^0 and for landfilled waste were: c'=12 kPa and ϕ'=32^0. These results show that the shear strengths of fresh waste and landfilled waste (slightly degraded in 15-19 months) were similar. It should be recognized that the test results are based on small-scale tests on shredded waste samples, and the effects of the test conditions on in-situ waste properties should be properly assessed and accounted in analysis and design of bioreactor landfills.

ACKNOWLEDGEMENTS

The authors are thankful to Veolia Environmental Services, particularly C. Peters, for supporting field sampling. The assistance of S. Grellier in waste sampling and testing is appreciated. Funding for this research is received from the National Science Foundation (Grant#0600441) and is gratefully acknowledged.

REFERENCES

Eid, H.T., Stark, T.D., Douglas, W.D., and Sherry, P.E. (2000). "Municipal solid waste slope failure waste and foundation properties:" *Journal of Geotechnical and Geoenvironmental Engineering*, 126(5), 397-407.

Gabr, M.A., and Valero, S.N. (1995). "Geotechnical properties of solid waste." *ASTM, Geotechnical Testing Journal,* 18(2), 241-251.

Grellier S., Reddy, K.R., Gangathulasi, J., Adib, R., and Peters, C. (2007). "US MSW and its biodegradation in a bioreactor landfill," *Proc., Sardinia '2007, Eleventh International Landfill Symposium,* Cagliari, Italy.

Hendron, D.M., Fernandez, G., Prommer, P.J., Giroud, J.P., and Orozco, L.F. (1999), "Investigation of the Cause of the 27 September 1997 Slope Failure at the Dona Juana Landfill," *Proc. Sardinia '99 - 7th International Waste Management and Landfill Symposium*, Cagliari, Italy, Vol. III, 545-567.

Hossain, M.S. (2002). "Mechanics of compressibility and strength of solid waste in bioreactor landfills." Ph.D. Thesis, Department of Civil Engineering, North Carolina State University at Raleigh, Raleigh, North Carolina, USA.

Kavazanjian, E., Jr. (2001). "Mechanical Properties of Municipal Solid Waste," *Proc. Sardinia '01 - 8th International Waste Management and Landfill Symposium, Cagliari, Italy, October, Vol. III,* 415-424.

Landva.A.O and Clark.J.I (1990). "Geotechnics of waste fill", *Geotechnics of waste fills – theory and practice, ASTM STP 1070*, 86-103.

Zekkos, D.P. (2005). "Evaluation of static and dynamic properties of municipal solid waste." Ph.D. Thesis, Department of Civil and Environmental Engineering, Univ. of California, Berkeley, California, USA.

Laboratory Tests on Creep and Shear Behavior of Municipal Solid Waste and Mitigation of Its Long-Term Subsidence

Ikuo Towhata[1] and Masaaki Uno[2]

[1]Professor, Department of Civil Engineering, University of Tokyo, 7-3-1, Bunkyo, Tokyo 113-8656; towhata@geot.t.u-tokyo.ac.jp
[2]Senior Research Engineer, Fudo-Tetra Corporation, Tokyo; masaaki.uno@fudotetra.co.jp

ABSTRACT: This paper addresses a series of laboratory tests in which the mechanical properties of municipal waste are investigated. In addition to the high shear strength of this material, a special emphasis is placed on the mitigation of long-term subsidence. This is because the practical usage of a closed landfill depends on the risk due to subsidence. It is shown that preloading is effective to achieve this goal. Moreover, model tests as well as numerical calculations were conducted on the behavior of a building foundation, which revealed that a piled-raft foundation combined with preloading is a good measure.

INTRODUCTION

FIG.1. Subsidence problem in foundation of building constructed on waste landfill by using end-bearing pile foundation.

The authors have been investigating the possibility to reuse the landfill area as a space resource for urban development. Evidently, there is a problem of long-term subsidence in a landfill. Fig.1 illustrates a building which was constructed upon a closed landfill. Being made of unprocessed waste in which all kinds of family waste were mixed without incineration, this landfill has been settling down since 1960s, and at present the foundation of a building, supported by end-bearing piles, is suffering differential settlement which has lasted for many years. Shimizu et al. (1989) studied the rate of subsidence of landfills in Tokyo Harbor area to show that more than 25 years is needed for the subsidence to be completed (Fig.2). Fig.3 illustrates another closed landfill near downtown Tokyo where waste disposal was stopped in 1950s and there are many buildings today without evidence of subsidence problem. Being older than 50 years, this former landfill ground appears stable; suggesting that more or less 50 years is needed for the long-term subsidence to be completed in landfills of Tokyo. Note that this information was obtained from landfills made mainly of unprocessed waste, while recent landfills in Japan are made of different materials such as incinerated ash and incombustible waste.

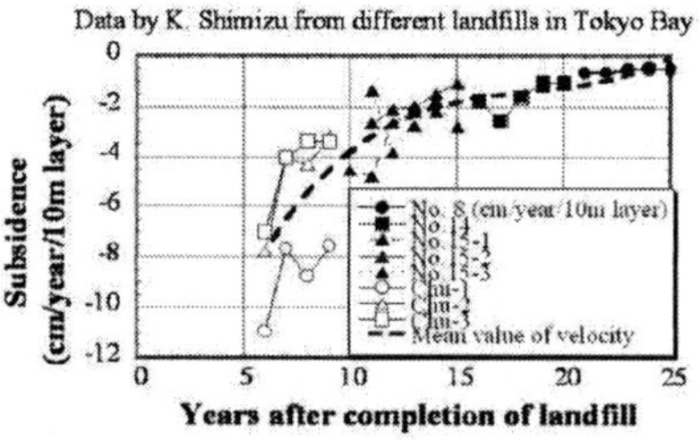

FIG.2. Rate of subsidence of landfills in Tokyo Bay (Shimizu et al., 1989).

FIG.3. Former landfill site in Shiomi of Tokyo.

RESULTS OF TRIAXIAL COMPRESSION TESTS ON WASTE SPECIMENS

Early interest in mechanical behavior of waste lay in consolidation. Sowers (1973) carried out consolidation tests and mentioned the possibility of preloading for mitigation of creep settlement. On the other hand, Gabr and Valero (1995) and Zekkos et al. (2007) reported high shear strength of waste. It seems therefore that subsidence problem should be further studied from the viewpoint of its mitigation. Note that Al-Yaqout and Hamoda (2007) applied high load to accelerate subsidence. It is not clear, however, whether or not they later removed this additional load.

A laboratory testing program has been conducted by using waste materials which was collected from landfills in Tokyo Harbor. Fig.4 illustrates the incombustible waste and incinerated ash. Since the former is much softer and more problematic than the latter, the present paper addresses those tests conducted on the incombustible waste. For its detailed composition, see Fig.5. Note that the incinerated ash, which constitutes more than 70% in weight, is of high rigidity and hence is mechanically very stable.

Triaxial compression tests were conducted in drained manner on incombustible waste which was compacted and consolidated isotropically under 15-45 kPa. The tested samples measured 30 cm in diameter and 60 cm in height. The test results in Fig.6 reveal that the incombustible waste has low rigidity when strain is small, being much softer than Toyoura sand. Conversely, the stress level in larger strain range is very high without yielding, indicating approximately linear elastic behavior. Note that both rigidity and stress level increased as the consolidation pressure became higher.

The high shear strength of waste accounts for the fact that a vertical cut in a landfill can maintain stability (Fig.7). This high strength is developed by plastic sheets and other fibers which produce tensile resistance in the horizontal direction when loading occurs in the vertical direction. Thus, high bearing capacity is developed in foundation resting on a waste landfill. The engineering problem, therefore, lies in the high compressibility of waste or subsidence of the foundation. It is important to note that such recent failures of landfill in the Philippines and Indonesia were preceded by fires (Merry et al., 2005; Kölsch et al., 2005) which probably destroyed the fibrous reinforcement unintentionally (Fig.8).

(a) Incombustible waste (b) Incinerated ash

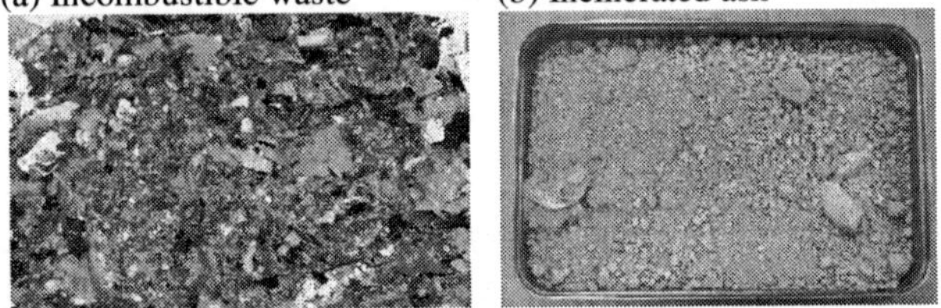

FIG.4. Two kinds of municipal waste in Tokyo Port Landfill.

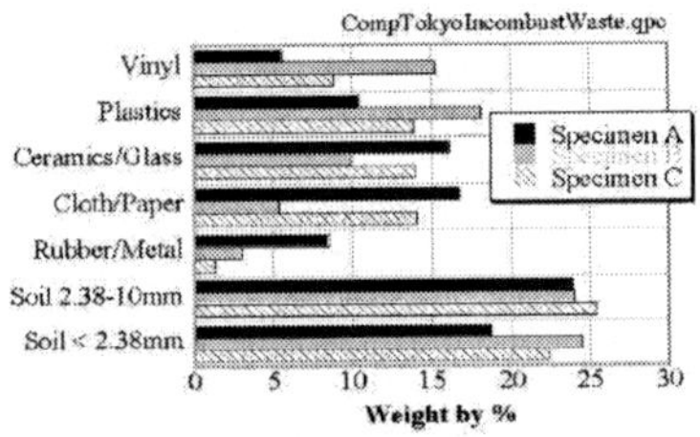

FIG.5. Composition of incombustible waste collected from Tokyo Port Landfill.

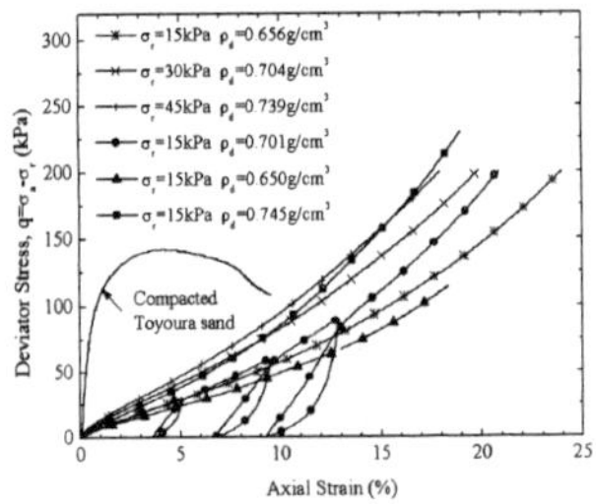

Fig.6 Drained triaxial compression tests on incombustible waste and Toyoura sand.

FIG.7. Vertical cliff cut in municipal waste landfill near Goettingen, Germany.

FIG.8. Cross section of burnt landfill after catastrophic failure in 2005 (Leuwigajah, Bandung, Indonesia).

ON VOLUME CONTRACTION DUE TO CREEP

Since triaxial tests revealed that shear failure is unlikely in waste, further studies were made of long-term subsidence. Since the specimens were air-dry, the measured volume contraction was caused not by consolidation but creep under constant effective stress. Fig.9 demonstrates the development of creep volume change of incombustible waste which was measured under isotropic stress state in the aforementioned large triaxial apparatus. The isotropic effective stress was raised from 10 to 15 kPa and then maintained constant in order to monitor creep deformation. As shown in the figure, creep deformation did not stop for one hour after raising the stress; the extent of creep deformation was greater when the waste was looser.

Preloading is a well-known technology in soft-ground engineering in which the overburden stress is raised and then removed so that ground subsidence is reduced or stabilized afterwards. Fig.10 conceptually shows the idea of long-term compression tests by which the effects of preloading on mitigation of creep volume change were examined. The feature of employed preloading is characterized by the initial high effective stress (σ_p), its duration time (t_p), and the stress level after unloading (σ_1).

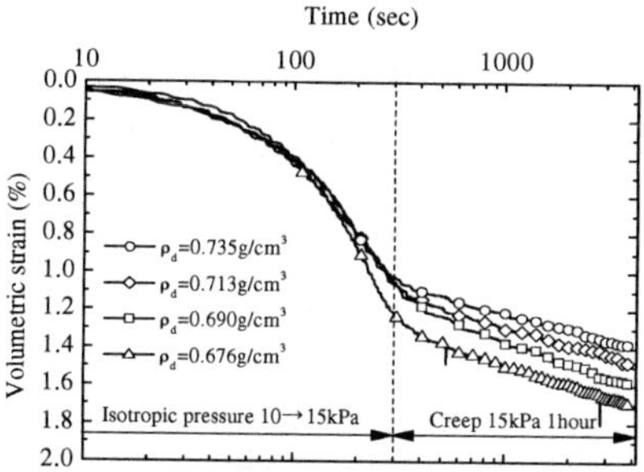

FIG.9. Volume contraction of incombustible waste with time under isotropic stress state.

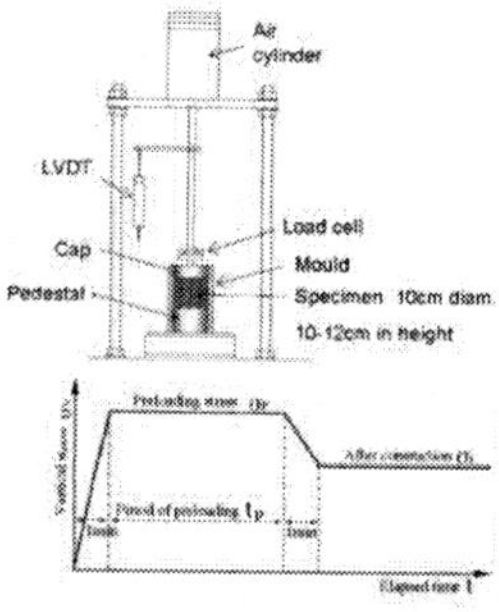

FIG.10. One-dimensional compression tests on verification of preloading effects.

Fig.11 examines the effects of preloading time (t_p) on creep deformation after stress unloading. Both stress levels during preloading and after unloading were held constant. The time coordinate in this figure starts at the time when preloading stress was unloaded completely. Since the extent of preloading was small (33 kPa unloaded to 30 kPa), no swelling occurred and volume contraction started after some stationary period. It is noteworthy in this particular set of data that the stationary period is more or less similar to t_p, implying that longer preloading time can maintain a landfill for a longer time. Even after initiation of creep subsidence, the rate of subsidence is lower as suggested by the logarithmic time scale in the figure.

Fig.12 shows the effects of preloading stress, while holding the preloading time constant. It is evident therein that the greater preloading stress can delay the initiation of creep subsidence, and even after that, the rate of subsidence is significantly reduced.

Similarly, the mitigation of long-term volume contraction by preloading was verified by the aforementioned large triaxial apparatus. Fig.13 indicates that volume change was stabilized after unloading from 45 kPa to 30 kPa or from 60 kPa to 30 kPa.

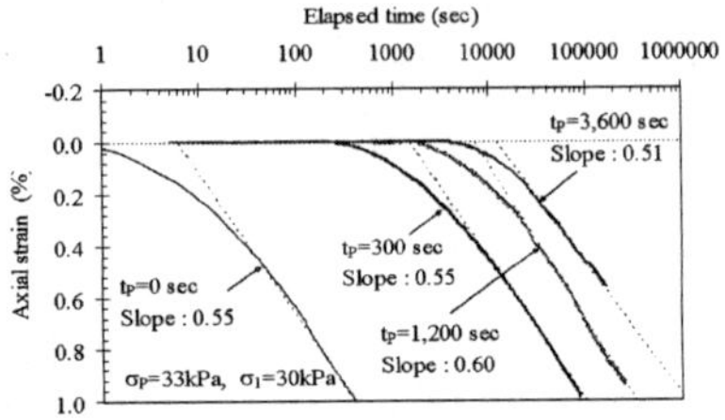

FIG.11. Effects of preloading time on mitigation of creep volume contraction.

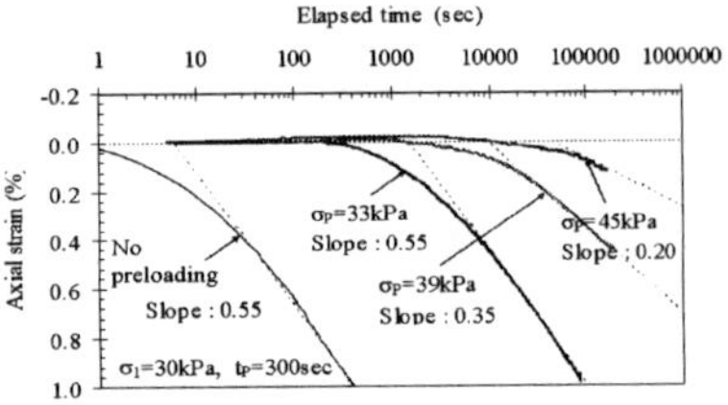

FIG.12. Effects of preloading stress on mitigation of creep volume contraction.

SKIN FRICTION OF PILE EMBEDDED IN INCOMBUSTIBLE WASTE

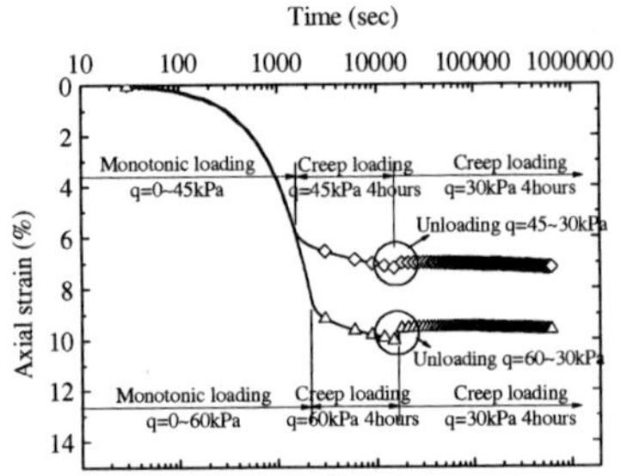

FIG.13. Validation of preloading by tests in large triaxial apparatus.

Buildings upon landfill used to be supported by end-bearing piles (Fig. 1). Accordingly, those buildings have been rather stable, although differential subsidence could be a problem. End-bearing piles are not practical nowadays because those piles penetrate through underlying clay barriers before reaching the bearing stratum, thus allowing leachate in waste landfill to come out along the penetration channel. To avoid the consequent environmental pollution, it is important to seek for the possibility of friction piles. It is certainly true that foundations supported by friction piles are less vulnerable to differential settlement.

Little is known about skin friction of piles embedded in waste. Hence, model tests were conducted in a model container (Fig.14) in which a waste deposit was compressed by controlled overburden pressure (air balloons) and a sheet of instrumented aluminum plate, which modeled the surface of a metal friction pile, was pushed downwards in order to mobilize the skin friction. Note that the aluminum plate penetrated through the bottom of the container so that any movement of the plate did not affect the net contact area between the plate and the waste. As mentioned later, it was attempted to increase the friction by attaching lateral plate(s) (or node) to the aluminum plate.

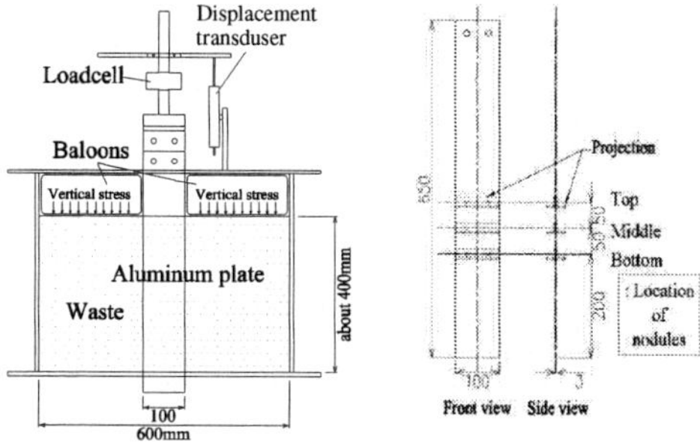

FIG.14. Model tests on skin friction of pile embedded in incombustible waste (container and metal plate as pile model).

Conditions in 7 model tests are summarized in Table 1. Effects of both preloading and attaching nodes were examined. Cases 1, 2, 3, and 4 were conducted on a flat pile without node. Fig.15 reveals that Cases 2 and 4 developed greater skin friction than Cases 1 and 3 with normally consolidated model. Cases 3 and 4 with denser waste developed greater skin friction than Cases 1 and 2 with looser material. It is important that preloading reduces creep effects as well. Moreover, the skin friction in these 4 cases achieved the ultimate value at a very small penetration; revealing what is called rigid-perfectly plastic behavior.

Table 1. Conditions in tests on skin friction.

Case	Mass density (g/cm^3)	Overburden Stress (kPa)	Node
1	0.533	20	None
2	0.577	40->20*	None
3	0.652	20	None
4	0.665	40->20*	None
5	0.547	20	Top
6	0.566	20	Top & Bottom
7	0.543	20	Top, Middle, Bottom

* Preloading

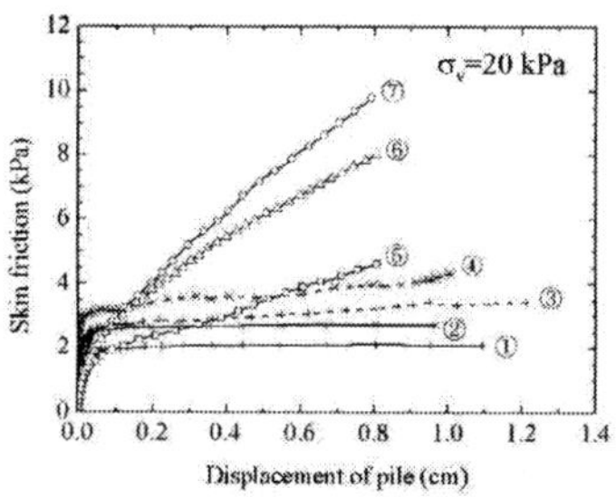

FIG.15. Variation of skin friction with progress of pile penetration.

FIG.16. Nodular piles in practice.

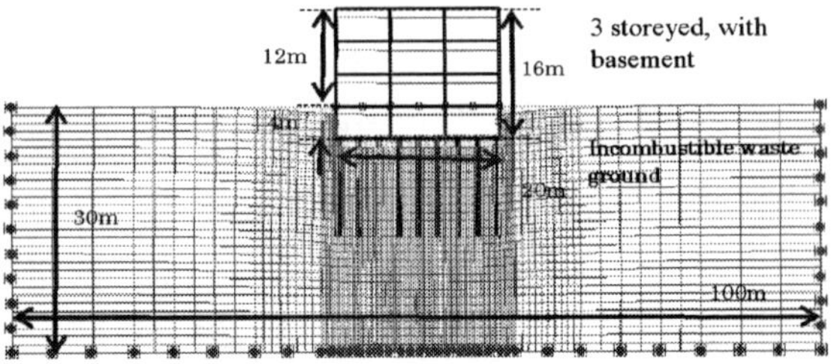

FIG.17. Model piles with and without node.

FIG.18. Numerical model of building foundation resting on waste deposit.

A nodular pile is a type of pile foundation which has an irregular shape and develops greater skin friction (Fig.16). Fig.17 illustrates a model pile with and without node.

Cases 5, 6, and 7 attempted to investigate the effects of a nodular pile by attaching horizontal wing(s) to the aluminum plate. It is supposed that those wings can transfer the pile load more efficiently to the surrounding ground. The load-displacement relationships in Fig.15 show that irregular wings made it possible to develop greater skin friction at large penetration, while, at smaller penetration, pile displaced slightly more easily due probably to deformation of surrounding ground due to load transfer.

NUMERICAL ANALYSES ON SUBSIDENCE OF BUILDING FOUNDATION

The experimental results so far described were employed in numerical analyses in which the mitigation effects of preloading and friction piles were examined. It is not necessary to consider negative skin friction because piles are floating in the subsiding waste. Material properties for linear elasticity (Fig.6) were obtained from triaxial tests on both normally-consolidated and preloaded specimens. Their dependency on confining stress was considered as well. Fig. 18 illustrates the model of a 3-storeyed building with basement; the width being 20 m. Note that the creep deformation of waste is not fully considered, leading to underestimation of deformation when preloading is not carried out. The thickness of incombustible waste is 30 m as is the case in reality in mind. Since the present analysis assumes good contact between the waste deposit and the base slab of the building, the interaction at the base slab was taken into account. This type of foundation is called piled-raft foundation.

The first stage of the analysis was conducted by considering only the weight of the waste. Afterwards, the weight of the building foundation was loaded numerically so that the differential subsidence between the building and the free field of landfill (at far distance from the building) was calculated. Results are shown in Table 2.

In Table 2, comparison of Cases 1 and 2 shows that piled raft foundation can significantly reduce the subsidence. Cases 1 and 3 further show that the effects of densification due to preloading (Fig.15) are even more substantial. Consequently, Case

4 with both piled raft and preloading is able to reduce the subsidence from 28.6 cm in Case 1 to only 3.1 cm which is around 0.15% of the width of the building.

CONCLUSIONS

The present study carried out laboratory tests on municipal waste material, aiming at mitigating its long-term creep subsidence. Triaxial tests showed that the mechanical strength of incombustible wastes is still higher than that of sand, although the rigidity is much lower. Hence, the research was focused on subsidence which in particular occurs after long term of creep. Tests revealed that preloading is an efficient measure to mitigate subsidence. Moreover, pile model tests indicated the advantage of piles with surface nodes. By combining all the results, numerical analyses were run on subsidence of building on landfill. It was found that a combination of densification / reloading and nodular piles can efficiently reduce the subsidence of a low-rise building.

Table 2. Calculated differential subsidence between landfill and building foundation.

Case	Preloading	Type of foundation	Subsidence cm
1	No	Shallow	28.6
2	No	Piled raft	9.9
3	Yes	Shallow	5.3
4	Yes	Piled raft	3.1

Acknowledgments

The authors appreciate the financial support of the MEXT (Ministry of Education, Culture, Sports, Science, and Technology) and the Japanese Iron and Steel Federation as well as the Tokyo Metropolitan Government for supplying tested waste materials.

REFERENCES

Al-Yaqout, A/F. and Hamoda, M.H. (2007) Movement of unlined landfill under preloading surcharge, Waste Management, 27(3), 448-458.

Gabr, M.A. and Valero, S.N. (1995) Geotechnical properties of municipal solid waste, Geotechnical Testing Journal, 18(2), 241-251.

Kölsch, F., Fricke, K., Mahler, C. and Damanhuri, E. (2005) Stability of landfills – The Bandung dumpsite disaster, Proc. 10th Lanfill Symp.

Merry, S.M., Kavazanjian, E. and Fritz, W.U. (2005) Reconnaissance of the July 10, 2000, Payatas landfill failure, J. Perform. Construct. Facilities, ASCE, 19(2), 100-107.

Shimizu, K., Ebina, S., Saito, S., and Mizukoshi, K. (1989) Geotechnical study of refuse landfill in Tokyo Port (No.10), Proc. 24th Japan Nat. Conf. Soil Mech. Found. Eng., JGS, Tokyo, 149-150.

Sowers, G.F. (1973) Settlement of waste disposal fills, Proc. 8th Int. Conf. Soil Mech. Found. Eng., 2.2, 207-210.

Zekkos, D., Bray, J.D., Athanasopoulos, G.A., Riemer, M.F., Kavazanjian Jr., E., Founta, P.A., and Grizi, A.F. (2007) Compositional and loading rate effects on the shear strength of municipal solid waste, Proc. 4th Int, Conf. Earthq. Geotech. Eng., Paper No. 1525.

Mechanical properties of municipal solid waste from Suzhou landfill in China

Yunmin Chen[1], Tony L. T. Zhan[2] and Wei-an Ling[3]

[1]Professor, MOE Key Laboratory of Soft Soils and Geoenvironmental Engineering, Zhejiang University, Hangzhou, 310027, China; Chenyunmin@zju.edu.cn
[2] Associate Professor, MOE Key Laboratory of Soft Soils and Geoenvironmental Engineering, Zhejiang University, Hangzhou, 310027, China; zhanlt@zju.edu.cn
[3] Research student, MOE Key Laboratory of Soft Soils and Geoenvironmental Engineering, Zhejiang University, Hangzhou, 310027, China.

ABSTRACT: Landfill expansion is taking place in many cities of China. Knowledge of compressibility and shear strength of MSW are required for the design of a vertical landfill expansion. This paper presents a laboratory study on the mechanical properties of MSW within the Qizhishan landfill in Suzhou of China, which is under design for a vertical and lateral expansion. The laboratory study included 31 one-dimensional compression tests and 34 triaxial compression tests on the borehole samples taken at different depths. The experimental results showed that both the mechanical compressibility and shear strength parameters of MSW significantly depended on the embedment depth. The primary compression index (C_c) was observed to decrease from 1.0 to 0.3 with an increase in depth from 0 to 37 m. The magnitude of friction angle increased with an increase in the embedment depth of MSW, and the magnitude of cohesion decreased with depth.

INTRODUCTION

In China, urban population and economy have increased significantly since 1990. The growth results in a rapid increase in the quantity of municipal solid waste (MSW). At present, the annual total generation is approximately 150 million tons. Over 90% of the amount of MSW is disposed in landfills. Most of the landfills in major cities of China were built in early 1990s, and now have reached their service life (i.e., about 15 years). Due to the social and political problems associated with identifying new landfill sites, expansion of the existing landfills is taking place in many cities of China.

The Qizhishan landfill in Suzhou, China, is under being designed for vertical and lateral expansion (Fig. 1). There are two major technical difficulties associated with the landfill expansion. One is the design of an intermediate liner system in between the existing and expanded waste body, the other is the slope stability of the landfill body subjected to a significant increase in height (Qian et al., 2002). The additional waste fill from a vertical expansion will cause substantial settlement of the existing landfill. The intermediate liner system constructed on the existing landfill may experience large differential settlements, and lose its effectiveness as a hydraulic barrier. The design of the intermediate liner system requires a reliable settlement prediction for the existing landfill. There is also a requirement of performing slope stability analyses for assessing the factor of safety of the higher waste fill.

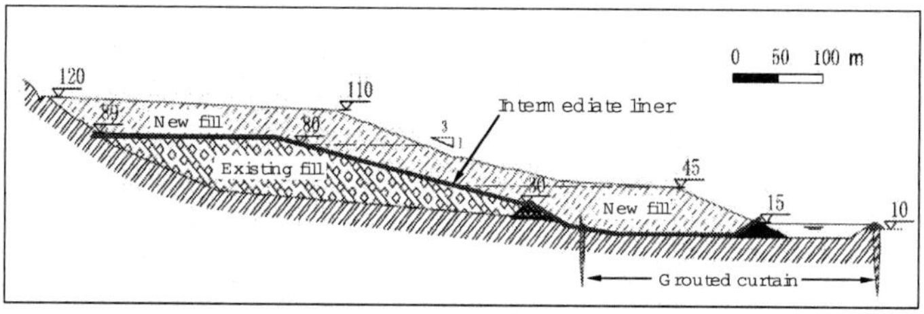

Fig. 1 A preliminary design of landfill expansion in Suzhou, China

To assist in the design of the landfill expansion project, field and laboratory studies were carried out on the existing landfill. The field study included borehole investigation, sampling, cone penetration tests and monitoring of pore fluid pressures. Waste samples obtained at different depths were sent to a laboratory to measure mechanical properties including compressibility and shear strength. This paper mainly presents some representative experimental results from the laboratory study.

BOREHOLE SAMPLING AND LABORATORY TEST PROGRAM

Fig. 2 shows a cross section of the Qizhishan landfill corresponding to the period when the field study was conducted. The landfill comprises of a number of waste benches set back at 3H/1V embankment slope, with a rockfill dam retaining the lowest bench. The existing landfill initiated operation in 1993. Its capacity and service life were designed to be 4.7 million m^3 and 15 years, respectively.

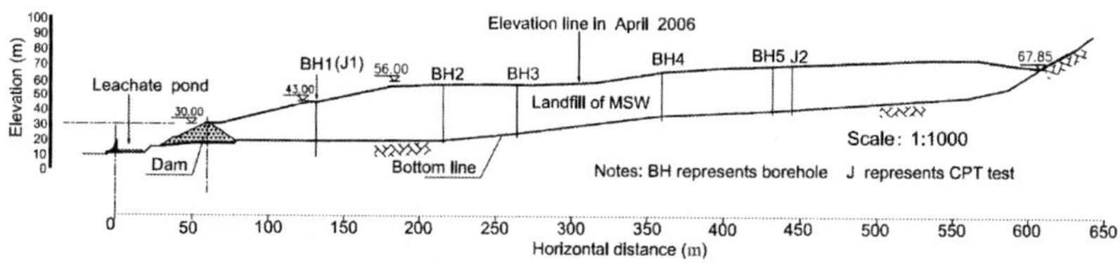

Fig. 2 Cross section of the Suzhou landfill in April 2006

Five boreholes (BH1 to BH5) were drilled to the bottom of the existing landfill (Fig. 2). The depths of the boreholes ranged from 26 to 38 m. Drilling was performed without introduction of drilling mud and liquids. To avoid a collapse of borehole wall, steel casings were installed in each borehole. Each borehole consisted of three vertical sections (i.e., from top to a depth of 10 m, from 10 m to 20 m and below 20 m) with different-diameter casings installed. The borehole diameters for the three sections were 130 mm, 110mm and 90mm, respectively. MSW samples were taken with thick-wall samplers at an interval of 1 or 2 m. More than 20 samples were obtained from each of the boreholes. The sample diameters were about 96 mm and 82 mm for those taken above and below a depth of 20 m, respectively. Each sample was about 200 mm in height.

In total, sixty-five borehole samples of MSW were used in this laboratory study. The following experimental procedure was adopted to handle each of the borehole samples: Firstly, the weight and dimensions (diameter and height) of the sample were measured to obtain its bulk density. Subsequently, the sample was either moved into a purposely-designed compression cell to perform 1-D compression test or installed in a triaxial cell to perform a triaxial compression test. After the completion of the compression test, all material was retrieved from the cell, and then dried in an oven with a temperature of 60°C. The water content of the sample can thus be calculated. The major components of the sample (plastic, paste, textiles, wood, metal, glass, ceramic ect.) were identified by optical investigation and weighted individually. Finally, all the materials were divided into two parts with the same weight as well as the same composition. For one half part, all the materials except the plastic matter were incinerated for 2 hours in an oven at a temperature of 300°C. Thus, the organic content of the paste and other waste materials can be evaluated by measuring the amount of incineration loss. For the other half part, all the materials were moved to a cylindrical container with a siphonal tube to measure its overall specific gravity by using water replacement method. Thus, the void ratio of the sample can be deduced from the measurements of overall specific gravity and dry density.

Thirty-one MSW samples taken at different depths were used for 1-D mechanical compression tests. The dimension of the samples is approximately 96 mm (or 82 mm) in diameter and 200 mm in height. Two sets of compression apparatuses were purposely designed to accommodate the borehole samples (Chen and Zhan, 2006). Step-loading method was adopted to measure the compression curve of each MSW sample. After each sample was placed in the compression cell, dead loads equivalent to vertical pressures of 25, 50, 100, 200, 500 and 1000 kPa were applied in stages. The duration for each step of loading was set at 8 hours. This duration was determined on the basis of five trial compression tests, which lasted 24 hours. The trial compression test results demonstrated that over 95% mechanical compression occurred within 8 hours.

Thirty-four MSW samples taken at different depths were selected to carry out triaxial compression tests by using two modified triaxial apparatuses (Chen and Zhan, 2006). The 34 MSW samples were classified into 8 groups corresponding to different ranges of depth. For each group, there were at least 4 samples, which were consolidated to effective confining pressures of 50, 100, 200 and 400 kPa. The borehole samples with a diameter of approximately 82 mm or 96 mm and a height of 200 mm were directly installed on the triaxial apparatuses. Drained consolidated strain-controlled compression tests were performed. The strain rate was set as 1.9×10^{-3} per min. Each specimen was sheared to reach a strain level beyond 20%. It should be noted that puncture of membrane due to a sharp matter in the specimen occurred to several samples when the effective pressure was higher than 200 kPa. The data from these tests were not considered.

EXPERIMENTAL RESULTS

Change of waste composition with depth

Knowledge of waste composition is helpful to evaluate the mechanical properties of MSW. MSW generally consists of putrescent organics (i.e., food and garden wastes), cinder, dust, paper, plastics, rubber, textiles, wood, glass, metal etc. Here it should be noted that the MSW in China generally contains much more food and garden wastes (between 50% and 60% by wet weight) than that in USA. This is attributed to the difference in the cooking style. After being disposed in landfills, waste composition inevitably changes with time due to a biological degradation of the organics. The wastes embedded at different depths generally have experienced different durations of fill, and hence pose different composition.

Fig. 3 shows variations of MSW composition with the embedment depth in borehole BH5. The data points were obtained from the composition analysis conducted on the borehole samples. It can be seen that the waste composition changed significantly with depth for the top 10 m of fill. The content of cinder and dust increased from 30% to 60%. The organic content decreased from 28% to 16%. The other components (plastic, paper, textiles, wood, glass, metal etc.) also showed a decreasing trend with depth. The significant changes in organic content and cinder content for the top 10 m were related to the degradation of the putrescent organics. It is noticed that the data points corresponding to a depth of 11.9 m showed an extremely high content of cinder and an extremely low content of plastic and fiber. This is because there was an intermediate soil cover existing at the depth. For the depths greater than 12 m, the content of cinder and dust in waste was over 60 %, and showed a slight increasing trend with depth. The organic content changed slightly around a mean value of 16%. It should be pointed out that the change in the waste composition observed in Fig.3 was partially attributed to the change in the composition of MSW generated over the last decade. In addition, the composition analysis results are expected to be affected to some extent by the relatively small-size borehole samples.

Change of void ratio with depth

The mechanical properties of MSW (e.g., compressibility) greatly depend on its initial void ratio. The void ratio of waste in a landfill is related to waste placement and compaction methods as well as the in situ overburden pressure. Fig. 4 shows variations of void ratio with the embedment depth obtained from all the samples taken from the five boreholes (BH1 to BH5). It can be seen that almost all the data points fall into the range from 4 to 1. There appears to be a decrease of void ratio with an increase in the embedment depth, as illustrated by the best-fitting trend line. The significant discrepancy of the data points against the trend line can be explained by the heterogeneous nature of MSW. Of course, the discrepancy may also be owed to the disturbance and localization effects associated with the borehole sampling. The decreasing trend of void ratio with depth shown in Fig. 4 is consistent with the experimental data reported by Zornberg et al. (1999). However, the data of void ratio reported by Zornberg et al. (1999) is located within a narrower range from 1.2 to 2. This indicates that the MSW in Qizhishan landfill is significantly looser than that in US landfills, particularly at shallow depths. The profiles of void ratio provide the initial state parameters required for landfill settlement analysis.

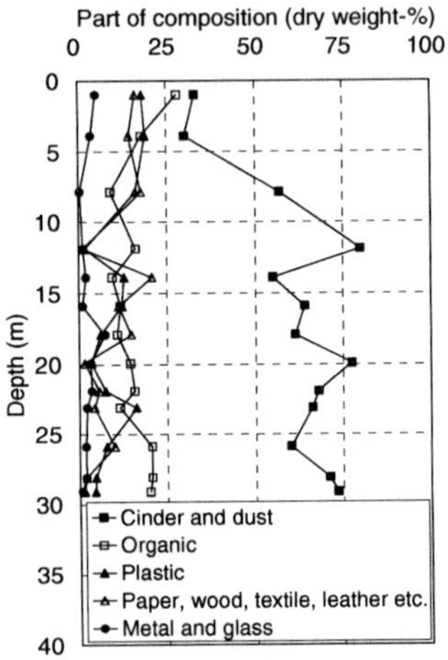

Fig. 3 Change of waste composition with depth in BH5

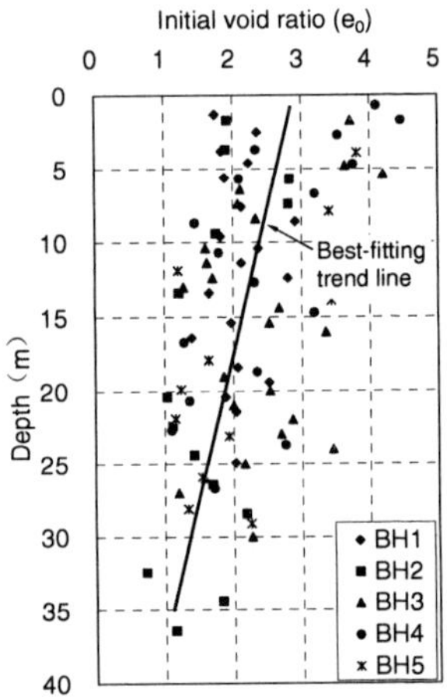

Fig. 4 Change of void ratio with the embedment depth

Change in compressibility of MSW with depth

Fig. 5 shows the mechanical compression curves obtained from the six samples taken at different depths in borehole BH5. The compression curves in a semi-logarithm coordinate exhibit a bi-linear shape. The pressure corresponding to the inflection point is defined as yielding stress (or pre-consolidation pressure). It is unexpected that the yielding stresses for the samples taken at different depths seem to fall into a narrow range from 25 to 50 kPa. The inclination of the post-yielding curve is defined as the primary compression index (C_c). It can be seen that the MSW sampled at shallow depths generally showed a higher compressibility (i.e., high value of C_c) than the deep waste.

Fig. 6 shows a plot of the primary compression index (C_c) as a function of the embedment depth for all the 31 samples taken from the five boreholes (BH1 to BH5). Almost all data points fall into the range between 0.3 and 1.0. Although the data points exhibit scatter, a decrease in C_c with the embedment depth is observed. The observed changes in compressibility with the embedment depth can be explained by considering the change in the MSW composition with depth (see Fig. 3). As the embedment depth increases, the MSW sample contains more cinder materials and less compressible components, becoming denser and uniform and hence exhibiting lower compressibility. With the experimental data shown in Fig. 6, engineers are able to carry out a rational settlement analysis for the existing landfill subjected to a vertical expansion. The above results indicate that the current practice of using a single value of C_c for settlement calculation may lead to inaccurate predictions.

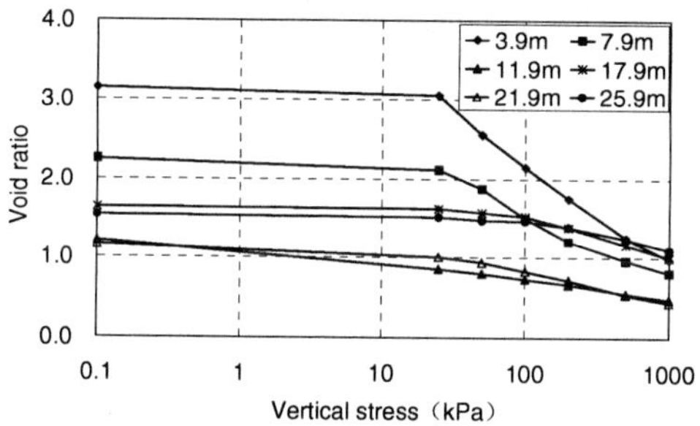

Fig. 5 Compression curves for the samples taken from BH5

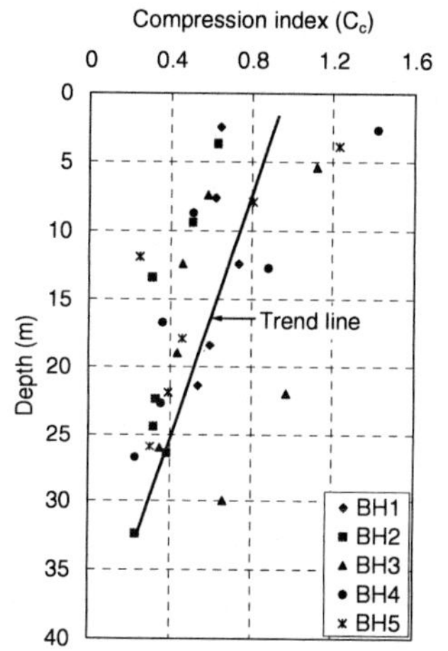

Fig. 6 Change of compressibility index (C_c) with the embedment

Change in shear strength property of MSW with depth

As mentioned previously, in total 8 groups of consolidated drained triaxial compression tests were carried out to investigate shear strength characteristics of the MSW samples taken from the Qizhishan landfill. One group of representative test results in terms of stress-strain relationship is shown in Fig. 7. The four stress-strain curves, corresponding to confining pressures of 50, 100, 200 and 400 kPa, were obtained from four samples collected from the top 5 m of fill. The waste sample age was about 5 years. As shown in Fig. 7(a), the stress-strain curves exhibit the typical strain-hardening behavior of MSW. It appears that the stress increase rate even becomes greater when axial strains exceed a value of 20%, particularly for a high confining pressure. This may indicate that the reinforcing effect contributed by the fiber materials becomes pronounced at a high strain level and under a high confining pressure. The volumetric strain versus axial strain curves demonstrate that all the specimens exhibited a contractive behavior during shearing. The volumetric strains measured at the end are all exceed 5%, being significantly greater than that of a soil. The above stress-strain behavior of MSW is generally consistent with that observed by other researchers (Machado et al., 2002; Grisolia et al., 1996; Jessberger and Kockel, 1993).

The MSW exhibited a strain-hardening behavior during triaxial shearing. Therefore, it would be appropriate to define strength in terms of the mobilized shear strength corresponding to a selected strain level. The shear strength mobilized at two strain levels (i.e., 10% and 20%) was investigated by plotting Mohr-Coulomb strength envelope. It was found that the strength envelope was basically linear on a τ verse σ plane. Shear strength parameters (c' and φ') were estimated for each of the 8 groups triaxial tests (Chen and Zhan, 2006).

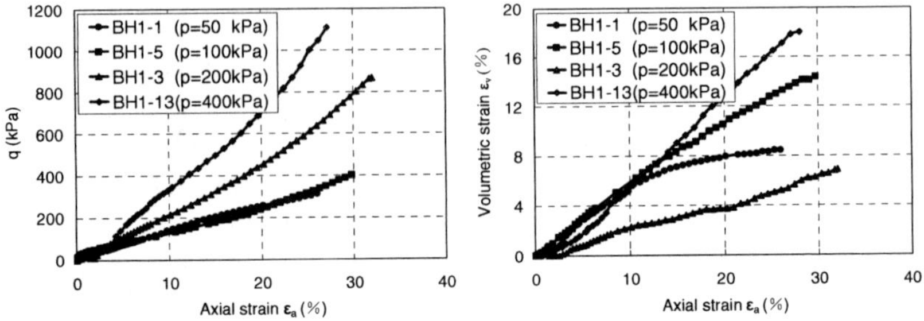

Fig. 7 Stress-strain curves for the samples taken from BH1

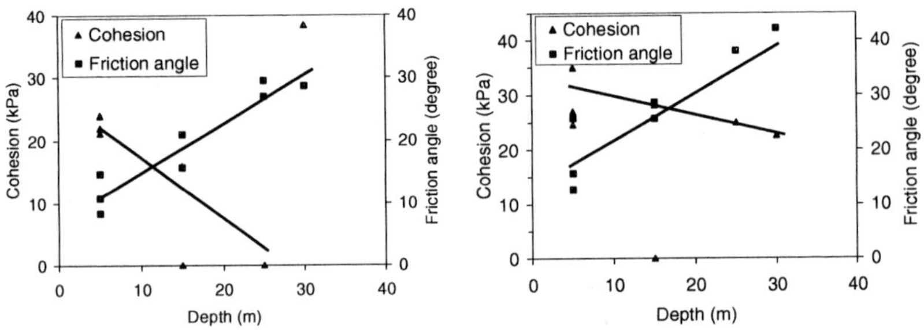

(a) $\varepsilon_a = 10\%$ (b) $\varepsilon_a = 20\%$

Fig. 8 Change of shear strength parameters with the embedment depth

Fig. 8 shows the relationships of shear strength parameters (c′ and φ′) to the embedment depth of MSW. Fig. 8 (a) and (b) correspond to the two strain levels of 10% and 20%, respectively. For a given strain level, it was found that the value of friction angle (φ′) increased with the embedment depth of MSW, and the value of cohesion (c′) decreased with depth. The trends for the two strain levels were consistent. The data show that the shallow MSW had a high friction angle up to 40° but a low cohesion close to zero, and the deep MSW had a low friction angle between 10° and 20° but a relatively high cohesion up to 30 kPa. As expected, both cohesion and friction angle increase with the strain level as a result of the strain-hardening behavior of the MSW.

The observed changes of shear strength with the embedment depth can be explained by considering the change in the MSW composition with depth (see Fig. 3). The shallow MSW contained more plastic, fiber and organic but less cinder and dust than the deep MSW. Thus, the plastic and fiber materials would dominate on the response of the shallow waste samples. This is understandable by considering the low density of plastic and fiber materials. The predominant plastic and fiber materials generally pose a low friction resistance, and hence lead to a low mobilized friction angle. On the other hand, the plastic and fiber materials provide a significant reinforcement effect, resulting in the relatively high mobilized cohesion for the shallow MSW. On the contrary, the deep MSW contained about 70% cinder. The cinder content dominated the material and resulted in a high mobilized angle of internal friction and a low mobilized cohesion. The above discussion indicated that the dependence of shear strength properties on the embedment depth of MSW should be taken into account when performing slope stability analyses on landfills.

SUMMARY

This paper presented experimental results from 1-D compression tests and triaxial compression tests on the MSW samples taken from the Qizhihan landfill. The experimental results showed that the mechanical properties of the MSW significantly depended on the embedment depth of waste. The primary compression index (C_c) was observed to decrease from 1.0 to 0.3 with an increase in depth from 0 to 37 m. The experimental results indicate the dependence of mechanical properties on the embedment depth. This effect should be taken into account when performing slope stability analyses or settlement analyses for landfills.

ACKNOWLEDGEMENTS

Financial support from research grants (50538080 and 50425825) provided National Natural Science Foundation of China (NSFC) is acknowledged. .

REFERENCES

Chen, Y.M., and Zhan L.T. (2006). Field and laboratory investigation on engineering properties of municipal solid wastes in Qizhishan landfill. Technical report, Zhejiang University, Hangzhou, China. (in Chinese).

Grisolia, M., and Napoleoni, X. (1996). "Geotechnical characterization of municipal solid waste: choice of design parameters. " *Proc. 2nd Int. Cong. on Environmental Geotechnics*, Osaka, Japan, Vol. 2, pp. 641-646.

Jessberger, H. L., and Kockel, R. (1993). "Determination and asessment of the mechanical properties of waste materials. " *Proc. Sardinia 93, 4th Int. landfill Symp.*, Cagliari, Italy, 1383-1392.

Machado S.L., Carvalho F.M., and Vilar O.M. (2002). "Constitutive Model for Municipal Solid Waste." *Journal of Geotechnical and Geoenvironmental Engineering*, Vol.128(11): 940 ~951.

Qian, X. D. Koerner, R.M., Gray, D.H. (2001). Geotechnical aspects of landfill design and construction. Prentice-Hall, Inc, New Jersey.

Zornberg, J.G, Jernigan, B.L., Sanglerat, T.R., Cooley, B. H. (1999). "Retention of free liquids in landfills undergoing vertical expansion." *Journal of Geotechnical and Geoenvironmental Engineering,* Vol. 125 (7): 583-594.

Municipal Solid Waste as a Reinforced Soil: Investigation Using Synthetic Waste

George Athanasopoulos[1], Athena Grizi[2], Dimitrios Zekkos[3], Polyxeni Founta[2], and Elli Zisimatou[4]

[1] Professor, Dept of Civil Engineering, University of Patras, Greece, M. ASCE; gaa@upatras.gr
[2] Graduate student, Dept of Civil Engineering, University of Patras, Greece;
[3] Engineer, Geosyntec Consultants, PhD, PE, Oakland, CA 94612, USA, M. ASCE; zekkos@geoengineer.org
[4] Undergraduate student, Dept of Civil Engineering, University of Patras, Greece;

Abstract

It is commonly accepted that the shear strength of Municipal Solid Waste (MSW) may be enhanced by the reinforcing effect of fibrous waste constituents. However, until recently no investigation has systematically evaluated the effects of the fibrous waste materials on the stress-strain-strength response of MSW. The present study is a continuation of a previous investigation conducted by the authors and aims at experimentally quantifying the effect of the fibrous materials on the stress-strain-strength response of MSW, as well as demonstrating the analogy of MSW to reinforced soil. Preliminary direct shear test results indicate that, similarly to tests performed on reinforced soils, the orientation of the fibrous materials with respect to the shear surface is critical and affects significantly the stress-displacement-strength response. In addition, the type of fibrous materials also affects the stress-displacement-strength response.

Introduction

The stress-strain-strength response of Municipal Solid Waste (MSW) is of interest for the design and operation of landfills as well as the post-closure behavior and re-use of the filled landfill areas. It is commonly stated that MSW is a relatively strong material due to the reinforcement effect of the fibrous materials (Landva and Clark, 1986, Kavazanjian et al. 1999). However, to the authors' knowledge, until recently no investigation has systematically evaluated the effects of the fibrous waste materials on the stress-strain-strength response of MSW. Recent investigations on MSW from the San Francisco Bay Area (Zekkos et al. 2007a, Zekkos et al. 2007b) have indicated that the effect of waste composition on the stress-strain-strength response of MSW is significant. This contribution presents the findings of additional work performed on synthetic waste from Greece, aiming to quantify the influence of the fibrous materials on the stress-strain-strength response of MSW.

The Influence of Fiber Reinforcement – Recent Studies

Zekkos et al. (2007a) recently presented results of a series of large-scale direct shear and triaxial compression tests on MSW collected from the Tri-Cities landfill located in the San Francisco Bay Area. The waste material was initially segregated, as part of the waste characterization procedures, and subsequently the various waste constituents were mixed to create a uniform sample. The mixed material was compacted in layers – resulting in an approximately horizontal orientation of fibrous constituents - to prepare the specimens. The results of direct shear tests indicated that the horizontal orientation of fibers produces the weakest response of MSW. Shearing in direct shear occurs parallel to the orientation of the fibrous materials within the waste, and thus, the contribution of the horizontal fibrous materials to the stress-displacement response is minimal. On the other hand, direct shear tests performed with waste fibers oriented perpendicular to the horizontal shear failure surface yield an upward curvature in the stress-displacement response that is attributed to the progressive mobilization of the fibrous materials within the waste matrix.

In triaxial compression tests an upward curvature was observed in the stress-strain response which can also be explained by the progressive mobilization of fibrous materials during shearing, similarly to the direct shear tests where fibers were forced to participate in the shearing mechanism by being oriented perpendicular to the horizontal shear surface. Because shearing in triaxial compression occurs at an angle from the horizontal orientation of the fibers, the shear resistance of MSW in triaxial compression is higher than that observed in direct shear where the fibers are typically oriented parallel to the shear surface.

There are some analogies between the findings of Zekkos et al. (2007a) on MSW and those made by other investigators on the shear response of reinforced soils (Michalowski and Cermak, 2002, Shewbridge and Sitar, 1989, Jewell and Wroth, 1987, Gray and Ohashi, 1983). For example, Gray and Ohashi (1983) presented direct shear test results from sands reinforced with stiff, rough reinforcement and #2 reed fibers. The authors varied the angle between the horizontal shear failure surface and the fiber orientation and found that it had a significant effect on the shear strength of the reinforced soil (Fig. 1). When the angle between fibers and the horizontal shear surface was equal to 90°, the strength of the reinforced soil was significantly higher than when the fibers were parallel to the horizontal shear surface. The highest contribution of the fibers was observed when the fibers were in tension during shearing and at an angle of 60° from the horizontal shear surface. Similar results are reported by Jewell and Wroth (1987) for direct shear tests in sand reinforced with a single plane of stiff, rough reinforcement.

An investigation is currently underway to quantify the effect of the fibrous materials on the stress-strain-strength response of MSW and evaluate if the trends observed in MSW are similar to those observed in reinforced soils. This contribution presents some first laboratory test results from this investigation. The motivation for this study is that, should such trends be confirmed, there may be a basis for increasing the design shear strength of MSW for the sections of the failure surfaces that intersect the waste mass at an angle to the horizontal.

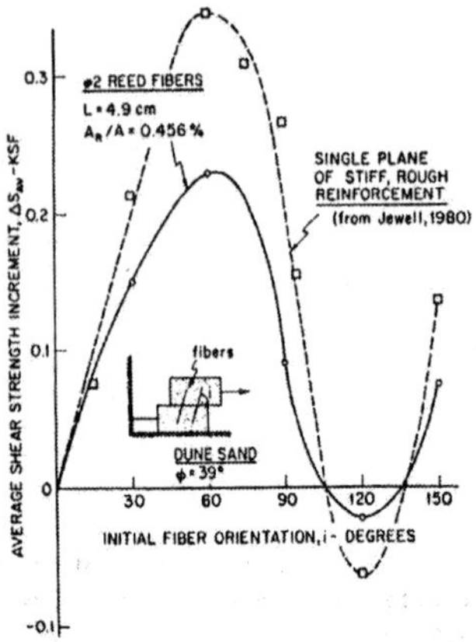

Figure 1. Direct shear test results on sand reinforced with reed fibers or stiff, rough reinforcement (from Gray and Ohashi, 1983)

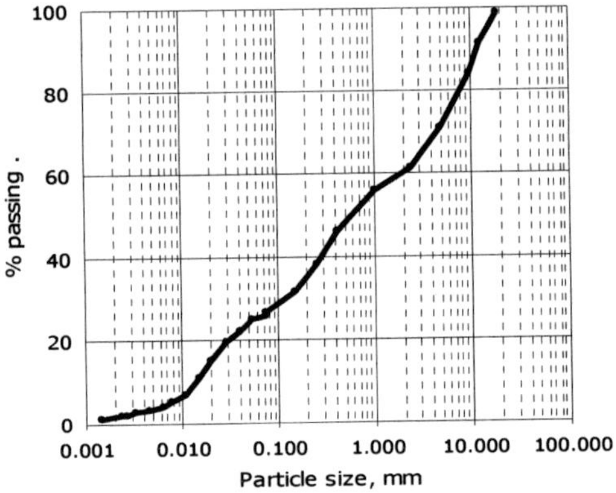

Figure 2. Grain size curve of the smaller than 20 mm material from Xerolaka landfill, Greece.

Specimen Preparation and Testing Program

Based on the waste characterization procedures established by Zekkos (2005), soil-like material with particles smaller than 20 mm - which is mostly representative of the daily soil cover - was collected from the Xerolaka landfill in Greece. Xerolaka landfill serves the city of Patras, the third largest city in Greece with a population of approximately 200,000. The smaller than 20 mm material is characterized as Silty Sand (SM) with 27% of material passing the #200 sieve (Fig. 2). The finer material has a Liquid Limit of 24, a Plasticity Index of 5, D_{10}=0.013mm, D_{60}=2mm and

C_u=154.

The coarser than 20 mm material, which was combined with the soil-like material, consisted of pieces of paper, plastic and wood, similar in nature to materials directed to the landfill. The above pieces of reinforcement were carefully cut from cardboard paper (60mmx50mm), plastic bags (130mmx60mm) and wood sheets (90mmx20mmx10mm).

To ensure that the orientation of the fibrous materials relative to the shearing plane is at a well-defined angle, specimens were prepared by laying successive layers of soil-like material (less than 20mm) and reinforcement (fibrous material) at specific angles and then compacting them manually as shown in Fig. 3. Efforts were made to apply the same compaction effort to all specimens. The angle between the orientation of the fibrous materials and the horizontal shear surface is referred herein as the "fiber orientation angle". Specimens with fiber orientations of 0° and 30° to the horizontal were prepared inside the direct shear box. Specimens with fiber orientations of 60° and 90° to the horizontal were prepared outside the direct shear box in a specially fabricated split-mold (Fig.4) and then transferred to the direct shear box. A completed specimen with 60° fiber orientation in the disassembled split-molt is shown in Fig.5.

Figure 3. Preparation process of synthetic MSW specimen in the split mold.

It should be noted that the test specimens were intentionally more structured than typical MSW placed in landfills, to systematically evaluate the effects of the orientation and type of fibrous waste on the stress-displacement response. MSW in landfills is also relatively structured. Geosyntec (1996) reports that waste particles in the field tend to be sub-horizontally oriented during compaction. Thus, a failure surface through the waste mass would be expected to shear the waste at different orientations relative to the sub-horizontal orientation of the fibrous materials.

Specimens were tested in a large-scale direct shear test device. The direct shear box has a square plan view with a side dimension of 300 mm and a total height of 180 mm. The composition (by weight) of specimens containing paper or wood reinforcement was approximately: 62% soil-like material, 38% reinforcement. Specimens containing plastic reinforcement, due to the difficulty of controlling the compaction effort had a somewhat different soil-like material content ranging from 60% to 75%. Because the unit weight of plastic is significantly lower than the unit weight of wood, significantly more plastic was necessary to reach the target 38% by weight composition compared to wood. All tests presented in this paper were performed at a normal stress of 50 kPa. For each specimen composition, tests were performed with reinforcement (i.e. fibrous materials) orientation of 0°, 30°, 60° and 90°, measured from the horizontal direction.

Figure 4. Split-mold used for specimen preparation.

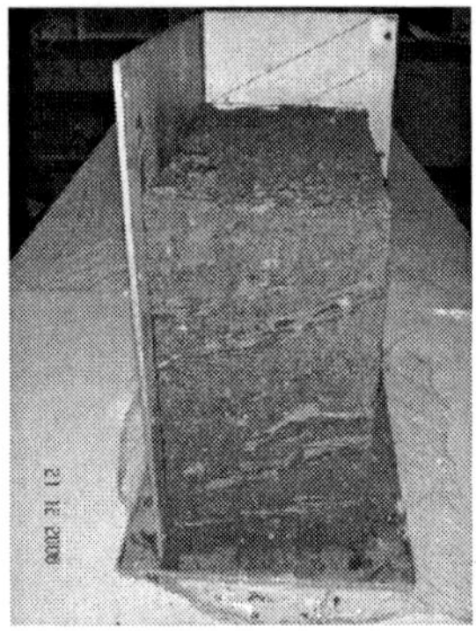

Figure 5. Completed specimen in disassembled split-mold.

Test Results and Discussion

Fig. 6 presents test results for specimens containing plastic reinforcement. The fiber orientation angle is equal to 0°, 30°, 60° and 90°. At small displacements (less than 10 mm), the shear resistance is not significantly affected by orientation. However, at larger displacements, the results indicate that the specimen response is highly anisotropic with the mobilized shear stress at 55 mm, varying greatly for the various fiber orientations. The mobilized shear stress at 55 mm is approximately 4-5 times higher for the specimen with a fiber orientation angle of 60° compared to the specimen with a fiber orientation angle of 0°.

The specimen with a fiber orientation angle of 0°, i.e. when the plastics are oriented parallel to the horizontal failure surface, has almost reached an ultimate shear stress condition and is the only specimen that does not exhibit an upward curvature in the stress-displacement response. This response is consistent with the observations made by Zekkos et al. (2007a) for MSW from the Tri-Cities landfill and indicates that the fibrous materials do not contribute significantly to the stress-displacement response of the specimen. The highest mobilized shear stress is observed at a fiber orientation angle of 60° and then gradually decreasing values of mobilized stress are observed for fiber orientation angles of 90° and 30°, which is still almost twice the mobilized shear stress for the specimen with a fiber orientation angle of 0°.

Fig. 7 presents a summary of all currently available testing results at a normal stress of 50 kPa. In Fig. 7, shear strength τ is defined as the mobilized shear stress at a displacement of 55 mm. The vertical axis is the ratio of the shear strength of any specimen to the shear strength of the specimen with the same composition, but with a fiber orientation angle of 0°.

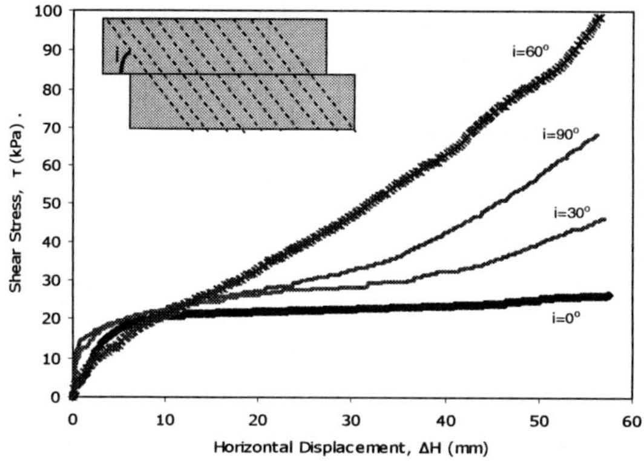

Figure 6. Stress-displacement response for specimens with plastic reinforcement.

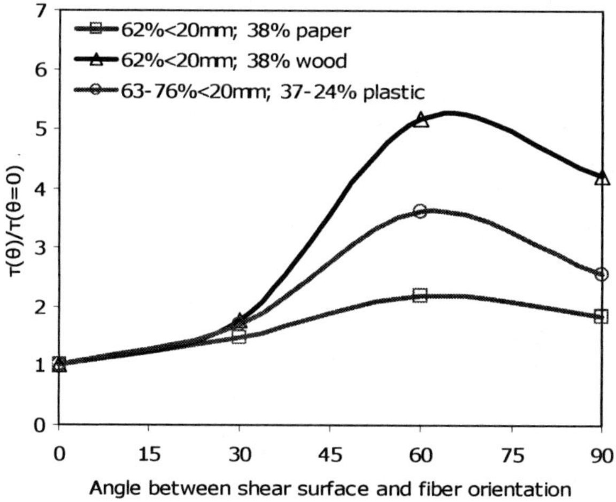

Figure 7. Effect of fiber orientation angle for all specimens at a normal stress of 50 kPa.

The results indicate that the shear strength of MSW is highly anisotropic and the mobilized shear stress depends greatly on the fiber orientation angle and the type of fibrous material. For specimens containing wood reinforcement, the mobilized shear stress for a fiber orientation angle of 60° is more than 5 times higher than the mobilized shear stress observed for specimens with an orientation angle of 0°. Specimens that include wood fibers are also stronger than specimens that include plastic fibers, even though the volumetric composition in plastic is significantly higher than the volumetric composition in wood. This may be attributed to the varying axial stiffness of the reinforcement. Wood fibers are significantly stiffer and stronger than plastic or paper fibers resulting in a more pronounced contribution of wood fibers to the stress-displacement response compared to paper or plastic fibers. Shewbridge and Sitar (1989) also observed a trend of increasing specimen strength with an increase of reinforcement stiffness.

For all specimens tested, the highest contribution of fibers is observed for specimens with a fiber orientation angle of approximately 60°, similarly to the trends observed in reinforced soils (Gray and Ohashi, 1983, Jewell and Wroth, 1987). The change in shear strength as a function of the fiber orientation angle for the synthetic waste (Fig. 7) is similar in shape to the change in shear strength as a function of the fiber orientation angle for the reinforced sands tested by Gray and Ohashi (1983), as observed in Fig. 1. However, the contribution of the fibers in the synthetic waste strength is significantly higher in magnitude compared to the contribution of the fibers in the reinforced sand strength. A reason may be the fact that the fibers content in the specimens tested by Gray and Ohashi (1983) is significantly lower than the fiber content in the waste specimens. This is because in reinforced soil research, emphasis is given in optimizing the benefits of the fibers without increasing significantly the fibers content, as these are typically more expensive to purchase and handle.

The results from this investigation are also generally consistent with the results from the Tri-Cities landfill investigation as reported by Zekkos et al. (2007a). For two specimens tested in the above investigation, the MSW composition was 62%<20 mm material, 18% paper, 5% wood, 5% plastics and 10% gravel. The specimens were tested at a normal stress of 2 kPa and two different fiber orientations: 0° and 90°. The test results indicated that the ratio of the shear strength corresponding to the orientation of 90° and 0° is equal to 2.1. This ratio is slightly higher than the ratio for paper fibers (equal to 1.85) in Fig. 7 (which is the predominant constituent larger than 20 mm), but lower than the ratio of plastic (equal to 2.6) and wood fibers (equal to 4.2), and is also consistent with the composition of the specimen.

It must be stressed that the aforementioned tests on Tri-Cities waste were performed at a normal stress of 2 kPa, whereas the tests from the present investigation were performed at a normal stress of 50 kPa. Further investigation is necessary before any recommendations are made for use in practice. The effects of the fiber orientation at different normal stresses and the effect of varying fiber compositions need to be further investigated. The authors are currently conducting additional tests to address these issues. An additional consideration, not addressed in this study, is the degradation potential of the various fiber constituents. It may be prudent to consider a reduced reinforcement effect to account for long-term degradation of the fibrous materials.

Conclusions

A series of direct shear tests on synthetic waste has been performed aiming to quantify the analogy between MSW and reinforced soils. Preliminary results indicate that, similarly to tests performed on reinforced soils, the orientation of the fibrous materials with respect to the horizontal shear surface is critical and affects significantly the stress-displacement-strength response. The largest contribution of the fibrous materials on the stress-displacement response is observed when the angle between the horizontal shear surface and the fibers orientation is approximately 60°, similarly to reinforced soils. The type of fibrous materials also affects the stress-displacement-strength response. The contribution of the fibrous materials to the MSW stress-displacement-strength response appears to be more important than that observed in reinforced soils and may warrant the use of enhanced values of MSW shear strength in landfill analysis and design.

Note

Additional information on this investigation and relevant studies on the properties of MSW is available from the Geoengineer website at: http://www.geoengineer.org .

References

Geosyntec (1996) "Waste Mass Field Investigation," Technical Report SWP-2, prepared for New Cure, Inc., Monterey Park, California for submission to the USEPA Region IX. GeoSyntec Consultants, Huntington Beach, California.

Gray, D. H., and Ohashi, H. (1983). "Mechanics of fiber reinforcement in sand." ASCE Journal of Geotechnical Engineering, Vol. 109 (3): 335-353.

Jewell, R. A. and Wroth, C. P. (1987). "Direct shear tests on reinforced sand." Geotechnique, Vol. 37 (1): 53-68.

Kavazanjian, E. Jr., Matasovic, N. and Bachus, R. C. (1999). "Large-diameter static and cyclic laboratory testing of Municipal Solid Waste." Proceedings, Sardinia 1999, 7th International Waste Management and Landfill Symposium, 4-8 October.

Landva, A.O. and Clark, J. I. (1986). "Geotechnical testing of wastefill." Proceedings, 39th Canadian Geotechnical Conference Ottawa, Ontario, 371–385.

Michalowski, R. L. and Cermak, J. (2002). "Strength anisotropy of fiber-reinforced sand." Computers and Geotechnics, Vol. 29: 279-299.

Shewbridge, S. E. and Sitar, N. (1989). "Deformation characteristics of reinforced sand in direct shear." ASCE Journal of Geotechnical Engineering, Vol. 115 (8): 1134-1147.

Zekkos (Zeccos), D. P. (2005), "Evaluation of static and dynamic properties of Municipal Solid-Waste", dissertation submitted in partial fulfillment of the requirements for the degree of Doctor of Philosophy, Department of Civil and Environmental Engineering, University of California, Berkeley, Fall. Also available at http://waste.geoengineer.org

Zekkos, D., Bray J. D., Athanasopoulos, G.A., Riemer, M. F., Kavazanjian, E., Founta, X., Grizi, A. (2007a). "Compositional and loading rate effects on the shear strength of Municipal Solid Waste." 4th International Conference on Earthquake Geotechnical Engineering, Thessaloniki, June 25-28, 2007, paper no. 1525. Also available at http://waste.geoengineer.org

Zekkos, D., Bray, J.D., Riemer, M., Kavazanjian, E., Athanasopoulos, G.A. (2007b). "Response of Municipal Solid-Waste from Tri-Cities landfill in Triaxial Compression." Proceedings, Sardinia 2007, 11th International Waste Management and Landfill Symposium.

Recent Findings on the Static and Dynamic Properties of Municipal Solid Waste

Zekkos, D.[1], Bray, J. D.[2], Stokoe, K.[3], Kavazanjian, E.[4], Rathje, E.[5], Athanasopoulos G. A.[6], Riemer, M.[7], Matasovic, N.[8], Lee, J. J.[9], Seos, B.[10]

[1] Engineer, Geosyntec Consultants, Oakland, California, USA;
[2] Professor, Dept. of Civil and Environmental Engineering, University of California at Berkeley, USA;
[3] Professor, Dept. of Civil, Architectural and Environmental Engineering, University of Texas at Austin, USA;
[4] Associate Professor, Dept. of Civil and Environmental Engineering, Arizona State University, Tempe, Arizona, USA;
[5] Associate Professor, Dept. of Civil, Architectural and Environmental Engineering, University of Texas at Austin, USA;
[6] Professor, Dept. of Civil Engineering, University of Patras, Greece;
[7] Adjunct Professor, Dept. of Civil and Environmental Engineering, University of California at Berkeley, USA;
[8] Associate, Geosyntec Consultants, Huntington Beach, California, USA.
[9] Graduate Student Researcher, Dept. of Civil, Architectural and Environmental Engineering, University of Texas at Austin, USA;
[10] Graduate Student Researcher, Arizona State University, Tempe, Arizona, USA;

ABSTRACT: The design of Municipal Solid Waste (MSW) landfills requires the selection of representative MSW material properties. The profession's understanding of the mechanical response of MSW has evolved significantly in recent years, but many aspects of the response of MSW remain uncertain. A research project involving five organizations was conducted to systematically evaluate the factors that affect the static and dynamic properties of MSW. A summary of major findings from the work performed to date by this team of investigators is briefly described in this paper.

INTRODUCTION

The selection of representative MSW properties is important for reliable landfill performance analyses, including static and seismic slope stability, seismic response, containment system design, and design of facilities for post-closure development. The profession's understanding of the response of MSW has evolved significantly since the early work performed by Sowers (1973) and Landva and Clark (1986). However, despite many advances in our understanding of MSW properties, many of the factors affecting the response of MSW remain largely uncertain. In 2002, the US National Science Foundation funded a collaborative research project that involved the University of California at Berkeley (UCB), the University of Texas at Austin (UT), and Geosyntec Consultants to evaluate systematically factors that affect the static and dynamic properties of MSW by means of in situ investigations and an extensive large-scale laboratory testing program. Subsequently, Arizona State University (ASU) and the University of Patras, Greece (UP), joined the project team. The fieldwork and laboratory testing performed by investigators is briefly described in this paper. Relevant findings are presented that provide insight regarding the mechanical response of MSW.

FIELD INVESTIGATION AND WASTE CHARACTERIZATION

Waste sampling was coordinated by Geosyntec Consultants. Two large-diameter (760 mm) borings were augered to depths of 10 m and 32 m using a bucket auger at the Tri-Cities landfill, located in the San Francisco Bay Area. Shallow and deep bulk samples of waste were retrieved and stored separately in 39 sealed 55-gallon drums. Bulk samples of similar visual composition from the same depth interval within a borehole were designated as sample groups. Sample groups included relatively young waste (waste placed within the past 2 years) and relatively old waste (waste placed for approximately 15 years) as well as waste of varying composition. In situ unit weight tests were performed using a gravel replacement procedure developed by Geosyntec Consultants (Matasovic and Kavazanjian, 1998) and described in Zekkos et al. (2006a). The MSW unit weight was found to range from 10 kN/m^3 near the surface to 16 kN/m^3 at depth. Shear wave velocity soundings were performed by UT at the boring locations using the Spectral Analyses of Surface Waves (SASW) method. The shear wave velocity was found to vary from 75 to 210 m/sec at the surface, reaching 250 m/sec at a depth of 25 m (Lin et al., 2004).

Waste material was characterized using a procedure that was developed to efficiently collect relevant information about the waste material. The procedure is described in Zekkos (2005) and included segregating the waste into material larger and smaller than 20 mm (0.75 in). This segregation is useful because: a) material <20 mm is composed predominantly of equidimensional particles, including soil from daily cover, organic materials, and some fine waste inclusions, whereas material >20 mm consists mostly of fibrous constituents; b) material <20 mm can be characterized using conventional soil mechanics tests and can be tested using conventional size geotechnical testing equipment. Evaluation of the influence of the fibrous >20 mm material on MSW was an important part of this investigation.

About 50-75% by weight of each waste sample was <20 mm material. This material contains a significant amount of soil (and soil-like) particles, but also contains organic material so that it is lighter and softer than many inorganic soils. The remaining coarser (>20 mm) material consisted primarily of paper, plastic, wood, and gravel. Constituents such as metals, glass, stiff plastics, and textiles, comprised a significantly lower percentage of the material by weight and by volume. Laboratory tests were performed on the three sample groups summarized in Table 1. Group A3 is older material sampled from a relatively large depth. Group C6 is younger material sampled from a relatively shallow depth. Group C3 was selected for testing as the most different sample group from sample groups A3 and C6. A coordinated laboratory testing program using both large-scale and small-scale equipment was conducted on material from these groups at the various institutions. A brief summary of the laboratory testing performed on the waste and some of the primary findings are presented in this paper.

Table 1. MSW sample groups tested

	A3	C6	C3
Borehole	BH-2	BH-1	BH-2
Depth, m	25.6-26.2	7.6-9.6	3.5-4.5
% moisture content[1]	12	13	23
% organic[1]	15-30	10-16	20-36
Age (years)	15	<1	2

[1]As measured on the smaller than 20 mm material at 55° C.

LABORATORY TESTING AT THE UNIVERSITY OF CALIFORNIA AT BERKELEY AND THE UNIVERSITY OF PATRAS

Large-scale cyclic triaxial (CTX) equipment (d=300 mm, h=600-630 mm upon specimen preparation) was used at UCB to perform cyclic triaxial, triaxial compression, and triaxial extension tests. Additional conventional-scale cyclic triaxial tests were performed on waste that was processed to make all particles less than 20 mm in size. A direct shear (DS) box, 300 mm by 300 mm by 180 mm, was used at UP to perform shear testing. One dimensional compression tests were performed prior to shearing. More information about the testing devices is presented in Zekkos (2005).

The laboratory testing program evaluated the effects of unit weight, compaction effort, confining stress, waste composition, and loading rate on the monotonic stress-strain response and shear strength, and the dynamic properties, i.e. the small-strain shear modulus, strain-dependent shear modulus and material damping, of MSW. Poisson's ratio was also measured during some of the triaxial tests.

A total of 23 large-scale direct shear and 27 large-scale monotonic triaxial tests were performed. As shown in Figure 1, direct shear test specimens prepared with the same compaction effort and with 100%, 62% and 12% <20 mm material, were found to have a conventional concave (e.g. hyperbolic) stress-displacement response and similar shear resistances. The similarity in shear resistance is attributed to the fact that shearing occurred parallel to the sub-horizontal orientation of fibrous materials within the waste and thus the contribution of the fibrous materials to the shear strength minimal. The tendency of the fibrous materials to be oriented horizontally was confirmed by inspection after testing. This tendency towards a horizontal orientation of the fibrous particles was also observed in the triaxial testing conducted for this investigation and was observed in the field by Geosyntec (1996). Subsequent direct shear tests performed with waste fibers oriented perpendicular to the shear failure surface yielded a convex, upward curvature in the stress-displacement response, which is attributed to the progressive mobilization of the fibrous materials within the waste matrix. Figure 2 illustrates this anisotropic aspect of MSW behavior, comparing two specimens with the same composition, prepared with the same compaction effort, and tested at the same normal stress, but with a different preferred orientation for the fibrous (>20 mm) materials. Zekkos et al. (2007a) provides details of these tests.

Consistent with the direct shear tests, the stress-strain response in triaxial compression was found to be strongly dependent on fibrous waste content. An upward curvature of the stress-strain curve was observed for all the triaxial compression tests, with the exception being specimens with 100% <20 mm material. Based upon the direct shear tests using different fiber orientations and on the triaxial test where the shear plane cuts across the preferred fiber orientation, the upwards curvature of the stress-strain curve may be attributed to the progressive mobilization of fibrous material reinforcement during shearing. This explanation is supported by the observation that the shear resistance in triaxial compression was higher than that observed in direct shear tests in which the preferred fiber orientation and shear plane were aligned.

The secant friction angle was found to decrease with confining stress both in direct shear and triaxial testing, indicating that a nonlinear strength envelope may be most appropriate for MSW. Shear resistance was found to increase with unit weight, compaction effort, and strain rate (for axial strain rates between 0.5%/min and 50%/min). Test results are presented in Zekkos et al. (2007a, 2007b) in greater detail.

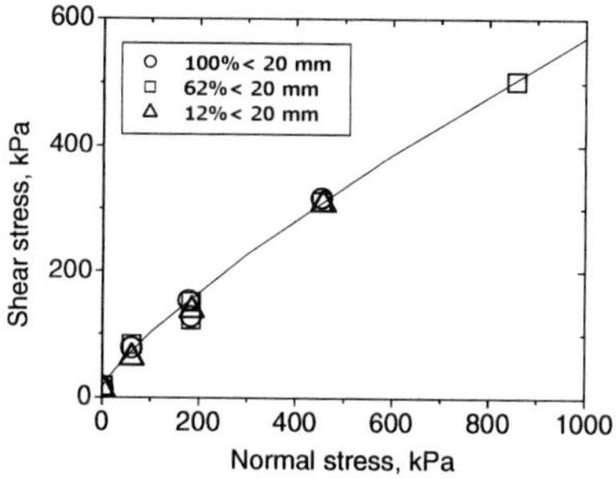

FIG. 1. Direct shear strength of MSW with horizontal fiber orientation (from Zekkos et al. 2007a).

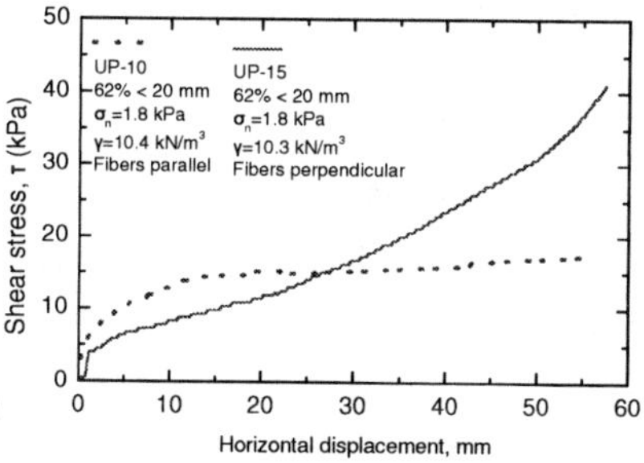

FIG. 2. Direct shear testing for MSW specimens with different fiber orientations (from Zekkos et al. 2007a).

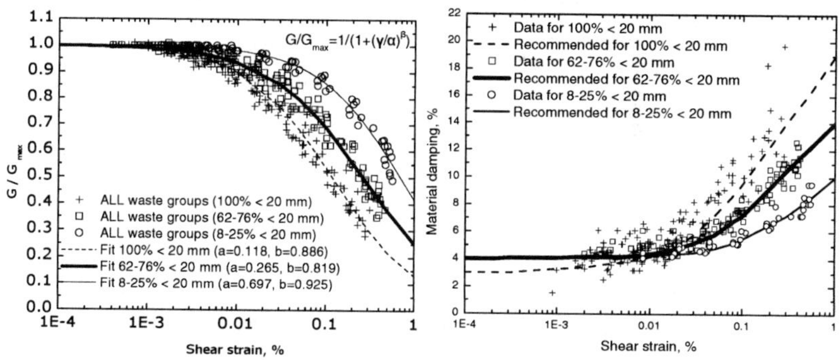

FIG. 3. Strain-dependent (a) normalized shear modulus reduction and (b) material damping curves for MSW based on results from all CTX tests at confining stress up to 125 kPa (from Zekkos et al. 2006b).

More than 90 cyclic triaxial test series were performed at UCB to evaluate the parameters that affect the dynamic properties of MSW. Of the parameters investigated, waste composition and unit weight were found to have the most important effect on the dynamic properties of MSW. As the amount of fibrous >20 mm material increases, the absolute value of G_{max} decreases, the normalized shear modulus reduction curve (G/G_{max}) shifts to the right (i.e., threshold strain increases), and the material damping (D) at larger strains reduces significantly. G_{max} is also significantly affected by confining stress, by unit weight, and by time under confinement, and to a lesser degree by loading frequency. The G/G_{max} and D relationships are affected primarily by waste composition (i.e., the amount of >20 mm material) and confining stress. Neither unit weight, time under confinement, nor loading frequency appear to significantly affect modulus reduction or damping. The G/G_{max} curves shift to the right as the amount of >20 mm material increases or as confining stress increases. Correspondingly, D reduces at larger strains as the amount of material larger than 20 mm increases or as confining stress increases. Figure 3 presents the G/G_{max} and D versus shear strain (γ) curves for all three sample groups for tests performed at confining stresses up to 125 kPa and for varying waste composition. No significant differences in the absolute value of the G_{max} G/G_{max}, or D relationships were observed among the three sample groups. The results of the cyclic triaxial investigation are presented in more detail in Zekkos et al. (2006b).

LABORATORY TESTING AT THE UNIVERSITY OF TEXAS AT AUSTIN

The linear and nonlinear dynamic properties of MSW were investigated at UT using reconstituted specimens with torsional resonant column devices. Two devices were employed. One device was a combined resonant column and torsional shear (RCTS) device. In this device, the specimen has a fixed-free configuration, with nominal dimensions of 71 mm in diameter and 142 mm in height. In the second device, called the large-scale resonant column (LSRC), the specimen has a free-free configuration, with nominal dimensions of 152 mm in diameter and 305 mm in height. The devices are described in Lee (2007).

The advantage of the LSRC device is that larger particle sizes can be used in constructing the specimens. Particle sizes as large as 38 mm were used to construct LSRC specimens. The specimens were reconstituted using the same procedure as used to construct the specimens tested by UCB and ASU. MSW specimens of two ages were tested; old MSW (A3 group) that was about 15 years old and fresh MSW (C6 group) that was less than one year old. For each age, three different waste composition groups were tested: (1) 100% of particles smaller than 20 mm, (2) 62% smaller than 20 mm, and (3) 14% smaller than 20 mm. The larger particles in groups #2 and #3 consisted of paper, soft plastic, wood and gravel. The total unit weights for these three groups were, on average, 12 kN/m^3, 10 kN/m^3, and 8 kN/m^3, respectively.

Staged testing was performed at five confining pressures. Typical results from testing old MSW with the LSRC are shown in Figures 4 through 7. Variations in the small-strain dynamic properties with total isotopic confining pressure, σ_o, are shown in Figures 4 and 5, in terms of shear wave velocity, V_s, and small strain damping ratio, D_{min}, respectively. As seen in Figure 4, the log V_s – log σ_o relationships for each MSW group are composed of two straight lines. The first straight line represents the waste in the overconsolidated state (induced by compacting the specimen) and the second straight line represents the waste in the normally consolidated state. At each test pressure, the values of V_s decrease as the percentage of particles <20 mm decreases. As a point of reference, the V_s values for specimens with 100% < 20 mm are about 50-75% of the V_s values of reconstituted loose sand at the same pressures.

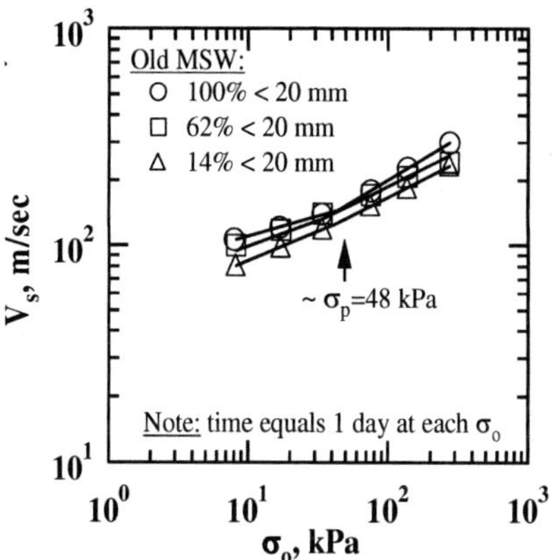

FIG. 4. Variation in V_s with σ_o for Old MSW (Group A3) with waste composition

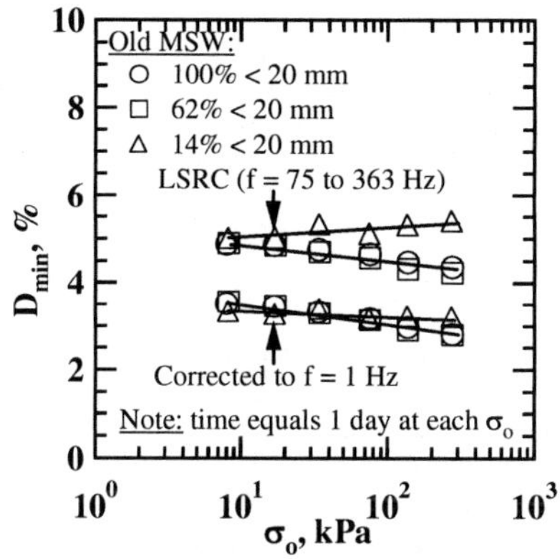

FIG. 5. Variation in D_{min} with σ_o for Old MSW (Group A3) with waste composition

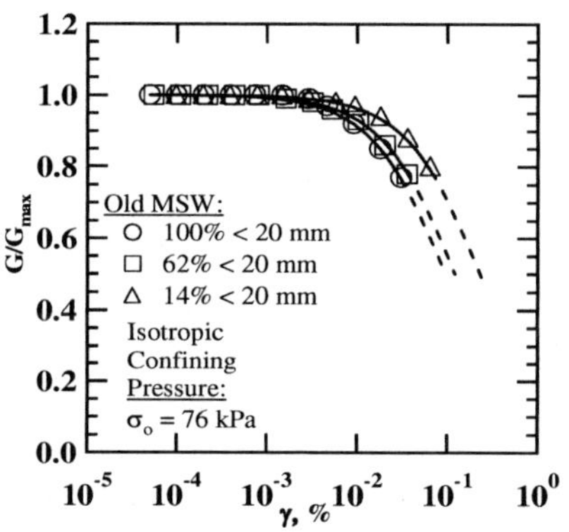

FIG. 6. Variation in G/G_{max} with Shear Strain at σ_o = 76 kPa for Old MSW (Group A 3) with waste composition

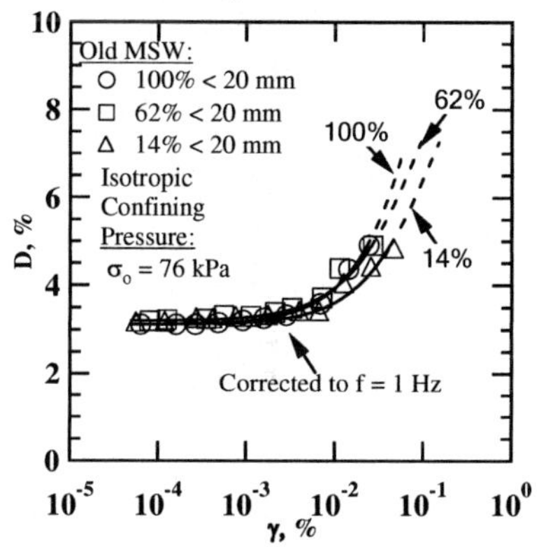

FIG. 7. Variation in D with Shear Strain at σ_o = 76 kPa for Old MSW (Group A3) with waste composition

The variation of D_{min} with σ_o is shown in Figure 5. The value of D_{min} increases significantly with excitation frequency. A frequency correction was applied to D using a second order polynomial relating D/D(1Hz) to log(f), developed based upon tests conducted on MSW over a range of frequencies from 0.03 Hz to approximately 200 Hz. Details are provided in Lee (2007). When corrected to an excitation frequency of 1 Hz, D_{min} varies only slightly with percentage of particles < 20 mm and decreases slightly with increasing σ_o. For these tests, the average value of D_{min} at f = 1 Hz is about 3%. This value is on the order of three times that of most granular soils, and is more similar to values for clay or organic soils.

Examples of the nonlinear response of old MSW are shown in Figure 6 for the variation of G/G_{max} with γ and in Figure 7 for the variation in D with γ. All results shown in Figures 6 and 7 were determined at a confining pressure of 76 kPa. The linearity in normalized shear modulus and material damping ratio increases as the percentage of particles < 20 mm decreases; that is, the G/G_{max} – log γ and D – log γ relationships shift to the right (to larger values of γ) as the percentage of larger particles increases in the waste. As a point of reference, all waste groups exhibit more linear G/G_{max} – log γ curves than sand. However, they exhibit higher damping than sand at shear strains below 0.02 % and lower values at strains above 0.05 %.

The fresh MSW was tested over the same test and material parameters as discussed above. In comparison with the old waste, the fresh waste generally has: (1) nearly the same values of V_s, (2) slightly higher values of D_{min}, (3) very similar nonlinear behavior in terms of G/G_{max}, and (5) slightly higher values of D.

LABORATORY TESTING AT ARIZONA STATE UNIVERSITY

Large-scale simple shear tests are being conducted at ASU using a stacked-plate direct shear apparatus. Teflon-coated steel plates 12 mm-thick with a rectangular opening approximately 400 mm-long (in the direction of the applied shear stress) and 300-mm wide are used to impose a zero lateral strain condition on the compacted specimen. Specimens approximately 150 mm-tall from A3 group are being tested under monotonic and cyclic loads. Most tests are being conducted at a normal stress of 150 kPa in order to achieve a mean normal stress of 76 kPa, based upon a K_0 value of between 0.25 and 0.3, as measured by tactile pressure sensors placed against the side walls of the stacked frames in the simple shear device prior to compaction of the specimen. Uniform cyclic testing is being conducted at shear strain amplitudes from 0.03% to 3%. Laboratory tests at ASU are still in progress and will be reported on subsequently. Initial test results, however, show a conventional concave stress-strain curve and similar modulus reduction characteristics as measured in the UCB tests.

CONCLUSIONS

As part of a collaborative research project, an extensive laboratory testing program of reconstituted MSW samples recovered from a landfill in the San Francisco Bay Area using several different testing devices has been performed. The results from the comprehensive testing program were consistent and indicated, among others, that the influence of waste composition and confining stress on the static and dynamic properties of MSW is critical. In particular, the influence of fibrous materials on the stress-strain and shear strength of MSW is highlighted by the test results.

ACKNOWLEDGMENTS

This material is based upon work supported by the National Science Foundation under Grant Nos. CMMI-022064, CMMI-0220159, CMMI-0219834, CMMI-04137572, and CMMI-0635435 as part of a collaborative study by the University of California at Berkeley, Arizona State University, Geosyntec Consultants, the University of Texas at Austin, and the University of Patras. Any opinions, findings, and conclusions or recommendations expressed in this material are those of the authors and do not necessarily reflect the views of the NSF. The authors would like also to thank Mr. Guy Petraborg of Waste Management for allowing the field activities at the Tri-Cities landfill. More information on this research project is available at: http://waste.geoengineer.org/ .

REFERENCES

Geosyntec Consultants (1996) "Pre-Design Analyses of the OII Landfill Technical Report SWP-3: Waste Mass Field investigation," *Technical Report* prepared for New Cure, Inc., Monterey Park, California for submission to the USEPA Region IX. GeoSyntec Consultants, Huntington Beach, California.

Landva, A.O. and Clark, J. I. (1986). "Geotechnical testing of waste fill." Proceedings, *39th Canadian Geotechnical Conference*, Ottawa, Ontario, 371–385.

Lee, J. J. (2007) "Dynamic Characteristics of Municipal Solid Waste (MSW) in the Linear and Nonlinear Strain Ranges." Ph.D. Dissertation in Progress, The University of Texas at Austin.

Lin, Y-C, Rosenblad, B., Stokoe, K. H. II. (2004). "Data Report on Shear Wave Velocity Profiles Determined by SASW Method at: Altamont Landfill, Redwood Landfill, and Tri-Cities Landfill." Geotechnical Engineering Report GR04-3, Geotechnical Engineering Center, Civil Engineering Department, The University of Texas at Austin, October 29, 2004.

Sowers. G. F. (1973). "Settlement of waste disposal fills." Proceedings of 8th *Int. Conf. on Soil Mechanics and Foundation Engineering*, Moscow, Vol. 2, pp. 207-210.

Zekkos (Zeccos), D. P. (2005). "Evaluation of static and dynamic properties of Municipal Solid-Waste", dissertation submitted in partial fulfillment of the requirements for the degree of Doctor of Philosophy, Department of Civil and Environmental Engineering, University of California, Berkeley, Fall.

Zekkos, D. Bray, J. D., Kavazanjian, E. Jr., Matasovic, N., Rathje, E. M., Riemer, M. F., Stokoe, K .H. (2006a). "Unit weight of Municipal Solid Waste", *ASCE Journal of Geotechnical and Geoenvironmental Engineering*, Vol. 132, No. 10, October.

Zekkos, D. P., Bray, J. D., Riemer, M. F. (2006b). "Shear modulus reduction and material damping relationships for Municipal Solid-Waste", *Proceedings of the 8th U.S. National Conference on Earthquake Engineering*, April 18-22, 2006, San Francisco, California, USA, Paper No. 1324 (in proceedings cd-rom).

Zekkos, D., Bray J. D., Athanasopoulos, G.A., Riemer, M. F., Kavazanjian, E., Founta, X., Grizi, A. (2007a). "Compositional and loading rate effects on the shear strength of Municipal Solid Waste." *4th International Conference on Earthquake Geotechnical Engineering*, Thessaloniki, June 25-28, 2007, paper no. 1525.

Zekkos, D., Bray, J.D., Riemer, M., Kavazanjian, E., Athanasopoulos, G.A. (2007b), "Response of Municipal Solid Waste from Tri-Cities landfill in Triaxial Compression", *11th International Waste Management and Landfill Symposium, Sardinia 2007*, 1-5 October 2007.

Stability Of Slopes For Closure Of Old Waste Dumps

Manoj Datta

Professor, Civil Engineering Department, Indian Institute of Technology Delhi, New Delhi -110016; mdatta@civil.iitd.ernet.in

ABSTRACT: Old waste dumps often have steep slopes because waste has been tipped over from the top of the dump. For the closure of such dumps, low permeability cover systems are to be provided to control leachate formation as well as to capture landfill gas. The stability of such cover systems is governed by the shearing resistance that develops at the interfaces of various components of the cover systems. This paper describes a case study for closure of an 18 m high municipal solid waste dump. It highlights the influence of slope angle, slope height, surface roughness of geomembrane and tensile strength of veneer reinforcement on the stability of slopes. The impact of seepage force and the earthquake loading on the stability is also discussed. The case study shows that use of textured geomembrane, berms at small heights and veneer reinforcement of high strength, all help in the improvement of slope stability.

INTRODUCTION

The cover and liner configurations for landfills in India, as per the guidelines issued by the regulatory authorities (CPCB 2000, CPCB 2002 and MUA 2000), are shown in Figs. 1 and 2 for municipal solid waste (MSW) and hazardous waste (HW) respectively.

The cover system for a MSW landfill has five components as shown in Fig. 1, including a single barrier of 0.6 m thick compacted clay. An HW landfill cover comprises of the same five components but includes a geomembrane as the sixth component which is a part of a composite barrier as shown in Fig. 2.

For closure of old waste dumps, no guidelines relating to the cover systems have been issued by regulatory authorities. The cover systems are expected to be similar to or more stringent than those adopted for new landfills because old waste dumps do not have a liner system at the base (Datta 2003).

Old waste dumps have steep side slopes along which cover systems are to be installed. The presence of multiple layers in the cover system gives rise to the phenomenon of slippage at the interface of various layers along sloping sides. The interface between the geomembrane and the layer above it, as well as the layer beneath it, are the two critical locations, which govern slope stability.

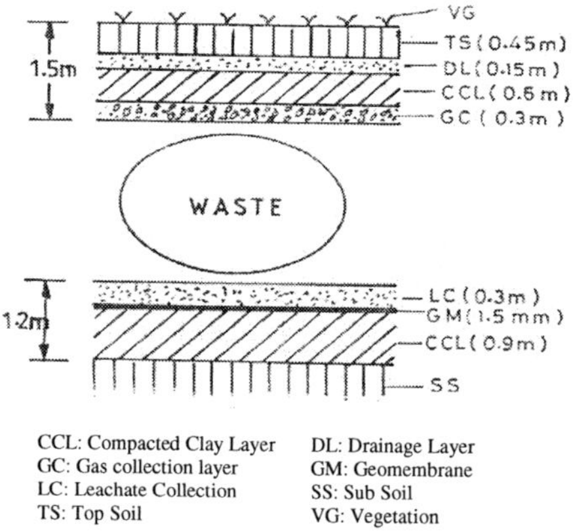

FIG. 1. Components of MSW landfill.

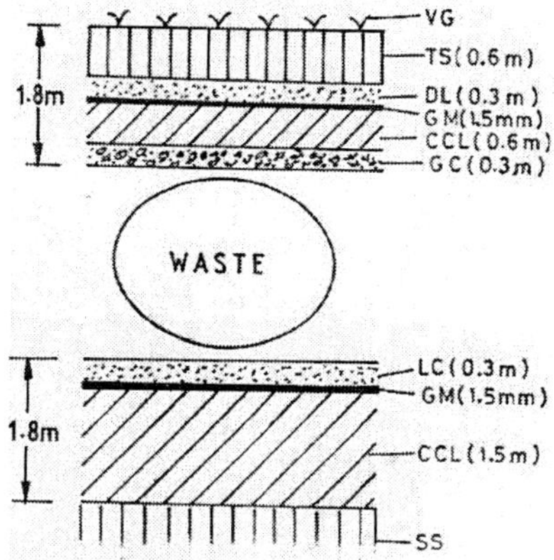

FIG. 2. Components of HW landfill.

One case study is presented which highlights the need to take great care in arriving at safe slopes for covers of waste dumps. The factors affecting the stability of cover systems along the slope are discussed.

CLOSURE OF AN OLD WASTE DUMP

Background

An old municipal solid waste dump occupies an area of 400 m x 500 m (approximately) in the suburb of a metropolitan city in western India (Fig. 3). The waste height is 18 m above the ground level (Fig. 4). On the northern side, the dump protrudes into a creek where the waste has been pushed forward to a depth of a few meters below the water level. On the eastern side, a few buildings are close by and on the southern side, a road runs parallel to the boundary at some distance away from the perimeter (Datta 2006).

Waste filling activities continue on top of the dump by the 'tipping forward' method and at many locations the slope of the waste along the side is of the order of 1 : 1 (horizontal : vertical). Uncontrolled gaseous emissions and foul odor emanate from the top of the waste dump. The base of the dump slopes towards the creek and dark colored leachate flows along the base into the creek. Site investigations reveal that the waste dump is underlain by clay followed by bedrock (Fig. 4).

Closure

From a long-term perspective, after closure, the waste dump must be stable, give an aesthetically pleasing appearance and have no harmful impact on the adjacent environment. This implies that the sides of the dump must be made stable such that the

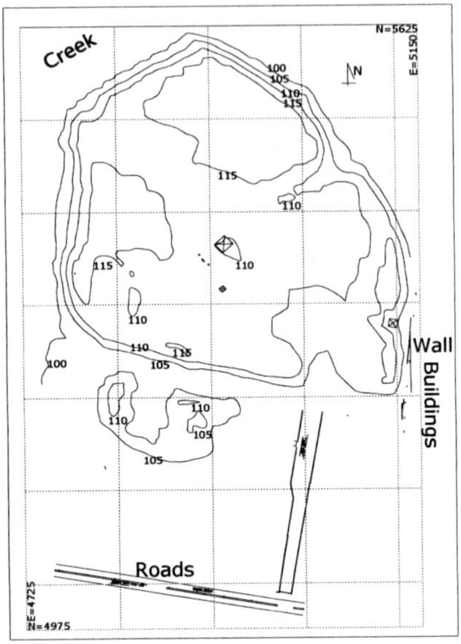

FIG. 3. Plan view of existing waste dump.

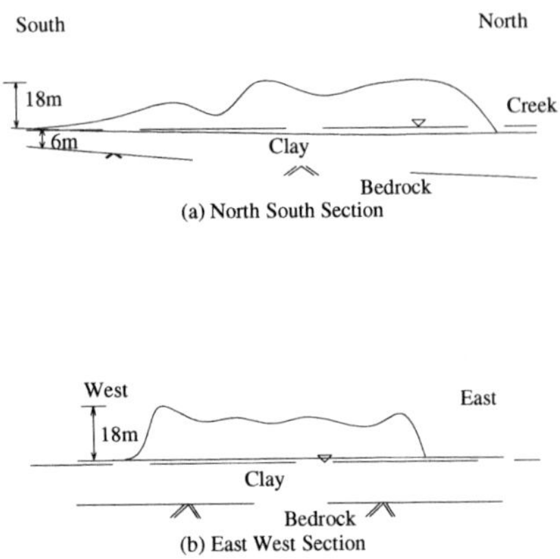

FIG. 4. Cross-section of the dump.

factor of safety against slope instability is of the order of 1.5 (as against 1.0 at present). It should be provided with a cover capable of supporting vegetative growth.

The harmful impact of the leachate on the aquatic life in the creek is to be minimized by reducing the infiltration and by collecting and treating the leachate. Further, the problem of foul odor and the green house gas emissions is to be minimized by collecting and flaring or utilizing the landfill gas.

Provision of an impermeable cover system along with a gas collection system is considered to be a suitable control measure to reduce the harmful impact on the environment. A cover system similar to that used for HW landfills (Fig. 2) is proposed as presence of geomembrane above the compacted clay prevents loss of landfill gas and facilitates efficient collection of gas. Such a cover system is to be checked for stability along the side slopes.

Stabilization of Existing Side Slopes

The most attractive option is to provide a cover system on the top and the side slopes after stabilizing the slopes without relocating the existing waste and without moving the toe of the slope. Three options were considered (Fig. 5), namely a nailed slope, an anchored wall and a reinforced wall. These options are discussed hereafter.

Steep slopes in soils are often stabilized by the techniques of anchoring or nailing (Hausmann 1990). Hence, an anchored or nailed slope with an impermeable cover and a vegetative mat was considered as an option (Fig. 5a). The vegetative mat would initiate and sustain vegetation along the slope. However, three uncertainties were associated with this option: (a) the longevity of the anchors embedded in waste was suspect because the waste products could react with and weaken the anchors; (b) the settlements in the waste due to continuous biodegradation could change the alignment of the anchors and thus affect their functioning; and (c) the self-sustainability of the vegetation on such steep slopes could not be assured over the long–term. As a consequence, this option was ruled out.

An anchored wall (Fig. 5b) was also considered as an alternative. It had the advantage of a concrete fascia on exterior which would require much lower maintenance in comparison to vegetation on a steep slope. However, this alternative also suffered from the disadvantage of uncertainty in the performance of the anchor buried in the waste. It was necessary to provide an anchor since a free standing wall would not be stable for a height in excess of 6 to 8 m. A reinforced earth wall (Fig. 5c) too was considered and ruled out on account of complexities involved in placement and anchorage of the reinforcement.

As a consequence, the focus shifted to stabilization of the slopes by re-grading and re-profiling of the waste.

Re-Profiling of Side Slopes

A simple method for stabilization of the steep side slopes is to re-profile and re-grade them to a gentle slope of the order of 4(horizontal) : 1(vertical). Such gentle slopes have adequate safety against sliding of components of the cover material over the waste. Re-profiling involves filling or excavating and re-locating the waste as shown

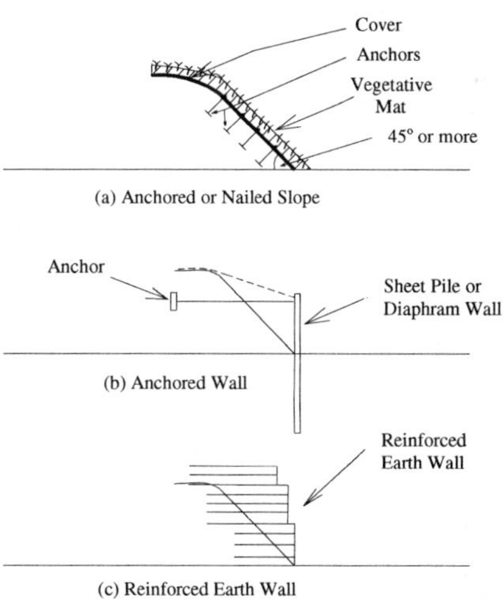

FIG. 5. Possible measures for stabilization of existing slope.

in Figs. 6 and 7, resulting in an increase in the height of waste to about 22 m. Fig. 7 shows how availability of space adjacent to the toe influences the adoption of this solution. Wherever space is available beyond the toe of the landfill, the slope can be re-graded to a gentler profile by filling waste or soil (Fig. 7a). Such a situation is valid for the southern side of the landfill. At other locations, the toe would have to stay at its existing location or move inwards due to shortage of space as shown in Figs. 7b and 7c. The process of excavation and re-location of waste would produce additional foul odor. This would be acceptable on the side of the creek but would cause discomfort to the residents of the buildings on the eastern side. Hence on the eastern side, use of special bio-sprays to control the odor would be made.

STABILITY OF COVER SYSTEM

Two cover systems (Fig. 8) were considered for the waste dump. Cover A is more stringent than the MSW landfill cover shown in Fig. 1 because of the presence of an additional component - geomembrane above the compacted clay. Cover B (Fig. 8b) does not meet the MSW landfill criteria for covers since the compacted clay is substituted by a geomembrane. For such a cover system, a more rigorous programme for leakage detection, maintenance and repair is expected to be adopted during the post – closure period.

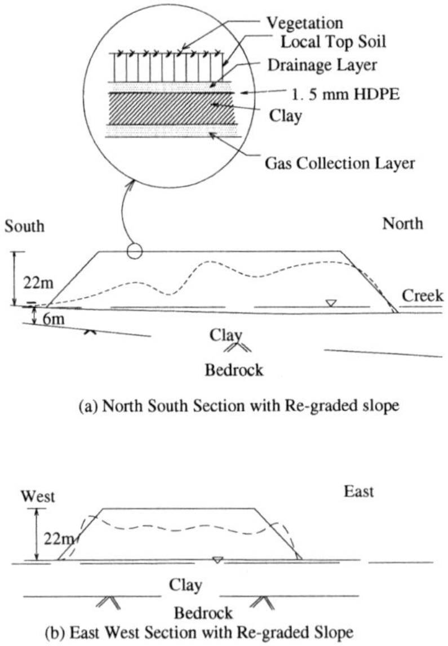

FIG. 6. Landfill with re-graded slope and cover (ver. : hor. scale is 5:1).

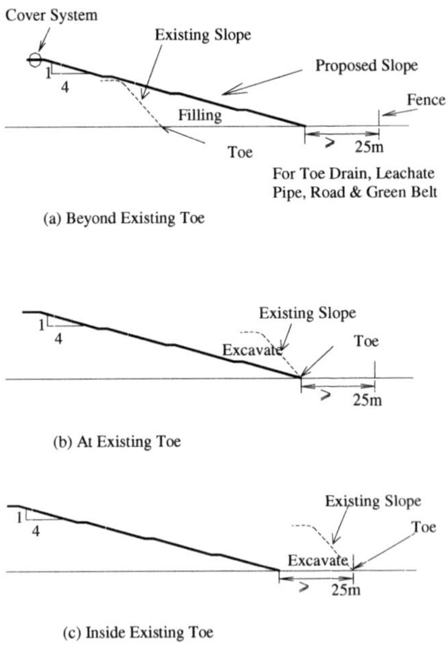

FIG. 7. Position of existing toe and re-gradation of slope.

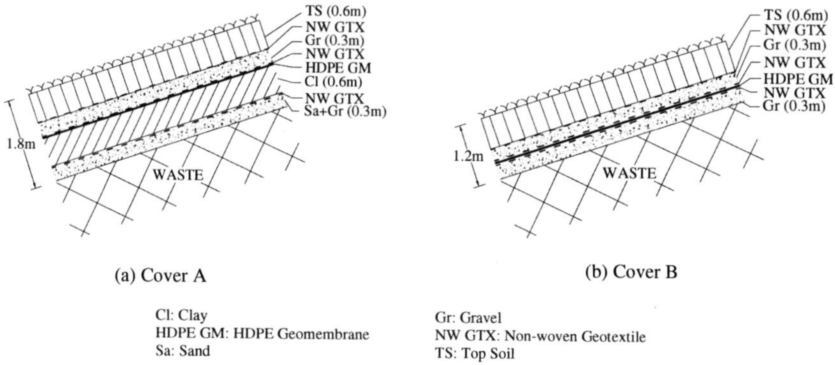

FIG. 8. Cover system.

An attempt was made to arrive at the steepest slope with adequate factor of safety for both types of covers, so that maximum volume of waste could be accommodated.

Four cases were considered critical for slope stability namely:

(a) long term case of dry slope under static loading;
(b) short term case, during monsoon, of slope with seepage flow in drainage layer parallel to the other slope (submergence ratio of 0.5);
(c) short duration case of slope under earthquake loading (pseudostatic approach with horizontal seismic coefficient of 0.1 (as per Bureau of Indian Standards));
(d) rare case of slope with seepage flow and earthquake loading occurring simultaneously.

Table 1 lists the minimum acceptable values of factor of safety adopted for the each of these critical cases. Stability analysis was performed for failure parallel to outer slope (Qian et al. 2002, Koerner and Daniel 1997).

The interface shear strength parameters were made available by the owner of the project. These were determined by performing modified direct shear tests under saturated conditions in a 300 x 300 mm shear box. The peak and residual angles of shearing resistance were reported as 18° and 14° respectively for the interface between clay and geomembrane (textured). At the interface between geotextile (non-woven, needle punched) and geomembrane (textured), the peak and residual values were reported to be 22° and 17° respectively. Adhesion was negligible. There was considerable debate on the choice of parameters - peak or residual - for the purpose of design. Keep in view the fact that sliding movement (pre-shearing) between various components could not be ruled out during the installation, it was decided to adopt residual parameters.

The results of the stability analysis for covers A and B are presented in Table 2 and Table 3 respectively for the static case, along the weakest interface. One notes from Table 2 that the geomembrane-clay interface in cover A has a low angle of shearing resistance (14°) and a factor of safety of 1.5 is achieved at a slope of 5:1 and slope height (between berms) of 5 m. This is so despite use of a textured geomembrane instead of a smooth one. For cover system B (Table 3), a similar slope is required

Table 1. Factor of Safety

	Condition	Acceptable factor of safety
1	Static case (long term)	1.5
2	Seepage flow during monsoon (short duration)	1.3
3	Earthquake loading (very short duration)	1.1
4	Earthquake loading + Seepage flow (rare)	1.0

Table 2. Factor of Safety for GM (textured) – Clay Interface (δ=14°) for Cover A

Slope angle	Height (between berms) (m)	Factor of safety
3 : 1 (18.4°)	10	0.86
	5	0.98
4 : 1 (14.0°)	10	1.11
	5	1.23
5 : 1 (11.3°)	10	1.36
	5	1.48

Table 3. Factor of Safety for GM (Textured) – Geotextile Interface (δ=17°) for Cover B

Slope angle	Height (between berms) (m)	Factor of Safety	
		Without reinforcement	With reinforcement
3 : 1	10	0.92	1.14
	5	0.92	1.53
4 : 1	10	1.22	1.53
	5	1.22	2.05
5 : 1	10	1.53	2.13
	5	1.53	3.52

Table 4. Factor of Safety for GM (Textured) – Geotextile Interface (δ=17°) for Cover B with Reinforcement

Slope Angle	Height (between berms) (m)	Factor of safety (with reinforcement)		
		EQ only	Seepage Only	EQ and Seepage
3 : 1	10	0.81	1.03	0.74
	5	0.99	1.37	0.91
4 : 1	10	1.00	1.37	0.92
	5	1.19	1.82	1.11
5 : 1	10	1.15	1.72	1.08
	5	1.38	2.28	1.27

without reinforcement as the residual angle of shearing resistance between geomembrane and geotextile is also low (17°). However when a veneer reinforcement (with allowable tension of 30 kN/m) is introduced, in the soil above the geomembrane, one can achieve a factor of safety of 1.5 for a slope inclination of 3 : 1 and height of 5.0 m between berms.

Table 4 lists the values of factor of safety obtained for cover B (with veneer reinforcement) at the weakest interface under earthquake and seepage loading. Cover slope with inclination of 3 : 1 is not observed to be stable under earthquake loading as the factor of safety falls below 1.0. However a slope of 4 : 1 is observed to be stable for a height of 5.0 m between berms for all conditions. Additional computations reveal that a veneer reinforcement, with allowable tension of 50 kN/m, results in a stable slope for all conditions at an inclination of 3.5 : 1 for slope height of 7.5 m between berms.

CONCLUSIONS

The present study leads to the following conclusions:

(a) The interfaces between the geomembrane and the clay beneath it or the geotextile above it are the weak locations at which slippage is likely to occur.

(b) Seepage force parallel to the geomembrane during monsoon as well as horizontal seismic loading during earthquakes causes the factor of safety to reduce significantly.

(c) Provision of veneer reinforcement in the soil above the geomembrane, and use of textured geomembrane, improves the stability of slope.

(d) Provision of berms at intervals of low heights also helps in increasing the stability of the cover system.

REFERENCES

CPCB, (2000). "Criteria for hazardous waste landfills." *Central Pollution Control Board*, New Delhi.

CPCB, (2002). "Manual for design, construction and quality control for liners and covers of hazardous waste landfills." *Central Pollution Control Board*, New Delhi.

Datta, M. (2003). "Geotechnical study for hydraulic barrier system at tailings pond." *ASCE Practice Periodical for Hazardous, Toxic and Radioactive Waste*, Vol. 7(3): 163-169.

Datta, M. (2006). "Geotechnical aspects of landfills and old waste dumps – some case studies." *Proceedings IGC 2006*, Chennai, India: 221-228.

Hausmann, M.R. (1990). "Engineering principles of ground modification." *McGraw Hill*, New York.

Koerner, R.M. and Daniel, D.E. (1997). "Final covers for solid waste landfills and abandoned dumps." *ASCE Press*, Virginia, USA.

MUA. (2000). "Manual for municipal solid waste management." *CPHEEO*, Ministry of Urban Affairs, New Delhi.

Qian, X., Koerner, R.M. and Gray, D.H. (2002). "Geotechnical aspects of landfill design and construction." *Prentice Hall*, New Jersey, USA.

Geomechanics and Long-term Landfill Settlement

John McDougall

School of Engineering & Built Environment, Napier University, Edinburgh, EH10 5DT, UK

ABSTRACT: Long term landfill settlement is commonly attributed to two phenomena: creep and biodegradation. Biodegradation, which may contribute the greater proportion of long-term settlement, is a function of a range of factors, many of which are not easily associated with conventional geomechanical analysis. A more fundamental treatment of biodegradation effects within the constitutive framework of a hydro-bio-mechanical (HBM) model is presented in this paper. The role of a decomposition-induced void volume change parameter Λ, and its ability to handle biodegradation-related settlement, are revealed through the simulation of a long-term, laboratory-scale waste settlement test.

INTRODUCTION

Long term landfill settlement is commonly attributed to two phenomena: creep and biodegradation. In the case of creep, physico-chemical and mechanical phenomena produce a long-term settlement that can be adequately interpreted using conventional secondary compression (or time-dependent) techniques. In contrast, biodegradation, which may contribute the greater proportion of long-term settlement, is a function of many factors. These factors relate to a number of waste attributes such as: waste type, organic content, leachate amount and composition, microbiological conditions, physical conditions namely density and conductivity; and landfill management procedures such as the filling sequence and leachate control opeations.

The evolution of the landfill environment depends on how the landfill is operated. In the case of dry entombment, biodegradation-related factors will change gradually (if at all) and the use of time-dependent methods for combined creep and biodegradation phenomena may well be expedient. However, failure of a dry containment system or deliberate manipulation of landfill hydraulics, biochemistry, or microbiology, as in the case of a bioreactor landfill, unleashes a system that is not easily reconciled with conventional geomechanics.

There is now considerable evidence to show qualitatively the influence of biodegradation factors on landfill settlement. A general acceleration in long-term landfill settlement rates attributable to biodegradation has been described by

Bjarngard & Edgers (1990) and Konig et al (1995). Wardwell & Nelson (1981) show that settlement is directly affected by organic content, whilst tests reported by Andersland & Al-Khafaji (1981) are more equivocal. McDougall et al (2004) indicated that the impact of mass loss may be episodic with phases of void enlargement during which little overall settlement occurs. The effect of moisture in the form of leachate recirculation is clearly demonstrated by Olivier & Gourc (2007) amongst others. Finally, Ivanova (2007) showed the inhibiting effects of acidification.

Whilst knowledge of the influence of individual factors is good, especially at the laboratory scale, their joint influence at field scale is conceptually more challenging. Time-dependent methods therefore continue to be attractive in the analysis of long-term landfill settlement. However, such an approach is constrained to a highly empirical and observational mode of operation.

MORE FUNDAMENTAL APPROACH TO WASTE MECHANICS

A more fundamental approach to the geomechanics of waste is one that allows factors other than time, namely those controlling the decomposition process, to determine long-term settlement. Time is important; creep and biodegradation are phenomena that progress at a certain rate but it is more revealing to interpret and model these processes, especially biodegradation, as rate-limited, rather than uniquely time-dependent. In this way, the decomposition process is free to accelerate, decelerate, even stop, within the rate-limit, according to conditions within the landfill milieu.

The resulting model will be more complex but offers a framework that accounts for a wider range of landfill and waste characteristics and thereby moves the analysis of long-term landfill settlement towards a more predictive mode of operation. However, a mechanism for determining volumetric strain (or other change in mechanical properties) due to decomposition is required. Only then can the biodegradation model influence landfill settlement.

Geomechanics of Decomposition

In the HBM model a constitutive relation between two volumetric state variables, the void phase volume V_V and solid phase volume V_S, handles the impact of decomposition on the mechanical state. This relation is given by

$$dV_V = \Lambda \, dV_S \tag{1}$$

where Λ is an mass-loss induced void change parameter. A range of responses can be accommodated, which are summarised in Table 1, from which it can be seen that either hardening or (more likely) softening with decomposition can result. For example, when $\Lambda = -1$, loss of solids results in a direct and equal increase in void volume, i.e. there is no change in overall volume although the solid skeleton is likely to be weaker. Notice in Table 1 that $\Lambda = e$ is particularly significant. When $\Lambda < e$ the void ratio increases and by implication the material becomes weaker, whereas when $\Lambda > e$ loss of solid matter leads to a more compact particle rearrangement implying a strength increase. In this latter case significant settlement would occur since both solid and void phases volumes reduce. A full derivation and description of Λ can be found in McDougall & Pyrah (2004).

Table 1: Decomposition-induced void change parameter - reference values and associated phase composition changes.

Λ	Void ratio	Overall volume	Phase composition
−1	Maximum increase	No change	Much looser & weaker
0	Increase	Reduction	Looser & weaker
e (= void ratio)	No change	Large reduction	No change
>e	Decrease	Maximum reduction	More compact & stronger

HBM MODEL

The HBM model is a three-part model specifically developed for the analysis of landfill settlement. It comprises hydraulic, biodegradation and mechanical system models, coupled in a sequential manner. The solution passes from model to model, through link routines, as shown in Fig.1. System variables are used within link routines to update the problem circumstances within which each model operates. In this way, a more representative condition at each stage of the simulation is obtained.

Operation of the model, i.e. preparation of simulation input data and interrogation of output data, is possible using a graphical user interface. Although space constraints allow only an outline here, a full description of the HBM model can be found in McDougall (2007) and at www.sbe.napier.ac.uk/HBM.

Hydraulic Model

The hydraulic model is an unsaturated flow model that requires the moisture retention properties of the waste to be defined. Initial dry density and a degradable: inert matter breakdown, which in conjunction with phase specific gravities enable void volume and saturated moisture contents to be determined, are also required. Residual moisture contents are calculated as a function (gravimetric moisture content) of the solid mass.

Saturated moisture contents are continually updated using changes in void volume; residual volumetric moisture contents are updated according to changes in solid mass. Absorptive capacity and associated discharge throughout the cell are then internally calculated.

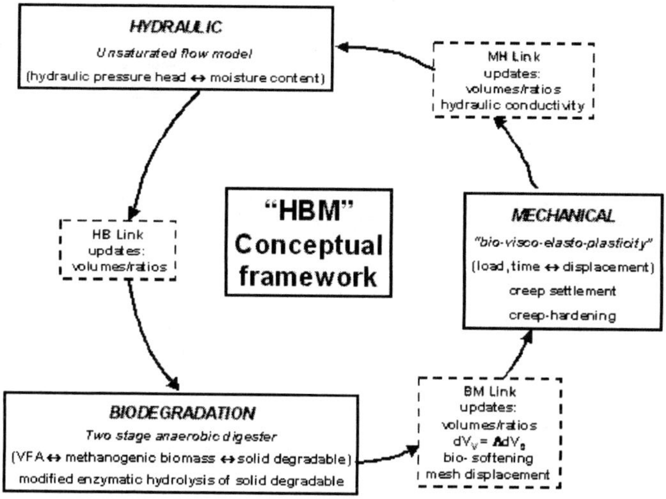

FIG. 1. Schematic diagram showing the main components of the HBM model.

Biodegradation Model

The biodegradation model is a two-stage anaerobic digestion model in which indicative volatile fatty acid and methanogenic biomass concentrations control the mineralisation of organic matter. Spatial and temporal distributions of all field variables and related properties are available. Some key modifications have been made to reflect more realistically the enzymatic hydrolysis of solid organic matter including sensitivity to moisture content, product inhibition and the changing digestibility of the decomposable fraction. A number of natural phenomena and operational treatments can then be simulated, e.g. souring due to acid accumulation, addition of methanogenic inoculum, clean water flushing or leachate recirculation. These modifications are described in detail in McDougall & Pyrah (2001).

Mechanical Model

The mechanical model combines identifiable load, creep and biodegradation-induced effects to analyse and predict landfill settlement. It is the distinction between time-dependent creep and rate-limited biodegradation effects that differentiates this mechanical model from many other landfill settlement models.

Load induced settlement is treated as an elastoplastic process with volumetric hardening – a plane strain formulation of modified Cam Clay. Creep strain is also an elastoplastic process based on an equivalent time method (Yin & Graham, 1993), which gives rise to volumetric hardening.

Implementation

The HBM model is implemented using the finite element method, with each system sharing a common mesh. In this way it is possible to address material and operational features such as complex section geometry, waste heterogeneity and simulation of the filling phase, see Fig. 2. By controlling element disclosure, compression, biodegradation, infiltration and moisture transfers within the waste body can be modelled to produce a waste profile that on completion of filling depicts a more realistic hydro-bio-mechanical condition.

ILLUSTRATIVE SIMULATION – LIRIGM WASTE COMPRESSION TEST

The aim of this simulation is to illustrate the performance and potential of the HBM model by comparison with a laboratory-scale long-term compression test on MSW undertaken at LIRIGM, University of Grenoble (Olivier & Gourc, 2007). The full range of parameters required by the HBM model has not been collected within a single test programme so only a subset, the mechanical parameters, are being evaluated here.

The general strategy of the simulation was: firstly to establish the as-placed conditions of the waste; secondly to determine mechanical compression parameters by fitting to the initial compression data; finally, without further adjustment of compression parameters, to explore secondary settlement under a range of decomposition-induced void change parameter values and in relation to a sequence of leachate recirculation.

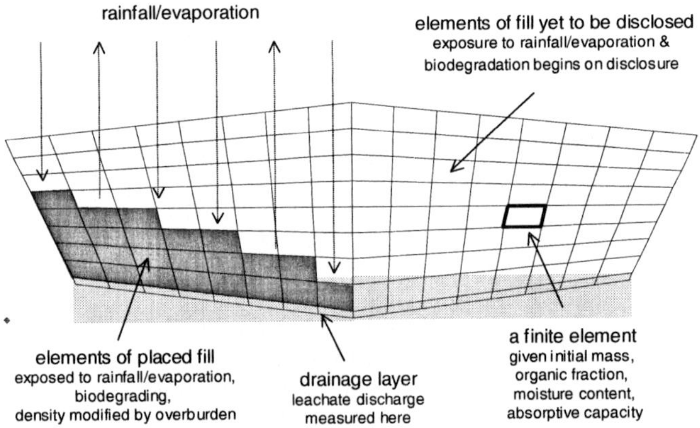

FIG. 2: Finite element discretisation illustrating control of filling.

Test conditions and waste properties

The LIRIGM test comprised a cell of approximately 1 m^2 plan area, which was filled with 420 kg of raw MSW giving an as-placed unit weight of 0.51 Mg/m^3. The as-placed gravimetric moisture content was 16% and degradable:inert ratio was 55:45, both by dry mass. After placement, 155litres of water were added to bring the waste up to field capacity. The waste was incrementally loaded to a total vertical stress of 130 kPa then left under this load for a period of 677 days, during which time leachate discharge from the bottom of the cell was collected and periodically recirculated at the upper surface. A full description of the test procedure and outputs can be found in Olivier & Gourc (2007).

Initial loading

After establishing initial conditions, the HBM model was run to simulate the first four days of the LIRIGM test. Mechanical model parameters were adjusted until predicted surface displacements matched the experimental data, as shown in Fig.3 (fitted parameter values: $\kappa = 0.072$, $\lambda = 0.23$, $\mu = 0.35$. $p_o = 30$ kPa). Of particular importance is the unload/reload cycle that occurred at about 70 hours into the test, which imposes a more stringent control on the selection of mechanical parameters. This unload/reload loop was produced solely by the removal and subsequent restitution of vertical load, i.e. there was no in-run manipulation of compression parameter values.

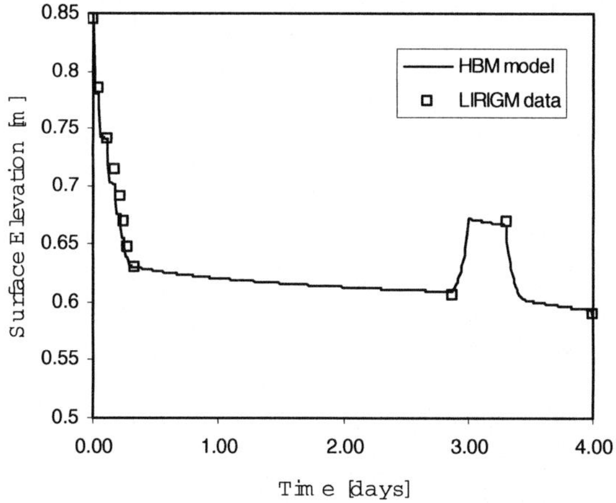

Fig. 3: Predicted and measured load/creep behaviout during first 4 days of the LIRIGM, University of Grenoble, experiment.

Long-term Behaviour

This part of the simulation captures secondary settlement over the entire 677 days of the experiment. Leachate recirculation, which can be classified into three distinct periods, as described in Table 2, was modelled as an applied surface infiltration rate. No changes in mechanical compression or viscosity parameters were made after the initial loading test.

Table 2. Periodic recirculation rates in LIRIGM test (Olivier & Gourc, 2007)

Period [days]	Leachate recirculation
0-200	Moderate & irregular [<0.8L/t/day]
200-400	Intensive [peak 6.8 L/t/day]
400-677	None

The results presented in Fig.4 compare measured and predicted long-term settlement for selected values of Λ (0.0, –0.2, –0.4). Best fit occurs with $\Lambda \approx -0.2$, which corresponds to a loosening of the waste with decomposition. The acceleration in settlement that occurs about 200 days is particularly interesting as it coincides with the commencement of a period of intensive recirculation and evinces the postulated influence of increasing moisture content on decomposition.

Since the decomposition-induced void change parameter controls biodegradation-induced settlement in conjunction with the progress of decomposition, its validation can be approached by comparing mass conversion in the test with that predicted by the model. According to Olivier & Gourc (2007), 37.1% of the degradable fraction has been lost during the test, whereas the corresponding figure is 39.7% the HBM model.

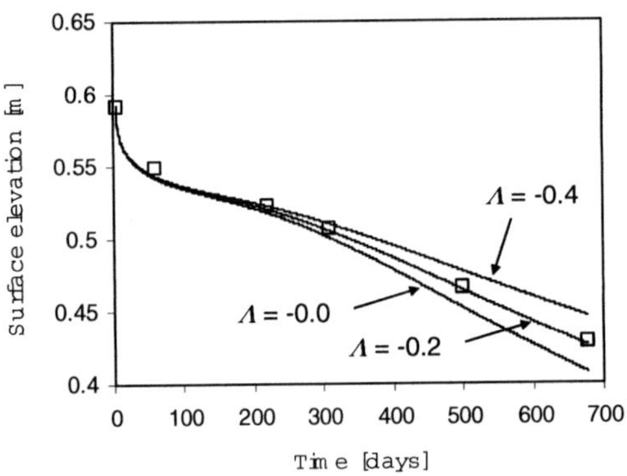

Fig. 4: Predicted and measured secondary settlement behaviour during the experiment for selected values of decomposition-induced void change parameter, Λ.

CONCLUSIONS

A more fundamental interpretation of long-term landfill settlement has been presented. Using a decomposition-induced void volume change parameter Λ within the framework of a hydro-bio-mechanical model, the long-term settlement behaviour of a large-scale laboratory test on decomposing waste has been modelled. Both qualitative and quantitative performance of the model is good, including the response to moisture addition, with $\Lambda = -0.2$. At the end of the test, the experimental and simulated amounts of degradable matter remaining were 37.1% and 39.7% respectively.

REFERENCES

Al Khafaji, A.W.N. and Andersland O.B. (1981) "Compressibility and strength of decomposing fibre." *Geotechnique*. 31 (4): 497-508.

Bjarngard, A. and Edgers, L. (1990) "Settlement of solid waste landfills." *Proc 13th Annual Madison Waste Conf., Wisconsin*: 192-205.

Ivanova, L. (2007) "Quantification of factors affecting rate and magnitude of secondary settlement of landfills." *PhD Thesis*, University of Southampton.

Konig, D., Kockel, R. and Jessberger, H.L. (1995) "Observations of the stability of municipal solid waste landfills." (in German) *Nurnberger Landfill Symposium '96*, PH Mullner.

McDougall, J.R. and Pyrah, I.C. (2001) "Settlement in landfilled waste: extending the geotechnical approach." *Sardinia 2001*, CISA Cagliari, Vol 3: 481-490.

McDougall, J.R. and Pyrah, I.C. (2004) "Phase relations for decomposable soils." *Geotechnique*, 54(7): 487-494.

McDougall, J.R., Pyrah, I.C., Yuen, S.T.S., Monteiro, V.E.D., Melo, M.C. & Juca, J.F.T. (2004) "Decomposition and settlement in landfilled waste & other soil-like materials." *Geotechnique*, 54(9): 605-610.

McDougall, J.R. (2007) "A hydro-bio-mechanical model for settlement and other behaviour in landfilled waste." *Computers & Geotechnics Special Issue: Chemo-Mechanical Interaction in Geomaterials*, 34(4): 229-246 (doi: 10.1016/j.compgeo.2007.02.004)

Olivier, F. and Gourc, J-P. (2007). "Hydromechanical behaviour of municipal solid waste subject to leachate recirculation in a large-scale compression reactor cell." *Waste Management*, 27(1): 44-58.

Wardwell, R.E. and Nelson, J.D. (1981) "Settlement of sludge landfills with fiber decomposition." *Proc 10th ICSMFE, Stockholm*, 2: 397 - 401

Yin, J.H. and Graham, J. (1989) "Viscous-elastic-plastic modelling of one-dimensional time-dependent behaviour of clays." *Canadian Geot. J.* 26: 199-209

ESTIMATION OF THE MECHANICAL PROPERTIES OF MSW DURING DEGRADATION IN A LABORATORY COMPRESSION CELL

M.K.Singh,[1] and I.R.Fleming[2]

[1]Research Asst. Dept Civil & Geol. Engineering, Univ. of Saskatchewan, Saskatoon, Canada S7N 5A9; Tel: (306)966-5013; FAX: (306)966-5427; manoj.singh@usask.ca

[2]Assoc. Professor, Dept Civil & Geol. Engineering, Univ. of Saskatchewan, Saskatoon, Canada S7N 5A9; Tel: (306)966-5341; FAX (306) 966-5427; ian.fleming@usask.ca

ABSTRACT

Significant advances have been made during the past two decades in the understanding of the mechanical behavior of municipal solid waste (MSW) landfills. Large post-closure deformations have been observed in landfills, often interfering with the buried infrastructure used to collect landfill gas. Slope failures have also been observed in landfills, in some cases with significant loss of life. Since a major component of MSW is degradable material, it is expected that the mechanical properties of waste likely change over time. It is therefore necessary to understand and quantify the evolution of mechanical properties of waste with time as it degrades. The present study is an effort in this direction intended to explore the long-term changes in the mechanical behavior of waste. Some preliminary results are presented in this paper.

INTRODUCTION

Increasing size and height of landfill sites have presented new challenges in terms of landfill stability. The popularity of vertical and "piggyback" expansions have been extensively documented in the literature, as have a number of major failures of high, steep landfill slopes (Hendron et al, 1999; Eid et al, 2000; Koelsch et al, 2005; Kavazanjian and Merry, 2005). Concurrent with the trend toward higher, steeper landfills has been an increase in the provision of engineered systems for landfill gas capture and utilization. The deformation behavior of landfilled waste may be an important consideration in the performance of these systems. Short of catastrophic failure of the waste slopes, there may be lateral deformations sufficient to impair function of gas collection and other engineered systems. These considerations have sparked interest and research into the mechanical behaviour of waste.

In the past two decades, the researchers around the world have contributed to a better understanding of waste mechanical behavior. However, the subject still continues to evolve with time and each new piece of information is a valuable contribution to our understanding of waste mechanical behavior. The current study, though in preliminary stage, is an effort in this direction intended to explore the long-term changes in the mechanical behavior of waste as it degrades.

The mechanical behavior of MSW may be described in terms of its unit weight, shear strength parameters (cohesion, c' and angle of internal shearing resistance ϕ') and the elastic properties (Young's modulus E', Poisson's Ratio ν', Shear Modulus G, and coefficient of at-rest lateral pressure K_o). The existing database of elastic properties is sparse compared with that for the shear strength parameters. A considerable scatter in the estimated mechanical properties of MSW has been observed which reflect the inherent heterogeneity of the material, difficulties in sampling, and the lack of universally accepted protocols for the testing of MSW.

TEST EQUIPMENT

Dual- purpose Laboratory Compression Cell

The equipment used in this study has been designed and built to function as a one-dimensional consolidometer as well as a landfill bioreactor. It was custom built at the University of Saskatchewan and is equipped with automated instrumentation for measuring average horizontal stress, vertical stress, pore pressures, vertical settlement and the quality and quantity of generated gas.

The Landfill Compression Cell (LCC) is a 442 mm internal diameter, 600 mm high stainless steel cylinder with wall thickness of 6.8mm. The cylinder is mounted on a 600 x 600 x 50 mm aluminum plate, fixed to the bottom rail of a 2 m steel girder frame (Figure 1). The dimensions of the LCC were selected to enable at least 4 cylindrical samples of 150 mm diameter and 300 mm height to be taken for triaxial compression tests subsequent to compression and degradation. The wall thickness of the cylinder was chosen so as to be able to measure hoop strains under vertical stress as low as 20 kPa while satisfying K_o conditions. The plunger used for vertical loading is 442mm in diameter and 38mm thick, constructed from two aluminum plates each 19mm thick. The bottom plate is provided with holes (5mm diameter) uniformly distributed to facilitate biogas escape into a cavity engraved between the two aluminum plates. An outlet is provided on the top aluminum plate for transporting the gas from the cavity to a gas collection system via a non-return valve. The plunger is sealed against the cylinder wall by two O-rings.

Four leachate injection ports are provided on the top aluminum plate of the plunger. The ports are spaced so as to provide uniform distribution of moisture within the waste sample. A peristaltic pump is used to inject leachate at a desired flow rate. Six quick-connect ports in two tiers on the lower middle outer surface of the cylinder are provided for pore pressure measurements. An airjack (bore size: 200mm; Mfr: Hydro-line, Inc) with a regulated supply of compressed nitrogen is used to apply the desired vertical load while allowing compression of the sample. A pancake type load cell (45kN) is rigidly fastened by means of a bracket assembly in the center of the top aluminum plate of the plunger for recording the applied vertical stress.

Instrumentation attached to the LCC

The instrumentation attached to the LCC includes strain gauges (for measuring hoop strains induced by the pressure of the waste sample against the cylinder wall), cable extension transducer (CET, for measuring the vertical settlement of the sample), load cell (for measuring the vertical stress), and pore pressure transducers. Six strain gauges are mounted on the outer surface of LCC in two tiers (75mm and 150mm from the base of the cylinder). The strain gauges at each level are 120° apart horizontally. They are 3.0mm foil type strain gauges (Tokyo Sokki Kenkyujo Co., Ltd.; Type: FCA-3-350-17-3L) with self-temperature compensation. The gauge resistance is 351.3 ± 1.5 ohms and has a gauge factor of 2.13 ± 1%. The data acquisition system used for strain gauge measurement is a National Instruments SCXI-1520, 8 channel signal conditioning system.

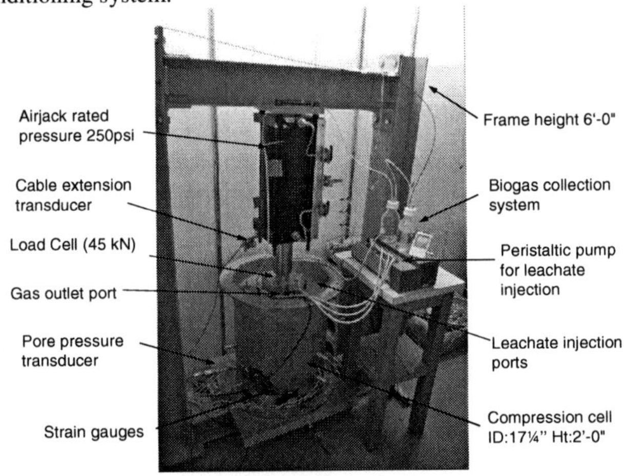

FIG. 1. Dual-purpose landfill compression cell (LCC) used in the present study

Calibration

The calibration of the load cell, CET, and pressure transducers was carried out using normal procedures followed in geotechnical engineering. The frictional force generated between the plunger O-rings and the stainless steel cylinder is subtracted to determine the actual vertical stress transferred to the waste sample. Calibration of the strain gauges was carried out as per the technique demonstrated by Edil and Dhowian (1981) for estimation of at-rest lateral pressure of peat soils. The compression cell was filled with de-aired water and vertical load were applied. For each increment of vertical load, the response of strain gauges and pressure transducer was recorded. Under these hydrostatic conditions in the cylinder, the vertical stress inside the cylinder is equal to the measured pressure and the horizontal stress. A calibration curve was developed to relate the horizontal stress and the corresponding response of the strain gauges. This was found to be a linear relationship (R^2 varying from 0.97 to 0.99) for the range of applied horizontal stresses (20 kPa to 120 kPa). This calibration curve is used to infer the horizontal stresses in the MSW sample.

THE TESTED WASTE

The MSW sample was collected from excavation and using auger cuttings at the Spadina landfill, Saskatoon, Canada at depths of 4-8 m. The waste was black in color and includes a significant fraction of organic matter with negligible fractions of metal, glass and inert material. Waste samples collected at the same depth had been previously tested for organic content and methane potential. These served as a reference for the extent of degradation of the waste (Singh and Fleming, 2004).

PROCEDURE

The LCC is placed in a temperature controlled room at 34-37°C. The unsorted waste was compacted in layers in the cylinder at its natural moisture content to achieve a desired unit weight. The plunger was mounted on the top of the cylinder and was moved down slowly to bring its bottom in contact with waste without compressing the waste. The system was left undisturbed for few hours for the instrumentation to get stabilized. At this stage the initial (baseline) values of the instrumentation attached to the LCC (strain gauges, load cell, pore pressure transducer, and CET) were recorded. An increment of vertical stress was then applied and the sample was allowed to compress and degrade at this constant stress for a pre-defined time. The data for each instrument was continuously logged using an automated data acquisition system. The leachate was injected at pre-defined rate typically followed for bioreactor landfills. Biogas was collected intermittently using tedler bags and its composition and production rate were estimated. The biogas analysis was done using a respirometer and its volume was estimated using water displacement method.

From the test data, effective horizontal stress (σ_h') and effective vertical stress (σ_v') were used to estimate $K_o (= \sigma_h' / \sigma_v')$ and $\nu' (= K_o / [1 + K_o])$. The constrained modulus, $E_o' (= \Delta\sigma_v' / \Delta\varepsilon_v)$ which is the reciprocal of m_v (coefficient of volume compressibility) was used to estimate E' using the relationship $E_o' = E'(1 - \nu')/(1 + \nu')(1 - 2\nu')$.

RESULTS & DISCUSSION

To date, two short-term compression tests have been conducted using LCC and a long-term test is currently in progress. The duration of each short-term test was 3 to 4 weeks. These short-term tests were conducted with a view to demonstrate the working of the LCC with respect to data acquisition and to check for leakages (if any) in the system. The procedure for the long-term test is similar to short-term test except the waste was allowed to consolidate and degrade under constant vertical stress for a longer duration.

For the long-term test, the waste was compacted to an initial bulk unit weight of approximately 10.5 kN/m^3. The response of strain gauges, pore pressure transducer, CET, and the load cell were recorded every 100 seconds using an automated data acquisition system. Six different quantities were recorded to document the complete stress history of the waste sample: time, horizontal and vertical stresses, vertical deformation, pore water pressure and quantity of biogas produced. For each instrument, the average of all the values logged each day was used for inferring the waste mechanical properties. Leachate and biogas sample were collected biweekly for analyses of pH and gas composition respectively. The leachate pumping rate was maintained at approximately 8ml/minute so as to prevent channelized flow inside the waste.

The results presented in this paper are from the short-term tests and also from the preliminary stage of the long-term experiment. So far, the waste sample in the long-term test has been subjected to the first increment of $\sigma_v^{'} = 26$ kPa which lasted for 55 days and the current increment of $\sigma_v^{'} = 42$ kPa. Figure 2 shows the change in bulk unit weight over time at $\sigma_v^{'} = 26$ kPa. Though the change in unit weight for this vertical stress is insignificant (given the short time to date), the pattern illustrates the creep behavior of MSW. This pattern is similar to the change in unit weight with overburden stress as suggested by Kavazanjian et al. (1996) and Zekkos et al. (2006).

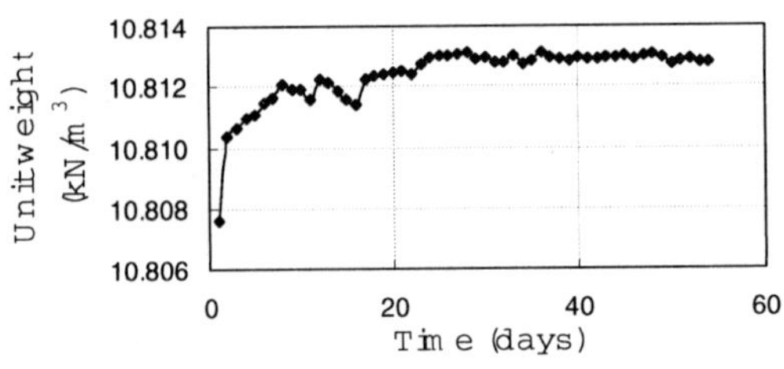

FIG. 2. Variation in Unit weight over time at constant vertical stress of 26 kPa (Creep behavior of MSW)

Figure 3 shows a plot of $\Delta\sigma_h^{'}$ vs. $\Delta\sigma_v^{'}$ obtained from two short-term tests. The K_o is obtained as the slope of the regression line fitted to the experimental data as shown in Figure 3. This value of $K_o (\nu = 0.29)$ is consistent with that obtained by other researchers (e.g. Landva et al, 2000 and Matasovic and Kavazanjian 1998). The variation of $\sigma_h^{'}$ with time at constant $\sigma_v^{'}$= 26 kPa has been shown in Figure 4 along with its moving average values. The observation suggests that the waste relaxes as it degrades. At this preliminary stage, although it is difficult to conclusively explain the decaying behavior of K_o; however, it might be possible that the degradation of waste progressively weakens its solid skeleton which might results in relaxation in K_o values.

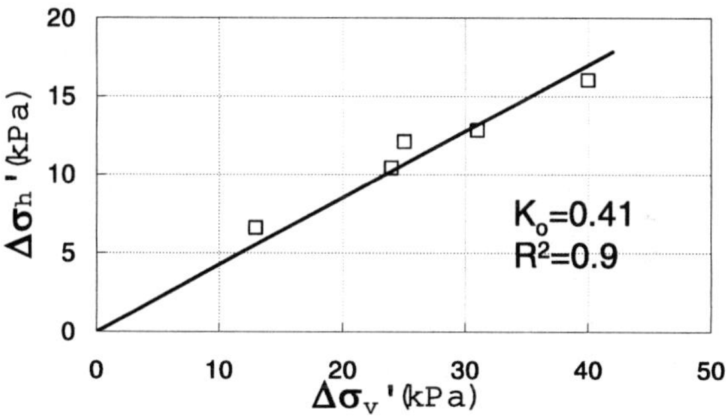

FIG.3. At-rest earth pressure coefficient estimated from short-term test.

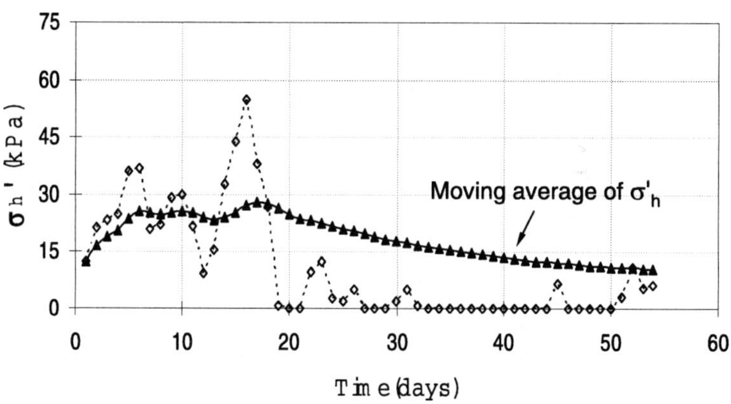

FIG. 4 Decay of lateral stress with degradation for $\sigma_v^{'} = 26kPa$

The process of weakening of waste skeleton is an episodic process of void enlargement and subsequent collapse which is expected at very low stress state (McDougall et al. 2004). The biochemical methane potential (BMP) of the waste from this approximate depth within the Spadina landfill had earlier been estimated to be 60 litres CH_4 per kg dry waste (Singh & Fleming 2004). In the present study, the cumulative CH_4 production recorded during the period for which the vertical effective stress was maintained constant at 26 kPa had reached 3.02 l CH_4 per kg dry waste. In terms of waste degradation, this represents approximately 5% degradation of waste during the period monitored in this study. Given, the huge potential for CH_4 generation, it is anticipated that K_o value might further decrease if the waste is allowed further to consolidate and degrade at $\sigma_v^{'}$ of 26 kPa. Further research is required to support this argument. The pH of the leachate varied consistently between 6.8 and 7.9. The average biogas composition consisted of 43% CO_2 and 51% CH_4. Both these parameters are indicative of optimum microbial activity inside the LCC.

The $E_o^{'}$ from two previous short-term tests is plotted along with the $E_o^{'}$ values from the first two steps of the current long-term test in Figure 5. The values for the long–term test shown in Figure 5 are the value of $E_o^{'}$ measured on day 55 for $\sigma_v^{'} = 26$ kPa and the current value for $\sigma_v^{'} = 42$ kPa. The $E_o^{'}$ values for the current test are higher than the estimated values from previous short-term tests for the same level of vertical stress. Also, for the current long-term test, the $E_o^{'}$ value at $\sigma_v^{'} = 26$ kPa is more than for $\sigma_v^{'} = 42$ kPa. From this observation it follows that at low vertical stress, $E_o^{'}$ is a function of initial compacted unit weight (high unit weight and therefore less displacement). It is worth pointing here that for the current long-term test, the waste was compacted at a higher density (10.5 kN/m^3) than the previous short-term tests (8.7 kN/m^3). At this preliminary stage, nothing can be conclusively said about the trend in $E_o^{'}$ due to degradation; however, the short term test shows a trend of increasing $E_o^{'}$ with increase in $\sigma_v^{'}$.

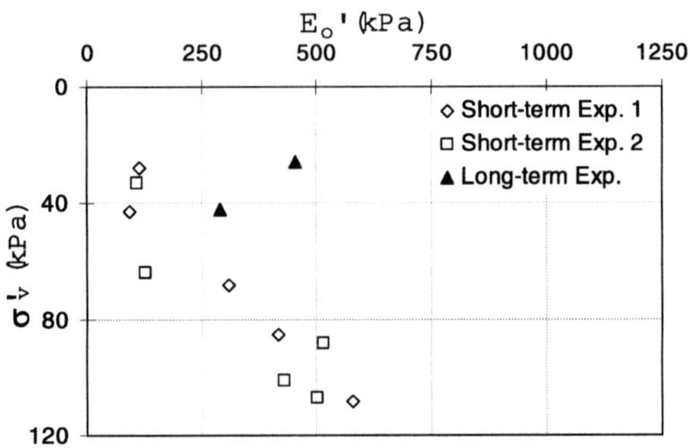

FIG. 5. Constrained Modulus obtained from short- term and long-term tests.

CONCLUSION

A considerable database is developing regarding the mechanical properties of MSW. However, little is known about the evolution of these mechanical properties as the waste degrades. Some preliminary results are presented in this paper for estimation of mechanical properties of MSW during degradation. The unit weight of waste at constant low vertical effective stress does not show significant change with time although it reflects the creep behavior of MSW. The K_o appears to decay with degradation. Though the possible decay in K_o can not be confirmed at this preliminary stage, it might be possible that the degradation of waste progressively weakens its solid skeleton which might results in relaxation in K_o values. The trend in E_o' due to degradation is not clear. However, E_o' appears to be a function of initial compacted unit weight particularly at low vertical stress.

REFERENCES

Beaven, R.P., and Powrie, W. (1995). "Hydrogeological and geotechnical properties of refuse using a large scale compression cell." *Proc., 5th International Waste management and Landfill Symposium,* Cagliari, Italy, 745-760.

Edil, T.B., and Dhowian, A.W. (1981). "At rest lateral pressure of peat soils." *J. Geotechnical Engineering*, ASCE, 107, GT2, 721-738.

Eid, H.T., Stark, T.D., Evans, W.D., and Sherry, P.E. (2000). "Municipal solid waste slope failure – I: Waste and foundation soil properties." *J. Geotech. Geoenviron. Eng.*, 126(5), 397-407.

Hendron, D.M., Fernandez, G., Prommer, P.J., Giroud, J.P., and Orozco, L.F. (1999). "Investigation of the cause of the 27 September 1997 slope failure at the Dona Juana landfill." 7th *Intl Waste Mgmt & Landfill Symposium,* Cagliari, Italy, 545-567.

Kavazanjian, E., Jr., Matasovic, N., Stokoe, K., and Bray, J.D. (1996). "In-situ shear wave velocity of solid waste from surface wave measurements."*Proc., 2nd Intl. Congress on Environmental Geotechnics*, Osaka, Japan, Balkema, vol. 1, 97-104.

Kavazanjian, Jr. E., and Merry, S.C. (2005). "The 10th July 2000 Payatas landfill failure." 10th *Intl Waste Mgmt & Landfill Symposium,* Cagliari, Italy (on CD ROM).

Koelsch, F., Fricke, K., Mahler, C., and Damanhuri, E. (2005). "Stability of landfills-The Bandung dumpsite disaster." *10th Intl Waste Mgmt & Landfill Symposium,* Cagliari, Italy (on CD ROM).

Landva, A.O., Valsangkar, A.J., and Pelkey, S.G. (2000). "Lateral earth pressure at rest and compressibility of municipal solid waste." *Can. Geotech. J*, 37, 1157-1165.

Matasovic, N., and Kavazanjian, E. Jr. (1998). " Cyclic characterization of OII landfill solid waste." *J. Geotech. and Geoenviron. Eng.*, 124(3), 197-210.

McDougall, J. R., Pyrah, I. C., Yuen, S. T. S., Monteiro, V. E. D., Melo, M. C., and Juca, J. F. T. (2004). "Decomposition and settlement in landfilled waste and other soil-like materials." *Geotechnique, 54(9), 605-609.*

Singh, M.K., Fleming, I.R. (2004). "Evaluation of existing and potential methane generation at the city of Saskatoon landfill". Draft report submitted to Saskatchewan Power Corporation, Regina, SK.

Zekkos, D.P., Bray, J.D., Kavazanjian, E. Jr., Matasovic, N., Rathje, E.M., Riemer, M.F., and Stokoe, K.H. (2006). "Unit weight of municipal solid waste", *J. Geotech. Geoenviron. Eng.*, 132(10), 1250-1261.

Experimental analysis of waste compressibility

Francesco Castelli[1] and Michele Maugeri[2], M. ASCE

[1]Researcher, Department of Civil and Environmental Engineering, University of Catania, Viale Andrea Doria no.6, 95125, Catania, Italy; fcastelli@dica.unict.it
[2]Professor, Department of Civil and Environmental Engineering, University of Catania, Viale Andrea Doria no.6, 95125, Catania, Italy; mmaugeri@dica.unict.it

ABSTRACT: Mechanical behavior of the waste body controls many aspects of landfill lining system design and performance, including stability issues and integrity of the geosynthetic components. The compression behavior of municipal waste is influenced by the properties of the individual constituents and on their composition, and particularly characterized by large amounts of settlements which are time-dependant. Mechanical properties of waste are often described by application of soil mechanics principles so that conventional shear or compression parameters can be determined. But data from literature are scarce and show large variations. In this paper test results on the determination of waste compressibility are reported and analyzed.

INTRODUCTION

In the geotechnical design of landfills attention must be given to the mechanical properties of the deposited waste materials as these define landfill stability and deformations which are usually assessed applying soil mechanics principles.

Often researchers and engineers have relied on their knowledge of the behavior of soils, even if the current understanding of waste behavior is far from being complete.

Although this has been helpful to some extent, there is an increasing realization that behavior of waste should be considered in the context of a separate discipline of soil and/or waste mechanics. Similarities and differences in experimental behavior can then lead to the development of laboratory and field tests specifically for obtaining engineering properties of waste.

The study of waste compressibility gives rise to important conceptual and application problems actually affecting the design and landscaping of a controlled landfill (Dixon et al. 2004; 2005).

The considerable deformations to which wastes will be subjected in the course of time greatly affect the landfill overall efficiency, the stability and effectiveness of final coating as well as the final recovery of the area.

Stability and deformability analyses require extensive knowledge on the mechanical behavior of waste materials which could be provided by laboratory or in situ testing. A comprehensive approach to the problem, based on the direct control of a settlement process is particularly useful and instructive. A better understanding of the mechanical properties of waste can be provided by conducting confined compression tests.

In the present paper, some data relating to compression tests carried out by the use of a large-scale one-dimensional compression cell are reported and discussed.

The analysis of the experimental results permit an upgrading of the knowledge on the mechanical behavior of waste materials and factors influencing deformation processes as: initial composition of waste, compaction techniques, conditions affecting the decomposition of the organic substances, etc.

GEOTECHNICAL WASTE PARAMETERS

The physical and mechanical characterization of waste is a very complex task given the testing difficulties and uncertainties adapting the theory and techniques used in soil mechanics (Grisolia et al. 1995; Jessberger and Kockel, 1993a). It is difficult to extrapolate the results obtained from one situation to another, because of the differences in initial composition, in the type of pre-treatment carried out and in the way in which the waste was dumped, etc.

Waste materials differ widely in type, composition, consistency and state of biochemical decay. Typical examples for different waste naturals are excavated soil, construction material debris, sludges or municipal waste showing large variations in type and properties of its constituents.

Generally waste types are divided into two groups: soil-like waste, defined as "*granular waste*", for which conventional soil mechanics principles are applicable, and "*other waste*".

For fine-grained, soil-like waste mechanical properties - compressibility, shrinkage and swelling behavior, shear strength- are determined using conventional geotechnical test methods. For mixed and coarse-grained soil-like waste some modifications of the testing equipment are necessary, i.e. compression tests on large scale samples. For other wastes, it is necessary to undertake in situ tests on trial areas.

The mechanical properties (strength and compressibility) are dependant on the individual composition of the waste material and on the mechanical properties of its constituents. In addition the mechanical parameter are time-dependent related to the state of decomposition. Taking account of this effect, parameters for fresh and for decomposed waste must be distinguished. In some cases to provide applicable parameters for stability or deformation analysis, it might be useful to conduct appropriate tests.

To characterize the deformation or compression behavior of waste, parameters like stiffness modulus or compression index can be used (Dixon et al. 2004). Comparable to the strength parameters, for mixed waste materials, compression parameters from literature also show large variations. The amount of settlement of waste materials is dependent on the properties of the individual constituents and on their composition. So the evaluation of compression parameters requires an extensive description and classification of the waste materials.

An appropriate method to classify "*other wastes*" may be the conduction of tests like grain-size analysis, determination of the main waste constituents, moisture content, consistency limits, organic content, etc..

Generally the tendency of increasing fine-grained material with increasing age of waste can be observed. This tendency probably can be explained by different states of decomposition.

CONFINED COMPRESSION TEST

The compression behavior of municipal waste is particularly characterized by large amounts of settlements which are time-dependant. For assessing landfill settlements, mechanical and degradation induced settlements should be distinguished (Manassero et al. 1996).

Considering an ideal compressibility test, different deformation phases can be pointed out: *A*) a first phase regulating the initial overall settlement of wastes and the big deformations of highly deformable components. These deformations take place very quickly, immediately after the application of external loads. *B*) a second longer phase connected to the release of overpressures of fluids absorbed. *C*) a last phase related to the viscous behavior of solid materials. In addition to the these processes, there is the decomposition of the organic substance implying, beside a decrease in volume and consistency of solid materials, the formation of liquid and gas.

Compression parameters of soil-like waste can be determined by using conventional geotechnical testing methods. For other wastes, appropriate modifications of the testing equipment or the testing procedures are necessary. In this last case testing methods must take account of several factors not comparable to the testing of soils: size of samples, high compressibility, simulation of in situ placement methods, simulation of in situ structure of waste. Thus laboratory testing of waste materials requires equipment that provides the possibility of testing large-sized samples in combination with large strains.

By the help of confined compression tests on waste materials load induced, mechanical settlements of a landfill can be assessed. In combination with a suitable approach for settlements caused by biochemical degradation processes, a total settlement provision may be possible (Jessberger and Kockel, 1993b).

At the Department of Civil and Environmental Engineering of the University of Catania (Italy), tests for the evaluation of the mechanical properties of wastes are conducted. The research are concentrates on compression testing and analyzing landfill settlement measurements.

Confined compression tests are conducted by the use of a large-scale one-dimensional compression cell as shown in Figure 1. The compression cell consists of a steel cylinder, 60 cm in diameter and 80 cm high, into which waste material is placed for testing. The whole compression cell is mounted on a support frame and the refuse in the cylinder is compressed by an upper platen, which can be moved vertically inside the testing cylinder. The upper platen is connected to, and moved by, a hydraulically operated pump. A drainage stratum constituted by 120 mm thick layer of gravel is placed at the base of the cell prior to refuse being loaded.

A maximum vertical stress of 600 kPa can be applied to the surface of the refuse, thus taking the bulk density of a landfill as approximately 10 kN/m^3, the compression cell is capable of simulating loads equivalent to a 60 meters depth of material.

As geotechnical parameter characterizing the settlement of municipal waste, the stiffness modulus E_o can be used. For different load intervals it is defined as follows:

$$E_0 = \frac{\Delta\sigma}{\Delta\varepsilon} \quad (1)$$

where $\Delta\varepsilon$ is the increase in vertical compression strain which results from an increase in vertical stress of $\Delta\sigma$ under conditions of zero lateral displacements. This use of the stiffness parameters is advantageous, in that it is readily determined from one-dimensional compression tests of the type reported in the paper, and directly applicable for the estimation of field vertical settlements (Beaven and Powrie, 1995).

It is important to point out that these results only describe the load-dependant settlement. Influences from waste decomposition are considered when using a stiffness modulus back-analyzed from settlement measurements. But extensive settlement measurements are scarce and therefore such data are difficult to provide.

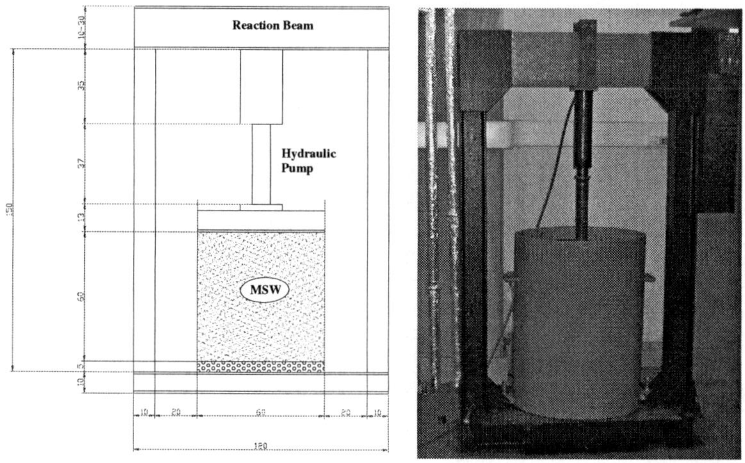

FIG. 1. Large-scale one-dimensional compression cell.

EXPERIMENTAL MEASUREMENTS

The tests were conducted on unsaturated, disturbed samples of municipal fresh waste and 5 years old pre-treated waste. In Figure 2 and 3 is reported the material classification of the fresh and old wastes tested respectively. A breakdown of the refuse by weight into 11 categories, such as plastics, paper, textiles, glass etc. was made (Figures 2 and 3). The grain-size distribution of the 5 years old pre-treated waste sample is reported in Figure 4.

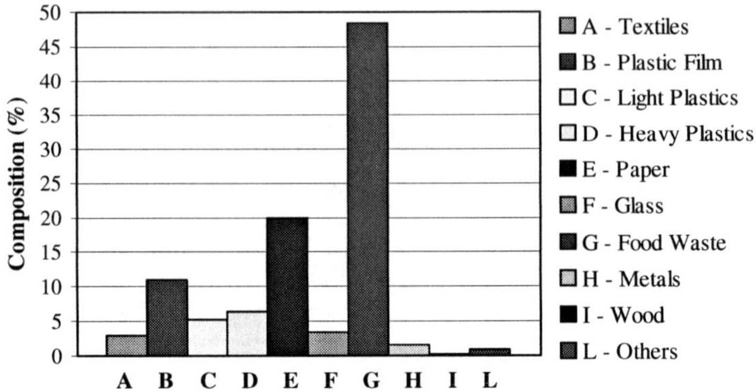

FIG. 2. Material classification of fresh waste tested.

At first, a consolidation pressure of about 20 kN/m^2 was applied. After that, different constant loads (40, 80,150, 300, 600 kPa) were applied for 24 hours and sample compression with time was measured. The time-vertical strain curves obtained for the fresh and 5 years old wastes tested are shown in Figures 5 and 6 respectively.

The bulk density of the refuse was determined as the ratio of the mass of the refuse at its original water content to the volume occupied in the compression cell. The initial density of the refuse varied from 2,6 kN/m^3 for the fresh waste to 7,8 kN/m^3 for the 5 years old waste. The final density at an applied stress of 600 kPa varied from 10,1 kN/m^3 for the fresh waste to 15,9 kN/m^3 for the 5 years old waste.

The maximum sample compression reached a considerable degree of more than 50 % of the original sample height.

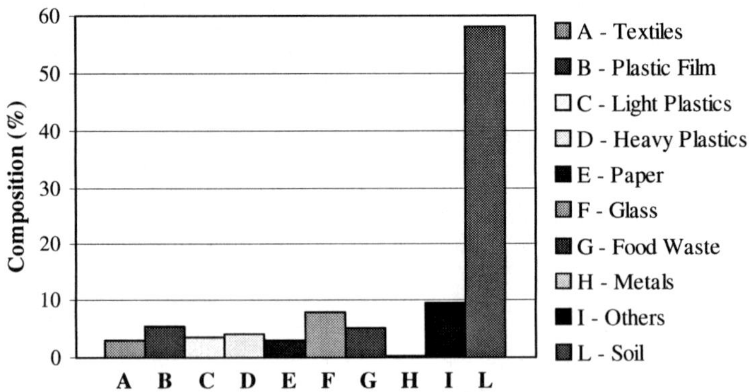

FIG. 3. Material classification of 5 years old waste tested.

During the first 10÷30 seconds after load application an initial settlement takes place, which is about 20÷30 % of the maximum settlement per load interval. After that the settlement rate per time decreases, but the amount of compression continuously increases. This may be defined as the primary compression (Jessberger and Kockel, 1993b).

Primary compression occur during waste placement. As the thickness of the waste increases, the stiffness of the waste also increase with depth. From equation (1) the stiffness modulus E_o is therefore not a constant but depends upon the level of mean stress in the layer under consideration. Table 1 and Figure 7 shows the variation in stiffness related to stress level of the 5 years old waste sample. As expected, the data indicate an increase in stiffness with applied stress and refuse density.

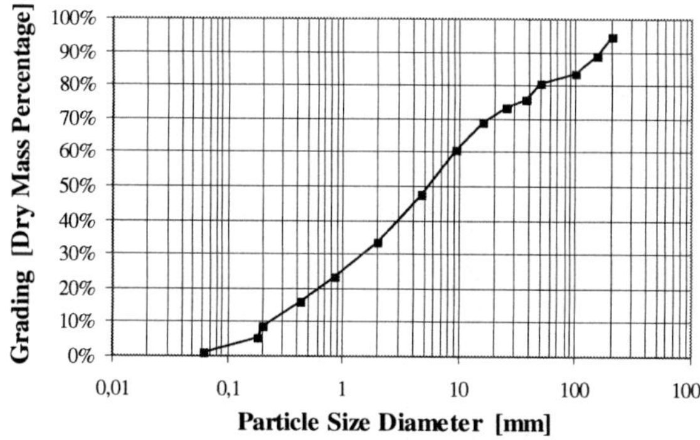

FIG. 4. Grain-size distribution of 5 years old municipal waste sample.

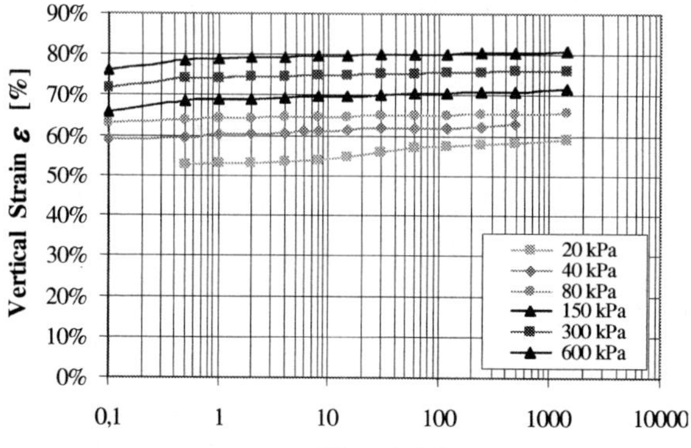

FIG. 5. Time-vertical strain curves of fresh waste tested.

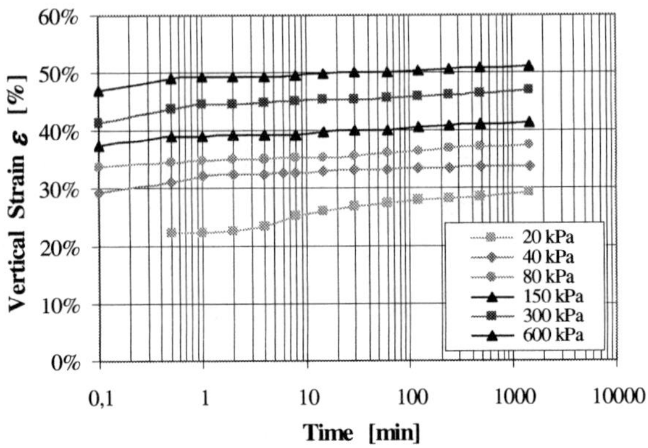

FIG. 6. Time-vertical strain curves of 5 years old waste tested.

Results obtained are in good agreement with similar results reported by Jessberger and Kockel (1993b). It should be emphasized that the present stiffness modulus derived from confined compression tests only represents the mechanical settlements. Therefore further research should aim at the determination of degradation dependent deformations caused by biochemical processes.

Table 1. Values of the stiffness modulus with applied stress.

σ (KPa)	ΔH (cm)	H_i (cm)	ε (%)	$m_v = 1/E_o$ (kPa^{-1})	E_o (kPa)	ρ (kN/m^3)
20	13,76	47,0	29,3	0,014654	68	11,1
40	15,85	33,2	33,7	0,002207	453	11,8
80	17,60	31,2	37,4	0,000931	1074	12,5
150	19,45	29,4	41,4	0,000562	1778	13,4
300	22,0	27,6	46,8	0,000362	2765	14,7
600	23,95	25,0	51,0	0,000138	7231	15,9

CONCLUSIONS

The geotechnical design of a waste requires extensive knowledge on the mechanical properties of municipal solid waste. In this paper results of confined compression tests are reported. Laboratory tests show that a significant mechanical material properties of municipal solid wastes is the high compressibility and the influence on it of consolidation or preloading.

The results of a number of compression tests have demonstrated the variability of the actual density and stiffness modulus E_o with vertical stress.

Moreover, in situ tests and extensive landfill monitoring are necessary for a better evaluation of the mechanical waste properties and waste body behavior.

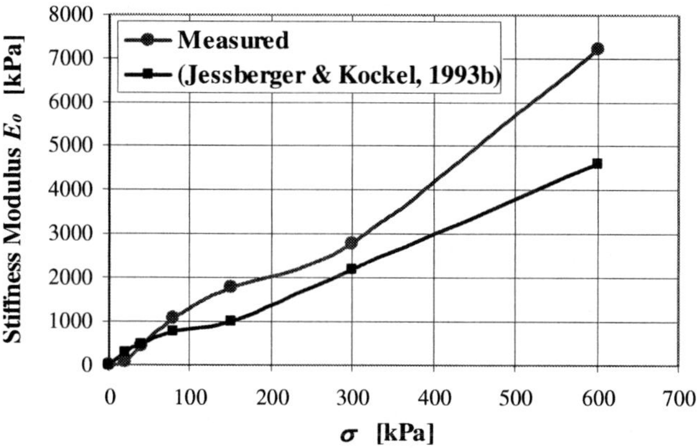

FIG. 7. Measured values of stiffness modulus of the 5 years old waste sample.

REFERENCES

Beaven, R.P. and Powrie, SW. (1995). "Hydrogeological and geotechnical properties of refuse using a large scale compression cell." *Proceedings Fifth International Landfill Symposium, Sardinia 95,* Cagliari, Italy: 746-760.

Dixon, N., Ng'ambi, S. and Jones, D.R.V. (2004). "Structural performance of a steep slope landfill lining system. *Proceedings of ICE*, Vol.157: 115-125.

Dixon, N., Russell, D. and Jones, V. (2005). "Engineering properties of municipal solid waste." *Geotextiles and Geomembranes*, Vol.23: 205-233.

Grisolia, M., Napoleoni, Q. and Tancredi, G. (1995). "Contribution to a technical classification of MSW." *Proceedings Fifth International Landfill Symposium, Sardinia 95,* Cagliari, Italy: 703-710.

Jessberger, H.L.. and Kockel, R. (1993a). "Determination and assessment of the mechanical properties of waste." *Waste Disposal by Landfill - Green '93.* R.W. Sarsby (ed). Balkema, Rotterdam NL: 313-322.

Jessberger, H.L.. and Kockel, R. (1993b). "Determination and assessment of the mechanical properties of waste materials." *Proceedings Fourth International Landfill Symposium, Sardinia 93,* Cagliari, Italy: 1383-1392.

Manassero, M., Van Impe, W.F. and Bouazza, A. (1996). "Waste disposal and containment." *Proceedings Second International Congress on Environmental Geotechnics*, Vol.3, Osaka: 193-242.

Deformation of MSW Bioreactor Landfills: Properties and Analysis Approach

MD Sahadat Hossain[1], Mohamed A. Gabr[2], M. ASCE and Mohamed A. Haque[3]

[1]Assistant Professor, Department of Civil and Environmental Engineering, University of Texas at Arlington, TX 76019, hossain@uta.edu
[2]Professor, Department of Civil and Environmental Engineering, North Carolina State University, Raleigh, NC , gabr@eos.ncsu.edu
[3]Graduate Engineer, Bryant Consultants, Inc., Carrolton, TX 75006, mhaque@geoneering.com

ABSTRACT: Bioreactor landfills are operated for rapid stabilization of waste, increased landfill gas generation for cost-effective energy recovery, increase in landfill space, enhanced leachate treatment, and reduced post closure maintenance period. Due to rapid stabilization and settlement of solid waste, bioreactor landfills are gaining popularity as an alternative to conventional Subtitle D landfills. However, the addition of leachate to accelerate waste decomposition changes the physical and engineering characteristics of Municipal Solid Waste (MSW), which affects the compressibility and shear strength behavior of MSW. Settlement during the active landfilling period is beneficial as it increases the landfill capacity, however, large differential settlement may cause serious damage to the existing leachate recirculation pipe system and interim covers. Also, due to accelerated decomposition and changes in shear strength properties, the stability of landfill slopes is expected to be affected. The objective of this paper is to analyze the compressibility of MSW in a bioreactor landfill as a function of construction sequence, time and waste placement using the finite element program PLAXIS. In this analysis, the layer properties are adjusted to account for extent of decomposition. The results from PLAXIS are compared with waste settlement data collected during the filling of a landfill cell at Calgary Biocell in Canada.

INTRODUCTION

Landfilling is the primary source of solid waste disposal in United States. About 54% of the generated solid waste goes in to the landfills (US EPA, 2005). In a conventional landfill that is designed and operated in accordance with Subtitle D of the Resource Conservation and Recovery Act (RCRA), efforts are typically made to minimize moisture infiltration. Due to the absence of moisture, refuse decomposition proceeds at suboptimal rates for decades or even centuries. Recently, a type of landfilling technology known as Bioreactors, in which generated leachate is re-circulated, has emerged. Bioreactor landfills are operated for rapid stabilization of waste, increased landfill gas generation for cost-effective energy recovery, gain in landfill space, enhanced leachate treatment, and reduced post closure maintenance period. However, the design and operation of landfills as bioreactors raise some concerns. First, the physical properties of MSW are expected to change with the increasing moisture content. These changes must be considered during the design phase to ensure slope and cover stability. Second, the enhanced rate of waste decomposition can result in an enhanced rate of waste settlement as refuse solids are converted to landfill gas (CH_4 and CO_2). Third, the compressibility parameters of the waste will vary with time, as decomposition takes place (Hossain et al 2003). Having these changes occur over a relatively short time can affect the design philosophy of landfills as well as operating practices as waste exhibits a highly compressible skeleton depending on its degradation phase (Hossain, 2002).

Landva & Clark (1990), Gabr & Valero (1995), Coumoulous & Koryalos (1999), Sharma et al. (1999), El Fadel and Al-Rashed (1998), Wall and Zeiss (1995), and Hossain et al. (2003) conducted laboratory and field studies to determine compressibility parameters and estimate settlement of landfills' waste. In general, their results showed high compressibility parameters for the waste. Hossain et al (2003) reported Compression Ratio (CR) in the range of 0.2 to 0.5 depending on the stress level, and the rate of creep index, $C\alpha e$ was found to be in the range of 0.2% to 3%.The secondary compression (creep and biological) index appeared to be less dependent on void ratio and more dependent on microbial activity conditions. The range of the secondary compression coefficient ranged from 0.02 to 0.32.

Crow-wing landfill is one of few landfills operated as a bioreactor. The secondary coefficient in this case is relatively higher than values computed from traditional landfills. Hossain et al (2003) reported data that correlated compressibility coefficients with the chemical composition of the waste in an attempt to explain the high values.

It is of interest to note that the time difference between waste placements at bottom versus the top of a landfill may be between 15 to 35 years depending on the active filling period. Therefore, the level of decomposition of MSW within such time period is expected to be different (Gabr et al., 2000) depending on waste location. Accordingly, considering properties to be uniform throughout a bioreactor landfill is not a reasonable assumption, and it may lead to erroneous computation of settlement.

The objective of this paper is to analyze the compressibility of MSW in a landfill as a function of construction sequence, time and waste decomposition using the finite element program PLAXIS. The results from PLAXIS are compared with waste settlement data collected during the landfilling at Calgary Biocell in Canada.

FIELD INSTRUMENTATION – CALGARY BIOCELL

The Calgary biocell is a field scale bioreactor project instrumented in Calgary, Canada. The biocell covers an area of approximately 100 square meters, with a waste foot print of 85 square meters and a maximum height of 18 meter. The conceptual design of the biocell is presented in Figure 1. The construction of this biocell started on August 4, 2004. However waste was not accepted until April 2005. By the end of July 2005 first lift was completed, the leachate collection systems and intermediate covers were placed. The second lift accepted waste between

July and August 2005, and the final lift of 8 meter was completed between January and May 2006.

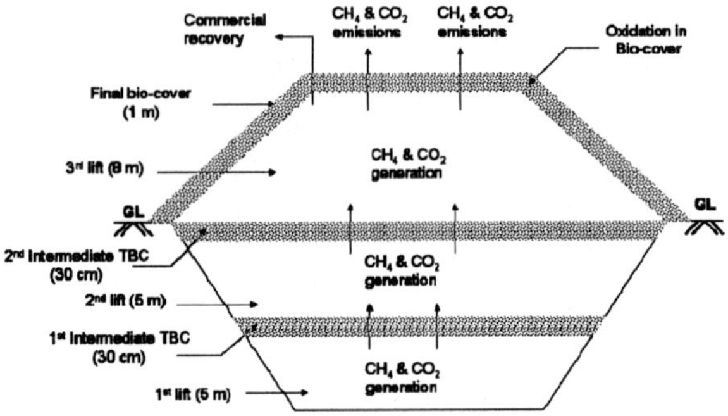

Figure 1. Schematic Diagram of Calgary Biocell (Hunte et al., 2006)

The waste was filled in three lift of approximately 5 to 8 meter as presented in Figure 1. A total of 20,752 tons of waste was placed in second lift between August and December 2005. The average daily disposal rate for second lift was approximately 340 tons. A total of 6,663 tons of waste was placed in third lift between January 2006 and March 2006. The cumulative waste placed in the second and third lifts of biocell is presented in Figure 2. Based on the amount of solid waste placed on the biocell Hunte et al. (2006) back calculated the unit weight of the waste as 5.6 kN/m^3.

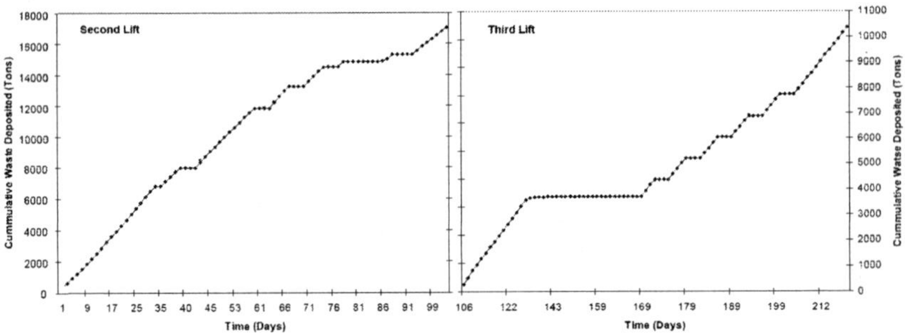

Figure 2. Waste Placement in Calgary Biocell (Hunte et al., 2006)

The soil settlement gauges were installed at each of the intermediate cover to monitor the settlement of landfill over time. With this arrangement the settlement of first intermediate cover due to the placement of second and third lift, and settlement of second intermediate cover due to the filling of third lift can be computed. However, in this paper the waste settlement data obtained from the first intermediate cover was used for comparison with the settlement results obtained from FEM analysis.

The settlement data between August 2005 and February 2006 obtained from the soil settlement sensors installed in the first intermediate cover is presented in Figure 3. Settlement data were not obtained between August 2005 and September 2005 due to the maintenance at the biocell. Increase in strain on the settlement gauge due to the filling of second and third lift can be clearly observed in Figure 3. Figure 3 presents the strains in the settlement gauge placed in the first intermediate cover increased with filling operation. The strain on the settlement gauge was approximately 8% during the filling of second lift and was 6% during the filling of third lift. Based on the settlement, Hunte et al. (2006) calculated a compression index of 0.21.

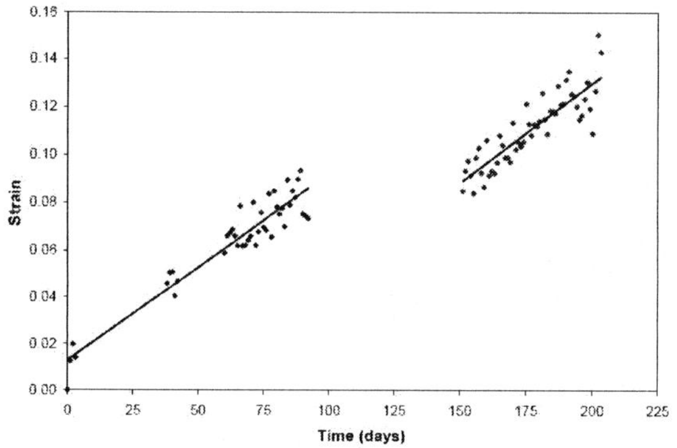

Figure 3. Strain over Time in Settlement Gauge (Hunte et al., 2006)

NUMERICAL MODELING

The numerical modeling is performed to analyze settlement of the landfill considering changing MSW properties with time due to decomposition. The settlement of Calgary bioreactor landfill is simulated (using two-dimensional program, PLAXIS) and the results are compared with field data collected during waste filling of one of the cells.

The 15-node triangle elements were used in the modeling. The unstructured mesh was generated using PLAXIS. The two vertical boundaries are simulated as "boundary free to move" in vertical direction and fixed in horizontal direction, whereas the horizontal bottom boundary is considered to be fixed in both directions. The foundation soil was considered to be stiff soil and the bottom boundary is considered fixed.

1. Settlement Modeling

The cross-section of the Calgary bioreactor landfill cell utilized in the numerical modeling of settlement is presented in Figure 4. The cell is modeled as a two dimensional plain strain domain.

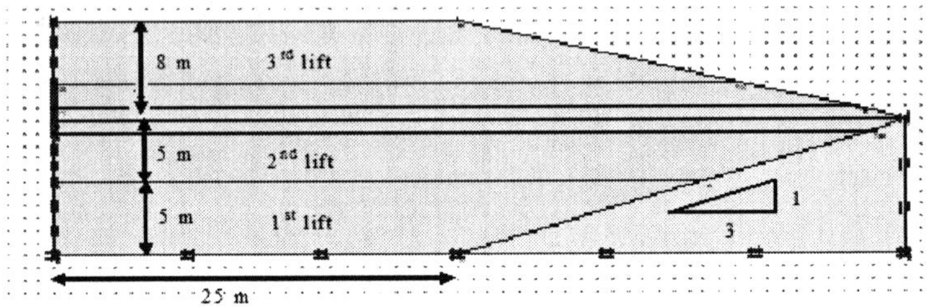

Figure 4. Cross-section of the Numerical Model for Settlement Analysis.

Material Model and Parameters

The Mohr-Coulomb model was assumed in the analyses. The parameters assumed in the analyses are presented in Table 1. The unit weight value was obtained from the field data from Calgary biocell. The permeability values at different stages of decomposition were obtained from Penmethsa (2007). The author reported a value of 8 cm/day for MSW at initial stage of decomposition. PLAXIS uses alternative stiffness moduli, such as shear modulus, G, and the oedometer modulus, E_{oed}. These stiffness moduli relate to Young's modulus according to Hooke's law of isotropic elasticity, which involves Poisson's ratio. The Young's modulus (E) value was back calculated using the compression index and unit weight data reported by Hunte et al. (2006) from Calgary landfill, the modulus was 350 kN/m^2.

Table 1. Parameters for M-C Model in FEM analysis

Material Set	Value
Drainage Condition	Drained
Unit Weight, γ_{moist} (kN/m^3)	5.6
Permeability, m/day	8×10^{-2}
Cohesion, c (kN/m^2)	3
Friction Angle, Φ°	26
Youngs Modulus, E (kN/m^2)	350

Analysis Type and Calculation Stages

The settlement in PLAXIS was computed using the consolidation calculation at each stage of filling. The governing equations of consolidation used in PLAXIS follows: (a) Biots theory; and (b) Darcy's law for fluid flow and elastic behavior of the soil skeleton.

The first lift was assumed to be there on day "0" and its construction time was not considered. The second lift of 4 m was constructed in 76 days, followed by 15 days of consolidation time. The final 1 meter of second lift was constructed in 15 days, followed by 2 days of consolidation before the start of next lift. The first one meter of the final lift constructed in 26 days and it took 37 days of consolidation time before the start of next 2 m of final lift. This layer took 46 days to get completed.

RESULTS AND DISCUSSIONS

Results from settlement analysis for a bioreactor cell in Calgary are presented in Fig. 5. The PLAXIS analysis predicted a total incremental displacement of 1.09 m. at the end of approximately 200 days. The total displacement results from the PLAXIS analyses and those from field instrumentation are plotted in Figure 6. The total displacement predicted by the PLAXIS analysis was similar to those observed in the Calgary landfill for the given properties assumed for the analysis.

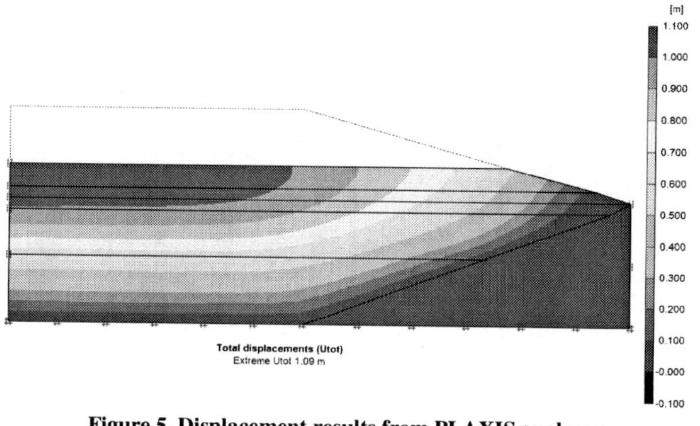

Figure 5. Displacement results from PLAXIS analyses

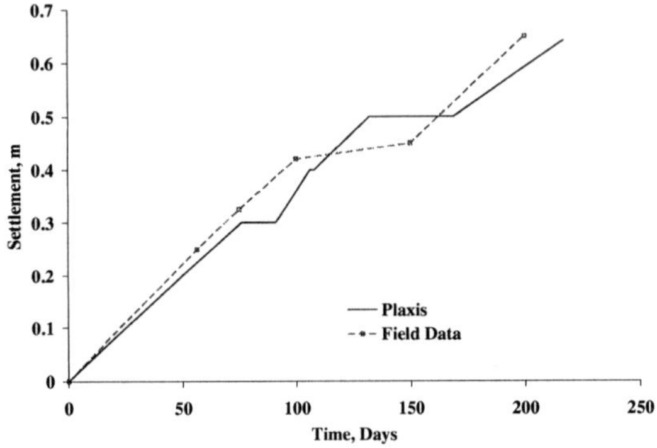

Figure 6. Comparison between displacement from PLAXIS and field instrumentation

SUMMARY AND CONCLUSIONS

The Finite Element Program PLAXIS was used to model and compute settlement and stability of MSW in a bioreactor landfill assuming various waste decomposition levels with depth. The results from deformation analysis are compared with field data. Based on the analysis conducted in this paper, the following findings are presented:

i. The total displacement (settlement) predicted by PLAXIS analysis was similar to those observed in the Calgary landfill given the magnitude of the assumed waste parameters.

ii. The results of this study suggest that the settlement of bioreactors waste should be evaluated using the compressibility parameters as function decomposition rather than using average values.

REFERENCE

Coumoulous, D.G., and Koryalos, T.P. (1999). "Prediction of long-term settlement behavior of landfill covers after closure." 7th International Waste Management and Landfill Symposium, Sardinia (Italy).

El-Fadel, M. and Al-Rashed,H. (1998). "Settlement in municipal solid waste landfills I. Field scale experiments." Journal of Solid Waste Technology and Management, Vol. 25, No.2, pp. 89-98.

Gabr, M.A., and Valero, S.N. (1995). "Geotechnical properties of solid waste." *Geotechnical Testing Journal*, ASTM 18(2), 241-251.

Gabr, M. A., Hossain, M. S., and Barlaz, M. A. (2000). "Solid Waste Settlement with Leachate Recirculation," Geotechnical News, 2 (18), 50 – 55.

Hossain, M. S. (2002). Mechanics of Compressibility and Strength of Solid Waste in Bioreactor Landfills, Ph. D Thesis, NC State University, Raleigh, NC.

Hossain, M.S., Gabr, M.A., and Barlaz, M.A. (2003), "Relationship of Compressibility Parameters to Municipal Solid Waste Decomposition", ASCE Journal of Geotechnical and Geoenvironmental Engineering, Vol. 129, No. 12, December 1, 2003, pg. 1151 –1158.

Hunte, C., Hettiaratchi, P., Meegoda, J. N., and Hettiarachchi, C. H. (2006). "Settlement of Bioreactor Landfills During Filling Operation," *Geotechnical Special Publication No. 163, Geoenvironmetal Enginering.*

Landva, A.O. and Clark, J.I. (1990). Geotechnics of Waste Fill – Theory and Practice, ASTM STP 1070, pp. 86-103.

Penmethsa, K. K. (2007). "Permeability of Solid Waste in Bioreactor Landfill with Degradation" Master Thesis, the University of Texas at Arlington.

Sharma, H. D., Dukes, M. T., and Olsen, D. M. (1990). "Field Measurements of Dynamic Moduli and Poisson's Ratio of Refuse and Underlying Soils at a Landfill Site," *Geotechnics of Waste Fills – Theory and Practice,* STP 1070, ASTM, Philadelphia, PA.

U.S. Environmental Protection Agency (EPA) (2005). ''Basic facts: Municipal solid waste.'' http://www.epa.gov/epaoswer/non-hw/muncpl/facts.htm

Wall, D.K., and Zeiss, C. (1995). "Municipal landfill biodegradation and settlement." Journal of Environmental Engineering., ASCE, 121 (3), 214-224.

The Impact of Degradation on MSW Shear Strength

Edward Kavazanjian, Jr.[1], M. ASCE

[1]Associate Professor, Department of Civil and Environmental Engineering, Arizona State University, Tempe, AZ 85287-5306 edkavy@asu.edu

ABSTRACT: Multiple lines of evidence are used to draw conclusions concerning the shear strength of degraded municipal solid waste (MSW). Interpretation of the accumulated evidence leads to the conclusion that for conventional "dry" landfills the shear strength envelopes commonly used in practice provide reasonable strength values for circular and non-horizontal failure surfaces even for old landfills where the waste may be in a relatively advanced state of degradation. However, for horizontal failure surfaces in conventional landfills that may not engage the fibrous waste components, for landfills where mechanical pre-processing has been used to reduce waste constituent size and facilitate degradation, and for bioreactor landfills, a strength envelope characterized by zero cohesion and an effective friction angle of 30 degrees appears to provide a conservative representation of degraded MSW shear strength.

INTRODUCTION

The impact of degradation on the shear strength of municipal solid waste (MSW) is an important consideration in the landfill design. Concerns that degradation reduces the shear strength of MSW must be addressed when considering the long term stability of a conventional landfill. Furthermore, degraded MSW shear strength may govern both short and long term stability of bioreactor landfills. Recent studies suggesting that the degraded shear strength of MSW may be lower than the shear strengths typically employed in landfill design practice give cause for a critical revaluation of available data on the shear strength of degraded MSW.

MSW SHEAR STRENGTH ENVELOPES USED IN PRACTICE

Two of the most common MSW shear strength envelopes used in practice are the envelopes recommended by Kavazanjian et al. (1995) and Eid et al. (2000). Kavazanjian et al. (1995) recommended a MSW shear strength envelope represented by a cohesion of 24 kPa at low normal stresses and by a friction angle of 33 degrees at high normal stresses. Eid at al. (2000) provided lower bound, median, and upper bound recommendations for MSW shear strength. The lower bound envelope was represented by a friction angle of 35 degrees, the median envelope was represented by a cohesion of 25 kPa and a friction angle of 35 degrees, and the upper bound envelope was represented by a cohesion of 50 kPa and a friction angle of 35 degrees. The Kavazanjian et al. (1995) bi-linear envelope, based in part on back analysis of stable slopes at existing landfills that are now over 30 year old, may be inferred to include at least some degraded waste and explicitly characterized as a lower bound envelope for all ages of waste. However, Kavazanjian et al. (2001b) suggest that it may not be prudent to include the low normal stress cohesion component of this bi-linear envelope when considering the strength of degraded MSW. As the Eid et al. (2000) back analysis of landfill slopes may also be inferred to include degraded waste, the Eid et al. (2000) lower bound envelope may similarly be considered an indicator of degraded waste shear strength. Thus based on the Kavazanjian et al. (1995) and the Eid et al. (2000) strength envelopes, the shear strength of degraded MSW may be interpreted to be characterized by a friction angle of 33 degrees.

OII LANDFILL SHEAR STRENGTH

The Operating Industries, Inc. (OII) landfill is a predominantly MSW landfill located approximately 20 km east of downtown Los Angeles. Waste was placed in the OII landfill between the late 1940's and 1983. The north slope of the south parcel (NSSP) at OII rises 60 m above grade at a typical inclination of 1.5 horizontal to 1 vertical (1.5H:1V) approximately and were as steep as 1.3H:1V prior to remedial construction in the late 1990s. Waste liquids, including organic solvents, were co-disposed of with MSW at the west end of the OII site from the mid 1970's until 1983. As a result of this co-disposal, much of the waste at the west end of the site is extremely degraded, with samples recovered from depths of 5 m to 30 m described as being composed of greater than 80 percent by weight black, highly organic soil and soil-like material (GeoSyntec, 1996). Based upon the angle of repose alone, the 1.5H:1V slopes on the NSSP suggest a secant friction angle of at least 33 degrees.

The results of direct shear tests on 454 mm-diameter reconstituted specimens reported by Kavazanjian, et al. (1999), conducted as part of the remedial action program at the site, are presented in Figure 1. The numbers next to the data points in Figure 1 represent the percentage of un-degraded waste ("refuse") in the corresponding specimen. The results in Figure 1 suggest that MSW shear strength decreases as the amount of un-degraded waste decreases (i.e. as waste degradation increases). However, the lower bound strength envelope developed using specimens with less than 16 percent and as little as 8 percent refuse, presumed to represent fully degraded MSW, corresponds to a cohesion of 43 kPa and a friction angle of 26 degrees. The secant friction angle (the friction angle calculated assuming zero cohesion) for the point with the highest normal stress on the lower bound envelope (approximately 950 kPa), corresponding to the specimen with 8% refuse and 92% soil and soil-like material, is equal to 28 degrees.

After including the results of direct shear tests and other available data, a shear strength represented by zero cohesion and a friction angle of 30 degrees was adopted for "planes of weakness" in the OII landfill, presumed to represent either continuous layers of daily and intermediate cover soil or soil-like material (e.g. degraded waste) or horizontal shear planes in the waste along which the strength was controlled by the soil and soil-like material because fibrous waste components tend to be aligned horizontally and thus are not engaged during shear (GeoSyntec, 1996). This "lower bound" shear strength was critically reviewed and approved by two "blue-ribbon" panels composed of eminent geotechnical engineers: an internal review panel assembled by the responsible parties for the OII remedial action program and an external review panel assembled by the U.S. Environmental Protection Agency.

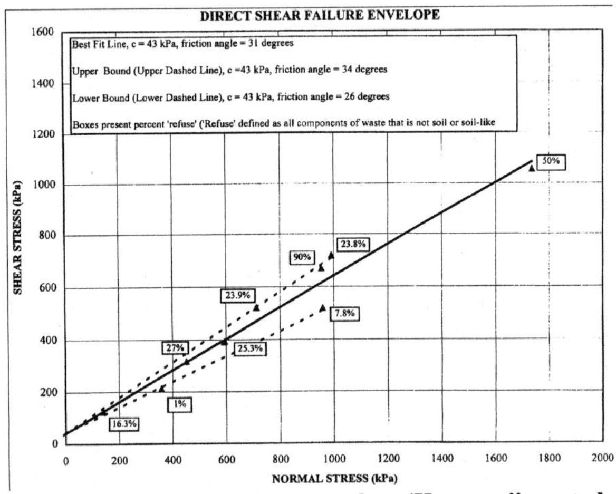

Figure 1. OII landfill direct shear test data (Kavazanjian, et al., 1999)

DOÑA JUANA AND PAYATAS LANDFILL FAILURES

Slope failures at the Doña Juana Landfill in Bogota, Columbia, in September 1997 (Hendron, et al., 1999) and at the Payatas Landfill in Quezon City (Metropolitan Manila)., Philippines, in July 2000 (Kavazanjian and Merry, 2005) provide additional insight on the shear strength of degraded waste. Both landfills were subject to introduction of large amounts of liquid into the waste mass, suggesting that waste degradation was enhanced at these facilities. Visual description of the waste at both landfills supports this suggestion. Back analysis of the shear strength of the waste at both of these facilities is complicated by the large amount of liquid in the waste mass which necessitates estimation of pore pressures within the waste and the assumption that MSW shear strength is governed by effective stresses. Estimation of excess pore pressures is further complicated by the potential for increase n the pore pressure above hydrostatic values due to gas-pore water interaction (Merry et al., 2006).

The Doña Juana Landfill Failure

Hendron et al. (1999) describe back analyses of the failure of the Doña Juana Landfill. While the landfill was not a bioreactor, liquid (leachate) were being reinjected into the landfill under pressure. Data collected prior to the failure showed high pore pressures in the waste from this leachate recirculation. Back analysis of the failure using these pore pressures indicated that the shear strength of the waste could be characterized by a cohesion of 19 kPa and a friction angle of 28 degrees.

The maximum overburden in the failure zone at Doña Juana was approximately 50 m, corresponding to a maximum overburden pressure of less than 1000 kPa and an estimated effective overburden pressure on the order of 600 kPa. For this maximum effective overburden pressure, the secant friction angle of the waste based upon strength envelope cited by Hendron et al. (1999) is 29 degrees. Furthermore, this analysis did not appear to consider the potential for increased pore water pressure due to landfill gas interaction (Merry et al., 2006). While Hendron et al. (1999) discuss the potential for landfill gas-pore water interaction to increase pore water pressure, they do not appear to have considered this in their analysis. If increased pore pressures due to landfill gas-pore water interaction in included in the stability assessment, the secant friction angle for the Doña Juana Landfill waste would be greater than 29 degrees.

The Payatas Landfill Failure

The Payatas Landfill failed following 10 consecutive days of heavy rain induced by two typhoons. As described by Merry et al., 2006, a large retention basin on the top of the landfill, adjacent to the failed section, virtually assured infiltration of large amounts of water into the waste fill. Waste scavenging removed most building materials, metals, and other recyclables from the waste, resulting in a waste mass high in organic material and smaller-sized constituents. Back analysis of the Payatas Landfill is described in Kavazanjian and Merry (2005). These analyses indicate that, when landfill gas-pore water interaction is taken into account, the factor of safety of the waste mass reduces to 1.0 for the Kavazanjian et al. (1995) strength envelope.

LABORATORY TESTING OF WASTE

WMI Bioreactor Waste Testing

Figure 2 presents the results of 454 mm-diameter direct shear tests run on reconstituted samples of waste from a Waste Management, Inc. (WMI) bioreactor in the southeastern United States reported by Kavazanjian et al. (2001b).

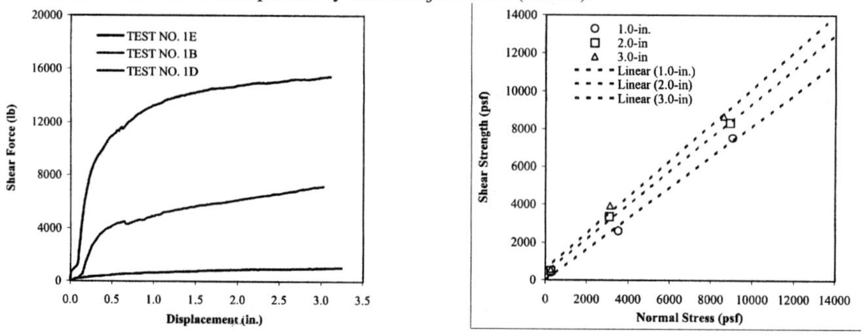

Figure 2. Direct shear results for bioreactor waste (Kavazanjian et al., 2001b)

Visual description of the waste constituents as "goo" and "slime" indicate the waste may be considered fully degraded. Reconstituted specimens of unprocessed waste from the site were consolidated to normal pressures of up to 430 kPa and sheared at a very slow rate to create a state of drained shear. Because the stress-displacement curves from these tests did not show a pronounced peak, shear strength envelopes were defined on the basis of a lateral displacement. These tests yielded a friction angle of 39 degrees for the waste at a shear deformation of 25 mm in the 454 mm diameter shear box, i.e. at a displacement of less than 6 percent of the diameter of the shear box.

Separated Waste Testing by Zekkos et al.

Zekkos et al. (2007a, 2007b) report on 300-mm direct shear (DS) and 300-mm diameter triaxial compression (TC) testing of MSW from a San Francisco area landfill. Particles less than 20 mm in dimension (<20 mm) were separated from the larger (>20 mm) particles and testing was conducted using specimens reconstituted with different proportions of <20 mm and >20 mm material. DS testing looked at the impact of the orientation of the long axis of >20 mm particles relative to the shear plane on shear strength (Zekkos et. al, 2007a). The DS tests, conducted on specimens compacted such that the preferred orientation of the >20 mm particles could be varied, indicated that conventional DS tests in which the preferred >20 mm particle orientation and the shear plane are both horizontal gave a lower bound strength for MSW, as illustrated in Figure 3. Analysis of the TC test data (Zekkos et. al, 2007b) showed that use of an axial strain-based failure criteria for isotropically consolidated triaxial tests led to unrealistically low shear strength values but that correction for at rest (K_0) field consolidation using K_0 equal to 0.3 led to shear strength values consistent with with the Kavazanjian et al. (1995) and Eid et al. (2000) strength envelopes. Figure 4 presents the secant friction angle from the TC tests calculated using a failure criterion of an axial strain of 5% past the K_0 condition.

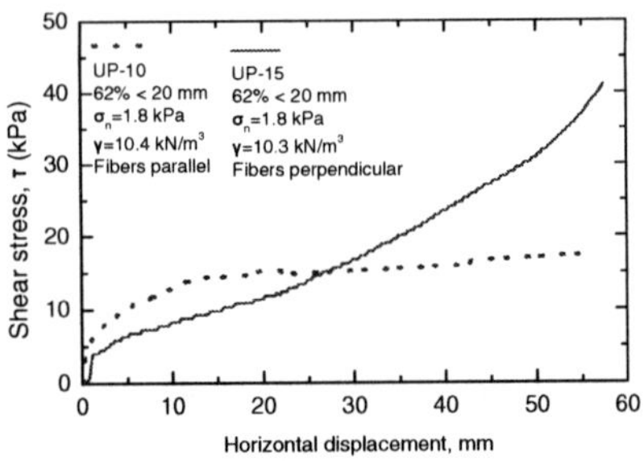

Figure 3. Effect of Particle Orientation Relative to the Shear Plane on Direct Shear Test Results (Zekkos et al., 2007a)

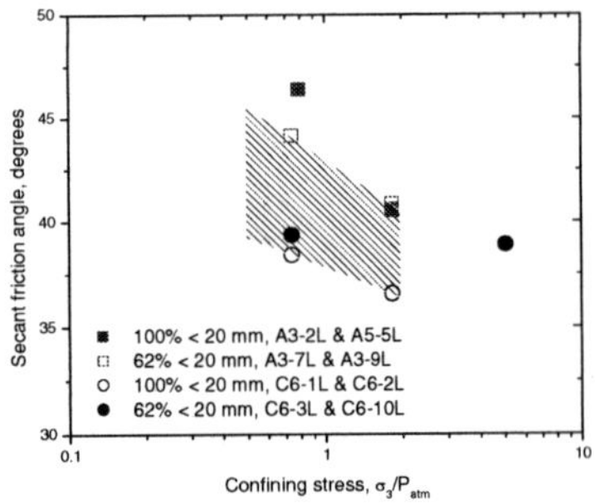

Figure 4. Influence of particle size and confining pressure on secant friction angle for K=0.3+5% (Zekkos et al., 2007b)

The data in Figure 4 indicate decrease with increasing confining pressure but, interestingly no systematic influence on the friction angle of the percentage of particles <20 mm on waste strength. The secant friction angle for MSW was in excess of 36 degrees for a confining pressure of approximately 200 kPa for all particle size distributions tested.

Harris et al. Degraded and Processed Waste Testing

Harris et al. (2006) report on simple shear and triaxial testing of reconstituted samples of MSW recovered from landfills in Utica, New York, and Louisville, Kentucky. The waste from the Utica landfill was recovered from a section of the landfill where the waste was below the water table and the waste from the Louisville landfill was recovered from a section of the landfill operated as a bioreactor. Hence the waste recovered from both landfills may be considered to be well-degraded. Harris et al. shredded and processed the waste prior to sample reconstitution, eliminating any fibrous material reinforcement. Based primarily upon direct simple shear (DSS) and isotropically consolidated TC tests (testing also included several multi-stage DS tests), these authors concluded that the shear strength of the degraded waste they tested could be represented by an effective cohesion of between 9 and 14 kPa and an effective friction angle of between 23 and 29 degrees. However, their interpretation of the DSS and TC tests is suspect, as in the following paragraphs.

DSS test results were interpreted by Harris et al. (2006) in the same manner as a DS test: the shear and normal stresses on the horizontal plane (i.e. the applied stresses) were assumed to represent the shear stresses on the failure plane. This is an exceedingly conservative method of interpretation as it assumes that the horizontal normal stress during the DSS test has increased from an initial K_0 state to a value significantly greater than the vertical (applied) normal stress. Numerical analysis by Prevost and Hoeg (1978) suggest that in the fully plastic state the horizontal normal stress in a DSS test may approach, but should not exceed, the vertical normal stress. Under this conservative limiting condition of a horizontal normal stress equal to the applied vertical normal stress, the stress points for the horizontal plane in the direct shear test become the peak points on the Mohr's circle and the effective cohesion (c') and effective friction angle (ϕ') values cited by Harris et al. (2006) become values of a' and α', the intercept and angular inclination of the line connecting the peak points on the Mohr's circle at failure. Converting values of a' and α' to values c' and ϕ' using the relationships c' = a'/cosϕ' and sinϕ' = tanα' (Lambe and Whitman, 1969) yields the following values for the effective cohesion and effective friction angles from the DSS tests conducted by Harris et al. (2006): c' = 11.5 kPa and ϕ' = 33.6 degrees for undrained tests and c' = 10.5 kPa and ϕ' = 30.6 degrees for drained tests on waste from the Utica landfill and c' = 15.1 kPa and ϕ' = 26.4 degrees for undrained tests and c' = 10.6 kPa and ϕ' = 25.1 degrees for drained tests on waste from the Louisville landfill. At the maximum normal stress of approximately 700 kPa employed in these tests, these values convert to secant friction angles of 34 degrees for undrained shear and 31 degrees for drained shear at the Utica landfill and 27 degrees for undrained shear and 26 degrees for drained shear at the Louisville landfill. It should be emphasized that these are still very conservative values for the shear strength as they are based on the assumption of fully plastic flow within the specimen and the waste was processed before testing.

Harris et al. (2006) interpret shear strength from consolidated undrained TC tests using the mobilized shear strength at an axial strain of 9% in one isotropically consolidated test and one test consolidated such that the ratio of the horizontal effective stress to vertical effective stress at the end of consolidation (K) was 0.65. This interpretation leads to an effective stress strength represented by an effective cohesion of 22 kPa and an effective friction angle of 20 degrees. If only the result of the test with K=0.65 is used, a secant friction angle on the order of 42 degrees is calculated from the Harris et al. (2006) data for a mean effective normal stress of approximately 60 kPa. Furthermore, results from a test with K=0.29, not employed by Harris et al. (2006) in developing their TC shear strength envelope, yield a secant friction angle of between 46 and 56 degrees for a mean effective normal stress of between 100 and 140 kPa. As discussed previously, Zekkos et al. (2007b) recommend interpreting shear strength from triaxial test data based upon K=0.3.

CONCLUSIONS

The multiple lines of evidence discussed above can be employed to draw conclusions on the shear strength of degraded MSW. Strength envelopes from Kavazanjian et al. (1995) and Eid et al. (2000), developed in part from back analyses of stable and failed landfill slopes, suggest that a friction angle of 33 degrees is adequate to characterize the shear strength of degraded MSW in conventional landfills. Back analysis of the failures of the Doña Juana and Payatas Landfill suggest that a friction angle of at least 30 degrees may adequately describe the shear strength of degraded MSW up to effective normal stresses of 600 kPa. Drained direct shear tests on unprocessed waste from a bioreactor yield a friction angle of 39 degrees for normal stresses of up to 430 kPa. Triaxial tests conducted by Zekkos et al. (2007b) on waste with all >20 mm material removed suggest secant friction angles in excess of 35 degrees for degraded waste for effective confining pressures of up to 200 kPa. A very conservative reinterpretation of direct simple shear tests on processed degraded waste from both a bioreactor landfill and a landfill where the waste was below the water table yield secant friction angles of 26 to 34 degrees for effective normal stresses of up to 700 kPa. Reinterpretation of this triaxial data yields secant friction angles of 42 degrees or greater at effective confining pressures of up to 140 kPa.

Based upon the above-cited evidence, the following recommendations are made concerning the shear strength of degraded MSW for normal stresses of up to 600 kPa. For conventional "dry" landfills, where some fibrous waste should still remain after degradation, shear strength envelopes commonly used in practice provide reasonable strength values for circular and non-horizontal failure surfaces, even for old landfills where the waste may be in a relatively advanced state of degradation. However, for horizontal failure surfaces in conventional landfills, i.e. failure surfaces that may not engage the fibrous waste components, for landfills where pre-processing has been used to reduce waste constituent size and facilitate degradation, and for bioreactor landfills, an effective stress strength envelope characterized by zero cohesion and a friction angle of 30 degrees may be an appropriate conservative representation of the shear strength until further data is available.

ACKNOWLEDGEMENT

Some of the material described in the paper was generated under National Science Foundation Grant No. CMMI - 0634435.

REFERENCES

Eid, H.T., Stark, T.D., Evans, W.D., and Sherry, P.E. 2000. Municipal Solid Waste Slope Failure: Waste and Foundation Soil Properties. *Journal of Geotechnical and Geoenvironmental Engineering,* ASCE. Vol. 126, No. 5, pp. 397-407

GeoSyntec (1996) "Waste Mass Field Investigation, Operating Industries, Inc. Landfill, Monterey Park, California," report prepared by GeoSyntec Consultants for submission to EPA Region IX, San Francisco, California

Harris, J.M., Shafer, A.L., DeGroff, W., Hater, G.R., Gabr, M., and Barlaz, M.A. (2006) "Shear Strength of Degraded Reconstituted Municipal Solid Waste," *Geotechnical Testing Journal,* ASTM, Vol. 29, No. 2.

Hendron, D.M., Fernandez, G., Prommer, P.J., Giroud, J.P., and Orozco, L.F. (1999) "Investigation of the Cause of the 27 September 1997 Slope Failure at the Dona Juana Landfill," Proc. *Sardinia '99, 7th Waste Management and Landfill Symposium,* Environmental and Sanitary Engineering Centre, Cagliari, Sardinia

Kavazanjian, E., Jr., Matasovic, N., Bonaparte, R., and Schmertmann, G.R. 1995. Evaluation of MSW Properties for Seismic Analysis. *Geoenvironment 2000.* ASCE Geotechnical Special Publication No. 46, Volume 2, pp. 1126-1141.

Kavazanjian, E., Jr. (2001a) "Mechanical Properties of Municipal Solid Waste," Proc. *Sardinia '01 - 8th International Waste Management and Landfill Symposium,* Environmental Sanitary Engineering Centre, Cagliari, Sardinia

Kavazanjian, E. Jr., Hendron, D. and Corcoran, G.T. (2001b). Strength and Stability of Bioreactor Landfills. Proc., *6th Annual Landfill Symposium,* Solid Waste Association of North America, Silver Springs, Maryland.

Kavazanjian, E., Jr. and Merry, S.M. (2005) "The July 10, 2000 Payatas Landfill Failure," Proc. *Sardinia '05 - 10th International Waste Management and Landfill Symposium,* Environmental Sanitary Engineering Centre, Cagliari, Italy.

Lambe, T.W. and Whitman, R.V. (1969) Soil Mechanics, John Wiley, New York, NY

Merry, S.M., Fritz, W.U., Budhu, M., and Jesionek, K. (2006). "Effect of Gas on Pore Pressures in Wet Landfills," *Journal of Geotechnical and Geoenvironmental Engineering,* ASCE, Vol. 32, No. 5, May.

Prevost, J-H, and Hoeg, K. (1978) "Reanalysis of simple shear soil testing," *Canadian Geotechnical Journal,* Vol. 15, pp. 419 – 429.

Zekkos, D., Bray J. D., Athanasopoulos, G.A., Riemer, M. F., Kavazanjian, E., Founta, X., Grizi, A. (2007a) "Compositional and Loading Rate Effects on the Shear Strength of Municipal Solid Waste," 4th International Conference on Earthquake Geotechnical Engineering, Thessaloniki, June 25-28, 2007,

Zekkos, D, Bray, J.D., Riemer, M., Kavazanjian, E., Jr., and Athanasopoulos, G.A. (2007b) "Response of Municipal Solid Waste From Tri-Cities Landfill in Triaxial Compression," Proc. *Sardinia '07 - 11th International Waste Management and Landfill Symposium,* Environmental Sanitary Engineering Centre, Cagliari, Italy.

Application of a Morphologic Classification of Brazilian MSW

A.V.A. Borgatto[1], R.L.S. Izzo[2] and C. F. Mahler[3]

[1]D.Sc. Student of Environmental Geotechnical of COPPE-UFRJ-Brazil and Leichtweiss Institute for Hydraulic Engineering, Department of Waste Management, Technical University of Braunschweig, Germany. Beethovenstraße 51A, 38106 - Braunschweig, Deutschland. E-mail: aborgatto@gmail.com

[2]D.Sc. Student of Environmental Geotechnical of COPPE-UFRJ-Brazil and Leichtweiss Institute for Hydraulic Engineering, Department of Waste Management, Technical University of Braunschweig, Germany. Cidade Universitária, Centro de Tecnologia, Bloco B, Sala 101, Ilha do Fundão, Caixa Postal 68506, Rio de Janeiro, RJ, Brasil, CEP. 21.941-972. E-mail: ronaldoizzo@gmail.com

[3]D.Sc. COPPE/UFRJ, Dr. Hab. FSP/USP, Professor of Geotechnical Engineering of the Civil Engineering Department and Coordinator of the Group of Waste Technology (GETRES) Federal University of Rio de Janeiro. Cidade Universitária, Centro de Tecnologia, Bloco B, Sala 101, Ilha do Fundão, Caixa Postal 68506, Rio de Janeiro, RJ, Brasil, CEP. 21.941-972.
E-mail: mahler0503@yahoo.com

ABSTRACT: The paper describes the classification of municipal solid waste (MSW) of Rio de Janeiro/RJ/Brazil according to the established criteria by the German recommendation GDA E1-7, DGGT (1994). This recommendation entails the identification and description of residues according to waste mechanics. These studies will provide to the environmental control agencies and the companies responsible for the final residues disposal a new approach about the geotechnical characteristics of the waste.

The residues are classified regarding the type, identification and its condition. From the determination of the residue type the indications for the analysis into groups of substances are obtained in a second step.

The dimensions of the waste particles were emphasized considering their size, length and volume. In this way, materials with one significant dimension (e.g. rope), two significant dimensions (e.g. foil), three significant dimensions (e.g. coconut) and small particles (e.g. grain) were identified and separated.

The results showed that in the Rio de Janeiro sample, the percentage by weight of waste with 1 significant dimension and 2 significant dimensions is (~34%), into the concentration level presented in GDA E2-35, DGGT (1994) – "Geotechnical Parameters for Sanitary Landfill" for fresh waste (>30%), inducing viability by including the fibre reinforcement effects.

INTRODUCTION

In the past few years the international social concern about the Municipal Solid Waste (MSW) issue is growing. In Metropolitan Centers, where the available space for waste disposal is becoming rare, the situation is even more problematic. As a consequence various measures have been taken to increase the lifespan of existing waste placement areas.

Current studies usually lack of knowledge about the characteristics and behavior of the MSW. Slope instability problems in MSW landfills have occurred all over the world, including Brazil, as for example the slide which occurred in 1991 at Bandeirantes sanitary landfill (São Paulo).

This paper objective is the application of a morphologic classification of Brazilian MSW based on the German recommendation GDA E 1-7, DGGT (1994). The results found using this classification provide information for an improved analysis of MSW landfill stability by including the fibre reinforcement effects.

TEST PROCEDURES

The first activity carried out was the identification of the residues where the samples had been collected. This identification included:

- Amount of residue received (t/day);
- Class of the residues;
- Origin of the residues;
- Supply type;
- Estimation of the homogeneity of the residues.

Sampling

For this research it was chosen a fresh waste before being sent to be landfilled. The samples were collected at the MSW storage and redirection shed of Municipal Company of Urban Cleaning (COMLURB) in Jacarepaguá, district of Rio de Janeiro. Therefore, this sample concerning about a fresh waste before being landfilled. Collected samples were transferred to an external area where they were deposited on a plastic blanket in the interior of a square formed by wood rulers with the 2.00 x 2.00 m dimension to homogenize and quarter.

The sampling procedure was repeated until the volume of a COMLURB's standard container of approximately 50 kg of MSW was obtained.

Physical Characterization Carried Out

At COMLURB's laboratory the following physical analyses had been made:

- Determination of water content;
- Distribution of the substances groups according to the recommendation established by the German DGGT (1994);
- Analysis of the pieces sizes that compose the groups of substances established by the German recommendation DGGT (1994);

- Morphological classification by groups of substances established by the German recommendation DGGT (1994).

Water Content

Upon collection, the buckets were weighed and transferred to the selection table for the residue separation and placement into trays. Trays were placed in the oven at a temperature of 105 ºC until a constant weight value had been reached (at least 48 to 72 hours).

Through this procedure the water content of the samples was determined. The water content was determined for each group of substances.

Segregation into Substances Groups

The segregation into groups of substances consists of the MSW samples separation as defined in German recommendation GDA E1-7, DGGT (1994). These samples are separated in such a way that each group of substances presents characteristics of similar materials with respect to the mechanical behavior and biochemistry stability. The groups of substances are:

- Big pieces: big residual substances, composed by diverse components as furniture, mattress, etc;
- Paper / Cardboard: residual substances basically composed by paper or paper similar fibres, like cardboards, paper packings, carpets, diaper, etc;
- Soft plastics: residues composed basically of soft synthetic substances or with similar characteristics, like soft plastic, packing, plastic foils, textiles, soft rubber, soft leather, etc.;
- Hard plastics: residues composed basically of hard synthetic substances, like rigid plastic packing, PET bottles, rigid plastics, rigid leather, hard rubber, etc;
- Metals: ferrous metal and non-ferrous metal;
- Minerals: residues that basically are composed by mineral substances or present similar mechanical or biological behavior (inert) like glass, ceramic, soil, etc;
- Wood;
- Organic: residues that have natural origin, organic, e.g. vegetables, grass cuttings, plants, dry leaves, etc.

Morphologic Classification of the MSW

The morphologic classification of the MSW was carried out based on the German recommendation GDA E1-7, DGGT (1994) intending to classify the residues regarding its form and size according to relevant mechanical characteristics. Each group of substance was submitted to this geometric description considering the parameters observed in Figure 1.

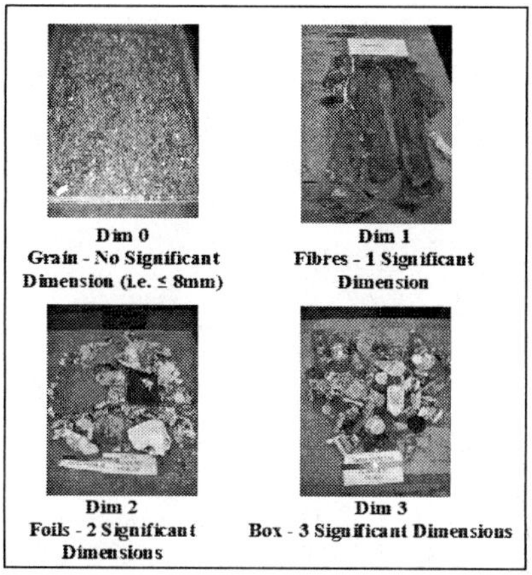

FIG. 1. Parameters considered in the morphologic classification of the MSW.

The MSW was also classified according to the pieces sizes. The fraction bigger than 120 mm was visually separated and again visually separated into 500 mm and 1000 mm sizes. The remaining fraction was sieved according to the procedures applied for soils following the Brazilian technical regulation NBR-7181. Large sieves were primary used, the 40 and 8.0 mm mesh. The fraction that passed was sieved into seven different diameter sizes (37.5, 25, 19, 9.5, 4.75, 2.36 and 2.0 mm).

RESULTS

Classification into Groups of Substances

The values of the classification into groups of substances are given in percentage of the total mass related to dry weight. The result is presented in Figure 2.

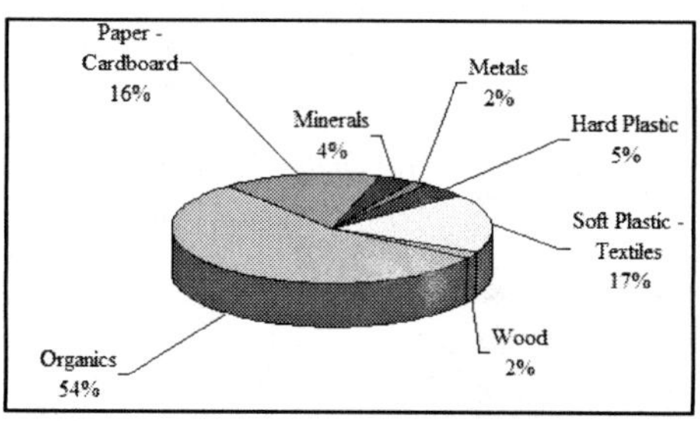

FIG. 2. Groups of MSW substances studied.

Classification according the Morphology

As described before, the classification according to the morphology is a combination of sieving and description of the geometrical dimensions. The morphologic classification of the substances groups is shown in Figure 3. The results are also given in relation to the total dry mass of the substance group.

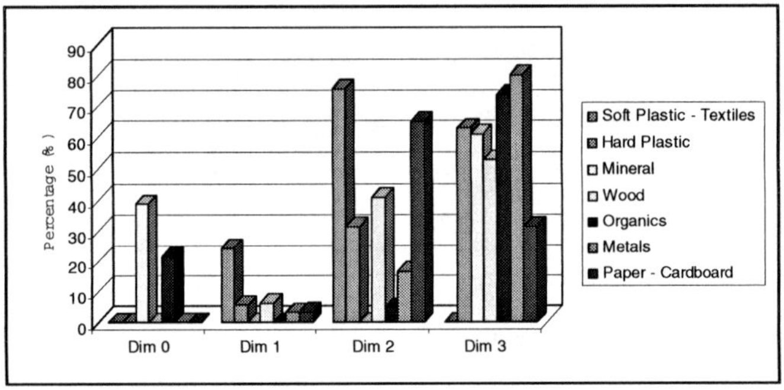

FIG. 3. Morphologic Classification of the Studied MSW Substances Groups.

Considering the waste mechanics, regarding the increments of shear resistance, the percentage of fibre materials with dimension 1 and 2 are the most interesting due to the reinforcement generated. In analogy with the reinforced soils, the amount of fibres in the mass influences the reinforcement gain.

For the soft plastics group the two-dimensional material (Dim 2) is composed mostly (~ 76 %) by plastic bags, plastic packing, textiles materials, etc. In the group of paper and cardboard, the dimension 2 is composed by paper sheets, cardboards, newspaper, smashed packing boxes type Tetra Pak, etc (~ 65 %). The dimension 3 is composed by cardboard boxes, packings, among other materials. Paper materials that compose the "dimension 3" group can, inside the landfill body, become dimension 2 materials due to collapse and compression caused by loading. In the hard plastics group, the high percentage of dimension 3 is explained by materials such as PETs bottles, different types of plastic packings, etc.

Again, as at the hard plastics group, the metal group shows a high percentage of dimension 3 materials, which is explained by the presence of food cans, vegetal oil cans, drink cans, etc. The mineral group presented materials with 0 dimension like small pieces of ceramic and glass. The percentage of dimension 3 is represented, in its majority, by glass pieces. The wood group presented materials like wood foils (dimension 2), boxes and packages (dimension 3). The organic group included in general food and organics that depending of its size and shape were placed at 0 and 3 group. The high percentage of dimension 3 is explained by the presence of voluminous materials such as coconuts, oranges, among others.

Summary results for all the groups of substances can be observed through Figure 4. The total percentage of each group of substances was calculated as the percentage of each size divided by the percentage of each group of substance present in the MSW sample.

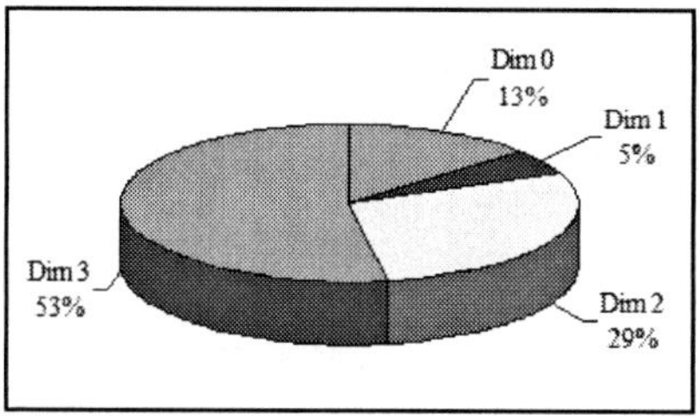

FIG. 4. Summary Results of the Morphologic Classification.

Application of the Morphology Classification in Landfill Stability Analysis

In order to give an example of the results produced by the MSW morphologic classification use, stability analyses have been carried out in a generic sanitary landfill cross section. For those analyses was used the software GGU-Stability, developed in Germany, that makes possible the inclusion of the fibre reinforcement effects in its calculations.

At the first analysis, the classic method based on the soil mechanics was used (without considering the fibre reinforcement) and at the second analysis was include the fibre reinforcement.

According KÖLSCH (1995), the fibrous materials present in the composition of MSW (textile, plastics, etc) induce tensile forces that are function of the normal stress. The shear strength parameters of the MSW would be composed of two different parts being the first regarding to the friction forces in the shear plan and the second one regarding to the tensile forces due to the fibrous material or also denominated fibres cohesion.

The fibres cohesion can be described by the internal angle of tensile forces (ζ). The size of ζ ranges between 14° for decomposed waste and 35° for fresh waste. Using the fibres cohesion in stability analysis, it has to be considered, that the size of active fibres cohesion is depending on the angle between the fibres direction and the sliding surface. Based in investigation on materials of reinforced earth it seems to be correct to calculate with completely active apparent cohesion ($\alpha = 1$) that means ζ if the angle between the sliding surface and fibre is bigger than 45°. For lower angles, it is reduced to ($\alpha = 0,5$) that means 0,5 ζ (KÖLSCH, 1997).

The mechanical properties of the waste used in that stability analysis was taken according to the established criteria by the German recommendation GDA E2-35, DGGT (1994) that means:

- Specific weight: 5 – 10 (kN/m^3) after compaction;
- Angle of friction: 30 – 35 (°) for anisotropic condition;
- Cohesion: 10 – 15 (kN/m^2) for anisotropic condition;
- Angle of internal tensile force: 20 – 35 (°) for fresh waste;

The fibres cohesion can be described by Eq.(1).:

$$\tau(\zeta) = z \cdot (1{,}5 \cdot \alpha) \cdot a_{\zeta} \qquad (1)$$

With:

- $\tau(\zeta)$ – fibres cohesion,
- z – tensile stress,
- a_{ζ} – maximum activation rate under optimal angle α (rage between 0,8 – 1,1),
- α – angle of surface failure with horizontal direction of fibres.

The analysis results are presented in Table 3.

Table 3. Comparison between Analyses Results

Typical Sanitary Landfill Cross Section						
Parameters	**Classic Methods (without fibre reinforcement)**			**Presented Method (with fibre reinforcement)**		
	$r_u = 0{,}0$	$r_u = 0{,}2$	$r_u = 0{,}5$	$r_u = 0{,}0$	$r_u = 0{,}2$	$r_u = 0{,}5$
Angle of Friction (°)	30,00	30,00	30,00	30,00	30,00	30,00
Cohesion (kN/m²)	15,00	15,00	15,00	15,00	15,00	15,00
Specific Weight (kN/m³)	10,00	10,00	10,00	10,00	10,00	10,00
Angle of Internal Tensile Forces ζ (°)	-	-	-	35,00	35,00	35,00
Correction Factor α	-	-	-	0,50	0,50	0,50
Tensile Stress (kN/m²)	-	-	-	125,00	125,00	125,00
Safety of Factor (SF)	**1,72**	**1,44**	**1,00**	**2,43**	**2,15**	**1,72**

As expected the fibre reinforcement leads to higher values for the safety factor. However, the fibre effect should be used for landfill planning very carefully since the collapse may occur due to other reasons as fire inside the landfill body, pore pressures caused by leachate recirculation, etc.

CONCLUSIONS

The morphologic classification of MSW is very valuable for waste mechanics analysis since it depends on the shape and size of the waste particles. By determining these new MSW characteristics it was possible to include in the slope stability analyses the effect of fibre reinforcement, because the materials presented dimension (dimension 1+2 – fibre material) and part size (40mm) with percentages compatible with recommendations in GDA E2-35, DGGT (1994) – Geotechnical Parameters for

Sanitary Landfill. When comparing fibre reinforced soil, the concentration and size of the fibres are also compatible, being able to apply the effect of reinforcement despite the mechanical behaviour difference between soil and MSW.

Percentages of materials with dimension 1 and 2 - the fibre materials responsible for the reinforcement effect on the MSW shear strength - had presented value (33.84%) into the concentration level for fresh waste (>30%), inducing viability by including the fibre reinforcement effects.

Practice shows that a landfill can collapse due many reasons such as the high pore pressure inside the mass, poor compaction, fire, new building operation in the landfill, etc. For this reason the fibre reinforcement should be considered for back analysis in case studies of accidents. For the planning of new landfill areas and the maximum final slope it should be used very carefully, due to the results may indicate factors of safety that do not represent the reality.

ACKNOWLEDGMENTS

The authors would like to register their acknowledgement to COMLURB for the research viabilization in its MSW unit of Jacarepaguá. To CAPES, CNPQ and FAPERJ and LWI/TU Braunschweig, Department of Waste Management for their support.

REFERENCES

Borgatto, A.V.A (2006). "Study of the Fibre Effect and Morphology in the Stability of MSW Landfills." *M.Sc. Dissertation,* Federal University of Rio De Janeiro, COPPE, Civil Engineering Programe, Rio de Janeiro, RJ, 157p.

Comlurb (2005). "Gravimetrical and Microbiological characterization of the MSW from Rio de Janeiro City." *Comlurb Publications*, 124-146.

DGGT, Deutsche Gesellschaft für Geotechnik (1994). "Empfehlungen des Arbeittskreises "Geotechnik der Deponien und Altlasten". E 1-7: Identifizierung und Beschreibung von Abfällen (Entwurf)." *Bautechnik 71*, Helf 9, Berlin. Wilhelm Ernst & Sohn.

DGGT, Deutsche Gesellschaft für Geotechnik (1994): "Empfehlungen des Arbeittskreises, Geotechnik der Deponien und Altlasten - E 2-35: Geotechnische Kennwerte für Siedlungsabfälle." *Bautechnik 71,* Helf 9, Berlin. Wilhelm Ernst & Sohn.

Kölsch, F. (1993). "The Bearing Behaviour of Domestic Waste and Related Consequences for Stability." *Proceedings Sardinia 93,* 4th International Landfill Symposium, S. M. di Pula, Cagliari, Italy, 11-15 October 1993, p.1393 - 1410.

Kölsch, F. (1995). "Material Values for some Mechanical Properties of Domestic Waste." *Proceedings Sardinia 95*, 5th International Landfill Symposium, S. M. di Pula, Cagliari, Italy, Volume II, p.711 - 722.

Kölsch, F. (1997). "Standsicherheitsberechnung fur Siedlungabfalldeponien unter Berucksichtigung anisotroper Stoffeigenschaften." *Geotechnik 20,* Heft 3, Essen: Gluckauf.

Relationship Between Classification and Shear Behavior of MSW

Neil Dixon[1] and Ulrich Langer[2]

[1] Professor, Department of Civil and Building Engineering, Loughborough University, Leicestershire, LE11 3TU, UK. Email: n.dixon@lboro.ac.uk, Tel: +44 1509 228542
[2] Project Engineer, Wasteconsult International, Robert-Koch-Str. 48b, 30853 Langenhagen, Germany. Email: langer@wasteconsult.de, Tel.: +49 511 3655718

ABSTRACT: Waste forms the largest structural element in a landfill, it stresses the lining system and often controls both stability and integrity of the landfill. A rigorous classification system is required to help explain mechanical behavior of waste bodies, to group wastes with similar mechanical properties and to facilitate the exchange and interpretation of measured properties. A classification system is outlined. A waste survey of MSW has been conducted and used to identify and select key components to produce a family of synthetic wastes based on this classification. Synthetic wastes have been tested in both a large and a small shear device. Tests in the small device used downscaled components. Results for similar wastes show comparable behavior in the two sizes of devices. They have also been compared to tests on real wastes and these have also shown comparable shear behavior, thus validating the approach of using synthetic wastes for such studies. It has been demonstrated that trends of shear behavior can be related to waste sample classification.

1 INTRODUCTION

In a landfill, waste forms the largest structural element and often controls both stability and integrity of the containment system through interaction and straining of lining elements. However, in spite of this critical role there is still a dearth of knowledge on behavior of waste as an engineering material. It is proposed that a rigorous classification system is required to help explain mechanical behavior of waste bodies, to group wastes with similar mechanical properties and to facilitate the exchange and interpretation of measured properties. Given the significant variation in waste materials, a classification system is deemed crucial to development of a unified framework for waste mechanics. A classification system has been developed and previously presented by Dixon and Langer (2006) to relate mechanical behavior and waste composition. A waste survey has been conducted on a representative sample of MSW from a UK landfill site and this has been used to identify and select key component types. Synthetic wastes have then been produced using different percentages of these components. These have been used to investigate the relationship between classification and shear behavior. Results from direct shear tests conducted at different scales are presented and compared to measurements for real MSW reported in the literature. The results are used to investigate the relationship between shear behavior and waste classification.

2 METHODOLGY

2.1 Direct Shear Devices

It is preferable to examine representative in situ MSW component size ranges in a shear device, and in addition to be able to generate large shear displacements so that the influence of reinforcing components can be investigated. This necessitates the use of a large device. The large shear device (1000x1000x800mm high) used in the laboratory tests was developed by LIRIGM, Joseph-Fourier-University, France, and has been utilized previously for research on real waste (e.g. Gotteland *et al.*, 2000). Shear displacements up to 200mm could be achieved. This device was used to test synthetic wastes formed using full size components identified in a waste sorting analysis. The large volumes of material and extended testing times required to conduct tests using the large direct shear device limits the number of tests, and hence variables, that can be investigated. Therefore, a series of tests were conducted in a relatively small standard soil direct shear apparatus (300x300mm) that could achieve 50mm shear displacements, and this required the use of scaled synthetic waste components.

2.2 Classification Framework

A waste classification has been presented in detail by Dixon and Langer (2006) and a brief summary is provided in this section. Key criteria for a classification include material groups based on typical component material properties, and component shape and size, in association with a certain mechanical function and biodegradability. A classification has to describe component characteristics and also reflect typical changes caused by biological and mechanical modification. Components in waste placed in a landfill have properties that change due to physical forces and degradation (i.e. size and shape). Biological degradation potential decreases over time mainly due to alteration of degradable organic into carbon dioxide and methane. Material groups, mechanical properties of the material groups, shape-related subdivision of waste components, size of waste components and the degradation potential within the material groups are used to define the waste classification. A key element of the classification is the categorization of components into one of the three shape-related subdivisions of: reinforcing components, compressible components (with high and low compressibility subdivisions) and incompressible components. The percentages of components in these subdivisions are used to investigate the relationship between waste composition and mechanical behaviour as discussed below.

2.3 Synthetic Waste

Use of synthetic wastes to assess mechanical behaviour of MSW has a number of advantages. By using a composition of well-defined components uncertainties like size, shape and material properties are eliminated. Minimisation of unknowns leads to a more controllable test material and hence aids interpretation and comparison of test results. It is possible to examine gradual changes of waste composition and of the factors influencing waste mechanical behaviour as discussed by Dixon *et al.* (2004). Knowledge of the synthetic waste component properties and percentages forming each sample allows reproducible results to be obtained. To engineer a family of synthetic wastes a reference waste had to be designated. An in situ sorting analysis was conducted to provide information on the composition of a representative sample of real UK MSW, including a thorough description of its components. This allowed the selection of representative synthetic waste components (i.e. considering material types, size and shape). It should be noted that in this study, the components, and hence wastes, were tested in an air dried condition. The waste analysis, selection of synthetic components and derivation of synthetic wastes are described in detail by Langer (2006). Tests in the large shear device used full size synthetic waste components and in the small shear box tests the synthetic waste components were downscaled to produce model synthetic wastes. The results for reference classifications of synthetic wastes tested in both the large and small direct shear devices are compared and discussed below to consider the influence of down-scaling.

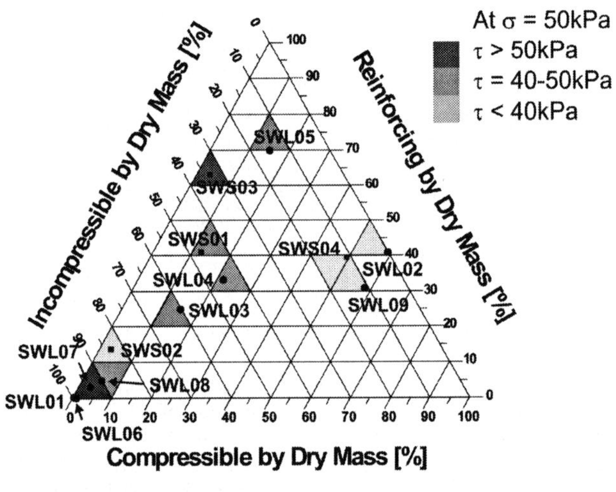

FIG. 1 Classification of synthetic waste samples used in the direct shear tests based on percentages by mass of the three shape-related subdivisions, with associated measured waste shear strengths for a reference normal stress of 50kPa.

2.4 Testing Programme

Nine synthetic wastes (SWL01 – SWL09) were tested in the large shear device. Only synthetic waste SWL09 was designed to represent 'real' UK waste (i.e. to replicate the sample from the waste sorting analysis). The remaining eight samples were selected to represent a range of classifications and hence potentially a range of mechanical behavior, although it is not intended that they represent real waste compositions. Sample SWL01 was designed to be a low compressible waste, comprising mainly sand with a small percentage of sheets of paper (i.e. reinforcing components). SWL02 had a high percentage of compressible and reinforcing components (e.g. aluminum cans and sheets of paper). Incompressible material (e.g. tire shreds and brick fractions) accounted for the highest percentage mass in SWL03, although there was also a high percentage of compressible components (e.g. cans) resulting in SWL03 having a compressible character. Samples SWL04 and SWL05 represented compositions with one component type dominating the composition (i.e. flattened cans). SWL05 was used to investigate the impact of changing the ratio between dominant components. In wastes SWL06, SWL07 and SWL08, the percentages of plastic and cans were tested in a sand matrix to assess the impact of relatively small changes in percentages by mass of components. In the small shear box four different waste compositions were tested. SWS01 was designed to be relatively compressible as it contained a high percentage of packaging and had a classification similar to SWL04. SWS02 was regarded as likely to be relatively incompressible as it contained a high percentage of incompressible components and had a classification similar to SWL08. SWS03 contained a relatively high percentage of reinforcing components such as paper and plastic sheets and had a classification similar to SWL05. SWS04 was produced to have the same classification as the 'real' waste sorting analysis sample (i.e. SWL09 and SWS04 are comparable). The classification of each of the samples is shown by their position on the ternary shape-related classification diagram in FIG. 1, based on percentage of components by dry mass.

3 RESULTS

FIG. 2 provides a comparison of Coulomb envelopes derived from the large shear tests on synthetic waste and from the small shear device on downscaled waste components. In addition, results of tests conducted by Gotteland *et al.* (2000) using the same large direct shear device on samples of real waste are also shown. The real wastes tested were non-hazardous industrial waste (NHIW) and domestic waste (DW). The envelopes are best fit straight line relationships through mobilized shear strengths determined at maximum displacements at each normal stress (i.e. 200mm in the large device tests and 50mm in the small device). The failure envelopes are defined by shear strength parameters as summarized in Table 1.

4 DISCUSSION

4.1 Influence of Downscaling

A key issue is whether the large shear tests using full size components and the small shear tests using scaled components give comparable results. As discussed in Section 2.4, samples with similar classifications were tested in the two devices. Samples SWL09 and SWS04 both had the same classification as the 'real' waste sorting sample. Table 1 and FIG. 2 show that the two devices have produced comparable shear strengths for this classification. However, comparisons for the other waste classifications are less clear. Mobilized friction angles for incompressible compositions (e.g. SWL01, SWL06, SWL07, SWL08 and SWS_02) exhibited high values in both test devices but mobilized cohesion values are variable with no clear trend. Friction angles for reinforcing (SWL05 and SWS03) and compressible compositions (SWL04 and SWS01) are significantly lower for the small device but mobilized cohesion values are larger. There is a consistent trend for the compressible and reinforced samples tested in a given device size. Significant challenges are posed by testing the compressible waste types in the small device due to the large compression of the sample under application of the normal stress. Given the nature of both the components and sample compositions tested, it is inevitable that scatter of results will occur and repeat tests are planned to investigate these issues. However, in general it can be concluded that only relatively small differences between mobilized shear strengths in the small and the large shear devices were observed.

Table 1 Mobilized shear parameters at maximum displacement (a) synthetic waste in large device, (b) scaled synthetic waste in small shear device and (c) real waste in the large device (Gotteland *et al.* 2000).

a) Synthetic/large	**SWL**	**01_1**	**01_2**	**02**	**03**	**04**	**05**	**06**	**07**	**08**	**09**
Mob. Friction Angle Φ	[°]	38.8	34.2	28.7	34.0	34.9	34.9	40.5	43.3	36.4	33.9
Mob. app. Cohesion c	[kPa]	28.5	30.8	11.5	12.1	5.8	6.2	12.3	16.0	15.0	0
b) Synthetic/small	**SWS**		**01**				**02**			**03**	**04**
Mob. Friction Angle Φ	[°]		24.4				31.3			23.7	35.8
Mob. Cohesion c	[kPa]		20.4				0			11.0	1.4
c) Real Waste/large		**NHIW_1**		**NHIW_2**		**DW_1**		**DW_2**		**DW_3**	
Mob. Friction Angle Φ	[°]	29.6		34.0		33.0		29.5		39.2	
Mob. Cohesion c	[kPa]	23.4		52.5		75.0		40.0		21.5	

4.2 Comparison of Real and Synthetic Waste Shear Behavior

An assessment is required of whether the synthetic wastes have comparable shear behavior to real wastes in order to validate the use of synthetic wastes to investigate relationships between classification and shear behavior. Shear strengths for the synthetic wastes are, apart from SWL01 and SWL07, lower than for the real waste samples. The differences in shear strength could be a function of the higher unit weights of the real waste samples leading to greater interlocking of particles and mobilization of tensile forces in reinforcing elements (see section 4.3). The synthetic and real wastes have comparable mobilized friction values but widely differing apparent cohesions. A key factor could be that the real wastes tested by Gotteland *et al.* (2000) are reported to contain a large amount of fine soil like components (e.g. up to 66% by mass of the samples) and to have a significant moisture content. In addition, the influence of water softening, and hence changing, component material properties may modify shear behavior of the real wastes and this is not investigated in the current study using air dried synthetic components. Although not presented here, the shear stress vs. shear displacement curves have a similar shape and magnitude for synthetic and real samples when tested in the large direct shear device. It is concluded that the shear behavior of the synthetic wastes is comparable to real MSW for similar test conditions, however shear strength is variable and is a function of both classification, unit weight and moisture content. A full comparison of synthetic and real wastes is given by Dixon *et al.* (2008).

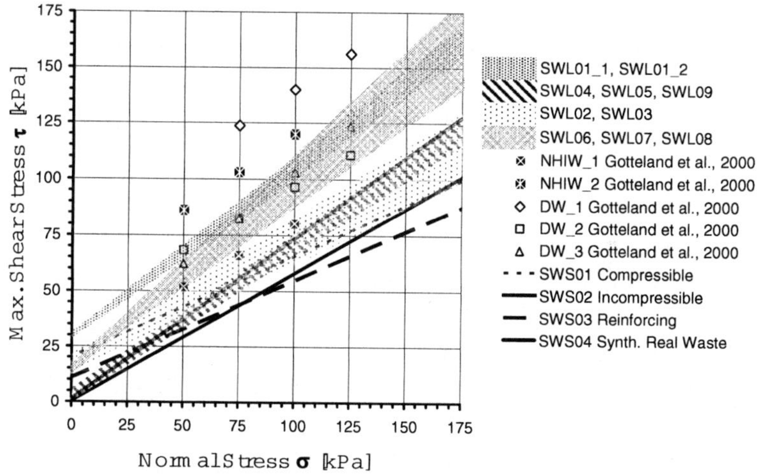

FIG. 2 Results from small (SWS) and large direct shear tests (SWL) on synthetic waste compositions and real waste values from Gotteland *et al.* (2000).

4.3 Influence of Unit Weight

FIG. 3 shows the unit weights measured for samples tested in the large and small shear tests. Also provided for comparison are profiles taken from Zekkos *et al.* (2005) based on values of unit weight measured in situ for different depths and conditions of compaction. The unit weight of the sample at the end of the shearing phase is plotted against the test normal stress. Apart from soil like samples (e.g. SWL01 and SWL06) that have a relatively high unit weight, the shear sample values appear to be low when compared with the results from the extensive programme of field measurements of unit weight conducted by Zekkos *et al.* (2005). However, it must be considered that the values for synthetic wastes are dry unit weights. Powrie and Beaven (1999) have shown that dry unit weights in the order of 2 to 4kN/m^3 equate to bulk unit weights in the range 6 to 8kN/m^3 for the same waste at field capacity. It can be concluded that the low unit weights are not unrealistic considering that the material is dry, although they are at the lower bound. This could be due to the use of dead load to compact the samples compared to modern site practice using a sheep's foot roller, and which may not replicate it. There is no clear link between these differences and the trends in shear strength discussed in Section 4.1.

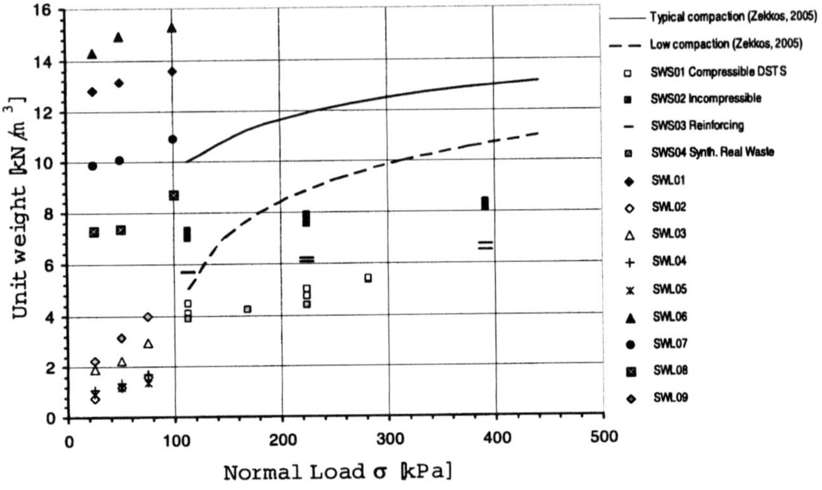

FIG. 3 Relationship between unit weight and normal load for test samples and equivalent field values.

4.4 Relationship between Shear Behavior and Waste Classification

The ternary diagram (FIG. 1) shows measured shear strengths for a reference normal stress of 50kPa from both the small and large shear tests related to the shape-related subdivisions. Waste samples with a relatively high percentage of compressible particles (SWL02, SWL09 and SWS04), including the synthetic 'real' waste samples, exhibit low shear strengths. Also shown are relatively high shear strengths for incompressible compositions tested in the large device (SWL01, SWL06 and SWL07). It also appears that an increase in percentage of reinforcing components results in higher shear strengths than those with a greater percentage of compressible components (e.g. SWS03, SWL05 and SWS04) but further testing is required to investigate this trend in more detail. The most obvious inconsistency in the data is the low shear strength for the small device sample with a high percentage of incompressible components (SWS02). This could be a function of the size of the particles used to form the sample. In SWS02 the particles were relatively large pieces of brick, tile, and rigid plastic with a small percentage of sand, while in SWL01 the predominant incompressible particles were sand. It is noticeable that the unit weight of SWS02 is considerably lower than SWL01 and SWL07. FIG. 1 shows general trends between shear strength and classification and demonstrates that the waste classification proposed by Dixon and Langer (2006) can be linked to shear behavior by using mass relationships of shape related component . A full description and discussion of the large direct shear tests and also of compression behavior is provided by Langer (2006) and Dixon *et al.* (2008).

5 CONCLUSIONS

A study has been conducted using synthetic wastes to investigate the relationship between waste classification and shear behavior. A classification based on component material properties, size and shape related sub-divisions has been outlined. Synthetic components were selected based on a sorting analysis of a representative sample of real MSW. Shear tests have been conducted using large and small direct shear devices. Measured shear strengths for samples with similar classifications have produced comparable shear behavior. Consistent mobilized friction values have been measured but there is greater variability in the mobilized cohesion parameters. A comparison of behavior measured using the large device for both synthetic and real wastes is presented. Although the synthetic wastes have lower unit weights, as a result of the components being in a dry state, it is concluded that comparable shear strengths are obtained, thus validating the approach taken. A ternary diagram has been used to investigate relationships between classification and shear strength. Lowest strengths are obtained for samples dominated by compressible components and highest strengths for those dominated by incompressible components. Increasing percentages of reinforcing components produced increased strengths. Further research is required to conduct repeatability tests and to add results for wastes in regions of the ternary diagram currently unpopulated. The classification could be used to compare mechanical behavior of different wastes and to assess implications of changes in waste composition.

6 ACKNOWLEDGEMENTS

The authors would like to express their gratitude to Mr Rikesh Shah for his work in studying downscaled synthetic waste materials.

7 REFERENCES

Dixon, N., Langer, U., Gourc, J.P. and Gotteland, P., 2004. "Factors influencing shear behaviour of MSW." *Waste 2004*, Stratford-Upon-Avon, UK. 591-600.

Dixon, N. and Langer, U., 2006. "Development of a MSW classification system for the evaluation of mechanical properties." *Waste Management*, Vol. 26 (3): 220-232.

Dixon, N., Langer, U. and Gotteland, P., 2008. "Classification and mechanical behaviour relationships for MSW: A study using synthetic wastes." *ASCE, J. Geotechnical & Geoenv. Engrg.*, 134,1-12.

Gotteland, P., Gourc, J.P., Aboura, A. and Thomas, S., 2000. "On site determination of geomechanical characteristics of waste". *International Conference on Geotechnical and Geological Engineering*, Melbourne, Australia, ICMS, (CD).

Langer, U., 2006. "Shear and compression behaviour of undegraded municipal solid waste." Ph.D. Thesis, Loughborough University, UK.

Powrie, W., and Beaven, R.P., 1999. "Hydraulic properties of household waste and implication for landfills". *Proceedings of the Institution of Civil Engineers*; Geotechnical Engineering, 137, 235-247.

Zekkos, D.P., Bray, J.D., Kavazanjian, E., Matasovic, N., Rathje, E., Riemer, M. and Stokoe II, K.H., 2005. "Framework for the estimation of MSW unit weight", *10th International Waste Management and Landfill Symposium,* Cagliari, Italy, CISA.

Hydrogeological characterization of Suzhou landfill of municipal solid wastes

Tony L. T. Zhan[1], Daosheng Ling [2], Wen-jie Zhang [3] and Yunmin Chen[2]

[1] Associate Professor, MOE Key laboratory of Soft Soils and Geoenvironmental Engineering under Ministry of Education, Zhejiang University, Hangzhou, 310027, China; zhanlt@zju.edu.cn
[2] Professor, MOE Key laboratory of Soft Soils and Geoenvironmental Engineering under Ministry of Education, Zhejiang University, Hangzhou, 310027, China; dsling@zju.edu.cn
[3] Research student, MOE Key laboratory of Soft Soils and Geoenvironmental Engineering under Ministry of Education, Zhejiang University, Hangzhou, 310027, China.

ABSTRACT: Municipal solid waste landfills in southern China generally possess a higher leachate level than expected. A representative landfill in the region, i.e., Suzhou landfill, was chosen to perform field and laboratory study to understand the hydrogeological conditions. This paper presents some preliminary findings from the study, including laboratory measurements of hydraulic properties of the MSW and field measurements of pore pressures within the landfill. It was found that the hydrogeological conditions within the landfill are closely related to the local hydrogeology of the valley. Laboratory test results show that municipal solid wastes embedded at different depths in the landfill have different values of field capacity and water permeability as a result of the change in waste composition with depth. The field measurements of pore pressures within the landfill revealed a perched leachate mound above the intermediate cover of soils and a substantial leachate mound at the bottom of the landfill.

INTRODUCTION

Leachate is the main vector of contamination from a municipal solid waste (MSW) landfill. One important design criterion for a landfill is that the leachate level within the landfill is kept below a specific elevation, so as to minimize the likelihood of leachate leaking into the surrounding ground. A high leachate level generally produces a high contaminant loading on the bottom engineering barrier, and reduces the factor of safety with regards to the slope stability of waste body (Rowe and Nadarajah, 1996). The development of high leachate mound also has adverse effect on landfill gas recovery.

In the southern cities of China, MSW landfills are usually located in valleys surrounded by hills. Field reconnaissance shows that the leachate levels in most of the landfills are higher than expected. It is believed that the high leachate level is related to the humid climate (i.e., annual precipitation greater than 1000 mm) and the inappropriate design of water management system associated with the landfills. It was found that the landfill's rainwater interception facilities were not as effective as expected. Clogging of leachate drainage facilities was observed after the landfills were operated for several years. To study in more detail the hydrogeological conditions associated with the landfills, a representative landfill in Suzhou, China, called Suzhou landfill, was chosen for a field and laboratory study. This paper presents some findings obtained from the study.

SITE DESCRIPTION AND FIELD INVESTIGATION

The Suzhou landfill is located in a valley surrounded by hills about 13 km to the south of Suzhou, China (Fig. 1). The average annual precipitation in this region is about 1100 mm. About 70% of the annual precipitation is distributed in the spring and summer seasons. The landfill started operation in 1993. Its design capacity and design service life were 4.7 million m^3 and 15 years, respectively. The current landfill body covers an area of 7.44 hectare. The landfill body consists of a number of filled platforms that are set back at a slope of 3H:1V. The thickness of the landfill body in April 2006 ranges from 15 to 40 m. A rockfill dam retains the lowest platform of the landfill. A toe drain was installed near the bottom of the rockfill dam for leachate drainage.

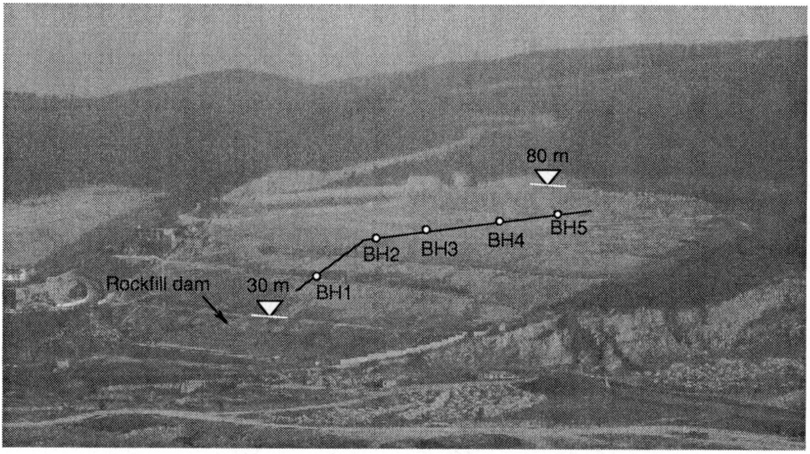

Fig. 1 An overview of Suzhou landfill

The bottom of the landfill was not lined with any engineering barrier. An injected grout curtain was installed downstream the rockfill dam in an attempt to limit downstream movement of leachate. The natural soil strata below the landfill bottom comprise of a layer of alluvial-colluvial deposits, highly-decomposed sandstone along with slightly-decomposed and fresh sandstone. The alluvial-colluvial deposits consist of gravelly clay with a thickness ranging from 5 to 27 m. The gravelly clay has a water permeability in the order of 10^{-6} m/s. Joints were well developed in the highly-decomposed sandstone, resulting in a high hydraulic conductivity. However, the fresh rock at the bottom is more intact and has a water permeability less than 1×10^{-9} m/s. The grout curtain extends to the underlying fresh rock. The grout curtain and the fresh bedrock were expected to constitute a closed barrier system against the leachate in the landfill. However, groundwater monitoring downstream of the grout curtain indicated that the barrier system was imperfect.

Around the existing landfill, there is an upstream catchment area of 54.4 hectare. The upstream catchment area has a well vegetation of shrub. The top soil layer in the catchment area consists of a loose colluvium soil with a thickness ranging from 1 to 3 m. The loose colluvium is composed of gravelly clay and has a high infiltrability (i.e., ranging from 10^{-5} to 10^{-6} m/s). An interception trench was constructed at an elevation of 80 m to collect the surface runoff from the upstream catchment area. The depth of the interception trench was about 1 m.

Borehole investigation and sampling was carried out in the landfill in April 2006. Five boreholes (BH1 to BH5) were drilled to the bottom of the existing landfill (see Fig. 1). MSW samples were taken using heavy-wall samplers (100 mm in diameter) at an interval of 1 or 2 m. After the completion of sampling, pore pressure transducers were installed in two boreholes (i.e., BH1 and BH3) to measure pore fluid pressures in the landfill. In each borehole, there were three transducers located at depths of 8 m, 17 m and 25 m. An intermediate cover soil layer was found at a depth of about 10 m into the borehole. It should be noted that the intermediate soil cover layer was relatively impermeable as compared with the MSW. After the installation of the pore pressure transducers, the boreholes were backfilled with gravel (3 to 6 mm in grain size) with the exception of the top 1 m and the section corresponding to the intermediate cover soil layer. These two exceptions were backfilled with a sealing clay column for a length of 1 m. It was expected that the two transducers separated by the sealing layer could register the leachate heads within different hydrogeological regimes.

HYDROGEOLOGY SYSTEM ASSOCIATED WITH SUZHOU LANDFILL

The height of leachate mount in a landfill is closely related to the site hydrogeology. Fig. 2 shows a schematic diagram of the hydrogeology system in the Suzhou landfill. The input of water to the landfill includes rainfall infiltration on the surface of the landfill, leachate generation caused by degradation and consolidation of waste, as well as surface and subsurface inflow from the upstream catchment zone. The output of water from the landfill mainly includes the actual evaporation from the surface of the landfill, and leachate discharge from the toe drain constructed near the bottom of the rockfill dam as well as leachate discharge through the bottom of the landfill. The leachate mound within the landfill was developed until the hydrogeology system reached a state of water balance.

As mentioned previously, the upstream catchment area around the landfill is well vegetated and is covered by a loose colluvium soil with a high infiltrability. It is estimated that a rainfall with a rate up to 10 mm/hr will completely infiltrate into the loose soil. It was found that little surface runoff was collected in the interception trench. Most of the infiltrated rainwater will be consumed by the evaporation and vegetation evapotranspiration. The remnant part becomes a down-slope subsurface water flow when the water reaches a relatively impermeable stratum (e.g., the underlying fresh bed rock). The subsurface flow is likely to exit at downstream and seep into the landfill body. Some exit points were observed on the excavation faces during the spring season.

The landfill has been operated with the majority of surface area being exposed to precipitation. As the municipal solid waste on the landfill surface is loose and has high permeability. It is believed that a high percentage of precipitation percolates the landfill and becomes leachate. In addition, gradual clogging was observed at the toe drain near the bottom of the rockfill dam. The above conditions resulted in a continual accumulation of water, and hence caused a substantial leachate mound in the landfill. Leachate was observed on the sloping surface of the landfill during a wet season.

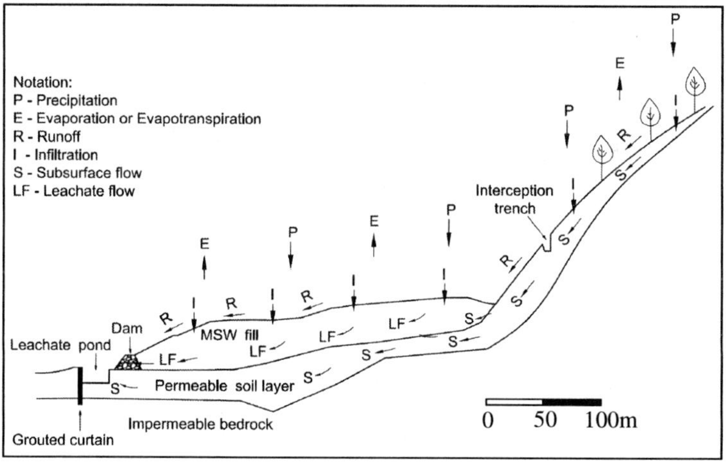

Figure 2 Hydrogeology system associated with Suzhou landfill

HYDRAULIC PROPERTIES OF MUNICIPAL SOLID WASTE

The distribution of leachate mound within the landfill is also affected by stratification of waste pile, field capacity of the MSW, vertical and horizontal hydraulic conductivity of the MSW, as well as the existence of daily cover soil or intermediate cover soil layer. The MSW as collected generally consists of putrescent organics (i.e., food and garden wastes), cinder, dust, paper, plastics, rubber, textiles, wood, glass, metal etc. After being placed in landfills, waste composition inevitably changes with time due to biological degradation of the organics. The waste embedded at different depths is generally at different stages of degradation, and hence poses different composition. Therefore, the waste samples taken at different depths into the landfill will have different hydraulic properties.

Pressure plate tests were carried out to investigate the water retention characteristic of the MSW samples taken from different depths into the Suzhou landfill. Three groups of re-compacted MSW specimens were prepared using the disturbed MSW samples taken from the shallow, middle and deep sections of the boreholes (i.e., from 3 m to 8 m, from 14 m to 16 m and from 24 to 28 m, respectively). The information, including waste composition and volume-mass properties, for each of the re-compacted MSW specimens is listed in Table 1. Any large-size waste material was shredded to a size less than 20 mm. Each of the specimens was compacted to a dry density close to the corresponding in-situ dry density by using static compaction method. Each of the specimens was 100 mm in diameter and 50 mm in height. The water retention curves were measured by the use of a 5-bar pressure plate extractor. Fig. 3 shows three water retention curves for the re-compacted MSW specimens prepared by the shallow, middle and deep MSW samples. For the purpose of comparison, the water retention curve of a fine sand reported by Elzeftawy et al. (1981) is also shown in Fig. 3. As compared with the fine sand, all the MSW specimens at a saturation state pose significantly higher volumetric water content. This is consistent with the high void ratio of the MSW. It is found that the residual water content of each MSW specimen is also higher than that of the fine sand. This is primarily attributed to the inherent water retained in organic component and fibrous materials. The air entry values of all the MSW specimens are less than 1 kPa. The desorption rate of the MSW specimens seems to be close to that of the fine sand when the applied suction exceeds the air-entry value.

Table 1 Information for the re-compacted specimens used for laboratory tests

Parameters		Shallow MSW	Middle MSW	Deep MSW
Corresponding embedment depth (m)		3~8	14 ~ 16	24 ~ 28
Corresponding fill age (year)		3	6	10
Dry density (g/cm^3)		0.41	0.60	0.81
Void ratio		2.7	2.2	1.60
Part of composition (dry weight-%)	Organics	16.1	16	15
	Plastics	11.5	24.3	17
	Paper, rubber, textiles, wood etc.	5.9	17	7.3
	Metal and glass	1.9	2.5	1.7
	Cinder and soil	64.6	40.2	59

Notes: All the above values were measured from the borehole samples

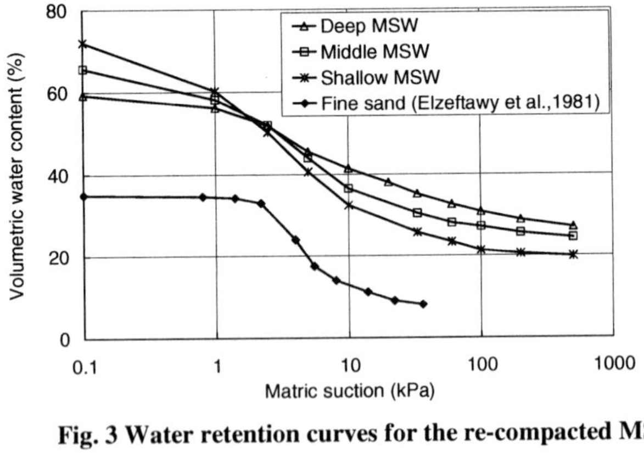

Fig. 3 Water retention curves for the re-compacted MSW samples

A comparison of the three water retention curves of the MSW specimens shows that the desorption rate decreases with an increase in the embedment depth of the waste. The MSW collected at a shallow depth has a higher saturated water content and a lower residual water content than the waste collected from the deep fill. The above difference in the water retention characteristic can be explained by considering the difference in the waste composition and density (see Table 1). The shallow MSW contains a higher content of coarse materials such as plastics and fibers, and the deep MSW contains a higher content of fine-grained cinder and soil. In addition, the void ratio of the shallow MSW is higher than the deep MSW. If the field capacity of MSW can be defined as the volumetric water content corresponding a suction of 10 kPa as suggested by Qian et al. (2001), the values of field capacity for the deep, middle and shallow MSW specimens are 41.3%, 36.5% and 32.4%, respectively. The magnitude of field capacity is greater than the data reported by Schroder et al. (1994) but less that the data reported by Zornberg et al. (1999).

Constant head permeability tests were also carried out to measure the saturated permeability of the MSW samples taken from different depths into the Suzhou landfill. Three groups of re-compacted MSW specimens, corresponding to the shallow, middle and deep waste layers, were prepared using the same way as those for the measurement of water retention curves. For each group, there were two duplicate re-compacted specimens compacted to a dry density close to the individual in-situ dry density (see Table 1). Each of the specimens was compacted in a rigid-wall permeater (100 mm in diameter and 200 mm in height). No vertical stress was applied on the specimens during the permeability tests. The test results are listed in Table 2. It can be seen that the magnitude of saturated permeability for the shallow, middle and deep MSW specimens are in the order of 10^{-4}, 10^{-5} and 10^{-6} m/s, respectively. In other words, the saturated permeability of the MSW shows a decreasing trend with an increase in the embedment depth. This is basically consistent with the fact that the void ratio of the MSW decreases with the embedment depth and that the content of fine-grained cinder and soil increases with the embedment depth. The magnitude of saturated permeability obtained in this study is within the ranges reported in the literature (Oweis et al., 1990).

Table 2 Results of constant-head permeability tests on the MSW specimens

Specimen #	Void ratio	Dry density (g/cm^3)	Average value of saturated permeability (m/s)
Deep MSW	1.78	0.72	3.56×10^{-6}
Middle MSW	2.16	0.61	3.50×10^{-5}
Shallow MSW	2.65	0.41	4.81×10^{-4}

PORE PRESSURES IN THE LANDFILL

As stated previously, pore pressure transducers were installed in two boreholes (i.e., BH1 and BH3) to measure pore fluid pressures in the landfill. Fig. 4 shows changes of pore pressures with time recorded by the three transducers installed in borehole BH1. The pore pressures represent the total pressures resulting from gas and leachate. As shown in Fig. 4, an increase in pore pressures is observed for all the three depths after the completion of instrumentation. This may be related to a recovery in the hydraulic conditions, which had been disturbed by the borehole drilling. After about 20 days passed, the changes in pore pressure with time at all the depths were relatively insignificant. The maximum pore pressures measured at depths of 8, 17 and 25 m are 50, 80 and 145 kPa, respectively. A slight decrease in pore pressures was registered after the summer season started. This may be due to a decrease in the subsurface water recharge from the upstream catchment area. The evaportranspiration of new green vegetation extracts more water from the soil in the summer.

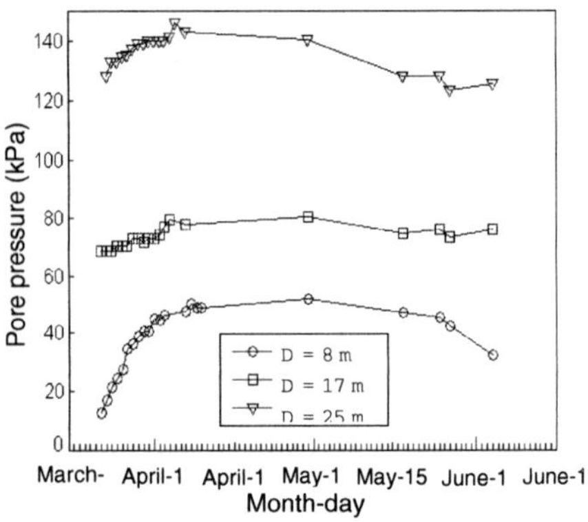

Fig. 4 Pore pressures measured in boreholes BH1

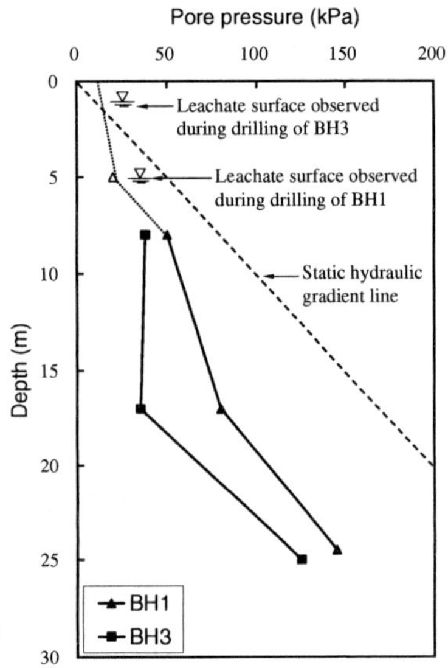

Fig. 5 Pore pressures profiles obtained from boreholes BH1 and BH3

Fig. 5 shows the changes of pore pressure with depth as recorded by the transducers installed in boreholes BH1 and BH3. The top levels of leachate observed while drilling the two boreholes were also shown in the figure to provide a reference. The pore pressures measured in borehole BH3 were obviously lower than those obtained from borehole BH1 for all three depths, regardless of the higher leachate surface observed while drilling. It should be noted that the bottom of the landfill is sloping and borehole BH1 is located downstream of borehole BH3 (see Figure 1). Between the depths of 8 m and 17 m, the gradients of pore pressure for each of the boreholes (i.e., BH1 and BH3) were found to be significantly greater than the static hydraulic gradients. This indicates that the waste within the region may be unsaturated and downward flow of leachate exists. The pore water pressure profiles also indicate that the leachate mound above a depth of 8 m is in a perched state. This finding is further supported by the intermediate cover soil layer observed at the depth of 10 m during the borehole drilling. By taking the observed leachate surface in borehole BH1 as a reference, the pore pressure profile above a depth of 8 m was approximated (see the dash line in Fig. 5). For the 8 m long section near the bottom of the landfill, the gradient of the pore pressure profile for each of the boreholes (i.e., BH1 and BH3) was found to be close to the static hydraulic gradient. This indicates that the MSW within the region was saturated or nearly saturated. In other words, there exists a saturated zone with a thickness of about 8 m just above the landfill bottom. The substantial leachate mound in the landfill must pose a threat to the landfill stability as well as to the engineering barrier.

SUMMARY

This paper presents some findings from the field and laboratory investigation on the hydrogeological conditions of Suzhou landfill. Being located in a valley, the landfill has a subsurface water recharge from the upstream catchment area which can not be ignored in water balance analyses. Laboratory test results shows that the municipal solid waste at different depths of the landfill have different hydraulic properties as a result of the change in waste composition with depth. The field capacity of the waste appears to increase with the embedment depth, and the water permeability shows a decreasing trend with the embedment depth of the wastes. The field measurements of pore pressures and water content revealed a perched leachate mound above the intermediate cover of soils and a substantial leachate mound at the bottom of the Suzhou landfill.

ACKNOWLEDGEMENTS

Financial support from research grants (50408023 and 50538080) provided National Natural Science Foundation of China (NSFC) is acknowledged.

REFERENCES

Elzeftawy A and Cartwright K. (1981). Evaluating the saturated and unsaturated hydraulic conductivity of soils, in *Permeability and Groundwater contaminant Transport, ASTM STP 746*, pp.168-181.

Oweis, I. S., Smith, D. A., Ellwood, R. B., Greene, D. S. (1990). Hydraulic characteristics of municipal refuse. *Journal of Geotechnical Engineering*. Vol. 116, No. 4, pp. 539-553.

Qian, X. D. Koerner, R.M., Gray, D.H., (2001). Geotechnical aspects of landfill design and construction. Prentice-Hall, Inc, New Jersey.

Schroeder, P. R., Dozier, T. S., Zappi, P. A. and Aziz, N. M. (1994). The Hydrologic Evaluation of Landfill Performance (HELP) Model: Engineering Documentation for Version 3. EPA/600/R- 94/168b, U.S. Environmental Protection Agency Office of Research and Development. Washington, DC, September.

Zornberg, J.G,, Jernigan, B.L., Sanglerat, T..R., Cooley, B. H., (1999). Retention of free liquids in landfills undergoing vertical expansion. *Journal of Geotechnical and Geoenvironmental Engineering*, Vol. 125 (7): 583-594.

MBT Waste Used as a Capillary Barrier in a Sanitary Landfill

Ronaldo L. Izzo[1], Claudio F. Mahler[2], Kai Munnich[3], Jan Bauer[4]

[1]PhD student, COPPE/UFRJ, Rio de Janeiro – Brazil. ronaldoizzo@gmail.com
[2]Prof. Dr. Civil Engineering Department, COPPE/UFRJ. Coordinator of the Waste Technology Study Group (GTRES) – Federal University of Rio de Janeiro. cfmahler@acd.ufrj.br
[3]Dr. Leichtweiß-Institute/Department of Waste Management, Technical University of Braunschweig, Braunschweig, Germany. k.muennich@tu-bs.de
[4]Civil Engineer. Leichtweiß-Institute/Department of Waste Management, Technical University of Braunschweig, Braunschweig, Germany. jan.bauer@tu-bs.de

ABSTRACT: A capillary barrier is an alternative technology for the final cover of sanitary landfills that has been very successfully studied in recent years on several landfills. Its functionality is based on the retention capacity and unsaturated hydraulic conductivity of the materials used. Although a more attractive alternative than the traditional use of geosynthetic materials and clay layers in the construction of the final cover system, a large amount of soil is still required for the capillary barrier, giving rise to the need to find appropriate soils and occupying a large volume in the landfill. A more sustainable solution would be to build the capillary barrier using mechanical biological pre-treated waste (MBT) as building material. This paper presents the results of laboratory tests to evaluate the main factors that influence a capillary barrier built with mechanical biological pre-treated waste.

INTRODUCTION

The main concern of the planet nowadays is the recent climate change and imminent scarcity of natural resources that may occur due to constant human aggression against the environment. This matter is quite alarming since the prospects are not so good and the tendency is for them to worsen with uncontrolled global population growth, absence of specific plans and policies to deal with the issue (especially in undeveloped countries), and the apparent indifference of some governments around the world.

In Brazil these problems are just as great as the others worldwide, in view of its continental size and status as a "developing country". There is a tremendous shortage, specifically in Brazil, of research and technology for production, treatment and final disposal of municipal solid waste (MSW).

Even today, according to official data, uncontrolled dumping is the main method of MSW disposal in Brazil. Although a sanitary landfill generates an environmental debt that could last for centuries, it is, nevertheless, still an interesting alternative to the MSW final disposal when done according to appropriate geotechnical criteria.

In all countries having directives concerning the waste disposal on landfills it is therefore regulated by law that landfills must be built as capsulate-construction with a base and surface sealing. The base liner has the function to avoid a spreading of contaminants into the subsoil and the groundwater. The main function of the top cover system is, beneath the collection of gaseous emissions and the recultivation for future use, to avoid an uncontrolled infiltration of rain water in the waste. The concequence of the uncontrolled water infiltration would be the generation of leachate which has to be purified over a long period and which might lead to stability problems of the landfill at all.

Compacted soils with low permeability are the most common system used to construct the final top cover system, but other alternatives exist that, depending on the situation, could be more interesting.

FUNDAMENTALS OF CAPILLARY BARRIERS

Working Principle

According to Weiss and Witsche (2005) the capillary barrier is a system consisting of two materials, one with finer granulometry called the capillary layer, and the other with coarse granulometry called the capillary block.

In soil science the intergranular pores are commonly described by a bundle of interconnected capillary tubes. To specify the working principle of a capillary barrier, the governing rules of hydraulics in tubes are therefore adopted to the soil material (Figure 1).

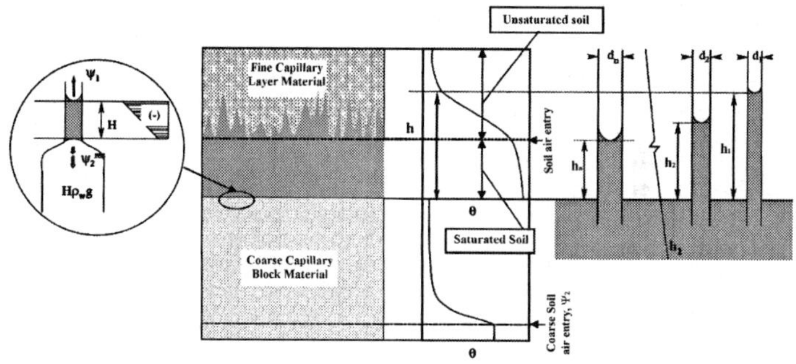

Figure 1 – Retention capacity and water distribution in a capillary barrier.

The most important factor in the capillary barrier is the distinct difference between the pore diameter of the fine capillary material and the underlying coarse capillary block material. This has the effect that on one hand the surface forces at the interface of the two materials prevent instantaneous break through of water and on the other hand at the same water content the degree of water saturation in the fine layer is much higher than in the coarse material. From the second effect results that the unsaturated hydraulic conductivity of the fine material is much higher than in the coarse material so that lateral flow occurs in this layer when a minimum incline of the layer is applied. The critical point of the capillary barrier is backwater in the fine material which may lead at a certain level to a breakthrough into the coarse material.

EXPERIMENTAL STUDY

The Technical University of Braunschweig and the Federal University of Rio de Janeiro have been working together in a cooperative research program involving many issues relating to environmental geotechnics, one of which is the development of a capillary barrier built with mechanical and biological pre-treated waste (MBT). The advantages of this new capillary barrier construction method are:

- Optimized use of landfill space, considering the reduction in soil use for the top cover system.
- Cutting costs with soil transportation to the landfill.

RESULTS AND DISCUSSION

Materials from Germany

Today in Germany all municipal solid waste (MSW) disposal in a sanitary landfill undergoes mechanical biological treatment, which has positive effects on the long-term gaseous and leachate emissions of the landfill and which greatly reduces the need for landfill space.

At the Technical University of Braunschweig three mechanical biological pre-treated MSW have been studied by Pahl (2006), the material collected from the sanitary landfills in Münster, Mansie and Rethmann. The particle size curves from the materials are given in Figure 2.

By examining Figure 2 it is possible to see that the Rethmann material has a finer grading than the Münster and Mansie material.

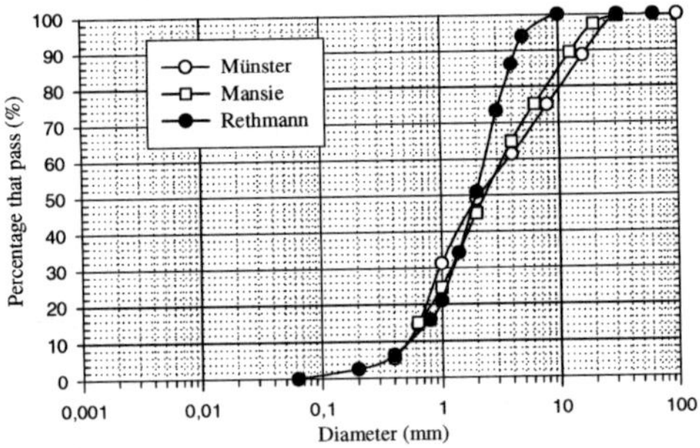

Figure 2 – Granulometric curves of the pre-treated MSW from Münster, Mansie and Rethmann.

Loss-on-ignition tests were conducted by Pahl (2006) using Mansie and Rethman materials. Mansie material had a loss of 38.24% while Rethmann lost 41.96%, indicating that the Rethmann material has more organic matter than the Mansie material.

Permeability tests were carried out by Izzo (2006) using Münster and Mansie material for different densities. As expected, the higher the density the lower the permeability. The results can be seen in Figure 3 (A).

The lowest permeability values were obtained for the Rethman material (the finer particle size material) by Pahl (2006) and can be seen in Figure 3 (B).

In order to study the capillary barrier in the laboratory a small box was built where the discharge from the capillary layer and the discharge from the capillary block could be collected and measured separately during simulated rainfall, with different inclination angles. Figure 4 shows the assembled apparatus.

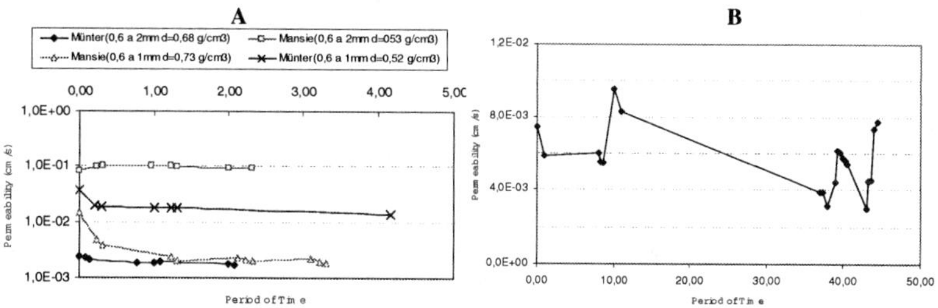

Figure 3 – (A) Constant head permeability test results for Münster and Mansie material at different densities (IZZO, 2006). (B) Constant head permeability test result for Rethmann material obtained by Pahl (2006).

Pahl (2006) tested the capillary barrier built using Rethmann material in the laboratory, varying the angle of inclination, and the results can be seen in Figure 5. The results show that in all cases the flux in the capillary block (CB) is negligible. The maximum flux in the capillary layer (CL) is reached at an incline of 10° to 14°. Larger angle of inclination reduces the flux as under these conditions of artificial rainfall surface runoff occurs, which has to be avoided in practice because of the risk of erosion of the soil material.

Figure 4 – Laboratory capillary barrier test assembled at TU Braunschweig.

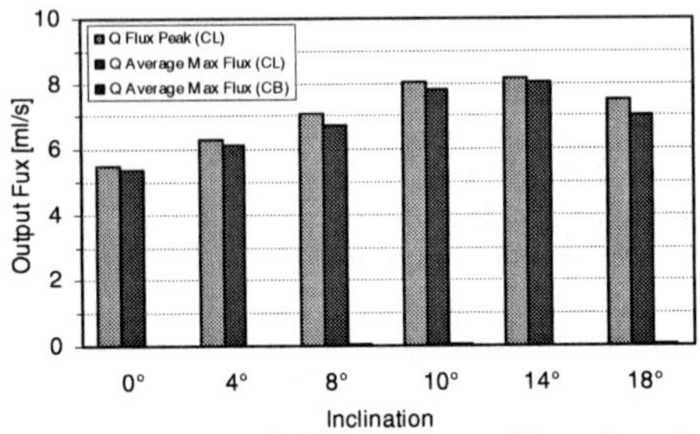

Figure 5 – Results obtained at a laboratory capillary barrier test varying the inclination (PAHL, 2006).

Materials from Brazil

In Brazil MSW disposal in landfills undergoes no treatment and its disposal is still a major problem. The two treatment pilot systems implanted in Brazil based on German technologies were shut down in 2005. For this reason compost produced at the Caju treatment plant in Rio de Janeiro, Brazil, was studied in order to find a material suitable for building the capillary layer for a capillary barrier. Two samples from the Caju compost were collected. The granulometric curves obtained for the Caju compost can be observed in Figure 6.

The samples were collected after previous screening, when the material that passed through a 10mm sieve and 22mm sieve was collected. Permeability tests were carried out using the Caju Sample 01 material. The results can be seen at Figure 7

The compost from Caju had a loss of 44.04% at the loss-on-ignition test, which shows that the organic content is practically the same as that in the Rethmann material.

Another test carried out in Brazil and in Germany was the capillary rise test. This test shows the capillary process and potential of a material and consists of putting a column filled with the material under study in contact with water, and then measure the water content every five centimeters of the column. The results obtained for the Caju samples with different granulometric range and the Rethmann sample are compared in Figure 8.

It can be seen that the Rethmann material performs better than the Caju compost since the humidification reaches the top of the test column, so that the water content value is almost constant from the base to the top. According to Pahl (2006) the capillary rise reached the top of the column within 24 hours.

Although the performance of the Brazilian compost was inferior to that of the Rethmann material, it is still not possible to give a conclusive explanation because of the many variables involved in the process. Even when not showing the same result as Rethmann's, it is very possible that the Caju material could be used as a capillary layer since it effectively developed suction.

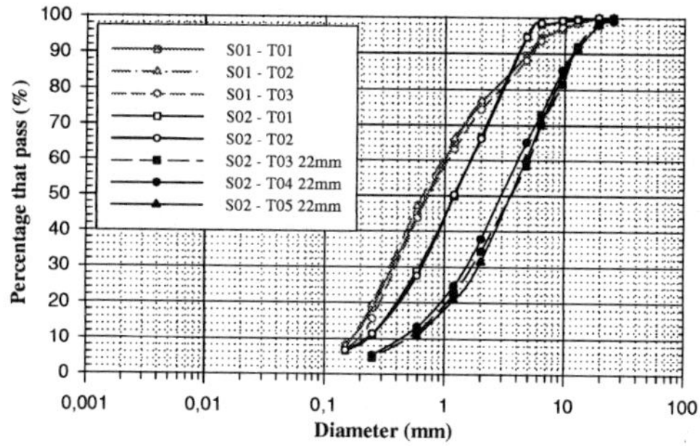

Figure 6 – Granulometric curves obtained for Sample 01 (S01) and Sample 02 (S02) collected from the Caju treatment plant, Rio de Janeiro, Brazil.

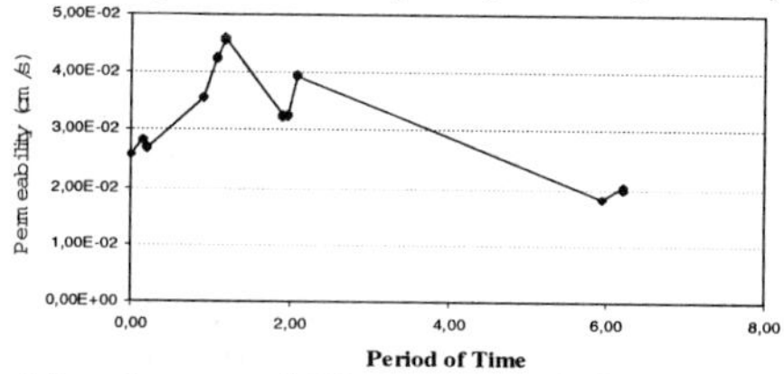

Figure 7 – Constant head permeability test result for Sample 01 collected from Caju treatment plant, Rio de Janeiro, Brazil.

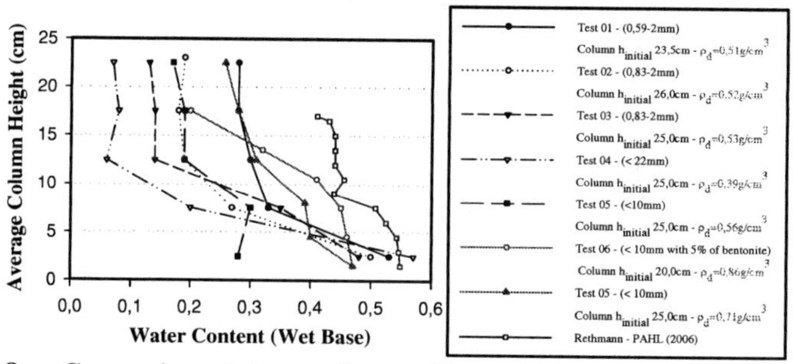

Figure 8 – Comparison between the capillary rise test results for the Caju samples and Rethman sample.

UNSATURATED COMPRESSION TESTS

The tests made so far were investigating the hydraulic behaviour of the capillary barrier when it is used as a final top cover system. In many countries it is required by the regulations that a daily soil cover of about 50 cm height has to be placed on the waste to avoid bad odor, insects etc.. In the extreme case depending on the conditions of the landfill the difference between the volume of disposed waste and soil is only small. Under these conditions the use of MBT material instead of soil would be of great advantage in view of the cost for the cover and also in view of a better utilization of the landfill volume for the waste.

Under these boundary conditions the unsaturated hydraulic and compression behavior is changing as now the system is compacted by the next deposits of waste. In order to study these changes, an unsaturated compression test was performed, measuring the settlement and suction of the material during the test by using a tensiometer developed by Mahler (2004) installed at the bottom of the sample. The assembly method can be seen through the Figure 9. The first tests results can be seem in the Figure 11.

At this first test, the compost sample was built with initial humidity of 29.22% (dry base) and wet density of 0.7 g/cm^3.

Figure 10 B shows that there is an increase in suction as the load increases to about 60 kPa and remains constant until reaching 800s$^{1/2}$ (load at this point 800 kPa). At this moment the suction begins to decrease until the sample starts to be unloaded.

This behavior is probably due to the faster decrease in voids than water drainage in the sample, and so the suction continues to increase.

Figure 9 – View of the assembly method of the compression test.

At this first test only drainage at the top of the sample was allowed. In future tests, if drainage is allowed at both top and bottom of the sample, it is believed that the development of further suction is possible.

From this test it may be believed that this kind of material and the capillary barrier technology can be used as daily cover in a sanitary landfill, although a final conclusion is still premature.

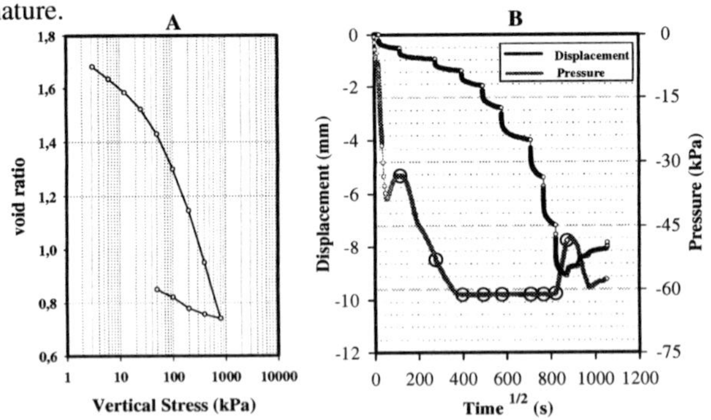

Figure 10 – First results of the unsaturated compression test.

CONCLUSIONS

Many variables can influence the performance of a capillary barrier built using mechanical biological treated MSW, such as density, granulometry, permeability, temperature and quantity of organic matter, but its principle construction is possible as shown from the results made in laboratory in Germany.

The material used in Brazil was coming from a composting plant, which means that the treatment of this material was quite different from the one in Germany, so that a different physical behavior is to be expected. Even though one of the major factors influencing the effectivity of the compost capillary barrier is lower than in the MBT material the prospect of using this material as a capillary barrier deserves more studies. The next test will be the determination of lateral flow and of the critical height of backwater in the capillary liner for the compost material. The installation of a test area on a landfill to show the functioning under practical conditions is required.

The tests made so far illustrate that materials coming from MBT can be used as a capillary barrier, although a final conclusion is still premature, because so far a long-term functioning in practice has not been proved. In any case an optimization of the material for the capillary liner and the capillary block has to be made depending of the waste and technology of mechanical biological treatment.

Beyond the economic advantage in a better exploitation of the limited landfill volume with waste, the use of MBT material may become an initial start-up for the landfill owner to install a mechanical biological treatment plant on his area.

ACKNOWLEDGEMENTS

The authors wish to thank CNPq, CAPES, DAAD and FAPERJ for their continuing support.

REFERENCES

Izzo, R. L. S. (2006). “Final report of PhD apprenticeship at the Technical University of Braunschweig.” CAPES BEX1997/05-0, Rio de Janeiro, RJ. (in Portuguese)

Mahler, C. F. and Gonçalves, H. (2004). (To COPPE/UFRJ Engineering Post-Graduation Program of the Federal University of Rio de Janeiro), “Tensiometer for High Suction Measurements.” Brazilian Patent BR n. PI 0403670-0, 01 September 2004 (in Portuguese).

Pahl, A. (2006). “Laboratory tests on the use of mechanical biological pre-treated waste in capillary barriers.” Leichtweiss-Institute for Hydraulic Engineering, Dept. of Waste Management, Technical University Braunschweig (in German)

Weiß, J. and Witzsche, A. (2005).“ Results of the Efficiency of the Capillary Barrier System at the Landfill Grix Offenbach-Germany.” *Proceedings Sardinia 2005, Tenth International Waste Management and Landfill Symposium.* S. Margherita di Pula, Cagliari, Italy; October 3-7, 2005.

The influence of landfill gas on the hydraulic conductivity of waste

William Powrie[1] Richard Beaven[2] and Andrew Hudson[3].

[1]Professor of Geotechnical Engineering and Head of School; School of Civil Engineering and the Environment, University of Southampton, UK. wp@soton.ac.uk
[2]Principal Research Fellow, School of Civil Engineering and the Environment, University of Southampton, UK. rpb2@soton.ac.uk
[3] Research Assistant, School of Civil Engineering and the Environment, University of Southampton, UK . a.p.hudson@soton.ac.uk

ABSTRACT: Knowledge of the hydraulic conductivity of waste is essential to the efficient and effective management of leachate in landfills. Over the past few years it has become apparent that the presence of gas in a waste landfill will affect the hydraulic conductivity. This paper reports the results of tests carried out in a large scale (2 m diameter) compression cell to investigate the influence of gas on the hydraulic conductivity of municipal solid wastes compressed to different initial stresses, at different pore water pressures.

INTRODUCTION

Landfill has been the primary means of solid waste disposal in many countries for the past hundred years or more. Sites containing both old and new wastes will often require active pumping of leachate for level control or treatment purposes, and/or to achieve long-term stabilisation through biodegradation and flushing. For these reasons, knowledge of the hydraulic conductivity of domestic wastes and an understanding of the factors that control it are essential. Previous research (e.g. Landva and Clarke, 1990; Powrie and Beaven, 1999; Hudson et al., 2000; Beaven, 2000) has demonstrated that one of the key factors controlling the hydraulic conductivity of municipal solid waste (MSW) is effective stress, through its impact on waste compression and density.

More recently, Hudson et al. (2001) and Powrie et al. (2005) reported data on the effect of landfill gas accumulation on the hydraulic conductivity of one waste sample tested in a large compression cell. The data indicated that landfill gas could have a significant impact on hydraulic conductivity. In this paper, data from new tests are presented and discussed, and used to quantify the effect of gas generation on MSW hydraulic conductivity.

APPARATUS AND EXPERIMENTAL PROCEDURES

Tests were carried out in the Pitsea compression cell (Fig. 1), a purpose-built apparatus that accommodates a sample of waste 2 m in diameter and up to 2.5 m high

FIG. 1. Pitsea compression cell

The size of the apparatus enables representative results to be obtained from samples of generally highly heterogeneous wastes.

Overburden pressures are simulated by applying a vertical stress via hydraulic rams acting on a platen on top of the waste. Typically, the applied stress is increased in five or six stages to a maximum of 600 kPa, representing landfill depths of up to 60 m. At the end of each compression stage, the bulk density, drainable porosity and saturated hydraulic conductivity of the waste are determined.

Detailed descriptions of the compression cell, including a correction to the applied stress to account for side wall friction are given by Powrie & Beaven (1999) and Beaven (2000). The general testing procedure adopted for the current tests was as follows. The waste sample was loaded into the cylinder, and its as-placed weight determined by means of the load cells on which the apparatus is mounted. Waste composition was determined by sorting sub-samples typically 200 kg in mass. As-placed water content was determined from the weight loss of 15-20 kg oven dried samples. When compression in response to the application of a vertical stress had substantially ceased (i.e. the sample thickness was changing at a rate of less than 1% in 24 hours), the waste was saturated by allowing water to flow into the sample through the lower platen. After the refuse has been saturated it was allowed to drain under gravity to field capacity (defined as the water content of the waste in conditions of free downward gravity drainage), and the drainable porosity was calculated from the volume of leachate drained, per unit total volume. The bulk density could be calculated from the known mass and volume of the waste at any stage.

Following initial compression under a given vertical load, the position of the top platen was fixed. This was necessary because movement of the top platen during subsequent hydraulic conductivity testing (which occurred over a period of several days or weeks), would have ruptured the inflatable seals between the top platen and the cylinder side wall. Consequently all hydraulic conductivity data were determined on a waste sample held at constant volume, following compression to a given initial effective stress.

The hydraulic conductivity of the refuse at each volume and with different degrees of gas accumulation was measured in a series of constant head flow tests. Water from header tanks was allowed to flow upward through the refuse – initially at a fast rate so as to purge the sample as far as possible of air and gas. The hydraulic gradient during testing was determined by means of piezometers inserted through ports in the side of the column. Piezometers at the same horizon indicating the same hydraulic head confirmed that flow was vertical and approximately uniform. The flow rate was measured using electromagnetic flowmeters, except at low flow rates when direct measurement of the outflow into a graduated container was found to be a more reliable method.

Each hydraulic conductivity test ran continuously for several weeks, during which time gas generated naturally by waste degradation, although able to vent to atmosphere through the top platen, began to accumulate within the sample. The volume of gas accumulated within the sample was deduced from the reduction in weight indicated by the load cells mounted under the compression cell framework. Each test was concluded when no further gas accumulation was apparent. For each volume, hydraulic conductivity tests were carried out at two different average pore water pressures of about 25 kPa and 60 kPa. The change in pore water pressure was effected by switching between two different sets of header tanks and outflow pipes located at different levels.

Once testing at both pore water pressures over the range of gas accumulation conditions at each volume had been completed, the top platen seal was deflated, the sample was compressed to an increased vertical stress, and the cycle of measurements was repeated.

WASTE TESTED

The initial data reported by Powrie et al. (2005) were carried out on a sample of fresh MSW, denoted DN1, that had been partly sorted and tumbled in a drum using the Dano system: a full characterization analysis for this sample is given by Hudson et al. (2001).

The new tests reported in this paper were carried out on a sample of shredded, fresh MSW denoted SW1. Particle size and composition analysis of sample SW1 was carried out for the University of Southampton by M·E·L Research. A total of 170 kg of waste was analysed, with the results summarized in Table 1. The dominant size band was 20 to 40 mm, and the dominant material type was "miscellaneous combustibles", which is likely to include items such as disposable nappies, sanitary items, shoes, wood, rubber and cork.

The total weight of the sample SW1 used in the test in the Pitsea Cell was 5580 kg. The initial water content expressed as a proportion of the total weight was 38%, giving a mass of dry solids of 3460 kg and a water content (defined as mass of water ÷ mass of dry solids) of 61.3%.

Table 1. Material characterization and size assay for waste SW1

Material		+80 mm Kg	60-80mm Kg	40-60mm kg	20-40mm Kg	10-20mm kg	<10 mm kg	TOTAL %
Paper and card	kg	2.3	4.0	12.1	5.8	0.0	0.0	14.2
	%	9.5	16.5	50.2	23.9	0.0	0.0	
Plastic film	kg	0.7	5.0	7.3	0.9	0.0	0.0	8.1
	%	5.1	36.1	52.6	6.2	0.0	0.0	
Dense plastic	kg	2.7	2.3	3.2	4.3	0.0	0.0	7.4
	%	21.4	18.4	25.5	34.8	0.0	0.0	
Textiles	kg	0.0	0.2	0.1	0.0	0.0	0.0	0.2
	%	12.1	51.5	36.4	0.0	0.0	0.0	
[1]Misc combustible	kg	0.8	2.2	7.1	20.1	4.4	0.0	20.3
	%	2.3	6.3	20.6	58.2	12.6	0.0	
[2]Misc non-combustible	kg	1.3	1.1	2.3	17.9	5.1	0.0	16.3
	%	4.5	3.8	8.2	64.9	18.5	0.0	
Glass	kg	0.7	2.1	5.8	5.2	3.2	0.0	10.0
	%	3.9	12.4	34.5	30.6	18.6	0.0	
Ferrous Metal	kg	2.2	1.0	0.6	1.2	0.0	0.0	2.9
	%	43.6	20.3	11.2	25.0	0.0	0.0	
Non ferrous Metal	kg	0.2	0.6	0.5	0.4	0.0	0.0	1.0
	%	13.2	34.1	28.1	24.6	0.0	0.0	
[3]Putrescible	kg	1.4	1.3	7.1	11.5	0.0	0.0	12.5
	%	6.5	6.2	33.2	54.1	0.0	0.0	
[4]Fines	kg	0.0	0.0	0.0	0.0	0.0	12.3	7.2
	%	0.0	0.0	0.0	0.0	0.0	100.0	
Total	kg	12.2	19.6	45.9	67.3	12.6	12.3	100.0
	%	7.2	11.6	27.0	39.6	7.4	7.2	

1: shoes, pet litter, foam, disposable nappies, wood etc. 2: rubble, soil, plasterboard, ceramics etc.
3: waste grass, leaves, twigs, roots, fruit and vegetable matter, prepared food etc; 4: particles <10 mm

RESULTS

Data presented by Hudson et al. (2001) and Powrie et al. (2005) from tests on a Dano-processed MSW waste sample (DN1) indicated a two orders of magnitude reduction in hydraulic conductivity in initially nominally saturated waste, and a decrease in water content (drainable porosity) of the waste, in response to *in situ* gas generation. Increasing the pore water pressure compressed the gas and reduced its effect on hydraulic conductivity.

Further investigations into the effect of gas accumulation and pore water pressure on hydraulic conductivity have been carried out at lower compression stresses, on the sample of fresh shredded domestic waste SW1. The tests were carried out at constant volume following initial compression under applied stresses of 40 kPa and 87 kPa, representing landfill depths of approximately 4 m and 9 m respectively.

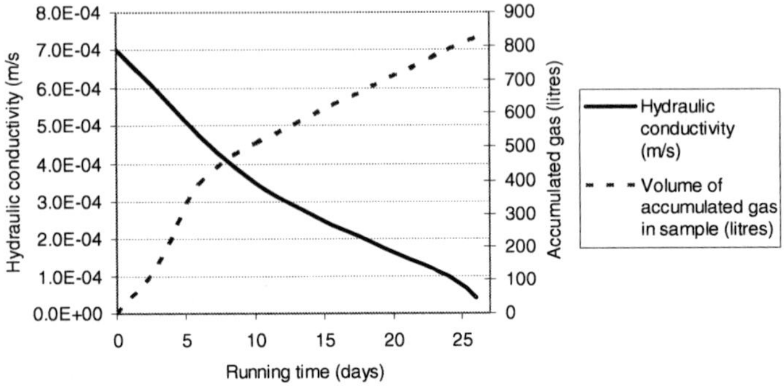

FIG. 2. Changes in hydraulic conductivity with time for sample SW1 at a constant volume corresponding to an initial applied stress of 40 kPa, with an average pore water pressure of 60 kPa

Fig. 2 shows the changes in average bulk hydraulic conductivity and the volume of accumulated gas over a 27 day period for a test conducted at constant volume (corresponding to an initial applied stress of 40 kPa) with a relatively high average pore water pressure of 60 kPa. The reduction in hydraulic conductivity in response to the increase in the volume of gas contained within the sample is clear.

Table 2 summarises the results of all tests undertaken at the two different initial stresses under both high (60 kPa) and low (25 kPa) average pore pressure conditions. At both stresses the hydraulic conductivity was reduced by gas accumulation, by approximately two orders of magnitude in low pore water pressure conditions and one order of magnitude in high pore water pressure conditions.

These large reductions are understandable when the volumes of leachate that must have been displaced from the sample by gas accumulation are considered. After compression to an applied effective stress of 40 kPa, the drainable pore volume in nominally saturated conditions was 1219 litres. Following compression to 87 kPa, the drainable pore volume of the sample had reduced to 496 litres. Some gas might have accumulated in the bottom gravel drainage layer, which had a drainable pore volume of 117 litres. In the last column of Table 2, an estimated range for the proportion of the readily drainable pore volume occupied by gas in gas accumulated conditions is given: the range depends on the amount of gas assumed to have accumulated in the lower gravel layer.

Although the tests lasted for periods of several weeks, gas accumulation could occur very quickly. In a test carried out at a pore water pressure of 60 kPa following compression to an applied stress of 87 kPa, 136 litres of gas accumulated within the first three hours of the test, causing a 76% reduction in hydraulic conductivity. This test was carried out during the summer in warm conditions. Rapid gas generation was not evident in earlier tests carried out at cooler ambient temperatures.

Table 2. Summary of test results, sample SW1

Applied Stress[1] (kPa)	Av. pwp (kPa)	Length of test (days)	Accumulated volume of gas (litres)	Initial K (m/s)	Final K (m/s)	% of drainable pore volume occupied by gas (range)
40	25	36	1064	1.0×10^{-3}	1.5×10^{-5}	78 – 87
40	60	27	820	7.0×10^{-4}	4.0×10^{-5}	58 – 67
87	25	37	297	1.1×10^{-4}	1.1×10^{-6}	37 – 60
87	60	34	571	1.5×10^{-4}	1.2×10^{-5}	93 – 100

[1]Initial stress at which constant volume was established

In Fig. 3 the data from the tests on waste samples compressed to comparatively low stresses has been combined with data from the tests on sample DN1 reported by Hudson et al. (2001) and Powrie et al. (2005) to give an indication of how the hydraulic conductivity of fresh processed MSW varies following compression to stresses between 40 and 228 kPa, in different conditions of gas accumulation and pore water pressure.

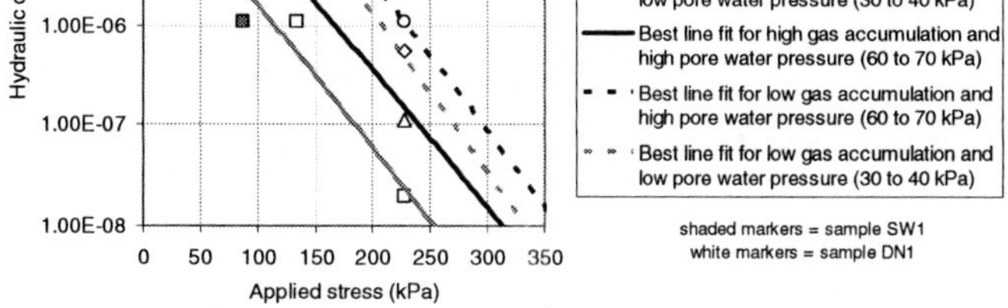

FIG. 3. Hydraulic conductivity for fresh processed MSW in different conditions of gas accumulation and pore water pressure

DISCUSSION AND IMPLICATIONS

The results indicate that gas accumulation in fresh, shredded MSW can significantly reduce the hydraulic conductivity, especially at lower pore water pressures. The reduction in hydraulic conductivity could be up to two orders of magnitude, compared with generally only a one order of magnitude reduction in the drainable porosity and hence the saturated area available for liquid flow. This implies that the intrinsic permeability for liquid flow is also reduced, for example because of a reduction in the size of the pore network through which flow occurs.

Although this paper has focused on fresh MSW, similar effects are likely in any biodegradable waste; even those that have been highly processed (e.g. by MBP) prior to landfilling (Danhamer et al., 1999). Consequently, gas must be expected to be present in many landfills, and a genuinely saturated zone is unlikely. If gas is present, leachate injection - which results in a localized increase in pore water pressure - would be expected to increase the hydraulic conductivity in the vicinity of the injection infrastructure. Conversely, in the vicinity of leachate extraction wells which cause a local reduction in the pore water pressure, the hydraulic conductivity of the waste would be expected to be reduced.

The generation of landfill gas below the leachate level (i.e., in the notionally saturated zone of the landfill), and its ability to vent (e.g. through open leachate monitoring wells) could both impact on measured leachate heads. If discrete bubbles of gas, unconnected to the atmosphere, are present within the notionally saturated zone, the initial response to pumping from a leachate well and the consequent reduction in pore water pressure could be an increase in leachate levels. This was observed during pumping tests by Burrows et al. (1997), and has also been detected in geophysical surveys on leachate pumping systems (e.g. Jolly et al., 2007).

CONCLUSIONS

The very significant (up to two orders of magnitude) decrease in hydraulic conductivity with degradation gas accumulation in a biodegradable MSW suggests that, in addition to the reduction in saturated area available for flow, the intrinsic permeability also changes as a result of a change in the geometry of the relevant pore network.

Hydraulic conductivity in a gassy waste must be expected to increase if the pore water pressure is increased (e.g. in the vicinity of leachate injection infrastructure), and to reduce if the pore water pressure is reduced (e.g. around a pumped leachate extraction well).

Variations in gas content and pore water pressure could affect measured leachate levels, hence care is needed in their interpretation.

ACKNOWLEDGMENTS

The research presented in this paper was undertaken with the financial support of the UK Engineering and Physical Sciences Research Council (EPSRC) and the former Category C Landfill Tax Credit Scheme.

REFERENCES

Beaven, R.P. (2000). "The hydrogeological and geotechnical properties of household waste in relation to sustainable landfilling." *PhD dissertation, University of London (Queen Mary and Westfield College)*

Burrows, M.R., Joseph, J.B. and Mather J.D. (1997). "The hydraulic properties of in-situ landfilled waste." *Proceedings of the 6th International Waste Management and Landfill Symposium, S. Margherita di Pula, Cagliari, Italy. pp 73-83*

Danhamer H., Dach J., Obermann I., Jager J. and Ostrowski M.W. (1999). "Simulation of Emissions from Landfills containing MBP Waste" *Proceedings of the 7th International Waste Management and Landfill Symposium, S. Margherita di Pula, Cagliari, Italy. Vol I, pp 519-532*

Hudson, A.P., Beaven, R.P and Powrie, W. (2000). "Current research into the hydrogeological properties of household waste using a large scale compression cell" *Proceedings of the Waste 2000 Conference, Stratford-upon-Avon, Warwickshire, UK. pp227-237*

Hudson, A.P., Beaven, R.P & Powrie, W. (2001). "Interaction of water and gas in saturated household waste in a large scale compression cell." *Proceedings of the 8th International Waste Management and Landfill Symposium, S. Margherita di Pula, Cagliari, Italy, Vol III, pp585-593*

Jolly, J., Barker, R.D., Beaven, R.P. and Herbert, A.W.(2007). "Time Lapse electrical imaging to study fluid movement within a landfill." *Proceedings of the 11th International Waste Management and Landfill Symposium. S Margherita di Pula, Cagliari, Sardinia, Italy.* (CD only)

Landva, A.O., and Clark, J.I. (1990). "Geotechnics of Waste Fills," *In: Knowles A., and Landva, G.D., Ed., Geotechnics of Waste Fills - Theory and practice. Baltimore: ASTM Special Technical Publication*

Powrie, W., and Beaven, R.P. (1999) "Hydraulic properties of household waste and implications for landfills." *Proceedings of the Institution of Civil Engineers (Geotechnical Engineering)* **137** (4), 235-247

Powrie, W., Beaven R.P., and Hudson, A.P. (2005) "Factors affecting the hydraulic conductivity of waste," *In: J-P. Gourc, Ed., International Workshop "Hydro-physico-mechanics of landfills". Grenoble.*

Variation of fluid conductivity with settlement of domestic waste

G. Stoltz[1,2], JP. Gourc[1]

[1] LTHE Laboratory, Grenoble University, BP 53 - 38041 Grenoble Cedex 9, France
[2] CRPE, Veolia Environnement, Limay, France

ABSTRACT: Domestic waste (MSW) can be considered as a gradually changing porous medium. This paper reports tests of a new type on a waste sample in order to get correlations between mechanical and hydraulic properties. Compression tests coupled with gas permeability tests have been made on waste sample taken from a hole drilled in a French landfill. The apparatus used for these tests is an oedo-permeameter developed in Lirigm-LTHE. In a second stage, the oedo-permeameter cell has been modified to carry out retention tests on a waste sample under compression. These tests allow to get water retention curve and to assess pores volumetric distribution of the waste.

INTRODUCTION

Land disposal of solid waste has changed dramatically because the "dry tomb" approach is not sustainable. Indeed optimal moisture is required for biological degradation. Municipal solid waste landfills can be operating as bioreactor which intends to accelerate waste degradation by circulating fluids through the waste in a controlled manner. One of the main uncertainties concerning the practicality of operating a landfill as a bioreactor is the control of fluids flows and of the water content of waste. The hydraulic permeability (leachate), the gas permeability (biogas) and the retention properties of the waste govern the ease with which fluids may be introduced into and extracted from the landfill. These hydrodynamics properties depend on the mechanical compression and the saturation degree of the fluids (liquid and gas) and also on the waste evolution upon the elapsed time characterized by biodegradation. Moreover these parameters depend on initial mechanical and biological pre-treatment of the waste (MBT).

In the context of bioreactor technique, these data could be specifically useful (Beaven 1995, Benson 2006, Durmusoglu 2006, Jain 2005 and 2006, Olivier 2007, Powrie, 1999)

PRESENTATION OF THE RESEARCH PROGRAM

Domestic waste is a complex porous medium. Its void structure is largely dependent of the components and the possible mechanical pre-treatment. Its evolution over time is dramatically conditioned by its settlement due to the surcharge of the column of waste in the landfill and by biodegradation.

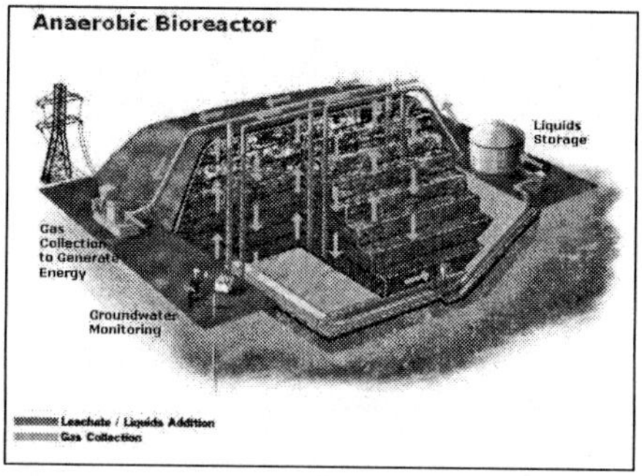

FIG 1. General schema of a bioreactor (Ademe source)

Characterization of the evolution of the void structure can be obtained through measurement of the fluid permeability and water retention. In addition, quantification of these properties is important for a better understanding of the bioreactor operation.

Following a large research programme about waste settlement and geomechanical properties of waste, a new programme is starting in Lirigm-LTHE about the correlation between mechanical and hydraulic properties. The programme attempts to use for waste, experimental tests previously used for soils.

Several specific prototype devices are used for coupling measurement of settlement of a refuse sample and measurement of the hydraulic or gas permeability.

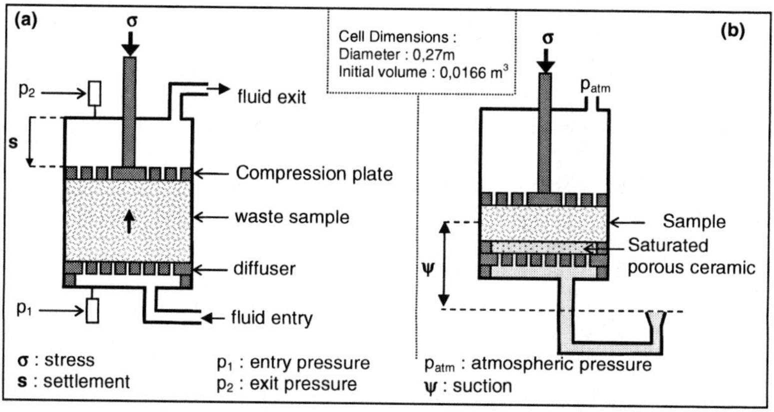

FIG 2. (a) Schematic section of the compression cell of the oedopermeameter equipped for the gas permeability test , (b) Schematic section of the modified oedopermeameter compression cell equipped for the suction test

GENERAL DESCRIPTION OF THE OEDOPERMEAMETER

The oedopermeameter (Figure 2a) consists of a rigid cylindrical cell (diameter 0,27m and height 0,29m) subject to vertical compression σ in oedometric conditions (lateral strains prevented) up to 200 kPa (equivalent to the self-weight of a 20 m waste column) by means of a load apparatus. The cell is equipped to perform gas permeability test. All electronic equipments are connected to an automatic data-acquisition system.

ORIGIN AND COMPOSITION OF THE TESTED WASTE

The studied waste was collected from a hole drilled (few meters under the surface) in a French landfill. This waste was stored one year ago. It was dumped in the landfill without any pre-treatment but the effect of drilling was to shred the waste particles.

Waste characterization of the material showed that it is composed of 20 % of organic waste (including kitchen and green waste, wood, paper and cardboard), 16 % of plastics, 8 % of glass, 3 % of textile, 7 % of metals, 29% of very small particles and 17 % of others. Its particle size distribution was also evaluated: The fraction < 50 mm is around 70% of the waste mass.

The waste was characterised first with regard to its initial water content (by oven-drying at 70°C during 72 h). The water content (w) expressed as the ratio of the mass of constitutive water to the mass of solid particles was hence evaluated to approximately 30%.

Only short duration tests are presently carried out, so the influence of the biodegradation is not taken in consideration.

COMPRESSION OF THE WASTE SAMPLES

The waste was prepared in the cell in thin layers and compacted with added water. After humidification for compaction in the cell, the water content (w) reached 56% and the corresponding volumetric water content (θ) (i.e. the ratio of the volume of water in the sample to the total volume of the sample) was around 27%.

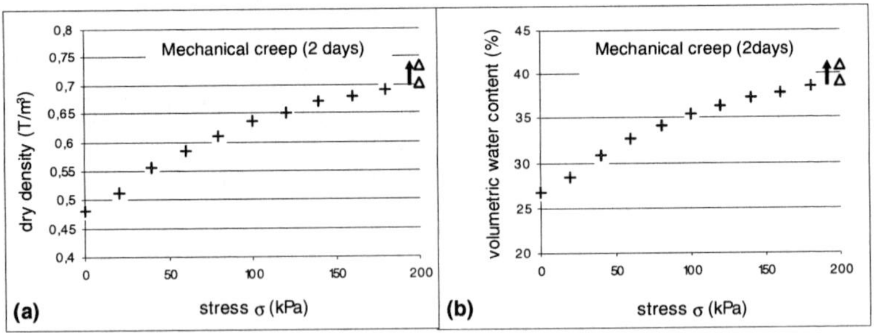

FIG 3. (a) Evolution of the dry density versus the compression vertical stress, (b) Evolution of the volumetric water content versus the compression vertical stress

The test procedure consisted of compressing the waste to 200 kPa (normal stress) in 20 kPa steps. Every step lasts 20 minutes. A diagram corresponding to the increase of the dry density due to the compression is presented Figure 3a and the increase of the volumetric water content is presented Figure 3b. At each compression step, a gas (nitrogen) permeability test was carried out.

At the last step, the load was held at 200 kPa for two days. During this phase at constant load, a mechanical creep is observed. A gas permeability test was done at the end of this phase.

EVOLUTION OF THE HYDRO-PHYSICAL CHARACTERISTICS OF THE WASTE VERSUS COMPRESSION STRESS

In order to perform the gas permeability tests, pressure sensors are installed at the entry and exit of the cell (Figure 2a). A mass flow meter is placed at the entry of the cell.

The gas permeability measures are realized in Darcy's conditions and the gas permeability coefficient k_g is given by the relation:

$$\frac{q_1}{A} = \frac{k_g}{\mu_g} \frac{1}{2L} \frac{p_1^2 - p_2^2}{p_1}$$

With A the section of the sample, L the height of the sample, μ_g the dynamic viscosity of gas (taken at 20°C, temperature of the permeability tests), q_1 the gas volumetric flow at the entry of the sample, p_1 the gas entry pressure and p_2 the gas exit pressure.

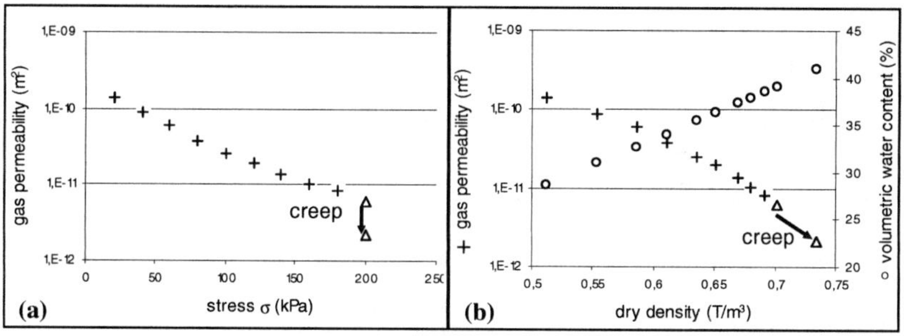

FIG 4. (a) Evolution of the gas permeability versus the compression vertical stress, (b) Evolution of the gas permeability versus the dry density

Figure 4a shows that the gas permeability decreases when compression stress and thus the dry density increases. The Figure 4b shows that the gas permeability is depending on the dry density and volumetric water content. The creep phase is identified on every diagram. It is worth noting on Figure 4b that the relationship between gas permeability and dry density seems independent of the method used to achieve settlement: increasing compression stress (points drawn with a cross on Figure 4b) or creep under constant compression stress (points drawn with a triangle on Figure 4b)

INFLUENCE OF THE SATURATION ON THE PERMEABILITY

A first characteristic of the porous structure is that intrinsic permeability is independent of the fluid. When the fluid saturation S of the porous media is 100%, the fluid permeability is marked $k_{g,wSat}$ and corresponds to the intrinsic permeability with $k_{gSat} = k_{wSat}$. At a constant density, the fluid (gas, water) permeability varies with the water saturation (Figure 5) and whether the system is wetting or drying (hysteresis). Fluid permeability ($k_{g,w}$ in m^2) is related to fluid conductivity ($K_{g,w}$ in m/s) as follows:

$$K_{g,w} = k_{g,w} \frac{\rho_{g,w} G}{\mu_{g,w}}$$

With $\mu_{g,w}$ the dynamic viscosity of fluid, $\rho_{g,w}$ the density of fluid and G the acceleration due to gravity.

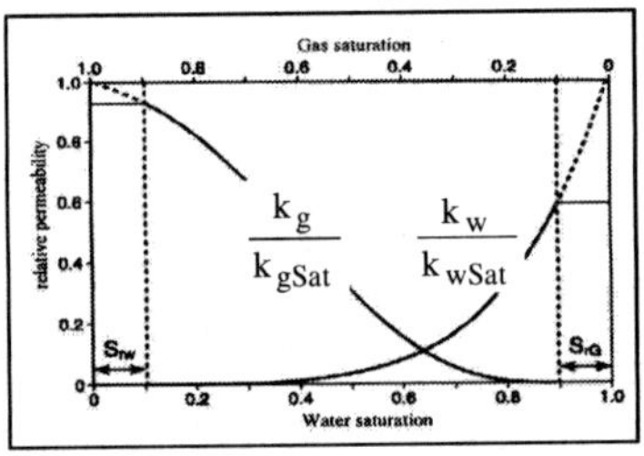

FIG 5.Schematic relative permeability with respect to fluid saturation (Warrick, 2002)

At the end of the compression process, the waste sample are initially fully saturated with water and subsequently drained to reach the field capacity. At field capacity, the volumetric water content is around 43,5%. A gas permeability test was done at this volumetric water content value (Figure 6a).

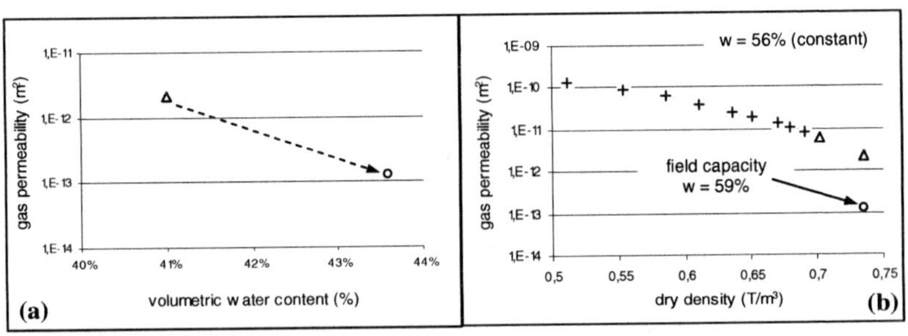

FIG 6. (a) Evolution of the gas permeability versus the volumetric water content after a saturation-desaturation cycle (field capacity state), (b) Evolution of the gas permeability during the phase of compression and after a saturation-desaturation cycle (field capacity state)

It is worth noting that for the same dry density, the gas permeability is decreasing with increasing volumetric water content in agreement with Figure 5. In the future, we intend to perform tests at a constant density and at different water content to obtain a diagram similar to Figure 5.

WATER RETENTION CURVE AND PORE VOLUMETRIC DISTRIBUTION

The oedopermeameter is modified so as to realize a retention test on a waste sample under compression. The waste sample is set above a porous ceramic plate which has an air entry pressure of 50kPa. The suction is given by the capillary pressure which is the difference between the pressure of the water and the atmospheric pressure (Figure 2b).

At suction equilibrium, the capillary pressure is plotted versus the corresponding volumetric water content. This is the conventional retention curve obtained from desaturation of the sample. The hysteresis effect is not studied. Results of Figure 7 correspond to two compression stresses and therefore to two dry densities.

The retention curve obtained shows that the porometric structure is modified by the compression.

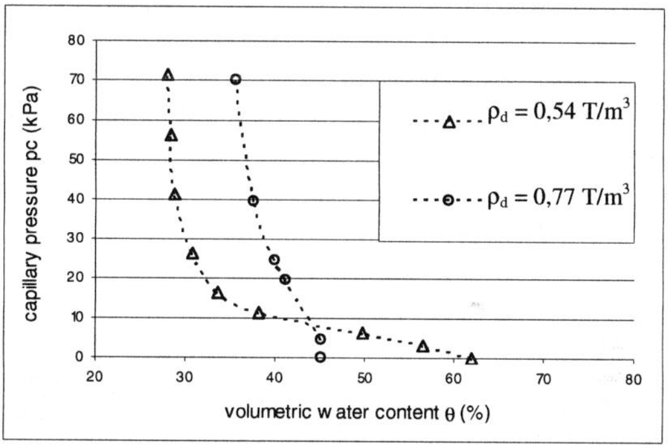

FIG 7. Water retention curve of the drilled domestic waste at two dry densities

The water retention curve can be interpreted in pores volumetric distribution. Indeed, the Laplace relationship gives a relation between the capillary pressure p_c and the equivalent pores diameter d_p:

$$p_c = 4\frac{T}{d_p}$$

With T the superficial tension of the fluid and given by the value of the water at 20°C ($T = 7{,}3.10^{-2}\,N.m^{-1}$). The solid surface is supposed perfectly wettable.

An increment of suction is correlated to a pore diameter d_p. This suction induces a drawdown of a porometric class corresponding to an increment of void volume.

Figure 8a and 8b show the pores volumetric distribution deduced from the Laplace relationship for the water retention curves presented on Figure 7.

The two suction tests at two dry densities are realized at constant solid dry mass (m_d=1970g). Figure 8a shows that for a given sample in the oedopermeameter the volume of void of each class decreases when the sample is compressed. The volume of void with pore diameter d_p < 40 μm is almost the same for the two tests when the largest pores are dramatically decreasing.

Figure 8b extrapolated from Figure 8a corresponds to two samples of the same overall volume: the total void volume (characterized by θ_{Sat}) is decreasing with increasing density, but the volume of smallest pores is increasing.

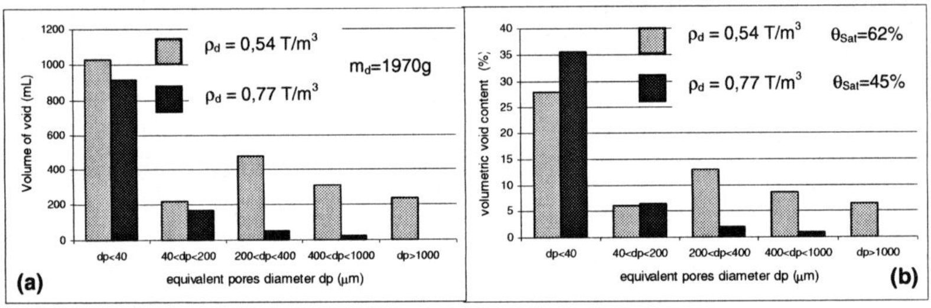

FIG 8. (a) Volume of void versus equivalent pores diameter of the drilled domestic waste at two dry densities, for a given mass of waste, (b) Volumetric void content versus equivalent pores diameter of the drilled domestic waste at two dry densities (corresponding to a given unit overall volume of waste)

CONCLUSION AND PERSPECTIVES

Bioreactor efficiency is depending on the optimal wetting of the waste body. As demonstrated, the porometric structure is dramatically changing with compression. Consequently the alteration of the hydraulic and gas conductivities will be taken into account in the management of a bioreactor. The use of the oedopermeameter to assess the hydro-mechanical properties of waste is promising.

AKNOWLEDGMENTS

This research programme is supported by the Environment French Agency (Ademe) and the Veolia Environment Research Center (Crpe).

REFERENCES

Beaven, R.P. et Powrie, W. (1995) Hydrogeological and geotechnical properties of refuse using a large scale compression cell. Proc. Sardinia 95, 5th International Landfill Symposium, Cagliari, Vol.II

Benson CH, Barlaz MA, Lane DT, et al. (2007) Practice review of five bioreactor/recirculation landfills Waste Management 27

Durmusoglu E, Sanchez IM, Corapcioglu MY (2006) Permeability and compression

characteristics of municipal solid waste samples Environmental Geology
Jain P, Powell J, Townsend TG, et al. (2005) Air permeability of waste in a municipal solid waste landfill Journal of Environmental enginneering
Jain P, Powell J, Townsend TG, et al. (2006) Estimating the hydraulic conductivity of landfilled municipal solid waste using the borehole permeameter test Journal of Environmental enginneering
Olivier, F. and Gourc, J.P. (2007) Hydro-mechanics of Municipal Solid Waste subject to leachate recirculation in a large-scale compression reactor cell. Waste Management 27
Olivier, F. Gourc, J.P. Achour, F. Morais, J. Bayard, R. (2005) Evolution of bio-physical and mechanical characteristics of MSW after 2 years incubation in a laboratory-scale bioreactor Proc. Sardinia 2005, 10th International Landfill Symposium, Cagliari, Italy
Olivier, F. (2003) Tassement des déchets en CSD de classe II : du site au modèle. PhD dissertation, University of Grenoble, 325 p.
Powrie, W. et Beaven, R. P. (1999) Hydraulic properties of household waste and implications for landfills. Proc. of the Institution of Civil Engineering, Vol. 137
Warrick. A.W. (2002) Soil Physics Companion CRC Press LLC 389 p

Evaluating Cr(VI) Leaching from Recycled Waste Concrete Aggregate Using Acceleration Tests

Toru Inui[1], Masashi Kamon[2], Takeshi Katsumi[3], M. ASCE and Akiko Kida[4]

[1] Assistant Professor, Graduate School of Global Environmental Studies, Kyoto University, Yoshida-honmachi, Sakyo, Kyoto 606-8501, Japan, inui@mbox.kudpc.kyoto-u.ac.jp
[2] Professor, Graduate School of Global Environmental Studies, Kyoto University, Yoshida-honmachi, Sakyo, Kyoto 606-8501, Japan, kamon@mbox.kudpc.kyoto-u.ac.jp
[3] Associate Professor, Graduate School of Global Environmental Studies, Kyoto University, Yoshida-honmachi, Sakyo, Kyoto 606-8501, Japan, tkatsumi@mbox.kudpc.kyoto-u.ac.jp
[4] Head of Laboratory, Research Centre for Material Cycles and Waste Management, National Institute for Environmental Studies, 16-2, Onogawa, Tsukuba, 305-8506, Ibaraki, Japan, akida@nies.go.jp.

ABSTRACT: Cement based secondary materials, such as waste concrete aggregates and cement stabilized wastes, are often utilized as geo-materials. Their physical and chemical properties may change in field due to the loss of soluble chemicals and/or the surface wearing. Thus, the laboratory testing protocol for evaluating their in situ leaching behavior needs to be developed. In this study, three different acceleration tests, wetting-drying test, freezing-thawing test and abrasion test, are conducted for several waste concrete aggregates (WCAs) to evaluate the effects of the exposure to these conditions, which may actually occur in field conditions, on the leaching behavior of hexavalent chromium (Cr(VI)) contained in the mortar of WCA. Leaching amounts obtained are compared and correlated with those in the conventional batch leaching test. Exposure to these accelerated conditions, particularly the wetting-drying process, promotes the leaching of Cr(VI). Leaching amounts from granular WCA exposed to these conditions are no more than that in the conventional batch leaching test, which is conducted for WCA crushed with < 2 mm in grain size. This finding supports that the Cr(VI) leaching amount of WCA exposed under environmental conditions can be conservatively estimated by the conventional batch leaching test.

INTRODUCTION

Annual discharge of waste concrete from demolition works has reached 32 million ton in Japan. Most of waste concrete is processed into waste concrete aggregate (WCA) by crushing and grain-size control (Fig. 1) and recycled as geo-materials (Kamon et al. 2000). However, a certain amount of chromium is contained in both raw materials for cement and cement products (Kamon 2000). Accordingly, the leaching of hexavalent chromium (Cr (VI)) with considerable concentrations from the mortar fraction of WCA has been observed (Kamon et al. 2004). Thus, characterization of the Cr(VI) leaching behavior is necessary for assuring the environmental suitability of the WCA.

Physical and chemical properties of cement based secondary materials, such as WCA and cement stabilized wastes, may gradually change in field due to the loss of soluble chemicals and/or the surface wearing. Thus, developing the testing method to evaluate the effect of these property changes on the leaching behavior has been one of important issues (e.g. Gervais et al. 2004).

In this study, leaching behavior of Cr(VI) is evaluated for the WCA exposed to three types of acceleration tests, i) wetting-drying, ii) freezing-thawing and iii) abrasion, to which the WCA may be subjected in field conditions. Leaching amounts obtained are compared and correlated with those in the conventional batch leaching test to discuss a possibility to estimate the long term leaching performance with a simple evaluation test.

FIG. 1. Waste concrete aggregate used in the experiment.

LABORATORY EXPERIMENTS

Waste Concrete Aggregate

Three different WCA samples, which are referred to WCA-1, 2 and 3 were used. They were collected from demolition works of the concrete structure. Table 1 shows the basic properties including the grain size distribution of three WCA samples. Water content and water absorption tests are carried out according to JIS A1203 and JIS A1110, respectively. WCA-2 is considered to contain a larger amount of mortar since the water absorption, which is a good index of mortar content, is greater than others.

Table 1. Basic properties of WCA samples.

Sample	WCA-1	WCA-2	WCA-3
Site	Tokyo, Japan	Kanagawa, Japan	Tokyo, Japan
Age of original structure	Unknown	40 years	40 years
Natural water content (%)	3.3	5.5	4.2
Water absorption capacity (%)	5.4	7.0	5.2
Grain size distribution (%)			
19 mm –	14	0	0
9.5 mm – 19 mm	41	13	23
4.75 mm – 9.5 mm	22	50	55
– 4.75 mm	23	37	22

Leaching Test

Both batch and tank leaching tests were conducted to determine the Cr(VI) leaching amount without any accelerated exposure. Table 2 indicates the testing conditions of these tests. Solid samples for the batch leaching test were prepared in three different forms; 1) original form, 2) sieving through a 13.2 mm-opening sieve and 3) crushed with the maximum grain size less than 2 mm. The third one is equivalent to the method for the official batch leaching test for the soil quality, regulated by the Japanese Environmental Agency notification No.46, 1991 (JLT46). In addition, acid solvent was employed to evaluate the pH effect. In tank leaching test, WCA was stored in the solvent for 28 days with and without the solvent renewal. In the former case, the solvent was renewed after 1, 3, 7 and 14 days. The leachate was filtered using a 0.45 μm-opening membrane filter and then analyzed with the diphenylcarbazide absorption method to determine the Cr(VI) concentration immediately after the collection.

Table 2. Testing conditions of batch and tank leaching tests.

Test type	Batch leaching test			Tank leaching test
	Original form	≤ 13.2 mm	≤ 2 mm	Original form
Sample weight (g)	200	80	50	200
Solvent	Distilled water (pH = 5.8 to 6.3) or Acid solvent			
Solvent volume (ml)	2,000	800	500	2,000
Liquid to solid ratio, L/S	10	10	10	10
Duration	6 hours	6 hours	6 hours	28 days
Remarks	Horizontal shaking			With and without the solvent renewal

Acceleration Test

Three types of acceleration tests, wetting-drying test, freezing-thawing test and abrasion test, were conducted. For the WCA subjected to each accelerated exposure, batch and tank leaching tests were conducted respectively to quantify the Cr(VI) leaching amount. Cr(VI) leaching during the acceleration test was also considered.

Abrasion Test

Assuming the situation where the surface of WCA was gradually worn out under environmental conditions, accelerated abrasion test was carried out. WCA was kept in a rotary mill until its abrasion loss (the ratio of 2 mm under fraction in dry weight) reached 5, 10 and 15%, respectively. Then, leaching test was conducted for both gravel fraction and ≤ 2 mm fraction.

Wetting-Drying Test

WCA was firstly placed in a sealed oven to provide the temperature at 60 ± 3 °C for 24 hours. After cooling the specimen in the chamber, it was immersed in the distilled water at 20 ± 3 °C for 23 hours. The volume of the water represented by L/S (liquid to solid ratio) was 10. The test was terminated after the number of this cycle reached 12. As a result, the cumulative volume of the water contacting the sample goes up to 120, represented by L/S, at the termination of the test.

Freezing-Thawing Test

In the cold district where the temperature changes around 0 °C, water fraction in WCA freezes up and expands, and with the temperature raising it melts. This freezing-thawing cycle can cause the physical damage to WCA. WCA specimen was stored in the water which can fully cover specimen (L/S = 1.5). Then WCA specimen in the water was exposed to the freezing-thawing cycle (4 hours/cycle, freezing at -18 °C, thawing at 5 °C) until the number of cycles reached 10, 20, 30, 60, 90, 120, 150, and 180, respectively.

RESULTS AND DISCUSSIONS

Batch and Tank Leaching Test

Cr(VI) leaching concentrations of WCAs with different forms are shown in Fig. 2. In this paper, Cr(VI) leaching concentration and mass are normally reported by the average value of three-duplicate tests and its variation (only two-time tests in the freezing-thawing test). Grain size effect on the Cr(VI) leaching concentration is so significant that leaching concentrations of ≤ 2 mm specimen are 1.7 to 3.8 times higher than those of original-form WCA specimen. This indicates that the regulatory leaching test (JLT46), which is conducted for ≤ 2 mm specimen, tends to overestimate the Cr(VI) leaching concentration from the WCA utilized in original form.

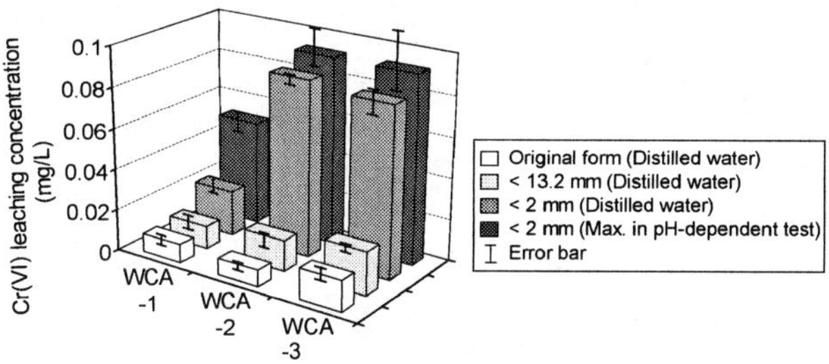

FIG 2. Cr(VI) leaching concentration in the batch leaching tests.

Figure 3 indicates the mass of Cr(VI) leached from the original-form WCA during the tank leaching test. Hereafter, results of the leaching tests are given by converting the leaching concentration into the mass leached from unit-weight WCA to compare the results of various testing methods. The variation of measured values is larger compared with in the batch leaching test. Renewal of the solvent promotes the Cr(VI) leaching; average leaching mass values after 28 days with the solvent renewal are larger than those without the solvent renewal for all WCA samples. This is because the fresh solvent could promote the diffusive leaching of Cr(VI). Compared with the results of batch leaching test, leaching amount is larger particularly for WCA-2. Judging from the fact that WCA-2 contains more mortar, this can be attributed to the diffusive leaching from mortar fraction, in which most Cr(VI) may exist.

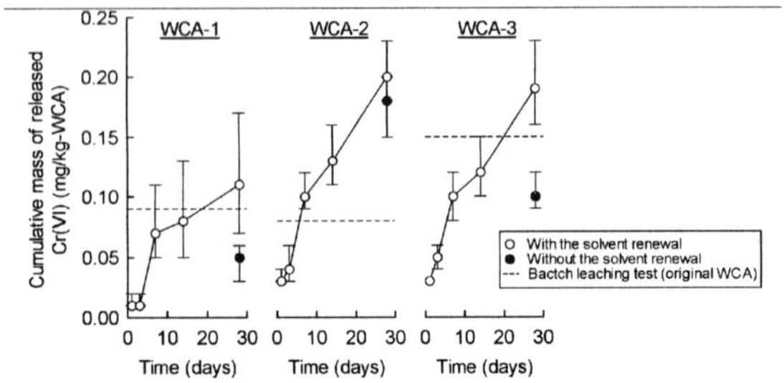

FIG.3. Mass of Cr(VI) leached in the tank leaching test.

Leaching Affected by the Accelerated Exposures

Abrasion

Figure 4 indicates the mass of Cr(VI) leached from unit weight of ≤ 2 mm fraction in the batch leaching test following the accelerated abrasion. For WCA-1 and 3, Cr(VI) leaching concentration is lowered as the abrasion loss becomes larger. For the lower abrasion loss, ≤ 2 mm fraction is mainly composed of the fragile mortar, which is considered to contain most of leachable Cr(VI). Thus, Cr(VI) leaching concentration is higher for the lower abrasion loss. For WCA-2, however, the leaching amount is lowest when the abrasion loss is 0 to 5 %. As shown in Fig. 5, which shows the sum of Cr(VI) leaching mass from gravel fraction and ≤ 2 mm fraction, Cr(VI) leaching from the gravel fraction is dominant for WCA-2 unlike others. Considering that WCA-2 contains more mortar than WCA-1 and 3, as described previously, it can be estimated that Cr(VI) at the surface of original-form WCA have already been leached for some reasons and Cr(VI) content in the fragile mortar of WCA-2 was relatively small consequently.

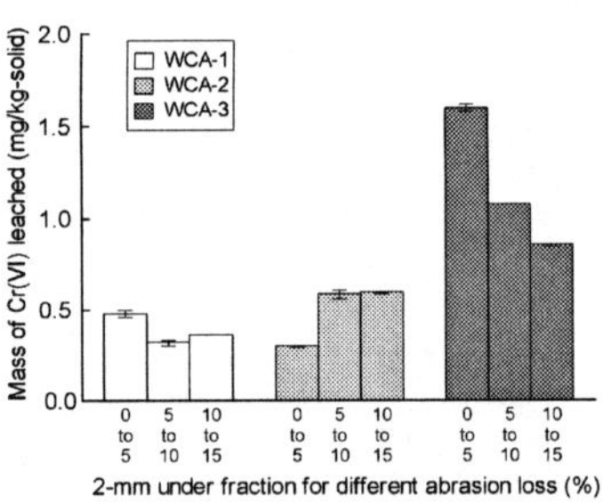

FIG.4. Mass of Cr(VI) leached from 2 mm under fraction after abrasion.

Figure 5 indicates the effect of the accelerated abrasion on the Cr(VI) leaching is more significant for WCA-2 than for WCA-1 and 3. When the abrasion proceeds to some extent, Cr(VI) leaching from the gravel fraction is decreased due to the detachment of mortar. Therefore, effect on the total leaching amount is not major for

WCA-1 and 3. On the other hand, for WCA-2 with the larger mortar content, Cr(VI) leaching from the gravel fraction was not decreased even though the abrasion loss got larger. As a result, the total Cr (VI) leaching amount was continuously increased. These observations support that the Cr (VI) leaching is promoted by the abrasion particularly for the WCA with the larger mortar content.

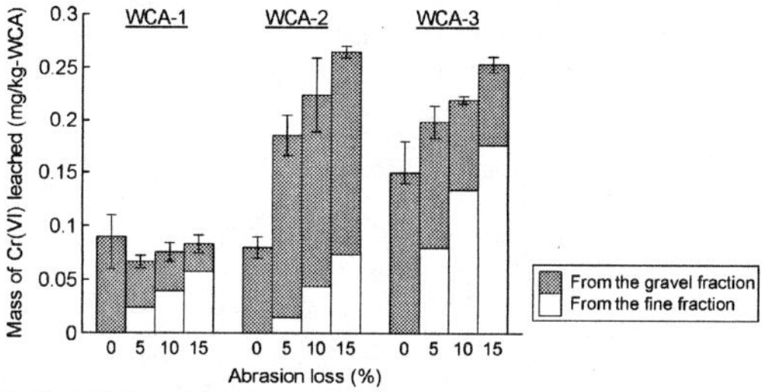

FIG.5. Total mass of Cr(VI) leached from the fine and the gravel fractions.

Wetting-Drying

Leaching behavior of Cr (VI) during and after the wetting-drying cycles is shown in Fig. 6, where mass of Cr(VI) leached in each cycle and their cumulative value are given. The largest leaching amount was observed at the first cycle for all WCAs. Thereafter, the Cr(VI) leaching was converged with cycles. Cumulative mass of Cr(VI) leached after 12 cycles reached approximately 0.6 to 0.9 mg/kg. WCA-2 and 3 exhibited the relatively larger leaching amount throughout the cycles. This observation is consistent with the testing results shown in Figs. 2, 3 and 5. Wetting-drying test can be regarded as the tank leaching test with drying process. Comparing with these two tests, the Cr(VI) leaching mass values after the 12 wetting-drying cycles (cumulative L/S = 120, duration of storage in water = 12 days), approximately 0.6 mg/kg for WCA-1 and 0.8 mg/kg for WCA-2 and 3, are much larger than those after the tank leaching test with the solvent renewal (cumulative L/S = 50, duration of storage in water = 28 days), approximately 0.1 mg/kg for WCA-1 and 0.2 mg/kg forWCA-2 and 3, although the testing conditions are not consistent in cumulative L/S and duration of storage. This comparison indicates that the exposure to the wetting-drying cycles increases the Cr(VI) solubility in the mortar fraction of the WCA.

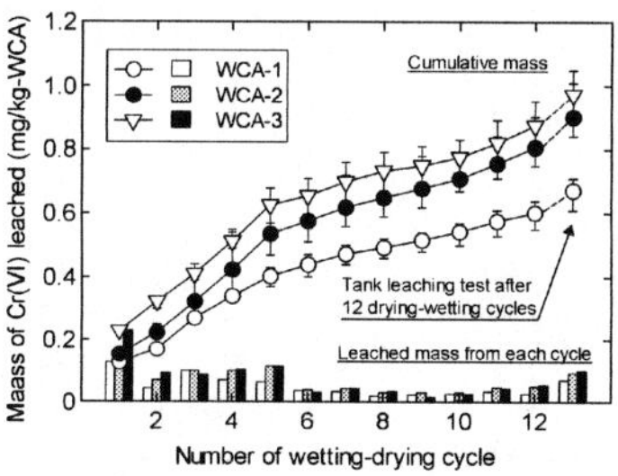

FIG.6. Mass of Cr(VI) leached during and after the wetting-drying cycles.

Freezing and Thawing

Figure 7 indicates the mass of Cr(VI) leached during and after the freezing-thawing with the number of freezing-thawing cycle. Bach leaching test was conducted on the WCA after the freezing-thawing cycles. Although variations in Cr(VI) leaching amount are relatively large, average leaching amount is getting larger with cycles. In particular, WCA-2 exhibits the largest leaching amount. According to the visual observation, mortar fraction of the WCA-2 was severely exfoliated and crushed after more than 20-cycle exposure. This is probably because WCA-2 having a higher water retention capacity is easily affected by the freezing-thawing cycles.

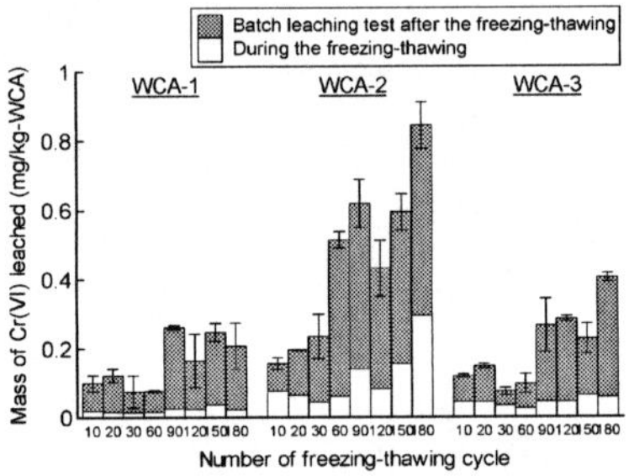

FIG.7. Mass of Cr(VI) leached during and after the freezing-thawing cycles.

DISCUSSIONS

Figure 8 shows the comparison of Cr (VI) leaching amounts among all the tests for WCA 1 to 3, including the testing results for other WCA (referred to WCA-4) reported by Kamon et al. (2007). WCA-4 has a coarser grain size distribution than others (≥ 19 mm: 24%, 9.5 – 19 mm: 54%, 4.75 – 9.5 mm: 11%, < 4.75 mm: 11%). The leaching amount of Cr(VI) are plotted with the cumulative volume of solvent contacting with WCA during the test, represented by the L/S. The wetting-drying test provided the largest leaching mass of three different acceleration tests. In addition, for the WCA with the larger mortar content like WCA-2, the freezing-thawing promotes the Cr(VI) leaching to the same degree.

Comparing the leaching amounts after the accelerated exposure with those in simple and/or standard leaching tests (batch and tank leaching tests, acid addition test), leaching amounts during and after the accelerated tests are larger than those in the batch leaching for WCA in original form. However, they are no more than those in the acid addition test for 2 mm under specimens for all WCA samples. Considering how simple and time-saving the batch leaching test is, it can be concluded that the acid addition test can be employed to estimate quantitatively the Cr(VI) leaching amount of WCA subjected to environmental conditions for the safe side. Of course, further studies is necessary to identify the mechanisms of promoting the Cr(VI) leaching influenced by testing procedures.

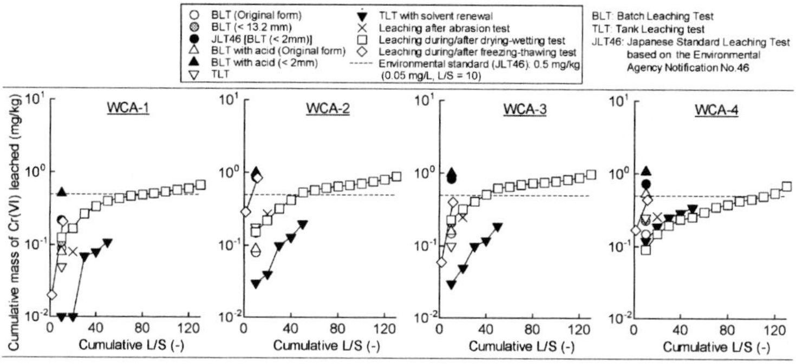

FIG.8. Comparison of Cr(VI) leaching amounts in various testing methods (Data for WCA-4 is collected from Kamon et al. (2007)).

CONCLUSIONS

Three different accelerated exposure tests were conducted for various WCAs to evaluate how much impact the exposure to environmental conditions has on the leaching behavior of Cr(VI) contained in the mortar of WCA. Exposure to these accelerated conditions increases the leaching amount of Cr(VI). Particularly, effect of the wetting-drying exposure is most significant. Leaching amounts of WCA subjected to these accelerated exposures are no more than that in the simple batch leaching test, which is conducted for the crushed WCA < 2 mm in grain size with acid solution, for various WCA samples. This finding supports that the Cr(VI) leaching amount exposed under environmental conditions can be conservatively estimated by the simple batch test using crushed samples.

REFERENCES

Gervais, C., Garrabrants, A.C., Sanchez, F., Barna, R., Moszkowicz, P., and Kosson, D.S. (2004). "The effects of carbonation and drying during intermittent leaching on the release of inorganic constituents from a cement-based matrix." *Cement and Concrete Research*, Vol.34: 119-131.

Kamon, M. (2000). "Remediation techniques by use of ground improvement." *Soft Ground Technology*, GSP No.112, J.L. Hanson and R.J. Termaat (eds.), ASCE: 374-387.

Kamon, M., Hartlén, J. and Katsumi, T. (2000). "Reuse of waste and its environmental impact." *GeoEng2000, Proc. Int. Conf. on Geotechnical and Geological Engineering*, Technomic Publishing Company, Pennsylvania: 1095-1123.

Kamon, M., Shoji, Y., Endo, K., Kida, A. and Sakai, S. (2004). "Leaching characteristics of hexavalent chromium from sludge stabilized with Portland cement and waste concrete powder." *Proc. 4th Japan-Korea Joint Seminar on Geoenvironmental Engneering*: 117-120.

Kamon, M., Inui, T., Shimada, H., Tanabe, M., Katsumi, T. and Kida, A. (2008). "Evaluating the leaching characteristics of waste concrete aggregate using acceleration tests", *Journal of the Society of Material Science, Japan*, (in printing).

Evaluation of Base Prepared from Road Surface Gravel Stabilized with Fly Ash

Bulent Hatipoglu[1], Tuncer B. Edil[2], Member Geo-Institute, P.E., Craig H. Benson[3], Member Geo-Institute, P.E.

[1]PhD , Faculty of Civil Engineering, İstanbul Technical University, Maslak, 80626, İstanbul, Turkey, bulent@ins.itu.edu.tr
[2]Professor, University of Wisconsin-Madison, 1415 Engineering Drive, Madison, 53706, WI, USA, tbedil@wisc.edu
[3]Wisconsin Distinguished Professor, University of Wisconsin-Madison, 1415 Engineering Drive, Madison, 53706, WI, USA, benson@engr.wisc.edu

ABSTRACT: This paper describes a study of sustainable roadway construction where cementitious fly ashes (Class C and off-specification) were used to stabilize road surface gravel (RSG) in situ to form base for asphalt paving during re-construction of a roadway in Minnesota. In situ stabilization of the existing RSG provided significant cost savings compared to traditional reconstruction where road surface materials are replaced with new aggregate, was more expedient than traditional construction, and enhanced sustainability by re-using existing roadway materials (RSG) and an industrial byproduct (fly ash). California bearing ratio (CBR) and resilient modulus (Mr) tests were conducted on the RSG alone and on fly-ash stabilized RSG (S-RSG) prepared in the field and laboratory. After 7 d of curing, CBR and Mr of the S-RSG were at least 2 times higher than those of the road surface gravel alone. Falling weight deflectometer surveys have shown that the in-situ elastic modulus of the S-RSG varies seasonally, with the lowest moduli occurring in spring when the pavement profile is wet and the highest occurring in late fall when the profile is drier. The findings indicate that fly ash stabilization of RSG renders it a suitable base material for paving.

INTRODUCTION

Use of industrial byproducts to improve the structural capacity of pavement materials is becoming more common, particularly when roads are being reconstructed or upgraded. This study describes an application of self-cementing fly ash to stabilize an existing road-surface gravel (RSG) to form a stable base for a hot-mix asphalt (HMA) layer when converting a gravel road to a paved road. Both the geomechanical and geoenvironmental performance of the road were evaluated in this study. This paper describes geomechanical aspects of the study.

The project covered a 3.5-km section of County Road 53 in Chisago County, MN, which is approximately 90 km north of Saint Paul, MN. Cementitious fly ash (10% by dry weight) was spread uniformly on the surface of the RSG using truck-mounted lay-down equipment similar to that described in Edil et al. (2002) and Trzebiatowski et al. (2004). RSG is a gravelly sand material spread over the subgrade and compacted to form an unpaved riding surface. Fly ash was blended into the RSG with a CMI RS-650-2 road reclaimer to a depth of 250 mm, with water added during mixing using a water truck. Within 1-2 hr of blending, the mixture was compacted by a tamping foot compactor followed by a vibratory steel drum compactor to form a stabilized RSG (S-RSG) base. The base material produced in this way is called Stabilized-Road Surface Gravel (S-RSG). The S-RSG was overlain with 90 mm of HMA within 7 d after compaction of the S-RSG.

In situ stabilization of the existing RSG provided significant cost savings (≈ 33%) compared to traditional reconstruction where road surface materials are replaced with new aggregate. Construction also was more expedient than is typical using traditional construction methods, and sustainability was enhanced by re-using existing roadway materials (RSG) and an industrial byproduct (fly ash) in lieu of virgin construction materials. This paper describes the strength and stiffness of the S-RSG in terms of California Bearing Ratio and resilient modulus, and provides comparisons to unstabilized RSG. Data collected from falling weight deflectometer (FWD) surveys are also presented. Geoenvironmental aspects of the project are described in Hatipoglu et al. (2006).

MATERIALS

Disturbed samples (≈ 20 kg) of subgrade soil, RSG, fly ash, and the S–RSG mixture were collected at ten stations along the alignment as shown in Fig. 1.

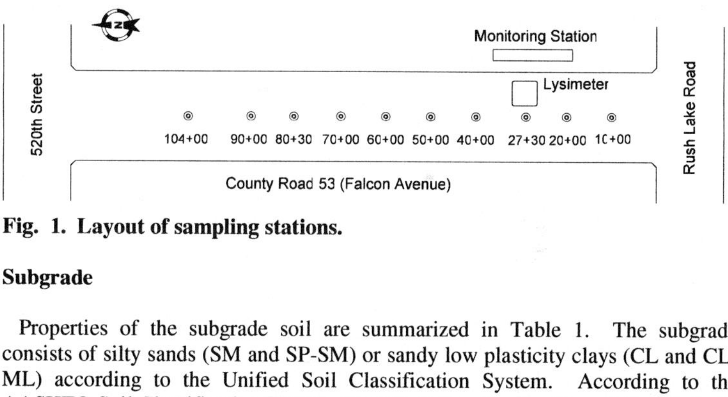

Fig. 1. Layout of sampling stations.

Subgrade

Properties of the subgrade soil are summarized in Table 1. The subgrade consists of silty sands (SM and SP-SM) or sandy low plasticity clays (CL and CL-ML) according to the Unified Soil Classification System. According to the AASHTO Soil Classification System, the subgrade soils classify as A-2-4, A-3, A-4, and A-5.

Road Surface Gravel (RSG)

Particle size distribution curves for the RSG are shown in Fig. 2. The RSG consists of well-graded gravelly sand with 11-14% fines (Station 90+00, which has a fines content of 22%, is an exception). The sand content is consistently around 60% and the gravel content is about 25%. Because the particle size distribution of the RSG did not vary considerably along the project alignment, a composite sample of RSG was prepared for laboratory testing. The composite RSG classifies as gravelly clayey sand according to Unified Soil Classification System. Compaction tests were performed on a sample of RSG from Station 27+30 (Fig. 1) and on the composite using the standard Proctor method. Optimum water content was 9% and the maximum dry unit weight was 20.0 kN/m^3 for both samples.

Table 1. Index properties and classification of subgrade soils.

Group	Stations	γ_d (kN/m^3)	w_n (%)	SSG Stiffness (MN/m)	DCP Index (DPI) (mm/blow)	LL	PI	Classification USCS	Classification AASHTO
A	20+00	17.0	3.0	8	31	NP	NP	SP-SM	A-3
	40+00	18.0	6.0	8	30				
B	27+30	18.0	15.0	14	19	44	28	CL	A-5
	60+00	19.0	12.0	17	19				
C	50+00	21.0	8.0	13	39	18	NP	SM	A-2-4
	70+00	22.0	7.0	12	7				
	80+30	20.0	8.0	9	9				
	104+00	19.0	5.0	9	20				
D	10+00	19.0	12.0	12	14	21	4	CL-ML	A-4
E	90+00	19.0	5.0	8	16	NP	NP	SM	A-2-4

Notes: γ_d = in situ dry unit weight and w_n = in situ water content (measured by nuclear gauge). SSG = Soil Stiffness Gauge, DCP = Dynamic Cone Penetrometer (DPI is the weighted average DPI over a depth of 150 mm), LL = liquid limit. PI = plasticity index. USCS = Unified Soil Classification System. AASHTO = American Association of State Highway and Transportation Officials.

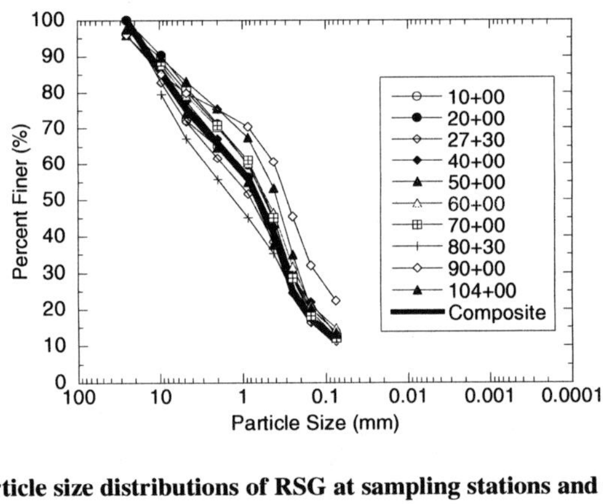

Fig. 2. Particle size distributions of RSG at sampling stations and composite of all stations.

Fly Ash

Fly ash from Units 7 and 8 of the Riverside Power Station in St Paul, MN was used for stabilization. Calcium oxide (CaO) contents of Unit 7 and Unit 8 fly ashes are 24% and 22% and silicon dioxide (SiO_2) contents are 32% and 19%, respectively. The CaO/SiO_2 ratios of the fly ashes, which are indicative of cementing potential, are 0.75 and 1.18 for the Unit 7 and 8 fly ashes. Loss on ignition (LOI), which is an indication of the amount of unburned coal in the fly ash, is 0.9% for the Unit 7 fly ash and 16.4% for the Unit 8 fly ash. According to ASTM C 618, Unit 7 fly ash is a Class C fly ash, whereas Unit 8 fly ash is an off-specification self-cementing fly ash (i.e., Unit 8 ash does not meet the Class C or F criteria).

Stabilized Road Surface Gravel (S-RSG)

S-RSG is prepared by pulverization of RSG to 255.0 mm depth and mix with 10% by weight fly-ash Water content and unit weight of the compacted S-RSG were measured at each station using a nuclear density gage (ASTM D 2922) immediately after compaction was completed. Samples of S-RSG were also collected immediately after mixing for preparation of test specimens for CBR and resilient modulus (Mr) testing. These specimens were compacted to the measured density at each station, sealed in plastic wrap, and cured for 7 d at 100% humidity.

Test specimens of S-RSG were also prepared in the laboratory using the composite RSG and 10% (dry weight) Unit 8 fly ash. This mixture, referred to as laboratory-mix RSG, was prepared at the mean water content (6.4%) and mean dry unit weight (19.3 kN/m^3) recorded in the field, and then cured using the same procedures employed for the S-RSG specimens prepared in the field. A specimen was also prepared with air-dry composite RSG (no fly ash) at a dry unit weight of 19.3 kN/m^3.

METHODS

California Bearing Ratio

CBR tests were conducted on the S-RSG and RSG specimens following the methods in ASTM D 1883. The specimens were un-soaked and were tested at a strain rate of 1.3 mm/min. The tests were conducted after 7-d curing period (S-RSG specimens) or immediately after compaction (RSG specimens).

Resilient Modulus

The Mr tests were performed in accordance with AASHTO T292 after 7 d of curing (S-RSG) or immediately after compaction (RSG). All of the Mr specimens had an aspect ratio of two. No standard loading protocol currently exists for resilient modulus testing of materials stabilized with fly ash. Thus, the S-RSG specimens were tested using the loading protocol for cohesive soils. The RSG specimens were tested using the protocol for granular soils.

Falling Weight Deflectometer

FWD surveys were conducted using a Dynatest 8002E FWD following the method described in ASTM D 4694. The FWD data were inverted using the program MODULUS v6.0 from the Texas Transportation Institute.

RESULTS

California Bearing Ratio

CBR of the field mix S-RSG along the alignment is shown in Fig. 3. Also shown is CBR of the S-RSG estimated from the DCP data and the CBR of the RSG and the S-RSG prepared with the composite sample. There is no systematic variation in CBR of the RSG or S-RSG along the alignment, suggesting that the variability in the CBR is more likely due to heterogeneity in the material rather than systematic variations in site conditions or construction methods. Similar variability was reported Li et al. (2007) for a project in Waseca, MN where reclaimed pavement material (RPM) was stabilized with fly ash.

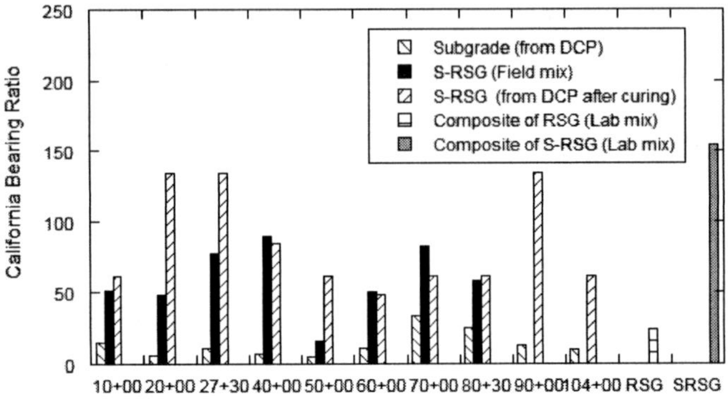

Fig. 3. California Bearing Ratio of S-RSG and RSG. Tests performed after 7 d of curing.

Type of fly ash (Unit 7 Class C fly ash vs. Unit 8 off-specification fly ash) had no apparent effect on the CBR. For instance, at Stations 27+30, 60, and 70 only Unit 8 was used, whereas only Unit 7 was used Station 40. Although the highest CBR for these stations corresponds to Station 40 (Unit 7 ash), the CBR is comparable at Stations 27+30, 40, and 70 (Unit 8 ash). Moreover, S-RSG prepared with both ashes had appreciably higher CBR than unstabilized RSG. In fact, the CBR of the unstabilized composite RSG was only 24, making the RSG unqualified as a base course material without stabilization.

The CBR of the S-RSG mixed in the field was as much as 60% lower than the CBR of the laboratory-mix S-RSG. This is consistent with observations made for stabilized recycled pavement material in Waseca, MN (Li et al. 2007) and for fine-grained subgrade soil stabilized with fly ash in Wisconsin (Bin-Shafique et al. 2004). This difference may reflect less uniform blending in the field relative to that achieved in the laboratory.

The CBR of the S-RSG estimated from the DCP ranges between 62 and 134 (mean = 85), which is lower than the CBR of the laboratory-mix S-RSG (154), but higher than the CBR of the S-RSG mixtures from the field (16-90, mean = 59). The reason for this difference has not been determined, but it may indicate that DCP-CBR calibrations need to be updated for use with fly ash-stabilized materials.

Resilient Modulus

The Mr along the alignment is shown in Fig. 4. These Mr correspond to a deviator stress of 21 kPa, which represents typical conditions within the base course of a pavement structure (Tanyu et al. 2003). As observed for CBR, there is no systematic variation in Mr along the alignment. The field-mix S-RSG specimens had Mr between 96 and 195 MPa (mean = 153 MPa), which is markedly higher than the typical resilient modulus of crushed rock aggregate (48 to 103

MPa) and more than two times higher, on average, than the Mr of the unstabilized composite RSG (51 MPa).

The Mr of the field-mix S-RSG is higher, on average, than the Mr of the laboratory-mix S-RSG (153 versus 112 MPa). This is inconsistent with the investigator's experience, which has shown that Mr of field mixtures of fly ash-stabilized materials typically lower than Mr of mixtures prepared in the laboratory (Bin-Shafique et al. 2004; Li et al. 2007). However, this discrepancy may be due to limited number of tests that were conducted on laboratory-mix S-RSG in this study (only one laboratory-mix specimen was tested, Fig. 4)

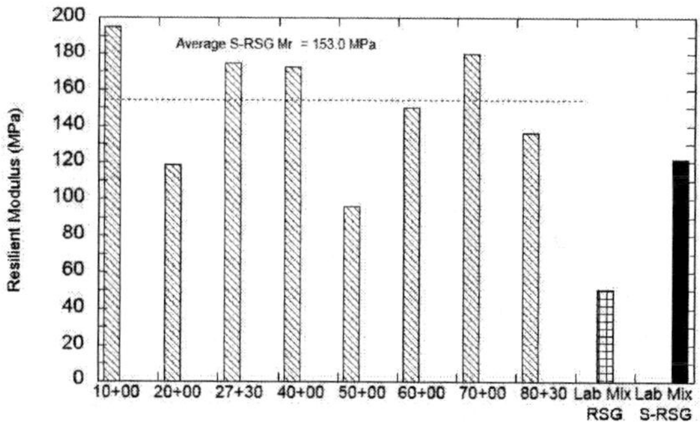

Fig. 4. Resilient modulus of S-RSG mixed in the field, S-RSG mixed in the laboratory, and RSG. Tests on S-RSG performed after 7 d of curing.

FWD Deflections and Moduli

Maximum deflections from FWD (falling weight deflectometer) tests conducted with a 40-kN load are shown in Fig. 5a. Tests were conducted in November 2005 (two months after construction), the following spring (May 2006), and the following fall (October 2006). Frozen conditions were absent during all three testing periods. However, temperatures and volumetric water contents within the pavement structure were lowest during the tests conducted in November 2005, highest during the tests conducted in May 2006, and intermediate for the tests conducted in October 2006 (Hatipoglu et al. 2006).

The deflections are fairly uniform along the alignment, except near Stations 70 and 80, where larger deflections occurred in May and October 2006 (Fig. 5a). Deflections at all stations increased markedly in May 2006, which most likely reflects thaw weakening of the subgrade (Jong et al. 1998) as well as the higher water content of the S-RSG. This is especially true near Stations 70 and 80, which have a silty subgrade that is more frost sensitive. By October 2006, however, the deflections diminished and were close to the deflections measured after construction (November 2006).

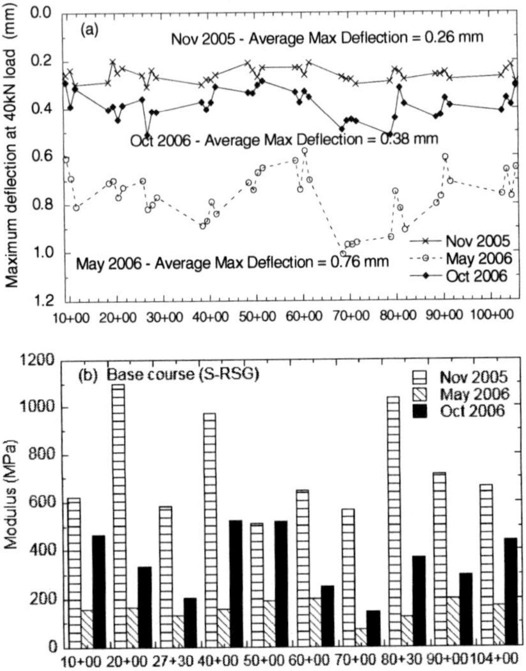

Fig. 5. Maximum FWD deflections for 40-kN load (a) and moduli of S-RSG obtained by inversion of FWD data (b).

Elastic moduli of the S-RSG obtained by inversion of the FWD data are shown in Fig. 5b. The moduli drop between November 2005 and May 2006, and then increase between May 2006 and October 2006. These changes in modulus probably reflect changes in the water content of the S-RSG, which was highest in the spring and lower in the summer and fall. Some of the moduli determined in November 2005 were much higher than anticipated (see stations 20, 40, and 80). The reason for these high moduli is unclear, but high CBRs were also obtained from the DCP data collected after construction near Stations 20, 40, and 80 (Fig. 3). However, the moduli at these stations were consistent with the other moduli in May and October 2006.

CONCLUSIONS

The fly ash stabilized road surface gravel (S-RSG) at this site had CBR and Mr at least 2 times higher than the CBR and Mr of unstabilized road surface gravel (RSG). The stabilized RSG also had higher CBR and Mr than typical base course materials. Thus, from the perspective of strength and stiffness, stabilizing in-place RSG during road re-construction and upgrading appears to be a viable means to create a sturdy base for paving with HMA.

Moduli back-calculated from falling weight deflectometer tests conducted along the alignment at various times since construction showed that the modulus of the S-RSG varies seasonally, with the lowest moduli occurring in the spring when the pavement has thawed and has higher water content. However, the moduli recover by the fall as the water content of the RSG diminishes.

ACKNOWLEDGEMENT

Financial support for this study was provided by the Minnesota Local Road Research Board (LRRB) under administration of the Minnesota Department of Transportation (Mn/DOT). The conclusions and recommendations in this report are solely those of the authors and do not reflect the opinions or policies of LRRB or Mn/DOT. Appreciation is expressed to the Chisago County Department of Public Works, X. Wang, M. Rosa, and J. Baugh. Financial support for B. Hatipoglu was provided by Istanbul Technical University, The Scientific and Technological Research Council of Turkey, and the University of Wisconsin-Madison.

REFERENCES

Bin-Shafique, S., Edil, T. B., Benson, C. H., and Senol, A. (2004). "Incorporating a fly-ash stabilised layer into pavement design." *Proceedings of the Institution of Civil Engineers-Geotechnical Engineering*, 157(4), 239-249.

Edil, T. B., Benson, C. H., Bin-Shafique, M. S., Tanyu, B. F., Kim, W. H., and Senol, A. (2002). "Field evaluation of construction alternatives for roadways over soft subgrade." *Transportation Research Record*, No. 1786, National Research Council, Washington D. C., 36-48.

Hatipoglu, B., Edil, T., and Benson, C. H. (2006). "Fly Ash Stabilization of Gravel Road as Base for Pavement at Chisago County Road 53, Minnesota." *GeoEngineering Report No 06-19*, University of Wisconsin-Madison, Madison, WI.

Jong, D., Bosscher, P.J., and Benson, C.H. (1998), "Field assessment of changes in pavement moduli caused by freezing and thawing," *J. Transportation Research Board*, 1615, 41-50.

Li, L., Edil, T. B., Benson, C. H., and Hatipoglu, B. (2007). "Evaluation of recycled asphalt pavement material stabilized with fly ash." *Soil and Material Inputs for Mechanictic-Emprical Pavement Design*, GSP No. 169, E. Tutumluer, L. Tashman, and H. Ceylan, eds., ASCE, Reston, VA.

Tanyu, B., Kim, W., Edil, T., and Benson, C. (2003). "Comparison of laboratory resilient modulus with back-calculated elastic modulus from large-scale model experiments and FWD tests on granular materials." *Resilient Modular Testing for Pavement Components*, STP 1437, ASTM International, West Conshohocken, PA, 191-208.

Trzebiatowski, B., Edil, T. B., and Benson, C. H. (2004). "Case study of subgrade stabilization using fly ash: State Highway 32, Port Washington, Wisconsin." *Beneficial Reuse of Waste Materials in Geotechnical and Transportation Applications*, GSP No. 127, A. Aydilek and J. Wartman, eds., ASCE, Reston, VA, 123-136.

Feasibility Study of Recycling Facilities in Brazil: The Case of Rio de Janeiro

K.R.A. Nunes[1], D.Sc.; C.F. Mahler[2], D.Sc.; R. Valle [1], D.Sc.

[1] Department of Production Engineering, Federal University of Rio de Janeiro, C.P. 68507 - CEP 21945-970, Rio de Janeiro – RJ - Brazil. knunes@pep.ufrj.br

[2] Department of Civil Engineering, Federal University of Rio de Janeiro, C.P. 68506 - CEP 21945-970, Rio de Janeiro – RJ - Brazil. mahler@coc.ufrj.br

SUMMARY: There are very few construction and demolition (C&D) waste recycling centers in Brazil and these use simple technology and process basically mineral inert material. To encourage the building and operation of new units with similar technology, data relating to C&D waste management and recycling in Brazil were collected and analyzed. Based on the results of this analysis, a conceptual model was proposed for conducting feasibility studies of future C&D waste recycling centers. Applying the model to verify the feasibility of private recycling centers, results show that C&D waste recycling centers in current market conditions in Brazil are not financially feasible based solely on revenue from the sale of processed products. A case study for the city of Rio de Janeiro is presented. The results of this application and the analysis can provide support to public authorities and private enterprises in their decision-making about investments in waste recycling centers in Brazil.

INTRODUCTION

Only eleven (0.2%) of the 5,507 Brazilian municipalities, besides the Federal District, have C&D waste recycling centers (Nunes, 2004). Thirteen centers are stationary plants and recycle part of the C&D waste produced in local communities. The technology adopted in Brazil for C&D recycling is simple and labor intensive. In these projects, the required equipments process only the mineral fractions and a magnetic component removes the metal fraction. The other fractions, excluding minerals and the metals, are previously removed by hand before beginning the crushing processes.

In Nunes (2007), a feasibility study for C&D waste recycling centers in Brazil was carried out without considering the geographic location. The main conclusions of the study were that C&D waste recycling centers could be economically feasible for public authorities in current market conditions depending on the particular circumstances of each county (landfill disposal costs, transportation costs from C&D waste to landfills and price of procuring natural products). The feasibility, however, also depends on continuity and the production volume achieved in recycling centers. Another main conclusion in Nunes (2007) was that the C&D waste recycling centers in Brazilian market conditions were not financially feasible for private enterprises. Revenue from the sale of the processed product alone was not enough to make these recycling centers feasible. It was suggested that other sources of revenue be sought, such as charging for the acceptance of C&D waste in recycling centers and the reduction of taxes and loans to recycling projects.

The study herein was carried out to combine the data relating to C&D waste in Brazil and, based on the conceptual model for financial feasibility studies of C&D waste recycling centers proposed by Nunes (2007), the feasibility of a private recycling center was verified through a case study for Rio de Janeiro. Data collection involved four activities. First, (1) a literature review was carried out. Next, (2) some professionals from the sector were consulted (such as equipment and material suppliers, businessmen from the mining and recycling sectors), who were qualified to provide essential data about C&D waste recycling. The data on waste recycling centers were obtained through (3) questionnaires sent to local governments and (4) visits to some of the counties and facilities. Most of the data was collected between January and November 2003, but there is no significant change in the situation presented in this paper by the end of 2006, according to other inquiries made to professional in the sector.

FEASIBILITY ANALYSIS: CONCEPTS AND STAGES

This phase of the study showed that there are no large-scale private C&D waste recycling centers in Brazil. Aiming to provide support to change this situation, this study is targeted towards private enterprises. The consultations of professionals in the sector showed that private projects with less than 20 tonne per hour (t/h) of waste processing flow will probably not be financially viable, due to low productivity and low prices of the processed product. It was also found that a center with a nominal processing capacity of 50t/h would have costs equivalent to those of a 100 t/h center (Sindibrita, 2003). Therefore, it was concluded that two different recycling center projects need to be analyzed: one on a small scale (20t/h) and the other midsize (100t/h). According to Sindibrita (2003), projects with a greater production than 100t/h would involve too many risks because of the necessary heavy investments and lack of tradition in consumption of recycled material in Brazil.

UNIDO (1987) and Kohler (1997) presented models for feasibility studies of complex projects involving heavy investments from various funding sources. By simplifying this structure, the key stages in pre-feasibility studies of C&D waste recycling centers were identified as follows: (a) market and competition analysis; (b) estimated generation of C&D waste; (c) estimated revenues and costs; and (d) investment analysis. These stages form a conceptual model for feasibility studies of C&D waste recycling centers. Stages (a) and (b) depend on the geographic location of the center. In this research the location was the city of Rio de Janeiro.

The comparison method between alternative investments used in the financial analysis of this research was NPV (Net Present Value), which is one of the most commonly used dynamic methods. The scenario technique is also applied in the analysis in addition to the NPV. This technique in a financial analysis permits an estimate of favorable and unfavorable results considering different hypotheses

(Motta & Caloba, 2002). An exchange rate of US$1 to R$ 3.00 was used in the financial analysis (Basis: November 2003). In November 2006, US$1 was equal to R$ 2.20 and the influences of this fluctuation in the analysis herein were not evaluated. Rio de Janeiro is the second largest Brazilian city with approximately six million inhabitants. The Rio de Janeiro city garbage collection company provides services to approximate 96% of the commercial and residential units in Rio de Janeiro (IBGE, 2000).

Market and Competition Analysis

In Brazil the resources for civil construction aggregates are abundant. The major consumer centers are generally located in regions with good quality reserves. Aggregate consumption in Brazil is estimated to be around 1.5 tonnes per inhabitant/ year (Sindipedras, 2004). The population of Brazil is around 180 million inhabitants (IBGE, 2000), therefore total aggregate consumption corresponds to 267 million tonnes (about 170 million m^3). The estimated aggregate consumption for the Metropolitan Region of Rio de Janeiro between 2004 and 2011 is approximately 11,250 tonnes a year. Current prices for civil construction aggregates have changed little since 1994 (Sindipedras, 2004) and the average price for sand, stone powder and gravel is 3.09 US$/t. Transportation costs are estimated to be approximately US$0.1/($m^3$.km) (Sindibrita, 2003).

Both aggregate prices and the number of new construction projects have been relatively low over the past ten years (DNPM, 2006). Consequently, today there is a large aggregate surplus. The recycling centers compete with the landfills for the receipt of C&D waste. According to Brazilian practices, the landfills require large amounts of inert material to permanently cover the landfill cells. Moreover, the landfill operating companies need C&D waste and in its absence they would use natural aggregates. However, according to some C&D estimates (Nunes, 2004) waste contributes to a significant fraction (20%-50%) of municipal solid waste. It must be considered that landfill operating companies often earn revenues based on per received tonne. This current form of income discourages the landfill to separate waste materials and recycle because it would result in lost revenue. Today Rio de Janeiro has a number of low-lying areas. With the spread of the city it may consider filling these low-lying areas (Nunes, 2004).

Demand Estimate

Use of recycled aggregates has been a little explored alternative to solve housing problems. There are some successful experiences in large-scale use of recycled aggregates as a road surface base in some Brazilian cities. In 2004, five ABNT (Brazilian Association of Technical Norms) standards were published about C&D waste, which, for example, regulate specifications about the use of recycled aggregates as road base. In the market for aggregates used by civil construction in Brazil, only 30% of produced gravel is found to be directly allocated to road surfacing. Seventy percent of produced gravel and 100% of produced sand are consumed by other markets (mixed with cement, production of concrete and pre-cast components and mortar) (DNPM, 2006). In a scenario where 30% of the present sand

and gravel market could be replaced by recycled aggregates, a market for at least 3.4 millions tonnes a year would be available for Rio de Janeiro in the next eight years[1]. The public authorities themselves are a large consumer of aggregates, for example, in the building of state-financed housing and infrastructure projects and street maintenance services. The public authorities are therefore able to encourage the consumption of recycled aggregates. Other incentives to use recycled aggregates could be obtained by tax discounts for production and sale of recycled aggregates or for building recycling centers. An additional option could also be public-private partnerships, where governments and the private sector work together to promote the use of recycled aggregates.

Estimated Production of C&D Waste

The estimated C&D waste production and collection in some Brazilian cities is presented in Table 1. COMLURB (2004) estimates a collection of 1,000 tonnes/day (0.17 kg/inhab.day), below the average of other Brazilian cities (0.51 kg/inhab.day). A reason for this may be that illegal disposals are considered public waste in the official statistics. According to COMLURB (2004), the average estimated total of domestic waste is 4,880 t and the average public waste collected in Rio de Janeiro is 2,848 t between 1994 and 2003.

Table 1: Estimated C&D waste in some Brazilian cities (Nunes, 2004)

Cities	Estimated C&D waste (tonne/day)		Year-Base	Population (IBGE, 2000)	Production per inhab. (kg/inh.d)	Collection per inhab. (kg/inh.d)
	Produced	Collected				
Rio de Janeiro	i.u.	1,000	2003	5,857,904	i.u.	0.17
Salvador	i.u.	2,746	2000	2,443,107	i.u.	1.12
São Paulo	16,000	3,360	2001	10,434,252	1.53	0.32
Ribeirão Preto	1,000	250	2003	504,923	1.98	0.50
S. J. dos Campos	733	i.u.	1995	539,313	1.36	i.u.
Piracicaba	620	i.u.	2003	329,158	1.88	i.u.
Vinhedo	i.u.	11	2003	47,215	i.u.	0.23
Guarulhos	i.u.	i.u.	-	1,072,717	i.u.	i.u.
Ribeirão Pires	i.u.	i.u.	-	104,508	i.u.	i.u.
S. J. Rio Preto	687	i.u.	1996	358,523	1.92	i.u.
Santo André	1,013	i.u.	1996	649,331	1.56	i.u.
Belo Horizonte	i.u.	2,220	2000	2,238,526	i.u.	0.99
Londrina	1,280	i.u.	2003	447,065	2.86	i.u.
Brasília	i.u.	i.u.	-	2,051,146	i.u.	i.u.
Macaé	-	35	2003	132,461	i.u.	0.26
Florianópolis	636	i.u.	2001	285,281	2.23	i.u.
Averages					1.92	0.51

i.u.: information unavailable

[1] Thirty percent of the smaller value was considered among the values between 2004 and 2011 presented in Table 3.

Table 2 shows the estimated total produced C&D waste (t/day) in the City of Rio de Janeiro made by Nunes (2004). According to the gravimetric composition of C&D waste presented by COMLURB (2004), 94.1% of the total C&D waste generated is inert mineral material. When the multiplication of this percentage by the total C&D waste generated (2,877 t/d - see Table 2) is carried out, it can be concluded that there is a potential production of inert mineral material of approximately 2,712 t/day (10^6 t/y), which can be recycled and sold in the city.

Table 2: Estimated total C&D waste produced (t/day) in Rio de Janeiro

Estimated total produced C&D waste	t / day
C&D waste produced in construction of new licensed buildings	940
C&D waste produced in licensed renovations	115
C&D waste produced from construction of non licensed buildings	222
Amount of C&D waste collected by the city in the residential units	1,000
Amount of C&D waste collected by the city in illegal disposals	600
Total C&D waste generated	2,877
Daily production per inhabitant	0.49 kg/inhab.day
Annual production per inhabitant	180 kg/inhab.year

Estimated revenues and costs

The key stages in pre-feasibility studies were presented in previous subsections. Stages *a* (market and competition analysis) and *b* (estimated C&D waste production) depend heavily on the center's location. Stage *c* (estimated revenues and costs) depends partially on location. Because of such dependences, the results of stages *d* and *e* are also affected by the definition of the center's location. In Nunes (2007), a feasibility study for C&D waste recycling centers in Brazil was carried out without considering the geographic location. This was done because the Nunes (2007) feasibility study was a preliminary study and a location for the recycling center had not still been defined. This paper continues this preliminary study. The location to be evaluated is a plot of land in the city of Rio de Janeiro, as described in the following subsections. The estimated cost presented by Nunes (2007) was divided in two parts: (1) estimated fixed capital investment and (2) estimated operational costs.

The equipment used in C&D waste recycling centers requires substantial investment. Since the used equipment market (based on the mining sector) is strong in Brazil (Sindibrita, 2003), this was taken into account in the analysis. In Nunes (2004) the fixed capital investments required for 20 t/h and 100 t/h recycling centers are illustrated in accordance with production volumes and the choice of new or used equipment. After defining the site for building the recycling center, it is possible to add to the previous cost estimates made by Nunes (2004) costs with land acquisition and for transportation and proper disposal of the recycling center rejects. The price of the plots of land in the industrial area would be at least US$ 14,500 (6,000 m² area) for 20 t/h plants and US$ 72,500 (30,000 m² area) for 100 t/h plants.

Having defined the recycling center site, it is possible to estimate the transportation costs. This is computed by multiplying the freight price (0.15 US$/(t.km) (Sindibrita, 2003) by the distance between the recycling center and the Rio de Janeiro landfill (around 10 km). Considering the 20% of refuse from C&D waste recycling (Nunes, 2007), the daily refuse production for a 20 t/h and a 100 t/h recycling center (eight hours' production) are 32t/d and 160t/d, respectively. For an amount over 18t/d the municipal landfill charges US$ 3.33/t for receiving C&D waste. Calculating the estimates, a 20t/h center would spend US$ 16,905 [(32 -18 t/d).(250 d).(3.33/t+ 10 km. 0.15 US$/(t.km)] a year, and for the 100 t/h center US$ 171,465 [(160 -18 t/d).(250 d).(3.33/t+ 10 km. 0.15 US$/(t.km)] for payment to local landfills alone.

INVESTMENT ANALYSIS

As already mentioned, the paper herein is a continuation of the preliminary study by Nunes (2007), in which eight scenarios were used. These scenarios were based on three hypotheses: (a) nominal capacity (20 or 100 t/h); (b) equipment: new or used; and (c) revenue from acceptance of waste (yes or no). When a scenario includes revenues from acceptance of waste, the estimated minimum value to be charged for financial feasibility becomes positive. This occurs when the financial analysis has a NPV higher than zero. The results of Nunes (2007) did not include the cost of land for the plant site. In Rio de Janeiro 24 scenarios were investigated. These scenarios were based on five hypotheses: (a) nominal capacity; (b) equipment; (c) revenue from acceptance of C&D waste; (d) the cost of land included; and (e) transportation and waste disposal costs.

Table 3 shows that scenarios, which do not consider revenues from C&D waste acceptance (Scenarios 12, 13, 14, 18, 19, 20, 24, 25, 26 and 30), are not economically feasible in present market conditions of the city of Rio de Janeiro. The revenues alone from the sale of recycled aggregates do not make plants feasible. A similar conclusion was also reached in the preliminary feasibility study by Nunes (2007) for market conditions in Brazil, but without considering a geographic location. The gate fees (revenues from the acceptance of waste) give the plants an additional revenue option. Some of the scenarios with gate fees are presented in Table 3 and the best results, considering location as defined, were scenario 23 (US$ 1.09/t with 100 t/h second-hand equipment) and scenario 29 (US$ 1.26/t with 100 t/h new equipment). It means that a C&D waste recycling plant, with second-hand equipment of 100 t/h capacity can be economically feasible if for each received tonne of waste a gate fee of US$ 1.09 must be paid by the user. For the feasibility of a recycling plant with new equipment with 100 t/h capacity, a gate fee of US$ 1.26 must be paid by the user.

Table 3: Evaluation of scenarios

Scenario	Nominal capacity	Equipment	Land included?	Revenue from acceptance of C&D waste?	Expense with waste disposal & transport?	Price charged with C&D waste acceptance NPV=0 (US$)	NPV value without C&D waste acceptance
09	20	Used	No	Yes	Yes	2.81	-
10	20	Used	Yes	Yes	No	2.52	-
11	20	Used	Yes	Yes	Yes	2.80	-
12	20	Used	No	No	Yes	-	- 553,423
13	20	Used	Yes	No	No	-	- 509,883
14	20	Used	Yes	No	Yes	-	- 566,420
15	20	New	No	Yes	Yes	3.18	-
16	20	New	Yes	Yes	No	2.96	-
17	20	New	Yes	Yes	Yes	3.24	-
18	20	New	No	No	Yes	-	- 625,533
19	20	New	Yes	No	No	-	-581,996
20	20	New	Yes	No	Yes	-	-638,530
21	100	Used	No	Yes	Yes	1.00	-
22	100	Used	Yes	Yes	No	0.49	-
23	100	Used	Yes	Yes	Yes	1.09	-
24	100	Used	No	No	Yes	-	-986,158
25	100	Used	Yes	No	No	-	- 484,193
26	100	Used	Yes	No	Yes	-	- 585,920
27	100	New	No	Yes	Yes	1.19	-
28	100	New	Yes	Yes	No	0.66	-
29	100	New	Yes	Yes	Yes	1.26	-
30	100	New	No	No	Yes	-	-392,070
31	100	New	Yes	No	No	-	-651,237
32	100	New	Yes	No	Yes	-	-1,241,196

For the feasibility of 20t/h plants, considering location as defined, the best results were scenario 11 (gate fee of US$ 2.80/t and second hand equipment) and scenario 17 (gate fee of US$ 3.24/t and new equipment). From these results it can be seen that gate fees for the feasibility of 20 t/h plants are higher than the gate fees for 100 t/h plants. This means that a larger production of recycled aggregates facilitates financial feasibility of recycling plants. Table 3 also shows that the scenarios that considered the waste disposal and transportation expenses showed even less favorable results than the scenarios that did not consider these expenses. This means that these expenses strongly impact the economical feasibility of the recycling plants. By shortening distances between the landfills and recycling centers, the refuse transportation costs will be reduced. The possibility of working on more than one shift was not considered, but perfectly possible to implement if a demand for the recycled material exists. Multiple shifts should lower the overall unit and cost of the recycled aggregates by distributing certain fixed costs across greater production, assuming that the market demand exists to support the increase.

CONCLUSION

The economical feasibility of C&D waste recycling centers in Rio de Janeiro was investigated by using multiple scenarios based on aspects such as nominal capacity, equipment, revenues from acceptance of C&D waste, cost of land included and waste disposal and transportation costs. It was found from the financial results of the scenarios that do not consider revenues from C&D waste acceptance, that private recycling centers in today's market conditions would not be economically feasible in the city. The revenue alone from the sales of recycled aggregates does not make the plants feasible. These results have confirmed the conclusions of a previous preliminary study that C&D waste recycling centers in current Brazilian market conditions are not financially viable for private enterprises without receipt of both C&D acceptance revenues and recycled aggregate sales revenues. It is suggested seeking other sources of revenue, such as charging for acceptance of C&D waste in recycling centers and reducing tax and lending at lower costs than practiced in the market. An option may also be public-private partnerships, where governments and private sector work together.

REFERENCES

COMLURB (Rio de Janeiro City Urban Garbage Collection Company) (2004). "Information about reception of C&D waste". Available at: http://www2.rio.rj.gov.br/comlurb. Access on: February 2004. Rio de Janeiro.

DNPM (National Department for Mineral Research) (2006) *Sumário Mineral 2006.* Available at: http://dnpm.gov.br/dnpm-legis. Access on: February 2007.

IBGE (Brazilian Institute of Geography and Statistics) (2000) *Population estimative and other information: Year 2000*). Rio de Janeiro, Brazil.

Kohler, G. (1997) *Recyclingspraxis Baustoffe* (Practice of Recycling: Building Materials). 3. ed., Köln: TÜV Rheinland.

Motta, R. R.; Caloba, G. M. (2002) Investment Analysis: Decision-Making in Industrial Projects). Atlas, SP, Brazil.

Nunes, K.R.A., (2004). *Investment and Performance Analysis in Construction and Demolition Waste Recycling Plants.* Ph.D. Thesis. Federal University of Rio de Janeiro, COPPE, Production Engineering, Rio de Janeiro, Brazil.

Nunes, K.R.A., et al. (2007). Evaluation of Investments in Recycling Centers for Construction and Demolition Waste. Article in press. *Waste Management* (International Journal of Integrated Waste Management), Elsevier, USA.

Sindibrita (Federation of the Gravel Mining Industry of the State of Rio de Janeiro) (2003) *Data provided between January and November 2003.*

Sindipedras (São Paulo State Federation of the Gravel Mining Industry) (2004). *Aspectos Institucionais* Available at: http://www.sindipedras.org.br/institucional Access on: January 2004.

UNIDO (United Nations Industrial Development Organisation) (1987). *Manual for the preparation of industrial feasibility studies.* Atlas, São Paulo, Brazil.

Use of Tire Chips in the Final Cover System of a Superfund Site Landfill

Majdi A. Othman[1], M. ASCE, Ph.D., P.E., and John Seymour[2], P.E.

[1]Associate, Geosyntec Consultants, 1255 Roberts Boulevard, NW, Suite 200, Kennesaw, GA 30144
Ph: 678 202 9500, Fax: 678 202 9501, Email: mothman@geosyntec.com
[2] Associate, Geosyntec Consultants, 55 W Wacker Dr, Suite 1100, Chicago, IL 60601
Ph: 312-658-0500, Fax: 312-658-0576, E-mail: jseymour@geosyntec.com

ABSTRACT: The final cover system of a Superfund Site landfill in Illinois was designed to include a tire chips gas-collection layer and a tire chips drainage layer. A design-phase performance demonstration was prepared in support of the use of the tire chips layers. The demonstration addressed the tire chips hydraulic conductivity, compressibility under anticipated loading, chemical stability and leachability, and combustion potential. Protection of the adjacent geomembrane from puncture was also addressed in the design. Prior to construction, a field test pad program was conducted to: (i) evaluate suitability of the proposed construction methods and equipment; (ii) select an appropriate geotextile cushion for protection of the geomembrane from puncture by wires in the tire chips; and (iii) assess if the clay barrier layer could be compacted to meet the project specifications despite the anticipated compressibility of the underlying tire chips gas-collection layer. In this paper, the data and analyses that were used for the performance demonstration are summarized and the results of the test pad program are presented.

INTRODUCTION

The United States currently generates approximately 300 million scrap tires each year. Recycling and beneficial use of scrap tires continue to grow, thus reducing the number of tires that are stockpiled or landfilled. Tire chips have been used in many civil engineering projects including, but not limited to: (i) subgrade fill in highway embankments; (ii) backfill for walls and bridge abutments; (iii) road subgrade insulation; and (iv) septic system drain fields. Tire chips have also been used for numerous landfill applications including, but not limited to: (1) leachate collection layers; (2) operations (protective) layers; (3) alternate daily covers; (4) landfill gas-collection layers; (5) foundation (grading) layers for final cover systems; and (6) drainage layers for final cover systems. Example landfills where tire chips were used include the DSI Superfund Site landfill in Vermont, the Lowry Landfill Site in Colorado, the Modern Landfill in Pennsylvania, the Modern Landfill in New York, and the Crossroads Landfill in Maine.

The Superfund Site landfill occupies approximately 16 hectares and includes waste that is 30 to 40 years old. The waste thickness is 9 m maximum and 4.5 m on average. A required component for the selected remedy for the Site was to construct a final cover system over the landfill. The remedy required that components of the final cover system include a gas collection layer with a hydraulic conductivity of at least 10^{-3} cm/sec and a geosynthetic infiltration drainage layer with a hydraulic conductivity of at least 20 cm/sec. The function of the gas-collection layer, located directly underneath the barrier components of the final cover, is to provide a highly transmissive path for landfill gas to be conveyed to either an active or passive gas-removal system. Conventionally, this layer consists of either granular soils (i.e., typically sand) or geosynthetics (i.e., geotextiles and geonets). The function of the drainage layer is to provide a path for removal of storm-water that percolates through the vegetative layer, thereby minimizing head build-up on top of the barrier layer. Conventionally, this layer consists of either a granular soil or a geonet.

Tire chips were proposed for the gas collection and drainage layers of the final cover system. A design-phase performance demonstration and a test pad program were performed to allow the use of tire chips for these two layers, as described in the remainder of this paper. It is noted that the terms "tire chips", "tired shreds", and "shredded tires" are used interchangeably in this paper.

FINAL COVER SYSTEM DESIGN AND DEMONSTRATIONS

The final cover system of the landfill was designed to consist of a 0.15-m thick gas-collection layer, overlain by a 0.6 m thick clay barrier/geomembrane composite liner, overlain by a 0.3-m thick drainage layer and a 0.45-m thick protective cover soil layer. Tire chips were proposed for the gas collection and drainage layers. Based on the design, geotextiles are installed directly above the tire chips gas collection layer, above the geomembrane, and above the tire chips drainage layer.

During the design stage, several design issues were identified relative to the use of tire chips in the gas collection and drainage layers. The design issues included the tire chips hydraulic conductivity, compressibility under anticipated loading, chemical stability and leachability, and combustion potential; and protection of the adjacent geomembrane from puncture. A design-stage performance demonstration was performed to justify the use of the tire-chip layers, as described below.

Hydraulic Conductivity

The selected remedy required that the 6-inch thick tire chips layer must be able to maintain the specified hydraulic conductivity (i.e., 1 x 10^{-3} cm/sec), when subjected to the confining pressure of the overlying final cover system components. The hydraulic conductivity of tire chips, in civil engineering applications, has been investigated by Humphrey and Sandford (1993), Donovan et al. (1996), Humphrey (1996), Humphrey (1997), and Reddy and Saichek (1998), amongst others. The range of hydraulic conductivity values was shown by Humphrey (1997) to be on the order of 0.6 to 24 cm/sec. The large range is a function of varying tire chip sizes, the initial density of the layer, hydraulic gradients, and confining pressures. Duffy (1995) showed the hydraulic conductivity of tire chips to be 0.7 cm/sec even when subjected to a confining pressure of 125 kPa. These ranges clearly exceed the required minimum hydraulic conductivity (1×10^{-3} cm/sec) by at least two orders of magnitude, even under loading conditions which are much higher than will occur at the Site (20 kPa).

The selected remedy required that the final cover system have a geosynthetic drainage layer with a hydraulic conductivity of at least 20 cm/sec overlain by a protective geotextile filter fabric to prevent clogging. To evaluate the proposed tire-chip layer relative to the hydraulic conductivity requirement stated above, it is necessary to perform an equivalency comparison based on hydraulic transmissivity. Transmissivity is defined as $\theta = k_p t$, where θ is the transmissivity (cm^2/sec), k_p is the hydraulic conductivity in the plane of the material (cm/sec), and t is the thickness of the layer (cm). Given a geonet drainage layer with a hydraulic conductivity of 20 cm/sec and assuming a typical thickness of 0.75 cm, the transmissivity of the geonet drainage layer would be approximately 15 cm^2/sec. The hydraulic conductivity of tire chips is in the range of 0.6 to 24 cm/sec (Humphrey, 1997). The proposed drainage layer is 30-cm (12-inch) thick, resulting in a range of transmissivity values of 18 to 730 cm^2/sec. Therefore, based on transmissivity, the use of tire chips will provide higher drainage capacity than the geonet drainage layer.

Additionally, modeling using the U.S. Environmental Protection Agency (EPA) Hydrologic Evaluation of Landfill Performance (HELP) model was used to determine the effectiveness of the tire chips in limiting hydraulic pressure ("head") build-up on top of the proposed geomembrane/clay barrier layer. These calculations demonstrate that as long as the tire chips have a hydraulic conductivity of greater than approximately 0.15 cm/sec, the maximum head in the drainage layer will not exceed the thickness of the layer. This criterion was considered acceptable because a head buildup of 30 cm would still provide an acceptable factor of safety for stability of the final cover system on the 2.5% landfill slopes. The range of hydraulic conductivity of tire chips (0.6 to 24 cm/s reported in the literature) is clearly above the engineering design criterion of minimum allowable hydraulic conductivity of 0.15 cm/s. In the project specifications, the tire chips were required to have a hydraulic conductivity that is greater than or equal to 1 cm/sec for both the gas collection layer and for the drainage layer, when tested in accordance with ASTM D 2434 at a confining pressure of 35 kPa and a hydraulic gradient of less than 0.10. The contractor submitted test data that met these specifications.

Compressibility under Anticipated Loading

As shown in Figure 1, the gas-collection layer will be subjected to an overburden of 4.5 feet of cover materials. The calculated overburden pressure from the overlying materials is approximately 20 kPa. Inspection of data included in Donovan (1997) reveals that an average vertical stress of 20 kPa would result in approximately 6 percent compressive strain of the tire chip layer. This would result in the thickness of a 15-cm layer being reduced by less than 1.3 cm. The corresponding reduction in transmissivity of the layer would be on the order of 6%, which, based on the range of tire-chip hydraulic conductivities and the thickness of the layer would still greatly exceed the design requirement.

Chemical Stability and Leachability

Surface-water infiltration through the vegetative cover layer will be conveyed via the drainage layer, eventually exiting along the landfill perimeter and rejoining the surface water. The use of tires for vehicles and multiple non-waste related applications indicates that tire chips do not exhibit high leachability or degradable properties. Research has been performed, however, to confirm this. For example, studies by Andrews and Guay (1996), and Donovan (1997), have shown that degradation of tire chips will not occur in landfill leachate environments. Most of these studies have been performed using permeating liquids (such as leachate or low-pH solutions) that are much harsher than storm water. These results are obviously conservative when compared with leachability using clean water. The Andrews and Guay (1996) study evaluated compounds produced by exposing tires to different leaching environments (i.e., pH of 3.5, 5.0, 7.0, and 8.0); toxicity characteristic leaching procedure (TCLP) threshold limits were not exceeded in any of the test conditions.

Combustion Potential

Tire chips have a flash point equal to 322°C, and require an external ignition source for combustion. Therefore, tire chips are not in and of themselves capable of spontaneous combustion. However, under certain conditions, an initial exothermic reaction may occur within a tire shred embankment or backfill that could eventually raise the temperature within the fill to a point where ignition could possibly occur.

In 1995, three shredded tire projects experienced combustion problems. Two of the projects were located in Washington, and the third project was located in Colorado. The combustion process may have been initiated by heat released either by the presence of soils and microbial degradation, the oxidation of exposed steel wires, or microbes consuming liquid petroleum products that could have been spilled on the chips during construction. Also, the presence of both sufficient air and crumb rubber particles in the embankment may have contributed to the process. The installations had a number of common features including free access to oxygen, thin soil cover, topsoil placed in direct contact with tire chips, chips contaminated with liquid petroleum, a large number of exposed steel wires, contact of chips with fertilizer, and concentrations of crumb rubber. The combustion process also occurred where the fill had a thickness greater than 6 m.

Spontaneous combustion is not expected to be a limiting factor for a landfill gas collection system because the typical thickness of such a system is on the order of 15 cm to 30 cm. A few guidelines were included in the project specifications to minimize the internal heating and combustion of tire chips. In particular, the project specifications included the following requirements:

- Tire chips shall be free of contaminants (i.e., grease, oil, gasoline, diesel, fuel, etc.), dirt, wood, fines, and remains of tires that have been subjected to a previous fire.
- A geotextile shall separate the chips and the surrounding soil.
- Common features from the installations that experienced combustion shall be avoided (i.e., free access to oxygen, thin soil cover, topsoil placed in direct contact

with tire chips, chips contaminated with liquid petroleum, a large number of exposed steel, contact of chips with fertilizer, and concentrations of crumb rubber).

Protection of the Adjacent Geomembrane from Puncture

All tires contain a bundle of high tensile strength wires surrounded by rubber that forms the bead of a tire to provide a firm contact with the rim. The individual wires, that compose this bundle, can be up to 3 mm in diameter and are relatively stiff. This type of wire is called "bead wire". Most tires also contain steel belt wire in the tread and sidewall areas. This wire is much smaller diameter than bead wire and is therefore more flexible. Bead wires pose a significant puncture hazard for geosynthetic barrier materials (GeoSyntec, 1998). Therefore, tire shreds containing bead wires should not be placed in direct contact with the underlying geosynthetic barrier layer. In contrast, the thinner belt wires in direct contact with a geomembrane in a field trial did create some minor damage (i.e., indentations, scratches, dents) but did not puncture the geomembrane (GeoSyntec, 1998).

The design required that for the gas collection layer, belt or bead wires protruding from the tire chips shall not extend more than 1.3 cm beyond the edge of the tire chips. For the drainage layer, the belt wires shall not extend more than 1.3 cm beyond the edge of the tire chips. The bead wires for the drainage layer shall not extend more than 0.6 cm beyond the edge of the tire chips if a 23.7 kg/m^2 geotextile cushion is used under the tire chips, and not more than 0.9 cm if a 35.6 kg/m^2 geotextile cushion is used. Other combinations of protrusion lengths of bead wires and geotextile thicknesses were allowed if field trials can demonstrate that these combinations do not significantly affect the hydraulic performance of the underlying geomembrane.

FINAL COVER SYSTEM CONSTRUCTION

Geotextile Cushion Test Pad

The project was specified to have the potential to utilize either a 23.7 kg/m^2 or 35.6 kg/m^2 geotextile cushions between the tire chip drainage layer and the geomembrane. The geotextile cushion was to protect the geomembrane from damage primarily from the bead wires in the tire chips. The final geotextile cushion selection was made based upon a full scale field test pad that was constructed using each of the geotextile cushions between the geomembrane and the tire chip drainage layer. The test pad was constructed during Contractor mobilization to allow enough time to order the selected geotextile cushion to be on site for cover construction.

The clay soil subgrade below the geomembrane was first prepared in accordance with the specifications for the low permeability clay barrier. The test pad was approximately 6-m wide by 9-m long. The geomembrane was placed on the subgrade and a sheet of each geotextile cushion (the Contractor selected 25.2 kg/m^2 and 35.6 kg/m^2 non-woven, needle-punched polypropylene fabrics) was placed next to each other on the geomembrane (Photo 1). The tire chips were placed in a 0.3-m thick layer followed by a separator fabric and a 0.4-m thick layer of vegetative cover soil layer over the tire chips. Tire chips were placed using a front end loader and spread using a Caterpillar low ground pressure (CAT D-RH LGP) dozer. Soil was placed using the same dozer and compacted with the tracks of the dozer. Finally, a loaded front end loader was driven across the soils to simulate final cover design conditions. Photo 2 depicts construction of the geotextiles cushion test pad.

Photo 1. Placement of tire chips over the cushion geotextiles.

Photo 2. Front end loader compacting clay over the tire chips drainage layer.

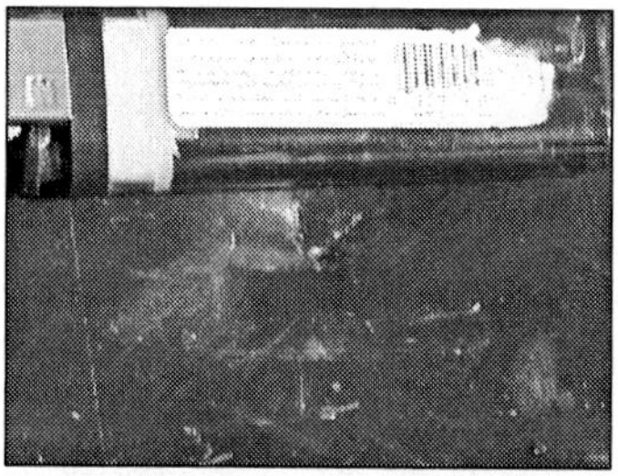

Photo 3. The tire chip wires caused gouges in the geomembrane protected by 25.2 kg/m^2 cushion geotextile.

Photo 4. The tire chip wires caused scratches in the geomembrane protected by 35.6 kg/m^2 cushion geotextile.

After the geotextile cushion test pad construction was complete, the vegetative layer, separator fabric, tire chips, and geotextile cushions were carefully removed and the geomembrane was inspected below the areas where the 25.2 kg/m^2 and 35.6 kg/m^2 geotextile cushions were placed. Photos 3 and 4 depict the conditions of the geomembrane below the 25.2 kg/m^2 and 35.6 kg/m^2 geotextile cushions, respectively. Examining the geomembranes indicated that the 25.2 kg/m^2 geotextile cushion did not provide acceptable performance because punctures and gouges were observed in the underlying geomembrane. However, the 35.6 kg/m^2 geotextile cushion performed with no punctures and limited scratching of the geomembrane. Further, the results indicated that the drainage layer design using tire chips could be constructed while maintaining the integrity of the geomembrane.

Clay Barrier Test Pads

Overview

A low-permeability soil barrier test pad was completed as a part of the full scale cover construction to identify the equipment and procedures to adequately construct the clay barrier over the compressible and "spongy" tire chip gas collection layer. Because there were two contractors, one for each of two phases of construction, both Contractors constructed a test pad using their own equipment and soil-borrow sources.

First Test Pad

The first test pad was comprised of a 30-m wide by 30-m long by 0.61-m thick section of clay barrier that was incorporated into the final cover. The pad was constructed in one 0.305-m and two 0.152-m thick lifts. In-place moisture content and density tests were performed on the lifts using a nuclear moisture-density gauge. The measured moisture contents were 0.3% to 1.6% above optimum moisture content, and the measured dry densities were a minimum of 92% and an average of 95% of the maximum dry density. The clay barrier had measured hydraulic conductivities of 1.2 x 10^{-8} to 8.5x 10^{-7} cm/sec. These results showed that three full coverage passes using a CAT 815F padded foot compactor provided adequate compaction of the clay barrier.

Second Test Pad

The second Contractor constructed a test pad that was 0.61-m thick, 60-m long, and 15-m wide. The clay in the pad was spread in one (1) 0.381 to 0.406-m thick initial lift and in two (2) additional lifts to make the full 0.61-m thickness. The clay was spread by low ground pressure dozers and compacted by a padded foot drum compactor making three full coverage passes. In-place moisture and density tests were conducted using a nuclear gauge to evaluate the entire thickness of the clay barrier. The measured moisture contents were from 0.4% below the optimum moisture content to 4.6% above the optimum moisture content. The measured dry densities were a minimum of 95.3% and an average of 97.5% of the maximum dry density. The clay barrier had a measured average hydraulic conductivity of 9.1 x 10^{-8}, which met the project specifications.

Drainage Layer Design Change

At the outset of construction, it was identified that an adequate supply of tire chips was not available to construct both the gas collection and infiltration control layers in accordance with the project schedule. Consequently, the drainage layer design was changed from tire chips with a geotextile cushion and overlying separator geotextile, to a drainage geocomposite layer consisting of a high density polyethylene (HDPE) geonet with a geotextile filter heat bonded to the top of the geonet.

PERFORMANCE OF THE FINAL COVER SYSTEM

The performance of the final cover system was assessed during construction by observing the cover response to compaction of the clay barrier and through long-term monitoring of groundwater and landfill gas (LFG) in the post-closure period. The clay barrier was successfully compacted without significant impacts to the underlying tire chip gas collection layer. Limited difficulties occurred that were associated with soft waste materials under the tire chip gas collection layer. It required removal and replacement of the waste and overlying cover. No repairs of the cover were necessary within three years of construction completion.

CONCLUSIONS

The case study described in this paper shows that tire chips can be successfully utilized in landfill final cover systems as drainage layers and gas collection layers. Tire chips have adequate transmissivity for these applications even after undergoing compression due to loading. Tire chips are durable, chemically stable, and do not contribute significant chemical concentrations to the environment. The potential for combustion and puncture of adjacent geosynthetic layers can be addressed through the design and construction phases of the project. The compacted clay barrier layer of the final cover system can be constructed above a tire chip layer and achieve typical compaction and hydraulic conductivity requirements. The final cover described in this study is functioning properly after more than three years of construction.

REFERENCES

Andrews, D.W. and Guay, M.A. (1996). "Tire Chips in a Superfund Landfill Cap: A Case History of the First Use of a Tire Chip Drain Layer." *Proceedings of the Nineteenth International Madison Waste Conference*, Madison/WI: 206-216.

Donovan, R., Dempsey, J., and Owen, S. (1996). "Scrap Tire Utilization in Landfill Applications." *Proceedings of SWANA's 34th Annual International Solid Waste Exposition*, Portland/OR: 353-383.

Donovan, R. (1997). "Processing Scrap Tires for Landfill Projects." *Solid Waste Technology*: 26-29.

Duffy, D.P. (1995). "Using Tire Chips as a Leachate Drainage Layer." *Waste Age*, Vol. 26 (9): 113-122.

GeoSyntec Consultants Inc. (1998). "Technical Memorandum: Application of Tire Shreds in the Phase 11, Cell A Landfill Lining System".

Humphrey, D.N., Sandford, T.C. (1993). "Tire Chips as Lightweight Subgrade Fill and Retaining Wall Backfill." *Proceedings of the Symposium on Recovery and Effective Reuse of Discarded Materials and By-Products for Construction of Highway Facilities,* Federal Highway Administration, Washington/D.C.: 5-87 to 5-99.

Humphrey, D.N. (1996). "Engineering Properties and Water Quality Effects of Tire Chips for Landfill Construction." *Proceedings of a Seminar on Landfill Design Using Tire Shreds*, Arlington, TX: 34-92.

Humphrey, D.N. (1997). "Civil Engineering Applications of Shredded Tires, A Short Course." *California Integrated Waste Management Board.*

Reddy, K.R. and Saichek, R.E. (1998). "Assessment of Damage to Geomembrane Liners by Shredded Scrap Tires." *Geotechnical Testing Journal*, Vol. 21(4): 307-316.

Evaluation of Cemented Quarry Fines as a Pavement Base Material

[1]Anand J. Puppala, [2]Sireesh Saride, [3]Sunil K. Sirigiripet, [4]Richard Williammee, [5]Venkata S. Dronamraju

[1]Professor, Department of Civil and Environmental Engineering, The University of Texas at Arlington, Texas 76019, anand@uta.edu
[2]Post Doctoral Fellow, Department of Civil and Environmental Engineering, The University of Texas at Arlington, Texas 76019, sireesh@uta.edu
[3]Masters Student, Department of Civil and Environmental Engineering, The University of Texas at Arlington, Texas 76019, sunil110@gmail.com
[4]Materials Engineer, Texas Department of Transportation, Fort Worth, Texas 76137, rwillia@dot.state.tx.us
[5]Doctoral Student Department of Civil and Environmental Engineering, The University of Texas at Arlington, Texas 76019, dronamraj@gmail.com

ABSTRACT: This paper discusses the laboratory test results and field monitored data to evaluate the efficacy of recycled material used as a pavement base material. In an on-going research, Quarry Fines (QF), a residue deposit produced from lime stone quarries, was used for enhancing recycling efforts. Since QF material exhibits low strength and resilient moduli, cement stabilization was attempted to enhance the material properties. A cement-treated QF material was then used as a pavement base to support a new pavement test section (3100 ft long x 70 ft wide) on SH-360 in Arlington, Texas. The test section was extensively instrumented with inclinometers and pressure cells to monitor the performance of this recycled material as a pavement base material.

INTRODUCTION

Recycled pavement materials have become a viable alternative to be considered in road construction. Conservation of resources, preservation of the environment, and retention of existing highway geometrics are some of the benefits obtained by reusing pavement materials (Taha et al. 1999). Using the recycled materials is probably the best choice for road construction and maintenance since it allows very large savings as compared to conventional materials.

In the United States, the recycled materials market comprising of recycled asphalt pavements, light weight aggregates, fly ash, and quarry byproducts is estimated to be between 352 million tons (320 million metric tons) and 859 million tons (780 million metric tons) per year. It is estimated that out of 41 million tons produced, 33 million tons (nearly 80%) are used effectively in the pavement and other geotechnical construction (Holtz and Eighmy, 2000). Collins and Ciesielski (1994) concludes that the annual production of quarry byproducts alone amounts to 175 million tons and these byproducts are used by highway agencies as cement-treated subbases, flowable fills, embankment fills and stabilized bases.

Limestone quarry fines are a by-product from crushed natural stone production for use as construction aggregate. The process consists of blasting, primary and secondary crushing, washing, screening, and stockpiling operations. Quarry by-products are produced during crushing and washing operations (Tepordei, 1992). Quarry fines are generally angular and fine materials (Okamura, 1997). Depending upon the gradation, larger sizes can be replaced as sand in concrete mixtures and also in lateritic soil to improve its properties (Soosan et al. 2001). The physical and chemical composition and mineralogy of quarry fines vary with parent rock type and production source, but are relatively consistent at each quarry location (Wood and Marek 1993). However, the variations gradation and compaction characteristics of the quarry fines obtained from different sources are marginal (Soosan et al. 2005). Sand occupies a larger volume in the mixture and replacement by larger quarry fines can be considered as an economical, high-volume use option (Naik and Kumar, 2003).

Lightweight aggregate is produced from the introduction of certain expansive shale materials into a rotary kiln and fired at temperatures greater than 1000°F. Once cooled, it is crushed into appropriate graded sizes for use in many different commercial applications. Using recycled materials in road construction offers opportunities to use huge quantities of recycled materials while extending the supply of traditional construction materials and ensuring good-quality roads.

The USGS calculates that even though the use of recycled crushed concrete aggregates increased by 170 percent between 1994 and 1996, it still constituted only 0.4 percent of the total aggregates consumed in 1995 (Gnanendran and Woodburn, 2003). The application of recycled quarry fines, which substituted the natural sand, reduced the requirement of expensive chemical admixtures, high-range water-reducing admixture (HRWRA) and viscosity-modifying admixture (VMA) without affecting the strength of self compacting concrete and proved to be an economical solution to concrete producers (Tarun et al., 2005).

The present paper discusses about, Quarry Fines (QF), a residue deposit produced from lime stone quarries, was used in this research for enhancing recycling efforts. Untreated QF material exhibits low strength and resilient moduli. To enhance these material properties 2.3 % of cement was added to these quarry fines. Then this Cemented Quarry Fines (CQF) used as a pavement base to support a new pavement test section (3100 ft long x 70 ft wide) on SH-360 in Arlington, Texas. A typical test section is shown in Figure 1.

EXPERIMENTAL TEST RESULTS AND ANALYSIS

Table 1 presents the details of preliminary laboratory tests performed on QF and CQF materials. It is to be noted that the percentage of cement added by dry weight to the QF was obtained priory from optimum strength criteria.

QF has 25% material retained on 2 mm sieve with coefficient of curvature (c_c) of 0.5 and uniformity coefficient (c_u) of 7. Figure 2 shows the grain size distribution curve of QF. The liquid limit, plastic limit and specific gravity of the QF were found to be 21.5%, 11.7% and 2.65 respectively using ASTM D 4318-00 method. As per Unified

Soil Classification System (USCS) the QF was classified as silty sand (SM). Compaction characteristics of both QF and CQF were determined as per ASTM methods to evaluate the strength properties at optimum moisture contents.

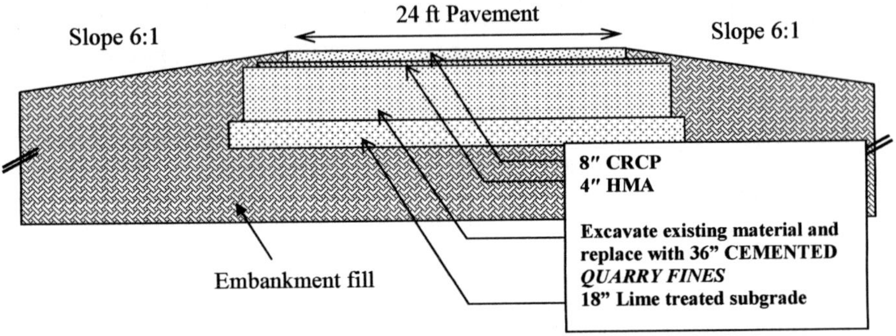

Figure 1 Typical test embankment with CQF as a base material supporting pavement (SH 360, Arlington, Texas. Section 750.10 to Section 750.60)

Table 1 Details of laboratory tests on Quarry Fine Materials

Material	Description of tests/parameters
Quarry Fines (QF)	Grain size analysis Atterberg limits (LL, PI) Compaction (OMC, MDD) Vertical free swell Unconfined Compressive Strength (UCS) Resilient modulus (M_R)
Cemented Quarry Fines (CQF)	Compaction (OMC, MDD) Vertical free swell Unconfined Compressive Strength (UCS) Resilient modulus (M_R)

Figure 3 shows compaction relationships between the moisture content and dry density of QF with and without cement treatment. QF material has an optimum moisture content of 11.2% and a corresponding maximum dry unit weight of 18.8 kN/m^3. With the addition of 2.3% cement to the QF, the maximum dry weight of the QF was slightly decreased to 17.9 kN/m^3 whereas the optimum moisture content has increased to 13.8%.

To understand swelling nature of this material, vertical free swell tests were conducted on both treated and untreated samples. These were performed on a 2.8 inch sample as per ASTM D-4546 method. Treated sample was subjected to curing for 7 days under controlled environment. Three samples were compacted at optimum moisture content (OMC), and were then subjected to testing. Figure 4 shows the average swell strain (%) variation with respect to time (hours) on logarithmic scale.

The vertical free swell strain of untreated quarry fines was close to 6% and the same for cement treated QF was close to 0%. This confirms that the QF is a moderately swelling material whereas the cement treatment of quarry fines has resulted in a non-swelling material. The non-swell behavior was attributed to cementing actions in the treated sample.

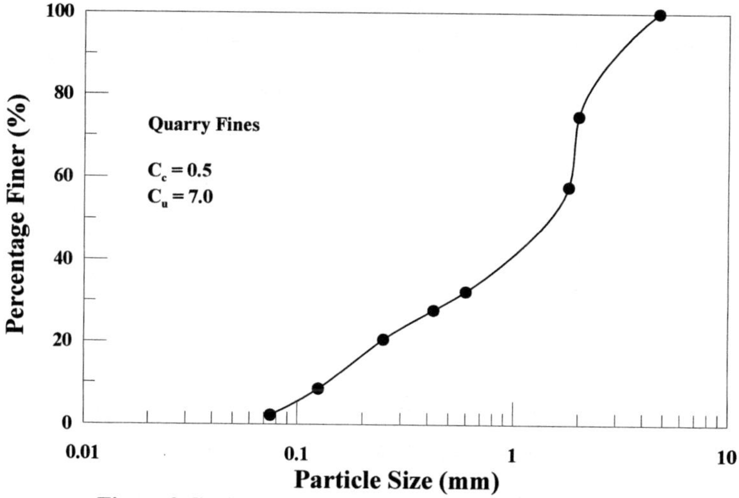

Figure 2 Grain size distribution curve of Quarry Fines

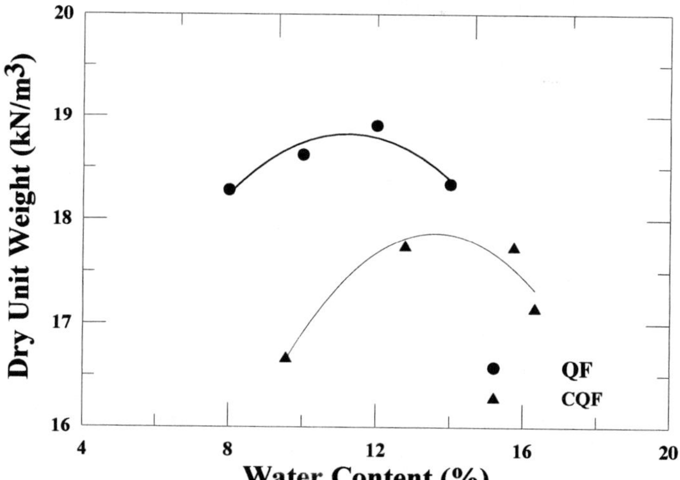

Figure 3 Compaction characteristics of untreated and treated QF

Unconfined Compressive Strength (UCS) tests were performed on both treated and untreated samples. The cylindrical samples were prepared of size 2.8 in. x 5.6 in. (diameter × height) and were compacted at the optimum moisture content of the respective compaction curve. The treated sample was tested after a curing period of 7 days in a controlled environment. Figure 5 shows UCS test results. Untreated sample showed a low strength close to 100 kPa. The cemented QF material showed a peak UCS value close to 1200 kPa, which was 12 times more than the strength of the untreated QF material. High strength of CQF is attributed to cementing reactions of the cement and fine sandy fraction of the QF material.

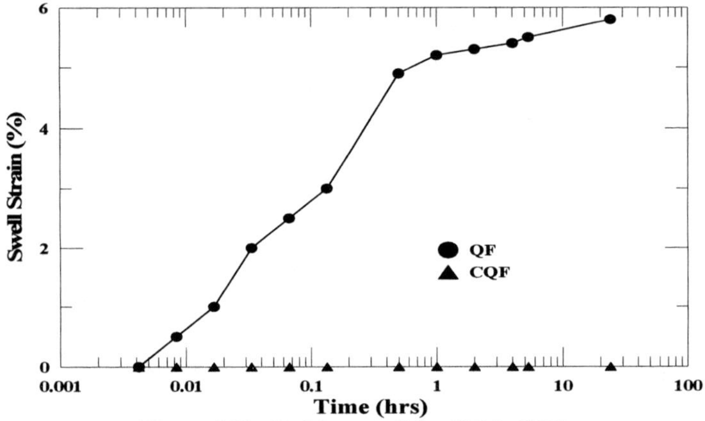

Figure 4 Vertical free swell for QF & CQF

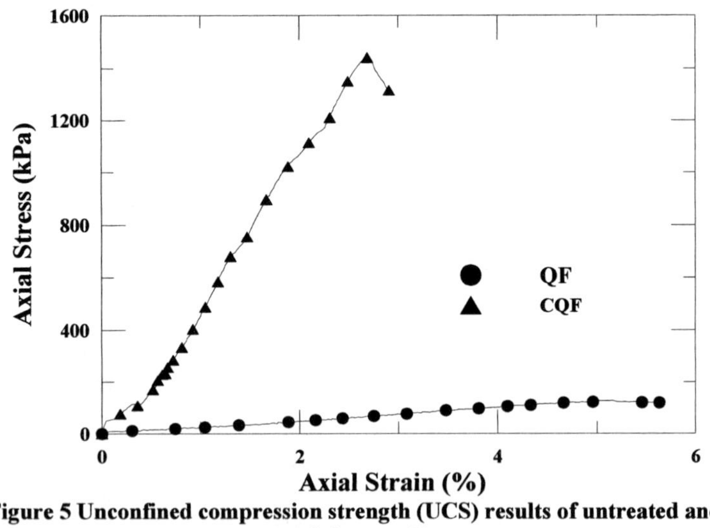

Figure 5 Unconfined compression strength (UCS) results of untreated and treated Quarry Fines

A repeated loading triaxial test was followed to determine the resilient modulus of treated QF material by following the AASHTO test procedure, T 307-99. The test was conducted on compacted soil specimens that were prepared in accordance with the standard procedure.

The Resilient Modulus (M_R) test applies test loads, which simulate the traffic wheel loads on the in situ soils by applying a sequence of repeated or cyclic loading on soil specimens. A haversine-shaped wave load pulse with a frequency of 10 Hz was applied. These loading features are in accordance with the resilient modulus test procedure outlined in AASHTO T 307-99. Both QF and CQF samples were compacted at the optimum moisture content in three layers. After compaction, the specimens have been extruded and cured for 7 days. The cylindrical specimens were then subjected to testing immediately after the curing period.

Table 3 presents the average results of resilient modulus tests carried out on both control and cemented quarry fines. As expected, cemented QF exhibited higher moduli than the control samples. The resilient modulus of QF and CQF increased with an increase in confining stress. This trend explains that the cemented granular material behaves as a strain hardening material and the M_R properties depend on the confining pressure conditions.

Table 3 Maximum resilient modulus values for cemented quarry fines at different confining stresses

Confining Pressure (σ_c), kPa	Resilient Modulus (M_R), MPa (QF)	Resilient Modulus (M_R), MPa (CQF)
20.7	65	152
34.5	118	216
68.9	228	317
103.4	232	351
137.9	230	369

FIELD MONITORING RESULTS

The field monitoring of the test section utilizing this CQF material as a base is currently under progress. Monitored data is generally collected twice a month from all field instruments (horizontal inclinometers, moisture probes, pressure cells and temperature probes) placed around the CQF layer. Inclinometers were used to monitor vertical deformations, pressure cells were used to calculate the pressure coming from the traffic. These monitored data are used here to evaluate the efficacy of the CQF as a base material for pavements. Based on the monitored data (Figure 6), small amounts of plastic deformations (less than 0.05 in.) were recorded, which are attributed to construction induced deformations since these values were recorded immediately after construction. No additional major deformations are recorded thereafter. However, further monitoring of the test site is necessary to assess whether the CQF can be used effectively as a base material in the pavements.

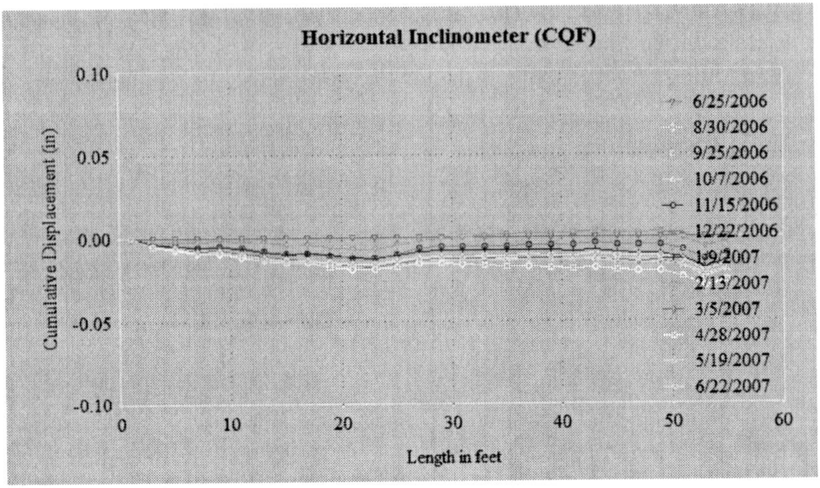

Figure 6 Horizontal inclinometer placed underneath pavement section with CQF as a base

SUMMARY AND CONCLUSIONS

This paper presented the results of a study conducted to evaluate the swell, strength and resilient characteristics of Cemented Quarry Fines (CQF). The Quarry Fines (QF) has characteristics similar to a sandy material, with very few amounts of fines. This material experienced a free swell strain close to 6%, which was reduced to 0% when treated with cement. Untreated sample exhibited low UCS value close to 100 kPa. The same material, when cemented, showed a peak UCS value close to 1200 kPa, which is a 12 fold improvement. High strength of CQF is attributed to cementing reactions of the cement and fine sandy fraction of the QF material. Field monitoring data using horizontal inclinometer showed that this recycled material experienced low amounts of permanent deformation during service. Further monitoring will be needed to address the permanent deformation of this material.

ACKNOWLEDGEMENTS

Authors would like to acknowledge the Fort Worth District and Jeff Seiders of the Construction Division of the Texas Department of Transportation (TxDOT) for providing financial support to this project under an interagency contract agreement.

REFERENCES

American Association of State Highway Transportation Officials (1993). "AASHTO Guide for Design of Pavement Structures." *American Association of State Highway and Transportation Officials*, Washington, D. C.

ASTM International (2003). "ASTM Book of Standards", vol. 04.08, ASTM, Philadelphia, Pennsylvania.

Collins, R. J. and Ciesilski, S. K. (1994). "Recycling and Use of Waste Materials and By-Products in Highway Construction." *Synthesis of Highway Practice 199*, National Academy Press, Washington, D.C., 1–77.

Gnanendran, C. T. and Woodburn, L. J. (2003). Recycled Aggregate for Pavement Construction and the Influence of Stabilization. Proceedings- Conference of the Australian Road Research Board. Vol. 21: 1755-1768.

Holtz, K. and Eighmy, T. (2000). "Scanning European Advances in the Use of Recycled Materials in Highway Construction." Vol. 64 (1).

Naik, T. R. and Kumar R. (2003). "Use of Limestone Quarry By-Products for Developing Economical Self-Compacting Concrete," Report No. CBU-2003-15, *UWM Center for By-Products Utilization*, University of Wisconsin - Milwaukee, USA.

Okamura, H. (1997). "Self-Compacting High Performance Concrete." *ACI Concrete International*. Vol. 19 (7): 50-54.

Soosan, T. G., Jose, B. T. and Abraham, B. M. (2001). "Use of Quarry Dust in Embankment and Highway Construction." *Proc. of Indian Geo-Technical Conference*, Indore.

Soosan, T. G., Sridharan, A., Jose, B. T. and Abraham, B. M. (2005). "Utilization of Quarry Dust to Improve Geotechnical Properties of Soils in Highway Construction." *Geotech. Testing J.*, ASTM. Vol. 28 (4): 1-10.

Taha, R., Ali, G., Basma, A. and Al-Turk, O. (1999). "Evaluation of Reclaimed Asphalt Pavement Aggregate in Road Bases and Subbases." *Transportation Research Record 1652*, 1, *7th Int. Conf. on Low-Volume Roads*, National Research Council, Washington, D.C., 264–269.

Tarun, R. N., Rudolph, N. K., Chun, Y. M., Fethullah, C. and Bruce, W. R. (2005). "Use of Fly Ash and Limestone Quarry By-Products for Developing Economical Self-Compacting Concrete". *Proc of Int. Cong. on Fly Ash Utilization*, 4th – 7th December 2005, New Delhi, India.

Tepordei, V.V. (1992). "Construction Sand and Gravel." Annual report 1990. *U.S. Department of Interior*, Bureau of Mines.

Gypsum Waste Reduction through Stabilization for Trench Backfill.

Md. Waliur Rahman[1], Gurmel. S. Ghataora[1], D. N. Chapman[1], Mark Tyrer[2], Peter Claisse[3], Essie Ganjian[3].

[1]Department of Civil Engineering, School of Engineering, University of Birmingham, B15 2TT, UK. mwr133@bham.ac.uk, g.s.ghataora@bham.ac.uk, d.n.chapman@bham.ac.uk. PH: (00) 44-121-4145153.

[2]Centre for Environmental Control and Waste Management, Imperial College, London, UK. m.tyrer@imperial.ac.uk.

[3]Department of Civil Engineering, Faculty of Engineering and computing, Coventry University, UK. cbx054@coventry.ac.uk, cbx111@coventry.ac.uk.

ABSTRACT

Conventional backfill for utility trenches often comprises of granular crashed rock, which is Type I subbase material as used in road pavement construction. However, the increasing need to preserve resources and to use recovered and recycled materials led to the consideration of using gypsum as a utility trench backfill material. Thus a range of mixtures containing waste gypsum were investigated. During this study it was found that mixes containing waste gypsum, limestone quarry dust and ordinary Portland cement complied with the existing Specifications for trench backfill in terms of unconfined compressive strength as well as durability in respect of loss of strength on soaking and freeze thaw. Leachate tests showed that sulphate levels were also within acceptable limits.

1. INTRODUCTION

Provision of suitable trench backfill had been a problem for many years before the Street Works Act was passed in 1991 (HMSO, 1991) in the UK, which placed the duty of care for performance of the reinstated trenches on utility companies. This was done to ensure that those who make the excavations are responsible for their proper reinstatement. Following the Act (HMSO, 1991), Highway Authority and Utility Committee (HAUC) Specifications were prepared (Department of Transport, 1991). Currently all reinstatement has to be carried out according to the revised 2nd edition of this Specifications. Unlike the 1st edition, the revised HAUC Specifications permit the use of alternative reinstatement materials (ARM). Procedures required for ARM are described in Appendix 9 of the Specifications.

In addition to replacing good quality granular material for trench backfill, the drivers of the research included the need to find uses for gypsum waste in order to reduce disposal costs. The aim was to recycle the waste or reduce its production. At present the Waste & Resources Action Programme (WRAP) has launched an initiative to encourage researchers and practitioners to find alternative uses for gypsum and other wastes (WRAP, 2007).

2. MATERIALS

Titanium oxide pigment production yields 250 Kt of 'Red' and 84 Kt of clean 'White' gypsum per year in the UK. World wide production of 'Red gypsum' is around 1.25 Mt from Huntsman Tioxide factories alone (Mini-Waste proposal, 2004). Titanium dioxide is a white pigment used in a wide range of products such as paints, plastics, cosmetics and food products. Titanium is extracted from a type of sand, called ilmenite, using sulphuric acid before being reformed as titanium dioxide. It is then coated and ground to a consistent particle size before being dispatched for use as a pigment. This process leads to waste iron sulphate. If this sulphate solution is neutralized with lime, it produces red gypsum for use as a soil conditioner on local farms but this has been restricted because of contamination, and therefore, red gypsum is being landfilled with high tax and requires an alternate use. Neutralizing waste acid with calcium carbonate (chalk) produces calcium sulphate, called white gypsum, which is sold as a raw material in the manufacture of plasterboard or cement. Waste gypsum also arises from plasterboard production. Plasterboard off-cuts from construction sites and plasterboard wastes from demolition projects amount more than 1 million tons per year in the UK (WRAP, 2006). This investigation aims at developing recycling techniques for both these wastes streams.

Quarry dust is another easily available material and likely to be an effective filler material for flowable backfill. Limestone quarry fines contain particle sizes that lie in the range 5mm to dust. Depending on the geological source of the rock quarried, the crushing equipment used, and the method used for coarse aggregate separation, the size distribution, particle shape, and other physical properties of quarry fines can be slightly different from one quarry location to another (Kalcheff and Machemehl, 1980). However, 'quarry fines' from same quarry was used in this study.

3 MIX DESIGN CRITERIA

According to the revised 2nd edition of HAUC Specification (Department of Transport, 1991), cemented material can be used as a backfill material. The mix of the material is restricted only by the need to comply with the 90 day compressive strengths from Appendix 9 of the HAUC Specifications (1991). That is minimum 4 MPa for Types 1 and 2 roads and minimum 2 MPa for Types 3 and 4 roads. The maximum permitted strength is 10 MPa for all of the road types. Care should be taken not to exceed the maximum values for two reasons. Firstly, above 10 MPa it starts to behave more like structural concrete and may crack and produce reflective cracking on the road surface. Secondly, re-excavation would be difficult. [The roads are classified in the UK according to their traffic volume. Type 1 roads carry 10 to 30 million standard axles (msa), Type 2 roads carry 2.5 to 10 msa, Type 3 carry 0.5 to 2.5 msa and Type 4 carry up to 0.5 msa].

4. METHODOLOGY

ARMs may offer significant environmental or practical advantages and cost benefits, compared to conventional materials which may include reduced usage of virgin materials by including a range of factors, such as, utilisation of recycled and secondary materials and use of self-levelling backfill to avoid the need for compaction. Source of the research on the use of secondary materials that require no compaction has been reported by Ghataora and Alobaidi (2000), Ghataora et al (2006), Alobaidi and Ghataora (2006). The materials considered by the above authors were sourced from utility excavations in the Midlands of England.

To reduce the gypsum wastes the methods adopted in this investigation for gypsum-based cementitious mixes were to introduce flowable, self-compacting controlled low strength materials in compliance with HAUC Specifications for strength properties, British Standard Specifications for durability characteristics and Building Research Establishment (BRE) of UK guidelines for leachable sulphate quantities.

Red gypsum (RG) is a sticky and lumpy material. Mixing of gypsum-based mixes proved difficult in the pilot mixes, however, addition of sandy 'quarry fines' (QF) soaked moisture readily, and consequently got rid off the stickiness of gypsum. Therefore, high amounts of QF improved mixability of the materials substantially. QF is also a well-graded material and provides mechanical stability to a mix. RG was replaced by Plaster-board gypsum (PG) off-cuts in the mixes with cement and QF. As PG waste contained stiffer lumps than RG waste, mixing was difficult in its available state, and therefore, it was necessary to grind PG in order to get a uniform mix. PG waste was available with and without paper. Both of the types were investigated.

5. LABORATORY RESULTS AND DISCUSSION

5.1 Mixes with Ordinary Portland Cement, Gypsum and Quarry Fines

A range of mixes were tested for UCS with a range of cement contents. Water-solid ratio of the mixes was kept constant at 0.3 because high flowability was achieved with this water content. Flow table test (British Standard, 1984) results showed a spread of more than 500 mm after 15 drops. The compressive strength of all the mixes increased with the increase in cement content. The mixes with cement (OPC), RG and QF with the ratio of 1:4:15 or with higher cement content complied with the HAUC Specifications as the 90 days strength of these mixes was more than 2 MPa (Fig. 1).

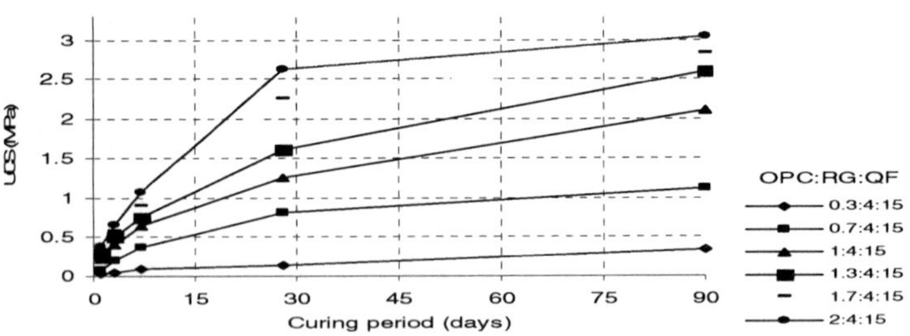

FIG. 1: UCS of mixes with a range of cement content

The mixes were cured by double wrapping with cling film at room temperature (18 to 20°C) and humidity (about 50%). Curing in high humidity (above 90%)

decreased the strength by only about 5%. The cementitious material was highly flowable and the water content was much higher than the optimum moisture content of the mix. Curing in high humidity increased the setting time of the mix, consequently decreasing the strength slightly.

The strength in case of PG-based mixes was lower than the mixes with RG probably due to the smaller particle size of PG, which increases the cement demand in the mix. Slightly higher cement (OPC:PG:QF=1.3:4:15) in the mix made it acceptable as a backfill material. PG 'without paper' performed better than 'with paper' one because the binding of the materials was obstructed by the paper with the latter material (Fig. 2).

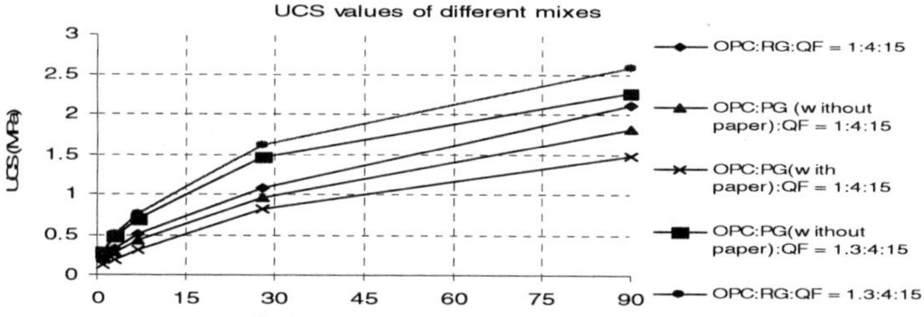

FIG. 2: UCS of the mixes with plasterboard and red gypsum.

5.2 Effect of Gypsum Contents on Strength

Investigation has been carried out with the mixes with OPC, QF and a range of RG content. Many researchers like Goswami et al. (1990) and Colak (2001) reported that gypsum normally retards the setting time and reduces the strength of cement paste. However, effect on strength due to the increase in gypsum contents has not been investigated. The ratios investigated in this study were OPC:RG:QF of 1:(0, 3, 4, 5, 6 or 7):15. The amount of OPC and QF were kept the same with a range of RG content in the mixes. The water-solid ratio of the mixes was also kept constant at 0.3. The compressive strength for curing periods of 1, 3, 7, 28 and 90 days increased with the increase in RG content up to 4 (i.e. OPC:RG:QF=1:4:15). This is probably due to the flowability effect, the flowability index is too high for RG of 0 and 3, and therefore had lower strengths than with RG of 4. However, when the amount of RG was increased from 4 to 5, strength dropped, this happened due to the retardation of strength by gypsum superseded the flowability effect at this point. For amounts of gypsum from 5 to 6 the strength increased again due to the flowability effect once more. The mix with RG of 6 was less flowable and had higher strength. The strength again dropped when RG was increased from 6 to 7 (Fig. 3). This happened due to one of the two reasons, firstly, the amount of water added might not be sufficient to allow proper mixing or permit full hydration of cement and secondly, the increased amount of gypsum (more than 30%) might have retarded the strength.

In this investigation two peak strengths were achieved, however, the mix with RG of 4 has better flowability than RG of 6. Therefore, for OPC, RG and QF mix the ratio of 1:4:15 (i.e. RG of 4) was likely to be most effective as a flowable backfill material. It can be predicted from this investigation that due to the effect of flowability and strength retardation of the mixes by gypsum, use of 20% red gypsum with OPC and QF was justified RG content for a flowable, self compacting trench backfill material.

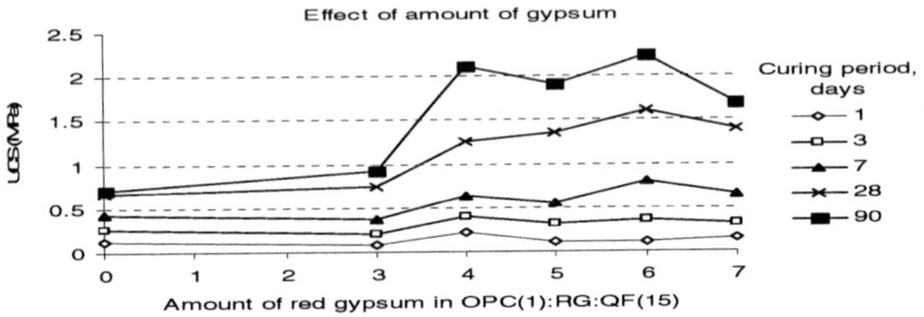

FIG. 3: Effect of gypsum contents on strength.

6. CEMENT HYDRATION AND GYPSUM

During the cement hydration, both C_2S and C_3S hydrate to produce $Ca(OH)_2$ and CSH-gel, the latter being responsible for the majority of the strength in hydrated OPC. C_3S is the most rapidly reacting silicate and contribute most of the strength development during the first few weeks, whilst C_2S influences the gain in strength at later ages. The hydrated silicates are much stronger than hydrated C_3A and C_4AF, although the latter compound exhibits a significant strength while C_3A has a negligible strength. C_3A is generally the fastest reacting phase in Portland cement. In fact, gypsum is specifically added to Portland cement to slow down this reaction, thus avoiding 'flash set' of the material (Neville 1995). If an additional amount of gypsum is added to a mix with cement, the strength development is expected to be slower; however, if the total aluminate content is low the addition of gypsum would not be expected to affect the strength.

When gypsum is not present in the system, C_3A reacts with water to form a variety of crystalline hydration products, with hydrogarnet, being the ultimately stable hydration product. Reactions are much more complex in the presence of gypsum. In this case, the C_3A will react with the gypsum to form ettringite, CsA_3SH_{32}, whose crystals are often observed to grow as needles within the cement paste. When all of the gypsum is consumed, the ettringite may decompose, reacting with more of the C_3A to form the Afm-SO_4 monosulphate phase, C_4ASH_{12}. Additionally, ettringite is unstable above about 6O°C, exemplifying the fact that cement reactions are a function of temperature as well as solution concentrations. The sulfate ions are generally more mobile than the aluminate ions so that ettringite forms either in solution or at the surfaces of the aluminate phases in cement paste. Formation of ettringite may reduce the strength of the materials, however, considerable amount of aluminium is required to form ettringite.

7. DURABILITY

Two types of durability tests were considered in this investigation. These were the effect of soaking (soaked UCS) and freeze/thaw resistance. Soaked durability test

was performed according to BS 1924 (British Standard, 1990a) and freeze/thaw test was carried out in accordance with BS 5075 (British Standard, 1982).
Two mixes were investigated for soaked durability, firstly, OPC, RG and QF at a ratio of 1.3, 4 and 15, and secondly, OPC, PG (without paper) and QF at the same ratio (Fig. 4). The mixes showed less than 20% strength reduction after soaking, which is acceptable (British Cement Association, 1998).

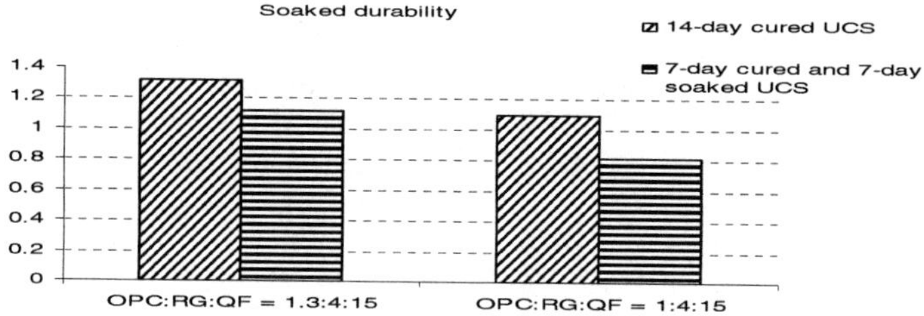

FIG. 4: Soaked durability test results.

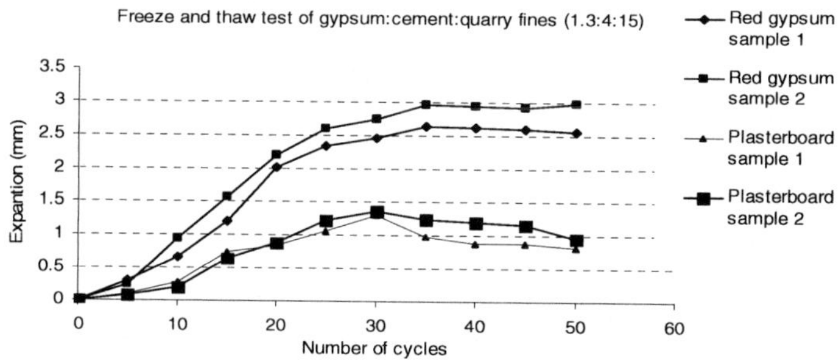

FIG. 5: Freeze and thaw test results of OPC, gypsum and QF mixes.

Freeze and thaw tests of the mixes with OPC, RG or PG and QF at a ratio of 1.3:4:15 were performed. Fig. 5 shows the expansion of standard 225 mm long prism for 50 cycles according to BS 5075. Two specimens of the same mixes were investigated. The maximum expansion of the specimens with RG was higher (about 3 mm) than the maximum expansion with PG (about 1.5 mm) (see Fig. 5 above). This happened probably due to the finer particles of PG, which resulted in reduction in void ratio and thereby produced discrete air cavities in the mix with the consequence of reduced overall frost susceptibility.

During and after freeze and thaw cycles the expansion of specimens made from both of the mixes was found to be low. According to Department of Transport (1991), frost susceptible material with a mean heave greater than 15 mm when tested in accordance with BS 812: Part 124 (British Standard, 1989) is considered to be acceptable. However, this test is for compacted granular materials. In this investigation the freeze and thaw test undertaken was more severe since it is normally carried out for concrete. The results showed maximum expansion of only 3 mm and the mixes did not disintegrate after 50 cycles. Therefore, both of the mixes were considered to be suitable.

8. SULPHATE CONTENT AND SPECIFICATION

Sulphate content of the mixes was a major concern due to two main reasons. Firstly, sulphate leaching from the mixes may contaminate ground water and secondly, sulphate emanating from the mixes can attack any surrounding concrete and any material stabilised with cement or lime. Building Research Establishment 363 digest of UK (Department of Trade and Industry, 1991) limits the sulphate content up to 1.2g/l. Sulphate test of cement, gypsum and quarry fines mixes has been carried out in accordance with BS 1377 (British Standards, 1990b). The results showed that the sulphate contents were lower than the BRE limits even after 28 days of mixing (Fig. 6).

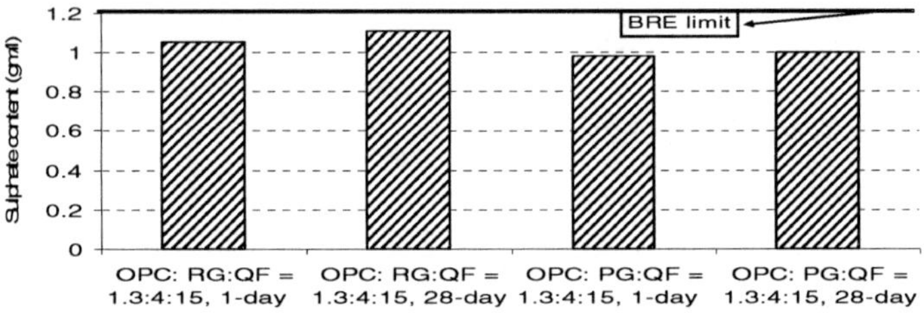

FIG. 6: Sulphate content of the mixes with cement, gypsum and quarry fines.

9. CONCLUSIONS

Ordinary Portland cement, red or plasterboard gypsum and quarry fines mixes at a ratio of 1.3, 4 and 15 proved to be effective for using as a backfill material in Type 3 and 4 roads of the UK in consideration of strength, durability and sulphate test results. The optimum gypsum content of these mixes was 20% for their effective use as flowable backfill materials. All of the mixes investigated were highly flowable and self compacting, therefore, no form of compaction will be required to place the mixes in trenches.

10. REFERENCES

Alobaidi, I.M. and Ghataora, G.S. (2006). "Performance of Trial Trenches in Roads". Ash Tech 2006. 15-17 May. Birmingham. UK.

British Cement Association (1998). "Cement-Bound Materials for Subbases and Roadbases, Material Selection and Mix Design, Construction and Control Testing". BCA Publications. Crowthorne, Berkshire, England.

British Standard (1982). "Concrete Admixtures, Part 2: Specification for Air-Entraining Admixtures". BS 5075-2:1982. BSI. London. UK.

British Standard (1984). "Testing Concrete, Part 105: Method for determination of flow". BS 1881-105:1984. BSI. London. UK.

British Standard (1989). "Testing Aggregates: BS 812 Part 124 Method for Determination of Frost- Heave". BSI, London, UK.

British Standard (1990a). "Stabilised Materials for Civil Engineering Purposes, Part 2: Methods of Tests for Cement-Stabilised and Lime-Stabilised Materials". BS 1924-2:1990. BSI. London. UK.

British Standard (1990b). "Methods of Tests for Soils for Civil Engineering Purposes". BS 1377-3:1990. BSI. London. UK.

Colak, A. (2001). "Physical, mechanical, and durability properties of gypsum–Portland cement–natural pozzolan blends". Canadian Journal of Civil Engineering 28. pp375-382.

Department of Trade and Industry (1991). "Sulfate and Acid Resistance of Concrete in the Ground". BRE Digest 363. Construction Research Communication Ltd. Watford. UK.

Department of Transport (1991). "Specification for the Reinstatement of Openings in Highways". Highway Authorities and Utility Committee (HAUC), HMSO, London.

Ghataora, G.S. and Alobaidi, I.M. (2000). "Assessment of the performance of Trial Trenches Backfilled with Cementious Materials". International Journal of Pavement Engineering. Issue 4. Vol. 1. pp297-316.

Ghataora, G.S. and Alobaidi, I.M. Faragher, E. and Grant, S. (2006). "Use of Recycled Granular Materials for Non-compaction Trench Backfill". ICE. Waste and Resource Management Journal. Vol 159. Issue Wr1. pp23-28.

Goswami, G., Mohapatra, B. and Panda, J.D. (1990). "Gypsum Dehydration During Comminution and Its Effect on Cement Properties". Journal of the American Ceramic Society. Volume 73. Issue 3. pp721-723.

Her Majesty's Stationery Office (HMSO) (1991). "Chapter 22". New Road and Street Works Act. First Published in 1991 and Reprinted in 1997 in the UK.

Kalcheff, I.V. and Machemehl, C.A. "Utilization of Crushed Stone Screenings in Highway Construction." Presented at the 59th Annual Meeting of the Transportation Research Board, Washington, DC, January, 1980. Sighted in http://www.rmrc.unh.edu. >June 30, 2007.

Mini-Waste Farady Research Proposal (2004). "Gypsum Waste Reduction". A Report Circulated by Mini-Waste Faraday in the United Kingdom.

Neville, A. M. (1995). "Properties of Concrete". Fourth Edition. Longman Group Limited. Essex. England.

WRAP (2006). "Review of Plasterboard Material Flaws and Barriers to Greater Use of Recycled Plasterboard, Technical Report, WRAP, AEA Technology Plc.

WRAP (2007). "Material Change for a Better Environment". The Waste & Resources Action Programme, www.wrap.org.uk >11th July 2007.

Performance of Biosurfactant Produced from Used Vegetable Oil

C. Vipulanandan[1], S.Harendra[2] and S.G.Hariharan[3]

1 Chairman, Professor and Director of Center for Innovative Grouting Materials and Technology (CIGMAT), Department of Civil and Environmental Engineering, University of Houston, Houston, Texas 77204-4003. Phone: (713) 743-4278; FAX (713) 743-4260; Email: cvipulanandan@uh.edu;

2 PhD Student, Department of Civil and Environmental Engineering, University of Houston, Houston, Texas 77204-4003. Email: ps_harendra@yahoo.com;

3 M.Sc, Department of Civil and Environmental Engineering, University of Houston, Houston, Texas 77204-4003.

ABSTRACT: Disposing of used vegetable oil into the sewer system is causing numerous problems for the cities. Hence there is interest in recycling the used vegetable oil for other applications. In this study, a nonaseptic method was used to produce biosurfactants from used vegetable oil. The biosurfactant produced was characterized by various physical and biochemical assays. The critical micelle concentration (CMC) of the produced biosurfactant was 0.7 g/L and the surface tension was 30 dynes/cm. Also the active groups in the biosurfactant were identified using the Fourier transform infrared spectroscopy (FTIR) and Thin Layer Chromatography (TLC).The ability of the biosurfactant to solubilize dense non aqueous phase liquids (DNAPLs) was studied using chlorinated organic compounds such as trichloroethylene (TCE).The efficiency of the biosurfactant produced from used vegetable oil was compared to other commercially available surfactants and the results were comparable.

INTRODUCTION

Traditionally, biosurfactants have been produced by growing a single strain of bacteria on glucose, under aseptic conditions. Large-scale production of biosurfactants under sterilized conditions could be expensive. Copper and Paddock (1984) showed that the yeast *Torulopsis bombicola*, when grown on vegetable oil, produced considerable amounts of biosurfactant. Kosaric et al. (1984) showed that the cost of producing biosurfactants could be reduced if waste material could be used as process feed-stock for biosurfactant production. Mattei & Bertrand (1985) proposed a continuous biosurfactant producing strategy by growing a mixed bacteria culture on crude oil. Ghurye et al. (1994) have shown that a mixed culture obtained from an activated sludge plant was able to produce surface active agents when grown on 2% molasses.

Chemical surfactants have been shown to be very effective in mobilizing organic and inorganic pollutants from soil subsurface (Volkering et al. 1995). They have also been shown to increase the apparent solubility of organic compounds (Dohse et al. 1994). But the use of chemical surfactant in soil remediation is rather limited due to the fear that these compounds themselves may be pollutants, as they are not biodegradable. Researchers have shown that surfactants that are produced biologically also posses similar properties, are less toxic and are generally biodegradable (Scheibenbogen et al. 1994). Since biosurfactants are produced from substrates of natural origin, their structures are amenable for biodegradation.

Although a notable amount of work has been done on biosurfactants, very few researchers have identified waste materials that could be used for producing biosurfactants. Kitamoto et al. (1990) used vegetable oils to produce their surfactant. Mercade and Manresa (1994) pioneered the work on producing biosurfactants from agro-industrial by products. They showed that a lot of industrial by products are potential substitutes for expensive glucose. Polman et al. (1995) have shown that potato processing wastes are viable food source for fermentation. If the entire waste material could be used for fermentation, it could considerably reduce the overall cost of production of biosurfactants. Biosurfactants have been used to solubilize naphthalene and remove Pb from wastewater.

This research investigated the possibility of producing biosurfactant from a recycled material, used vegetable oil, which has negative cost and hence, decrease the cost of production of biosurfactants. Biosurfactant produced under non sterilzed batch fermentation was characterized based on various physical and biochemical assays.

OBJECTIVE

The overall objective of this study was to produce the biosurfactant using used vegetable oil and to investigate its potential for solubilizing chlorinated solvents.
The specific objectives are as follows.

1) To characterization of biosurfactant produced from used vegetable oil.
2) To investigate the ability of the biosurfactant to solubilize trichloroethylene and compared to the performance to other commercially available surfactants.

MATERIALS AND METHODS

Microorganisms: Two different microbial strains were used for producing the biosurfactants. One of them was isolated from a crude-oil contaminated soil, in Houston, Texas.. The microorganisms were sustained on 1 % crude oil (w/v), with adequate amounts of mineral salts. The different types of microbes present in the isolate were not ascertained. After the batch reactor runs were over, biomass samples were taken out and streaked on agar plates. The single colony that grew on the agar plate was identified as a *Pseudomonas* sp. using the Oxi/Ferm Tube (II) test kit from Roche. The other strain used was from a sample of activated sludge from a sewage treatment plant in Houston. The bacterial strain was identified to be *Flavobacterium* sp. using the Oxi/Ferm Tube (II) test kit.

Chemicals: The mineral salts used were as described by Cooper, et al (1979). The salts used were K_2HPO_4 0.1%, KH_2PO_4 0.05%, KCl 0.01%, $MgSO_4$ 0.05%, $NaNO_3$ 0.2%. All chemicals were procured from Sigma, NJ, and were reagent grade. Based on literature review, SDS, Triton X-100 and CTAB were selected for the solubility study to represent anionic, non ionic and cationic respectively. Except for the UH-biosurfactant, other surfactants were purchased from Sigma-Aldrich, St.Louis, Missouri.

Substrates: Used vegetable oil was obtained from a domestic restaurants used as the sole carbon source for biosurfactant production.

Experimental Methods:

The biosurfactants were produced in 2 L glass bottles, with 1 L volume. 10 g of substrate was mixed in 1 L of de-ionized water, along with the mineral salts. Four mL of acclimated micro-organism was added to the reactor and stirred at a moderate speed. Samples were taken out at different times for analysis.

Reactors were run under both aseptic and nonaseptic conditions. For the aseptic reactors, mineral salts medium and glassware used were sterilized by autoclaving at 250 °C, at 15 psig for 20 minutes. The used-vegetable-oil substrate was sterilized by filtering through a 45 μm filter. The reactors were covered by a sponge stopper to keep the reactor aerobic, and the broth was stirred using magnetic stirrers.

The surfactant solutions of TCE were prepared in 40 ml vials. The vials were continuously agitated using an Innova 2000 platform shaker (New Brunswick Scientific, N.J) at 200 rpm held in a horizontal position. A Gas chromatography (Shimadzu GC-14A) with a flame ionization detector and a CR501 data processor was used for determining the chlorinated compounds concentration. The stainless steel column (Alltech) was 6-ft long × 1/8-in outer diameter × 0.085-in inner diameter and detection temperature was set at 250°C. The injection volume of sample used for GC analysis was 2 μL. The detection limit for TCE was 1 mg/L.

<u>Characterization of UH biosurfactant</u>

Surface Tension: The surface tension of the whole broth was measured using a De-Nuoy surface tensiometer. All above mentioned surfactants including UH biosurfactant were then diluted with de-ionized water, and the surface tension measured. This was done to estimate the critical micelle dilution (CMD) of the broth (Fig. 1).

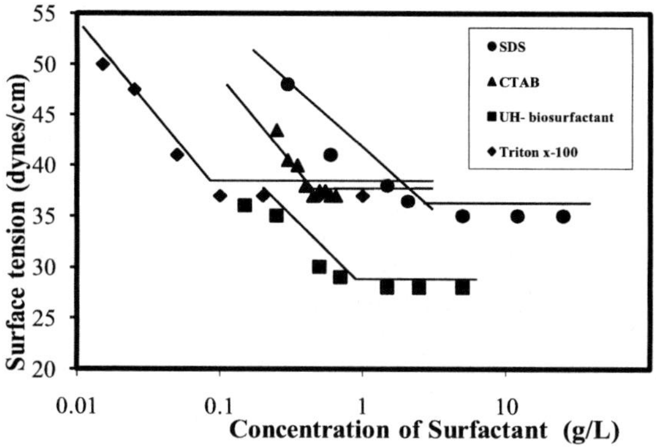

Fig. 1: Variation of surface tension with surfactant concentration

Thin Layer Chromatography (TLC): The components of the biosurfactant were analyzed using the TLC technique. One mL samples were taken from the broth and centrifuged at 7000 g for 10 minutes. The supernatant was separated from the pelletized biomass and used as the sample for the TLC analysis. A solvent mixture of chloroform: methanol (81.8: 18.2) was used to separate the components present in the sample. Precoated silica gel 60 Fisher Rediplates were used as the chromatographic medium. The developed plates were sprayed with different reagents (Rhodamine 6G, Molybdenum Blue, Bial's Orcinol and Diphenyl amine). The plates sprayed with Rhodamine 6G were viewed under UV light (240 nm). The plates sprayed with Molybdenum Blue and Bial's Orcinol reagents were heated in an oven at 105° C, for 5 minutes before viewing (Table 1).

Table 1 Components present in biosurfactant produced from used vegetable oil (TLC Assay)

No.	R_f (%)	Component	Spray Reagent Used
1	0	Lipid / Hydrocarbon/ Fatty Acid	Rhodamine 6G
2	18	Glycolipid	Bial's Orcinol Solution
3	26	Phospholipid	Molybdenum Blue
4	35	Lipid / Hydrocarbon/ Fatty Acid	Rhodamine 6G
5	45	Glycolipid	Bial's Orcinol Solution, Diphenyl Amine
6	55	Sulfolipid / Sugar	Bial's Orcinol Solution
7	71	Glycolipid	Bial's Orcinol Solution, Diphenyl Amine
8	94	Lipid / Hydrocarbon/ Fatty Acid	Rhodamine 6G
9	100	Glycolipid	Bial's Orcinol Solution, Diphenyl Amine

The Rf values of each component were calculated as the ratio of distance moved by the spot from the origin to the total distance moved by the developing front.

Infrared analysis

Changes in the chemical structure of the surfactants were investigated in AgCl pellets using the Perkin- Elmer Fourier transform infrared spectroscopy (FTIR) in the range of 4000-400 cm^{-1} (wave number) at a 4 cm^{-1} resolution and the intensities were expressed in terms of transmittance (Fig. 2).

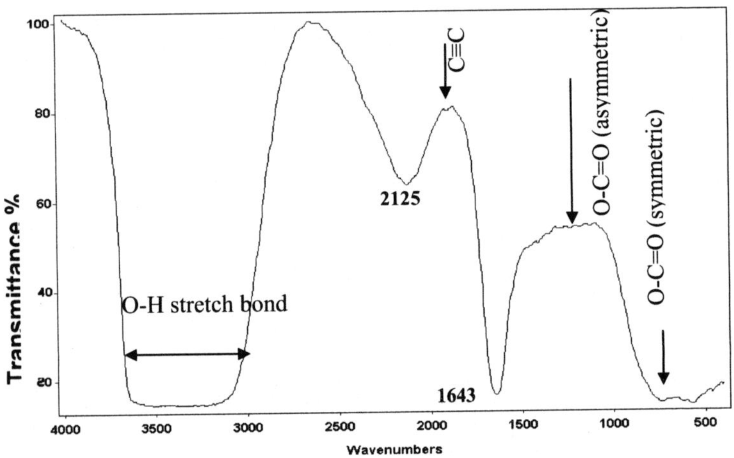

Fig. 2: FTIR Analysis of 10 g/L UH biosurfactant solution

RESULTS AND DISCUSSIONS

The biosurfactant produced from used vegetable oil was able to reduce the surface tension of water to less than 28 dynes/cm. The biosurfactant had a CMC of 0.7 g/L. The surfactant was found to be mixture of atleast nine compounds, as was seen from the TLC assay (Table 1). Of the nine components, five were extractable using a solvent system. Enzymatic hydrolysis of the biosurfactant using protease and β-glucosidase destroyed the surface active components of the surfactant, and hence increased the surface tension of the sample. This shows that the surface activity of the biosurfactant is mainly due to a protein fraction, and partially due to a polysaccharide component.

Fig.2 shows the FTIR analysis of UH biosurfactant. Functional groups identified were very broad O-H stretch bond (3160-3660 cm^{-1}) which overlaps the C-H absorption, C≡C (2095 cm^{-1}), O-C≡O asymmetric stretch (near 1643 cm^{-1}), O-C≡O symmetric stretch (near 1433 cm^{-1}) [Kim et al., 2006].

Fig. 3 shows the effect of pH on the surface activity of the biosurfactant. The surface tension values measured at different pH showed that there was not much loss in surface activity due to pH. At very low pH (<4.5) and at very high pH (>10.5), there was a slight loss in surface activity. This shows that the surfactant produced from used vegetable oil is stable in a pH range of 4.5 to 10.5.

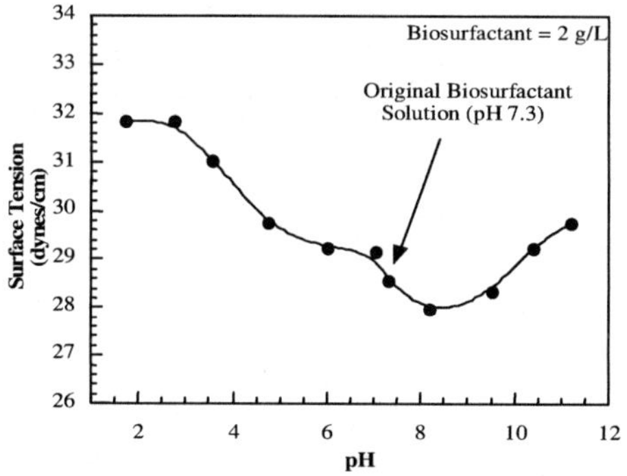

Fig. 3: Effect of pH on surface activity of biosurfactant produced from used vegetable oil.

Solubility

Surfactants increased the solubility of TCE in water. Of the surfactants used the UH biosurfactant solubilized TCE as equally well or much better compared to the other surfactants used in this study.

Based on preliminary investigation (Where S > 0, $\frac{dS}{dt}$>0 and $\frac{d^2S}{dt^2}$<0 (t→∞ S→ Limiting value)) solubilization kinetics of chlorinated compounds was represented by hyperbolic relationships than exponential or bilinear relationships as follows:

$$S=\frac{t}{A+Bt} \tag{1}$$

Where S is the concentration of chlorinated compound solubilized in time t and parameters A and B depend on the type and concentration of surfactant. Parameter A represents the initial rate solubilization and parameter 1/B represented the ultimate solubilization. The hyperbolic relationship can be rearranged to a linear form as follows.

$$\frac{t}{S} = A + Bt \quad (2)$$

Solubilization kinetics are shown in Fig. 4. If the data satisfied Eq (2) then the assumption of hyperbolic is acceptable and the parameters A and B can be determined. In Fig. 5, the relationship in Eq. (2) is verified for the various surfactants.

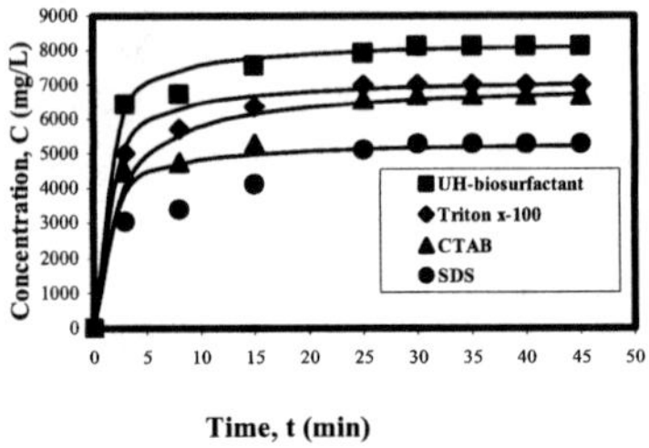

Fig. 4: Solubilization Kinetics of TCE in Various Surfactant Solutions using 10 g/L of Surfactants

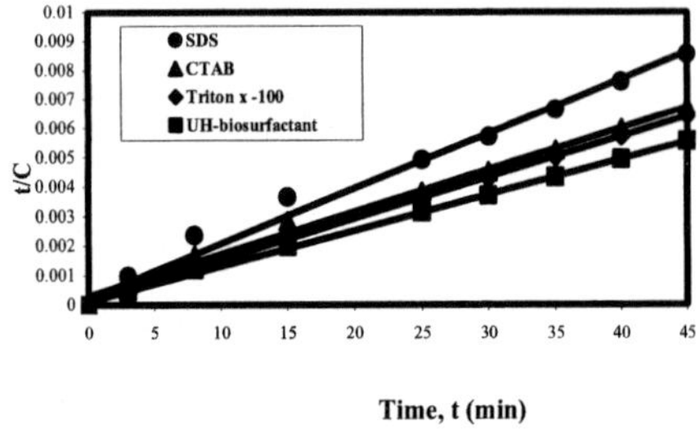

Fig. 5: Relationship of t/c, (time/concentration) vs. t, (time)

Linear relationship with coefficient of R^2=0.97 or higher indicated that hyperbolic relationships can be used to represent the solubilization kinetics of chlorinated compound in various surfactant solutions (Table 2).

Table 2 Solubilization kinetic parameters A and B for TCE

Surfactant	A	B	R^2
UH Bio surfactant	0.0011	0.0012	0.99
CTAB	0.00031	0.00014	0.99
SDS	0.00021	0.00019	0.99
Triton	0.00016	0.00014	0.99

CONCLUSIONS

Based on the experimental study following can be concluded:

1) Biosurfactant was produced from used vegetable oil which had carboxyl and hydroxide groups in its structure
2) Biosurfactant solubility potential of chlorinated solvent TCE were compared to commercially available surfactants.
3) The rate of solubilization of TCE by various surfactants have been quantified and represented by a hyperbolic relationship.

ACKNOWLEDEMENT

The study was supported by the Center for Innovative Grouting Materials and Technology with funding from the Advanced Research Program of the Texas Higher Education Coordinating Board (THECB) and the Texas Hazardous Waste Research Center.

REFERENCES

Copper D.G. and Paddock D.A., "Production of a biosurfactant from Torulopsis bombicola" *Journal of Applied and Environmental Microbiology (1984)*" Vol 47, No 1, pp 173-176

Dohse, D.M. and Lion, L.W; "Effect of Microbial Polymers on the Sorption and Transport of Phenanthrene in a Low-Carbon Sand", Environmental Science and Technology, 28 (4), 541-548.

Ghurye, G.L, Vipulanandan, C. and Wilson, R.C; 1994; "A Practical Approach to Biosurfactant Production Using Nonaseptic Fermentation of Mixed Cultures", Biotechnology and Bioengineering, 44, 661-666.

Kim, J and Vipulanandan, C. (2006). "Removal of Lead from Contaminated Water and Clay Soil Using a Biosurfactant," Journal of Environmental Engineering, Vol. 132, No. 7, pp.857-865.

Mercade, M.E. and Manresa, M.A; 1994; "The Use of Agroindustrial By-Products for Biosurfactant Production", Journal of American Oil Chemists Society, 71 (1), pp 61-64.

Kitamoto, D., Haneishi, K., Nakahara, T. and Tabuchi, T. (1990); "Production of Mannosylerythritol Lipids by *Candida antarctica* from Vegetable Oils", Agricultural Biological Chemistry, 54 (1), 37-40..

Mettei, G. and Bertrand, J-C, 1985; "Production of Biosurfactants by a mixed Bacteria Population grown in continuous culture on Crude Oil", Biotechnology Letters, 7 (4), 217-222.

Scheibenbogen, K; Zytner, R.G; Lee, H. & Trevors, J.T; 1994; "Enhanced Removal of Selected Hydrocarbons from Soil by *Pseudomonas aeruginosa UG2* Biosurfactants and Some Chemical Surfactants", Journal of Chemical Technology and Biotechnology, 59 , 53-59.

Volkering, F; Breure, A.M; Van Andel, J.G. & Rulkens, W.H; 1995; "Influence of Nonionic Surfactants on Bioavailability and Biodegradation of Polycyclic Aromatic Hydrocarbons", Applied and Environmental Microbiology, 61 (5), 1699-1705.

Zajic J.E., Guignard H. and Gerson D.F., 1977a, "Emulsifying and surface active agents from Corynebacterium hydrocarboclastus", *Journal of biotechnology and bioengineering,* Vol 19, pp 1285-1301.

Soil Fused with Recycled Plastic Bottles for Various Geo-Engineering Applications

Naji Khoury[1], A.M. ASCE, Charbel Khoury[2], S.M. ASCE and Younane Abousleiman[3], M.ASCE

[1]Research Associate, School of Civil Engineering and Environmental Science, University of Oklahoma, 202 W. Boyd, Room 334, Norman, OK, 73019; Phone: 405-325-4236; nkhoury@ou.edu.
[2]Doctoral Candidate, School of Civil Engineering and Environmental Science, University of Oklahoma, 202 W. Boyd, Room 334, Norman, OK, 73019; Phone: 405-325-9244; ckhoury@ou.edu.
[3]Larry W. Brummett/ONEOK Chair and Professor, PoroMechanics Institute, 100 E. Boyd Street, Suite P119, Norman, Oklahoma 73019, USA, Email: Phone: 405-325-2900; yabousle@ou.edu

ABSTRACT: The emphasis of the present work is to study the mechanical properties of a new generation of geo-plastic products that can be used in various aspects of geo-engineering applications. The plastic-soil based geoplastic-material (PS) is produced by heating and mixing simultaneously recycled plastic bottles with soil at a specified temperature to melt the thermoplastic material so that a uniform fused mix is attained. This study is also directed to assess the potential of PS as an additive to stabilize open graded aggregate base. Laboratory results showed that PS properties varied with the types of soil. PS produced with sand exhibited unconfined compressive strength and indirect tensile strength values higher than PS produced with lean clay. Generally, specimens exhibited unconfined compressive strengths as high as ordinary concrete specimens. In addition, permeability and strength results of open-graded aggregates stabilized with PS indicated that such technology produces a drainable and a stable base for a better pavement performance with ecological and economic implications.

INTRODUCTION

Plastic waste has become a visible and significant component in municipal solid waste and continues to increase due to the increase of plastic sales and plastic packaging over the last decades. Our continuous dependence of plastics stresses the need of handling them after their initial use. A recent report by Municipal Solid Waste (MSW, 2006) indicated that 223 million tonnes of wastes were generated in the USA in 2005, of which approximately 26 million tonnes are plastic waste. Based on the information provided in the report, approximately 6% of the plastics are recovered and the others are discarded mainly in landfills. Recently, the Container Recycling Institute (CRI, 2007) reported that the number of beverage bottles that have been landfilled, littered and incinerated in the U.S. is more than 50 billion as of July 2007.

The development of new technologies and methods help divert a large amount of plastic waste from being landfilled and incinerated with tremendous ecological and economic implications to our nation.

To this end, the primary emphasis of the proposed research is to study the mechanical properties of a new generation of an environmentally sound plastic-soil based geoplastic-material (PS) developed by Khoury et al. (2006). The product is formed by heating and mixing simultaneously disposed plastic bottles and plastic containers with soil at the melting point of the thermoplastic material so that a uniform fused mix is attained. The manufactured product could be revolutionary in geo-engineering applications such as:

(1) Pavement and transportation application; PS as a new additive to stabilize open-graded aggregate base and enhance performance in sulfate bearing soils.
(2) Geotechnical application, stabilizing expansive soils.
(3) Manufacturing building material, such as plasti-tiles and bricks with a higher tensile and compressive strength.
(4) Finally, to rapidly construct military based airport runways in remote areas on challenging soils.

The secondary focus of the study is the fundamental behavior of manufactured product (PS) as a new additive to stabilize open-graded aggregate base using the measurements of permeability and strength. A free-draining and rapid drained aggregate base is preferable for a better overall pavement performance, as recommended by the FHWA and as revealed in the literature (Khoury and Zaman, 2006; Cedergren, 1994; Cedergren, 1988; Baldwin, 1987). The stabilized open-graded aggregate base has high permeability and adequate strength, as revealed by the results presented in this study.

BACKGROUND

The increase of plastic municipal wastes has led to extensive and innovative technologies on the integration of recycled plastics in various applications. Department of Transportation, highway agencies and researchers have been involved in various studies concerning the feasibility, economic and ecological impact and the overall performance of recycled plastic in related engineering projects. Relevant technologies and techniques to the proposed study are described in the following sections.

Recycled Plastic Flakes as an Aggregate Replacement in Pavement Materials

Recycled thermoplastics, namely, PET, HDPE and LDPE, have been used in AC (asphalt concrete) and PCC (Portland Cement Concrete) mixtures to replace aggregates with specified diameters (see e.g., Hassani et al., 2005; Hinislioglu and Agar, 2004; Flynn, 1993; Zoorob and Suparman, 2000; Zoorob, 2000; Conigliaro and Watson, 2000; Mallick and Teto, 2000; among others). Findings from these studies showed enhancement in the durability, strength and fatigue life. However, the degree of enhancement is a function of the plastic types and amount.

Mechanical Stabilization Using Plastic Strips

Mechanical stabilization has been widely used as a remediation to enhance the engineering properties of unsuitable subgrade soils. The method consists mainly of blending aggregates or sand with soils prior to compaction to overcome deficiencies in grading and decreasing the plasticity index. For the last few decades recycled plastic strips have been used in mechanical stabilization as revealed by a number of studies (Sobhan and Mashnad, 2003; Sobhan, 1997; Cavey et al., 1995; Benson and Khire, 1994). The degree of improvement is influenced by many factors such as strips type and amount, soil and aggregate types.

Recycled Plastic Waste as New Composite Materials

Innovative and creative research has lead to the evolution of new composite materials using recycled plastic waste for various applications. The common process for preparing the composite consists of heating and mixing the intake materials, namely, recycled plastic (unprocessed, flakes, or pellets), granulates and by-products, among others, to a specified temperature. The heated mixture is compressed into a specific mold to form a final product. The properties of the composite depends on the pressure, type of recycled plastic and granulates, as revealed by numerous studies (Schmidt, 1993; Meyers, III et al., 2006; Malloy et al., 2003; Balkum, 2002; Kashi et al., 2002; Kashi et al., 2001; Krishnaswamy and Lampo, 2001; among others).

MATERIAL PROPERTIES

Type of Soils

Two types of soils, i.e., sand and clay, were used to produce different types of PS, in this study. The Atterberg limits were determined in accordance with the ASTM D 4318 test methods. For clay the average plasticity index and liquid limit values are 15 and 35, respectively. The sand is non-plastic. The grain size distribution for each specimen was determined by conducting gradation tests according to the ASTM D 422 test method. The clay soil has 80% passing No. 200 sieve and the sand soil had a value of 25%. In view of these test results, the clay is classified as lean clay (CL), while the sand as SW, according to the Unified Soil Classification System.

Selected Aggregate

A limestone-type aggregate was selected to assess the effect of PS stabilization on aggregate base. The aggregate has an average Los Angeles abrasion value of approximately 30 and a bulk specific gravity of 2.64. Bulk samples were collected and shipped to the laboratory for testing. The bulk samples for each aggregate were sieved in accordance with the sieve sizes for AASHTO No. 57 gradation. Aggregates retained on each sieve were washed, dried for 24 hours and then placed in different batches. It is important to note that materials retained on the US Standard No. 200 sieve were only dried and placed in a separate bucket. This method helped reconstitute specimens to meet AASHTO No. 57 gradation. It also gave a better control over the gradation of each specimen, because gradation has significant influence on the engineering and physical properties of an aggregate mix (Wright and Dixon, 2004).

Plastic Waste

Discarded plastic bottles (# 1 PET, i.e., Polyethylene Terephthalate) were collected and only used in this study. Bottles were emptied from any leftover liquid and the paper around each bottle was removed as well as the cap. Bottles were dried in an oven at a temperature of approximately 110°C to assure no moisture prior to mixing with soil at an elevated temperature. PET properties are listed in Table 1.

Table 1 PET Properties (After Wikipedia, 2007)

Density	1370 kg/m^3
Young modulus (E)	2800–3100 MPa
Tensile strength(σ_t)	55–75 MPa
Elongation @ break	50–150%
Melting point	260 °C
Thermal conductivity	0.24 W/m.K
linear expansion coefficient (α)	7×10^{-5}/K
Specific heat (c)	1.0 kJ/kg.K
Water absorption	0.16

SPECIMEN PREPARATION

Plastic-Soil Based Geoplastic-Material (PS)

PS was formed by heating and mixing simultaneously the plastic bottles and soil at a temperature of approximately 275°C, for approximately one hour so that a uniform fused mix is attained. A plastic to soil ratio (PSR) of one (by weight) was selected to produce PS from the clay soil (i.e., PS(C)) and from the sand (i.e., PS(S)). Fig. 1 shows a photographic view of soil and plastic mixed together prior and after reaching a uniform mixture. It is important to note that various PSR value produce PS with different properties; PSR effect on PS is beyond the scope of this study.

FIG. 1. Mixing of Plastic and Soil

In its liquid form, the mixture was poured in a mold to prepare specimens for unconfined compressive strength (UCS) and indirect tensile strength (ITS). UCS specimens were molded in a Harvard miniature mold (i.e., 33.5 mm x 68.6 mm) while a 101.6 mm x 85.0 mm mold was used for ITS specimens. After 5 minutes, specimens were extruded and then placed at room temperature for one hour prior to placing them in a water bath for approximately 24 hours. Fig. 2 shows Harvard miniature specimens prepared for UCS testing.

(a) (b)
FIG. 2. PS specimens: (a) at room temperature and (b) and in water

Aggregate-Base Stabilized Specimens

PS(C) produced from the clay soil with a PSR of one was used to evaluate its effects on the properties of an open-graded aggregate. A total of eight specimens were prepared with 10% and 20% PS(C) and tested for permeability and UCS. PS(C) was added to aggregates (oven dried at 110°C for 24 hours) and mixed uniformly together at a temperature of approximately 275°C, for one hour. The mix was poured in a split mold and compacted using a steel rod by hand. Fig. 3 shows aggregate base stabilized with PS(C) for permeability and strength.

(a) (b)
FIG. 3. Aggregate base stabilized with PS(C): (a) for UCS tests and (b) for Permeability tests

LABORATORY TESTS

PS specimens were tested for UCS and ITS. UCS and ITS values were determined using ASTM D 2166 and ASTM D 3967 test methods, respectively. The ASTM D 2166 test methods was also used to assess the UCS values of aggregates specimens stabilized with 10% and 20% PS(C). Permeability tests were performed using a falling head approach in accordance with a laboratory procedure detailed in Fwa et al. (1998).

PRESENTATION AND DISCUSSION OF RESULTS

Mechanical and physical properties of PS

The laboratory results revealed that PS(S) specimens (i.e., produced with sand) had an average UCS value of approximately 35.5 MPa compared to PS(C) (i.e., produced with lean clay) with an average UCS value of 18.3 MPa. It is clear that such products produce UCS values as high as ordinary concrete specimens cured for 28 days. Fig. 4 shows a typical failure of PS(S) specimens. The ITS values exhibited the same behavior of UCS values. The average ITS value of specimens with PS(S) was approximately 1240 kPa, which is 25% higher than the corresponding values of PS(C). From these results, it can be concluded that changes in soil properties influence the engineering properties of PS.

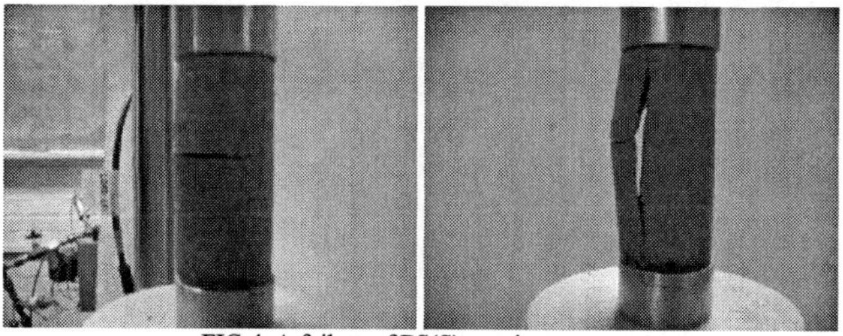

FIG.4. A failure of PS(S) specimens.

Aggregate Base Stabilized with PS

The effect of PS stabilization on the performance, using measurements of permeability and strength, of open-graded aggregate specimens was examined. Results showed that specimens with 10% PS(C) produced an average permeability value of approximately 0.49 cm/s compared to 0.42 cm/s for specimens with 20% PS(C). The compressive strength values of specimens with 10% and 20% of PS(C) were 1,895 kPa and 2,585 kPa, respectively. From these results, it is concluded that PS stabilization technique produce a permeable and stable aggregate base.

CONCLUSIONS

This study was undertaken to assess the engineering properties, namely, UCS and ITS, of PS for various geo-engineering applications. Results showed that PS produces a UCS value as high as ordinary concrete specimens. The effect of different types of soil on the engineering behavior of PS was also observed. PS produced with sand exhibited higher UCS and ITS values than the corresponding values of PS with lean clay. In addition, the study examined the use of PS as a new generation of additive to stabilize open-aggregate base. It was found that PS stabilized aggregate specimens exhibited high permeability and strength values, resulting in a better pavement performance and increases pavement longevity. Such a revolutionary technique, in addition to its superior performance, helps upscale the recycling effort currently underway and provides tremendous ecological and economic benefits to our nation and society. On the basis of the data obtained in this study, it is concluded that the PS mix has attained an acceptable level of sustainability which can be further advanced and verified by long-term performance tests.

ACKNOWLEDGEMENT

The authors are thankful to Dr. Gerald Miller for his support and help.

REFERENCES

Baldwin, J.S. (1987) " Use of Open-Graded, Free-Draining Layers in Pavement Systems: A National Synthesis Report," Transportation Research Record, No. 1121, pp 86-89.

Balkum, E (2002) "Aggregate Using Recycled Plastics" U.S. Patent No. 6,488,766.

Benson, C. H., and Khire, M. V. (1994) "Reinforcing sand with strips of reclaimed high-density polyethylene."' *J. Geotech. Eng.*, 120 (5), 838–855.

Cavey J. K., Krizek, R. J., Sobhan, K., and Baker, W. H. (1995). "Waste fibers in cement-stabilized recycled aggregate base course material." *Transportation Research Record 1486*, Transportation Research Board, Washington D.C., 97–106.

Cedergren, H.R., (1994) "America's Pavements: World's Longest Bathtubs, " Civil Engineering, Vol. 64, September 1994, pp 56-58.

Cedergren, H.R. (1988) "Why All Important Pavements Should be Well Drained, " Transportation Research Record 1188. Transportation Research Board, Washington D.C., pp 56-61.

Conigliaro, A and Watson, P. (2000) "Determining the Best Formulation for a Unique Asphalt Cold Patch Product Made with # 3-7 Rigid Plastic Aggregate," Chelsea Center for Recycling and Economic Development Technical Research Program, January 2000.

Container Recycling Institute (CRI) (2007), "http://www.container-recycling.org/," accessed, March 2007.

Flynn, L. (1993) "Recycled Plastic Finds Home in Asphalt Binder", Journal, Roads and Bridges, March 1993.

Fwa, T.F., Tan, S.A., and Chuai, C.T., (1998) "Permeability Measurement of Base Materials using Falling-Head Test Apparatus," Transportation Research Record, No. 1615, 1998, pp 94-99.

Hassani, A., Ganjidoust, H. and Abedin Maghanaki, A (2005) "Use of plastic waste (poly-ethylene terephthalate) in asphalt concrete mixture as aggregate replacement," *Waste Manage Res 2005: 23: 322–327*

Hinislioglu, S. & Agar, E. (2004) Use of waste high density polyethylene as bitumen modifier in asphalt concrete mix. *Materials Letters*, **58**, 267–271.

Kashi, M., Malloy, R. and Swan, C. (2002) "Fly Ash/Plastic Synthetic Aggregate for Construction Material," Chelsea Center for Recycling and Economic Development Technical Research Program, August 2002.

Kashi, M., Malloy, R. and Swan, C. (2001) " Development of Synthetic Lightweight Aggregate for Construction Material," Chelsea Center for Recycling and Economic Development Technical Research Program, July 2001.

Khoury, N.N., Khoury, C.K., Abousleiman, Y. and Miller, G., (2006) Office of Technology and Development, i.e., OTD, Disclosure No. 07NOR020, The University of Oklahoma, Norman, Oklahoma. December, 2006.

Khoury, N.N. and Zaman, M.M. (2006) "Permeability and Resilient modulus of Different Aggregate Bases Commonly used in Oklahoma," Technical report submitted to Oklahoma Transportation Center, HWY04-10 - OTC # 3, Nov. 2006.

Krishnaswamy, P. and Lampo, R. (2001) "Recycled-Plastic Lumber Standards: From Waste Plastics to Markets for Plastic-Lumber Bridges," *Standardization News* (December 2001), http://www.astm.org/SNEWS/DECEMBER_2001/wsd_dec01.html., accessed March 2007.

Mallick, R.B. and Teto, M (2000) ""Evaluation of Use of Manufactured Waste Asphalt Shingles in Hot Mix Asphalt, 2000" Chelsea Center for Recycling and Economic Development Technical Research Program, July 2000.

Malloy, R. and Kashi, M. and Swan, C. (2003) "Fly Ash/Mixed Plastic Aggregate and Products Made Therefrom" U.S. Patent No., 6,669,773.

Meyers, III, Swartz, J, Nathaniel, Kurczewski, N and Kurczewski, M. (2006) "Recyclable Composite Materials Artciles of Manufacture and Structures and Method of using Composite Materials" U.S. Patent No. 6,984,670.

Municipal Solid Waste (MSW) (2006). "Municipal Solid Waste in the United States: 2005 Facts and Figures," U.S. Environmental Protection Agency, October, 2006.

Schmidt, E. (1993) "Methods for Preparation of Composite" United States Patent, No. 5,217,655.

Sobhan, K. and Mashnad, M (2003) "Mechanical Stabilization of Cemented Soil–Fly Ash Mixtures with Recycled Plastic Strips," Journal of Environmental Engineering, ASCE, October 2003

Sobhan, K. (1997), Stabilized Fiber-Reinforced Pavement Base Course With Recycled Aggregate, Ph.D. Dissertation, Northwestern University, Evanston, Illinois, June, 1997, 303 pages.

Wikipedia (2007) "Polyethylene Terephthalate," http://en.wikipedia.org/wiki/Polyethylene_terephthalate, accessed March 2007.

Wright, P.H. and Dixon, K.K. (2004) "Highway Engineering," 7th edition, ISBN 0-471-45258-0, John Wiley & Sons, Inc., New Jersey, United States.

Zoorob S.E. and Suparma, L.B. (2000) "Laboratory Design and Investigation of Proportion of Bituminous Composite Containing Waste Recycled Plastics Aggregate Replacement (Plastiphalt), CIB Symposiumon Construction and Environment Theory into Practice, Sao Paulo, Brazil (November, 2000)

Zoorob, S.E. (2000) "Laboratory Design and Performance of Improved Bituminous Composites Utilising Plastics Packaging Waste", Conference on Technology Watch and Innovation in Construction Industry, Belgium, Building Research Institute, Brussels, Belgium (April, 2000).

Towards the Development of Sustainable Landfills

H. N. Hsieh[1], M. ASCE, J. N. Meegoda[2], M. ASCE, J. P. A. Hettiarachi[3], S. El Haggar[4], and R.I. Stressel[5]

[1]Professor, Department of Civil & Env. Eng., New Jersey Institute of Technology, University Heights, Newark, NJ 07102; hsieh@njit.edu
[2]Professor, Department of Civil & Env. Eng., New Jersey Institute of Technology, University Heights, Newark, NJ 07102; meegoda@njit.edu
[3]Professor, Department of Civil Engineering, and CEERE, University of Calgary, Calgary, Alberta, Canada; jhettiar@ucalgary.ca
[4]Professor of Energy & Environment, Department of Mechanical Engineering, the American University in Cairo, 113 Kasr El-Aini St. Cairo, 11511, Egypt; elhaggar@aucegypt.edu
[5]Deceased, Department of Civil and Environmental Engineering, University of South Florida, Tampa, USA

ABSTRACT: Historically, municipal solid waste is mostly disposed in landfills. However, with the more stringent environmental regulations and difficulties in locating new waste disposal facilities, the low cost advantage of landfills is no longer true. This manuscript discusses the development of a new generation of landfills to achieve sustainable management of municipal solid waste. A sustainable landfill involves sequential application of anaerobic degradation, aerobic decomposition, and landfill mining. Biodegradation of waste in landfills is enhanced through leachate re-circulation, landfill gas collection, air injection, and environmental monitoring. Practical waste mining and recycling processes are also presented. This new generation of landfills will eliminate the commonly encountered problems of ground/surface water contamination and landfill gas emissions, as well as the need for new land for waste disposal and associated truck routing issues for new landfills.

Keywords: Aerobic, Anaerobic, Landfill Mining, Recycling, Sustainable, Management, Municipal Solid Waste, Bioreactor Landfill

INTRODUCTION

Landfills have been favored due to their low capital and operation/maintenance costs. However, there are four primary problems associated with landfills; aesthetics of operation, ground/surface water contamination, landfill gas emissions, and space requirements. Because of these, stringent environmental regulations have been promulgated to promote sanitary landfills and associated operation/maintenance and monitoring programs. This, however, results in engineering cost increase. Additionally, many developed countries have already exhausted landfill capacities and locating and permitting a new facility is a challenge. Hence, development of alternative/innovative disposal methods is in great need.

Recently, sanitary landfills have managed to solve, to a large extent, the first three problems mentioned above. Nevertheless, the fourth one, searching for a large piece of land for waste disposal is not an easy task, as it needs to be done in a rather short time. Therefore, the introduction of the concept of a sustainable landfill with perpetual use of the same land has become essential. Sustainable landfill is defined as a landfill that has the ability to provide the needs of waste disposal for future generation without exhausting its capacity. This can be accomplished through the use of modified bioreactor landfill and subsequent mining.

SUSTAINABLE LANDFILL

Research on sanitary landfills during the past decade indicated that the operation of landfills as bioreactors is viable (Reinhart and Townsend 1998). A bioreactor landfill is a sanitary landfill with enhanced biodegradation (Pacey 2001). Sustainable landfill is a further advancement of bioreactor technology (Hettiarachi 2006). It involves three stage landfill operations. The first stage incorporates the cell as an anaerobic bioreactor with leachate recirculation to recover the full energy content of organic waste. During the second stage, the landfill operates in an aerobic mode to produce compost-like material. The last stage involves cell mining for resource and space recovery, thus making the landfill operation sustainable.

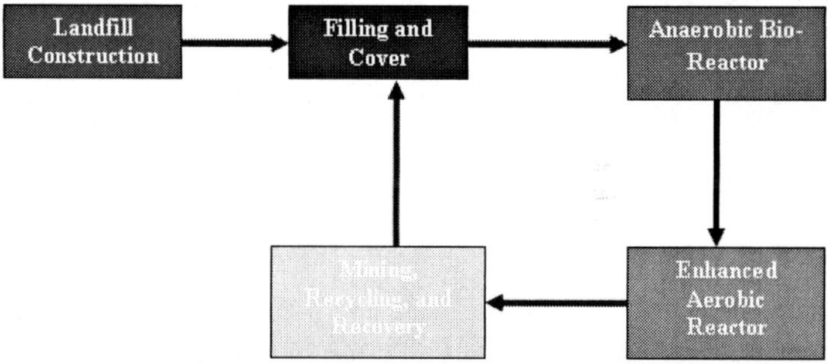

Stage of Landfill	Timeline (Hettiarachi 2006)
Stage I: Anaerobic Bio-Reactor	4 to 10 years
Stage II: Enhanced Aerobic Reactor	1 to 5 years
Stage III: Mining, Recycling, and Recovery	½ to 1 year
Total Expected Life Span of the Landfill	5½ to 16 years

Figure 1 Conceptual Framework of Proposed Sustainable Landfill.

This concept of the sustainable landfill is illustrated in Figure 1. Please note that landfill liner and leachate collection systems are used to minimize groundwater contamination and maximize recovery of leachate. Daily, intermediate and final covers are provided to manage gas emissions and liquid entry. A gas collection system is installed for landfill gas collection and emission control. In addition to these typical sanitary landfill installations, pipe systems are added to re-circulate the collected leachate, as well pump air into the waste matrix during the second stage aerobic operation to accelerate biological degradation. Re-circulation of leachate will add the much needed moisture and microbial mass during both aerobic and anaerobic modes of cell operation. The landfill is fully instrumented with moisture and temperature monitoring sensors and the leachate composition is chemically analyzed. The final process will be the recovery and recycling of various products. Landfill mining has been attempted as an enhancement technique in last few years (Murphy, 1993). Sequential application of all these techniques will materialize the concept of the sustainable landfill (Yedla, 2005).

In a conventional dry-tomb type landfill, many factors such as low moisture content of incoming waste, landfill operation to keep moisture out, insufficient separation of waste, and inappropriate conditions for microbial growth, may contribute to low biodegradation rates. The sustainable landfill is operated differently, and its operation promotes proliferation of microbes. In addition, the waste can be pre-sorted and mixed during disposal to ensure optimum carbon (C) to nitrogen (N) ratio for a maximum biodegradation rate. Mixing waste with sludge with high nitrogen and phosphorus content will also help achieve optimum C: N ratio and maximum biodegradation (Tchobanoglous et al. 1993).

The following sections describe the three stages of sustainable landfill in detail.

ANAEROBIC DECOMPOSITION WITH GAS EXTRACTION

Biodegradation of waste occurs in stages in a sanitary landfill, first aerobic reaction, until the depletion of oxygen, then anaerobic. For a conventional dry-tomb sanitary landfill, it may take as long as 50 to 100 years to degrade the majority of biodegradable organics (Crawford and Smith, 1985). Hence leachate is collected and disposed during this time.

In a sustainable landfill, a controlled anaerobic condition is introduced first and prevails until the majority of the biodegradable organics are exhausted. A highly efficient leachate re-circulation network is installed and instead of treating and/or disposing of the leachate, it is re-circulated back into the landfill. Because of the leachate re-circulation, the waste is maintained at optimum moisture content (50-60%) for waste decomposition. It also serves as a medium for bacteria and nutrients transfers. Due to the rapid degradation of waste large amounts of gases are produced in a short time. This gas production will increase rapidly from early stages of operation and will attain its peak within one year. After which, the gas production will start to decrease exponentially and reach a stage where the energy recovery is not economically worthwhile to continue. A suitable gas extraction system with gauge measurement is installed for such collection and monitoring of the quantity of landfill gas released.

In order to effectively implement the technology, moisture content, temperature, and gas production rate in the landfill should be monitored along with the measurement of pH and nutrients level in the leachate. Optimum microbial growth conditions depend on the presence of nutrients such as nitrogen and phosphate; and proper temperature and pH (neutral); and absence of toxic materials.

Once methane production decrease to a critical value, the next stage of sustainable landfill will initiate, that is to use aerobic degradation to rapidly decompose the remaining organic waste. Since aerobic degradation rate is usually higher for simple organic compounds, this stage may take only a year or two to complete.

ENHANCED AEROBIC DECOMPOSITION

To convert from anaerobic to aerobic conditions in the landfill, air has to be introduced to the landfill and maintained to enhance the rate of decomposition of waste (Stessel and Murphy, 1992; Hettiaratchi, 2006). The aerobic bioreactor is not a new concept and was first proposed by Merz and Stone (1962) and there are more than 20 operating aerobic bioreactor landfills in North America. The aerobic bioreactor is quite efficient, where the biodegradable waste is rapidly converted to compost like material. However, the energy content of the biodegradable waste is lost contributing to global warming. Once the landfill stop producing extractable amounts of landfill gases during the anaerobic stage of the sustainable landfill, aerobic stage is initiated to expedite waste degradation. The starting point of this stage depends on land value and cost of operating the gas extraction and the pumping systems. The gas extraction system used during the anaerobic stage will be used to pump air into the landfill to create aerobic conditions, which accelerates and completes biological degradation of organic waste. The recirculation of appropriately adjusted leachate for aerobic degradation is also required for the same reason stated in the first stage of operation. Towards the end of aerobic decomposition, periodic testing of waste composition from boreholes and the analysis of the leachate would ensure the complete biodegradation. When the organic waste is stabilized, the landfill is ready for mining.

MINING OF USEFUL/RECYCLABLE PRODUCTS

After biodegradation, mining will ensure recovery and recycling of various products (Murphy, 1993; Zee et al., 2003). Figure 2 shows the process flow diagram of landfill mining and recovery of resources.

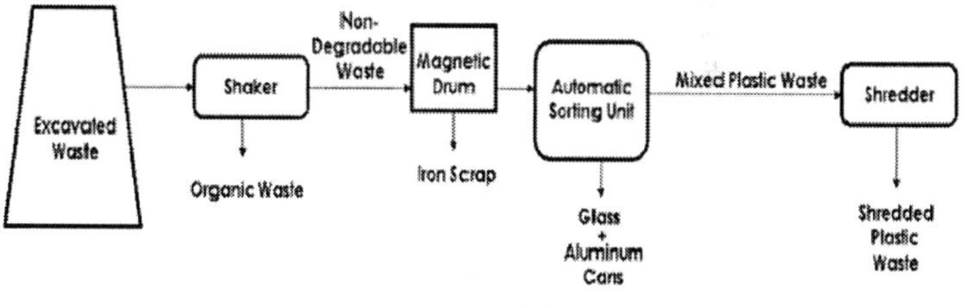

Figure 2 Process Flow for Mining the Landfill.

At first, the whole cell is excavated. To reduce the cost of mining process, EPA recommends the use the same technology and equipment that has been used in the mining industry (USEPA 2007). Bulky waste, if any present, should be removed first. Recycling a landfill involves a series of activities that includes separating, sorting, and processing recyclables. Separation of waste based on size can be performed through the use of various types of screens (trommel, vibrating, rotating, and disc).

The excavated waste, as shown in Figure 2, is stockpiled and fed to a shaker with sieves to separate the leftover biodegradable organic waste or by-products of biodegradation. The organics that passes through 1 cm – 5 cm sieve size can be used as compost in agricultural applications (Hudgins and March 1998). Typically the recovered material is first sieved in a 5 cm sieve shaker and then in a 1 cm sieve shaker as a second stage with 3-4 weeks of maturing before final use to enhance the quality of compost (Hudgins and March 1998).

The collected non-degradable fraction will be sent to an automatic sorting unit with metal separation. A feeding belt with a magnetic drum installed on the belt is used to remove the ferrous metals. The magnetic drum can be an overhead magnet whereby ferrous metals are separated from other waste material by utilizing their magnetic properties. A further separation is conducted through the difference in waste density. Light materials such as plastics and non-degraded packaging can be separated from heavy materials such as metal and glass because of their weight difference in an air stream or a cyclone. An Eddy Current Separator is often used for separating non-ferrous metals, such as aluminum. The mixed plastic waste could be shredded into small pieces for further processing. At the site, recovered materials can be crushed, bailed or shredded for the convenience of transportation.

The mined separated/recovered waste can be sold to companies that manufacture new products. Many recycle them back into the same product process line, such as aluminum, and glass. The majority of the compost-like product resulting from biodegradation can be used as soil conditioner with various agricultural applications. Thermoplastics, such as PETE, PVC, and HDPE, soften when heated and can be recycled and remolded. Cullet, the industrial term for the furnace-ready scrap glass, is one of the most important ingredients of the glass manufacturing process. Recycled paper has many applications in newspaper industry and packaging. Other than these versatile uses, recyclables can be used as feedstock for a number of products, such as plastic lumber park benches or decks made from plastic milk jugs or soda bottles. Also, cardboard containers such as juice boxes and plastics have been used to produce construction materials (El Hagger, 2007). Table 1 provides a detailed account of potential recycling options for all waste components, where stages I, II and III are as defined in Figure 1.

Recycled plastics are the least recycled material in developed nations (Tchobanoglous et al. 1993). If sorted and cleaned economically and efficiently, plastic waste can be recycled according to Figure 3. Figure 3 illustrates a mechanical recycling process for plastic recycling according to physical property, namely, hard plastic such as bottles and containers, and thin film plastic such as plastic bags. The thin film plastic can be agglomerated through an agglomerator machine and converted into small granules, while hard plastic can be shredded into small pieces. Both types of plastics can be pelletized through a pelleting machine to remove any impurities and convert it into regular shapes for further processing using blowing "film" machine, injection molding machine or extruder machine depending on the type of product desired.

If cleaning the recovered plastic is not economically feasible, the mixed plastics can be processed. In this case, the mixed plastic is heated indirectly to a temperature below the melting temperature with a heating furnace. The hot paste is then transferred to a mold and pressed with a hydraulic press to required shapes according to the desired application. Additives can be used with the mixed plastic waste to adjust the required properties. One of the very successful applications (El Haggar, 2007) is the production of manhole cover and base shown in Figure 4.

Even with enhanced biodegradation and recycling processes described above there will be some residual waste which are non-recoverable, but with high caloric value. These wastes may include plastics, textiles, rubber, waste wood, and a number of other organic waste types. These can be used to produce refuse derived fuel (RDF) and co-incinerated in cement kilns.

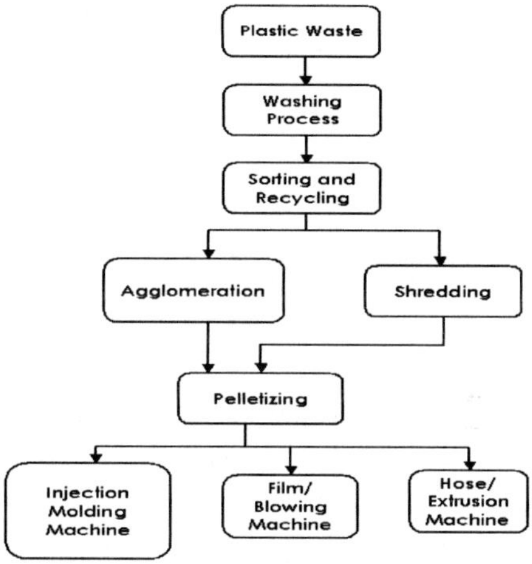

Figure 3 Recycling of Plastic Waste Flow Diagram

Figure 4 Manhole Cover Made from Recycled Plastics

SUMMARY AND CONCLUSION

There are four main problems with landfill operation, namely aesthetic issues, ground/surface water contamination with leachate, emission of landfill gases, and extensive space requirements. Over the last three decades, the design and operation of sanitary landfills have managed to solve, to a large extent, the first three problems. However, the fourth problem has not been adequately addressed. A sustainable landfill is a concept proposed in this manuscript to solve all four problems at the same time. The approach involves sequential operation of a landfill cell to produce methane gas during the first stage of anaerobic degradation, in-situ composting within the cell footprint during aerobic degradation is the second stage and landfill mining for resources and space recovery during the third stage. The resources that can be recovered include compost-like material and recyclables such as plastics, metal, and glass. Non-recovered waste but with high energy content can be used as refuse derived fuel. The practice of this approach will no longer require the need to allocate valuable land for new landfills on an on-going basis and hence will revolutionize the management of municipal solid waste.

ACKNOWLEDGMENTS
The work described in this manuscript was partially supported by a grant from the National Science Foundation (grant # OISE-0405900) where the program manager is Dr. Osman Shinaishin and the Natural Science and Engineering Research Council of Canada.

REFERENCES

Crawford, J.F. and Smith, P.G. (1985) *Landfill Technology*. London: Butterworth.
El Haggar, S. (2007) *Sustainable Industrial Design and Waste Management: Cradle-to-Cradle for Sustainable Development*, Academic Press.
Hettiaratchi, J. P. A. (2006) "Bio-Cell Project," www.eng.ucalgary.ca/resrch_civil/bio-cell-project/Hettiaratchi-bio-cell-project.htm
Hettiarachchi, C. H., Tavantzis, J., Meegoda, J, N. and Hettiaratchi, J. P. A., "Numerical Model to Predict Settlements Coupled with Landfill Gas Pressure in Bioreactor Landfills," Journal of Hazardous Materials, B139 (2007) 514–522
Hudgins, M. and March, J., "In Situ Municipal Solid Waste Composting Using an Aerobic Landfill System", http://www.p2pays.org/ref/12/11566.pdf, American Technologies, Inc., 1998
Merz, R.C. and Stone, R. (1962) Landfill Settlement Rates. *Public Works*, 93(9): 103-106, 210-212.
Murphy, R.J. (1993) *Optimization of Landfill Mining*, Technical Report, Center for Solid and Hazardous Waste Management and the Collier County Government of Naples, Florida.
Rathje, W.L. and Murphy, C. (2001), *Rubbish! The Archaeology of Garbage*, University of Arizona Press.
Reinhart, D.R. and Townsend, T. G. (1998) *Landfill Bioreactor Design and Operation*, Lewis Publisher, New York
Pacey, J. (2001) *Bioreactor Landfill: An Overview Perspective*, MSW Management, http://www.forester.net/msw_0011_bioreactor.html#anaerobic
Stessel, R.I. and Murphy, R. J. (1992) A Lysimeter Study of the Aerobic Landfill Concept, *Waste Management & Research*, Vol. 10 (6): 485-503
Tchobanoglous G, Theisen H, and Vigil S. (1993) *Integrated Solid Waste Management, Engineering Principles and Management Issues*. New York: McGraw-Hill, Inc. Chapter 11, 361-540.
USEPA (2007) *Mining Waste*, www.epa.gov/epaoswer/other/mining/index.htm
Yedla, S. (2005) .Modified Landfill Design for Sustainable Waste Management., *Int. J. Global Energy Issues*, Vol. 23 (1): 93 - 105.
Zee, D.J. van der, Achterkamp, M.C. and de Visser, B.J. (2003) *Assessing the Opportunities of Landfill Mining*. Research Report 03B39, University of Groningen, Research Institute SOM (Systems, Organizations and Management).

Table 1 Possible Recycling Options for Different Wastes

Waste Type	Remainder			Alternatives for Final Usage
	Stage I	Stage II	Stage III	
Paper and Paperboard	Not completely degraded	Higher percentage would have been decomposed.	Degraded percentage will pass through 0.5" sieve and rest can be collected for recovery	Paper has a very good recycle value. Recycled paper can be used for various purposes such as newspaper, packaging, tissue paper, printing and writing paper.
Glass	Very little or No change	No further change	Should be separated with respect to the color	Recycled glass is mainly used as synthetic aggregate and cullet for production of bottles, containers, fiberglass, and other products. Glass may also be recycled and separated based on color and type.
Ferrous	No significant change	Possible loss of value due to corrosion	Should be separated and processed	Aluminum and tin have one of the highest recovery revenues. Recycled metals can be used to make cans and several other products using much less energy (70-90%). Ferrous material is likely to be corroded, but it can be processed and sent for recovery. Recycled metal can be mixed with fresh metal to gain additional strength.
Aluminum	No quantifiable change	No major damage	High recycle value and should be collected for recovery	
Other Nonferrous	Not much degradation	No further change	Collected and inspected before processing	
Plastics	No change	No significant degradation, but chances of contamination	Should be separated from metals and other items and checked for recovery worth	There are many types of plastics and they need to be separated before recycling. Recycled plastic can be used to make different products or can be burnt in an incinerator.
Rubber and Leather	Not completely degraded will cause odor to an extent	Leather will mostly be decomposed and rubber decomposition depend on the quality and prior usage	Rubber gets spoiled and contaminated especially tires. Leather if at all present will be separated and processed	Recycled rubber is not only cheaper, but the recycled products use less energy. Recycled rubber can be used for erosion control, in fences, as tree guard etc.
Textiles	Not completely degraded	Considerable amount will be decomposed	Textile products are difficult to separate.	The pre-consumer textile waste can be recycled into new raw materials.
Wood	Bulky wood products will require more time to decompose	Small items decomposed, while the large ones may still be intact	Bulky wood items can be separated easily. They can be crushed before recovery	Wood products can be separated, processed and sent for recovery.
Food Scraps	High percentage is decomposed	Most should have been turned into compost	It will mostly be composted	Food waste generally decomposes completely. Compost has several agricultural applications. It can also be used as daily cover in subsequent cells.
Yard Trimmings	High percentage is decomposed	Most should have been turned into compost	It will mostly be composted	

Use of Sewage Sludge and Copper Slag for Land Reclamation

Jian Chu[1] and Teik Thye Lim[2]

[1] School of Civil and Environmental Engineering, Nanyang Technological University, Blk N1, 50 Nanyang Ave, Singapore 639798, Singapore cjchu@ntu.edu.sg.
[2] School of Civil and Environmental Engineering, Nanyang Technological University, Blk N1, 50 Nanyang Ave, Singapore 639798, Singapore cttlim@ntu.edu.sg.

ABSTRACT: A study on the use of sewage sludge and copper slag for land reclamation is presented in this paper. The physical, chemical and geotechnical properties of sewage sludge and copper slag are characterized. Methods to mix sludge with binders, copper slag and other waste materials before disposal and consolidate the mixture on site using fill or vacuum surcharge and prefabricated vertical drains (PVDs) are proposed. The geotechnical aspects of the proposed methods are studied by conducting model tests using a fully instrumented consolidation tank. Some of the test results are presented in this paper.

INTRODUCTION

The disposal of sewage sludge and other waste materials such as copper slag in Singapore has become increasingly a problem due to a shortage of disposal space. On the other hand, there is a great demand for fill materials for land reclamation. Therefore, it would be ideal if sewage sludge and other waste materials can be used as fill materials for land reclamation. Toward this purpose, a study on the feasibility of using binder treated sewage sludge and copper slag for land reclamation was carried out. One of the methods studied is to mix sludge, copper slag and other waste materials with binders and then consolidate the mixture on site after it is disposed. Depending on the usage of the reclaimed land, the sludge mixture needs not to be converted into a hard granular material before it can be used for land reclamation. The sludge only needs to be treated to gain a shear strength comparable to the seabed marine clay on which the sludge mixture is to be placed. In this way, the costs involved in treating and disposing sludge can be much lower compared with other disposing methods. The demand for granular fill can also be reduced.

In this study, Portland cement and/or hydrated lime[1] were used as binders. Copper slag was used as a filler. A preliminary study (Chu et al. 2005) shows that sewage

[1] Quicklime was not used as it was not practical to use it on a large scale due to the weather condition and the health regulations in Singapore.

sludge cannot be treated by binders or by consolidation alone and a combined chemical and mechanical method has to be used in order to improve the geotechnical properties of sludge in a cost effective way. To establish a cost-effective method to treat the sludge, some laboratory model tests were conducted using a fully instrumented consolidation tank. As the consolidation behavior of the binder treated sludge and copper slag mixtures can be quite different from that of soils, the model tests also serve to study the consolidation behaviour of the binder treated sludge mixture around vertical drains. The testing arrangement of the model tests is described. Some results of the model tests are presented. This paper discusses only the geotechnical aspects. The environmental aspects of the proposed methods were also studied and presented elsewhere (Goi 2004; Chu et al. 2006; Lim et al. 2004; 2006).

MATERIALS TESTED

The anaerobically-digested sewage sludge used in this study was taken from a wastewater treatment plant in Singapore. The sludge had been filter-pressed for dewatering at the plant. Nevertheless, the so-called dewatered sludge was still very soft, very high in water content and had little shear strength. It contained a considerable proportion of organic matters and some heavy metals. The physical and geochemical properties of the sludge are summarized in Tables 1 and 2 respectively. The heavy metal contents in the sludge were determined by microwave-assisted acid digestion of their oven-dried sub-samples followed by analysis of the digestates with the Inductively Coupled Plasma Optical Emission Spectrometer (ICP-OES). The results are presented in Table 3. Calcium and Silicon existed in high concentrations because quartz and calcite or calcium carbonate was the main mineral found in sludge, as revealed by the X-ray diffraction (XRD) tests conducted on the sludge sample (Goi 2004). Concentrations of Zn and Cu were relatively high. Zn could be introduced by factories producing galvanized zinc, whereas Cu might come from copper production plants and corroded copper pipe system. The coefficients of variations (COV) for Zn and Cu concentrations in the sludge were 56% and 36% respectively. These high COV values suggest that there were considerable spatial variations in the heavy metal concentrations in the sludge. Sulphate should not present as the sludge is anaerobically-digested. Owing to the potential environmental concerns, the sludge needs to be treated to meet both the geotechnical and environmental requirements.

Table 1 Physical properties of sewage sludge

Property	Value
Moisture Content (%)	400-514
Total Solids (TS) Content (%)	16.3-20
Bulk Density (Mg/m^3)	1.02
Specific Gravity	1.634
Liquid Limit, LL (%)	397
Plastic Limit, PL (%)	63
Plasticity Index, PI (%)	334
Liquidity Index, LI	1.35

Table 2 Geochemical properties of sewage sludge

Property	Value
pH at 25°C ± 2 (1:6 solid: water)	8.42
Electric Conductivity (EC) at 25°C ± 2 (1:6 solid: water) (mS/cm)	39.4
Redox Potential (Eh) at 25°C ± 2 (1:6 solid: water) (mV)	323
Dissolved Organic Carbon (DOC)* (mg/L)	1581
Phosphates* (mg/L)	66

* Based on the extraction of 1:5 solid: water ratio

Table 3 Total heavy metal concentrations in sewage sludge

Element	Concentration (mg/kg)
Sodium – Na	2,070 [705][a]
Calcium – Ca	32,600 [5,000]
Magnesium – Mg	4,820 [540]
Potassium – K	2,960 [290]
Aluminum – Al	12,700 [2490]
Iron – Fe	22,300 [2,670]
Silicon – Si	66,200 [4,440]
Barium – Ba	427 [76]
Arsenic – As	< 36
Selenium – Se	< 24
Cadmium – Cd	< 1
Chromium – Cr	694 (102)[b]
Copper – Cu	1,491 (533)
Lead – Pb	146 (24)
Nickel – Ni	314 (111)
Zinc – Zn	1,844 (1,032)

[a] Values in [ ] are standard deviations based on test results from 6 samples
[b] Values in ( ) are standard deviations based on test results from 12 samples

Table 4 Physical and geotechnical properties of copper slag

Property	Value
Bulk Density (Mg/m^3)	2.0
Specific Gravity	3.57
Void Ratio	0.264 - 0.651
D_{50} (mm)	0.5
Particle Distribution	Well graded
Particle Shape	Subrounded to
Permeability (m/s)	$10^{-4} \sim 10^{-5}$
Friction Angle (°)	29 ~ 47

Copper slag is a non-toxic waste product after it has been used as a blast-cleaning agent to remove rust and marine deposits on ships. It is generated in substantial quantity every year in Singapore and its disposal also becomes increasingly a problem. It is well-graded with a mean grain size of 0.5 mm. The geotechnical and geochemical properties of copper slag is given in Table 4. As copper slag contains heavy metals such as Pb, Zn, Cu, Cr and Ni, it cannot be used directly as a fill material for land reclamation. Ordinary Portland cement or/and hydrated lime were used as binders.

MODEL TESTS

A consolidation tank as shown in Fig. 1 was used to conduct model tests. The sewage sludge was first mixed with binders and the mixture was then placed in the consolidate tank for consolidation. The settlement measured during consolidation is shown in Fig. 2. Using the ultimate settlement estimated by the Asaoka (1978) method (Goi 2004), the average degree of consolidation achieved at the end of the test was estimated to be 89.1%.

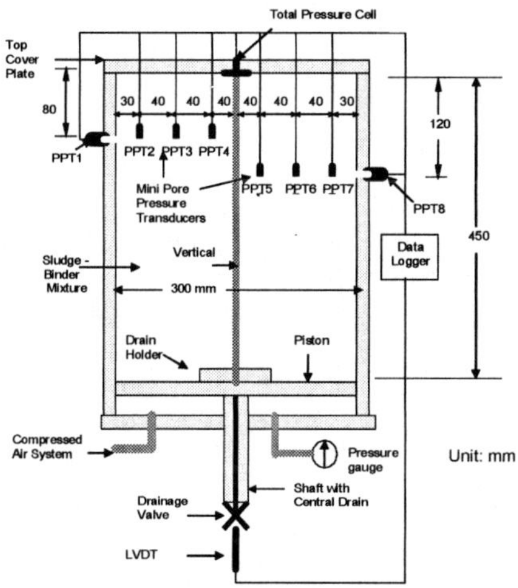

FIG. 1 Schematic arrangement of consolidation test

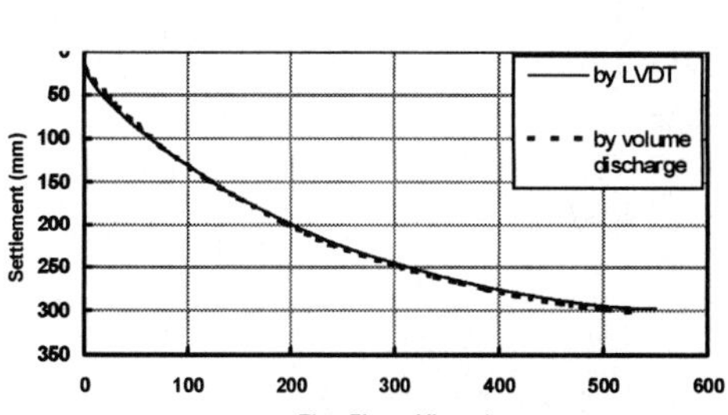

FIG 2 Settlement versus time curve measured during consolidation

The distribution of excess pore pressures along the radial direction at different time intervals are shown in Fig. 3. The left-hand side was the pore pressure measured at 80 mm depth, while the right-hand side was the pore pressure measured at 120 mm depth. The initial excess pore pressures at the various locations were normalized to 80 kPa. Assuming that the excess pore water pressures at the 2 depths are the same, a family of isochrones are formed. In general, the pore pressure dissipations were generally faster at locations closer to the drain. The difference in the pore pressure dissipations at both sides was largely attributed to the lateral distortion and buckling of the vertical drain during the process of consolidation. Based on the isochrones shown in Fig. 3, the average degree of consolidation, U, can also be calculated (Chu et al. 2005). The U at the end of the test calculated based on pore water pressure is 79%, which is smaller than the U calculated based on settlement.

At the end of the model test, the sample was extruded and cut open along a vertical plane. The moisture content at the tip of each PPT was determined and plotted in Fig. 4. It can be seen that the moisture content had reduced from the initial 434% to the final 57 to 94%. Therefore, the combined chemical and mechanical treatment method is effective.

The UC strength obtained from the consolidated sludge varied from 45 to 160 kPa (Chu *et al.* 2005), which corresponds to an undrained shear strength to the effective vertical stress ratio, c_u/σ_{v0}', of 0.36 to 1.14. This range of value is higher than the typical c_u/σ_{v0}' ratio for normally consolidated remoulded soils, which reflects the effect of cementation due to the use of cement. It is also shown (Chu *et al.* 2005) that the microstructures of the treated sludge mixtures are more densely packed and have much lesser pore spaces as compared to that of the raw sludge.

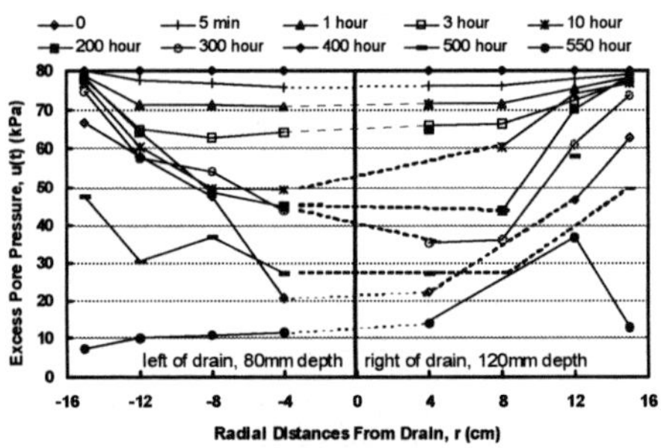

FIG 3 Isochrones of excess pore pressure distribution along the radial distance from the drain at different time intervals.

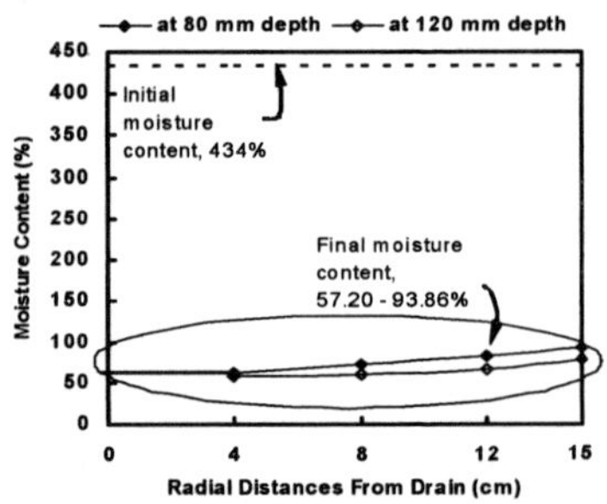

FIG 4 Comparison of the initial and final moisture content distributions.

SUGGESTED DISPOSAL METHODS

As the water treatment plants in Singapore are near the sea, it will be convenient to use the sludge for land reclamation. Two such methods are suggested. In the suggested methods, the sludge will not be pressed. Therefore the energy for pressing can be saved. The sludge will be mixed using a mixer with other wastes, such as copper slag. The optimum ratios for different waste can be established easily by conducting laboratory tests. Based on the present study, a ratio of 77% of sludge, 15% of copper slag and 8% of cement by weight or 54% of sludge, 37.5% of marine clay and 8.5% of cement appears to be suitable. No segregation was found during the tests.

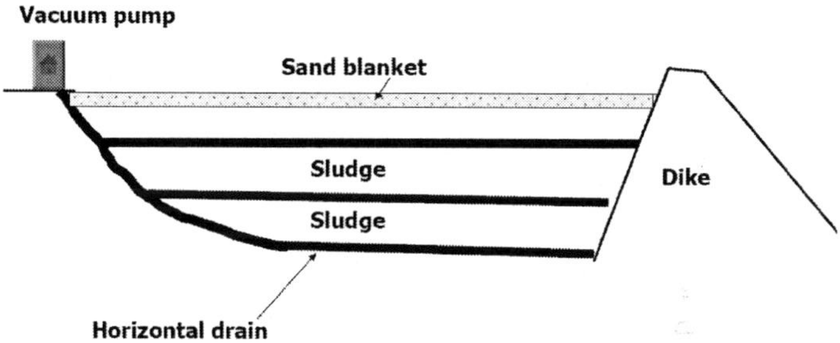

FIG 5 Schematic illustration of Method 1

The first method can be adopted when sludge and waste materials are used as the main fill materials. This method is suitable if the water table in the reclamation area is relatively shallow and the location is not far from the water reclamation plant. In this method, the mixed sludge can be pumped directly through pumping pipes into the designated reclamation area. A dike is required to contain the fill along the boundary of the reclamation area, as shown in Fig. 5. The dike does not have to be impervious as in the case of a land fill site. This is because the mixed sludge is low in permeability and the water inside the sludge mixture will not flow freely. This is particularly the case if the water inside the mixed sludge is to be dissipated using vacuum pressure via installed drainage layers. Therefore, leaching of heavy metals or any other contaminants into the sea is unlikely. Horizontal drains will be placed on the seabed before the pumping of sludge. Prefabricated drain panels such as those shown in Fig. 6 can be used as horizontal drainage layers. When the sludge fill has reached 3 to 4 m thick, another layer of horizontal drains can be installed. The drainage layers will be connected to vacuum pumps so the mixed sludge can be consolidated under vacuum pressure after it is disposed. After the sludge fill reaches the mean sea water level, a sand layer can be used to cap the sludge layer, as shown in Fig. 5.

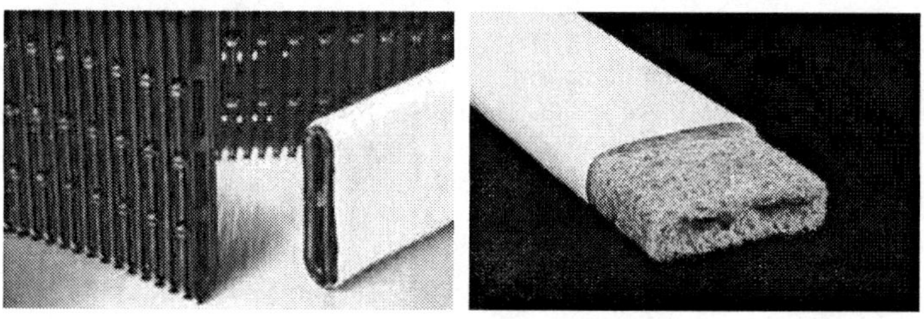

FIG 6 Examples of prefabricated drainage panels for horizontal drains

The advantage of this method is that there will be little environmental risks as the water from the sludge is centrally collected. The mixed sludge treated in this way will have geotechnical properties that are much better than the seabed marine clay. However, this method is only workable if the main purpose is to dispose sludge and other waste materials as the amount of sludge produced each year is not sufficient for the purpose of land reclamation.

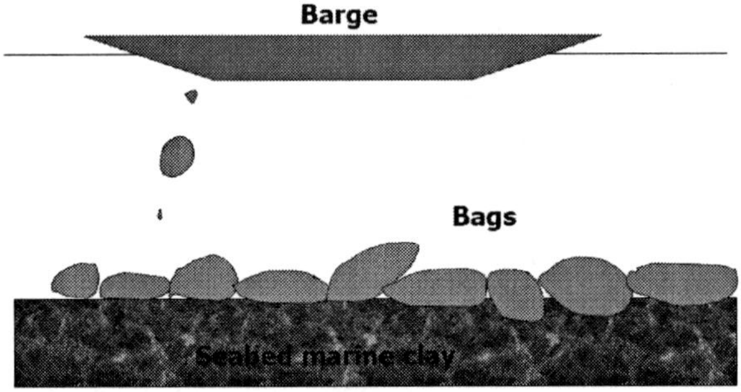

(a)

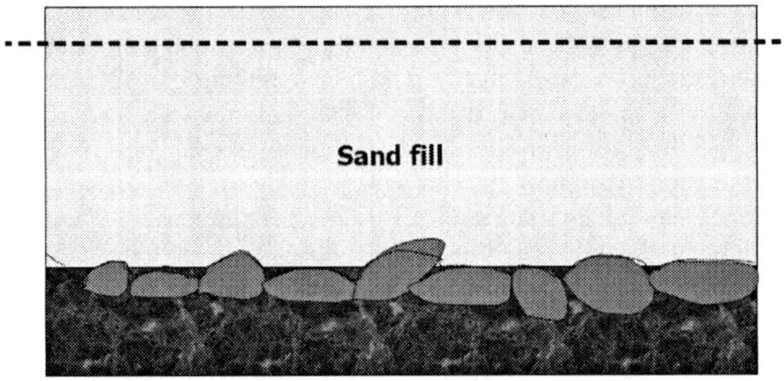

(b)

FIG 7 Schematic illustration of Method 2: (a) dumping of sludge filled geotextile bags (b) Consolidation of the sludge under sand fill

The second method is to put the mixed sludge into geotextile bags, and then dump the bags from a barge on to seabed before sandfill or other fill material is placed as part of the normal land reclamation process, as shown in Fig. 7. Under the load of the sandfill, the sludge filled will consolidate through the drainage provided by the geotextile used for the bags. The geotextile bags may also sink into the marine clay under the sandfill. This method is suitable when the water depth of the reclamation area is deep. The advantage of this method is that no dike needs to be built. As the sludge mixture is bagged, it will not draft or cause siltation problem. The disadvantage is that there is less control in the mobility of the water discharged out from the sludge.

However, as shown by previous studies (Lim et al. 2004, 2006), the leaching of heavy metals from the sludge can be effectively contained after it is mixed with copper slag or marine clay and bound by binders. Furthermore, as majority of the bagged sludge mixture is buried in marine clay, the discharged from the sludge will not be too small to cause an environmental problem even if a small amount of leaching might occur. It should be pointed out that as the sludge matrix is highly porous, the formation of a small amount of ettringite will only fill the void space without causing volume expansion.

CONCLUSIONS

The following conclusions can be drawn from this study:

1) The geotechnical properties of sewage sludge cannot be improved effectively by using either chemical or mechanical method alone. A combined chemical and mechanical method is required in order to stabilise the sludge in a cost-effective way.
2) By mixing sludge with cement and copper slag and consolidating the mixture under a pressure of 80 kPa, the physical and geotechnical properties of the sludge can be significantly improved. The moisture content of the sludge can be reduced from more than 500% to a level of 57 to 94%. The cementation effect can be clearly identified.
3) Based on the model tests, two methods of using the sludge and copper slag for land reclamation are suggested. In both methods, the sludge will be mixed with binders and waste materials before disposal and disposed sludge mixture will be consolidated on site.

ACKNOWLEDGMENTS

The model tests presented in this paper were conducted by Mr M.H. Goi. His contribution to this study is gratefully acknowledged.

REFERENCES

Chu, J., Goi, M.H. and Lim, T.T. (2005). "Consolidation of cement treated sewage sludge using vertical drains." *Canadian Geoteh.l Journal,* Vol. 42, No. 2, 528-540.

Chu, J., Lim, T.T. and Goi, M.H. (2006). "Stabilization of Sewage Sludge by a Combined Chemical and Mechanical Method." *Proc. 5th International Congress on Environmental Geotechnics, 26-30 June, Cardiff, UK.*

Goi. M.H. (2005). "Use of stabilised sewage sludge for land reclamation." *MEng Thesis*, Nanyang Technological University, Singapore.

Lim, T.T., Chu, J., and Goi, M.H. (2004). "Assessment of heavy metals leachability in clay-amended sewage sludge stabilized with cement for use as fill material." *Journal of Residuals Science & Technology*, Vol. 1, No. 3, 157-164.

Lim, T.T., Chu, J., and Goi, M.H. (2006). "Effects of cement on redistribution of trace metals and dissolution of organics in sewage sludge and its inorganic waste-amended products" *Waste Management, Vol.* 26, No. 11, 1294-1304.

Innovative Use of Geomembranes and Recycled Materials in the Closure of a Landfill in the Environmentally Sensitive New Jersey Pinelands Protection Area

Authors: Ernest J. Kuhlwein, Jr., PE[1] ; Joseph J. Koehler, PE[2]; M. Brendan Mullen, PE[3]

1 Director of Solid Waste, Ocean County, New Jersey 08754; ekuhlwein@co.ocean.nj.us
2 Senior Associate, Hatch Mott MacDonald, Millburn, New Jersey 07041; joseph.Koehler@hatchmott.com
3 Vice President, Hatch Mott MacDonald, Millburn, New Jersey 07041; Brendan.mullen@hatchmott.com <

ABSTRACT: The Southern Ocean Landfill (SOLF) is located in Ocean County, New Jersey. The privately owned landfill comprised two distinct disposal areas - a 50-acre unlined landfill area and a 12-acre lined landfill. Due to insufficient funds, neither landfill was closed in compliance with the prevailing New Jersey Department of Environmental (NJDEP) regulations. In an effort to achieve proper closure, Ocean County entered into a partnership agreement with the NJDEP to assume responsibility for the closure work. The NJDEP provided a grant in the amount of $15 million to the Ocean County Board of Chosen Freeholders to fund the proper closure and long-term post-closure operation, maintenance and monitoring of the 50-acre unlined landfill area.

The County retained Hatch Mott MacDonald (HMM) to complete permit application packages, prepare the engineering design and provide construction management to close the 50-acre unlined landfill.

The closure design included consolidation of waste areas within the landfill footprint and regrading to reduce steep sideslopes. The innovative cap design incorporated reuse of recycled materials as part of a 12-inch landfill gas collection and bedding layer overlain. A 40-mil thick synthetic geomembrane liner, an 18-inch sand drainage layer and a 6-inch vegetated topsoil layer completed the cap components. HMM developed an innovative stormwater management design that accomplished a zero discharge under compounded 100-year and subsequent 10-year storm events.

INTRODUCTION

The Southern Ocean Landfill (SOLF) is located on a 283-acre tract in Ocean Township, Ocean County, New Jersey. The SOLF is an unlined 50-acre landfill that preceded a 12-

acre contiguous lined landfill on the southern third of the property. The Southern Ocean Landfill, Inc. and its predecessor company operated the Southern Ocean Landfill (SOLF) from approximately 1950 through the cessation of disposal activities in 1988. Originally, the site was owned by Ocean Township and was used as a waste disposal site in the 1950's and 1960's. In 1969, Southern Ocean Landfill, Inc., acquired the site and continued acceptance of solid waste through 1988, developing the 12-acre lined area in the mid1980's.

Neither landfill was properly closed following cessation of filling activities; the leachate collected in the lined 12-acre site was pumped to an on-site lined lagoon and subsequently tankered off-site to a POTW for treatment. The management of the collected leachate exhausted the escrowed closure funds; ultimately the lined portion of the site was closed using revenue generated from the importation of grading materials including crushed glass and kaofin. With no funding available and no

PHOTOGRAPH 1. Pre-Existing Site

opportunity to place additional fill on an unlined landfill, the State of New Jersey Department of Environmental Protection ultimately funded the closure of the 50-acre unlined site.

The closure of the 50-acre unlined SOLF, in particular, the innovative use of geosynthetic and recycled materials, is the focus of this paper. Hatch Mott MacDonald (HMM) completed the design and construction management of the closure directly for Ocean County, the contract administrator. Photograph 1 depicts the site prior to the commencement of closure activities.

RESOLUTION OF INITIAL ISSUES

Prior to the implementation of the closure plan, several issues had to be resolved. These issues included the excavation and consolidation of outlying portions of the landfill footprint. In addition, several portions of sideslope that exhibited sideslope grades up to 1 horizontal: 1 vertical had to be excavated and regraded to the approvable design grade of 3 horizontal: 1 vertical. A third concern was the amount of broken glass that had accumulated at the site that had to be accommodated in the closure. This glass was not appropriate for use as an alternative construction material as it was comprised of large portions of glass bottles and shards and did not have the same properties as crushed glass. The design and construction ultimately called for the placement and covering of the broken glass in a five-acre area of the topslope. Compaction of the excavated and consolidated waste areas was completed by means of a landfill compactor. Dynamic compaction was deemed unsuitable for this situation, however, the landfill compactor was effective in consolidating the landfill material.

DESIGN ELEMENTS

Following the excavation and consolidation of the outlying landfill materials, the regrading of the steep sideslopes and the placement and covering of the broken glass, the entire 50-acre site was grubbed and regraded to produce an undulating surface. The overall top of slope was approximately 60 feet above the original grade. An initial six-inch layer of crushed glass was placed on the regraded landfill surface. The crushed glass

PHOTOGRAPH 2. During Construction

represented a key element of recycled material reuse and significantly assisted in the project economics. The crushed glass was an important and integral component of the gas collection system. The decision to beneficially reuse crushed glass in the gas collection system was based upon practical reasoning. The local availability of crushed glass from the adjacent County-owned recyclables processing facility allowed for the stockpiling of the crushed glass. Sufficient quantity of the material for project needs was stockpiled on-site and its use was accordingly beneficial in construction scheduling. A grid of four-inch diameter perforated ADS drainage pipe was laid in the crushed glass trench. At the longitudinal and lateral intersections of the ADS piping vertical gas wells were advanced. The vertical gas wells comprised a three-foot diameter borehole with a six-inch perforated PVC riser; the annual space through the entire waste column was filled with gravel up to the liner. A six-inch cushion layer of sand was placed on top of the crushed glass and beneath the liner. This layer also assisted in the gas collection and conveyance.

The liner employed was a GSE 40-mil textured VFPE. The liner was placed on the entire 50-acre site. An 18-inch sand drainage layer was placed immediately above the liner. The cap was completed with a six-inch layer of vegetated topsoil. Photograph 2 depicts the site with the majority of the liner in place.

INNOVATIVE USE OF GEOSYNTHETIC MATERIALS

In the design of landfill caps, one of the most critical components is the management of stormwater. In addition to the precipitation fall on the surface of the landfill, it is also imperative that the rainfall infiltrating the vegetative cover is transmitted effectively and expeditiously to the drainage structures. Many landfill caps that were otherwise well designed and constructed failed due to improper material selection and design of the stormwater controls. Consequently, this aspect of the design required careful attention, and innovative and creative use of geomembrane to isolate and convey stormwater off the site as effectively as possible.

The design of the stormwater management structures incorporated three miles of longitudinal swales that discharged into three gabion-lined downchutes. The swales were located at 100-foot travel intervals along the sideslope. Therefore, in effect, no precipitation that remained as surface runoff on the landfill cover had a travel distance of more than 100 feet before it encountered a swale. The failure of many caps, however, is related to the movement of infiltrating precipitation and its lateral movement in the layers beneath the topsoil and above the geomembrane. The longer the travel distance, the greater will be the destructive power of this infiltrating precipitation through liquefaction, erosion and washout. In designing the cap for this site, HMM incorporated a key element that directs all infiltrating precipitation to daylight in the swales. This is accomplished by welding a flap of geomembrane to the geomembrane on the landfill sideslope at each of the swale locations; the flap follows the contour of the swale invert. The flap extends from the interface weld on the sideslope geomembrane through the successive cap layer and daylights in the swale. A separate flap corresponding to each successive swale along the sideslope from top to bottom. Consequently, as is the case in the surface runoff, no

infiltrating precipitation has a travel distance greater than 100 feet before it is directed via the welded geomembrane flap to daylight in the swale and ultimately be conveyed downslope via the downchutes. The downchutes were also similarly lined to ensure that there is no potential for washout beneath the structure or at the interface with the swales. Figure 1 is a detail of the typical downchute profile elements, and Figure 2 is a corresponding cross-section.

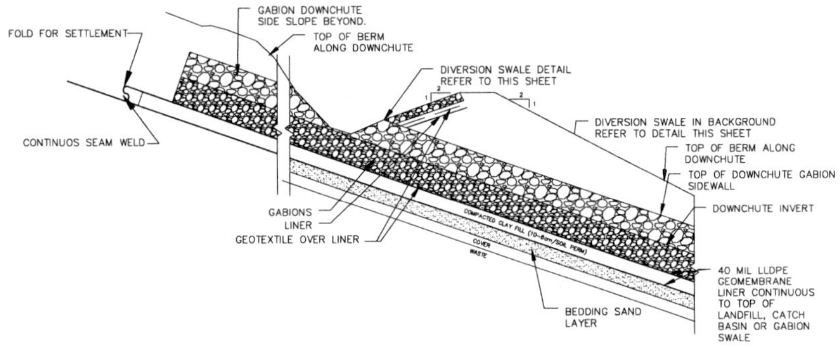

FIG. 1: Typical Downchute Profile

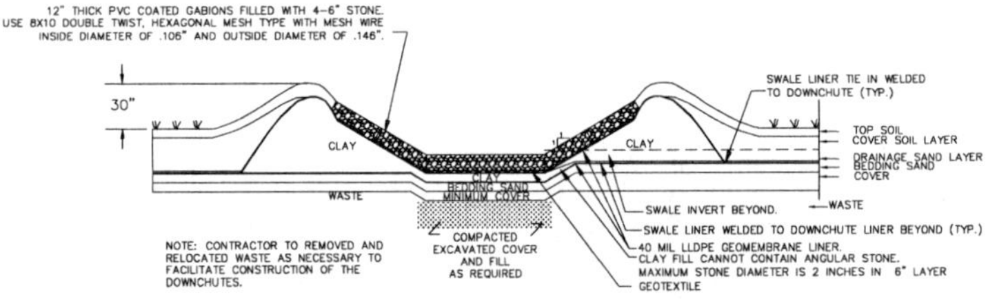

FIG. 2. Typical Downchute Cross Section

Figure 3 is a cross-section depicting a typical swale configuration, and Figure 4 is an enlargement depicting the typical arrangement.

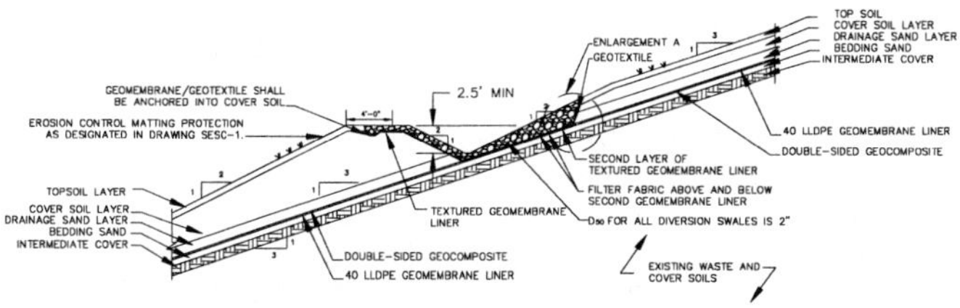

FIG. 3. Diversion Swale And Side Slope Berm Detail

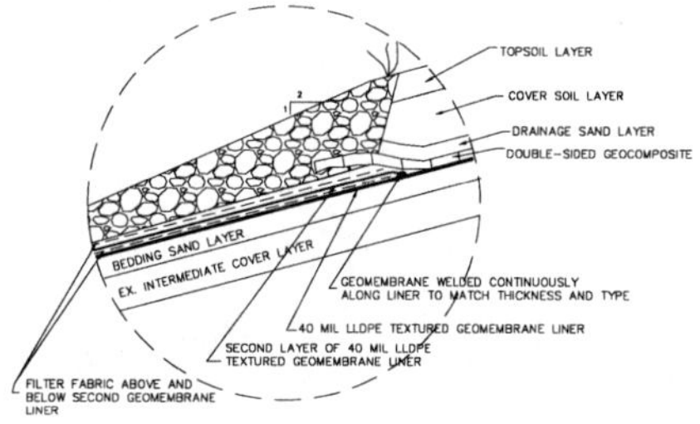

FIG. 4. Enlargement A

The construction was completed in 2003 and, despite having experienced several hurricanes, the stormwater features have performed effectively to date. The landfill cover has remained intact and has required standard post-closure maintenance. Photograph 3 depicts the completed closure.

CONCLUSION

The closure of this site was a significant win for the State of New Jersey, Ocean County and the host municipality, Ocean Township. The State removed a problem unclosed landfill from its books while from the perspective of the County and host municipality an environmental eyesore has been pleasantly replaced. Without the availability of the State funding the site could not have been closed as the owner was bankrupt. The innovative utilization of crushed glass in combination with geosynthetic materials in the closure not only helped accelerate the project but also helped keep project costs in check.

PHOTOGRAPH 3. After Construction Completion

Characterization of Water Treatment Residuals and Their Beneficial Uses

Hsin-Neng Hsieh[1], M. ASCE, and Dorairaja Raghu[2], M. ASCE

[1]Professor, Dept. of Civil & Env. Eng., New Jersey Institute of Technology, University Heights, Newark, NJ 07102; hsieh@njit.edu
[2]Professor, Dept. of Civil & Env. Eng., New Jersey Institute of Technology, University Heights, Newark, NJ 07102; raghu@njit.edu

ABSTRACT: Technologies have been developed and tested for the beneficial use of water treatment residuals. Application of these techniques depends on the understanding of the property of the residuals. Ten samples were collected, tested in the laboratory. The properties of the residuals and beneficial uses of residuals are also provided. Lime residuals can be used in agricultural soil conditioning, reclamation of acidic soils and cement production. Iron residuals are employed for iron recovery and desulphurization. Alum residuals are utilized in brick making and composting.

INTRODUCTION

Handling and disposal of water treatment residuals have always been important consideration and a common problem for water industry. Because of increasing environmental constraints, regulations governing residuals disposal have become more restrictive. Also, the costs of landfilling are increasing because of the limited land availability, strict design and operation standards, and political pressure. For these reasons, water industry has an increasing interest in alternative methods for residuals disposal. Beneficial application has gained wide attention.

A number of methods and technologies have been developed and tested in the past. These include agricultural soil conditioning with lime softening residuals, mining uses, hydrogen sulfide binding with iron residuals, brick making, iron recovery, land application, composting with alum residuals, land reclamation, lime residue for emission control, eutrophication control, cement production, and some other innovative methods. Application of these technologies is summarized in Table 1.

Proper application of beneficial techniques depends on the understanding the properties of the residuals. In this study, water treatment plant residuals were collected from ten facilities. Environmental and geotechnical tests were conducted on these samples. Data are presented to correlate the physical and chemical properties of

Table 1 Beneficial Uses of Water Treatment Plant Residuals

Beneficial Uses	Residual	Usage/Purpose	Recommended Amount	Considerations	Reference
Agricultural soil conditioning	Lime	The application of nitrogen fertilizer causes a reduction in soil pH. If optimum pH conditions do not exists, crop yield will be reduced.	For 100 lb (45.5 kg) of ammonia fertilizer, 3 to 4 lb (1.4 to 1.8 kg) of limestone be applied	Farmers are unfamiliar with its use and logistics of its transportation and application. The trace metals in the slurry do not appear to cause any adverse effects on the crop	Russel 1980; Reeves 1980
Mining uses	Lime	Aids in the reclamation of unusable lands by neutralizing acid soils	10 tons of dry solids/acre (2.2 kg of dry solids/m^2)	Check if runoff quality meets stream standards	Ohio Department of Health 1969
Hydrogen sulfide binding (Desulfurization)	Iron	H_2S is toxic gas. NIOSH recommends 10 ppm to be the 10-min max. in 8-h working environment. Cause Crown corrosion in sewer and sludge digester.	Overall average iron content in digestion sludge must be at least 18 g Fe/kg solids	No adverse effects on digestion process for iron conc. over 160 g Fe/kg solids. May cause problems due to the presence of $CaCO_3$ (reducing the CO_2 content)	Koppers 1980; Shannon et al. 1997

Table 1 Beneficial Uses of Water Treatment Plant Residuals (Continued)

Beneficial Use	Residual	Usage/Purpose	Recommended Amount	Considerations	Reference
Brick making	Alum	First used to replace the water required in the brick making process	Test bricks containing 1 to 15 percent WTP sludge were made.	Color change. No structural problem encountered.	Roland 1976
Iron recovery	Iron	Reuse as a raw material in the iron-producing and processing industry		Iron content cannot be too low because problems may arise in processing	ven der Veer 1987
Land application	Alum	Improve the aeration and available moisture-holding capacity of potting soil		Plant growth was adversely affected - aluminum's ability to fix phosphorus; Non-hazardous	Bugbee and Frink 1985; Permit – CWA Part 503 WTP residuals management guideline
Composting	Alum	Composting an alum sludge with leaves	A mix of 4:1 (leaves to alum sludge on a volume basis)		E and A Environmental Consultants 1988
Land reclamation	All	Establish vegetative growth or restore soil productivity lands		GW monitoring, trace metals, soil pH	Criteria used in Pennsylvania
Emission control	Lime	Clean emissions (scrubber system)			Engineering News Record 1988
Eutrophication control	Alum	Aluminum to fixate phosphorus reservoirs or lakes			Harper and Yousef 1988
Cement production	Lime			No pollutants & sufficiently dewatered	Kreft 1988

the residuals to the possible beneficial use of the residuals and to find if any difficulties may be encountered in such applications.

BENEFICIAL USES OF WTP RESIDUALS

A survey of beneficial uses of residuals was made and listed in Table 1. Lime residuals can be used in cement production and in emission control. They are also used in neutralizing the effects of acids in situations such as reclamation of acidic soils and in agricultural conditioning acidic soils to be able to grow plants and crops. Alum residuals are employed for brick making, composting and in land applications. Iron residuals are used to recover iron and hydrogen sulphide binding. However, some of the applications were performed 20 years ago and properties of some of the residuals were not reported. This study would provide useful additional information and confirm studies conducted earlier.

SAMPLE AND SAMPLE ANALYSIS

The basis for selecting the types of Water Treatment Plant (WTP) samples and the plants from which the samples were obtained involved several factors, which were (1) major types of WTP residuals (based on chemicals added); (2) different raw water sources; (3) treatment processes; (4) types of dewatering methods; and (5) geographical location. Samples were collected from ten water treatment plants and shipped for testing at NJIT. Samples were selected to reflect the consideration of the above factors. Methods of sample analysis were in accordance with US EPA methods (USEPA 1986).

Major types of WTP dewatered residuals produced in the United State are alum, ferric, and lime. Most WTP residuals are generated from coagulation, water softening, sedimentation and filtration processes. The raw water sources of the samples are reservoirs and rivers. Dewatering processes utilized for the samples tested were mechanical and non-mechanical (lagoon and drying bed).

COMPOSITION OF THE RESIDUALS

Chemical addition during the dewatering process and the method of dewatering affects the composition of WTP residual. Composition tests include determination of solids concentration, pH, primary metals (calcium, iron, and aluminum) extracted from the WTP residual, and TCLP (Toxicity Characteristics Leaching procedure) Analysis.

Composition and pH of the residuals are presented in Table 2. Metals detected in dewatered residuals are related to the types and quantities of chemicals added. For example, in the dewatered cakes collected from Jersey City WTP, NJ (JCD), Little Falls WTP, NJ (PVC), and Minneapolis Water Works (MWD), where lime was used for softening or as dewatering conditioning reagent, calcium concentrations were very high (10.06% - 17.97%). For iron residuals such as those from West River WTP, CT (RWA and RWAF), iron concentrations in residuals range from 2.81% to 4.41%, which are much higher than those for other residuals. Concentrations of aluminum in alum residuals Wanaque WTP, NJ (WQD), Haworth WTP, NJ (HWD), Manasquan WTP, NJ

Table 2 Relationship with Primary Metals and Addition of Chemical Agents of Residuals

Sample	Type	pH	Solids content	Concentrations of primary metals (dry basis)			Concentration of chemical added in WTP process		
				Aluminum (%)	Iron (%)	Calcium (%)	Alum (ppm)	Ferric (ppm)	Lime (ppm)
JCD	Lime	9.94	27.56	3.92	0.06	12.86	10	-	1
PVD	Lime	11.15	29.28	1.26	0.04	17.97	9.53	-	50
MWD	Lime	7.46	70.94	0.01	0.08	10.06	20	-	170
SLD	Lime	9.15	74.49	0.04	0.11	10.02	-	12.8	94
ELD	Lime/Ferric	6.25	41.31	0.014	2.81	7.58	-	20	
RWA	Ferric	7.15	20.56	0.34	3.87	0.27	-	7.2	-
RWAF	Ferric	6.71	45.48	0.07	4.41	0.35	-	7.2	-
WQD	Alum	6.55	14.83	5.80	0.32	1.16	10-12	-	-
HWD	Alum	6.82	60.06	0.82	0.006	0.100	5	-	-
MQD	Alum	7.83	46.71	2.02	0.16	0.98		-	-
FLD	Alum	6.62	28.74	3.03	0.16	0.13	75	-	25
FLDM	Alum	6.45	81.95	0.75	0.04	0.12	75	-	25

-: Not applicable

(MQD), Bradenton WTP, FL (FLD and FLDM) were 0.75% to 3.30%, which are the highest among all samples tested.

Generally, the pH of raw WTP residuals was in the range of 6.6-8.2, depending on the types of coagulants used. The pH of dewatered residuals was also related to the type and amount of chemicals used. The lime residual had high pH ranging from 7.46 to 11.15. In places where large quantities of lime are added as conditioning agent during mechanical dewatering, such as the case in Jersey City (JCD) and Passaic Valley water treatment plants (PVD), the pH values increased after conditioning (pH = 9.94 - 11.82). The pH of ferric and alum residuals were from 6.25 to 7.81, as observed from the test results from samples of other treatment plants.

The paint filter tests were conducted on all WTP residuals before shipping to NJIT. There was no liquid passing through the filter for all the dewatered residuals because of their high solids content, which are 14.8 to 82 percent as shown in Table 2.

In TCLP analysis, none of the TCLP regulated constituents exceeded the TCLP regulatory limits for all the ten residuals tested. Most of these constituents were either not detected or below the minimum detection limit.

DEWATERING CHARACTERISTICS OF WTP RESIDUALS

Two types of dewatering methods are commonly used for WTP residuals, air drying and mechanical dewatering. Sand drying beds and lagoons are most commonly used for air drying. Mechanical devices used include belt filter presses, centrifuges, pressure filter presses, and vacuum filters. Among the ten water treatment plants studied, six plants used drying beds, three plants utilize filter presses, and one used centrifuges to dewater residuals. Related information is presented in Table 3, which shows that higher solids content can be obtained using drying beds. This could be due to trapped water that can be removed by dewatering machines only when the residual floc changes.

Table 3 Relationship of solids contents with dewatering processes

Sample	Dewater Process	Solids Content (%)	Volatile Solids Content (%)
FLDM	Drying Bed	81.95	63.41
SLD	Drying Bed	74.49	3.62
HWD	Drying Bed	60.06	14.33
MQD	Drying Bed	46.71	3.55
RWAF	Drying Bed	45.48	35.62
ELD	Drying Bed	41.31	17.37
MWD	Centrifuge	70.94	15.02
PVD	Filter Press	29.28	9.47
JCD	Filter Press	27.56	34.45
WQD	Filter Press	14.83	45.99

Similar results were obtained from laboratory tests. Dewatering characteristics of WTP residuals can be determined by various methods by which the water from residuals was removed in the laboratory, such as pressure filtration, air-drying, and oven drying. The paint filter test can be used as an indicator for determining free water content in WTP residuals. In dewatering test, laboratory mechanical drying, air drying, and oven drying were conducted. In each experiment, five samples were tested. Fresh WQD sample was taken from lagoon, and solids content was about 0.25-1%. RWA sample was taken from the lagoon, and solids content was about 2%. The results are shown in Table 4.

Table 4 Solids Contents of Three WTP Residuals under Different Dewatering Conditions

Sample	Type	Filter (%)	Mechanical (%)	Air-Dried (%)	Oven Dried (%)
WQD	Alum	3.0	30.05	83.86	100
RWA	Ferric	8.5	12.08	86.61	100
PVD	Lime	8.0	22.38	88.65	100

However, it should be pointed out that these observations were based on tests on small quantities of samples in the laboratory. Under field conditions, where large quantities are involved, the residual will have to be spread in relatively thin layer for drying. Solids content of three types of WTP residuals through mechanical dewatering were from 12.08% to 23.82%, which are very close to the dewatered samples from WTP filter presses as shown in Table 4, except for five-year old sample WQD, which has the solids content of 44.78% because of the effects of freeze/thaw. The solids content of WTP residuals of the laboratory air drying samples can reach 83.66% to 88.65%, whereas the solids contents of WTP residuals obtained from drying beds were from 41.31% to 81.95% depended on the weathering conditions.

CONCLUSIONS

Based on the data, the following conclusions may be drawn:

1. The pH of dewatered residuals is related to the type and amount of coagulant and conditioning agent added. Lime residual had a higher pH (7.46 to 11.82). The pH of ferric and alum residuals were from 6.25 to 7.81. The lime residual is a good candidate for use in farmland.
2. The major cations found in residuals were closely related to the type and quantities of chemicals added. Iron concentrations in iron WTP residuals vary from 7 to 20 ppm. Iron recovery for iron WTP residue may depends on cost effectiveness of the process.
3. None of the TCLP regulated constituents exceeded the TCLP regulatory limits for all the ten residuals tested. Most of these constituents were either not detected or below the minimum detection limit. Therefore, land application would not be a problem as far as toxicity is concerned.
4. Dewatering properties of WTP residuals were more likely to be related to solids composition, dewatering characteristics of residuals, and dewatering processes. If dewatered residue is required for beneficial use, proper dewatering technique should be considered.

ACKNOWLEDGMENTS

Authors wish to acknowledge American Waterworks Association Research Foundation (AWWARF) who funded part of this study.

REFERENCES

1. Bugbee, G.J. and Frink, C.R. (1985). *Alum Sludge as Soil Amendment: Effect on Soil properties and Plant Growth.* The Connecticut Agricultural Experiment Station. Bull. 827.
2. E and A Environmental Consultants (1988). *Connecticut-American Sludge Composting Pilot Study Report.* E and A Environmental Consultants Inc., Greenwich, Conn.
3. Engineering News Record (1988). Lime Sludge Put to Good Use. *ENR*, 220:4:19.
4. Grabarek, R.J. and Krug, E.C. (1987). Silvicultural Application of Alum Sludge, *Jour. AWWA*, 79:6:84.
5. Harper, M. and Yousef, Y. (1988). Reuse of Water Treatment Sludges for Improvement of Reservoir Water Quality. *Proc. AWWA Ann. Conf.*, Las Vegas, Nev.
6. Hsieh, H.N. and Raghu, D. (1997) *Criteria Development for Water Treatment Plant Residual Monofills*, American Water Works Association Research Foundation, 1P-5C-90707-3/97-CG, Denver, CO, Chapter 4.
7. Koppers, H.M.M. (1980). Treatment and Disposal of Wastes from Water Treatment Plants, *H_2O*, 13:25:627.
8. Kreft, W. (1988). Entsorgung im Zementwerk. Verwertung von Rest- und Abfallstoffen durch die Zementindustrie. *Entsorgung Praxis*, 9/88:360.
9. Ohio Department of Health. (1969). *Supplement to Report on Waste Sludge and Filter Washwater Disposal from Water Softening Plants.*
10. Reeves, T. (1980). The Mahoning Valley Sanitary District Plans to Spread Slurry on Farmland. *Farm and Dairy*, Salem, Ohio.
11. Roland, A.T. (1976). Evaluation of Water Plants Disposal Methods. *Proc. Ann. Meeting. Jour. North Carolina Section*, 12:1:56.
12. Russel, G.A. (1980). Agricultural Application of Lime Softening Residue. Illinois AWWA Section Meeting (Mar 1980) and the Missouri AWWA Section Meeting.
13. Shannon, L.D., Yu, C.H., Wleklinski, R. E., Jarvis, J. B., and DeKraker, D. P. (1997). *Beneficial Reuse of Lime Softening Residuals for Flue Gas Desulfurization.* AWWARF Report.
14. US EPA. (1986). Test Methods for Evaluating Solid Waste Physical/Chemical Methods - SW 846, 3rd Ed. Office of Solid Waste and Emergency Response. United States Environmental Protection Agency, Washington, D.C.
15. ven der Veer, A.J. (1987). Appropriate Sludge Disposal. KIWA Report 98 *Sludge Problems at Waterworks.* KIWA, Nieuwegein, The Netherlands. 112

Assessment of brownmillerite and periclase hydration in chromite ore processing residue at elevated temperature

Deok Hyun Moon[1], Dimitris Dermatas[2], M. ASCE, Adriana M. Sanchez[3], Maria Chrysochoou[4], Mahmoud Wazne[5] and Dennis G. Grubb[2], M. ASCE.

[1]Research Assistant Professor, Center for Environmental Systems, Stevens Institute of Technology, Hoboken, NJ 07030, USA; dmoon@stevens.edu
[2]Senior Research Associate, Center for Environmental Systems, Stevens Institute of Technology, Hoboken, NJ 07030, USA.
[3]Research Assistant, Center for Environmental Systems, Stevens Institute of Technology, Hoboken, NJ 07030, USA.
[4]Assistant Professor, Department of Civil and Environmental Engineering, University of Connecticut, Storrs, CT, 06929-2037, USA.
[5]Assistant Professor, Center for Environmental Systems, Stevens Institute of Technology, Hoboken, NJ 07030, USA.

ABSTRACT: Chromite ore processing residue (COPR) is a solid waste that was generated by the high temperature process of chromium extraction from chromite ore using soda ash and lime. The purpose of this study was to evaluate whether brownmillerite and periclase could be transformed into other minerals to understand and predict phase transformation during weathering in COPR. In this study, brownmillerite hydration to hydrogarnets and periclase to brucite in COPR materials were respectively assessed at elevated temperatures (100°C and 200°C) for 30 days because these reactions are very slow at ambient temperatures. The results showed that no apparent phase transformation was observed at 100°C. However, brownmillerite and periclase started dissolving the first day at 200°C, forming hydrogarnets and brucite, and completely dissolving after 15 days. Moreover, hydrogarnets were the main phase after 30 days of reaction. This indicated that the phase transformation could be possible when the initial COPR materials with high content of brownmillerite are exposed to long-term weathering.

INTRODUCTION

Chromite ore processing residue (COPR) was generated during the hexavalent chromium (Cr (VI)) extraction processes (Hillier et al., 2003). Specifically, chromite ore was mixed with soda ash and slaked lime and roasted at approximately 1200°C in a rotary kiln to oxidize chromium from the trivalent to hexavalent state and sequester impurities (Hillier et al., 2003). COPR was deposited as fill material in different areas including the Dundalk Marine Terminal (DMT) area, located in Baltimore, Maryland.

The COPR "parent" phases are brownmillerite (Ca_2FeAlO_5), periclase (MgO) and quicklime (CaO). COPR also contains hydration products such as hydrogarnets, brucite, and portlandite (brownmillerite, periclase and quicklime hydration products, respectively), as well as pozzolanic compounds such as calcium aluminum chromium oxide hydrates (CAC), and carbonation products like calcite.

In the DMT areas, two different types of COPR were observed. One type of COPR is a gray-black granular sand-sized (4.75 mm-75 μm) material (designated as GB). The other type of COPR is a fine-grained (> 75 μm) material of orange to reddish-brown color with varying degrees of lithification (designated as HB) (CES, 2004). The GB and HB materials essentially constitute the same mineral phases, but they differ quantitatively from each other with respect to their mineralogical composition. GB materials contain higher amounts of brownmillerite and periclase (approximately 60% and 4%, respectively), while HB materials contain much higher percentages of the transformation products of these materials.

Two major types of phase transformations are believed to occur in COPR. The first can occur through the hydration of brownmillerite, periclase, and quicklime to hydrogarnets, brucite and portlandite, respectively. Hydration reactions are believed to occur under hydrothermal conditions. The second involves the introduction of external chemicals, like sulfate, to form products such as ettringite (Moon et al., 2004).

The purpose of this study was to assess the transformation of the GB to the HB material, focusing on the hydration of brownmillerite to hydrogarnets and periclase to brucite to understand phase transformation in COPR during long-term weathering. Hydration kinetics seem to be strongly associated with temperature, and brownmillerite hydration in Portland cements has been reported to be slow at temperatures below 70°C (Jupe et al., 2001). Similarly, brownmillerite hydration is extremely slow in the COPR material at ambient temperature. The evidence for that is the presence of high amounts of brownmillerite that still exits as a crystalline phase along with several phases with very small amounts in GB material even though the COPR was deposited decades ago (Jupe et al., 2001). In the present study, brownmillerite hydration was investigated under hydrothermal conditions, which cause a preferred acceleration of the hydration reactions (Jupe et al., 2001).

MATERIALS AND METHODS

Summary of Hydration Experiments

Two GB materials from two trenches at DMT were tested in this study, designated as GB1 and GB2. The total metal analyses of GB1 and GB2 are presented in Table 1. Moreover, based on the X-ray powder diffraction (XRPD) analyses, the samples were rich in brownmillerite and contained low amounts of transformation products, as presented in Table 2. The amorphous content of GB1 and GB2 at 20°C was 41.6 wt% and 38.9 wt%, respectively.

Specifically, 7 g of each sample was mixed at a liquid to solid ratio of 2:1 (weight/weight) with deionized water. Each sample was placed in closed vessels (digestion bomb, Fisher Scientific, GA), heated to 200°C, and cured for 30 days. Another GB1 sample was heated at 100°C for 30 days to observe the potential for hydration reactions at lower temperatures. The samples were subjected to rapid (2 hours) cooling to room temperature, before analysis.

Table 1. Total Metal Analyses (mg/kg)

Sample	pH (2:1)	W/C	Al	Ca	Total Cr	Cr(VI)
GB1	12.1	24.29	60,900	231,000	17,400	3,710
GB2	13	17.97	56,800	244,400	19,100	3,315
	Fe	**Mg**	**Mn**	**K**	**Na**	**Si**
GB1	140,000	50,900	1,380	116	4,690	2,070
GB2	142,150	49,050	1,315	111	4,355	2,160

Table 2. Quantitative XRPD Results (wt%) of Samples GB1 and GB2 at 20°C

pH	GB1 Initial 12.3	GB2 Initial 12.75
Brownmillerite	62.7	51.5
Periclase	4	4.1
Brucite	6.2	4.4
Portlandite	-	2.6
Hydrogarnets	18.4	19.4
Hydrotalcites	3.9	5.4
Calcite	-	5.3
Dolomite	-	1.3
Quartz	-	1.5
AFMs	4.8	4.5

note: "-" indicates no-detection (< 0-5 wt%)

X-ray Powder Diffraction (XRPD) Analyses

XRPD was used to evaluate the phase transformations of the samples. Sub-XRPD samples were collected at 1, 7, 14 and 30 days, air-dried and then pulverized to pass through a U.S. standard #400 sieve (38 μm). Step-scanned x-ray patterns were collected in a Rigaku DXR-3000 diffractometer using Bragg-Brentano geometry. XRPD analyses were conducted at 40 kV and 40 mA using a diffracted-beam graphite monochromator with CuKα radiation. XRPD data were collected between two-theta values of 5° to 65°, with a step size of 0.02° and a count time of 3 seconds per step. XRPD patterns were analyzed by the Jade software version 7.1 (MDI, 2004) with reference to the patterns of the International Centre for Diffraction Data database (ICDD) (ICDD, 2002) and the Inorganic Crystal Structure Database (ICSD) (ICSD, 2004). Quantitative phase analysis by the Rietveld method (Rietveld, 1969) was conducted using the Whole Pattern Fitting function of Jade.

Scanning Electron Microscopy and X-ray Dispersive Spectroscopy (SEM-EDX) Analyses

SEM-EDX analyses were performed on air-dried samples mounted on a carbon coating using a LEO 982 field emission scanning electron microscope with an Oxford energy dispersive X-ray analyzer. EDX includes the qualitative analysis of crystal composition, as well as a semi-quantitative analysis of the element percentages. These were performed in order to confirm the phase transformations and to assess the elemental composition of the newly formed phases.

RESULTS AND DISCUSSION

XRPD Analyses

The XRPD results obtained from the sample GB at 100°C are presented in Table 3. No significant dissolution of brownmillerite was observed. However, the XRPD results obtained after 1, 7, 14 and 30 days at 200°C show that considerable transformations have occurred, as illustrated in Table 4 and FIG. 1. The amorphous content of GB1 and GB2 at 200°C after 30 days of reaction was 40.3 wt% and 37.7 wt%, respectively, indicating minor variability in amorphous quantity throughout the experiments.

Table 3. Quantitative XRPD Results (wt%) – Samples at 100°C

	GB1	
Days	**Initial**	**30 days**
pH	**12.3**	**12.2**
Brownmillerite	53.3	48.5
Periclase	5.2	4.5
Brucite	6.5	7.8
Portlandite	-	0.3
Hydrogarnets	15.8	18.6
Hydrotalcites	3.7	8.1
Calcite	12.2	8.6
AFMs	3.3	3.5

note: "-" indicates no-detection (< 0-5 wt%)

The XRPD results (FIG. 1) indicated the dissolution of brownmillerite in both samples after 1 day, approximately 17.38% and 67.38%, in GB1 and GB2, respectively. Moreover, approximately 90% of the brownmillerite was dissolved after 7 days and brownmillerite was not detected in GB2 after 14 days but persisted at 2.9% in GB1.

A change in color and a higher degree of lithification were noted for both samples at 7 days. The different rates of dissolution may be a result of differences in sample morphology and the initial mineralogy, as GB1 contained a greater quantity of parent materials than GB2, indicating less extensive weathering.

In both samples, hydrogarnets started to form immediately, reaching their maximum between 7 and 14 days (Table 3 and FIG. 1). Thereafter, these minerals became unstable and their subsequent transformations likely provided a source of minerals (Ca and Al) for the formation of CACs, portlandite, and calcite. Calcite formation seems to be a major reaction, which takes place despite limited access to external CO_2 due to sealing of the vessel (Hillier et al., 2003; Meller et al., 2004).

Dissolution of periclase was also observed. The periclase dissolved completely after seven days in GB1, while it was 0.5% in GB2; however, a material balance of magnesium showed that not all the magnesium was converted to brucite, suggesting that magnesium may have been incorporated in other minerals or was transformed to the amorphous phase.

Table 4. Quantitative XRPD Results (wt%) – Samples at 200°C

	GB 1				GB 2			
Days	**1**	**7**	**14**	**30**	**1**	**7**	**14**	**30**
pH	12.94	11.99	12.36	12.25	13.08	11.96	12.35	12.2
Brownmillerite	51.8	6.5	2.9	-	16.8	4.5	-	-
Periclase	1.3	-	-	-	0.9	0.5	-	-
Brucite	10	10.1	10.9	8.7	11.1	9.5	9.4	9.3
Portlandite	2.2	13.9	12.3	18.3	7.1	15.3	17.9	19.5
Hydrogarnets	23.8	40.5	37.7	29.2	36.9	37.3	33.1	29
Hydrotalcites	-	-	-	-	-	-	-	-
Hematite	2.7	19.4	22.3	20.2	13.4	19.4	21	19.8
Calcite	8.2	4.4	5.9	11.4	8.4	8.3	9	14.3
Quartz	-	-	-	-	1.0	0.8	0.9	0.8
AFMs	-	4	6.7	3.7	4.4	3	6.1	4.9
Ca Al Oxide Hydrate	-	1.1	1.3	4.6	-	1.5	2.5	2.4
Green Rust	-	-	-	3.8	-	-	-	-

note: "-" indicates no-detection (< 0-5 wt%)

Small amounts of hydrotalcites were present initially, but they were not observed after 1 day of curing. According to the literature (Hutson et al., 2004), the weakly bonded water in the interlayer of hydrotalcites is liberated around 200°C, causing the interlayer spacing to collapse. The loss of the water in the interlayer decreases the well-defined structure of the crystals.

Hematite (Fe_2O_3) formed during the first 7 days, presumably incorporating iron derived from brownmillerite dissolution. The amount of hematite present remained stable after 7 days. Contrary to the average COPR mineralogy in the field, hematite emerged as one of the main phases (CES, 2004). It is believed that during hydration, an iron-based network is left to form amorphous iron oxide/hydroxide at low temperatures, but at higher temperatures promotes the conditions which favor hematite (Hematite, 2007; Meller et al., 2004).

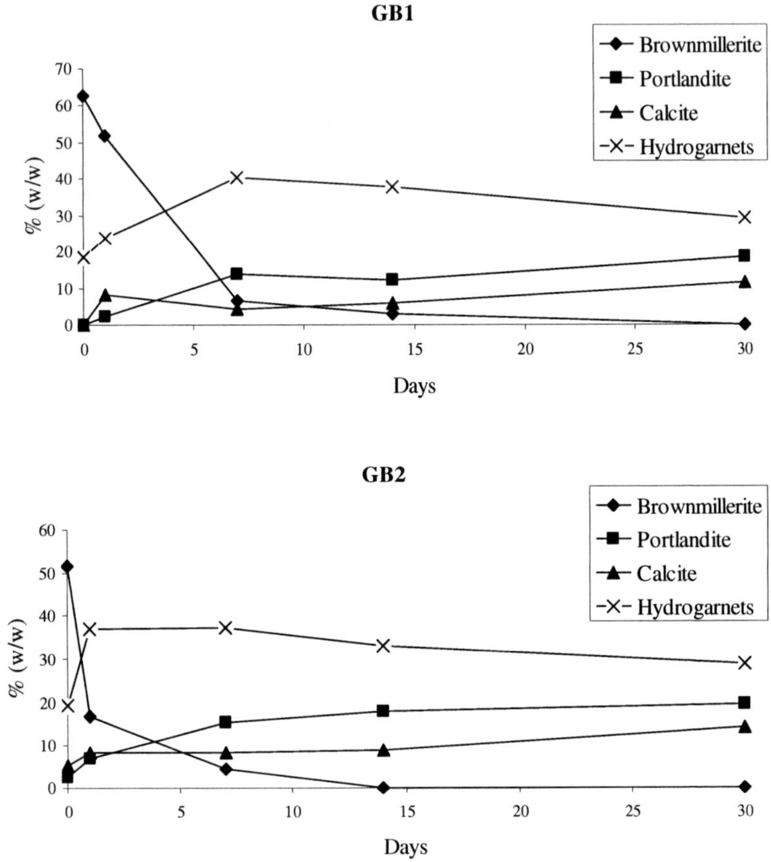

FIG 1. Phase transformations of COPR samples GB1 and GB2 at 200°C

Overall, the brownmillerite and periclase hydration at 200°C indicated that they could be transformed into hydrogarnets and brucite, respectively. Likewise, it appears that GB materials may be transformed into HB materials when the GB materials are exposed to weathering for long-term (more than 50 years) via Arrhenius modeling principles. If the rate of hydration at ambient temperature is known, the time for complete phase transformation can be calculated. This observation may help predicting COPR dissolution. Further investigation should be performed.

SEM-EDX Analyses

Hydrogarnets are crystals with a hexagonal shape containing calcium and aluminum at a 1.5:1 ratio (CES, 2004). SEM–EDX analyses of both GB1 and GB2 samples identified hexagonal shape crystals containing magnesium in addition to calcium and aluminum, potentially indicating partial magnesium substitution of hydrogarnets. These crystals in sample GB2 shown in FIG. 2 have a ratio (Ca+Mg)/Al of 1.29:1, which is close to the theoretical value for hydrogarnets.

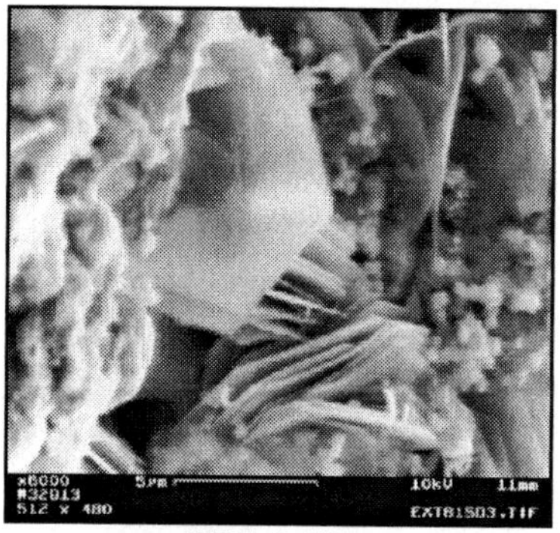

ELEMENTAL COMPOSITION (%)			
	Ca	Al	Mg
Theoretical (hydrogarnets)	60%	40%	
GB2	24.31%	43.73%	31.96%

FIG. 2. Hexagonal crystal in sample GB2 after 14 days

Another crystal found in GB1 at 7 days. This portlandite is very rich in calcium with only a little magnesium substitution, shown in FIG. 3. The crystal has a ratio Ca/Mg of 10.2.

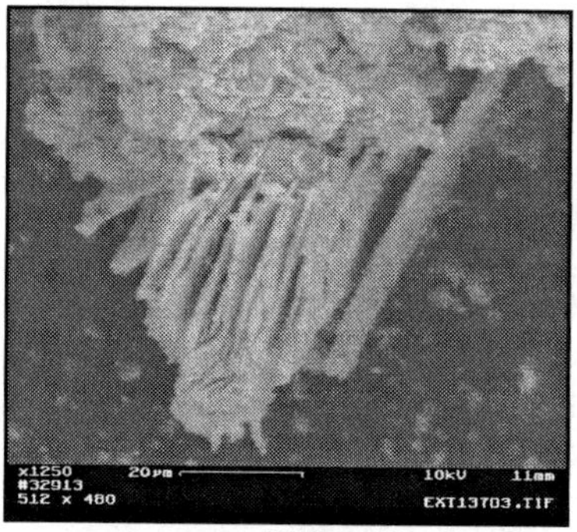

ELEMENTAL COMPOSITION (%)		
	Ca	Mg
Theoretical (Portlandite)	100%	
GB1	91.1%	8.9%

FIG. 3. Portlandite crystal in GB1 after 7 days

CONCLUSIONS

In this study brownmillerite hydration in COPR was investigated at both 100°C and 200°C. No apparent phase transformation was observed at 100°C after 30 days. However, immediate dissolution of the brownmillerite was observed in all COPR samples tested at 200°C. Brownmillerite was completely dissolved in GB1 and GB2 after 30 days and 14 days, respectively. Periclase was significantly dissolved after 1 day and completely dissolved after 7 days, forming brucite, but the mass balance for magnesium suggested that all of the magnesium was not converted into brucite, but may have included other phases such as hydrogarnets. Hydrogarnet formed immediately and reached its maximum after 14 days both in GB1 and GB2 and thereafter likely dissolved to provide a source of minerals for the formation of calcite, portlandite and other minor phases. Upon brownmillerite dissolution, hematite became a main phase, reaching values as high as 22% for some of the samples and remaining stable throughout the experiment. Overall, the transformation from the GB-type COPR, rich in brownmillerite and periclase, to the HB-type of COPR rich in hydrogarnets, was evident for both samples at 200°C after 30 days of reaction.

ACKNOWLEDGMENTS

This research was sponsored by Honeywell International Inc.

REFERENCES

Center for Environmental Systems, Stevens Institute of Technology. (2004) "Final Report on Trench Investigation and COPR analyses at Dunkalk Marine Terminal". Baltimore, Maryland.

Center for Environmental Systems, Stevens Institute of Technology. (2005) "Chromium Contaminated Study Area (SA7) and Dunkdalk Marine Terminal Swell Testing. Phase –II"August 2005.

Hematite (2007) http://www.webmineral.com/data/Hematite.html

Hillier, S., Roe, M.J., Geelhoed, J.S., Fraser, A.R., Farmer, J.G. and Paterson, E. (2003) "Role of quantitative mineralogical analysis in the investigation of sites contaminated by chromite ore processing residue." *Science of the Total Environment*, Vol. 308: 195-210.

Hutson, N. D., Speakman, S. A. and Payzant, E. A. (2004) "Structutral Effects on the High Temperature adsorption of CO_2 on a synthetic hydrotalcites." *Chem. Mater.* Vol. 16 (21): 4135-4143.

Inorganic Crystal Structure Database (ICSD), Fachinformationszentrum Karlsruhe, Germany, 2004.

International Centre for Diffraction Data (ICDD), 2002. Powder Diffraction File, PDF-2 Database Release.

Jupe, A. C., Cockcroft, J. K., Barnes, P., Colston, S. L, Sankar, G. and Hall, C. (2001) "The site occupancy of Mg in the Brownmillerite structure and its effect on hydration properties: and X-rat /neutron diffraction and EXAFS study." *Journal of Applied Crystallography*, Vol. 34: 55-61

Material's Data Inc. (2004) "Jade Version 7.0 User's Manual." CA, USA.

Meller, N., Hall, C., Jupe, A., Colston, S., Jacques, S., Barnes, P. and Phipps J. (2004) "The paste hydration of brownmillerite with and without gypsum: a time resolved synchrotron diffraction study at 30, 70, 100 and 150°C." *Journal of Materials Chemistry*, Vol. 14: 428 – 435.

Moon, D.H., Chrysochoou, M., Dermatas, D., Christodoulatos, C., Kaouris, M., Morris, J. (2005). "Investigation of ettringite formation in chrome ore processing residue." *Proceedings of the 8th International In Situ and On-Site Bioremediation Symposium*, Baltimore, Maryland, USA.

Rietveld, H.M., 1969. A profile refinement method for nuclear and magnetic structures. Journal of Applied Crystallography, 2, 65-71.

Conversion of Chromium Ore processing Residue to Chrome Steel

Jay N. Meegoda[1], M. ASCE, Wiwat Kamolpornwijit[2], Zhengbo Hu[2], Robert Mueller[3] and Ravi Patraju[3]

[1]Professor, Department of Civil & Env. Eng., New Jersey Institute of Technology, University Heights, Newark, NJ 07102; meegoda@njit.edu
[2]Post Doctoral Researchers, Department of Civil & Env. Eng., New Jersey Institute of Technology, University Heights, Newark, NJ 07102;
[3]New Jersey Dept of Environmental Protection, Division of Science, Research and Technology, 401 E. State St., P. O. Box 409, Trenton, NJ 08625

ABSTRACT: In Hudson County, New Jersey, there are more than two million tons of leftover chromite ore processing residues (COPR). Part of COPR was used as construction fill materials spreading the problem to a larger area. High solubility of some chromate compounds along with their toxicity is threatening the environment as well as human health. The primary objective of this manuscript is to demonstrate an efficient and optimum way to recover chromium and iron from COPR in the production of chrome steel or stainless steel. In this research, COPR was thermally treated to recover iron with chromium. The research applied techniques used in steel manufacturing to extract metallic iron and chromium from COPR. An extensive experimental program was performed using a Thermo-Gravimetric Analyzer (TGA) to thermally treat the processed chromium contaminated soils with additives at varying temperatures and under reducing environment. The optimum chemical composition of COPR and additives to be used in the melts was evaluated based upon the thermodynamic properties of the mixture to ensure good phase separation, least amounts of iron and chromium oxides in the slag and minimum variability of final product (steel or iron with chromium). The impact of other oxides on the steel making process was evaluated to minimize the adverse impact on the process. The research demonstrated the feasibility of the proposed process.

Keywords: Chromium Ore Processing Residue, chrome steel, Reduction, thermal treatment, Beneficial Use, and Spinel

INTRODUCTION

In the early 20th century more than 2 million tons of chromite ore processing residues (COPR) were left over from chromate extraction in Hudson County, New Jersey. In the chromate manufacturing, chromite ore was pulverized to a size smaller than #100 mesh, mixed with soda ash (Na_2CO_3), and lime (CaO), and heated to a temperature between 1100°C and 1500°C to produce sodium chromate (Na_2CrO_4). The soluble chromate compounds was recovered for beneficial use while the slowly dissolving chromate (such as $CaCrO_4$) left in the COPR. Chromium exists in COPR mostly in two oxidation states, trivalent and hexavalent. The hexavalent chromium is well known as group "A" human carcinogen with sufficient evidence. The extent of chromium contamination is now estimated to be 10 million tons making it a major environmental problem in New Jersey.

The vitrification process, making glass out of waste, has proven to be a very effective treatment technology for COPR and produces by-products with reusable values (Meegoda et al., 1999; 2000). In the bench-scale and pilot-scale vitrification study of COPR from several sites, the separation of metal was observed. The separation of metallic iron in the above study prompted the need for a further investigation on the possibility of converting the hazardous waste into a reusable material, chromium steel.

Meegoda et al., (1999) showed that COPR samples obtained from nine contaminated sites on average had 29% iron and 7% chromium. Due to the original chromium extraction process, a large concentration of sodium and calcium, and small quantities of silica, aluminum and magnesium are also present in COPR. The XRD characterization showed that metal oxides exist in the COPR in the form of spinel ($AO.B_2O_3$) or other mineral compounds. Generally, the substances which are considered basic oxides are those which are compounds of the elements forming basic compounds in ordinary chemical reactions in water solution, i.e. CaO, MgO, and Na_2O. The SiO_2 is an example of acidic flux. There are compounds of elements, for example alumina, that act both as acids and bases depending on the slag composition.

Chromium plays an important role when present in the iron by providing corrosion resistance, and reduced oxidation at elevated temperatures. The chromium concentration in the metal alloy steel varies depending upon the application. Chromium percentage in stainless steel is higher than 12%. Low chromium steel and structural steel for tool making contain chromium in the range of 0.5-3%. The reduction of chromium oxide to metallic chromium along with iron can potentially form an alloy of higher value provided it has similar composition to stainless steel.

Both metal oxide reduction and metal separation are important in the iron and steelmaking process. Chemical compositions of COPR determine the material properties such as basicity and viscosity. It will influence the kinetics rate for the metal reduction, and also play an important role in the separation of metal from slag. Most of the impurities, existing as oxides have very high melting temperatures for example SiO_2 at 1600°C. The separation of impurities requires the formation of intermediate compounds having lower melting point, sometimes by the addition of compounds to change the chemical composition of the mixture.

The primary objective of this research is to determine the feasibility of recovering iron and chromium from COPR. It also provides the basic understanding of the process, the optimum operating conditions, applicability and limitations. The remediation of the COPR is also a major requirement. The by-product of the melting process, slag, will be no longer toxic after the treatment and extraction of chromium at high temperatures. The slag will have similar properties to the vitrified product, which satisfies requirements as a construction aggregate set forth by American Society for Testing and Materials (ASTM) and New Jersey Department of Transportation (NJDOT).

EXPERIMENTAL PROCEDURE

COPR samples were collected from three sites in Jersey City, NJ. Three 25-gallon COPR samples from three sites were collected. One COPR sample (sample-1) was obtained from NJDEP site 131 and the two other samples (Sample 2-S and 2-N) from NJDEP site 115. Approximately 25 gallons of COPR from NJDEP site 131 was collected from depths less than 15 cm, since chromium is typically concentrated at shallow depths due to the capillary rise (Meegoda et al., 1999). Other two samples were supplied to NJIT form NJDEP site 115. All samples were homogenized before measurement or testing. Then approximately 5 gallons of COPR from each site was dried overnight in oven at 80°C and ground to fine powder and mixed homogenously. The fine powder that was ground to size 100-mesh was used in this research.

The characteristics of concern are chemical compositions, phases, and morphology of COPR samples, the reduced products (metal) and byproducts (slag). The homogenous samples were analyzed by X-Ray Diffraction spectrometer (XRD) and X-Ray Fluorescence spectrometer (XRF). The chemical compositions of the treated products were determined by X-ray Fluorescence (XRF) and scanning electron microscope (SEM) with Energy Dispersive X-ray (EDX) detector.

X-Ray Fluorescence spectrometer (XRF): Wavelength dispersive XRF (Philips, PW2400), was used in this study. All standards and COPR samples were prepared as pressed powder and analyzed under vacuum. Several calibration standards were also prepared.

X-Ray Diffraction spectrometer (XRD): XRD (Philips, X'Pert) was used to investigate phase assemblages of the sample. The incident X-ray Cukα was diffracted by the sample of different phases from 5-70 2θ angles with a step size of 0.01°. Samples were ground to smaller than 75 micron and placed in inverse sample holders in such a way to minimize preferred orientation.

Thermo-Gravimetric Analyzer (TGA): The reduction study was performed in a Perkin-Elmer, model TGA7, where samples were flushed with nitrogen to create an oxygen-free environment. The particle size of soils used was smaller than 100 mesh. Carbon was added as a reducing agent up to 15% by weight relative to the COPR. Sand was also added up to 50% by weight of COPR to neutralize the high CaO content in COPR which resulted in lower melting temperatures. The standard TGA test condition was a 2-hour flushing with nitrogen at a rate of 100 ml/min at 1 atm and room temperature, then ramping at a rate of 40°C/min to a set temperature (1500°C) with N_2 flushing rate at 20 ml/min, maintaining in isothermal state for 1 hr, and cooling at a rate of 20°C/min down to room temperature. COPR samples weighing approximately 70-100 mg were placed in graphite crucible which was then placed on a standard platinum pan for TGA experiments.

Morphology and spatial composition: Scanning electron microscope (SEM) with Energy Dispersive X-ray (EDX) detector was used for the morphology studies and analysis of local composition of the reaction mixtures after TGA experiments.

The SEM is LEO 1530 VP FE-SEM. The EDX is from Oxford instruments with INCA software.

EXPERIMENTAL RESULTS AND DISCUSSION

Mineral phases: FIG. 1 shows XRD diffractogram of COPR for Sample 2-S. The phase identification in COPR is not trivial and very often not conclusive due to the heterogeneity. It is not possible to identify all phases present in COPR. However, other researchers including Geelhoed et al., (2003) and Dermatas et al., (2006) have performed quantitative analysis of COPR. Some major phases that could be identified included $CaCO_3$ (or $MgCO_3$), $Ca_2(Fe_{2-x}Al_x)O_5$ (Brownmillerite), $Ca(OH)_2$ and SiO_2.

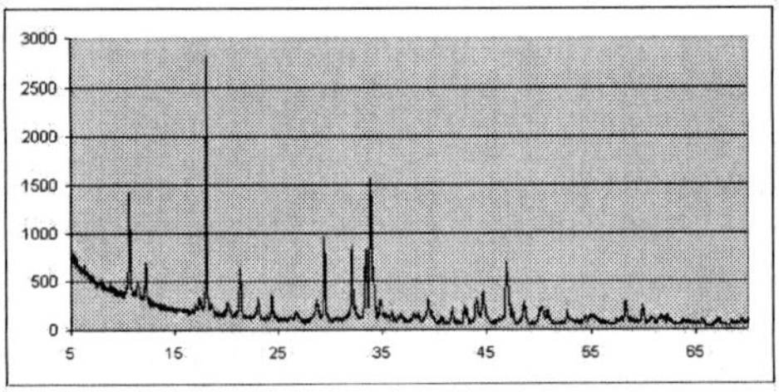

FIG. 1 XRD diffractograms of COPR Sample 2-S.

While previous XRD studies (Meegoda et al, 1999) on COPR samples from nine sites in Hudson County showed the major phase for iron and chromium is Spinel (Mg, Fe).(Fe, Cr, Al)$_2$O$_3$, the XRD pattern of these 3 samples indicated the major phase for Fe content as Brownmillerite, $Ca_2(Fe_{2-x}Al_x)O_5$. Also the 2-S and 2-N samples had significant peaks of $Ca(OH)_2$ (portlandite), along with that from $Ca(Mg)CO_3$ (calcite). The XRF results also showed large Ca content (~46% weight in the form of CaO). These results are consistent with the fact that high weight percentage of calcium was added during chromate extraction process.

The Cr content is relatively small (2-4% by weight in the form of Cr_2O_3 by XRF). The minor minerals containing Cr is identified as $Mg_6(Al,Cr,Fe)_2(CO_3)(OH)_{16}\bullet 4H_2O$ (hydrotalcite), and $Mg_5(Al,Cr,Fe)(Si_3Al)O_{18}H_{18}$ (Cr-Clinochlore). The existence of hexavalent chromium in soil is a concern, but the phase of hexavalent chromium could not be identified. Other major mineral phases were also not identified due to low concentrations. $CaCrO_4$ (calcium chromate) could be a possible mineral phase for chromium, although its peaks overlapped with brownmillerite, which is one of the most dominant phases in COPR. The yellow taints in COPR suggested the presence of hexavalent chromium. Water extraction of COPR produced yellowish filtrate. The precipitate from filtrate was subsequently identified using XRD as calcium carbonate and $CaCrO_4.2H_2O$ (chromatite hydrate). Although speculative there is a possibility that the hexavalent chromium present in COPR as $CaCrO_4.2H_2O$ (Kamolpornwijit et al., 2007).

Graphite weight loss from TGA analysis: The graphite crucible was very sensitive to oxidation by oxygen at high temperature. Hence the furnace chamber was flushed with nitrogen at a rate of 100 CC/min for 2 hours before heating up to desired temperature for isothermal period. It shows that under standard operation conditions for metal oxide reduction approximately 3 mg of carbon weight loss occurred. This information will be useful for the calculation of the reaction yield and later systematic kinetics studies.

Phase changes, i.e. from $CaCO_3$ to CaO, and metal reduction, i.e. Fe_2O_3 to Fe, proceed at certain temperatures. This information allow the identification of chemical phases and their kinetic of reaction in COPR when weight loss of COPR was monitored during TGA experiment.

Weight loss Curve and weight differences versus temperatures for Sample-2S are shown in FIG. 2 and FIG. 3. Weight difference is simply another way to present weight changes during TGA experiment, it is in fact a weight change of COPR between 2 consecutively monitored temperatures. The peaks in FIG. 3 represented weight losses from moistures and organic parts in sample at 150~400°C, from $CaCO_3$ at ~650°C, and from metal oxide at 1000~1400°C. Weight losses increased drastically at above 1000°C. Figure 3 shows that there are two peaks in the range of 1000-1400°C. It might imply that an easily reducible fraction reduced leaving hardy reduced phases with modified composition. Ignoring other factors, the stability of oxides is in the order of SiO_2> Cr_2O_3> FeO> Fe_3O_4> Fe_2O_3. The highly reducible type, possibly Fe_3O_4 and Fe_2O_3, was reduced to FeO at 1000°C. At 1100°C, FeO was reduced to metal. Cr_2O_3 and some amount of SiO_2 can be reduced to the metal at higher temperature.

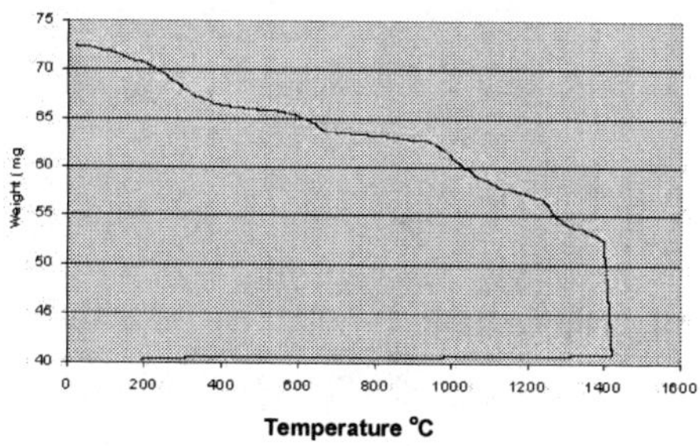

FIG. 2 Weight Loss Curve from the TGA experiment.

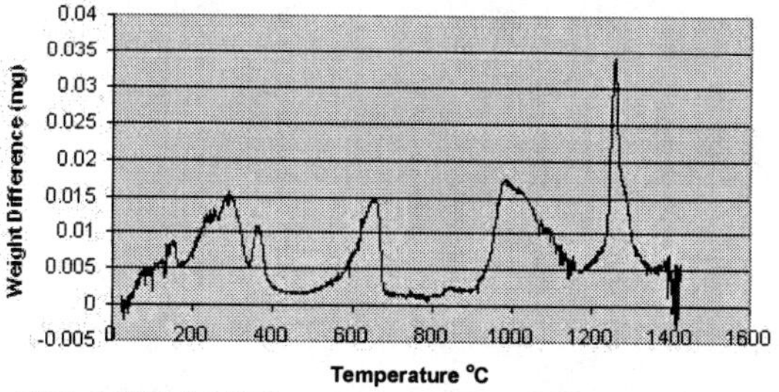

FIG. 3 Weight difference pattern from the TGA experiment.

The weight difference peaks also implied the reduction of iron oxide proceeded very fast and readily finished before the isothermal period at 1400°C. Previous studies and literatures suggested the reduction of chromium oxide (which is in smaller amount than iron oxide) proceeded at around 1400°C. The reduction of some SiO_2 to Si in the isothermal step of high temperature has also been reported and confirmed by the EDX results later in this manuscript. The reaction mixtures yielded a dust and some glasslike droplets. There was neither the metal separation nor the presence of tiny shiny specks. The metal separation was not achievable without sand addition.

The reduction of iron oxide in COPR with sand was rapid; therefore the limiting step for the iron droplet formation would depend on the grain growth and metal separation. Metal separation is easier in the liquid state. It is proposed that metal separation could be achieved by adjusting the melt composition to make COPR melt at low temperature and to have low viscosity. The solid state diffusion of acidic and basic oxides at high temperature may yield intermediate compounds with lower melting temperatures, and subsequently melt and form a liquid pool to which the high melting temperature compounds can be further dissolved.

TGA experiments were performed on sample 2S with 15% carbon and 50% sand. The addition of sand, an acid oxide to COPR rich with CaO, a basic oxide, should bring the melting temperature down and ease the agglomeration of the metallic iron. Some shiny metal balls were found in the final melted product, as shown in the FIG. 4 of the SEM-EDX results from morphology study.

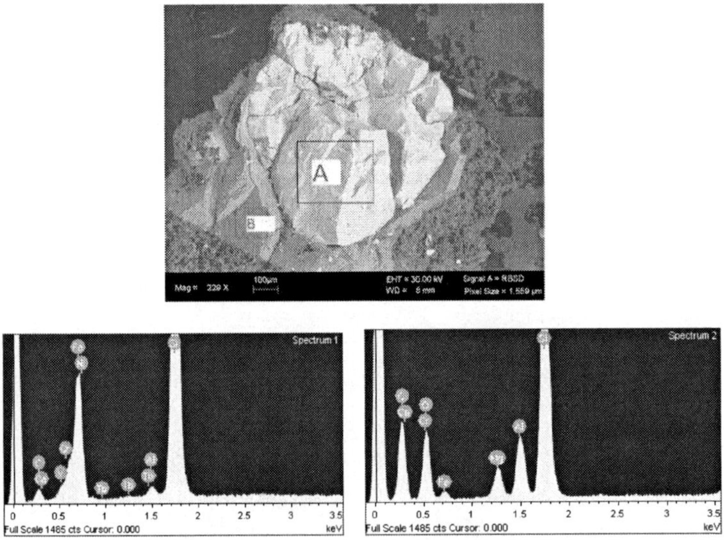

(a) Region A (b) Region B

FIG 4 SEM-EDX Showing separation of Metal and Oxide Phases

Table 1 shows the composition of the metal phase (Region A) was mainly Fe (28.1%) and Cr (3.6%). The Oxygen peak was not observed, and the other major element carbon comes from the graphite crucible to which the product was attached, and silicon from the reduction of SiO_2. The oxide region contained large percentages of CaO and SiO2, and a small faction of Fe and Cr. Please note for the semi-quantitative nature of EDX results.

Table 1 EDX Results for Regions A and B of SEM shown in FIG. 4

	A		B	
Elements	**Weight %**	**Atomic%**	**Weight %**	**Atomic%**
Fe	50.81	28.11	3.63	1.07
Cr	6.03	3.58	0.81	0.26
Si	24.23	26.66	9.37	5.50
Ca	0.23	0.18	11.32	4.66
C	15.42	39.67	37.70	51.78
O	nd	nd	33.12	34.15
Al	0.88	1.00	2.40	1.47
Mg	nd	nd	1.64	1.11

nd: not detected by EDX.

The factors governing the metal separation include but are not limited to the melting temperature and the slag viscosity which depends on slag compositions. The behavior of slag or oxide system can be partly described through phase diagrams. The slag components of interest are CaO, MgO, SiO_2, and Al_2O_3. FIG. 5 shows pictures of metal balls with different amount of sand addition to COPR Sample 2S. Preliminary data suggested that the possibility for metal separation with 40% or more sand addition, but not with 20% or lesser amount of sand. Systematic studies can use the phase diagram of quaternary system CaO-MgO-SiO_2-Al_2O_3 because phase diagrams provide information on thermodynamically stable states, i.e. as liquid or liquid in equilibrium with solids, depending on sample compositions.

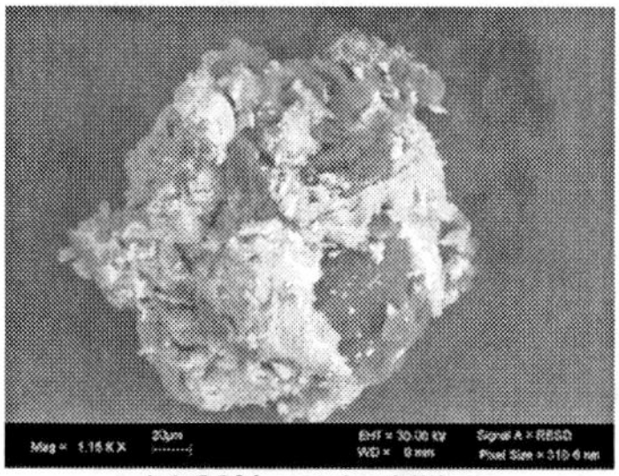

(a) 20% sand addition

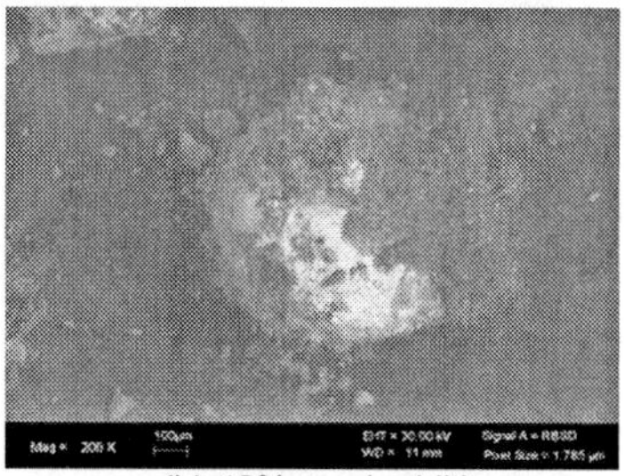

(b) 40% sand addition

FIG 5 SEM Pictures of Metal Balls with Different Amounts of Sand

SUMMARY AND CONCLUSIONS

Chromium contamination by chromite ore processing residues (COPR) is a major environmental issue in New Jersey. Characterization of soil samples obtained from nine contaminated sites on average had 29% iron and 7% chromium. Other major metal elements were Ca, Mg, Al, Si and Na. Due to the early chromate extraction process, COPR is basic with relatively high Ca/Mg contents and low Si content. Separation of metallic iron from the previous vitrification process of COPR prompted the need for a further investigation on the possibility of converting the hazardous waste into a reusable material, chromium steel. The remediation of COPR could be accomplished by recovering almost all Cr in the form of metal alloy.

The reduction study of COPR was performed with graphite under a reduced environment (N_2). The temperature for reduction went up to 1450-1500°C to ensure the full recovery of iron and chromium. Metal separation from the reduction of COPR could be achieved with correct chemical composition and high temperature up to 1450-1500°C for 1 hour. The reduced metallic and non-metallic products were characterized by SEM with EDX.

The XRD analysis indicated that the major phase of iron was Brownmillerite, and the major phases of Ca were $Ca(OH)_2$ and $CaCO_3$. The chemical compositions of mineral phases helped to optimize the operating conditions (additive, temperature, and reaction time) for the metal oxide reduction and metal separation. The reduction of iron oxide with chromium in COPR to a metal was achievable under the oxygen free environment with sufficient amount of reducing agent. The reaction occurred rapidly at high temperatures and completed during the heating period from 1000°C to 1400°C. Metal separation occurred when 40-50% of sand was added to balance the large amount of CaO in COPR. The morphology study by SEM-EDX confirmed that the metal phase contained mainly Fe/Cr, the oxide phase had intermediate compounds, such as $CaO\text{-}SiO_2$, and the Cr was well separated.

ACKNOWLEDGMENTS

The work described in this manuscript was supported by a research contract from the New Jersey Dept of Environmental Protection. The contents of this manuscript reflect the views of authors. The contents do not necessarily reflect views or policies of NJIT or NJDEP.

REFERENCES

Burke, T., Fagliano, J., Goldoft, M., Hazen, R. E., Tglewicz, R. and Mckee T., (1991), "Chromite Ore Processing Residue in Hudson County, New Jersey." Environ Health. Pres., 92, 131-137.

Dermatas, D., et al., (2006), Ettringite-induced heave in chromite ore processing residue (COPR) upon ferrous sulfate treatment, Environmental Science and Technology, Volume 40, Issue 18, 15 September 2006, Pages 5786-5792

Ding, Y. L.; Merchant, A. J. Ironmaking and Steelmaking (1999), 26(4), 247-253. Kinetics and mechanism of smelting reduction of fluxed chromite. Part 1. Carbon - chromite -flux composite pellets in Fe-Cr-C-Si melts.

Geelhoed, J. S., et al., (2003), Chromium remediation or release? Effect of iron(II) sulfate addition on chromium(VI) leaching from columns of chromite ore processing residue, Env. Science and Technology, 37 (14), pp. 3206-3213

Meegoda, J. N., et al., (1999), "Remediation of Chromium Contaminated Soils - A Bench-scale Investigation," ASCE Practice Periodical of Hazardous, Toxic, and Radioactive Waste Management, 3(3), 124-131.

Meegoda, J. N., et al., (2000), "Remediation of Chromium Contaminated Soils - A Pilot Scale Investigation," ASCE Practice Periodical of Hazardous, Toxic, and Radioactive Waste Management, 4(1), 7-15.

Kamolpornwijit, K., Meegoda, J. N. and Hu, Z. (2007) Characterization of Chromite Ore Processing Residue, ASCE Practice Periodical of Hazardous, Toxic, and Radioactive Waste Management, Vol. 11#4, pp. 234-239

Current Knowledge on Heaving Mechanisms of Chromite Ore Processing Residue

Dimitris Dermatas, M. ASCE[1], Maria Chrysochoou[2], Deok Hyun Moon[3], Maria Kaouris[4], John Morris[4], Chris French[4]

[1]Senior Associate, Center for Environmental Systems, Stevens Institute of Technology, Hoboken, NJ 07030; dermatas@stevens.edu
[2]Assistant Professor, Department of Civil and Environmental Engineering, University of Connecticut, Storrs, CT, 06929-2037;
[3]Research Professor, Center for Environmental Systems, Stevens Institute of Technology, Hoboken, NJ 07030;
[4] Honeywell International Inc., 101 Columbia Road, Morristown, NJ 07962;

ABSTRACT: Chromite Ore Processing Residue (COPR) has in certain cases been associated with heaving phenomena. To understand the COPR heave mechanism, the site owner sponsored research that included extensive field investigation, site characterization and laboratory programs. Based on the results, this paper presents the current knowledge on COPR heave mechanisms. Among the different heaving mechanisms considered, ettringite formation had previously been reported as an expansion mechanism in cements and lime-treated soils. While site investigation revealed significant, localized presence of ettringite, it could not be correlated to the observed heave features. However, observations from test-trench activities and heave phenomena in ferrous sulfate treated COPR, confirmed that ettringite induced expansion constitutes a viable COPR heave mechanism. From other potential mechanisms that were evaluated, brownmillerite hydration to hydrogarnets remains the most viable. Although brownmillerite hydration proceeds extremely slowly at ambient temperature, hydrogarnet-rich material has been found in discrete layers in trenches at different sites, and has lately been associated with a major heave feature at SA-7. Accordingly, the conditions for rapid brownmillerite hydration and its associated potential for heave manifestation remain open to further investigation.

INTRODUCTION

Study Area 7 (SA-7) is a 34-acre site located in Hudson County, N.J., between Route 440 and the Hackensack River. About 1.5 million tons of COPR were deposited at SA7 in the period 1905-1954 (Dermatas et al. 2005). After hydraulic deposition, the COPR was spread mechanically across the site, graded, and covered with soil for the purpose of commercial site development. The major structure at the site had to be demolished due to pronounced surface heave. Heaving phenomena were also observed at another COPR site known as Dundalk Marine Terminal (DMT).

In the past, research on COPR focused mainly on Cr(VI) abundance, leaching and treatment (Meegoda et al. 2000; Graham et al. 2006; Farmer et al. 2006; Cao and Zhang 2006; Hillier et al. 2007) and not on COPR-related heave. COPR volumetric expansion was mentioned only by one group in early publications (Raghu et al. 1989). Consequently, there was limited background information with regard to COPR properties and treatment, and no background with regard to COPR heave, prior to the initiation of the investigations at the two sites, SA-7 and DMT, starting in July 2003. The research involved a thorough literature review, extensive field investigation and site characterization programs and laboratory-based experiments. Based on these results, this paper presents the current knowledge on COPR heave mechanisms.

SCOPE OF COPR RESEARCH INVESTIGATION

An extensive subsurface investigation was conducted at SA-7 starting in 2003 and continuing through 2005. A total of 92 discrete disturbed samples were obtained from different depths in 38 soil borings and 25 relatively undisturbed samples were obtained from tube samplers. One block sample was obtained from a trench excavation conducted in January 2004. Further sampling via excavated test trenches was pursued following pilot treatment of COPR during the summer of 2005. Three trenches were excavated in Area 1800 at DMT in 2004. Trenches A and B traversed major heave ridges, located in the center of the trenches. A total of 63 samples were obtained from distinct positions of the three trenches.

The testing program for both COPR sites included both physical and chemical characterization. COPR characterization showed that macroscopic heave resulted from the microscopic volume expansion induced by mineralogical changes (Dermatas et al. 2005a; Chrysochoou et al. 2005; Dermatas et al. 2006). Thus, COPR characterization mainly relied on mineralogical testing by X-ray Powder Diffraction (XRPD) followed by qualitative/quantitative Rietveld-based analyses; selective Scanning Electron Microscopy/ Energy Dispersive X-ray (SEM/EDX) used to study/confirm specific mineralogical features as identified by XRPD Rietveld-based analyses (Chrysochoou 2006; Chrysochoou and Dermatas 2007).

These analyses, along with a close look at site histories and a targeted literature review lead to the identification of several possible heaving mechanisms; that is the formulation of possible mineralogical transformations that lead to volume expansion and may induce heave. In order to verify the different heaving mechanisms hypothesized, laboratory slurry and swell tests were also conducted (Moon et al. 2006a; Moon et al. 2006b). Slurry tests involved the reaction of small amounts of COPR with water and different reagent chemicals to investigate mineralogical transformations by XRPD at select time intervals, whereas swell tests constituted a quantitative measure of monitoring COPR expansion upon its reaction with water and different chemicals.

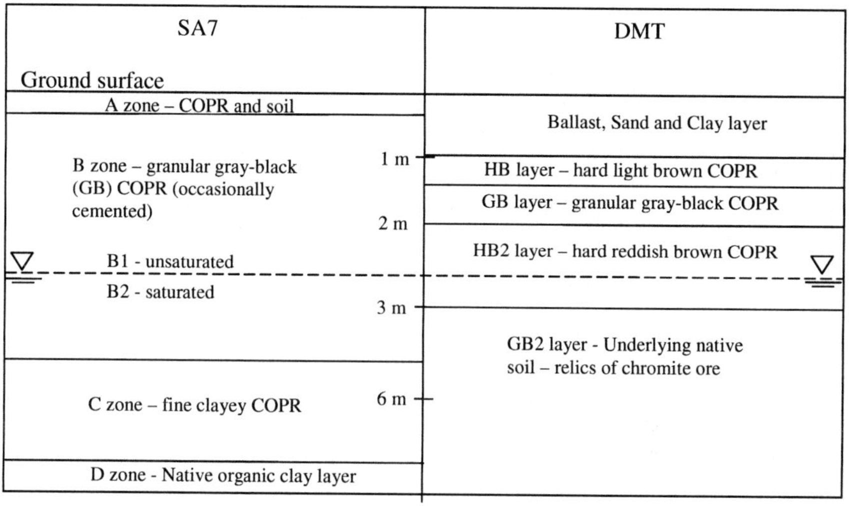

Figure 1: General outline of COPR zones at SA7 and DMT

COPR HEAVING MECHANISMS

Figure 1 shows the outline of COPR stratigraphy at the two sites. It should be noted that this outline is only a representative depiction, as COPR stratigraphy is not uniform across the sites. In summary, the properties of the two COPR types that were found relevant to heave were:

a) GB: grey-black granular COPR resembling a sandy soil. This type of COPR was encountered as the top massive layer (denoted as zone B in Figure 1) at SA7. Zone B was subdivided into two zones, B1 above and B2 below the groundwater table. Distinct horizontal GB layers of 0.5 m thickness were also identified in all three DMT trenches. GB2 COPR was found underneath the HB2 layer at DMT.

b) HB: orange to reddish brown, very fine and highly lithified/cemented material. This type of COPR was identified at isolated locations as of thin layer interbeds at SA7 that seemed to be associated with the phreatic surface (and is thus not shown in Figure 1) and as distinct horizontal layers at DMT. It should be noted that subsequent work to ours has uncovered the presence of substantial HB masses.

The main mineralogical transformations occurring in COPR are hydration reactions of the parent minerals (brownmillerite and periclase), and carbonation. A list of the common crystalline COPR phases identified at both sites is shown in Table 1. Mineralogically-induced volume changes can be described in terms of changes to specific gravity (Gs). As evidenced by their Gs values (Table 1), both brownmillerite and periclase are compact oxide structures of high density; when they hydrate, they produce phases of considerably lower density; this implies that hydration reactions lead to an overall volume increase, the extent of which depends on the ΔGs and the amount of the newly formed phases. Also, kinetics of mineral formation affect the rate of volume expansion and thus the rate of heaving phenomena development. Considering the relatively low amount of periclase in COPR (Table 1), the literature-reported slow rate of hydration and the difference in porosity between cements and COPR, and also based on laboratory slurry and swell tests, this mechanism was ruled out (Moon et al. 2006a; Moon et al. 2006b). Based on Gs, CAC and hydrotalcite monophase formation can be accompanied by significant volume increase (Table 1); however, their platy morphology is generally not conducive to expansion in high porosity systems as their crystals can be accommodated in the existing voids. For this reason, although monophase formation is a very common reaction in cements (Hewlett 1998), it has not been considered to cause expansion. Carbonation reactions, on the other hand, lead to increase of mass due to CO_2 uptake, but calcite density is higher compared to portlandite and most hydration products potentially resulting in a decrease in volume. Based on previous literature, mineralogical considerations and laboratory tests, only two of the considered heaving mechanisms seem viable: (a) Brownmillerite hydration to form hydrogarnets and (b) Ettringite formation.

Table 1: Common crystalline phases identified in most SA7 and DMT COPR samples

COPR Mineral name	Gs[1]	Chemical formula	SA7-GB[2]	DMT-GB[2]	DMT-HB[2]	DMT-HB2[2]
Brownmillerite	3.76	Ca_2FeAlO_5	29	59	16	2
Periclase	3.78	MgO	2	4	2	0
Portlandite	2.23	$Ca(OH)_2$	4	2	0	1
Brucite	2.39	$Mg(OH)_2$	8	4	3	8
Calcite	2.71	$CaCO_3$	16	5	11	7
CAC	1.33	$Ca_4Al_2O_6(CrO_4)\cdot xH_2O$	7	4	7	33
Ettringite	1.80	$Ca_6Al_2(SO_4)_3(OH)_{12}\cdot 26H_2O$	6	1	12	0
Hydrogarnets	2.96	$Ca_3Al_2(OH)_{12}$	12	15	14	31
Hydrotalcites	2.13	$Al_xMg_yFe_z(OH)_n(CO_3)(H_2O)_m$	7	6	10	18
Afwillite	2.62	$Ca_3(HSiO_4)_2(H_2O)_2$	5	0	0	0
Soil minerals	2.67	-	4	0	25	0

[1]Source: http://www.webmineral.com [2]Rietveld-based average as percent of dry weight of crystalline phases (Chrysochoou, 2006) [3]CAC: Calcium Aluminum Chromium Oxide Hydrate

Brownmillerite is encountered in larger quantities in oil/well cements and sulfate-resisting Portland cements. Although its hydration behavior has been well studied, there is no mention of expansive behavior of ferrite-rich cements. The lack of volume expansion phenomena in ferrite-rich cements can be attributed to the reported brownmillerite hydration rate being extremely slow (Chrysochoou 2006). This apparent kinetic inhibition was traced to the early formation of pozzolanic product and ferric oxihydroxide passivation layers around COPR brownmillerite nodules (Chrysochoou et al. 2005). Interestingly, as seen in Table 1, the qualitative mineralogy for GB versus HB is quite similar, but quantitatively it is at the opposite ends of the brownmillerite hydration reaction pathway; DMT-GB resembles almost unreacted COPR whereas HB2-DMT resembles almost fully hydrated COPR, with SA7-GB being somewhere between the COPR reaction pathway and DMT-HB being a carbonated version of HB2. Moreover, the fact that HB layers are found as interbedded with GB layers, predominately at DMT but also at SA7 and Glasgow, UK (Manzur et al. 2001), clearly points to a post-deposition, in-situ transformation of GB to HB. As already mentioned, based on Gs, GB to HB (brownmillerite to hydrogarnet) transformation may be a viable heaving mechanism, provided that the rate of transformation is somehow dramatically increased over the currently extremely slow brownmillerite hydration rate. The question is then what does it take for the brownmillerite rate to substantially increase so that rapid HB formation occurs and associated heave develops?

Sulfate attack or Delayed Ettringite Formation (DEF), which is the prevalent term for ettringite-induced concrete expansion, is considered one of the major mechanisms of concrete deterioration (Taylor et al. 2001). Ettringite has also been reported in the literature as the cause of swell in lime-stabilized clays (Mitchell 1984; Dermatas 1995). Mineralogical analyses of the SA-7 soil boring samples showed that ettringite presence at the site was significant; ettringite was present in 78 out of 120 XRPD patterns (65%), its concentration (without considering amorphous phases) ranging from 1% to 33% (Dermatas et al. 2005a; Chrysochoou 2006). Despite those findings, as well as laboratory swell tests (Moon et al. 2006a) that showed ettringite formation as the most viable heave mechanism, ettringite presence could not be correlated to the observed heave features at SA-7. Conversely, ettringite was indeed correlated with the observed heave features at DMT (Chrysochoou 2006; Dermatas et al. 2005a), without though ruling out the brownmilerite to hydrogarnet (GB to HB) mechanism. Moreover, during field pilot testing of COPR treatment with ferrous sulfate, ettringite formation was clearly identified as the observed heave mechanism, being responsible for dramatic vertical heave (up to 3 feet) (Dermatas et al. 2006).

The fact that ettringite formation does not seem to account for the reported heave at SA-7, coupled with the fact that during more recent investigations at the SA-7 site, a massive layer of HB-like COPR was linked to a major heave feature at the site, point again to the viability of the GB-to-HB transformation serving as a heave mechanism. However, one cannot easily account for the field conditions that would make this possible under realistic time frames, commensurate with observed heave development at either SA7 or DMT. At DMT for instance, where massive layers (and not thin interbeds like at SA-7) of HB COPR were encountered, the GB to HB transformation seems to have taken place in-situ sometime in the past, and it is unclear whether it is still the operative mechanism. The lack of well documented historical data would rule out making any reliable assessments to this end. The documented GB-to-HB transformation time frame is extremely slow at ambient temperature, as evidenced both by the mineralogy of the DMT GB COPR material, the mineralogy of COPR deposited at SA7 and laboratory testing (Moon et al. 2006a; Moon 2006b). Thus it was hypothesized that the only viable mechanism for a preferred acceleration of the hydration reactions in the HB material could have been its exposure to a local high temperature environment (Chrysochoou et al. 2005). How or why such hydrothermal conditions occurred is not clear; one possible explanation would be that the hydration of brownmillerite and residual quicklime (CaO), both strong exothermic reactions (Hewlett 1998), lead to a significant increase in the interior of deposited COPR near the phreatic surface.

To elucidate the GB-to-HB transformation, slurry tests (Moon et al. 2006b) and high temperature experiments (Sanchez 2006) were conducted. Slurry tests using both ultrafine (10 μm) fresh synthetic brownmillerite and actual GB COPR showed that the transformation from brownmillerite to hydrogarnet was very fast (less than 1 hour) and was complete within 7 days when using synthetic brownmillerite (Moon et al. 2006b). CAC formation was also thermodynamically favored upon chromate influx, as it was observed in the XRPD pattern within one hour of reaction. However, hydrogarnet and CAC formation were not observed at all with actual COPR; no major phase transformation was observed in these samples (Moon et al. 2006b). This confirmed that the GB-to-HB transformation appears to be kinetically inhibited. To overcome this kinetic inhibition at ambient temperatures, COPR grinding to a much finer gradation was thought to be necessary. During high temperature (200°C) testing, brownmillerite dissolution was clearly observed in all GB COPR samples tested following 24 hours; and for most of the samples, the mineral dissolved completely after 14 days (Sanchez 2006). This confirmed the viability of the hydrothermal hypothesis (Chrysochoou et al. 2005). The latter was also confirmed in the field during a pilot COPR study, where brownmillerite dissolution of up to 80% was observed in a treatment-induced high temperature environment (Dermatas et al. 2006). Consequently, both lab and field observations clearly point to high temperature as overcoming the apparent kinetic inhibition of GB hydration. However, more research is needed in order to isolate the mechanism of the kinetic inhibition in GB COPR hydration and to address the conditions under which it can be effectively overcome.

Moreover, the fact that GB to HB transformation appears to occur at or near the phreatic surface needs some explanation. A working hypothesis is that the heat of hydration of "fresh", largely unreacted COPR in the unsaturated zone can lead to a temperature increase which, in-turn, would effectively accelerate hydration and further temperature build-up, until brownmillerite is mostly consumed and no more "fresh" surface is available for further reaction. Conversely, hydration of COPR in the saturated zone, where heat can be effectively dissipated, precludes internal heat build up and thus brownmillerite hydration ceases once the formation of a passivation layer/matrix of pozzolanic products and/or ferric oxyhydroxides takes place around the COPR nodules. Clearly, again, more research is needed in order to establish the validity of the aforementioned hypothesis.

In summary, research to date has suggested that ettringite-induced heave is a viable COPR heaving mechanism. Another viable COPR heave mechanism appears to be possibly linked to GB-to-HB transformation, albeit the underlying mechanisms remain obscured. Complete brownmillerite hydration has to occur relatively fast for volume expansion to take place (i.e., from hours to days to less than a year or so based on the observations from the SA-7 pilot study and DMT heave development observations, respectively), otherwise slow expansion can be easily accommodated within the high void ratio of GB COPR. Aside from the observed heave, there seems to be no conclusive evidence that GB-to-HB transformation is an on-going reaction in otherwise undisturbed COPR deposits. DMT HB layers seemed to have majorly formed long ago (it may be, to a certain extend, that HB layers continue to form today but there is no conclusive evidence for this either), and brownmillerite hydration (in the lab) could not be reproduced unless temperature was substantially increased or COPR particle size substantially reduced. Similarly, in the field, brownmillerite was effectively hydrated within a month under a substantially increased temperature regime. It follows that the exact conditions that may accelerate GB-to-HB transformations in the field would have to be further investigated and ultimately defined, as this may better explain the ongoing heave at DMT. On the other hand, ettringite formation most clearly constituted the main heave culprit during COPR pilot scale treatment with ferrous sulfate. It also effectively contributed to the heave observed at the DMT trenches where the relative role of GB to HB volumetric contribution, if any, still remains unclear. Ettringite formation is clearly thermodynamically favored in the presence of sulfates, especially in HB COPR and should thus be carefully considered during COPR treatment applications (Dermatas et al. 2006).

ACKNOWLEDGMENT

The authors wish to thank Honeywell International Inc. for the financial support.

REFERENCES

Cao, J. and Zhang, W-X. (2006). "Stabilization of COPR with nanoscale iron particles." *Journal of Hazardous Materials*, B132: 213-219.

Chrysochoou, M.; Moon, D. H.; Dermatas, D.; Wazne, M., Christodoulatos, C.; Meng, X.; Kaouris, M.; Morris, J.; French, C. and Sass, B. (2005), "Mineralogical Analysis of Chromite Ore Processing Residue (COPR) by X-Ray Powder Diffraction (XRPD)", *Proc. of the 8th Battelle In-situ and On-site Bioremediation International Symposium*, Baltimore, Maryland, June 6-9, 2005.

Chrysochoou, M. (2006), "Investigation of the Environmental and Geotechnical Implications Associated with Deposition of COPR", *Ph.D. Thesis*, Stevens Institute of Technology, May 2006.

Chrysochoou, M. and Dermatas, D. (2007). "Application of the Rietveld Method to Assess Chromium (VI) Speciation in Chromite Ore Processing Residue", *Journal of Hazardous Materials*, Elsevier, Vol. 141, No. 2, pp 370–377, March 15, 2007.

Dermatas, D. (1995). "Ettringite-induced heaving in soils: State-of-the-art", *Appl. Mech. Rev.*, 38(10), 659-673.

Dermatas, D.; Chrysochoou, M.; Moon, D. H.; Pardali, S.; Christodoulatos, C.; Lazarte, C. A.; Pendleton, C. H.; Bonaparte, R.; Briggs, R.; Myers, M.; French, C.; Morris, J. and Kaouris, M. (2005a). "Mineralogical Characterization of Chromite Ore Processing Residue (COPR) at Dundalk Marine Terminal Area 1800", *Proc. of the 8th Battelle In-situ and On-site Bioremediation International Symposium*, Baltimore, Maryland, June 6-9, 2005.

Dermatas, D.; Bonaparte, R.; Chrysochoou, M. and Moon, D.H. (2005b). "Chromite Ore Processing Residue (COPR): Hazardous contaminated soil or solid waste?", *Journal of ASTM International*, Vol. 2, No.10, November/December 2005.

Dermatas, D.; Chrysochoou, M.; Moon, D.H.; Grubb, D.G.; Wazne, M. and Christodoulatos, C. (2006). "Ettringite-Induced Heave in Chromite Ore Processing

Residue (COPR) upon Ferrous Sulfate Treatment", *Environ. Sci. Technol.*, 40, Issue 18, 5786-5792.

Farmer, J.G.; Paterson, E.; Bewley, R.J.F.; Geelhoed, J.S.; Hillier, S.; Meeussen, J.C.L.; Lumsdon, D.G.; Thomas, R.P. and Graham, M.C. (2006). "The implications of integrated assessment and modelling studies for the future remediation of chromite ore processing residue disposal sites", *The Science of the Total Environment*, 360, 90– 97.

Graham, M.C.; Farmer, J.G.; Anderson, P.; Paterson, E.; Hillier, S.; Lumsdon, D.G. and Bewley R.J.F. (2006). "Calcium polysulfide remediation of hexavalent chromium contamination from chromite ore processing residue", *The Science of the Total Environment*, 364(1-3), 32-44.

Hewlett, P.C., (Ed.) (1998). "Lea's chemistry of cement and concrete", Elsevier Butterworth Heinemann, Oxford, UK.

Hillier, S.; Lumsdon, D . G .; Brydson, R . and Paterson, E . (2007). "Hydrogarnet: A Host Phase for Cr(VI) in Chromite Ore Processing Residue (COPR) and Other High pH Wastes", *Environ. Sci. Technol.*, 41, 1921-1927.

Manzur, L.C.G.; Thomas, R.P.; Geelhoed, J.S.; Farmer, J.G. and Campbell, C.D. (2001). "Variability in chemical properties of a COPR disposal site and assessment of potential toxicity using a biosensor", *Environmental Geochemistry and Health*, 23, 213-217.

Meegoda, J.N.; Partymiller, K.; Tichards, M.K.; Kamopornvijit, W.; Librizzi W.; Tate, T.; Noval, B.A.; Mueller, R.T. and Santora, S., (2000). "Remediation of chromium-contaminated soils: pilot-scale investigation", *Practice Periodical of Hazardous, Toxic and Radioactive Waste Management*, 4(1), 7-15.

Mitchell, J.K. (1984). "Practical problems from surprising soil behavior", *J. of Geotech. Eng.*, 112(3), 250-289.

Moon, D. H.; Dermatas, D.; Chrysochoou, M. and Shen, G. (2006a). "An Investigation of the Heaving Mechanism Related to Chromite Ore Processing Residue", *Journal of ASTM International*, ASTM, Vol. 3, No. 6.

Moon, D.H.; Dermatas, D.; Chrysochoou, M. and Grubb, D.G. (2006b). "An Investigation of the Phase Transformation of Brownmillerite in Chromite Ore Processing Residue", *Proc. of the International Conference: Protection and Restoration of the Environment VIII*, Chania, Greece, July 2006.

Raghu, Dorairaja and Hsieh, H.-N. (1989). "Performance of some structures constructed on chromium ore fills", *J. Perform. Construct. Facil.*, 3(2), 113-120.

Sanchez A. M. (2006), "Brownmillerite Hydration in COPR at Elevated Temperatures", *MS Thesis*, Stevens Institute of Technology, Hoboken, NJ, May 2006.

Taylor, H.F.W.; Famy, C. and Scrivener, K.L. (2001), "Review Delayed Ettringite Formation", *Cement and Concrete Research*, 31, 683-693.

Diagenesis of Buried Chrome Ore Processing Residue

Bruce M. Sass[1], Daniel T. Kremser[2], Mohit Bhargava[3], Jody Lipps[4]

[1]Research Leader, Battelle, 505 King Ave., Columbus, OH; sassb@battelle.org
[2]Principal Research Scientist, Battelle, 505 King Ave., Columbus, OH; kremserd@battelle.org
[3]Researcher, Battelle, 505 King Ave., Columbus, OH; bhargavam@battelle.org
[4]Researcher, Battelle, 505 King Ave., Columbus, OH; lippsj@battelle.org

ABSTRACT: The purpose of this study was to investigate mineralogical changes that have taken place in buried COPR due to hydration of mineral phases that were produced during ore processing. Along with the high-temperature phases, hydration products have been identified and quantified using x-ray powder diffraction (XRD) with whole pattern (Rietveld) fitting. Detailed chemical compositions of all phases were determined using scanning electron microscopy (SEM) and electron probe microanalysis (EPMA) of particulate (3D imaging) and polished sections (2D imaging). Laboratory experiments were conducted using a synthetic brownmillerite to determine reaction pathways and rates of mineral transformation. When the experiments were run for 200 days and longer with calcium chromate, lithified pellets were observed. Electron micrographs showed that the pellets are composed of hydrogarnet with residual brownmillerite fragments, which were cemented by calcium monochromate. These results prove that when chromate availability is high, calcium monochromate is the favored reaction product.

INTRODUCTION

Chromite Ore Processing Residue (COPR) was produced in Hudson County, New Jersey by milling and roasting operations that involved heating chromite ore with soda ash and unslaked lime. This process converted trivalent chromium in the ore to hexavalent sodium chromate that was separated from the residue by water leaching. However, residual hexavalent chromium remained in the residue at levels of 0.5%, on average, based on numerous chemical analyses. Due to its granular form, the residue was frequently used as fill material, which later resulted in groundwater contamination by semi-soluble chromate.

Over the decades that the COPR has been buried and capped with asphalt, mineralogical changes have taken place due to hydration of mineral phases that were produced during ore processing. Irregular heave patterns have been noted at sites where COPR has been buried for several decades, resulting in pavement, foundation, and utility damage. Heaving is thought to be due to expansion at the grain scale.

RESEARCH OBJECTIVE

This study is an investigation into the mineralogical changes that have taken place in buried COPR due to hydration of mineral phases that were formed in the high-temperature kiln process. Extensive characterization of COPR mineralogy and chemistry has been conducted and reported elsewhere (Kremser et al., 2005; Chrysochoou and Dermatas, 2007). A key finding among all of the field investigations to date is that two morphologically and mineralogically distinct forms of COPR exist: one is loose, granular, and gray-black in color; the other is consolidated, weakly-to-strongly lithified, and dark brown or brick-red in color (Figure 1). We refer to the former as GB (gray-black) and the latter as HB (hard brown). The major element chemistry (Ca, Al, Fe, and Mg) of both types is very similar, on average, suggesting that the chromite ore and the chemical process that produced these products were the same. However, mineralogical transformations that have taken place since the COPR was deposited differ considerably in terms of reaction progress. That is, both contain the same minerals, but the extent of reaction is much greater in the HB than in GB. For example, the GB typically contains more than 60% brownmillerite, which is a major product of the kiln process. The remainder is composed of a small amount of periclase (MgO, another high temperature phase), plus hydration products that include hydrogarnet, Al_2O_3-Fe_2O_3-mono (AFm) phases, brucite, calcite (as a result of carbonation), and <1% chromite. Conversely, the HB COPR typically contains less than 10% brownmillerite. The majority of compounds in HB are the hydration products that are identified in GB, with an abundance of chromate- and carbonate-substituted AFm phases and the mineral hydrotalcite.

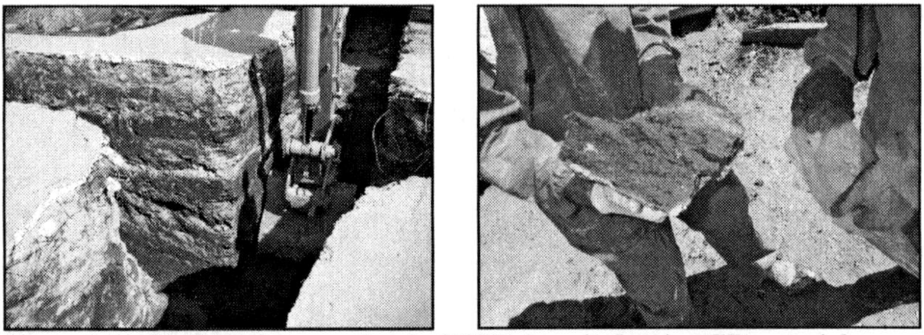

Figure 1. (Left) Trench excavation in COPR deposit showing GB material resting above HB; (right) sample of HB with yellow chromate-rich layer at boundary with upper GB.

A chemical distinction between the two types of COPR can be based on the concentration of divalent anionic species consisting of chromate, carbonate, and sulfate. Among this group of anions, lithification is most consistently associated with elevated chromate, which typically is above 1%, and as much as 5% or more, in HB, whereas Cr(VI) tends to be less than 0.5% in GB (Figure 2, left). In addition, total inorganic carbon (Figure 2, right) and sulfate (not shown) are higher in HB than in GB.

It is hypothesized that the fundamental mechanism responsible for lithification of COPR (i.e., conversion of GB into HB) is the reaction between divalent anions and

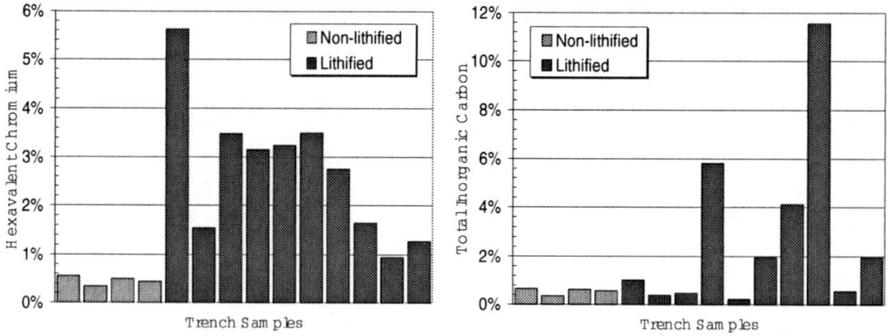

Figure 2. Typical chromate and carbonate concentrations in GB and HB COPR

AFm compounds that are produced by hydration of brownmillerite. The following equation describes brownmillerite hydration:

$$C_4AF + 22H \rightarrow C_4AH_{19} + FH_3 \quad (1)$$
(Brownmillerite + Water → AFm + Ferric Hydroxide)

where cement chemistry nomenclature has been used (C=CaO; A=Al_2O_3; F=Al_2O_3; H=H_2O). Eq. 1 is an irreversible hydration reaction (at ambient temperature) which states that under proper conditions brownmillerite becomes converted to one mole of the AFm compound C_4AH_{19} (alternatively written $C_3A{\cdot}Ca(OH)_2{\cdot}H_{18}$) and ferric hydroxide $Fe(OH)_3$. In Eq. 1, the form of the AFm compound was assumed so that the equation could be written with definite stoichiometry. Because the hydration product is dependent on a number of factors, a more generic description is calcium-aluminum-hydrate, or C-A-H. Based on thermodynamic calculations, the stable form of C-A-H is hydrogarnet, C_3AH_6. However, formation of hydrogarnet in brownmillerite pastes can take several days or longer at ambient temperature (Meller et al., 2004). In comparison, hydration of C_3A leads to C_3AH_6 in only minutes. It has been proposed that metastable C-A-H passivates the brownmillerite surface and limits further dissolution and hydration (Meller et al., 2004). Possible mechanisms for increasing brownmillerite reactivity include increasing temperature and incorporating other species that react with C-A-H.

Reaction of the AFm phase with divalent anions has been most studied in the case of sulfate (Taylor, 1990) where the product of brownmillerite reacting with gypsum ($C\underline{S}H_2$) is the trisulfoaluminate phase (AFt) ettringite:

$$C_4AF + 3C\underline{S}H_2 + 30H \rightarrow C_6A\underline{S}_3H_{32} + CH + FH_3 \quad (2)$$
(Brownmillerite + 3Gypsum + Water → Ettringite + Portlandite + Ferric Hydroxide)

If the availability of sulfate ($\underline{S}$=SO_3) is low, ettringite may decompose to an AFm monosulfoaluminate hydrate. The overall reaction for sulfate converting brownmillerite to a stable AFm can be written as

$$C_4AF + C\underline{S}H_2 + (n+2)H \rightarrow C_4A\underline{S}H_n + CH + FH_3 \quad (3)$$

(Brownmillerite + Gypsum + Water → AFm-sulfate + Portlandite + Ferric Hydroxide)

(Meller et al., 2004), where n is variable, but typically the number is 10, 12, 14, or 16, depending on temperature and relative humidity conditions.

Based on information about the COPR chemical system, it is obvious that some factor other than sulfate is facilitating brownmillerite reaction in the HB material. Considering that chromate may be providing the impetus for reactions leading to the almost complete conversion of brownmillerite and lithification of COPR, we explore some possible mechanisms using published information about sulfate as a basis.

Because there is no evidence for a chromate AFt phase (i.e., chromate-ettringite) in any COPR deposits that we are aware exist, it is assumed that either chromate-ettringite formed at an early stage and then converted to AFm monochromate hydrate, or that the AFm monochromate hydrate formed directly from hydration of brownmillerite.

$$C_4AF + CaCrO_4 + (n+4)H \rightarrow C_3A[CaCrO_4]H_n + CH + FH_3 \quad (4)$$

(Brownmillerite + Calcium Chromate + Water → AFm-chromate + Portlandite + Ferric Hydroxide)

where n can be 9, 12, or 14. Rates for synthesizing calcium monochromates and x-ray diffraction data have been published by Göske et al. (2004). In Eq. 4, calcium chromate was chosen to be the source of chromate because is has moderate solubility, and likely was an initial precipitate in COPR after soluble chromate was removed by leaching.

Reactions similar to Eq. 4 could be written to illustrate the potential for carbonate to react with brownmillerite. However, the chemical determination of carbonate is ambiguous because it can be distributed among two AFm compounds $C_4A\underline{C}_{0.5}H_{12}$ and $C_4A\underline{C}H_{11}$ ($\underline{C}$=CO_2), and calcite ($CaCO_3$). Typically, $C_4A\underline{C}H_{11}$ and calcite are present in HB COPR samples, based on x-ray diffraction. Thus, the presence of carbonate doesn't imply that all of it exists in the COPR hydration product. Conversely, chromate is almost exclusively contained in calcium monochromate, based on mineralogical analyses by SEM and electron microprobe. In addition, hydrogarnets and other phases are thought to be additional sinks for hexavalent chromium (Hillier et al., 2003). Compared to calcite, $CaCrO_4$ is too soluble to remain a solid phase in a water-rich environment.

EXPERIMENTAL APPROACH

Experiments were conducted using pure synthetic brownmillerite (Ca_2AlFeO_5) to determine reaction pathways and rates of mineral transformation under different sets of conditions. In contrast, brownmillerite in COPR ranges from the ideal composition (same as pure synthetic form) to one where trivalent iron and trivalent chromium partially substitute for aluminum. Calcium chromate and hydrated lime (portlandite) were added to a system containing synthetic brownmillerite which was previously reacted with water to form AFm. Chromate was added in either the exact stoichiometric requirement to convert all of the brownmillerite to calcium monochromate, according to Eq. 4, or exactly half the stoichiometric requirement.

All samples were analyzed on a Rigaku wide angle powder x-ray diffractometer (XRD) running Materials Data, Inc. (MDI) automation and software (JADE 7+). The radiation source was a copper, normal focus x-ray tube operated at 40 kV and 30 mA. Several grams of the dried material were ground with an agate mortar and pestle to pass through a 400-mesh sieve (< 38 μm). A portion of this material was carefully packed into a standard 0.5 mm recessed glass holder and scanned in the diffractometer from 5° to 70° 2θ, using a step size of 0.024 and a speed of 0.24°/minute. An external standard (NBS 640b Silicon powder) was run routinely to ensure that the instrument's mechanical alignment met the manufacturer's specifications.

Scanning electron microscopy (SEM) was used to view and map polished sections (2-dimensional imaging) and to acquire 3-dimensional images of intact samples to inspect their morphology and probe individual minerals. In addition, energy dispersive spectroscopy (EDS) on the SEM was engaged to obtain reconnaissance chemistry on these phases. Additional details about the XRD instrument and electron microscope are given in Kremser et al. (2005).

RESULTS

Solid phase products of the experiments were examined for each of the three conditions that were tested. The results of initial 14-day experiments are summarized as follows:

Experiments with Brownmillerite and Pure Water

Less than 2% brownmillerite remained after 14 days of testing. With the exception of calcium chromate which formed as a result of oversaturated conditions, all of the remaining brownmillerite converted to AFm phases, which included C_4AH_{19} and $C_4A\underline{C}H_{11}$ (the carbonate phase exists because atmospheric CO_2 was not controlled in this set of experiments).

Experiments with Brownmillerite, Water, and Half-Stoichiometric Chromate

As with the water-only test, less than 2% brownmillerite remained after 14 days of testing with half the stoichiometric requirement for chromate. Approximately 20% of the crystalline mass consisted of calcium chromate. The remainder included: C_4AH_{19} (17%); calcium monochromate (22%); a chromate equivalent of kuzelite (15%); $C_4A\underline{C}H_{11}$ (19%); and excess portlandite (7%). Figure 3 shows the texture of the brownmillerite before use and after 200 days of experiments.

Experiments with Brownmillerite, Water, and Full-Stoichiometric Chromate

Also in this test, less than 2% brownmillerite remained after 14 days of testing with the full stoichiometric requirement for chromate. However, there are several differences between these results and those for the test with half-stoichiometric chromate: only 4% of the solid consisted of calcium chromate and only 9% C_4AH_{19} had formed; calcium monochromate was very high (66%), while the chromate equivalent of kuzelite was low (5%); excess portlandite was again 7% and calcite (8%) was formed only in this test. No carbonated AFm phases were detected. These results indicate that when chromate availability is high, calcium monochromate is the favored reaction product.

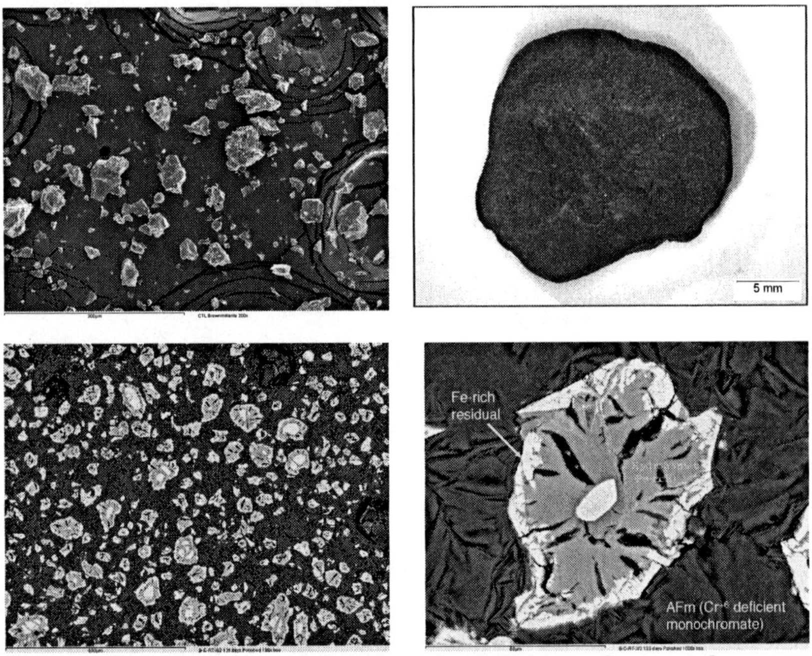

Figure 3. Clockwise from top left: 3-dimensional SE image of unreacted synthetic brownmillerite on carbon tape at low magnification (200X); light photograph of lithified pellet after 200 days in half-stoichiometric calcium chromate solution; BE image of pellet in cross-section at high magnification (1000X); BE image of pellet in cross-section at low magnification (100X).

CONCLUSIONS

COPR diagenesis involves a number of complex mineral changes that take place over a long time due to slow hydration of brownmillerite and periclase. The mineralogy of COPR is well understood, in the sense that all minerals and amorphous substances have been identified. The experimental study was conducted to test the hypothesis that chromate greatly increases the rate of brownmillerite hydration and subsequent reaction with the anion that 'activates' the brownmillerite phase. When a large amount of chromate was available to the experimental system, the transformation of AFm to calcium monochromate was very high after only 14 days. In fact, this reaction seemed to inhibit formation of AFm carbonates, which were able to form when chromate levels were half the stoichiometric amount or in pure water systems.

When the experiments were run for 200 days and longer, lithified pellets were observed in half- and full-stoichiometric calcium chromate solution. Electron micrographs showed that the pellets were composed of residual brownmillerite fragments that are cemented by calcium monochromate (Figure 3). The interior of most fragments contains residual brownmillerite (with no change in composition) and an iron-rich hydrogarnet. The exterior portion of the grains contains an amorphous iron-rich residual compound (see compositions of individual phases in Figure 4). These results show that aluminum is highly mobile in the COPR system. In addition, the AFm compound inside the mineral grains that was formed by reaction of brownmillerite with water was converted to hydrogarnet after chromate and hydrated lime were added to the system. Apparently, the elevated pH caused AFm inside the mineral grains to convert to hydrogarnet, while chromate caused AFm outside the mineral grains to become converted to calcium monochromate.

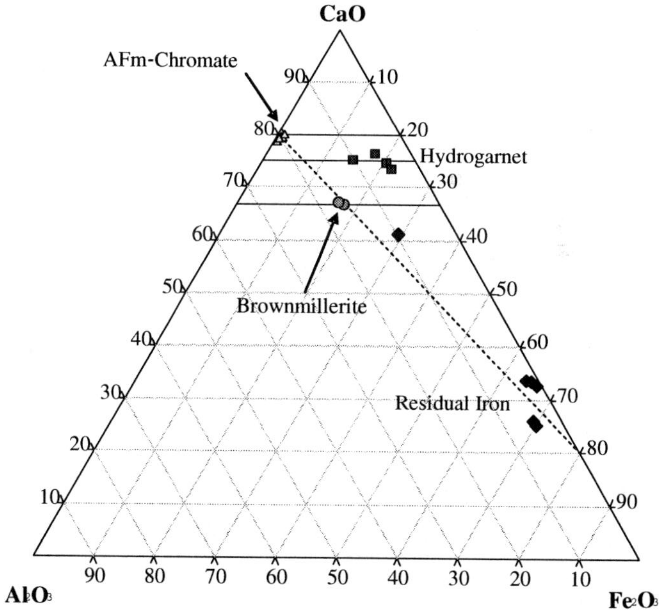

Figure 4. Figure 4. Ternary diagram showing compositions of individual phases based on SEM/EDS analysis

ACKNOWLEDGMENTS

This work was sponsored by Honeywell International, Inc. The authors would like to thank John Morris, Chris French, and Maria Kaouris of Honeywell and Ted Toskos of MACTEC for their stewardship and guidance throughout this project.

REFERENCES

Chrysochoou, M. and Dermatas, D. (2007). "Application of the Rietveld method to assess chromium(VI) speciation in chromite ore processing residue." *J. Hazard. Mater.* Vol. 141: 370-377.

Göske, J., U. König, and H. Pöllmann. (2004). "Kinetical Investigations of Lamellar Calcium AluminateChromateHydrates with High Speed Detector." *Mater. Sci. Forum*. Vol. 443-444: 299-302.

Hillier, S., M.J. Roe, J.S. Geelhoed, A.R. Fraser, J.G. Farmer, and E. Patterson. (2003). "Role of quantitative mineralogical analysis in the investigation of sites contaminated by chromite ore processing residue." *Science of the Total Environment.* Vol. 308: 195-210.

Kremser, D.T, B.M. Sass, G.I. Clark, M. Bhargava, and C. French. (2005). "Environmental Forensics Investigation of Buried Chromite Ore Processing Residue." Proceedings of the *Eighth International In Situ and On-Site Bioremediation Symposium,* Baltimore, MD, June 5-8.

Meller, N., C. Hall, A.C. Jupe, S.L. Colston, S.D.K. Jacques, P. Barnes, and J. Phipps. (2004). "The paste hydration of brownmillerite with and without gypsum: a time resolved synchrotron diffraction study at 30, 50, 70, and 150°C." *J. Mater. Chem.* Vol. 14: 428-435

Taylor, H.F.W. (1990). *Cement Chemistry*. Academic Press.

Field Investigation Techniques for Characterization and Delineation of COPR

James Tinjum[1], M. ASCE, P.E., Christopher Houck[2], P.G., and Chris French[3], P.G.

[1]Project Geotechnical Engineer, CH2M HILL, 1717 Arch Street, Suite 4400, Philadelphia, PA 19103; jtinjum@ch2m.com
[2]Staff Geologist, CH2M HILL, 15010 Conference Center Drive, Suite 200, Chantilly, VA 20151; christopher.houck@ch2m.com
[3]Remediation Manager, Honeywell International Inc., 101 Columbia Road, Morristown, NJ 07962-1139; Chris.French@Honeywell.com

ABSTRACT: A field investigation was conducted at a chromium ore processing residue (COPR) fill site in the Baltimore area. The purpose of the field investigation was to delineate the vertical and horizontal boundaries of the COPR fill area in and assess the unique variations in the physical properties within COPR. Two types of COPR were identified at the site: a weakly consolidated, granular, gray-black (GB) material; and a moderately to strongly lithified/weathered, hard-brown (HB) material. Multiple field techniques were used to characterize and sample the COPR at the site, including down-hole geophysics; cone penetrometer testing (CPT) with resistivity and piezo cones; and direct drilling with split-spoon, direct push macro cores, and Pitcher tube–sampling techniques. Correlations were developed between the outputs from the CPT, direct drilling, and geophysical approaches to characterize the COPR stratigraphy over a large area.

INTRODUCTION

At many COPR sites in the Mid-Atlantic region of the U.S., COPR consists of either unweathered granular or weathered lithified material (Dermatas et al., 2007). Although both types of COPR are derived from the same source material, unique differences exist in their physical properties. Over time, the parent minerals in COPR hydrate owing to interactions with surface water infiltration, fluctuations in groundwater levels, and other interactions in the vadose zone (Dermatas et al., 2005; Sass et al., 2005). Some COPR deposits have undergone changes in their mineral matter, resulting in the material becoming rocklike in hardness by a process similar to lithification. Since the two forms of COPR have different physical properties (such as tip resistance, bulk density, and textural/visual features), different investigation techniques and field tests may be used to differentiate COPR from fill and native soil.

The Dundalk Marine Terminal (DMT) is located in Baltimore, MD, and is an active marine shipping terminal (Figure 1). The southern portion of the terminal was constructed on lands reclaimed from prior marshlands and the Patapsco River by means of fill placement. The fill material at DMT includes mixtures of COPR, sand dredging, and other locally available fill materials such as clay and borrow sand.

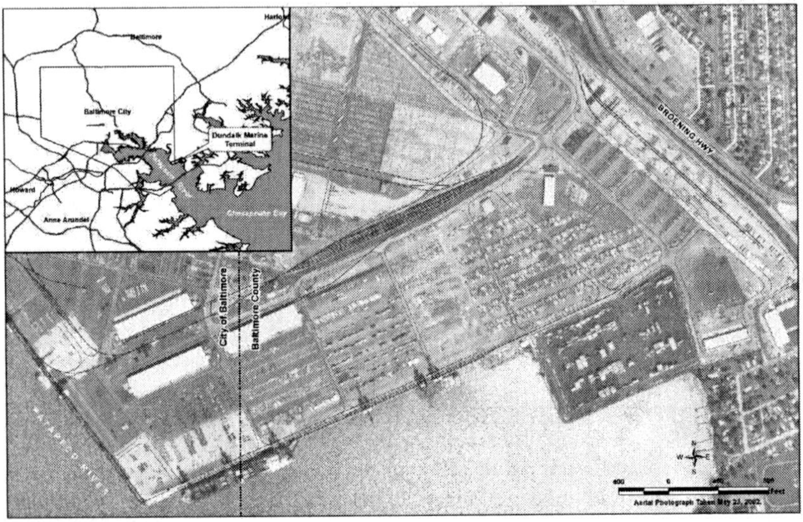

FIG. 1. Site location and map.

CPT-SOIL BORING CORRELATIONS

CPT soil classifications are empirically based and generally rely on tip resistance and friction ratio measurements collected from naturally occurring subsurface materials (Robertson, 1989). As such, there is no CPT classification scheme readily available for COPR. Therefore, to address the extent and stratification of COPR using CPT data, the CPT signatures for the various subsurface materials that are encountered at the DMT were interpreted by correlating CPT measurements with soil boring data at paired borehole/CPT locations.

CPT measurements were correlated with lithologic data from 36 boring pairs installed during this investigation. The correlations were performed by comparing CPT measurements with lithologic information provided in core and split-spoon samples from hollow-stem auger (HSA) boring logs. The correlations provide the means to interpret COPR and non-COPR stratigraphy solely from CPT data.

Correlation Trends

The primary goal of the CPT correlations was to define a characteristic signature for the soil types that are encountered at the DMT, which can be classified into three general soil types. Any soil designated as "fill" in this classification scheme refers to fill materials that do not include COPR. The fill is further subdivided into Types I, II, and

III. COPR is subdivided into HB COPR and GB COPR. The underlying alluvial sediments include stratified silt, clay, and sand deposits.

Fill Material

Type I fill includes dense, gravel base fill that is present below asphalt or concrete pavement to depths of up to 1 m beneath the site. The CPT characteristics of Type I fill include tip resistance values that exceed 50 MPa and resistivity values that exceed 50 ohm-m. The relatively high tip resistance values tend to decrease abruptly in subsurface materials that underlie Type I fill.

Type II fill is encountered across much of the site and is typically classified as homogeneous, poorly-graded sand. Type II fill is non-COPR that is characterized by tip resistance values that range between 10 and 40 MPa at shallow depths. The high tip resistance values in this material tend to decrease steadily to a minimum range of 5–10 MPa at deeper depth. Type II fill is relatively homogeneous, so variations in tip resistance occur over relatively long depth intervals, which creates a smooth pattern in the tip resistance data. Resistivity values in Type II fill generally exceed 40 ohm-m and tend to slowly and steadily decrease with depth.

Type III fill is composed predominately of stratified fine-grained soils, including silty sand, clayey sand, silt, or clay. Consequently, tip resistance values are generally lower (< 5 MPa) and tend to be more variable than those of Type II fill. Resistivity values generally exceed 25 ohm-m and tend to have a "noisy" pattern (i.e., the data change abruptly over short depth intervals). Pore pressure data also tend to be noisy in Type III fill, and the values are generally less than 6 m. This unit is classified as clay, clayey silt, or organic soil.

Chromium Ore Processing Residue

HB COPR is typically described on HSA logs as dense, poorly to well-graded sand with gravel, silty sand, or gravelly sand that is weakly to strongly indurated. It was noted in many of the HSA logs that the gravel component of HB material is composed of indurated COPR fragments that were disaggregated by the action of the direct push sampling device. HB COPR is characterized by tip resistance values that range from 25 to > 50 MPa. Relatively thin layers of HB COPR tend to produce a high tip resistance "spike" and typically also produce a spike in friction sleeve resistance and resistivity. Positive or negative pore pressure data spikes are common in intervals of HB COPR. Positive spikes indicate that the HB COPR is responding to CPT cone advancement much like a normally consolidated, fine-grained material or loose sand below the water table. Negative spikes indicate that the HB COPR is responding like an overconsolidated or fissured material or a compact granular stratum.

GB COPR is described on HSA logs as poorly-graded, medium- to fine-grained sand or silty sand. Tip resistance values in GB COPR generally range between 5 to 25 MPa and tend to change abruptly with depth, which creates a noisy pattern in the tip resistance data. Resistivity values in GB COPR within the vadose zone range from 5 ohm-m to greater than 50 ohm-m, and the pattern of the data also tends to be very noisy. The noisy patterns are attributed to differences in density and zones of partial cementation. Resistivity values in saturated GB COPR are commonly less than

5 ohm-m and are generally the defining characteristic of this material. Low-resistivity values are commonly maintained throughout the saturated GB COPR interval. Pore pressure values in GB COPR tend to be negative or fluctuate around an approximate hydrostatic pressure line. The pattern of pore pressure data in GB COPR also tends to be noisy as compared to other materials encountered at the site.

Alluvial Sediments

Alluvial sediments immediately underlie fill and COPR across the site. The alluvial sediments include silt and sand layers. Silt and clay portions of the alluvial sediments have tip resistance values that are typically < 1 MPa. Materials above the alluvial sediments tend to have much higher tip resistance values, so the contact between the alluvial silt and overlying material is abrupt and apparent on most of the CPT logs. Resistivity and pore pressure data in the alluvial sediments are fairly uniform across the site. Resistivity values tend to be low, typically slightly greater than or equal to 5 ohm-m. Pore pressures in the alluvial sediments typically show a gradual increase with depth and range up to 15 m, although higher pore pressures are common. The alluvial sediments are generally classified as silt or clay, which is consistent with lithologic descriptions.

Use of CPT-Boring Log Correlations to Delineate COPR

Soil types encountered at the DMT generally have characteristic CPT signatures that allow them to be identified on an unpaired CPT log. However, there is a significant overlap of CPT response and values for some parameters, which make certain interpretations ambiguous. Therefore, when applying the correlations to unpaired CPT logs, multiple lines of evidence using a variety of CPT parameters and field test results were used to define the various soil types.

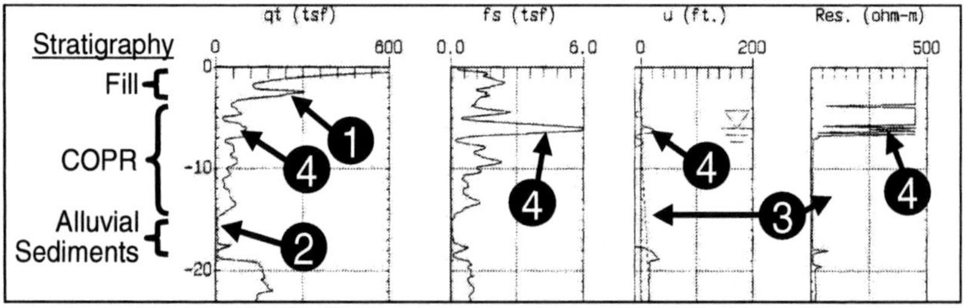

FIG. 2. Unpaired CPT log used to define COPR: (1) base of Type 1 Fill; (2) base of Fill/COPR; (3) interval of low resistivity and negative pore pressure typical in COPR; (4) spikes in q_t, f_s, resistivity, and pore pressure that identify HB COPR.

Figure 2 is an example of a CPT log in which the soil types discussed above are discernable based on the CPT correlations. Fill (Type I) and alluvial sediments are first identified on the log since these horizons have unique tip resistance and resistivity signatures. The tip resistance from 1.2–4.3 m (4–12 ft) is indicative of either Type III fill or COPR but is identified as COPR based on the low resistivity and negative pore pressure. HB COPR is separated from GB COPR based on the data spikes in tip resistance, friction sleeve resistance, and pore pressure.

Multiple CPT parameters are often needed to distinguish between HB and GB COPR. HB COPR often occurs above or at the water table, at the base of fill material, or in thin layers, so it is difficult to ascertain whether increases in tip resistance values are due to the presence of HB COPR or are related to other lithologic variations. Fortunately, intervals of HB COPR tend to produce spikes in tip resistance, friction sleeve resistance, resistivity, and pore pressure. HB COPR was identified by first isolating depth intervals within COPR that contained relatively high tip and friction sleeve resistance values. If data spikes in pore pressure and/or resistivity were encountered over the same depth interval, then HB COPR was ascertained.

DOWNHOLE GEOPHYSICS

Downhole gamma and seismic data were collected at several CPT locations. These data were collected in order to assess the effectiveness of these geophysical techniques for the delineation of COPR. Radioactive particles tend to be associated with fine-grained soils, and therefore these materials typically produce higher gamma counts than deposits of sand and gravel. Additionally, radioactive elements are generally associated with granitic igneous rocks (Sharma, 1997), whereas chromium is generally associated with gabbroic igneous rocks. Therefore, COPR should theoretically contain very low gamma counts since it is usually classified as a sandy material and its composition resembles that of a gabbroic igneous rock.

Six of the 10 CPT locations where gamma data were evaluated were paired with HSA boreholes, and COPR was encountered in all six of the paired CPT logs. A distinct interval of very low gamma counts coincides with the COPR in three of the six paired logs. Gamma counts within the COPR are generally fewer than 2–3 counts per minute and are lower than any of the other subsurface materials encountered at the site. The gamma count data appear to be a reliable indicator of the upper and lower surface of COPR as abrupt changes in the data occur near these contacts.

Seismic waves travel faster in very dense or hard materials. COPR tends to be a dense material relative to other stratigraphic horizons at the site and, portions of the COPR are known to be indurated. Therefore, it was hypothesized that the COPR should have high shear velocity relative to other subsurface materials. Nine of the 11 CPT locations where seismic data were collected were paired with HSA boreholes, and COPR was encountered in seven of the paired locations.

High shear velocities were observed within COPR in five of the nine paired locations. Shear velocity within the COPR interval ranged from 400 m/s to approximately 700 m/s. On most seismic logs, the transition from COPR to underlying material is marked by an abrupt decrease in seismic velocity. The shear velocities observed in other subsurface materials at the site fall within a range of approximately 150–250 m/s (note that high-velocity layers above low-velocity layers may tend to mask the lower layers). In the two remaining paired locations, an abrupt change in shear velocity was observed in the depth interval immediately below the COPR. Elevated shear velocities were also observed in unpaired seismic logs from CPT locations in known COPR areas.

PHYSICAL CHARACTERISTICS OF GB AND HB COPR

COPR Classification Based on Field Findings

There is no single physical classification criterion that distinguishes HB COPR from GB COPR. However, various descriptor characteristics can be used collectively to classify the COPR type. Table 1 summarizes the characteristics of HB and GB COPR.

Table 1. Physical Classification Differences between COPR Types

Descriptor	HB COPR	GB COPR
DPC Reactivity[1]	Predominantly ++, sometimes +	Predominantly ++, sometimes +
Group Symbol and Name	Disturbed samples commonly described as WELL-GRADED SAND WITH GRAVEL (SW)	Predominantly POORLY-GRADED SAND (SP) or SILTY SAND (SM)
Color	• Strong brown (7.5 YR 5/8, 4/6) • Yellowish brown (5 YR 5/8) • Dark yellowish brown (10 YR 4/6) • Brown (7.5 YR 4/4) • Dark brown (10 YR 3/3)	• Black (10 YR 2/1, 7.5 YR 2.5/1) • Very dark gray (10 YR 3/1, 7.5 YR 3/1) • Dark gray (7.5 YR 3/1)
Moisture Content	Commonly described as dry or damp in vadose zone, moist below water table	Commonly described as damp or moist in vadose zone, wet or saturated below water table
Relative Density or Consistency	Typically medium dense to very dense	Typically loose to medium dense
Degree of Lithification	Typically lightly to strongly indurated	Typically particulate to slightly indurated
Major Constituent	Undisturbed – Blocky Structure Disturbed – FINE or COARSE GRAVEL	FINE or MEDIUM SAND
Minor Constituent	Fine to coarse sand or silt	Coarse sand or fine gravel
Other Descriptors	Commonly interspersed with nodules of yellow, light greenish, or whitish material and nodules of GB COPR	Often contains particles of whitish, light green, or pale yellow coloration
Specific Gravity	Mean = 2.78 (2.53–3.00)	Mean = 3.05 (2.67–3.41)

[1]A solution of diphenylcarbazide (DPC) was used to test for the presence of Cr(VI). The procedure consists of applying drops of DPC to a sample. The result is interpreted as a colorimetric reaction as follows.

- None—no visible reaction; no Cr(VI). Symbol: (-)
- Strong—some light purple coloration; presence or some contamination with Cr(VI). Symbol: (+).
- Very Strong—dark-purple reaction; definitive presence of Cr(VI). Symbol: (++)

Example Classifications:

DPC	SPT	COPR	Soil Description
++	3-4-5-4 (9)	GB	POORLY-GRADED SAND (SP), Dark gray (7.5 YR 3/1), moist, loose, particulate, FINE SAND with some medium sand.
++	14-28-40-38 (68)	HB	WELL-GRADED SAND WITH GRAVEL (SW), Strong brown (7.5 YR 5/8), damp, very dense, strongly indurated, FINE TO COARSE SAND with some course gravel.

Geotechnical Characterization of COPR Types

Particle-Size Distribution

Typical results from particle-size analyses of GB COPR samples are graphically presented in Figure 3. According to the Unified Soil Classification System (USCS), GB COPR classifies as either poorly-graded sand or silty sand. The majority of GB COPR samples contain less than 15 percent fines (i.e., particles passing a 75-μm, no. 200 sieve) by mass. A majority of the coarse component divides roughly equally between fine- and medium-sized sand fractions, according to USCS criteria. Particle-size distributions were not conducted on HB COPR due to its cemented, rock-like nature. Neither HB nor GB COPR samples were plastic according to USCS criteria.

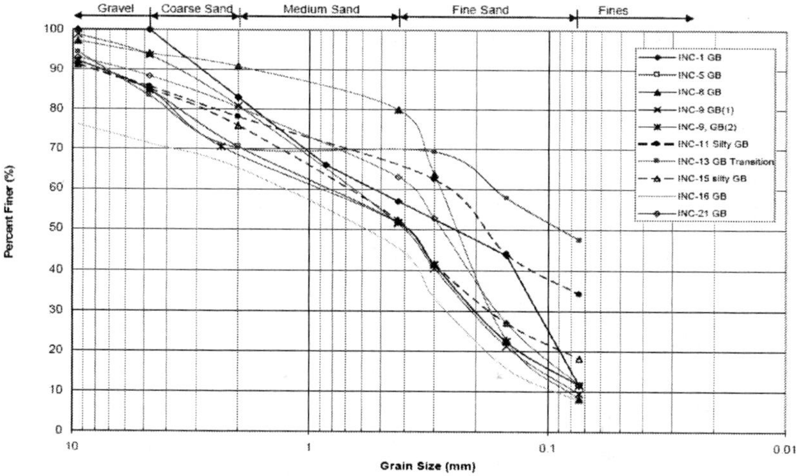

FIG. 3. Particle-size distribution curve for GB COPR samples.

Specific Gravity

The specific gravity of HB COPR ranged from 2.53 to 3.00, with a mean of 2.78 (standard deviation of 0.10). Specific gravities of COPR encountered in GB layers averaged 9% higher than those for the HB COPR, having values ranging from 2.67 to 3.41, with a mean of 3.05. The GB samples had a relatively higher spread (standard deviation of 0.18) in specific gravity, which may be the result of material variability or partial weathering/hydration.

Moisture Content

COPR moisture contents (gravimetric basis, laboratory oven-drying methods) averaged 18 and 22 percent in unsaturated HB and GB COPR, respectively. COPR moisture contents for saturated HB and GB COPR were much higher, with averages of 28 and 33 percent, respectively. For saturated COPR, the range and variability of the water contents was greater. Inspection of the moisture contents did not identify any obvious trends with respect to depth below ground surface or location.

CONCLUSIONS

A characteristic CPT signature was defined for each shallow soil type encountered at DMT by correlating CPT response with soil characteristics as observed at 36 paired CPT and HSA borehole locations. Correlation trends show that COPR has a CPT signature that is distinct from other soil types. COPR is generally defined by relatively high tip resistance that is coupled with very low resistivity and/or cone dynamic pore pressure data values. Within the COPR interval, it also appears that HB COPR can be distinguished based on the presence of spikes in tip resistance, friction sleeve resistance, resistivity, and cone dynamic pore pressure values. In addition, seismic and gamma CPT tests may be used to support the definition of COPR.

Two types of COPR were identified at DMT based on the physical characteristics. HB COPR is moderately to strongly lithified or indurated. GB COPR is particulate to weakly cemented. According to the USCS, GB COPR classifies as either poorly-graded sand or silty sand, containing less than 15 percent fines. HB COPR does not classify as a soil because of its cemented state.

ACKNOWLEDGMENTS

The authors wish to thank Honeywell International Inc. for the financial support of the field investigation. Parratt-Wolff, Inc., of Hillsborough, NC, conducted the drilling; ConeTec of West Berlin, NJ, conducted the CPT effort.

REFERENCES

Dermatas, D., Chrysochoou, M., Moon, D.H., Pardali, S., Christodoulatos, C., Lazarte, C.A., Pendleton, C., Bonaparte, R., Briggs, R., Myers, M., French, C., Morris, J., and Kaouris, M. (2005). "Mineralogical characterization of chromite ore processing residue at Dundalk Marine Terminal Area 1800." *Proc. 8th Intl. In Situ and On-Site Bioremediation Symposium*, Battelle Press, Columbus, OH.

Dermatas, D., Chrysochoou, M., Pardali, S., and Grubb, D.G. (2007). "Influence of X-ray diffraction sample preparation on quantitative mineralogy: implications for chromate waste treatment." *J. Environ. Qual.* 36: 487-497.

GeoSyntec Consultants. 2005. Subsurface Investigation Data Summary Report—Area 1800, Dundalk Marine Terminal. July 22.

Robertson, P.K. 1989. Soil Classification Using the Cone Penetration Test. *Can. Geotech. J.*, *27*, 151–158.

Sass, B.M., Kremser, D.T., Clark, G.I., and French, C. (2005). "Hydration-induced expansion of chromite ore processing residue." *Proc. 8th Intl. In Situ and On-Site Bioremediation Symposium*, Battelle Press, Columbus, OH.

Sharma, P.V. (1997). *Environmental and Engineering Geophysics*. Cambridge University Press, Cambridge, UK, 475 p.

Field-Scale Evaluation of In-situ and Ex-situ Treatment Technologies for Chromite Ore Processing Residue (COPR)

Cuneyt Gokmen[1], M. ASCE, P.E., Tamara Hebeler[2], M. ASCE, P.E., Maria Kaouris[3], Rudolph Bonaparte[4], M. ASCE, Ph.D., P.E., and Lina Kodjo-Wayo[5], Ph.D.

[1]Senior Engineer, Geosyntec Consultants, 1255 Roberts Blvd., Ste 200, Kennesaw, GA 30144; CGokmen@geosyntec.com
[2]Project Engineer, Geosyntec Consultants, 1255 Roberts Blvd., Ste 200, Kennesaw, GA 30144; THebeler@geosyntec.com
[3]Project Manager, Honeywell International, Inc., 101 Columbia Road, Morristown, NJ 07962; Maria.Kaouris@Honeywell.com
[4]Principal, Geosyntec Consultants, 2002 Summit Blvd NE, Suite 885, Atlanta, GA 30319; RBonaparte@geosyntec.com
[5]Senior Staff Geoscientist, Geosyntec Consultants, 1255 Roberts Blvd., Ste 200, Kennesaw, GA 30144; LKodjowayo@geosyntec.com

ABSTRACT: A field-scale pilot program was conducted to evaluate treatment options to reduce hexavalent chromium [Cr(VI)] concentrations and pH levels in chromite ore processing residue (COPR). The treatment performance objectives were to: (i) reduce Cr(VI) concentrations to less than 240 mg/kg; and (ii) reduce the pH of the COPR to less than 9.5. Various types and dosages of chemical reductants and pH reducing reagents were utilized. Implementation methods included: (i) pugmill; (ii) large-diameter vertical auger; and (iii) horizontal rotary mixer. Reductions in Cr(VI) concentrations were observed immediately after treatment, regardless of the reagent or implementation technology. Cr(VI) concentrations increased in some samples within 600 days post-treatment. Results for the pilot program are presented and discussed, and implementation challenges are identified.

INTRODUCTION

COPR, a byproduct of the sodium dichromate manufacturing process, was placed as fill between 1905 and 1954 at a tidal marshland site adjacent to the Hackensack River in New Jersey. COPR at the site is a slag-like material with a pH above 12, large acid neutralization capacity, and average Cr(VI) concentrations of approximately 4,000 mg/kg (typical total Cr concentrations of 15,000 to 40,000 mg/kg). In its natural state, the material leaches chromium in the U.S. EPA TCLP test in excess of the RCRA characteristic hazardous waste threshold concentration of 5 mg/L (Dermatas, et al., 2005). Cr(VI) leaching from COPR may potentially impact groundwater quality. Cr(VI) from COPR is also a concern with respect to human inhalation, ingestion, and dermal contact exposure pathways. Site COPR can be expansive upon weathering, creating unfavorable geotechnical/structural engineering foundation conditions.

The mineralogy of COPR at the site is known to include primarily brownmillerite (a ferrite compound), hydrogarnet, periclase, and hydrated lime, with lesser and varying amounts of calcium aluminum chromate hydrate (CAC), calcite, hydrocalcite, ettringite, calcium silicate hydrate, and other compounds (Kremser, et al., 2005). Much of the Cr(VI) is contained in the CACs, which becomes thermodynamically unstable at a pH below approximately 9.5.

In 2005, a field-scale pilot program was conducted to evaluate treatment options to reduce Cr(VI) concentrations and pH levels in site COPR. Established treatability criteria were Cr(VI) concentrations less than 240 mg/kg and pH values below 9.5 s.u. The field program treatment approaches were based on the results of laboratory studies conducted at the Stevens Institute of Technology (SIT) in Hoboken, New Jersey, which showed the potential for: (i) chemical reduction of Cr(VI) to essentially non-toxic and insoluble Cr(III) species using several commercially-available reagents; and (ii) acid neutralization of COPR to eliminate expansive characteristics while at the same time accelerating the release of Cr(VI) from the mineral matrix.

At the test site, several different COPR strata are typically encountered (Figure 1). B1 and B2 COPR are commonly described as a black, fine to coarse sand size materials with trace silt and gravel size particles. C COPR is commonly described as a lighter-colored, finer grained material with silt and clay size particles and trace fine to medium sand size particles.

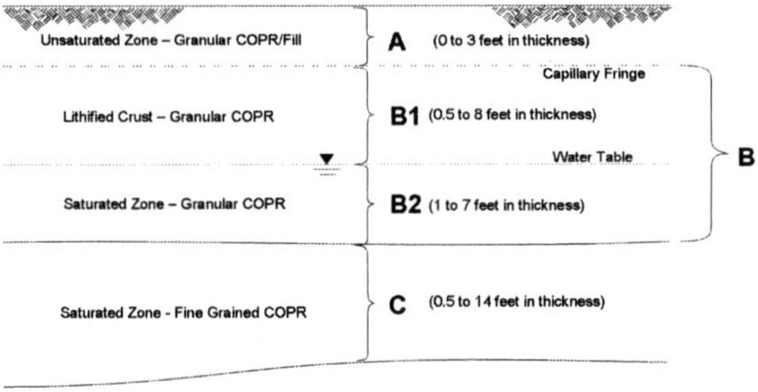

FIG 1. Generalized Subsurface Stratigraphy

PILOT PROGRAM

Overview

Both organic (e.g., lactate, molasses) and inorganic (e.g., reduced sulfur- or iron-containing compounds) reagents are known to be effective electron donors for aqueous Cr(VI) reduction. For example, sulfide [S(-II)] or ferrous iron [Fe(II)] have been used to reduce oxidized Cr(VI) to Cr(III). The laboratory studies performed at SIT showed that reduction of Cr(VI) to Cr(III) in COPR could be achieved using commercially-available reagents, including: (i) ferrous sulfate heptahydrate ($FeSO_4 \cdot 7H_2O$) (referred to herein as $FeSO_4$); (ii) calcium polysulfide (CaS_x); (iii) ferrous chloride ($FeCl_2$); and (iv) a proprietary calcium/sodium polysulfide (proprietary polysulfide). In addition to the reductants, several acids, including liquid sulfuric acid (H_2SO_4) and granular sodium bisulfate ($NaHSO_4$), were evaluated for pH reduction.

The goals of the pilot program were to not only evaluate reagent effectiveness, but also field implementation methods, including: (i) ex-situ treatment in a pugmill; (ii) in-situ treatment using a large-diameter vertical auger (VA); and (iii) in-situ treatment using a horizontal rotary mixer (HRM). A summary of the pilot program testing matrix is presented in Table 1, and as can be noted, ferrous sulfate was more extensively studied than the other reagents. This decision was based on the SIT laboratory test results and the fact that ferrous sulfate is an acid generating reductant that would itself help to reduce COPR pH.

Table 1. Treatment Reagent and Dosages

Technology	Chemical Reductant[1] (Stoichiometric Dosage)	pH Reducing Reagent (% based on dry weight of COPR)
Pugmill	Ferrous Sulfate (3X, 5X, 8X)	None
	Ferrous Sulfate (3X, 3.5X, 5X, 8X)	Sulfuric Acid (10%, 20%, 30%)[2]
	Ferrous Sulfate (5X)	Anhydrous Sodium Bisulfate (25%)
	Ferrous Chloride (3X, 5X, 8X)	None
	Calcium Polysulfide (1.5X, 2X, 2.25X)	None
	Proprietary Polysulfide	None
VA	Ferrous Sulfate (3X, 5X)	None
	Ferrous Sulfate (3X)	Sulfuric Acid (22%)[1]
HRM	Ferrous Sulfate (3X, 5X, 8X)	None
	Ferrous Sulfate (3X)	Sulfuric Acid (23%)[1]
	Ferrous Sulfate (5X)	Anhydrous Sodium Bisulfate (25%)

Note:
[1]"X" reports the reagent dosage as a multiple of Cr(VI) stoichiometric demand.
[2]The sulfuric acid was a 66° Baume solution (93% concentrated) for the pugmill tests and a 35° Baume solution (40% concentrated) for the VA and HRM tests.

Ex-situ Pugmill Pilot Program

Pugmill pilot program activities commenced with excavation of COPR and transportation of the material to an off-site permitted RCRA TSDF. The excavated material was segregated by COPR horizon and prepared for treatment. Particle size reduction (PSR), which increases the surface area of the COPR material and improves contact with treatment chemicals, was investigated for its effect on treatment effectiveness. The PSR was accomplished using a plate mill disc grinder with a delumper attachment (Models GP-140/DL-140, Modern Process Equipment Corporation, Chicago, IL). The ex-situ tests (i.e., mixing of COPR and reagents) were conducted using a 1.5 cubic meter (m^3) capacity batch-type pugmill mixer (i.e., Mini-MaxCrete, Maxon Industries, Milwaukee, WI) with dual directional mixing capabilities and an eight paddle agitator shaft (Figure 2A).

Approximately 450 kg of COPR (dry weight) was treated in each of the 27 tests; 22 using B1/B2 COPR, four using C COPR, and one using B1/B2/C COPR. Seven samples from the B1/B2 were "fine" ground, three samples were "coarse" ground, and the remaining 15 were mixed in the "as excavated" particle size condition. The grinding process reduced the size of the larger COPR particles (e.g., average d_{90}

decreased from 2.3 mm for as excavated to 0.9 mm for fine ground), but the d_{50} (from average of 0.34 mm to 0.31 mm) and percent fines (from average of 14% to 18%) did not change significantly. The C material is naturally fine-grained and PSR was not used with this material. Moisture contents were adjusted in several tests to evaluate the effect of moisture on treatment; target moisture contents ranged from "as-excavated" to 100 percent. Following treatment, each test batch was transferred to a 208-liter (55-gal) drum for archiving (Figure 2B), from which post-treatment samples were obtained at 0, 30, 60, 120 and 600 days (Figure 2C).

FIG 2. Pre-treatment and Long-term Sampling (Ex-situ Pugmill Program)

In-situ Vertical Auger Pilot Program

The VA pilot program consisted of three test plots (i.e., VA1, VA2, and VA3) at the Hackensack River test site, each approximately 6 m wide by 9 m long in plan view. The vertical auger rig was a Manitowoc 4000W Series crane mounted with a Hain 450 turntable capable of applying 42,900 m-kg of rotary torque (Figure 3A). The turn table was connected to a hollow 0.3 m square Kelly Bar that rotated a 3 m diameter single flight auger with a 0.6 m long pilot point (Figure 3B).

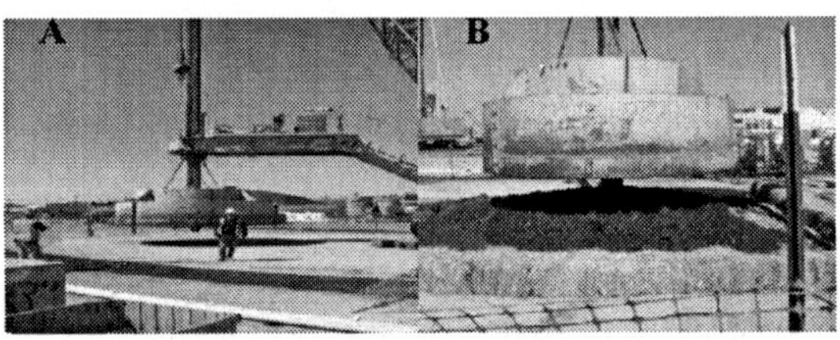

FIG 3. In-Situ Mixing of the VA Test Plot

Reagents were prepared as a slurry in a nearby batch plant and pumped through the Kelly Bar into the auger, which was equipped with two injection ports and a surface shroud to minimize dust, splashing, and/or gas emissions. The COPR was mixed to a depth of 5 m. Post-treatment samples were obtained using direct-push sampling techniques at 0, 30, 60, 120, and 600 days.

In-situ Horizontal Rotary Mixer Pilot Program

The HRM pilot program consisted of five test plots. The first three plots (i.e., HRM1, HRM2, and HRM3) were designed for shallow treatment (i.e., 1 m depth) of B1 COPR using HRM equipment with a wide mixing head (i.e., 3.3 m wide) (Figure

4A-B). HRM4 and HRM5 were designed for deeper treatment (i.e., 3 m depth) of B1/B2 COPR using HRM equipment with a narrower mixing head (i.e., 0.6 m wide). Both systems were equipped with an environmental hood (not shown in Figure 4) to minimize dust and splashing. Post-treatment samples were obtained using direct-push sampling techniques at 0 and 30 days at each test plot and at 600 days at HRM3.

FIG 4. In-Situ Mixing using a Horizontal Rotary Mixer

RESULTS

Post-treatment samples were analyzed for Cr(VI) using US EPA Methods 3060A and 7196A. Other analyses included pH, Cr(total), TCLP(Cr), ferrous iron, redox potential, and moisture content. Select Cr(VI) and pH results from pugmill pilot program are presented in Figures 5 and 6, respectively, and discussed below.

Reagent Type and Dosage

As shown in Figure 5, CaS_x, proprietary polysulfide, and $FeCl_2$ were the tested reagents that achieved Cr(VI) concentrations less than 240 mg/kg based on the 600 day test results, with no evidence of increasing Cr(VI) concentrations in the polysufide based systems. Cr(VI) concentrations in the $FeCl_2$ based systems began to show increased Cr(VI) concentrations in the 120 day results, but were still below the 240 mg/kg threshold in the 600 day results. Polysulfide-based treatments did not reduce the COPR pH due to the alkaline nature of the reagent (Figure 6).

Results from the $FeSO_4$ tests show increasing Cr(VI) concentrations and pH values with time, even at high stoichiometric dosages, which may be due to scavenging of the reductant by oxygen or another oxidizer coupled with the rate-limited release of Cr(VI) from the COPR mineral matrix. This affect was seen with the $FeCl_2$ treatment to a lesser degree, and not at all in the CaS_x treatment within the 600 day analysis interval. As these results show that Cr(VI) release from the mineral matrix is rate limited, two conclusions can be drawn: (i) the observed treatment levels in the test results reflect a combination of Cr(VI) reduction in the treated mass prior to laboratory analysis and Cr(VI) reduction in the laboratory results from alkaline digestion of the sample in the presence of excess sample reductant; and (ii) for COPR based treatment systems, it is important to select reductants that will remain effective for as long as necessary given the rate-limited release of Cr(VI) from the mineral matrix.

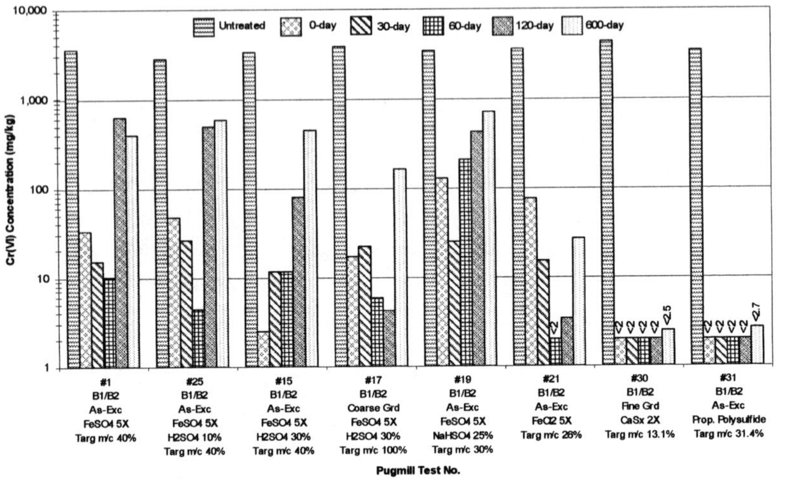

FIG 5. Effect of Chemical Treatment on Cr(VI) Concentrations in Pugmill Tests

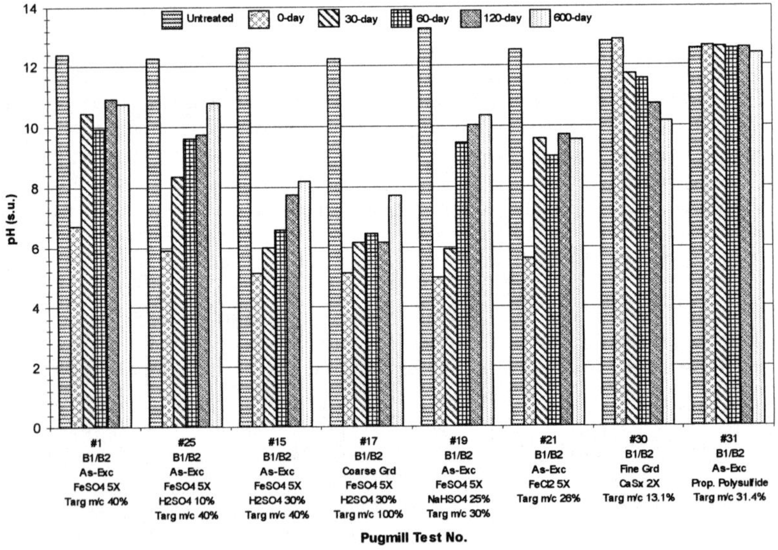

FIG 6. Effect of Chemical Treatment on pH values in Pugmill Tests

$FeSO_4$ with acid was initially more effective at reducing Cr(VI) concentrations than was $FeSO_4$ alone. With time, however, post-treatment concentrations tended to rebound to similar levels, regardless of acid addition. This was true even when 600 day pH levels remained below 9.5 s.u.. Although only one test was conducted with anhydrous sodium bisulfate, it appears that the two pH reducing reagents tested were comparable in terms of treatment effectiveness at similar acid equivalencies. Higher dosages of ferrous sulfate resulted in lower Cr(VI) concentrations and pH values.

Particle Size and Moisture Content

No noticeable improvement in treatment effectiveness was observed due to pre-treatment PSR for the level of PSR achieved during the pugmill pilot program.

There was some indication that higher target moisture contents in B1/B2 COPR resulted in lower post-treatment Cr(VI) concentrations, possibly due to better distribution of the reductant at higher moisture content, or lower levels of pore oxygen to scavenge the $FeSO_4$. Higher moisture contents, however, result in poorer post-treatment geotechnical characteristics.

Implementation Methods

The pilot program demonstrated that the treatment reagents and technologies described herein can be implemented at a large-scale. However, differences in treatment effectiveness were observed between technologies. The pugmill provided the highest level of process flexibility and quality control, allowed moisture adjustment based on the feed conditions, and produced more homogenous mixing conditions. Although the VA did not provide the same level of flexibility, results were comparable to the pugmill. Reagents must be applied in liquid or slurried form with the VA, which can be a limiting factor in certain conditions. Additionally, a higher degree of variability was observed from samples obtained from the VA test plots compared to samples obtained from pugmill samples. Cr(VI) concentration results from VA2 are presented in Figure 7, along with the results from a pugmill test with a similar stoichiometric dosage. HRM technology was found to have the least amount of process flexibility. The technology is limited to shallow depth applications and difficult to implement when mixing liquids (or for those conditions with high moisture content).

In the pugmill, VA, and HRM tests, the mixing of acid with the alkaline COPR resulted in significant heat generation. However, in the pugmill, engineering controls were successfully implemented to dissipate the excess heat and to prevent any damage to the pugmill and its components. In the VA and HRM, it was demonstrated that the heat generation does not adversely impact the operation of the implementation equipment. Corrosivity concerns associated with using strong acids and $FeCl_2$ can be overcome during full-scale implementation with proper selection of equipment materials and coating types. During the pilot program, the relatively short exposure period encountered did not result in any noticeable corrosion problems. Health concerns associated with the potential generation of hydrogen sulfide gas during mixing of polysulfide with COPR were addressed through air monitoring. No hydrogen sulfide gas was detected. However, it is important to note that the potential for gas generation increases in neutral and acidic environments and this may be an important application consideration.

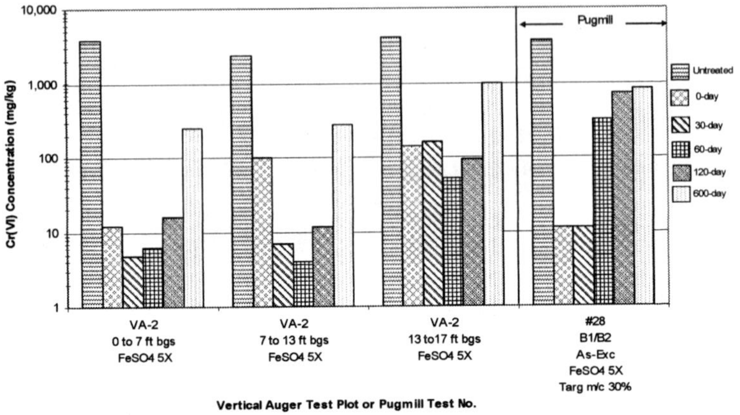

FIG 7. Effect of Chemical Treatment on Cr(VI) Concentrations at VA2

Geotechnical Characteristics

COPR treated with ferrous sulfate was observed to undergo mineral bulking, especially when the pH value could not be maintained below 9.5 s.u. This bulking is attributed to the formation of ettringite resulting from the addition of sulfate. For CaS_x and proprietary polysulfide based systems, post-treatment mineral bulking was not observed, but mineralogical analysis would be required to evaluate whether ettringite or other potentially problematic secondary minerals were formed.

The liquid nature of the CaS_x, proprietary polysulfide, and $FeCl_2$ resulted in initial post-treatment consistencies ranging from a thin slurry to a wet concrete. Although some improvement in post-treatment consistency was observed, the consistency observations after 600 days were still indicating low compressive strengths (e.g., remolded clay at or near its liquid limit, and some with free liquids on top). Further study is required to evaluate the long-term geotechnical characteristics of material treated with these reagents.

ACKNOWLEDGEMENTS

The authors appreciate the support of Honeywell International, Inc. and the contributions of Stevens Institute of Technology in conducting this pilot program.

REFERENCES

Dermatas, D., Bonaparte, R., Chrysochoou, M., and Moon, H. (2005) "Chromite Ore Processing Residue (COPR): Hazardous Contaminated Soil or Solid Waste," *Journal of ASTM International*, Vol. 3, No. 7, pp. 1-10.

Kremser, D.T., Sass, B.M., Clark, GI, and Bhargava, M. (2005) "Environmental Forensics Investigation of Buried Chromite Ore Processing Residue", *Eighth International In Situ and On-Site Bioremediation Symposium*, Baltimore, Maryland.

Degradation of Naphthalene in Aqueous Phase of Saturated Ottawa Sand using Alternating and Direct Currents

Milind V. Khire[1], Ph.D., P.E., M. ASCE and Emmanuel Pepprah[2]

[1]Assistant Professor, Dept. of Civil and Environmental Engineering, Michigan State University, A135 Research Complex Engineering, E. Lansing, MI 48824, U.S.A., Email: khire@egr.msu.edu.

[2]Graduate Research Assistant, Dept. of Civil and Environmental Engineering, Michigan State University, A30 Research Complex Engineering, E. Lansing, MI 48824, U.S.A.

ABSTRACT: The objectives of this study were to assess the effect of direct current (DC) and alternating current (AC) on the degradation rates of naphthalene in the aqueous phase of spiked Ottawa sand. Square wave AC at frequency and peak current density equal to 0.1 Hz and 3 mA/cm^2, respectively, was used. DC was applied in a separate test at current density equal to 3 mA/cm^2. The key findings of this study are: (1) both DC and square wave AC degraded naphthalene in the aqueous phase of the sand; (2) at an equivalent current density, the rate of degradation of naphthalene in the aqueous phase when sand was present was less than when only the aqueous solution were subjected to AC or DC; and (3) rate of degradation for AC was less than that for DC at an equivalent current density.

INTRODUCTION

Polycyclic aromatic hydrocarbons (PAHs) have been found in at least 600 of the 1,430 National Priorities List (NPL) sites identified by the United States Environmental Protection Agency (USEPA) (ASTDR 1995) and are classified as potential carcinogens. Lageman *et al.* (1990) has demonstrated that removal of many aromatic organic compounds from soils is feasible using electrical methods.

Most studies on the electroremediation of organic compounds have been focused on using DC to transport them by electromigration and/or electroosmosis with subsequent removal onto electrodes or into solutions surrounding the electrodes (catholyte or anolyte). The objective of this study was to use alternating and direct currents (AC, DC) to assess the degradation of a PAH (naphthalene) in a clean poorly graded sand with spiked naphthalene solution. AC is readily available from the power grid and can be potentially used minimal hardware. Use of DC requires power supplies which are relatively expensive for field-scale applications. Naphthalene was used as the model PAH compound because it has relatively high solubility in water (~30 mg/L) as compared to other higher order PAHs. Ottawa sand was selected as the porous media because it would not significantly adsorb naphthalene from the spiked solution used for saturating the sand.

MATERIALS AND EXPERIMENTAL METHODOLOGY

Reaction Chamber

The experimental setup is illustrated in Fig. 1. The test cell, which was an undivided cell, consisted of a 1 liter Pyrex glass beaker with a Teflon cap. These materials of the cell were selected to eliminate or reduce sorption of the contaminants onto the walls of the reaction vessel and the cap.

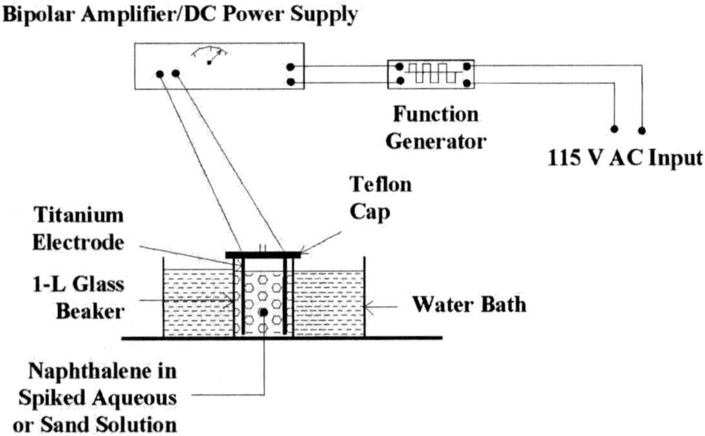

FIG. 1: Schematic of the experimental setup.

Titanium plates were used as electrodes in electrochemical tests of this study. Alshawabkeh and Sarahney (2005), Li *et al.* (2005), and Goel *et al.* (2003) have used titanium as a base metal and coated it with mixed metal oxides such as RuO_2, IrO_2, and SnO_2. The titanium plates used in this study were not coated. Uncoated titanium is cheaper than commonly used electrodes such as platinum or titanium coated electrodes. Also, titanium offers resistance to corrosion in a wide range of aggressive conditions. The titanium sheets used as electrodes in this study were 12.0-cm-long by 5.0-cm-wide by 1.72-mm-thick and the distance between them was maintained at 8.0 cm. The purity of the titanium was about 99%.

Electrical Equipment

A 0.1 Hz to 2 MHz Sweep Function Generator was used to generate the AC signal. The power supply consisted of a bipolar operational amplifier/power supply capable of producing up to 200 V AC/DC and 1 A AC/DC output or a maximum electrical power of 200 W. Voltage and current measurements were obtained with the aid of an oscilloscope with accessories.

Spiked Sand

Aqueous solutions of naphthalene were prepared by adding naphthalene (98%, purity) crystals to one-liter of deionized (DI) water in a capped Pyrex bottle and allowing the suspension to stand for up to three days. This was followed by the addition of 0.5 g of anhydrous sodium sulfate (Na_2SO_4) to each suspension, to increase its electrical conductivity. Sodium sulfate was selected as the supporting electrolyte because all aqueous solution tests without sand were carried out using Na_2SO_4 as the supporting electrolyte. The target concentration of Na_2SO_4 was approximately 500 mg/L (~3.5 mM). This concentration was selected to meet USEPA's maximum contaminant level goal for sulfate equal to 500 mg/L (~3.5 mM). The resulting spiked aqueous solution was added to 1,000 cm^3 of dry Ottawa sand placed in a 1-liter glass beaker until the sand was fully saturated.

Testing Procedure

Tests were conducted in batch mode in 1-liter undivided cells. Prior to applying AC or DC to the cells, samples were collected for high performance liquid chromatography (HPLC) analyses. For each sampling event for the tests with sand, three aqueous samples: one from the middle of the test cell and one each close to the surface of the electrodes, were collected with the aid of a glass syringe and needle. For tests without sand (aqueous solution only), samples were collected only from the center of the cell. During the tests, samples were collected for HPLC analyses at elapsed time intervals of 1, 2, 4, 8, 24, and 48 hours. HPLC analyses were performed using a diode array detector (detection at 255 nm), a Waters C-18 PAH, 250 mm x 4.6 mm x 5 αm column, an automated gradient controller, and deionized (DI) water/acetonitrile mobile phase. The flow rate of the mobile phase was 1.2 mL/min and the column was flushed with 10:90 (v/v) acetonitrile/DI water for 5 minutes followed by 60:40 (v/v) acetonitrile/ DI water for 10 minutes, and by 100:0 (v/v) acetonitrile/ DI water for 5 minutes. This gradient was used for the analysis in order to obtain appropriate separation of the degradation product peaks on the chromatograms during the tests. Standard solutions were prepared to bracket the expected concentration level of the analytes. Calibration curves with an average R^2 value of 0.998 were used to determine the concentration of naphthalene in the solutions as the tests progressed.

RESULTS AND DISCUSSION

Comparison of Degradation Rates

Fig. 2 shows a comparison of the concentration of naphthalene in spiked aqueous solution with or without Ottawa sand using square wave AC. The peak current density (j) and frequency (f) were equal to 3 mA/cm^2 and 0.1 Hz, respectively. The control tests were carried out without the passage of AC or DC while the rest of the experimental setup was similar to other tests where AC or DC was passed. In addition, the control cell for sand was placed in a static temperature chamber at 44 °C to simulate a conservative scenario of heating. When AC or DC was applied, while the water bath kept the temperature rise to a minimum, it reached 32.9 °C when DC was applied for the cell containing sand. For other tests, the temperature was less than 21 °C. DC test produced higher temperature because the DC resistance of the system was greater than the impedance when AC was applied. Hence, greater electrical power was utilized in the DC cell for a fixed current.

Aqueous phase samples collected close to the surface of the electrodes generally exhibited faster degradation rate than samples collected from the center of the cell when sand was present. This result is consistent with the general belief that electrochemical reactions occur on the surface or in the vicinity of the electrodes. Pepprah and Khire (2007) reported that in AC electrolysis, constant polarity reversal of the electrodes ensures that both electrodes participate in the anodic oxidation process. By applying an AC, which resulted in constant polarity reversal of the electrodes at a rate dependent upon the AC frequency, anodic oxidation of naphthalene in solution occurred on both electrodes (Pepprah and Khire 2008). Pepprah (2007) analyzed the degradation products in undivided as well as divided cells and confirmed that similar to DC, AC also resulted in hydroxylation of naphthalene.

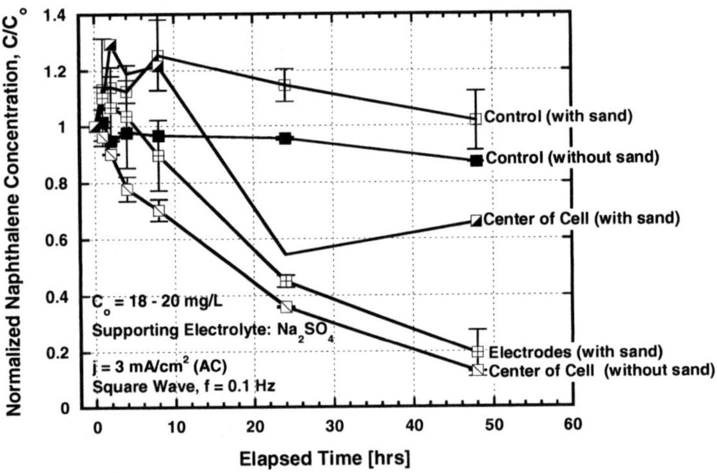

FIG. 2: Concentration of naphthalene in spiked solution with and without sand using AC.

Previous studies on electrochemical degradation have focused on the use of DC where the cathode acts as a "passive" electrode in the oxidation process. However, when an AC is used, the constant polarity reversal creates anodic oxidation reactions on both electrodes alternately, depending on their polarity in a given AC cycle. During half cycle when one of the electrodes has a positive electrical potential, it instantaneously acts as the anode (instantaneous anode) while the other electrode, at a negative electrical potential instantaneously acts as the cathode (instantaneous cathode). In the subsequent half cycle, polarities are reversed and the electrodes acquire electrical potentials opposite to what they had in the previous cycle and this continues throughout the period of application of the AC.

Fig. 3 shows a comparison of the normalized concentration of naphthalene with and without sand subjected to DC having current density equal to 3 mA/cm^2. Aqueous phase samples collected in the vicinity of the anode of the cell with sand exhibited faster degradation of naphthalene compared to samples collected from the middle or near the cathode. Samples collected in the vicinity of the cathode showed less degradation because in DC electrolysis, anodic oxidation occurs only on one electrode - the anode (Pepprah and Khire 2008). Hence, unlike in an AC test, the cathode in DC test does not participate in the oxidation of the organics in the aqueous phase. However, because it is an undivided cell, the reduction in the concentration at anode also decreased the concentration throughout the cell. Pepprah (2007) have shown via modeling using the Nernst-Planck equation that electromigration of naphthalene with the supporting electrolyte results in naphthalene reaching the electrodes where it is converted. In addition, sparging of hydrogen gas at the cathode also contributed to a reduction in the concentration of naphthalene at cathode (Goel at al. 2003).

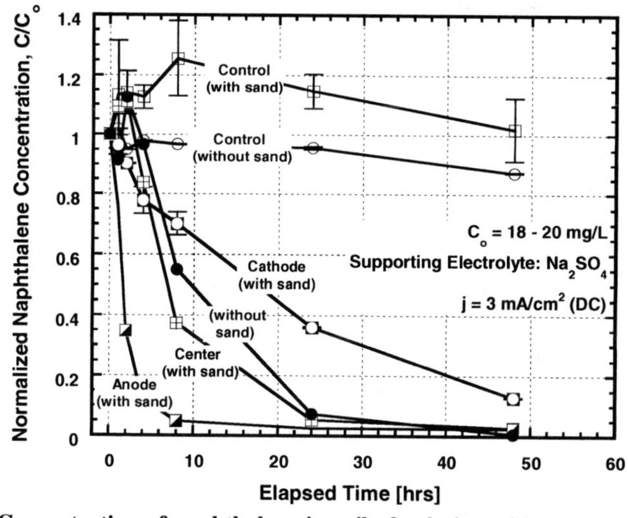

FIG. 3: Concentration of naphthalene in spiked solution with and without sand using DC.

Comparison of Degradation Rates in Spiked Sand Using AC and DC

Fig. 4 shows the normalized concentration of naphthalene for cells containing sand where AC and DC having current densities equal to 3 mA/cm^2 were passed. Samples collected from the vicinity of both electrodes in the AC tests exhibited degradation. However, samples collected from the vicinity of the anode showed much faster degradation than those collected in the vicinity of the cathode in the DC test. Based on the observed concentration profile, it is likely that anodic oxidation occurred only at the anode, and as naphthalene was being oxidized, more molecules moved from the center of the cell as well as in the vicinity of the cathode, towards the anode where they were subsequently oxidized. The pH of the samples collected at the end of the test (Table 1) indicated acidic/oxidizing conditions at anode and alkaline/reducing conditions at cathode. This observation is consistent with a review presented by Page and Page (2002) who reported that in DC electrolysis of moist soils, the predominant process is the electrolysis of water molecules, which results in the formation of H^+ and oxygen at the anode and OH^- and hydrogen at the cathode. Therefore, a local change in the pH of the pore solution occurs so that acidity increases with time at anode and alkalinity increases at cathode.

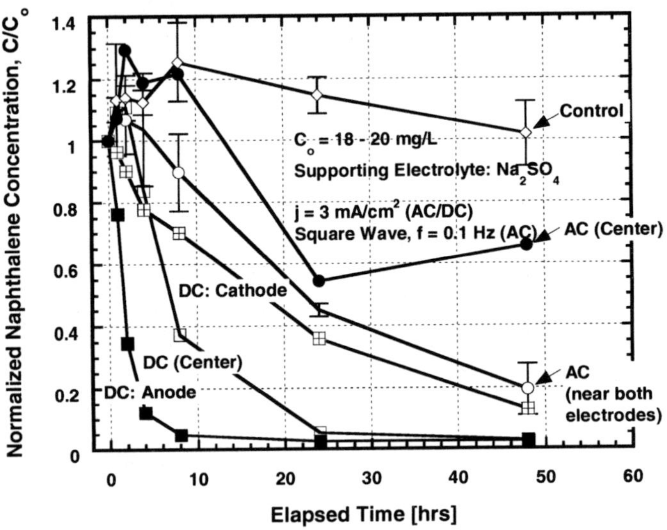

FIG. 4: Concentration of naphthalene in spiked Ottawa sand using AC and DC.

Experimental Parameters Monitored

Apart from monitoring the aqueous phase concentration of naphthalene over time, other experimental parameters monitored included the initial and final values of pH, temperature, redox potential, and electrical conductivity. A portable dissolved oxygen/pH meter and oxidation-reduction probe, were used to measure pH and redox potential of the test solutions. Table 1 lists these measurements.

SUMMARY AND CONCLUSIONS

AC and DC were separately applied to spiked naphthalene solutions with and without Ottawa sand to assess degradation. Potential application of this technology

Table 1: Measured initial and final values of temperature, pH, standard redox potential, and electrical conductivity.

Medium	Peak Current Density [mA/cm^2]	Temperature (°C)		pH		Standard redox potential (mV)		Electrical conductivity (μS/cm)	
		Initial	Final	Initial	Final	Initial	Final	Initial	Final
Without Sand (AC)	3	19.2 (C)	18.0 (C)	6.5 (C)	4.5 (C)	394 (C)	415 (C)	811 (C)	818 (C)
With Sand (AC)	3	20.1 (C)	20.8 (C)	6.8 (C)	11.5 (C) 6.4 (E1) 6.3 (E2)	339 (C)	184 (C)	835 (C)	1297 (C)
With Sand (DC)	3	21.5 (C)	31.9 (C)	6.7 (C)	3.1 (C) 2.9 (AN) 12.2 (CT)	359 (C)	771 (C) 804 (AN) 235 (CT)	874 (C)	1044 (C) 1555 (AN) 355 (CT)
With Sand (Control)	-	20.0 (C)	44.1 (C)	7.2 (C)	8.0 (C)	420 (C)	394 (C)	867 (C)	1052 (C)
Without Sand (Control)	-	19.9 (C)	19.6 (C)	6.7 (C)	6.7 (C)	458 (C)	469 (C)	811 (C)	815 (C)

Notes:

1. AN = Sample collected in the vicinity of anode (DC);
2. C = Sample collected from the center of the test cell (AC or DC);
3. CT = Sample collected in the vicinity of the cathode (DC);
4. E 1 = Sample collected in the vicinity of electrode 1 (AC); and
5. E 2 = Sample collected in the vicinity of electrode 2 (AC);

would be for cleanup of PAH contaminated ground water in sandy or rocky aquifers. Square wave AC having frequency equal to 0.1 Hz was used. The current density for both AC and DC was 3 mA/cm^2. The key findings of this study are: (1) both DC and square wave AC degraded naphthalene in the aqueous phase of the sand; (2) at an equivalent current density, the rate of degradation of naphthalene in the aqueous phase when sand was present was less than when only the aqueous solution were subjected to AC or DC; and (3) rate of degradation for AC was less than that for DC at an equivalent current density.

ACKNOWLEDGEMENT

Financial support for this project has been provided by the National Science Foundation (Grant No. BES-0402772). Donggeun Li assisted in the experiments on cells containing sand and James Brenton assisted in the fabrication of the electrochemical cells. We are thankful to Prof. Satish Udpa and Mr. Michael Shiu C. Chan for their help with the selection of the function generator, oscilloscope, and the amplifiers. Any opinions, findings, and conclusions or recommendations expressed in this paper are those of the writers and do not reflect the views of the National Science Foundation.

REFERENCES

Agency for Toxic Substances and Disease Registry (ATSDR) (1995). "Toxicological profile for polycyclic aromatic hydrocarbons." Atlanta, GA: U.S. Department of Health and Human Services, Public Health Service.

Alshawabkeh A. N. and Sarahney (2005). "Effect of current density on enhanced transformation of naphthalene." *Environ. Sci. & Technol.* 39: 5837-5843.

Goel, R. K., Flora, J. R. V., and Ferry, J. (2003). "Mechanisms of naphthalene removal during electrolytic aeration." *Water Res.*, 37: 891-901.

Lageman, R., Pool, W., and Seffinga, G. A. (1990). "Electro-reclamation: State-of-the-art and future developments." Contaminated Soil '90, F. Arendt, M. Hinsenveld and W. J. van den Brink, eds., Kluwer Academic, Dordrecht, The Netherlands, 1071-1078.

Li X-Y., Cui, Y-H., Feng, Y-J., Xie, Z-M., and Gu, J-D. (2005). "Reaction pathways and mechanisms of electrochemical degradation of phenol on different electrodes." *Water Res.*, 39: 1972-1981.

Page, M. M. and Page, C. L. (2002). "Electroremediation of contaminated soils." *J. Environ. Engr.*, 128(3): 208-219.

Pepprah, E (2007). "Degradation of Polycyclic Aromatic Hydrocarbons (PAHs) in Aqueous Media using Alternating Current." Ph.D. Dissertation, Dept. of Civil & Environmental Engineering, Michigan State University.

Pepprah, E. and Khire, M. V. (2008). "Electroremediation of naphthalene in aqueous solution using alternating and direct currents." *J. Environ. Engr.*, ASCE, 134-1, in press.

Electrolytic Alkaline Decomposition of a Munition Constituent (RDX) Contaminated Groundwater

David B. Gent[1], Altaf Wani[2], Akram Alshawabkeh[3], Jeffrey L. Davis[1]

[1]Environmental Laboratory, USACE Engineer Research and Development Center, 3909 Halls Ferry Road, Vicksburg, MS 39180, Tel. 601 634 4822 Fax. 601 634 3518, David.B.Gent@erdc.usace.amry.mil, Jeffrey.L.Davis@ erdc.usace.amry.mil
[2]Environmental Resources Management (ERM) 15810 Park Ten Place Tel: 281- 600-1133, Suite 300 Fax: 281-600-1001, Houston, TX 77084, Altaf.Wani@erm.com
[3]Dept. of Civil and Environmental Engineering, 400 Snell Engineering Center, Northeastern University, 360 Huntington Avenue, Boston, MA 02115, aalsha@neu.edu

ABSTRACT: Combined use of electrolysis and alkaline hydrolysis is explored for in-situ decomposition of hexahydro-1,3,5-trinitro-1,3,5-triazine (RDX) in laboratory experiments with simulated groundwater. Alkaline medium generated by water electrolysis at the cathode under direct electric currents will develop a permeable alkaline barrier for in-situ decomposition of RDX. Pseudo-first order transformation rate coefficients were developed from alkaline hydrolysis and electrolytic batch experiments for RDX decomposition with time. The results provided a target pH of 12 and a current density of 1 mA/cm^2 for treatment of RDX under groundwater flow column experiment. The results from the one-dimensional sand-filled alkaline hydrolysis columns (5-cm ID) were used to develop reaction rate coefficients used in designing a large scale-up one-dimensional column to test the electrolytic generation of hydroxide (10-cm ID). The rate coefficient from the 5-cm alkaline columns (0.46 per hr) was used to calculate a column length (160 cm) for complete removal of RDX and its nitroso-substituted products under groundwater flow rate of 30 cm/day. Effluent RDX concentrations from the 10-cm scale-up column (4,000 µg/L influent) were less than 0.1 µg/L for 36 days of treatment. The study concludes that cathodes placed at the down-gradient of groundwater RDX plume can perform as an effective permeable alkaline hydrolysis barrier for decomposition of RDX to levels below EPA drinking water advisory limits.

INTRODUCTION

The use, manufacturing, and storage of nitroaromatic and nitramine explosive compounds resulted in contamination of soil and groundwater. The U.S. EPA lists hexahydro,1-3-5-trinitro-1,3,5-triazine (RDX) under the Unregulated Contaminant Monitoring Regulation for Public in List 2. RDX groundwater contamination disrupts the use of military training areas and impacts local drinking water.

Base hydrolysis decomposes energetic material into small organic and inorganic molecules. The high pH attacks the functional groups of the energetic molecule. Energetic nitramines decompose to form nitrate, nitrites, ammonia, nitrogen, hydrogen, organic acids and formaldehyde. Base hydrolysis treatment has been used with scrap material from demilitarization of munitions and pink water (EPA, 2005). Davis et al. (2006) used microcosms with lime to treat explosive contaminated soil from munition plants and firing ranges. Bonin et al. (2004) treated RDX in a divided flow through electrochemical cell and determined that the rate of degradation increases with current density.

The objective of this study is to explore in-situ groundwater treatment by combining electrolytic and alkaline hydrolysis processes. The specific research objectives were to: assess RDX decomposition under alkaline hydrolysis and electrolytic processes in continuously stirred batch reactors, investigate RDX decomposition under one-dimensional (1-D) flow conditions (using columns filled with sand) by alkaline hydrolysis processes, and evaluate the up-scaling of electrolytic alkaline hydrolysis treatment process in one dimensional flow systems (Fig. 1).

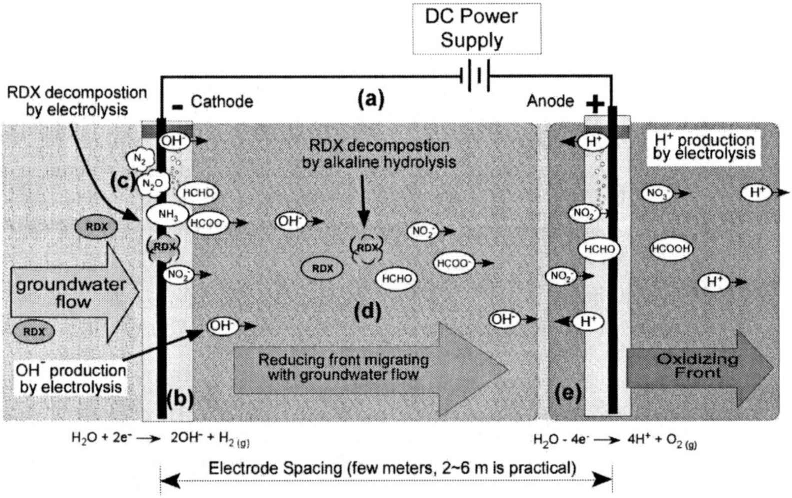

Fig. 1. Concept of the proposed electrolytic remediation barrier (a) in-situ application of direct current across cathodes and anodes, (b) cathode well upstream of anode wells, (c) RDX decomposition direct electrolysis, (d) alkaline front downstream of cathode toward the anode, and (e) anode oxidation.

METHODS AND MATERIALS

Batch Experiments: Five triplicate batch alkaline hydrolysis experiments were conducted with pH ranging from 11 to 12.5 using sodium hydroxide at 25 C. Six triplicate electrolytic experiments were performed with current ranging from 4 to 20 mA. The initial RDX concentrations of the alkaline and electrolytic batch experiments was 1.0 mg/L. Aspirator bottles (500 mL) were used as batch reactors for both the base and electrolytic studies. The electrode material for the electrolytic batch experiments consisted of mixed metal oxide catalyst sintered to an expanded titanium mesh (Elgard™ 85 mesh, Corrpro Companies, Inc., Medina, OH). The anode and cathode electrodes (5 x 1 cm) were inserted into the aspirator bottles with 5-cm spacing. The electrode dimensions and spacing were maximized to fit within the narrow bottle opening. An Agilent Model E3612A dc power supply (60 – 120 V / 0.5 - 0.25 A Dual) operated in constant current mode supplied the electric power to each electrolytic reactor. The results were fitted by non-linear regression pseudo-first order decay equations to determine RDX reaction rate coefficients.

Column Experiments: All columns were prepared by installing polyethylene (PE) porous filter (6.4 mm thickness) material (Porex Technologies, Fairburn, GA) with a neoprene gasket (3.2 mm thickness) and bolting a blind flange to the column. The column was placed in a vertical position and wet filled by pouring sand into the water to prevent air bubble entrapment. After sealing the column, all remaining fittings were installed and the columns were laid horizontal. Clean reverse osmosis water (RO) was pumped through the column for several days to calibrate the volumetric flow rate.

Small diameter sand column (5-cm ID)*:* Three column experiments were conducted to separate the effect of downstream alkaline hydrolysis of RDX from electrolytic decomposition at the electrode (Fig. 2). These experiments include sodium hydroxide (31.6 mM) dripped at a rate of 40 mL/day with a target pH of 12.5 into the cathode well (no electric current) while RDX was pumped into the column from the influent reservoir at concentrations of 0.5 and 1.0 mg/L. In the three experiments the column flow was 0.175 mL/min (seepage velocity equal to 30.5 cm/day) plus the additional hydroxide solution pumped into the cathode well (Fig.2). Port samples were removed every three days for RDX analysis. The RDX results from column experiments were modeled by nonlinear regression to determine pseudo-first-order reaction rate constants using a steady-state solution to the convective-dispersive first-order decay equation (van Genuchten and Alves, 1982) and were compare with the batch results.

Up-Scaled Column (10-cm ID)*:* The up-scaled column setup and operation were similar to those used with the 5-cm column experiments (Fig. 3). After the column was sealed, it was lowered to the horizontal position. The inlet and outlet fitting, in-line pH cells with pH electrodes, and voltage gradient probes were installed and water flowed through the column at a steady rate for 4 days. Six sampling ports were place along the column with one port before the cathode and five ports place at 30.5-cm intervals on center between the cathode and anode (Fig. 3). After the setup was complete the cross circulation pumps were turned on and allowed to equilibrate. A chloride tracer test was conducted to determine the column dispersivity and porosity.

The results from the tracer study analysis were used to calibrate the column flow to a target seepage velocity of 30.5 cm/day.

The Elgard™ 85 mesh electrodes (12.7 by 2.54 cm) and #10 slotted PVC electrode wells (2.54 cm I.D) were installed in the column (65 cm^2 wetted surface area). The column was filled with solution of 0.5 mg/L concentration of RDX by constant hydraulic head gradient. Two pore volumes of 0.5 mg/L concentration RDX solution were passed through the column and samples were collected from all ports and effluent to ensure a consistent concentration throughout the column. After the power supplies were turned on, effluent and port samples were collected twice a week for pH, voltage, current, and RDX analysis.

Effluent samples were collected in 40 mL vials and 1 liter jars. The 1 liter jars were used to collect the effluent when the RDX concentration was below the 20 μg/L HPLC detection limit. Approximately 800 mL of effluent was collected each day and used in solid phase extraction (SPE). The 800 mL sample size lowered the RDX laboratory reporting limits from 20 to less than 0.100 μg/L. Influent concentrations of 0.5, 1.0, 2.0, and 45 mg/L were used during this experiment. The three lower concentrations were used to approach a target concentration of 0.5 mg/L at Port 1. The Port 1 concentrations were used as a performance indicator for the up-scale column since most contaminated field sites have less that 0.5 mg/L RDX in the groundwater.

RDX and its known nitroso-transformation products (MNX, DNX, and TNX) aqueous phase concentrations were measured from the influent and effluent streams, electrode wells, and the sampling ports along the column length. MNX, DNX, and TNX calibration standards were obtained from Supelco® (Boston, MA). Analysis using a modified EPA Method 8330 with a Dionex Summit HPLC system (Sunnyvale, CA) with chemical separation with a Supelco® CN reverse-phase HPLC column.

Inorganic and organic aqueous phase concentrations of chloride, nitrite, nitrate, and formate were analyzed with a DIONEX (Sunnyvale, CA) Ion Chromatograph. Chemical separation and detection were achieved using an Ionpac® AS11 analytical column (250 x 4 mm) and a Dionex conductivity detector (1.25 μL internal volume). The mobile phase consisted of 18.2 MΩ water mixed with concentrated KOH at a flow of 1.5 mL/min. The sample volume was 100 μL of filtered (0.45 μm) sample.

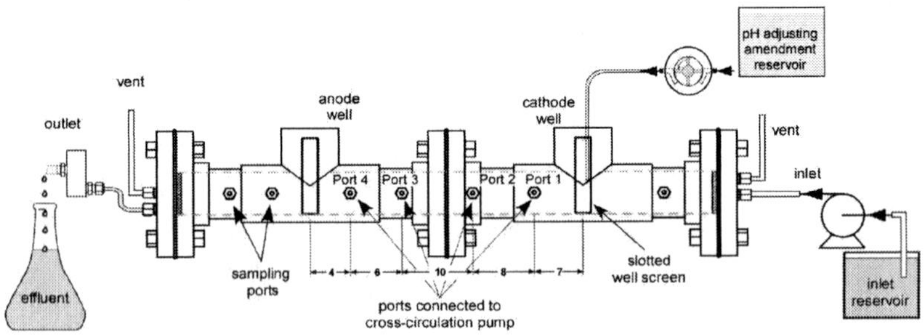

Fig. 2 Alkaline Hydrolysis Column (5-cm ID) by flow

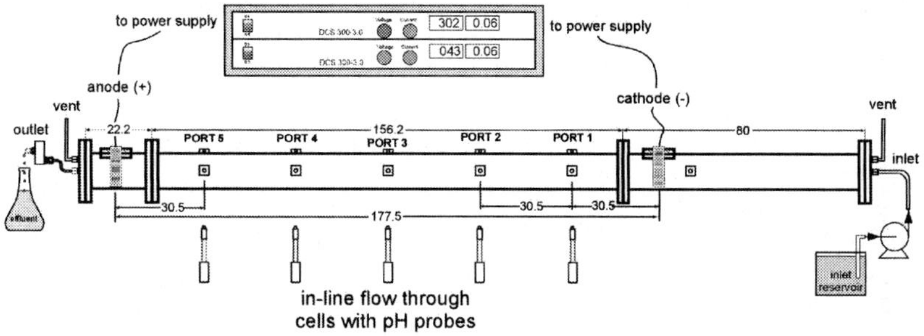

Fig. 3. Up-Scale Column (10-cm ID) Alkaline Hydrolysis by Electrolysis

RESULTS

Batch studies: The results from the alkaline hydrolysis batch experiments confirm that excess hydroxide concentrations increase the rate of RDX destruction. Hydrolysis of RDX shows first-order dependence with hydroxide concentration. The results indicate that alkaline hydrolysis of RDX is not practical below hydroxide concentrations of 3.16 mM (pH 11.5) because of slow reaction kinetics. A practical hydroxide concentration appears to be near 0.01 mM (pH 12), with a 4 to 5.6 hour half-life. The results from the alkaline hydrolysis experiments also provided a target pH (12) for use in the sand filled column studies (Fig. 4a).

The results of the batch electrolytic hydrolysis studies are consistent with those of Bonin et al. (2004) and Doppalapudi et al. (2003) and conclude that higher electric currents results in accelerations of the reaction rate. It was determined that 10 and 20 mA were the best currents to use with the sand filled column experiments because the reaction rates were the highest in these experiments and are optimum for the seepage velocity used in the sand filled columns experiments (Fig. 4b).

Alkaline hydrolysis column by flow: A steady-state solution to the convective-dispersive-reactive transport equation was used to develop reaction rate coefficients for the 0.5 and 1.0 mg/L RDX inlet concentrations. By rearranging the equation and using target influent and effluent concentrations (0.5 and 0.0002 mg/L), an up-scale column length was calculated. An alkaline hydrolysis first order rate coefficient (0.046 hr^{-1}) from the 5-cm alkaline column was used to estimate the optimized scale-up column length (160-cm) required for complete destruction of RDX and its nitroso-substituted products.

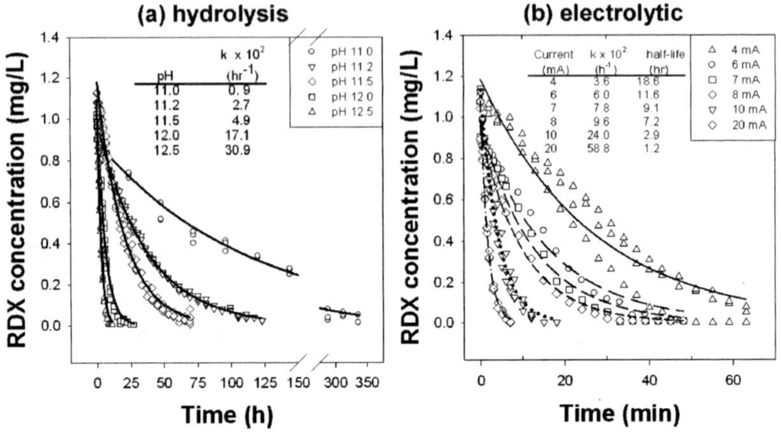

Figure 4. Batch Results (a) alkaline hydrolysis and (b) electrolytic

Up-Scaled alkaline hydrolysis by electrolysis: The initial current was adjusted from 40 to 60 mA after 60 days to increase the decomposition of RDX by increasing the hydroxide concentration through the column. Effluent RDX concentrations from the 10-cm diameter scale up column (500 to 4,000 µg/L influent) were less than the 2 µg/L health advisory standard established by the US EPA (Fig. 5). The alkaline hydrolysis first order reaction rate coefficient from the 10-cm diameter column was 5.4×10^{-2} h^{-1} with a half life of 13.1 h. This rate is similar to batch alkaline hydrolysis at pH 11.5. The decline in RDX concentration in Port 1 was caused by the increase in current at Day 60. After the inlet RDX concentration was increased to 4,000 µg/L (Day 68), the Port 1 RDX concentration increased to target levels (Day 78). The lag time for the concentration changes to reach each ports was because of the long hydraulic retention time in the column (8.3 days).

The RDX removal percentage for 32 days of operation of 10-cm diameter column was 99.9 percent with an electrode current density of 7.2 A/m^2 (Table 1). Nitroso-substituted products (MNX, DNX, and TNX) were below laboratory reporting limits in the column effluent. The effluent end products of RDX ring cleavage detected were formate, nitrite, and nitrate.

Table. 1. Influent concentrations and percent removal (10-cm column)

influent level □	Port A RDX × 10^3 (mg/L)	Port 1 RDX × 10^3 (mg/L)	Effluent RDX × 10^3 (mg/L)	% removal EC†	AH‡	EC + AH§
1	494. ± 10.8	128 ± 22.5	3.4 ± 4.0	70.4 ± 9.88	99.5 ± 0.2	99.5 ± 0.8
2	994 ± 11.4	232 ± 22.0	0.8 ± 0.5	60.0 ± 16.9	99.7 ± 0.3	99.90 ± 0
3	4,000 ± 110	478 ± 170	0.2 ± 0.2	87.8 ± 4.3	99.9 ± 0.1	99.99 ± 0.02

†Electrochemical treatment only
‡Alkaline Hydrolysis
§Electrochemical combined with alkaline hydrolysis

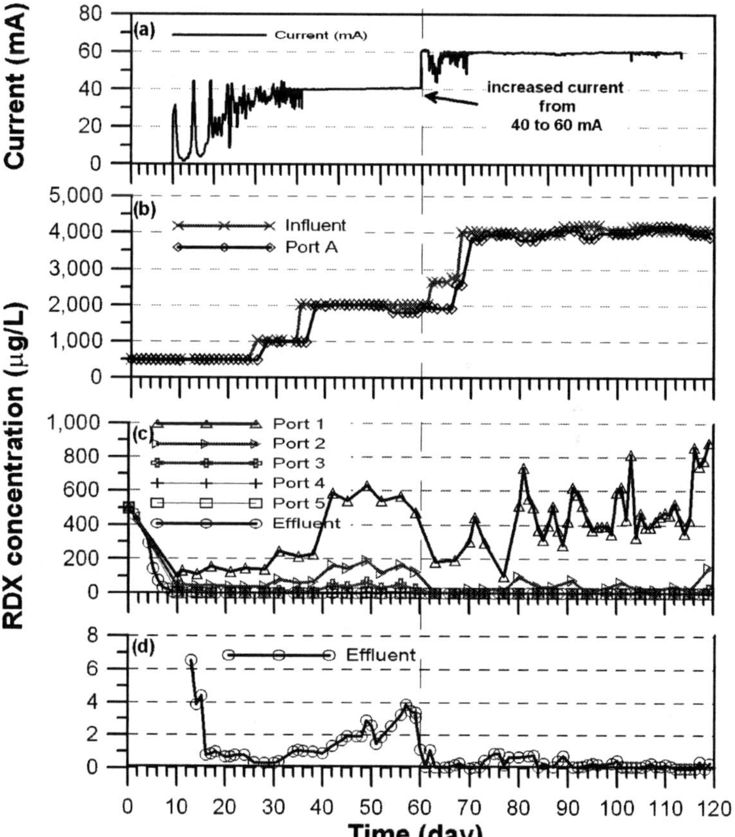

Fig. 5 Up-Scaled Column results vs. time (a) current, (b) influent and Port A RDX (µg/L), (c) Port RDX (µg/L), (d) effluent RDX (µg/L)

CONCLUSIONS

For the up-scale column a specific groundwater flow of 30.5 cm/day (1 ft/day), a current density of 7.4 A/m^2 is sufficient to completely (>99%) destroy 4 mg/L of RDX in flowing groundwater. Up-scaling the decomposition of RDX by electrolytic reactive barriers can be based on the reaction rate coefficients from RDX decomposition by alkaline hydrolysis

The significance of this in-situ electrochemical barrier is that the contaminant can be removed from the groundwater without the addition of external chemical amendments and it can be applied below trenching depths. Electrodes placed at the leading edge of a groundwater plume can act as a barrier to RDX transport.

REFERENCES

Bonin, P.M., Bejan, D., Schutt, L., Hawari, J., Bunce, N.J., (2004), "Electrochemical reduction of hexahydro-1,3,5-trinitro-1,3,5-triazine in aqueous solutions," *Environ. Sci. Technol.*, 38, 1595-1599.

Davis, J. L., Brooks, M/ C., Larson, S. L., Nestler C. C., Felt, D.R. (2006), "Lime Treatment of Explosives-Contaminated Soil from Munitions Plants and Firing Ranges," Soil and Sediment Contamination, Volume 15, Number 6, 565 – 580.

Doppalapudi Rajesh B., George A. Sorial, and Stephen W. Maloney (2003), "Electrochemical Reduction of 2,4-Dinitrotoluene in a Continuous Flow Laboratory Scale Reactor," *J. Envir. Engrg.* 129, 192.

U. S. Environmental Protection Agency, "2006 Edition of the Drinking Water Standards and Health Advisories," EPA 822-R-06-013, accessed 27 July 2007, http://www.epa.gov/waterscience/criteria/drinking/dwstandards.html#advisory

USEPA (2005), Handbook on the Management of Munition Response Actions, Interim Final Report, EPA 5005-B-01-001.

Van Genuchten, M. Th. and W. J. Alves (1982), Analytical Solutions of the One-Dimensional Convective-Dispersive Solute Transport Equation. United States Department of Agriculture, Agricultural Research Service, Technical Bulletin 1661.

Enhanced Electrokinetic Remediation of Soil Contaminated with Heavy Metals

Bala Ramudu Paramkusam[1], Ram Prakash Tiwari[2] and Rakesh Kumar Srivastava[3]

[1]Research Scholar, Department of Civil Engineering, Motilal Nehru National Institute of Technology, Allahabad 211 004, INDIA; balaramudu_p@yahoo.com
[2]Asst. Professor, Department of Civil Engineering, Motilal Nehru National Institute of Technology, Allahabad 211 004, INDIA; rpt_ced@yahoo.co.in
[3]Professor, Department of Civil Engineering, Motilal Nehru National Institute of Technology, Allahabad 211 004, INDIA; rksciv@yahoo.com

ABSTRACT: The experimental set-up was designed to examine the removal of heavy metals cadmium and copper from clayey soil by electrokinetic (EK) remediation process. Different enhancement agents such as chelating (ammonium citrate) and surface active agent (tween-20) have been tested for better removal of the metallic species by electrokinetic process. Applied voltage was 35 V DC and time duration was 120 h. A large pH influence has been observed over removal of metals. The results showed that the ammonium citrate during remediation process enhanced the removal rate of cadmium (48.8%) and copper (30.0%) compared to experiments with tween-20. EK treated soil samples were analysed by X-ray diffraction (XRD) and X-ray fluorescence (XRF) to understand the mineralogical changes during electrokinetic process.

INTRODUCTION

The assessment of heavy metal contaminated soil has received much attention in the last few decades. Copper (Cu) and cadmium (Cd) can be considered as the most extensive pollutants. They are toxic trace metals, which can be introduced into and accumulate in soil through agricultural application of sewage sludge and fertilizers, and/or through land disposal of metal-contaminated municipal and industrial wastes (Wang and Xing, 2002). They have been included on the US Environmental Protection Agency's (EPA) list of priority pollutants (Mulligan et al., 2001). Thus, it is imperative to develop innovative technologies to clean the contaminated soils. This paper presents an emerging in-situ cleanup technology, i.e., electrokinetic remediation (EKR). Use of chelating agents is unavoidable for enhancement of electrokinetic extraction sometimes due to increase in pH (Puppala et al., 1997). A number of studies on metal extraction using HNO_3 and HCl (Tampouris et al., 2001) and the most commonly used chelater like Ethylenediaminetetraacetic acid (EDTA) (Chaiyaraksa and Sriwiriyanuphap, 2004) have been reported. EDTA forms very stable soluble complexes that are highly effective in remediating heavy-metal-contaminated soils (Kedziorek and Bourg, 2000; Kim et al., 2003). Despite satisfactory removal ratios by these extractants, many studies have observed that metal removal is associated with a number of disturbing physical, chemical and biological impacts (Bassi et al., 2000; Papassiopi et al., 1999). The present study is to develop an effective EKR system for the removal of Cd and Cu from the spiked Cd(II) contaminated (CdS) and Cu(II) contaminated (CuS) soil. In particular, several flushing solutions were tried and specifically in this paper results of one chelating agent (ammonium citrate) and one surfactant (Tween-20) have been discussed for their potential use in the removal of cadmium and copper from the field soil.

MATERIALS AND METHODS

For the present study, field soil was obtained from a non-polluted site, Meja village, Allahabad district (Uttar Pradesh, India) from an approximate depth of 2 m. Soil was air-dried and washed through 0.425 mm sieve to obtain finegrained soil. The soil was classified as clayey soil with low compressibility (CL) as per USCS and has the following properties: organic content 6.25%; soil pH 8.36; cation exchange capacity 22 meq/100g; $CaCO_3$ content of 2.5%; Cd below detection limit and Cu 0.15 mg g^{-1}. Qualitative XRD was performed for mineral identification and the minerals found were Quartz, Feldspar, Calcite, Dolomite, Illite and Kaolinite. The artificial Cd(II) and Cu(II) contaminated soils were prepared by adding an appropriate amount of cadmium nitrate ($Cd(NO_3)_2.4H_2O$), copper sulphate ($CuSO_4.5H_2O$) solution and soil into a plastic container individually. The slurry was mixed for 3 days to reach equilibrium adsorption. At the end of the aging period the soil was analysed for its Cd(II) and Cu(II) concentrations individually. It was determined by acid digestion with concentrated HNO_3 and using Atomic Absorption Spectrometry (AAS) at a wavelength on 228.8 nm for Cd and 324.8 nm for Cu. An average concentration of 10,044 mg kg^{-1} for Cd(II) and 15,880 mg kg^{-1} for Cu(II) was measured, which was below the maximum adsorption capacity (13,300 mg kg^{-1} and 19,970 mg kg^{-1} for Cd(II) and Cu(II) respectively) of the soil as determined from Langmuir isotherm. A series of electrokinetic experiments (Table 1) were conducted in an experimental set up which comprised of a perspex cylindrical cell shown in Fig. 1. The cell has an inner diameter of 6.7 cm and consists of three compartments: anode electrode chamber (5 cm length), soil cell (35 cm length) and cathode electrode (graphite) chamber (5 cm length). A fliter paper reinforced with cork and rubber sheet was arranged between the electrode chamber and soil chamber to separate the soil from procesing fluid. An anode reservoir was attached to anode electrode chamber so that it can supply the fluid whenever reduction of processing fluid in electrode chamber at a hydraulic gradient of 0.3. The soil cell contained approximately 1800 g of the spiked Cd(II) and a Cu(II) contaminated soil with a moisture content of 40.0% considering the fully saturated level at liquid limit value of the soil. Electrodes were connected with a DC power supply which provides a constant electric potential gradient to induce the movement of electrolyte solution within the soil cell. During the EK process, current across the soil specimen, the amount of electro osmotic (EO) flow collected at the cathode end, the pH of processing fluid at anode and cathode chambers and the amount of migrated Cd(II) and Cu(II) with processing fluid were monitored. At the end of the each test, the soil was removed from the cell and sliced into 6 segements. The soil pH and residual Cd(II) and Cu(II) in the soil were analyzed. XRF technique was followed in analyzing these residual concentrations. The mineralogical varaions were identified by XRD technique.

Table 1. The Test Program for EK Remediation

Test	Heavy metal	Concentration (mg g^{-1})	Soil Saturation	Anode Flushing Agent
EKCdD	Cd	10.044	DI Water	DI Water
EKCdF	Cd	10.044	DI Water	Tween – 20 (0.2 % v/v) (6.4 pH)
EKCdAC	Cd	10.044	DI Water	AC (1M) (10 pH)
EKCdAA	Cd	10.044	AC (1M) (5 pH)	AC (1M) (10 pH)
EKCuD	Cu	15.880	DI Water	DI Water
EKCuF	Cu	15.880	DI Water	Tween – 20 (0.2 % v/v)
EKCuAC	Cu	15.880	DI Water	AC (1M) (10 pH)
EKCuAA	Cu	15.880	AC (1M) (5 pH)	AC (1M) (10 pH)

EKCdD: EK remediation of Cd(II) contaminated soil using DI water.
EKCdF: EK remediation of Cd(II) contaminated soil using tween-20.
EKCdAC: EK remediation of Cd(II) contaminated soil using ammonium citrate.
EKCdAA: EK remediation of Cd(II) contaminated soil saturated with ammonium citrate and same used as flushing agent.
EKCuD: EK remediation of Cu(II) contaminated soil using DI water.
EKCuF: EK remediation of Cu(II) contaminated soil using tween-20.
EKCuAC: EK remediation of Cu(II) contaminated soil using ammonium citrate.
EKCuAA: EK remediation of Cu(II) contaminated soil saturated with ammonium citrate and same used as flushing agent.

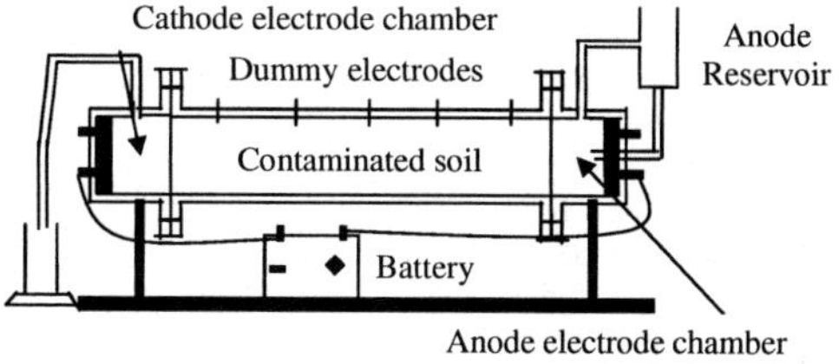

FIG. 1. Schematic diagram of overall electrokinetic test setup

RESULTS AND DISCUSSION

Current Variation

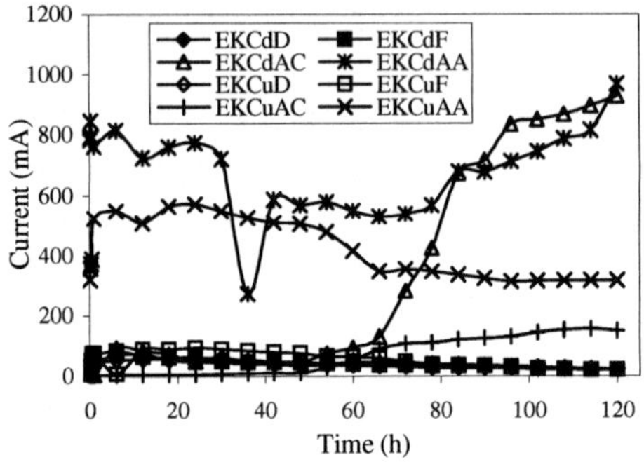

FIG. 2. Current with time during EK process

The variations of current through out the duration for all EK experiments are shown in Fig. 2. It has been reported that the measured electric current is proportional to the dissolved species present in the solution (Acar and Alshawabkeh, 1993). In EKCdD, EKCuD, EKCdF and EKCuF the electrical conductance decreased. This sharp decrease of current with time reflects that the total electric resistance increased in the system. Possibly as a result of decreased dissolved species due to contaminant migration or precipitation and consequently, decreasing electric current. High average current values were observed in case of EKCdAA and EKCuAA. Initial explanation for the two tests EKCdAA and EKCuAA where current was quite high throughout the experiments is that it may be due to the presence of dissolved species and less contaminant migration in the system.

Electroosmotic Flow

The direction of EO flow was observed to be from anode to cathode in all the experiments. Fig. 3 shows the plot for the cumulative volume of effluent versus time. When DI water was used for purging, the flow rate was relatively low in EKCdD and EKCuD. It was observed that flow ceased within one day in case of DI water and surfactant purged experiments and EO flow continued until almost the end of the experiment for EKCdAA and EKCuAA. But the volumes collected in case of copper are much less than volumes collected in cadmium contaminated soil when comparing EKCdAC & EKCdAA with EKCuAC & EKCuAA respectively. EO flow values of EKCdF and EKCuF are observed to be low compared to AC purged experiments and high in comparison to DI water purged experiments. The calculated EO permeability (k_e) values for EO flow are presented in Table 2.

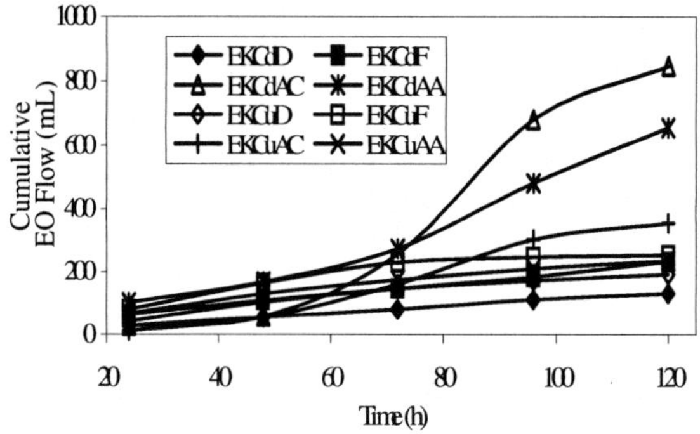

FIG. 3. Cumulative EO Flow vs Time

Table 2. Electroosmotic Permeability Values (k_e)

Test	k_e (cm^2/V-s)	Test	k_e (cm^2/V-s)
EKCdD	3.17×10^{-05}	EKCuD	4.86×10^{-05}
EKCdF	5.79×10^{-05}	EKCuF	7.32×10^{-05}
EKCdAC	1.44×10^{-04}	EKCuAC	6.09×10^{-05}
EKCdAA	1.32×10^{-04}	EKCuAA	6.31×10^{-05}

The EO permeability, k_e (cm^2/V-s) was calculated using the equation

$$k_e = Q_e/(i_e.A), \qquad (1)$$

Where, Q_e is EO flow (mL/d); i_e=Electric gradient (V/cm); A= Cross-sectional area of soil cell (cm^2).

Soil pH vs Concentration Profiles

pH Variations and distribution of Cd(II) and Cu(II) concentrations of the EK treated soils are shown in Figs. 4-7. In test EKCdD, about 60% of cadmium was removed from the anode to section 4 but this amount precipitated in the next slice, and thus no cadmium was removed further. In the case of EKCdAC, EKCdAA, EKCuAC and EKCuAA all values of C/C_0 were below 1, which implies that heavy metals were partly removed from soil. In comparison with EKCdAC (869 mL of EO Flow), EKCdAA (657 mL of EO Flow) achieved less cumulative EO flow and high removal of cadmium was observed. Similarly EO flow in EKCuAA was also observed less than EKCuAC but the removal efficiency is high. Under these experimental conditions, removal efficiencies for EKCdD, EKCdF, EKCdAC, EKCdAA, EKCuD, EKCuF, EKCuAC, and EKCuAA were 4.15%, 6.9%, 36.3%, 48.9%, 0%, 0%, 25.3%, and 30.0% respectively.

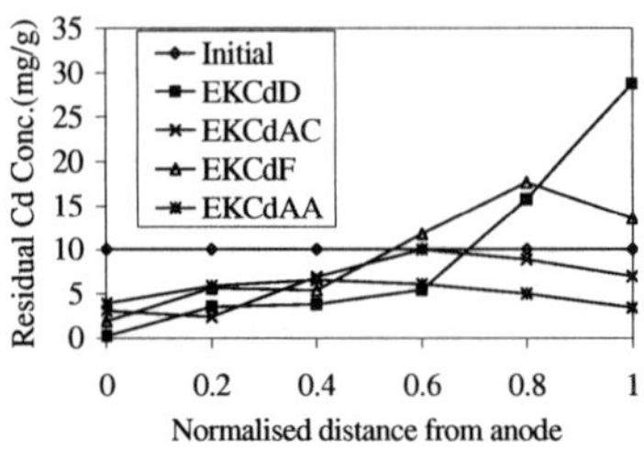

FIG. 4. Residual Cd(II) concentrations profiles of EK treated soils

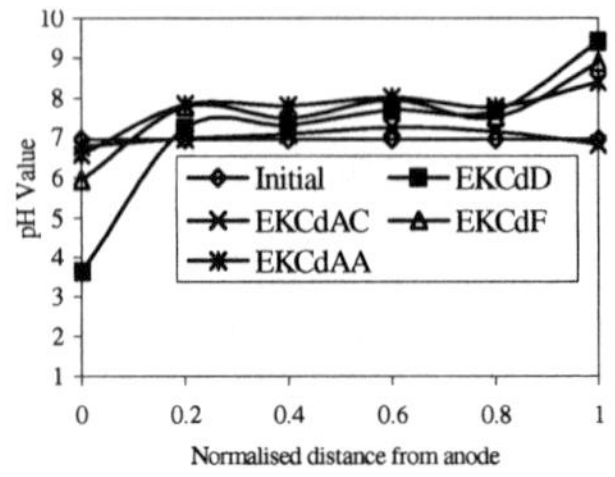

Fig. 5. pH distribution of EK treated Cd(II) contaminated soils

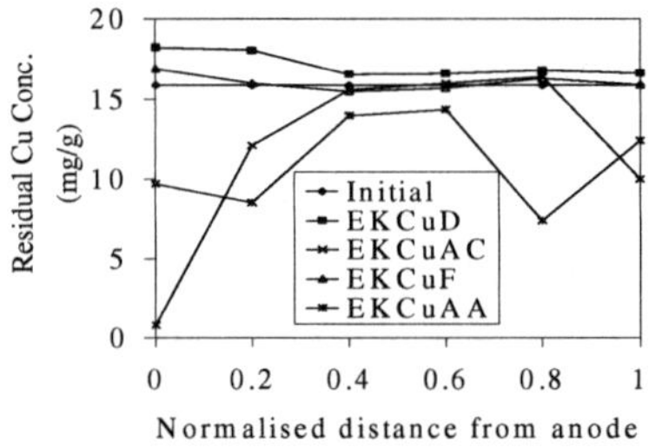

FIG. 6. Residual Cu(II) concentrations profiles of EK treated soils

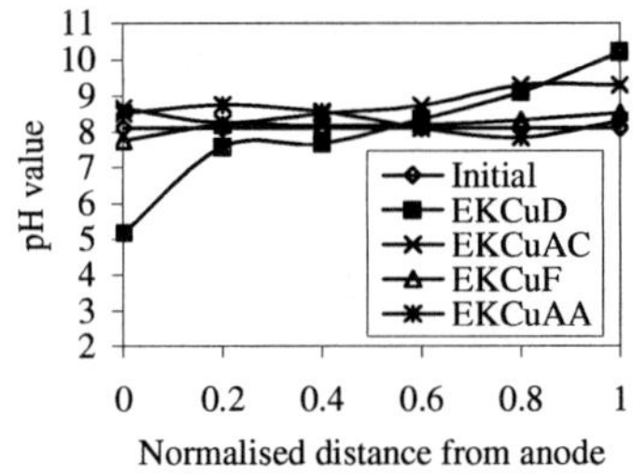

Fig. 7. pH distribution of EK treated Cu(II) contaminated soils

Moisture Content

Moisture content of the soil was measured and is presented in Fig. 8. The precipitation of cadmium ions caused the reduction in electroosmotic flow, which led to significant changes of moisture content with variation in contaminant concentration. There was no significant change in moisture content in case of EKCdF. It indicates that no significant physico-chemical changes occur with the use of Tween-20 as flushing agent. The overall moisture content was decreased in EKCdAC and EKCdAA and volume changes were observed. As a result, consolidation had occurred and crack formation was observed in EKCdAA. High moisture content has been observed in EKCuAC and the lowest moisture content was observed in case of test run with surfactant purged copper contaminated soil (EKCuF) amongst all. The moisture content in case of EKCuD and EKCuAA are observed to be at the level of initial moisture content.

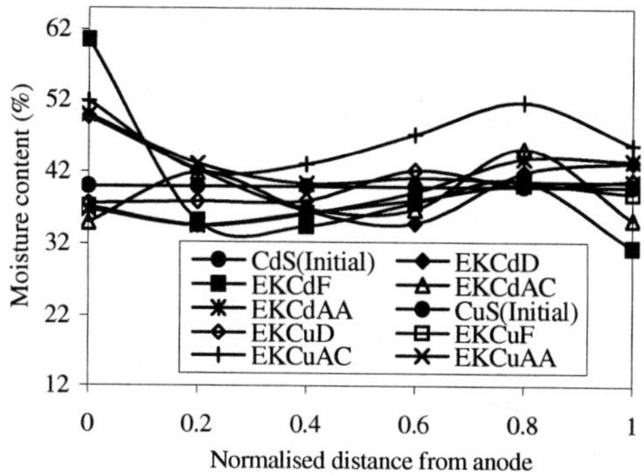

FIG. 8. Moisture content variations in EK treated soils

Mineralogical and Microstructural Analysis

Considering importance of heavy metal – mineral association in high sorption capacity soils, in the present study, the mineralogical and microstructural analytical methods including XRD was used to characterize EK treated soils for correlating metal removal efficiencies with respect to soil mineral changes. X-ray diffraction (XRD) patterns measured in a Rigaku DMAX 2000/JADE 6.0 X-ray diffractometer. The qualitative XRD results are presented in Fig. 9.

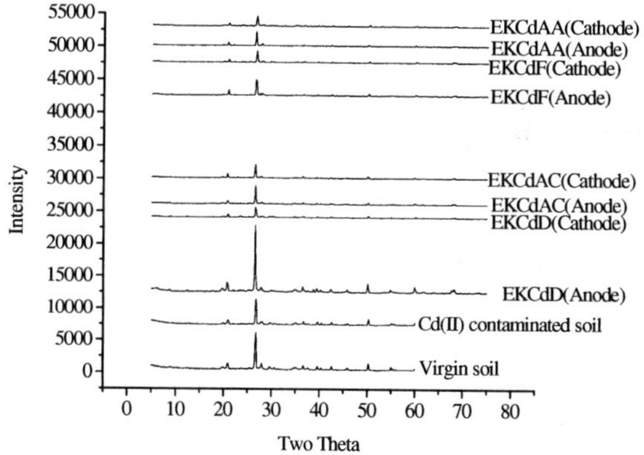

FIG. 9. XRD of the virgin, cadmium contaminated and EK treated soils

The over all mineralogical composition changes are observed to be insignificant in all the cases of both cadmium and copper contaminated soils. Remarkable variations have been observed with respect to the peak heights of the various minerals. Especially in case of EKCdAA and EKCuAA the leaching of minerals has occurred, which was confirmed from the XRD peak data.

CONCLUSIONS

Better results are observed in the case of soil saturated with ammonium citrate for enhancing the Cd & Cu transport. The soil can be fully remediated, if the remediation times are longer. The use of ammonium citrate, whose cost is relatively low, is attractive because its metal complexes are easily biodegradable and therefore, do not have long lasting effects. The results of present experimental study are supportive of investigation on the Cd & Cu transport in low hydraulic conductive field contaminated sites for saturation with ammonium citrate before applying electric fields.

REFERENCES

Acar, Y.B., and Alshawabkeh, A. (1993). "Principles of electrokinetic remediation." *Environ. Sci. Technol.,* Vol. 27(13): 2638-2647.

Bassi, R., Prasher, S.O., and Simpson, B.K. (2000). "Extraction of metals from a contaminated sandy soil using citric acid." *Envi. Prog.,* Vol.19(4): 275–282.

Chaiyaraksa, C., and Sriwiriyanuphap, N., (2004). "Batch washing of cadmium from soil and sludge by a mixture of $Na_2S_2O_5$ and Na_2EDTA." *Chemosphere*, Vol (56): 1129–1135.

Kedziorek, M.A.M., and Bourg, A.C.M., (2000). "Solubilization of lead and cadmium during the percolation of EDTA through a soil polluted by smelting activities." *J. Contaminant Hydro.*, Vol.40 (4): 381–392.

Kim, C., Lee, Y., and Ong, S.K., (2003). "Factors affecting EDTA extraction of lead from lead-contaminated soils". *Chemosphere*, Vol.51 (9): 845–853.

Mulligan, C.N., Yong, R.N., and Gibbs, B.F. (2001). "Remediation technologies for metal contaminated soils and groundwater: an evaluation." *Eng. Geo.,* Vol. 60: 193–207.

Papassiopi, N., Tambouris, S., and Kontopoulos, A. (1999): Removal of heavy metals from calcareous contaminated soils by EDTA leaching, *Water, Air, Soil Pollu,*. Vol.109: 1–15.

Puppala, S.K., Alshawabkeh, A.N., Acar, Y.B., Gale, R.J., and Bricka, M. (1997). "Enhanced electrokinetic remediation of high sorption capacity soils." *J. Haz. Mat.,* Vol.55: 203–220.

Tampouris, S., Papassiopi, N., and Paspaliaris, I. (2001). "Removal of contaminant metals from fine grained soils, using agglomeration, chloride solutions and pile leaching techniques." *J. Haz. Mat.*, Vol.B(84): 297–319.

Wang, K.J., and Xing, B.S. (2002). "Adsorption and desorption of cadmium by goethite pretreated with phosphate." *Chemosphere*, Vol. 48: 665–670.

Transport and Speciation of Heavy Metals in Soils during Electrokinetic Remediation: Influence of Soil Type and Electric Potential

Ashraf Z. Al-Hamdan[1] and Krishna R. Reddy[2]

[1]University of Alabama in Huntsville, Department of Civil and Environmental Engineering, 301 Sparkman Drive, Huntsville, Alabama 35899, USA; alhamdan@eng.uah.edu
[2]University of Illinois at Chicago, Department of Civil and Materials Engineering, 842 West Taylor Street, Chicago, Illinois 60607, USA; kreddy@uic.edu

ABSTRACT: This paper presents transport and speciation of heavy metals in soils during electrokinetic remediation. A series of laboratory electrokinetic experiments was conducted using two different clayey soils, kaolin and glacial till, to investigate the migration and removal of heavy metals Cr(VI), Ni(II), and Cd(II). In kaolin, the acid front advances towards the cathode faster than the base front towards the anode, leading to a development of an acidic condition in most of the soil. In glacial till, however, because of its high buffering capacity, the development of the acid front is hindered, which results in alkaline conditions throughout the soil during electrokinetic remediation. In both kaolin and glacial till, Cr(VI) migration towards the anode increases as the applied voltage gradient increases. However, Cr(VI) migration was higher in glacial till as compared to kaolin because of the high pH conditions that exist in glacial till. As the voltage gradient increases, greater migration of Cr(VI) was observed in glacial till as compared in kaolin. Some Cr(VI) was reduced to Cr(III) in both soils. The Cr(VI) reduction rate to Cr(III) as well as the Cr(III) migration are significantly affected by the applied voltage gradient, leading to different chromium distributions throughout the soil. In kaolin, the applied voltage gradient has a pronounced effect on the migration of nickel and cadmium. The Ni(II) and Cd(II) migration towards the cathode increases in kaolin as the applied voltage gradient increases. Unlike kaolin, in glacial till the applied voltage gradient has no effect on nickel and cadmium migration because of its high acid buffering capacity.

INTRODUCTION

Electrokinetic remediation has received increased attention to remediate contaminated sites, and it involves installing electrodes in the contaminated soil and applying a low DC electric potential. As a result of transport processes such as electromigration, electroosmosis, and diffusion, contaminants are transported into the electrodes where they are removed by pumping out water from the electrodes or treated using electroplating and adsorption methods (Acar and Alshawabkeh 1996; Reddy and Parupudi 1997). Numerous laboratory experimental studies have been performed to assess electrokinetic remediation on soils contaminated with heavy metals (Acar and Alshawabkeh 1996; Eykholt and Daniel 1994; Pamukcu and Wittle 1992; Reddy and Parupudi, 1997). Most of these studies involved conducting experiments for a specific duration and voltage gradient conditions and assessing the extent of metal migration and removal. The objective of this study is to systematically evaluate the effects of soil type and voltage gradient on the migration and speciation of heavy metals, specifically chromium, nickel, and cadmium, during electrokinetics. An assessment of increasing voltage gradient effects will help determine the optimal voltage gradient that produces maximum contaminant migration and removal, yet does not increase energy costs.

EXPERIMENTAL METHODOLOGY

Electrokinetic Test Setup

Figure 1 shows a schematic of the electrokinetic test setup used for all of the tests performed in this study. The detailed description of this setup is given by Reddy and Parupudi. (1997). The setup consisted of an electrokinetic cell, two electrode compartments, two electrode reservoirs, a power source, and a multimeter. The electrokinetic cell had an inside diameter of 6.2 cm and a total length of 19.1 cm. This cell was made of plexiglass. Each electrode compartment consisted of a valve to control the flow into the cell, a slotted graphite electrode, a porous stone, and a filter paper. Small holes were drilled in the electrode compartment for insertion of the electrode pins. The graphite electrode and the porous stone were separated with filter paper. The electrode compartment was connected to either end of the cell using four screws. The soil specimen in the electrokinetic cell and the porous stone were separated by filter paper. Electrode reservoirs were made of 3.8 cm inner diameter plexiglass tubes and were connected to the electrode compartments using tygon tubing. Exit ports were made in the electrode compartments and thin tubes were then inserted into these ports to allow gases generated from the electrolysis of water to escape. The other end of each gas tube was connected into the reservoir to collect any liquid that was removed along with the gases. A power source was used to apply a constant voltage to the electrodes and a multimeter was used to monitor voltage and measure the current flow through the soil sample during the testing.

Materials

Two different clayey soils were selected for this study: kaolin, which is a typical low buffering soil, and glacial till, which is a typical high buffering soil. Kaolin consists of kaolinite mineral, while glacial till consists of quartz, feldspar, carbonates, illite, chlorite, vermiculite, and smectite. The kaolin is free of any organic matter and it possesses low hydraulic conductivity of 1.3×10^{-7} cm/s under effective confining pressure of 69 kPa. The glacial till consists of 2.8% organic content and its hydraulic conductivity is 4.1×10^{-8} cm/s under effective confining pressure of 69 kPa. The kaolin is slightly acidic (pH 4.9), while glacial till is slightly basic (pH 8.2). A detailed characterization of these soils is given by and Reddy and Parupudi (1997).

Hexavalent chromium, Cr(VI), nickel, Ni(II), and cadmium , Cd(II) were used to represent heavy metal contaminants for this research. The source chemicals used were potassium dichromate (K_2CrO_4, ACS certified) for Cr(VI); nickel chloride

($NiCl_2.6H_2O$, technical fisher chemical) for Ni(II), and cadmium chloride ($CdCl_2.2.5H_2O$, ACS certified) for Cd(II).

Testing Program

The testing program consisted of four experiments to investigate the effect of the induced voltage gradient on enhanced migration of heavy metals in kaolin and glacial till soils during electrokinetics. These tests were conducted with different applied voltage gradients, 0.5 and 2 VDC/cm. These experiments were conducted by spiking both soils with 1000 mg/kg of Cr(VI), 500 mg/kg of Ni and 250 mg/kg of Cd. All of these tests were run for 4 days, and deionized water was used as the electrode solution. The deionized water used for all experiments in this study had pH=5.7, redox=199 mV, and electrical conductivity = 0 μS/cm.

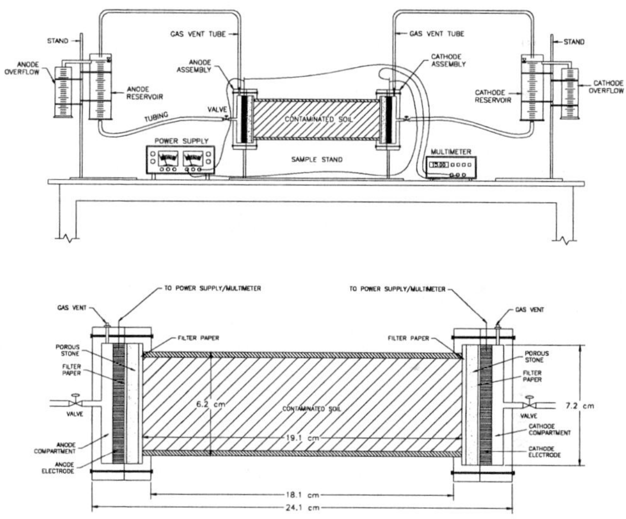

FIG.1. Bench-scale electrokinetic test setup

Testing Procedure

Approximately 1100 g of dry soil were used for each test. The required amounts of the chemicals that would yield the desired concentrations of Cr(VI), Ni(II), and Cd(II) were weighed and then dissolved individually in deionized water. These contaminant solutions were then added to the soil and mixed thoroughly with a stainless steel spatula in a HDPE container. A total of 375 mL of deionized water (to yield 35% moisture content) was used for kaolin, while 285 mL of deionized water (to yield 25% moisture content) was used for the glacial till.

The contaminated soil was placed in the electrokinetic cell in layers and compacted uniformly using a hand compactor. The soil was then allowed to equilibrate for a week. The exact weight of the soil used in the cell was determined. The pH, redox potential, and electrical conductivity (EC) of the remaining contaminated soil in the HDPE container were measured both before and after equilibration. The electrode compartments were then connected to the electrokinetic cell. These electrokinetic compartments consisted of graphite electrodes, porous stones with filter papers inserted between the electrode and porous stone as well as between the porous stone and the soil. The electrode compartments were connected to the anode and cathode reservoirs with tygon tubing. The reservoirs were then filled with the desired electrode solution. The elevations of the electrode solution in both reservoirs were kept the same so that no hydraulic gradient existed across the specimen. The electrokinetic cell was then connected to the power supply and a constant voltage gradient of 0.5 VDC/cm, 1VDC/cm, or 2 VDC/cm for was applied to the soil sample. The electric current across the soil sample as well as the water flow, pH, redox potential, and electrical conductivity in both the anode and cathode reservoirs were measured at different time periods throughout the duration of the experiment.

At the end of the test duration, aqueous solutions from the anode and cathode reservoirs and the electrode assemblies were collected and volume measurements were made. Then, the reservoirs and the electrode assemblies were disconnected and the soil specimen was extruded from the cell using a mechanical extruder. The soil specimen was sectioned into five parts and each part was weighed and subsequently preserved in glass bottles. For each section, about 10 g of soil was mixed with 10 mL of $CaCl_2$ solution in a glass vial. The mixture was shaken using a mechanical shaker and the solids were allowed to settle. The pH, redox potential, and the electrical conductivity of the soil-water mixture as well as that of the anolyte and catholyte were measured (ASTM D4972). The moisture content of each soil section was also determined in accordance with the ASTM Standard D2216.

Chemical Analysis

Contaminants in different soil sections were extracted by performing acid digestion in accordance with USEPA 3050 procedure and analyzed using an atomic absorption spectrophotometer (AAS) to determine the concentrations of total chromium, nickel and cadmium in accordance with the USEPA methods 7190, 7520, and 7130, respectively (USEPA, 1986). Aqueous samples from the electrode reservoirs were directly tested using AAS for the contaminant concentrations. Alkaline digestion was performed on soil sections in accordance with the USEPA 3060A procedure which extracts only Cr(VI) into the solution and then analyzed using AAS to determine Cr(VI) concentrations. Cr(III) concentrations were calculated by mass balance, i.e., subtracting Cr(VI) from the total chromium.

RESULTS AND ANALYSIS

Tests performed using kaolin showed that the current reached a peak current values of 238 mA, 82 mA, and 34 mA in 14, 20, and 49 hours for the tests with 2, 1, and 0.5 VDC/cm, respectively, then, it decreased and finally stabilized at approximately 8 mA after 20, 30, and 80 hours for the tests with 2, 1, and 0.5 VDC/cm tests, respectively. It is clear that higher current is generated initially with higher voltage gradient, but the current decreases and then stabilizes at a low constant value. This is attributed to the accelerated generation of the acid and base fronts at the anode and cathode reservoirs under a high voltage gradient. Glacial till also had a current variation similar to that for kaolin under different voltage gradient application in that the current reached its highest value after a certain time and then decreased and stabilized. Since glacial till contains many dissolved species other than the spiked contaminants, it has higher final stabilized current densities. Overall, the peak current generated is higher voltage gradient, but long-term stabilized current is the same for all voltage gradients.

The electroosmotic flow increased as the induced potential increased. However, the increase between 0.5 VDC/cm and 1.0 VDC/cm tests was not significant as compared to 2 VDC/cm test. The flow rate was high during the initial stages of testing then the flow rate decreased at later stages. This was attributed to the zeta potential of the soil particles, which is significantly affected by the liquid phase properties (i.e., pH and ionic strength of the fluid). As a result, the electroosmotic coefficient K_e varies during electrokinetic remediation, causing a change with time in the flow volume. During the first 60 hours of the tests, the coefficient of electroosmotic conductivity (K_e) was calculated to be around $2.2x10^{-4}$-$7.4x10^{-4}$ cm^2/s-volt and $9.2x10^{-5}$–$4.6x10^{-4}$ cm^2/s-volt for kaolin and glacial till, respectively. After the first 60 hours, the flow rate reduced to a lower K_e value of $8.8x10^{-5}$- $2.4x10^{-4}$ cm^2/s-volt and $3.9x10^{-5}$ - $1.5x10^{-4}$ cm^2/s-volt for kaolin and glacial till, respectively.

The generation of the acid and base in the anode and cathode reservoirs increased as the applied voltage gradient increased. This was attributed to the fact that the greater the potential difference, the greater will be the amount of current carried in the electrochemical cell, hence, the greater will be the amount of H^+ and OH^- ions produced electrochemically. As shown in Figure 2(a), as the voltage gradient increased the net acid front advanced towards the cathode. For the glacial till, however, which had a high buffering capacity, Figure 2(b) shows clearly that the net base front advances from the cathode region to the middle of the soil sample.

Figures 3(a) and 3(b) show the normalized total chromium profiles for kaolin and glacial till, respectively, after electrokinetics under different applied voltage gradients. In kaolin, the total chromium ions transported towards the anode increased slightly with an increase in the applied voltage gradient. Low pH conditions that existed through most of the soil caused Cr(VI) to adsorb, leading to insignificant effect of voltage gradient. However, in glacial till, significant chromium migration occurred towards the anode. This is attributed to the suitable conditions in glacial till (i.e., high pH) for the hexavalent chromium to exist as aqueous form.

Speciation of residual chromium in terms of Cr(VI) and Cr(III) is shown in Figures 4 and 5 for kaolin and glacial till, respectively. Higher voltage gradient application caused increased electromigration of Cr(VI) towards the anode. A partial reduction of Cr(VI) to Cr(III) is observed in both kaolin and glacial till. Moreover, the distribution profile of Cr(III) in kaolin indicates that as the applied voltage gradient increased, the produced Cr(III) migrated and accumulates (precipitates and/or adsorbed) further towards the cathode. In glacial till, the generated Cr(III) mostly accumulated near the anode region because of the high pH conditions in the soil that hindered its migration.

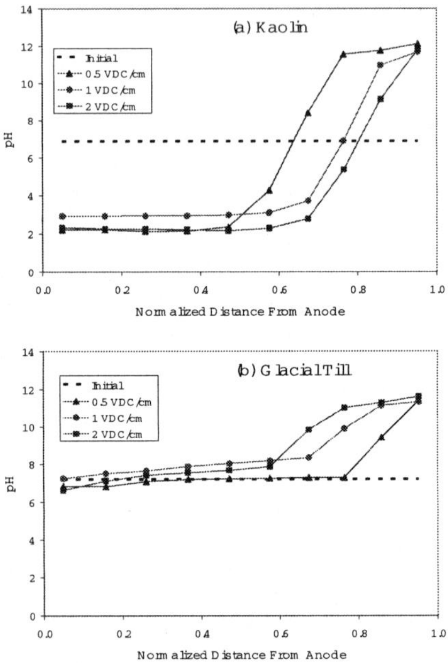

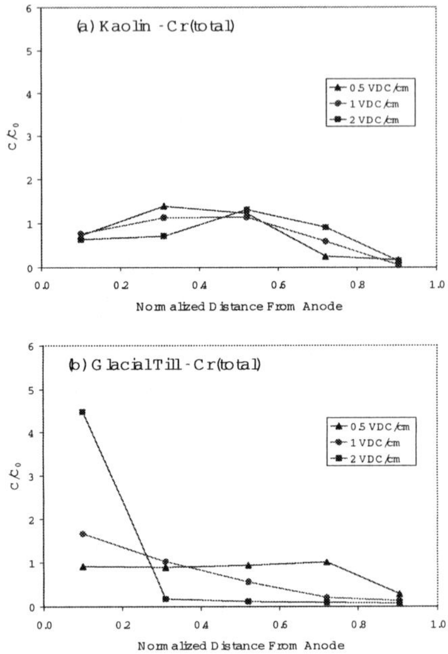

FIG2. pH distribution **FIG.3. Total chromium distribution**

Figures 6(a) and 6(b) show the extent of nickel migration under different voltage gradient conditions in kaolin and glacial till, respectively. In kaolin, nickel migrated toward the cathode as the voltage gradient increased (Figure 6(a)). In the test with 2 VDC/cm gradient, nickel accumulated and the maximum value existed at the region close to the cathode. The effect of the voltage gradient on the contaminant movement happens by affecting the advance of pH profile in the soil, the electroosmotic flow (i.e., osmotic velocity increases as voltage gradient increases), and the migrational flux of the ion (i.e., as the voltage increases, the migrational flux increases).

As shown in Figures 6(b), nickel was retained in the glacial till and none or very little migration occurred. This is due to the high pH of the soil, leading the nickel to precipitate throughout the soil sample. The high buffering capacity of glacial till hindered the acid front in the soil close to the anode region even during the high voltage gradient of 2VDC/cm. This high pH condition caused the nickel and cadmium to exist in the adsorbed or precipitated form.

Figures 7(a) and 7(b) show the extent of Cd(II) migration under different voltage gradient conditions in kaolin and glacial till, respectively. The migration of Cd(II) and the effects of increasing voltage gradient on enhanced cadmium migration were similar to that observed for nickel. Cadmium migration towards the cathode increased as the voltage gradient increased. In glacial till, however, no migration of Cd(II) occurred irrespective of the magnitude of voltage gradient. Cadmium adsorbed/precipitated due to high pH resulting from the high carbonate content of this soil.

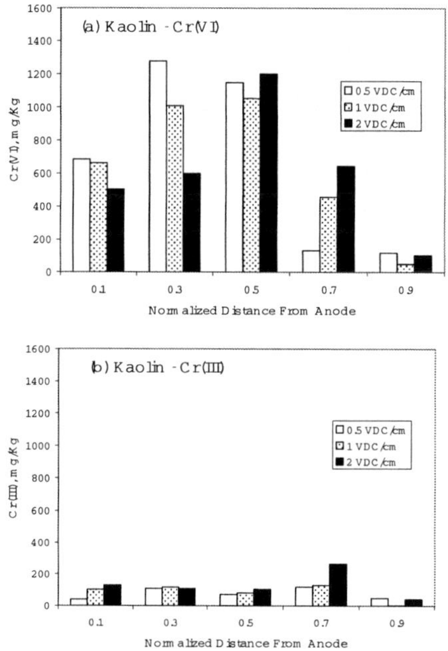

FIG.4. Chromium species in kaolin

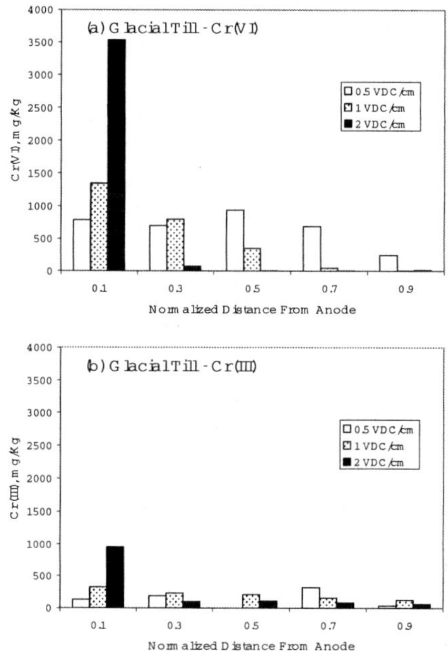

FIG.5. Chromium species in glacial till

CONCLUSIONS

In both kaolin and glacial till, Cr(VI) migration towards the anode increased as the applied voltage gradient increased. However, Cr(VI) migration was higher in glacial till as compared to kaolin because of the high pH conditions that existed in glacial till. As the voltage gradient increased, greater migration of Cr(VI) was observed in glacial till as compared in kaolin. Some Cr(VI) was reduced to Cr(III) in both soils. The Cr(VI) reduction rate to Cr(III) as well as the Cr(III) migration are significantly affected by the applied voltage gradient, leading to different chromium distributions throughout the soil. In kaolin, the applied voltage gradient has a pronounced effect on the migration of nickel and cadmium. The Ni(II) and Cd(II) migration towards the cathode increased in kaolin as the applied voltage gradient. However, the time effect was noticeable to a certain extent after which it diminished because of the generated soil conditions by the cathode that hindered the nickel and cadmium migration. Unlike kaolin, in glacial till the applied voltage gradient had no effect on nickel and cadmium migration because of its high buffering capacity.

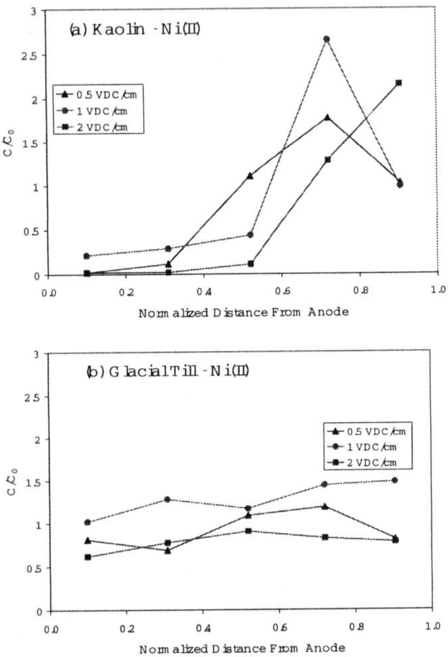

FIG.6. Nickel distribution

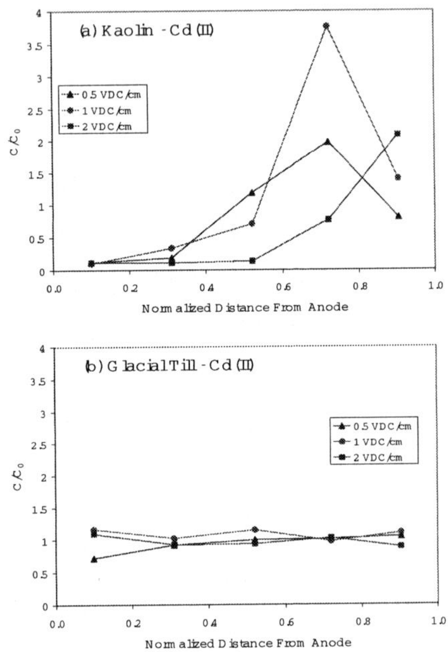

FIG.7. Cadmium distribution

REFERENCES

Acar, Y.B., Gale, R.J., Alshawabkeh, A.N., Marks, R.E., Puppula, S., Bricka, M., and Parker, R. (1995). "Electrokinetic remediation: Basics and technology status." *J. Hazard. Mater.* 40, 117–137.

Eykholt, G.R., and Daniel, D.E. (1994). "Impact of system chemistry on electroosmosis in contaminated soil." *Journal of Geotechnical Engineering* 120(5): 797-815.

Pamukcu, S., and Wittle, J.K. (1992). "Electrokinetic removal of selected heavy metals from soil." *Environmental Progress* 11(3): 241-250.

Reddy, K.R., Parupudi, U.S., Devulapalli, S.N., and Xu, C.Y. (1997). "Effect of soil composition on the removal of chromium by electrokinetics." *Journal of Hazardous Materials* 55, 135-158.

USEPA: Test Methods for Evaluating Solid Waste, Vol. 1A-1C: Laboratory Manual, Physical/Chemical Methods. *SW-846*. Washington, D.C. 1986.

Remediation of PCE Contaminated Soil Using Nanoparticles

C. Vipulanandan[1], S.Harendra[2]

[1] Chairman, Professor and Director of Center for Innovative Grouting Materials and Technology (CIGMAT), Department of Civil and Environmental Engineering, University of Houston, Houston, Texas 77204-4003. Phone: (713) 743-4278; FAX (713) 743-4260; email: cvipulanandan@uh.edu;

[2] PhD Student, Department of Civil and Environmental Engineering, University of Houston, Houston, Texas 77204-4003. Email: ps_harendra@yahoo.com;

ABSTRACT: Contamination of soils by dense nonaqueous phase liquids (DNAPLs) such as chlorinated hydrocarbons is an environmental concern in industrial and hazardous waste sites. The purpose of this study was to evaluate the effect of different solvents and surfactants to extract tetrachloroethylene (PCE) from contaminated soils and then degrade the solubilized PCE using fine metallic nano-size particles. Solvents such as water, methanol, iso propyl alcohol, anionic surfactant (Sodium dodecyl sulfate (SDS)), cationic surfactant and UH biosurfactant were used to extract PCE from 200,000 mg/kg contaminated soils. Fe/Ni nano-bimetallic particles were used to degrade the extracted PCE in batch reactors. Nanoparticles were characterized using SEM and EDS. Desorption kinetics and isotherms for various solvents and the surfactant in extracting PCE from contaminated soils have been quantified. Rate and order of degradation of PCE by 100g/L Fe-Ni nanoparticles in various extraction solutions have been quantified.

INTRODUCTION

Synthetic organic compounds account for approximately one third of the chemical production in the U.S. Many are intentionally or inadvertently released into the environment. Evidence has mounted that nation's ground water resource, which supplies over 50% of the drinking water in the U.S. is threatened by contamination caused by past and present industrial, agricultural and commercial activities [USGS, 2000]. Chlorinated solvents such as tetrachloroethylene (PCE), trichloroethylene (TCE), dichloroethene (DCE) and vinyl chloride (VC) are among the most prevalent contaminant. Many of these compounds are known or potential threats to public health and the environment, so there is an urgent need to understand their transport and fate in the environment and develop effective control methods.

The removal of chlorinated solvents$_s$ from contaminated soil is becoming a major concern, as their water insoluble characteristics may retain those compounds in solid phase and eventually impair human health through bioaccumulation, bioconcentration and food chain systems.

Surfactants have recently been studied by several investigators for their ability to act as soil washing agents and as facilitators of subsurface remediation of hydrocarbon spills [Chu et al., 2003]. Solvents such as iso propyl alcohol, ethanol have been used to enhance the dissolution of PCE system [Michael et al., 2004]. Both co solvent and surfactant solutions can be used to accelerate the removal of residually trapped NAPLs from porous media through either mobilization or enhanced solubilization. Mobilization of NAPL trapped within the porous media by capillary forces is facilitated by a reduction in the interfacial tension (IFT) between the NAPL and the flushing solution as either a cosolvent or surfactant solution. Similarly, solubilization of the NAPL is enhanced due to the reduced polarity of the flushing solution as compared to that of the resident groundwater [James et al., 1998].

Nanosized zerovalent iron with high surface area has been synthesized using conventional approaches such as solution method and the micro emulsion method to degrade chlorinated compounds such as PCE. These nano particles can increase degradation rates and enhance the formation of ethylene and ethane from chlorinated compounds [Yit et al., 2005]. Nanosized bimetallic particles such as Pd/Fe, Ni/Fe, Pt/Fe, Ag/Fe, Cu/Fe, Pd/Zn, Ni/Zn and Cu/Zn have been synthesized to achieve higher degradation rates and to prevent the production of chlorinated intermediates that are even more harmful than the chlorinated parent compounds [Li et al., 2003 and Lee. et al, 1998]

In this study methanol and iso propyl alcohol were selected as solvents to extract PCE from contaminated soils. Cetyl trimethyl ammonium bromide (CTAB) and SDS have been selected as surfactants to extract PCE from contaminated soils.

OBJECTIVE

The overall objective of this study was to investigate the extraction of PCE in different aged contaminated soils and degrade it using Fe/Ni particles.

1) Extraction efficiency of different solvents, surfactants in different aged PCE contaminated soil.
2) Quantify the degradation of extracted PCE using fine Fe/Ni particles

MATERIALS AND METHODS

Based on literature review, SDS, CTAB and UH-biosurfactant were selected for the solubility study to represent anionic, cationic and biosurfactant respectively. Methanol and Iso propyl alcohol were selected as solvents to extract PCE. Except for the UH-biosurfactant, other surfactants were purchased from Sigma-Aldrich, St.Louis, Missouri. Magnesium sulfate, yeast extract, potassium chloride, potassium phosphate, potassium phosphate and sodium nitrate used for the production of biosurfactant [Hariharan 1996 and Ren 1998]. Nickel (II) chloride hexa hydrate (Reagent plus), Iron (II) sulfate heptahydrate (99%, certified A.C.S), ferric chloride and sodium borohyride powder (98%) were also purchased from Sigma Aldrich company for the preparation of iron and Fe-Ni bi-metallic particle using the solution method.

Synthesis of fine particles

Fe-Ni particles: To produce the bi-metallic particles 6.15 g of $FeSO_4.7H_20$ and 1.5 g of $NiCl_2.6H_2O$ were mixed in 50 ml water using magnetic stirrer in 200 ml bottles for 30 minutes and 1.3g $NaBH_4$ was added into the solution for preparation of Fe-Ni bi-metallic particles. The entire system was kept under nitrogen atmosphere in order to prevent the oxidation of Fe-Ni bi-metallic particles and the proposed reaction is as follows

$$3\ FeSO_4 + NiCl_2 + 8\ NaBH_4 + 24\ H_2O \rightarrow 3\ Fe\text{-}Ni \uparrow + 28\ H_2 \uparrow + 8\ B(OH)_3 + 2\ NaCl + 3\ Na_2SO_4$$

Microanalysis of metals

SEM, EDS were used to verify the production and composition of Fe-Ni particles. The particles synthesized using the solution method were characterized using the JEOL 2000FX scanning electron microscope (SEM) .The bi-metallic particles were deposited on carbon grid for SEM characterization. Morphology of the particles was determined using the SEM image. The SEM micrographs of Fe-Ni bi-metallic particles are shown in Fig. 1. The particles were spherical and the size varied from 100 nm to 3 μm. The EDS analysis is shown in Fig 2.

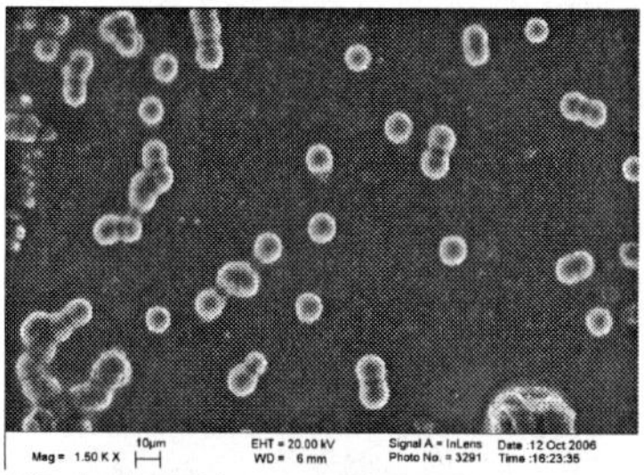

Fig 1: SEM image of Fe-Ni particles produced by solution method

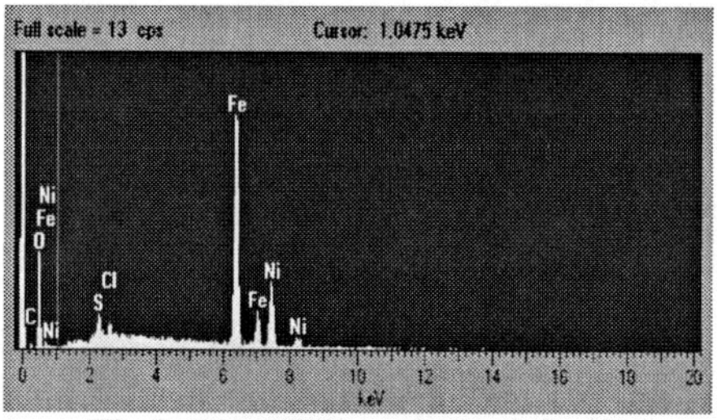

Fig 2: The EDX analysis of Fe-Ni bimetallic particles

Experiment

Soil with 82.5% sand with 17.5 % kaolinite clay was contaminated with PCE. Extraction experiments were conducted at room temperature. Surfactants and solvents were used either separately or as a mix. Those solutions were transferred to 40 ml EPA vials with contaminated soil. After end over end rotation of the tumbler the incubation vials were centrifuged and subsequently the clear supernatant was analysed for the total extracted PCE by Gas chromatography. A Gas chromatography

(Shimadzu GC-14A) with a flame ionization detector and a CR501 data processor was used for determining the PCE concentration. The stainless steel column (Alltech) was 6-ft long × 1/8-in outer diameter × 0.085-in inner diameter and detection temperature was set at 250°C, respectively. The injection volume of sample used for GC analysis was 2 μL. The detection limit for PCE was 1 ppm. The extracted PCE was degraded using 100g/L Fe/Ni particles.

RESULTS AND DISCUSSIONS

The soil (82.5 % sand and 17.5% kaolinite) was aged with PCE (200000 mg/kg) for one month and 3 month respectively. PCE was extracted using pure solvents and surfactant solutions and mixtures of these extraction solutions. The extraction mixtures were prepared by mixing equal volumes of the two components.

1 month

The extraction efficiencies of water, CTAB, SDS, Methanol and ISP were 4%, 32%, 32%, 50% and 51% respectively. The efficiency of water and CTAB, ISP and methanol, CTAB and methanol, CTAB and ISP, ISP and water, CTAB and SDS was 14%, 60%, 35%, 40%, 49% and 48% respectively.

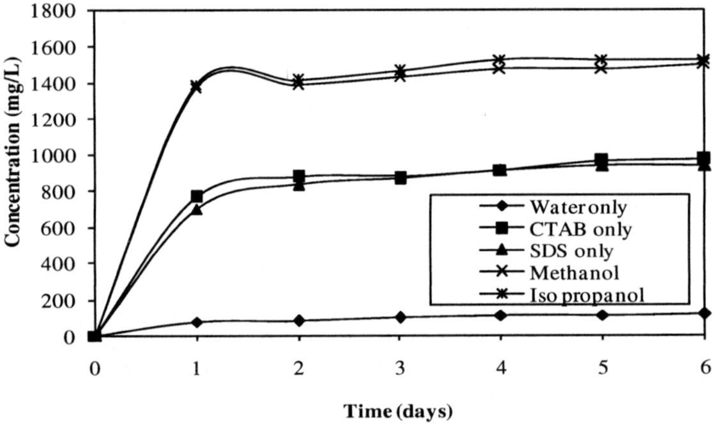

Fig 3: PCE extraction from contaminated soil- One month

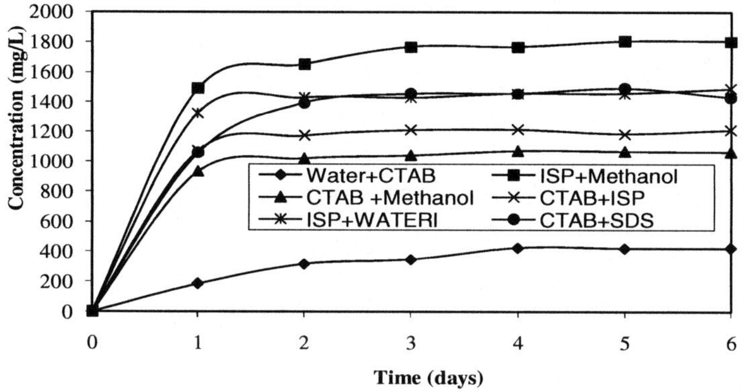

Fig. 4: PCE extraction from contaminated soil from different mixtures- One month

3 month

The extraction efficiency of water, CTAB, SDS, Methanol, ISP and UH biosurfactant was 5%, 10%, 8%, 12%, 10% and 7% respectively. The efficiency of water and CTAB, ISP and methanol, CTAB and methanol, CTAB and ISP, ISP and water, CTAB and SDS was 4%, 10%, 8%, 7%, 7% and 5% respectively.

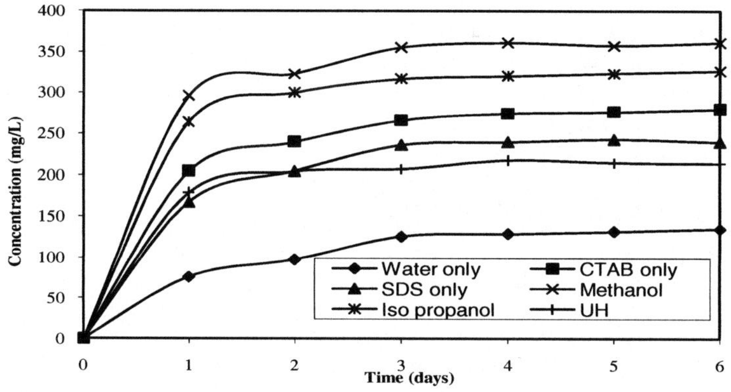

Fig. 5: PCE extraction from contaminated soil–Three month

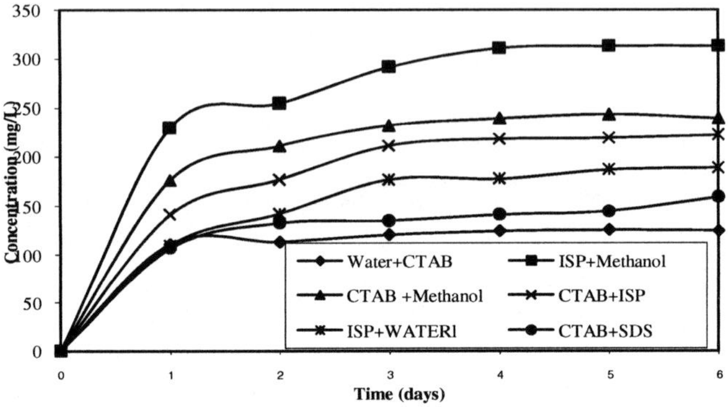

Fig. 6: PCE extraction from contaminated soil-Different mixtures–Three month

Degradation

The extracted PCE from 3 month aged samples were degraded using 100g/L Fe/Ni particles in batch reactors. The extracted PCE using water from contaminated soil was 135mg/L. The complete degradation was observed in 45 minutes. The extracted PCE using CTAB from contaminated soil was 280mg/L. The complete degradation was observed in 45 minutes. The extracted PCE using SDS from contaminated soil was 240 mg/L. The complete degradation was observed in 45 minutes. The extracted PCE using methanol from contaminated soil was 361 mg/L. The complete degradation was observed in 480 minutes. The extracted PCE using ISP from contaminated soil was 327 mg/L. The complete degradation was observed in 1500 minutes. The extracted PCE using UH biosurfactant from contaminated soil was 214 mg/L. The complete degradation was observed in 45 minutes.

CONCLUSIONS

Different solvents and surfactant solutions were used to extract the PCE from 200,000 mg/kg contaminated soil in batch studies. Based on the experimental study, following conclusions are advanced.

1) Solvents and surfactants were effective to varying degree in extracting PCE from contaminated soil. PCE extracted from one month aged soil was more compared to that of 3 month aged using solvents and surfactant solutions.

2) The Fe/Ni particles were effective in degrading PCE but the rate of degradation was influenced by the solvent and surfactant type in reactor.

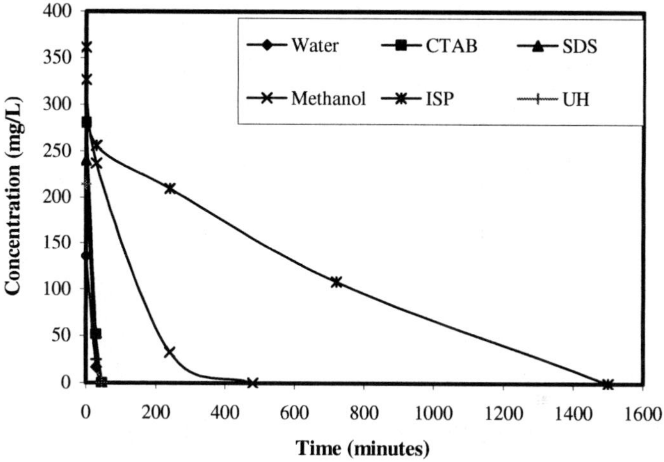

Fig 7: Degradation of extracted PCE using Fe/Ni particles

ACKNOWLEDEMENT

The study was supported by the Center for Innovative Grouting Materials and Technology with funding from the Advanced Research Program of the Texas Higher Education Coordinating Board (THECB) and the Texas Hazardous Waste Research Center.

REFERENCES

Wei X., Chuan-Bao W. and Hsing-Lung L. "Treatment of chlorinated organic contaminants with nanoscale bimetallic particles" *Journal of Catalysis Today* (1998), Vol 40, pp 387-395

Michael C. K." Alternatives for ground water cleanup" National Academic Press, Washington DC, 1994

Chu W. "Remediation of contaminated soils by surfactant aided soil washing" *Journal of Practical Periodical of Hazardous, Toxic and Radioactive Waste Management (2003)*, Vol 7, No 1, pp 19-24

Michael C.B., Michael D.A., Suresh P.C., Kirk H. and James W.J. "Controlled release, blind test of DNAPL remediation by ethanol flushing" *Journal of Contaminant Hydrology (2004)* ,Vol 69, pp 281-297

James W.J. et al., " Field implementation of a winsor type 1 surfactant/alcohol mixture for in situ solubilization of a complex LNAPL as a single phase micro emulsion" *Journal of Environ. Sci.Technol,* Vol 32, pp 523-530.

Yit H.T., Eric G. and Dibakar B. "Role of Ni/Fe nano particle composition on the degradation of TCE from water" *Journal of Ind.Eng.Chem.Res (2005),* Vol 44, pp 7062-7070.

Li, F., Vipulanandan, C., and Mohanty, K.M.," Microemulsion solution approaches to nanoscale iron production for degradation of TCE", *Colloids and Surfaces A: Physiochemical Eng. Aspects*", 2003, Vol. 223, pp. 103-112.

Lee H.S., Lee W. and Furubayashi T., "A comparison of coprecipitation with microemulsion methods in the preparation of magnetite" *Journal of Appl.Phys (1999),* Vol 85, pp 5231-5233

Hariharan S. "Production of biosurfactant from used vegetable oil and use for lead precipitation" Thesis-Department of Civil and Environmental Engineering, University of Houston, Texas, 1996.

Ren X. "Application of a biosurfactant in naphthalene, phenanthrene and TCE remediation" Doctoral dissertation –University of Houston, TX, 1998.

Removal and Degradation of Pentachlorophenol in Clayey Soil using Nanoscale Iron Particles

Krishna R. Reddy [1], and Madhusudhana R. Karri[2]

[1]Professor, University of Illinois at Chicago, Department of Civil and Materials Engineering, Chicago, Illinois 60607, USA; kreddy@uic.edu
[2]Graduate Research Assistant, University of Illinois at Chicago, Department of Civil and Materials Engineering, Chicago, Illinois 60607, USA

ABSTRACT: The main objective of this study was to investigate the efficiency of nanoscale iron particles (NIPs) to promote the reductive degradation of pentachlorophenol (PCP) in clayey soils. A series of batch experiments was conducted using kaolin as a low permeability clayey soil that was spiked with PCP at 1000 mg/kg and NIPs at different concentrations. Specifically, different contact time (1, 2, 8, 12, 24 and 48h) and different NIPs concentrations (1, 2, 4, 5, 8, 10, 20, 40, 60, 80 and 100 g/L) were investigated. PCP concentrations in supernatant and the residual soil for each test were measured. These results showed that 80 to 98% PCP was removed from the soil within an hour, and PCP reduction was increased from 50 to 78% at 1h contact time to 40 to 90% at 24 h contact time for different NIP concentrations. There was no significant effect of NIPs concentration on the PCP removal; however, the amount of PCP reduction increased with increase in concentration of NIPs with 30% at 1 g/L to 98% at 100 g/L. There appears to be an optimal NIPs concentration beyond which benefits are diminished.

INTRODUCTION

Pentachlorophenol is a general biocide and has been used extensively for a variety of purposes such as agriculture and timber preservation. Worldwide use of PCP has led to severe contamination problems particularly around former timber treatment plants (Dahlgren et al. 2003; Khodadoust et al. 1999; Weiss et al. 1982). Most of the readily available literature data show a PCP aqueous solubility between 10 to 20 mg of PCP per kg of water and its solubility increases drastically with increasing pH (Arcand et al. 1995). PCP is toxic to humans and may target and/or damage the liver, kidney, hematopoietic system, pulmonary system and central nervous system (Samokyszyn et al. 1995). Thus known toxicity and persistence of PCP in contaminated soils poses a long-term threat to the environment and has been listed as a priority pollutant in many countries including the USA (Keith et al. 1979; USEPA 1999). There is, therefore, considerable interest in the remediation of sites contaminated by PCP.

Various methods that have been employed to remediate PCP from contaminated soils are soil washing, chemical oxidation, and bioremediation, but they are either ineffective or expensive. One method to degrade PCP is via its reaction with metallic iron (Kim and Carraway 2000; Morales et al. 2002). Kim and Carraway (2000) used zero valent iron and their bimetals for the dechlorination of PCP. They concluded that zero valent iron showed the maximum affinity for the dechlorination of PCP as compared to other modified zero-valent iron used in their study. Nearly 50% of PCP was removed in few hours. Morales et al. (2002) reported the feasibility of using zero valent metals (Fe and Mg) for the dehalogenation of various chlorinated phenols. Marshall et al. (2002) investigated the effect of temperature and catalyst on the dehalogenation of PCP. Many laboratory and field scale studies used zero valent iron (Fe^0) as an effective media to remediate certain groundwater pollutants (Grittini 1995; Fennelly 1998; Xu 2000; Lien 2002; Chuang 1995).

The Fe^0-water system has been studied in detail and more than 73 potential reactions have been reported (Biernat et al. 1972). Under aerobic conditions, the following reactions are expected to occur:

$$2Fe^0 + O_2 + H_2O \leftrightarrow 2Fe^{2+} + 4OH^- \quad (1)$$

$$4Fe^{2+} + O_2 + 2H^+ \leftrightarrow 4Fe^{3+} + 2OH^- \quad (2)$$

Under anaerobic conditions, Fe^0 can undergo corrosion according to either of the following reactions depending upon the solution pH:

$$Fe^0 + 2H_2O \leftrightarrow Fe^{2+} + H_2 + 2OH^- \quad (3)$$

$$Fe^0 + 2H^+ \leftrightarrow Fe^{2+} + H_2 \quad (4)$$

In the presence of reducible substrate such as organohalide (R-Cl), Fe^0 can be oxidized:

$$Fe^0 + R\text{-}Cl + H^+ \leftrightarrow Fe^{2+} + R\text{-}H + Cl^- \quad (5)$$

Thus the pH of the reaction solution is important in these reactions probably because a low solution pH allows more of the iron surface to remain available for reaction with the halogenated molecule or at least assist in the corrosion of the iron by halogenated molecule (equation 5). At higher pH values oxide and hydroxide coatings unquestionably hinder access to the Fe^0 surface. Pretreatment of iron with mineral acids, or increasing the available iron surface area generally increases the rate of metal-contaminant dechlorination reactions (Orth and Gillham 1996) confirming the importance of iron surface availability.

The use of zero valent iron has many drawbacks in practical applications. Firstly, after a short period of reactions Fe^0 is liable to form an oxide film on the surface, which subsequently reduces the reaction activity (Wang and Zhang 1997). Secondly, the retention of the surface activity of Fe^0 is difficult to maintain. Once Fe^0 is in contact with air, even under proper storage condition, its reactivity towards chlorinated organic compounds is inevitably reduced (Cheng and Wu 2000). Thirdly, there is considerable variation in the reactivity towards chlorinated organic compounds of Fe^0 of different origins. The reaction rates can differ by up to three orders of magnitude (Su and Puls 2001). The above factors cause the zero valent iron technique to be largely restricted in its application to in-situ remediation.

In recent years, a great deal of research has concentrated on the improvement of zero-valent iron based remediation technique. This research included the use of ultra-sonic waves to remove oxide film on the Fe^0 surface and the use of external voltage to maintain the surface activity of Fe^0. Cantrell et al. studied the use of shear-thinning fluids as a carrier solution for the injection of micrometer-size Fe^0 particles within the sand column for its uniform distribution. Use of bimetals as well as palladized iron also showed significant promotional effect for the degradation of chlorinated organic compounds. In the most recent studies, nanoscale iron and palladized iron synthesized in the laboratory have been used for the degradation of PCB, TCE and other organic contaminants (Zhang et al. 1998; Elliott and Zhang 2001; Schrick et al. 2002). These studies showed that nanoscale particles provide enormous flexibility for in-situ application.

The main objective of this study was to investigate the efficiency of nanoscale iron particles to promote the reductive degradation of PCP in kaolin- a low permeability clayey soil. Specifically, the parameters such as amount of nanoscale iron (mechanism) and contact time (kinetics) were investigated to understand the extent of removal and reduction of PCP in the soil.

MATERIALS AND METHODS

Spiking Soils

The initial set of tests was conducted on kaolin spiked with PCP. PCP was chosen only because it is one of the commonly detected contaminants at former wood preserving contaminated site soils. Kaolin was spiked with PCP with a target concentration of 1000 mg/Kg. To spike the soil, 100% hexane (500 ml) was used to dissolve solid PCP (1.5 g). This hexane-PCP mixture was slowly added to 1500 g of kaolin soil. Additional hexane was added so that the soil-hexane-PCP mixture could be easily stirred and blended homogeneously. The mixtures were stirred with stainless steel spoons, and all mixing operations were performed within glass beakers. The purpose of this mixing technique was to distribute PCP evenly throughout the soil. The soil-hexane-PCP mixture was then placed beneath a ventilation hood for nearly a week until the hexane completely evaporated and the contaminated soil was dry.

Nanoscale Iron Particles

Nanoscale iron particles used in this study were obtained from Toda Kogyo (Japan). The particles had an average diameter of 70 nm (50-300 nm), pH of 10.7, and BET surface area of 37.1 m^2/g.

Testing Variables and Procedure

Batch tests were designed to optimize the mechanism and kinetics of the PCP reduction by nanoscale iron. A number of 40-mL stoppered glass vials containing 25 mL of nanoscale iron solution of known concentration (1, 2, 4, 5, 8, 10, 20, 40, 60, 80 and 100 g/L) were placed in a shaking assembly. 5 g of PCP spiked kaolin soil (concentration 1000 mg/Kg) was added into each vial and the solutions were agitated intermittently for a maximum period of 48 h.

The kinetic studies were also performed at different NIP (5, 10, 40 and 100 g/L) concentrations. At pre-determined intervals of time (1, 2, 8, 12, 24 and 48h), the test solutions in the specified vials were centrifuged at 4000 rpm for 30 min and separated from the soil. The supernatant was analyzed for PCP concentration and tetrachlorophenol isomers. The residual soil was also analyzed to determine the concentration of unreacted PCP. Control experiments without NIP solution were also performed with double distilled water.

Chemical Analysis

Two types of extractions were used in this study. The first method was a typical liquid-liquid extraction method using methylene chloride as a solvent. While the second type of liquid-liquid extraction was modified by the extraction of soil residue by ethyl alcohol and HCl. The percent removal and percent reduction of PCP were calculated using the following two equations:

% Total PCP removal = [Initial mass in soil – final mass in soil] / Initial mass in soil x 100

% Total PCP reduction= [Initial mass in soil – (final mass in soil + final mass in filtrate)] / Initial mass in soil x 100

Liquid-Liquid Extraction

This extraction procedure was followed to determine PCP concentration in liquid samples. It consisted of placing 1 mL of the supernatant in a conical flask with the help of a syringe. The syringe was thoroughly rinsed with ethyl alcohol after use. Then, the sample was diluted in the ratio of 1:10 with water. The conical flask was shaken thoroughly before transferring the diluted sample into a test tube. Then, 200 μL of tribromo phenol was added. After that, 2.5 mL of methylene chloride was added in to the test tube. The test tube was stirred at least for two minutes. Then, the two phases, methylene chloride phase and the aqueous phase, were allowed to separate. Approximately, one to two mL of the methylene chloride phase was taken with the help of a syringe into a two mL target autosampler vial. Then, the vial was closed with a cap containing septa. This sample was then injected into gas chromatography (GC).

Water-Ethanol Extraction

This extraction procedure was used to determine PCP in residual soil samples and it consisted of placing 1 g of soil sample in a conical flask and shaking it with 50% water-ethanol solution (9 ml) and 1 ml HCl for 1 hr at 18 rpm speed at 25 C. water-ethanol solution was used to extract the remaining unreacted PCP from the residual soil while HCl was used to completely desorb the protonated phenolate ions from the soil as well as from the surface of iron. The mixture was then centrifuged and the supernatant was analyzed for PCP by liquid-liquid extraction procedure as described above.

RESULTS AND DISCUSSION

Effect of Contact Time

Figure 1 shows the effect of contact time for the total removal of PCP from the soil using 5, 10, 40 and 100 g/L of NIP solution. All the tests were performed in duplicates and are presented as such. These results for PCP removal showed that at 1 hr contact time, higher concentration of NIP solution was required to attain 98% removal efficiency. But as the duration of contact time was increased from 2 to 24 hr almost similar removal efficiency was observed with very slight difference. Not much difference was observed in the removal efficiency with increased NIP concentration.

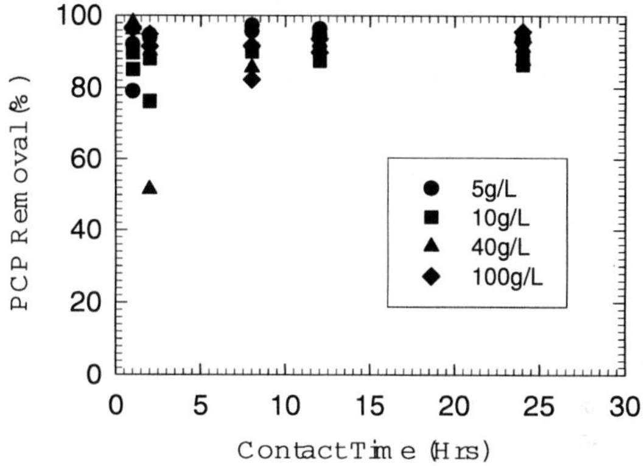

FIG.1. Effect of contact time on PCP removal

Figure 2 presents the results of total PCP reduction (reductive degradation) with contact time and reveal the efficiency of the nanoiron particles for the mineralization of PCP in the contaminated soil. It is clear from these results that with low NIP concentration, the amount of PCP reduction ranges from 8 to 50% and increases with increasing time. Maximum destruction efficiency was achieved with 100 g/L NIP concentration and found to increase with increasing concentration. The variation in the reduction efficiencies among the duplicates may be due to the heterogeneous distribution of PCP within the soil as well as high chemical reactivity of the kaolin clay.

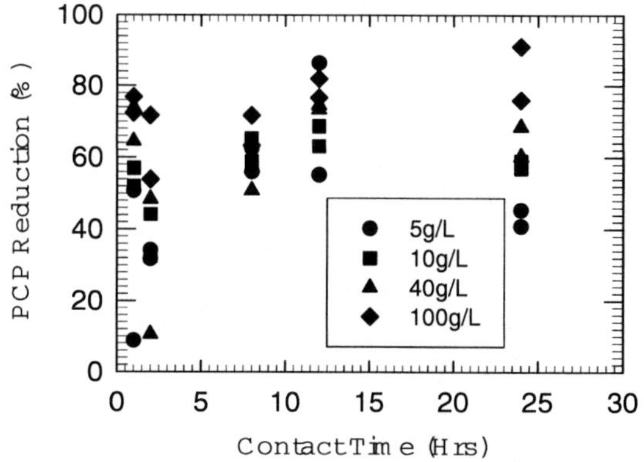

FIG.2. Effect of contact time on PCP reduction

Effect of Concentration

Figure 3 shows the total PCP removal from the soil using different NIP concentrations ranging from 1 g/L to 100 g/L. All the tests were run for a contact time of 48 hr. It is found that nearly 85 to 100% of PCP was removed from the kaolin soil in all the concentration range. However, the removal was higher with higher NIP concentration. The sorption of NIP particles may have changed the surface reactivity of the kaolin soil that might had promoted the enhanced desorption of protonated PCP. The variation in duplicate test results may be attributed to heterogeneous distribution of PCP in the soil.

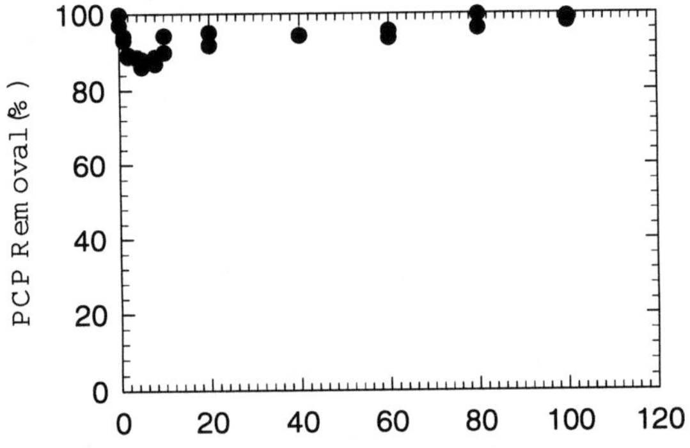

FIG.3. Effect of NIP concentration on PCP removal

Figure 4 clearly shows that though the PCP removal at low NIP concentration was high, but it causes only low amount of PCP reduction. It is clearly seen that the reduction of PCP increases with the increased NIP concentration, and maximum PCP reduction was found with 100 g/L NIP concentration. Further investigations are warranted to investigate the mechanisms responsible for the observed reduction efficiencies and enhancement strategies.

CONCLUSIONS

The objective of this study was to determine the efficiency of using nanoscale iron particles for the remediation of PCP in kaolin soil. Based on a series of batch experiments, high removal of PCP from the soil was observed even with a low NIP concentration, but the extent of PCP reduction was found to depend on the concentration (amount) of NIPs and the contact time. The PCP reduction was 98% for 100 g/L NIP concentration; however, the use of such high NIP concentrations may not be economical and not conducive for transport during in-situ treatment. Overall, this study indicates that the remediation of PCP using NIPs is a complex process and requires advanced surface-specific investigation to predict the exact mechanism for the removal/reduction of PCP.

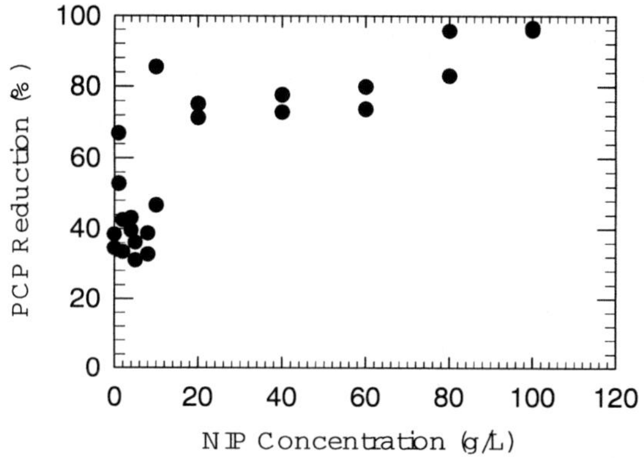

FIG.4. Effect of NIP concentration on PCP reduction

ACKNOWLEDGEMENTS

This project was funded by the U.S. National Science Foundation Grant CMMI #0727569 and the support of this agency is gratefully acknowledged. The authors are thankful to Saurabh Sharma, Craig Chawla, Naveen Parakalla, and Amid Khodadoust for their assistance during this study.

REFERENCE

Arcand, Y., Hawari, J., and Guiot, S.R. (1995). "Solubility of pentachlorophenol in aqueous solutions: The pH effect." *Water Res.*, 29, 131-136.

Biernat, R. J. and Robins, R. G. (1972). "High-temperature potential/pH diagrams for the iron-water and iron-water-sulphur systems." *Electrochim. Acta*, 17, 1261-1283.

Cheng, S.-F. and Wu, S.-H. (2000). "The enhancement methods for the degradation of TCE by zero-valent metals" *Chemosphere,*41, 1263-1270.

Chuang, F., Larson, R., and Wessman, M., (1995). "Iron-promoted dechlorination of polychlorinated biphenyls." *Environ Sci. Technol.*, 29, 2460-2463.

Dahlgren, J., Warshaw, R., Horsak, R.D., Parker, F. M. and Takhar, H. (2003). "Exposure assessment of residents living near a wood treatment plant." *Environ. Res.*, 92, 99-109.

Elliott, D.W., and Zhang, W. (2001). "Field assessment of nanoscale bimetallic particles for groundwater treatment." *Environ. Sci. Technol.*, 35, 4922-4926.

Fennelly, J.P., and Roberts, A.L. (1998). "Reaction of 1,1,1-Trichloroethane with zero-valent metals and bimetallic reductants." *Environ. Sci. Technol.*, 32, 1980-1988.

Grittini, C., Malcomson, M., Quintus, F. and Korte, N. (1995). "Rapid dechlorination of polychlorinated biphenyls on the surface of a Pd/Fe bimetallic system." *Environ. Sci. Technol.* 29, 2898-2900.

Keith, L. H. and Telliard, W. A. (1979). "Priority pollutants: I—A perspective view." *Environ. Sci. Technol.* 13, 416–423.

Khodadoust, A.P., Suidan, M. T., Acheson, C. M. and Brenner, R. C. (1999). "Solvent extraction of pentachlorophenol from contaminated soils using water-ethanol mixtures." *Chemosphere*, 38, 2681-2693.

Kim, Y. H. and Carraway, E. R. (2000). "Dechlorination of pentachlorophenol by zero valent iron and modified zero valent irons." *Environ. Sci. Technol.*, 34, 2014-2017.

Lien, H. L. and Zhang, W. (2002). "Enhanced dehalogenation of halogenated methanes by bimetallic Cu/Al." *Chemosphere,* 49, 371-378.

Marshall, W.D., Kubatova, A., Lagadec, A. J. M., Miller, D. J. and Hawthorne, S.B. (2002). "Zero-valent metal accelerators for the dechlorination of pentachlorophenol (PCP) in subcritical water." *Green Chem.*, 4, 17-23.

Morales, J., Hutcheson, R. and Cheng, I.F. (2002). "Dechlorination of chlorinated phenols by catalyzed and uncatalyzed Fe(0) and Mg(0) particles." *J. Haz. Mater.*, B90, 97-108.

Orth, W.S. and Gillham, R.W. (1996). "Dechlorination of trichloroethene in aqueous solution using Fe^{0}", *Environ. Sci. Technol.*, 30, 66-71.

Samokyszyn, V.M., Freeman, J.P., Maddipati, K.R. and Lloyd, R.V. (1995). "Peroxidase-catalyzed oxidation of pentachlorophenol." *Chem. Res. Toxicol.*, 8, 349-355.

Schrick, B., Blough, J.L., Jones, A.D. and Mallouk, T.E. (2002). "Hydrodechlorination of trichloroethylene to hydrocarbons using bimetallic nickel-iron nanoparticles." *Chem Mater.* 14, 5140-514

Su, C. and Puls, R.W. (2001). "Arsenate and arsenite removal by zero-valent iron: Kinetics, redox transformation, and implications for in-situ groundwater remediation." *Environ. Sci. Technol.,* 35, 1487 -1492.

USEPA (1999). EPA*Monthly Reporting of CDD/CDF Content in Technical Pentachlorophenol manufactured by KMG-Bermuth, Inc. and Vulcan Chemicals, 1987 to 1999*, Office of Pesticide Programs, Washington, DC.

Wang, C.G. and Zhang, W. (1997). "Nanoscale metals particles for dechlorination of TCE and PCBs." *Environ. Sci.Technol.,* 31, 2154-2156.

Weiss, U. M., Scheunert, I., Klein, W. and Korte, F. (1982). "Fate of pentachlorophenol-14C in soil under controlled conditions." *J. Agric. Food Chem.*, 30,1191-1194.

Xu, Y. and Zhang, W. (2000). "Subcolloidal Fe/Ag particles for reductive dehalogenation of chlorinated benzenes." *Ind. Eng. Chem. Res.*, 39, 2238-2244.

Zhang, W., Wang, C., and Lien, H. (1998). "Treatment of chlorinated organic contaminants with nanoscale bimetallic particles." *Catal. Today* 40, 387-395.

Transport of Fe/Ni Bimetallic Fine Particles through PCE Contaminated Clayey Soil

C. Vipulanandan[1], S.Harendra[2]

[1] Chairman, Professor and Director of Center for Innovative Grouting Materials and Technology (CIGMAT), Department of Civil and Environmental Engineering, University of Houston, Houston, Texas 77204-4003. Phone: (713) 743-4278; Fax: (713) 743-4260; Email: cvipulanandan@uh.edu;

[2] PhD Student, Department of Civil and Environmental Engineering, University of Houston, Houston, Texas 77204-4003. Email: ps_harendra@yahoo.com;

ABSTRACT: In this study, transport of Fe/Ni bimetallic particles produced from solution method was investigated. Soil was packed in 5.1 cm diameter column to a height of 6 cm and the permeability of the soil was 10^{-4}cm/s. Breakthrough curve for bimetallic particles was determined from the effluent from a series of column studies. The bimetallic particles were produced at room temperature using the solution method. The produced particles were kept under nitrogen atmosphere in order to prevent the oxidation of Fe-Ni bimetallic particle. The SEM micrograph of the Fe/Ni particles showed that particles were < 10 μm and were spherical in shape. The breakthrough curve for a pulse input of 100 g/L of Fe/Ni was bell shaped, typical response observed for pulse input of colloidal solution as reported in the literature. The variation of Fe/Ni particle in the soil column was investigated with depth. The transports of the Fe/Ni particles were modeled using the advective-diffusion equation and the transport parameters were quantified for the test conditions.

INTRODUCTION

The transport and deposition of colloidal particles during flow through porous media have been studied extensively using well characterized model systems [Ruben et al.1997, Terri et al. 1999]. Early research was driven by the need to understand the performance of deep bed filters used in chemical engineering and wastewater treatment [Elimelech et al. 1995]. Filtration theories have been developed from theoretical considerations and experimental results for the transport of monodisperse microspheres through columns packed with spherical grains. More recently, the focus of interest has shifted to the transport of mobile colloids in natural subsurface porous media.

Colloid deposition kinetics in natural and model porous media have been studied as a function of colloid size, colloid type, surface properties, flow velocity, water content, pore size, and solution chemistry (Daniel et al., 1996; Elimelech and O'Melia, 1990-a, 1990-b; Liuzong et al., 2001; Jie et al. 2003 & 2004 and Lakshmi 1997). Although great advancements have been made in the study of colloid reaction and transport in porous media, both theoretically and experimentally, understanding of colloid–soil interactions and the ability to predict transport of colloids in natural subsurface media are limited.

The primary mechanisms controlling the transport of colloidal particles in subsurface porous media are particle advection, dispersion and deposition (filtration). These are influenced by the surface chemical characteristics of the natural porous media and colloids, size and morphology of colloidal particles and granular porous media, solution chemistry and the fluid flow field. Moreover, some of these mechanisms are closely interrelated. For instance, particle deposition is influenced by particle advection where as particle dispersion in porous media is directly influenced by particle advection in porous media is directly related to fluid velocity and particle advection.

Fine Fe/Ni bimetallic particles represent a new generation of environmental remediation technologies that could provide cost effective solutions to some of the most challenging environmental clean up problems [Li et al. 2003, Daniel et al. 2001, Wei et al., 2003 and Yit et al. 1998]. Those particles have large surface areas and high surface reactivity. This leads to greater remediation of chlorinated compounds present in soil and aquifer. Newer techniques such as the injection of reactive slurries or suspended solids may serve to overcome some of the limitations so that contaminants at greater depths can be reached and more directly targeted [Bettina et al., 2004]. So in this study Fe/Ni particles transport was studied in sandy clayey soil. Bi metallic particles were transported through contaminated (Perchloroethylene (PCE)) and uncontaminated soil. The retardation factor (R), dispersion coefficient (D) and velocity (v) of particles were determined.

OBJECTIVES

The overall objective of this study was to determine the transport parameters of Fe/Ni bimetallic particles in packed contaminated soil column. The specific objectives were as follows:

1) Determination of transport parameters of Fe/Ni particles in clayey soil
2) Verify the concentration profile of Fe/Ni particles along soil column studies .
3) Determine the effectiveness of Fe/Ni particles in degrading PCE in contaminated soil.

THEORY

Consider a soil column with saturated flow condition and constant solution composition as pulse input. Transport of Fe/Ni particles can be described by accounting for particle advection, hydrodynamic dispersion and deposition (filtration). The concentration of suspended Fe/Ni particles C(x,t) at a column depth x and time t follows the one dimensional advection dispersion equation with a sink term for particle deposition (Daniel et al. 2001).

$$R\frac{dC}{dt} = D\frac{d^2C}{dx^2} - v\frac{dC}{dx} - kC \text{ ---------------------------------- [1]}$$

where ν is the interstitial velocity of the colloidal particles, D is the hydrodynamic dispersion coefficient and k is the particle deposition rate coefficient. The transport equation assumes particle deposition to follow first order kinetics and to be irreversible. Both assumptions are justified at sufficiently low particle concentration and for moderate to high ionic strengths where particle release is negligible compared to particle deposition.

For steady state flow in a soil transport of a linearly adsorbed solute is given by two site non equilibrium transport model which makes a distinction between type 1 (equilibrium) and type 2 (first order kinetics) adsorption sites [van Genuchten and wagenet, 1989]. This model was verified and used in many past research works. f is the fraction of exchange sites (mobile) that are always at equilibrium and (1-f) represents the immobile sites which are first order kinetics. If f=0, it will fall into one site kinetic model. The relationship for the mobile phase is represented as follows [van Genuchten and wagenet, 1989]:

$$(1+\frac{\rho f k}{\theta})\frac{dC}{dt} = D\frac{d^2C}{dx^2} - v\frac{dC}{dx^2} - \frac{\alpha\rho}{\theta}[(1-f)kc - S_2] - \mu_l C - \frac{f\rho k\mu_{sl}C}{\theta} \text{ --- [2]}$$

where **θ** porosity, ρ density, α mass transfer filter coefficient, μ degradation constant, k partition coefficient, x distance. One and two site kinetic models have been applied to fit and predict breakthrough curves from column experiments. The two site model fitted all breakthrough curves satisfactorily, accounting for the skewness of the rising limb as well as for the smooth transition of the declining limb to the tail of the breakthrough curve. The one site kinetic model does not follow the curvature of the breakthrough tail, leading to an overestimation of the inactivation site. The resulting two site kinetic model has been successfully used to describe the transport of a number of solutes (Nubuo et al. 1993) that interact with a solid phase composed of such different constituents as soil materials, organic matter and various oxides. This model is widely used in experiments such as viruses, bacteria, protozoa and colloidal particles. The particles often exhibit a typical skewness. Typical response of such breakthrough has a slowly rising limb and smooth transition of a declining limb to a very long tail. One site kinetic models fail to fit the rising and declining limbs together with the tail satisfactorily.

MATERIALS AND METHODS

Cationic surfactant CTAB was purchased from Sigma-Aldrich, St.Louis, Missouri. Iron (II) chloride (98%) and sodium borohyride powder (98%) were also purchased from Sigma-Aldrich (St.Louis, Missouri) for producing the iron nano particles using the solution method. Nickel (II) chloride hexa hydrate (Reagent plus), Iron (II) sulfate heptahydrate (99%, certified A.C.S), ferric chloride were also purchased from Sigma Aldrich company for the preparation of Fe-Ni bimetallic particle using the solution method.

Synthesis of fine particles

Fe-Ni particles (solution method): To produce the bimetallic particles 6.15 g of $FeSO_4.7H_20$ and 1.5 g of $NiCl_2.6H_2O$ were mixed in 50 ml water using magnetic stirrer in 200 ml bottles for 30 minutes and 1.3g $NaBH_4$ was added into the solution for preparation of Fe-Ni bimetallic particles. The entire system was kept under nitrogen atmosphere in order to prevent the oxidation of Fe-Ni bimetallic particles and the proposed reaction is as follows. The reaction can be represented as follows [Li et al., 2003]:

$$3\ FeSO_4 + NiCl_2 + 8\ NaBH_4 + 24\ H_2O \rightarrow 3\ Fe\text{-}Ni\downarrow + 28\ H_2\uparrow + 8\ B(OH)_3 + 2\ NaCl + 3\ Na_2SO_4 \quad \text{—[3]}$$

Microanalysis of Particles

SEM and EDX methods were used to verify the production and composition of Fe-Ni particles. The particles synthesized using the solution method were characterized using the JEOL 2000FX scanning electron microscope (SEM). The bimetallic particles were deposited on carbon grid for SEM characterization. Morphology of the particles was determined using the SEM image. The SEM analysis for Fe-Ni bimetallic particles are shown in Fig 1. The EDS pattern of Fe-Ni particles are shown in Fig 2.

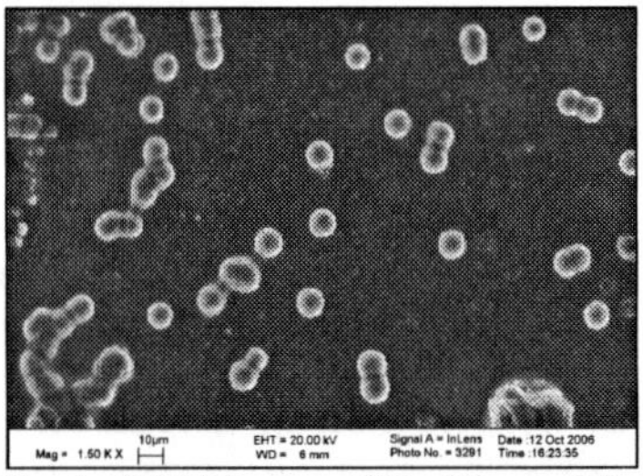

Fig. 1: SEM image of Fe-Ni particles produced by solution method

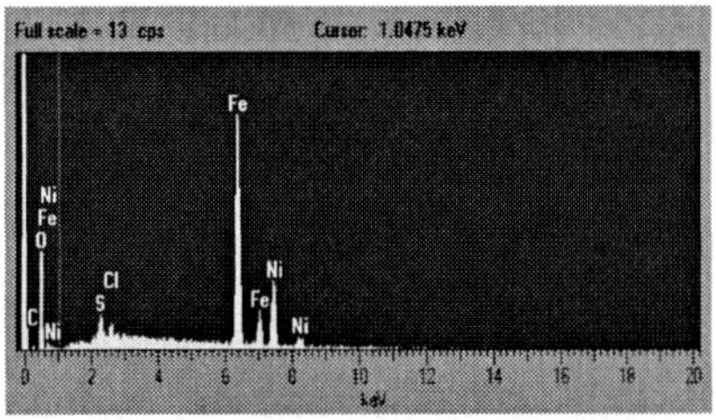

Fig. 2: The EDX analysis of Fe-Ni bimetallic particle before PCE degradation

EXPERIMENT

Column study

The experimental setup for the column studies is the column having 5.1 cm diameter and 50 cm long made out of glass which is equipped with air compressor and pressure regulator unit which is used to pump the solutions. The column is packed with 17.5 % kaolinite and 82.5 % blasting sand. The soil was packed about 6 cm height. The Fe/Ni particle slurry was transported through clayey soil with and without PCE contamination (Fig. 3).

Tracer experiments

Tracer breakthrough curves were analyzed to determine Peclect number and column porosity. From the first and second moments of the Br^- breakthrough curve (Fig. 4), the average travel time of the tracer and the dispersion coefficient were calculated. Fitting the convective-dispersive transport equation to the Br^- breakthrough data using non linear least square procedure gave very similar results. The porosity varied from 0.4 to 0.5.

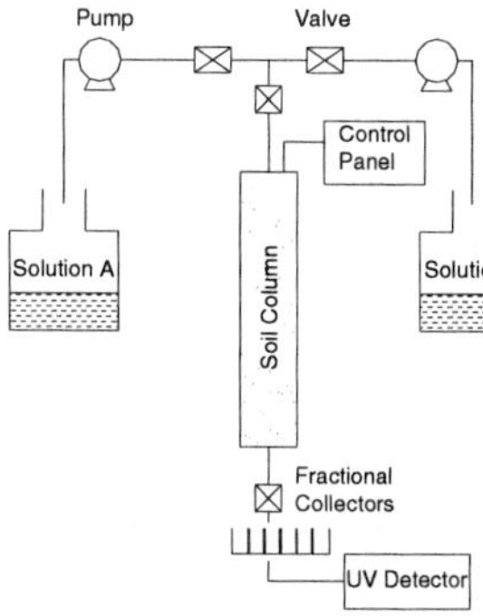

Fig. 3: Schematic drawing of the experimental setup used for breakthrough experiments

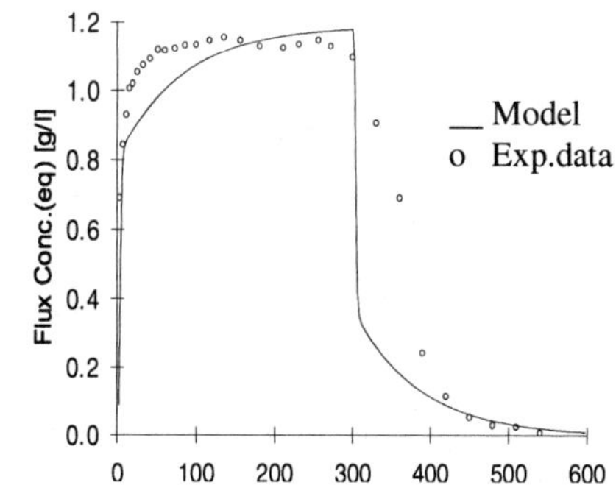

Fig. 4: Breakthrough of Tracer (Br^-) in Soil column

The UV spectrometer was used to measure the Fe/Ni particle concentration in the solution. The Fe/Ni particle solution was oxidized using H_2SO_4 and mixed with 1, 10 phenanthroline which is the chelating agents used for the spectrophoto metric determination of Fe/Ni particles. The formed Fe/Ni -1, 10 phenanthroline complex will be anaylsed using UV spectrophotometer at 510 nm.

The soil (82.5% blasting sand and 17.5% kaolinite) was contaminated to 100000 mg/kg and aged for 3 months. Two columns were packed with 3 month aged soil in columns. The surface tension of CTAB was 37 dynes/cm at CMC. The critical micelle concentration (CMC) of CTAB was 0.4 g/L. The breakthrough studies were done using 5 g/L CTAB surfactant solutions as continuous input. After passing four pore-volumes of surfactant solutions, 100 g/L Fe/Ni particles was injected as a pulse input to one soil column. After the test, soil samples were obtained from various depths to determine the Fe/Ni concentration.

RESULTS AND DISCUSSIONS

The variation of effluent concentration of Fe/Ni particles with time (breakthrough curve) is shown in Fig. 5.The response can be described as bell shaped curve. The fraction of the solid phase that is in equilibrium with the nanoparticles (*f*) was 0.43.

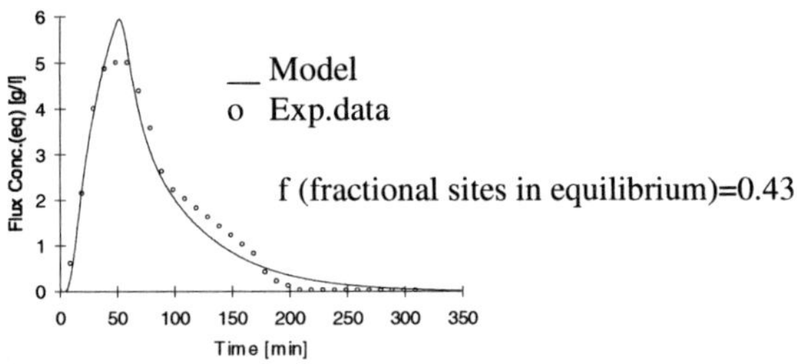

Fig. 5: Breakthrough curve of Fe-Ni in Soil column

The pH of the effluent increased from 4.5 to 6.33 where as ORP reduced from 189 mV to 90.3 mV (Fig. 6). The remaining Fe/Ni particle concentration profile from top layer to bottom layer varied from 0.29 mg Fe-Ni/ g Soil to 0.0012 mg Fe-Ni/ g Soil (Fig. 7). The retardation factor (R), velocity (v) and diffusion coefficient (D) were 17.38, 1.77 cm/min and 2.45 cm/min^2 respectively.

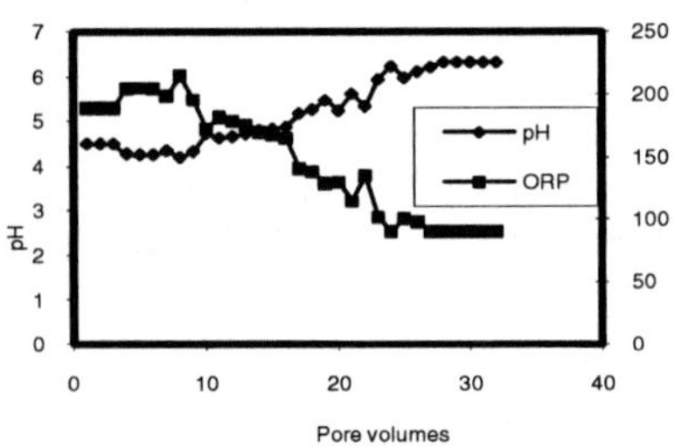

Fig. 6: pH and ORP with of Fe-Ni nano particle transportation

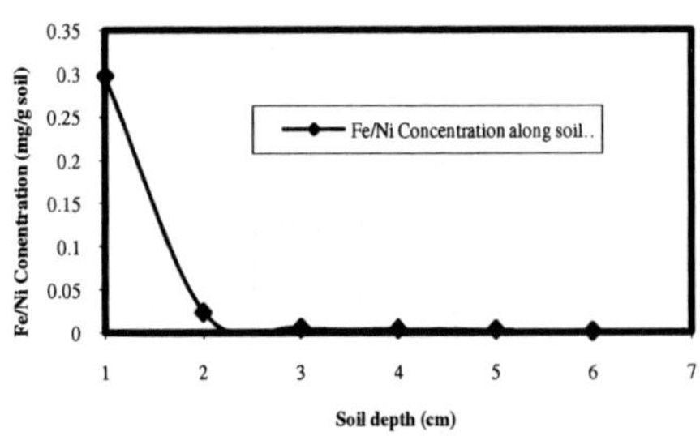

Fig. 7: Measured Fe-Ni concentration along the soil depth

The effluent PCE concentration was changed from 11-12 mg/L to 5-6 mg/L due to the injection of 100g/L Fe-Ni particle solutions (Fig. 8.).

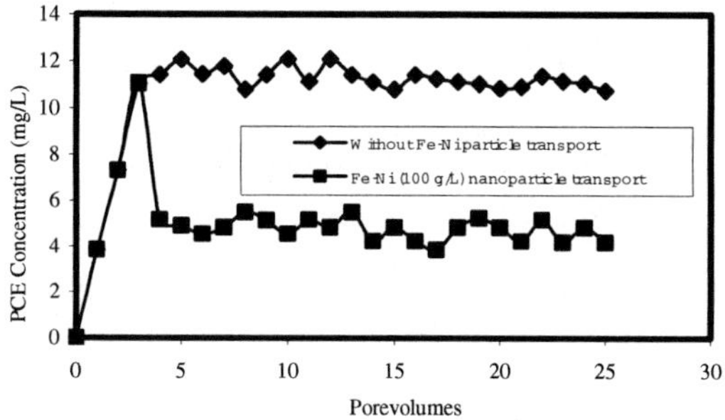

Fig. 8: Effluent PCE concentration with and without Fe-Ni Nanoparticles

CONCLUSIONS

Transportation of Fe/Ni particles in PCE contaminated soil was investigated in packed clayey soil column. Based on the modeling and experimental study, following conclusions are advanced.

1) Transport of 100 g/L of Fe/Ni nanoparticles in the size range of 10 μm were filtered out of the solution within the top 33% of the soil column. Transportation was modeled using the CDE. The pH and ORP of the effluent changed from 4.5 and 189 mV to 6.33 and 90.3 mV respectively.

2) The effluent concentration of PCE was reduced from 12 mg/L to 5 mg/L with the transport of Fe-Ni particles. The Fe/Ni particles degrade the PCE in the contaminated soil column.

ACKNOWLEDEMENT

The study was supported by the Center for Innovative Grouting Materials and Technology with funding from the Advanced Research Program of the Texas Higher Education Coordinating Board (THECB) and the Texas Hazardous Waste Research Center.

REFERENCES

Bettina S., Bianca W.H, Jennifer L.B. and Thomas E.M " Delivery vehicle for zerovalent metal nanoparticles in soil and groundwater" Journal of Chem. Mater.(2004),Vol 16, pp 2187-2193.

Daniel G., Michal .B. and Kurt .B "Colloid facilitated transport of strongly sorbing contaminants in natural porous media: A laboratory column study" Journal of Environ. Sci. and Technol. (1996), Vol 30, No 10, pp 3118-3123.

Daniel W.E. and Wei X.Z." Field Assessment of nanoscale bimetallic particles for groundwater treatment" Journal of Environ. Sci. and Technol. (2001), Vol 35, pp 4922-4926.

Daniel G., Menachem E., Michal B., Kurt B., Ruben K. and Hans S. " Transport in situ mobilized colloidal particles in packed soil columns" Journal of Environ. Sci. and Technol. (1998), Vol 32, pp 3562-3569.

Elimelech M. and O'Melia, C.R."Kinetics of deposition of colloidal particles in porous media" Journal of Environ.Sci.Technol. 1990-b, Vol 6, pp 1153 -1163.

Jie Z., Yan J. and Markus F. "Comparison of Hanford colloids and Kaolinite transport in porous media" Vadoze zone journal (2004),Vol 3, pp 395-402

Lakshmi N.R "Particle transport in soils: Review of significant processes in infrastructure systems" Journal of Infrastructure Systems (1997), Vol 3, No. 2, pp 78-86.

Li, F., Vipulanandan, C., and Mohanty, K.M.," Microemulsion solution approaches to nanoscale iron production for degradation of TCE", Colloids and Surfaces A: Physiochemical Eng. Aspects", 2003, Vol. 223, pp. 103-112.

Nubuo T., Feike J.L. and Van genuchten " A comprehensive set of analytical solutions for non equilibrium solute transport with first order decay and zero order production" Journal of Water Resources Research (1993),Vol 29, No 7, pp 2167-2182.

Ruben K., Kurt B., Daniel G., Yao Y., Michal B. and Hans S. " Experimental determination of colloid deposition rates and collision efficiencies in natural porous media" Journal of water resources research (1997),Vol 33, No 5, pp 1129-1137.

Terri A.C., Kenneth M.U. and Bruce E.L." Blocking and ripening of colloids in porous media and their implications for bacterial transport" Journal of Colloids and surfaces (1999), Vol 160, pp 291-308.

Wei, Z. "Nanoscale iron particles for environmental remediation: An overview "Journal of Nanoparticle research (2003), Vol 5, pp 323-332.

Transport of Lactate-Modified Nanoscale Iron Particles in Sand Columns

Amid P. Khodadoust[1], Krishna R. Reddy[2], and Srinivasa Varadhan[3]

[1]Associate Professor, Department of Civil & Materials Engineering, University of Illinois at Chicago, 842 W. Taylor Street, Chicago, IL 60607, USA; akhodado@uic.edu, kreddy@uic.edu
[2]Professor, Department of Civil & Materials Engineering, University of Illinois at Chicago, 842 W. Taylor Street, Chicago, IL 60607, USA; kreddy@uic.edu
[3]Graduate Research Assisatnt, Department of Civil & Materials Engineering, University of Illinois at Chicago, 842 W. Taylor Street, Chicago, IL 60607, USA

ABSTRACT: Nanoscale iron particles have great potential for in-situ remediation of subsurface soils.The transport of reactive nanoscale iron particles (RNIPs) into the contaminated subsurface is essential for the success of this remediation technology. RNIP cannot be transported through porous media effectively without surface modification. In this study, the transport of RNIP modified with lactate was investigated in 1-D sand column experiments under pressurized conditions using a natural sandy soil and an electrolyte solution. Sodium lactate and aluminum lactate were considered as lactate species for modification of RNIP. Aluminum lactate was mixed with bare RNIP at concentrations of 2, 4, 6, 8, 10 and 12%, and introduced at the top of the sand column. The elution of RNIP from the sand column increased with increasing concentration of aluminum lactate up to 10% using a constant air pressure of 30 psig. Elution of RNIP slug through the sand column ranged from 55% for bare RNIP to 90% for RNIP modified with 10% aluminum lactate using an RNIP dosage of 4.8 g/L. Using a constant aluminum lactate concentration of 10%, the elution of RNIP through the sand column decreased with increasing dosage of RNIP, ranging from 90% for 4.8 g/L to 65% for 12 g/L. Using a dosage of 4.8 g/L of RNIP, 10% sodium lactate performed worse than 10% aluminum lactate, resulting in an elution value of 80% versus 90%, respectively.

INTRODUCTION

The transport of iron-based nanoparticles in most environmental media such as soil is not possible without surface modification (Schrick et al. 2004). For efficient subsurface transport, nanoiron with controllable surface properties may be used. The objective of the coating or surface modification is primarily to provide better transport through water-saturated porous media. The goal of this study was to determine the transport behavior of surface-modified Reactive Nanoscale Iron Particles (RNIP). The mobility of iron-based nanoparticles in soils is a critical parameter for the success of this remediation technology. However, experimental data can be obtained to determine the mobility of different RNIP slurries in a given soil, thereby providing valuable data for assessing the potential effectiveness of this remediation technology using modified RNIP. For this study, a natural sandy soil with a hydraulic conductivity (K) between 0.005 and 0.01 cm/s was used.

By developing a reliable soil preparation and RNIP addition procedure along with an appropriate eluted iron mass determination method, it is possible to generate reliable data that can be used to estimate the mobility of modified RNIP slurries in soils. The effective transport of RNIP through any given soil depends on the proper surface modification of RNIP. Recently, RNIP modified with polymers and polyelectrolytes have shown to be superior to bare RNIP for enhanced transport of RNIP through porous media (Saleh et al. 2007). This paper investigates the use of lactate species as surface-modifying agents for enhanced transport of RNIP. Lactate species such as sodium and aluminum lactate are environmentally benign and considered green compounds for in-situ applications. The transport behavior and effectiveness of lactate-modified RNIP were assessed through visual observations and quantitative determination of effluent RNIP.

MATERIALS AND METHODS

Sand Columns and Pressurized Flow Testing Apparatus

Transport tests were performed in one-dimensional bench scale sand columns. The experimental apparatus consisted of 30-cm long borosilicate glass columns with a 2.5-cm inside diameter. The bottom of the glass columns were fitted with a 100 mesh HDPE screen and a 20 um HPDE filter. The solution reservoir was connected to the glass column and a source of compressed air. The solution was fed to the glass column using pressurized flow under a head of compressed air. The apparatus setup is illustrated in Figure 1. Pressurized flow tests were conducted using air at an applied air pressure of 30 psi; this pressure is a typical pressure applied in field applications for in-situ injection of solid slurries such as RNIP.

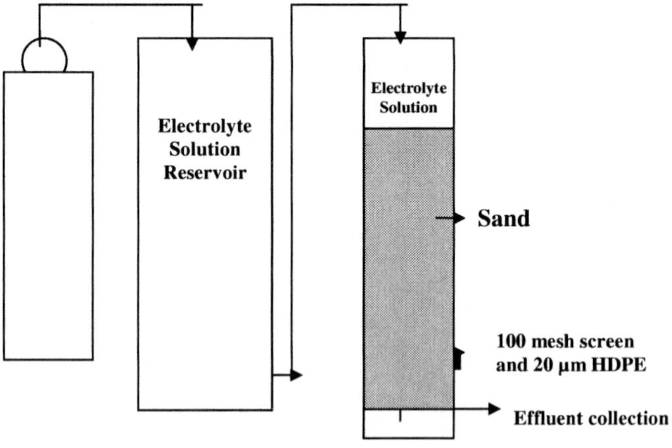

FIG. 1. Pressurized flow and soil column testing apparatus

Soil

The soil used in these experiments was a sandy field soil from the Chicago area. The soil was characterized for particle size distribution, hydraulic conductivity and pH. The absolute permeability (hydraulic conductivity) of the soil was determined by measuring the quantity of water flowing out of the column within a given time interval under a constant head and using Darcy's law for a fully saturated porous medium containing a single fluid. The soil had a hydraulic conductivity of 0.0069 cm/s, pH of 7.9, a porosity of 39.6 percent and an organic content of 0.98 percent, while containing 99.1 percent sand. The soil was added in several 5-cm increments inside the 30-cm long glass column until two-third of the column was filled with soil. The side of the columns was tapped with a rubber mallet during the addition of the soil to ensure uniform packing. The soil column was saturated with simulated groundwater by flushing the solution through the top of the columns first by gravity flow followed by pressurized flow.

RNIP Slurry Addition

A buffered electrolyte solution was used as simulated groundwater for transport of RNIP through the sand columns. The electrolyte contained 0.006 M of sodium bicarbonate, 0.002 M of calcium chloride and 0.002 M of magnesium chloride. The pH, total dissolved solids (TDS) and conductivity of the electrolyte solution were 7.76, 500 mg/L and 1020 μS, respectively. The RNIP was obtained from Toda Kogyo (Japan). The RNIP had an average particle diameter of 70 nm (50-300 nm), pH of 10.7, a BET surface area of 37.1 m^2/g. For pressurized column flow tests, a concentration of 4.8 g/L of RNIP was applied by the addition of a slug of 2 g of RNIP stock solution in 12 pore volumes of flow (approximately 420 mL). For RNIP modified with lactate, the aluminum or sodium lactate was added to 2 g of RNIP. Pressurized flow was used to induce transport of the RNIP slurry through the columns. One slug of RNIP slurry was passed through each column. Twelve pore volumes of simulated groundwater were passed through each column following the RNIP injection. In order to assess the effectiveness of transport through the porous medium, a mass balance was conducted to determine the mass of iron that eluted from the columns. The RNIP/lactate slurry with the highest RNIP elution was considered the optimum slurry. The pore water velocity under gravity flow or induced flow was calculated.

RESULTS AND DISCUSSION

The performance of bare RNIP and RNIP modified with lactate was evaluated by comparison of the recovery of solid mass of RNIP in the effluent from the soil column versus the mass of RNIP input to the system. The recovery of RNIP in the effluent from the soil column is indicative of the degree of elution of RNIP through the soil column. Soil column tests were conducted for 4.8 g/L of RNIP modified with aluminum lactate and sodium lactate as function of lactate added to determine the optimum lactate concentration for each lactate compound. The results for addition of aluminum lactate in increments of 2 percent lactate are shown in Figure 2. The recovery data from Figure 2 show that there was a substantial increase in recovery from nearly 51 percent for bare RNIP to 71 percent for RNIP with 2 percent aluminum lactate, followed by an eventual increase to 91 percent for RNIP with 10 percent aluminum lactate. There was no further increase in recovery for RNIP with 12 percent aluminum lactate. Similar results were obtained for sodium lactate. Using RNIP with 10 percent aluminum lactate, the recovery of RNIP was evaluated as function of RNIP concentration from 2.4 to 12 g/L RNIP (as shown in Figure 3). The RNIP recovery data from Figure 3 show that the elution of RNIP modified with 10 percent aluminum lactate decreased with increasing concentration of RNIP.

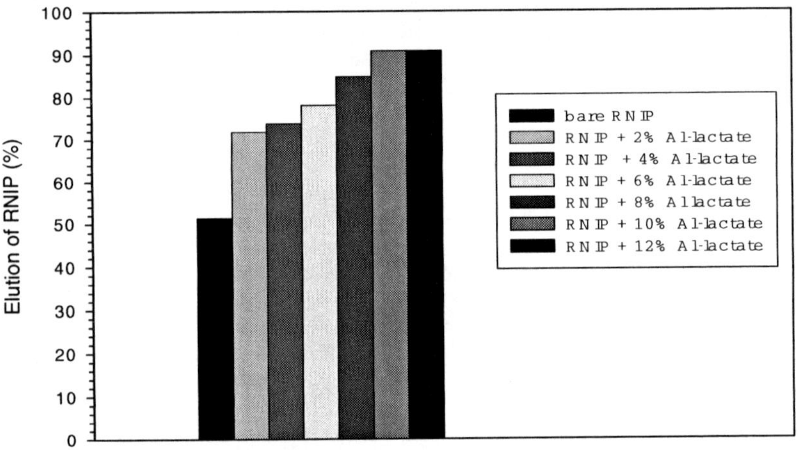

FIG.2. Elution of modified RNIP as function of aluminum lactate concentration

The results for the elution of RNIP from the sand column are shown in Figure 4. The average elution of solid RNIP were nearly 50, 80 and 90 percent for bare RNIP, RNIP with sodium lactate and RNIP with aluminum lactate, respectively. In all three cases, the elution of RNIP was obtained within three pore volumes of flow through the sand column. The data from Figure 4 show that RNIP modified with aluminum lactate and bare RNIP exhibited the greatest and the smallest recoveries, respectively. The RNIP modified with aluminum lactate performed best, indicating that modification with aluminum lactate effected the greatest elution of RNIP through the sand column. The greater elution of RNIP through the sand column occurred mainly due to the effectiveness of aluminum lactate as dispersant for RNIP particles. While RNIP modified with sodium lactate was more effective than bare RNIP, sodium lactate was observed as the less effective dispersant versus aluminum lactate, when added to RNIP on an equivalent mass basis (10 percent). The main hindrance to transport of bare RNIP in porous media such as sand is the ineffective dispersion of colloidal or semi-colloidal particles of RNIP, which results in the agglomeration (clumping) of particles and the adsorption of RNIP onto soil surfaces.

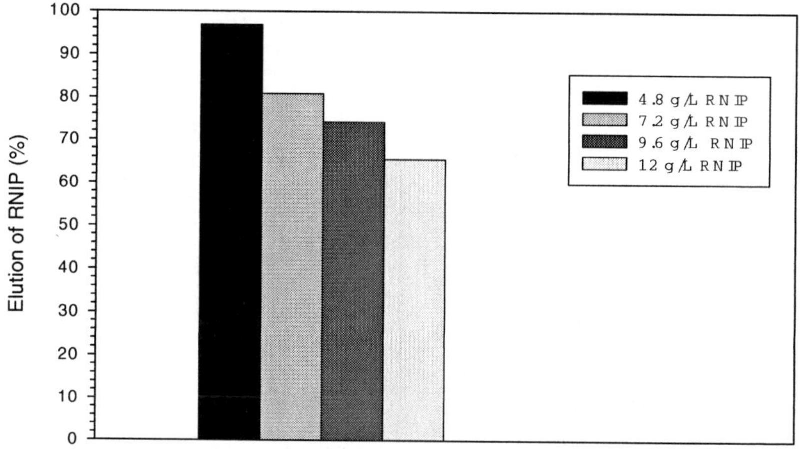

FIG.3. Elution of modified RNIP as function of RNIP concentration

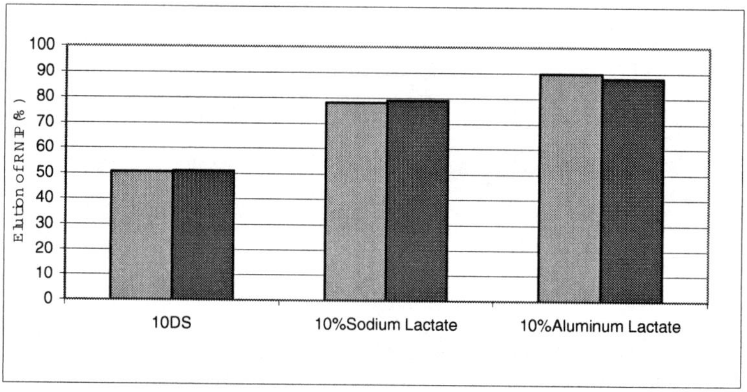

FIG.4. Elution of lactate-modified RNIP from soil column

The elution of RNIP through the sand column was observed visually; this was feasible due to the distinct color difference between the lighter sand and the black RNIP. The visual observations show that the top of the sand column was clear and free of RNIP in the case of RNIP modified with aluminum lactate, whereas the top of the sand column was covered with a layer of RNIP when bare RNIP was used and was relatively dark when RNIP modified with sodium lactate was used. This supports the results for greater elution and recovery of RNIP modified with aluminum lactate, which evidenced better performance.

The data for effluent quality are presented in Figures 5, 6 and 7 for pH, conductivity and soluble iron, respectively. The pH data from Figure 5 show that the effluent pH was acidic as low as 5.5 for RNIP modified with aluminum lactate, while the effluent pH was slightly higher for RNIP modified with sodium lactate that for bare RNIP. Although the electrolyte solution was buffered, the variation in effluent pH occurred for RNIP modified with 10 percent aluminum lactate. The effluent pH stabilized to the levels for bare RNIP within four to five pore volumes of flow after the elution of RNIP modified with aluminum lactate from the sand column. The data for effluent conductivity in Figure 6 show that conductivities for RNIP modified with lactate were greater than the conductivity for bare RNIP. Sodium lactate exhibited a larger conductivity than aluminum lactate due to the larger concentration of soluble salt for 10% sodium lactate than for 10% aluminum lactate. Similar to effluent pH, the effluent conductivities decreased to the levels for the bare RNIP after the elution of RNIP modified with lactate from the sand column during the first four to five pore volumes of flow. The effluent soluble iron data in Figure 7 show that RNIP modified with aluminum lactate evidenced the largest concentration of soluble iron in effluent, whereas little soluble iron was found in the other two systems; the larger soluble iron concentration in effluent was due to the lower pH witnessed by the RNIP modified with aluminum lactate. Again, similar to effluent pH, the soluble iron concentration subsided after the elution of RNIP modified with lactate from the sand column within four to five pore volumes of flow.

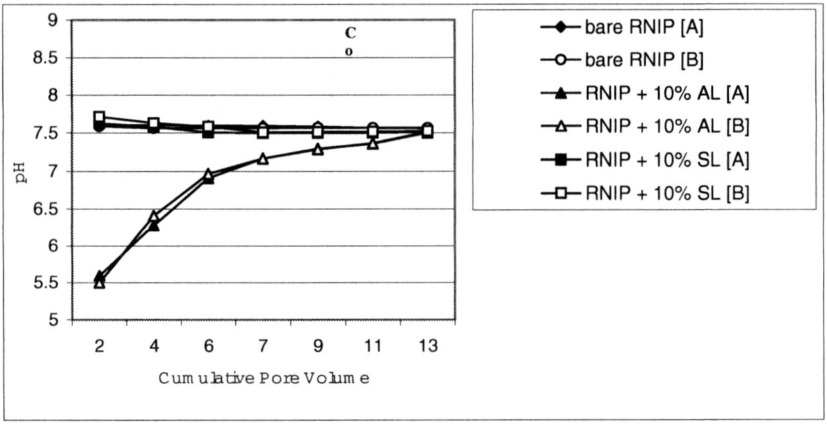

FIG.5. Effluent pH for lactate-modified RNIP [AL=aluminum; SL=sodium]

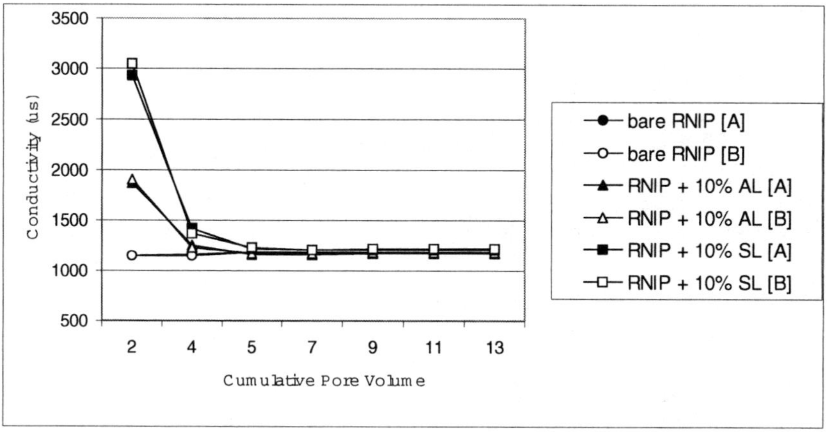

FIG.6. Effluent conductivity for lactate-modified RNIP [AL=aluminum; SL=sodium]

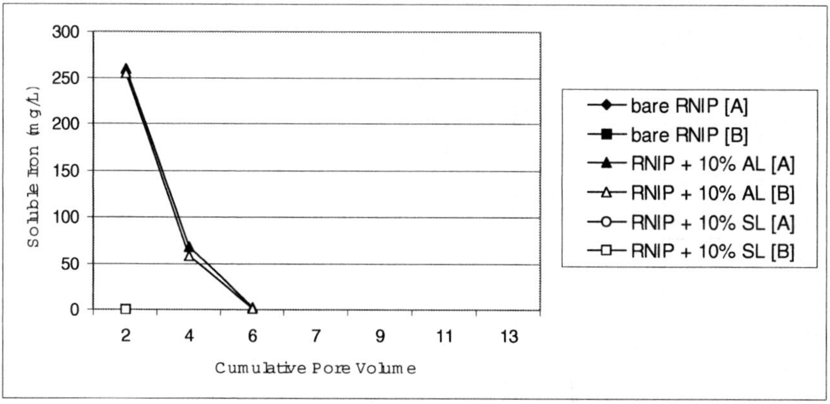

FIG.7. Effluent soluble iron for lactate-modified RNIP [AL=aluminum; SL=sodium]

The permeability (hydraulic conductivity) of sand during flow through the sand column was determined (based on Darcy's Law) for all three systems. The results for permeability as function of pore volume of flow are presented in Figure 8. The results show that the system for RNIP modified with sodium lactate evidenced slightly lesser permeability. Overall there was no significant difference between the permeabilities in all three systems (0.005 to 0.02 cm/s), indicating that the transport of RNIP at a concentration of 4.8 g/L with and without modification was not conducive to appreciable decrease in permeability inside the sand column; in other words, no clogging occurred and flow through the sand column was not impeded.

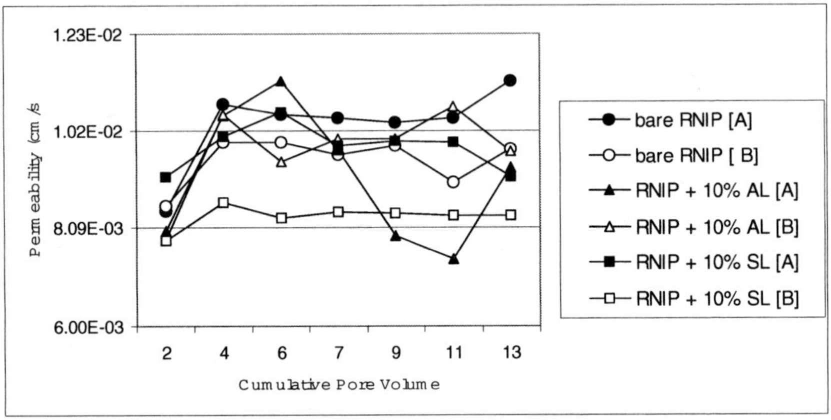

FIG.8 Permeability for lactate-modified RNIP systems [AL=aluminum; SL=sodium]

CONCLUSIONS

This study results show that RNIP modified with aluminum or sodium lactate can be effectively transported through the tested soil under pressurized conditions. Detailed investigations into transport and reactivity of lactate-modified RNIP in contaminated subsurface soils are being explored in on-going studies.

ACKNOWLEDGEMENTS

The funding for this project was received from the U.S. National Science Foundation and is gratefully acknowledged.

REFERENCES

Saleh, N., Sark, K., Liu, Y., Phenrat, T., Dufour, B., Matyjaszewski, K., Tilton, R.D., and Lowry, G. V. (2007). 'Surface modifications enhance nanoiron transport and NAPL targeting in saturated porous media.' *Environ. Engr. Sci.*, 24(1), 45-57.

Schrick, B., Hydutsky, B., Blough, J., and Mallouk, T. (2004). "Delivery vehicles for zerovalent metal nanoparticles in soil and groundwater." *Chemistry of Materials*, 16, 2187-2193.

Modeling of Cr (VI) Transport and In-situ Remediation with Nano scale Irons

Kaliappan S[1], Subbulakshmi M[2], Saisubhashini S[2]

[1]Professor of Civil Engineering, [2]Research Scholar
Dept. of Civil Engg., College of Engineering, Guindy, Anna University, Chennai, India

ABSTRACT: About 1,500,000 tons of chromites ore processing residues (COPR) containing 2 to 5% chromium, were generated from a chromites ore processing plant in Ranipet, Tamilnadu, INDIA, between 1975 to 1995. Since zero valent iron is a strong reducing agent it readily immobilize to Cr(III). In the present study, a 3D model, using finite element code, FEFLOW was used to evaluate the effect of Cr(VI) reduction, resulting from injection of nano scale zero valent iron (NZVI) particles. Initially the solute transport model was simulated to predict the distribution of the Cr(VI) in groundwater and calibrated. The NZVI injection was centered over the source zone and area of highest concentrations were divided into number of grids and the dosage of NZVI was determined. Column studies were undertaken to determine the rate of Cr(VI) reduction from the aqueous phase, by NZVI (RNIP), under laboratory conditions. The results illustrate the percentage reduction of Cr (VI) to Cr (III) and their spatial and temporal distribution in groundwater.

INTRODUCTION

In natural conditions, chromium is present as chromites ore and from which chromium salts are produced. Approximately 1,500,000 tons of chromium ore processing residue, containing 2 to 5% chromium was generated from "Tamil Nadu Chromites and Chemicals Ltd" Ranipet, India between 1975 and 1995. The residues were stacked in an open yard (three to five meters high and on 2 hectares of land) on the facility premises. The slow dissolution kinetics of the Cr(VI) bearing minerals in the residue makes it a continuing source of Cr(VI) to the environment (Chih-Huang Weng et al.2002).

The chromium leaching processes is mainly governed by the adsorption/desorption process and redox reactions. As the surface of most natural soil is negatively charged, the CrO_4^{2-} get adsorbed onto the soil as weak outer- sphere complex, (*Liyuan Liang et al*, 2005). Under alkaline conditions, CrO_4^{2-} is quite soluble and mobile and it may become hazardous as rain or irrigation water percolates into the ground unless it is reduced to an insoluble form of Cr (III) (*Bruce R. James*, 2002).

The study area was shown in Figure 1. The objective of this study is to investigate the application of NZVI for Cr(VI) reduction in the study area through Reactive Transport Modeling. To achieve this the solute(*Cr(VI)*) transport in groundwater was predicted and the effectiveness of NZVI in Cr(VI) reduction was assessed in laboratotary.

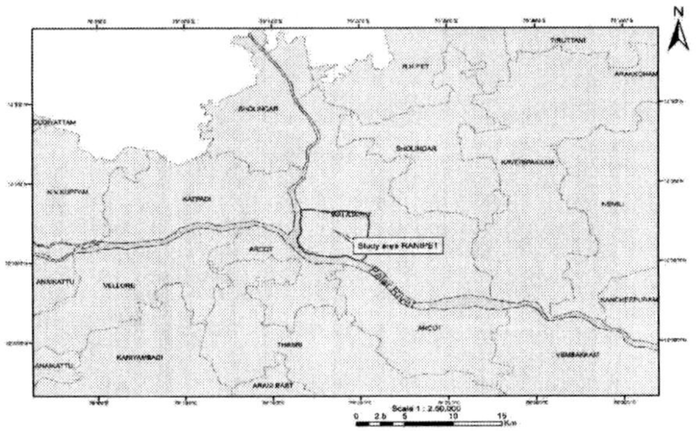

Figure 1. Study Area Map

Remedial Measure

Nanoscale iron particles have large surface areas, high surface reactivity and better transport property in groundwater and are able to remain in suspension for extended periods due to their colloidal nature. Therefore they are very effective for the transformation of contaminants *(Jiasheng Cao et al, 20 Climent et al, 1998; Steven Bryant et al 2001)*. Nanoiron acts as a reductant by supplying electrons directly from the neutral surface (*Taeyoon Lee 2003)*.

Site Description

The study area Ranipet (79°29´E / 12°55´N) is located in the coarse of *Poini* tributary and *Palar* river and about 160 km from Chennai city. The subsurface conditions at the site contain sandy loamy soil, followed by weathered granite gneiss rock zone that extend to the hard granitic gneiss bed rock at depth of 20 to 26 m below ground surface(*bgs*).

Groundwater samples were collected from different locations around the Cr (VI) residue disposal site and downstream. Cr(VI) concentrations were determined by Diphenylcarbohydrazide (DPC) method and analyzed by the 1,5-diphenylcarbohydrazide chromate complex in an acidic medium at 540 nm.

Table 1. Typical Characteristics of Groundwater

Sl.No	Parameters	Groundwater at Landfill Site	Groundwater Upstream side of Landfill Site
1	pH	11.8	7.7
2	TSS	40	44
3	TDS	2460	1200
4	Chloride	257	125
5	Sulphate	820	420
6	Cr(VI)	380	BDL
7	Total Chromium	470	BDL

Solute Transport Modeling

Advection and dispersion are the two major transport processes that determine the maximum extent of the leachate plume spread and the geometric character of the concentration distribution. Transport and reaction of the contaminant in the porous medium is described as shown in equation 1.

$$\frac{\partial c}{\partial t}=\left[\frac{\partial}{\partial x}\left(Dx\frac{\partial c}{\partial x}\right)+\frac{\partial}{\partial y}\left(Dy\frac{\partial c}{\partial y}\right)+\frac{\partial}{\partial z}\left(Dz\frac{\partial c}{\partial z}\right)\right]-\left[\frac{\partial}{\partial x}(VxC)+\frac{\partial}{\partial y}(VyC)+\frac{\partial}{\partial z}(VzC)\right]\pm\sum_{1}^{m=n}Rm \quad - \quad (1)$$

Where, C is solute concentration, [mg/L], t is time, [t], V is to represent velocity in three dimensions, [m/s], D is the dispersion coefficient, [m^2/s] and Rm is physical, chemical, and biological reaction rates, [$mg/m^3.s$].

Model Simulation:

The study was georectified using Arc View / Arc GIS and used as a base map, over which Super element mesh was constructed. The model domain was subdivided for Nodal classification and Finite Element Mesh was done. The problem was defined as transient flow with mass transport for an unconfined aquifer and set to run over 10 years up to a depth of about 25 m *bgl*. The geological stratum comprise of top soil (about 1.5m to 6.0m thick) being sandy loamy nature followed by moderately weathered rock (about 12 to 18m thick). Thus the conceptual model was created.

The ground water surface at the beginning of simulation was assigned with the water level data in year 1996 with the Dirichlet boundary condition. The western and southern part of the model domain was set with Palar River and a constant hydraulic head of 21m was assigned. Hydraulic conductivity (0.381×10^{-4} to 3.657×10^{-4} m/s), Porosity (0.36 to 0.42), and the dispersion coefficient (1.375×10^{-3} m^2/min to 1.593×10^{-3} m^2/min) as determined experimentally were used for modelling. Initial water heads for 16 wells were observed and with Akima inter/extrapolation module extrapolated for the study area. A continuous release of leachate with 500mg/L was assumed. The mass of Cr (VI) leached is about 1% of the total Cr (12.5 mg/g) in the soil sample. The model output of velocity distribution and the ground water flow are shown in figure 2.

The transient state simulation of the solute transport was carried out using automatic time step control via predictor-corrector schemes, with initial time step length as 0.001 day and final time as 3650 days (10 years) to reach steady state conditions. The simulated result of contaminant movement in the model domain was shown in figure 3.

Model was calibrated by comparing the model-simulated conditions and field conditions for hydraulic head and contaminant concentration for January 2006. The RMSE was observed to be 0.042. The comparison was shown in figure.4 and figure.5.

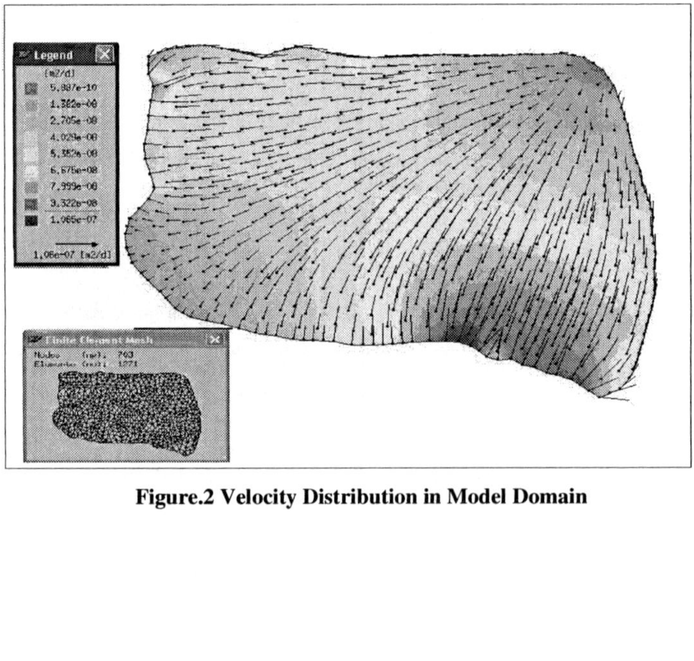

Figure.2 Velocity Distribution in Model Domain

Figure.3 Simulated Cr(VI) Spread from the

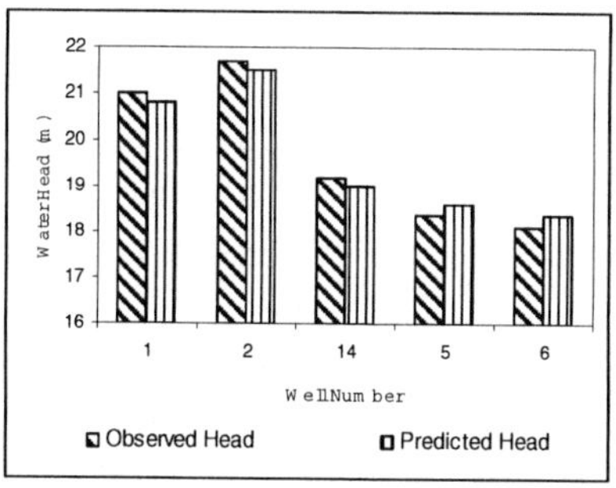

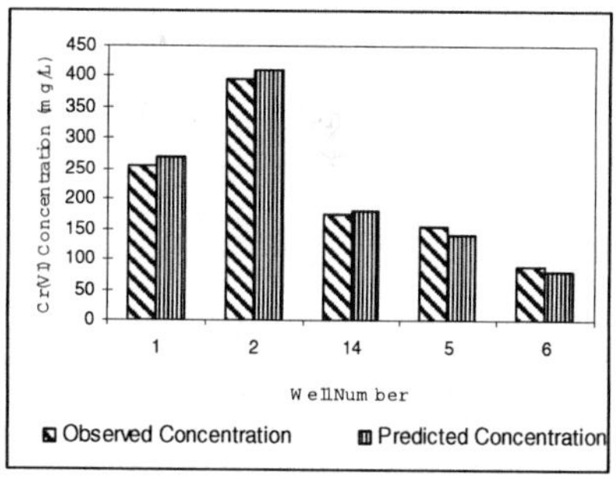

Figure.4 Observed Water Head Vs Predicted Water Head

Figure.5 Observed Cr(VI) Concentration Vs Predicted Cr(VI) Concentration

Experimental Study on NZVI Reaction:

From experimental study, Continuous flow column experiment is ideal to simulate in situ injections, transport and reactions of nanosclae iron particles in sub surface. (*Alexandre Duarte Gusmao et al* 2004, *Amos et al* 2004). An acrylic column of length 30cm and internal diameter 5cm with 5 numbers of ports along its length, equally positioned, was used. The groundwater collected from the site was made to flow through the column packed with reactive material (sand and RNIP 10) at a flow rate of about 4-5 ml/min with varying concentration of nano irons and the reaction rate was calculated (*Kaplan D. L.,* 2002). From the column study, reaction rate was found as $2.843x10^{-3}$ s^{-1}

Reduction-Corrosion Reactions

Reduction-Corrosion Reactions was assumed to be irreversible and the stoichiometries reaction were normalized with zero valent iron*(Mayer et al 2001)* as shown

$$CrO^{2-} + Fe^{0} + 8H^{+} \rightarrow Fe^{3+} + Cr^{3+} + 4H_2O$$

$$(1\text{-}x)Fe^{3+} + (x)Cr^{3+} + 2HO \rightarrow Fe_{(1-x)}Cr_x\,OOH_{(s)} + 3H^{+}$$

Reaction precipitates has indicated a one-to-one correspondence of the locations of Cr(III) and Fe(III) hydroxide phases on the surface of reacted iron filings, indicating co-precipitation and the likelihood of the formation of a solid solution phase of the general formula $(Cr_x, Fe_{1-x})(OH)_3$ *(Salas et al 2004, Peter Bayer et al 2005).*

The concentration of NZVI of above 200 mg/L was used for injection. The residue disposal site and it's environs of about 500m down stream was simulated. Since the groundwater table in that area is at a height of about 3m *bgl*, the injection at a depth of about 3.5 to 4.5m *bgl* was done. About 10 injection wells were simulated to provide sufficient coverage of treatment zone (5 at Treatment Zone and 5 in the Downstream Side). With the calibrated results of solute transport modeling as a base, multi species transport was carried out. The transport in 3D for three-species, Cr(VI), Fe(0) and Fe.Cr(III) was simulated. Cr(VI) and Fe(0) were assigned as fluid phase where Cr(III) was assigned as solid phase which is to be precipitated *(Sherman et al 2000)*. The Chemical reaction kinetics was specified in FEMATHED module as follows.

$R_1 = -k_1C_1$ ------------- (2)

$R_2 = -k_2C_2$ ------------------ (3)

$R_3 = k_1C_1 + k_2C_2 - k_3C_3$ -------------------- (4)

Where R_1, R_2, R_3 represent the rate constant, k_1, k_2, k_3 are first order rate constant and C_1, C_2, C_3 are concentrations correspond to three-species, Cr(VI), Fe(0) and Fe.Cr(III). The simulation of Cr(VI) decay for 30 days was carried out and the Figure 6 shows the concentration at two days interval for various observation wells. The results provide a clear-cut idea of about how the degradation will take place in the field. Due to the injection of nano iron in the field, the iron content might increase and this may be cleared off from the site by natural attenuation.

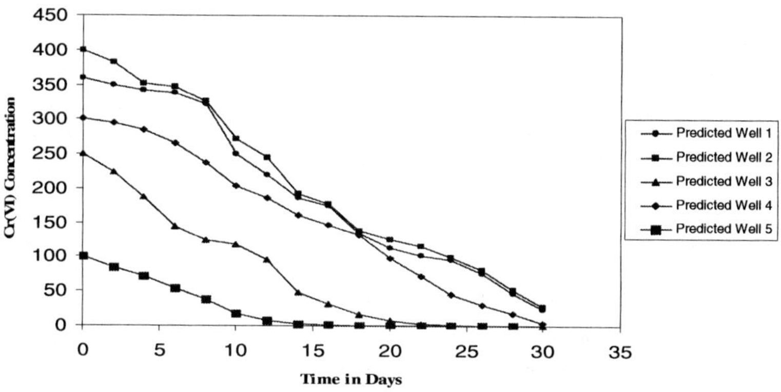

Figure 6. Decay Curve of Cr(VI)

CONCLUSION

In the current study, the spatial distribution of Cr(VI) in the groundwater was modeled using FEFLOW and calibrated with accuracy of RMSE = 0.042. The Cr (VI) reduction rate was found to be 2.843x10-3 s-1 from the column study. The model reveals that about 92% Cr(VI) reduction can be achieved in about 30 days at the study area. Consequently Reactive Transport Model can be effectively used as management tool for initiating, control and remedial measure of Pollutants.

REFERENCES

Alexandre Duarte Gusmao, Tacio Mauro Pereira de Campos, Manoel de Melo Maia Nobre, Euripedes do Amaral Vargas Jr (2004), 'Laboratory tests for reactive barrier design', J. Hazardous Materials, Vol.110: 105-112.

Amos, R. T., Mayer, K. U., Blowes, D. W., and Ptacek, C. J (2004), 'Reactive transport modeling of column experiments for the remediation of acid mine drainage', *Environmental science and Technology*, Vol.38: 3131-3138.

Bruce R. James (2002), 'Chemical Transformations of Chromium in Soils: Relevance to Mobility, Bio-availability and Remediation', The Chromium file; No-8

Chih-Huang Weng1; C. P. Huang; and Paul F. Sanders, (2002) 'Transport of Cr (VI) in Soils Contaminated with Chromite Ore Processing Residue (COPR)', *Practice Periodical of Hazardous, Toxic, and Radioactive Waste Management,* Vol. 6 : 6- 13.

Climent.T.P, Sun.Y, Hooker B.S. and J.N. Petersen (1998), 'Modeling Multispecies Reactive Transport in Ground Water',*Ground Water Monitoring & Remediation*, vol.18 (2) : 79-92.

Kaplan. D. L (2002), 'Removal Rates of Aqueous Cr(VI) by Zero-Valent Iron Measured Under Flow Conditions', *Water Resources Research*, Vol.15 : 63-72.

Liyuan Liang , Gerilynn R.Moline, Wiwat Kamolpornwijit, Olivia R.West (2005), 'Influence of hydrogeochemical processes on Zero_Valent iron barrier performance: A field investigation', *J. contaminant Hydrology*, Vol.38 : 291-312.

Mayer, K. U., Blowes, D. W., and Frind, E. O.(2001), 'Reactive transport modelling of groundwater remediation by an in-situ reactive barrier for the treatment of hexavalent chromium and trichloroethylene', *Water Resources Research*, Vol.37: 3091-3103.

Peter Bayer, Michael finkel (2005), 'Modeling of sequential Groundwater treatment with Zerovalent iron and Granular activated carbon', J. of contaminant hydrology, Vol. 78 : 129-146.

Salas.J, C. Ayora (2004), 'Groundwater chemistry of the Oke´lobondo uraninite deposit area (Oklo, Gabon): two-dimensional reactive transport modelling', J. Contaminant Hydrology, Vol- 69 : 115– 137

Sherman M.Ponder, John G.Darab and Thomas E.Mallouk (2000), 'Remediation of Cr (VI) and Pb (II) aqueous solutions using supported Nanoscale Zero Valent Iron', *Environmental science and Technology*, Vol. 34 : 2564-2569.

Steven Bryant L., Karsten Thompson E. (2001), 'Theory, modeling and experiment in reactive transport in porous media', *Current opinion in colloid & interface science*, Vol. 6 : 217-222.

Taeyoon Lee, Hyunjung Lim, Yonghun Lee, Jae-Woo Park (2003), 'Use of waste iron metal for removal of Cr(VI) from water', *Chemosphere*, Vol. 53, pp. 479–485.

Biodegradation of Phenol-Pb Contaminated Clay Soil

C. Vipulanandan, Ph.D., P.E.[1], and B. Mamidi, M.S.[2]

[1]Chairman, Professor and Director of Center For Innovative Grouting Materials And Technology, Dept. of Civil and Environmental Engineering, University of Houston, Houston, TX 77204-4003
[2]Former Student, Dept. of Civil and Environmental Engineering. University of Houston, Houston, TX, 77204-4003.

ABSTRACT: Various methods are being investigated to treat soils contaminated with organic and metal wastes. Hence the effect of metal wastes on the biodegradation of organic waste must be investigated. Since higher levels of phenol contaminated soils are encountered with metals in industrial and hazardous waste sites, in this study, acclimate sludge was used to treat up to 10,000 mg/kg of phenol contaminated soils. Synthetically contaminated soil consisting of 80% sand and 20% clay was used in this study. The effect of lead on the biodegradation of phenol was found to be significant at concentration higher than 30 mg/L lead in the aqueous phase. The growth of biomass was not affected up to a lead concentration of 30 mg/L but was reduced by about 99% when the lead concentration was 120 mg/L.

INTRODUCTION

Application of technology to effectively treat the hazardous wastes generated by the industrial activity and safe disposal of these wastes has gained major attention in the past few years. With regard to wastes, the basic strategies to be adopted are to minimize the amount of waste generated, to reduce the toxicity of wastes that are generated, and to find more satisfactory ways of disposing of and degrading wastes which have accumulated. In order to achieve the above objectives it is very important to select the appropriate treatment method to treat the contaminant or the contaminated site. Before the selection of the treatment method it is necessary to evaluate the potential applicability of the treatment process and to demonstrate the feasibility of the approach.

There are many physical, chemical and biological treatment methods being employed for the treatment of hazardous wastes (Wang and Vipulanandan 2001). In this laboratory research bioremediation was selected and tested for its feasibility to apply the process in the field. Bioremediation is a treatment technology involving the application of microbial processes to convert environmental contaminants into harmless substances. The use of these microbial processes for the treatment of the waste materials has a long history, the area of sewage and wastewater treatment systems like activated sludge, aerated lagoons, digesters, trickling filters and fixed film reactors being among the most widely known applications of biological treatment (Brenner et al. 1992; Wang et al. 2001).

OBJECTIVES

The objective of this research was to study the feasibility of remediating phenol-contaminated soil using bioremediation treatment processes. The effect of heavy metals on the biodegradation process was evaluated.

MATERIALS AND METHODS

CONTAMINANT

Phenol was selected as the contaminant for biodegradation as it is listed among the top ten priority pollutants. Lead is one of the most commonly found contaminants in the environment. Of the 1200 superfund sites all over the United States over a third are contaminated by lead and of these about half have a documented release of lead to air, surface water or ground water. Hence, lead was selected to represent the heavy metals.

MICROORGANISMS

The microorganisms used in the biodegradation process were obtained from activated sludge consisting of a mixed population of cultures (Vipulanandan et al. 1993). Thickened activated sludge from the City of Houston waste water treatment plant was used. The treatment plant is used for treating both industrial and domestic wastes and the sludge was believed to have a mixed population of microbes that are already adapted to wastes containing moderate concentrations of organic compounds. Identification of this pure culture was done based on the oxi/ferm tube test and the morphological characteristics of the bacteria. This bacteria was identified as *Pseudomonas Putida*. The SEM micrograph (Fig. 1) of this bacteria indicates that the bacteria is sperical in shape.

EXPERIMENTAL PROGRAM

This section explains the various experiments that were designed for the biodegradation of phenol and the effect of two parameters (soil, Pb). The objective was to compare the performance of bioremediation on two types of wastes: (i) liquid waste and (ii) contaminated soil.

METHOD OF ANALYSIS

The reactors were monitored periodically for phenol and optical density (for biomass). Lead was also monitored for the experiments involving the study of the effect of lead on biodegradation of phenol.

Optical density was measured and expressed in terms of % transmittance of light at wavelength of 540 nm. A Baush & Lomb spectrophotometer (Model Spectronic 20) was used to measure the transmittance. The phenol concentration was measured using a Perkin-Elmer GC equipped with a flame ionization detector (FID).

The samples were filtered using a 0.45 μm gelman sciences filter paper to remove the biomass before analyzing for phenol.

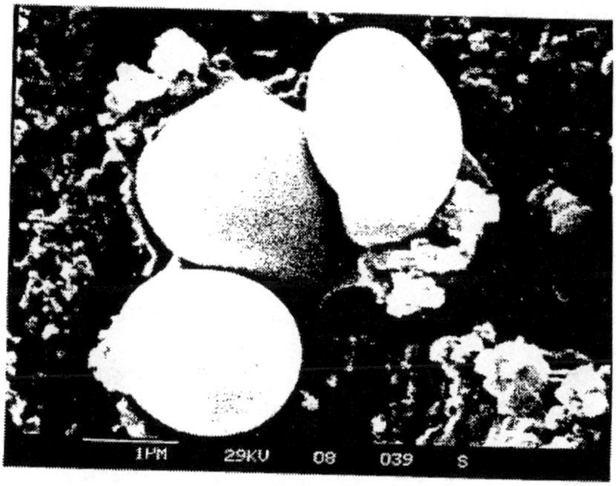

Figure 1. SEM Micrograph of *Pseudomonas Putida* Used for Degrading Phenol

Lead concentration was measured according to the standard method SW846-6010 using the Inductively Coupled Plasma (ICP). Lead concentrations were monitored at two different pHs, i.e. at pH 6.5 and pH 2. Acidification of the samples to pH 2 was done using nitric acid (HNO_3).

RESULTS AND DISCUSSIONS

Phenol was the target contaminant for all the studies conducted. Biodegradation of phenol was performed in the aqueous phase and in the presence of soil. The effect of metals (lead) on the biodegradation of phenol in the aqueous phase and in the soil slurry phase was also investigated and quantified.

(a) LIQUID WASTE

(i) Phenol

In this study, the biodegradation of phenol at a concentration of 1800 mg/l has been evaluated. A lag time of 120 hours was observed before the growth of the bacteria occurred, for an initial phenol concentration of 1800 mg/L. The total time taken for the complete biodegradation of phenol was 300 hours. This is the highest concentration of phenol degraded by bacteria reported in the literature. The degradation was rapid till the phenol reached a concentration of 200 mg/L. At lower concentrations the degradation was slower. This could be due to the fact that since the bacteria is acclimated to high concentrations of phenol it fails to recognize very low concentrations of phenol. The degradation rate of phenol with time is shown in Fig 1. The maximum growth rate obtained is 0.028 /h and the yield was 0.8.

The biodegradation of phenol was much faster when 0.5 g/l yeast extract was added to the reactor. The yeast extract basically reduced the lag period from 120 hours to 24 hours. The total time taken for the biodegradation of phenol was 170 hours. The growth rate of the bacteria is much higher when yeast extract was used. The final yield when yeast extract was used as the nutrient was 0.94 indicating that the addition of yeast extract increased the yield of the biomass. The pH was monitored continuously during the biodegradation of phenol. A drop in pH of 0.2 units was observed during the course of the complete biodegradation of phenol.

(ii) Phenol and Lead

Five concentrations of lead (6, 12, 30, 60, 120 mg/L) were used to study the effect of lead on acclimated bacteria. It was observed that this bacteria was resistant to lead concentrations up to 30 mg/L lead as there was no change in the time taken for biodegradation when compared to the control reactor (0 mg/l lead). For the lead concentrations from 0 to 30 mg/l, the time taken for biodegradation of phenol varied from 230 to 260 hours and for lead concentrations of 60 and 120 mg/L there was significant increase in the time taken for biodegradation which was 744 and 870 hours respectively. The specific growth rate was reduced by 40% when the lead concentration was 30 mg/L, by 83 % when the lead concentration was 60 mg/L and by 97% for a lead concentration of 120 mg/L (Fig. 3). Table 1 summarizes the results of the biodegradation of phenol in the presence of lead in the aqueous phase.

(b) CONTAMINATED SOIL

The biodegradation of contaminated soil containing phenol (10,000 mg/kg) and lead (300 mg/kg and 600 mg/kg) was investigated..

(i) Phenol

Mostly, contamination is encountered in the form of contaminated soil where the contaminant is adsorbed or physically attached to the soil. In such cases, the contaminant should be desorbed from the solid phase and be soluble in water in order to be available for biodegradation. This desorption process might slow down the biodegradation rate.

In order to study the effect of soil on the biodegradation of phenol tests were conducted on spiked soils. Though spiked soils do not represent the field soils as field soils are exposed to the contaminant for long periods which makes it difficult to remove the contaminant from these soils, but since phenol is a highly soluble substance these results can be extrapolated for field soils as well.

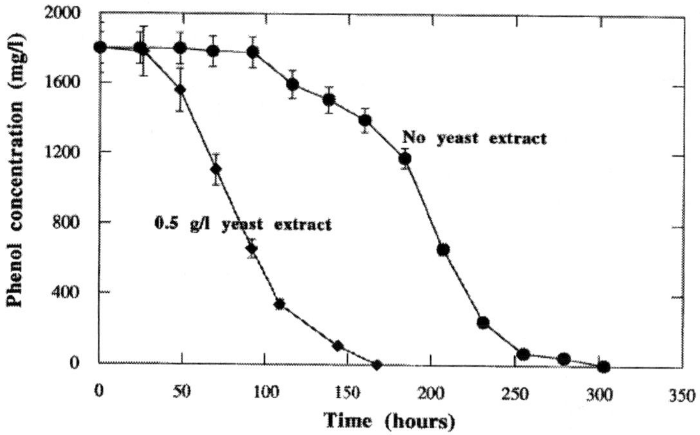

Figure 2 Biodegradation of 1800 mg/L of Phenol with and without yeast.

Table 1. Summary of Results for the Biodegradation of Phenol in the Presence of Lead.

Pb (mg/l)	Time of biodegradation (hours)	Specific growth rate (h^{-1})	$\frac{\mu'_m}{\mu_m}$*
0	210	0.08875	1
6	220	0.07875	0.8876
12	250	0.06040	0.6805
30	225	0.05380	0.6062
60	744	0.01	0.1144
120	840	0.0023	0.0259

* μ'm indicates the specific growth rate at a given metal concentration

Initial Phenol Concentration 10,000 mg/kg

For a concentration of 10,000 mg/kg, the biodegradation in the aqueous phase and in the soil slurry reactors was almost the same which was around 50 hours (Fig 4). This could be due to the fact that there was a lag time of 30 hours for the biodegradation of phenol in the aqueous phase. By the end of the lag time all the phenol was desorbed from the soil. Once the phenol was desorbed from the soil, the rate of biodegradation was the same in the two cases. The effect of soil was seen for lower concentrations of phenol as the time taken for the biodegradation of phenol was longer. This slow biodegradation could be attributed to the fact that phenol was bound on to the soil and in order for the phenol to be bioavailable, it was necessary to be desorbed from the solid phase in to the aqueous phase.

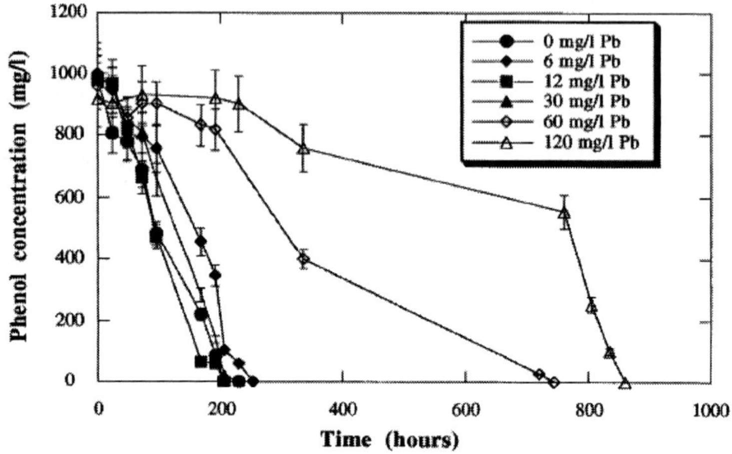

Figure 3 Biodegradation of 1000 mg/L of phenol in the presence of lead.

(ii) Phenol and Lead

The results of the bioremediation of contaminated soil containing mixed wastes (phenol and lead) are presented in Fig. 5. When a reactor loading of 50 g/l was used, the maximum phenol concentration in the aqueous phase measured was 500 mg/L. The time taken to degrade this concentration of phenol was about 90 hours when there was no lead present in the reactors and the biodegradation of phenol in the presence of lead also took about 90 hours.

The concentrations of lead measured in the liquid phase in the reactors is indicated in Fig. 6. Here again the lead was monitored at pH of 6.5 and 2. At pH 6.5 the samples were centrifuged and filtered and the filtrate was analyzed for the lead concentration. In order to determine the total lead present in the reactors, the samples were acidified using nitric acid to pH 2, then centrifuged and filtered and the filtrate was analyzed for lead. The results indicate that there is an initial increase in the lead concentration, in the aqueous phase, within 24 hours. This could be due to the removal of lead which was bound to the soil. The maximum concentration of lead attained in the aqueous phase at neutral pH was 7 mg/L for both the concentrations of lead and phenol contaminated soil studied. The maximum concentration of lead attained in the aqueous phase at pH 2 was 21 mg/L for 600 mg/kg of lead contaminated soil and 35 mg/L for 1200 mg/kg of lead contaminated soil. Some of the lead in these reactors could not be recovered even after acidification and this could be due to the possibility of irreversible adsorption of the lead on to the clay or due to the uneven distribution of lead in the soil, which lead to higher predictions of lead concentrations in the aqueous phase.

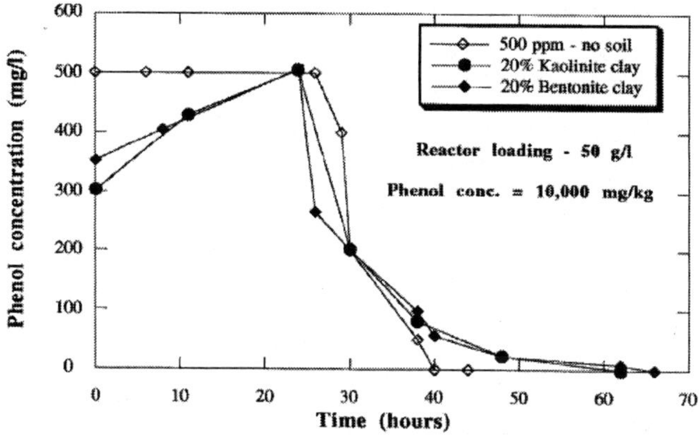

Figure 4. Biodegradation of phenol from contaminated soils

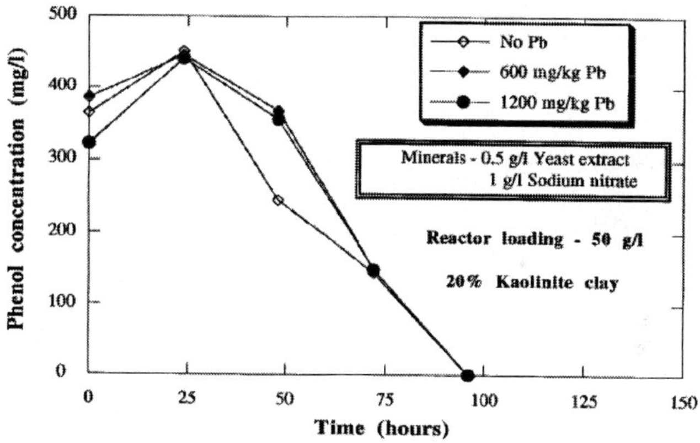

Figure 5. Degradation of Phenol – Pb Contaminated Soil

CONCLUSIONS

Based on the experimental study following are concluded:

(1) Biodegradation of 1800 mg/L phenol was successfully completed using acclimated bacteria. Presence of 0.5 g/L yeast extract reduced the lag time from 120 hours to about 24 hours.

(2). From the results of these studies, it was observed that up to a lead concentration of 30 mg/L the time taken for the biodegradation was not significantly affected.. The specific growth rate of bacteria was reduced by 40 fold when the lead concentration was increased to 120 mg/l.

(3) Biodegradation of phenol was observed to be slow down in the presence of soil. Lead which was present in relatively high concentrations (600 mg/kg, 1200 mg/kg) in the soil does not seem to affect the biodegradation process of phenol.

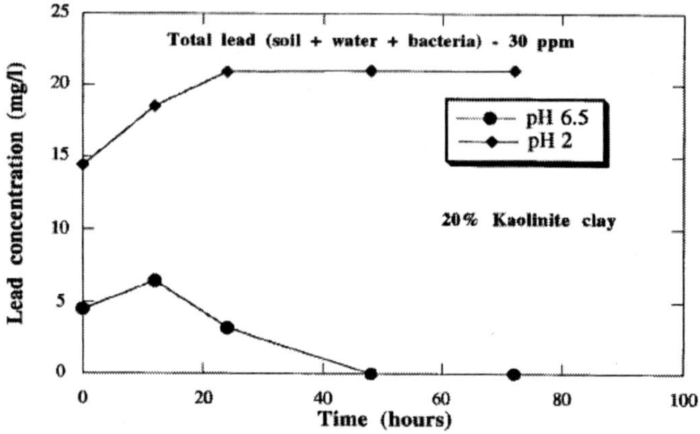

Figure 6. Variation of Lead Concentration in the Reactor with pH.

ACKNOWLEDGEMENT

The study was supported by the Center for Innovative Grouting Materials and Technology (CIGMAT) at the University of Houston.

REFERENCES

Brenner, A., Chozick, R. and Irvine, R. L.. (1992), "Treatment of High-Strength Mixed Phenolic Wate in an SBR, Water Environmet Research, Vol. 64, No. 2, pp. 128-133.

Vipulanandan, C. and Krishnan, S. (1993), "Leachability and Biodegradation of High Concentration of Phenol and O-Chlorophenol, Hazardous Waste and hazardous Materials, Vol. 61, No. 3, pp. 320-326.

Vipulanandan, C. (1995), "Effect of Clays and Cement on the Solidification/Stabilization of Phenol contaminated Soils," Waste Management, Vol. 15, No. 5 pp. 309-406.

Wang, S. and Vipulanandan, C. (2001) "Biodegradation of Naphthalene-Contaminated Soils in Slurry Bioreactors," Journal of Environmental Engineering, ASCE, Vol. 127, No. 8, pp. 748-754.

Bioremediation of military area contaminated by petroleum products

Barbara Kołwzan[1], Kazimierz Grabas[2], Adam Pawełczyk[3] and Mieczysław Steininger[4]

[1] Dr hab., Wrocław University of Technology, Wybrzeże Wyspianskiego 27, 50-370 Wrocław, Poland; barbara.kolwzan@pwr.wroc.pl
[2] Dr hab., Wrocław University of Technology, Wybrzeże Wyspianskiego 27, 50-370 Wrocław, Poland; kazimierz.grabas@pwr.wroc.pl
[3] Dr Wrocław University of Technology, Wybrzeże Wyspianskiego 27, 50-370 Wrocław, Poland; adam.pawelczyk@pwr.wroc.pl
[4] Dr Wrocław University of Technology, Wybrzeże Wyspianskiego 27, 50-370 Wrocław, Poland; mieczyslaw.steininger@pwr.wroc.pl

ABSTRACT: Examples of the bioremediation method applied at the former Soviet air bases in Poland polluted with the petroleum products are presented in this paper. The bioremediation of polluted soils was conducted to remove aviation fuel and heavy fractions of diesel oil. Polluted clay-based grounds were cleaned by the *ex situ* method, which utilized specially built installations. Indigenous microorganisms carried out the biodegradation. The polluted soil was situated in special reactor for the treatment process. The reactor included systems to provide irrigation water and nutrients, a liner at the bottom of soil, and a means to collect leachates. A closed circle water system was created to eliminate the possibility of polluting environment. The efficiency of that process was very high. At the Bagicz airbase, where the original concentration of the petroleum products amounted up to 17,400 mg/kg a 90 % reduction of the petroleum in soil was achieved in a few months. Similar results were obtained at the former Russian airbase "Brzeg" in South-West Poland where obout 1,000 m3 of polluted ground has been cleaned.

INTRODUCTION

High demand for oil and its by-products has resulted in an increasing amount of oil pollution entering ecosystems every year (Surygała and Śliwka 2000). In Poland an especially high devastation of the environment was observed on the terrain occupied by the Russian Federation Military Troops. A serious contamination of the water-ground environment at the former Soviet military airfields has been revealed in Poland, which was caused as a result of reprehensible carelessness. The Soviet air bases were located in western and north-western part of Poland (fig.1). The General

Inspector for Environmental Protection has prepared reports giving full data on damages in the environment (General Inspector 1994). According to a report published in 1994, costs of grounds remediation in the area of 406 hectares was estimated at 2,3 billion US $. The remediation of the former Soviet military airfields was recognized as one of the priorities in environment protection in Poland.

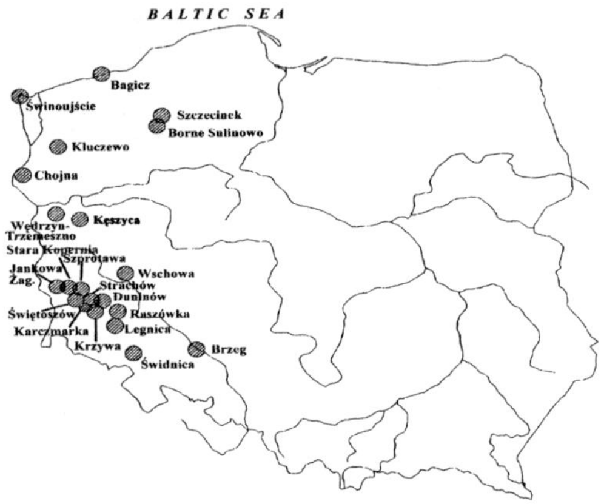

FIG.1. Location of the Soviet Army airfields in Poland (General Inspector 1994).

The choice of the method of soil restoration requires knowledge of many parameters such as the quantity, type and placement of soil pollutants, geological build-up of the terrain, aquiferous layer distribution, and direction of the underground water flow. The effectiveness of the purification processes depends on grain size analysis, chemical and the geological origin and morphology of the grounds, amount, kind and distribution of the pollutant (Kolwzan 2005).

In recent years the remediation methods based on biological treatment have found a wider application. The main role in the biological process of purification is played by the microorganisms capable of utilizing xenobiotics as the source of carbon and energy. Studying the properties of microorganisms and pathways of microbiological break-down allows the effective use of biodegradation (Lebkowska 1996 and Kolwzan 2005). Various technologies base on the biodegradation activities of microorganisms and concentrate on the increase of productivity in already existing, however slow, self-cleanup taking part in nature (Leahy and Colwell 1990).

Biological methods of self-cleanup are usually assisting techniques. This means that within the polluted areas, the negative effects of contamination should be limited by the following additional means:

- Elimination of the contamination source
- Securing the contaminated area
- Collecting petroleum products from ground surface and the surface waters

- Limiting or bringing to a stop of pollution migration
- Pumping out the petroleum products from the aquiferous layer.

The authors developed two methods of soil remediation. Depending on the contamination level and the character of the environment to be treated the, bioremediation process can be carried out using *in situ* or *ex situ* method (Pawełczyk, Kołwzan and Grabas 2006). The first one consists of treating the ground directly in site without moving the contaminated soil. In this method bacteria strains, biogens and air are introduced directly into the soil through a network of pipes and piesometers inserted into holes drilled in the area. *In situ* method is usually applied in areas where the danger of remediation agents' migration to non-polluted locations has to be excluded.

The second method (*ex situ*) refers to the excavating and treatment of the soil at a specially prepared place isolated from the subsoil by a foil layer. After completing the process the purified ground has to be transported to the original site (Lebkowska 1996).

Two examples of the *ex situ* bioremediation method developed by the authors and applied at former Soviet air bases in Poland polluted with the petroleum products are presented. *Ex situ* method was used in the area where full control of ground parameters was needed and where a risk of any secondary pollution of ground with mineral fertilizers and microorganisms used could not be tolerated. The method appeared to be effective and enabled restoration biological life in the area.

The main task of the remediation treatment of grounds is reducing the petroleum products' concentration in the contaminated ground to the level acceptable by obligatory standards.

SITE DESCRIPTION

Bioremediation of polluted grounds was the last step in the recultivation of two former Soviet airports:

- Former Bagicz airport located at the Baltic Sea east from Kołobrzeg. Due to the recreation ground located in the nearest neighborhood at the sea side, there was a necessity to develop such a remediation method that would eliminate danger of secondary pollution of the waters and grounds in that area.
- Former Soviet Army base located close to Skarbimierz south-west from Brzeg. Due to possibility of petroleum products' migration, the airport can be a threat for the water-bearing terrains for Wroclaw city.

The source of environment pollution was aviation fuel leakages from damaged tanks and pipes. Also, heavy ends of diesel oil were observed in the ground. Content of petroleum hydrocarbons in the soil was high, for the Bagicz airport 2732 mg/kg d..m. and in Brzeg: 5540 mg/kg d.m.(table 1).

The bioremediation process was carried out in three stages.

I stage – Selection of bioactive microorganism strains in relation to the pollutants present in the ground.

II Stage - Real bioremediation process carried out in a windrow reactor.

III stage – Displacing the ground from the windrow to the original site after completing the bioremediation process.

Table 1. Physicochemical characteristics of polluted soil.

Soil parameters:	***Airfield Bagicz***	***Airfield Brzeg***
Cubic capacity of polluted soil (m^3)	500	1000
Pile dimensions (m) (length x width x height)	80m x 4.2m x 1.5m	50m x 6.2m x 3.1m
Soil texture (% by mass)	clay-loam (42% clay, 33% silt, 25% sand)	sandy-loam (44% sand, 30% silt, 26% clay)
pH	7.1 – 7.4	6.8-7.5
Moisture (% by mass)	13-25	11-19
Total Petroleum Hydrocarbons (TPH) concentration range before soil remediation (mg/kg d.m.)	142-8 771	1716 – 14 891
Average of TPH concentration in soil before remediation (mg/kg d.m.)	2732	5541
TPH concentration range after soil remediation (mg/kg d.m.)	18-103	77 -1788
Average TPH concentration in soil after remediation (mg/kg d.m)	55	367
Effluent parameters:		
BOD at the beginning of bioremediation (gO_2/m^3)	620	810
BOD after bioremediation (gO_2/m^3)	24	31
COD at the beginning of bioremediation (gO_2/m^3)	1450	1840
COD after bioremediation (gO_2/m^3)	362	460

MICROORGANISMS USED IN THE BIOREMEDIATION PROCESS

The bioremediation of polluted grounds was based on bio-preparations obtained from an autochthonous microflora base. For the purpose of isolating bacteria from polluted ground, a mineral medium containing diesel oil was used as the only source of carbon and energy. The selection of microorganisms was done under natural temperature conditions, with the aim of obtaining microorganisms which effectively degrade petroleum products under field conditions. ID 32 GN test from Biomerieux Company was used to determine if pathogens were present. Before starting multiplication, the pathogenic strains were eliminated. The remaining strains were used for obtaining the biopreparation. Taxonomic affiliation of microorganisms included in the inoculants differed. The strains collected from Bagicz belonged to following genera: *Pseudomonas fluorescens, Acinetobacter calcoaceticus, Pseudomonas putida I, Pseudomonas putida II, Pseudomonas stutzeri, Rhodococcus maris, Rhodococcus erythropolis, Candida spp.* Biopreparation made based on autochthonic bacteria from Brzeg soil consisted of only 3 bacteria strains (*Aeromonas caviae, Alcaligenes xylosooxidans, Pseudomonas fluorescens*).

BIOREMEDIATION SYSTEM DESIGN

Bioremediation of grounds polluted by fuel oil was carried out by a biopile method (fig.2). Due to the texture of the soil the prisms varied by height. Due to clay character of the soil-which prevented from proper aeration of the soil-the Bagicz prism had to be lower. This facilitated handling the pile and migration of air in the interior. Growth of hydrocarbons-degrading microorganisms in polluted soil was stimulated by:

- inoculation the prisms with bio-preparations containing autochthonous bacteria multiplied earlier in a field bioreactor,
- introduction of atmospheric oxygen,
- regulation of the nutrients concentration, pH level and humidity.

Excess water from pile was drained out to tanks of 1m^3 volume each. The collected effluent was used for wetting the pile, multiplication of the bacteria as well as the dissolution of the mineral fertilizers (sources of nitrogen and phosphorus). The originality of this method was using effluent from the pile as a cultivation medium (after assuring proper pH level and adding biogenic compounds). Applied technology lets us avoid contamination of surrounding area with toxic hydrocarbons and their metabolites as well as microorganisms released form prism. Biopreparations were introduced in the amount enough to obtain the number of bacteria in the soil on the level of 10^6-10^8 cells/g s.m. soil.

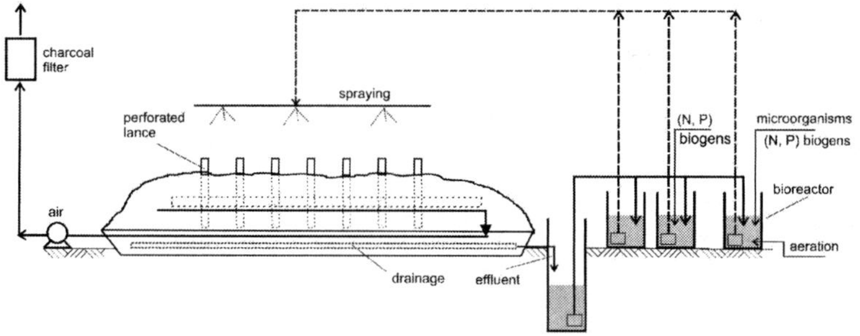

FIG. 2. Scheme of biopile method.

The soil was enriched in additional sources of nitrogen and phosphorus (ammonium nitrate, ammonium phosphate) in the amount that assured their right proportion to carbon according to the molar ratio C:N:P=10:1:0,1 (Liebeg and Cutright 1999). Fertilizers were introduced to the soil in a few steps in order to avoid toxic influence of heavily concentrated salts (Manilal and Alexander 1991). The selection of nutrients and choice of their appropriate dosages was performed in laboratory tests. The type of missing nutrients was established by growing soil microorganisms in the presence of various sources and concentrations of biogenic elements and the careful supervision of the number of microorganisms and their dehydrogenase activity (Kołwzan 2005).

Proper aeration was maintained through the drainage system. The air from the prism was directed to filters of activated carbon (about 30 kg) in order to protect surrounding atmosphere by volatile compounds emanating from the prism during aeration. Moreover, additional perforated vertical air lances were inserted into the soil. They assured access to the air plus they were used as a way to introduce multiplied bacteria and nutrients to the prism. The pH of soil was maintained at 6-8. The wetness of the soil wavered between 10–20% depending on the atmospheric conditions.

BIODEGRADATION MONITORING

The bioremediation process in field conditions was controlled by constant monitoring soil parameters in selected measuring points. The basic parameters were: petroleum products concentration, the number of hydrocarbons degrading bacteria, soil dehydrogenase activity, nutrients concentration, pH level and humidity. Soil samples were collected every month at 10 (Bagicz) or 16 (Brzeg) locations in the soil pile.

The speed of biodegradation of petroleum products in the polluted soil constantly wavered. The decrease of hydrocarbons in each measuring point varied (figure 3). It depended on their beginning concentration in soil and local environment. Biopreparations were introduced in the amount enough to obtain the number of bacteria in the soil on the level of 10^6-10^8 cells/g d.m. of the soil. The used technology assured an appropriate number of hydrocarbon-degrading bacteria in polluted soil. The highest intensity of the degradation process was observed in the first 3 months of bioremediation. The fastest loss of petroleum products occurred in the surface layer of the prism (figure 4). Bioremediation was conducted during the summer (from March to September) in natural weather conditions. In spite of unfavorable properties of polluted soil and technical problems, the applied technology of bioremediation succeeded the achievement of significant reductions in a short period of time. The final average level of petroleum hydrocarbons in polluted grounds was a lot lower than concentrations specified by obligatory standards. A good quality of effluents from prisms was also observed (tab. 1).

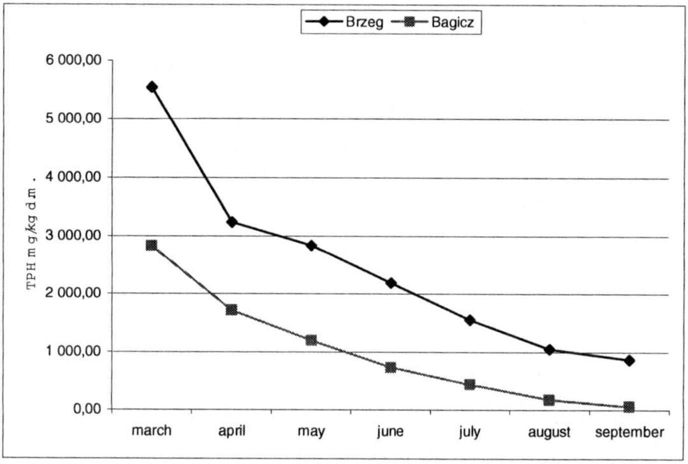

FIG.3. Petroleum products concentration' (average) in soil during bioremediation process.

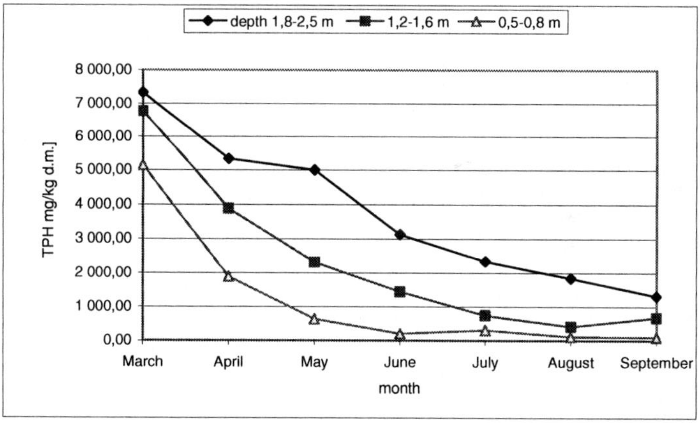

FIG.4. Petroleum products concentration' in soil during bioremediation process – airfield Brzeg, measuring point nr 1.

CONCLUSIONS

Results from this study suggested that the polluted soil can be a source of active bacteria strains able to degrade the petroleum hydrocarbons. Biopreparations based on autochthonous microflora can be applied in various bioremediation processes.

The bioremediation of soil by biopile method conducted with closed water circulation system was very successful thanks to using autochtonous strains of bacteria already adapted to the type of pollutants present in the soil. Also, closing the circle of pile effluents and using them to bacteria multiplication sped up the process. The cultivation of microorganisms as well as their introduction into the ground could be conducted continuously throughout the cleaning process. Moreover, in the field bioreactor, only those cells were multiplied which weren't sensitive to metabolites created during petroleum products biodegradation.

The bioremediation of soil polluted by petroleum products carried out by biopile method allowed elimination of over 90% of pollutants during only few months. The final concentration of pollutants was lowest in the surface layer of piles. This can be associated with better accessibility of air to bacteria capable of degrading hydrocarbons; thus they had better growing conditions.

REFERENCES

General Inspector for the Environmental Protection. (1994). „Identyfikacja i wycena szkód spowodowanych przez stacjonujące w Polsce wojska Federacji Rosyjskiej". Wydawnictwo ELWOJ-TRIO: 9-42.

Kołwzan, B. (2005). „Bioremediacja gleb skażonych produktami naftowymi wraz oceną ekotoksykologiczną. Oficyna Wydawnicza Pol. Wr., Wrocław

Leahy, J.G. and Colwell, R.R. (1990). "Microbial degradation of hydrocarbons in the environment". *Microbiological Reviews*, 54 (3): 305-315

Lebkowska, M. (1996). „Wykorzystanie mikroorganizmów do biodegradacji produktów naftowych w środowisku glebowym". *Gaz, Woda i Technika Sanitarna*, (3): 117-118.

Liebeg, E.W. and Cutright, T.J. (1999) "The investigation of enhanced bioremediation through the addition of macro and micro nutrients in a PAH contaminated soil". *Int. Biodeterior. Biodegradation*, (44): 55-64

Manilal, V.B. and Alexander, M. (1991). "Factors affecting the microbial degradation of phenanthrene in soil". *Appl. Microbiol Biotechnol*, 35, 401-405.

Pawełczyk, A. Kołwzan, B. and Grabas, K.(2006) "Bioremediation of grounds polluted with petroleum products". In: *Sustainable chemistry and biotechnology - a contribution to rivers management*. W. L. Filho, G. Greif, B. Delakowitz (Eds). Frankfurt am Main: Peter Lang, 165-194.

Surygała, J. and Śliwka, E. (2000). „Wycieki ropy i produktów naftowych". (In:) *Zanieczyszczenia naftowe w gruncie*, Oficyna Wydawnicza Pol. Wr.,Wrocław: 130-207.

Biosurfactant Flushing of PCE Contaminated Clayey Soil

S. Harendra M.Sc.[1] and C. Vipulanandan[2] Ph.D., P.E, MASCE

[1] PhD Student, Department of Civil and Environmental Engineering, University of Houston, Houston, Texas 77204-4003. Email: ps_harendra@yahoo.com

[2] Chairman, Professor and Director of Center for Innovative Grouting Materials and Technology (CIGMAT), Director of Texas Hurricane Center for Innovative Technologies, Department of Civil and Environmental Engineering, University of Houston, Houston, Texas 77204-4003 (Phone: (713) 743-4278; Fax (713) 743-4260; Email: cvipulanandan@uh.edu)

ABSTRACT: In this study biosurfactant (UH–biosurfactant) was used to flush the PCE contaminated soil. The adsorption and desorption isotherms for PCE and biosurfactant with sand and kaolinite clay have been quantified. Transport parameters such as seepage velocity (v), diffusion (D) and retardation factor (R) of the above mentioned surfactant solution was determined in clayey soils (sand (82.5%) with 17.5% kaolinite clay mixture). In this study 0.1 g/L NaCl was used as the electrolyte solution. Convection-Diffusion equation (CDE) was used to model the breakthrough curves of the surfactant. Bromide ion was chosen as the tracer material in order to characterize the flow. The effectiveness of biosurfactant solution in flushing the PCE from the contaminated soil was quantified.

INTRODUCTION

Perchloroethylene (PCE) has been used in industrial and commercial applications for metal degreasing, cleaning of electronic components and dry cleaning. PCE represents a long term source of soil and aquifer contamination because PCE has low solubility and high interfacial tensions. Surfactants have been studied by several investigators to determine their ability to act as soil washing agents and as facilitators of subsurface remediation of hydrocarbon spills (Lee et al. 2004). Hence understanding of surfactant adsorption and desorption is of importance for the application such as surfactant enhanced soil remediation.

Surfactant solutions can be used to accelerate the removal of trapped non-aqueous phase liquids (NAPLs) from porous media through either mobilization or enhanced solubilization. The degree of surfactant sorption in an aqueous system depends largely on the nature of the surfactant monomers hydrophilic head group. In general, cationic surfactants have the potential for significant sorption on mineral surfaces. Anionic surfactants generally adsorb less than nonionic surfactants and much less than cationic surfactants. The degree of sorption of cationic surfactants can be so great that certain studies have proposed creating in-situ sorption zones through targeted injection of these compounds in the path of organic contaminant plumes (Lee et al. 2004).

Biosurfactants are compounds that will have low environmental impact compared to synthetic surfactants. Thus, complete removal after treatment may not be necessary [Wouter et al. 2000]. To be effective for enhancing the removal of sorbed contaminants, a biosurfactant should have high solubilizing properties, should be water soluble and should not strongly adsorb to the soil [Torrens et al. 2000].

OBJECTIVE

The overall objective of this study was to flush the PCE contaminated soil using the UH-biosurfactant.
The specific objectives are as follows.

1) Determine the sorption parameters of UH- biosurfactant with the sandy clayey soil.
2) Investigate the effectiveness of UH-biosurfactant in flushing the PCE from the contaminated soil.
3) Determine the transport parameters of UH- biosurfactant in the contaminated clayey soil.

MATERIALS AND METHODS

The blasting sand was obtained from TEC minerals of Eagle Lake, TX. Sand particles are made mostly of quartz and feldspar and have a very stable tetrahedral structure. The particle size ranged from 0.07 mm to 5 mm. The specific gravity was 2.65. Hydrite-R kaolinite clay was purchased from D & F Distributing Inc. The properties of the sand and kaolinite clay are summarized in Table 1.

Table 1. Properties of materials from manufacturer's data sheets

Properties	Sand	Kaolinite
Specific gravity (g/cm^3)	2.65	2.58
Natural water content (%)	0.6	0.4
Hydraulic conductivity (cm/s) (current study)	5.0E-02	1.2E-06
Porosity (%) (1 bag)	44.4	81.8
Organic carbon % (current study)	0.35	1.9

Experimental Methods:

The UH-biosurfactant was produced in 2 L glass bottles, with 1 L broth volume. About 10 g of used vegetable oil was mixed in 1 L of de-ionized water, along with the mineral salts. About 4 mL of acclimated micro-organism was added to the reactor and stirred at a moderate speed [Kim et al. 2006; Wang et al, 2001; Vipulanandan et al. 2000].

A Gas chromatography (Shimadzu GC-14A) with a flame ionization detector and a CR501 data processor was used to determine the chlorinated compounds concentration. The stainless steel GC column (Alltech) was 6-ft long × 1/8-in outer diameter 0.085-in inner diameter and detection temperature was set at 250°C. The injection volume of sample used for GC analysis was 2 μL. The detection limit for PCE was 1 ppm.

Batch study

Clean sand and kaolinite clay were used for adsorption and desorption study of surfactant solutions. Batch sorption experiments were conducted at room temperature. Biosurfactant solutions were prepared in different concentrations and transferred to 40 mL EPA vials with sand and clay mixture. After 21 days in the tumbler (rotating at 4 rotation/min) the incubation vials were centrifuged and subsequently the clear supernatant was analyzed for the total aqueous surfactant by Gas chromatography using the back titration method (Fig. 1).

Column study

The experimental setup for the column studies is the column having 5.1 cm diameter and 50 cm long made out of glass which is equipped with air compressor and pressure regulator unit which is used to pump the solutions (Fig. 2).

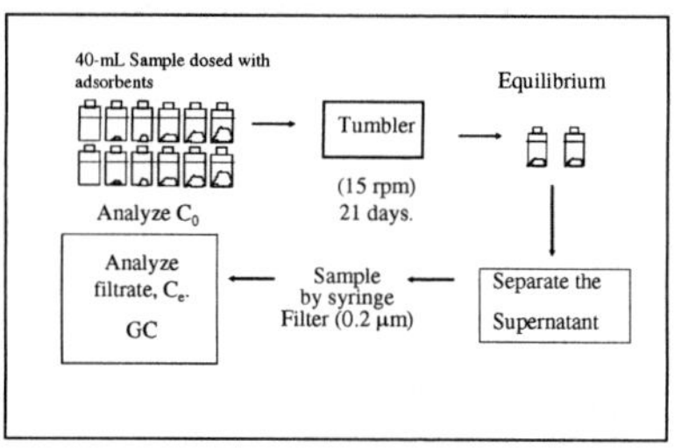

Fig. 1: Schematic diagram for sorption experiments in batch studies

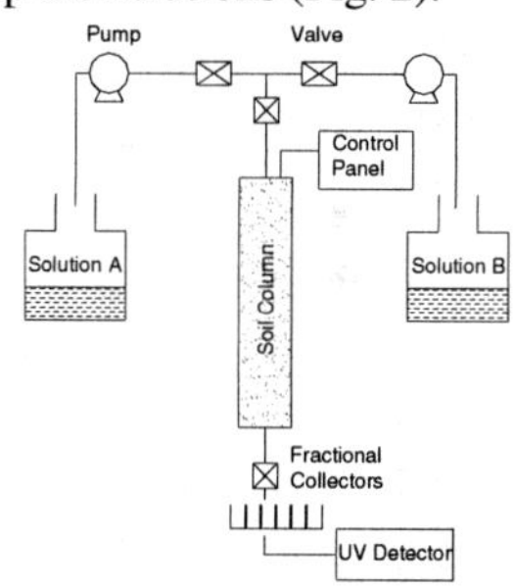

Fig. 2: Schematic drawing of the experimental setup used for breakthrough experiments

Surfactant experiment

The UH biosurfactant solutions were prepared in 1 L volumetric flask and were stirred for atleast 2 hours for proper mixing before injecting it through the column. The surfactant solutions were injected through the soil column and the effluent was collected. The effluent surfactant concentration was measured by determining the solubility of PCE and then back calculating the surfactant concentration.

PCE (300 g) was used to contaminate 1200 g of sand (82.5%) with kaolinite (17.5%) for 2 months. During the two month period the soil was sealed in a container. The contaminated soil was used to pack the column. Then water was flushed through the column until there was no measurable PCE in the effluent. Then 5 g/L UH biosurfactant was flushed through the column to further remove PCE.

Surfactants transport can be modeled as two region model. The two region model assumes that the liquid phase can be partitioned into mobile (*f*-flowing) and immobile ((1-*f*) stagnant) regions [Van Genuchten et al, 1982]. Solute exchange between the two regions is modeled as a first order process.

For the mobile phase [Van Genuchten et al, 1982]

$$(1+\frac{\rho_b f k_d}{\theta_m})\frac{\partial C_m}{\partial t} = D_m \frac{\partial^2 C_m}{\partial x^2} - v_m \frac{\partial C_m}{\partial x^2} - \frac{\alpha}{\theta_m}[C_m - C_{im}] \quad (1)$$

For the immobile phase [Van Genuchten et al, 1982]

$$(1+\frac{\rho_b (1-f) k_d}{\theta_{im}})\frac{\partial C_{im}}{\partial t} = \alpha[C_m - C_{im}] \quad (2)$$

where the subscripts m and im refer to the mobile and immobile liquid regions respectively. The following initial and boundary conditions are applied for the transport of surfactants in the column

Initial condition : $C(x,0) = C_i$
Boundary conditions : $C(0,t) = C_0 \quad 0<t<t_0$
$0 \quad t>t_0$

$$\frac{\partial C}{\partial x}(\infty, t) = 0$$

Version 1.0 of STANMOD (CXTFIT 2.0 [*Toride et al.* 1995]) was used to model the breakthrough curves of surfactants. Bromide ion was chosen as a tracer material in order to characterize the flow. The retardation factor, diffusion coefficient and seepage velocity were determined by fitting the two region model with experimental data.

RESULTS AND DISCUSSIONS

The adsorption of the UH surfactant followed the Freundlich isotherm model for clay and a linear isotherm model for sand (Tables 2, 3, and 4, and Figure 3). Desorption experiments suggested that the Langmuir isotherm was the best-fit model for sand, except for kaolinite which was best modeled using the Freundlich isotherm (Tables 5 and Figure 4).

Table 2 Overall adsorption isotherms for sand and kaolinite

Soil	Equation (No salt)	Equation (salt)
Sand	$Y=\dfrac{0.015X}{1+0.0015X}$	$Y=\dfrac{0.17X}{1+0.73X}$
Kaolinite	$Y=\dfrac{0.23X}{1+0.0239X}$	$Y=\dfrac{1.34X}{1+1.43X}$

Table 3 Adsorption of UH-biosurfactant to the soil below CMC

Soil	Equation (R^2)	Remarks
Sand	$Y=0.88X^{0.51}$(0.94)	Freundlich
Kaolinite	$Y=0.84X^{1.06}$(0.35)	Freundlich

Table 4 Adsorption of biosurfactant to the soil below CMC with ionic strength 0.1g/L

Soil	Equation (R^2)	Remarks
Sand	Y=1.25X (0.92)	Linear
Kaolinite	$Y=2.37X^{2.99}$ (0.88)	Freundlich

Table 5 Desorption of UH biosurfactant from soil

Soil	Equation (R^2)	Remarks
Sand	$Y=\dfrac{0.30X}{1+0.18X}$ (0.95	Langmuir
Kaolinite	$Y=1.71X^{0.60}$(0.91)	Freundlich

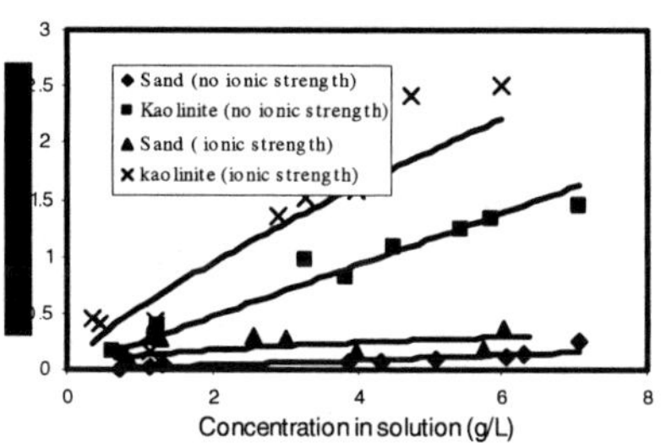

Fig. 3. UH Biosurfactant adsorption in soil

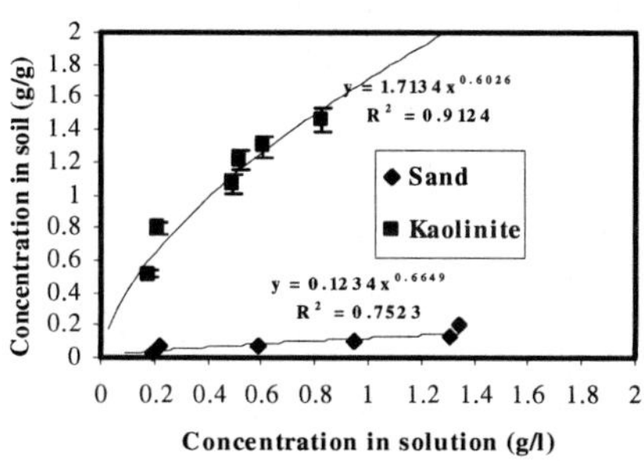

Fig. 4. UH Biosurfactant desorption in soil

Ionic strength of solution may impact the sorption kinetics [Derjaguin et al. 1941]. In natural sediments, repulsive double-layer interactions resulting from the negative surface charges of the colloidal particles and matrix grains must be considered. These interactions are qualitatively described by the DLVO theory [Derjaguin et al. 1941] and play an important role in the deposition of colloidal particles on sediment matrices. Laboratory experiments concerning the deposition rates of colloidal particles have been observed to be several orders of magnitudes larger than predictions based on the DLVO theory [Elimelech et al. 1990]. This discrepancy may be a consequence of local roughness and variation of charge distribution at the surface of colloidal particles and sediment matrices. Thus, the interaction energy of a given particle may vary widely with position across these surfaces. Due to the influence of parameters such as the ionic strength, the valence of the dominant counterion and the pH on colloidal particles and matrix grains must be considered, too [Elimelech et al. 1990].

With the increase in ionic strength (addition of salt) in the biosurfactant solution the retardation factor (R) increased from 7.20 to 41.08 (Figs. 5 and 6). Similar trends were observed with the convection (v_m), diffusion (D_m) and mobile (*f*) parameters (Table 6). The filtering parameter (α) was reduced with the increase in the ionic strength. The observed trends indicated that the electrolyte solution had an impact on the transportation of the surfactant solution in soil (Table 6).

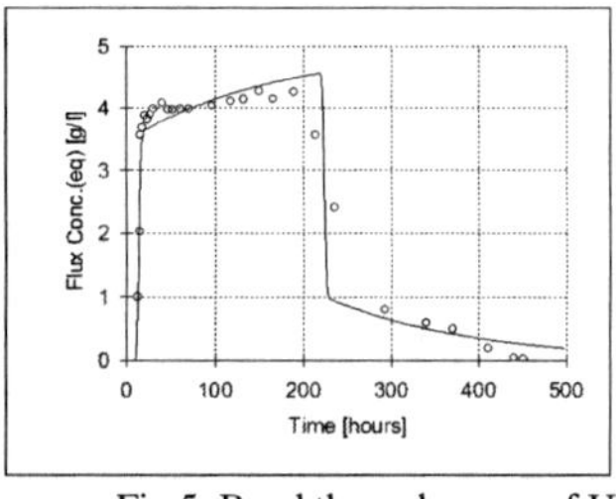

Fig 5: Breakthrough curve of UH biosurfactant in sand (82.5%) and kaolinite (17.5%)

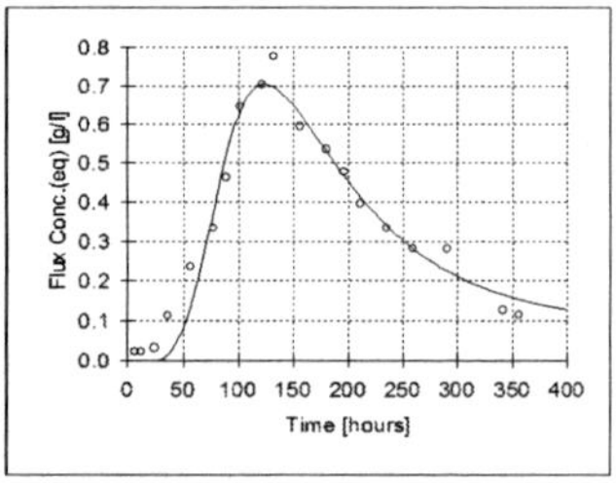

Fig. 6: Breakthrough curve of UH biosurfactant in sand (82.5%) and kaolinite (17.5%) with ionic strength 0.1 g/L

Table 6 Transport parameters of UH Biosurfactant in soil

Soil type	**Parameters**					**RMS (Root mean square)**
	$\mathbf{v_m}$ (cm /min)	$\mathbf{D_m}$ (cm^2/min)	**R**	**f**	$\boldsymbol{\alpha}$ (min^{-1})	
Sand (82.5%) and Kaolinite (17.5%)	0.00028	0.0016	7.19	0.16	0.64	0.1944
Sand (82.5%) and Kaolinite (17.5%) with ionic strength	0.0073	0.020	41.08	0.23	0.05	0.003147

Initially PCE from the contaminated soil was removed by flushing the column with water. UH Biosurfactant was used to remove the remaining PCE (Fig.7). The water removed 1.18% of the estimated PCE from the column. The biosurfactant flushing further removed 0.27% of PCE.

Based on very limited data, the variation of the retardation factor (R) (which was obtained from the column studies) with the partition coefficient (K_d) (which was obtained from the batch study) is shown in Fig 8. With the increase in ionic strength,

K_d increased and so did the retardation factor (R). But the percentage increase in K_d was much higher the increase in the retardation factor (R).

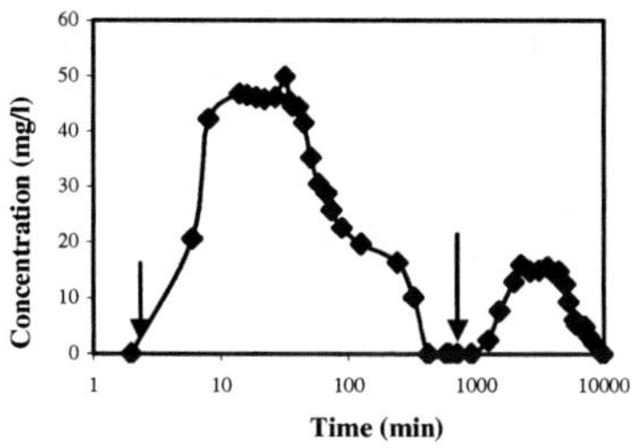

Fig. 7: PCE contaminated soil -water-UH biosurfactant solutions breakthrough in Sand (82.5%) and Kaolinite (17.5%) mixture

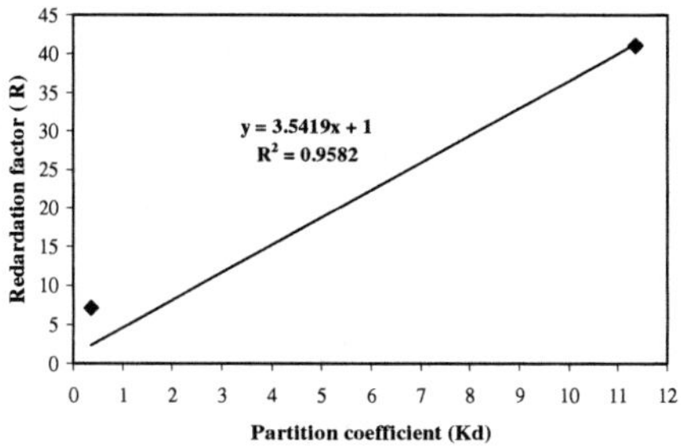

Fig. 8: Relationship of Retardation factor (R) vs Partition coefficient (Kd)

CONCLUSIONS

The effect of a biosurfactant in flush PCE from a column of contaminated soil was investigated. Based on the experimental and modeling study, following conclusions may be made.

1) Adsorption and desorption isotherms for the UH-biosurfactant with the kaolinite clay was represented by the Freundlich isotherm in adsorption.
2) Adsorption isotherm for the UH-biosurfactant with sand, with and without salt, was linear. The desorption was represented by the Langmuir isotherm.
3) Ionic strength of 0.1 g/L in the surfactant solution increased the convection, diffusion and retardation factor.
4) UH biosurfactant flushed an additional 0.27 % of PCE after water removed 1.18 % of PCE from contaminated soil.

ACKNOWLEDEMENT

The study was supported by the Center for Innovative Grouting Materials and Technology with funding from the Advanced Research Program of the Texas Higher Education Coordinating Board (THECB) and the Texas Hazardous Waste Research Center.

REFERENCES

Chu W. (2003) "Remediation of contaminated soils by surfactant aided soil washing" Journal of Practical Periodical of Hazardous, Toxic and Radioactive waste management , Vol 7, No 1, pp 19-24.

Derjaguin, B.V.; Landau, L.D. (1941)" Theory of the stability of strongly charged lyophobic sols and the adhesion of strongly charged particles in solutions of electrolytes" Journal of Acta Physicochim.USRR, Vol 14, pp 733-762

Elimelech M. and O'Melia, C.R. (1990) "Kinetics of deposition of colloidal particles in porous media" Journal of Environ.Sci.Technol., Vol 6, pp 1153 -1163

Grolimind D., Elimelch M., Michal B., Kurt B., Ruben H. and Hans S. (1998)" Transport of in situ mobilized colloidal particles in packed soil columns" Journal of Environ.Sci.Technology, Vol 32, pp 3662-3569.

Kim, J and Vipulanandan, C. (2006). "Removal of Lead from Contaminated Water and Clay Soil Using a Biosurfactant," Journal of Environmental Engineering, Vol. 132, No. 7, pp.857-865.

Lee D.H., Chang H.W. and Cody R.D. (2004) " Synergism effect of mixed surfactant solutions in remediation of soil contaminated with PCE" Journal of Geo Sciences, Vol 8, No 3, pp 319-323.

Torrens, J. L.; Herman, D. C.; Miller-M., R. M.; "Biosurfactant (Rhamnolipid) Sorption and the Impact on Rhamnolipid-Facilitated Removal of Cadmium from Various Soils under Saturated Flow Conditions (1998)" Journal of Environ.Sci.Techno. (2000) Vol 32, pp 776-781

Toride N., Leij F.J. and Van Genuchten M.T. (1999) " The CXTFIT code for estimating transport parameters from laboratory or field tracer experiments" Research report 137, U.S Salinity Report, Riverside, California

Van Genuchten M.T. and Parker J.C. "Boundary conditions for displacement experiments through short laboratory soil columns" Journal of Soil.Sci.Soc.America, Vol 48, pp 703-708.

Vipulanandan, C. and Ren, X. P. (2000). "Enhanced solubility and biodegradation of naphthalene with biosurfactant" Journal of Environmental Engineering, Vol. 126, No. 7, pp. 629-634.

Wang, S. and Vipulanandan, C. (2001) "Biodegradation of Naphthalene-Contaminated Soils in Slurry Bioreactors," Journal of Environmental Engineering (ASCE), Vol. 127, No. 8, pp. 748-754.

Wouter H. N., Mark.L.B. and Dick B.J. (2000) "Adsorption of a multicomponent Rhaminolipid surfactant to soil" Vol 34, pp 832-838

Contributions of Xenobiotic-Degrading Bacterial Endophytes to the Field of Phytoremediation

Sarah L. Schmidt[1] and Jong-In Han, Ph.D.[2]

[1] Rensselaer Polytechnic Institute, Department of Civil and Environmental Engineering
110 8th St. Troy NY, 12180; schmis2@rpi.edu
[2] Professor, Rensselaer Polytechnic Institute, Department of Civil and Environmental Engineering
110 8th St. Troy NY, 12180; jihan@rpi.edu

ABSTRACT: Phytoremediation of organic contaminants occurs as a result of a complex set of interactions between plants and their associated microorganisms. However, the technique has failed to produce consistent and satisfactory results due to incomplete degradation and phytotoxicity of intermediate and parent compounds. In the case of the phytoremediation of recalcitrant organic pollutants, such as polycyclic aromatic hydrocarbons (PAHs), discrepancies between desired and actual results become more pronounced due to reduced mobility of the compounds in soils. Addition of bacterial endophytes, a novel group of microorganisms that reside within plants, may prove to be the missing ingredient that both degrades the chemical compounds and protects plants from their toxicity. It is our conjecture that these microorganisms are a key part of a set of interactions between the host plant and all of its associated microorganisms (endophytic and rhizospheric) that coordinate to mobilize and break down the soil pollutants. We have isolated a species of endophytic bacteria and shown it to use the model PAH phenanthrene as a sole carbon source. Greater understanding of the interactions between plants and endophytic bacteria in remediation systems may serve to optimize the technique of phytoremediation.

INTRODUCTION

Phytoremediation, the use of plants for the removal of soil contaminants, has been established as a feasible treatment option for sites contaminated with recalcitrant and potentially toxic petroleum hydrocarbons such as polycyclic aromatic hydrocarbons (PAHs) [1,2]. This technique can be greatly enhanced by plant-associated microorganisms existing in the rhizosphere (termed rhizoremediation) [1,2]. Plants play multiple roles in this process by drawing compounds into their rhizosphere via the transpiration stream, excreting peroxidases able to oxidize organic pollutants, and enhancing bioavailability of less soluble compounds [1,2]. In the case of soil contamination by highly hydrophobic organic compounds such as PAHs, mass transfer of parent compounds into plant tissue, and hence biodegradation inside the plant, is limited. Therefore, initial biodegradation of these compounds occurs largely in the rhizosphere [3]. This initial biodegradation is often incomplete and results in intermediates that are more soluble, and therefore more assimilable, to plants [4,5]. However, intermediates are generally more toxic, and when taken up by the plant can exert potentially detrimental effects [6]. This potential toxicity of the intermediates to plants, along with inefficient plant uptake of parent compounds and the plant's innate inability to completely mineralize the compounds, has to date provided limited utility of phytoremediation for highly hydrophobic organic contaminants including PAHs.

Endophytic bacteria exist in the interior of plants and can have a great impact on the growth of their hosts by producing growth promoting chemicals and antibiotics [7]. Recent research has raised the possibility that certain endophytic bacteria can further degrade or completely mineralize toxic intermediates taken up by the plant to prevent detrimental effects otherwise exerted on the host. For example, Van Aken et al., reported that an endophytic methylotroph was able to degrade explosives such as 2,4,6-trinitrotoluene (TNT) [8]. Barac et al. also successfully demonstrated that endophytic bacteria can be genetically engineered with a toluene-degrading gene to effectively reduce toluene concentration in a plant [9]. This unique yet relatively uncharacterized type of degrading bacteria may provide the key to acheiving the successful application of phytoremediation to recalcitrant organic contamination by serving as one component of a phytoremediation tripod which encompasses plants and their endophytic and rhizospheric microorganisms. Fundamental understanding of how these three general groups of organisms interact can lead to a variety of applications, most notably for the remediation of PAHs.

PAHs, which consist of multiple fused aromatic rings, are widespread as a result of incomplete combustion of organic matter such as in forest fires, waste incineration and the use of fossil fuel. PAH contamination has long reaching effects on human and environmental health: fifteen different PAH compounds are classified as probable human carcinogens by the National Institutes of Health and pollution can cause a drop in biodiversity and large scale trophic changes in certain environments [10,11].

The following research is the beginning of a project aimed at the discovery of an optimal combination of endophytic pollutant-degrading bacteria, rhizospheric bacteria and host plant to be used for improvement of the efficiency and feasibility of the technology of phytoremediation. For this purpose, PAHs were used as model contaminants as they are ubiquitous and recalcitrant to biodegradation. Phenanthrene was further chosen as a model PAH because of its relatively high aqueous solubility and representative structure consisting of three fused aromatic rings.

METHODS

Isolation of Endophytic Bacteria

Multiple species of grasses were harvested from the Water Street Manufactured Gas Plant (MGP) site in Troy, New York, known to be PAH-polluted. The plants were manually washed with tap water and antibacterial soap to remove bound soil from the root systems and then cut into sections of root, stem and leaf. Sections of root were soaked in 100mL of 2% NaOCl for 30 minutes and rinsed three times with sterile distilled water. They were then dipped three times in pure ethanol and the ethanol was shaken off between dips. The sections were then placed under an ultraviolet lamp inside a laminar flow hood for a total of one hour (30 minutes per side). Sections of stem and leaf were sterilized similarly except for the soaking in 2% NaOCl which was omitted. All sections were pressed onto R2A agar (Difco) plates to test for surface sterility. The samples were then cut into smaller pieces and macerated with a pestle and mortar containing 3 mL of MOPS mineral medium. This medium, (pH 7.2) was prepared according to Huang *et al.* [12] with some modification, containing: 5 mM morpholinepropanesulfonic acid (MOPS), 4.20g Na_2HPO_4, 2.92 g $NaH_2PO_4{\bullet}H_2O$ 1g NaCl, 0.4 g $MgCl_2{\bullet}6H_2O$, 0.1g $CaCl_2{\bullet}2H_2O$, 0.5g KCl, 0.267g NH_4Cl, 0.322g Na_2SO_4, 1mL trace element stock and 100µL vitamin solution. All tools and reagents were sterilized before use. The resulting pulp was serial diluted in MOPS mineral media and spread on MOPS mineral medium plates containing 1.5% agar supplemented with 100µL L^{-1} grass extract. Grass extract was made by macerating root, stem and leaf sections separately in a juicer. Extracts were filter sterilized and added to the agar at approximately 60°C before pouring the plates. Inoculated plates were sprayed with 1.5% phenanthrene dissolved in acetone. The acetone was allowed to evaporate leaving a thin layer of phenanthrene crystals on the surface of the plate. Plates were incubated in the dark at 28°C. Colonies that displayed a surrounding zone of cleared phenanthrene were serially diluted or successively plated on MOPS plates with phenanthrene to isolate species of bacteria.

Identification and Characterization of Endophytic Bacteria

Genomic DNA was extracted using the QIAGEN® DNeasy Blood & Tissue Kit and the 16S rDNA was amplified using the QIAGEN® Taq PCR Master Mix Kit and an Eppendorf Mastercycler Gradient Thermocycler. The amplified DNA was then purified using the QIAGEN® QIAprep Spin Miniprep Kit. Sequencing was done by the University of Maine DNA Sequencing Facility (Orono, ME) using 27F and 1492R primers. Sequences were edited using BioEdit Sequence Alignment Editor software and aligned using the NCBI/BLAST search tool located at http://www.ncbi.nlm.nih.gov/BLAST [13]. Growth curves were obtained by measuring the optical density of liquid cultures of nutrient broth (Remel, Lenexa, KS) and MOPS minimal media with solid phenanthrene crystals as the sole carbon source. Optical density was measured on a Thermo Electron Corporation Spectrophotometer 20D+ set to a 600nm wavelength.

One of the eleven strains isolated was selected for further study because of its fast growth kinetics. Oxidase and Catalase reactions were performed using standard reagents (Fisher). Urease production was tested according to Steadham 1979 [14]. Tween 80 hydrolysis, niacin production, iron uptake and photochromogenesis were tested according to Wayne and Doubek, 1968 [15]. Nitrate reduction was tested according to Tsukamura, 1968 [16].

Fatty acid methyl ester (FAME) analysis of the cell membrane was done by Microbial ID, Inc. (Newark, DE) which uses the MIDI Sherlock® Microbial Identification System for comparison to known strains. Serial plating on R2A agar followed by MOPS agar with phenanthrene was done to determine if the bacteria lost its degrading capability when grown on a rich medium.

Confirmation of Endophytic Origin

Soybean seeds were obtained from Johnny's Selected Seeds (Winslow, ME). Seeds were surface sterilized in 100mL of 1% NaOCl with 10µL of Triton® X-100 Solution (Fluka) for 30 minutes. Seeds were then rinsed three times with sterile distilled water and pressed onto R2A agar plates to test for surface sterility. The seeds were then put on wet paper towels in Petri dishes, inoculated with one milliliter of log phase liquid culture and allowed to germinate in the dark at 30°C for one to two days. Germinated seeds were placed in large test tubes (25X150 mm) containing 1 gram of perlite (Miracle-Gro Lawn Products, Inc. Marysville, OH) and 6 mL of Murashige and Skoog (MS) basal salt media with minimal organics (Sigma). Perlite was rinsed once with nitric acid and three times with distilled water to remove bound organics then dried and sterilized before use. The pH was 3.9. The germinated seeds and tubes were placed in an incubator at 30°C with a 12 hour light/dark cycle from a New Wave T5 fluorescent growth light. The relative humidity was kept at 40%. All reagents and equipment were sterilized before use.

After one week of growth, the plantlets were surface-sterilized as described above. The whole plant was then macerated with a mortar and pestle and plated on both R2A and MOPS agar sprayed with phenanthrene. Presence of clearing zones on the MOPS agar was taken as evidence of the colonization of the inside of the plant by the endophytic phenanthrene degrader.

Phytotoxicity Effects

One gram of perlite was placed in a test tube along with 1.2 mL of 0.05% phenanthrene in acetone and the acetone was allowed to evaporate off. Five milliliters of MS media was added, creating a final phenanthrene concentration of 100ppm. The pH was adjusted to 3.9 ± 0.2 and the tubes were autoclaved. Both inoculated and uninoculated seeds, surface sterilized and germinated as before, were placed in the tubes to grow. Two weeks after germination, heights of the plantlets were recorded.

RESULTS

Identification and Characterization of Bacteria

A total of eleven phenanthrene degrading strains were extracted from the plant tissue. One strain, strain U1A, was chosen for further study because it exhibited fast growth kinetics and did not lose its degrading capabilities when grown on rich media.

Strain U1A is a medium sized acid fast rod which displays medium sized yellow mucoid colonies on R2A agar and small off-white mucoid colonies on MOPS mineral medium agar with phenanthrene. It has a doubling time of 9 hours on nutrient broth and its growth rate on phenanthrene as a sole carbon source is constant (as shown in Figure 2) and dictated by diffusion of the compound into the liquid media.

Strain U1A is non-motile, non-sporulating, oxidase negative and catalase positive. It can also use pyrene, 9-phenanthrol, phthalate and 2-carboxybenzaldehyde as a sole carbon source. 16S rDNA analysis of U1A shows a 99.6% similarity to *Mycobacterium sp. KMS.* FAME analysis (data not shown) shows a predominance of unsaturated fatty acids and the presence of tuberculostearic acid, a common component in Mycobacteria. MIDI Sherlock® comparison shows a similarity index of 0.372 to a *Mycobacterium* species, which is only sufficient for identification of genus.

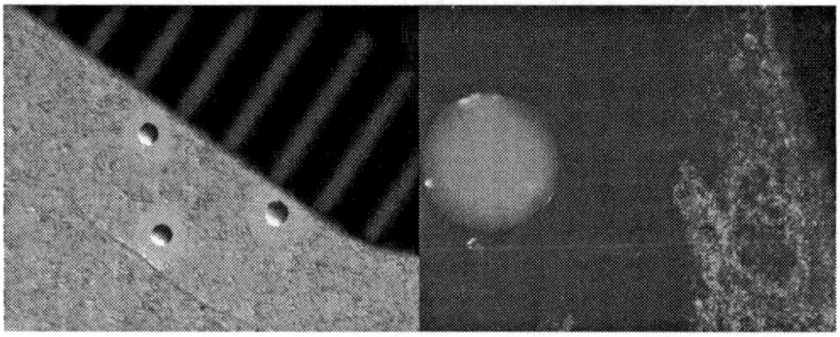

FIG. 1. Colonies of strain U1A growing on MOPS agar plates with phenanthrene sprayed on the surface. Colonies have metabolized the phenanthrene crystals immediately adjacent, forming circular clearing zones. Left: white lines represent 1 mm.

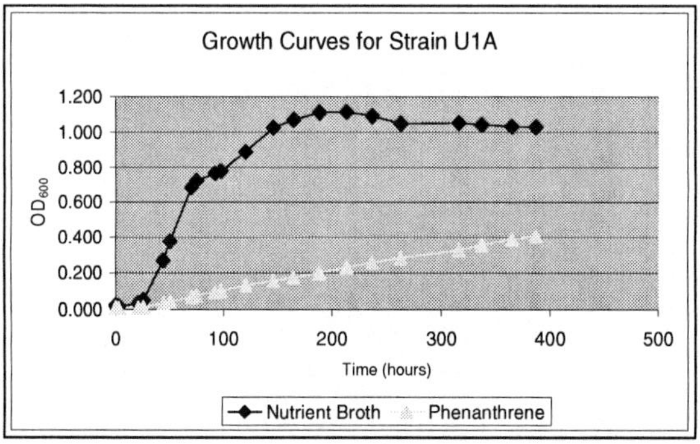

FIG. 2. Growth of strain U1A on MOPS mineral media supplemented with phenanthrene crystals versus nutrient broth. Growth rate on phenanthrene is constant and governed by dissolution of the compound into the media.

Strain U1A was positive for urease production, Tween 80 hydrolysis, scotochromogenesis, iron uptake and nitrate reduction. It was negative for niacin production. U1A did not lose its phenanthrene degrading capability when successively plated 15 times on R2A agar.

Confirmation of Endophytic Origin

Despite successful surface sterilization of both the seed and grown plantlet, many natural endophytes were observed growing on the R2A plates spread with the macerated soy plantlets not inoculated with bacteria. MOPS plates spread with extract from the control plants and sprayed with phenanthrene also displayed some growth, but did not display clearing of phenanthrene and were thus considered a successful negative control. Further characterization of these natural seed endophytes was not done as they were not observed to have a direct effect on colonization of the soybean seed by strain U1A. Preliminary results show clearing zones on MOPS plates spread with surface sterilized plant extract, indicating that U1A has successfully re-colonized soybean plant stems and roots. Confirmatory repetitions are still pending.

Phytotoxicity Effects

After two weeks, there was a large difference in the heights of the soybean plants with and without strain U1A. The average height of the uninoculated plants (4 replicates) was 3.7cm while the average height of the inoculated plants (7 replicates) was 19.4cm (data not shown). Pictures of the plants are shown in Figure 3. Differences between inoculated and uninoculated plants were not noted in plants grown without phenanthrene.

FIG. 3. Soybean plants grown on 100ppm phenanthrene in MS media with perlite. Left: Uninoculated plants. Right: Plants inoculated with strain U1A.

DISCUSSION

While strain U1A almost definitely belongs to the genus *Mycobacterium*, differences in biochemical reactions, 16S rDNA sequences and FAME profiles make it difficult to identify U1A as a known species. Its closest matches in 16S rDNA sequence, the three soil bacteria *Mycobacterium sp. KMS, MCS* and *JLS*, were also isolated from a PAH polluted site [17]. These strains are fast growing like U1A and degrade the same types of PAHs; however, while their genomes have been sequenced, there is no biochemical data or FAME profile to compare with strain

U1A. Its next closest match by 16S rDNA alignment (99.6%), *Mycobacterium monacense*, is morphologically and biochemically similar but has a significantly different FAME profile [18]. Further tests, such as the sequencing of the gene *hsp65*, will reveal if U1A is indeed a novel species.

There are several PAH degrading microorganisms within the genus *Mycobacterium*, many of them soil and/or rhizospheric bacteria [17]. This is not surprising as endophytic and rhizospheric bacteria frequently are closely related, if not the same organism, because the sites of endophytic colonization are often surface tears in the root system made by growing root hairs [7]. Strain U1A's close association with the host plant is evident from the data presented in Figure 3, suggesting, at the very least, colonization of areas within the transpiration stream where phenanthrene would enter the plant. Protection of the soybean plant from phenanthrene's evidently harmful effects may be aided by U1A's display of diffusion-mediated growth on phenanthrene, meaning that the pollutant is metabolized as quickly as it can reach the bacterial cell. Further experiments will explore whether this protective effect translates to the other compounds strain U1A was shown to degrade.

While characterization and testing are still in the beginning phases, strain U1A presented here represents a step toward utilization of a new type of remediation inoculum. Further research will attempt to elucidate the relationship between bacteria and host plant and explore degradation capabilities of the pair in field and laboratory-prepared soil samples containing PAHs. As a next step, further interaction between an endophyte/host pair and a strain of known rhizospheric PAH degrader will be explored. An optimized combination of the correct host plant and associated bacteria has the potential to tackle the major areas where phytoremediation has, to date, been lacking. Increased solubilization and initial degradation by rhizospheric bacteria coupled with increased degradation by endophytes and, in turn, increased development of the host plant root system creates a mutually beneficial, closed-loop system for removal of these pollutants.

Discovery of an endophytic PAH degrader - a virtually uncharacterized type of bacteria - is an exciting advance in the knowledge base behind the technology of phytoremediation. Once fully characterized, correct utilization of this type of bacteria, in concert with plants and rhizospheric microorganisms, has the potential to bring phytoremediation technology to the forefront of remediation techniques aimed at recalcitrant organic soil pollutants.

REFERENCES

1 - McCutcheon, S.C. & Schnoor, J.L. (2003). "Phytoremediation transformation and control of contaminants." John Wiley & Sons, Inc.: New York.

2 - Shaw, L.J. & Burns, R.G. (2003). "Biodegradation of organic pollutants in the rhizosphere." *Adv. in Appl. Microbiol.,* 53:1-60.

3 - Hutchinson, S.L., Schwab, A.P., & Banks, M.K.. (2003). "Biodegradation of petroleum hydrocarbons in the rhizosphere." In *Phytoremediation: Transformation and Control of Contaminants*, ed. SC McCutcheon, JL Schnoor, pp. 355–86. John Wiley & Sons, Inc.: New York.

4 - Meulenberg, R., Rijnaarts, H.H.M., Doddema, H.J. & Field, J.A. (1997). "Partially oxidized polycyclic aromatic hydrocarbons show an increased bioavailability and biodegradability." *FEMS Microbiol. Lett.,* 152:45-49.
5 - Harvey, P. J., et al. (2002). "Phytoremediation of polyaromatic hydrocarbons, anilines and phenols." *Environ. Sci. Pollut. Res. Int.,* 9: 29–47.
6 - Harms, H., Bokern, M., Kolb, M. & Bock, C. (2003). "Transformation of organic contaminants by different plant systems." In *Phytoremediation: Transformation and Control of Contaminants,* ed. SC McCutcheon, JL Schnoor, pp. 285–316. John Wiley & Sons, Inc.: New York.
7 - Sturz, A.V., Christie, B.R. & Nowak, J. (2000). "Bacterial endophytes: Potential role in developing sustainable systems of crop production." *Crit. Rev. Plant Sci.,* 19:1-30.
8 - Van Aken, B., Yoon, J.M., & Schnoor, J.L. (2004). "Biodegradation of nitro-substituted explosives 2,4,6-trinitrotoluene, hexahydro-1,3,5-trinitro-1,3,5-triazine, and octahydro-1,3,5,7-tetranitro-1,3,5-tetrazocine by a phytosymbiotic *Methylobacterium sp.* associated with poplar tissues (*Populus deltoides x nigra* DN34)." *Appl. Environ. Microbiol.*, 70:508-517.
9 - Barac, T., et al. (2004). "Engineered endophytic bacteria improve phytoremediation of water-soluble, volatile, organic pollutants." *Nature Biotech.,* 22:583-588.
10 - Website: http://toxnet.nlm.nih.gov
11 - Venturini, N. & Tommasi, L.R. (2004). "Polycyclic aromatic hydrocarbons and changes in the trophic structure of polychaete assemblages in sediments of Todos los Santos Bay, Northeastern, Brazil." *Marine Pollut. Bull.,* 48: 97–107.
12 – Huang, J.J., Han, J.-I., Zhang, L.-H. & Leadbetter, J.R. (2003). "Utilization of acyl-homoserine lactone quorum signals for growth by a soil pseudomonad and *Pseudomonas aeruginosa* PAO1." *Appl. Environ. Microb.*, 69: 5941-5949.
13 - Hall, T.A. (1999). "BioEdit: a user-friendly biological sequence alignment editor and analysis program for Windows 95/98/NT." *Nucl. Acids. Symp. Ser.*, 41:95-98.
14 - Steadham, J.E. (1979). "Reliable urease test for identification of mycobacteria." *J Clin. Microb.*, 10(2): 134-137.
15 - Wayne, L.G. & Doubek, J.R. (1968). "Diagnostic key to Mycobacteria encountered in clinical laboratories." *Appl. Microb.*, 16(6): 925-931.
16 - Tsukamura, M. (1967). "Identification of Mycobacteria." *Tubercle*, 48(4): 311-338.
17 - Miller, C.D. et al. (2004). "Isolation and characterization of polycyclic aromatic hydrocarbon-degrading mycobacterium isolates from soil." *Microb. Ecol.* 48: 230-238.
18 - Reischl, U. et al. (2006). "*Mycobacterium monacense* sp. nov." *IJSEM*, 56: 2575-2578.
19 - Kiyohara, H. & Nagao, K. (1978). "The catabolism of phenanthrene and naphthalene by bacteria." *J. Gen. Microb.,* 105: 69-75.

Biomass, Remediation, re-Generation (BioReGen Life Project): Reusing brownfield sites for renewable energy crops

Richard Lord[1], Janet Atkinson[2] and Andy Lane[3], Jonathan Scurlock[4], Graham Street[5]

[1] Bioremediation Programme Manager, Clean Environment Management Centre (CLEMANCE), School of Science & Technology, University of Teesside, Middlesbrough, TS1 3BA, UK; r.lord@tees.ac.uk

[2] BioReGen Project Officer, Clean Environment Management Centre (CLEMANCE), School of Science & Technology, University of Teesside, Middlesbrough, TS1 3BA, UK; j.atkinson@tees.ac.uk

[3] Tutor in Applied Geology, Centre for Lifelong Learning, Newcastle upon Tyne, NE1 7RU, UK; andy.lane@sunderland.ac.uk

[4] Now Chief Adviser, Renewable Energy and Non-Food Crops, National Farmers' Union, Agriculture House, Stoneleigh Park, Warwickshire, CV8 2TZ, UK; jonathan.scurlock@nfu.org.uk

[5] Director, Clean Environment Management Centre (CLEMANCE), School of Science & Technology, University of Teesside, Middlesbrough, TS1 3BA, UK; g.street@tees.ac.uk

ABSTRACT: Biomass fuel composition is compared to host soil contamination for energy crops grown on five contrasting sites in NE England. These include three contaminated brownfield sites and control sites in both urban and rural settings. Fuel quality is compared for willow (*Salix* spp.) short rotation coppice (SRC), miscanthus (*Miscanthus* spp.), reed canarygrass (*Phalaris arundinacea*) and switchgrass (*Panicum virgatum*). The information is used to assess the potential for long-term remediation of contaminated land during energy crop growth. Concentrations of Zn and Cd are consistently higher in SRC willow for a given site, whereas the grasses have higher ash contents, which are richer in SiO_2 but lower in K_2O. Initial actions of the full-scale demonstration plantings carried out under the Life III Environment Programme are described together with an analysis of the wider economic, environmental and social benefits of this sustainable type of reuse of derelict brownfield land and carbon neutral approach to remediation.

INTRODUCTION

In the UK the term "brownfield" refers to land that has been previously developed for any non-agricultural purpose (i.e. the opposite of "greenfield" land). Current estimates suggest that there are 63,500 ha of brownfield land in the UK, of which some 36,000 ha are currently vacant or derelict (National Land Use Database, 2006). This definition is independent of any potential or known contamination, environmental hazard or risk. Indeed some undeveloped greenfield sites may also be contaminated. In many areas of the UK, however, brownfield land left by the retraction of traditional heavy industries is indeed contaminated to the degree that it would require treatment prior to being suitable for use. Much of this land area is either too expensive to remediate or simply surplus to current requirements for any appropriate new industrial use. As a result derelict industrial sites are often cleared of above ground structures, secured and left to await future redevelopment.

The Tees Valley, NE England, contains one of the largest heavy industrial clusters in W Europe, including petrochemicals, iron and steel, bulk inorganic and specialty chemicals. Recent industrial history included extensive coal-based heavy industry, while mining, smelting and metallurgy of ferrous and non-ferrous metals has been carried out over two millennia. The result is a legacy of contaminated land amounting to an estimated 20,000 ha (2.3%) of NE England (Environment Agency, 2002). Due to decline and restructuring of remaining heavy industries there is also now 1155 ha (1.5%) of available derelict or vacant brownfield land in the Tees Valley alone.

Development of the 30 MWe Wilton 10 biomass power station by SembCorp Utilities has created a local market opportunity for energy crop production in Teesside. In 2003 a consortium led by CLEMANCE was formed to investigate the potential of reusing derelict industrial sites for energy crop production to supplement production from agricultural land.

The aim of the BioReGen Life project (www.bioregen.eu) is to demonstrate at full commercial scale on a variety of contaminated sites the feasibility of reusing brownfield land for biomass production (Lord et al. 2007). Five 1 ha demonstration sites were planted in April-May 2007 included a former shipyard site, a steel slag heap, coal and coke oven yard, a former sewage treatment works, and a landfill site. The purpose of this paper is to compare the results of soil contamination and fuel properties for energy crops grown at five earlier pilot scale field trials.

PILOT SCALE FIELD TRIALS

Field trials commenced in 2004 with the establishment of duplicate hand-planted plots of four candidate energy crop species on a former landfill site at Fyland's Bridge near to Bishop Auckland, County Durham (site (a) Table 1). Desk study of historical mapping and initial site investigation showed that this former glacial brick clay pit had been back-filled largely with domestic coal ash and incineration residues, resulting in potentially phytotoxic heavy metal contamination and levels of As and Pb of concern for redevelopment. Individual plots were planned to be 6.9 x 7.5 m, the area needed to accommodate 100 short-rotation coppice willow plants using conventional spacing and agricultural equipment. No amendment was used, although the plants were mulched with compost to aid weed suppression in August 2004. Lord et al. (2007) gives further details of planting methods, contamination assessment, bio-accessibility testing and energy crop survival rates for this site.

Hand planting was also employed at site (d) which was located within an eco-park, a visitor attraction specializing in educational facilities, which demonstrate sustainable living (www.naturesworld.org.uk/). The aim was to provide examples of energy crops in a publicly accessible location, coupled with a control site. The site is in an urban location a few km from the heavy industrial areas of Teesside, which

Table 1. Details of pilot scale sites

	Site (a)	Site (b)	Site (c)	Site (d)	Site (e)
	Clay pit and coal ash landfill	Oil foam land farm, near to Pb-Zn smelter	Agricultural (arable)	Agricultural, (formerly licensed wastes management site)	Shipyard, reclaimed estuary
Current land use	Derelict	Within industrial compound	Agricultural (biomass)	Ecopark (visitor attraction)	Derelict, partially cleared site
Site type	Brownfield	Brownfield	Rural control	Urban control	Brownfield
Planting date	April 2004	May 2004	May 2004	June 2004	(May) June 2005
Area, ha	0.04	0.4	4	0.01	(0.5 failed), 0.01
Coppice planting details	Hand-planted, 20 cm sticks	Mechanized step-planter, 20 cm sticks	Mechanized step-planter, 20 cm sticks	Hand-planted, 20 cm sticks	(Mechanized) hand-planted 2 m sticks with tree-guards
SRC age (growth), years	3 (2)	3 (2)	3 (2)	3 (2)	2 (2)
Grasses	MC, RC, SG	MC & RC (failed)	MC, RC	MC, RC	None – Future trials

Note: SRC = short rotation coppice, MC = miscanthus, RC = reed canarygrass, SG = switchgrass.

include an integrated steel works, coke works, power station and other potential sources of airborne heavy metal contamination. The site was originally part of a small horticultural farm and had also formerly included licensed waste management but no contamination was suspected. The site was pretreated with glyphosate but no fertilizer amendment was used. The plants were mulched in late summer with wood chip to aid weed suppression.

Conventional tractor-towed step-planters were used to plant SRC at three sites with varying success. Site (b) was part of an oil refinery complex within the Seal Sands petrochemicals complex. This industrial area is built within former salt marshes of the Tees Estuary on a raised platform typically composed of blast-furnace slag with some thin imported soils. The North Tees area includes the site of a former Zn and acid works that had associated widespread heavy metal soil contamination. The pilot site had previously been used for land-farming of dissolved-air flotation scum, a foam of crude oil, air and formation water generated by the initial treatment of crude oil. Land-farming, a type of bioremediation, involved regular ploughing of soil within shallow bunded depressions, onto which oil-water mixtures were occasional discharged or covered with imported soil. As a result the land was already in a cultivated state.

Site (c) was an agricultural site on heavy clay soil with no known contamination. The site is in a rural area but still only 2 km from heavy industrial facilities, including a chromium metallurgical plant. The main area of the 4 ha site was planted with a mixture of six clone varieties, including Scandinavian and UK examples. No amendments were made. Smaller single clone plots were also planted for direct comparison of performance. An area of miscanthus rhizomes were planted semi-mechanically by dropping into the slit made by a sub-soiler. Reed canarygrass and switchgrass were broadcast, although the latter failed to establish.

The following year site (e) was used for a broader trial of the technical and regulatory issues of establishing crops on non-agricultural land. The former shipyard had been cleared in preparation as an industrial estate. The soil was essentially made ground on former salt marshes, comprising coal ash, crushed brick, concrete and slag, mixed with clay. Site preparation included amendment at 250T/ha of composted source-segregated green waste prepared to BSI PAS100 (2005) standard. At the time this material was still treated as a "waste" by the UK Environment Agency until fully "recovered" by combination with the receiving soil, and was therefore subject to UK Wastes Management Licensing Regulation (1994). This amendment rate was the maximum annual application to land permitted under a Paragraph 7A Exemption for "Land Treatment for Agricultural Benefit or Ecological Improvement", a formal derogation under specific circumstances from the usual regulatory controls. It was only equivalent to adding a layer of compost < 5 cm thick across the site. During site preparation with a sub-soiler plough a variety of obstructions were encountered, including steel plate, wire rope, bricks, boulder-sized pieces of slag and concrete. SRC was step-planted but due to remaining ground obstructions and the stony made ground, poor penetration was achieved with cuttings left protruding from the soil. Due to difficulties of site preparation and the regulatory issues surrounding compost use, the planting was not completed until late May and was followed by a period of drought. As no perimeter rabbit fencing was erected, the substantial rabbit population in the surrounding site nibbled shoots and all willow cuttings later died.

To establish if soil composition and contamination had contributed to this failure limited replanting was carried out by hand using 2 m willow whips protected by spiral plastic tree-guards. They were planted by inserting into a 0.5 m hole created using a crowbar and then watered in thoroughly. These willows survived where tree guards remained intact. It is this growth which is used to compare fuel characteristic to the other sites.

SOIL CONTAMINATION

Surface soil samples were collected during walkover site survey, typically at depths of 10-20 cm. Pots of > 1kg were submitted to commercial laboratories (Seven Trent or NRM) for contaminant suite analysis (Table 2). Current UK soil guideline values are set as triggers to indicate levels of contamination requiring further investigation and site-specific assessment (DEFRA-EA, 2002). The values vary according to the sensitivity of the end-use, reflecting the pollution linkage from source, though pathways, to the receptors that it introduces. The sites can be ranked in order of the severity of contamination as follows: site e (shipyard) > site a (coal ash landfill) > site b (land farm) > site d (urban control) > site c (rural control). Sites a and e show average concentrations of As and Ni respectively that exceed the SGV for allotments or residential uses with plant uptake, and both have Cu and Zn levels exceeding the ICRCL (1987) thresholds for the onset of phytotoxic effects at low pH.

Table 2. Concentrations of potentially toxic elements in soils (ppm).

Site code	(a)	(b)	(c)	(d)	(e)	SGV[a]	ICRCL[b]
No. samples	8	3	2	2	6		
As	31	10	4.7	8.0	18	20	n/a
Cr	29	28	33	28	45	130	n/a
Cu	130	30	12	23	1153	n/a	130
Pb	271	73	37	60	425	450	n/a
Ni	51	21	21	22	161	50	70
Zn	365	168	68	111	541	n/a	300
Cd	0.86	0.34	0.17	0.28	0.62	1, 2, 8	n/a
Hg	0.42	2.49	0.06	0.15	0.48	8	n/a

[a]soil guideline values, [b]Inter-Departmental Committee on the Redevelopment of Contaminated Land

BIOMASS FUEL ANALYSIS

All crop samples were collected between mid February and late March 2007. Representative examples of willow SRC were cut using a pruning chain saw whereas the smaller areas of miscanthus and reed canarygrass were completely harvested by hand using gardening secateurs or edging shears. All material was hand supported during sampling and collected with extreme care, so as to avoid any additional soil contamination. The material was not washed, as the intention was to assess the contribution to the fuel composition that any adhering contaminated soil dust would have after conventional harvesting, together with any plant uptake. All samples, including the grasses, were processed on site using a petrol-driven garden chipper. After homogenization 25L sub-samples were sealed and dispatched to a specialist commercial laboratory (Knight Energy Services Ltd) for fuel and contaminant suite analysis (Table 3).

Contamination by potentially toxic elements is at sub-ppm levels for elements other than Zn and Cu. The Zn contents found in SRC willow are consistently higher than for other crops grown at the same site (Fig. 1). Cd concentrations are lower but are only detected in SRC. The highest recorded Cu concentration also occurs in SRC from the most Cu contaminated site.

Ash contents are higher for the grasses than for the SRC willow, particularly for reed canarygrass. The ash compositions are calculated as weight percent oxides for an ashed sample, so reflect the compositions of the ash component, rather than the actual concentration in the biomass. The relative proportions of alkali metals show consistent differences for each crop. The ash of willow SRC is richest in K_2O, in miscanthus this is accompanied by higher concentrations of SiO_2, whereas in reed canarygrass K_2O and Na_2O are lower but SiO_2 is higher still. The different elemental ratios indicate that this is not merely due to different degrees of soil adhesion related to the different physical properties of the plants. Levels of fluorine are uniformly low, whereas chlorine shows wider variation.

Table 3. Biomass fuel quality

Site	(a)	(a)	(a)	(b)	(c)	(c)	(c)	(d)	(d)	(d)	(e)
Crop	SRC	MC	RC	SRC	SRC	MC	RC	SRC	MC	RC	SRC
Contaminants (ppm)											
As	< 0.1	< 0.1	< 0.1	< 0.1	<0.1	< 0.1	< 0.1	< 0.1	< 0.1	< 0.1	< 0.1
Cr	0.18	0.4	0.47	0.41	0.97	1.3	0.28	< 0.1	< 0.1	0.34	0.57
Cu	5.2	2.3	5.8	6.2	3.9	2.1	2.5	3.3	0.86	3.2	35
Pb	<0.30	0.36	2.08	0.47	0.36	<0.30	<0.30	<0.30	<0.30	0.4	3.73
Ni	0.33	0.36	0.54	0.61	0.58	0.25	0.34	0.22	< 0.1	0.47	0.46
Zn	160	48	100	180	132	58	28	74	8.1	55	200
Cd	0.26	<0.05	<0.05	0.95	0.41	<0.05	<0.05	0.34	<0.05	<0.05	0.15
Hg	<0.02	<0.02	<0.02	<0.02	<0.02	<0.02	<0.02	<0.02	<0.02	<0.02	<0.02
Halogens											
Cl (%)	0.13	0.31	0.31	0.28	0.03	0.07	0.69	0.10	0.22	0.25	0.48
F (ppm)	220	210	230	220	220	210	220	220	230	220	220
Ash content											
(%)	1.5	4.0	6.7	1.8	1.8	4.1	5.6	2.0	2.6	6.7	2.3
Ash composition (weight % oxides)											
Na_2O	1.59	1.89	0.40	3.54	1.49	1.90	0.49	0.60	0.79	0.45	3.34
K_2O	18.75	10.33	2.44	18.27	21.25	6.46	4.15	17.96	10.71	3.38	18.68
SiO_2	2.19	68.42	83.36	3.74	1.72	73.69	82.99	4.42	77.24	84.01	5.77

Note: SRC = short rotation coppice, MC = miscanthus, RC = reed canarygrass, SG = switchgrass.

DISCUSSION

An excellent environmental case can be made for the activities and results that the full scale BioReGen project will demonstrate: Climate change is addressed both by providing carbon-neutral biomass and by diverting biodegradable material from landfill as compost and thereby reducing methane emissions. In situ remediation of organic contaminants is anticipated due to amendment and cultivation while sequestration and phytoremediation may have a long-term impact on available metal contamination. Not only are organic wastes diverted but this activity also eliminates the need for disposal of contaminated soils as hazardous wastes. Instead the land is reused for a purpose for which it is suitable and without displacement of land otherwise used for food production. This is a low energy passive treatment or management strategy for a contaminated brownfield site, without the energy intensive, resource utilization or waste generation that typifies many remediation methods. Wider environmental benefits include the improved aesthetic and amenity value, coupled with habitat creation and ecological value that exceed those of more intensive farming on agricultural land. In summary, the type of brownfield land use envisaged following application of the BioReGen Project results appears to meet the economic, environmental and social aspirations of sustainable development.

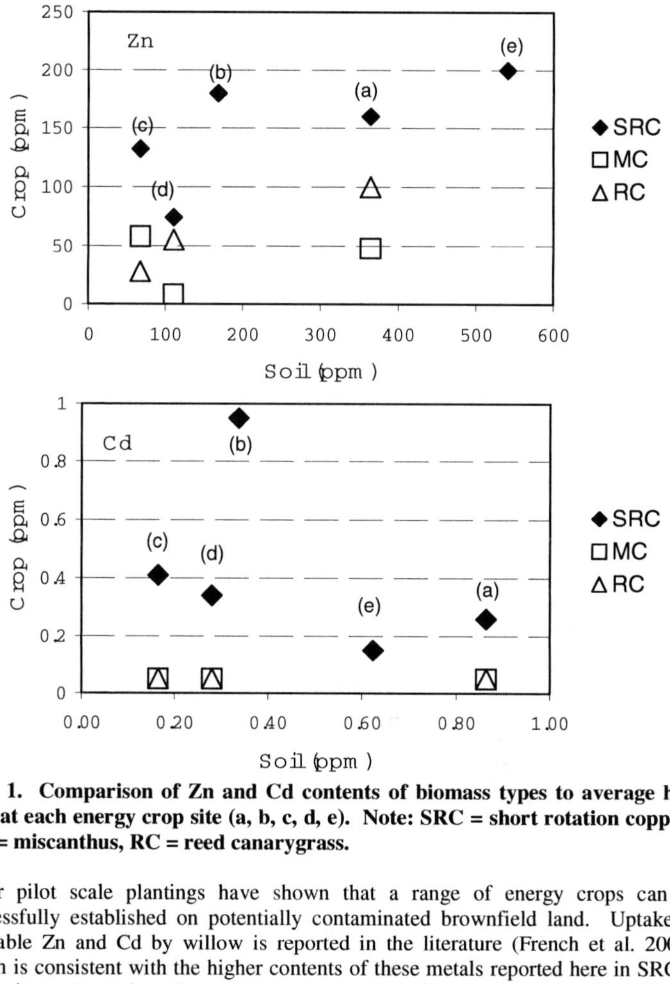

FIG. 1. Comparison of Zn and Cd contents of biomass types to average host soils at each energy crop site (a, b, c, d, e). Note: SRC = short rotation coppice, MC = miscanthus, RC = reed canarygrass.

Our pilot scale plantings have shown that a range of energy crops can be successfully established on potentially contaminated brownfield land. Uptake of available Zn and Cd by willow is reported in the literature (French et al. 2006), which is consistent with the higher contents of these metals reported here in SRC at each site. Assuming the metals are extracted from a 0.3 m soil profile, a conservative harvest of 10 oven-dry tonnes per hectare, biomass concentrations of 1 ppm Cd and 200 ppm Zn equate to annual removal of 2 ppb Cd and 40 ppb Zn from the average soil concentration. Critically, it is the most available (and hence most ecotoxic) fraction of the contaminant that is removed by phytoremediation (Lord et al. 2007). Higher metal uptake can be expected on more contaminated sites. Conversely, planting other energy crop species will produce clean fuel while reusing a dirty site. Differences in ash contents and compositions between biomass types indicates that any undesirable characteristic of one fuel can be offset by blending with another to gain an intermediate composition between those given in Table 3.

In northern UK 1 ha of energy crop can produce fuel for 1-1.5 KWe of continuous generation. The scale of available derelict land would allow immediate widespread implementation without displacing food production. Furthermore, if this carbon-neutral renewable energy is used to reduce use of fossil fuels, the BioReGen approach to phytoremediation probably also has the smallest carbon footprint of any currently available remediation technique.

CONCLUSION

Field scale trials confirm the technical feasibility of reusing brownfield sites to grow energy crops. The higher relative Cd and Zn content of SRC biomass offers the prospect of additional long-term phytoremediation during biomass production. Future trials will demonstrate the commercial viability of these fuels.

ACKNOWLEDGMENTS

The European Regional Development Fund, County Durham Environment Trust, European Life III Environment Programme and the UK Waste and Resources Action Programme have supported the BioReGen Project and the pilot projects before it.

REFERENCES

DEFRA-Environment Agency (2002). *Soil Guideline Value (SGV) Reports* No.1 As, No. 4 Cr, No. 7 Ni & No. 10 Pb (2002), www.environment-agency.gov.uk/.

Environment Agency (2002). "Dealing with Contaminated Land in England." Environment Agency, Bristol UK, 40 pp.

French, C.J., Dickinson N.M. and Putwain, P.D. (2007). "Woody biomass phytoremediation of contaminated brownfield land", *Environmental Pollution* 141, 387-395.

ICRCL (1987) *Inter-Departmental Committee on the Redevelopment of Contaminated Land Guidance Note 59/83* 2nd ed. HMSO, London.

Lord, R. A., Atkinson, J., Scurlock, M.O, Lane, A.N., Rahman, P.K.S.M., Connolly, H.E. and Street, G. (2007). "Biomass, Remediation, re-Generation (BioReGen Life Project): Reusing brownfield sites for renewable energy crops, *Proceedings 15th European Biomass Conference & Exhibition (Berlin 7-11 May 2007),* ETA-Renewable Energies, Florence, Italy.

National Land Use Database (2006). "Previously developed land that may be available for development: England 2005", Department for Communities & Local Government, www.communities.gov.uk/

BSI PAS100 (2005) "Publicly available standard for composted material", Waste & Resources Action Programme, Banbury, UK, www.wrap.org.uk/

Wastes Management Licensing Regulation (1994), www.opsi.gov.uk/

Evaluation of Fe(III) Reducing Microorganisms for the Biostabilisation of Chromium in Contaminated Soils

Nymphodora Papassiopi[1], Athina Kontoyianni[2], Katerina Vaxevanidou[3] and Anthimos Xenidis[4]

[1] National Technical University of Athens (NTUA), School of Mining and Metallurgical Engineering, Zografos Campus, GR-157 80, Athens, Greece; papasiop@metal.ntua.gr
[2] NTUA, S. of Mining and Metallurgical Engineering, GR-15780, Greece, akonto@mail.ntua.gr
[3] NTUA, S. of Mining and Metallurgical Engineering, GR-15780, Greece, vaxevank@central.ntua.gr
[4] NTUA, S. of Mining and Metallurgical Engineering, GR-15780, Greece, axen@central.ntua.gr

ABSTRACT: Hexavalent chromium (Cr(VI)) is a common and highly toxic contaminant in soils. Most remediation strategies developed for Cr(VI) contaminated soils aim at reducing the hexavalent form to the less toxic trivalent state, Cr(III), which can be efficiently immobilised in situ due to the very low solubility of Cr(III) compounds. This conversion can be achieved by conventional chemical methods. Another alternative is the exploitation of microbial processes, using microorganisms that exhibit the appropriate metal bioreduction potential. Current study investigates the possibility of exploiting the metabolic activity of an iron reducing microorganism, *Desulfuromonas palmitatis*. During the initial experiments Cr(VI) was added in the growth medium of the microorganism at three levels, i.e. 5, 15 and 30 mg/l. Complete reduction was obtained at all three concentrations, but growth of the bacteria was observed only in the medium with the lowest initial Cr(VI) content, i.e. 5 mg/l. In the subsequent experiments *D. palmitatis* were used for the treatment of soil samples artificially contaminated with Cr(VI) at two levels, i.e. 200 and 500 mg/kg. Experimental results with soils confirmed the bioreducing potential of *D. palmitatis*.

INTRODUCTION

Chromium contamination of soil and groundwater is a significant problem worldwide. The extensive distribution of this pollutant is due primarily to its use in numerous industrial processes such as metal plating and alloying, leather tanning, and wood treatment. Chromium persists in the environment in two stable oxidation states, Cr(VI) and Cr(III), which have widely contrasting toxicity and transport characteristics. Hexavalent chromium, Cr(VI), which typically exists as the oxyanion chromate (CrO_4^{2-}) has a high solubility in soils and waters and consequently tends to be mobile in the environment. Additionally, Cr(VI) is acutely toxic, subject to biological uptake, and is teratogenic and carcinogenic. In contrast, trivalent chromium, Cr(III), has a limited hydroxide solubility and forms strong complexes with soil minerals making it relatively immobile and less available for biological uptake. Therefore, reduction of Cr(VI) to Cr(III) is considered an important remediation strategy for reducing the deleterious effects of this pollutant.

Methods for remediation of chromate contaminated soils include a wide variety of techniques, such as in situ vitrification, chemical treatment with a reductant, removal by elektrokinetic techniques etc. (Tai and Jou, 1999; Sawada et al., 2004; Ignatiadis et al., 2005). Bioremediation is recently receiving attention as an alternative remediation strategy supported by the discovery that a wide variety of microorganisms possess chromate reductase activity. Several species amongst the sulphate, the sulphur and the Fe(III) reducing bacteria have been demonstrated to possess a high potential for obtaining the direct (enzymatic) or indirect (through biologically produced reductants, e.g. S(-II) or Fe(II)) reduction of chromate (Michel et al., 2001; Wielinga et al., 2001). It is thus a particular technological challenge to exploit the bioreducing potential of these microorganisms and develop appropriate remediation techniques. A great research effort has been devoted in the development of water treatment techniques, mainly based on the use of sulphate reducing microorganisms, and some processes have been already developed up to a pre-industrial scale (Battaglia et al., 2004, 2006). The development of bioremediation techniques for Cr(VI) contaminated soils is a field where intense research is currently in progress (Tseng and Bielefeldt, 2002; Jeyasing and Philip, 2005).

Present study investigates the possibility of exploiting the iron reducing microorganism *Desulfuromonas palmitatis* for obtaining the reduction and immobilization of chrome in contaminated soils. The reducing potential of the microorganism was initially evaluated in tests with soluble chromate. In subsequent experiments *D. palmitatis* were used for the treatment of soils artificially contaminated with Cr(VI) at two levels, 200 and 500mg Cr(VI) per kg soil.

MATERIALS AND METHODS

Preparation of Soil Samples with Artificial Cr(VI) Contamination

The initial soil is a calcareous material containing 4.42% of total C, which is almost all inorganic, and exhibiting a pH value of 8.16. Chemical analyses indicated that the soil contained low levels of some important trace elements of environmental concern. As far as chrome is concerned, the measured concentration was 43.5 mg/kg, a value which is lower compared to most soil quality guidelines.

Artificial contamination with Cr(VI) was applied in three representative subsamples of this soil material, weighing approximately 2 kg each. Each sample was mixed with an aqueous solution containing the appropriate concentration of potassium dichromate ($K_2Cr_2O_7$) in order to obtain three levels of contamination, i.e. 20, 200 and 500 mg Cr(VI) per dry kg of soil. The amount of solution corresponded to 1 litre per kg of soil. Wet soils were periodically stirred for 3 days to obtain homogenous distribution of Cr(VI) and left at room temperature for air drying. After the initial natural drying, the soils were sprayed with deionized water every 48 hours and stirred up daily for a period of approximately three months.

After the completion of artificial contamination the soils were subjected to the standard EN12457-2 leaching test to evaluate the amount of Cr(VI) which remains water soluble and mobile.

EN12457-2 leaching test

Standard test EN12457-2 is a batch leaching procedure, which is applied as compliance test in order to assess whether a waste material can be accepted for disposal in a landfill site. The European Council Decision 2003/33/EC established criteria for acceptance of wastes in three main classes of landfill sites, i.e. landfills for inert wastes, non hazardous wastes and hazardous wastes, based on the leachability of contaminants in this particular test. The procedure involves mixing of the solid material with deionised water at a liquid to solid ratio (L:S) 10:1, agitation of the mixture for 24 hours, filtration and determination of dissolved species in the filtrate.

Microorganisms and Media

Desulfuromonas palmitatis (DSMZ 12391) was purchased from the German Collection of Microorganisms (DSMZ, Branschweig, Germany). Standard anaerobic culturing techniques were used throughout. All incubations were carried out at 30 °C using 125 mL serum bottles sealed with butyl rubbers and aluminium crimps. The initial growth of the organism was conducted in the medium suggested by the DSMZ with fumarate (50 mM) as electron acceptor and acetate (10 mM) as electron donor (DSMZ Medium 837). Experiments with soluble Cr(VI) and soil were carried out using the same medium.

Bioreduction experiments with soluble chrome (VI)

Initial bioreduction experiments were conducted with soluble Cr(VI). The tests were carried out in 125 ml serum bottles containing 80 ml of nutrient broth medium. Chrome (VI) was added from a stock $K_2Cr_2O_7$ solution to obtain three levels of Cr(VI) concentration, i.e. 5, 15 and 30 mg/l. Oxygen was removed by bubbling a N_2/CO_2 (80:20) gas mixture for 20 min and the serum bottles were sealed and autoclaved. After cooling, the serum bottles were inoculated with 8 ml from a *D. palmitatis* culture and incubated horizontally on a shaking orbital incubator at 30°C.

During the tests aliquots of 200μl were withdrawn with syringes and analyzed for Cr(VI) using the diphenyl carbazide method (Lovely and Philips, 1994). Total chrome was measured with atomic absorption spectroscopy (AAS). Growth of bacteria was determined by measuring the corresponding total protein with Bradford assay (Bradford 1976). All tests were carried out in replicate and the mean values were taken into account. Similar tests, without Cr(VI) addition, were also conducted to obtain comparative results regarding *D. palmitatis* growth in the absence of Cr(VI).

Biostabilization of chrome (VI) in soils

Biostabilization of chrome in soils was evaluated only for the soil samples contaminated with 200 and 500 mg Cr(VI)/kg. Three tests (in replicates) were carried out for each soil: (a) the soil was mixed with nutrient broth solution without addition of microorganisms; (b) the soil was mixed with nutrient broth solution and inoculated with a 10% v/v *D. palmitatis* inoculum; (c) the soil was added in a fully developed culture of *D. palmitatis.* Concentration of microorganisms in the two latter tests corresponded to 7 and 70 mg/l proteins respectively. Tests were carried out in the 125 ml serum bottles mixing 8 g of soil with 80 ml of solution and applying standard anaerobic techniques as previously described.

After completion of the tests, the aqueous solution was separated from the soil by filtration and the solid residues were subjected to the EN12457-2 leaching test.

RESULTS AND DISCUSSION

Evaluation of chrome mobility in the artificially contaminated soils

The results of standard EN12457-2 test, regarding leachability of Cr from the three artificially contaminated soils, as well as from the initial soil, are presented in Figure 1. Compared with the established criteria, the soil with the lower level of contamination is characterized as inert, while the soil with medium contamination level could be accepted only in a landfill site for hazardous wastes. Finally, the soil with the highest chrome contamination is not acceptable for disposal in any class of landfill site. Based on the provisions of EC Decision 2003/33/EC, if the criteria for acceptance at a hazardous waste landfill are not met, the waste must be subjected to a stabilization treatment and tested again against the criteria, until they are met, before obtaining permitting for disposal.

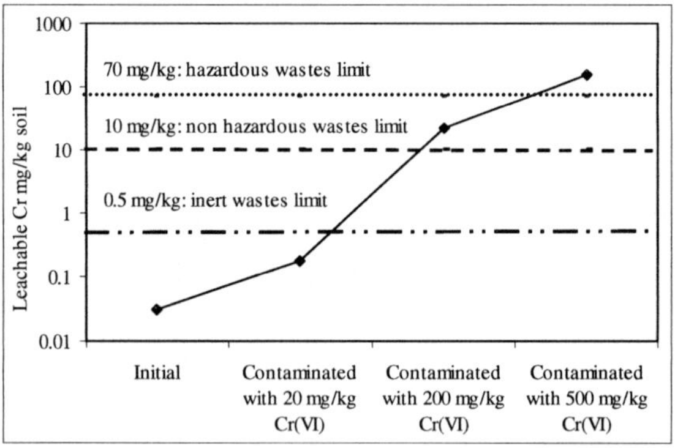

FIG.1. Leachability of chrome according to the standard EN 12457-2 test

EN leachability results are also indicative about the capacity of this particular soil to retain hexavalent chromium. At the highest level of contamination, 500 mg/kg Cr(VI), a considerable percentage of Cr(VI), about 35%, is mobilized with this simple water extraction treatment, indicating indirectly that the maximum retention capacity of this soil towards Cr(VI) has been exhausted. When contamination decreases to 200 and 20 mg/kg, the percentage of water soluble chrome is reduced to 10 and 1% respectively.

Bioreduction experiments with soluble chrome (VI)

The results of bioreduction experiments that were carried out with soluble Cr(VI) are presented in Figure 2. As seen in the figure reduction was observed in all three levels of Cr(VI) concentrations, but with different kinetics. At the lower concentration of 5 mg/l, Cr(VI) reduction was completed within 30 days. Reduction kinetics was much more slower in the solutions containing initial Cr(VI) concentrations of 15 and 30 mg/l. Complete Cr(VI) reduction was observed after approximately 120 days of treatment.

Analyses of total Cr in the aqueous solutions by AAS gave values close to the concentrations of Cr(VI) as determined by the diphenyl carbazide method. It was thus concluded that reduced Cr(III) is removed from the solution, probably as $Cr(OH)_3$, which is known to have a very low solubility at the neutral pH region 6-9 (K_{sp} =6.7 $\times 10^{-31}$). Formation of a white precipitate was also observed supporting this assumption.

It is worthwhile to mention that in the serum bottle with the low initial concentration of Cr(VI), 5 mg/l, there was also a clearly visible growth of *D. palmitatis.* Certainly growth took place with considerable delay compared to the tests that were carried out without chrome (VI) addition. More particularly growth became visible only after the initial 20 days of treatment, when Cr(VI) reduction and precipitation was complete. The amount of proteins that were measured in this test was about 32 mg/l, a concentration which is approximately 50% of the proteins measured in the control growth test conducted without Cr(VI).

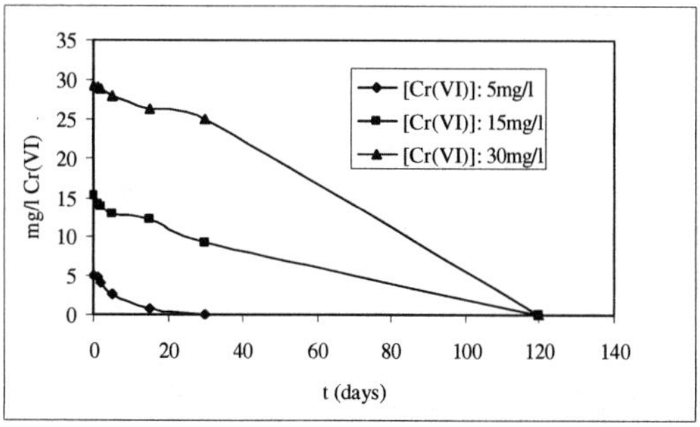

FIG. 2. Reduction of aqueous Cr(VI) by *D. palmitatis*.

Biostabilization of chrome (VI) in soils

The concentration of Cr(VI) in the aqueous solution during the biostabilization treatment of soil with the higher level of contamination (500 mg Cr(VI)/kg soil) is presented in Figure 3. As seen in the figure, treatment of this soil with the elevated concentration of microorganisms, corresponding to 70 mg proteins per liter, resulted in the relatively rapid reduction of Cr(VI). Approximately 90% of reduction was achieved within the first 7 days of treatment. Reduction kinetics was slower during the test carried out with the lower concentration of *D. palmitatis*, i.e. 7 mg proteins per liter. The results indicate a specific reduction rate in the order of 1.0-1.5 mg of Cr(VI) reduced per gram of protein per hour. Chromate was not reduced during the control test that was conducted without addition of *D. palmitatis,* indicating that reduction is due to the microbial activity.

The concentration of aqueous Cr(VI) was below the detection limit of the diphenyl carbazide method in all the kinetic samples, that were taken during the treatment of soil contaminated with 200 mg Cr(VI) per kg soil. The concentration of chrome was measured in the final solutions by AAS and was found to be 0.9 mg/l in the control experiments that were carried out without *D. palmitatis* and lower than 0.04 mg/l in the tests that were conducted with 7 and 70 mg of microbial proteins per liter.

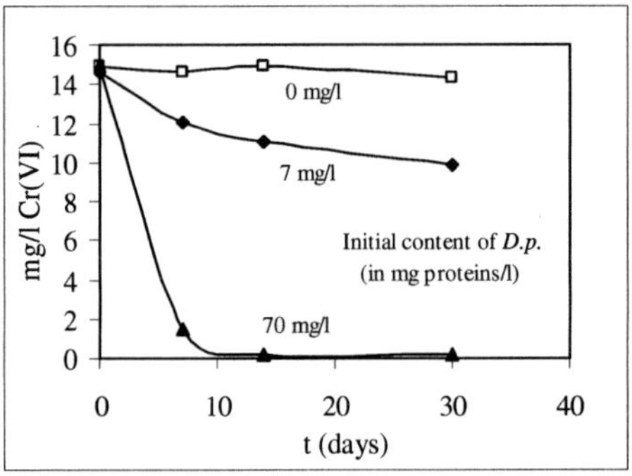

FIG. 3. Concentration of aqueous Cr(VI) during the treatment of soil contaminated with 500 mg Cr(VI) per kg soil.

Evaluation of biostabilization treatment with the EN12457-2 leachability test

The EN 12457-2 leaching test was applied to evaluate the mobility of chrome following the biostabilization treatment of soils. The results are presented in Figures 4a and 4b. As seen in the figures, the concentration of Cr that was measured in the EN solutions was low for all the solid residues, regardless of the efficiency of stabilization. This is due to the fact that in the samples where chrome was not stabilized, the water soluble fraction of the element was removed with the treatment solution. The real performance of stabilization can be evaluated taking into account the sum of chrome recovered in both solutions, i.e. treatment solution and EN leachate. The results presented in Figure 4a indicate that biostabilization was very efficient in the case of soil contaminated with 200 mg Cr(VI) per kg. The water soluble fraction of chrome was reduced from the initial 13 mg/kg to less than 0.5 mg/kg, which is the upper permitted level for characterizing a solid material as inert. Stabilization was obtained treating the soil either with the low or the high microbial population. On the contrary treatment of the soil contaminated with 500 mg/kg was efficient only using the higher microbial concentration (Fig. 4b). With this treatment the water soluble Cr was reduced from 122 mg/kg in the initial soil to approximately 2.9 mg/kg, which is the sum of chrome recovered in the treatment solution and in the EN leachate.

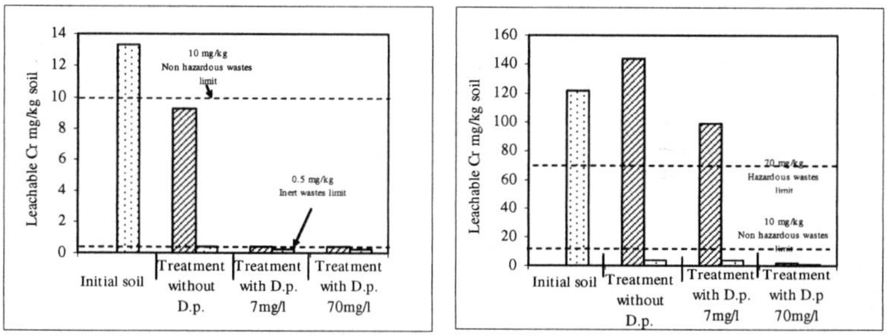

⬚ : Cr in the biotreatment solution ▨ : Cr in the EN12457 solution

(a) Soil contaminated with 200 mg Cr(VI) per kg

(b) Soil contaminated with 500 mg Cr(VI) per kg

FIG.4. Water soluble chrome recovered in the biotreatment solutions and in the solutions of the EN12457-2 test.

CONCLUSIONS

The iron reducing microorganism *D. palmitatis* was evaluated as potential biostabilization agent for the remediation of chromate contaminated soils. During the initial experiments Cr(VI) was added in the growth medium of the microorganism at three levels, i.e. 5, 15 and 30 mg/l. Complete reduction was obtained at all three concentrations, but growth of the bacteria was observed only in the medium with the lowest initial Cr(VI) content, i.e. 5 mg/l. In the subsequent experiments *D. palmitatis* were used for the treatment of soil samples artificially contaminated with Cr(VI) at two levels, i.e. 200 and 500 mg/kg. The kinetics and efficiency of bioreduction was seen to be strongly affected by the initial bacterial concentration. When the soils were treated with a bacteria population corresponding to 70mg proteins/l, reduction and precipitation of aqueous chrome (VI) was completed within approximately 10 days. With this treatment the water soluble Cr(VI) was reduced by more than 95% in both soil samples.

ACKNOWLEDGEMENT

This research was supported by the European Social Fund (75%) and Greek National Resources (25%) - Operational Program for Educational and Vocational Training II (EPEAEK II) and particularly the Program PYTHAGORAS.

REFERENCES

Bradford, M.M. (1976). "A rapid and sensitive method for the quantification of microgram quantities of proteins utilizing the principle of protein dye binding". *Anal. Biochem.*, Vol . 72: 248-254.

Battaglia-Brunet, F., Touze, S., Michel, C., and Ignatiadis, I. (2006). Treatment of chromate polluted groundwater in a 200 dm3 pilot bioreactor fed with hydrogen. *Journal of Chemical Technology and Biotechnology*, 81(9): 1506-1513.

EN 12457-2 (2002). "Characterisation of waste. Compliance test for leaching of granular waste materials and sludges. Part 2.", *European Committee of Standardization CEN/TC 292, 12/02/2002*, 28 pp.

EC Decision 2003/33/EC: "Council Decision of 19 December 2002 establishing criteria and procedures for the acceptance of waste at landfills pursuant to Article 16 of and Annex II to Directive 1999/31/EC." *Official Journal L 011, 16/01/2003*, pp. 0027–0049.

Ignatiadis, I., Morin, D. Ragot, C., Oudin, J.C., and Dueso, N. (2005) "Development and in situ implementation of a chemical process for immobilisation of the chromate contained in an industrial ground", *Proc. 9th Int. Conference on Contaminated Soil (ConSoil 2005)*, Bordeau , pp. 2555-2564.

Jeyasingh J. and Philip L. "Bioremediation of chromium contaminated soil: optimization of operating parazmeters under laboratory conditions", *J. of Hazardous Materials*, B118: 113-120.

Lovley, D.R. and Phillips E.J.P. (1994). "Reduction of chromate by Desulfovibrio vulgaris and its c3 cytochrome." *Appl. Environ. Microbiol.*, 60: 726-728.

Michel C., Brugna M., Aubert C., Bernadac A. and Bruschi M. (2001) "Enzymatic reduction of chromate: comparative studies using sulfate-reducing bacteria." *Appl Microbiol. Biotechnol.*, 55: 95–100.

Sawada A., Mori K., Tanaka S., Fukushima M. and Tatsumi K. (2004) "Removal of Cr(VI) from contaminated soil by electrokinetic remediation." *Waste Management*, 24: 483-490.

Tai H-S and Jou C-J G. (1999) "Immobilization of chromium-contaminated soil by means of microwave energy", *J. of Hazardous Materials*, B65: 265-275.

Tseng, J.K. and Bielefeldt, A.R. (2002). "Low temperature chromium(VI) biotransformationin soil with varying electron acceptors", *J. Environ. Qual.* 31: 1831–1841.

Wielinga B., Mizuba M., Hansel C. and Fendorf S., (2001) "Iron promoted reduction of chromate by dissimilatory iron-reducing bacteria", *Eniron. Sci. Technol.*, 35: 522-527.

Fracturing Mechanism in Soils with Three Dimension Stress State

By Dingwen Zhang[1], Songyu Liu[2] M. ASCE and Lei Chen[3]

ABSTRACT

Hydraulic and pneumatic fracturing has been used to improve the permeability of the soil and rock and has been widely used in geotechnical practices. Although two soil fracturing mechanisms (i.e., tensile failure mechanism and shear failure mechanism) were proposed in literatures, the soil fracturing mechanism is still not yet fully understood. A comprehensive review was conducted on the available approaches existing in the literature to calculate the fracture-initiation pressures around a borehole. And then, two fracture-initiation pressure calculating formulas with three dimension stress state were put forward in this paper, based on tensile failure mechanism and shear failure mechanism, respectively. In addition, the suitability of those two calculating formulas was assessed. The results imply that the possibility of shear failure mechanism is increased with the increasing of the minor principal stress and decreasing of undrained shear strength of soil. A hydraulic fracturing test results in literatures were also verified the suitability of proposed methods.

INTRODUCTION

Hydraulic fracturing is the injection of pressurized water and pneumatic fracturing is the injection of pressurized air through boreholes to develop fractures or extend existing fractures in soil and rock (Alfaro et al., 2001). The concept of generating fractures in soil or rock by liquid or gas has been recognized firstly by the oil industry and has been utilized for many years to increase productivity of oil and gas wells (Massarsch, 1978). Fracturing has also been recognized as important when considering the maximum allowable pressure in connection with pressure grouting, field

[1] Research Assistance, Institute of Geotechnical Engineering, Southeast University, 2 Sipailou, Nanjing, 210096, People's Republic of China
[2] Professor, Institute of Geotechnical Engineering, Southeast University, 2 Sipailou, Nanjing, 210096, People's Republic of China
[3] Research Assistance, Institute of Geotechnical Engineering, Southeast University, 2 Sipailou, Nanjing, 210096, People's Republic of China

permeability measurements in soils (Bjerrum et al., 1972), pile driving and installation of sand drains (Massarsch, 1978), and drilling mud pressure (Xia et al., 2006). Fracturing has also been proposed as a way to measure the in situ minor principle stress by recording the closing pressure of a crack opened by fracturing (Bjerrum et al., 1972). Some researchers also drew attention to the effect of soil fracturing in the dam design. Another aspect of fracturing that has recently gained increasing attention lies in the area of enhancement the effectiveness of most in situ remediation methods for contaminated sites underlain with unfavorable low-permeability soils (Alfaro et al., 2001).

In despite of the wide utilization of fracturing, the failure mechanism during soil fracturing is not clear. For example, some researchers indicated that the initiation of fracture in soils is governed by a shear failure mechanism, whereas others reported a tensile failure mechanism. A comprehensive review was conducted on the available approaches existing to calculate the fracture-initiation pressure around a borehole. And then, two fracture-initiation pressure calculating formulas with three dimension stress state were put forward in this paper, based on tensile failure mechanism and shear failure mechanism, respectively. At last, a hydraulic fracturing test results in literatures were also verified the suitability of proposed formulas.

REVIEW OF PAST RESEARCH

Andersen et al. (1994) and Alfaro et al. (2001) conducted reviews concentrating on the fracturing theories. Based on their reviews, this paper examined different approaches existing in the literature to calculate the fracture-initiation pressure for situations in low permeability soils under undrained conditions. The various approaches that have been proposed may be grouped into the following categories: (a) cavity expansion theory, (b) initial yielding, (c) tensile failure mechanism, (d) empirical formulas, and (e) linear elastic fracture mechanism.

Cavity Expansion Theory

Massarsch (1978) used the cavity expansion theory to analyze the hydraulic fracturing around driven piles and sand drains. The fracture pressure P_f was taken as the ultimate pressure when large radial displacement or plastic expansion occurred. For a generalized cohesive-frictional material, an expression to determine P_f can be found in Carter et al. (1986). In a purely cohesive material, P_f is given in Eq. (1).

$$P_f = \sigma_h + \tau[1 + \ln(G/\tau)] \tag{1}$$

where σ_h is the horizontal confining pressure, G is the shear modulus of soil and τ is the shear strength of soil.

The cavity expansion theory gives the excess borehole pressure when large radial displacements occur. However, fracturing may occur before large radial displacements occur, and the fracture pressure will in most cases be lower than that calculated by cavity expansion theory (Andersen et al. 1994). This was been substantiated by the comparison of fracture pressure predicted by cavity expansion theory and those observed by Overy et al. (1986), Andersen et al. (1994) and Alfaro et al. (2001).

Initial Yielding

Initial yielding implies that fracturing in soil occurs when the stress of a soil element around the borehole intercepts the strength envelope such as Mohr-Coulomb criterion. Initial yielding in the soils was used by Overy et al. (1986), Mori et al. (1987), Panah et al. (1989) and Yanagisawa et al. (1994) as a criterion to determine the fracture-initiation pressure.

Panah et al. (1989) used the elastic theory to calculate the stress changes around a borehole and the fracture pressure was set equal to the borehole pressure when the initial yielding occurs. In a thick-walled cylinder of homogeneous, isotropic, elastic material with Mohr-Coulomb criterion, the fracture pressure is given by Eq. (2).

$$P_{\mathrm{f}} = \frac{(1+\sin\varphi)\sigma_{\mathrm{h}} + c\cos\varphi(1-r_{\mathrm{i}}^2/r_{\mathrm{h}}^2)}{1+(r_{\mathrm{i}}^2/r_{\mathrm{h}}^2)\sin\varphi} \quad (2)$$

where φ and c are the angle of internal friction and cohesion of soil, respectively; r_{i} is the radius of the borehole and r_{h} is the radius of the cylinder being considered.

With the very fundamental theory and the simplicity of obtaining the soil parameters from conventional laboratory tests, this approach is very attractive for practical use (Alfaro et al. 2001).

Tensile Failure Mechanism

Tensile failure mechanism means that fracturing occurs when the minor principle stress become negative (i.e. tensile) with the magnitude equal to or greater than the tensile strength of soil. Elasticity theory can be used to calculate the stress changes round a borehole in an elastic medium. Elasticity theory assumes that the soil is linearly elastic and neglect the shear induced pore pressure. The total pressure inside the borehole when fracturing occurs will then be

$$P_{\mathrm{f}}=2\sigma'_{\mathrm{h0}}+u_0+\sigma_{\mathrm{t}} \quad (3)$$

where σ'_{h0} is the lateral effective stress prior to drilling of the borehole, equal in all horizontal directions; u_0 is the pore pressure prior to drilling of the borehole, and σ_{t} is the tensile strength of the soil.

Andersen et al. (1994) proposed an approach to calculating fracture pressure based on tensile failure mechanism. In their approach, they emphasized the non-linearity of the stress-strain properties of soil, and the pore pressure changes in the soil induced by changes in total mean normal stress and shear stress.

Empirical Formulas

Many empirical formulas in determining the fracture pressure are available, which have slight differences in the form. A common empirical formula has the form of Eq.

(4).

$$P_f = m\sigma_3 + n \tag{4}$$

where m and n are an empirical factor; and σ_3 is the minor principle stress.

Jaworski et al.(1981) suggested that σ_t can vary between zero and the ultimate pressure from cavity expansion theory, and that m varies between one and two. In their tests, Jaworski et al. (1981) found m to be between 1.5 and 1.8 and n varied from 20~265 kN/m^2. Mori et al. (1987) found good agreement between fracture pressures from laboratory experiments and those calculated from Eq.(4) with m equal to 1 and n equal to the unconfined compressive strength. Panah et al. (1989) gave values of m equal to 1.05 to 1.085. The empirical approach seems to be very simple but lacks theoretical basis.

Linear Elastic Fracture Mechanism

Murdoch (1992) established the fracture criteria based on the conventional linear fracture mechanics and fracture toughness was introduced as a factor to estimate the fracture pressure. However, fracture toughness is very difficult to measure in practice. This made the method not so attractive for practical use, but it seemed to be the only method available to describe the fracture mechanism for wells with initial fracture slots.

As shown in the above literature reviews, there seems to be no general consensus among researchers concerning failure mechanisms of soil fracturing by pressure injection through borehole. The first two approaches (i.e., those based on cavity expansion and initial yielding) indicate that the initiation of fracture in the soil is governed by a shear failure mechanism as reflected by the Mohr-Coulomb failure criterion. The other three approaches (i.e., empirical formulas, tensile strength approach and linear elastic fracture mechanism) indicate that initiation of fracture is a tensile failure mechanism.

DETERMINING THE FRACTURE PRESSURE WITH THREE DIMENSION STRESS STATE

Fracture Pressure Based on Tensile Failure Mechanism

The initial radius of the borehole is R_0.The pressure inside the borehole increased to a value P_i and the borehole radius is increased to R_i. In the initial part of loading it is assumed that soil exhibits a linear elastic response up to a certain pressure limit, defining using the Mohr-Coulomb failure criterion. After initial yielding, the soil is assumed to exhibit perfectly plastic behavior, and a plastic zone develops around the borehole with a plastic radius R_e that travels outwards as the pressure applied within the borehole increase.

Before the commencement of shear failure (initial yield), the stress distribution in the elastic region is expressed as (Yu, 2000)

$$\sigma_r = \frac{\sigma_2+\sigma_3}{2}(1-\frac{R_0^2}{R^2}) + \frac{\sigma_2-\sigma_3}{2}\cdot(1-\frac{4R_0^2}{R^2}+3\frac{R_0^4}{R^4})\cos 2\theta + P_i\frac{R_0^2}{R^2} \quad (5)$$

$$\sigma_\theta = \frac{\sigma_2+\sigma_3}{2}(1+\frac{R_0^2}{R^2}) - \frac{\sigma_2-\sigma_3}{2}\cdot(1+3\frac{R_0^4}{R^4})\cos 2\theta - P_i\frac{R_0^2}{R^2} \quad (6)$$

where σ_r is the radial stress; σ_θ is the circumferential stress; σ_3 is the minor principle stress; σ_2 is the middle principle stress; R_0 is the initial radius; θ is the angle from the horizontal line; P_i is the expansion pressure inside the borehole; and R is the distance from the center of the borehole.

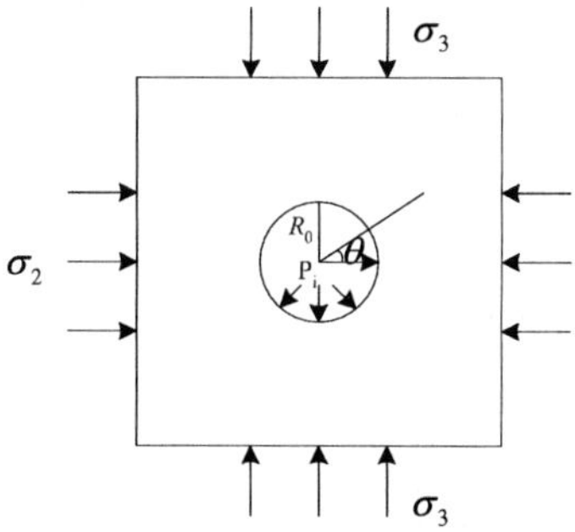

Fig.1. Cylindrical cavity expansion under anisotropic stress.

The critical circumferential stress at the borehole boundary would reach when R equals to R_0 and θ equals to zero or π. Hence,

$$\sigma_r = P_i \quad (7)$$

$$\sigma_\theta = 3\sigma_3 - \sigma_2 - P_i \quad (8)$$

It is assumed that the fracturing occurs when the minor principle effective stress become negative with a magnitude exceeding the tensile strength, σ_t, of the soil. Hence, the fracture pressure would be

$$P_f = 3\sigma_3 - \sigma_2 + \sigma_t \quad (9)$$

If the original minor principle effective stress is large and (or) the shear strength of soil is small, the soil around the borehole may reach undrained shear failure before the minor principle effective stress become negative or reach the tensile strength of soil. Therefore, the minor principle effective stress never reaches zero. Upon shear failure, the difference between the radial and circumferential stresses remains equal to $2S_u$, (i.e., σ_r-σ_θ=$2S_u$, S_u is the undrained shear strength). So the maximal fracture pressure based on tensile failure mechanism is given by

$$P_{\mathrm{f,\,max}} = S_{\mathrm{u}} + \frac{1}{2}(3\sigma_3 - \sigma_2) \tag{10}$$

Eq.(10) implies that if the calculating fracture pressure based on tensile failure mechanism is lager than $P_{f,max}$, the fracturing is controlled by shear failure mechanism.

Fracture Pressure Based on Shear Failure Mechanism

Some researches indicated that the initiation of fracturing in soil is governed by a shear failure mechanism and fracturing occurs when the stress of a soil element around the borehole intercepts the Mohr–Coulomb strength envelope.

$$(\sigma_r - \sigma_\theta) = (\sigma_r + \sigma_\theta)\sin\varphi + 2C\cos\varphi \tag{11}$$

The stress at the point of interface between the plastic zone and the elastic region satisfied the Mohr-Coulomb strength envelope. Substituting Eq. (7) and Eq. (8) into Eq. (11) and considering the tensile strength of soil, it is found that

$$P_{\mathrm{f}} = \frac{1}{2}(3\sigma_3 - \sigma_2)(1+\sin\varphi) + C\cos\varphi + \frac{\sigma_t}{2}(1+\sin\varphi) \tag{12}$$

It is difficult to understand why fracturing should occur because of yielding in the soil. A possible explanation is that the increase of the plastic shear failure zone created shear bands or an unstable state around the cavity. This leads to a localized micro-scale crack and the injected fluid or gas can penetrate into the crack to produce local tensile stresses at the crack tips (Mitchell et al. 2006).

Comparison of Measured and Calculating Fracture Pressure

A hydraulic fracturing test results performed by Yanagisawa et al. (1994) are presented here to verify the suitability of proposed methods. Yanagisawa et al. (1994) performed hydraulic fracturing test on compacted cohesive silt clay. The properties of the soil sample are as follow: φ'= 29°, LL = 49.82%, PL = 25.54%, PI = 24.28, Gs = 2.71, clay = 43%, silt = 41%, sand = 16%, optimum water content = 36%. Also from a series of unconsolidated undrained triaxial tests, φ_u = 6.5° and c_u = 10.8 kPa were found for compacted specimens with the same density. Other tests details can be found in the original paper.

Test results and calculated fracture pressure were presented in Table 1 and Fig.2. The tensile strength of the soil was neglected. The results indicate that the calculated fracture pressure based on tensile failure is greater than the maximal fracture pressure based on tensile failure and it implied that shear failure occurs before the tensile failure. In fact, it can be observed that the experimental data are in good agreement with calculating fracture pressure based on shear failure mechanism. Moreover, the discrepancy between the measured P_f and calculated P_f based on shear failure mechanism decreases with the increasing of the minor principal stress. Considering the tensile strength of soil, the discrepancy between the measured P_f and calculated P_f based on shear failure mechanism would decreases. The results indicates that soil fracturing based on shear failure occurs easily than the tensile failure in the case of high minor principal stress and (or) small soil shear strength, whereas fracturing is governed by the tensile failure mechanism in the case of lower minor principal stress and (or) lager soil shear strength.

Table 1. Calculated fracturing pressure and measured test results

σ_3 kPa	σ_2 kPa	Measured P_f kPa	Tensile failure mechanism		Shear failure mechanism
			Calculated P_f kPa	Maximal P_f kPa	Calculated P_f kPa
29	29	49.0	58	43.1	43.0
29	34	47.9	53	40.6	40.2
29	39	40.7	48	38.1	37.4
29	44	38.9	43	35.6	34.7
29	49	37.1	38	33.1	31.9
49	49	66.4	98	65.4	65.3
49	54	63.9	93	62.9	62.5
49	59	61.7	88	60.4	59.7
49	64	56.7	83	57.9	56.9
49	69	55.6	78	55.4	54.1

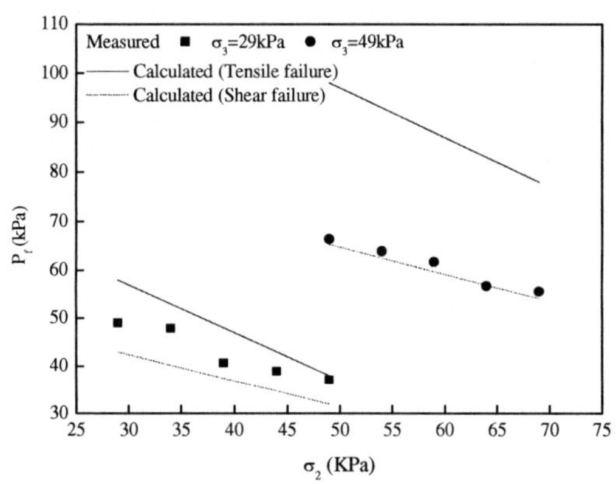

Fig.2. Comparison of measured and calculated fracture pressure.

CONCLUSIONS

A comprehensive review was conducted concentrating on the available approaches existing in the literature to calculate the fracture-initiation pressures around a borehole. Two mechanisms (i.e., tensile failure mechanism and shear failure mechanism) have been identified as having the potential to response for the soil fracturing. And then, two initiation fracturing pressure calculating formulas with three dimension stress state were put forward in this paper, based on tensile failure mechanism and shear failure mechanism, respectively. And the suitability of those two calculating formulas was assessed. The results indicates that soil fracturing based on shear failure occurs easily than the tensile failure in the case of high minor principal stress and (or) small soil shear strength, whereas fracturing is governed by the tensile failure mechanism in the case of lower minor principal stress and (or) lager soil shear strength.

ACKNOWLEDGMENTS

The authors appreciate the support of Excellent Doctor Thesis Fund of Southeast University (No. YBJJ0603).

REFERENCES

Alfaro, M.C. and Wong, R. C. K. (2001). Laboratory studies on fracturing of low-permeability soils, *Canadian Geotech. J.*, Vol. 38(3): 303-315.

Andersen, K. H., Rawlings C. G., Lunne, T. A. and By, T. H. (1994). Estimation of hydraulic fracture pressure in clay, *Canadian Geotech. J.*, Vol. 31(6): 817-828.

Bjerrum, L., Nash, J. K. T. L., Kennard, R. M., and Gibson, R. E. (1972). Hydraulic fracturing in field permeability testing, *Geotechnique*, Vol. 22(2): 319-332.

Carter, J P, Booker J R, Yeung S K. (1986). Cavity expansion in cohesive frictional soils. *Geotechnique*, Vol. 36(3) : 349-358.

Jaworski, G. W., Duncan, J. M., and Seed, H. B. (1981). Laboratory study of hydraulic fracturing, *J. Geotechnical & Geoenv. Engrg.*, Vol. 107(GT6): 713-732.

Massarsch, K. R. (1978). New aspects of soil fracturing in clay, *J. Geotechnical & Geoenv. Engrg.*, 104(GT8): 1109-1123.

Mitchell, J. K. and Soga, K. (2006). *Fundamentals of Soil Behavior* (3rd Edition), John Wiley and Sons., 577pp.

Mori, A. and Tamura, M. (1987). Hydrofracturing pressure of cohesive soils, *Soils and Foundations*, 27(1): 14-22.

Murdoch, L. C. (1992). Hydraulic fracturing of soil during laboratory experiments, Part 3: Theoretical analysis, *Geotechnique*, 43(2): 277-287.

Overy, R. F. and Dean, A. R. (1986). Hydraulic fracturing testing of cohesive soil. In proceeding of the 18th offshore technology conference. Houston, Vol. 3: 45-53.

Panah, A.K., and Yanagisawa, E. (1989). Laboratory studies on hydraulic fracturing criteria in soil. *Soils and Foundations*, 29(4): 14–22.

Xia, H. W. and Moore I. D. (2006). Estimation of maximum mud pressure in purely cohesive material during directional drilling, *Gomechanics and Geoengineering*, Vol. 1(1): 3-11.

Yanagisawa, E. and Panah, A. K. (1994). Two-dimensional study of hydraulic fracturing criteria in cohesive soils. *Soils and Foundations*, 34(1): 1-9.

Yu, H. S. (2000). *Cavity expansion methods in geomechanics*, AH Dordrecht: Kluwer Academic, 385pp.

Remediation of Petroleum-Contaminated Groundwater Using High Carbon Content Fly Ash

M. Melih Demirkan[1], Doina Morar[1], Ahmet H. Aydilek[2], Eric A. Seagren[2], Members, ASCE, and Alice Tsai[3]

[1, 2, 3] Graduate Research Assistant, Associate Professor, and Undergraduate Research Assistant, respectively, University of Maryland, Department of Civil and Environmental Engr., College Park, MD, 20742, aydilek@eng.umd.edu

ABSTRACT: Approximately 90% of the coal used in the United States is burned to produce electricity. As a result, the power plants produce vast quantities of Class F fly ash. The amount of unburned carbon in these fly ashes has been increasing in the last decade due to introduction of low nitrogen oxide burners to coal-burning power plants. Such fly ashes have no value as a concrete additive and require high land-filling costs. The objective of this study was to investigate potential use of these high carbon content fly ashes (HCCFA) in sorptive barrier applications. Naphthalene and o-xylene sorption onto fly ash was studied in a series of batch adsorption tests. Column sorption tests were also conducted on sand-fly ash mixtures with naphthalene and o-xylene. The results indicated that HCCFA has significant capacity for retardation of petroleum contaminants in barrier applications. Sorbed amounts measured in batch scale tests and breakthrough curves from column tests suggested that the amount and type of carbon has a significant effect on remediation efficiency.

INTRODUCTION

Remediation of groundwater contaminated with petroleum-based products has been an important task for engineers and scientists in recent years. As a result, the difficulty of reducing subsurface contamination levels has led to research and development of several innovative in-situ treatment technologies. One of the passive remediation technologies gaining wide acceptance is the use of permeable reactive barriers (PRB). In this technology, the pollutants are immobilized permanently or their levels are reduced to the Maximum Contamination Limits (MCL) while the plume is passing through an underground barrier system.

One relatively new variation on the PRB concept is to use an immobilization process, in which the organic pollutants are adsorbed onto sorptive surface of the barrier material. The reaction mechanism in these PRBs is often adsorption, and a term "permeable sorptive barriers (PSB)" has been recently introduced (Woinarski et al. 2003). Typical compounds of interest treated with such systems are trichloroethylene (TCE), petroleum hydrocarbons as well as trace elements and heavy metals.

Laboratory studies have demonstrated the effectiveness of various natural and synthetic sorbents as potential reactive/sorptive medium for treatment of groundwater containing both organic and inorganic pollutants. Specific materials tested include such as wood chips, limestone, manure (USEPA 2006), peat (Guerin et al. 2002), and lignitic coal (Jenk et al. 2003) have been investigated. For example, Schad and Gratwohl (1998) reported a successful field application of sorptive barrier using activated carbon as sorbent for a groundwater contaminated with petroleum hydrocarbons. Furthermore, there is growing interest in the utilization of recycled materials for remediation of contaminated groundwater as a part of sorptive barrier investigations. Recycled materials, such as tire chips (Kim et al 1997) and foundry sand (Lee et al. 2004) have also been studied to investigate their feasibility as sorptive medium in these barrier systems.

Approximately 90% of the coal used in United States is burned to produce electricity. As a result, the power plants produce vast quantities of coal combustion by-products (CCB) that present another environmental challenge. One of these CCB, Class F fly ash, is generated in large quantities and occasionally contains significant amounts of unburned carbon (i.e., high loss on ignition) as a result of equipping the power plants with the low NOx burners. In certain cases, the unburned carbon content exceeds the ASTM limit of 6%, and the fly ashes cannot be used in concrete production. Recent data indicate that approximately 68% of this high-carbon content fly ash (HCCFA) is placed in landfills, thereby consuming valuable land space and creating the potential to impact aquatic resources (Petzrick 2001). HCCFA is ideally suited as a sorptive medium in PSBs. Previous research has showed that high carbon content fly ashes can be effective in removal of herbicides (Konstantinou and Albanis 2000), PCBs (Nollet et al. 2003), and dyes (Jonas et al. 2003, Wang et al. 2005) via adsorption. Fly ash was also found to be capable of removing o-xylene, a commonly encountered pollutant in the environment (Banerjee et al. 1997).

The objective of this study is to investigate the adsorption chacteristics of HCCFA for its potential use as a PSB medium for removal of petroleum residues from groundwater. To meet this objective, a series of batch adsorption tests and column sorption experiments were conducted with three HCCFA and two potential groundwater contaminants.

MATERIALS

Fly ashes

Three different fly ashes obtained from three different electrical power plants located in Maryland were employed in the testing program. Fly ashes were named after their plant of origin: Morgantown (MT), Paul Smith (PS), and Dickerson Precipitator (DP) ashes. All three were produced as a result of burning pulverized bituminous coal and classified as Class F fly ash according to ASTM C 610. Specific gravity tests conducted per ASTM D 854 indicated that PS, DP and MT had specific gravities of 2.20, 2.37 and 2.45, respectively. Loss on ignition (e.g., unburned carbon content) values for MT, PS, and DP were 3.1 %, 10.7 %, and 20.5 %, respectively. Specific surface area (SSA) measurements conducted following five point BET analysis by using a Micromeritics Gemini 2360 indicated that PS, DP and MT had SSAs of 5.142, 11.078 and 1.357 m^2/g, respectively.

Contaminants

Petroleum products (e.g., diesel fuel) typically contain more than 100 chemical compounds (Lee et al. 1992). Due to this complex nature of diesel fuel, naphthalene, a representative polycyclic aromatic hydrocarbon (PAH), and o-xylene, a benzene-toluene- ethylbenzene-xylene (BTEX) compound, were selected as the target model contaminants for the batch sorption tests and column studies. The naphthalene and o-xylene used in this research were purchased from VWR Scientific, Inc. (99% purity).

METHODS

Batch Adsorption Experiments

The batch adsorption technique is one of the most commonly used laboratory test methods for determining adsorption isotherms and estimating partitioning coefficients of geologic materials. The details of batch adsorption tests performed in this study are provided in Demirkan et al. (2007). So, the methodology followed is only briefly summarized here. A series of preliminary batch sorption tests were conducted on fly ash at different solid-to-solution ratios, and a mass ratio of 1/120 that resulted in 50% sorption of naphthalene and *o*-xylene onto the ash was selected. Batch kinetic test results suggested that an equilibrium time of 24-hrs was appropriate for testing of both naphthalene and *o*-xylene. Three 0.5 g-fly ash specimens were placed in 60 mL centrifuge tubes and the naphthalene solution prepared 24 hrs before the test was mixed with the fly ash. Two control specimens (containing naphthalene or o-xylene solution only) were prepared as well. The centrifuge tubes were sealed with Teflon caps and were rotated for 24 hrs using an end-over-end rotator. The agitated samples were then separated by centrifuge at 3000 rpm for 15 min and the supernatant was transferred to a clean vial for analysis.

Column Sorption Experiments

Column sorption tests were performed to study the reaction mechanisms in PSBs constructed with HCCFA. The tests were performed in CHROMAFLEX® glass columns (48 mm in inner diameter and 300 mm in height) to simulate the one dimensional flow of a petroleum hydrocarbon plume through a PSB medium composed of HCCFA and uniform sand. The columns were operated in an up-flow mode. The lower 80 mm of the column was packed with sand only in order to maintain a uniform flow. The 190 mm sorptive medium located above this sand layer was a mixture of 60% sand and 40% HCC fly ash by weight overlain by another 30 mm layer of uniform sand. The medium was tamped into the glass column in 0.5 cm layers to a dry density typical of that expected under field placement conditions following the procedures outlined by Oliviera et al. (1996). Three sampling ports were placed along the column in order to monitor the concentrations during the test. The concentrations at the effluent and influent ports were also periodically monitored. Outer surface of the columns was covered with aluminum foil to eliminate photodecomposition. The top and bottom caps were fabricated from PTFE and all tubing and fittings were Teflon. After placement of the materials, the columns were exposed to gamma irradiation at a dose of 30 kGw for complete sterilization, i.e., to eliminate the bacteriological effects on remediation. A constant flow rate of 60 mL/hr was selected based on literature to simulate a typical groundwater flow and was sent from the bottom using a peristaltic pump.

The influent for the column sorption experiments was synthetic groundwater solution prepared with 10 mg/L naphthalene or 40 mg/L o-xylene.. The synthetic groundwater included 0.1 mg/L $FeSO_47H_2O$, 2 mg/L $MgSO_47H_2O$, 3 mg/L NH_4Cl, and 0.6 mg/L $NaH_2PO_4H_2O$ and 5 μg/L each of $MnCl_2$, Na_2SeO_3, H_3BO_3, $Na_2MoO_42H_2O$, $CoCl_26H_2O$, $NiSO_46H_2O$, $CaSO_45H_2O$, and $ZnSO_47H_2O$ in deionized water. The PIPES (Sigma Chemical Co., 99%) stock buffer solution was prepared in 10 mM, which was adjusted to pH = 6.8 with 4 N NaOH. The synthetic groundwater was autoclaved before spiking with filter sterilized contaminants stock solution. . Before starting the sorption tests, columns were saturated by applying 10 mL/hr flow for four pore volume equivalents.

A series of non-reactive tracer tests were performed in the same columns to define the porosity and hydrodynamic dispersion coefficient for each column. The influent for the tracer tests was 1000 mg/L sodium bromide (NaBr) solution prepared using synthetic groundwater at a concentration approximating the ionic strength of infiltrating water in the field. The breakthrough curves from tracer tests were evaluated using an absolute non-linear least-square regression to obtain the best fit estimates for the pore water velocity and hydrodynamic dispersion coefficient values. Upon completion of the tracer tests, the column specimens were subjected to constant head hydraulic conductivity tests.

A Shimadzu 5301 Fluorescence Spectrophotometer was utilized to measure the contaminant concentrations from batch adsorption and column sorption experiments. The excitation and emission wavelengths for naphthalene and o-xylene in aqueous solution were 273 and 336 nm, and 267 and 289 nm, respectively. The samples from tracer tests were analyzed using bromide electrode (Cole Parmer C-27502-05).

RESULTS

Batch Adsorption Tests Results

Batch adsorption tests were conducted on three HCCFA at three different carbon contents (3.1 %, 10.7 %, and 20.5 % for MT, PS, and DP fly ashes respectively) using naphthalene and o-xylene. In analyzing batch adsorption test data, it is common to relate the mass of sorbed contaminant per mass of sorbent (q_i, mg/kg) to the final aqueous equilibrium concentration of the contaminant in solution (C_f, mg/L) by using sorption isotherm models. The adsorption test data collected in this study were analyzed using a nonlinear regression analysis to fit various isotherm models to the data. The best fit was determined through observing the highest value of R^2 and the lowest standard error of the regression.

The best fit adsorption isotherm models are summarized in Table 1. Adsorption of naphthalene onto the all fly ashes is best described by the Freundlich isotherm, as evidenced by the relatively high R^2 values and low standard errors of regression. Similarly, the tests with o-xylene yielded a Freundlich-type isotherm for the MT fly ash; however, a Langmuir-type was observed for PS and DP fly ashes. The Langmuir sorption isotherm is generally the preferred model for describing the sorption of chemicals on a solid surface that contains a fixed number of identical active sorption sites (Bartelt-Hunt et al. 2005).

Table 1. Adsorption isotherms parameters from batch adsorption tests

Fly ash	Naphthalene		*o*-xylene	
MT	K_F (L/kg)[a]	102	K_F (L/kg)[a]	46.41
	n [a]	0.3062	n [a]	0.5275
	R^2	0.99	R^2	0.88
	RMSE [c]	4.58	RMSE [c]	40.95
DP	K_F (L/kg)[a]	696.1	Q_{max} (g/kg)[b]	875.8
	n [a]	0.3389	K_L (L/kg) [b]	0.6442
	R^2	0.97	R^2	0.81
	RMSE [c]	70.16	RMSE [c]	76.89
PS	K_F (L/kg)[a b]	1506	Q_{max} (g/kg)[b]	1983
	n [a]	0.2073	K_L (L/kg) [b]	0.3957
	R^2	0.98	R^2	0.92
	RMSE [c]	77.53	RMSE [c]	155.6

[a] Freundlich Isotherm: ($q_i=K_F\,[C_f]^n$) K_F = Freundlich equilibrium isotherm constant (L/kg), n = dimensionless empirical, C_f = final concentration of naphthalene (mg/L). [b] Langmuir Isotherm: ($q_i=(Q_{max}K_LC_F)/(1+K_LC_F)$) Q_{max} = sorption capacity of particular solid (mg/kg), K_L = Langmuir isotherm coefficient (L/kg), [c] RMSE: Root mean squared error PS=Paul Smith, DP=Dickerson Precipitator, MT=Morgantown fly ash.

The batch adsorption tests were also used to investigate the effect of carbon content on the sorption capacity of the particular fly ash. Based on the isotherm analysis it is observed that the fly ashes with high carbon content have a higher Freundlich isotherm coefficient (K_F) for naphthalene.

Column Sorption Experiments

Results of tracer analysis from the column experiments are given in Table 2. The values listed in Table 2 are calculated by averaging the data determined from the ports located in the fly ash medium (ports B and C in Figure 1).

Lee et al. (2004) reported similar dispersion coefficients when tested excess foundry systems sands for their possible use as a PRB medium. The measured pore water velocities are also comparable with the reported field velocities (Mercer and Cohen 1990). Relatively high hydraulic conductivity values suggest that flow is not impeded in the fly ash medium, and the PSBs are likely to maintain their high permeability in field applications.

Table 2. Advection-dispersion parameters from non-reactive tracer tests

Fly ash	Porosity	Hydrodynamic dispersion coefficient (cm^2/hr)	Pore water velocity (cm/hr)	Hydraulic conductivity (cm/s)
MT	0.31±0.01	2.02±0.64	6.53±1.23	1.1×10^{-4}
PS	0.34±0.02	0.58±0.44	6.56±0.39	1.8×10^{-4}
DP	0.42±0.02	1.02±0.51	5.27±0.53	1.5×10^{-4}

PS=Paul Smith, DP=Dickerson Precipitator, MT=Morgantown fly ash

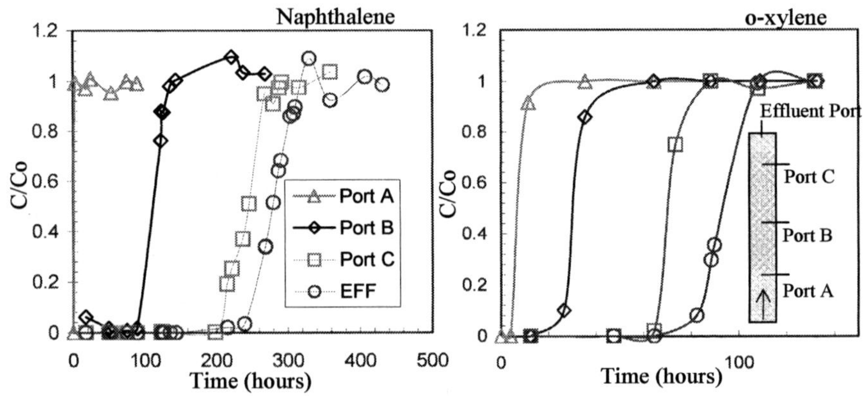

Figure 1 Breakthrough curves from column sorption test with PS fly ash sorptive medium ▨ :Sand ▨ : Fly Ash Medium (60% sand-40% fly ash)

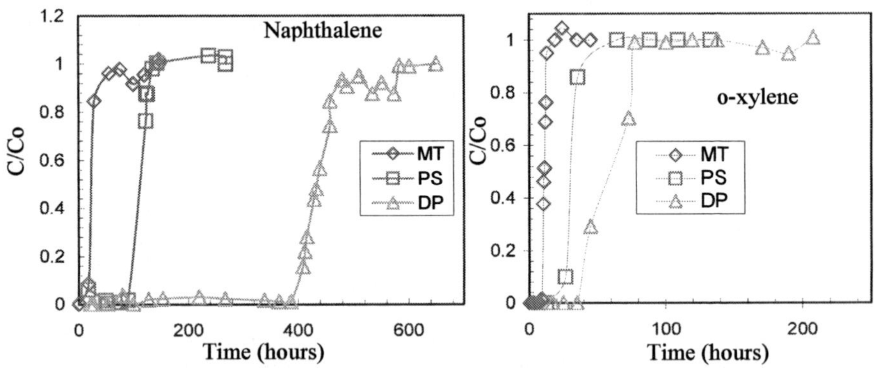

Figure 2 Naphthalene sorption desorption tests breakthrough curves from Port B. MT: Morgantown, PS: Paul Smith, DP: Dickerson Precipitator fly ash.

A series of breakthrough curves are plotted for PS in Figure 1. As evidenced from measurements at port A, the sand exhibits plug flow type behavior due to its very low naphthalene adsorption capacity. It is well known that the hydrophobicity of the contaminant has a great influence on its environmental fate (Luthy et al 1999). This effect can be seen in Figure 1, when the breakthrough curves for the two contaminants are compared. Relatively greater hydrophobicity (log K_{ow}) of naphthalene (3.29 versus 3.12) resulted in longer retention times for the contaminant.

One of the goals of this study was to investigate the effect of carbon content on sorption during column tests. This effect can be seen by comparing the contaminant concentrations at the same sampling locations. Figure 2 depicts the breakthrough curves based on naphthalene and o-xylene concentrations at Port B. Relatively higher unburned carbon content of Dickerson Precipitator fly ash (20.5%) has resulted in longer retention times due to its higher naphthalene or o-xylene adsorption capacity.

CONCLUSIONS

There is a growing interest in using sustainable technologies in geoenvironmental remediation and as a result, recycled materials are increasingly being more incorporated into design. One of these recycled materials is high carbon content fly ash (HCCFA). Laboratory batch adsorption and column sorption tests conducted on three HCCFAs collected from Maryland showed that the material has significant organic contaminant sorption capacity at both equilibrium and non-equilibrium conditions. The collected data indicated that the HCCFA provides good hydraulic and adsorption properties which may justify its potential use as a permeable sorptive barrier material in remediation of groundwater contaminated with petroleum residues. However, additional research is required to study the rate and amount of desorption of the contaminants, leaching of inorganic compounds from the HCCFA, and effect of biodegradation on the overall process. Moreover, future tests with commercially available powder activated carbon will provide a benchmark to evaluate the effectiveness of HCCFA as a sorbent in these applications.

REFERENCES

Bartelt-Hunt, S.L., Smith, J.A., Burns, S.E., and Rabideau, A.J. (2005). "Evaluation of granular activated carbon, shale, and two organoclays for use as sorptive amendments in clay landfill liners." *J. of Geotechnical and Geoenvironmental Eng.*, 131 (7): 848-856.

Banerjee, K., Cheremininoff, P.N., and Cheng, S.L. (1997). "Adsorption kinetics of o-xylene by fly ash." *Water Research*, 31(2):249-261.

Demirkan, M.M., Seagren, E., and Aydilek, A.H. (2007). "Reuse of fly ash-amended contaminated soils in highway embankments." *Journal of Transportation Research Board*, 1975:104-111.

Guerin, T.F., Horner, S., McGovern, T., and Davey, B. (2002). "An application of permeable reactive barrier technology to petroleum hydrocarbon contaminated groundwater." *Water Research*, 36: 15-24.

Janos, P., Buchova, H., and Ryznarova, M. (2003). "Sorption of dyes from aqueous solutions onto fly ash." *Water Research*, 37: 4938-4944.

Jenk, U., Schreyer, J., and Klinger, C. (2003). "Fe(0)/lignitic coal: an efficient and mechanically stable reactive material for purification of water containing heavy metals, radionuclides, and nitroaromatics." *Environmental Science and Technology*, 37(3):644-651.

Kim, J.Y., Edil, T.B., and Park, J.K. (1997). "Sorption of organic compounds in the aqueous phase onto tire rubber." *J. of Environmental Management*, 123(9):827-835.

Konstantinou, I. K. and Albanis, T. A. (2000). "Adsorption – desorption studies of selected herbicides in soil – fly ash mixtures." *J. Agric. Food Chem.,* 48: 4780-4790.

Lee, S. L., Hagwall, M., Delfino, J. J., and Rao, P.S.C. (1992). "Partitioning of polycyclic aromatic hydrocarbons from diesel fuel into water." *Environmental Science and Technology,* 26: 2104-2110.

Lee, T., Benson, C.H., and Eykholt G.R. (2004). "Waste green sands as reactive media for groundwater contaminated with trichloroethylene (TCE)." *Journal of Hazardous Materials*, 109(1-3):25-36.

Luthy, R.G., Aiken, G.R., Brusseau, M.L., Cunningham, S.D., Gschwend, P.M., Pignatella, J.J., Reinhard, M., Traina, S.J., Weber, Jr.W.J., and Westall, J.C. (1997). "Sequestration of hydrophobic organic contaminants by geosorbents." *Environmental Science and Technology*, 31(12):3341-3347.

Mercer, J.W. and Cohen. R.M. (1990). "A review of immiscible fluids in the subsurface properties, models, characterization and remediation." *J. of Contaminant Hydrology*, 6: 107 – 163.

Nollet, H., Roels, M., Lutgen, P., Van der Meeren, P., and Verstraete W. (2003). "Removal of PCBs from wastewater using fly ash." *Chemosphere*, 53: 655-665.

Oliviera, I.B. Demond A.H. and Salehzadeh A. (1996). "Packing of sands for production of homogenous porous media" *Soil Science Society of America. Journal*, 60: 49-53.

Petzrick, P. (2001). "The use power plant coal combustion products in Maryland." *Proceedings of International Ash Utilization Symposium*, Paper #51, pp. 7, Lexington, Kentucky.

Schad, H. and Growohl, P. (1998). "Funnel and gate systems for in-situ treatment of contaminated groundwater at former manufactured gas sites." *Proceedings of NATO/CMS Phase III Meeting*,Vienna, Austria

USEPA, (2006). Performance evaluation of organic carbon + limestone based PRB for treatment of Pb and acidity, (http://www.epa.gov/ada/research/waste/research_02.pdf).

Wang, S., Li, L., Wu, H., and Zhu, Z. (2005). "Unburned carbon as a low-cost adsorbent for treatment of methylene blue-containing wastewater." *Journal of Colloid and Interface Science*, 292: 336-343.

Woinarski, A.Z., Snape, I., Stevens, G.W., and Stark, S.C. (2003). "The effects of cold temperature on copper ion exchange by natural zeolite for use in a permeable reactive barrier in Antarctica." *Cold Regions Science and Technology*, 37: 159-168.

A LABORATORY COLUMN STUDY FOR ADSORPTIVE REMOVAL OF CADMIUM BY FLY ASH

K. Bandyopadhyay
Department of Construction Engineering, Jadavpur University (2nd Campus), Kolkata, India, 700098,kb@const.jusl.ac.in,ph-(033)2335-5211,fax-(033)2335-7254
B.Thakur
Department of Civil Engineering, Meghnad Saha Institute of Technology, Kolkata, India, 700107.
A. K. Misra
Department of Civil Engineering, Jadavpur University, Kolkata, India-700032.
C.Goswami and D.Chaudhuri
Department of Construction Engineering, Jadavpur University (2nd Campus), Kolkata, India, 700098,chandrima_goswami@const.jusl.ac.in, devaleena_chaudhuri@const.jusl.ac.in,
ph-(033)2335-5211,fax-(033)2335-7254

ABSTRACT: The use of heavy metals and the concomitant adverse effects has increased to a great extent over the recent times. Cadmium is one of the most harmful toxic metals. This metal is being increasingly consumed in the industries like nickel-cadmium batteries, pigments, and stabilizers for PVC, electroplating and other coatings, metallurgical, electronics, petroleum and coal, pulp and paper, fertilizers, photography, etc. Its pollution potential, has therefore given rise to the necessity of removing the metal from wastewater.

On the other hand fly ash, which is produced in huge quantities even in a medium sized coal-fired thermal power plant, is a major cause of concern with regard to its safe disposal.

Findings of a few earlier investigations in this regard have lent support to the belief that cadmium-bearing wastewater could possibly be safely disposed of on a land filled-up with fly ash. The fly ash would adsorb cadmium and hence reduce the problem of ground water pollution due to leaching. In the present investigation a column study experiments has been carried out to evaluate performance of fixed bed adsorption columns made of fly ash exposing them to cadmium bearing wastewaters. The gathered characteristic information from the pertinent breakthrough curve can be used as a design rationale for a fly ash bed as a liner overlying a clay mantle for safe disposal of cadmium bearing wastewater. The leaching behaviour of exchangeable cations from the fly ash column has also been studied.

INTRODUCTION

The use of heavy metals in different industrial processes as well as the risk of metal pollution has increased to a large extent over the last few decades.

Several noxious effects have been attributed to cadmium, the target metal of the present study, in living systems, e.g. isomorphic substitution of zinc (II) (in enzymes, with their consequent damage) and of cadmium (in bone tissues, making them fragile). Cadmium is known to get accumulated in the kidneys causing its malfunctioning at overdoses spilling proteins in the urine and disrupting protein metabolism. Consumption of rice containing high concentrations of cadmium led to a surge in the Itai-Itai disease in Japan in 1955 (Srivastava et al., 2006).

The options available to reduce this metal pollution problem are either removal and/or recovery methods, or the use of substitutes of metallic compounds. Conventional removal methods may be ineffective or extremely expensive for treating effluents with low heavy metal concentrations. The continual search for a new economically viable but effective contaminant removal method has revealed in its wake that adsorption of metal ions present in low levels on to the solid can prove to be a cost-effective and better option for treating industrial effluents and wastewater streams (Agarwal et al., 2006).

Low cost adsorbents including industrial and agricultural wastes, solid wastes, fly ash, leaves, bones, shells, hairs, soils, etc. have been identified as economical substitutes to the activated carbon (Rao et al., 2006).

On the other hand, the disposal of huge quantity of fly ash generated from thermal power plants is a burning problem nowadays. The accumulated fly ash in the year 2003 over India was about 100 million tons which is expected to be more than 150 million tons by the year 2010. Thus fly ash management would remain an important source of natural concern. Researchers have been attempting at conversion of this waste into wealth by exploring viable avenues for fly ash management. One of such possible uses can be its use as a low cost adsorbent material to remove various kinds of pollutants from water and wastewater (Kumar et al., 2004).

Findings of a few earlier investigations by the present authors and others in this regard have lent support to the belief that cadmium- bearing wastewater could possibly be safely disposed of on a land filled-up with fly ash. The fly ash would adsorb cadmium and hence reduce the problem of groundwater pollution due to leaching. (Thakur et al. 2004)

In the present investigation a column study experiment was carried out to evaluate performance of a fixed bed adsorption column, and also to gather some characteristic information from the pertinent breakthrough curve.

MATERIALS AND METHODS

In the present investigation, fly ash was procured from National Thermal Power Corporation (NTPC) thermal power plant, Farakka, West Bengal. The study was divided into two main parts. In the first part some characteristics of the fly ash were determined. In the second part the column adsorption study was carried out to evaluate performance of a fixed bed adsorption column, and also plot the pertinent breakthrough curve. Here the column study experiment is discussed.

The column study was carried out using the experimental set up as shown in Figure 1. At first 2.50 kg of unexposed, oven-dried fly ash was compacted in the permeability mould at optimum moisture content (O.M.C.) of 20.00% and maximum dry density (d_{max}) of 1.467 g/cc. This acted as the adsorbent column.

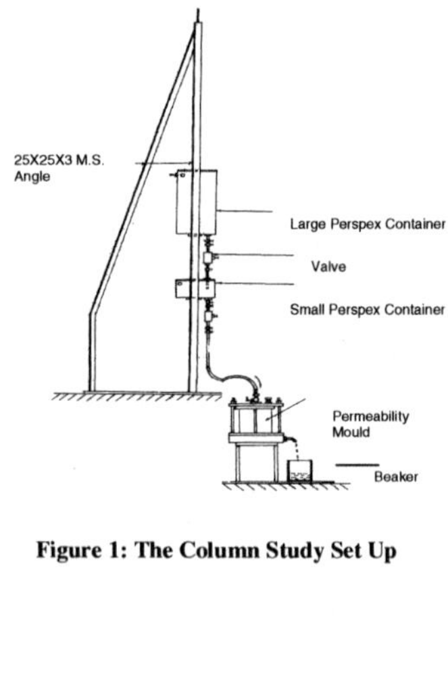

Figure 1: The Column Study Set Up

In an earlier column study by the same authors, the cadmium concentration of the adsorbate solution was chosen to be 35 mg/L and consequently very high adsorption was observed with the exhaustion point of the fly ash column been far from reach (Thakur et al. 2004). Therefore, in the present study the cadmium concentration of the adsorbate solution has been increased to 350 mg/L. pH of this solution was 1.03. Sufficient volume of this solution was placed in the large perspex container and with the valve it was allowed to fill the small perspex container upto a definite level. The level of adsorbate solution in the latter was maintained by a by-pass pipe so that any excess solution would overflow through it. The top of the permeability mould was connected with the small perspex container with a polythene tube. By placing the permeability mould at a definite lower level a constant hydraulic gradient (i=h/L) of 12 was maintained throughout the experiment.

The experiment was started by opening the top valve and bottom tap of the permeability mould so that the adsorbate solution can percolate through the adsorption column. A beaker was placed underneath the tap to collect the effluent coming out of the adsorption column. Once the percolation started, at first the bed got saturated with the contaminant solution and then the effluent started coming out of the column. This was collected and the collection volume was measured periodically at an interval of two hours. Five such collections were made during the first day. After that the percolation was stopped for 13 hours by closing the top valve of the permeability mould. During the second day four such effluent collections were made. The percolation was again stopped for 13 hours. During the third day three such effluent collections were made and then the experiment was stopped. A total of twelve periodic collections were made during the entire experiment for a total period of seventy five hours. The collected samples were measured for the collection volume and pH. The coefficients of permeability of the column for these eighteen collections were also calculated. The effluent samples were analysed for the concentration of three cations viz. cadmium (Cd^{2+}, by Atomic Absorption Spectrometer), Sodium (Na^{+}, by Flame Photometer) and Potassium (K^{+}, by Flame Photometer).

RESULTS AND DISCUSSIONS

Geotechnical Properties of the Unexposed Fly Ash

The specific gravity of the unexposed fly ash sample was found to be 2.06.

The grain size distribution study shows that the fly ash contains 12.75% of clay, 80.50% of silt, and 6.75% of fine sand.

From proctor compaction test, the optimum moisture content (O.M.C.) and maximum dry density (d_{max}) of the unexposed fly ash sample was found to be 20.00% and 1.467 g/cc respectively.
The permeability of the unexposed fly ash sample was found to be 6.63 X 10^{-6} cm/s.

Chemical Properties of the Unexposed Fly Ash
The pH of the unexposed fly ash sample was found to be 7.35.

The Cation Exchange Capacity (CEC) of the unexposed fly ash was found to be 6.25 meq/100g of fly ash. The Exchangeable Cations are Sodium, Potassium and Calcium and their amounts are 2.03 meq/100g of fly ash, 1.44 meq/100g of fly ash and 2.61 meq/100g of fly ash respectively.

Results of the Column Study Experiment
Five parameters viz. column permeability, effluent pH, and Cd^{2+}, Na^{+}, K^{+} were determined and tabulated with the cumulative volume of outflow through the adsorption column and shown in Table 1.

Table 1: Results of the Column Adsorption Study

Day Of Expt.	Cum. Vol. Of Outflow Σ(L)	Cd^{2+} mg/L	Na^{+} mg/L	K^{+} mg/L	pH	Co-Eff. Of Permea Bility *10X^{-6} cm/s
1st Day	0.0000	350	0	0	-	-
	0.0610	BDL	90.17	33.67	7.84	8.9893
	0.1160	0.03	72.14	56.12	8.31	8.1051
	0.1700	0.06	72.14	33.67	8.37	7.9577
	0.2210	10.00	72.14	56.12	8.40	7.5156
	0.2637	31.00	90.17	56.12	8.37	6.2881
Experiment was Stopped for 13 Hours						
2nd Day	0.3137	70.00	180.34	112.25	7.51	7.3683
	0.3357	100.00	306.58	145.92	5.71	3.2420
	0.3507	100.00	342.65	179.59	5.11	1.1052
	0.3673	120.00	324.62	190.82	4.81	1.6308
Experiment was Stopped for 13 Hours						
3rd Day	0.3735	150.00	324.62	190.82	4.84	0.4568
	0.3791	160.00	324.62	202.43	4.88	0.2751
	0.4815	250.00	324.62	213.27	2.80	1.7753

Variation in Cd^{2+} concentration
The effluent Cd^{2+} concentrations were plotted against the cumulative volume of outflow as in Figure 2.

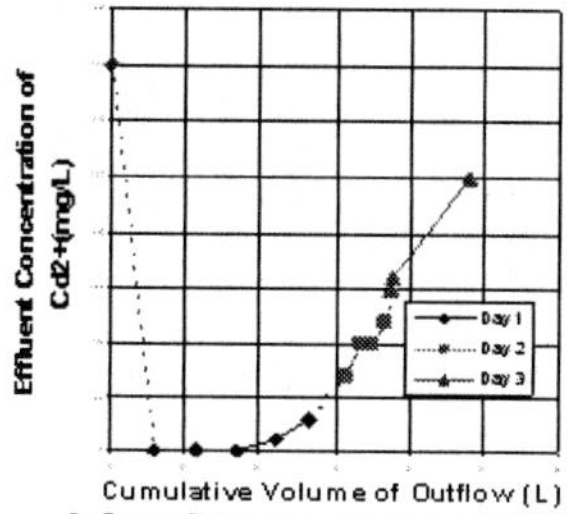

Influent Cd+2 concentration = 350.00mg/l
Figure 2: Effluent Cd^{2+} Concentration vs. Cumulative Volume of Outflow

It is evident from the plot in Figure 2 that high adsorption of Cd^{2+} has taken place throughout the entire period of the experiment The initial breakpoint can be located at a cumulative volume of outflow of 0.1700 L. The satur5ation break point is yet to reach and to locate it properly experiments with longer duration is required.

Variation in Na^{+} and K^{+} Concentration
The effluent Na^{+} concentrations were plotted against the cumulative volume of outflow as in Figure 3.

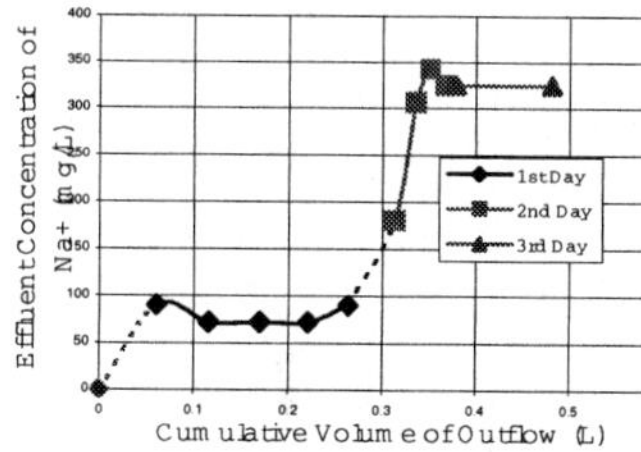

Figure 3: Effluent Na^{+} Concentration vs. Cumulative Volume of Outflow (Influent Na^{+} concentration=0.00 mg/L)

The effluent K^{+} concentrations were plotted against the cumulative volume of outflow as in Figure 4.

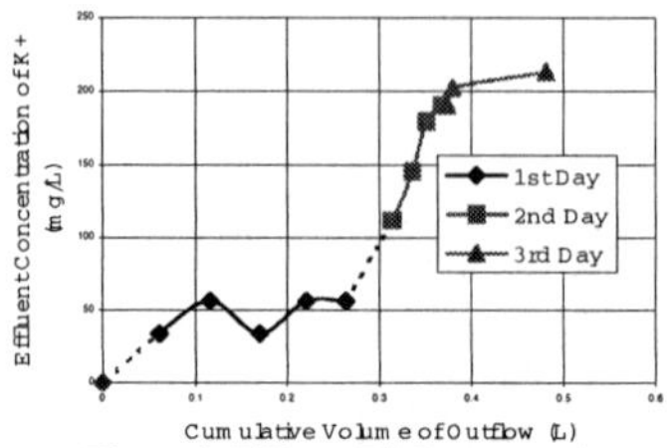

Figure 4: Effluent K^+ Concentration vs. Cumulative Volume of Outflow(Influent K+ Concentration = 0.00mg/L)

Both of the above plots show a common feature of sudden increase in the effluent concentrations of Na^+ and K^+ ions at the start of the experiment, from their zero influent concentrations. As the experiment progressed there was a gradual increase in both concentrations coming to almost constant values in third day. This can be explained by the constant exchange of these exchangeable cations of the fly ash mass with the Cd^{2+} ions of the adsorbate solution by adsorption during the experiment. As time progressed the fly ash column neared exhaustion of these exchangeable cations and ultimately came to a constant value.

Variation in pH

The effluent pH values were plotted against the cumulative volume of outflow as in Figure 5.

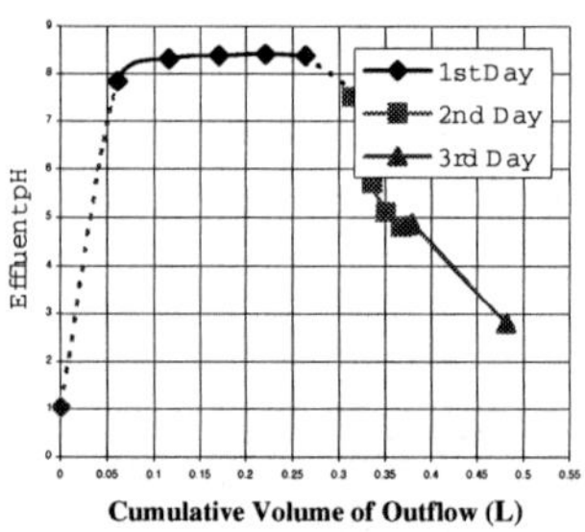

Figure 5: Effluent pH vs. Cumulative Volume of Outflow(Influent pH=1.03)

The pattern of variation in the pH values of the effluents can be explained similarly. Sudden initial increase from the influent pH of 1.03, more or less constant pH values round about 8 during the middle and again sharp decrease at the end can be explained by the constant leaching of alkaline cations like Na^+ and K^+ in the effluent from the fly ash mass of the adsorption column during the experiment. Towards the end the column gets exhausted of these alkaline cations and the pH value falls back towards the initial level.

Variation in Co-efficient of Permeability of the Adsorption Column

The co-efficient of permeability (K) values of the fly ash column were plotted against the cumulative volume of outflow as in Figure 6. The initial increase can be explained by the washing out of some fly ash from the column causing some increase in the K value for a short period. Then the gradual decrease in the K values during the experiment can be explained by the gradual filling of the pore spaces of the fly ash mass of the adsorption column by the adsorbed Cd^{2+} ions.

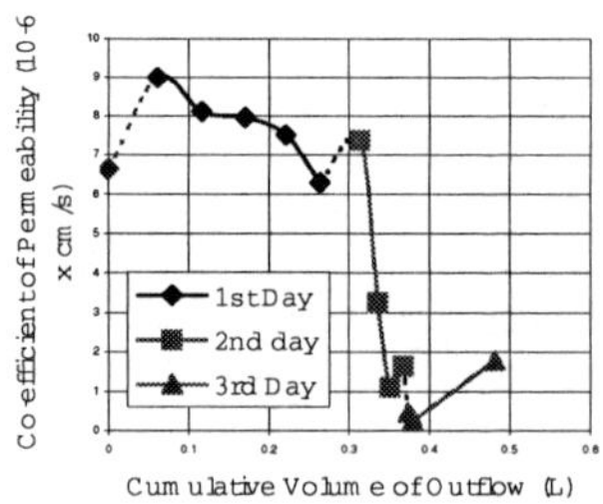

Figure 6: Co-efficient of Permeability vs. Cumulative Volume of Outflow

CONCLUSION

From the present study, it can be concluded that fly ash bed may be very effectively utilized for renovation of cadmium bearing wastes. However actual field experiments should be conducted to design the bed to attain desired level of cadmium in the effluent for a considerable duration of time.

REFERENCES:

Rao, M.M., Ramesh, A., Rao, G.P.C., Seshaiah, K., (2006). "Removal of Copper and Cadmium from the Aqueous Solutions by Activated Carbon Derived from *Ceiba Pentranda* Hulls", Journal of Hazardous Materials, B129, pp 123-129.

Thakur, B., Bandyopadhyay, K., Misra A.K., and Som, N.N., (2004). "Fixed Bed Column Study with Fly Ash as Adsorbent for Evaluation of its Cadmium Removal Potential", Proceedings, IGC-2004, pp 549-552.

Remediation of Lead Contaminated Gypsum Sludge

Raju G.C[1], Sudhakar M Rao[2] and B.V. Venkatarama Reddy[3]

[1]Research Student, Dept. of Civil Engineering, Indian Institute of Science, Bangalore-560 012, India; gcr@civil.iisc.ernet.in
[2]Professor, Dept. of Civil Engineering & Associate Faculty, Center for Sustainable Technologies, Indian Institute of Science, Bangalore-560 012, India; msrao@civil.iisc.ernet.in
[3]Associate Professor, Dept. of Civil Engineering & Associate Faculty, Center for Sustainable Technologies, Indian Institute of Science, Bangalore-560 012, India; venket@civil.iisc.ernet.in

ABSTRACT: Characterization of industrial waste materials and by-products is a crucial step in their assessment for reuse or for their safe disposal in environmental friendly and economical manner. The objective of the present study is to examine the possibility of using washing technique to remediate lead contaminated gypsum sludge. The sludge is produced by neutralization of acidific effluent by lime addition in the effluent treatment plant (ETP) of lead-acid storage battery. Chemical forms of the lead contaminants are determined by sequential extraction. Since, acids and chelating agents are recognized to the most effective washing agents, the present study attempts remediation of lead contaminated gypsum sludge by washing the hazardous waste with hydrochloric acid (HCl) and ethylenediaminetetraacetic acid (EDTA) solutions. The results of washing experiments showed that EDTA is more efficient than HCl in lead extraction. Further, based on its high lead removal efficiency and lesser dissolution of solids, 0.1 M EDTA is considered as a most effective washing agent for removing lead from contaminated gypsum sludge.

INTRODUCTION

Hazardous wastes are solid, semi-solid or non-aqueous liquids that exhibit characteristics of corrosivity, reactivity, ignitability, toxicity and infectious property (Granados et al., 1995; Reinhart, 1993). Industries are major sources of generation of hazardous wastes (Reinhart, 1993; Polprasert and Liyanage, 1996; Misra et al., 2005; Sandya et al., 2006). Electroplating units, petroleum refineries, tanneries, thermal power plants, pulp and paper, chemical, textile, pesticides, paints, and storage batteries industries are some major producers of hazardous waste substances (Misra et al., 2005; Polprasert and Liyanage, 1996). These industries generate hazardous wastes containing unsafe levels of inorganic and organic contaminants. Recycling of hazardous wastes appears in top of the hierarchy of priorities in hazardous waste management. Common recycling methods include, re-use and treatment. Hazardous wastes are commonly treated by solidification/stabilization process. This technique uses lime, cement, fly ash, clay and blast furnace slag to bind the contaminant metals in the cemented matrix (Leist et al., 2003; Singh & Pant, 2006). Soil washing is another common technique for treatment/remediation of soils contaminated with organics and heavy metals. In conventional soil washing process, the excavated soil is vigorously mixed with a solution that leaches contaminants from the waste matrix. Washing usually employs, acids, bases, chelating agents, alcohols or other additives (Griffiths et al., 1995; Sandhya et al., 2006). However, the most efficient washing agents are acids and chelating agents (Neilson et al., 2003; Peters, 1999). Solids content, reaction time and pH are key factors in acid washing (Lo and Chen, 1990). Acid washing involves solubilization of heavy metals in an acid environment as shown in Eq.1 (Lo and Chen, 1990)

$$\text{Sludge-M} + \text{Acid} \longrightarrow \text{Sludge} + M^{z+} + \text{Acid} \qquad (1)$$

Strong chelating agents such as EDTA, function by complexing heavy metals to form EDTA- metal complexes. Theoretically, one mole EDTA can remove one mole of heavy metals and the reaction can be expressed as in Eq.2 (Lo and Chen, 1990).

$$\text{Sludge-M} + \text{EDTA} \longrightarrow \text{Sludge} + \text{EDTA-Metal Complex} \qquad (2)$$

The present study employs the washing technique to remediate lead contaminated gypsum sludge produced by neutralization of acidific effluent by lime addition in the effluent treatment plant (ETP) of lead-acid storage battery. Chemical forms of the lead contaminants are determined by sequential extraction. Since, acids and chelating agents are recognized to the most effective washing agents, the present study attempts remediation of lead contaminated gypsum sludge by washing the hazardous waste with HCl and EDTA solutions.

MATERIALS AND METHODS

Preparation of Sample

Lead contaminated gypsum sludge was supplied by a lead-acid storage battery industry near Bangalore, India. The sludge sample had average moisture content of 135%. The samples were homogenized and air-dried at 50° C prior to characterization tests. The air-dried mass was pulverized to pass through 1.18 mm Indian Standard sieve and stored in air-tight polythene bags.

Water Leach Test

In the water leach test (ASTM D 3987-85), the air-dried sample is agitated with distilled water for 18 hours (solid to solution ratio = 1:20). After 18 hours, the slurry is filtered and the filtrate is analyzed for dissolved ions such as lead, calcium, magnesium, sodium, potassium, sulfate, chloride and bicarbonate ions using standard test methods.

TCLP Test

Toxicity characteristic leaching procedure (TCLP-Method 1311) test of USEPA, 1996 was performed with air-dried sludge. The mobilization of toxic metals released from wastes is measured using a variety of risk assessment procedures including the Toxicity Characteristic Leaching Procedure (TCLP), among others. The TCLP test is designed to simulate leaching of heavy metals and organics from industrial wastes to be disposed in landfills. The TCLP test measures the possible leachable substances of certain toxic constituents from solid waste under a specified set of laboratory conditions. In the TCLP test, the lead contaminated gypsum sludge was agitated with an acetate buffer (pH= 4.93) solution for 18 hours (solids: solution ratio=1:20). After 18 hours, the slurry is filtered. The leachate is acidified to a pH< 2 (Al-Abed et al., 2006) and analyzed for lead concentration using atomic absorption spectrophotometer.

Estimation of Total Lead Content in Contaminated Gypsum Sludge

Nitric acid digestion was employed to estimate total lead content in the air- dried sludge as per ASTM D 5198-92. In this procedure, 5 g of air-dried sludge was mixed with 25 ml of 1:1 nitric acid: distilled water solution and heated at 90ºC for 2 hours. The mix was cooled to room temperature, filtered and diluted with distilled water to a volume of 200ml. The diluted solution was analyzed for total lead concentration in the sludge sample using atomic absorption spectrophotometer.

Chemical Analysis of Lead Contaminated Gypsum Sludge

The chemical analysis of lead contaminated gypsum sludge was performed using IS; 1288; 1982 procedure. Known mass of sludge (0.2g) is thoroughly mixed for 10 minutes with 50ml of dilute hydrochloric acid (1:250, solid: solution ratio) and filtered. The filtrate was diluted to 200ml. The diluted filtrate was boiled for 2 hours, cooled to room temperature and precipitated with 12% barium chloride solution. The precipitate was filtered and the filter paper was ignited. The residue was weighed and is used to calculate gypsum [(Calcium Sulphate Dihydrate ($CaSO_4.2H_2O$)] present in the sludge sample.

The amounts of silica and acid insolubles present in the sludge were determined by the following procedure. 50 ml of diluted hydrochloric acid was added to 1g of air-dried sludge sample in a porcelain dish and evaporated on a sand bath. Concentrated hydrochloric acid solution was added to wet the residue followed by addition of distilled water. The acidified mix was boiled and filtered. The precipitate contained in filter paper was ignited and weighed to calculate the amount of silica present in the air-dried sludge.

The filtrate from the previous step was used to estimate the amount of iron and aluminium oxides present in sludge sample. The filtrate was acidified with concentrated nitric acid and boiled to ensure oxidation of iron. The boiled solution was cooled to room temperature. Ammonium chloride, methyl red indicator and dilute ammonium hydroxide solution were added to the cooled solution. The solution was once again boiled for a few minutes to facilitate precipitate coagulation. The solution was filtered and the precipitate was washed with hot ammonium nitrate solution until the washings were free from chlorides. The filter paper was ignited and the mass of the residue was used to calculate the amount of iron and aluminium oxides in the sludge sample. The filtrate from previous step was analyzed for dissolved calcium concentration by the EDTA titration method. The calcium concentration in the filtrate was used to calculate the amount of calcium oxide present in the sludge sample.

Sequential Extraction Test

In the sequential extraction process, the sludge was agitated with consecutive extractants for a fixed time interval (Figure1). A solid: liquid ratio of 1:8 was used in the extraction tests. After extraction, the slurry was filtered and the filtrate was acidified (< 2 pH) and analyzed for lead concentration using atomic absorption spectrophotometer. The sludge retained on the filter paper at the end of step 1 was subjected to the next extraction (step 2) using extractant number 2. The procedure outlined in step 1 was repeated till the last extrantant was reached.

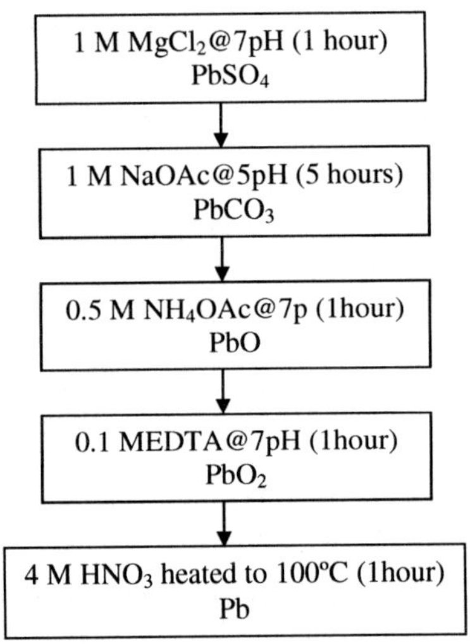

FIG.1 Sequential extraction protocol.

Sludge Washing Experiments

The sludge samples were washed using HCl and EDTA in batch experiments. 0.01 M, 0.1 M and 0.5 M HCl solutions and 0.01 M, 0.1 M and 0.2 M EDTA solutions were used as washing agents in the batch experiments. Preliminary experiments revealed that a solid: liquid ratio of 1:20 was most effective in lead extraction by HCl solutions; comparatively, a solid: liquid ratio of 1:10 was most effective in lead extraction by EDTA solutions. In a typical batch experiment, the sludge was agitated with required concentration of HCl solution (S/L ratio 1:20) or EDTA solution (S/L ratio 1:10) for 18 hours using a mechanical shaker at 120 revolutions/minute. At the end of equilibration, the slurry was filtered and filtrate was acidified and analyzed for lead concentration. Concomitantly residue retained on filter paper was dried at 50°C and weighed to calculate the amount of sludge dissolution by the washing procedure. The weighed mass was used to calculate the percentage dissolution of sludge by the washing procedure using the formula:

$$\frac{Initialmas\ sofsludge - massofslud\ geafterwas\ hing}{Initialmas\ sofsludge} \times 100\% \qquad (3)$$

RESULTS AND DISCUSSION

Chemical & Mineralogical Characterization of Lead Contaminated Gypsum Sludge

Table 1. Chemical Composition of Air-Dried Sludge

Constituent	% Weight
Calcium sulfate dihydrate	90
Silica	0.9
Iron & Aluminium oxide	1.4
Calcium oxide	5.7
Magnesium oxide	2.7
pH	9.01
Total Dissolved Salts (mg/l)	2616

Table 2. Water Leach Test on Air-Dried Sludge

Constituent	Conc. (mg/l)
Lead	0
Sulfate	908
Calcium	720
Magnesium	49
Potassium	2
Sodium	39
Chloride	840
Bicarbonates	58

Table 1 presents chemical compositions of the sludge sample. The sludge is predominantly (90%) constituted by calcium sulfate dihydrate ($CaSO_4.2H_2O$) and lesser amounts of calcium oxide (5.7%), magnesium oxide (2.7%), iron and aluminium oxide (1.4%), and silica (0.9%) respectively. Data in Table 1 also shows that the sludge has an alkaline pH of 9.01. Ordinarily calcium sulfate dihydrate ($CaSO_4.2H_2O$) has slightly acidic pH of 5.7. The alkaline nature of the lead contaminated gypsum sludge arises due to the presence of unreacted lime in the sludge (Table 1). The sludge sample has total dissolved salt concentration of 2616 mg/l that is mainly contributed by calcium, magnesium, potassium, sodium, sulfate, chloride and bicarbonate ions.

Lead Leachability of Contaminated Sludge

Table 3 presents the results of water leachability test, TCLP test and total lead content estimation in the sludge sample. Leaching with distilled water did not extract any lead from the sludge sample. However, the TCLP leachate was characterized by lead concentration of 10 mg/l that exceeds the TCLP limit of 5 mg/l (USEPA, SW 846). Nitric acid digestion of lead contaminated sludge sample revealed total lead content of 2960 mg/ kg of air-dried sludge. The total lead content in the gypsum sludge is less than the prescribed limit (5000 mg/kg) for hazardous wastes (Ministry of Environment and Forests, New Delhi in HWRules-2003). Calculation showed that the acetate buffer (TCLP extraction fluid #1) could only leach 6% of the total lead content from the sludge.

Table 3. Lead Leachability Test Results

Test	Lead Concentration in Leachate mg/l	Lead Concentration in Sludge mg/kg
Water leach test	0	0
TCLP test	10	200
Total lead	----	2960

Fractionation of Lead

Results of the sequential extraction procedure are provided in Table 4. The major fraction of lead in the contaminated sludge occurs in the carbonate form (40%). Oxide forms of lead-lead oxide and lead dioxide constitute 15% and 24% of the total lead in the contaminated sludge. Exchangeable lead (in the form of lead sulfate) is present in minor amount corresponding to 2.8%. The remaining lead is constituted by residual (metallic) lead (14.3%).

Table 4. Sequential Extraction Procedures (Tessier et al., 1979; Clevenger et al., 1991)

Reagent	Time (Hr)	Extracted Form	% of Total Pb
1M $MgCl_2$ (pH 7)	1.0	$PbSO_4$(exchangeable)	2.84
1M NaOAc (pH 5)	5.0	$PbCO_3$(carbonate)	40
0.5 M NH_4OAc (pH 7)	1.0	PbO	15.24
0.1M EDTA (pH 7)	1.0	PbO_2	24.40
4M HNO_3	1.0 (heated)	Pb (residual)	14.26

Acid and EDTA Washing

The results of batch washing experiments are summarized in Table 5. The results in Table 5 show that EDTA is more efficient than HCl in lead extraction. For example 0.01 M EDTA extracts 67% of the total lead in the sludge, while 0.01 M HCl has zero efficiency for lead removal. Like wise 0.1 M EDTA extract 96% of the total lead in the contaminated sludge while 0.1 M HCL extracts only 10% of the available lead in the contaminated sludge. The zero efficiency of 0.01 M HCl in extracting lead is attributed to neutralization of the acid solution by free lime present in the sludge material. Neutralization reaction increases the pH of 0.01 M HCl extractant from an initial value of 1.97 to 8.0 (Table 5); at this altered pH the solubility of lead is minimal (Al-Abed et al., 2006). Examining the efficiency of EDTA solutions it is seen that increasing the strength of EDTA solutions from 0.01M to 0.2 M increases the lead extraction efficiency of the chelating agent from 67% (at 0.01M EDTA) to 100% (at 0.2 M EDTA). Besides increasing molar strength, the final slurry pH values of the EDTA samples may also have influenced the extraction efficiency. The EDTA ligand is most effective in chelating metals from an aqueous system in the pH range of 3 to 6 (Cline & Reed, 1995). The final slurry pH of 0.1 M and 0.2 M EDTA samples correspond to 4.98 and 4.52 that lies in the favorable pH range for chelating heavy metal ions. The final slurry pH value of 0.01 M EDTA sample is 7.98 which does not lie in the favorable pH range.

Table 5. Sludge Washing

Reagent	pH of Reagent	pH of Slurry	Extracted Pb mg/g	% Removal of Pb	% Loss of Solid
0.5 M HCl	0.3	0.48	2.3	77.7	29
0.1 M HCl	1.06	1.6	0.29	9.8	11
0.01 M HCl	1.97	8.0	0.00	0.0	3.5
0.01 M EDTA	---	7.98	1.97	66.6	0.3
0.1 M EDTA	---	4.98	2.83	95.6	5
0.2 M EDTA	---	4.52	2.96	100	17

Table 5 also includes the percentage loss of solids experienced by the sludge sample on extraction with various HCL and EDTA solutions. Data in Table 5 shows that increasing EDTA concentrations 0.01 M to 0.2 M progressively increases the percentage loss of solids from 0.3% to 17%. Although 0.01 M EDTA causes minimum loss of solids, it has the least removal efficiency (67%). 0.1 M EDTA solution has a much higher efficiency of lead removal and also causes lesser dissolution of solids (5%) in comparative to 0.2 M EDTA solution (17%). Based on its high lead removal efficiency and lesser dissolution of solids, 0.1 M EDTA is considered as a most effective washing agent for removing lead from contaminated gypsum sludge.

CONCLUSIONS

The lead contaminated gypsum sludge was characterized for chemical, mineralogical and toxicity properties. Chemical analysis revealed the predominant presence of calcium sulfate dihydrate ($CaSO_4.2H_2O$) in the lead contaminated sludge. TCLP test showed that lead concentration in the leachate exceeded the critical level of 5 mg/l, characterizing it as hazardous waste based on toxicity characteristics. Interestingly, the total lead content in the gypsum sludge (2960 mg/ kg) was below the prescribed limit (5000 mg/kg) for hazardous wastes. Results of the sequential extraction procedure revealed that the major fraction of lead in the contaminated sludge occurs in the carbonate form; lead oxide, lead dioxide, exchangeable lead and residual lead constitute the other forms of lead. The results of washing experiments showed that

EDTA is more efficient than HCl in lead extraction. Further, based on its high lead removal efficiency and lesser dissolution of solids, 0.1 M EDTA is considered as a most effective washing agent for removing lead from contaminated gypsum sludge.

REFERENCES

Al-Abed, S.R; Hageman, P.L; Jegadeesan, G; Madhavan, N and Allen, D (2006). "Comparative evaluation of short- term leach tests for heavy metal release from mineral processing waste." *Science of the Total Environment,* 364, 14-23.

Clevenger, T.E; Salwan, C and Koirtyohann, S.R (1991). "Lead speciation of particles on air filters collected in the vicinity of a lead smelter". *Environ. Sci. Technol.,* Vol. 25, No 6, 1128-1133.

Cline, S.R & Reed, B.E (1995). "Lead removal from soils via bench-scale soil washing techniques". *Jl of Environ. Engg.* Vol.121, No.10, Oct, 700-705.

Granados, A. J and Peterson, P.J (1995). "Hazardous waste indicators for national decision makers." *Journal of Environmental Management,* 55, 249-263.

Griffiths, R A. (1995). "Soil-washing technology and practice" *J. Haz. Mater.* Vol. 40,175-189.

Leist, M; Casey, R.J and Cardi, D (2003). "The fixation and leaching of cement stabilized arsenic". *Waste Management.* 23, 353-359.

Lo, K.S.L and Chen, Y.H. (1990). "Extracting heavy metals from municipal and industrial sludges." *Sci. of the Total Environ.* 90, 99-166.

Misra, V. and Pandey, S.D. (2005). "Hazardous waste, impact on health and environment for development of better waste management strategies in future in India." *Environment international.* No.31, 417-431.

Neilson, J W; Artiola, J F, and Maier, R M. (2003). "Characterization of lead removal from contaminated soils by nontoxic soil-washing agents" *J. Environ. Qual.* 32:899-908.

Peters, R.W. (1999). "Chelant extraction of heavy metals from contaminated soils." *Jl of Hazardous Materials,* 66: 151-210.

Polprasert, C and Liyanage, L.R.G, (1996). "Hazardous waste generation & Processing." *Resource, Conservation and Recycling.* 16, 213-226.

Reinhart, D.R (1993). "A review of recent studies on the sources of hazardous compounds emitted from solid waste landfills: A U.S. experience." *Waste Management & Research,* 11, 257-268.

Sandhya B. and Dominica del M D. (2006). "Heavy metal removal from contaminated sludge for land application: a review." *Waste Management.* 6, 988-1004.

Singh, T.S and Pant, K.K (2006). "Solidification/stabilization of arsenic containing solid wastes using Portland cement, fly ash and polymeric materials". *Jl of Hazardous Materials,* 131, 29-36.

Tessier, A.; Campbell, P.G.C. and Bisson, M (1979). "Sequential extraction procedure for the speciation of particulate trace metals." *Analytical Chemistry.* No. 51, 844-851.

USEPA, 1996. Toxicity characteristic leaching procedure, Method 1311 in Test methods for evaluating solid waste, Physical/chemical methods (SW-846).

Synergistic Coupling of ISCO with SEAR and Bioremediation

M. Crimi[1], R.L. Siegrist[2], J. Munakata-Marr[2], J. Sahl[2], P. Dugan[2]

[1]Dept of Environ. Health, East Tennessee State University, PO Box 70682, Johnson City, TN 37614, PH (423) 439-7066; FAX (423) 439-5230; email: crimi@etsu.edu
[2]Division of Environmental Science and Engineering, Colorado School of Mines, 1500 Illinois Street, Golden, CO 80401, PH (303) 273-3467; FAX (303) 273-3413

ABSTRACT: Treatment of organic contaminants at hazardous waste sites may best be achieved by coupling technologies for improved treatment effectiveness. In particular, sites with high heterogeneity and high contaminant mass may be difficult to treat with a single approach due to difficulties associated with contacting remedial agents with contaminant, the cost of remedial amendments for complete contaminant degradation/removal, and the performance limitations of individual technologies. In situ chemical oxidation (ISCO) is one technology that may be applied prior to, simultaneous with, or after application of other active or passive techniques, in the same or adjacent treatment location. For example, bulk contaminant (e.g., non-aqueous phase liquid, or NAPL) removal, such as provided by surfactant-enhanced aquifer remediation (SEAR), may be necessary prior to ISCO treatment. Passive technologies that provide for a "polishing step", such as natural attenuation, may be required following source zone removal using ISCO. Laboratory studies were conducted to investigate the important synergies and challenges associated with coupling ISCO with SEAR and with bioprocesses (e.g., natural attenuation). Results demonstrate the promise of these coupling strategies and demonstrate the importance of characterizing reactants and byproducts associated with each individual process in order to exploit positive effects and avoid negative effects in system design.

INTRODUCTION

The subsurface environment is highly complex with media heterogeneities and varied contaminant mass distributions, challenging the applicability of any one individual remediation technology. Also, individual treatment technologies have performance and cost limitations (e.g., transport in different media, amenability of contaminants, contaminant mass or mass distribution, etc.) that may challenge their site-specific applicability. Because of these challenges, the use of coupled remediation technologies to improve treatment and cost effectiveness is receiving increased attention. The focus herein is the coupling of in situ chemical oxidation (ISCO) with surfactant-enhanced aquifer remediation (SEAR) or with bioprocesses. There are both synergies and challenges associated with coupling technologies that must be understood to apply them effectively.

ISCO is a remediation technique where organic contaminants are degraded in place by the delivery of oxidants to the subsurface. Effective treatment relies on effectively distributing/dispersing oxidant and contacting contaminant. Reaction can occur via electron transfer or free radical oxidation, depending on the oxidant employed. Common oxidants include permanganate, persulfate, catalyzed hydrogen peroxide propagations (CHP), and ozone. Byproducts of reaction also depend on the oxidant utilized and may include manganese dioxide solids (with permanganate), gas and heat (with CHP), sulfate (with persulfate), oxygen (CHP and persulfate), and increased dissolved metals (any oxidant). SEAR is the use of surfactants and/or cosolvents (such as water-miscible alcohols, ketones, and sugars) to solubilize or mobilize contaminants via micellar solubilization or due to interfacial tension reductions. Bioremediation is the use of organisms, primarily microorganisms, to degrade contaminants into less toxic forms under either aerobic or anaerobic subsurface conditions. Treatment effectiveness depends on microbial populations present, nutrient availability, moisture, temperature, and pH.

There are two potential approaches for coupling ISCO and SEAR, a co-injection approach where the treatment agents are delivered together (enhanced contaminant dissolution/desorption with concurrent contaminant destruction), and a sequential application of SEAR followed by ISCO to first remove significant contaminant mass followed by a chemical treatment polishing step. Table 1 presents the synergies and challenges associated with coupling ISCO and SEAR described below.

If properly designed, an ISCO/SEAR coupled treatment system can reduce chemical costs for remediation. For example, as more contaminant is removed from a site via SEAR, the remaining mass in the subsurface decreases and less contaminant is removed using a given volume of surfactant/cosolvent due to mass transfer limitations. The cost-benefit (i.e., mass of contaminant removed vs. mass of surfactant/cosolvent introduced) decreases as treatment continues. Treatment of this remaining mass of contaminant may be more cost-effective through a post-SEAR process such as ISCO (Dugan et al., 2007). Furthermore, this sequential approach of SEAR followed by ISCO may improve contact of ISCO oxidants with contaminant remaining post-SEAR. For example, if contaminant is initially present as a NAPL pool at a site where an oxidant is introduced, contact of the oxidant with the contaminant will occur only at the interface of the pool and the treatment solution delivered, then rely on slower contaminant dissolution (which will be enhanced due to destruction of dissolved phase contaminant) (e.g., Schnarr et al. 1998; Hood 2000). If significant NAPL mass is removed via SEAR first, the remaining contaminant may be more evenly distributed in the subsurface, allowing for greater contact of oxidant with contaminant on a mass to mass basis than treating the original NAPL with ISCO alone. An additional advantage is realized in coupling ISCO and SEAR using a co-injection approach. SEAR alone will remove contaminant from the subsurface via enhanced solubilization and/or mobilization, but a post-flush above ground process is necessary to then separate and manage the contaminant. Coupling SEAR with ISCO allows for simultaneous destruction of the contaminant at high concentrations, which increases the efficiency of oxidant use (Li 2004, Zhai et al. 2006).

Table 1. Synergy and Challenges of Coupling ISCO with SEAR

Synergy
Potentially reduced chemical costs if properly designed
Improved contact of oxidant with remaining contaminant post-SEAR (sequential application)
Enhanced contaminant dissolution/desorption concurrent with contaminant destruction (co-injection)
Challenges
Compatibility of oxidants with surfactants/cosolvents to avoid nonproductive oxidant consumption and/or byproduct generation
Compatible *concentrations* of oxidant and surfactant/cosolvent
Managing solids, gas, byproducts, and/or mobilized NAPL
Characterizing remaining contamination post-SEAR (sequential application)

There are challenges, however, in coupling ISCO and SEAR. The primary challenge is in identifying a compatible surfactant/cosolvent and oxidant solution. The ideally compatible solution is one where the surfactant/cosolvent exerts no inherent demand for the oxidant (i.e., nonreactive) that would result in nonproductive oxidant consumption and extensive byproduct generation. This issue of compatibility is important with respect to both sequential application of SEAR and ISCO (where residual surfactant/cosolvent will remain in the subsurface post-SEAR) and co-injection. Laboratory studies conducted to evaluate coupled ISCO and SEAR demonstrate a variety of issues associated with byproduct generation using incompatible surfactant/cosolvent and oxidant mixtures (e.g., Conrad et al. 2002). Another challenge in coupling these technologies relates to the sequential approach of SEAR followed by ISCO. Extensive efforts may be made to characterize the site pre-treatment, however it is essential to characterize the site to an extent post-SEAR so ISCO may be applied appropriately. This involves additional time, effort, and funds, however the cost-benefit can be realized in chemical cost savings.

Typically the coupling of ISCO with *bioprocesses* involves applying ISCO as a first step for significant contaminant mass treatment, followed by biotreatment (enhanced or monitored natural attenuation) as a polishing step for residual contamination. Table 2 presents the synergies and challenges associated with this coupled approach.

If properly designed, an ISCO/bioprocesses coupled approach can lead to reduced chemical costs, particularly at sites with low permeability layers or lenses into which contaminant has diffused and is not readily accessible for oxidation. Rather than continuing to apply oxidants to a site where diffusion processes are significantly slower than oxidation reactions, it is more cost-effective to rely on enhanced or natural bioprocesses for treating residual contaminant. The application of ISCO prior to biotreatment may offer an advantage to post-ISCO biotreatment by generating advantageous byproducts (Azadpour-Keeley et al., 2004; Droste et al., 2002, Sahl et al. 2007). For example, oxidants react with natural organic matter in the subsurface, generating dissolved organic carbon that can serve as a "food" source for microorganisms and enhance contaminant degradation (Sahl, 2005).

Byproducts generated during ISCO, however, can also challenge post-ISCO biotreatment. For example, metals potentially mobilized, along with pH changes, can affect microbial biotreatment, and these impacts are currently not well understood. Temporary perturbations in microbial activity post-ISCO that have been observed in some instances (e.g., Kastner et al., 2000, Klens et al., 2001, Bui and Cotton, 2002) are likely due to these changes in geochemistry. Field studies of coupled ISCO and bioremediation, however, indicate a rebound of biological activity and/or biomass post-ISCO (Gardner et al., 1996; Miller et al., 1996; Allen and Reardon, 2000; Hazen et al., 2000; Kastner et al., 2000; Klens et al., 2001; Droste et al., 2002; Hrapovic et al., 2005; Macbeth et al., 2005).

Table 2. Synergy and Challenges of Coupling ISCO with Bioprocesses

Synergy
Potentially reduced chemical costs if properly designed
Viable approach for sites with low permeability layers/lenses
Potentially improved biotreatment effectiveness due to ISCO byproducts generated (e.g., dissolved organic carbon, oxygen, sulfate, etc.)
Challenges
Unclear impacts of ISCO byproducts (e.g., metals, pH changes, etc.) on microbial activity
Potential for temporary decline in microbial activity post-ISCO

LABORATORY EVALUATIONS OF ISCO COUPLED APPROACHES

To improve the understanding of both the synergies and challenges of coupling ISCO with SEAR or with bioprocesses, laboratory studies were conducted at the Colorado School of Mines at the batch experimental and 1-dimensional (1-D) and 2-dimensional (2-D) transport scales (Sahl 2005; Dugan et al. 2007). Highlights of experimental systems and results are presented below.

Experimental Studies: Coupling ISCO with SEAR

The first stage of experimentation to evaluate the coupling of ISCO with SEAR was to identify a compatible surfactant/cosolvent and oxidant system. Batch screening studies were performed with a range of concentrations of 72 surfactants and 7 cosolvents in the presence of a range of concentrations of potassium permanganate ($KMnO_4$) and catalyzed hydrogen peroxide (CHP or H_2O_2). Full experimental design and results are described in Dugan et al. (2007). Oxidant concentrations were measured over a two week reaction period. Figure 2 shows example results for 5000 mg/L $KMnO_4$ and H_2O_2 reaction systems for varied concentrations of representative surfactants. These results demonstrate that many surfactants/cosolvents were incompatible with the oxidants as evidenced by the significant decrease in oxidant concentrations. These incompatible systems also resulted in significant byproduct (manganese dioxides in permanganate systems and gas/heat in CHP systems) generation. However, several compatible mixtures were identified (e.g., anionic surfactant Dowfax 8390 with permanganate in Figure 2), establishing the proof-of-concept for the ISCO and SEAR coupled approach.

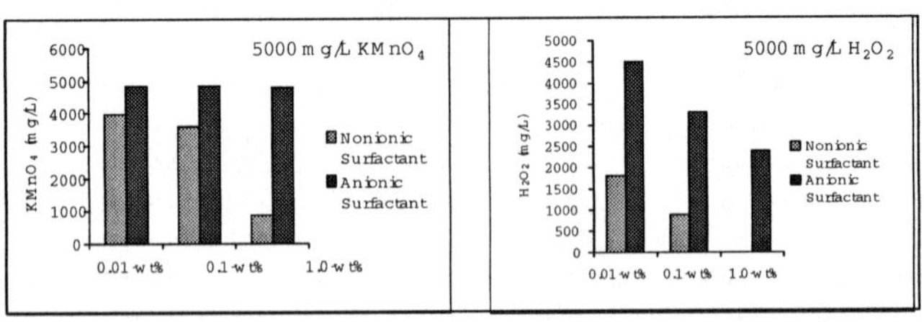

Figure 2. Representative results from ISCO/SEAR screening tests.

After identifying compatible mixtures of surfactant/cosolvent and oxidant, 2-D transport studies were conducted using a co-injection approach, where surfactant and oxidant were introduced simultaneously, to treat tetrachloroethylene (PCE). The primary goals were to (1) compare treatment with non-coupled ISCO controls with the same mass/concentration of permanganate delivered (not presented here due to space limitations, and (2) characterize pre-, during, and post-coupled surfactant/ISCO contaminant distribution and flux. The resultsbelow focus on this latter objective.

The transport cells (Figure 3) were comprised of a 3-sided Delrin box (30 cm x 30 cm x 2.5 cm) with a glass face for observing results. The back side of the Delrin box contained multiple sampling ports from which samples could be periodically withdrawn and analyzed. The configuration of sampling ports allowed for assessment of PCE and oxidant concentrations in the vertical (transect 1 in Figure 3) and horizontal dimensions (transect 2 in Figure 3). The cells were packed with media to contain (1) gravel at the influent and effluent ends of the cell to facilitate vertical transport of introduced solution, (2) a fine-grained media as the primary component, and (3) a coarse-grained lens embedded within the fine-grained media. PCE was introduced to the coarse lens to allow for pooled architecture within the cell (~20% saturation). A simulated groundwater was first passed through the cell (ambient flow of 12 cm/d) for several pore volumes to establish an equilibrium PCE solubilization. Samples were periodically withdrawn from the back ports and analyzed ("pre-flush" condition in Figure 3). Next, the compatible surfactant/permanganate mixture (3.5 wt% of surfactant Dowfax 8390 and 4 wt% potassium permanganate) was introduced for one pore volume of flow at four times the rate of ambient flow to simulate field injection. The delivery scheme was NOT designed to completely oxidize all PCE present in the system, but to determine the extent of treatment within one pore volume of compatible surfactant/oxidant solution. The left side of Figure 3 depicts transport through the cell over time, with permanganate first breaking through the coarse lens zone, as anticipated. Again, samples were withdrawn from the back ports periodically ("during flush" condition in Figure 3). Finally, multiple pore volumes of simulated groundwater were flushed post-treatment at the ambient 12 cm/day rate until a steady concentration of PCE was observed in cell effluent. Samples were also withdrawn from back ports during this flow phase ("8 days post" results in Figure 3).

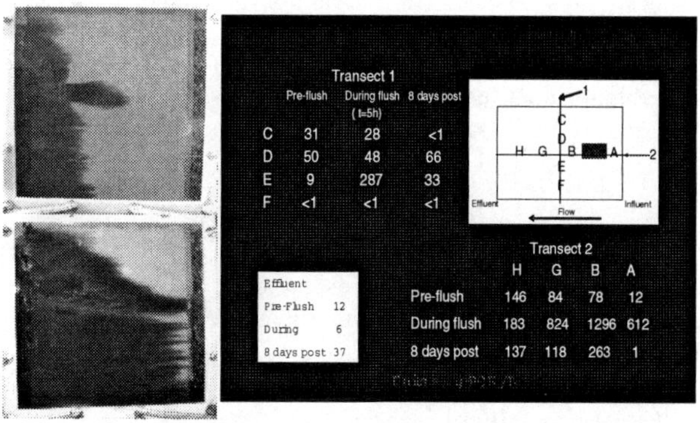

Figure 3. 2-D cell results for SEAR/ISCO co-injection.

Representative pre-, during, and post-flush results for PCE concentrations are shown in Figure 3 as a function of distance from the PCE source located in the coarse lens (sampling port transects labeled A-H in Figure 3), along with effluent concentrations. Vertical transect results demonstrate increased PCE concentrations during solution flush relative to the pre- and post-flush as a result of enhanced apparent solubility of the PCE due to the presence of the surfactant/cosolvent. These results also indicate no "uncontrolled" mobilization of the PCE as concentrations are centered near the lens (i.e., highest in D and E ports). The horizontal transect results also demonstrate increased PCE concentrations during the solution flush, significantly greater than the solubility limit for PCE of ~160 mg/L. In moving from port A to port H in the direction of flow, concentrations decrease with distance (ports approximately 5 cm apart) during the flush, indicating destruction of the PCE due to the presence of the permanganate. At port H, the "during flush" results are similar to those at the solubility limit both pre- and post-flush. Post-flush results are somewhat higher than pre-flush likely due to the presence of low residual concentrations of surfactant. Again, 100% treatment of the PCE was intentionally not planned so post-treatment contaminant distribution/flux could be assessed.

Experimental Studies: Coupling ISCO with Bioprocesses

A variety of experimental studies at the batch and 1-D transport scale were conducted to evaluate the coupling of ISCO with natural bioprocesses. The results presented here focus on the impact of the byproducts from oxidation of organic carbon naturally present in media on microbial degradation of PCE (Sahl 2005). For these experiments, a 1-D column was packed with a high organic carbon content natural porous media (characteristics in Sahl 2005). A solution of $KMnO_4$ was passed through the column (no contaminant present), and effluent from the column was collected up to the point of permanganate breakthrough. The goal was to collect dissolved organic carbon containing permanganate-oxidized organic matter. Next, microcosm studies were conducted to compare PCE degradation by a KB-1 culture obtained from a Southern Ontario TCE-contaminated site in the oxidized water and in the same water passing through the column without permanganate. Results are presented as "post-oxidation groundwater" and "pre-oxidation groundwater" in Figure 4, along with results of several control systems, and indicate a significantly enhanced rate of PCE degradation in the oxidized groundwater. Note that no residual oxidant is present in these reaction systems. It is postulated that the presence of low molecular weight organic acids, the likely byproduct(s) of permanganate oxidation of soil organic matter, facilitated microbial degradation of PCE (Sahl 2005).

SUMMARY

In summary, there are key frequently asked questions (FAQs) practitioners ask when considering a coupled approach for contaminant treatment. Specific to coupling ISCO with SEAR or bioprocesses, Table 3 presents key questions, along with brief responses.

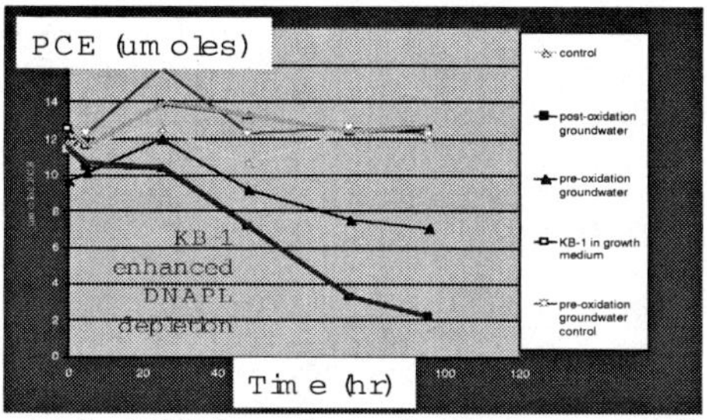

Figure 4. PCE degradation in microcosm studies

Table 3. Frequently Asked Questions About ISCO Coupled Treatment

FAQ	Response
When should coupling ISCO with another technique be considered?	Coupled approaches should be considered for EVERY site as part of the technology screening process.
Can ISCO be synergistically coupled with other remediation techniques?	YES, a wide variety of approaches are compatible.
How can a coupled strategy best be implemented with respect to space and time?	First consider advantages and limitations of technologies individually, then determine the appropriate coupled accordingly (e.g., Fig 1). The key factors for effective coupling include comprehensive site characterization, treatability testing, and cost-benefit analysis.
At what sites may ISCO be cost-effective with a SEAR pre-treatment step?	This approach is appropriate for sites with high NAPL mass, including NAPL pools.
At what sites is ISCO cost-effective as a treatment train with a bioremediation polishing step?	This approach is appropriate for highly heterogeneous sites having the site-specific potential for favorable post-ISCO conditions (e.g., generation of dissolved organic carbon).
Will the oxidant sterilize the soil and prevent future microbial degradation of contaminants?	No! There may be temporary decreases in microbial activity, but populations and activity will rebound and potentially be enhanced.
Can ISCO byproducts inhibit or enhance bioprocesses? Are potential geochemical effects of ISCO a concern with respect to coupled approaches?	ISCO byproducts can potentially both inhibit or enhance bioprocesses. It is important to understand site-specific ISCO conditions and potential effects on microbial activity.
What is the process for selecting the proper oxidant mass and concentration for an ISCO coupled approach?	Laboratory-scale treatability testing with field materials and/or pilot tests should be conducted to evaluate appropriate oxidant mass/concentration.
What laboratory tests are necessary to conduct?	Laboratory tests should evaluate the compatibility of the ISCO system with the technology to be coupled. Inputs and outputs of each process must be understood.

REFERENCES

Allen, S. A.; Reardon, K. F. (2000). "Remediation of Contaminated Soils by Combined Chemical and Biological Treatments." In *Physical and Thermal Technologies: Remediation of Chlorinated and Recalcitrant Compounds*, Wickamanayake, G. B., Gavaskar, A. R. (Eds.); Battelle Press: Columbus, OH.

Azadpour-Keeley, A.; Wood, L. A.; Ue, T. R.; Mravik, S. C. (2004). "Microbial Responses to In Situ Chemical Oxidation, Six-Phase Heating, and Steam Injection Remediation Technologies in Groundwater." *Remediation*, 14 (4), 5-17.

Bui, C. T.; Cotton, R. G. H. (2002). "Comparative Study of Permanganate Oxidation Reactions of Nucleotide Bases by Spectroscopy." *Bioorg. Chem.*, 30 (2), 133-137.

Conrad, S. H., Glass, R. J., and Peplinski, W. J. (2002). "Bench-scale visualization of DNAPL remediation processes in analog heterogeneous aquifers: Surfactant floods and in situ oxidation using permanganate." *J. Contam. Hydrol.*, 58, 13-49.

Droste, E. X.; Marley, M. C.; Parikh, J. M.; Lee, A. M. (2002). "Observed Enhanced Reductive Dechlorination After In Situ Chemical Oxidation Pilot Test." In *Remediation of Chlorinated and Recalcitrant Compounds*; Battelle Press: Monterey, California.

Dugan P.J., R.L. Siegrist, M. Crimi (2007). "Compatibility of Surfactants and Cosolvents with Chemical Oxidants for Groundwater Remediation." *Unpublished, to be submitted 2007.*

Gardner, F. G.; Korte, N.; Strong-Gunderson, J.; Siegrist, R. L.; West, O. R.; Cline, S. R.; Baker, J. L. (1996). "Implementation of Deep Soil Mixing at the Kansas City Plant." *ORNL/TM-13552*; Oak Ridge National Laboratory: Oak Ridge, Tennessee.

Hazen, T. C.; Sewell, G.; Gavaskar, A. R. (2000). "The Effect of Source Remediation Methods on the Presence and Activity of Indigenous Subsurface Bacteria at Launch Complex 34, Cape Canaveral Air Station, Florida." Battelle: Columbus, Ohio.

Heiderscheidt J. (2005). "DNAPL Source Zone Depletion during In Situ Chemical Oxidation (ISCO): Experimental and Modeling Studies." *Doctoral Dissertation, Colorado School of Mines*, 2005.

Hood E.D., and N.R. Thomson (2000). "Numerical Simulation of In Situ Chemical Oxidation." *Proceedings of the Second International Conference on Remediation of Chlorinated and Recalcitrant Compounds*, Monterey, CA, May 2000.

Hrapovic, L.; Sleep, B. E.; Major, D. J.; Hood, E. (2005). "Laboratory Study of Treatment of Trichloroethene by Chemical Oxidation Followed by Bioremediation." *Environ. Sci. Technol.*, 39:2888-2897.

Kastner, J. R.; Santo Domingo, J.; Denham, M.; Molina, M.; Brigmon, R. (2000). "Effect of Chemical Oxidation on Subsurface Microbiology and Trichloroethene Biodegradation." *Bioremediation*, 4 (3):219-236.

Klens, J.; Scarborough, S.; Graves, D. (2001). "The Effects of Permanganate Oxidation on Subsurface Microbial Populations." In *Natural Attenuation of Environmental Contaminants*. Leeson, A., Kelley, M. E., Rifai, H. S., Magar, V. S. (Eds.); Battelle Press: Columbus, Ohio, 253-259.

Li, X. D., and Schwartz, F. W. (2004). "DNAPL mass transfer and permeability reduction during in situ chemical oxidation with permanganate." *Geophys. Res. Lett.*, 31.

Macbeth, T. W.; Peterson, L. N.; Starr, R. C.; Sorenson, K. S.; Goehlert, R.; Moor, K. S. (2005). "ISCO Impacts on Indigenous Microbes in a PCE DNAPL Contaminated Aquifer." In *In Situ and On-Site Bioremediation Symposium; Battelle Press: Baltimore, Maryland.*

Miller, C. M.; Valentine, R. L.; Roehl, M. E.; Alvarez, P. (1996). "Chemical and Microbiological Assessment of Pendimethalin-Contaminated Soil after Treatment with Fenton's Reagent." *Water Res.*, 30 (11), 2579-2586.

Sahl J., J. Munakata-Marr, M. Crimi, R.L. Siegrist (2007). "Coupling Permanganate Oxidation with Microbial Dechlorination of Tetrachloroethene." *Water Environment Research*, WEF, 79(1):5-12

Sahl, J. (2005). "Coupling In situ Chemical Oxidation (ISCO) with Bioremediation Processes in the Treatment of Dense Non-Aqueous Phase Liquids (DNAPLs)," *Masters Thesis,* Colorado School of Mines, Golden, Colorado.

Schnarr M.J., C.L. Truax, G.J. Farquhar, E.D. Hood, T. Gonullu, and B. Stickney (1998). "Laboratory and Controlled Field Experiments Using Potassium Permanganate to Remediate Trichloroethene and Perchloroethene DNAPLs in Porous Media." *J. Contam. Hydrol.*, 29(3):205-224.

Zhai, X., Hua, I., Rao, P. S. C., and Lee, L. S. (2006). "Cosolvent-enhanced chemical oxidation of perchloroethylene by potassium permanganate." *J. Contam. Hydrol.*, 82(1-2), 61-74.

Treating PCB/Petrochemical-Contaminated Soil With Humic Mineral Concentrates

Alexander I.Shulgin[1] and D. Joseph Hagerty[2], F. ASCE, PE

[1]Research Professor, Chemical Engineering Dept., U. Louisville, Louisville, KY 40292
a.shulgin@insightbb.com
[2]Professor, Civil & Environmental Engineering Dept., U. Louisville, Louisville, KY 40292
hagerty@louisville.edu

ABSTRACT: Remediation of soil contaminated by petroleum and/or chlorinated hydrocarbons [e.g., PCBs] using Humic Mineral Concentrate (HMC) was demonstrated as efficient and effective, using laboratory and field tests. HMC molecules present dense populations of easily hydrated sorption sites, coupled with equally numerous hydrophobic sites. Thus, HMC products possess properties of both insoluble humic acids and soluble humates, in an optimal ratio, with pH of 6.5 to 7.5.

Sorption on HMC combines hydrophobic interaction with covalent bonding, followed by decomposition of petroleum hydrocarbons and dechlorination and decomposition of chlorinated hydrocarbons. Refractory organic pollutants become much more available to microbiological destruction. HMC is a natural substrate on which microorganisms flourish; HMC application supports higher microbial activity, and increases soil moisture retention and fertility.

In tests, HMC completely bound petroleum present at 5 to 7 percent, within two months. Residual petroleum concentration was 0.2 percent. The petroleum residue was distributed uniformly throughout the soil volume; after treatment the soil was much more hydrophilic. Average PCB content changed from 134 mg/kg to 10.8 mg/kg after 5 months and 1.7 mg/kg after 15 months. HMC immediately eliminated odors characteristic of chlorinated organics. Both sorption and decomposition of PCBs occurred in the presence of humic acids.

INTRODUCTION

Humic substances are extremely widespread in nature, existing as the major or even dominant proportion of many organic substances such as oil shale, peat, and brown and oxidized bituminous coal. They are also major components in fluvial, lacustrine and marine clastic deposits. The most important parts of these humic substances are humic acids (Orlov 1990)

Humic acids are high-molecular weight (5 to 100 kDa) organic compounds. Each molecule of humic acid incorporates tens of acidic groups, so those groups can be termed polybasic. In order to isolate humic acids from natural humified organic masses (carbon-series caustobiolites) where they are in a water-insoluble state as free acids or water-insoluble salts of such acids, an aqueous-alkaline extraction method is used. As a rule, a surplus amount of an alkaline reagent is used. Potassium, sodium, or ammonium hydroxides are used most commonly, but organic bases are also used to form water-soluble salts with the acids. The extract has a high pH and contains humic acid salts, with a residual excess of alkali. The excellent solubility and high dispersion properties of the products are due to the significant negative charge carried by the humic acid ions; those charges produce mutual repulsion between macromolecules, with consequent hydration. However, application of metal humates in the field is restricted to low concentrations because they are toxic to both microorganisms and plants at high concentrations. Therefore, the aforementioned high negative charges adversely affect sorption, flocculation and structural formation properties.(Lishtvan and Abramec 1993)

Conversely, free humic acids formed in aqueous systems at pH less than 3 aggregate because their macromolecules do not carry a significant negative charge; the aggregates expel water and settle out of aqueous solutions as flocculated precipitates. Consequent, the specific surface of the acid aggregates with respect to the liquid phase is drastically reduced, with detrimental effects on their sorptive and reactive capacities. From a practical standpoint, interest must be restricted to systems in which pH ranges between 3.5 and 8 because in such systems the humic acid molecules are partly dissociated.

The valuable sorptive and reactive properties of humic acids favor their use in agriculture, animal husbandry, and medicine, and in solving environmental problems. However, to be useful, the humic acids must be available in forms having high physicochemical and biological activity, and efficient production technology for such forms has not been available. Thus, not humic acids proper, but diverse natural materials such as humic soils, peat, clastic deposits (e.g., sapropels) and brown coal (e.g., disintegrated lignite and leonardite) have been used because they contain humic substances. The presence of the humic substances in such materials has made them attractive in remediation of contaminated soils, reclaimed land, disposal areas and waste dumps. Such materials contain only a relatively low proportion of humic acids, in only a slightly active state, so large amounts of the materials are needed. The desired positive effects sometimes are lost because of detrimental effects of such massive applications of materials that may contain small concentrations of harmful substances. At the same time, many potential applications for humic acids exist, and many environmental pollution problems may be solved.

POTENTIAL FOR APPLICATIONS OF HUMICS

Many methods are available for treatment of contaminated surface and ground waters, and domestic and industrial wastewaters; those methods remove inorganic, organic and microbiological pollutants. However, contamination of soils is a more difficult problem to remedy. A myriad of waste products have been released onto surface areas or buried beneath those surfaces. Some of those waste products are toxic, carcinogenic and/or mutagenic; especially problematic are petroleum hydrocarbons and petroleum products such as polycyclic aromatics (anthracene, benzopyrene, dibenz(a,h)anthracene, etc.), polychlorinated aromatics (polychloro-dibenzofurans, polychloro-biphenyls, dioxins, etc.), pesticides and organic chemicals of exogenic origin.

Several methods can be used to detoxify such organic compounds. For instance, biphenyls can be detoxified by decomposing them by photolysis using solar radiation or ultraviolet radiation. Another way includes dioxin decomposition by alkali in polyethilenglycol (Dioxin 87 1987). Organic pollutants can bind with soil organic matter and soil clay minerals (Bailey and White 1964). Furthermore, soil organic matter binds organic pollutants much more completely than soil clay minerals, and the bond "soil organic matter – organic pollutant" is much stronger than the bond "soil clay minerals – organic pollutants". Also, the hydrophobic interaction of soil humic acids with polychlorinated biphenyls is a principal mechanism of PCB immobilization in soil (Landrum 1985). "Humic acid– PCB" complexes were formed in prior research and these complexes reduced significantly the toxicity of PCB and their availability to plants (Shulgin et al. 2000). But hydrophobic interaction may be only the first step in the "humic acid—PCB" complexing, which may include covalent binding of PCB by humic acids (Bollag and Myers 1992). Such types of chemical interaction would form polymer structures very stable even to acid and alkali attack.

Toxic organic chemicals in soils and plants can be detoxified by applying humic substances to the contaminated zones; the humus substances used are salts of humic acids (cf. Detoxifying Agent SU Russian Patent A # 460,037, 1986). In a concentration above 0.01%, sodium humate used in this method is a phytotoxic substance that inhibits not only plants but also soil microflora. However, at concentrations between 0.005 % and 0.01% (dry weight), sodium humate stimulates growth and development of plants. Moreover, the toxic chemical concentrations in the plants in the treated area are reduced, not only because of direct decomposition of the toxic chemicals, but also because their concentration is decreased owing to the increased cover density of plants over the treated area. Unfortunately, this method is virtually inapplicable for detoxifying soils and other media featuring high levels of contamination with persistent organic pollutants because the degree of detoxification is extremely low and cannot therefore be demonstrated experimentally. Additionally, the danger of salinization of natural soils is very high.

HUMIC-MINERAL CONCENTRATES

A method was required for detoxifying organic compounds that was efficient in restoring a polluted natural environment. For this task, humic-mineral concentrates (HMC) were selected for binding and subsequent complete decomposition of organic impurities. HMC exhibits high efficiency as an ion-exchange, sorptive, and flocculent material that promotes formation of molecular structures. In HMC, molecules of humic acids are present in a transitional hydrated state, wherein only some of their acid groups are dissociated (A Method (and system) for Obtaining Humic-Mineral Concentrates, Patent Ru, 2175651, 2000). Such partial dissociation is characteristic of many bio-organic molecules and is demonstrated in many humic compounds by different ratios between humic acids and humates in those compounds. Typically the pH in such systems varies over a range safe for biota, that is, from 6.5 to 7.5. Depending on the degree of dissociation of humic acid molecules in the concentrate, the product may be insoluble in both dry and humid states. On the other hand, humic acids in HMC have a considerable number of non-dissociated acid groups and a lower negative charge than do humates. Typically, they form spatial structures in aqueous systems that impart economically useful properties to those systems. (Shulgin 2003).

HMC can be used for detoxifying organic compounds by efficiently binding diverse toxic, mutagenic, carcinogenic, and other toxic organic compounds, whereupon they undergo accelerated decomposition. The adsorption occurs after the pattern of a hydrophobic interaction and/or covalent binding. Formation of a complex is followed by dechlorination of polychlorinated organic compounds with concurrent-consecutive reactions of oxidation-hydroxylation of polychlorinated and polycyclic compounds.

Such a transformation of organic xenobiotic compounds converts them from refractory compounds that resist degradation to compounds amenable to microbiological destruction. Where pollution with organic compounds is intense, soil microflora exist in a suppressed state and thus are incapable of exerting a substantial degrading influence on xenobiotics. For as much as the humic concentrate is a natural soil component (i.e., that substrate on which microorganisms live and develop well), its application promotes higher microflora activity. Moreover, microorganisms, while sorbing on humic acids, actively destroy organic xenobiotics concentrated in polluted soils. Simultaneous application of humic concentrates and microorganisms, particularly those that oxidize hydrocarbons (bacteria of the genus B. Pseudomonas and Azotobacter in dosages of 10^5 to 10^7 cells/gram), achieves accelerated decomposition of organic pollutants. Such microorganisms also can be applied effectively to contaminated zones that have already been treated with humic concentrate.

Apart from detoxifying organic pollutants, the humic concentrate that has been developed provides increased soil moisture capacity and fertility, and enhanced soil structure. Application of humic concentrates has improved the cohesion and resistance to water and wind erosion of diverse waste products, and has improved the effectiveness of traditional methods of treatment to remove various contaminants from surface and ground waters, and wastewaters from an array of sources.

LABORATORY TESTING PROGRAM

A program of laboratory testing was carried out to investigate the process by which humic concentrates could be used to detoxify a sandy loam soil polluted with crude oil (carbon content 83%, hydrogen 14 %, nitrogen, sulfur and oxygen 3%). The research was directed toward regenerating soil fertility by incorporating petroleum hydrocarbons into the soil structure, followed by decomposition of those hydrocarbons. The moisture retention capacity of the sandy loam soil in these tests, free from petroleum pollution, was about 33.2 percent and had been determined by standard method; the moisture retention capacity of the same sandy loam after petroleum pollution could not be determined, since soil samples could not be dried completely because of the high petroleum content.

Humic concentrate was applied to and distributed within petroleum-polluted soil samples. Application of humic concentrate to the soil completely bound the petroleum present therein in an amount of 5 to 7 percent and caused formation of soil structure within two months. The selected soil samples dried well, which indicated a uniform petroleum distribution throughout the soil volume and an increase in the hydrophilic character of the soil. Moisture retention capacity of such samples was approximately 60 percent.

To assess the suitability of the soil for growth and development of plants, wheat and mustard seeds were planted in soil specimens. The experiment was carried out in 0.5-liter vessels placed in a controlled-climate chamber with an illumination of 7,000 units and a light duration of 6 hours per day. Inorganic supplements including nitrogen, phosphorus, potassium, and microelements were applied at the beginning of the experiment to all the vessels, including the control. Triplicate specimens were used. The control specimens consisted of plants grown on podzolic soil. The wet weights of the plant tops were determined three weeks after seed planting. Results of the experiment on sandy loam soil samples are presented in Table 1; they show that use of humic concentrate on this sandy loam soil proved effective as a means to remedy the petroleum hydrocarbon contamination.

TABLE 1. Results of Growth Experiments

Test Variant	**Wet Weight of Plant Tops (percent of control)**	
----	Wheat	Mustard
Podzolic Control	100	100
Sandy loam soil	56	42
Sandy loam soil with 5 % crude oil	1	0
Sandy loam soil with humic concentrate	110	101
Sandy loam soil with 5 % crude oil and humic concentrate	100	96

FIELD TESTING PROGRAM

Field experiments with this soil detoxication process were carried out on acreage available adjacent to a plant where condenser were manufactured; the area had been contaminated with polychlorinated biphenyls (PCB). The average PCB content on the experimental plot was 102.2 mg/kg. Plots were selected to provide an approximately equal level of PCB contamination, and then were divided into two subplots, with humic concentrate applied to the soil on one subplots in each pair of plots. Samples were obtained from the subplots at five months, and at fifteen months after the humic concentrate application.

PCB concentration in soil samples was determined using the modified technique (United States Environmental Protection Agency. Environmental Monitoring and Support Laboratory. EPA Test Method: Base/Neutral and Acids-

Method 625, July 1982) on a chromatic mass spectrometer (Hewlett Packard model MSD 5972). Results of the tests are presented in Table 2.

Experimentation clearly showed that application of humic acids to the soil immediately eliminated the odor characteristic of chlorinated organic matter. The observed changes in the content of isomers demonstrated that both sorption and decomposition of PCB occur in the presence of humic acids. In this case, more highly chlorinated isomers were converted into less chlorinated isomers. Subsequently, the concentration of the less chlorinated forms first rose, and then dropped; that is, a process of progressive dechlorination was observed.

In addition to determining the PCB content of the soil samples, the degree of reduction of their total toxicity was assessed by examining the condition of microorganisms of the genus Azotobacter, microalgae, and other soil organisms The results of the soil bioassay test are presented in Table 3, and show that

TABLE 2. Results of Field Dechlorination Experiments

Time from Date of Application (months)	Total concentration of tri-, tetra-, and penta-chlorinated biphenyls (mg/kg)	
	Control	Soil with 1 percent HMC added
0	134	116
5	134	10.8
15	121	1.7

application of humic concentrate to this PCB-polluted soil virtually eliminated the toxicity of the tested soil to microorganisms.

TABLE 3. Results of Bioassay for Growth of Azotobacter Microorganisms in Soil Samples

Test Variant	Index of soil microbiological activity in points (1 to 10)
Original soil	1
Soil five months after HMC application	6
Soil fifteen months after HMC application	10

In further tests, soil microorganisms that had been isolated earlier from the same soil and cultivated according to a routine laboratory maintenance technique were applied to one of the subplots in each pair of plots, in a dosage of about 10^5 to 10^7 cells/gram. The microorganisms that were applied belong to the genus *Pseudomonas* and are not classified as pathogens. Then, the surface was tilled on each of the two subplots to a depth of five to eight cm with a view to intermixing the soil and humic concentrate applied to the first subplot portion, and mixing the soil, humic concentrate and the microorganisms applied to the second subplot.

Soil samples were taken before the subplots were tilled and at periods of 25 and 60 days, respectively, after application of humic concentrate, or humic concentrates together with the microorganisms. The test results are given in Table 4; data on mono-chlorinated biphenyls, are expressed in absolute values (mg/kg) in view of the absence of that isomer in the control (initial) samples.

These changes in the content of isomers demonstrates that both sorption and decomposition of polychlorinated biphenyls occurred in the presence of humic concentrate. In this case, more highly chlorinated isomers were converted into less highly chlorinated isomers whose concentrations first rose, and then dropped, i.e., the data indicate a high-efficiency process of progressive dechlorination so that the concentration of polychlorinated (that is, most toxic) biphenyls is reduced by 14 to 20 times in a period as short as 60 days. Similarly, the data show more complete decomposition of polychlorinated biphenyls in samples where microorganisms were applied to the soil with the humic concentrates.

TABLE 4. Degradation of PCB Isomers with Time

Isomers	**Isomer Content (percent of control sample)**					
	Soil Samples with Humic Concentrate			**Soil Samples with Humic Concentrate and with Microorganisms**		
	Initial	**After 25 Days**	**After 60 days**	**Initial**	**After 25 days**	**After 60 days**
Mono-chlorinated	0	0	2.2	0	6	10
Dichlorinated	100	60	49	100	106	6
Trichlorinated	100	6	5	100	10	12
Polychlorinated	100	10	5	100	13	7

In addition to determination of the content of polychlorinated biphenyls in the soil samples, the relative reduction in total toxicity was assessed by measuring the growth of microflora such as the genus *Azotobacter*, a typical soil microbe. *Azotobacter* is highly sensitive to the presence of toxicants in the soil. The results of a bioassay in which the soil samples were maintained under natural conditions for 25 and 60 days, are presented in Table 5. The results show the effectiveness of the application of HMC in reducing total toxicity in soil polluted with PCB and the even greater effectiveness of application of HMC with microorganisms.

CONCLUSIONS

1. Humic-mineral concentrates (HMC) are effective at remediation of petroleum contaminations in soils, as shown in laboratory and field tests.
2. HMC can be used effectively to remedy PCB contamination in soils, as shown in field tests and demonstrated in results of bioassays.
3. The effectiveness of HMC can be amplified by simultaneous application of appropriate microorganisms to contaminated soils.

TABLE 5. Results of Bioassay for Growth of Azotobacter Microorganisms in Soil Samples

Test Variant	Degree of Accretion of *Azotobacter* to Soil Crumbs (%)
Original soil	58
Soil 25 days after HMC application	80
Soil 60 days after HMC application	91
Soil 25 days after application of HMC with microorganisms	99
Soil 60 days after application of HMC with microorganisms	99

REFERENCES

Orlov D.S. (1990) *Humic Acids of the Soils and the General theory of Humification.* Moscow State University, Moscow. 324 pp.

Lishtvan, I.I. and Abramec, A.M. (1993) "Humic preparations and an environmental protection." *Humic Substances in the Biosphere*. Proceeding of the First International Conference in Moscow. Science [Publisher], 126-138

Shulgin A.I., Bobovnikova C.I. and Pleskachevskaya, G.A. (2000) "Effective technology for detoxifying soil contaminated by polychlorinated bi-phenils with the use of humic acids." *Polychlorinated Bi-phenils: Super-toxins of the 21st Century; Information Bulletin No. 5*, VINTIN [All Russia Research Institute of Science and Technical Information], 182-187.

Shulgin A.I. (2003) "Effective technologies for detoxification of contaminated soils with the use of humates and humate-mineral substances." *Humic Substances in the Biosphere*. General Report at the II International Conference in Moscow, 132-133.

Bailey C.W.and White, J.L., (1964) "Soil-pesticide relationships: review of adsorption and desorption of organic pesticides by soil colloids, with implications concerning pesticide bioactivity." *J.Agric. Food Chem.*, 12 (4) 324-332

Bollag J.M. and Myers, C. (1992) "Decontamination of aquatic and terrestrial sites through binding of pollutants to humic substances." *Sci. Total Environ.*, 117-118, 357-66

Dioxin 87: Program and Abstracts. (1987) 7th Int. Symp.on Chlorinated Dioxin and Related Compounds. Las Vegas. Nevada, 20

Landrum P.F. (1985) "Predicting bioavailability of organic xenobiotics to Pontoporeia hoyi in the presence of humic and fulvic materials and natural dissolved organic matter." *Environ. Toxicol. Chem.*, 4, 459-467

Detoxifying Agent (1984) Soviet Union Patent A # 460,037

A Method (and system) for Obtaining Humic-Mineral Concentrates. (2000) Russian Patent 2175651

Evaluation of Two Strategies to Enhance the Long-Term Hydraulic Performance of Permeable Reactive Barriers

Lin Li[1], M. ASCE, P.E. and Craig H. Benson[2], M. ASCE, P.E.

[1]Assistant Professor, Department of Civil and Environmental Engineering, Jackson State University, Jackson, MS; lin.li@jsums.edu
[2]Wisconsin Distinguished Professor, Department of Civil and Environmental Engineering, University of Wisconsin-Madison, Madison, WI; benson@engr.wisc.edu

ABSTRACT: Groundwater flow and geochemical reactive transport models were used to assess the effectiveness of two strategies that are used to enhance the long-term hydraulic behavior of continuous-wall permeable reactive barriers (PRBs) employing granular zero valent iron (ZVI). The flow model accounted for geological heterogeneity and the reactive transport model included a geochemical algorithm for simulating iron corrosion and mineral precipitation reactions that have been observed in ZVI PRBs. The two strategies that were evaluated are (i) adding pea gravel equalization zones up gradient and down gradient of the reactive zone and (ii) placement of an up gradient sacrificial pre-treatment zone containing a mixture of pea gravel and ZVI. Results of simulations show that installation of pea gravel equalization zones results in a more uniform distribution of residence times within the PRB, slightly greater average porosity reductions, and smaller maximum porosity reductions. Sacrificial pre-treatment zones elevate the groundwater pH and consume many of the mineral forming ions that form secondary minerals, reducing mineral precipitation in the reactive zone. However, mineral fouling by $Fe(OH)_2$ still occurs within the PRB in response to iron corrosion reactions.

INTRODUCTION

The permeable reactive barrier (PRB) is an in situ remediation technology where contaminated groundwater is treated passively as it flows through a reactive medium. Most PRBs use granular zero valent iron (ZVI) to create redox conditions that result in the degradation or immobilization of chlorinated solvents and herbicides, heavy metals, and radionuclides (Naftz et al. 2002). These redox conditions also promote the precipitation of secondary minerals from ions typically in groundwater (and some contaminants as well), a process referred to as mineral fouling (Mackenzie et al. 1999, McMahon et al. 1999, Wilkin et al. 2003, Li et al. 2006). Accumulation of secondary minerals in the pore space can reduce the reactivity, porosity, and hydraulic conductivity of the ZVI, resulting in reorientation of flow paths, changes in flow rate, seepage velocity, and residence times, and deterioration of treatment efficiency (Kamolpornwijit et al. 2003, Li et al. 2005, Jeen et al. 2007). Mineral fouling causes annual porosity reductions in PRBs ranging between 0.0007 to 0.03 and annual reductions in hydraulic conductivity ranging between a factor of 1.5 to 7.0 (Li et al. 2005). Variability in flow velocities due to geological heterogeneity can exacerbate the effects of mineral fouling.

In many applications, PRBs employing ZVI are expected to last decades. Thus, considerable interest exists in strategies that will enhance the long-term performance of PRBs (Sarr 2001). Two approaches have been proposed: (i) pea gravel zones up gradient and down gradient of the reactive zone to equalize flows and (ii) placement of a sacrificial pre-treatment zone upstream of the reactive medium. Pea gravel equalization zones have been used in field PRBs (McMahon et al. 1999, Yabusaki et al. 2001), but their effectiveness is largely undocumented. They are intended to equalize flows, resulting in more uniform residence times and distribution of minerals within the PRB. Sacrificial pre-treatment zones consist of a mixture of granular material (e.g., gravel) mixed with ZVI to form a region where pH and redox conditions can change and secondary minerals can form before groundwater enters the reactive zone. In a field study at Monticello, UT, USA, a pre-treatment zone containing 13% ZVI and 87% gravel was installed along the upstream face of 100% ZVI reactive zone (Morrison 2003). Morrison (2003) has shown that most of the minerals form in this pre-treatment zone.

In this study, the effectiveness of these two strategies was evaluated in terms of their ability to minimize the impacts of flow variability and secondary mineral precipitation on the long-term hydraulic behavior of PRBs. The study made use of a three-dimensional groundwater flow model incorporating geological heterogeneity coupled with a reactive transport model simulating geochemical reactions and mineral precipitation in the PRB. The model was used to simulate flow and mineral fouling in a PRB located in an aquifer with a natural level of heterogeneity to represent conditions that may occur during the service life of a typical PRB.

FLOW AND TRANSPORT MODELING

The conceptual model is a horizontal PRB and a portion of the surrounding aquifer (Fig. 1). The hydraulic conductivity of the aquifer is spatially variable to represent the heterogeneity inherent in natural systems. The PRB is oriented perpendicular to the primary direction of groundwater flow and is composed of ZVI particles.

Groundwater flowing into the PRB is assumed to be in chemical equilibrium and includes Fe^{2+}, Ca^{2+}, OH^-, HCO_3^-, CO_3^{2-}, and dissolved O_2. These dissolved species are assumed to form three minerals in the PRB: $CaCO_3$, $FeCO_3$, and $Fe(OH)_2$. These ion species and minerals were selected based on the results of a sensitivity analysis conducted by Li et al. (2006) using the same flow and geochemical transport model employed in this study. The analysis showed that these minerals compose the majority of the secondary mineral mass that precipitates in the pore space of PRBs under a broad variety of conditions. Additionally, these minerals represent the most common minerals observed in column and field studies (Mackenzie et al. 1999, Wilkin et al. 2003, Li et al. 2006, Jeen et al. 2007).

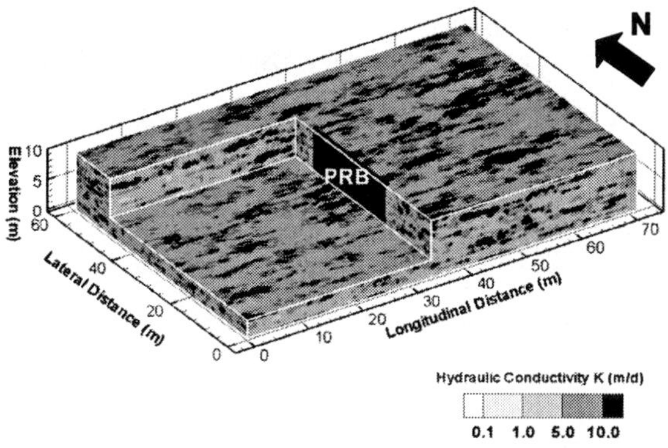

FIG. 1. Heterogeneous aquifer containing PRB.

A three-dimensional domain was used to simulate flow in the surrounding aquifer materials and the PRB. Aquifer heterogeneity was simulated using the turning bands method described in Elder et al. (2002). The aquifer hydraulic conductivity is assumed to be log-normally distributed and is characterized by the geometric mean hydraulic conductivity (K_g), the standard deviation of the logarithm of hydraulic conductivity (σ_{lnK}), and the correlation lengths (λ) along the principle axes (Hemsi and Shackelford 2006). The PRB is located in the middle of the aquifer, oriented perpendicular to the groundwater flow, and is 10-m deep, 25-m wide, and 1-m thick. The PRB was assigned a uniform hydraulic conductivity of 216 m/d and porosity of 0.6. For the simulation described in this paper, K_g = 3.9 m/d, σ_{lnK} = 1.0, λ_l = 3 m, λ_t = 1 m, and λ_v = 0.5 m.

Steady-state flow was simulated in the heterogeneous aquifer and PRB using MODFLOW (McDonald and Harbaugh 1988). The model domain consisted of 250 columns, 120 rows, and 20 layers, which represent a flow domain of 71 x 60 x 10 m. The grid spacing was 0.5 m in the vertical and lateral directions and varied from 0.3 m to 0.1 m in the longitudinal direction, with the smaller spacing (0.1 m) being used within the PRB. Specified-head boundaries were set along the east and west ends of the aquifer to establish an average hydraulic gradient of 0.01 in the longitudinal direction. No-flow boundaries were used along the lateral, top, and bottom boundaries. The flow field was updated annually to account for changes in hydraulic conductivity of the PRB caused by mineral fouling. The updating scheme employed an algorithm based on the Kozeny-Carman equation to relate changes in porosity to changes in hydraulic conductivity (Li et al. 2005).

Reactive transport was simulated with RT3D (Clement 1997), which uses the head solution from MODFLOW as input. The initial concentration of each species in the PRB was assumed to be zero. The upstream boundary was assigned the concentration of ions in groundwater, and was assumed to be spatially uniform and time invariant. The bottom boundary was assigned as no flux. All other boundaries were assigned the Cauchy boundary condition with no dispersive flux. Groundwater entering the PRB was assumed to be anaerobic and have the following composition: DO = 10^{-8} M, Fe^{2+} = 10^{-10} M, Ca^{2+} = 10^{-3} M, OH^- = 10^{-7} M, HCO_3^- = 10^{-3} M, and CO_3^{2-} = 10^{-7} M.

The geochemical algorithm used in RT3D was developed by Li et al. (2006) for simulating geochemical reactions in PRBs. Redox reactions related to anaerobic iron corrosion and three mineral precipitation-dissolution reactions were included (Table 1). The reaction rate for anaerobic iron corrosion was assumed to be proportional only to the reactive surface area of the ZVI (Reardon 1995). The kinetics of mineral precipitation-dissolution were assumed to follow transition state theory (Lichtner et al. 1996). Rate coefficients obtained by calibration to field studies were used as input (Li et al. 2006). Li et al. (2006) show that predictions made with the geochemical algorithm are in general agreement with field data.

Table 1. Geochemical reactions in the PRB.

Reactions	Kinetic Rate[a]	Solubility Constant Log(K_{eq})
$Fe^0 + 2H_2O \rightarrow Fe^{2+} + H_2(aq) + 2OH^-$	$k_{an,Fe}S$	-
$CaCO_3(s) \leftrightarrow Ca^{2+} + CO_3^{2-}$	$-k_{Ca}\left(1 - a_{Ca^{2+}} a_{CO_3^{2-}} / K_{eq}\right)$	-8.1
$FeCO_3(s) \leftrightarrow Fe^{2+} + CO_3^{2-}$	$-k_{Fe1}\left(1 - a_{Fe^{2+}} a_{CO_3^{2-}} / K_{eq}\right)$	-10.5
$Fe(OH)_2(am) \leftrightarrow Fe^{2+} + 2OH^-$	$-k_{Fe}\left(1 - a_{Fe^{2+}} a_{OH^-}^2 / K_{eq}\right)$	-15.2

[a]k = rate coefficient, S = reactive surface area of ZVI, a = activity of aqueous species.

Minerals precipitates formed in the iron media were assumed to be immobile and the pore space occupied by these minerals was estimated from the molar volume of each mineral. The pore volume reduction at a given location was computed as the total pore volume occupied by the mineral precipitates at the location less the pore volume gained by dissolution of iron.

Residence times in the PRB were defined using the particle tracking code Path3D (Zheng 1991) using the head solution from MODFLOW as input. One thousand particles were released from a 25 m wide by 10 m deep source located 20 m up gradient from the PRB. Path3D calculates the position and travel time of each particle across the domain by moving particles through a series of time steps. The residence time in the PRB was calculated for each particle as the difference between the times at which the particle entered and left the PRB.

EFFECTIVENESS

A series of simulations were conducted to evaluate the two strategies for limiting mineral fouling and its impact on hydraulic behavior of PRBs. The impact of these two strategies on hydraulic properties and behavior of PRBs are evaluated in terms of porosity reductions in the ZVI and residence times. Simulations were conducted for 30 yr of operation, which is a common upper bound for the service life of PRBs (Li et al. 2005).

Pea Gravel Equalization Zones

Simulations were conducted to assess the effectiveness of pea gravel equalization zones by adding 0.5-m-thick zones having a hydraulic conductivity of 864 m/d and porosity of 0.4 directly up gradient and down gradient of the 1-m-thick PRB with 100% ZVI. The pea gravel zones had the same lateral and vertical dimensions as the PRB (25-m wide and 10-m deep).

Box plots showing the distribution of residence time in the PRB with and without pea gravel zones at the onset of operation and after 30 yr are shown in Fig. 2. Installation of pea gravel zones results in more uniform residence times (4.4-16.0 d compared vs. 5.9-12.7 d with pea gravel zones), but no change in the median residence time remains (8.8 d without pea gravel zones, 8.9 d with pea gravel zones). More importantly, addition of pea gravel zones reduces the frequency of very short residence times, which can have the greatest impact on mineral fouling as well as quality of groundwater downstream of the PRB. This effect is realized initially and after 30 yrs, with more complete equalization the PRB ages.

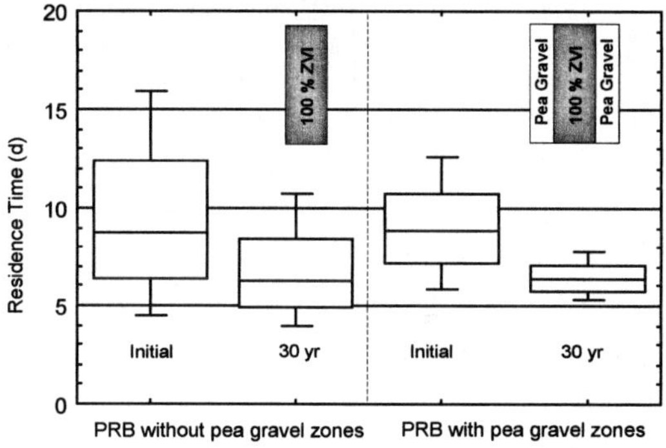

FIG. 2. Box plots of residence time in the PRB initially and after 30 yr of operation with and without pea gravel zones.

Average and maximum porosity reductions in the PRB (100% ZVI reactive zone) with and without the pea gravel zones are shown in Fig. 3 as a function of distance from the entrance face of the PRB at 30 yr. The average porosity reduction reaches a peak near the entrance face (≈ 0.1 m), which is followed by a decrease and then leveling off (Fig. 3). The maximum porosity reduction persists farther into the PRB, and diminishes less rapidly than the average porosity reduction, because mineral forming ions are being transported along preferential flow paths due to flow heterogeneity (Li et al. 2005).

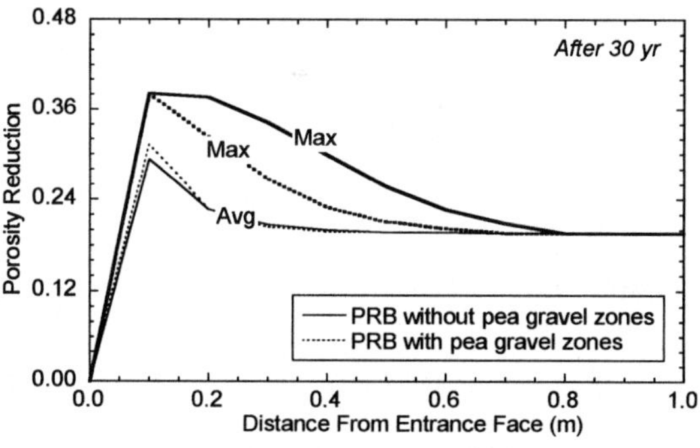

FIG. 3. Maximum and average porosity reduction as a function of distance from the entrance face of the PRB with and without pea

Pea gravel equalization zones result in a slight increase in average porosity reduction, and have no effect on the peak maximum porosity reduction (Fig. 3). Additionally, at greater distances in the PRB, the maximum porosity is smaller. Flow equalization by the pea gravel zones reduces the prevalence and impact of preferential flow paths, which are known to control maximum porosity reductions deeper in the reactive zone of PRBs (Li et al. 2005).

Sacrificial Pre-treatment Zone

Simulations to evaluate sacrificial pre-treatment zones were conducted with a 0.5-m-thick gravel-ZVI mixture placed directly up gradient of a 1-m-thick reactive zone containing 100% ZVI. The sacrificial pre-treatment zone had the same lateral and vertical dimensions as the 100% ZVI zone. The percentage of ZVI (13% by volume) and the porosity (0.4) of the pre-treatment zone at Monticello, UT (Morrison 2003) were used as input. The hydraulic conductivity of the pre-treatment zone was set at 640 m/d, which falls between typical hydraulic conductivities of gravel and ZVI.

Maximum porosity reductions in the reactive zone after 30 yr with and without a sacrificial pre-treatment zone are shown in Fig. 4a as a function of distance from the entrance face. Including a pre-treatment zone reduces the maximum porosity reduction within the PRB by shifting precipitation into the pre-treatment zone. The effect of mineral precipitation on the pre-treatment zone is modest, with a maximum porosity reduction of only 0.06.

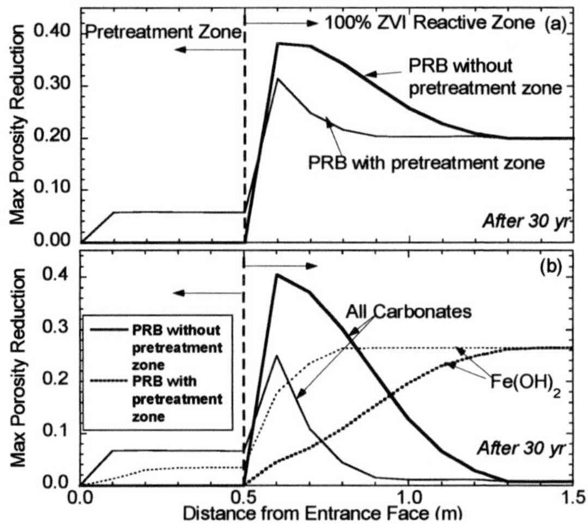

FIG. 4. Maximum porosity reductions after 30 yr as a function of distance from the entrance face with and without a pre-treatment zone: (a) total due to all minerals and (b) maximum porosity reductions due to carbonates and $Fe(OH)_2$.

The shift in porosity reductions is caused by precipitation of carbonate minerals in the pre-treatment zone due to elevated pH (pH) (Fig. 5). HCO_3^- is consumed when minerals precipitate in the pre-treatment zone, thereby limiting the amount of HCO_3^- available for precipitation of carbonate minerals in the reactive zone (Figs. 4b, 5). This effect could be enhanced by

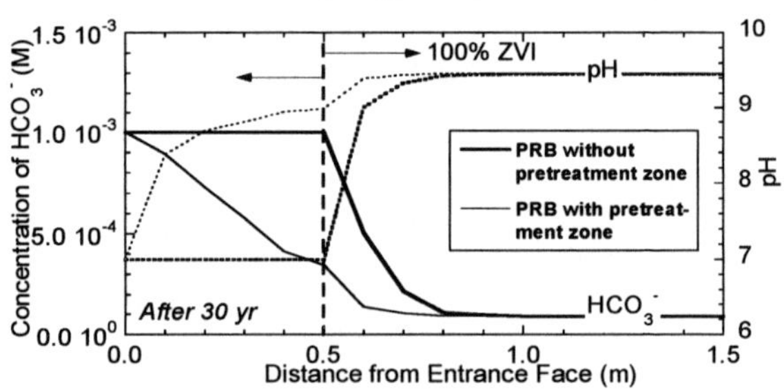

FIG. 5. pH and HCO_3^- concentration at 30 yr as a function of distance from the entrance face of the PRB with and without a pre-treatment zone.

adding more ZVI to the pre-treatment zone. However, even if the ZVI content was increased, precipitation within the reactive zone would not be eliminated, as $Fe(OH)_2$ will form in response to corrosion of iron within the reactive zone (Fig. 5).

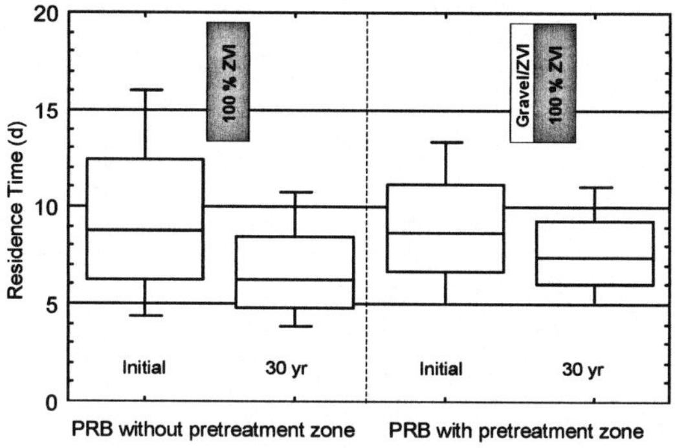

FIG. 6. Box plots of residence time in the PRB initially and after 30 yr of operation with and without pre-treatment zone.

Box plots of residence times in the PRB with and without the sacrificial pre-treatment zone are shown in Fig. 6 for conditions initially and after 30 yr of operation. Because the pre-treatment zone also provides for flow equalization, residence times within the PRB are more uniform when a pre-treatment zone is included. Longer and slightly more uniform residence times are also achieved with a pre-treatment zone after 30 yrs due to less mineral fouling within the reactive zone.

CONCLUSIONS

A series of simulations were conducted with the groundwater flow model MODFLOW and reactive transport model RT3D to evaluate the effectiveness of two strategies intended to enhance the long-term hydraulic behavior of PRBs: (i) adding pea gravel equalization zones up gradient and down gradient of the reactive zone and (ii) placement of a sacrificial pre-treatment zone up gradient of the reactive zone. For both strategies, porosity reductions in the PRB over time were determined and the impact on residence times was investigated.

Installation of pea gravel zones results in flow equalization and a more uniform distribution of residence times within the PRB, but has negligible impact on the median residence time in the near term and in the long term (30 yr). Average porosity reductions increase slightly when pea gravel zones are added, but maximum porosity reductions are reduced, particularly in deeper portions of the reactive zone.

Sacrificial pre-treatment zones provide a region where the groundwater pH is elevated and mineral forming ions are consumed, resulting in less precipitation of secondary minerals within the reactive zone. Consequently, residence times within the PRB are less affected by mineral precipitation when a pre-treatment zone is employed. However, pre-treatment zones do not eliminate porosity reductions completely, as secondary minerals (e.g., $Fe(OH)_2$) will still form within the reactive zone in response to iron corrosion.

REFERENCES

Clement, T.P. (1997). *RT3D: A Modular Computer Code for Simulating Reactive Multi-Species Transport in 3-Dimensional Groundwater Aquifers.* PNNL-SA-11720, Pacific Northwest National Laboratory, Richland, Washington.

Elder, C.R., Benson, C.H., and Eykholt. G.R. (2002). "Effects of heterogeneity on influent and effluent concentrations from horizontal permeable reactive barriers". *Water Resources Research,* Vol. 38(8): Art. No. 1152.

Hemsi, P.S., and Shackelford, C.D. (2006). "An evaluation of the influence of aquifer heterogeneity on permeable reactive barrier design". *Water Resources Research,* Vol. 42, W03402, doi: 10.1029/2005WR004629.

Jeen, S., Mayer, K.U., Gillham, R.W., and Blowes, D.W. (2007). "Reactive transport modeling of trichloroethene treatment with declining reactivity of iron". *Environmental Science & Technology,* Vol. 41 (4): 1432-1438.

Kamolpornwijit, W., Liang, L., West, O.R., Moline, G.R., and Sullivan, A.B. (2003). "Preferential flow path development and its influence on long-term PRB performance: column study". *J. of Contaminant Hydrology,* Vol. 66: 161-178.

Li, L., Benson, C.H., and Lawson, E.M. (2005). "Impact of mineral fouling on hydraulic behavior of permeable reactive barriers". *Groundwater,* Vol. 43 (4), 582-596.

Li, L., Benson, C.H., and Lawson, E.M. (2006). "Modeling porosity reduction caused by mineral fouling in continuous-wall permeable reactive barriers." *Journal of Contaminant Hydrology,* Vol. 83, 89-121.

Lichtner, P.C., Steefel, C.I. and Oelkers, E.H. (1996). *Reactive Transport in Porous Media.* Mineralogical Society of America, Washington, DC.

Mackenzie, P.D., Horney, D.P., and Sivavec, T.M. (1999). "Mineral precipitation and porosity losses in granular iron columns". *J. of Hazardous Materials,* Vol. 68 (1-2):1-17.

Mayer, K.U., Blowes, D.W., and Frind, E.O. (2001). "Reactive transport modeling of an in situ reactive barrier for the treatment of hexavalent chromium and trichloroethylene in groundwater". *Water Resour. Res.,* Vol. 37(12): 3091-3103.

McDonald, M.G., and Harbaugh, A.W. (1988). *MODFLOW: A modular three-dimensional finite-difference ground-water flow model.* USGS, Washington, DC.

McMahon, P.B., Dennehy, K.F., and Sandstrom, M.W. (1999). "Hydraulic and geochemical performance of a permeable reactive barrier containing zero-valent iron, Denver Federal Center". *Groundwater,* Vol. 37 (3): 396-404.

Morrison, S.J. (2003). "Performance evaluation of a permeable reactive barrier using reaction products as tracers". *Environmental Science & Technology,* Vol. 37 (10): 2302-2309.

Naftz, D.L., Morrison, S.J., Davis, J.D., and Fuller, C.C. (2002). *Handbook of Groundwater Remediation Using Permeable Reactive Barriers: Applications to Radionuclides, Trace metals, and Nutrients.* Academic Press, San Diego, CA.

Reardon, E.J. (1995). "Anaerobic corrosion of granular iron: Measurement and interpretation of hydrogen evolution rates". *Environmental Science & Tech.,* Vol. 29 (12): 2936-2945.

Sarr, D. (2001). "Zero-valent iron permeable reactive barriers-How long will they last?" *Remediation,* Vol. 11 (2): 1-18.

Wilkin, R.T., Puls, R., and Sewell, G.W. (2003). "Long-term performance of permeable reactive barriers using zero-valent iron: geochemical and microbiological effects". *Groundwater,* Vol. 41 (4): 493-503.

Yabusaki, S., Cantrell, K., Sass, B., and Steefel, C. (2001). "Multicomponent reactive transport in an in situ zero-valent iron cell". *Enviro. Sci. & Tech.,* Vol. 35 (7): 1493-1503.

Zheng, C. (1991). *Path3D: A ground-water path and travel-time simulator.* S.S.Papadopulus & Associates, Inc., Bethesda, Maryland.

Modeling Geochemical and Reactivity Changes of Different Iron Materials

Sung-Wook Jeen[1], Jin suk O[2] and Robert W. Gillham[3]

[1]Post Doctoral Fellow, Department of Earth and Environmental Sciences, University of Waterloo, Waterloo, Ontario, Canada N2L 3G1; sjeen@sciborg.uwaterloo.ca
[2]Geoenvironmental Scientist, Peto MacCallum Ltd., 165 Cartwright Avenue, Toronto, Ontario, Canada M6A 1V5; ojs692000@yahoo.ca
[3]Professor, Department of Earth and Environmental Sciences, University of Waterloo, Waterloo, Ontario, Canada N2L 3G1; rwgillha@sciborg.uwaterloo.ca

ABSTRACT: Including declining reactivity of iron, resulting from precipitation of secondary minerals, into reactive transport models is a key component for better estimation of longevity of iron permeable reactive barriers (PRBs). In this study, the accumulation of secondary minerals and reactivity loss were coupled using an empirically-derived relationship that was incorporated into an existing multi-component reactive transport code. The simulation results were compared to the observations from column experiments, which were designed to evaluate the changes of the reactivity of different iron materials for cis-dichloroethene (cis-DCE) treatment in the presence of dissolved $CaCO_3$. The model provided a reasonable representation of the evolution of iron reactivity toward cis-DCE treatment and the changes in geochemical conditions for each material. The modeling results suggest that iron material having a high corrosion rate is not beneficial in the presence of a high concentration of dissolved $CaCO_3$ because of a faster migration of cis-DCE profiles and greater porosity loss closer to the influent end. This study shows that declining reactivity of iron due to mineral precipitation should be considered at the design stage of iron PRB construction.

INTRODUCTION

In granular iron permeable reactive barriers (PRBs) receiving high concentrations of inorganic constituents in groundwater, accumulation of secondary minerals has raised concerns regarding the long-term performance in terms of reactivity and permeability (e.g., Mackenzie et al. 1999; Agrawal et al. 2002). In particular, carbonate mineral precipitation has been found to be a common cause of passivation of iron, resulting in reduced treatment efficiency and progressive migration of the mineral precipitation front (Zhang and Gillham 2005; Jeen et al. 2006). Therefore, including declining reactivity of iron, resulting from precipitation of secondary minerals, into predictive models is a key component for estimation of longevity of PRBs.

Because the chemical reactions and transport processes within reactive barriers are complex and interactive, to quantify the effects of mineral precipitation, or to predict future behavior of PRBs requires coupled transport and reaction models. In only a few studies, reactive transport and geochemical processes have been combined within predictive models (Mayer et al. 2001; Li et al. 2005). However, these studies did not include the change of the iron reactivity resulting from mineral precipitation, and thus predictions based on these types of models may be limited to the performance of iron PRBs for relatively short periods of time. More recently, Jeen et al. (2007) incorporated the reactivity change into the reactive transport model MIN3P (Mayer et al. 2002). In the model, the accumulation of secondary minerals and reactivity loss were coupled using an empirically-derived relationship by modifying the kinetic expressions of MIN3P. The simulation results reproduced the observations from the column experiments of Jeen et al. (2006), for the treatment of trichloroethene (TCE), reasonably well and demonstrated the potential predictive capability of the model. The model was the first attempt to directly incorporate the effects of mineral precipitation on the reactivity of iron in a reactive transport model.

One of the important implications from the results of Jeen et al. (2007) is that iron having moderate initial reactivity may be advantageous compared to iron having higher reactivity for the long-term performance of PRBs in the presence of a high concentration of dissolved $CaCO_3$. It was hypothesized that the high initial precipitation rate in iron having higher reactivity will result in faster accumulation of precipitates, a greater reactivity loss, and thus, faster migration of the contaminant removal front. In contrast, an iron PRB containing less reactive material will show a slower accumulation of precipitates, and thus slower migration of the contaminant removal front.

The objective of this study was to evaluate the applicability of the modified MIN3P model to the data of column experiments, which used different sources of iron materials for the treatment of cis-dichloroethene (cis-DCE) in the presence of dissolved $CaCO_3$. The particular purpose was to test the hypothesis that iron materials of moderate reactivity may give improved long-term performance relative to materials of high reactivity. In addition to supporting the predictive capability of the model, the results of this study have significant implications in selecting materials to be used in construction of PRBs.

EXPERIMENTAL SECTION

The column experiments were designed to evaluate the changes in reactivity of different iron materials (Connelly, Gotthart-Maier, Peerless, and ISPAT irons) for cis-DCE (10 mgL^{-1}) treatment in the presence of dissolved $CaCO_3$ (300 mgL^{-1} $CaCO_3$). The column design was similar to that described in Jeen et al. (2006), except larger in diameter (3.81 cm). Each column consisted of a 50 cm long clear Plexiglas™ tube. Nine sampling ports were located at 5 cm intervals along the column with two additional ports at 2.5 and 7.5 cm from the influent end. The identical feed solution was pumped to each column at a flow rate of approximately 0.4 $mLmin^{-1}$. The flow velocity varied between the columns as a consequence of differences in porosity, but the initial values were close to 1.0 md^{-1}. Samples were collected periodically for analyses of organic and inorganic parameters. Sampling was carried out from every week to 2 months for a total of 13 sampling events for Connelly and Gotthart-Maier irons and for a total of 7 sampling events for Peerless and ISPAT irons. Columns were operated for approximately 8 months.

MODELING APPROACH

In the modified MIN3P (Jeen et al. 2007), to incorporate the effect of mineral precipitation on reactivity of iron, which is variable in time and space, various kinetic reaction expressions were modified to update the reactive surface area along the flow path at different times using Eq. 1:

$$S(\mathrm{x,t}) = S_0 \exp\left(-\sum_i \alpha_i \varphi_i(\mathrm{x,t})\right) \tag{1}$$

where $S(\mathrm{x,t})$ is the reactive surface area of iron at a specific location along the flow path and time (m^2 iron L^- bulk), S_0 is the initial reactive surface area of the iron (m^2 iron L^- bulk), α_i is the proportionality constant for mineral phase *i*, which represents the extent to which mineral phase *i* contributes to the reactivity loss of the iron, and $\varphi_i(\mathrm{x,t})$ is the volume fraction of mineral phase *i* at a specific location and time (-). Eq. 1 was proposed from observations from a previous column study (Jeen et al., 2006) and is an empirical equation that relates the reactivity of iron to the amount of secondary minerals. The reactive surface area in Eq. 1 is a mathematical representation of the reactivity of iron. During the simulation, as secondary minerals precipitate and iron dissolves, the porosity and hydraulic conductivity are also updated to reflect the changes. In this study, the same modified MIN3P code as that presented in Jeen et al. (2007) was applied to the results of the column experiments.

For the simulation, cis-DCE degradation, iron corrosion, and secondary mineral precipitation were included as the important chemical reactions for the reaction network. The initial cis-DCE rate constants were taken to fit the initial cis-DCE profiles for each column. The initial rate constants were similar for Connelly, Gotthart-Maier, and Peerless irons, with the value for Connelly iron slightly higher than those for Gotthart-Maier and Peerless, whereas ISPAT iron had the lowest degradation rate (the first-order rate constants of 2.14×10^{-3}, 8.82×10^{-4}, 7.82×10^{-4}, and 5.77×10^{-5} s^{-1}, for Connelly, Gotthart-Maier, Peerless, and ISPAT, respectively). The initial iron corrosion rate used for the simulation was the lowest for Connelly iron, Gotthart-Maier and Peerless irons were similar and intermediate between Connelly and ISPAT irons, and ISPAT iron had the highest value (the corrosion rates of 6.63×10^{-7}, 7.00×10^{-7}, 7.13×10^{-7}, and 5.64×10^{-6} mol L^{-1} bulk s^{-1}, for Connelly, Gotthart-Maier, Peerless, and ISPAT, respectively). The differences in corrosion rates could be caused by differences in composition of iron surfaces and geometric factors such as packing and geometry of the particles. It was assumed that the two main carbonate mineral precipitates were aragonite and iron hydroxy carbonate, as identified in Jeen et al. (2006). $Fe(OH)_2$(am) was included as the initial phase for representative iron (hydr)oxide. The precipitation rate constants for the carbonate minerals and their proportionality constants were fixed to one set of values for all four columns. Once a set of parameters was determined, they were not calibrated further for each column.

RESULTS AND DISCUSSION

Simulation for the Column Experiments

Fig. 1 shows the measured and simulated cis-DCE profiles at selected times for each column. Consistent with the experimental results, the simulated trends for Connelly, Gotthart-Maier, and Peerless irons are rather similar, although Gotthart-Maier and Peerless irons have slightly more passivation near the influent end and thus lower degradation rates in that region compared to Connelly. ISPAT shows distinct features compared to the other columns. The initial degradation rate is significantly lower than the other columns, and almost complete reactivity loss is observed staring from the region close to the influent end because of greater accumulation of carbonates in that region. However, the initial degradation rate is generally maintained where there is no significant accumulation of carbonates, i.e., the region close to the effluent end.

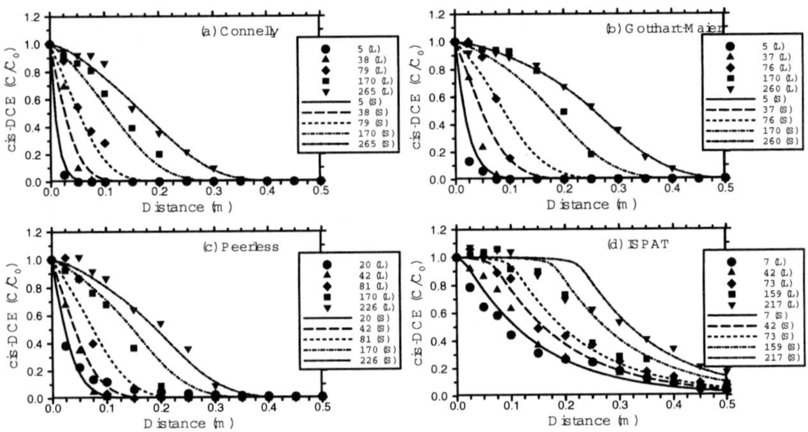

FIG. 1. The measured and simulated cis-DCE profiles for (a) Connelly, (b) Gotthart-Maier, (c) Peerless, and (d) ISPAT iron. Figures in legend represent the time periods as day, and L and S represent the laboratory and simulated results, respectively.

The simulated cis-DCE profiles do not match the measured profiles precisely. For example, the simulated profiles deviate from the measured profiles, particularly at intermediate time periods for Gotthart-Maier and Peerless irons (Fig. 1b and 1c). Nevertheless, the simulated profiles for all columns give a good representation of the major trends in the data, considering that there was no calibration of the model parameters for each column. More importantly, the significant difference in the pattern of passivation is successfully demonstrated. Consistent with the laboratory observations, for Connelly iron, there is no expanding (highly passivating) zone where there appears to be little degradation. In contrast, the expanding zone adjacent to the influent end for ISPAT iron, resulting from the higher iron corrosion rate, is well reflected in the simulation results. The simulations also show that the leading edge of the degradation profiles is somewhat advanced in Gotthart-Maier iron compared to Connelly iron.

It is important to note that except for the initial cis-DCE rate constants and the iron corrosion rates, indicative of reactivities of the irons, all other model parameters are the same for all columns. Thus, the differences in the simulated performance of the columns were due solely to the differences in the initial cis-DCE rate constants, which are intrinsic degradation characteristics of each material toward degradation of specific chemicals, and to the differences in iron corrosion rates, which govern the carbonate precipitation and thus passivation of iron materials over time. This indicates that matching the data was not simply a random curve fitting process. It would appear that changing the reactivity resulted in changes in profile behavior that were consistent with the data, showing that the proposed hypothesis can be represented by the model.

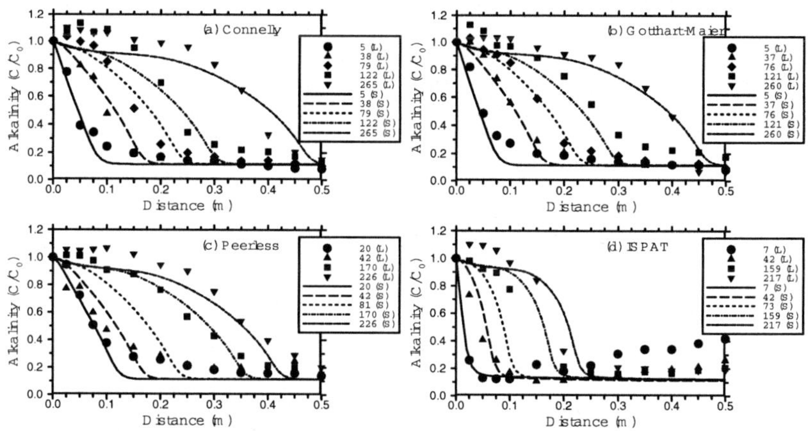

FIG. 2. The measured and simulated alkalinity profiles for (a) Connelly, (b) Gotthart-Maier, (c) Peerless, and (d) ISPAT iron. Note that the data at day 81 and 73 for Peerless and ISPAT iron, respectively, were not measured. Figures in legend represent the time periods as day, and L and S represent the laboratory and simulated results, respectively.

The matching of the alkalinity profiles between the measured and simulated results was also relatively successful (Fig. 2). The alkalinity profiles for Connelly, Gotthart-Maier, and Peerless irons are similar. The differences in migration of the alkalinity profiles between these three columns were less obvious than for the cis-DCE profiles, because the migration of the alkalinity profiles are governed only by the iron corrosion rate, whereas the migration of the cis-DCE profiles are affected both by the cis-DCE degradation rate and iron corrosion rate.

ISPAT iron has a much higher corrosion rate compared to the other irons. As a result, the first simulated alkalinity profile for ISPAT (Fig. 2d) is much steeper than for the other irons, indicating a higher alkalinity removal rate (higher carbonate precipitation rate) at early time, consistent with the laboratory results. However, the gradual increase in alkalinity from 15 cm from the influent end toward the effluent end of the ISPAT column on days 7 and 42 (Fig. 2d), which is due to a significant contribution of hydroxide alkalinity to the measured total alkalinity resulting from a very sharp pH increase, could not be reproduced by the simulation. Over time, both the measured and simulated alkalinity profiles for ISPAT iron did not advance much further toward the effluent end compared to the other columns for similar time periods (Fig. 2d), resulting in greater accumulation of carbonates near the influent end, which in turn resulted in greater passivaiton of the iron in that region. This pattern of accumulation of carbonates in the simulation was generally consistent with those observed in the acid digestion analyses for solid samples conducted after completion of the experiment.

In summary, the model provided a reasonable representation of the differences in the performance of the four columns, demonstrated to be consistent with the differences in iron corrosion rates. Although there are relatively small differences in reactivity and thus relatively small differences in the behavior of the various profiles over time between Connelly, Gotthart-Maier, and Peerless irons, the behavior for ISPAT iron confirms the proposed hypothesis. Thus, the simulation results appear to support the use of the model as a tool for predicting long-term performance of PRBs containing iron materials of differing reactivities.

Long-Term Prediction

It will be useful to test the effects of different reactivities on the performance of a PRB at a typical groundwater flowing condition. Thus, further simulations were performed to predict the differences in performance over longer periods of time for different iron materials. The longevities of each material were simulated under the following conditions. The groundwater velocity was assumed to be 0.1 md^{-1}. The cis-DCE and $CaCO_3$ concentrations were the same as in the column experiments (10 mgL^{-1} cis-DCE + 300 mgL^{-1} $CaCO_3$). The pH of the inflowing water was assumed to be 7.0, and the porosity and hydraulic conductivity of the PRB were assumed to be 0.5 and 6.57×10^{-5} ms^{-1}, respectively. The thickness of the PRB was 0.5 m and the period of operation was 40 years. For the four cases tested, the characteristics of the iron materials, such as the initial cis-DCE degradation rates and iron corrosion rates, were selected from the values representative of Connelly, Gotthart-Maier, Peerless, and ISPAT irons, as measured in the column experiments. All other parameters were taken from the values used in the simulations for the column experiments.

Fig. 3 shows the predicted breakthrough curves for cis-DCE at the effluent face of the PRB for the four iron materials. Under the imposed conditions, cis-DCE begins to breakthrough the PRB after about 10, 15, 16 and 23 years of operation, for ISPAT, Peerless, Gotthart-Maier, and Connelly irons, respectively. ISPAT iron, which has the highest corrosion rate and the lowest cis-DCE degradation rate, shows the earliest breakthrough, representing the worst performance. On the other hand, Connelly iron, which has the lowest corrosion rate and the highest cis-DCE degradation rate, shows the latest breakthrough. The steepness of the breakthrough curves is also dependent on the corrosion rate; i.e., the materials with higher corrosion rates have steeper curves than those of lower rates (Fig. 3). Depending on the incoming cis-DCE concentration and the remediation goal (e.g., maximum contaminant level (MCL)), this could have important consequences. For example, if the MCL corresponds to a C/C_0 value of 0.1, the Connelly iron would perform adequately for a period of more than 33 years, while ISPAT iron would fail at about 13 years.

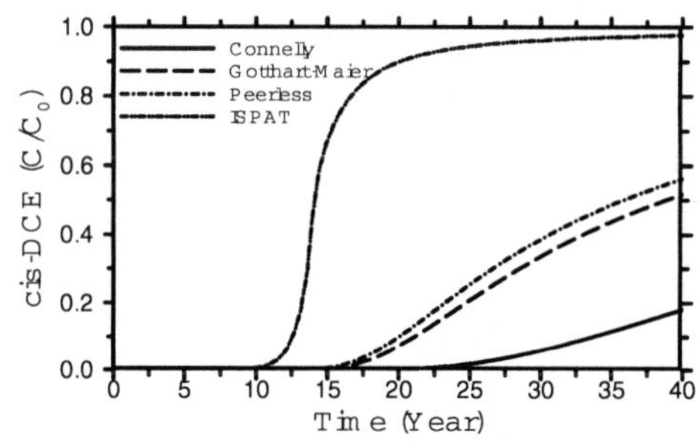

FIG. 3. The predicted cis-DCE breakthrough curves for each material at the effluent face of the PRB over a period of 40 years.

Because of more complex water chemistry and possibly other factors, the simulation results may not be directly applicable to particular field conditions. In particular, different mixtures of pollutants may develop different passivating secondary precipitates. For example, reduction of strong oxidants, such as chromate and nitrate, will result in precipitation of highly passivating iron (oxy)hydroxides. Modeling then should include those passivating minerals, based on the extent to which they affect iron reactivity. However, the predictions show how iron materials having different reactivitites affect the behavior of the cis-DCE degradation. The simulations thus provide valuable insight for selecting iron materials to be used in construction of PRBs.

CONCLUSIONS

The previous modeling study of Jeen et al. (2007) hypothesized that iron material having moderate reactivity may be more beneficial than material having a higher reactivity in terms of long-term performance because of less overall accumulation of secondary precipitates. One of the objectives of this study was to evaluate the validity of this hypothesis. The results of the numerical simulations, along with the results of the column experiments, generally support this hypothesis.

The difference in the pattern of accumulation of carbonate precipitates for each material was governed by the initial iron corrosion rate, and it resulted in a significant difference in the pattern of iron passivation. Whereas the iron with high corrosion rate (e.g., ISPAT iron) showed a highly passivated zone adjacent to the influent end, passivation of the iron with low corrosion rate (e.g., Connelly iron) appeared more as a decline in slope of the degradation profile, starting from the influent end, with minor evidence of a highly passivated zone. These patterns resulted in differences in onset of cis-DCE breakthrough and relative steepness of the breakthrough curves for the predictive simulations, which affected significantly the estimated longevities.

The pattern of accumulation of the precipitates also affected the pattern of permeability changes. It suggests that iron materials with high corrosion rates may experience local clogging problems, particularly near the influent face of the PRBs. Iron corrosion rate in the presence of dissolved inorganics should thus be accounted for as an important factor at the design stage of PRB installation to determine longevity of the PRB.

ACKNOWLEDGMENTS

Funding for this research was provided from the NSERC/DuPont/EnviroMetal Industrial Research Chair held by R.W. Gillham.

REFERENCES

Agrawal, A., Ferguson, W.J., Gardner, B.O., Christ, J.A., Bandstra, J.Z., and Tratnyek, P.G. (2002). "Effects of carbonate species on the kinetics of dechlorination of 1,1,1-trichloroethane by zero-valent iron." *Environ. Sci. Technol.*, 36 (20): 4326-4333.

Jeen, S.-W., Gillham, R.W., and Blowes, D.W. (2006). "Effects of carbonate precipitates on long-term performance of granular iron for reductive dechlorination of TCE." *Environ. Sci. Technol.*, 40 (20): 6432-6437.

Jeen, S.-W., Mayer, K.U., Gillham, R.W., and Blowes, D.W. (2007). "Reactive transport modeling of trichloroethene treatment with declining reactivity of iron." *Environ. Sci. Technol.*, 41 (4): 1432-1438.

Li, L., Benson, C.H., and Lawson, E.M. (2005). "Impact of mineral fouling on hydraulic behavior of permeable reactive barriers." *Ground Water*, 43 (4): 582-596.

Mackenzie, P.D., Horney, D.P., and Sivavec, T.M. (1999). "Mineral precipitation and porosity losses in granular iron columns." *J. Hazard. Mater.*, 68 (1-2): 1-17.

Mayer, K.U., Blowes, D.W., and Frind, E.O. (2001). "Reactive transport modeling of an in situ reactive barrier for the treatment of hexavalent chromium and trichloroethylene in groundwater." *Water Resour. Res.*, 37 (12): 3091-3103.

Mayer, K.U., Frind, E.O., and Blowes, D.W. (2002). "Multicomponent reactive transport modeling in variably saturated porous media using a generalized formulation for kinetically controlled reactions." *Water Resour. Res.*, 38 (9): 1174, doi:10.1029/2001WR000862.

Zhang, Y., and Gillham, R.W. (2005). "Effects of gas generation and precipitates on performance of Fe^0 PRBs." *Ground Water*, 43 (1): 113-121.

Passive Reactive Berm to Provide Low Maintenance Lead Containment

Gregory O'Connor[1], Steve Larson[2], PhD, W. Andy Martin[2], John Niles[1]

[1] US Army RDECOM-ARDEC, Building 355, Picatinny Arsenal, NJ 07806-5000

[2] US Army Corps of Engineers, Engineer Research and Development Center, 3909 Halls Ferry Rd, Vicksburg, MS 39180

ABSTRACT: A program was conducted to determine the viability of containing metallic residue from firing ranges by treating the range area chemically to allow passive blocking of the contaminants in sand, a technique called passive reactive berm (PRBerm™). Materials used in the PRBerm™ maintain near neutral soil conditions and promoting immobilization of metals. The PRBerm™ has the potential to provide a very useful, low cost tool for defense forces to maintain active firing ranges while managing the inevitable effects of munition residue in a low cost and effective manner.

INTRODUCTION

Live fire training at Small Arms Firing Ranges (SAFRs) is necessary to maintain mission readiness for our nations' war fighters. The U.S. Department of Defense (DoD) oversees more than 3,000 active SAFRs (ITRC 2003). In addition to these facilities, it is estimated that 9,000 non-military outdoor ranges are now active in the U.S. as private firing ranges, largely unregulated despite the fact that their operations can result in soil, ground water, or surface-water contamination at levels that would trigger major remediation efforts at industrial sites (Houlihan andWiles 2002). There have been several Best Management Practices (BMPs) suggested to control lead migration from active SAFRs (Fabian and Watts 2005, ITRC 2005, United States Environmental Protection Agency (USEPA) 2001). Systematic development of SAFR sustainment technologies is necessary to address the feasibility and functionality of such technologies. In order to demonstrate that a BMP such as PRBerm, the passive blocking of contaminants in amended sand, will work under field conditions, and at a reasonable cost, extensive batch, pilot, and full-scale research efforts are needed. Small arms ammunition is made up largely of metallic lead with smaller amounts of alloying materials such as antimony (a hardening agent) along with copper and zinc, which are the primary components of shell casings and jackets (HQDA 1996;Weiss 2005). The USEPA estimate that 4 percent of the 2 million tons of all lead produced in the U.S. in the late 1990s was made into bullets or lead shot (USEPA 2001). All of the lead entering a berm on a firing range is initially present as metallic lead in either particulate or bullet form. As the berm system ages, the metallic lead undergoes corrosion processes that result in the release of lead cations. Depending on the environment within the berm, these lead ions can transport to groundwater as soluble lead, transport to surface water as soluble lead, ion exchange onto clay surfaces, react with carbonates to produce lead carbonates, associate with iron and manganese oxides, react with sulfides, associate with organic matter, or react with compounds such as phosphates, sulfates, and aluminum. Lead contamination can escape from SAFR sites through weathering effects, storm water runoff, and through leachate (ground water) transport. If not properly maintained, storm water runoff can erode the lead-contaminated berms and carry the lead and contaminated soil particles away from the SAFR. The threshold toxicity characterization concentration for lead in contaminated soils established by the Toxicity Characteristic Leachate Procedure

(TCLP) is 5.0 mg L^{-1} and the universal treatment standard (UTS) for lead is 0.75 mg L^{-1} (ITRC 2003).

Lead is an amphoteric metal that exhibits its greatest solubility in acidic (pH < 4) and alkaline (pH > 11) solutions. Under acidic conditions, elemental lead will oxidize, releasing a hydrated cation, Pb^{2+}. Under alkaline conditions, elemental lead will oxidize under most circumstances to form a lead hydroxide complex $Pb(OH)_5^{3-}$. Lead that exists in the dissolved state can be sorbed to negatively-charged clay particle surfaces. Thus, the erosion and surface water transport of contaminated clays can provide a major source of lead in the environment; environmental transport can be either attenuated or increased depending upon the mobility of the soil particles. In some cases, local soil characteristics will be adequate to retain Pb within the firing range and maintain surface water or leachate water Pb concentrations within applicable regulatory permit limits and/or action levels. In cases where the local soils have undesirable properties for range management purposes, such as high acidity, low sorption capacity, K_d, for lead, or easy erosion/transport of soil particles from the berm surfaces, partial replacement of the berm soil with a more suitable ballistic material may provide an economical alternative to steel bullet-trap type systems. Naturally-occurring silicate (e.g., quartz, feldspar) mineral sands are relatively low cost materials that are widely available, due to their extensive use in construction. Impact berms made from sand are able to decelerate fired projectiles safely, with minimal bullet fragmentation and lead-dust generation, as compared to compacted or coarse-grained soils (Larson et al. 2007). Sand as a berm material provides a number of advantages such as: hydraulic conductivity, chemical inertness, biological inertness, consistent texture, density, and particle size, low water retention capacity, and the refractory nature of silicate sands minimizes the potential fire hazards associated with tracer ammunition. A disadvantage to using sand as a deceleration medium in SAFR berms is the lack of buffering capacity of most commercially-available silicate sands. For this reason, unamended sand berms subject to acid rain conditions could see a progressive reduction in berm pH, to levels where both the rate of lead corrosion and extent of lead solubility in water are significantly increased. Because of this inherent limitation of sand for containing lead following firing and under long term weathering conditions, the technology proposed incorporates berm amendments that will provide the capability to buffer the sand berm pH and react with lead ions produced during the inevitable metallic lead corrosion. This study was initiated to determine the appropriate sand and amendment for use in the field demonstration at an active SAFR. The treatment goals to meet the regulatory requirements of the range were lead and copper concentrations below 0.150 mg L^{-1} and 0.500 mg L^{-1}, respectively, a pH between 6 and 9 in the leachate and runoff waters, and a TOC of less than 50 mg L^{-1} for the leachate and runoff waters.

EXPERIMENTAL DESIGN

The experimental design for this effort was conducted in two main efforts: 1) static lysimeter studies that tested sand and amendments by using berm material collected from actual firing ranges and obtaining the runoff and leachate from the lysimeters and 2) live fire lysimeters (LFL) studies that used sand and amendments that were exposed to a known number of rounds fired into them and then collecting the leachate from them. The static lysimeters (Figure 1) were constructed and loaded with contaminated soil according to the specifications of Larson et al. (2005). The appropriate amount of rainfall was introduced and controlled using a rainfall simulator constructed using porous polyethylene. Several amendments were tested in both of the lysimeter studies, but the study has not investigated all of the possible or potential amendments that may or may not work in the PRBerm concept at immobilizing the metals within the berm. There were several amendments used in the study; these amendments were composed of various compounds that provide metal stabilization properties such as phosphate, aluminum, and Fe-Mn based compounds. These amendments include several forms of Apatite II™, Buffer Blocks™, and TRAPPS™.

RESULTS

The average Pb, Cu, and DOC leachate concentrations leaving the Mechanical and Enzymatic Apatite II amended berm sand exceeded the Barksdale permit limits. The leachate concentrations for the SulfiTech and Buffer Block™ #5 amended berm sand performed better than the control (with no amendments), when examination of the

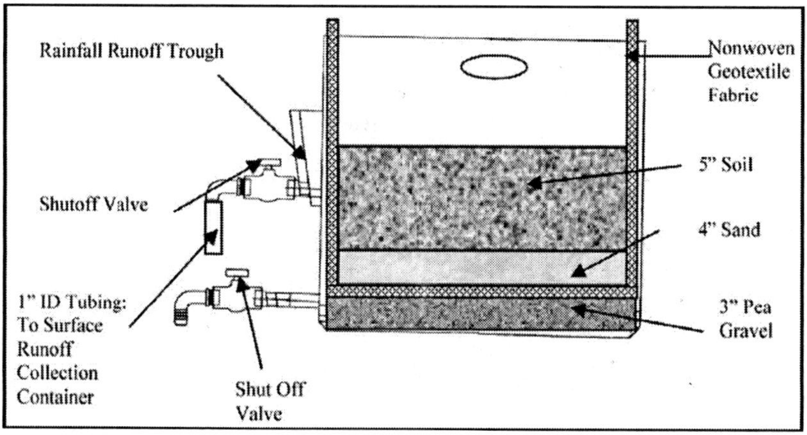

Figure 1. Cross section of a static lysimeter.

average leachate DOC is taken into account. Essentially, the SulfiTech A/T™ and Buffer Block™ #5 amendments created a biologically inert matrix within the lysimeter cells, which prevented significant DOC generation, while maintaining acceptable pH and leachate metals concentrations. The static lysimeter study is designed to provide a large volume of water in a short period of time. Under these simulated rainfall conditions, the ability to collect equal volumes of runoff and leachate exists; however, slow rainfall events are not simulated that allow the rainwater to percolate into the sand. These observations may help to explain why not all of the cells achieved the permit limit for total runoff lead concentrations.

There is a direct correlation between the amount of TSS leaving the lysimeter as runoff and leachate and the total Pb leaving the lysimeter as runoff and leachate. In general as more solids left the lysimeter, the Pb associated with the solids also left the lysimeter. The correlation coefficient for the lead associated with the TSS in the runoff is 0.6976, while in the leachate, it is 0.9484.

The live fire lysimeter (LFL) results indicate that the addition of the amendments in the sand berm media placed in the LFL's reduced the average concentration of Pb leaving the cells. There was a substantial decrease in the TSS leaving the cells with the addition of the different amendments. While there was a greater than 40 percent reduction of TSS for the TRAPPS™ amended cells as compared to the control cell, there was approximately 40 percent less leachate water collected for the TRAPPS™ amended cells than for the control and Buffer Block™ #5 amended cells. There was a substantial increase in DOC from the TRAPPS™ amendments than from the other two amendments, but this does not correspond with the TSS loss. The pH for all four systems remained within acceptable range.

CONCLUSIONS

The Passive Reactive Berm (PRBerm™) can be used as a practical, low cost technique to contain metals within four regulatory drivers, depending upon local conditions. The results from this study will be used to develop and improve SAFR berm technology through the incorporation of non-reactive ballistic sand with amendments for use during the field demonstration. However, the PRBerm™ may not be suitable for all range backstops. The range operations, climate conditions, and range soil conditions dictate what type of application, if any, is needed in order to reduce metals migration off the SAFRs. An alternative to the PRBerm™ is a fully contained bullet trap system. Although a steel bullet trap will reduce the potential metals migration into the environment, there is a significant cost and loss to training value associated with such a system as compared to the PRBerm™. Maintaining readiness of our armed forces is a perpetual challenge. Maintaining the facilities used by our forces for readiness is also a significant challenge. The results presented have the potential to provide a very useful, low cost tool for defense forces to maintain active firing ranges while managing the inevitable effects of munition residue in a low cost and effective manner. In addition there is the potential for application of the PRBerm™ at private shooting ranges.

ACKNOWLEDGEMENTS

We gratefully acknowledge the technical assistance provided by Ms. Haley Parsons

(ERDC-GSL), Mr. Chris Griggs, Ms. Michelle Thompson (Applied Research Associates, Inc.), Mr. Casey Trest (Mississippi State University), Ms. Allison Hearn (University of Southern Mississippi), Ms. Tarmiko Graham and Mr. Kerry Taylor (Jackson State University). We also wish to acknowledge the gracious participation of Ms. Barbara Legan, Barksdale Air Force Base (AFB) LA.

DISCLAIMER: The contents of this paper are not to be used for advertising, publication, or promotional purposes. Citation of trade names does not constitute an official endorsement or approval of the use of such commercial products. All product names and trademarks cited are the property of their respective owners. The findings of this report are not to be construed as an official Department of the Army position unless so designated by other authorized documents.

REFERENCES

Conca, J. L., Lu, N., Parker, G., Moore, B., Adams, A., Wright, J., Heller, P. (2000). PIMS-Remediation of Metal Contaminated Waters and Soils. Remed. Chlorinated Recalcitrant Compounds 7, 319-326.

Fabian, G. and Watts, K. (2005). Army Small Arms Training Range Environmental Best Management Practices (BMPs) Manual. SFIM-AEC-AT-CR-2006007. US Army Environmental Center, Aberdeen Proving Ground, MD.

Headquarters, Department of the Army (HQDA). (1996). "Army Ammunition Data Sheets Small Caliber Ammunition FSC 1305," Technical Manual (TM) 43-0001-27, Washington D.C.

Houlihan, J. and Wiles, R. (2002). Lead pollution at outdoor firing ranges. Environmental Working Group, Washington, D.C. Available at: www.ewg.org/reports/poisonouspastime/leadpoll.pdf

Interstate Technology and Regulatory Council (ITRC). (2003). Characterization and Remediation of Soils at Closed Small Arms Firing Ranges.

Interstate Technology and Regulatory Council (ITRC). (2005). Environmental Management at Operating Outdoor Small Arms Firing Ranges.

Larson, S.L., Tardy, B., Rainwater, K., and Tingle, J.S. (2005). Rainfall lysimeter evaluation of leachability and surface transport of heavy metals from six soils with and without phosphate amendment. ERDC-EL TR-05-9, U.S. Army Corps of Engineers, Engineer Research and Development Center, Vicksburg, MS.

Larson, S. L, Medina, V. F., Teeter, C., Martin, W. A. (2007). Treatment & Management of Closed or Inactive Small Arms Firing Ranges. ERDC-EL TR-07-6, U.S. Army Corps of Engineers, Engineer Research and Development Center, Vicksburg, MS.

Lu, N., Xu, J., Wright, J., Conca, J. L. (2001). PIMS-Remediation of Metal-Contaminated Groundwater and Soil using a Special Reactive Form of the Mineral Apatite. Appl. Miner. Res. Econ. Technol., Ecol. Culture. 2, 603-606.

Ma, Q. Y., Traina, S. J., Logan, T. J., Ryan, J. A. (1994). "Effects of aqueous Al, Cd, Cu, Fe(II), Ni, and Zn on Pb Immobilization by Hydroxyapatite," Environ. Sci. Technol. 1994, 28, 1219-1228.

W. A. Martin, S. Larson, D. Felt, J. Wright, C. Griggs, M. Thompson, J.L. Conca, and C. Nestler. (2007). The Effect of Organics on Lead Sorption onto Apatite II™. Applied Geochemistry, in press.

U.S. Environmental Protection Agency (USEPA). (2001b). "Best management practices for lead at outdoor shooting ranges," EPA/902-B-01-001, RCRA Compliance
Branch, New York.

United States Patent and Trademark Office (USPTO). (2001). "Low-lead leaching foamed concrete bullet barrier." U.S. Patent No. 6,264,735, Washington, DC.

United States Patent and Trademark Office (USPTO). (2003). "Material, and method of producing it, for immobilizing heavy metals later entrained therein." U.S. Patent No. 6,620,236, Washington, DC.

Weiss, C. A., Malone, P. G., Tom, J. R., Fransen, E. J., Peiten, J. (2005). "Modeling physical and chemical changes during bullet trapping and recovery," 6th Annual Environmental Technology Symposium and Workshop, Portland, OR.

Wright, J., Rice, K. R., Murphy, B., Conca, J. L. (2004). PIMS Using Apatite II: Remediation of Pb-contaminated range soil at Camp Stanley storage activity, TX. Proceedings of the Conference on Sustainable Range Management. New Orleans, LA.

Arrangement and Performance of Permeable Reactive Well (PRW) through Modeling

Gwanghun Lee[1], Jaeshik Chung[2], Yongju Choi[2], Junboum Park[3], Kyoungphil Nam[4]

[1]Ph D Candidate, Dept. of Civil, Urban and Geosystem Eng., Seoul National University, Republic of Korea; monkey@snu.ac.kr
[2]Graduate Student, Dept. of Civil, Urban and Geosystem Eng., Seoul National University
[3]Associate professor, Dept. of Civil, Urban and Geosystem Eng., Seoul National University, ASCE Member; junbpark@snu.ac.kr
[4]Associate professor, Dept. of Civil, Urban and Geosystem Eng., Seoul National University; kpnam@snu.ac.kr

ABSTRACT: Permeable Reactive Wells (PRWs) installed by boring have many advantages. It can reduce the construction time and it also can treat deep contamination compared to continuous wall or funnel and gate systems. In this paper, the ground water flows though PRWs was estimated by a numerical analysis. 2-dimensional, steady state flow, homogeneous and isotropic aquifer was assumed. Simulation results show that; (1) Capture width is largely affected by the transverse interval between wells. (2) In case of multiple wells, lead contaminant transport was considered for determining well arrangement. Transverse interval was increased, capture width was increased. When transverse interval was leached to 0.8m, capture width was not increased above 1.85 times of well diameter. Lead transport was affected by arrangement of wells. PRW performance was affected by longitudinal interval. And transverse interval was resulted in contaminant spread to the side of PRW.

Key words: PRW, Visual Modflow, capture width, arrangement, contaminant transport

1 INTRODUCTION

Permeable reactive barrier (PRB) is an in-situ technology for remediation of contaminated ground water. PRBs treat contaminated ground water by inducing chemical, physical, biological reactions between contaminants and reactive materials placed in the subsurface. The reactive wall is therefore emplaced to capture the contaminated ground water flow.

PRB technique has been studied extensively since the middle of 1990s. Currently, continuous wall and, funnel and gate systems are the most widely used. Both types of PRBs show good economical and remedial efficiency in shallow aquifer contamination. But in case of contaminants that lie within deep aquifers, there are limitations related to economical efficiency and difficulties in installation.

One alternative to overcome these disadvantages is the Permeable Reactive Well (PRW) system that is installed by boring rather than trench excavation. The permeable reactive well has many advantages. For example, it takes short time to be constructed and it allows treatment of deep contamination compared to a continuous wall or a funnel and gate system. Economical effects are expected due to relatively small amounts of reactive material necessary. Thus, most of permeable reactive barriers constructed in South Korea are well-type permeable reactive barriers. However, despite these advantages, examples of field construction are internationally rare, because research about the behavior of ground water flow though the PRW has been limited. In this paper, the ground water flow though the PRW will be estimated by a numerical analysis.

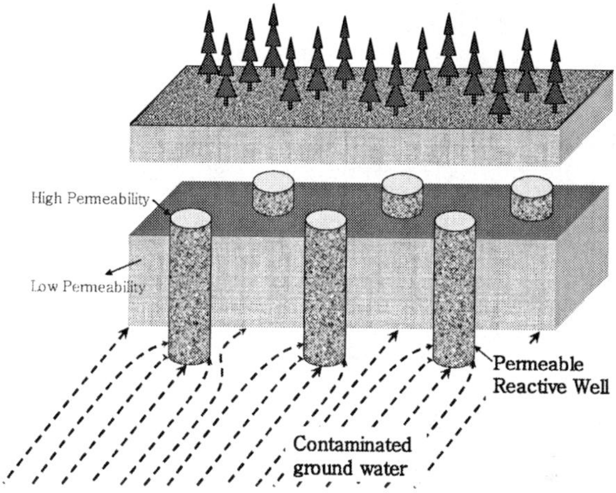

Figure 1. Concept of Permeable Reactive Well

The design and effective parameters for PRW system are the well diameter, the interval between each well, the interval between each line included wells, number of lines, depth of PRW, hydraulic conductivity difference between the well and aquifer, ground water velocity and retention time, etc. First, the relationship between transverse interval between wells and capture width; second, contaminant transport due to arrangement of wells.

2 COMPUTER MODELING

2.1 Method

To estimate the behavior of ground water flow through PRW, Visual MODFLOW (version 4.0), including MODFLOW and MODPATH was used. Finite difference method is employed in MODFLOW to solve partial differential equation related to ground water flow. MODPATH that allows particle tracking is used to visualize the ground flow path.

The simulated area is assumed two-dimensional with steady-state ground water flow in homogeneous and isotropic aquifer. About 20,000 cells are used to model the area of simulation. For a higher resolution, reactive wells were divided into cells that are smaller than those of the surrounding soil. Approximately 1,500 cells are used for each well.

The particles that are used for particle-tracking are set to be located at 1 centimeter intervals to check the capture width with the highest resolution.

The lateral boundaries are set sufficiently away from the wells, so that flow through the wells is not affected by the boundaries. WHS method, which obtains the solution using Bi-conjugate gradient stabilized (CGSTAB) was selected as the solver. To decrease errors caused by mesh dimension and operating process, head change criterion and residual criterion were set under $1 \times e^{-10}$ and iterations were set to be conducted sufficiently.

Numerical models used for simulations approximate solutions. The simulation results differ from each other according to the grid configuration (Han, 1999). The cause of difference is as follows;

(1) Discretization in continuous space domain
(2) Discretization in time-continuous domain
(3) Calculation using convergence values
(4) Simulated boundary conditions

2.2 Parameters

The hydraulic properties of aquifer used are based on field investigation data from abandoned mine site in Gosung city of South Korea. Additional properties that were not investigated in the field were used from references in technical literature. Aquifer consists mostly of residual soil. Hydraulic conductivity is 7.89 x 10^{-3} cm/sec, effective porosity (n_e) is 0.15 and hydraulic gradient (i) is 0.002. And to consider contaminant transport, sorption parameter was chosen by batch experiment. Sorption capacity, Q_0, is 153.8 mg/g, sorption energy, b, is 1.41 L/mg.

3 RESULT

3.1 General

There are two key objectives to accomplish in treatment and remediation of contaminated ground water using the PRB system. (1) PRB must capture the entire plume. (2) PRB must have enough retention time to remediate the contamination to environmentally acceptable concentrations. Therefore, the two most important factors in the design of PRW are the capture width (W_c) and the retention time (t_r) given by;

According to previous studies, the capture width is proportional to the well diameter and is changed by the difference in hydraulic conductivity between the aquifer and well. Generally, the capture width is assumed to be two times the diameter of the well. If hydraulic ratio between conductivities of the aquifer and well is smaller than fifty, the capture width becomes extremely small. However, if the ratio between the hydraulic conductivity of the aquifer and well is larger than fifty, the capture width becomes almost constant (Christ, 2006).

The capture width was irrelevant to hydraulic gradient and effective porosity of the well. Although the effective void ratio of the well affected the groundwater velocity, it did not bring about a change to the capture width.

Former results are based on a single well behavior. To reflect these results to design, simulation using multiple wells should be conducted. The factors that cannot be considered by single well simulation may be estimated through multiple well simulation.

3.2 Effect of transverse interval between wells (about Capture width)

Fig. 2 illustrates simulated flow line for single well. Well diameter is 20-centimeter, Hydraulic conductivity is $3.945 \times 10^{-1} cm/\sec$, total porosity is 0.3, effective porosity (n_e) is 0.15, and hydraulic gradient (i) is 0.002. Ground flow line is visualized by setting particles in locations within the aquifer that are far away from the well and running the particle-tracking model. The capture width was found to be about 1.85 times the well diameter. Well diameter is 20-centimeter, Hydraulic conductivity is $3.945 \times 10^{-1} cm/\sec$, total porosity is 0.5, effective porosity (n_e) is 0.3, and hydraulic gradient (i) is 0.002.

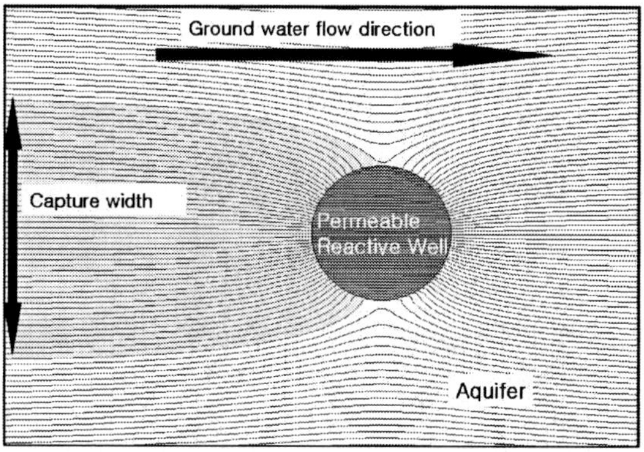

Figure 2. Simulation result of single well

Fig. 3 shows the changes in capture width according to d_t. The capture width from simulation result of single well that have same hydraulic properties is about 1.85 times the well diameter. When the transverse interval between wells is smaller than 1.85 times the well diameter, the capture width shows anticipated result that capture width decrease. But, when wells are set in transverse interval between wells wider than two times the capture width, the capture width was found to become smaller than 1.85 times the well diameter. The decrease of capture width becomes larger, when the transverse interval between wells is smaller. And when transverse interval between wells is wider than four times the well diameter, capture width becomes constant to 1.85 times the well diameter.

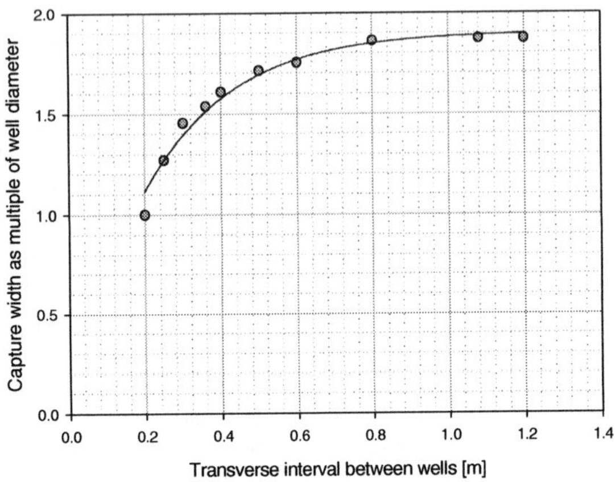

Figure 3. Effect of capture width on transverse interval between wells

3.4 Contaminant Transport

To determine the well arrangement, lead transport was considered. Figure 4 shows lead transport after about 4 years. The concentration contour is considered by well arrangement. When longitudinal interval is decreased, width of treatment area is decreased. And when transverse interval is decreased, width of treatment area is increased.

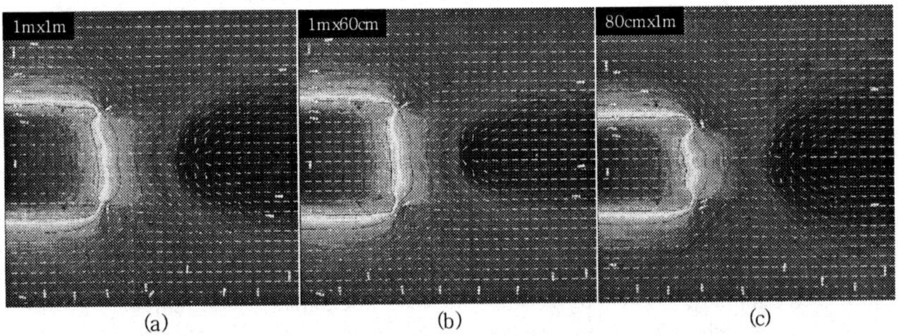

Figure 4. Pb (1 mg/L) transport modeling; (a) transverse interval 1m and longitudinal interval 1m; (b) transverse interval 1m and longitudinal interval 60cm; (c) transverse interval 80cm and longitudinal interval 1m

Fig 5 shows lead concentration in the center of PRW back-side through monitoring well. Fig 6 shows lead concentration in the side of PRW.

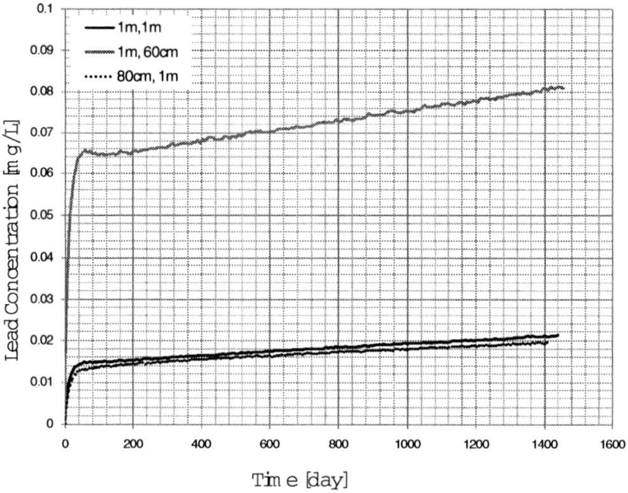

Figure 5. Lead concentration monitored in the center of PRW back-side

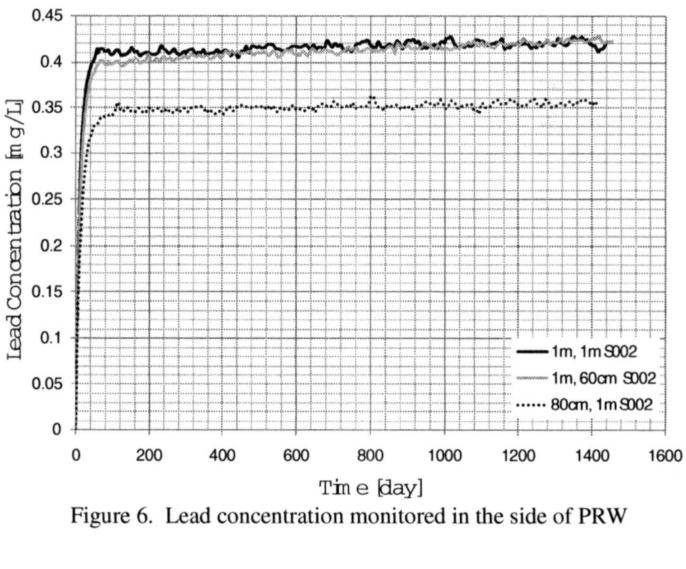

Figure 6. Lead concentration monitored in the side of PRW

4 SUMMARY AND CONCLUSION

The most common PRB configurations applied in the field are continuous wall and funnel and gate system. These two configurations of PRB work well to treat contaminated ground water. However, they face significant limitation in treatment of contaminants within deep aquifers due to problems related to installation and structural form.

PRW is easy to construct into deep aquifer because it can be simply installed through boring procedure. PRW can be viewed as a more innovative form of PRB because it allows treatment of deep aquifer contamination in addition to the advantage of ease of construction.

In this research, design parameters of the PRW have been evaluated, which can be summarized as follows;

(1) The dominant factor to determine the capture width is hydraulic conductivity difference between well and aquifer. The capture width is largely affected by transverse interval between wells. The capture width decreases extremely with the transverse interval between wells. The capture width becomes almost the same with the capture width of a single well when transverse interval between wells is larger than four times the well diameter. It was not found to be affected by the effective porosity of well and hydraulic gradient.

(2) When transverse interval between wells is larger than 1.25 times the well diameter, there were flow paths that develop between two wells. Some flow paths pass through the center of second well line. The flow path that passes through the outer edge of second line well was estimated to have passed through a zone that has 0.08 to 0.1 times well diameter interval from center of well. This zone has similar retention time with retention time of center, and the difference is very small.

Thus, when the ground water flows through staggered PRW system, the flow path that has the shortest retention time pass through the center of well.

(3) When transverse interval is more decreased, the performance of PRW is better. Longitudinal interval is decreased, longevity of PRW is also decreased.

The result of this paper is not exactly same with field condition because 2-dimensional, steady state flow, homogeneous and isotropic aquifer was assumed. Additionally, only the advection term was considered to obtain the retention time. In case of a sandy aquifer, the ground water flow has sufficiently high velocity. So, advection is the dominant behavior of contaminant. Thus, considering only advection may be acceptable. However, if the ground water velocity is low like in clay or silty aquifers, dispersion and diffusion must be considered to obtain more exact retention time.

ACKNOWLEDGEMENT

This subject is supported by Korea Ministry of Environment

REFERENCES

Gavaskar, A.R. (1998), "Design and construction techniques for Permeable Reactive Barriers", *Journal of Hazadous Materials* 68, pp. 41-71

Christ, M. and Park, J.B. (2006), "Capture width of permeable reactive wells as function of hydraulic conductivity", *Korea-Japan-France Joint Seminar on Geoenvironmental engineering*.

Freethey, G. W., Naftz, D. L., Rowland, R. C. and Davis, J. A. (2002), Deep Aquifer Remediation Tools: Theory, Design, and Performance Modeling, *Handbook of Groundwater, Remediation Using Permeable Reactive Barriers*, pp. 133-161

Scott, K.C. (2000), "Groundwater modeling of a permeable reactive barrier to enhance system performance", *Proceedings of the 2000 Conference on Hazardous Waste Research.*

Starr, R.C. and Cherry, J.A. (1994), "In situ remediation of contaminated Ground water", *GROUND WATER* Vol. 32, No.3

Wilson, R. D., Mackay, D. M. and Cherry, J. A.(1997), "Arrays of unpumped wells for plume migration control by Semi-Passive In-situ remediation", *J. Ground water Monitoring and remediation,* Vol. 17, No. 3, pp. 185-193

Durability Study of Eleven Years Old Cement-Bentonite Cut-off Wall Material

Kaushal Joshi[1], Kenichi Soga[2], Man Yin Albert Ng[3] and Cedric Kechavarzi[4]

[1] Ph.D. student, University of Cambridge, UK, CB2 1PZ; kdj22@cam.ac.uk
[2] Professor, University of Cambridge, UK, CB2 1PZ; ks207@cam.ac.uk
[3] Ph.D student, University of Cambridge, UK, CB2 1PZ; myan2@cam.ac.uk
[4] Lecturer, Cranfield University, UK, MK43 0AL; c.kechavarzi@Cranfield.ac.uk

ABSTRACT: Comprehensive understanding of the long-term performance of cement-bentonite slurry trench cut-off walls is essential as these mixes may degrade when exposed to aggressive environments or when subjected to prolonged drying. A series of wetting-drying and immersion experiments was carried out to evaluate the durability characteristics of laboratory mixed samples and block field samples from 40 days to 11 years of age. For the wetting-drying tests, the samples buried in medium graded sand were subjected to periodical flooding and drying cycles. They were then used for permeability testing and unconfined compressive strength (UCS) testing. For the immersion tests, the samples confined in perforated molds were submerged in magnesium sulfate solution for 16 weeks and their microstructures were then analyzed using X-ray diffraction (XRD) technique. This paper identifies the effects of contaminant exposure on durability of cement-bentonite and the effects of aging by comparing 11 years old samples to younger samples. Test results showed that young or previously contaminated cement-bentonite mixes are more susceptible to sulfate attack than old or less contaminated mixes.

INTRODUCTION

Containment technologies are cost effective control measure for contaminated land and hazardous waste sites. Slurry trench cut-off walls act as in-situ vertical barriers to confine the polluted ground in an enclosed area. These barrier walls are often constructed in-situ using cement-bentonite or soil-bentonite backfill slurries. The use of cement-bentonite cut-off walls has become more widespread in the last two decades in the UK, Europe and Japan due to substantial increase in the understanding of construction techniques. However, there is a need to investigate their durability because cement-bentonite mixes have been reported to degrade and even disintegrate when exposed to aggressive environments or when subjected to prolonged drying.

When cement-bentonite samples are allowed to dry in air, they shrink and crack quickly as shown in Figure 1 (Ratnam, 2002). Such drying shrinkage is irreversible and causes permanent loss of containment capability of the material. However, as in situ cut-off walls are always surrounded by moist soil, air drying of samples is not representative of field conditions. Field observation made by Tedd (2001) on a 3-year-old cement-bentonite wall suggested that it was intact throughout except for a very tight fissured area near the groundwater table.

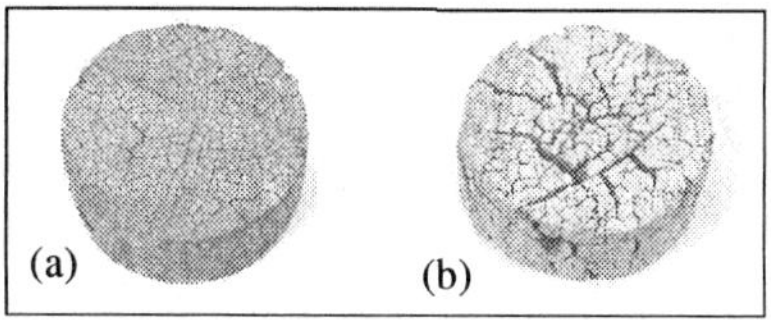

(a) After 1 day of drying
(b) After 5 days of drying

Fig. 1. Air dried cement-bentonite samples (Ratnam, 2002).

Another issue with regard to durability of cement-bentonite material is their susceptibility against aggressive chemicals. Like most of the cemented products, cement-bentonite material has been identified to have poor resistance against sulfate and acidic environments (Garvin & Hayles, 1999; Fratalocchi et. al., 2006).

Since most of the specifications for cement-bentonite cut-off wall only require compliance testing at the age of 90 days (ICE, 1999), very little literature reports on the evolution of cement-bentonite mixes properties for more than 90 days. However, the properties of wall material may continue to change with time due to hydration of cement and slag. Hence, the aim of this study is to evaluate the resistance of cement-bentonite wall material against chemical (sulfate) attack and wetting-drying cycles at various stages of its life. Test were conducted on 11 years old field samples, which were exhumed in the form of undisturbed blocks from cement-bentonite walls constructed at a disused gasworks site in the UK, and also on laboratory cast samples with age varying from 40 days to 11 years. The durability study of these samples can bridge the gap between young laboratory samples and the old field samples.

TESTED MATERIALS

A series of laboratory tests was carried out to assess the durability of cement-bentonite material on six types of samples, as summarized in Table 1. The first four types of samples, called 'mixer' samples, are uncontaminated homogeneous samples cast from fresh cement-bentonite slurry under laboratory conditions. These samples were tested at the age of 40 days, 90 days, 2.5 years and 11 years. The fifth and sixth type of sample, named 'block' samples, were obtained from the cut-off walls built on a disused gasworks test site in April 1996 (Fig. 2 (a)). Ground conditions at the test site consist of approximately 3 m of made ground, containing spent oxide, coal residues, carbon black and foul lime, overlying stiff clay. Further details of the test site are reported by Tedd et al. (1997, 2001) and Soga et al. (2005).

Table 1. Properties of cement-bentonite samples used for the experiment

Sample Type	Age of sample	Permeability (m/sec)†	UCS (kPa)	XRD Characteristic*	Test done
Mixer	40 days		415		Immersion
	90 days	1.90×10^{-10} to 6.70×10^{-11}	603 to 970	Calcite, Gypsum, Unreacted slag, Bentonite	Wet-dry
	2.5 years	7.00×10^{-11} to 1.38×10^{-12}	550 to 640	Calcite, Gypsum, Unreacted slag, Bentonite	Wet-dry Immersion
	11 years	7.97×10^{-11} to 1.06×10^{-11}	660 to 910	Calcite, Gypsum	Immersion
Block	11 years	1.98×10^{-10} to 1.56×10^{-10}	410 to 810	Calcite, Quartz, Gypsum	Wet-dry Immersion
Heavily contaminated block	11 years	1.94×10^{-10} to 4.97×10^{-11}	480 to 530	Calcite, Ettringite, Quartz, Gypsum	Immersion

† Tested in triplicates according to ICE (1999) specification * See Fig. 2 (b)

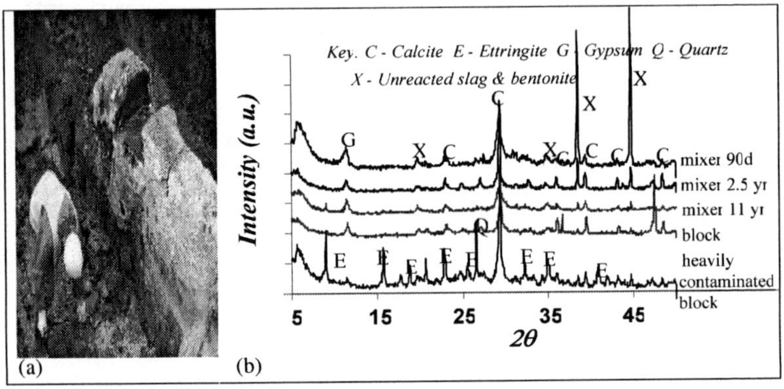

Fig. 2. (a) 'Block' sample exhumation, and (b) XRD diffractogram of 90 days, 2.5 years and 11 years 'mixer', 'block' and 'heavily contaminated block' sample.

The composition of slurry in all the samples was the same: 40 kg of sodium activated bentonite, 30 kg of cement and 120 kg of GGBS (ground granulated blast furnace slag) per 1000 liters of water. This mix design was developed by the UK Building Research Establishment (BRE) and the slurry complies with the specification of ICE (1999). Figure 2 (b) shows XRD analysis on dried powder (pestle-mortar crushed) of the tested samples to study their microstructure and to co-relate the changes in their physical properties. For example, Table 1 and Fig. 2 (b) depict that, as the samples age, their permeabilities decrease since unreacted slag and bentonite diminish from their microstructure due to slow hydration.

Unlike laboratory mixed samples, the field 'block' samples had been subjected to different environmental conditions in contaminated ground. In May 2004, the 8 years old walls were excavated down to the bottom and block samples (1 m x 0.6 m x 0.25 m) were retrieved at different depths as shown in Fig. 2 (a). While some blocks did not get much contaminated, some blocks were exhumed from walls at heavily sulfate and spent oxide contaminated ground. These types of field samples were very dark in color and are referred as 'heavily contaminated block' samples. Both types of 'block' field samples were stored underwater in laboratory and tested at the age of 11 years. Compared with the 11 years old 'mixer' sample, these 'block' samples were heterogeneous and characterized by the presence of surrounding soil, stones, contaminants and air voids. As evident from Table 1, both 'block' sample types have smaller UCS and greater permeability compared to 11 years old 'mixer' sample, mainly due to their heterogeneous state. Presence of quartz in diffractogram of both 'block' sample types implies the presence of stone or soil (Fig. 2 (b)). The 'heavily contaminated block' sample showed the presence of ettringite in their microstructure, which is one of the products of sulfate attack with hydrated cement paste.

DURABILITY TESTS – IMMERSION AND DRYING TESTS

The durability of cement-bentonite wall material was assessed in terms of its resistance to chemical attack and drying. Immersion tests were performed on all samples, except 90 days old 'mixer', to evaluate their resistance to chemical attack. One of the main limitations of immersion tests is that they allow free expansion of the samples, which is different from confined field conditions. In this study hollow perforated molds tied with plastic cable ties (Fig. 3) were used to confine the samples. Cemented soils confined in perforated molds during immersion do not loose much strength as compared to free expandable experimental conditions and such confinement may be representative of the field environment (Osman, 2007). All tested samples were 100-mm in height and 50-mm in diameter and were tested in duplication (Fig. 3(c)). The specimens were immersed in 5 liters of magnesium sulfate solution, which is reported as the most aggressive form of sulfate for cement-bentonite mixes (Garvin & Hayles, 1999). The concentration of magnesium sulfate solution was 27.2 g/l and the corresponding sulfate concentration was 10.8 g/l. This value of sulfate concentration is equivalent to the average amount of sulfate concentration found at the test site. The samples were immersed for 16 weeks at 21°C and the pH of the solution was measured periodically. After the immersion period, each specimen was unmolded and their chemical composition was examined by XRD analysis. For XRD test the specimens were firstly vacuum dried in a desiccator, which were then crushed with a pestle-mortar and sieved (< 425 μm) to produce fine powder.

Although it is acknowledged that cement-bentonite wall material hardens over a period of time, it is unclear whether the resistance to drying shrinkage increases with age. Wetting-drying tests were conducted on 90 days old 'mixer' sample, 2.5 years old 'mixer' sample and 11 years old 'block' sample to study the effect of age and heterogeneity. 11 years old 'mixer' sample was not used in the experiment because their general properties were similar to 2.5 years old 'mixer' sample. Each type of sample had two specimens of 100 mm diameter by 100 mm high for permeability test and four specimens of 50 mm diameter by 100 mm high for UCS test.

Fig. 3. Immersion test experiment on cement-bentonite material.

As shown in Figure 4(a) and 4(b), the samples were buried in British Standard fraction C sand (300 μm to 600 μm) to simulate ground condition. The sand tub was designed so as to control the water level inside the tub. Each specimen was first enclosed in a tight rubber membrane. As one end of the specimen was sealed, there is only one end of the specimen exposed to wetting-drying cycle. Such arrangement could replicate field situation where the interface between cut-off wall and soil is subjected to groundwater fluctuation while the interior of the wall is unaffected. The experiments were conducted at 21°C for 12 and 24 cycles of wetting and drying. During each cycle, the sand tub with specimens were saturated with water for 3 days followed by a draining period of 2 days and then the sand was left under residual moisture conditions for 13 days. Figure 4(c) shows the gravimetric moisture content of the sand in the tub during one wetting-drying cycle. After 12 cycles, two specimens were tested for permeability and another two for UCS. The remaining two specimens were subjected to 12 more cycles and were tested for UCS. Permeability tests were done in a flexible wall permeameter with effective confining pressure of 100 kPa and hydraulic gradient between 5 and 30. UCS test was carried out at a strain rate of 1.55%/min.

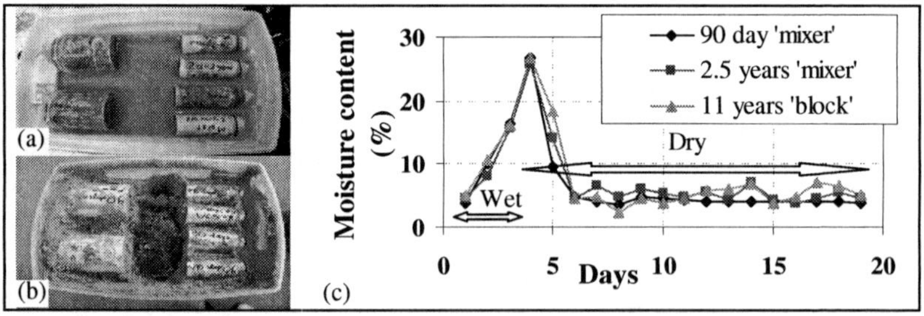

Fig. 4. Wet-dry experiment of cement-bentonite material.

EFFECTS OF CONTAMINANT EXPOSURE ON 11 YEARS OLD SAMPLES

At the end of immersion period, all the samples were found physically similar. They became touch-soft and it was not possible to conduct UCS test on these samples. All the samples were observed to have increased diameter, height and weight and had a vertical crack running mainly along the slit of the perforated mold (Fig. 5(c)) due to such expansion. Also the original dark grey color (Fig. 5(a)) of samples changed to white on the surface as shown in Fig. 5(b), (c) and (d). However, the original color was still present in the core of sample as shown in Fig. 5 (d). Therefore XRD test was performed on the outer white surface as well as the central dark core.

On the outer white surface, there were shinny needle type crystals as shown in Fig. 5(b) and (c). XRD test showed that they were pure gypsum. Similarly XRD analysis of outer white surface of all the specimens depicted only gypsum. The formation of gypsum suggests that the sulfate has reacted with hydrated cement product like calcium hydroxides and calcium silicate hydrate causing expansion of the specimen and loss of strength. The original minerals except quartz were not detected after reaction. This could indicate that chemical reaction near the exposed surface has been completed.

Fig. 5. 11 years old cement-bentonite sample before and after immersion test.

The 11 years old 'mixer' sample had a hard and brittle core like the original wall material before immersion (see Fig. 6 (b)). While the 'block' sample had a very small unaffected (hard) core, 'heavily contaminated block' did not have any hard core at all. The presence of unaffected core implies that the reaction front had not penetrated beyond that depth. The XRD analysis of these cores confirmed that their chemical composition is different from that of the white surface. Fig. 6(a) shows that the 11 years 'mixer' sample has calcite, gypsum and most notably ettringite, while the 'block' sample has gypsum, calcite and quartz. The 'heavily contaminated block' sample, however, is left with gypsum and quartz only without any calcite. These results imply that samples exposed to previous levels of contamination are less resistant to chemical attack as no solid cores were found in the 'heavily contaminated block' samples. Thus a wall that is not affected by previous contamination can have better resistance to sulfate attack.

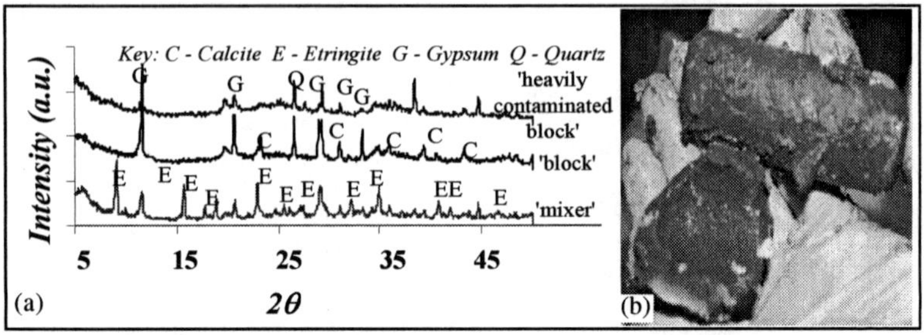

Fig. 6. XRD diffractogram of core of 11 years old samples.

EFFECTS OF AGING ON DURABILITY

Results from immersion tests on 'mixer' sample of 40 days, 2.5 years and 11 years show that the post-immersion samples were soft and cracked. Crystals formed on the exposed surface and color of the sample changed from grey to white. The XRD analysis identified that the crystals and the outer white surface were gypsum. The 40 days old sample disintegrated completely, whereas the 2.5 years and 11 years sample had similar sized non-disintegrated core and their XRD diffractogram were identical. The absence of core in 40 days old sample implies that the sample had not hydrated enough to resist reaction against sulfate. This implies that a cut-off wall should be built at a certain distance away from moving plume front so that the cement-bentonite slurry has enough time to set in order to be resistant to chemical attacks.

On exhumation of the samples after 12 cycles of wetting-drying, there were no visible changes, except the exposed surface became whitish. Permeability tests results on these samples show limited change (within the variation of measurements) after the wetting-drying cycles. Figure 7 shows UCS values for three samples before and after 12, and 24 wetting-drying cycles. Although their strengths reduced possibly due to development of discontinuous micro cracks, the samples did not crack as shown in Fig. 1. Such results comply with field observation, wherein the walls were found to be intact despite of groundwater fluctuation. This is because the specimens were surrounded by moist sand during the drying periods.

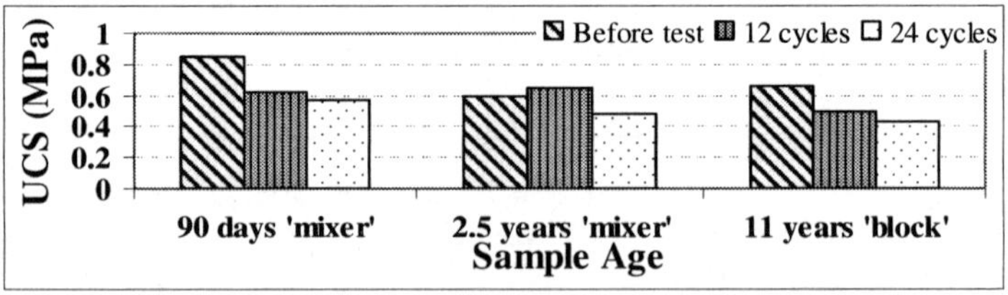

FIG. 7. UCS test results before and after wetting-drying cycles.

CONCLUSIONS

Cement-bentonite material has sufficient resistance to wetting drying cycles when surrounded by moist soils at residual saturation. However, sulfates have a detrimental effect on the properties of cement-bentonite. The XRD analysis confirmed that such deterioration was due to the formation of expansive products like gypsum. Although the samples were subjected to extremely aggressive conditions, it was possible to identify the effects of aging and contaminant exposure against chemical attack. Solid cores were not found in the young sample of 40 days of age (i.e. partially hydrated), which implies that sulfate has fully penetrated through the sample. It was also noticed that the wall material previously exposed to contamination (i.e. 11 years old heavily contaminated block sample) was more susceptible to chemical attack. Therefore a cement bentonite containment wall should be constructed away from contaminant hot spots in order to allow the material to age and harden before being exposed to any chemical contaminants.

ACKNOWLEDGEMENT

The authors thank Dr Paul Tedd of the BRE, JP Muriuki, S. Iyengar and M. Liska of the University of Cambridge for their help and advice on this work. The work was funded by the EPSRC and the Cambridge Commonwealth/Overseas Trust.

REFERENCES

Garvin, S. L. and Hayles, C. S. (1999). "The chemical compatibility of cement-bentonite cut-off wall material." *Construction and Building Materials*, Vol. 13: 329-341.

Fratalocchi, E., Pasqualini, E. and Balboni, P. (2006). "Performance of a cement-bentonite cut-off wall in an acidic sulphate environment." *Proc. 5th Int. Conf. Environmental Geotechnics*, Vol. 1, Cardiff: 133-139.

ICE (1999). "Specification for the construction of slurry trench cut-off walls as barriers to pollution migration". Thomas Telford.

Osman, A. (2007). "Durability and mechanical properties of deep-mixed clays." *PhD. Thesis*, Dept. of Engineering, Cambridge, UK.

Ratnam, S. (2002). "Development of a novel self-boring permeability measurement technique." *PhD. Thesis*, Dept. of Engineering, Cambridge, UK, 7-34 p.

Soga, K., Sutherland, K. J., Kechavarzi, C. and Whittle, R. W. (2005) "In situ Permeability Measurement of a Contaminant Containment Wall." *The 16th Int. Conf. on Soil Mechanics and Geotechnical Eng.*, Osaka, Vol. 4, 2445-2448.

Tedd, P., Holton, I. R., Butcher, A. P., Wallace, S. and Daly, P. J. (1997). "Investigation of the performance of cement-bentonite cut-off walls in aggressive ground at a disused gasworks site." *Proc. 1st Int. Containment Technology Conf.*, St Petersburg, Florida, USA: 125-132.

Tedd, P. (2001). "Field investigations of slurry trench cut-off walls to control pollution migration." *Proc. 3rd Geoenvirnomental Engineering Conf.*, Edinburgh. Thomas Telford, 575-582.

Hydraulic Barrier Performance of SBM Cut-off Wall constructed by the Trench Cutting and Re-mixing Deep Wall Method

Takeshi Katsumi[1], M. ASCE, Masashi Kamon[2], Toru Inui[3] and Susumu Araki[4]

[1] Associate Professor, Graduate School of Global Environmental Studies, Kyoto University, Yoshida-honmachi, Sakyo, Kyoto 606-8501, Japan, tkatsumi@mbox.kudpc.kyoto-u.ac.jp
[2] Professor, Graduate School of Global Environmental Studies, Kyoto University, Yoshida-honmachi, Sakyo, Kyoto 606-8501, Japan, kamon@mbox.kudpc.kyoto-u.ac.jp
[3] Assistant Professor, Graduate School of Global Environmental Studies, Kyoto University, Yoshida-honmachi, Sakyo, Kyoto 606-8501, Japan, inui@mbox.kudpc.kyoto-u.ac.jp
[4] Director, Raito Kogyo Co., Ltd., 4-2-35 Kudan-Kita, Chiyoda-ku, Tokyo 102-0073, Japan, s-araki@raito.co.jp

ABSTRACT: Containment using a soil-bentonite mixture (SBM) cut-off wall is a valid method to prevent the contaminants in soil from migrating in the aquifer. A new construction technique for the SBM vertical cut-off wall has been developed by employing the trench cutting and re-mixing deep wall method to achieve excellent homogeneity of the wall. This paper addresses this construction method and the laboratory testing results on the hydraulic barrier performance of SBM focusing on its chemical compatibility. Hydraulic conductivity values were measured for SBM specimens, which were processed from various types of soil with different concentrations of divalent cation (Ca^{2+}) in the permeant or in the pore water of original soil. For the Ca^{2+} concentrations lower than 0.01 M in soil pore water, there is no significant effect on the hydraulic conductivity of SBM. For the higher Ca^{2+} concentrations, more than one order of magnitude higher hydraulic conductivity values were observed. The correlation between the hydraulic conductivity and the void ratio change in SBM specimen implies that the void ratio change could be a good indicator of the hydraulic conductivity for different types of soil.

INTRODUCTION

Vertical cut-off walls are constructed at contaminated sites to contain the contaminants and prevent their transport in the aquifer. A hydraulic barrier material in the cut-off wall system needs to have a low hydraulic conductivity. Soil-bentonite mixture (SBM) is the most widely used material since it can provide extremely low hydraulic conductivity values as well as sufficient deformability (e.g. Grube 1992). A new method for constructing the SBM vertical cut-off wall has been developed by employing the trench cutting and re-mixing deep wall (TRD) method (Fig. 1).

It is well known that the hydraulic conductivity of bentonite-based material strongly depends on the chemical component of the permeants and/or the pore water (e.g. Katsumi et al. 2007, Lee and Shackelford 2005, Shackelford et al. 2000), as well as the effective vertical stress (Evans 1994), since the bentonite does not swell and/or the swelling is reversed in the presence of inorganic solutions or some organic liquids (Norrish and Quirk 1954). Particularly, inorganic solution containing polyvalent cations increased hydraulic conductivity with relatively lower concentration (~ 0.1 mol/L). This is because the existence of the polyvalent cations results in the development of thinner diffuse double layers and the less osmotic swelling. Accordingly, the first wetting liquid dominantly affects the hydraulic performance of the clay barriers. When the SBM is employed as a barrier material, its chemical compatibility should be assessed under the given conditions in the field. This paper addresses the experimental results of hydraulic conductivity tests for the SBM specimens, which were tested with soil containing different concentrations of the divalent cation in the permeants as well as in the pore water. Based on the test results, the effect of the divalent cation on the hydraulic conductivity of SBM was clarified. In addition, the void ratio change of the SBM specimen was correlated to the hydraulic conductivity of SBM affected by the divalent cation, regardless of the soil type.

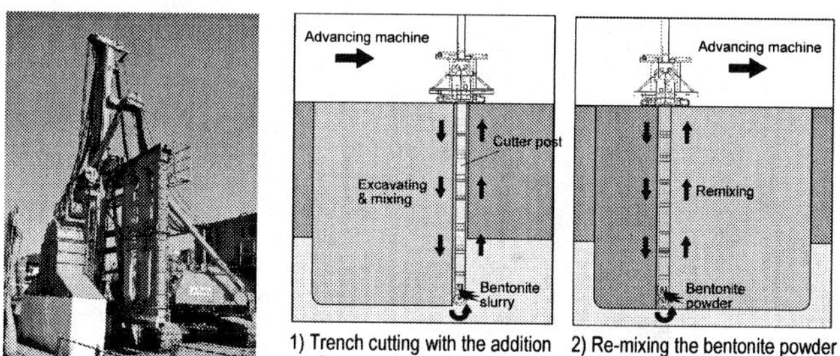

1) Trench cutting with the addition of bentonite slurry
2) Re-mixing the bentonite powder in the trench

FIG.1. Construction process of SBM vertical cut-off wall with TRD method.

WALL CONSTRUCTION USING TRD METHODS

Figure 1 shows the construction sequence of the SBM vertical cut-off wall employing the TRD method. First, the trench cutting is conducted supplying bentonite slurry to maintain the workability of the soil inside the trench. Then, re-mixing with bentonite powder is conduced in the trench.

The advantage of employing the TRD method is to construct the cut-off wall of high homogeneity even in the layered ground, since the cutter post which advances horizontally through the layers excavates, injects slurry or powder from its head, and mixes the soil, forming a continuous wall (e.g. Kamon et al. 1998). Homogeneity is an important factor since the larger variability in the hydraulic conductivity leads to the higher flux of contaminant out of the barrier system even though the average hydraulic conductivity values are equal (Britton and Filz 2007). Figure 2 shows the SBM cut-off wall (550 mm in thickness, 14.2 m in depth) constructed in the pilot test. Figure 3 shows the vertical distribution of earth pressure inside the SBM wall. This was monitored in a pilot test. Vertical stresses in the SBM wall was proved to be lower than the overburden pressures due to the arching effect (Evans et al. 1995). When the hydraulic conductivity of SBM is tested in the laboratory, the stress conditions must be considered so as not to underestimate the hydraulic conductivity.

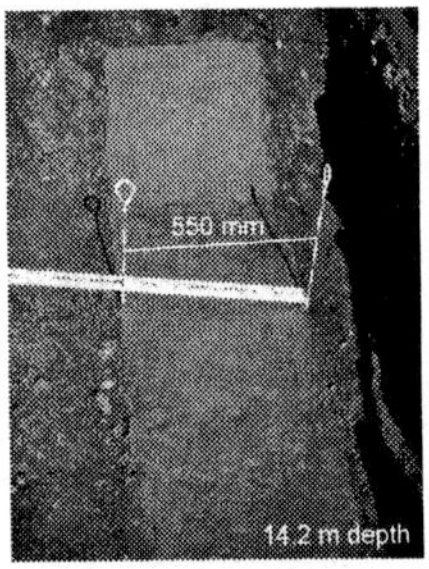

FIG.2. SBM wall constructed in pilot scale test (Kamon et al. 2006).

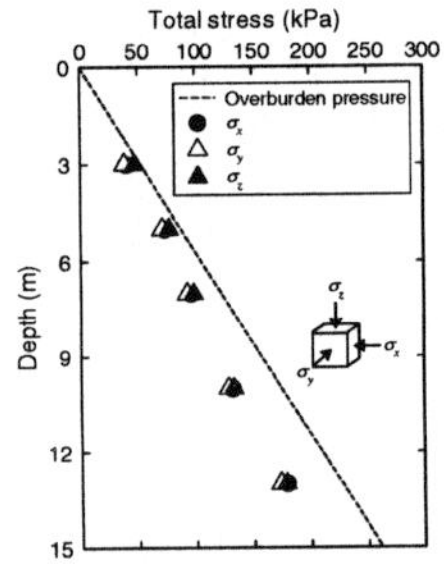

FIG.3. Total stress distribution in SBM wall vs. overburden pressure (Kamon et al. 2006).

MATERIALS AND METHODS

Materials

Three different soils were used for these studies. These soils included 1) a mixture of volcanic cohesive soil and sandy gravel collected at a pilot scale test site (referred to composite soil in the following), 2) silty clay and 3) fine sand. These soils were sieved through a 4.75mm-opening screen. The composite soil was prepared by mixing sandy gravel and volcanic cohesive soil at their natural water contents. The mixing ratio of 25:4 by dry weight (sandy gravel : volcanic cohesive soil) was determined based on the thickness of each layer obtained from the boring log on the pilot test site. Table 1 lists the grain size distributions of these soil samples. Ca^{2+} concentration in the pore water was adjusted to the target value by using the calcium chloride ($CaCl_2$) solution. Ca^{2+} was selected since it is commonly found in natural aqueous system as well as in water discharged from the industrial process or leached from the waste.

Simulating the process of the wall construction by the TRD method, 10%-concentration hydrated bentonite slurry was blended with the soil. The additive content of the bentonite slurry was determined based on the flowability of the soil-slurry mixture, approximately 150-mm flow value according to JIS R 5201. Once a mixture of suitable flowability was established, 50 to 150 kg/m^3 powder bentonite was added and mixed using the soil mixer.

Table 1. Grain size distribution of three different soils used

		Composite soil	Silty clay	Fine sand
Gravel fraction [2 mm–]	(%)	5.6	0.9	0
Sand fraction [75 μm – 2mm]	(%)	70.8	65.2	87.9
Silt fraction [5 μm – 75 μm]	(%)	15.8	19.1	7.9
Clay fraction [– 5 μm]	(%)	7.8	14.8	4.2

Hydraulic Conductivity Test

After consolidation at 30 kPa in the oedometer, SBM specimens having a 30 mm in height and 60 mm in diameter were subjected to the hydraulic conductivity tests. A flexible-wall permeameter with a fallen-head system according to ASTM D5084 was employed. A confining pressure of 30 kPa and a hydraulic gradient of approximately 30-40 were applied during the permeation. Permeation continued until the following requirements were confirmed: 1) the volume of the effluent and the influent were the same, 2) the change in hydraulic conductivity values with time was negligible, 3) pore volumes of flow were greater than 3, and 4) the electrical conductivity of the effluent was equal to that of the influent. Testing conditions were summarized in Table 2. P-series was designed to assess the chemical compatibility of SBM attacked by the solution containing $CaCl_2$. In this series, bentonite in SBM has been firstly wetted with pore water of original soil. In the N-series, the expected detrimental effect of the $CaCl_2$ concentrations in original soil on the hydraulic conductivity was verified. In this case,

Table 2. Experimental conditions for hydraulic conductivity tests

Test No.	Bentonite powder content, C_B (kg/m³)	Soil type[a]	$CaCl_2$ concentration in original soil, $c_{c,sw}$ (M[b])	$CaCl_2$ concentration in permeant, c_p (M[b])	Hydraulic conductivity (m/s)
P-00	100	P	0	0	5.0×10^{-11}
P-10	100	P	0	0.1	1.9×10^{-10}
P-20	100	P	0	0.25	2.2×10^{-10}
P-30	100	P	0	1.0	1.4×10^{-10}
N-00	100	P	0	0.1	1.2×10^{-10}
N-10	100	P	0.01	0.1	2.2×10^{-10}
N-20	100	P	0.025	0.1	5.6×10^{-10}
N-30	100	P	0.05	0.1	1.0×10^{-9}
N-40	100	P	0.1	0.1	1.3×10^{-9}
N-01	50	P	0	0.1	2.3×10^{-10}
N-21	50	P	0.025	0.1	1.5×10^{-9}
N-22	75	P	0.025	0.1	6.8×10^{-10}
N-31	75	P	0.05	0.1	1.1×10^{-9}
N-32	125	P	0.05	0.1	7.4×10^{-10}
N-41	125	P	0.1	0.1	1.1×10^{-9}
N-42	150	P	0.1	0.1	7.0×10^{-10}
S-00	100	P	0	0	2.8×10^{-11}
S-01	100	P	0.01	0.01	1.0×10^{-10}
S-02	100	P	0.1	0.1	1.3×10^{-9}
S-10	100	S	0	0	2.1×10^{-11}
S-11	100	S	0.01	0.01	2.9×10^{-11}
S-12	100	S	0.1	0.1	6.3×10^{-10}
S-20	100	F	0	0	3.1×10^{-11}
S-21	100	F	0.01	0.01	4.0×10^{-11}
S-22	100	F	0.1	0.1	2.5×10^{-10}

[a]Soil type; P: Soil from the pilot-test site, S: Silty clay, F: Fine sand [b]M represents mol/L

swelling of the bentonite in SBM is impeded, due to the concentration of inorganic chemicals in the soil and/or the groundwater at a contaminated site. In the S-series, chemical compatibility of the SBM, processed from three different types of soil, was tested to evaluate how the void ratio change during the permeation affects the hydraulic conductivity, regardless of the soil type.

RESULTS AND DISCUSSIONS

Effect of Inorganic Chemical

Figure 4 (a) illustrates the hydraulic conductivity values for the content of bentonite power (C_B) of 100 kg/m^3, affected by the $CaCl_2$ concentration in permeant, c_p (Kamon et al. 2006). Hydraulic conductivity for c_p = 0.1 M is 3.5 times higher than that for distilled water. Another finding is that the difference in hydraulic conductivity is negligible for c_p higher than 0.1 M. For c_p = 1.0 M, hydraulic conductivity is slightly lower than those for c_p = 0.1 and 0.25 M probably due to the higher viscosity of the permeant. These observations confirm that hydraulic conductivity of SBM is not significantly increased even against the permeant containing the 1.0 M polyvalent cation if the bentonite in SBM is well hydrated with soil pore water.

Figure 4 (b) shows the hydraulic conductivity values of SBM (C_B = 100 kg/m^3) for different $CaCl_2$ concentrations in pore water of the original soil, $c_{c,sw}$. The higher hydraulic conductivity values were obtained for the higher $c_{c,sw}$. Although the pore water containing calcium ions is diluted by the water fraction of the bentonite slurry, there was a significant increase in hydraulic conductivity. Hydraulic conductivity values were increased linearly for $c_{c,sw}$ lower than 0.05 M. However, for $c_{c,sw}$ higher than 0.05 M, they still increased but by the smaller rate. For 0.1 M, hydraulic conductivity reached higher than 1 x 10^{-9} m/s.

By comparing the effects of divalent cations on the hydraulic conductivity when they exist in permeant and in pore water of the original soil, it can be concluded that the cation in the pore water causes more significant increase in the hydraulic conductivity. The hydraulic conductivity for 0.1 M in permeant was 3.5 times as high as that for 0 M in permeant. In contrast, the increase in the $CaCl_2$ concentration in pore

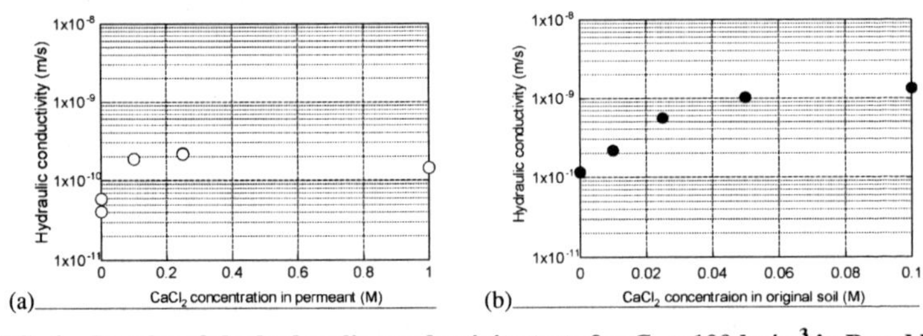

FIG. 4. Results of the hydraulic conductivity tests for C_B = 100 kg/m^3 in P and N series: (a) effects of $CaCl_2$ concentration in permeant (Kamon et al. 2006), (b) effects of $CaCl_2$ concentration in original soil.

water from 0 to 0.1 M resulted in the increase in hydraulic conductivity by more than one order of magnitude. This difference indicates that the prehydration of bentonite is an important factor for the chemical compatibility of the SBM. Its effect is not minor even in the case that the divalent cation concentration in pore water is lower than 0.05 M as shown in Fig. 4 (b). Thus, the concentration of the divalent cation and its variation in groundwater at the site of concern should be considered in evaluating the hydraulic barrier performance of SBM.

Effect of Bentonite Content

Figure 5 shows the hydraulic conductivity values in N-series for various C_B. Adding the larger content of bentonite was able to lower the hydraulic conductivity, even when the $CaCl_2$ concentrations in original soil were relatively high (0.05 or 0.1 M). Even for the highest $CaCl_2$ concentration of 0.1 M, hydraulic conductivity was lower than 1.0×10^{-9} m/s in case of C_B = 150 kg/m^3. From these observations, when the SBM vertical cut-off wall is constructed at the contaminated site where the inorganic chemical concentration in soil/groundwater is high, its hydraulic barrier performance can be enhanced by adding the larger amount of bentonite. In the case that the SBM was prehydrated ($CaCl_2$ concentration in pore water = 0 M), hydraulic conductivity of lower 1.0×10^{-9} m/s was achieved by adding 50 kg/m^3 of powder bentonite for the composite soil used in this study.

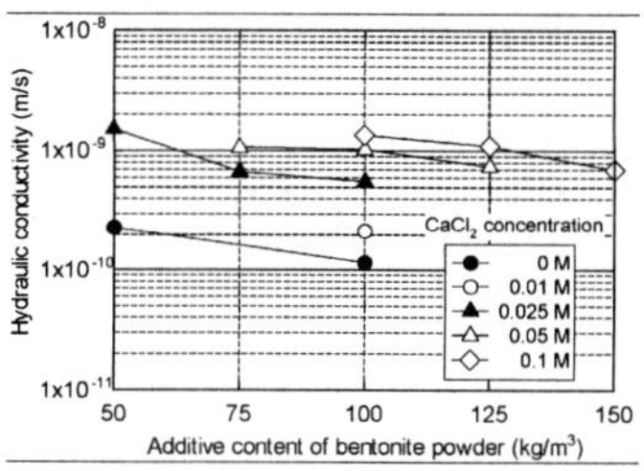

FIG.5. Hydraulic conductivity vs. content of bentonite powder.

Effect of Soil Type

Figure 6 (a) shows the relationship between hydraulic conductivity values and $CaCl_2$ concentrations in original soil and permeant for three types of soil in S-series, where the initial $CaCl_2$ concentration in pore water and permeant were set equal. For each soil, hydraulic conductivity increased with the concentration of $CaCl_2$. As $CaCl_2$ concentration in original soil and permeant increased from 0 to 0.1 M, the hydraulic conductivity of SBMs processed from composite soil, silty clay, and fine sand became 48 times higher, 30 times higher, and 8 times higher, respectively. Figure 6 (b) shows the relationship between the void ratio after permeation and the $CaCl_2$ concentration. Void ratio of SBM was lowered more by the higher concentration of $CaCl_2$. For the constant $CaCl_2$ concentration, the void ratio of composite soil-based SBM was the largest, and those of silty clay-based and fine sand-based SBMs were almost equal.

Comparing the values for $CaCl_2$ concentration of 0.1 M with 0 M, the void ratio of composite soil-based and silty clay-based SBMs became about 20% of magnitude lower. In contrast, decrease in the void ratio of fine sand-based SBM was only 10% of magnitude. This finding is consistent with the influence of the $CaCl_2$ concentration on the hydraulic barrier performance shown in Fig. 6 (a)

Figure 7 plots the relationship between the void ratio and the hydraulic conductivity, both of which were normalized with the value for the $CaCl_2$ concentration of 0 M. Correlation between them was observed regardless of the soil type. The lower concentration resulted in the larger void ratio of SBM. As a result, larger volume of water is retained as immovable water, which does not contribute to the permeation. This is why the lower hydraulic conductivity is achieved with the larger void ratio of SBM attacked by $CaCl_2$. This finding supports that the void ratio change can become a good indicator for the hydraulic barrier performance of SBM attacked by the divalent cation for various types of soil.

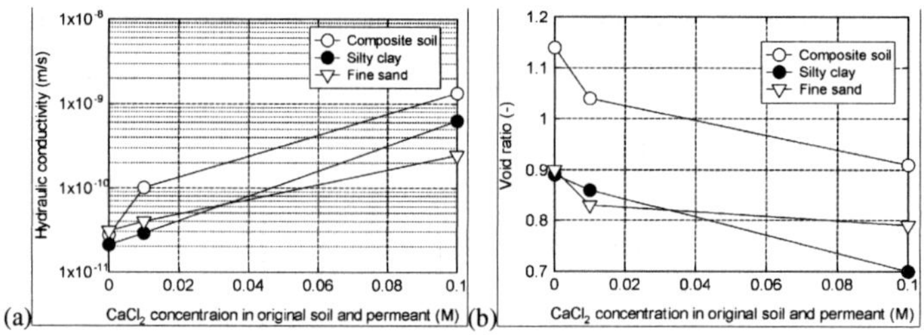

FIG. 6. Results of the hydraulic conductivity tests in S series: effects of $CaCl_2$ concentration on (a) hydraulic conductivity and (b) void ratio.

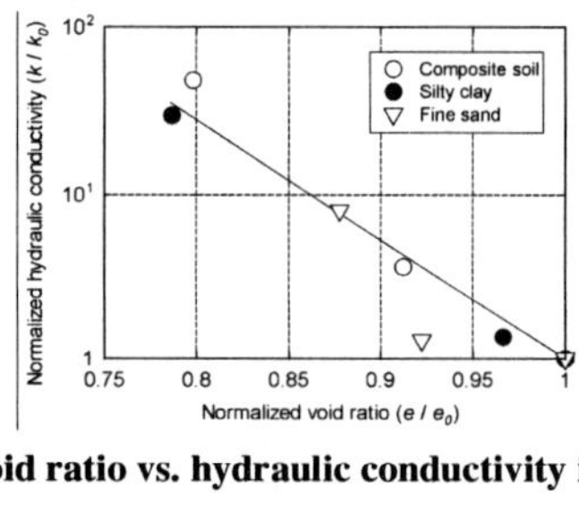

FIG.7. Void ratio vs. hydraulic conductivity in S-series.

CONCLUSIONS

This paper presents the results of hydraulic conductivity tests for the SBM focusing on its chemical compatibility. The conclusions can be summarized as follows:

1) A new technique for constructing the SBM vertical cut-off wall has been developed by employing the TRD (Trench cutting and re-mixing deep wall) method to attain the higher homogeneity of SBM.

2) When no divalent cation exists in soil pore water, the hydraulic conductivity of the SBM permeated with 1.0 M $CaCl_2$ solution is only 3.5 times as high as that with the distilled water. Sufficient hydration of the bentonite will maintain the hydraulic barrier performance permeated with the water in which the divalent cations exist.
3) Calcium ions in soil pore water have more significant effect on the hydraulic conductivity than those in permeant. However, hydraulic barrier performance can be enhanced by mixing the larger content of bentonite powder.
4) There is a strong negative correlation between the void ratio and the hydraulic conductivity of the SBM, for three different types of soil used. This observation suggests that the void ratio change can become a good indicator for the effect of the hydraulic barrier performance of SBM attacked by the divalent cation.

ACKNOWLEDGEMENT

The laboratory experiments were conducted with the cooperation of Mr. Y. Ogawa, Mr. D. Matsuhashi and Mr. A. Takai, former or current graduate students at Kyoto University.

REFERENCES

Britton, J.P. and Filz, G.M. (2007). "High uniformity versus low hydraulic conductivity for vertical barriers in contaminant containment applications." *Geoenvironmental Engineering*, GSP No.163, S.E. Burns, P.J. Culligan, J.C. Evans, P.J. Fox, K.R. Reddy, N. Yesiller (eds.), ASCE: on CD.

Evans, J.C. (1994). "Hydraulic conductivity of vertical cutoff walls." *Hydraulic Conductivity and Waste Contaminant Transport in Soils*, ASTM STP 1142, D.E. Daniel and S.J. Trautwein, eds., ASTM, Philadelphia: 79-94.

Evans, J.C., Costa, M., and Cooley B. (1995). "The state-of-stress in soil-bentonite slurry trench cutoff walls." *Geoenvironment 2000*, GSP No.46, Y.B. Acar and D.E. Daniel (eds.), ASCE: 1173-1191.

Grube, W.E. (1992). "Slurry trench cut-off walls for environmental pollution control." *Slurry Walls: Design, Construction, and Quality Control*, ASTM STP 1129, D.B. Paul et al., eds., ASTM, Philadelphia: 69-77.

Kamon, M., Ohta, H., Aoi, M. and Asada, S. (1998). "Development of new river protection method by continuous inclined diaphragm wall." *Environmental Geotechnics*, P.S. Sêco e Pinto (ed.), Balkema, Rotterdam: 35-38.

Kamon, M., Katsumi, T., Inui, T., Ogawa, Y. and Araki, S. (2006). "Hydraulic performance of soil-bentonite mixture barrier." *5ICEG Environmental Geotechnics*, H.R. Thomas (ed.), Thomas Telford Publishing, London: 733-740.

Katsumi, T., Ishimori, H., Ogawa, A., Yoshikawa, K., Hanamoto, K. and Fukagawa, R. (2007). "Hydraulic conductivity of nonprehydrated geosynthetic clay liners permeated with inorganic solutions and waste leachates." *Soils and Foundations*, Vol.47, No.1: 79-96.

Lee, J.M. and Shackelford, C.D. (2005). "Concentration dependency of the prehydration effect for a geosynthetic clay liner." *Soils and Foundations*, Vol.45, No.4: 27-41.

Norrish, K. and Quirk, J.(1954). "Cristalline swelling of montmorillonite, Use of electrolytes to control swelling." *Nature*, Vol.173: 255-257.

Shackelford, C. D., Benson, C. H., Katsumi, T., Edil, T. B and Lin, L. (2000). "Evaluating the hydraulic conductivity of GCLs permeated with non-standard liquids." *Geotextiles and Geomembranes*, Vol.18: 133-161.

Slump Evaluation of Soil-Bentonite Backfill Amended with Activated Carbon

Edward J. Barben[1], S.M. ASCE; Michael A. Malusis[2], A.M.ASCE, P.E.; and Jeffrey C. Evans[3], M.ASCE, P.E.

[1]Department of Civil and Environmental Engineering, Bucknell University, Lewisburg, PA 17837; ebarben@bucknell.edu
[2]Department of Civil and Environmental Engineering, Bucknell University, Lewisburg, PA 17837; mam028@bucknell.edu
[3]Department of Civil and Environmental Engineering, Bucknell University, Lewisburg, PA 17837; evans@bucknell.edu

ABSTRACT

This paper presents the results of slump tests to evaluate the constructability of a model soil-bentonite (SB) slurry trench cutoff wall backfill comprised of a fine sand and dry sodium bentonite (4% of total solids mass), amended with different amounts of either powdered activated carbon (PAC) or granular activated carbon (GAC) (2, 5, and 10% of total solids mass). The activated carbon (AC) products are added to provide enhanced attenuation capacity of the backfill for miscible organic contaminants. The results indicate that backfill slump at a given water content decreases as the activated carbon content increases. This decrease in slump correlates with an increase in liquid limit of the backfill, suggesting that the activated carbon provides a degree of water adsorption capacity. Thus, a greater water content is necessary for the AC-amended backfills to exhibit a similar slump relative to the control backfill containing no carbon. However, the slump can be adjusted to meet typical field requirements with an appropriate addition of slurry or reduction in dry bentonite amendment, if such a reduction does not compromise the hydraulic conductivity of the backfill.

INTRODUCTION

In situ vertical barriers (i.e., cutoff walls) have been employed for more than 40 years to control ground water flow and contaminant transport in the subsurface. While there are numerous methods and materials used to construct such barriers (e.g., Jefferis 1981, Millet et al. 1992, Larsson 2005, Opdyke and Evans 2005, Evans 2007), the most common type of vertical barrier utilized in the United States is the *soil-bentonite slurry trench* cutoff wall that takes its name from the nature of the final barrier materials (soil-bentonite) and the method of construction (slurry trench). This type of cutoff wall is constructed by excavating into an impermeable stratum while simultaneously filling the trench with bentonite-water slurry to maintain trench stability. The trench spoils are mixed with bentonite-water slurry and other amendments (e.g., dry bentonite) as necessary to achieve the desired properties, and the resulting mixture is backfilled into the trench to create the finished barrier.

In situ containment with soil-bentonite cutoff walls generally is advantageous because the backfill provides a barrier with a low hydraulic conductivity (i.e., 10^{-8} to 10^{-10} m/s) that impedes groundwater flow and, thus, impedes advective contaminant transport (Xanthakos 1979, U.S.EPA 1984, Evans 1994). In many applications, however, a vertical barrier utilized for *in situ* containment may be expected to perform over a long time span (e.g., years to decades). In such cases, there is increased opportunity for significant diffusive migration of contaminants through the barrier, a factor that has given rise to the consideration of barrier amendments to provide more sustainable containment performance (e.g., Inyang and Galvao 2004). For example, since most clay barriers have little intrinsic attenuation capacity for miscible organic pollutants, enhancing this attenuation capacity through the use of sorption amendments, such as high carbon fly ash, activated carbon, organoclay, and tire chips, has been proposed (e.g., Mott and Weber 1992, Bierck and Chang 1994, Smith and Jaffe 1994, Gray 1995, Smith et al. 1995, Gullick et al. 1996, Park et al. 1996, Park et al. 1997, Lo 2001, Bartelt-Hunt et al. 2005).

Soil-bentonite cutoff walls amended with materials possessing enhanced attenuation capacity for organic pollutants are attractive, given the increasing reliance on vertical barriers for long-term *in situ* containment and the prevalence of organic pollutants at contaminated sites. However, a viable amendment must provide sufficient sorption capacity without altering substantially the overall integrity of the soil-bentonite backfill, i.e., hydraulic conductivity and constructability cannot be compromised. Results of previous studies on amended clay barriers suggest that activated carbon appears to be a viable candidate amendment in terms of providing superior sorption capacity while limiting detrimental effects on barrier hydraulic conductivity (Bierck and Chang 1994, Baek et al. 2002, Bartelt-Hunt 2005, Lake and Rowe 2005, Malusis and Scalia 2007). However, all but one of these studies (i.e. Bierck and Chang 1994) was performed on compacted clay barriers or geosynthetic clay liners (GCLs), and the results of these studies may not be directly applicable to soil-bentonite backfills. In addition, limited information is available regarding constructability issues such as the influence of activated carbon amendment on backfill consistency and slump.

To ensure constructability of a soil-bentonite cutoff wall, the backfill should be a homogeneous mixture that exhibits a slump of 100 to 150 mm (4 to 6 in). Such a consistency is necessary to minimize the potential for heterogeneities in a completed cutoff wall that may impact the hydraulic conductivity and, thus, the containment performance of the wall. Within this slump range, the backfill is sufficiently fluid to fill the trench completely without creating large voids, yet sufficiently viscous that the solids will not segregate easily and the backfill will remain reasonably homogeneous during placement. Thus, the objective of this study is to evaluate the influence of activated carbon amendment on the slump of a model soil-bentonite backfill. This evaluation includes slump tests on the model backfill amended with different percentages of either granular activated carbon (GAC) or powdered activated carbon (PAC) to understand the potential implications of carbon content and carbon particle size on backfill consistency.

MATERIALS AND METHODS

Component Materials

The model backfill mixture considered in this study was composed of a mortar sand base soil and a powdered sodium bentonite (4% by dry mass). The mortar sand is a fine sand excavated from a local quarry and provided by Central Builders Supply (Lewisburg, PA). The natural gravimetric water content of the mortar sand, as received, was 9.0%. This sand was chosen to represent construction of a slurry trench cutoff wall though a sandy aquifer. The sodium bentonite is commercially available under the trade name Naturalgel (Wyo-Ben, Inc., Billings, MT) and is used commonly for SB cutoff wall applications. Two different activated carbon (AC) products were used as backfill amendment materials: (1) a fine GAC (type V8645; size 0.297-0.074 mm) primarily used for decolorizing oils, colored solutions, and syrups was obtained from Fisher Scientific (Fair Lawn, NJ); and (2) a PAC (type WPH; size 0.149-0.044 mm) commonly used for the treatment of potable water was acquired from Calgon Corporation (Pittsburgh, PA).

Classification of the component materials was based on grain-size distribution (ASTM D422-02), Atterberg Limits (ASTM D4318-05), and the Unified Soil Classification System (USCS; ASTM D2487-00). Measured grain-size distributions for all materials are shown in Fig. 1. The liquid limit (LL) and plasticity index (PI) for the bentonite are 488 and 443, respectively, signifying a high-plasticity clay (USCS CH). Both the mortar sand and GAC classify as poorly graded sands (USCS SP), whereas the PAC (LL = 113, PI = 5.6) classifies as a high-plasticity silt (USCS MH).

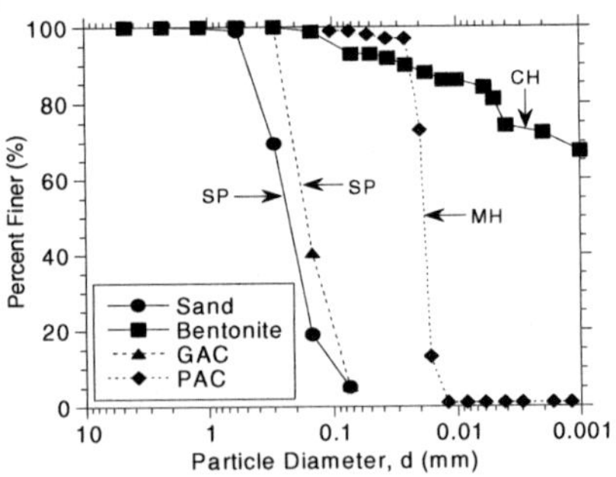

Fig. 1. Measured grain-size distributions for component materials used in study.

Backfill Mixtures

The mortar sand was air dried and re-adjusted to the as-received gravimetric water content of 9.0% for uniformity, and a bentonite-water slurry (5% bentonite by mass) was prepared by blending in a high-speed colloidal shear mixer (i.e., a Hamilton Beach 7-speed blender) for 5 min. The slurry was allowed to hydrate for a minimum of 24 h prior to use. After hydration, the density and Marsh cone viscosity of the slurry were 1.033 g/cm^3 and 38 s, respectively, and the pH and electrical conductivity (EC) of the slurry were 8.90 and 116.4 mS/m, respectively. The backfill mixtures then were prepared by combining the sand, dry bentonite, slurry, and GAC or PAC at 0, 2.0, 5.0, and 10% (by dry mass) in a Hobart mixer to ensure homogeneity. Thus, seven backfill mixtures (i.e., one control backfill and six carbon-amended backfills) were tested in this study. Liquid limits (ASTM D 4318-05) were determined for all seven backfills as an additional means to compare overall consistency.

Slump Tests

All slump tests were performed using a standard 300-mm slump cone, in accordance with ASTM C143-00. The influence of AC amendment on backfill slump was investigated by first preparing the control backfill (containing 0% AC) to a target slump within the range of 125 ± 12.5 mm, consistent with typical field requirements for SB cutoff walls. Bentonite-water slurry was added incrementally until the control backfill exhibited a final slump (S_f) within the target range based on triplicate measurements. Once the final slump was achieved, the corresponding final water content (w_f) also was measured in triplicate and the total bentonite content (i.e., dry bentonite plus bentonite added via slurry) was determined to be 5.8%.

Since the effect of AC amendment on backfill slump was not known *a priori*, all AC-amended backfills were prepared by adding enough slurry to obtain an initial water content (w_i) similar to w_f for the control backfill. Once w_i and the corresponding initial slump (S_i) were measured, the slump of each AC-amended backfill was adjusted to within the target range (i.e., 125 ± 12.5 mm) by adding slurry as needed. However, addition of slurry also causes an increase in the total bentonite content relative to the control backfill, confounding an evaluation of the sole influence of AC amendment on slump. Thus, more sand and AC were added with the additional slurry to achieve S_f within the target range while also maintaining a total bentonite content of 5.8% and maintaining the desired activated carbon content (i.e., 2, 5, or 10%). The final proportions of each backfill mixture satisfied the following set of equations:

$$X_{AC} = \left(\frac{M_{AC}}{M_S + M_B + M_{AC}} \right) \cdot 100 \quad (1a)$$

$$X_B = \left(\frac{M_B}{M_S + M_B + M_{AC}} \right) \cdot 100 \quad (1b)$$

where X_{AC} is the AC content (i.e., 2, 5, or 10%), X_B is the total bentonite content (i.e., 5.8%), and M_{AC}, M_B, and M_S are the total dry masses of AC, bentonite, and sand, respectively. Thus, as M_B is increased by a known amount due to addition of slurry, the required masses of sand and AC in the backfill were computed by solving Eqs. 1a and 1b simultaneously. Once S_f was obtained in triplicate, the final moisture content (w_f) of each AC-amended backfill was measured.

RESULTS AND DISCUSSION

All measured values of slump and corresponding moisture content are listed for each backfill in Table 1 and are plotted as a function of AC content in Fig. 2. The results indicate that amendment of the control backfill with AC increases the backfill stiffness considerably. For example, the data in Fig. 2a illustrate that the slump values for the AC-amended backfill mixtures are lower than the slump of the control backfill and decrease to zero with increasing AC content up to 10% when the mixtures are prepared to an initial water content, w_i, similar to the water content of the control backfill. These results suggest that the activated carbon provides a degree of adsorption capacity for water, as also evinced by the higher liquid limits observed

Table 1. Liquid limit and slump test results for SB backfills (total bentonite content = 5.8% by dry mass).

Backfill	Liquid Limit, LL	w_i (%)	S_i (mm)	w_f (%)	S_f (mm)			
					Trial 1	Trial 2	Trial 3	Mean
Control	27	43.2	128	43.2	130	124	130	128
2% GAC	34	43.0	92	43.5	127	124	133	128
5% GAC	35	43.2	83	46.3	133	127	121	127
10% GAC	42	44.5	0	52.1	127	121	121	123
2% PAC	30	44.2	89	44.9	117	114	127	119
5% PAC	32	44.0	38	48.3	127	121	114	121
10% PAC	36	43.6	0	53.1	127	140	127	131

w_i = initial water content; w_f = final water content; S_i = initial slump; S_f = final slump.

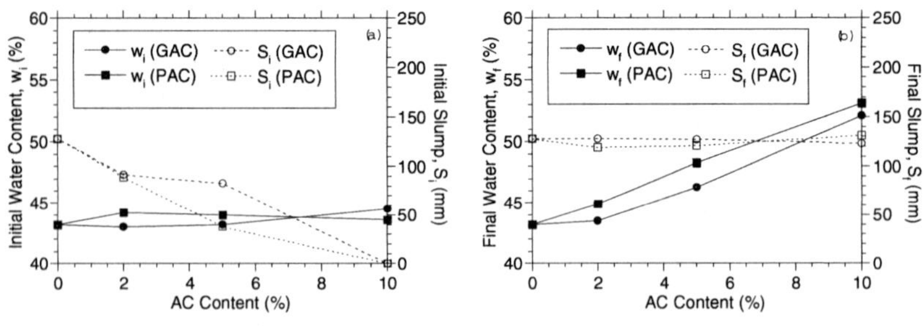

Fig. 2. Slump and water content of SB backfills containing 5.8% total bentonite (by dry mass) as a function of activated carbon (AC) content: (a) initial slump (S_i) and moisture content (w_i); (b) final slump (S_f) and moisture content (w_f).

for the AC-amended backfills (see Table 1). The results also indicate that the water content of the carbon-amended backfills must be increased to achieve the target slump (i.e., 125 ± 12.5 mm). The final slump and moisture content data, illustrated in Fig. 2b, show that as the activated carbon content increases, the moisture content must also increase to maintain a slump within the target range. The results in Fig. 2b also show that the PAC-amended backfills require a slightly higher moisture content (i.e., by ~1-2%) to achieve a similar slump as backfills amended with the same percentage of GAC. The increase in backfill stiffness caused by AC amendment should be easily overcome during typical field operations simply by adding more slurry. However, as stated previously, additional slurry also provides additional bentonite such that the final backfill mixture ultimately may contain more bentonite than is necessary to achieve an acceptable hydraulic conductivity. Thus, the relationship between AC content, total bentonite content, and hydraulic conductivity for a given backfill mixture also must be determined. Testing currently is underway to evaluate this relationship for the model backfill considered in this study.

CONCLUSIONS

The results of this study suggest that the addition of activated carbon to a sandy SB cutoff wall backfill material will not be detrimental to the overall constructability of the barrier on the basis of standard slump measurements. Although amendment with activated carbon may yield a backfill material with stiffer consistency, the slump can be adjusted to meet typical field requirements with appropriate addition of slurry or a reduction in dry bentonite amendment, if such a reduction is possible without compromising hydraulic conductivity. Additional research is required to evaluate the influence of activated carbon amendment on hydraulic conductivity of SB backfill and the potential differences in contaminant attenuation (sorption) capacity provided by the GAC and PAC.

ACKNOWLEDGMENTS

Financial support for this work was provided by the National Science Foundation (NSF) through Grant No. CMS-0625159. The opinions and recommendations provided in this paper are solely those of the authors and are not necessarily consistent with the policies of NSF. The authors thank Central Builders Supply (Lewisburg, PA) and Wyo-Ben (Billings, MT) for donating materials for this study.

REFERENCES

ASTM C143-00, "Standard test method for slump of hydraulic cement concrete." ASTM International, West Conshohocken, PA.

ASTM D422-63, "Standard test method for particle-size analysis of soils." ASTM International, West Conshohocken, PA.

ASTM D2487-00, "Standard classification of soils for engineering purposes (Unified Soil Classification System)." ASTM International, West Conshohocken, PA.

ASTM D4318-05, "Standard test methods for liquid limit, plastic limit, and plasticity index of soils." ASTM International, West Conshohocken, PA.

Baek, D.S., Kim, S.B. and Kim, D.J. (2002). "Irreversible sorption of benzene in sandy aquifer materials." *Hydrol. Process.* Vol. 17: 1239-1251.

Bartelt-Hunt, S. L., Smith, J.A., Burns, S. E., and Rabideau, A. J. (2005). "Evaluation of granular activated carbon, shale and two organoclays for use as sorptive amendments in clay landfill liners." *J. Geotechnical and Geoenv. Engrg.*, Vol. 131, (7): 848-859.

Bierck, B.R. and Chang, W.C. (1994). "Contaminant transport through soil-bentonite slurry walls: attenuation by activated carbon." *Proc. Water Env. Fed. Specialty Conf.*, Alexandra/Va. 461-472.

Evans, J. C., (1994). "Hydraulic conductivity of vertical cutoff walls." *Hydraulic Conductivity and Waste Contaminant Transport in Soil.* ASTM STP No. 1142, American Society for Testing and Materials, Philadelphia/PA: 79-93.

Evans, J.C. (2007). "The TRD method: slag-cement materials for *in situ* mixed vertical barriers." *Proc., Geo-Denver 2007, New Peaks in Geotechnics Geotechnical Special Publication 172* (CD-ROM), ASCE Reston, VA.

Gray, D.H. (1995). Containment strategies for landfill wastes. *Geoenvironment 2000*, GSP No. 46, ASCE, Reston/VA: 484-498.

Gullick, R.W., Weber, W.J., and Gray, D.H. (1996). "Organic Transport Through Clay Liners and Slurry Walls." *CMS Workshop Lectures, Vol. 8, Organic Pollutants in the Environment.* Boulder/CO: 95-136.

Inyang, H.I. and Galvao, T.C. (2004). "The application of innovative materials in waste containment barriers." *J. Environmental. Engrg.*, Vol. 130 (8): 834-835.

Jefferis, S.A. (1981). Bentonite-cement slurries for hydraulic cut-offs. *Proc. 10th Int. Conf.*, International Society for Soil Mechanics and Foundation Engineering, Vol. 1, Stockholm: 425-440.

Lake, C.B. and Rowe, R.K. (2005). "A comparative assessment of volatile organic compound (VOC) sorption to various types of potential GCL bentonites." *Geotex. Geomem.*, Vol. 23 (4): 323-347.

Larsson, S. (2005). "State of practice report - execution, monitoring and quality control." *Proc.,Int. Conf. on Deep Mixing. Best Practice and Recent Advances.* Vol. 2, Stockholm: 732-785.

Lo, I.M.-C. (2001). "Organoclay with soil-bentonite admixture as waste containment barriers." *J. Environmental Engrg.*, Vol. 127, (8): 756-759.

Malusis, M.A., and Scalia, J. "Hydraulic conductivity of an activated carbon-amended geosynthetic clay liner." *Proc., Geo-Denver 2007, New Peaks in Geotechnics, Geotechnical Special Publication 163* (CD-ROM), ASCE, Reston, VA.

Millet, R.A., Perez, J.-Y., Davidson, R.R. (1992). "USA practice slurry wall specifications 10 years later." *Slurry Walls: Design, Construction and Quality Control,* ASTM STP No. 1129, ASTM, Philadelphia/PA: 42-67.

Mott, H.V. and Weber, W.J. (1992). "Sorption of low molecular weight organic contaminants by fly ash: considerations for the enhancement of cutoff barrier performance." *Env. Sci. and Tech..* Vol. 26 (6): 1234-1242.

Opdyke, S.M. and Evans, J.C. (2005). "Slag-cement-bentonite slurry walls." *J. Geotechnical and Geoenv. Engrg.*, Vol. 131, (6): 673-681.

Park, J.K., Kim, J.Y., and Edil, T.B. (1996). "Mitigation of organic compound movement in landfills by shredded tires." *Water Env. Res*, Vol. 68, (1): 4-10.

Park, J.K., Kim, J.Y., Madsen, C.D., and Edil, T.B. (1997). "Retardation of volatile organic compound movement by a soil-bentonite slurry cutoff wall amended with ground tires." *Water Env. Res.*, Vol. 69, (5): 1022-1031.

Smith, J.A., Li, J., and Galan, A. (1995). "Organobentonites as components of earthen landfill liners to minimize contaminant transport." *Geoenvironment 2000*, GSP No. 46, ASCE, Reston/VA, 806-814.

Smith, J.A. and Jaffe, P.R. (1994). "Benzene transport through landfill liners containing organophillic bentonite." *J. Env. Engrg.* Vol. 120 (6): 1559-1577.

U.S. EPA (1984). "Slurry trench construction for pollution migration control." Office of Emergency and Remedial Response (Washington, D.C.) and Office of Research and Development (Cincinnati, OH), EPA-540/2-84-001. 268 p.

Xanthankos, P.P. (1979). *Slurry Walls.* McGraw-Hill, New York, NY, 622 p.

Strength and Permeability of a Deep Soil Bentonite Slurry Wall

Christopher R. Ryan P.E. M A.S.C.E. [1]
Charles A. Spaulding P.E. M A.S.C.E. [2]

[1]President, Geo-Solutions Inc., 1250 Fifth Ave, New Kensington PA 15068; cryan@geo-solutions.com
[2]Regional Manager, Austress Menard, 13-15 Lyonpark Rd. – Macquarie Park, NSW 2113, Australia; cspaulding@menard.com.au

ABSTRACT: In 2006, a Soil Bentonite (SB) slurry wall was constructed at a brownfield redevelopment of a former steel mill site in Mayfield, NSW Australia. At this site, the slurry wall is designed to block groundwater flow that might contribute to the contamination of an adjacent waterway, the Hunter River. The wall was approximately 1500 m long and up to 49 m deep, constituting an apparent depth record for walls of this type.

As a part of the construction QC, there was an extensive amount of testing done, including an unusual amount of in situ strength testing using both a static cone penetrometer and field vane shear measurements. These latter measurements offer a unique opportunity to determine the strength gain of SB backfill material.

Results show a moderate stiffening of the SB material after it has been in the trench. This is consistent with field observations which show that, while SB backfill is placed in a semi-fluid condition, after some weeks it can be excavated with a vertical face. Results also show that the wall does not achieve a full static state of stress over its full depth. Rather, as the material "sets", arching occurs, in effect holding some of the weight of the backfill on the sides of the trench.

Extensive permeability testing of field-mixed SB backfill samples also provides a basis for design of future walls. In an earlier design mix program, a good correlation between percentage of fines and reduced permeability was established. It is clear that the fines content must be at a certain minimum to achieve stability in the backfill and to permit the blending of a low permeability backfill mix. Data are also presented showing the effects of permeation over an unusually long test period with contaminated groundwater.

INTRODUCTION

In 1999, operations ceased at a steelworks site on the south bank of the Hunter River in Mayfield, a suburb of Newcastle in New South Wales, Australia. The development and operation of the site over a period of approximately 85 years resulted in significant contamination, and in its declaration as a remediation site in 2001.

The original ground consisted of river channels and low lying marshlands. During the progressive development of the site as a steelworks, river sediments and by-products from the steelmaking process were used to fill much of the site. Underlying the fill there is typically a layer of marine clay, and then a thick layer of dense sand with a less pervious bedrock at a depth that varied from 25-50 meters.

The most polluted part of the site was previously occupied by coke ovens, gas holders, by-products treatment, and other processes associated with steelmaking. This area represented approximately 90% of the hazard to the environment via offsite migration of contaminated groundwater flows.

The proposed remediation strategy focused on containment rather than treatment or removal, and was arrived at following consideration of a range of technological and logistical options. It addressed the risk of onsite exposure to contaminants and offsite migration of contaminants through groundwater flows. Also, given the strategic location of the site, it made the site compatible with future industrial and port-related land-uses.

The remediation strategy had four key elements: 1) Improve drainage infrastructure, 2) Install an up-gradient subterranean barrier around the contaminated area, 3) Contour the entire site, and 4) Cap the entire site.

The effect of the proposed barrier wall is to practically eliminate upgradient groundwater flows from entering the contaminated area known as Area 1. Groundwater modeling shows that this stops the movement of contamination towards the river.

A deep barrier wall was selected as the most appropriate method of managing site groundwater after extensive investigation and review of alternatives due to its passivity and a low level of encumbrance to future site development.

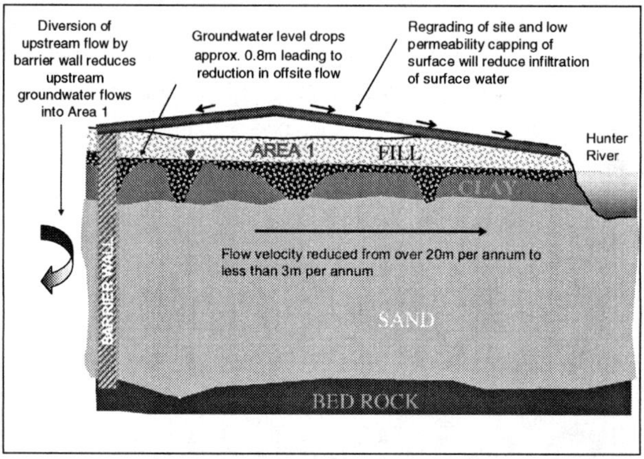

FIG. 1. Design Concept of Containment System

The wall was constructed using a long reach excavator to a depth of approximately 24 m, followed by a special slurry clamshell that excavated to the bedrock. The final depths varied from 23 m to 49 m, with the 49 m depth being an apparent new record depth for SB slurry walls. Stability of the trench was maintained with bentonite slurry during excavation. Once final depth was attained, the trench was backfilled with a blend of excavated spoils, select imported soils, and bentonite slurry. The permeability specification was a maximum of 1 x 10^{-8} m/sec. Additional information on the project can be found in Jones et al. (2007).

An unusual aspect of this project was that there were numerous static cone readings taken for the full depth of the wall. Since these types of readings can be used to infer shear strengths and since there are relatively few measurements of shear strength in situ in SB walls, this was a significant contribution to the state of knowledge. For additional information on in situ strengths, a deep vane shear test was run. The cone and vane data were correlated to each other and to in-situ measurements from other projects.

STRENGTH OF SOIL-BENTONITE BACKFILL

SB backfill is initially placed in a semi-fluid state with essentially zero shear strength. In the field, it has been observed that the backfill gains strength to the point where vertical cuts can be made through it. SB slurry walls generally do not consolidate vertically. Settlement plates installed at this site on top of the backfill, over a period of 4 months after construction, only settled from 0-10mm. The authors believe that the strength gain is actually due to thixotropic behavior of the backfill materials.

On this project, the strength of the in-situ SB backfill was measured with a series of 24 piezocone (CPTu) tests performed through the full depth of the wall. In addition, a vane test was performed to a depth of 18m at one of the cone locations. The CPTu tests were performed 4 to 67 days after backfill placement, with an average of 35 days. A review of the cone resistance corrected for pore pressure and area ratio, q_t, averaged over depth showed that consolidation time before testing had no measurable influence on measured q_t resistance.

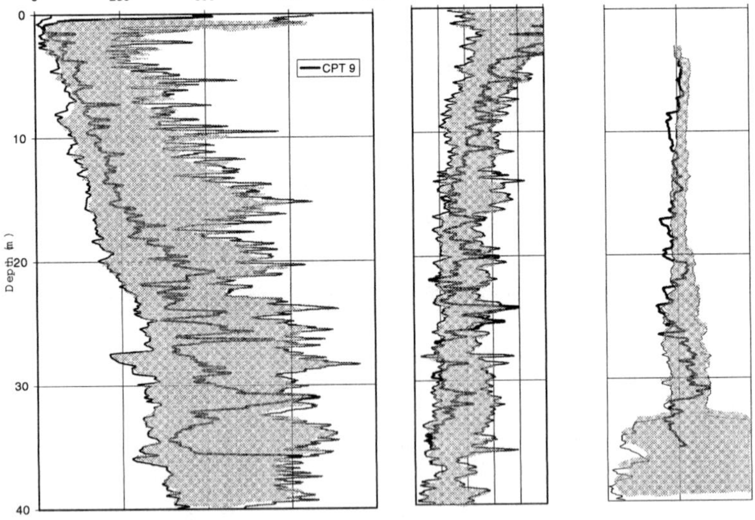

FIG. 2. Summary of CPTu results in SB wall

Figure 2 shows the minimum and maximum corrected cone resistance values, q_t, for the 24 tests. The cone resistance was first filtered to remove any set of values that exceeded 1MPa while at the same time showing a sudden increase within a depth increment of 2 or 4 cm, to eliminate the influence of small obstructions, such as gravel or unmixed clay clods left in the backfill.

The average value plus and minus one standard deviation of the Friction Ratio (ratio of cone friction to corrected cone resistance, fs/q_t) and of the pore pressure (u2) parameters are also displayed on Figure 2. In addition, one typical CPTu test, CPT-9 is also shown as an example for q_t, FR and u2 (thick blue line).

The undrained shear strength, s_u is calculated based on the corrected cone resistance and total vertical stress, σ_{vo}, as follows:

$$s_u = (q_t - \sigma_{vo}) / N_{kt} \qquad (1)$$

where N_{kt} is a correction factor; N_{kt}=15 is a recommended average, Lunne (1997)

The distribution of vertical stress in the wall is not well known since a combination of factors including arching have an influence (Evans and Ryan, 2005). However, assuming the backfill was normally consolidated at the time of the CPTu test, the following relationship would apply, p being the effective consolidation pressure, σ'_v, the effective vertical stress and u_o, the static pore pressure:

$$p = \sigma'_v = \sigma_v - u_o \qquad (2)$$

Hence, the total vertical stress, σ_{vo}, in equation (1) can be expressed as a function of u_o and p, which results in a direct expression of s_u as a function of q_t in the following formula:

$$s_u = (q_t - u_o) / (p/s_u + 15) \qquad (3)$$

In absence of site specific triaxial testing, the authors relied on values proposed by in Baxter et al. (2005) for determination of the ratio s_u/p (0.19 for a 35% plastic fines backfill) corrected to take into account the typical fines content of the wall of 25 to 30%:

$$s_u/p = 0.18. \qquad (4)$$

Hence the following relationship:

$$s_u = 0.0486 * (q_t - u_o) \qquad (5)$$

This relationship could be verified by comparing s_u derived from qt with peak shear strength (PSS) values measured in-situ using a deep vane test as shown in Figure 3. The vane blades were 75mm in diameter and 75mm in height and were equipped with a coupling to measure separately the friction developed by the rotation of the rods; the calculated PSS values were not corrected for plasticity index; note also that the CPTu was performed after 36 days, while the vane test was done after 83 days. The average s_u profile v. depth derived from the 24 CPTu tests is also plotted on Figure 3, the peak q_t values having been filtered as previously explained.

The increase of s_u with depth is nearly linear in the top 15 meters and then uneven to 40 meters but with a peak at around 25m. The values of the shear strength, s_u generally vary between 5 and 15kPa, which is consistent with values at other projects Evans and Ryan (2005), Baxter et al. (2005).

The backfill density as measured before placement was quite uniform during the whole construction project with an average value of 1.95g/cm3. Using the proposed s_u/p constant, and adopting an average Water Table depth of 4m, the s_u increment between 5 and 15m depth would hence be 17kPa as compared to an increment of only 3-4kPa in average from the Fig 3 plot, confirming that a predominant part of the weight of the backfill is transferred onto the surrounding dense native sands, illustrating how the arching effect plays a significant role in the stress distribution in the backfill.

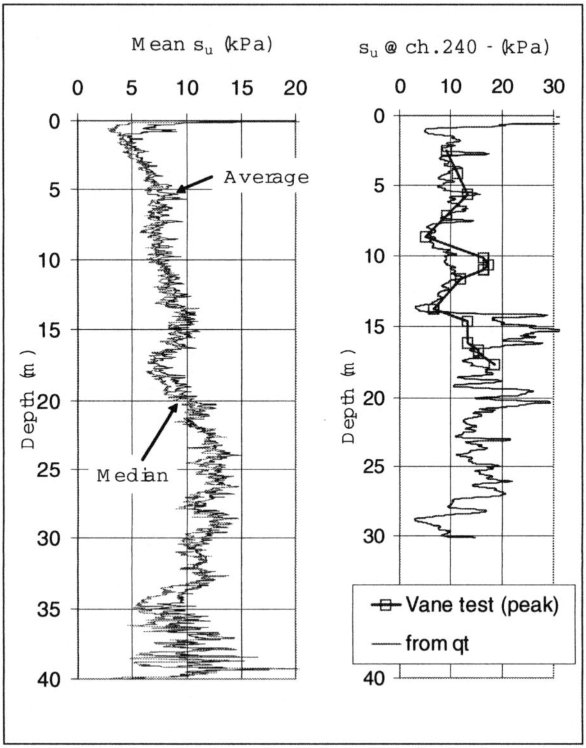

FIG. 3 – Left, Average S_u derived from CPTu using $s_u = 0.0486 * (q_t - u_o)$ and Right, Comparison of Field Vane data and S_u(CPTu)

PERMEABILITY OF SOIL-BENTONITE BACKFILL

The soil profile at this site consisted mainly of a very clean medium sand, with D_{50}=0.3mm and a Uniformity Coefficient close to 3. While there were layers of fine-grained material, it was readily apparent that there would not be enough fines to provide a backfill blend that would meet the requirements of the project (1 x 10^{-8} m/sec) with the excavated spoils alone. The option of adding additional dry bentonite was considered but was held in reserve because of the high cost and the relatively easy-to-meet performance standard. A search for sources of fines was launched and eventually encompassed sources at distances of up to 100 km away from the site.

All of the potential fines sources were tested in various additive ratios to the site base soils. The results, summarized on Fig 5 below, show that a minimum fines content of 20% in the backfill blend was required to reliably achieve the required permeability.

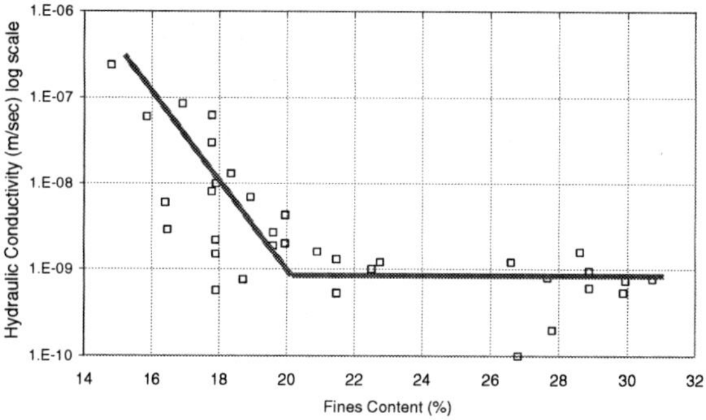

FIG. 4. Design Mix Summary Plot

An unusual aspect of this project was that there were several samples exposed for a very long term, up to six months, to the most contaminated water found at the site. Figure 6, shows a test, in which a volume of contaminants corresponding to 4.5 times the final pore volume was exchanged over a period of 3,100 hours, without variation of the permeability. The contaminated water was high in coal tars and had a very strong odor. The long-term exposure tests were run on the same samples and in the same triaxial test cells where the earlier permeability tests had been run. There was no significant degradation of any of the samples due to the contaminated materials.

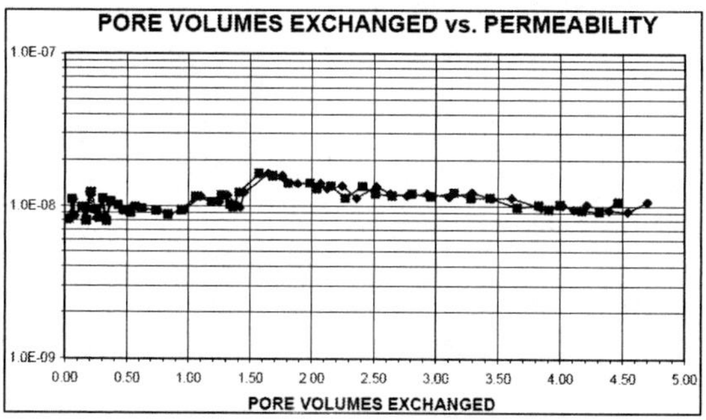

FIG. 5. Long Term Permeability Test (in cm/sec)

As a matter of practice, fines content in the field was generally higher than the minimum standard of 20% established for the site. In the figure below, the test results from grab samples taken at the surface from the backfill as it was placed are summarized. It can be seen that the results are comparable to the ones obtained in the pre-job design mix program. All results met the required permeability standards.

In an effort to understand better how the fines content of the backfill samples contributed to the impervious characteristic of the backfill, a number of backfill samples were selected for further analysis. For these samples, the fines portion of the backfill, less than 74microns in size (#200 sieve) was further analyzed with the hydrometer. As can be seen in the figure below, while there is a general correlation of decreasing permeability to increasing fines content in each size range, that correlation is very strong and clear for clay size fines. This suggests strongly that, as previously surmised, the best fines to use in an SB backfill blend are those that have good content of plastic, clay-sized fines.

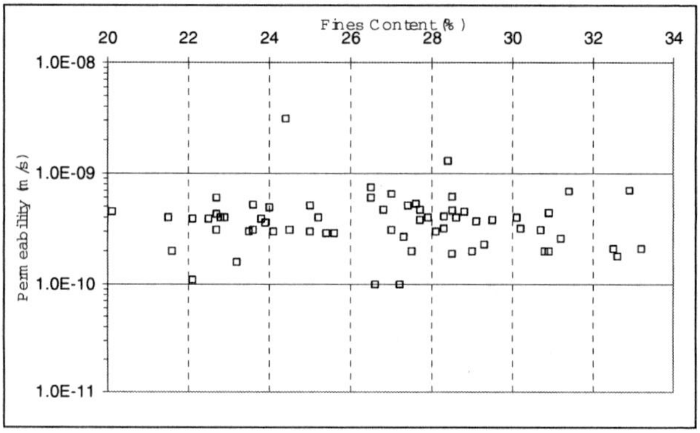

FIG. 6. Field Samples Tested for Permeability

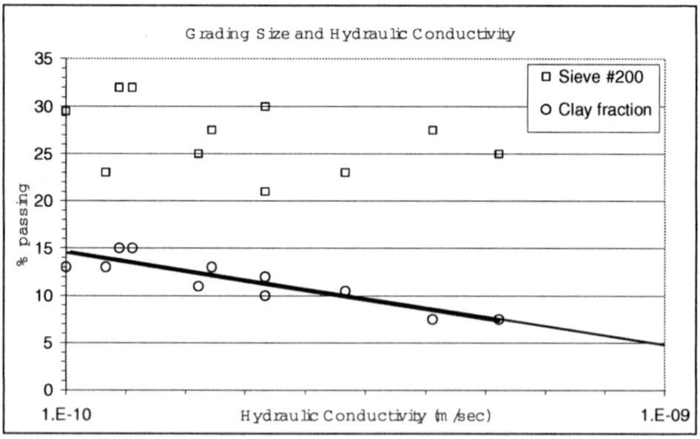

FIG. 7. Influence of Fines Size on Measured Permeability

CONCLUSIONS

The Mayfield project presented an unusual opportunity to gain a lot of performance data on an unusually deep soil bentonite slurry wall. Based on the data presented in this paper, as well as previous relevant experience, the authors conclude:

- The mechanism for strength gain in SB backfill is likely thixotropy and is not related to vertical consolidation of the backfill.
- A relationship was proposed to correlate corrected cone resistance q_t (CPTu) to the in-situ undrained shear strength of the backfill assuming a constant value for s_u/p; a good correlation was found between $s_{u\ (q_t)}$ and peak shear resistance from vane test in one test carried out to 17m depth;
- 24 CPTu tests, performed to the full extent of the SB wall, were used to obtain an average value of s_u v. depth, with mean and median values generally between 5 and 15kPa between 0 and 40m depth; these relatively low values of su illustrate the arching effect occurring in the trench through a progressive transfer of vertical loads on the native soils;
- A certain amount of fines is needed to establish an impervious matrix. In this case the criterion was set at 20%, which is a typical number for SB backfill.
- Permeabilities of field mixed samples were predicted well by the pre-job lab design mix study.
- Increasing percentages of fines, particularly clay-sized fines, have a good correlation to decreasing permeability.

ACKNOWLEDGMENTS

The authors appreciate the support of Stephen Jones of Douglas Partners, Greg Taylor from Austress Menard and Jay Santa of Geo-Solutions Inc. in helping to assemble the data in this paper. We also acknowledge the Regional Land Management Corporation and their representatives, Brad Foot and Patrick Smyth, for their continued support.

REFERENCES

Baxter, D. Y., Filz G. M., and Heslin G. M. (2005) "Strength and Compressibility of Soil-Bentonite Mixtures for Cutoff Walls" *Waste Containment and Remediation: Proceedings of the Geo-Frontiers 2005 Congress, GSP No. 142*, ASCE

Evans, J. C. and Ryan, C. (2005) "Time-Dependent Strength Behavior of Soil-Bentonite Slurry Wall Backfill" *Waste Containment and Remediation: Proceedings of the Geo-Frontiers 2005 Congress, GSP No. 142*, ASCE.

Jones, S., Spaulding, C. and Smyth, P. (2007). " Design and construction of a deep soil-bentonite groundwater barrier wall at Newcastle, Australia " *10th Australian New Zealand Conference on Geomechanics Common Ground* Brisbane, October 2007

Lunne, T., Robertson, P.K., Powell, J.J.M. (1997) Cone Penetration Testing in Geotechnical Practice - *Spon Press, UK*

Reactive Transport in Cut-Off Walls and Implications for Wall Durability

Stephan A Jefferis[1]

[1]Director, Environmental Geotechnics Ltd, Talbot Lodge, Middleton Stoney, OXON, OX25 4AD, England and Visiting Professor, University of Oxford, Email: egl@environmentalgeotechnics.com

ABSTRACT: Permeability is a key parameter for any containment system including cut-off walls but it tends to be regarded as a fixed property such as strength, sensitive principally to changes in the effective stress or chemical damage. Furthermore chemical attack, if it occurs, tends to be treated as a 'one-off' phenomenon. However, a barrier material such as a cement-bentonite is actually chemically reactive system with a dynamic permeability. Flow through this reactor is controlled by the overall permeability of all the reaction zones that develop within it. High permeability reactive barriers (PRBs) are well recognized for the management of contaminated groundwater but low permeability barriers are less often considered as reactors and opportunities for barrier protection and maintenance therefore can be overlooked. This paper is focused on the permeation of water through cement-bentonite materials: much can be learned from this about the dynamic permeability of the reactive matrix that a cement-bentonite material represents.

INTRODUCTION

Slurry trench cut-off walls are well accepted for the control of groundwater flow for dams and many other geotechnical structures. They are also regularly used for the control of pollution migration. These cut-off walls may be formed by (a) the single phase process whereby a trench is excavated under a clay-cement slurry which is then left in the trench to set and harden to form the barrier or (b) the two-phase method in which a trench is excavated under a bentonite slurry and then backfilled with a soil-bentonite mix or a soil-cement-bentonite mix. This paper considers cement based materials and therefore is applicable to single-phase cut-off wall and two-phase walls backfilled with soil-cement-bentonite mixes. For single-phase cut-off walls and especially in the UK and Europe, the cementitious material is often a blend of ground granulated blastfurnace slag (GBS) and a small amount of Portland cement (PC) as slag blends give good barrier properties at low total cementitious contents. However, for pollution migration control, there can be concerns about the durability of the barrier if the groundwater contains concentrated or unusual contaminants. For such situations cut-off materials containing large quantities of pulverised fuel ash (pfa) may be used – though this can result in poorer permeabilities and complications during the construction works as the resulting slurries may be rather thick.

THE CEMENT-BENTONITE SYSTEM

Before considering the durability of cement-bentonite slurries it is useful to consider some general features of cement-clay slurries. To enable excavation under a slurry to form a slurry trench by the single-phase method, it is important that the slurry should start as a free-flowing fluid. Furthermore for economy, the system should have as high a water: solids ratio as possible consistent with the hardened material achieving the required properties (permeability, durability, strength, etc.). However, in the absence of a dispersible clay, a cement or slag-cement slurry of high water: solids ratio will show excessive bleed leading to substantial loss of useful cut-off volume. In high bleed systems the escaping bleed water may force tracks through the slurry which become a focus for bleed water escape and appear at the surface as small 'volcanic vents' surrounded by fines elutriated from the slurry. A 'suspending' agent therefore is necessary for the cementitious material and for this a dispersible clay such as bentonite typically is used. The clay does not have to be a swelling clay and, for example, attapulgite clay can be used. The requirement is that the clay itself should form a stable but free-flowing suspension at an economic concentration – perhaps 30 to 70 kg per cubic metre of water. It also may have to be cement reactive as discussed below.

The Role of the Bentonite and the Slag

To prepare a cement-bentonite slurry, a sodium bentonite is dispersed into fresh water and usually allowed to hydrate for some hours before the cementitious materials are added. This hydration time may be reduced by the use of high shear mixing plant. However, even after high shear mixing the properties of a bentonite slurry will continue to develop with hydration time (Jefferis, 1982).

After mixing with water and a period of hydration, sodium bentonite will be in the form of finely dispersed clay sized particles of dimensions from less than a micron to a few microns (10^{-6} m) and it will provide a stable non-settling suspension. In contrast, the particles of typical Portland cement or ground granulated blastfurnace slag will range from a few microns up to about 150 microns. Dispersed bentonite particles thus are of considerably finer dimensions than those of PC or GBS and their gel system, if not compromised, helps to provide a support structure for the cement grains until a self-supporting cement hydrate structure develops. The dispersed bentonite particles also can act as nucleation sites for the precipitation of cement hydrates and thus they can promote a hydrate with a fine and low permeability structure. For optimum effect the bentonite must not be flocculated (edge to face and edge to edge bonding) nor aggregated (face to face bonding) into coarser units before the cementitious material has undergone some hydration.

However, Portland cement compromises the suspension and causes first flocculation and then aggregation of the dispersed bentonite particles. This can be observed from the flash stiffening (due to flocculation) followed by thinning (due to aggregation) which occurs when PC is added to dispersed bentonite slurry (Jefferis, 1991). As a result of this aggregation, PC-bentonite mixes give relatively high bleeding slurries (Guner, 1978), a hydrate structure containing many large pores and a set material of relatively high permeability. Under the electron microscope, hardened PC-bentonite mixes materials can appear to be much more porous than GBS-PC-bentonite – because of the evident large pores. However, the average porosity of both systems, if they are of the same bentonite content and total cementitious content, must be approximately equal as the density of the hydrates will be similar. Thus, the higher permeability of PC-bentonite mixes compared to GBS-PC-bentonite mixes is a result of their heterogeneous structure containing more coarse pores despite a similar average porosity to their GBS counterparts.

In contrast to PC-bentonite mixes, GBS-PC-bentonite mixes show a finer and more homogeneous structure (by electron microscopy and mercury intrusion porosimetry). The reason for this is that GBS does not cause the flocculation of a dispersed bentonite slurry. Indeed slag and bentonite can be added to a mixer at the same time with little adverse effect on the bentonite hydration and dispersion – a feature of slag that is exploited in some one-bag cement-bentonite mixes. The limited impact of GBS on bentonite is confirmed by the low bleed shown by cement-bentonite mixes containing a high proportion of GBS and a low level of PC (Guner, 1978). However, pure slag-cement mixes are slow setting unless accelerated, for example, by the addition of caustic alkalis (Card, 1981). Effectively in GBS-PC-bentonite mixes, the PC acts as a trigger for set but in so doing damages the permeability that can be achieved – hence the need to use slag contents over about 65% (see Guner, 1978; Card, 1981 and Jefferis, 1997 for the effect of GBS content on strength and permeability of mixes – using GBS available in the UK. Opdyke and Evans, 2005 confirmed the results with GBS available in the USA). Other setting agents than PC can be used (see for example, Taylor, 1997) – some achieving spectacularly low permeabilities – lower than can be measured with conventional triaxial permeability equipment.

A query that is often raised in connection with cement-bentonite slurries is: will the calcium from the cement 'damage' the bentonite by calcium ion exchange leading to aggregation of the clay particles – effectively the inverse of swelling? However, this type of question reflects an overly simplistic approach to the system. Clay particles are not stable in the high pH environment of a hydrating PC or GBS-PC system. They react to form calcium aluminate silicate hydrates probably comparable to those from normal PC or PC-GBS hydration (further research is needed on this). Thus the bentonite is not damaged by calcium ions, rather it is removed. An xrd study commissioned by the author in 1994 (unpublished) on a GBS-PC-bentonite system found only traces of bentonite (as calcium smectite) in the hardened material at six months from mixing. The initial mix contained 21% bentonite by dry weight. The xrd data were not quantitated but it is estimated that the residual clay content was less than 5% by dry weight. The removal of bentonite from cement-bentonite mixes, by reaction has been confirmed by the fuller and more detailed work of Yeborah and Evans (Evans, 2006).

It would seem that the bentonite is important in developing a fine microstructure in the set material as the author has found that the permeability of slag-cement-bentonite slurries is more sensitive to increase in bentonite content than in cementitious content (to reduce the permeability of a cement-bentonite material it is more effective to add an extra in kilogram of bentonite than of cementitious material).

As discussed above, in a slag-cement-bentonite slurry, the fine clay particles appear to act as nucleation sites for the slag-cement hydration promoting slag hydration and cement hydrates of fine microstructure and low permeability. It follows that if the bentonite is not properly hydrated before the slag and cement are added the slurry will show significant bleed and the set material will be of much higher permeability. This requirement for hydration is a disadvantage in the field as bentonite hydration facilities must be provided. However, in the laboratory, advantage can be taken of the feature to produce materials of high permeability to enable more rapid cement-bentonite – contaminant interaction. Sample A, discussed below, was prepared in this way with minimum bentonite hydration.

Without the reaction between the clay and the cementitious material it is doubtful whether a slag-cement-bentonite would be of low permeability as the flakes of bentonite can provide no particular sealing action if dispersed within a cementitious matrix. The key issue is to produce an overall system with a fine microstructure.

Further evidence that the persistence of bentonite particles in the set material is not significant can be adduced from the work by Thomas and Tallard (2001). They report the results of permeability tests on attapulgite clay-GBS slurries with no added PC, i.e. a slurry with slag as 100% of the cementitious material. These mixes showed permeabilities dropping to below 10^{-12} m/s. Now attapulgite is a lath-shaped clay mineral and the gel of attapulgite slurries is not created by electrostatic interactions of the particles (as in bentonite) but by tangling of the laths. The filtrate loss of attapulgite clay slurries (a measure of the permeability of a filter cake of attapulgite) tends to be much higher than that bentonite. Indeed Darley and Gray (1988) state that, in relation to the use of attapulgite in drilling fluids, "attapulgite does not afford filtration control" and yet despite this it gives a very low permeability slag-attapulgite slurry.

THE EFFECT OF CHEMICALS ON CEMENT-BENTONITE MATERIALS

A significant amount of work has been reported on the effect of contained chemicals on the physical properties of barrier materials such as clay liners, cement-bentonite and soil-bentonite walls. The prime focus of much of this work has been the assessment of the damage to the impermeability of a barrier that may be caused by contained chemicals. Much less work has been reported on the beneficial effects of chemicals although an examination of the literature shows that some work focused on the damage has actually demonstrated beneficial effects. Furthermore too often there has been a focus on investigating the effects of concentrated chemicals without first considering the effects of water alone. This is curious as relatively pure water will be the most common challenge for many barrier materials and it has a substantial dissolving power which may be much greater than the potential for reaction presented by chemical species dissolved in the water. Also, almost invariably when modelling chemical migration the 'chemistry' is described by a single parameter, a distribution coefficient, K_d – an approach that is entirely inappropriate to describe chemical reactions and the associated boundary conditions.

The work described in this paper is part of a long-term program designed to investigate both damaging and beneficial effects of chemicals on the behaviour of cement-bentonites. The work presently reported is limited to the results of tests involving the permeation by water as this sets a baseline for chemical investigations.

THE EFFECT OF PERMEATION BY WATER

The effects of extended water permeation on three cement-bentonite samples, A, B and C are considered in the following sections. Details of the sample mix designs and permeability test parameters are given in Table 1.

Table 1: Water permeation test details

Sample Mix proportions GBS/PC/bentonite kg/m^3	Age at start of test	Sample dimensions, diameter x height mm	Bentonite hydration time	Gradient and sample confining pressure, kPa	Periods of permeation, Days from start of test
A 105/45/45	6 days	36.5 x 69.5	Minimal	22 50	0-85, 103-110, 369-408, 1103-1119
B 105/45/45	23 days	104 x 209	24 hours	19 50	0-318
C 67/29/54	17 years	36.4 x 82	24 hours	9.3 100	0-71

Sample A was specifically designed to be of high permeability to enable more rapid water permeation and was 6 days old at the start of permeation. Samples B and C were taken from construction sites during cut-off works and were respectively 31 days and 17 years old at the start of permeation. Sample C was required to have a low strength and had an unusually low total cementitious content of 96 kg/m^3. The test sample had been continuously stored under water in a plastic tube, sealed at one end but open to the water at the other end. This curing regime allowed some softening of the exposed end of the sample as there was no confining stress. Over the 17 years this softening had penetrated some 20 mm into the sample. Many similar samples were stored together in the same closed curing box and the solids ratio in the curing box was about 4:1 by volume. The water was changed infrequently – about once per year. The regime will have allowed some leaching of soluble species such as alkali hydroxides and lime. However, carbonation may have been limited.

All three samples were permeated with tap water in flexible wall permeameters. The test program has been in progress for many years and is on-going. The samples have not been removed from the permeameters during the tests and thus no data on final dimensions, physical properties, mineralogy, etc. is available.

The behaviour of Samples A, B and C under water permeation is shown in Figure 1 as a function of time and in Figure 2 as a function of volume permeated. Unfortunately, long-term permeability tests present a challenge to the researcher, especially in the UK where laboratory closures are a regular event. The permeation of Sample A suffered a number of interruptions including, at 408 days, a stoppage of almost two years as a result of a laboratory closure. At each stoppage the sample was retained in the permeameter under a slight confining pressure. The other two Samples, B and C were permeated continuously for the times shown in Table 1. The sample confining pressures and test gradients are also shown in Table 1. Flow measurements generally were made daily. The laboratories were not temperature controlled and permeabilities were corrected to 20°C on the basis of the variation of water viscosity with temperature.

For Sample A, during the first period of permeation to 85 days, the permeability showed a steady decline from 3.5×10^{-7} to 6.2×10^{-8} m/s. This drop in permeability was mirrored by a drop in the pH of the effluent from about 12.9 reaching 11.8 at about 50 days at which time the pH began to drop more steeply and the permeability showed a slight increase. Thereafter the permeability although showing some significant oscillation began to plateau. Unfortunately the tests had to be stopped at 85 days – though usefully as it turned out as this provided further information. When the test was re-started for at 103 days (for 7 days) the permeability showed a declining trend. The first permeant sample after the stoppage had a pH of 8.83 which may be compared with pH 8.24, the last result before the stoppage at 85 days. From Figure 1 it can be seen that after the re-start at 103 days, the pH dropped very rapidly (from 8.83 to below 8.0). The same trend was observed when the flow was re-started on each of the two subsequent occasions (at 369 days for 39 days and 1103 days for 16 days). These increases and rapid declines in pH show that for the relatively high flow rate through this 'designer' high permeability sample, the permeating fluid was not fully in equilibrium with the matrix of the sample (demonstrating that there was some matrix diffusion resistance).

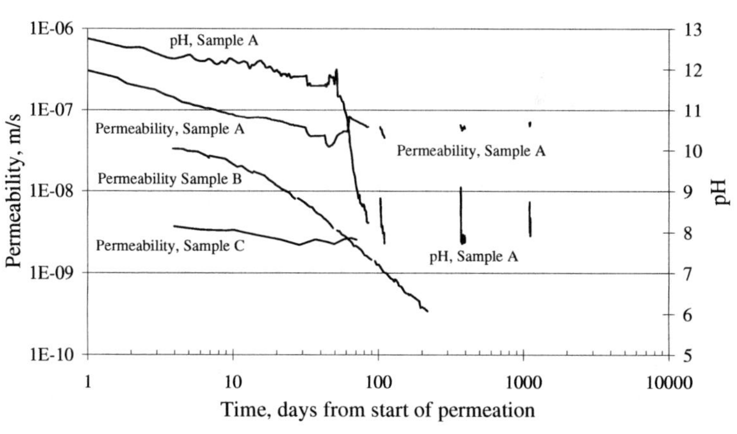

FIG. 1. Permeabilities as a function of time

However, the permeability showed no further decline suggesting that some degree of physical equilibrium had been achieved and there was no further decline in permeability as a result of continuing cement hydration reactions, reaction with bicarbonate in the permeating water or mineral changes due to continued leaching.

For Sample B, a typical GBS-PC-bentonite mix with 24 hr hydrated bentonite, the permeability dropped steadily throughout the test period. This well-recognised behaviour raises one of the key issues for a cut-off wall in service: for how long will the permeability reduction continue? For this it is necessary to consider permeability as a function of volume permeated (Figure 2).

Sample C, again a mix with 24 hr hydrated bentonite, had a low total cementitious content as it was required to be of low strength. The sample was 17 years old at the start of the test. From Figure 1 it can be seen that the permeability showed a slight reduction with time under permeation (from 3.8×10^{-9} to 2.6×10^{-9} m/s). It was interesting to find that there was some slight, reduction in permeability for such an old sample. This cannot have been due to further cement hydration as the test duration, 71 days, was just 1.1% of the age of the sample. The most likely reason was carbonation as the permeating water was deliberately left open to the atmosphere and hence atmospheric carbon dioxide (most groundwaters will be in equilibrium with carbon dioxide gas at least atmospheric levels and often significantly higher).

Figure 2 shows the same data as Figure 1 but the permeabilities are plotted against the volume of water permeated through the samples (normalised by the initial volume of the sample). From Figure 2 it can be seen that for Sample A, the steep pH drop and slight permeability increase started at a permeated volume of about 80 sample volumes. To achieve such a number of volumes permeated with a sample of permeability 10^{-9} m/s would have taken 9 years of continuous permeation – demonstrating the need for the development of a high permeability test sample.

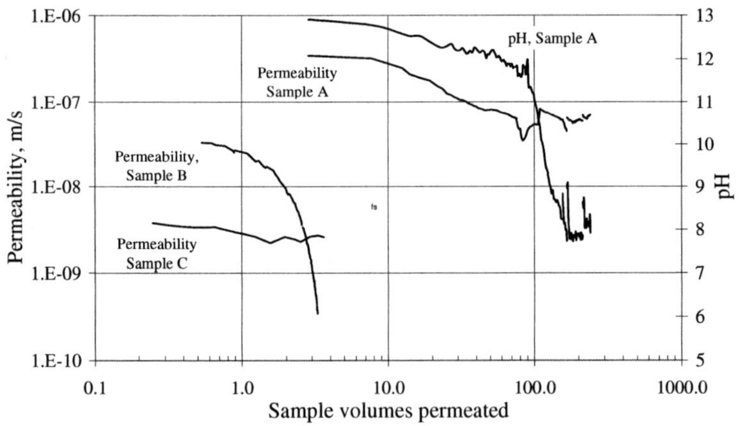

FIG. 2. Permeabilities as a function of volume of water permeated

If the permeability of Sample B continued to reduce until 80 sample volumes had passed, i.e. until the same number of volumes as Sample A, then the time to minimum permeability could be many tens of years, for as can be seen from Figure 2, the permeability drops below 10^{-9} m/s at about 3 sample volumes and is still declining very steeply. Thus the ultimate permeability of Sample B could be much below 10^{-10} m/s before reaction with and/or leaching by the permeating water reaches equilibrium. Clearly further long-term permeability investigations are required on these cut-off materials but the extrapolation from the high permeability sample (A) is promising. Sample C shows that permeability reduction will be small once a sample is fully or near fully hydrated and largely equilibrated with water by immersion.

CONCLUSIONS

These few but long-term permeation tests have been designed to confirm the hypothesis that although cement-bentonite materials show reduction in permeability with time there is a limit to the reduction that occurs. The tests are on-going and the samples are being subjected to selected chemicals to show that barrier performance can be enhanced by the introduction of selected chemicals to the groundwater.

Finally, it is important to remember that the greatest challenge to the integrity of any in-situ barrier is likely to be the volume changes that occur as a result of interaction with groundwater – volume changes that may cause cracking. Research is urgently needed on this at field scale.

REFERENCES

Card, G.B. (1981). “Properties and performance of bentonite-cement slurries for use as hydraulic cut-offs”, *University of London PhD thesis*, pp. 334.

Darley, H.C.H. and Gray G.R. (1988). *Composition and properties of drilling and completion fluids,* 5th Edition, Gulf Publishing.

Yeboah, N. and Evans, J. C. "The Role of Bentonite in Slag-Cement-Bentonite Slurry Wall Performance, in prep for *ASCE J. of Geotech and Geoenv Engg*, 2007.

Guner, A. (1978). “Properties and behaviour of bentonite-cement slurries.” *University of London PhD thesis*, pp. 279.

Jefferis, S.A. (1982). “Effects of mixing on bentonite slurries and grouts”. *ASCE Conference on Grouting in Geotechnical Engineering*, New Orleans, pp. 62-76.

Jefferis, S.A. (1991). “Grouts and slurries.” in *The Construction Materials Reference Book*, Doran, D.K., Editor, Butterworths (2nd edition in prep.).

Jefferis, S.A (1997). “The origins of the slurry trench cut-off and a review of cement-bentonite cut-off walls in the UK.” *International Containment Technology Conference and Exhibition*, Florida, pp52-61.

Opdyke, S. M. and Evans, J. C. (2006). “Slag-Cement-Bentonite Slurry Walls,” *ASCE J. of Geotech and Geoenv Engg,* Vol. 131, No. 6, pp. 673-681.

Taylor, H. F. W. *Cement chemistry*, 2nd Edition, Thomas Telford.

Thomas, G. and Tallard, G. (2001) “Unique hydraulic conductivity test results from a self-hardening slurry.” *International Containment and remediation technology Conference and Exhibition*, Florida

Geoenvironmental Characterization to Assess Waste Stabilization/Solidification Treatment Performance and Sustainability

Dimitris Dermatas, M. ASCE[1], Maria Chrysochoou[2], Deok Hyun Moon[3]

[1]Senior Associate, Center for Environmental Systems, Stevens Institute of Technology, Hoboken, NJ 07030; dermatas@stevens.edu
[2]Assistant Professor, Department of Civil and Environmental Engineering, University of Connecticut, Storrs, CT, 06929-2037;
[3]Research Professor, Center for Environmental Systems, Stevens Institute of Technology, Hoboken, NJ 07030;

ABSTRACT: Stabilized/Solidified (S/S) waste, whether a contaminated soil, sediment or industrial waste, is usually evaluated on the sole basis of regulatory leaching tests. These tests provide point-in-time performance data and cannot alone provide a solid foundation for a reliable assessment of long term performance and overall treatment sustainability. Instead, the use of state-of-the-art geoenvironmental characterization techniques constitutes a more reliable approach to effectively evaluate S/S treatment alternatives and develop a sustainable S/S treatment design and field implementation. This involves a combination of traditional geotechnical analyses (particle size distribution, unit weight, water content), physicochemical analyses (pH, total metal contents), a subset of controlled pH extraction tests (Toxicity Characteristic Leaching Procedure, TCLP, Synthetic Precipitation Leaching Procedure, SPLP, etc.) coupled with non-destructive quantitative mineralogical analyses (X-ray Powder Diffraction, XRPD combined with Rietveld Quantitative Analysis, RQA). Additionally, micromorphological (Scanning Electron Microscopy-Energy Dispersive Spectroscopy, SEM-EDS and/or Electron Probe Micro-Analysis, EPMA) and other contaminant speciation analyses (X-Ray Absorption Near-Edge Spectroscopy, XANES and/or Extended X-ray Absorption Fine Structure, EXAFS or other newly developed techniques) should be used once a more in-depth assessment is required. In this paper a general framework along with case studies involving such state-of-the-art geoenvironmental characterization of Pb and hexavalent Cr(VI) contaminated, stabilized waste are presented.

INTRODUCTION

Protecting the environment from hazardous pollutants associated with waste generation and disposal is a major concern in today's heavily industrialized world. Thus far, various technologies have been developed which transform hazardous wastes to non-toxic, or reduce the potential release of toxic species into the environment. One such promising technology is stabilization/solidification (S/S) of solid wastes by means of adding cementitious binders, such as lime and cement (Conner 1990; Dermatas and Meng 1996). S/S treatment aims to reduce the mobility and/or toxicity of hazardous compounds, by chemical transformation into less toxic and/or soluble forms, or by physical encapsulation in a low-permeability, high-strength media (USEPA 1999a). Moreover, the stabilized wastes may attain adequate stress-strain properties to enable their utilization in construction applications, such as engineering fill, road or pavement subgrade, backfill, and base material (Mitchell and Dermatas 1992; USEPA 1999a; Dermatas and Meng 2003; Dermatas et al. 2006d).

Customarily, S/S performance and overall sustainability, whether the waste is a contaminated soil, sediment or industrial waste, is evaluated solely on the basis of regulatory leaching tests. In the United States (US) and other parts of the world, this predominately involves treatability bench scale studies where the toxicity characteristic leaching procedure (TCLP) and total analyses (TA) form the basis for S/S performance evaluation. These tests and other similar leaching tests, used in Europe and elsewhere, provide point-in-time performance data and cannot solely provide a solid foundation for a reliable assessment of long term performance and treatment overall sustainability.

When such controlled pH extraction tests (TCLP, Synthetic Precipitation Leaching Procedure, SPLP, Sequential Extraction Testing, SET, etc.) are combined with traditional geotechnical analyses (particle size distribution, unit weight, water content) and physicochemical analyses (pH, total metal contents), useful information on contaminant speciation and release can be obtained. However, only when these more traditional geoenvironmental tests are coupled with non-destructive mineralogical quantitative analyses (X-ray Powder Diffraction, XRPD combined with Rietveld Quantitative Analysis, RQA), micromorphological testing (Scanning Electron Microscopy-Energy Dispersive Spectroscopy, SEM-EDS and/or Electron Microprobe Analysis, EPMA) and contaminant speciation analyses (X-Ray Absorption Near-Edge Spectroscopy, XANES and/or Extended X-ray Absorption Fine Structure, EXAFS or other newly developed techniques) a thorough and relevant picture of contaminant speciation can be obtained. The most recent global shift to remediation approaches based on risk assessment, makes it clear that TA, TCLP and other conventional geoenvironmental tests are no longer sufficient. This underscores the significance of using state-of-the-art geoenvironmental characterization aiming predominately in obtaining a deeper understanding of the solid state speciation of contaminants, and thus their bioavailability, in the geoenvironment.

In this paper, besides a general framework, case studies involving such state-of-the-art geoenvironmental characterization of S/S heavy metal waste, focusing on hexavalent Cr(VI) and Pb are presented. These demonstrate that the use of state-of-the-art geoenvironmental characterization constitutes a more comprehensive and reliable approach to: effectively evaluate S/S treatment alternatives, project long term leaching and structural integrity of the S/S waste and develop a sustainable S/S treatment design and field implementation.

TCLP AND SET LEACHING/EXTRACTION BASED S/S PERFORMANCE ASSESSMENT TOOLS

The TCLP was developed by the United States Environmental Pollution Agency (USEPA 1992) to mimic the worst case scenario for co-disposal of waste with municipal solid waste (MSW). The relevance of TCLP to evaluating and regulating waste materials has been repeatedly scrutinized. TCLP may overestimate the leaching of contaminants from wastes that are not co-disposed with MSW wastes (USEPA 1999b). Moreover, Hooper et al. (1998) reported that the TCLP fails to extract oxyanion-forming elements that are extracted by municipal solid waste leachates. Halim et al (2004) found that TCLP underestimates the leaching of Pb and Cd from cementitious wastes compared to MSW leachates. Ghosh et al. (2004) indicated that TCLP underestimated arsenic leaching from arsenic-containing solids under landfill conditions and equilibrium concentrations were not achieved within the 18h standard exposure time. Cao and Dermatas (2007), concluded that the standard one-point TCLP test would either over- or under-estimate Pb leachability in firing range soils and that the SPLP is a more appropriate alternative. Similarly, Kosson et al. (2002) and most recently van der Sloot et al. (2007), also point out to the inadequacies of regulatory leaching tests.

Similar to TCLP, SET procedures employed as investigatory testing tools, have also been scrutinized, mostly for failing to reliable assess contaminant release and speciation mechanisms. Dermatas et al (2006a) have pointed to the inadequacy of SET procedures in evaluating Pb speciation and release. Similarly, Scheckel et al. (2004) found that in phosphate-treated Pb contaminated soils, a significant shift of extractable Pb to the residual phase occurred due to the formation of pyromorphite during SET, artificially overestimating the presence of pyromorphite. Without definitive knowledge of contaminant speciation and release mechanisms, S/S treatment design becomes unreliable in attaining long-term treatment goals in terms of performance and overall sustainability.

Thus, it appears that treatability bench scale screening studies, which are mainly based on regulatory or investigatory extraction-based testing, may lead to the wrong treatment alternative selection. To this end, there have been many cases where S/S treatment that appeared to be initially successful, ended up failing in the near or long term (Dermatas et al. 2003). Moreover, this approach often leads to S/S over-treatment along with the associated waste of energy and resources.

THE ROLE OF CONTAMINANT SPECIATION IN S/S PERFORMANCE ASSESSMENT

What is a common thread in this growing list of studies that effectively scrutinize the use of extraction-based testing to evaluate S/S treatment performance and overall sustainability, is that testing inadequacies are mainly traced to a failure in qualitatively and quantitatively recognizing the contaminant speciation regime. Contaminant speciation, i.e., the physicochemical state at which the contaminant exists, ultimately dictates the degree of contaminant immobilization and thus its release and bioavailabilty into the geoenvironment (USEPA 1999a; Dermatas et al.

2006b). The use of contaminant speciation analyses is quite common for the evaluation of treatment approaches for soil or hazardous waste. For example, sieve analyses, specifically the presence of a high fines content, point to the availability of soil sorption sites for contaminant immobilization. However, the determination of soil mineralogy by XRPD is essential to verify and quantify the presence of active clays and amorphous and/or poorly crystallized metal oxides and hydroxides that would contribute to contaminant immobilization (Dermatas et al., 2006a). Feldspars and other non-clay and non-reactive fines would likely not significantly affect contaminant leaching. Interestingly, the amount of fines in a soil does not necessarily correspond to the amount of clay present, as other minerals (micas and feldspars) may contribute to the fine fraction (Mitchell, 1993), necessitating the use of XRPD, besides conventional particle size distribution testing. Similarly, the amount of carbonate species (calcite, dolomite) may also provide an indication of the buffering capacity of a soil or solid waste, that pH measurements alone cannot reliably provide (Dermatas et al. 2006c). Contaminant speciation analyses can be, and usually are, much more intricate then the simple examples provided. They may involve sophisticated quantitative and/or qualitative tools such as XRPD with RQA, SEM-EDS, XANES, EXAFS, etc. Overall, accurate knowledge of contaminant speciation can be ultimately used as reliable input to geochemical models (e.g., PHREEQC and MINTEQ), to evaluate whether the system is at equilibrium or transient and to create different scenarios and project how contaminant speciation and availability, will change with time.

STATE-OF-THE-ART GEOENVIRONMENTAL CHARACTERIZATION OVERVIEW AND CASE STUDIES

Besides the more traditional geoenvironmental characterization testing, microscopic and spectroscopic methods are deemed necessary to provide accurate information on contaminant speciation. Complimentary to XRPD, optical microscopy, SEM-EDS and EPMA are commonly applied to investigate the composition of crystalline phases (Hillier et al. 2003). Among these, XRPD presents the advantage that it allows a statistically defensible (global) analysis of a solid matrix (Chrysochoou and Dermatas, 2007). It is also a widely applied method with extensive databases of XRPD data that facilitate the identification of crystalline phases. Bulk mineralogy as determined by XRPD may reveal contaminant species only if contaminant concentrations are high enough (i.e., above the XRPD detection limit for the specific contaminant species). XRPD of discrete soil features, guided by optical microscopy observations and complemented by SEM observations, are critical in further elucidating aspects of contaminant speciation that cannot be readily discerned by bulk XRPD (Dermatas et al. 2006a). Finally, the use of advanced contaminant speciation analyses (XANES and/or EXAFS or other newly developed techniques) should be used once a more in-depth assessment is required.

Several studies have utilized XRPD to investigate heavy metal speciation in S/S matrices (Dermatas and Meng 2003; Duchesne and Reardon 1999; Mollah et al. 1998; Moon et al. 2004). The XRPD analysis in mainstream environmental applications is, however, to date mostly qualitative. Although XRPD quantitative analysis methods date back to 1969, with the publication of the Rietveld method (Rietveld 1969), quantitative phase analysis (QPA) for environmental applications has been limited. One reason for this is that the available software was previously difficult to manipulate. However, the progress in computerized powder diffraction and crystal structure data and the advances in XRPD analysis software in the past decade now allow for a wider application of QPA. Training in this regard is therefore essential to understand the physical meaning of quantitative analysis methods, as well as the intricacies of analysis strategies. Furthermore, the application of QPA on XRPD data of contaminated soils and waste is complicated by the nature of these matrices; the number of crystalline phases is usually high, mixtures are commonly non-uniform and solid phases are not necessarily chemically and structurally well defined, as is the case in materials science applications. Hiller et al. (2003) and Chrysochoou and Dermatas (2007), clearly demonstrate how the challenges presented by complicated solid waste matrices were successfully overcome to provide first order estimates of hexavalent Cr(VI) speciation.

The intricacies of XRPD sample preparation and its influence on contaminated media S/S remediation assessment are discussed in great detail in Dermatas et al. (2007). Sample preparation determines XRPD results. Erroneous sample preparation significantly alters conclusions drawn from qualitative and quantitative XRPD analyses of environmental samples, likely resulting in non-effective remediation treatment design. Dermatas et al. (2007), report that mechanical pulverization using non-polar fluids, namely cyclohexane, was the optimal method to determine the reactive potential of chromate waste, by reliably quantifying the brownmillerite content, the major unstable mineral phase in this waste. The use of water during milling caused extensive hydration reactions and yielded non-representative mineralogical results. Complementary hand pulverization proved necessary to accurately quantify softer, Cr(VI)-bearing minerals (Dermatas et al. 2007). Similar effects may take place in most reactive media, contaminated or otherwise, thus the choice of sample preparation is crucial to the validity of XRPD studies and the associated remediation treatment decisions.

To further illustrate the use of state-of-the-art geoenvironmental characterization, another one of our studies (Dermatas et al. 2006c) where several firing range soils were evaluated for different S/S treatment alternatives, is next considered. This study investigated the effects of Pb speciation on Pb leachability by evaluating TCLP and TA data from samples taken out of six different firing ranges. In all the soils tested, TCLP Pb leachability levels and TA Pb concentrations were found to be above the regulatory limits set by the USEPA and state agencies. However, TCLP Pb leachability levels did not always correlate well with TA Pb concentrations. Through the use of XRPD, QPA, optical microscopy and SEM-EDS, formation of Pb carbonates was found to play an important role in controlling Pb leachability and precipitation was found to be an important Pb immobilization mechanism in all soils tested. The effect of post-tumbling TCLP (leaching) pH, soil buffering capacity, grain size and mineralogy on Pb leachability was also considered. The leaching pH effect was most pronounced in the range that showed the lowest Pb leachability and was attributed to the presence of buffering compounds in this soil. Pb leachability was found to decrease with decreasing soil grain size, where lowered Pb leachability levels were attributed to clay, Fe-Mn oxide and aluminum hydroxide sorption mechanisms. The effectiveness of soil S/S treatment with phosphates and pozzolans (cement or lime) in achieving Pb immobilization was also evaluated. Formations of calcite and ettringite were identified in both cement and lime treated soils. $PbHPO_4$ formations were identified in soils treated by phosphate addition. Effective Pb immobilization in treated soils was attributed to the formation of these minerals and to an increase in the soil's buffering capacity in soils treated with pozzolans (Dermatas et al. 2006c).

In another study (Dermatas et al. 2006d), XANES analyses of S/S chromate waste clearly indicated that alkaline digestion (AD) results underestimated the total Cr(VI) by a factor on the order of 1.5 to 4.0. The reasons for this discrepancy may be that residual ferrous reduced the liberated Cr(VI) during AD and/or that Cr(VI) was not 100% released from the solid during AD. This discrepancy between XANES and AD results renders the latter unreliable for regulatory purposes, when applied to ferrous-treated chromate waste (Dermatas et al. 2006d). Similar findings are also reported elsewhere (Moon et al. 2007).

In yet another study (Chrysochoou et al. 2007), the use of EXAFS clearly demonstrated that the success and sustainability of applying phosphate as a Best Management Practice (BMP) (USEPA, 2001) in firing range soils and other Pb contaminated solid media remain questionable. Phosphate-based S/S treatment has emerged as a widely accepted approach to immobilize Pb in contaminated soils and waste media, relying on the formation of the highly insoluble mineral pyromorphite as solubility-controlling phase for Pb. As such, phosphate treatment has been proposed by EPA as a BMP for firing ranges where Pb occurs in its metallic forms and several other phases (carbonates, oxides). Treatability studies have shown that excess quantities of soluble and acidic phosphate sources, such as phosphoric acid, are necessary for successful in situ S/S treatment of Pb contaminated soils and sediments. Interestingly, even under these conditions, EXAFS, the only reliable method to identify and quantify Pb speciation in this case, showed that Pb conversion to pyromorphite in in-situ treated soils was less than 45% after 32 months (Chrysochoou et al. 2007).

In summary, the speciation of heavy metals and other contaminants will significantly affect their availability to participate in transformation and/or immobilization reactions. The accurate determination of contaminant speciation prior to, as well as following treatment, is essential to ensure the success of the treatment. Traditional geoenvironmental analyses (particle size distribution, unit weight, water content, pH, total metal contents) and extraction-based regulatory and investigatory tests (TCLP, SPLP, SET) are deemed necessary and can provide useful information. However, only when traditional geoenvironmental testing is accompanied by predominately non-destructive microscopic and spectroscopic methods a complete picture of contaminant speciation can be obtained. On this basis, especially when aided by geochemical modeling, a reliable S/S treatment design can be pursued. Reliable S/S treatment design is, in turn, a necessary precondition to ensure long-term S/S treatment performance and overall treatment sustainability.

REFERENCES

Cao, X. and Dermatas, D. (2007). "Evaluating the Applicability of Regulatory Leaching Tests for Assessing Lead Leachability in Contaminated Shooting Range Soils", *Environmental Monitoring and Assessment*, available online Feb 1 2007.

Chrysochoou, M. and, Dermatas, D. (2007). "Application of the Rietveld Method to Assess Chromium (VI) Speciation in Chromite Ore Processing Residue", *Journal of Hazardous Materials*, 141: 2, pp 370–377.

Chrysochoou, M., Dermatas, D. and, Grubb, D.G. (2007). "Phosphate Application to Firing Range Soils for Pb Immobilization: The Unclear Role of Phosphate", *Journal of Hazardous Materials*, 144: 1-2, pp 1-14.

Conner, J.R. (1990). "Chemical Fixation and Solidification of Hazardous Wastes", Van Nostrand Reinhold, New York, 1990.

Dermatas, D. and Meng, X. (1996). "Stabilization/Solidification (S/S) of Heavy Metal Contaminated Soils by Means of a Quicklime-Based Treatment Approach", *ASTM STP 1240*, ASTM, pp. 499-513.

Dermatas, D. and Meng, X. (2003). "Utilization of fly-ash for stabilization/ solidification of heavy metal contaminated soils", *Eng. Geol.,* 70, 377–394.

Dermatas, D.; Dadachov, M.; Mirabito, M. and Meng, X. (2003). "Strength Development of Stabilized/Solidified Organic Waste and Optimum Treatment Design", *Journal of the Air & Waste Management Association*, 53: 1363-1372.

Dermatas, D.; Shen, G.; Chrysochoou, M.; Grubb, D. G. and Dutko, P. (2006a). "Pb Speciation versus TCLP Release in Army Firing Range Soils", *Journal of Hazardous Materials*, 136:1, pp 34-46.

Dermatas, D.; Menounou, N. and Meng, X. (2006b). "Mechanisms of Pb Immobilization in Treated Soils", *Journal of Land Contamination and Reclamation*, 14 (1), pp. 43-56.

Dermatas, D.; Menounou, N.; Dadachov, M.; Dutko, P.; Shen, G.; Xu, X. and Tsaneva, V. (2006c). "Lead Leachability in Firing Range Soils", *Environ. Engr. Sci.*, 23: 1, pp. 86-99.

Dermatas, D.; Chrysochoou, M.; Moon, D.H.; Grubb, D.G.; Wazne, M. and, Christodoulatos, C. (2006d). "Ettringite-Induced Heave in Chromite Ore Processing Residue (COPR) upon Ferrous Sulfate Treatment", *Environ. Sci. Technol.*, 40 (18), 5786-5792.

Dermatas, D.; Chrysochoou, M.; Pardali, S. and Grubb, D.G. (2007). "Influence of X-Ray Powder Diffraction (XRPD) sample preparation on contaminated media remediation assessment: Mineralogical evaluation of Chromite Ore Processing Residue (COPR)", *Journal of Environmental Quality*, 36, pp 487-497.

Duchesne, J. and Reardon, E.J. (1999). "Lime treatment of fly ash: characterization of leachate composition and solid/water reactions", *Waste Manag.* 19, 221–231.

Ghosh, A.; Mukiibi, M. and Ela, W. (2004). "TCLP underestimates leaching of arsenic from solid residuals under landfill conditions", *Environ. Sci. Technol.*, 38, 4677-4682.

Halim, C. E.; Amal, R.; Beydoun, D.; Scott, J. A. and Low, G. (2004). "Evaluating the applicability of a modified toxicity characteristic leaching procedure (TCLP) for the classification of cementitious wastes containing lead and cadmium". *Journal of Hazardous Materials*, 103, 125-140.

Hillier, S.; Roe, M.J.; Geelhoed, J.S.; Fraser, A.R.; Farmer, J.G. and Paterson, E (2003). "Role of quantitative mineralogical analysis in the investigation of sites contaminated by chromite ore processing residue", *Sci. Total Environ.*, 308, 195-210.

Hooper, K.; Iskander, M.; Sivia, G.; Hussein, F.; Hsu, J.; Deguzman, M.; Odion, Z.; Ilejay, Z.; Sy, F.; Petreas, M. and Simmons, B. (1998). "Toxicity characteristic leaching procedure fails to extract oxoanion-forming elements that are extracted by municipal solid waste leachates", *Environ. Sci. Technol.*, 32, 3825-3830.

Kosson, D.S.; van der Sloot, H.A.; Sanchez, F. and Garrabrants, A.C (2002). « An integrated framework for evaluating leaching in waste management and utilization of secondary materials", *Environ. Engr. Sci.*, 19, 159-204.

Mitchell, J. K. (1993). "Fundamentals of soil behavior", John Wiley & Sons, Inc., New York.

Mitchell, J. K. and Dermatas, D. (1992). "Clay Soil Heave Caused by Lime-Sulfate Reactions", *ASTM STP 1135*, ASTM, pp. 41-64.

Mollah, M.Y.A.; Lu, F. and Cocke D.L. (1998). "An X-ray diffraction XRD and Fourier transform infrared spectroscopic FT-IR characterization of the speciation of arsenic V in Portland cement type-V", *Sci. Total Environ.* 224, 57–68.

Moon, D.H.; Dermatas, D. and Menounou, N. (2004). "Arsenic immobilization by calcium–arsenic precipitates in lime treated soils", *Sci. Total Environ.* 330, 171–185.

Moon, D. H.; Wazne, M.; Dermatas, D.; Christodoulatos, C.; Sanchez, A.M.; Grubb, D.G.; Chrysochoou, M. and Kim M.G. (2007). "Long-term treatment issues with chromite ore processing residue (COPR): Cr6+ reduction and heave", *Journal of Hazardous Materials*,143:3, pp 629-635.

Rietveld, H.M. (1969). "A profile refinement method for nuclear and magnetic structures", *J. Appl. Crystallogr.*, 2, 65–71.

Scheckel, K.G.; Impellitteri, C.A.; Ryan, J.A. and Mcevoy, T. (2004). "Assessment of a sequential extraction procedure for perturbed lead-contaminated samples with & without phosphorus amendments", *Environ. Sci. Technol.*, 37, 5296-5304.

USEPA (1992). "Test Methods for Evaluating Solid Waste, Physical/Chemical Methods", *SW-846 3rd Ed.*, Method 1311, Washington, DC.

USEPA (1999a). "Solidification/Stabilization Resource Guide", *EPA 549-B-99-002*, Washington, DC.

USEPA (1999b). "Waste leachability: The need for reviewing of current agency procedures", Washington, DC.

USEPA (2001). "Best Management Practices for Lead at Outdoor Shooting Ranges", *EPA-902-B-01-001*, Washington D.C., 2001

van der Sloot, H.A.; van Zomeren, A.; Meeuwsen, H.C.L.; Seignette, P. and Bleyerveld, R. (2007). "Test method selection, validation against field data, and predictive modelling for impact evaluation of stabilised waste disposal", *Journal of Hazardous Materials,* 141:2, 354-369.

In-Situ deep Soil Mixing for Solidification of Soft Estuarine Sediments – Shear Strength

Ali Maher[1], M. ASCE, Vernon Schaefer[2], M. ASCE and David Yang[3], M. ASCE

[1] Prof. of Civil Engineering, Dept. of Civil and Environmental Engineering, Rutgers University, 623 Bowser Road, Piscataway, NJ 08854-8014; mmaher@rci.rutgers.edu
[2] Prof. of Civil Engineering, Dept. of Civil, Construction and Environmental Engineering, Iowa State University, Ames, Iowa 50011; vern@iastae.edu
[3] V.P. for Technology, Raito, Inc., San Leandro, CA 94577, (510) 346-0840, email: dsyang@raitoinc.com

ABSTRACT: The paper reviews the use of Cement Deep Soil Mixing CDSM for solidification and stabilization of contaminated sediments and describes the shear strength behavior of solidified soft estuarine sediments from Newark Bay, near the mouth of the Passaic River, New Jersey. The objective of the study was to evaluate the shear strength gain with time for samples of solidified soft sediments following Cement Deep Soil Mixing (CDSM). An understanding of shear strength gain with time helps to determine the feasibility of conventional dredging/extraction methods once the sediments have been stabilized and allowed to cure. The results of the study show CDSM slurry mixtures, as low as 7% in cement content, result in significant solidification and strength gain of in-situ sediments under ambient conditions. Although the strength gain was significant, it was not so high that sediment dredging would become problematic or would require unconventional dredging equipment

INTRODUCTION and BACKGROUND

Solidification and Stabilization of Contaminated Soils (S/S)

Solidification/Stabilization (SS) is a widely used method for remediation of contaminated sites particularly those contaminated with hazardous waste. In S/S applications, the treatment immobilizes the contaminant in the media by mixing it with a binding reagent which prevents its migration to the receptors. The USEPA has identified SS as Best Demonstrated Available Technology (BDAT) for 57 RCRA-listed hazardous wastes (USEPA 1993). SS is the most frequently selected treatment for controlling the sources of environmental contamination at superfund program remediation sites (USEPA 2001).

In-Situ Techniques

The primary method of delivery of S/S treatment to contaminated sites is the use of in-situ mixing which is usually divided into two categories of Deep Soil Mixing (DSM) for depths greater than 40ft and Shallow Soil Mixing (SSM) for depths less than 40 ft. DSM equipment consists of a multiple auger mixing tool and SSM equipment has a single auger (Nicholson et. al, 1997).

The most frequently used reagent for treatment in in-situ operations is Portland cement. Cement is used for its ability to (a) chemically bind free liquids, (b) reduce the permeability of the waste form, (c) encapsulate waste particles surrounding them with an impermeable coating, (d) chemically fix hazardous constituents by reducing their solubility, and (e) facilitate the reduction of the toxicity of some contaminants. (Bone et al, 2005). Mix designs often include byproducts or additives in addition to Portland cement (Corner, 1997). These include Lime Kiln Dust (CLK) to modify pH and bulking, Fly ash to enhance pozzolonic reactions, and Cement Kiln Dust (CKD) and slag to produce amore economical mix (Wilk, 2003).

Considering the application of in-situ techniques for S/S of contaminants, which is primarily, based on short term results. A better understanding of the long-term effectiveness of S/S with cement is needed. Further research is needed in addressing issues such as: the uncertainties in test methods, observed deficiencies in the process application, observed lack of chemical binding in some crushed samples of treated waste, suggesting contaminants might leach under certain conditions, and uncertainties of performance arising from anticipated behavioral degradation of the material over time (Al-Tabbaa et al. 2005).

Use of Deep Soil Mixing (DSM) in S/S applications

The deep soil mixing technique is an effective method for solidification and stabilization (S/S) of contaminant sites. This is especially the case when treatment depths exceed 40 ft. A summary of a number of recent projects with focus on Deep Soil Mixing as a delivery system for Solidification/Stabilization of contaminates soils is given in Table 1. As shown in the table, cement-based reagents have been used successfully in these projects for SS of contaminants ranging from organics to PCBs, heavy metals, oil and grease. Design criteria presented in Tables by Bates and Mallot 2005, and Andromalos and Ameel, 1994, are examples of the basic framework for design guidelines in such applications. Furthermore, a comprehensive review of literature reveals that in most cases the in-situ applications have been primarily on upland sites and only in a few cases the efficiency of the method have been evaluated on marine and estuarine environments.

Table 1. Recent Examples of In-Situ S/S with Deep Soil Mixing

Cases	Contaminants	Additive	Volume (yd^3)	Criteria	Result	Notes
USA – In-situ S/S of Hazardous Acid Waste Oil Sludge (Bates and Mallot, 2005)	Hydrocarbons Sulfuric acid lead	Cement, Fly ash and Lime	39600 yr3/ DSM	Refer to Bates and Mallot, 2005	28-days and 5-year monitoring indicate that remediation meet cleanup objectives	Deep Soil mixing cost increased due to increasing the depth of SS from initial estimate, the pre-treatment of low pH oily sludge, time extension. Initial cost estimate at $135/yd3 but increased to $212 after pretreatment with lime.
USA - Use of DSM for Remediation of a Metal Site (Yang, 1995)	Arsenic Lead Cadmium Selenium Mercury	C(trade names: S-3 and P-1/P-1A)	14000 yr3/ DSM	Arsenic/Lead< 5 mg/l Cadmium/Selenium<1 mg/l Mercury<0.2 mg/l UC> 50 psi	Total of 484 samples for leach ability test- Criteria met	%50 of contaminated soil is below ground water table would make excavation and disposing of the groundwater very expensive if not using in-situ remediation.
USA – In situ Soil Treatment Using Deep Mixing for a Waterfront Site Contaminated with VOC and PAH (Fleri, et. al. 2005)	VOCs PAHs Oil and grease	Cement	80000 yr3/ DSM	UCS > 60 psi Permeability <1*10-5 cm/sec PAH content of TCLP(mg/l)<10	333 samples, all testes for UCS and the criteria met. 30 samples tested for PAH analyses and criteria met.	Former Manufactured Gas Plant contaminated with PAHs and VOCs and oil. Based on the treatability study, a design mix of 10% cement was chosen too meet the criteria and W:C ratio of the grout is 10%. Volume increase was 20% and the max depth is 38ft. Long term monitoring for a 5 year period shows no statistically significant levels of MGP related contaminants.
USA – In-situ stabilization of GEIGER Superfund site (Andromalosa nd Ameel	Chromium Lead PCB's Toluene Benzene Organic compounds	Cement	23000/ In-site mixing plant		Criteria met- 45 days monitoring	The first US army Corps of Engineers soil remediation project. Two pass injection process.
England- The use of deep soil mixing as part of the remediation strategy for contaminated sites (Essler , Kalin 2005)	DNAPL	Cement Nubmer of additives	Full- scale field demonstration	Permeability< $1*10^{-8}$ m/s UCS> 250 kpa Nitrate < 50 mg/l Phosphate< 1 mg/l ORC< 4mg/l	Criteria met	Combined implementation of soil-mixing for enhanced natural biodegradation of DNAPL(source treatment) and biologic permeable reactive barrier(pathway interception)

England- The use of deep soil mixing as part of the remediation strategy for contaminated sites (Essler , Kalin 2005)	DNAPL	Cement Zero Valent Iron (ZVI)	Full- scale field demonstration	N/A	out in November 2003 and the sampling and testing carried out in 2004 showed that the DNAPL had been sufficiently degraded to prove the application of ZVI soil mix technology as a potentially viable source term remediation option	Queen's University Belfast, Environmental Engineering Research Centre had shown that the DNAPL was effectively degraded by the ZVI and the trial was to investigate whether soil mixing would provide an acceptable delivery system.
Finland - Stabilization and solidification of contaminated ground – A preliminary study (Holm et. al., 2005)	Heavy metals PCBs	Cement	NA/ DSM	UCS > 50 psi	UCS criteria met. Leaching tests, 2 years after the completion of the project showed that oil pollutions were lower than before the stabilization. Heavy metals were higher than amounts for clean soil based on Finish environmental authorities but well below the amounts for contaminated soil.	The shore line was gradually developing by putting fills above clay. The soil found to be contaminated with PCB and heavy metals. The polluted soil could not be disposed on land or in the sea. It was decided to place the dredging masses between a supporting embankment toward the sea and the existing shore line. The stabilized soil was covered with 2m of till. Stabilization was made by rapid hardening Portland cement with an amount of 110 kg/m3.
Brazil - Solidification/ stabilization of a residual soil contaminated by diesel oil (Knop et. al., 2005)	Diesel oil	Cement	Lab experiments	Evaluating SS efficiency by measuring the % of contaminant retained in soil matrix	UCS depends on cement content and curing time. In leachiate tests, SS is efficient to immobilize diesel oil in for small oil amounts. For high oil amounts, a major part of the leachate is released indistinctly of cement amount. See fig 11-3 and table 11-4	Unconfined compressive strength (UCS) and column leaching (ASTM D4874) tests were carried out to evaluate the efficiency of remediation of a residual soil contaminated by diesel oil, through stabilization/solidification (S/S) method by means of Portland cement addition.

IN-SITU SOLIDIFICATION OF SOFT ESTUARINE SEDIMENTS

In the period from 2004 to 2005 the authors conducted a pilot study to evaluate the use of cement deep soil mixing for solidification and potential stabilization of soft sediments in the Newark Bay in the vicinity of the NY/NJ Harbor. The NY/NJ harbor and many of the surrounding waterways have been highly industrialized and commercialized for decades. As a result, the water quality in the area has long been degraded, the sediments contaminated, and many of the wetland areas have been completely destroyed or significantly reduced. Removal of toxic sediments by dredging is one way to remediate environmentally impacted areas, and has been suggested by some as a potential step toward remediating the most contaminated parts of the NY/NJ Harbor and surrounding waterways. One difficulty associated with dredging, however, is the potential for dispersion of sediments, and their associated contaminants, into the surrounding media. This is especially true for fine grained sediments. Unfortunately, fine grained sediments also tend to be those most likely to retain contaminants of concern such as dioxins/furans, PCBs, pesticides, organomercury and other semi-volatile toxins and metals. In fact, the dredging of some navigation channels and berths in hot spots like the lower Passaic River, Gowanus Canal, Newtown Creek and upper Newark Bay has not been performed in decades because of a combination of factors including the high dispersion potential, high levels of contamination and limited disposal/management opportunities (Maher, et. al. 2006).

The initial results from the study reported by Maher, et.al., (2006) by altering the characteristics of the estuarine sediment by Cement Deep Soil Mixing (CDSM), significant solidification of contaminated sediments could be achieved. The study also showed that the adverse dispersion from the in-situ treatment process was limited to a radius of 50 ft. from the mixing point. Solidification has the benefit of reducing dispersion potential, as well as improving the handling characteristics of the sediment for transportation and placement or treatment. Even if the sediments were never removed, the benefit of increased cohesion could reduce the potential for sediment (and associated contaminant) loss during storm events.

THE SHEAR STRENGTH OF IN-SITU SOLIDIFIED SEDIMENT

The Shear strength of DSM solidified sediment was evaluated by performing unconfined compression (ASTM D2216) and triaxial shear (ASTM D2850) tests on relatively undisturbed samples collected subsequent to deep soil mixing. Samples collected prior to mixing could not be tested for strength due to the very soft nature of the material. For the cemented soil samples, significant strength gain was measured as a result of mixing. The test results are summarized in Table 2 below. Soil samples were not saturated prior to testing in the triaxial chamber; therefore friction angle (φ) and cohesion were measured in an unconsolidated undrained shear test.

The overall test results indicate that CDSM significantly improved the strength characteristics of the sediment. Samples were collected in November and December of 2004, and were tested in November of 2004, and in June of 2005. The initial strength gain was considerable, but the strength gain over time, from November to June, was insignificant according to the above strength test results. This is typical of dredged material with high moisture contents.

Table 2. Shear Strength of Cemented Sediments

	Cement Content	Date Tested	Cohesion (psf)	Friction Angle(φ)
Unconfined Compression				
Sample 1	10.5%	11/23/2004	1870	---
Sample 2	14%	11/23/2004	1278	---
Sample 3	7%	11/23/2004	771	---
Triaxial				
Sample 1	10.5 %	6/10/2005	965	22.5
Sample 2	14%	6/10/2005	1282	24

In general, the long-term strength gain in cement treated soils is primarily a function of cement type, soil type, mixing and curing conditions. The most dominant soil characteristics affecting strength gain in cement treated soils are the organic content, pH, and the grain size distribution of the soil (Niina et al., 1977). The result of the laboratory investigation by Niina on more than 21 different types of soils showed that strength gain for a given amount of cement is more pronounced with higher pH, lower organic content and an average sand fraction of 60% in clay.

Past research by Saitoh (1988) indicates that knowing the pozzolanic reaction of soil to cement beforehand can lead to selection of appropriate cement type for a given site condition. In evaluating the unconfined compressive strength of two cemented soils from the Ports of Yokohama and Osaka, Saitho (1988) concludes that Portland cement is more effective in treatment of Yokohama clay which has high pozzolanic reaction while blast furnace slag cement type B is more effective in treatment of soils with low pozzolanic reaction such as Osaka clay. In both of the soil types tested the ratio of strength gain at 2000 days to that of the 28 days strength did not exceed 5 and averaged around 3.

The mixing time used in the preparation of the cemented soil has a direct impact on its strength. The temperature during curing also affects the strength gain. Saitho (1980) has shown that higher strength can be obtained under higher curing temperatures. This effect of curing temperature, however, is more pronounced for short-term strength and diminishes with time.

The strength gain with curing time occurs irrespective of soil type (Kawasaki et al., 1981). The strength gain with time is more pronounced with increasing cement content. Similar to the work by Saitho (1988), the results from Kawasaki shows that the strength gain ratio to that of the 28 days strength does not exceed beyond 2 for all soils tested.

It should also be noted that strength gain with time can be effectively controlled during the in-situ deep mixing process by changing the cement content and the amount of entrained air on the fly. Similar controls have been implemented effectively for long-term strength control of controlled low strength materials (CLSM) as referenced in the American Concrete Institute's document **ACI 229R-05.** In this specification the long term excavability of CLSM is controlled through entrained air and cement content and standard excavation tools can be used up for materials with unconfined compressive strengths of up to 2000 KN/m^2.

CONLUSIONS

The following conclusions can be made regarding the shear strength of in-situ solidified soft estuarine sediments: 1) Significant strength gain was observed for cemented sediments that were cured under water, and a significant increase in the workability of the material should be expected. And 2) although the strength gain was significant, it was not so high that sediment dredging would become problematic or would require unconventional dredging equipment, and

If the sediments were left in-place and capped, the potential for future dredging would not be affected. The strength developed in long term in average is 3 times the initial strength and will be within the ranges of strength that allows for removal of sediments by conventional dredging equipment

.

ACKNOWLEDGMENTS

The authors appreciate the support of the New Jersey Department of Transportation, and the Center for Advanced Infrastructure & Transportation (CAIT) at Rutgers University.

REFERENCES

Al-Tabbaa and A. and Perera, A. S. R. (2005), "State of practice report, UK stabilization/solidification treatment and remediation", Proc. of the International Conference on Stabilization/Solidification Treatment and Remediation, April, Cambridge, UK, Balkema, pp 365-385.

Andromalos, K.B. and M.E. Ameel. (1995). "Evaluation of Field Methods for the Solidification/Stabilization of Contaminated Soils and Sludges," Federal Environmental Restoration IV and Defense Cleanup Southeast, Hazardous Materials Control Resources Institute, Atlanta, Georgia, March 14-15.

Bates E.R. and Malott V. (2005), "Case study: In-situ Solidification/Stabilization of Hazardous Acid Waste Ail Sludge and Lessons Learned", International Conference on Deep Mixing – Best Practice and Recent Advances, May, Stockholm

Bone B.D., Barnard L.H. and Hills C.D., (2004), "Guidance on the use of Stabilization/ solidification for the Treatment of Contaminated Soil", Science Report: SC980003/SR1 Published by UK Environment Agency.

Conner, J. R.(1997). "Guide to Improving the Effectiveness of Cement-based Stabilization/Solidification", EB211, Portland Cement Association, Skokie, IL.

Kawasaki, T., Niina, A., Saitho, S., Suzuki, Y., Honjyo, Y. (1981), " Deep mixing method using cement hardening agent", Proc. Of the 10th Intl. Conference on Soil Mechanics and Foundation Engineering, 3:721-724.

Nicholson, P.J., B.H. Jasperse, and Fisher, M. J. (1997), "Economic Alternatives for Containment Barriers," International Containment Technology Conference, St. Petersburg, Florida, February 9-12.

Maher, A., Douglas, W.S., Jafari, F. and Yang, D. (2006), "Strategies for Handling of Contaminated River Sediments," Proc. Geo-Congress 06, ASCE.

Niina, A. S., Saitho, S., Babasaki, R., Tsutumi, I., and Kawasaki, T., 1977. Study on DMM using cement hardening agent (Part O). Proceedings of the 12th Japan National Conference on Soil Mechanics and Foundation Engineering: 1325-1328.

Saitho, S. 1988. Experimental Study of engineering properties of cement improved ground by deep mixing method. PhD. Thesis, Nihon University

Wilk C.M.(2003), " Solidification/Stabilization Treatment and Examples of Use at Prot Facilities", Published by Portland Cement Association

Waste characterization by leaching and extraction procedures

Franz G. Simon[1], Ute Kalbe[2] and Wolfgang Berger[2]

[1]Head of Division Waste Treatment and Remedial Engineering, Federal Institute for Materials Research and Testing, 12200 Berlin, Germany, franz-georg.simon@bam.de
[2]Working group Management of Contaminated Land and Environmental Engineering, Federal Institute for Materials Research and Testing, 12200 Berlin, Germany, ute.kalbe@bam.de, wolfgang.berger@bam.de

ABSTRACT: Laboratory leaching tests may be used for source term determination as a basis for risk assessment for soil-groundwater pathway. From the results the decision can be taken whether a material is suitable for utilization or not. In order to evaluate different leaching procedures, batch extraction tests and percolation tests were performed using three reference materials produced from contaminated soil, demolition waste and municipal solid waste incinerator bottom ash. Emphasis was placed on the investigation of the leachability of the heavy metals copper and chromium, polycyclic aromatic hydrocarbons (PAHs) and the anions chloride and sulfate. Significant discrepancies between column experiments and batch extraction tests were found for the release of PAHs and to a lesser extent for the heavy metals Cu and Cr.

The results from the experimental work will help establish standardized and feasible laboratory procedures as basis for substance specific risk assessment of contaminated sites and will be applied for execution of the planned German Ordinance on Utilization of Mineral Waste Materials.

INTRODUCTION

The utilization of mineral waste materials helps saving natural resources (Simon et al. 2005). However, residual contaminants in the waste can lead to detrimental effects on the environment, especially on soil and groundwater. The quantities of mineral waste are huge. In Germany 340 million tons of waste have been generated in 2004. More than 200 million tons were derived from the construction industry (64 % excavation residues and 36 % construction and demolition waste) (Heuser et al. 2007). According to the German Federal Closed Substance Cycle and Waste Management Act utilization has priority over disposal, but utilization has to be performed without harm to the environment. There is a variety of mineral materials from industrial processes for which different utilization options exist. Good examples are fly ash from coal fired power plant or slag from iron and steel production as cement clinker substitute. The proportion of construction and demolition waste (C&D waste) which is already utilized is almost 70 % in Germany for road construction, soil engineering and to a lesser extent as aggregate. Frequently occurring hazardous components within C&D waste are polyaromatic hydrocarbons (PAHs) and sulfate. Bottom ash (BA) from municipal solid waste incineration (MSWI) amounts to 3.2 million tons per year in Germany (Zwahr 2005). Here the hazardous components are chloride, sulfate and heavy metals. BA is mainly used for road construction.

The relevant pathway for risk assessment of utilization of mineral waste is the pathway soil-groundwater rather than soil-plant or soil-human. Therefore the total content of contaminants is less crucial than the mobility. It is therefore necessary to characterize materials which are suitable for utilization with regard to leachability of contaminants. The planned German Ordinance on Utilization of Mineral Waste Materials will implement a practicable approach for the risk assessment of mineral waste utilization. Leaching of solid materials is performed either by a batch test at a liquid to solid ratio (L/S) of 2 l/kg for the examination of the leaching behavior of inorganic substances or by a percolation test for the joint examination of inorganic and organic substances. Both test are designated for materials with a particle size up to 32 mm. The percolation test can be performed as comprehensive column test with leachate fractions at certain L/S ratios from 0.25 up to 4 l/kg or as compliance test where the leachate is collected as a whole up to an L/S ratio of 2 l/kg and analyzed subsequently. The batch test is executed as compliance test to assess the conformity with reference values of the specific and already well characterized materials such as blast furnace slag, fly ash, bottom ash or demolition debris. The standardization procedure is still in progress and shall be finished until the release of the ordinance. The present work was undertaken to compare results from both batch test and percolation test with columns.

MATERIALS AND METHODS

Batch and column test have been performed with three types of reference materials (RM) to provide a comparison of various source term determination methods (Berger et al. 2004):

- RM CS: Contaminated soil of grain size < 2 mm from a former tar works site with relevant PAHs contamination, bulk density 1.74 kg/dm^3
- RM BA: Bottom ash of grain size < 4 mm from MSWI with relevant Cu and Cr concentrations after ageing for more than three months, bulk density 1.72 kg/dm^3
- RM DW: Construction and demolition waste of grain size < 4 mm with PAHs, sulfate and heavy metals contamination, bulk density 1.76 kg/dm^3.

Characteristic parameters of the three reference materials are displayed in Table 1. Batch tests were carried out at L/S ratios of 2 and 10 l/kg with sample net weights of 0.25 and 0.125 kg, respectively. A test at L/S ratio of 0.25 l/kg was performed as saturation extract following the German Ordinance on Soil Protection. The test duration was 24 h. The leachates were decanted and filtrated through a 0.45 μm filter. Column tests were performed in glass columns of approximately 675 cm^3 capacities. The columns were filled with dry samples by placing them into the columns in five sub-layers which were compacted slightly with the aid of a rammer. Thus the average bulk density was very similar in all assembled columns. The bottom and top sections of the columns were equipped with a 1-2 cm layer of 1-2 mm grain size quartz sand. At the beginning of the tests the columns were rapidly saturated with the leachant using a higher pump velocity. Tests designated for determination of inorganic and organic constituents were performed separately. For the latter sodium azide (Na_3N) was used to reduce microbial activity. The L/S ratios of the collected fractions were increasing from 0.25 up to approximately 5 l/kg which was reached after 18 days. The contact time between the solid sample and the leachant in the column tests was between 15 and 19 hours. The samples were neither filtrated nor centrifugated prior analysis.

Table 1. Selected parameters of the reference materials

Parameter	Dimension	RM CS	RM DW	RM BA
Residual moisture content	% by mass	0.3	2.4	3.7
TOC (total organic carbon) [a]	% by mass	0.3	0.6	0.5
Carbonate content	% by mass	3.6	5.3	3.4
Loss on ignition	% by mass	0.6	1.8	2.0
Particle density [b]	g/cm^3	2.66	2.57	2.60
Specific surface	m^2/g	0.6	5.2	5.4
k_f value (water permeability)	m/s	$9.3 \cdot 10^{-6}$	$3.5 \cdot 10^{-6}$	$1.7 \cdot 10^{-6}$
Particle size distribution				
4 – 2 mm	% by mass	-	14.9	19.5
2 - 0.63 mm	% by mass	33.0	29.0	26.8
0.63 - 0.2 mm	% by mass	60.4	36.5	31.0
0.2 - 0.063 mm	% by mass	5.7	17.5	17.6
< 0.063 mm	% by mass	0.9	1.9	5.0
Chemical composition				
Sum PAH_{16} in solid matter	mg/kg DM [c]	83.8	31.8	-
Sulfate in solid matter	% by mass		1.37	0.58
Chloride in solid matter	% by mass		0.09	0.27
As	mg/kg DM	-	2.8	8.0
Pb	mg/kg DM	-	109	727
Cd	mg/kg DM	-	0.3	5.8
Cr	mg/kg DM	-	16.3	67
Cu	mg/kg DM	-	29.2	1657
Ni	mg/kg DM	-	14.5	68
Zn	mg/kg DM	-	66	2073

[a] TOC = TC – TIC (TIC calculated from carbonate content)
[b] He pycnometer Accu Pyc 1330, Micromeritics
[c] DM Dry matter
[d] TPH Total petroleum hydrocarbons

PAHs in the percolates were generally measured by High Performance Liquid Chromatography (HPLC) according to DIN EN ISO 17993:2004-03. Sulfate and chloride concentration in the leachates were analyzed by anion chromatography according to EN ISO 10304-2:1996-11. The concentration of the heavy metals copper and chromium were measured using Inductively Coupled Plasma Optical Emission Spectrometry (ICP OES). Electrical conductivity dissolved organic carbon (DOC) and pH value and turbidity were additionally determined.

RESULTS AND DISCUSSION

Concentration (in µg/l or mg/l) and release (in mg/kg) of the substances investigated are plotted against the L/S ratio (batch tests: points, column tests: curves, see Figures 1-3). The cumulative released amounts of contaminants in the column tests (Figures 2 and 3) were calculated by multiplying the concentrations by the L/S ratio and subsequently summing up in order to compare these results to those from batch/extraction tests.

In the column tests, relatively high initial concentrations were observed for all substances investigated (PAHs, anions, heavy metals) in the three RMs. Initially the curves drop steeply to a low level which then remains relatively constant for the remaining time (Fig. 1).

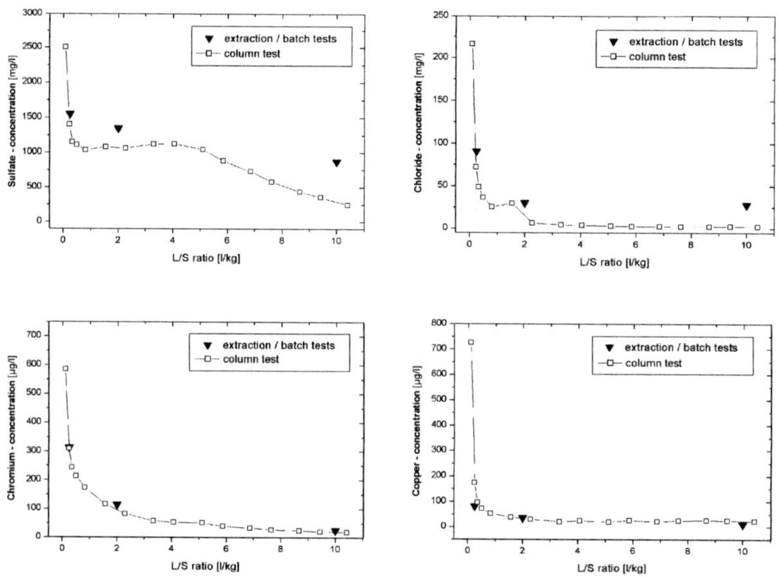

FIG. 1. Demolition waste (DW) reference material. Concentration of chloride, sulfate, chromium and copper in batch tests and column tests

The different batch tests and column tests for heavy metals as well as chloride and sulfate concentrations yielded results in the same order of magnitude at specific L/S ratios (Figures 1 and 2). This is directly applicable to the L/S ratios of 0.25 and 2 l/kg and can be anticipated from the progression of the column test curves for L/S 10. The comparison of the cumulative release of the column tests and the batch tests only yields a good agreement for sulfate (Figure 2) whereas for copper and chromium considerable differences were found at L/S 2 and 10 l/kg. The measured release of Cr and Cu from batch tests (especially L/S 10 l/kg) is considerably lower than the release measured in column tests. Therefore the results from the different test procedures cannot be easily compared with one another due to fundamental differences in the approach (e.g. contact time, static/dynamic conditions, etc.).

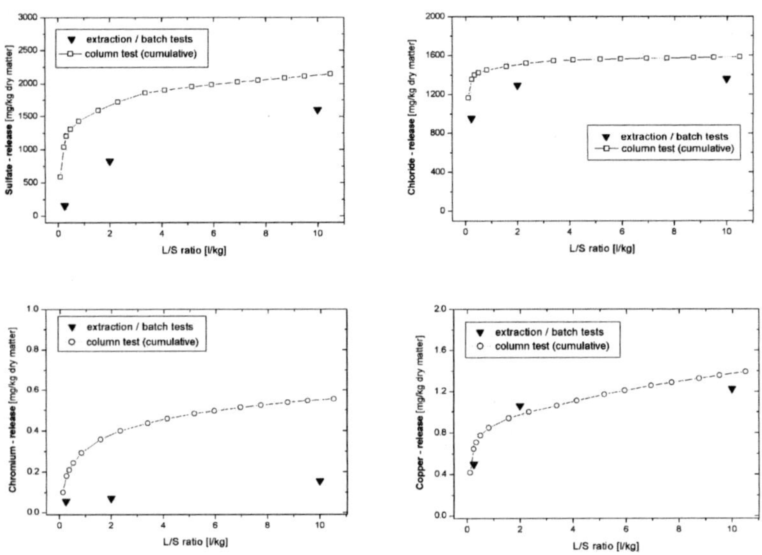

FIG. 2. MSWI bottom ash (BA) reference material. Release of chloride, sulfate, chromium, and copper. Note that y-axis has the dimension mg/kg (release) rather than mg/l (concentration) in Fig. 1.

Agreement of the estimated release of different leaching tests is better for anions because leaching is controlled mainly by their solubility in water and/or their availability. Leaching processes of heavy metals in bottom ash and demolition waste are controlled by other factors such as redox potential and pH value (Dijkstra et al. 2006). For the demolition waste material, high levels of sulfate concentrations were observed for all tests even after extended leaching periods in the case of column tests. It can be assumed that the concentration will remain high because less than 50 % of the sulfate potential (1.37 %, see Table 1) have been mobilized up to an L/S ratio of about 5 l/kg.

In contrast, chloride concentrations decrease rapidly to a low level as a result of their high water solubility (see Figure 1) and much lower content in the solid material (see Table 1). The results are similar for the RM BA. However, the values for chloride are higher due to the higher chloride content in MSWI bottom ash (see Table 1).

Release of PAHs (15 EPA PAHs) in the batch and column tests from the RM CS and DW is shown in Figure 3. The batch tests provide much lower concentrations and releases of PAHs than the corresponding column tests.

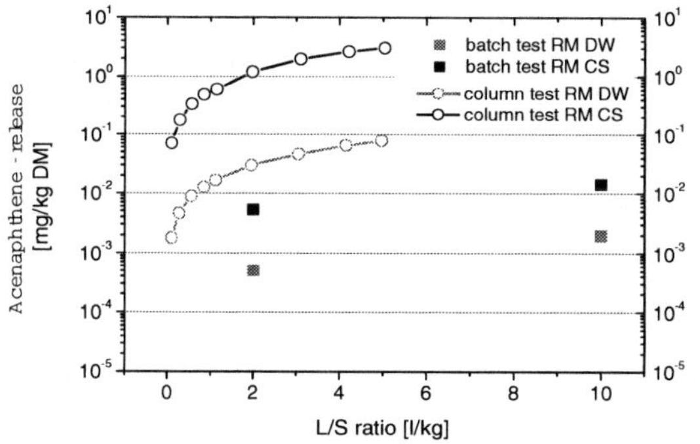

FIG.3. Cumulative release of PAHs for batch and column tests with reference materials DW and CS

The conclusion was drawn that the filtration step, i.e. the separation of the leachate from the solid sample, generally seems to be the critical step in the sample preparation procedure for the subsequent analytical measurement and was responsible for the reduction of the measured PAH concentrations (Kalbe et al. 2006). The porosity of the filter material (0.45 µm) was probably too low so that colloids containing PAHs were eliminated from the leachate whereas PAHs in solution remained in the leachate (Reemtsma and Mehrtens 1997). However, colloid facilitated transport of PAHs takes place within the unsaturated soil zone and thus PAHs associated with colloids should be taken into account. Further investigations are needed concerning the liquid/solid separation step in order to allow alternative recommendations. In contrast to the losses due to filtration, an overestimation of PAH concentrations from batch tests with a subsequent centrifugation for liquid/solid separation due to colloid facilitated release is reported by (Gadia and Raga 2005). The suggested recommendation of the ongoing standardization procedure will be a centrifugation for the batch test (E DIN 19529). In case of the percolation test (E DIN 19528) it is intended to stipulate a centrifugation of the leachate only if the turbidity scales to more than 20 FNU.

Although the total PAH content in solid matter of the RM CS and DW is in the same order of magnitude (see Table 1), leachate concentrations are considerably lower for the RM DW (see Figure 3). This can be explained by the PAH distribution patterns of both RMs which were totally different (Kalbe et al. 2007b). The solubility of the individual PAHs differ considerably (Mackay and Shiu 1977, de Maagd et al. 1998). Additionally biological decay is of great importance especially for soil where biological activity is higher than in man-made waste materials with high content of alkaline material (e.g. MSWI bottom ash or demolition waste) resulting in high pH values. Those PAHs with low solubility and low biodegradability offer a potential for leaching over a very long time. Even though the decay of pollutants can be described by models it is difficult to perform a proper leachate forecast based on leaching experiments with such interfering effects. Biodegradation mechanisms of organic pollutants in soil are discussed elsewhere (Hatzinger and Alexander 1995, Holden and Fierer 2005, Stieber et al. 2005).

In the column tests, PAH leaching is still not completed at a L/S ratio of approximately 5, equivalent to approximately 30 pore volumes in column tests (Figure 3), because after the initial mobilization of the more soluble PAHs the pollutant release is governed by diffusion (Grathwohl 1992, Grathwohl 1998).

CONCLUSIONS

In order to evaluate different leaching procedures, batch extraction tests and percolation tests were performed based on three reference materials containing organic and inorganic contaminants. Since interfering processes influence leaching behavior and status of equilibrium conditions can vary, the various leaching tests are difficult to compare. In particular, the determined concentrations cannot be directly correlated because of the different boundary conditions (e.g. contact time, L/S ratio, and agitation).

Sulfate and chloride releases exhibited a good agreement between column experiments and batch extraction tests. This is due to the fact that leaching kinetics of sulfates and chlorides is primarily controlled by their solubility in water and/or their availability. However, Cu and Cr release exhibited significant discrepancies and the difference between PAH results exceeded an order of magnitude.

Batch and extraction tests are generally appropriate for the investigation of leaching behavior of inorganic compounds. This has been confirmed also in interlaboratory comparisons (Kalbe et al. 2007a). For organic compounds batch tests are not applicable because due to the filtration step the analytical results underestimate the mobilized fraction. This fact has been considered for the planned German Ordinance on Utilization of Mineral Waste Materials.

ACKNOWLEDGMENTS

Part of the work was conducted within a joint project funded by the German Federal Ministry of Education and Research.

REFERENCES

Berger, W., Kalbe, U., Eckardt, J., Fischer, H. and Jansky, H.J. (2004). "Aufbereitung von Referenzmaterialien zur Untersuchung der Eluierbarkeit von Schadstoffen - Processing of reference materials for examination of contaminant leaching behaviour". *Aufbereitungstechnik - Mineral Processing* Vol. 45(11): 37-43

de Maagd, P.G.J., Ten Hulscher, D.T.E.M., van den Heuve, H., Opperhuizen, A. and Sijm, D.T.H.M. (1998). "Physicochemical properties of polycyclic aromatic hydrocarbons: aqueous solubilities, n-octanol/water partitition coefficients, and Henry's law constants". *Environmental Toxicology and Chemistry* Vol. 17(2): 251-257

Dijkstra, J.J., van der Sloot, H.A. and Comans, R.N.J. (2006). "The leaching of major and trace elements from MSWI bottom ash as a function of pH and time". *Applied Geochemistry* Vol. 21: 335-351

Gadia, R. and Raga, R. (2005). "Leaching of organic compounds from waste and soils", Proceedings 10th International Waste Management and Landfill Symposium - Sardinia 2005, Cagliari, CISA, Environmental Sanitary Engineering Centre, Italy

Grathwohl, P. (1992). "Die molekulare Diffusion als limitierender Faktor bei der Sanierung von Boden- und Grundwasserkontaminationen". *Zeitschrift für Umweltchemie und Ökotoxikologie* Vol. 4(4): 231-236

Grathwohl, P. (1998). "Diffusion in natural porous media: contaminant transport, sorption/desorption and dissolution kinetics", Kluwer Academic Publishers, Boston/Dordrecht/London

Hatzinger, P.B. and Alexander, M. (1995). "Effect of aging of chemicals in soil on their biodegradability and extractability". *Environmental Science and Technology* Vol. 29: 537-545

Heuser, B., Kohler, G., Ortleb, H., Pahl, G. and Schäfer, B. (2007). "5. Monitoring-Bericht Bauabfälle ". . AG Kreislaufwirtschaftsträger Bau (www.arge-kwtb.de)

Holden, P.A. and Fierer, N. (2005). "Microbial processes in the vadose zone". *Vadose Zone Journal* Vol. 4: 1-21

Kalbe, U., Berger, W., Eckardt, J. and Christoph, G. (2006). "Laborvergleichsuntersuchungen Sickerwasserprognose (Teil 1), Abschlussbericht innerhalb des UBA-Projektes "Evaluierung von Methoden zur Sickerwasserprognose" FKZ 20374278". Bundesanstalt für Materialforschung und -prüfung (BAM), Berlin

Kalbe, U., Berger, W., Eckardt, J., Christoph, G. and Simon, F.G. (2007a). "Results of interlaboratory comparisons on the evaluation of the reproducibility of column percolation tests". *Journal of Hazardous Materials*: in press (doi:10:1016/j.jhazmat.2007.03.039)

Kalbe, U., Berger, W., Eckardt, J. and Simon, F.G. (2007b). "Evaluation of leaching and extraction procedures for soil and waste". *Waste Management*: in press (doi:10.1016/j.wasman.2007.03.008)

Mackay, D. and Shiu, W.Y. (1977). "Aqueous solubility of polynuclear aromatic hydrocarbons". *Journal of Chemical Engineering Data* Vol. 22(4): 399-402

Reemtsma, T. and Mehrtens, J. (1997). "Determination of polycyclic aromatic hydrocarbons (PAH) leaching from contaminated soil by a column test with online solid phase extraction". *Chemosphere* Vol. 35(11): 2491-2501

Simon, F.G., Adam, C., Adamczyk, B. and Weimann, K. (2005). "Recovery of non-renewable resources from waste", in: White, T., Ferraris, C., Yu, L., Halada, K. and Umezawa, O. (Ed.), Advances in Ecomaterials, Vol. 2: 148-155, Stallion Press Pte Ltd, Singapore

Stieber, M., Brost, I. and Tiehm, A. (2005). "Dynamics of PAH biodegradation in unsaturated soil columns", 9th International FZK/TNO Conference on Soil-Water-Systems ConSoil 2005, Bordeaux, Uhlmann, O., Annokee, G.J. and Arendt, F. (Ed.): 804-808

Zwahr, H. (2005). "MV-Schlacke - mehr als nur ein ungeliebter Baustoff". *Müll und Abfall* (3): 114-122

Field scale characterization of fly ash stabilized with lime and FGD gypsum

John L. Daniels, P.E., M.ASCE,[1] and Gautham P. Das[2]

[1]Associate Professor, Department of Civil and Environmental Engineering and Global Institute for Energy and Environmental Systems, The University of North Carolina at Charlotte, Charlotte, NC 28223
[2]Ph.D. Candidate, Department of Civil and Environmental Engineering and Global Institute for Energy and Environmental Systems, The University of North Carolina at Charlotte, Charlotte, NC 28223

ABSTRACT: The use of lime and FGD gypsum material to stabilize coal combustion fly ash was evaluated. Five test pads were constructed at a scale that allowed use of field equipment, measuring approximately 15 m in length, 5 m in width and 0.6 m in height. The primary objective of the test pads was to evaluate the influence of lime and FGD gypsum on fly ash leachability. Secondary considerations included permeability, durability, strength and constructability. Measurements were made in the field and also with remolded laboratory specimens. Results suggest that lime influences fly ash leachability, although this varies according to mixing condition, as well as the contaminant of concern. Lime amendment reduced the mobility of cadmium, however, arsenic, chromium and selenium were all present at increased concentrations relative to the test pad constructed of fly ash alone. The hydraulic conductivity of all mixtures tested fell within the 10^{-4} - 10^{-3} cm/s range. Quality control over field mixing is an important concern. This is particularly sensitive when trying to use the least amount of lime to achieve solidification. Small localized variations in lime or moisture may lead to mixed results.

INTRODUCTION

More than half of the electricity in the U.S. and 37% in the world is generated from coal combustion. Notwithstanding vigorous interest in renewable resources and increased energy efficiency, U.S. coal consumption is projected to increase by 57% from 1128 million short tons in 2005 to 1772 million short tons to 2030 (EIA, 2007). As a result, more than 100 million metric tons of coal combustion by-products (CCBs) are produced annually, less than half of which is re-used in construction, highway or cement production applications (ACAA 2005).With a fluid regulatory climate, there is great concern over the handling and storage of this material. In particular, requirements have the potential to vary across a continuum that ranges from a full cap, liner and leachate collection system to direct placement on grade with no liner or cover system. Additionally, there is a need to evaluate CCB performance under field-relevant conditions (Bin Shafique et al. 2002; Daniels et al. 2006; Daniels and Das, 2005, 2006). To that end, a study was conducted to provide site-specific data regarding the potential use of lime as an additive to reduce ash leachability. The objective of this paper is to present elements of the testing program, as well as selected study results from the first 64 days after construction.

TEST PAD CONSTRUCTION

Five test pads were constructed at a scale that allowed use of field equipment, measuring approximately 15 m in length, 5 m in width and 0.6 m in height. The test pad mix designs and naming conventions are as follows:

- Test Pad #1 (TP1): 0.5% Lime + Fly Ash

- Test Pad #2 (TP2): 1.0% Lime + Fly Ash
- Test Pad #3 (TP3): 3.0% Lime + Fly Ash
- Test Pad #4 (TP4): 3.0% Lime + 2:1 Fly Ash:FGD Gypsum
- Test Pad #5 (TP5): 100% Fly Ash (No additives)

In general, material was end-dumped over the test pad location with a tandem truck, followed by multiple passes with a track-mounted dozer and steel drum roller (vibrator turned off). After the desired lift thickness was achieved (usually 15 cm, although 30 cm in some cases), lime was added using either a walking drop spreader (TP1) or a bobcat loader (TP1,2,3,4) and mixed through the use of rotary mixer typical of roadway stabilization. In general, the amount of lime added for a given test pad was based on dry weight computations. In particular, the lime content was defined in this project as:

$$\text{Lime Content}(\%) = \frac{M_{\text{Lime}}}{M_{\text{flyash}}} \quad (1)$$

During construction, the mass of lime to add was calculated on the basis of anticipated compacted ash dry density and ash thickness. The thickness was generally measured through the use of a surveying level and rod, although intermittent measurements were made by driving reference stakes and/or using tape measures. The reliability of these measurements was questionable given the difficulty in having the rod positioned in the same place as the height of the ash pad was increased. The thickness at a given location was also determined by augering through the compacted ash, particularly when measurements by the rod or markers were uncertain. After lime addition, the rotary mixer was driven over the material to mix the fly ash and lime together. After mixing, the material was watered, generally with four passes of the water truck, and then mixed again, tracked, and roller-compacted. In some cases the material was roller compacted and then watered, remixed and re-compacted. The achieved in place density was determined through a standard drive tube method whereby samples were recovered in steel cylinders, as presented in Table 1. The pH was also measured in the field as a check on lime addition. Samples were generally taken after mixing and before compaction. The pH was determined by mixing 10 grams of lime/fly ash with 40 grams of deionized water. Figure 1 shows the calibration curve developed in the laboratory prior to field testing, while Table 2 provides the measured pH of the test pads during construction.

Table 1. Moisture Density Relationship and Relative Compaction (Average)

Test Pad	ASTM D698 Maximum Dry Unit Weight (kN/m^3)	ASTM D698 Optimum Moisture Content (%)	Actual In Situ Dry Unit Weight (kN/m^3)	Actual In Situ Moisture Content (%)	Relative Compaction (%)
TP1	12.92	22.8	12.73	21.8	98.5
TP2	13.19	22.0	13.17	22.5	99.8
TP3	12.88	23.0	12.16	16.3	94.4
TP4	13.06	24.2	13.03	22.0	99.8
TP5	13.43	20.5	12.95	17.3	96.4

FIELD TESTING PROGRAM AND INSTRUMENTATION

The overall objective of the instrumentation was to capture information relevant to the quality and quantity of moisture that passes through the various test pads. Key components included gravity lysimeters, suction lysimeters, Boutwell permeameters, double ring infiltrometers and mini-disk infiltrometers. Also, an air temperature sensor, moisture sensor and rain gauge were installed on site to monitor weather conditions and assist in completing a hydrologic balance.

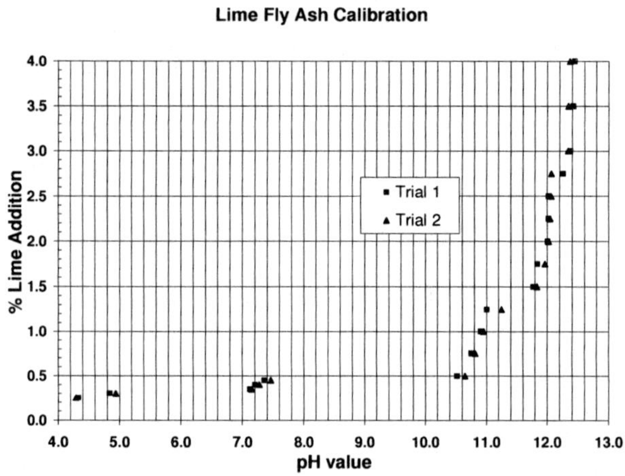

FIG. 1. Lime Fly Ash Calibration for use in Field Evaluation of Mixing Efficiency

Table 2. Test Pad pH Measurements (during construction)

Test Pad	Average Observed pH
TP1	5.08
TP2	10.56
TP3	12.74
TP4	12.47
TP5	4.50

Gravity Drain Lysimeters

The objective of the gravity drain lysimeters was to quantify the amount of infiltration through the various test pads. The gravity drain lysimeters were fabricated by hand, constructed of 40 mil PVC sheeting, and 6 mm thick geonet (Ambernet), double bonded with geotextile, from CFP, Inc. The PVC was wrapped around one side of the geonet and folded over the sides by about 8 cm. Schedule 40 PVC pipe (2.5 cm diameter) was slotted with 13 mm holes and inserted at one side of the lysimeter. The corners and pipe were sealed with duct tape and silicone glue. The geonet was fastened to the top of the slotted pipe with cable ties. The pipe was then connected to 13 mm OD flexible polyethylene tubing through threaded couplings and compression fittings to 19 liter collection bottles.

Suction Lysimeters

The objective of the suction lysimeters was to allow for pore fluid sampling over a wide range of moisture conditions. The gravity lysimeters only provide a sample when the moisture content of the test pads exceeds field capacity and moisture may freely infiltrate under the force of gravity. As such, if the prevailing infiltration is less than the available storage and evaporation capacity, then no samples will be recovered in the gravity collection bottles. However, in the case of suction lysimeter testing, the pore fluid may be sampled under drier conditions, even if the moisture content is less than field capacity. The suction lysimeter is similar in construction to a tensiometer, and consists of a porous ceramic probe with two 6 mm tubing lines that extend to the surface. One of the lines (the sampling line) extends to a lower sampling chamber while the other line (the pressure/vacuum line) extends to an upper chamber. A suction is applied to the upper chamber in the probe and pore fluid is drawn into the lower chamber. When enough fluid has been collected, a pressure is applied to the upper chamber to force the sample out to the surface. A check valve prevents fluid collected in the lower chamber from being forced back into the surrounding test pad when pressurizing for sample recovery. The type of suction lysimeter used was Model 1923, obtained from SoilMoisture Corporation, Goleta, CA. This device is designed for trace contaminant sampling and has a capacity of 560 mL. The suction lysimeters were installed immediately after construction. Samples, however, were not taken until after the designated cure time, taken as approximately 28 days. The choice not to sample until after curing was made because of the concern of removing moisture (by applying suction as described above) that would otherwise contribute to lime/fly ash reactions, and thus perhaps change the material behavior in the vicinity of the sampling device.

Hydraulic Conductivity

Hydraulic conductivity measurements are subject to considerable variation. Moreover, no one method is perfectly suited for all types of material. As such, two different techniques were used to determine the vertical hydraulic conductivity of the test pads, namely Boutwell Permeameters, as per Trautwein and Boutwell (1994) and Mini-Disk Infiltrometers, as per Zhang (1997). In addition, infiltration capacity was determined with open Double Ring Infiltrometers, as per ASTM D3385 (ASTM 2000). Infiltration capacity may be related to hydraulic conductivity, given a depth of saturation to define the hydraulic gradient. A summary of the three field-based techniques is given in Table 3.

RESULTS AND DISCUSSION

Gravity Lysimeter Observations

No samples were collected from any of the gravity lysimeters buried beneath the test pads for the first 64 days of evaluation. This appears consistent with the thickness and permeability of the test pad material, the time of study and prevailing weather conditions. Nonetheless it is reasonable to inquire whether these results are in agreement with hydrologic modeling. Therefore, Visual HELP 2.2 from Waterloo Hydrogeologic, was used to answer these questions. This HELP model is a widely accepted tool for conducting quasi two-dimensional water balance analyses on various soils, industrial byproducts and landfill systems. Details on this model can be found elsewhere (Schroeder et al. 1994).

Model predictions of leachate volumes are sensitive to the initial volumetric moisture content, relative to its field capacity and total porosity. Efforts to match this accurately are frustrated for all mixtures containing lime, since the initial hydration will change the volumetric moisture content and thus the amount of leachate produced. Moreover, the formation of cementitious products over time clearly influences key parameters, including the field capacity and porosity, which were assumed constant throughout time and mix design. Nevertheless, all of the test pads were modeled with HELP for the first 64 days using data collected from the on-site rain gauge and temperature sensor. Unless the initial moisture content exceeds field capacity, the HELP model does not predict any significant leakage from any of the test pads for the first 64 days and indeed none was observed in the field.

Table 3. Comparison of the Three In Situ Hydraulic Conductivity/Infiltrometer Tests

Item	**Boutwell Permeameters**	**Double Ring Infiltrometers**	**Mini-disk Infiltrometers**
Hydraulic Head	Falling Head	Constant Head	Falling Head
Testing Area (cm^2)	81.3	2916	6.5
Applicable hydraulic conductivity (cm/s)	$10^{-2} - 10^{-9}$	$10^{-2} - 10^{-6}$	$10^{-2} - 10^{-9}$
Relative Setup/Installation Time	Slow	Intermediate	Fast
Relative Testing Time	Intermediate	Slow	Fast
Relative Data Reduction Time	Intermediate	Fast	Slow

Suction Lysimeter Observations

Suction lysimeter measurements were made at 42, 49 and 57 days after construction. The sample pH was measured prior to acidification and analysis by a third party analytical laboratory. The pH results appear to correspond with the amount of lime added. For example, 42 days after construction, the measured pH for 0% (TP5), 0.5% (TP1), 1.0% (TP2), 3.0% (TP3) lime contents was found to be 7.32, 8.23, 10.54 and 11.68, respectively. It is noted that consistent samples were only available from TP3 and TP5. In general, the inability to retrieve a suction sample is connected with a lack of available moisture.

In terms of trace contaminants, Figures 2 and 3 show the results of various sampling events for TP5 (0% lime) and TP3 (3% lime), respectively. Lime amendment has reduced the mobility of cadmium, however, arsenic, chromium and selenium are all present at increased concentrations relative to the test pad constructed of fly ash alone (TP5). Remarkably, samples collected from TP3 (3% lime) reveal increasing concentrations with time for arsenic and chromium, with similarly high concentrations for selenium. These concentrations would not be predicted on the basis of laboratory sequential, wet/dry or freeze/thaw leaching tests (Daniels and Das 2006), which all imply reduced concentrations at the 3% level. Discrepancies may exist since the actual lime content is uncertain in the test pad. However, indications (from pH measurements and observed strength) are that enough lime was mixed to form cementitious reactions thought to be responsible for contaminant immobilization. As such, further investigation would be required to establish the optimum lime dose and mixing method.

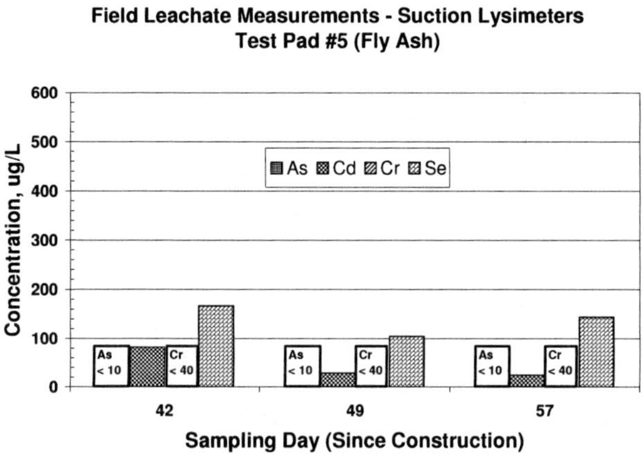

FIG. 2. Field Leachate Measurements of TP5

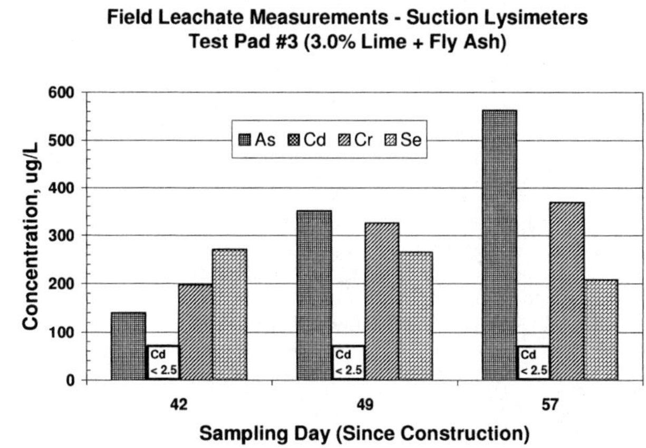

FIG 3. Field Leachate Measurements of TP3

Hydraulic Conductivity Observations

Figure 4 shows the relationship between hydraulic conductivity and lime content of the tested ash. In general, the hydraulic conductivity decreases in response to lime amendment at the 0.5% and 1.0% contents, while increasing the lime content to 3.0% results in an increase in hydraulic conductivity for the field tests.

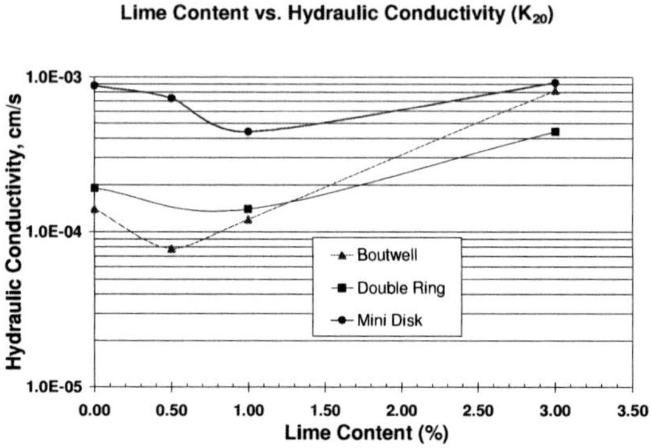

FIG. 4. Influence of Lime Content on Field-Measured Hydraulic Conductivity.

Typically, investigators observe modest decreases in hydraulic conductivity in response to lime amendment. Bowders et al. (1990) tested two types of fly ash with initial hydraulic conductivities of 7.2 x 10^{-6} and 1.9 x 10^{-5} cm/s. These values were reduced to 2.2 x 10^{-6} and 6.4 x 10^{-6}, respectively, in response to 3% lime amendment. In another example, the permeability of lime sludge amended Class F fly ash was tested by Maher et al. (1993). The authors measured hydraulic conductivities of 4.3 x 10^{-7}, 2.2 x 10^{-7} and 1.6 x 10^{-7} cm/s for amendment with 1.0, 1.2 and 1.4% lime, respectively. Notwithstanding these examples from the literature, the naturally broad variation in hydraulic conductivity values is such that the extent to which lime truly affects hydraulic conductivity, particularly at levels at or below 3%, remains uncertain. The author's assessment is that hydraulic conductivity is not in fact significantly affected by lime amendment at the 0.5 and 1.0% level. However, the uncertainty at 3% may be explained by considering the process by which lime stabilization occurs. For example, at very low lime contents (0.5%, 1.0%), it does not appear that cementitious reactions were initiated on any appreciable scale. This has been borne out by physically examining the various test pads, where the level of difficulty in retrieving Shelby tubes was not significantly different between TP1, TP2 and TP5 (0.5%, 1.0% and 0% lime contents, respectively). By contrast, TP3 and TP4 have shown considerably greater strength. This gain in strength occurs as silica and alumina from the fly ash react with calcium from the lime under conditions of high pH. The formation of these new products will, by definition, change the pore-size distribution and physical geometry of the particle assemblage. Initially, the average pore dimension may have actually increased through particle enlargement and overall volume expansion. Of course, one would expect that eventually this pore dimension would decrease (as would hydraulic conductivity) when enough lime is added. It would then seem that at the 3% lime content, enough lime was present to increase strength, but not enough to sufficiently or significantly fill pore spaces. In any case, the overall conclusion is that the hydraulic conductivity for the test pads is generally between 10^{-4} and 10^{-3} cm/s.

CONCLUSIONS AND RECOMMENDATIONS

Lime stabilization of the fly ash, at the levels tested and with the methods employed, is not completely effective. Notwithstanding the need for cementitious reactions, the appearance of strength at the field scale does not ensure reduced leachability. TP3 exhibited superior strength yet recorded some of the highest concentrations of arsenic, chromium and selenium. In terms of specific recommendations, efforts to use lime in the stabilization of fly ash should be focused on increasing the content beyond 3% and/or improving the efficacy (i.e., mixing and curing conditions) of lower lime contents. Lime should be studied with a range of fly ash types, ranging according to their leachability as defined by some metric (SPLP, column, etc.). For effective lime stabilization, one either needs to invest capital into monitoring the process (change in source contaminant concentrations, lime mixing, etc.) or simply adding enough "extra" lime to ensure that full contaminant fixation is achieved. Alternatively, other or low-cost cementitious products (e.g., kiln dusts) or surface modifiers (e.g., organo-silanes) may be investigated for comparison.

REFERENCES

ACAA (2005). 2004 Coal Combustion Product (CCP) Production and Use Survey, American Coal Ash Association, available from www.acaa-usa.org, last accessed 7/11/07.

ASTM (2000). "Standard test method for infiltration rate of soils in field using double ring infiltrometer" ASTM D3385, *Vol. 4.08 Soil and Rock (1)*, American Standards for Testing of Materials, West Conshohocken, PA.

Bin-Shafique, M., Benson, C., Edil, T., (2002). *Geoenvironmental Assessment of Fly Ash Stabilized Subbases*, Geo Engineering Report 02-03, Dept. of Civil and Environmental Engineering, University of Wisconsin-Madison.

Bowders, J.J., Gidley, J.S. and Usmen, M.A. (1990). "Permeability and leachate characteristics of stabilized Class F fly ash" *Transportation Research Record* No. 1288, National Research Council. Washington, D.C., pp. 70-77.

Daniels, J.L. and Das, G.P. (2006). "Leaching behavior of lime-fly ash mixtures" *Journal of Environmental Engineering Science* Vol. 23, No. 1, pp. 42-52.

Daniels, J.L., Hill, M.E., Das, G.P. and Ogunro, V.O. (2006). "Influence of residence time on fly ash leachability: long-term implications" 23rd Annual International Pittsburgh Coal Conference, Pittsburgh, PA, USA, September 25-28, 2006 (on CD)

Daniels, J.L. and Das, G.P. (2005). "The need for field-relevant leachability mass transfer coefficients" Twenty-Second Annual International Pittsburgh Coal Conference, Pittsburgh, PA, USA, September 12-15, 2005 (on CD).

EIA (2007). Annual Energy Outlook 2007 with Projections to 2030.Energy Information Administration, Report #:DOE/EIA-0383

Maher, M.H., Butziger, J.M., DiSalvo, D.L. and Oweis, I.S. (1993). "Lime sludge amended fly ash for utilization as an engineering material" *Fly Ash for Soil Improvement*, ASCE Geotechnical Special Publication No. 36, K. Sharp, Editor, New York, pp. 73-88.

Schroeder, P.R., Dozier, T.S., Zappi, P.A., McEnroe, B.M., Sjostrom, J.W. and Peyton, R.L. (1994). "The hydrologic evaluation of landfill performance (HELP) model: Engineering documentation for version 3," EPA/600/R-94/168b, U.S. Environmental Protection Agency Office of Research and Development, Washington, D.C.

Trautwein, S.J. and Boutwell, G.P. (1994). In situ hydraulic conductivity tests for compacted soil liners and caps" *Hydraulic Conductivity and Waste Contaminant Transport in Soil,* ASTM STP 1142, David E. Daniel, Stephen J. Trautwein, Eds., American Society for Testing and Materials

Zhang, R. (1997). "Determination of soil sorptivity and hydraulic conductivity from the disk infiltrometer" *Soil Science Society of America Journal*, 61, 1024-1030.

LEACHABILITY OF COMPACTED AGED AND FRESH COAL COMBUSTION FLY ASH UNDER HYDRAULIC FLOW CONDITIONS

Mutiu G.Ayoola[1] and Vincent O. Ogunro[2]

[1] Civil/Structural Engineer, AREVA NP INC., 7207 IBM Drive, Charlotte, NC 28262. Phone: 704-805-2198, Fax: 704-805-2370, Email: mutiu.ayoola@areva.com (formerly Doctoral Student, Civil and Environmental Dept., University of North Carolina at Charlotte)
[2] Associate Professor, Civil and Environmental Engineering Dept., University of North Carolina at Charlotte, 9201 University City Blvd., Charlotte, NC 28223. Phone: 704-687-3101, fax: 704-687-6953, Email: vogunro@uncc.edu

Abstract: Being that coal combustion byproducts (CCBs) such as fly ash are composed of trace and heavy metals that are known to pose great threat to human, ecology, and general environmental systems, addressing the long-term environmental concerns under field scenarios is paramount in any proposed management option. Leaching characteristics of trace metals under hydraulic flow conditions when subjected to isotropic confining stresses (35 kPa, 210 kPa and 420 kPa) are investigated. Leaching pattern of Zinc (ZN) and Copper (Cu) under this flow and stress conditions are similar. Concentrations of Cu and ZN are higher in leachates from fresh fly ash than aged fly ash. Cu and Zn are observed to be most leachable at 35 kPa and 420 kPa confining stresses. Leaching of Se is almost independent of the placement density and is higher in leachates from aged fly ash. While leaching of SE decreases with confining stress in the case of fresh fly ash, release of SE is approximately independent of confining stress in aged fly ash. Release of Cd is very minimal, dependent on the placement density, and not detectable in leachates from medium-dense aged fly ash.

INTRODUCTION

Extensive studies are been performed to understand and characterize leaching behaviors of reused coal combustion byproducts (CCBs) particularly fly ash, however, little data is available on the long-term performance of these materials. Without fully understanding the time dependent behaviors of these materials, tendency for high volume utilization in future geotechnical infrastructures is bleak. The metastable nature of fly ash requires that comprehensive investigation into the environmental characteristics including their potentials for their evolution over long period of time. From environmental and human safety perspectives, utilization of waste and waste-amended materials in civil engineering construction is limited by the maximum allowable concentration of contaminants such as heavy metals that is specified by regulatory agencies. Depending on the source, stress-strain condition, placement density, physico-chemical interactions, particles surface and leachant chemistry, and specific surface area, contaminants in recycled waste materials may be released over a long period of time during and after the service life of the infrastructure, thereby exposing nearby human and environmental systems to excessive concentration of the contaminants. This is particularly a serious problem if the CCBs contain metals or compounds that are categorized as hazardous by the U.S. Environmental Protection Agency (U.S. EPA).

The mineralogy composition of fly ash depends on the source of the coal, handling and storage methods, cleaning and pulverization, type and combustion operation at the thermal utility plants, and method of ash collection (Trivedi and Sud, 2004). However, silica (SiO_2), ferric oxide (Fe_2O_3), and alumina (Al_2O_3) typically constitute major components of most fly ash materials (especially bituminous coal), with other minerals such as calcium oxide (CaO), potassium oxide (K_2O), sodium oxide (Na_2O), magnesium oxide (MgO), sulfur oxide (SO_3), and phosphorous pentoxide (P_2O_5) occurring in trace quantities. Anthracite, lignite and sub-bituminous coal ashes contain lesser percentage of silica, ferric oxide, and alumina compared with bituminous coal ashes.

Recycled waste materials including fly ash are composed of soluble minerals that may be released (leached) under favorable conditions. Several laboratory-based methods that can be used to simulate leaching characteristics of waste material under field condition have been reported in literature (Sorini, 1997 and Kim, 2003). Generally, the most widely used leaching methods are in the form of column, batch, or bulk flow tests. Leaching under saturated and unsaturated condition can be modeled by using column leaching method. For the case of unsaturated condition, the leachant solution is intermittently added (Sanchez et al. 2003). Long-term leaching scenarios are typically simulated using serial batch method (ASTM D-5284 and ASTM D-4793) wherein same specimen is successfully leached with fresh but same leachant solution. The main limitations of this long-term leaching protocol include poor representation of field scenarios and lack of time dependent geochemical changes of the waste materials.

Among several factors that affect leachability potential of recycled waste materials include but not limited to pH and leachant chemistry, solubility of the minerals in the material, available surface area, leaching period, and leachant flow rate. Leachant solutions that are typically used include sulfuric acid, acetic acid, de-ionized water, and sodium carbonate.

To address the effect of time on geochemistry of CCBs, several studies have been conducted on weathering of these materials. Results from some studies have shown that weathering of volcanic ash which occurs over more than 250 years lead to formation of minerals that are typically found in clay soil (Parfitt and Wilson, 1985). Crystalline and/or non crystalline clay minerals can be formed depending on the percentage of Al and Si, magnitude of which are influenced by the geochemical process and weathering environment (Ugolini and Dahlgrn, 1991). However, most studies on weathering of CCBs have focused primarily on bottom ash with limited studies on coal combustion fly ash. Zevenbergen et al. (1999) investigated weathering mechanism of coal fly ash in terms of the time required for weathering and the magnitude of change in their original chemical composition. The results were compared with those from volcanic ash. The authors utilized coal fly ash from eight (8) years old ash deposit in India and a 14 years old ash in Denmark. Fresh fly ashes were also obtained from the same utility plants that produced the aged ashes for weathering analyses. Aged specimens were collected at a depth of 15-30 cm measured from top of the ash deposit. By comparing these results with those obtained from volcanic ash published by Dahlgren et al. (1997), the authors made following conclusions: (1) weathering of coal fly ash leads to formation of clay minerals, the process occurs at a rapid rate compared with similar formation from volcanic ash; (2) for the range of pH considered in the study, leachability of Cu, Ni, and Zn appears to be independent of pH of the leachant; and (3) extractable concentrations of the metals from weathered ash are much lower than those from fresh ash. The authors related this observation to the higher ability of weathered ash to retain metal compared with fresh ash.

Coupled with weathering, CCBs are subjected to different load regimes in most civil and environmental engineering applications. The number of studies on effects of stress conditions encountered in the field on contaminant release from CCBs is very limited. The exposed surface area is one of the most significant factors that influence contaminant release from waste material. Under stress conditions, the inter-particle porosity and hydraulic conductivity of the bulk material decreases but particle-particle contact increases (Ogunro et al., 2006). These changes in material properties affect the available volumetric surface area and hence the leachability behavior of the bulk material. Various loading scenarios that may influence contaminant release from recycled waste materials utilized in civil engineering constructions are presented in Inyang and Ogunro (2006). However, these studies limited to effective of stress on fresh fly ash. This paper explores the effect of confining loads on both fresh and aged fly ash.

MATERIALS AND TEST METHODS

In this study, fly ash materials used were produced by a utility plant in North Carolina. Fresh material was obtained directly from the coal combustion chamber while the aged material was from a closed ash monofill operated by the same utility plant that produced the fresh fly ash. Records indicate that the monofill facility was operated for fifteen (15) years and has been closed for five (5) years. The monofill facility is about 60 feet in depth and the aged material was obtained at a depth interval of 55 ft to 60 ft. Based on the number of years of operation and closure, and discussion with the facility operator/manger, aged fly ash material obtained at a depth of 60ft from this monofill is approximately twenty (20) years old.

Series of analysis including elemental composition and mineralogy of the materials, batch and column leaching tests performed on these materials are presented in Ayoola (2006). For this paper, leachability of several heavy metals (Selenium (Se); Cadmium (Ca); Zinc (Zn); Copper (Cu)) were analyzed under hydraulic flow condition. The flexible wall equipment, illustrated in Figure 1, allows the evaluation of the influence of confining stress on leaching pattern. Leachates were collected from specimens under confining stresses of 35, 210, 420 kPa. The dense specimens were compacted to 95% of maximum dry density at optimum moisture content determined from the standard proctor tests. Medium-dense specimens were compacted to relative density of 50% also at optimum moisture content. The maximum dry density of the fresh and aged fly ash materials are 12.3 kN/m^3 and 15.2 kN/m^3, respectively. The optimum moisture content of the fresh fly ash (28%) is significantly greater than that of aged fly ash (18%). The average specific gravities of the fresh and aged fly ash were computed to be 2.14 and 2.55, respectively. Mean diameter (D_{50}) of the fresh and aged fly ash materials are approximated to be 0.09mm and 0.05mm, respectively.

In addition to change in grain size distribution characterized by the mean diameter, consistency tests performed on the fresh fly ash indicate non-plastic behavior. However, in the case of the aged fly ash, the liquid limit and plasticity index were obtained to be 22 and 6, respectively. Based on the grain size distribution and Atterberg limits analysis on the materials, the fresh and aged fly can be classified as silt and silty clay materials, respectively.

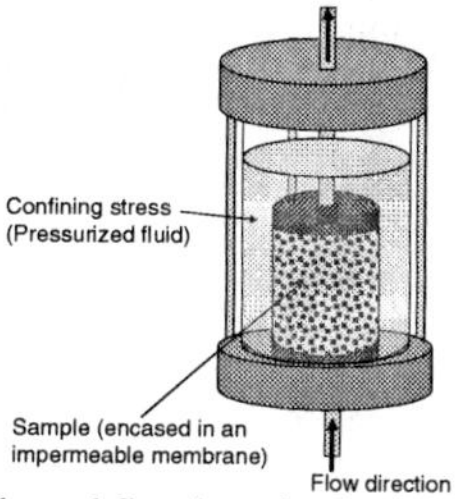

Figure 1: Leaching of fresh and aged flyash under hydraulic flow condition (isotropic confining stress

RESULTS

Using the atomic absorption spectrometer (AAS), the concentration of Se, vCd, Cu and Zn collected from both dense and medium-dense specimens of fresh and aged fly ash materials under hydraulic flow condition were analyzed. The results are presented in form of cumulative mass of metals leached expressed as percentage of total mass of fly ash solid plotted as a function of liquid to solid ratio (L/S). Detailed results of leaching data are presented in Ayoola (2006).

Cadmium: Figures 2: (a), (b), (c) and (d) present typical leaching pattern of Cd under hydraulic flow condition subjected to isotropic confining stress. Leaching is generally dependent on the placement density and confining stress. The influence of confining stress appears to vary with placement density and age of the material. Very low concentration of Cd was removed from aged material under low confining stress. The concentration further drops as the placement density decreases. In this study, no Cd was detected in leachates collected from medium-dense specimens under confining stress of 35 kPa (Figure 2 (d)).

Selenium: In this study, typical leaching pattern of Se from the specimens under hydraulic flow condition is presented in Figures 3: (a) and (b). From these figures leaching of Se is approximately independent of the placement density. Probably because of long-term chemical reactions that favor dissolution of Se, aged fly ash material shows stronger tendency to release Se. Leaching of Se from fresh fly ash is slightly dependent on the confining stress level. Generally, the concentration of Se released from fresh fly ash appears to decrease with increase in confining stress.

Copper: Leaching characteristics of Cu from both fresh and aged fly ash materials are shown in figures Figure 4. Regardless of the type of material, leaching is dependent on the placement density, with specimens compacted to high density showing higher leachability potential. The concentration of Cu is higher in leachates from fresh than aged fly ash. At low (35 kPa) and high (420 kPa) confining stresses, difference between concentration of Cu in leachates from dense and medium-dense specimens from fresh fly ash is very significant. However, at confining stress of 210 kPa, concentration is approximately equal. Leachability of Cu from fresh fly ash is more rapid at low confining stress regardless of placement density. For the case of aged dense specimen, leachability continues to decrease with increase in confining stress. Leachability of Cu in aged medium-dense specimen increases with confining stress.

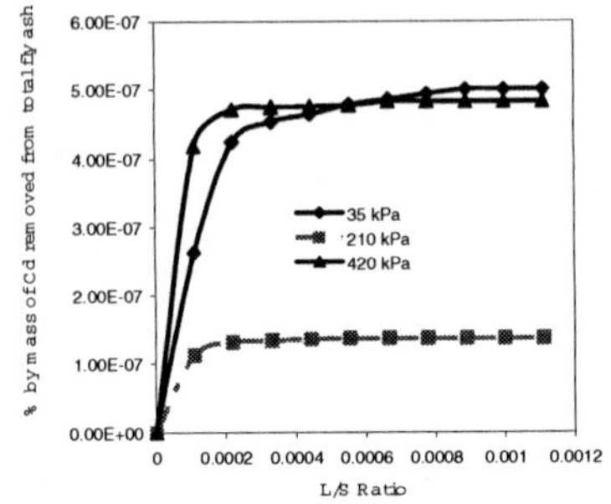

(a): Dense specimen of fresh fly ash

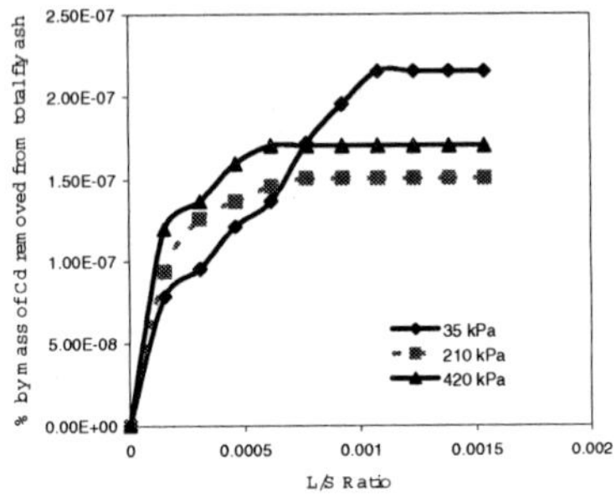

(b): Dense specimen of aged fly ash

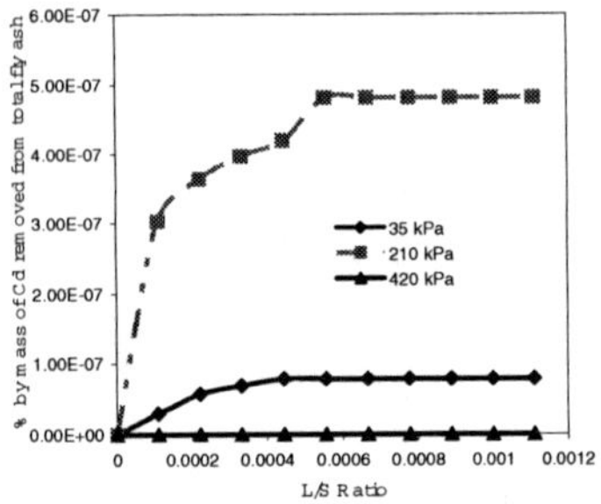

(c): Medium-dense specimen of fresh fly ash

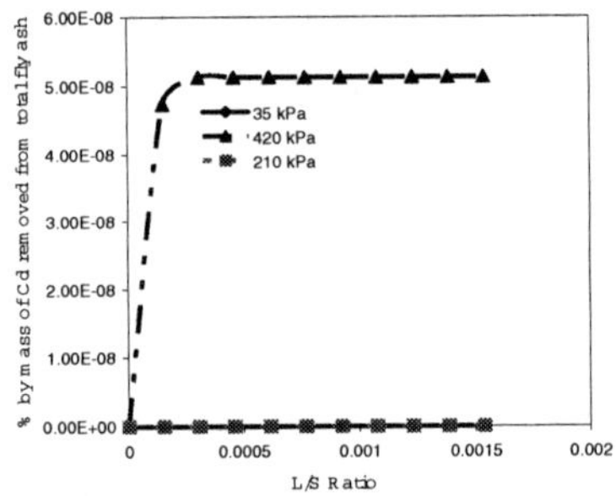

(d): Medium-dense specimen of aged fly ash

Figure 2: Leachability of Cd from total mass fly ash under hydraulic flow condition

Zinc: Leaching characteristics of Zn are similar to those of Cu (Figures 5: (a) and (b)). Release is dependent on the placement density and the effect of confining stress differs and depends on the age of the material. At all confining stress considered in this study, the concentration of Cu is higher in leachates generated by fresh fly ash. For the case of fresh fly ash, leachability of Zn increases significantly with placement density regardless of the confining stress level. On the contrary, release of Zn from aged fly ash is almost independent of the placement density at confining stress of 35 and 420 kPa. However, at confining stress of 210 kPa, medium-dense state appears to favor leachability. The results also indicate that while leachability of Zn from fresh fly ash occurs more rapidly at 35 and 420 kPa, release from aged fly ash decreases continuously with increase in confining stress.

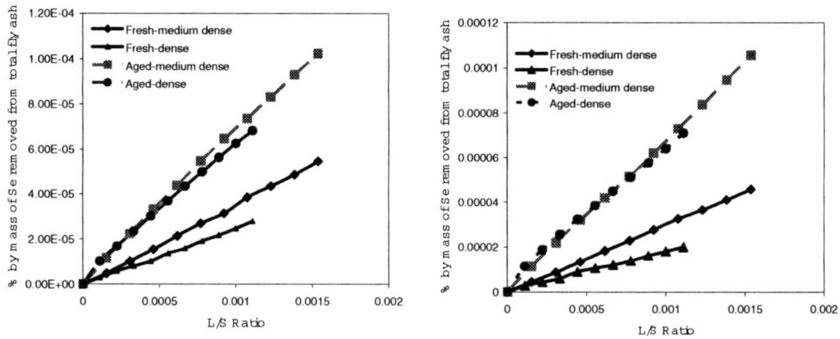

(a): At confining stress of 35 kPa

(b): At confining stress of 420 kPa

Figure 3: Leachability of Se from total mass fly ash under hydraulic flow condition

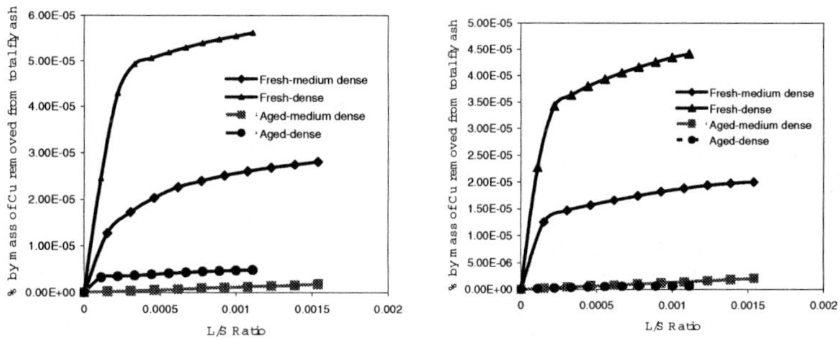

(a): Confining stress of 35 kPa

(b): Confining stress of 420 kPa

Figure 4: Leachability of Cu from total mass fly ash under hydraulic flow condition

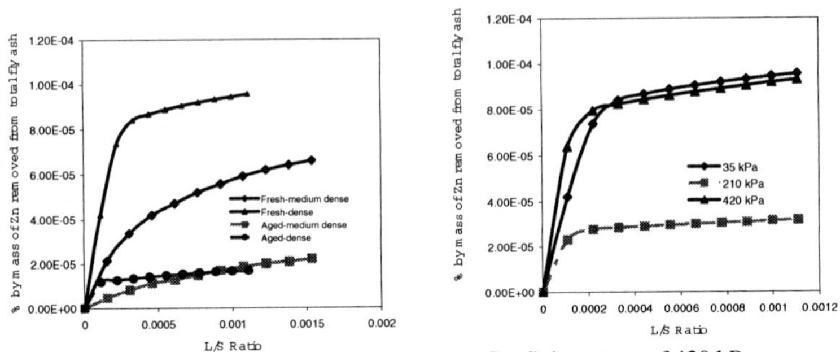

(b): Confining stress of 420 kPa

(a): Confining stress of 35 kPa

Figure 5: Leachability of Zn from total mass fly ash under hydraulic flow condition

CONCLUSIONS:

Leaching characteristics of trace metals from compacted fly ash under hydraulic flow conditions are quite complex and differ for most of the metals analyzed. Contaminant release is dependent on the age of the material, placement density, and magnitude of confining stress. Generally, release pattern of Zn and Cu under hydraulic flow condition are similar. Concentrations of Cu and Zn are higher in leachates from fresh fly ash than aged fly ash. Cu and Zn are observed to be most leachable at low (35 kPa) and high (420 kPa) confining stresses (Figure 6).

Leaching of Se is almost independent of the placement density and is higher in leachates from aged fly ash. Perhaps, a chemical dissolution reaction which involves Se compound occurs as the material ages. While leaching of Se decreases with confining stress in the case of fresh fly ash, release of Se is approximately independent of the confining stress in aged fly ash. Release of Cd is very minimal, dependent on the placement density, and is not detectable in leachates from medium-dense aged fly ash. The only scenario in which considerable concentration of Cd was detected is dense fresh fly ash at high confining stress. According to Ogunro et al. (2006), the inter-particle porosity and the hydraulic conductivity of bulk granular materials decrease under stress condition, while the particle-particle contact increases. It appears that these geometric and physical phenomena to be more dominant in leachability of Cd from fly ash.

For most metals considered in this study, after collection of leachate quantity which is in excess of two pore volume of the specimens, the tendency for further release is still very strong and will be purse in future experiments. It can be concluded therefore, that age of the fly ash and the loading regime should be considered in developing a long-term protocol for evaluation leachability and environmental impact of reuse of CCBs in civil engineering.

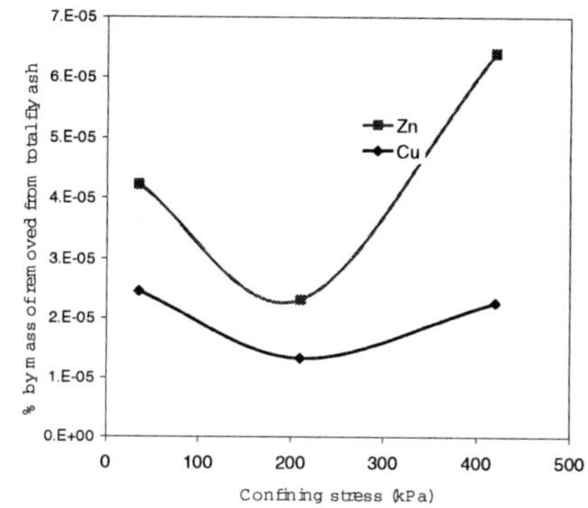

Figure 6: Leachability of Zn and Cu subjected under hydraulic flow condition subjected to isotropic confining stress (2 pore volume leached).

REFERENCES

Ayoola, M.G. (2006). "Ageing Effects on Stress-Strain Characteristics and Leachability of Recycled Granular Waste Materials". Ph.D. Dissertation, University of North Carolina at Charlotte, 284 p.

Dahlgren, R.A., Dragoo, J.P., and Ugolini, F.C. (1997). "Weathering of Mt. St. Helens Tephra under a Cryic-Udic Climatic Regime". *Soil Science Society of America Journal*, Vol. 61, 1519-1525.

Inyang, H.I., and Ogunro, V.O. (2006). "Chemodynamics of Contaminant Leachability from Particulate and Monolithic Media under Load". *Environmental Engineering Science (EES)*, Vol. 23 (1), 143 - 155.

Kim, A.G. (2002). "Fluid Extraction of Metals from Coal Fly Ash: Geochemical Simulation of Natural Leaching". PhD Dissertation, University of Pittsburgh, 217 p.

Ogunro, V.O., Long, J.M., and Inyang, H.I. (2006). "Modeling of Leaching Surface Evolution in Stressed Particulate Media". *Environmental Engineering Science (EES)*, Vol. 23(1), 171-186

Sanchez , F., Garrabrants, A. C. and D. S. Kosson (2003). "Effects of intermittent wetting on concentration profiles and release from a cement-based waste matrix". *Environmental Engineering Science (EES)*, Vol. 20(2), 135-153.

Sorini, S.S. (1997). "A Summary of Leaching Methods". *Prepared for American Coal Ash Association*, 73 p.

Trivedi, A. and Sud, V.K. (2004),"Collapse Behavior of Coal Ash". *Journal of Geotechnical and Geoenvironmental Engineering*, Vol. 130(4), 403-415.

Ugolini, F.C. and Dahlgrn, R.A. (1991)."Weathering Environment and Occurrence of Imogolite/Alophane in Selected Andisols and Spodosols" *Soil Sci. Soc. Am. J.*, Vol. 55, 1166-1171.

Zevenberg, C., Bradley, J.P., Reeuwijk, L.P.V., Shyam, A.K., Hjelmar, O., and Comans, R.N.J. (1999). "Clay Formation and Metal Fixation During Weathering of Coal Fly Ash", *Environmental Science and Technology*, Vol. 33, 3405-3409

Mitigation of Alkali Induced Heave in Rectorite Soil with Fly Ash

P. V. Sivapullaiah[1], P. Hari Prasad Reddy[2] and H.N. Ramesh[3]

[1]Professor, Department of Civil Engineering, Indian Institute of Science, Bangalore, India; siva@civil.iisc.ernet.in
[2]Research Scholar, Department of Civil Engineering, Indian Institute of Science, Bangalore, India; hariprasad@civil.iisc.ernet.in
[3]Professor, Department of Civil Engineering, Bangalore University, Bangalore, India; rheddur@yahoo.com

ABSTRACT: Mechanisms that control the volume changes behavior of foundation soils are well understood. The changes that occur in the behavior of soil due to migration of pollutants are not well understood. The extent of changes that occur in the presence of small concentration of contaminants can be predicted based on changes in the thickness of double layer and associated fabric changes. Interactions that occur with strong contaminants depends on the type of soil, type and concentration of contamination and duration of interaction etc It has been shown that different concentrations (1N and 4N) of sodium hydroxide solution causes abnormal changes on volume change behaviour of soil due to mineralogical changes. An attempt is made in this paper to stabilize contaminated soil using fly ash, after establishing its stability in alkali solutions. It was found that the effectiveness of fly ash to control the alkali induced heave increases with fly ash content incorporated into the soil. X-ray diffraction studies reveal that the mineralogical changes that occur in soil due to alkali interaction are inhibited by the presence of fly ash.

INTRODUCTION

Geotechnical behaviour of soils containing mixed layered minerals in natural water itself is not known, much less in polluted environment. Mixed-layer clay minerals are intermediate products of reactions involving pure end-member clays. Most mixed-layer clays contain smectite or vermiculte as a swelling component. Illite - smectite is the most abundant and wide spread of the mixed layered clay minerals in sedimentary rocks and soils and also more common than either discrete illite or smectite. Mixed layered minerals are more susceptible to environmental changes as the structural linkages between the layer minerals are weak compared to normal layered phyllosilicates and hence have been chosen for the study. One important pollutant that can have considerable effect on the behaviour of soils is the caustic alkali contamination released from various industries. Studies have shown that the behaviour of even stable minerals is affected by alkali contamination (Rao and Rao 1994; Sinha et al 2003; Sivapullaiah and Manju, 2005; and Dilek Turer, 2007). In the Recent past the effect of caustic alkali contamination on the behavior of soils containing mixed layered minerals has also been studied and is established that very high swelling can occur depending upon the concentration of alkali. However, no viable method is known to control the alkali-induced changes in soils.

Fly ash is the finely divided mineral residue resulting from the combustion of coal in thermal power plants. It consists mostly of silicon dioxide (SiO_2), aluminium oxide (Al_2O_3) and iron oxide (Fe_2O_3). The silica present in fly ash is generally in reactive form and hence fly ashes are pozzolanic in nature and produce cementitious compounds. Fly ash, which is normally used to stabilize expansive soils, can be attempted to stabilize alkali-contaminated soils. It is known that most pozzolanic materials show enhanced cementing effect in presence of alkali. This paper investigates the use fly ash to control the alkali induced heave in a natural soil containing mixed layered mineral.

EXPERIMENTAL INVESTIGATION

Black Cotton Soil

The natural soil black cotton soil (BC) used in this study was collected from Belgaum (Latitude: 15°52'N & Longitude 74°31'E), India. Belgaum is situated in northern Karnataka Plateau, a part of Deccan Plateau of India. The soil was collected by open excavation, from a depth of one meter from natural ground level. The soil was air dried and used after sieving through Indian Standard 425-micron sieve. The properties of soil are summarized in Table 1. Based on the index properties and the grain-size distribution, soils are classified. X-ray diffraction studies revealed the presence of a mixed layered illite-smectite (rectorite) mineral as the dominant component.

Table 1. Properties of Soil and Fly Ash

Properties	Black Cotton Soil	Fly Ash
Liquid limit (LL), %	82	-
Plastic limit (PL), %	35	Non Plastic
Plasticity index (PI), %	47	-
Specific gravity	2.79	2.60
Free swell index (ml/g)	2.2	Non-swelling
Clay (%)	40	2
Silt (%)	54	73
Fine sand (%)	6	25
Classification (USCS)	CH	Non-plastic
Max. dry unit weight (g/cc)	1.29	1.25
Optimum water content (%)	35	31.0

Fly Ash

The fly ash used in the investigation is obtained from Neyveli Thermal Power Plant, Tamil Nadu, which is a southern state of India. Neyveli fly ash (NFA) is typically obtained from lignite coal. The fly ash is dark grey in color. The physical properties and chemical composition of Neyveli fly ash are given in Table 1 and Table 2 respectively. Based on the chemical composition, the fly ash can be classified as ASTM class F fly ash.

Table 2. Chemical Composition of Neyveli Fly Ash

Constituent	Percentage
Total silica (SiO_2)	50.40
Reactive silica	5.61
Alumina (Al_2O_3)	18.81
Ferric oxide (Fe_2O_3)	16.61
Total calcium (CaO)	9.00
Free lime	3.92
Magnesium (MgO)	1.41
Titanium(TiO_2)	0.28
Potassium (K_2O)	0.23
Sodium (Na_2O)	0.18
Loss of ignition	2.60

Fluid Used

The contaminants used in the study are 1N and 4N sodium hydroxide solution. The solutions were prepared by dissolving the chemically pure salt in distilled water.

Oedometer Test

The volume changes in the form of swell and compression were measured by testing a compacted soil specimen in a consolidation apparatus following ASTM D-2435 method. Required amount of soil was statically compacted in the consolidation ring of 60 mm diameter and 20 mm height to Proctors maximum density at optimum moisture content conditions to a height of 14 mm and subjected to swell fully and reach equilibrium at seating load before subsequent loading and unloading. The changes in thickness of the specimen are recorded against time to obtain the swelling at seating load and the compression at the end of each load stage.

X-ray Diffraction Studies

X - ray diffraction pattern of all specimens were studied after consolidation test by standard method.

RESULTS AND DISCUSSION

Before studying the effectiveness of fly ash to control alkali induced heave, the volume stability of fly ash in alkali solutions has been studied.

Effect of NaOH solutions on volume change behavior of fly ash

Fly ash compacted with water and inundated with 1N and 4N NaOH solutions showed no swelling at nominal surcharge of 6.25kPa. As expected for material with no cation exchange capacity and silt like particles, there is no swelling. Because of this and because of cementitious nature fly ash can be used to control the swelling in soils.

The compressibility for fly ash is also very less in the alkali solution. Since there is no swelling either by double layer expansion or by soil alkali interaction, no compression is expected

Controlling the alkali induced Volume Changes in Soils Using Fly Ash

It is clear that the volume change behavior of fly ash is not affected by the presence of alkali solution. Hence it is proposed to study the effect of different percentages of fly ash on the volume change behavior of soil with alkali solutions so as to know the amount of fly ash needed to be incorporated into the soil to minimize the alkali induced volume changes.

Effect of 1N NaOH Solution on Black Cotton Soil and Fly Ash Mixture

Fig. 1 shows the effect of different percentages of fly ash (25% and 50%) mixed with black cotton soil and compacted with water and inundated with 1N alkali solution on the swelling at seating load. From Fig. 1, 2% of swelling of soil with water is observed which is due to incorporation of water molecules on the clay structure in between the clay plates to form diffuse double layer near the negative surface of clay mineral particles. A very high swelling of about 17% is observed for the soil with 1N alkali solution. It is known that with increase in the pH of the soil the charges on clay particles increase (Brady, 1995) which in turn increases the thickness of diffused double layer. Hence the increase in negative charges on soil particles after interaction with soil alkali interactions are responsible for larger swelling in sample inundated with 1N alkali solution compared to sample inundated with water. The swelling of the soil in alkali solutions is considerable reduced when the soil is mixed with fly ash. The reduction increases with increase in the percentage of fly ash in the soil. With 25% fly ash the swell of soil is reduced from 17 % to 4 % and with 50 % fly ash the swell is totally controlled and is about 0.5%. The high reduction, which is more than the proportion of the fly ash, indicates that it not just by decrease in clay content alone. The reduction in swelling is essentially due to binding of soil particles by cementitious compounds produced by fly ash upon hydration. Use of any non pozzolanic sandy material can not reduce this type of alkali induced swelling.

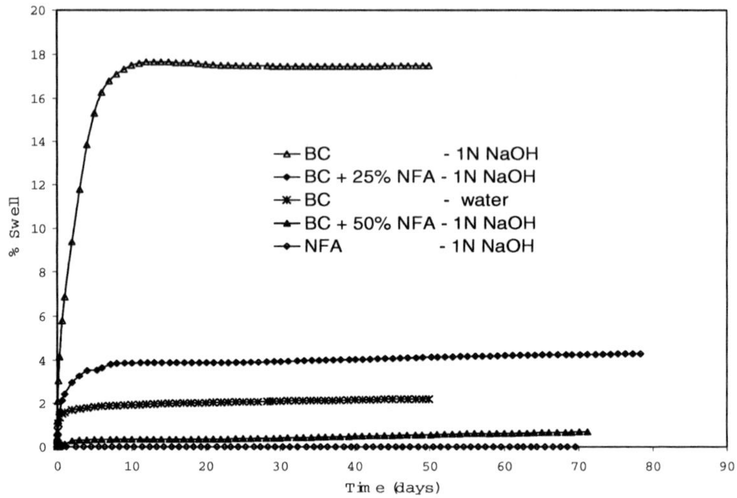

FIG. 1 Variation of swell with time for BC soil mixed with different percentages of fly ash inundated with 1N NaOH solution

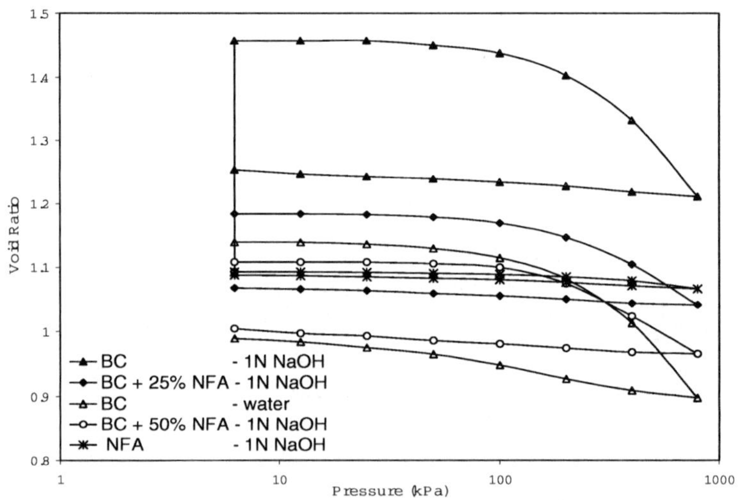

FIG. 2 Volume change behavior of BC soil mixed with different percentages of fly ash inundated with 1N NaOH solution

Fig. 2, represented in the form of void ratio versus logarithmic pressure relationship, shows the effect of fly ash on the compressibility of the soil when contaminated with alkali solution. The results indicate that fly ash treatment decreases the compressibility of the soil. This reduction in the compressibility of soil is due to decreased clay content and increased cementation of particles of soil. The reduction in the compressibility is seen in the compression index (C_c) values given in Table 3

Effect of 4N NaOH Solution on Black Cotton Soil and Fly Ash Mixture

The effect of different percentages of fly ash (25% and 50%) mixed with black cotton soil and compacted with water and inundated with 4N alkali solution on the swelling at seating load is shown in Fig. 3. From figure it can be noticed that a very high swelling of about 36% has occurred when the soil is contaminated with 4N NaOH. Further it is observed that the swelling with 4N alkali solution occurs in two stages, viz., steep initial swelling (of about 25 %) for about 4 days and gradual swelling (of about 11 %) beyond this period. There is some amount of settlement before the start of the second stage of swelling. Increase in negative charges on soil particles and mineralogical changes after interaction with soil respectively are responsible for the swelling in these two stages. The first stage of swelling is rapid like in most swelling soils. The second stage of swelling which is due to mineralogical changes occurs after considerable delay and proceeds for long period due to slow mineral dissociation and formation of new minerals by interaction of soil with 4N sodium hydroxide solution. From Fig. 3 it is observed that while the first stage of swelling has decreased, the second stage of swelling is totally absent in the presence of fly ash. Thus addition of fly ash effectively controlled swelling in alkali solution. The reduction in initial swelling depends on the amount of fly ash. With 25% fly ash the swell of soil is reduced from 25 % to 8 % and with 50 % fly ash the initial swell of soil is reduced from 25 % to 1.5 %. The reduction in initial swelling is due to replacement of swelling soil by fly ash and cementation of soil particles by pozzolanic fly ash. Elimination of swelling of soil in 4N alkali solution when fly ash is present is due to prevention of mineralogical changes that occur in soil with 4N solution. X-ray diffraction patterns on samples after consolidation test showed that the peaks at 6.30, 3.65, 2.58, and 2.11 A^0 due to sodium aluminium silicate hydroxide hydrate is not seen. This indicates that the soil particles are bounded by pozzolanic compounds and probably resist the attack by alkali. The variations in the swelling values with different percentage of fly ash are given in Table 3.

The effect of fly ash on the compressibility of the soil when contaminated with alkali solution is shown in Fig. 4. Fly ash effectively reduced the compressibility in alkali contaminated soil. The reduced compression is represented in the form compression index values and is given in Table 3. It can be seen that the reduction in compressibility of soil with fly ash is significant in 4N solution than in 1N solution. This indicates that effectiveness of fly ash is more significant in prevention of mineralogical changes than in reducing the swelling by reduced clay content and binding of soil particles.

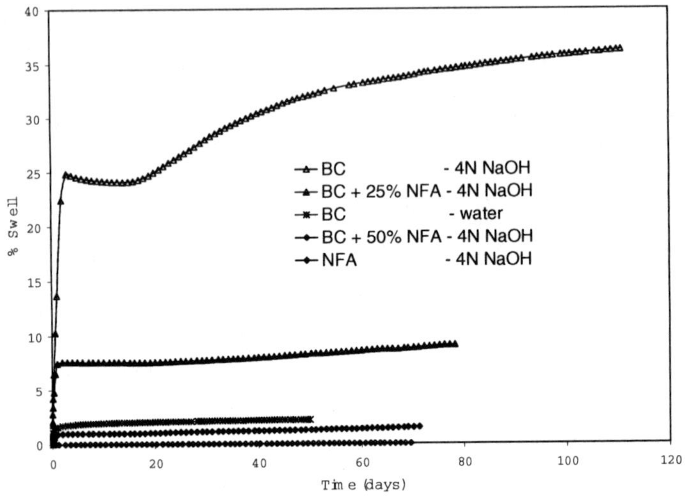

FIG. 3 Variation of swell with time for BC soil mixed with different percentages of fly ash inundated with 4N NaOH solution

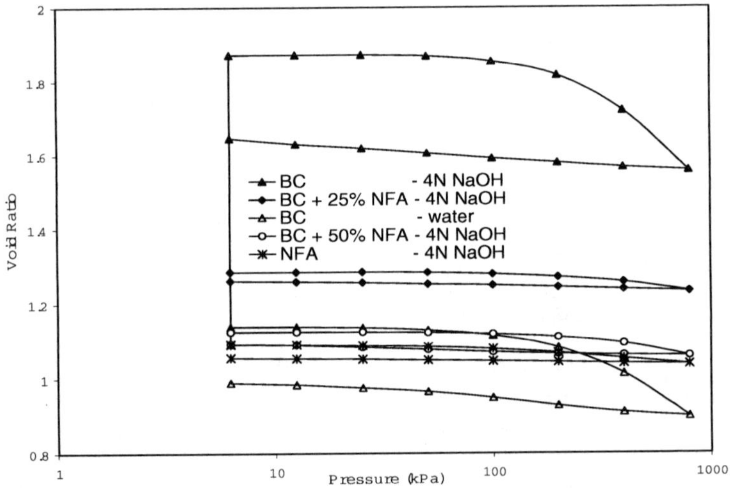

FIG. 4 Volume change behavior of BC soil mixed with different percentages of fly ash inundated with 4N NaOH solution

Table 3. Effect of Alkali on Volume Change Behavior

Inundating fluid	Soil	Maximum swell at 6.25 kPa, %	Compression index (Cc)
water	BC soil	2	0.24
1N NaOH	BC soil	18	0.25
	NFA	0	0.02
	BC soil + 25% NFA	4	0.14
	BC soil + 50% NFA	0.7	0.15
4N NaOH	BC soil	36	0.33
	NFA	0	0.04
	BC soil + 25% NFA	9	0.05
	BC soil + 50% NFA	1.5	0.06

CONCLUSIONS

Based on analysis of results obtained in this study, the following conclusions are drawn:

1. Addition of fly ash effectively reduces alkali induced swell in soil containing interstratified mineral. The effectiveness of fly ash increases with its content.
2. The reduction in swelling of soil with low concentration of alkali is by replacement of swelling soil by fly ash and cementation of soil particles of the soil by pozzolanic compound produced. Addition of fly ash also decreases the compressibility of the soil.
3. The second stage of swelling which occur in soil with high alkali concentration is absent in presence of fly ash. This is by preventing undesirable mineralogical changes in soil with alkali. This also leads to lower compressibility of soil with high alkali solution.

REFERENCES

ASTM D2435 (2003). "*Standard test method for one-dimensional consolidation properties of soils*." ASTM, WC, PA.

Brady, N.C. (1995). "*The nature and properties of soils.*", X Edition, Prentice Hall of India Private Limited, New Delhi.

Dilek Turer (2005). "Effect of heavy metal and alkali contamination on the swelling properties of kaolinite." *Environmental Geology,* Vol.52:421–425

Rao, S.M. and Rao, K.S. S. (1994). "Ground heaving from caustic soda solution spillage – a case study." *Soils and Foundations*, Japanese Society of Soil Mechanics and Foundation Engineering, 34(2), 13-18.

Sinha, U.N., Sharma, A.K., Bhargava, S.N., Minocha, A.K. and Pradeep Kumar (2003). "Effect of seepage of caustic soda on foundation and remedial measure in alumina plant", *Indian Geotechnical Conference*, Roorkee, India, 229-234.

Sivapullaiah, P.V., and Manju (2005). "Kaolinite – alkali interaction and effects on basic properties", *Geotechnical and Geological Engineering*, 23(5): 601-614

Stabilization of inorganic contaminants in lead crystal polishing sludge

Harald Weigand[1], Tina Gretzbach[2], Rupert Bäumler[2], and Clemens Marb[1]

[1]Bavarian Environment Agency, Josef-Vogl-Technology-Center, Am Mittleren Moos 46, 86167 Augsburg, Germany; josef-vogl-technikum@lfu.bayern.de
[2]Friedrich-Alexander-Universität Erlangen-Nürnberg, Department of Geography, Kochstr. 4/4, 91054 Erlangen, Germany; bäumler@geographie.uni-erlangen.de

ABSTRACT: Neutralization sludge from the lead crystal polishing process has a high leaching potential for a series of inorganic contaminants. Cost-effective treatment options would be desirable both for soil contaminated by this residue and for waste from current manufacturing. The feasibility of contaminant stabilization was studied in batch experiments using five phosphatic amendments. All additives reduced lead concentrations relative to the control presumably by formation of sparingly soluble lead phosphates. Arsenic and zinc concentrations were above the control in some treatments due to either competitive displacement or additive-borne secondary contamination. Both stabilization and mobilization were dose-dependent. Equilibration velocities were contaminant-specific and varied between treatments. Slowest kinetics and lowest stabilization efficiencies were observed for fluoride. Treatment optimization should therefore focus on combined amendments which address both lead and fluoride.

1. INTRODUCTION

Lead crystal production for high-quality tableware and chandelier glass is among the traditional industries of NE Bavaria, Germany. Resources of the region include quartz and lignite used as fuel for the melting process (CORIS, 2007). Lead crystal consists of at least 24 % wt lead oxide which is added to the melting batches to improve brilliance and reflectivity of the material (Nölle, 1977). Further additives include antimony oxide (Sb_2O_3) and arsenic oxide (As_2O_3) used as refining agents to eliminate bubbles from the melt (Glass Online, 2007). The finishing of lead glass products typically involves a series of acid etching and polishing steps. Therefore, workpieces are dipped into a bath holding a mixture of hydrofluoric and sulfuric acid. To avoid polishing flaws, the abraded material has to be separated from the bath and is further neutralized by lime addition. The resulting residue is referred to as neutralization sludge. Its main constituents are lead sulfate and alkali hexafluorosilicate (Pocham, 1977). Arsenic and antimony may be present depending on the use of refining agents. Improper handling of the material including its on-site disposal in some cases poses significant risks to groundwater quality at abandoned lead crystal production sites.

As a treatment option for lead contaminated soils, stabilization by phosphatic amendments to form sparingly minerals has been proposed (McGowen et al., 2001; Cao et al., 2002; Cao et al., 2005). Due to their low solubility and geochemical stability pyromorphites are regarded effective mineral phases controlling lead concentrations in treated residues (Eighmy et al., 1998; McGowen et al., 2001). Their formation has been demonstrated to reduce lead mobility in contaminated soils and incinerator bottom ash (Crannell et al., 2000; Ryan et al., 2001). Pb sorption on apatite has been reported as a further stabilizing process (Chrysochoou et al., 2007).

So far, no attempts have been made to stabilize lead crystal polishing sludge although this may be a cost-effective treatment option both for soil contaminated by this residue and for current production wastes. This study focuses the use of different phosphate sources with respect to dose dependence and reaction kinetics. In view of the questionable sustainability of phosphate treatments (Chrysochoou et al., 2007) it seemed particularly interesting to include biogenic and waste-derived P sources.

2. MATERIALS AND METHODS

2.1 Site Description and Sampling

The sample site is situated on an alluvial terrace in NE Bavaria, Germany, and covers an area of 9.5 ha. From the 1920s to the 1990s lead crystal was produced at the site. Soils are highly disturbed especially by infill with different types of technogenic material, e.g. demolition debris, glass melting slag, polishing residues, etc. A site survey revealed elevated arsenic and fluoride groundwater levels as well as contamination by lead and other heavy metals (Hartmann, 2002). Since insolvency of the owner, structures have been partially removed and redevelopment of the site is underway.

The polishing sludge was collected from a former neutralization basin. Ten kilogram of the material were sampled with a spade, stored in a plastic container and transferred to the laboratory. The white, paste-like material was air-dried, ground in a mortar and passed through a 2 mm sieve. The oversized grain was discarded. Total metal contents were determined in aqua regia extracts. Results are shown in Table 1. X-ray powder diffraction revealed that the crystalline portion of the neutralization sludge dominantly consists of anglesite. No further lead minerals were identified.

Table 1: Selected Metal Contents of Neutralization Sludge (mg/kg dry matter)

Al	As	Ca	Fe	Mg	Mn	Na	Pb	Zn
2,600	13	2,900	1,500	300	54	350	27,000	45

2.2 Additives

Five additives were tested with respect to their contaminant stabilization efficiency. Diammonium phosphate (DAP) is a commercial phosphate fertilizer with a slightly alkaline reaction. Due to its solubility DAP should provide a readily available phosphate source. Nanograde quality DAP was purchased from Merck KGaA, Darmstadt, FRG. Sewage sludge ash (SSA) was provided by Zweckverband Klärwerk Steinhäule, Neu-Ulm, FRG. At the plant, 10,000 Mg/a of municipal sewage sludge are treated in a fluidized bed incinerator. The residue contains up to 10 % wt phosphorus. TRAPPS, an apatite-based, granulated stabilizing agent was developed by Slater UK Ltd., Buckinghamshire, UK. References of the provider include successful treatment of a series of heavy metal contaminated materials, e.g. firing berms and soils from brownfield sites. Apatite II (AP) is a biogenic additive manufactured on the basis of ground fishbones by PIMS NW Inc., Richland, USA (Conca and Wright, 2001). Asides a slow-release phosphate source the material is an effective sorbent for a series of heavy metals due to its microcrystalline structure. Fe-rich Apatite II (FeAP) is a follow-up product developed by PIMS NW Inc., and Slater UK Ltd. To improve the efficiency for mixed cationic/anionic contaminations Apatite II was modified by adding a zerovalent iron source.

2.3 Experimental

Additive efficiency was tested in batch experiments. Tests were performed in 300 mL polyethylene flasks with 20 g of neutralization sludge (dry matter) and 200 mL distilled water. Triplicate batches were used (equilibrium). The additives were added such to yield amendment levels of 2.5 and 5 % dry matter. Blanks without amendments and samples containing only the additives were included. Batch kinetic experiments were performed without replication at an amendment level of 2.5 %. The batches were reacted on an end-over-end shaker at 2.5 rpm. Residence time was fixed to 24 h in the equilibrium study and ranged from 0.3 to 48 h in the kinetic experiments. Phases were separated by centrifugation and membrane filtration using 0.45 µm cellulose nitrate filters (Sartorius, Göttingen, FRG).

2.4 Analytical

After determination of pH (SenTix 81 electrode), electric conductivity (TetraCon 325 conductivity cell), and redox potential (SenTix ORP electrode, all WTW GmbH, Weilheim, FRG) samples were analyzed for major anions and metals/metalloids. Fluoride was measured with an ion selective electrode (Orion Ionplus fluoride electrode 96-09, Thermo Fisher Scientific Inc., Waltham, USA). To quantify total fluoride (free plus complex-bound), TISAB buffer was added to the samples. Sulfate and phosphate were determined by ion chromatography (Metrohm 733 IC Separation Center with an 732 IC conductivity detector, both Metrohm AG, Herisau, CH). Metals and metalloids were measured by inductively coupled plasma–mass spectroscopy (ICP–MS, ELAN 6000, Perkin Elmer Sciex, Ontario, CND).

3. RESULTS AND DISCUSSION

3.1 Efficiency of Additives

Electric conductivity, pH and redox potential after 24 h of treatment are shown in Fig. 1.

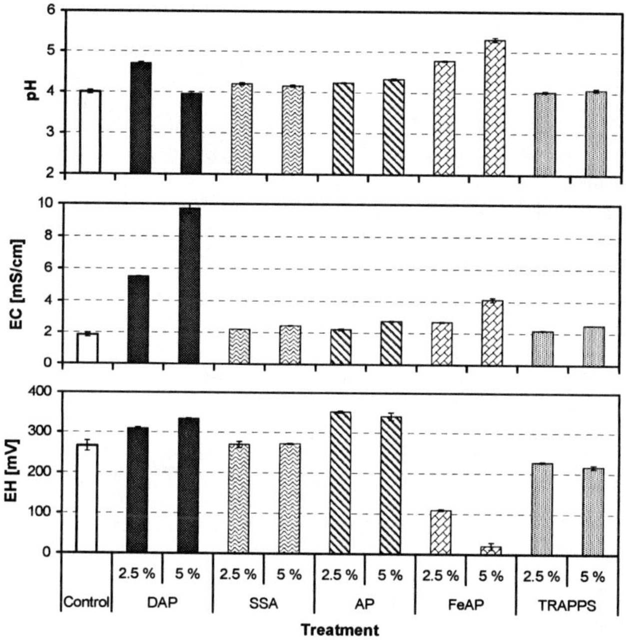

FIG. 1: Hydrochemical parameters in batches of treated and untreated neutralization sludge. Error bars show the standard deviation of three replicates

Although the pure additives were characterized by slightly to highly alkaline reaction, the pH of most treated samples was close to the control. A slight pH increase in the FeAP treatment is a common side effect of zerovalent iron oxidation. Overall, the pH behavior indicates high base buffering capacity of the neutralization sludge. Electric conductivity changes imposed by the additives were moderate with exception of the DAP treatment. Contrasting the other additives, DAP dissolved completely thereby increasing the total ion concentration roughly proportional to the dosage applied. The redox potential of the batches showed minor changes compared to the control except for the FeAP amended samples. Moderately reducing conditions are consistent with the reductive capacity of zerovalent iron. Equilibrium phosphate concentration of the batches decreased in the order DAP (1.6 g/L), AP (1.1 g/L), SSA (0.7 g/L), and TRAPPS (0.06 g/L) for the 5% amendmend level, respectively. High ferrous iron concentrations impeded phosphate measurement in the FeAP treatment.

The response of selected contaminants to the different additives is shown in Fig. 2. Arsenic concentrations in the supernatant of the untreated batches were between 75 μg/L and 80 μg/L. In the DAP, SSA, and AP treatments arsenic concentrations were clearly increased but still below the German waste acceptance criteria (WAC; AbfAblV, 2001) for non-hazardous waste to landfill (0.2 mg/L). This observation agrees with the competitive displacement of oxyanions like arsenate and antimonate by phosphate (Smith et al., 2002; Spuller et al. 2007). FeAP yielded a pronounced arsenic stabilization. Concentrations were reduced by 87 % relative to the control. Iron oxides are known as powerful arsenic sorbents (e.g. Bowell, 1994). Thus, effective stabilization by the FeAP treatment may be attributed to surface complexation by fresh oxidation products derived from zerovalent iron.

Lead concentrations of the control were between 7.9 mg/L and 9.8 mg/L. These values are clearly above the WAC (1 mg/L Pb). For all additives, stabilization was dose-dependent. Lowest concentrations were achieved by the 5 % amendment level of the SSA and TRAPPS treatments. However, to meet the WAC for non-hazardous waste a higher amendment level would be necessary.

Zinc concentrations of the control were around 1 mg/L which is below the WAC (5 mg/L Zn) for non-hazardous waste. The additives yielded non-systematic effects ranging from moderate stabilization to net zinc mobilization. The latter was the case in the DAP and SSA treatments. For the DAP treatment this is consistent with the lower pH at the higher additive dosage (cf. Fig. 1). In the SSA treatments this is con-

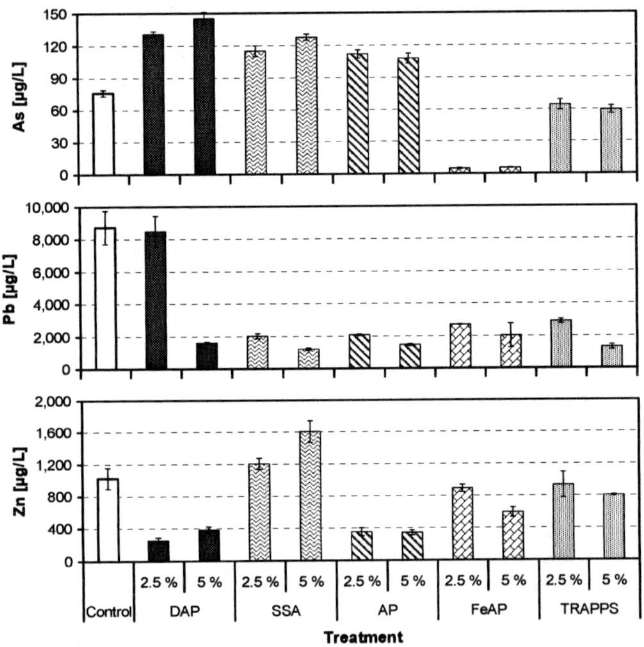

FIG. 2: Selected contaminants in batches of treated and untreated neutralization sludge. Error bars show the standard deviation of three replicates

sistent with the release of additive-borne zinc under acidic conditions (Levlin et al., 2004). The fact that increasing the dosage of DAP and SSA increases zinc concentrations limits the permissible amendment levels.

3.2 Kinetics of Contaminant Stabilization and Release

Kinetics of arsenic, lead, and zinc stabilization and total fluoride release in the SSA and TRAPPS treatments compared to the control are shown in Fig. 3. For clarity, the presentation is limited to the control, the SSA, and TRAPPS treatments, since these showed the highest lead stabilization efficiency in the equilibrium study. Rate equations and nonlinear functions were fit to the data using the software package TableCurve 2D (SYSTAT, 2002). Arsenic concentrations in the control and the TRAPPS-amended neutralization sludge approached equilibrium within 8 h. In the SSA treatment no equilibrium was reached and concentrations clearly exceeded the control. Phosphate concentrations were around 320 mg/L compared to 37 mg/L in the TRAPPS batches corroborating competitive displacement of arsenic by excess phosphate from SSA. Lead concentrations decreasing over time indicate rate-limited stabilization mechanisms. Surprisingly, this holds for the control suggesting a natural attenuation process which may be related to an unexplained pH increase from 3.7 to 4.2. Lead equilibration was somewhat delayed in the control and TRAPPS treatment. A fast drop to steady state was observed in the SSA-amended samples. Irrespective

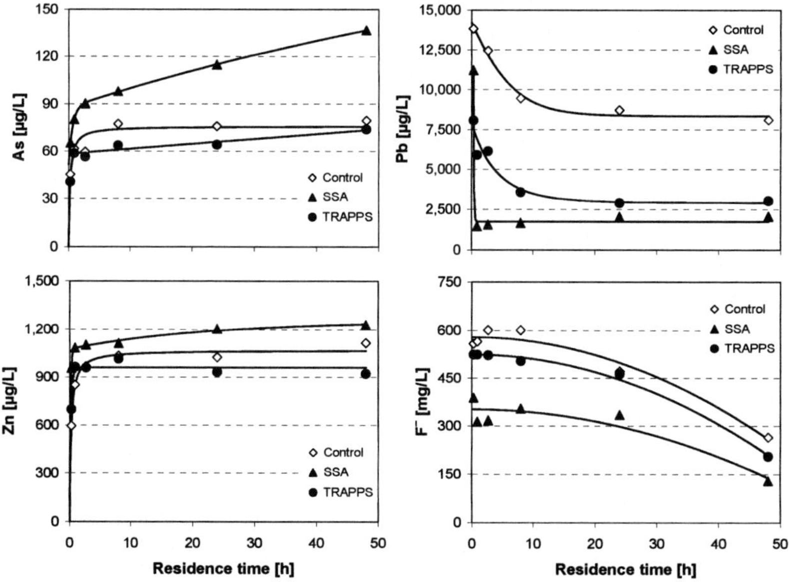

FIG. 3: Time-dependence of arsenic, lead, zinc, and total fluoride

of treatment type fast release and stabilization kinetics were also observed for zinc.
Also fluoride levels declined during the experiment possibly due to sorption to variable charge surfaces under acidic conditions and fluoride vs. hydroxide exchange in Apatite-type minerals of the treatments. Unlike the other contaminants, fluoride concentrations decreased slowly and exhibited a parabolic characteristic. Due to slow stabilization kinetics the fluoride WAC of 25 mg/L was not met.

CONCLUSIONS

Phosphatic additives show promise for lead stabilization in a highly contaminated glass polishing residue. Although in no case the waste acceptance criteria for non-hazardous waste were met, dose-dependence of lead levels suggests that amendment levels can be adjusted as long as a concomittant increase of phosphate leachate concentration does not pose an environmental problem (e. g. when the treated material is landfilled). In addition, further adverse side-effects may respond to increasing treatment intensity. This needs to be taken into account when customizing the amendment level. Fluoride seems to pose the greatest challenge for an effective treatment of the polishing sludge especially with respect to the slow stabilization kinetics. Future research should therefore consider the use of combined amendments capable of reducing both lead and fluoride concentrations.

ACKNOWLEDGEMENT

This study was funded by the Bavarian State Ministry of the Environment, Public Health and Consumer Protection within the R&D project "Chemical immobilization of inorganic contaminants in soils". We thank Zweckverband Neu-Ulm Steinhäule and Slater UK Ltd. for providing samples of the stabilizing agents.

REFERENCES

AbfAblV (2001). "Verordnung über die umweltverträgliche Ablagerung von Siedlungsabfällen und über biologische Abfallbehandlungsanlagen vom 20.02.2001 (Abfallablagerungsverordnung, AbfAblV) – Ordinance on environmentally compatible storage of waste from human settlements and on biological waste-treatment facilities of 02/20/2001." BGbl I, No 10, 305-324.

Bowell, R.J. (1994). "Sorption of arsenic by iron oxides and oxyhydroxides in soils." *Appl. Geochem.*, 9: 279-286.

Cao, X., Ma, L.Q,. Chen, M., Singh, S.P., Harris, W.G. (2002). "Impacts of phosphate amendments on lead biogeochemistry at a contaminated site." *Environ. Sci. Technol.*, 36: 5296-5304.

Cao, X., Dermatas, D., Shen, G., Ma, L.Q. (2005). "Lead contamination and immobilization at shooting range sites." In: A. Al-Tabbaa, J.A. Stegemann (Eds.), *Stabilisation/Solidification Treatment and Remediation*, Taylor & Francis Group, London, 315-321.

Conca, J., Wright, J. (2001). "Treatment of metal-contaminated leachates utilizing fish bones and fish hard parts." *United States Patent* 6,217,775, 8 pp.

CORIS (2007). "Clusterorientiertes regionales Informationssystem." *http://www.coris-online.de/app/primaer04_04.html*, last accessed 07/09/2007.

Crannell, B.S., Eighmy, T.T., Krzanowski, J.E., Eusden, J.D., Shaw E.L., Francis C.A. (2000). "Heavy metal stabilization in municipal solid waste combustion bottom ash using soluble phosphate." *Waste Management*, 20: 135-148.

Chrysochoou, M., Dermatas, D., Grubb, D.G. (2007). "Phosphate application to firing range soils for Pb immobilization: The unclear role of phosphate." *J. Haz. Mat.*, 144: 1-14.

Eighmy, T.T., Shaw, E.L., Eusden jr. J.D., Francis, C.A. (1998). "Chloropyromorphite ($Pb_5(PO_4)_3Cl$) by XPS: An environmentally important secondary mineral." *Surface Sci. Spectra*, 5: 122-129.

Glass Online (2007). "Illustrated Glass Dictionary." *http://www.glassonline.com-/infoserv/dictionary*, last access 07/05/2007.

Hartmann, R. (2002). "Erkundung einer ehemaligen Bleikristallglasfabrik in Nordostbayern." *GAB im Dialog*, 2: 4.

Levlin, E., Löwén, M., Stark, K. (2004): "Phosphorus recovery from sludge incineration ash and supercritical water oxidation residues with use of acids and bases." In: Plaza, E., Levlin, E., Hultman, B. (Eds.), *Integration and optimisation of urban sanitation systems*, TRITA-LWR.REPORT 3007, Report No 11, 19-28.

McGowen, S.L., Basta, N.T., Brown, G.O. (2001). "Use of diammonium phosphate to reduce heavy metal solubility and transport in smelter-contaminated soil." *J. Environ. Qual.*, 30: 493-500.

Nölle, G. (1977). "Technik der Glasherstellung." VEB Deutscher Verlag für Grundstoffindustrie, Leipzig, 200 pp.

Pocham, W. (1977). "Method of treating the fluorine containing mud produced by the acid polishing of lead glass so as to form basic lead carbonate." *United States Patent* 4,041,127, 4 pp.

Ryan, J.A., Zhang, P., Hesterberg, D., Chou, J., Sayers, D.E. (2001). "Formation of chloropyromorphite in a lead-contaminated soil amended with hydroxyapatite." *Environ. Sci. Technol.*, 35: 3798 -3803.

Smith, E., Naidu, R., Alston, A. M. (2002). "Chemistry of inorganic arsenic in soils: II. Effect of phosphorus, sodium, and calcium on arsenic sorption." *J. Environ. Qual.* 31: 557-563.

Spuller, C., Weigand, H., Marb, C. (2007). "Trace metal stabilisation in a shooting range soil: Mobility and phytotoxicity." *J. Haz. Mat.* 141: 378-387.

SYSTAT (2002). "TableCurve 2D 5.01 for Windows User's Manual." SYSTAT Software Inc., Richmond, CA, US, 613 pp.

Application of Two Novel Magnesia-based Cements in the Stabilization/Solidification of Contaminated Soils

Srinath Iyengar[1] and Abir Al-Tabbaa[2]

[1]PhD student in Engineering, University of Cambridge, Cambridge, UK; sri26@eng.cam.ac.uk
[2]Reader in Geotechnical Engineering, University of Cambridge, Cambridge, UK; aa22@eng.cam.ac.uk

ABSTRACT: This paper presents a preliminary investigation of the effectiveness of two novel magnesia-based cements in the stabilization/solidification of contaminated soils. The first is a reactive magnesia cement, which includes significant quantities of fly ash and in some cases also contains Portland cement. Magnesia cements are a recent invention and are claimed to have significant sustainability advantages over Portland cement. The second is a magnesia phosphate cement which consists of an acid phosphate source, such as potassium dihydrogen orthophosphate or triple super phosphate, and dead burned magnesia. It has mainly been used as refractory material, rapid-setting cement in a wide range structural repair works and has had limited application in stabilization/solidification treatments. Due to its low pH, it could have possible applications in facilitating biodegradation of organic contaminants in parallel to the stabilization/solidification of heavy metals in the form of very low solubility phosphates. The results presented here look specifically at the immobilization of zinc and lead in a model sand spiked with zinc chloride and lead nitrate respectively.

INTRODUCTION

Most stabilization/solidification (S/S) treatments use cementitious binders, most commonly Portland cement, in which the resulting high pH of the treated material, of around 11-13, leads to the immobilization of heavy metals in the form of metal hydroxides. However, there are a number of disadvantages to the use of Portland cement (PC). Firstly, the manufacture of PC is one of the most energy intensive of all industrial manufacturing processes and is responsible for up to 10% of anthropogenic CO_2 emissions worldwide (Sharp et. al., 2003). Hence there are continuous efforts to develop less environmentally damaging cements and although the quantities of cement used in S/S treatments would be small in terms of the global consumption of cement, there will be significant pressures placed on all sectors which use cement to adopt less environmentally damaging types. In addition, the high pH of cementitious S/S systems minimizes all biological activities leading to hindering of biodegradation activities of any organic contaminants present leading to limited suitability of those systems in the fixation of organic contaminants. Some organics have been reported to interfere with the hydration processes of PC resulting in reduced final strength and impaired stabilization, and hence facilitation of biodegradation of organics could be beneficial to the success of S/S treatments. In addition the facilitation of biodegradation would make S/S treatments a more sustainable and acceptable remediation option. Hence the paper considers two novel cements, one which addresses the first disadvantage and the other the second disadvantage, these are namely reactive magnesia cements and low pH phosphate cements respectively.

Reactive magnesia cements are a recently developed range (Harrison, 2001) in which the fundamental ingredient is reactive magnesia (MgO). They span from a modified PC range, which mainly includes PC and a small quantity of reactive magnesia, claimed to offer improved durability, to blends of MgO, a pozzolan (usually an industrial by-product e.g. pulverized fuel ash (pfa)) and, if necessary, a small quantity of PC, claimed to be a more sustainable than PC. This is because reactive magnesia is manufactured (from the calcining of magnesite) at much lower temperatures than PC (less than 850°C) and could potentially absorb large quantities of CO_2 making it carbon neutral (Harrison, 2003). Reactive magnesia cements are claimed to have many technical advantages relevant to S/S including reduced leachability, higher resistance to chemical attack and higher propensity for binding with waste inclusions (Sharp et. al., 2003).

Magnesia phosphate cements (MPCs) are formed by acid-base reaction, between dead burned magnesia and an acid phosphate source. The composition of these cements can be optimized so that the resulting pH is in the range of 5.5-8.5, i.e. the range required for biological activities. Although, metal hydroxides display different solubility profiles with pH and some reach solubility minima at this near neutral pH, there is a potential for the leaching of most cement-stabilized metals due to increased metal solubility (Rao et. al., 2000). However, MPCs immobilize the metals converting them into phosphates which have been reported to be insoluble under these pH conditions (Wagh, 2004). Also, MPCs have many relevant technical advantages including higher strength, higher durability, lower permeability and higher resistance to sulfate attack (Wagh, 2004).

This paper presents preliminary results from work investigating the effectiveness of those two magnesia-based cementitious binder systems, in comparison to PC, in the immobilization of a range of soils contaminated with heavy metals and organics. The specific aspect of the work being presented here is the investigation of the effectiveness of those two cements in the S/S of zinc and lead within a sand spiked with zinc chloride and lead nitrate respectively.

Reactive Magnesia Cement Binder Systems

The chemistry of reactive magnesia cements is fundamentally different from that of Portland cement (PC). Reactive magnesia hydrates to form brucite and carbonates to form nesquehonite (hydrated magnesium carbonate) (Vandeperre et. al., 2006) according to the following two reactions respectively:

$$MgO + H_2O \longrightarrow Mg(OH)_2 \quad (1)$$
$$Mg(OH)_2 + H_2O + H_2CO_3 \longrightarrow MgCO_3 \cdot 3(H_2O) \quad (2)$$

Brucite is far less soluble and reactive than Portlandite resulting in much reduced leachability and reactions with anions. Brucite is fibrous with a layered structure and is able to trap sulfate and chloride ions leading to high chemical resistance. This also results in a high propensity for binding with waste inclusions making it far less sensitive than PC to material impurities enabling it to mix more effectively with a range and higher content of ashes (Harrison, 2003). In the relatively porous high ash content magnesia cement, brucite can readily carbonate, under the correct curing conditions, to form mainly nesquehonite and gain significant strength (Vandeperre et. al., 2006). Nesquehonite has also been shown to be much more resistant to acid than the hydration products of PC (Liska, 2007).

Low pH Magnesia Phosphate Cement Binder Systems

Based on the phosphate source used, different variants of magnesium phosphate cements (MPCs) can be formed. One common phosphate source is potassium dihydrogen orthophosphate (KH_2PO_4) which forms magnesium potassium phosphate cements. The main reaction product of MgO with KH_2PO_4 is a hard, dense ceramic of magnesium potassium phosphate hexahydrate ($MgKPO_4.6H_2O$) (Wagh, 2004). For any small volume works, so as to reduce the reaction rate, the magnesia used is generally dead burned magnesia, which is a low reactivity magnesia form produced by calcining of magnesite ($MgCO_3$) at around 1300°C. However, the resulting reduced solubility of the calcined magnesia would still be too high for mixing larger volumes (Wagh, 2004). Therefore, pretreatment of MgO by retarder additions becomes necessary in order to decelerate the acid-base reaction and to lower the consequent heat generated. Whilst a reorganization of the MgO surface has been reported to occur due to calcination lowering its reactivity (Soudée and Péra, 2002), the addition of boric acid retarder has been reported to form a polymeric coating of low-solubility magnesium-boron-phosphate compound, called 'Lünebergite' ($Mg_3B_2(PO_4)_2.(OH)_6.6H_2O$) on the MgO particles prolonging its reaction with the acid phosphates in solution (Wagh, 2004).

MPCs have been reported to possess improved traits compared to PC such as higher early-age and long-term strengths (Abdelrazig et. al., 1989), higher freeze-thaw and wet-dry durability (Li et. al., 2004), lower permeability (El–Jazairi, 1987) and higher resistance to sulfate attack (Li et. al., 2004) which have made MPCs an attractive option for special applications. All these advantages are beneficial in the development of a more sustainable containment system and in particular in response to aggressive environmental conditions such as those likely to be imposed by future climate change scenarios. MPCs are used for a wide range of applications including refractory materials, dental cements, structural repair works and waste management. In waste management, MPCs have been reported to not only convert most of the hazardous contaminants into nonleachable phosphate reaction products, but also encapsulate these insoluble reaction products into a dense and durable matrix (Singh et al, 1998). Phosphates of these contaminants have a much lower solubility than their corresponding oxides or other salts (Rao et. al., 2000) at near-neutral pHs. Thus MPCs provide a very effective chemical immobilization and physical encapsulation of contaminants and, unlike PC, MPCs have also been reported to set even in the presence of hydrocarbons (Wagh, 2004).

In the work reported here, potassium dihydrogen orthophosphate, denoted as P, was used as the phosphate source. A much more cost-effective source of phosphate was also investigated for comparison. This is triple superphosphate, denoted as TSP, which is predominantly calcium dihydrogen phosphate ($Ca(H_2PO_4)_2.H_2O$) and also contains some gypsum ($CaSO_4.2H_2O$) and is a common fertilizer. One type of dead burned magnesia, DBM90 (Richard Baker Harrison, UK) with a citric acid reactivity of 1170mins, denoted as M, was used throughout and boric acid (H_3BO_3), denoted as B, was used as the retarder. Following some trials, ratios of P:M or TSP:M ranging from 8:1 and 1:5 were tested with B contents ranging from 0 to 1.5%. A range of cement paste mixes at their standard consistence, where appropriate, (water content 11-23%) were tested. Where it was not possible to test the standard consistence, primarily because of rapid setting of the mixes along with the hindrance caused to the penetration of Vicat needle by the excessive presence of TSP granules, a water content of 50% was used which was the value used when applying those cement mixes as grouts to the soil. Those mixes ranged from dense solid monoliths at the high M content end to highly fractured structures at the high P content end. Further details and properties of the cement pastes can be found elsewhere (Iyengar and Al-Tabbaa, 2007). The pH of the cement pastes for some of the mixes at 28 days curing time is shown in Figure 1 which clearly shows the lower pH produced by the MPC mixes, using both phosphate sources, compared to PC. The figure shows how the pH decreases as the quantity of the phosphate increases.

MATERIALS AND METHODS

Sand ($D_{50} \approx 0.8$mm), with a water content of 10% was used and was spiked with 3000mg/kg of zinc chloride as well as lead nitrate resulting in a theoretical concentration of Zn and Pb of 2310 and 2570mg/kg respectively. The reactive magnesia used was XLM (citric acid reactivity of 0.27mins), supplied by Causmag Ltd, Australia, and the dead burned magnesia was DBM90, supplied by Richard Baker Harrison Ltd, UK. The potassium dihydrogen orthophosphate and boric acid were of analytical grade supplied by Fisher Scientific, UK and the triple superphosphate granules was supplied by RS Minerals, UK. The PC used was CEM-I, supplied by Blue Circle, Lafarge, UK. The pfa was run-of-station obtained from Ratcliffe on Soar power station, Nottingham, UK. The ratio of the ingredient of the MPCs and that of the magnesia cement mixes can be see in Figure 4. Both cements were prepared as a grout with a water:cement ratio of 0.5:1 and mixed with the soil in a ratio of 3:7. The S/S treated contaminated sand were tested in

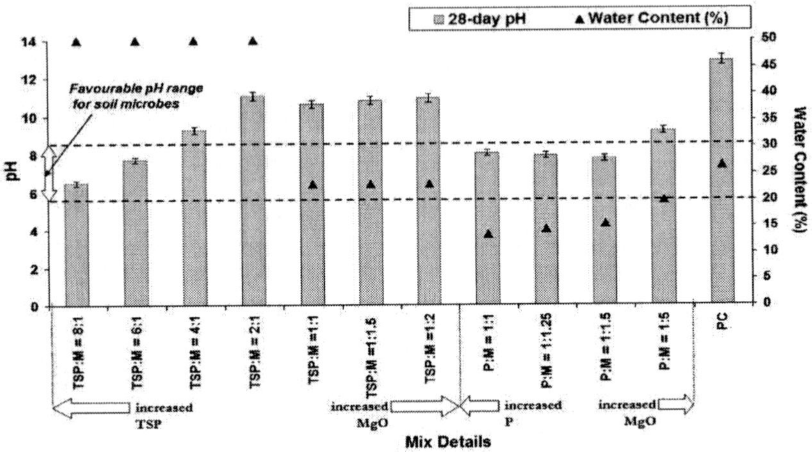

FIG. 1. 28-day pH of a range of the PC and the MPC pastes (all with B=1.5%).

duplicates using the BS EN 12457 batch leaching at 28 days (BSI, 2002). The test entails agitating 100g of the crushed stabilized/solidified soil sample (down to a size of <4mm) in a litre of water at pH of 5.6, on a rotating table for 24 hours and then analyzing the leachate for zinc and lead using an atomic absorption spectrophotometer.

RESULTS AND DISCUSSION

Figure 2 shows the leachate concentrations and leachate pH values of the Zn and Pb at 28 days in the MPC mixes, together with the PC mix. Figure 2 shows that the leached concentrations when using the MPC mixes is much lower than that produced by the PC mix for the pH range of 10.5-6.5. Although, the results show that stabilization was effective with the MPC mixes and was unaffected by the pH, it was difficult to compare the performance of individual mixes, since all the concentrations were at detection limit. This could further suggest the need to test higher initial concentrations of the contaminants to investigate these effects.

To develop more sustainable binders, industrial by-product inclusions in the form of pfa was incorporated into some of the MPC mixes. The 28-day leachate concentrations and pH of Zn and Pb from those mixes and those from the magnesia cement mixes are presented in Figure 3 as well as those of the PC-pfa mix for comparison. The figure shows very low concentrations of Zn and Pb for the MPC mixes, compared to the PC mix; which, are again at its detection limit. Moreover, the Zn and Pb leachability of PC-pfa mix was observed to increase which is in agreement with the findings which reported lower resistance of pfa-blended cement-based waste forms to heavy metal leaching under acidic medium (Poon et. al., 2001). This has been further attributed to their increased structural porosity. Conversely, the 50% replacement of PC with reactive MgO reduced the Zn leachability to its detection limit and nearly halved the Pb leaching value. When the PC was fully replaced with reactive MgO, the leachability of both the heavy metals dropped to its detection limits. It is thus evident that both MPCs and reactive MgO cements have a great potential for heavy metal immobilization in stabilization/solidification systems. Further work is currently being undertaken to develop a better understanding of the mechanisms involved and to assess the effectiveness of both cements in addressing organic as well as more complex type of wastes and their long-term related performance in comparison to PC-mixes.

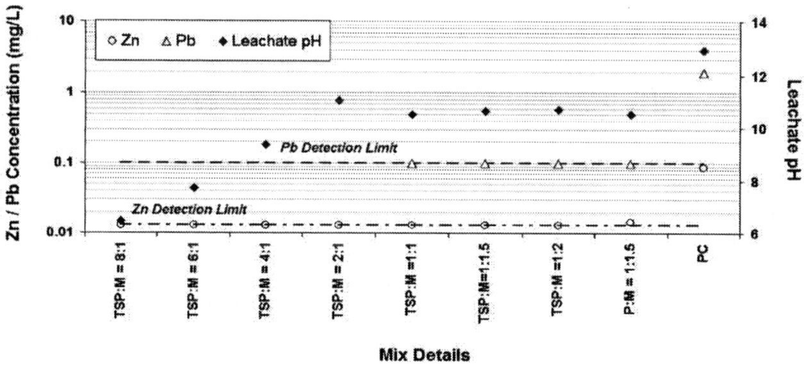

FIG 2. 28-day batch leaching results for Zn and Pb of the MPC mixes (all with B=1.5%) together with the PC mix.

The above results form a preliminary investigation into the applicability of the two magnesia-based cements in the S/S of contaminated soils. They form part of an extensive study into both cements which includes a detailed investigation of the S/S of a wide range of contaminants, the longevity of the treatment and issues related to field application in terms of relevant intrinsic properties and a detailed investigation of the wider technical and commercial advantage and disadvantages of both systems. For example, the metal phosphates formed by MPCs have a low solubilities at near-neutral pHs but their stability under acidic pH ranges could be a concern since lower pHs have been reported to favor dissolution of phosphates used in the stabilization of metals (Chrysochoou et al, 2007). Furthermore, the commercial use of reactive magnesia cements is still in its infancy, which reflects in their higher costs compared to PC.

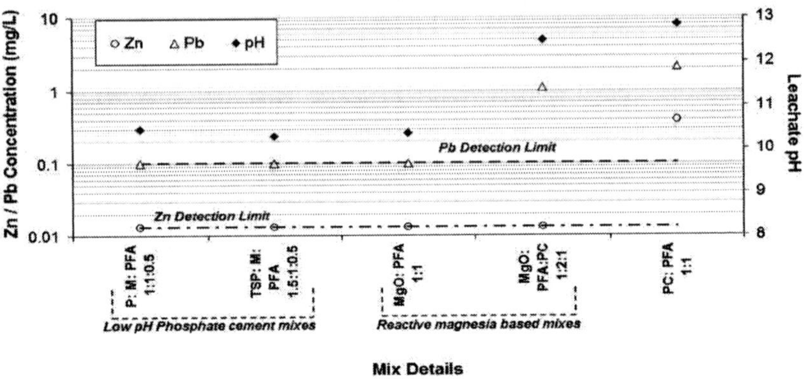

FIG 3. 28-day batch leaching results for Zn and Pb of some MPC mixes (all with B=1.5%), the reactive MgO mixes as well as a PC-pfa mix.

CONCLUSIONS

The paper presented details of two novel cement types, a reactive magnesia cement and a magnesia phosphate cement, which have the potential to offer a more effective and sustainable stabilization/solidification binder system compared to PC. By employing different proportions of two types of phosphate sources and dead burned magnesia, a broad range of low pHs between 6.5 and 9.2 was achieved by MPCs pastes in contrast to the high pH of ~13 developed by PC paste. Furthermore, the decrement in the pH as the phosphate content in the MPC mixes increased was quantified. The investigation of the 28-day leachability of Zn and Pb from a sand spiked with zinc chloride and lead nitrate respectively showed that both magnesia cements have the potential to significantly reduce the leached concentrations of both heavy metals down to their respective detection levels of 0.01 and 0.1 mg/L for Zn and Pb respectively. In contrast, the PC mix leached out 0.087 and 1.96 mg/L for Zn and Pb respectively and its performance further deteriorated with the inclusion of pfa.

ACKNOWLEDGMENTS

The authors are grateful for the funding provided by the UK Engineering and Physical Sciences Research Council and the Cambridge Trust. The authors are also thankful to Luc Vandeperre and Martin Liska for their assistance with some of the work presented.

REFERENCES

Abdelrazig, B. E. I., Sharp, J. H. and El-Jazairi, B. (1989). "The Microstructure and Mechanical properties of Mortars made from Magnesia-Phosphate Cement." *Cem. Concr. Res.*, Vol. 19: 247-258.

BSI. (2002). "Characterisation of waste leaching - Compliance test for leaching of granular waste materials and sludges." *BS EN 12457: Parts 1 to 4*, British Standards Institution, London, UK.

Chrysochoou, M., Dermatas,D. and Grubb, D. G. (2007). "Phosphate application to firing range soils for Pb immobilization:The unclear role of phosphate." *J. Hazard. Mater.*, Vol. 144: 1-14.

El–Jazairi, B. (1987). "The properties of hardened MPC mortar and concrete relevant to the requirements of rapid repair of concrete pavements." *Concr.*, Vol. 21: 25-31.

Harrison, J. (2001). "Reactive Magnesium oxide cements." *Patent WO 01/55049 A1*, W.I.P.O.

Harrison, J. (2003). "The Case for and Ramifications of Blending Reactive Magnesia with Portland Cement." *Paper presented on the 28th Conf. on Our World in Concrete and Structures*, Singapore.

Iyengar, S.R. and Al-Tabbaa, A. (2007). "Developmental Study of a low-pH Magnesium Phosphate Cement for Environmental Applications." *Environ. Technol.* (accepted).

Li, Z., Zhu, D. and Zhang, Y. (2004). "Development of Sustainable Cementituous Materials." *Proc. Int. Workshop on Sustainable Development and Concr. Technol.*, Beijing, China: 55-76.

Liska, M. (2007). Cambridge University Engineering Department, Personal Communication.

Poon, C. S., Lio, K.W. and Tang, C.I. (2001) "A Systematic Study of Cement/PFA Chemical Stabilisation/Solidification Process for the Treatment of Heavy Metal Waste." *Waste Manage. Res.*, Vol. 19: 276-283.

Rao, A.J., Pagilla, K.R. and Wagh, A.S. (2000). "Stabilization and Solidification of Metal-Laden Wastes by Compaction and Magnesium Phosphate-Based Binder." *J. Air Waste Manage. Assoc.*, Vol. 50: 1623-1631.

Sharp, J. H., Hill, J., Milestone, N.B. and Miller, E.W. (2003). "Cementitious Systems for Encapsulation of Intermediate level Waste." *Proc. ICEM '03: The 9th Int. Conf. on radioactive Waste management and Environmental Remediation.*

Singh, D., Wagh, S., Tlustochowicz, M. and Jeong, S.Y. (1998). "Phosphate ceramic process for macroencapsulation and stabilization of low-level debris wastes." *Waste Manage.*, Vol. 18: 135-143.

Soudée, E. and Péra, J. (2002). "Influence of magnesia surface on the setting time of magnesia-phosphate cement." *Cem. Concr. Res.*, Vol. 32: 153-157.

Vandeperre, L.J. and Al-Tabbaa, A. (2007). "Accelerated carbonation of reactive magnesia cements." *ICE J. Adv. Cem. Res.*, Vol. 19: 67-79.

Wagh, A.S. (2004). "Chemically Bonded Phosphate Ceramics." Elsevier Ltd, UK.

Solidification/Stabilization of PCB-Contaminated Wastewater Treatment Sludges

Ahmet H. Aydilek[1] and Tuncer B. Edil[2], Members, ASCE.

[1]Associate Professor, University of Maryland-College Park, Department of Civil and Environmental Engineering, College Park, MD, 20742, USA, aydilek@umd.edu
[2]Professor, University of Wisconsin-Madison, Department of Civil and Environmental Engineering, Madison, WI 53706, USA, edil@engr.wisc.edu

ABSTRACT: Treatment of polychlorinated biphenyl (PCB) contaminated sediments and sludges has been a challenge to the geotechnical and environmental community. Among various treatment options, solidification/stabilization (S/S) technology has often been considered. The technique can particularly be of interest for remediation of PCB-contaminated geomaterials as PCBs are quite insoluble and do not transport with water. The S/S approach requires knowledge of geotechnical properties of the treated waste and leaching of contaminants into the environment. The objective of this study was to investigate the applicability of the S/S technology to PCB-contaminated municipal sewage sludges at a U.S. EPA Superfund site located in Madison, Wisconsin. A battery of unconfined compression tests were conducted on the specimens prepared with various ratios of sludge and selected stabilizers (cement, fly ash, and lime) to study solidification. The effects of climatic stresses were observed by means of wet-dry and freeze-thaw tests on the solidified specimens. Consolidation behavior of one of the selected stabilized samples was also investigated. Moreover, PCB leaching from selected stabilized blocks was studied. The results indicated that successful encapsulation of the PCBs in the stabilized block of sewage sludges is possible by use of traditional binders.

INTRODUCTION

The retirement of large industrial waste storage facilities in accordance with environmental regulations has become a critical cost issue for industry and a challenge to geotechnical community. Many facilities were constructed prior to the emergence of modern environmental regulations and contain variety of contaminated high water content geomaterials. One of such facilities is located in Madison, Wisconsin. The Madison Metropolitan Sewerage District (MMSD) uses an activated sludge process to treat its sewage, and the main by-product of this process is sludge. This sludge has been spread onto the farmlands until U.S. EPA restricted its usage citing that it contains high concentrations of PCBs (polychlorinated biphenyls), which are detrimental to human health. Various remediation alternatives, including solidification/stabilization (S/S) have been evaluated for these sludges. The S/S is a remediation technology for the treatment of the hazardous wastes, and numerous studies have been conducted about its application to waste materials (e.g., Conner 1990, Means et al. 1995). However, it still remains as a waste and additive-specific process needs to be assessed for the MMSD sludges. The objective of this study was to determine the engineering properties of municipal sewage sludge relevant for the remediation alternative being considered. To meet this objective, a battery of unconfined compression tests was conducted on the specimens prepared with various ratios of sludge and selected admixtures to study solidification. The effects of the climatic stresses were observed by means of wet-dry and freeze-thaw tests on the solidified specimens. The settlement characteristics of one of the solidified mixtures were investigated in conventional consolidation tests. Leaching tests were conducted on one of the mixtures as well as raw sludge to see the effect of the stabilization process on PCB leaching.

MATERIALS AND METHODS

Two different types of PCB-contaminated wastewater treatment sludges obtained from the Nine Springs Wastewater Treatment Plant located in Madison, Wisconsin, were used in the study. The gravimetric water contents of the sludges A and B were 1300 and 300%, respectively. These water contents corresponded to solids content of 7 and 25%, respectively. The sludges contained 56-63% total volatile solids by weight. The nature of the sludge prevents systematic changes in the behavior of the sludge with changes in its water content and, measurements of the liquid and plastic limits were not possible. Similar difficulties with Atterberg limit tests have been reported by Genthe (1993) and Kraus (1994) for papermill sludges. The solid phase of the sludge contains 60-75% fines as determined by wash sieving the sludge on the U.S. No. 200 (75 μm) sieve. Specific gravity tests and total organic carbon (TOC) measurements indicated that the sludges have a mean specific gravity value of 1.6 and a TOC of 22%. Four different types of additives were used in the solidification process: Type I Portland cement, slaked lime, Class F fly ash, and Class C fly ash.

One of the common concerns in S/S is significant increase in volume of the treated waste as a result of curing. This issue becomes particularly important if the solidified waste is capped in situ. In order to quantify the volume changes, the sludge was mixed with the appropriate additive in a glass beaker and transferred into four 85-mm diameter PVC molds. Approximately two thirds of the molds were filled with the mixture of interest. One of the specimens was cured at 100% relative humidity and controlled temperature ($21 \pm 2\ ^{o}C$) for 28 days and its final volume was measured. The other three remained as control specimens and average of their volumes was calculated and compared with that of the cured mixture for volume change calculations.

A useful property to define the success of the S/S process is strength of the solidified block. This parameter can be critical in evaluating the stability of the landfill that sludge is placed in, or to determine the loads that can be successfully placed onto the solidified sludge in a capping application for insitu treatment. For strength tests, the sludge samples were mixed with the appropriate additive and transferred into PVC molds 76.2 mm in diameter and 152.4 mm in height in 3 layers of approximately equal volume. Each layer was rodded with a 10-mm diameter, 30.5-long steel rod using 25 strokes per layer. The molds were covered with plastic sheets that had several holes in them, and cured for 28 days at 100% relative humidity and controlled temperature ($23 \pm 1\ ^{O}C$) before testing. The specimen preparation and unconfined compression testing generally followed the procedures listed in ASTM C 192 and D 2166, respectively. In order to evaluate the effect of climatic stresses on the treated sludges, three mixtures-- sludge/Class F fly ash (1/1), sludge/Class F fly ash + cement (1/0.75/0.25), and sludge/cement (1/0.5)-- were subjected to unconfined compression tests after a series of wet-dry and freeze-thaw cycles. The durability testing generally confirmed with the procedures outlined in ASTM D 559 and D 560. The sludge-to-binder ratios utilized in the current study were chosen based on preliminary laboratory tests that included vicat needle (ASTM C 191), pocket penetrometer and density measurements.

The solidified wastes are often placed in hazardous waste containment facilities, or capped in situ in the lagoon where the waste has originally existed. Such a cap would invariably induce stresses in the sludges and cause consolidation. To evaluate the settlement characteristics of the solidified sludge under vertical loads, two conventional consolidation tests were conducted on 28-day cured specimens of sludge/Class F fly ash (1/1) mixtures, following the procedures outlined in ASTM D 2435. Sludge B was used in all tests. This mixture was chosen for consolidation testing because it was expected to have the highest compressibility among the feasible mixtures determined from a consideration of strength, volume increase and cost. The tests were conducted in fixed-ring consolidometers by loading the specimens from 12.5 kPa to 800 kPa in every 24 hours with a load increment ratio of 1. Sequential unloading of the specimens to 25 kPa was undertaken after consolidation under 800 kPa.

Finally, TCLP (Toxicity Characteristic Leaching Procedure) tests were conducted at Wisconsin State Hygene Laboratories to study the effect of stabilization activity on leaching of PCBs. Tests were conducted on 28-day cured specimens of sludge/Class F fly ash/cement (1/0.75/0.25) mixture. This mixture was chosen due to its cost-effectiveness as it included large amounts of fly ash and small amounts of cement and exhibited sufficient strength (>345 kPa) during laboratory tests (see Results and Discussion section). TCLP tests were also conducted on raw sludge for comparison purposes. All tests were performed following the testing protocols outlined in EPA method 1311.

RESULTS AND DISCUSSION

The volume increases which occurred as a result of the solidification activity are presented in Figure 1. Both types of fly ashes cause large volume increases. The use of the two fly ashes in the same mixture also increases the volume, which is inevitable, but does not cause a significant change on the solids content because of the lack of strong reactions between the two (i.e., Class C +cement 0.5/0.5 versus Class F and C 0.5/0.5). Another significant volume increase is observed for the Class

F fly ash/lime mixtures. The volume increase experienced by mixing fly ash and lime with Sludge A at ratios of 1/1 is 22% more than the one experienced when these additives were mixed with sludge at a ratio of 0.5/0.5. Similar observations, with a 25% more volume increase, can be made for sludge B mixtures. In spite of having wide applications in solidification of waste materials due to its cost-effectiveness, the greatest advantage of the cement-fly ash additives is the volume increase. The ranges of increase observed in this study with cement-fly ash are 38-54% and 47-65% for sludges A and B, respectively. These ranges are highly comparable with the ones reported by Conner (1990).

The data in Figure 1 also suggests that solids content of the mixtures prepared with Class F fly ash/lime, lime, and cement increases with increasing additive-to-sludge ratio. For instance, when the additive ratio is doubled, the solids content of these mixtures prepared with sludge A increased by 8, 36, and 77%, respectively. These increases are 6, 35, and 13%, respectively, for the mixtures prepared with sludge B. Table 1 shows that both type of the additive as well as the solids content play an important role in strength. The U.S. EPA requires an unconfined compressive strength of 345 kPa for solidified/stabilized materials so that the material can

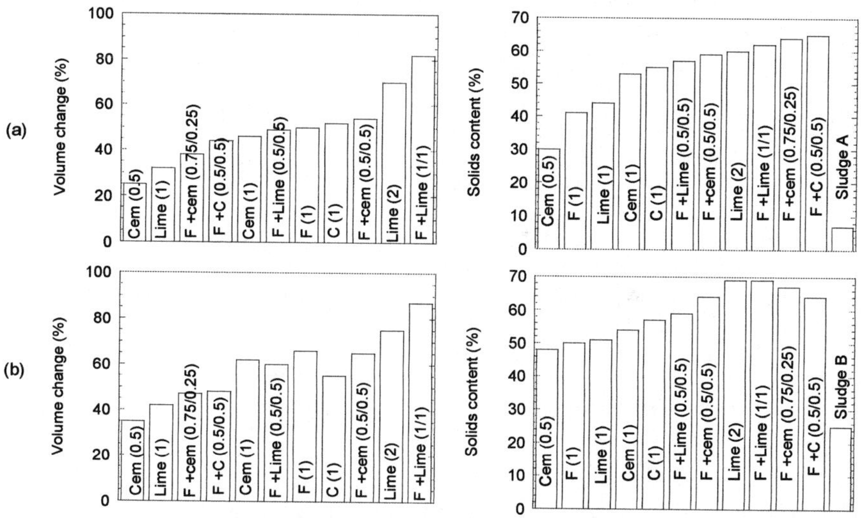

Figure 1. Contribution of additives to the increase in solids content and volume of (a) sludge A (water content = 1300%), and (b) sludge B (water content = 300%). The additive-to-sludge ratios are given in the parentheses. Cem=Cement, F=Class F fly ash, C=Class C fly ash.

withstand external loads during construction (e.g., construction equipment), and can act as a stable base material for the permanent loads placed over it (i.e., caps and covers). The data in Table 1 suggest that only the specimens containing cement had strength above this limit, consistent with the findings of previous researchers (Forslund et al. 1989, Conner 1990, Knox and Najar 1992).

Table 1. Unconfined compression test results

Additive (Binder-to-sludge ratio)	Sludge A (kPa)	Sludge B (kPa)
Class F fly ash + Lime (0.5/0.5)	10	15
Class F fly ash (1)	33	83
Class F fly ash + Lime (1/1)	39	54
Class F fly ash +Class C fly ash (0.5/0.5)	55	100
Class C fly ash (1)	103	91
Cement (0.5)	395	223
Class F fly ash + Cement (0.75/0.25)	572	570
Class F fly ash + Cement (0.5/0.5)	1479	1003
Cement (1)	2730	1686

Class F fly ash/lime reactions are analogous to those of cement, and form crystalline structures that result in a hardened product. However, these reactions are generally slower and the strength values are much lower than those of other processes (Conner 1990). Sharma and Lewis (1994) reported strengths of 140 to 7,000 kPa for sludge/lime/Class F fly ash mixtures when the lime-to-pozzolan ratio varies between 1 and 6. A ratio of 1.0 was used in the current study. Desired strength gain may eventually be achieved by increasing the lime (and fly ash) content; however, this may result in volume increases that may not be appropriate for in situ capping or landfilling.

Climatic stresses have been shown to have detrimental effects on strength of stabilized soils and sludges. An approach subjecting specimens to strength tests after freeze-thaw and wet-dry cycles and recording the change in unconfined compressive strength (q_u) has been used as indicator of durability (Means et al. 1995, Arora and Aydilek 2005). The durability test specimens in this study were prepared with sludge B. Among the specimens resulted in a q_u of higher than 345 kPa meeting the EPA rule and requiring the least cement contents, i.e., Class F fly ash + cement (0.75/0.25), was chosen for the durability tests. A sludge/cement (0.5) specimen was also tested to provide a comparison. Furthermore, a sludge/Class F fly ash mixture was prepared at a ratio of 1/1, as it may be cost-effective to use this mixture in applications where the sludge is subjected to low stresses, e.g., light-weight caps.

The data in Table 2 indicates that the sludge/cement (1/0.5) mixtures withstood best over the twelve wet-dry or freeze-thaw cycles. The control specimens experienced comparable strength loss with their sludge pairs, indicating that the sludge has been successfully solidified and turned into monolith blocks. It is of interest that the sludge/Class F fly ash (1/1) specimens seemed to lose only 13% of their initial strength over twelve cycles of wetting and drying. The sludge/Class F fly ash/cement (1/0.75/0.25) specimen experienced the greatest strength loss; however, its final strength is still higher than the strength of the sludge/Class F fly ash (1/1) specimen (73 kPa versus 83 kPa). The durability test results suggest that the solidified materials would be affected by climatic stresses, and it may be necessary to protect them during in situ remediation. However, it should be noted that the laboratory 24-hour cycles of wetting-drying and freezing-thawing represent much harsher conditions than the ones in the field.

Table 2. Effect of freeze-thaw and wet-dry cycles on unconfined compressive strengths

Additive (Binder-to-sludge B ratio)	%decrease in q_u after freeze-thaw cycles		%decrease in q_u after wet-dry cycles	
	After 3 cycles	After 12 cycles	After 3 cycles	After 12 cycles
Class F fly ash (1)	58	86	NA	13
Class F fly ash (Control)	NA	57 (after 6 cycles)	NA	78 (after 5 cycles)
Class F fly ash + Cement (0.75/0.25)	48	88	75	85
Class F fly ash + Cement (Control)	NA	88	NA	80
Cement (0.5)	24	51	35	48
Cement (Control)	42 (after 6 cycles)	84	NA	58

Note: NA=Not analyzed

Figure 2 shows that void ratio decreases for the sludge B/Class F fly ash (1/1) mixtures with increasing effective stress, and the shape of the trend is similar to the one observed for natural clays. The rate of compression decreased with time under a given load (not shown herein) as observed in clays; however, the coefficient of compressibility (a_v) was generally lower at high consolidation stresses and ranged from 7 x 10^{-5} to 8.6 x 10^{-4} m^2/kN (Table 3).

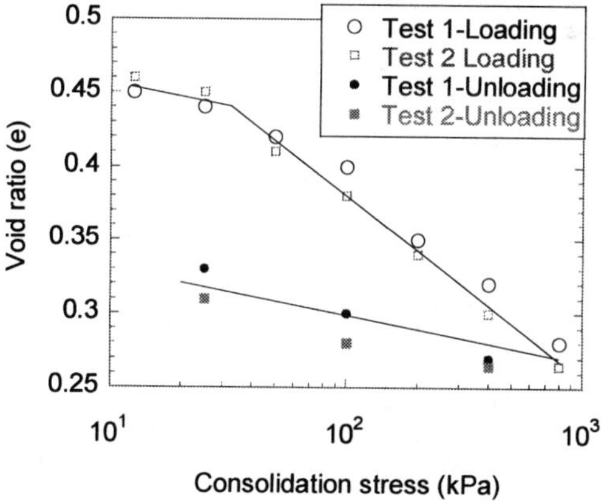

Figure 2. Void ratio-consolidation relationship for the sludge/Class F fly ash (1/1) mixture

While the mechanisms of compression may not be the same for solidified sludge and clay, the similarity in observed consolidation trends is noteworthy and it was felt that the use of one-dimensional consolidation analysis would be appropriate. Table 3

Table 3. Summary of the consolidation test results

Test No.	Consolidation stress (kPa)	c_{vT} (m^2/year)	c_{vC} (m^2/year)	c_{vT}/c_{vC}	a_v x10^{-4} (m^2/kN)
1	25	1.85	1.91	0.97	8.6
	50	5.42	5.4	1.0	5
	100	4.75	3.73	1.27	4.6
	200	1.14	1.14	1.0	0.9
	400	3.71	3.25	0.91	1.7
	800	7.7	7.32	1.12	0.6
2	25	1.6	1.74	0.93	3.1
	50	1.62	1.64	1.0	4.6
	100	3.49	3.39	1.1	2.3
	200	3.02	1.97	1.54	2.2
	400	3.39	3.15	1.1	0.8
	800	7.0	6.63	2.04	0.7

Note: c_{vC} and c_{vT} are coefficient of consolidation based on Casagrande and Taylor methods, respectively.

summarizes the coefficients of consolidation (c_v) corresponding to applied loads calculated in accordance with the Casagrande and Taylor methods (Duncan 1993). c_v varies from 1.14 to 7.7 m^2/year for the solidified sludge, and these values are higher than the ones generally observed for natural clays. Holtz and Kovacs (1981) reported that c_v values range from 0.1 to 3 m^2/year for most clays; however, they also indicated that it is not uncommon to see c_v values as high as 12 m^2/year for high plasticity (CH) clays. Duncan (1993) analyzed the results of 33 consolidation tests conducted on CH soils and reported c_v values in a range of 0.1 to 23.3 m^2/year. The c_{vT}/c_{vC} ratio ranged from 0.8 to 3.43, comparable with the range of data observed in the current study.

The TCLP tests suggested that there is approximately ten times reduction in the amount of leached PCBs as a result of the stabilization process. The leached PCB amount (dry basis) was 16, 420 nanograms/g for the raw sludge whereas it decreased to167 nanograms/g for the sludge/Class F fly ash/cement (1/0.75/0.25) mixture. TCLP tests have been so far conducted on one type of mixture. It is possible that higher PCB amounts can be observed in testing of other mixtures, particularly the ones that do not include cement or lime. However, the observed difference is noteworthy and highly encouraging for future stabilization activities. The authors are planning to conduct a similar array of leaching tests on the remaining mixtures for comparison purposes.

CONCLUSIONS

A study was conducted to investigate the feasibility of the solidification/stabilization of PCB-contaminated wastewater treatment sludges in

Madison, Wisconsin. A concept of solidification was considered in which the stiffness and strength of the sludge is enhanced to a desired level to allow construction on it using the most economical means. The results indicated that volume increase as a result of the solidification activity can be significant and ranges from 25 to 88%. Among the mixtures tested, the sludge/Class F fly ash/cement (1/0.75/0.25) mixture seemed to provide the most cost-effective solution. This mixture included the minimum amount of cement as an additive, and was one of the few ones that had strength above the limiting value of 345 kPa set by the U.S. EPA for solidified products. Even though the mixture experienced the greatest strength loss, its final strength was still higher than the strength of a mixture prepared solely with Class F fly ash, i.e., sludge/Class F fly ash (1/1). Furthermore, the mixture exhibited ten times lower PCB concentrations than raw sludge in TCLP tests. Based on visual observations and strength data, sludge/Class F fly ash (1/1) was chosen for consolidation testing as it was expected to have the highest compressibility among the mixtures. Moreover, low strength mixtures that are prepared with waste materials, such as fly ash, can often be useful especially if the solidified materials will later be capped in situ. The consolidation tests conducted to evaluate the settlement characteristics of the solidified sludge for capping applications showed that the mixture behaved like a CH soil, and the predictive tools commonly used for natural clays may be applicable for such soft solidified sludges.

REFERENCES

Arora, S. and Aydilek, A.H. (2005). "Class F fly ash amended soils as highway base materials." *J. of Materials in Civil Engrg.*, 17(6):640-649.

Conner, J.R. (1990). Chemical fixation and solidification of hazardous wastes. Van Nostrald Reinhold: 280-247.

Duncan, J.M. (1993). "Limitations of conventional analysis of consolidation settlement." *J. Geotechnical and Geoenvironmental Engrg.*, 119 (9): 1333-1359.

Forslund, B.L., Smith, L.J.S., and Bergstrom, W.R. (1989). "A physical testing program for stabilized metal hydroxide sludges." *Environmental Aspects of Stabilization and Solidification of Hazardous and Radioactive Wastes, ASTM STP 1033*, P.L. Cote and T.M. Gilliam Eds.:214-226.

Genthe, D.R. (1993). Shear strength of two pulp and paper mill sludges with low solids content. *M.S. Thesis*, Dept. of Civil Engineering, University of Wisconsin, Madison, Wisconsin.

Holtz, R.D. and Kovacs, W. (1981) An introduction to geotechnical engineering, Prentice-Hall, New Jersey

Knox, D.P. and Najar, R.A. (1992). "Solidification and stabilization of lagoon sludge and dredge spoils." *Proc. 1st Intl. Congress on Environmental Geotechnics, Edmonton, Canada*: 41-47.

Kraus, J.R. (1994). Effect of freeze-thaw on the hydraulic conductivity of barrier materials: laboratory and field evaluation. *M.S. Thesis*, Dept. of Civil Engineering, University of Wisconsin, Madison, Wisconsin.

Means, J.L., Smith, L.A., Nehring, K.W., Brauning, S.E., Gavaskar, A.R., Sass, B.M., Wiles, C.C., and Mashni, C.I. (1995). The application of solidification and stabilization to waste materials. Lewis Publishers.

Ferrite-Induced Immobilization of Pb-Contaminated Soil and Application of Magnetic Separation

Kiyoshi Omine[1], Hidetoshi Ochiai[1], Yuta Tajiri[1], Noriyuki Hirota[2] and Hidehiko Okada[2]

[1] Department of Civil Engineering, Kyushu University, 744 Motooka, Nishi-ku, Fukuoka 819-0395, Japan; oomine@civil.kyushu-u.ac.jp
[2] National Institute for Materials Science, 1-2-1 Sengen, Tsukuba, Ibaraki 305-0047, Japan

ABSTRACT: The ferritization method is applied to Pb-contaminated soil. In order to confirm the immobilization effect by ferritization, artificial Pb-contaminated soil was prepared and batch tests were performed. From the test results, it was confirmed that the insoluble effect of the ferritization-treated soil for Pb-contaminated soil is much higher over a wide range of pH levels. Furthermore, the superconducting magnetic separation was applied to feirritization-treated soil with magnetite. It is suggested that the combination of ferritization treatment and magnetic separation may be one of the techniques for the remediation of contaminated soil containing heavy metals.

INTRODUCTION

Several techniques are used for the removal of heavy metals. The ferrite-method has been widely used for the disposal of effluents containing heavy metals in Japan (Okamoto 1976 and Takada 1977). Ferrite is a compound oxide comprising iron and heavy metals and it exhibits the property of ferromagnetism. It is known that the solubility of ferrite is very low and the ferrite method is applicable to effluents containing many kinds of heavy metals.

In this study, the ferrite method is applied to Pb-contaminated soil (Omine et al. 2006). The immobilization effect of ferritization treatment for the contaminated soil is investigated by an elution test under different pH conditions. The applicability of the ferrite method and magnetic separation for the remediation of the contaminated soil is discussed based on the test results.

FUNDAMENTAL CONSIDERATION OF THE FERRITE METHOD

Ferrite is a compound oxide comprising iron and metal. It is known that the solubility of ferrite is very low and it has ferromagnetism. Ferrite can be formed by two methods, namely, oxidation and coprecipitation. In the oxidation method, generally, white deposits of a hexagonal tabular crystallite $Fe(OH)_2$ are formed by adding an alkali to an aqueous solution of Fe^{2+}. If there exists a heavy metal ion, M, then a compound hydroxide containing Fe and M is deposited during the deposition process. In other words, the compound hydroxide is deposited by the following process:

$$xM^{2+} + (3-x)Fe^{2+} + 6OH^{-} \rightarrow M_xFe_{3-x}(OH)_6 \tag{1}$$

The compound hydroxide containing Fe and M exhibits weak magnetism. The compound hydroxide transforms into a ferrite during the oxidation process by introducing air bubbles at a temperature of more than 60°C. This chemical reaction is represented as follows:

$$M_xFe_{3-x}(OH)_6 + \frac{1}{2}O_2 \rightarrow M_xFe_{3-x}O_4 \tag{2}$$

On the other hand, a coprecipitation method does not involve oxidation and heat processes has been proposed. Namely, when adding an alkali to a solution comprising heavy metal ions M^{2+} and Fe^{3+}, another type of ferrite is formed by the following process:

$$M^{2+} + 2Fe^{3+} + OH^{-} \rightarrow MO \cdot Fe_2O_3 \tag{3}$$

To conduct a fundamental study on the remediation of contaminated soil with heavy metal, the insoluble effect of the ferrite method is investigated in the following section.

IMMOBILIZATION OF LEAD- CONTAMINATED WATER AND SOIL

a) Preparation and Experimental Method

The process of ferrite formation by the oxidation method is shown in Fig.1. Pb-contaminated soil was prepared by adding a standard solution of Pb for atomic absorption analysis (1000 mg/l) into Kaolin clay with a Pb content of 300 mg/kg-dry soil. The equipment used for ferrite formation is shown in Fig.2. First, a prescribed volume of 0.2 mol/l iron sulfate, $FeSO_4$, was added to the container, and sodium hydroxide, NaOH, was mixed with a pH level greater than 10. The compound hydroxide containing Fe and Pb can be producted by agitating the mixtures. Furthermore, ferrite is produced by heating the compound hydroxide to a temperature of 60~70°C and bringing about oxidation by slowly introducing air bubbles for maintaining high alkalinity. The sample of soil obtained from these processes is referred to as the "ferritization-treated soil" in this study. In addition, an ideal ferrite sample was also prepared in an aqueous solution without clay for the purpose of comparison. As indicated in the reactions represented by Eqs. (2) and (3), when the heavy metal M is lead, $2Fe^{2+}$ and Pb^{2+} combine. Therefore, we use the ratio $2Fe^{2+}/Pb^{2+}$ in mol herein. In this test, the ferritization treatment was performed under the additive condition of $2Fe^{2+}/Pb^{2+}=40$. Pb-contaminated water was treated in the same manner.

X-ray diffraction was used for the analysis of the Kaolin clay, ferritization-treated soil, and the ferrite formation. The elution test was performed for analyzing the contaminated soil and the ferritization-treated soils.

In order to investigate the insolubility property of these samples, a batch test was performed under different pH conditions ranging from strongly acid to strongly alkaline and a liquid-solid ratio of 10:1. The specimens were agitated in a reciprocating shaker for 6 h, and the supernatant was separated by centrifugation. The concentration of Pb in the filtrates after passing through 0.45 μm membrane pore size filters was analyzed by atomic absorption photometry.

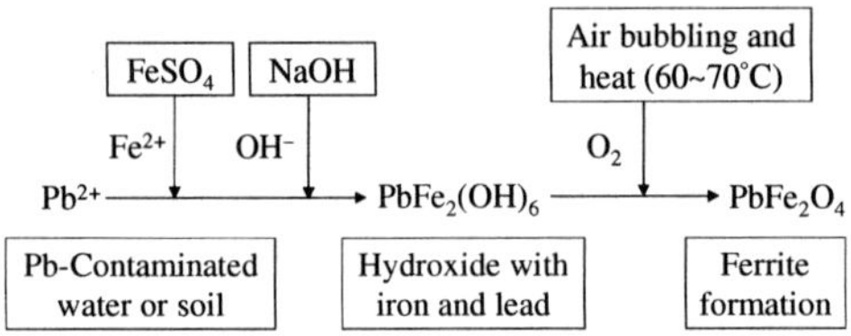

FIG. 1. Process of ferrite formation by oxidation method.

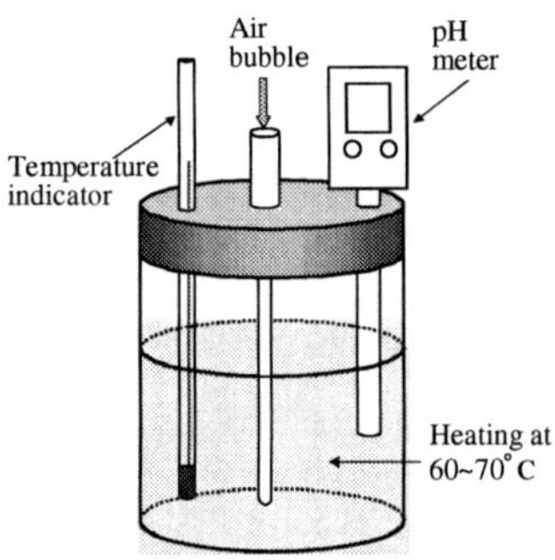

FIG. 2. Equipment for ferrite formation.

b) Confirmation of Ferrite Formation

Since ferrite is a ferromagnetic material, the application of magnetic separation is considered. For investigating the magnetism, we observed the state of the ferritization-treated soil when placed close to a permanent magnet, as shown in Fig.3. The ferritization-treated soil exhibits ferromagnetism and this indicates the existence of ferrite formation in the soil.

Figure 4 shows the results of the X-ray diffraction of each sample by ICDD-PDF2 database. The ideal ferrite formation prepared in the aqueous solution without clay has several peaks corresponding to the compound oxide comprising iron and lead as well as sodium sulfate Na_2SO_4. In fact, it is found that not only the ferrite, $PbFe_2O_4$, but also several other compound oxides comprising iron and lead are produced. On the other hand, it is indicated that the main components of Kaolin clay are silica, SiO_2, and pyrophyllite, $Al_2Si_4O_{10}(OH)_2$. The peak values of ferritization-treated soil are similar to those of Kaolin clay. This is because the ferritization-treated soil contains many constituents of the Kaolin clay. However, there are a few small peak values that correspond to the compound oxide comprising iron and lead.

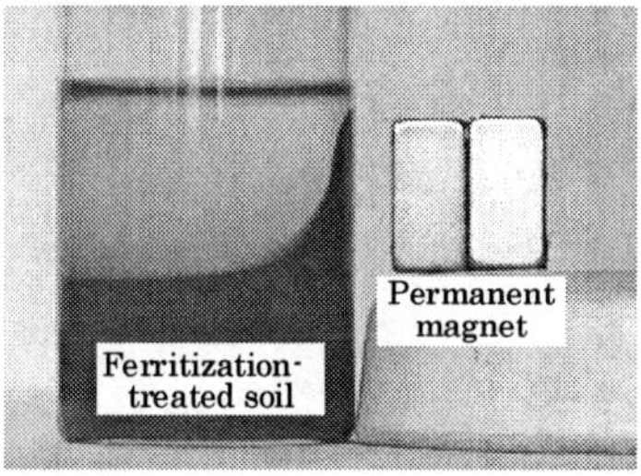

FIG. 3. The ferritization-treated soil when kept close to a permanent magnet.

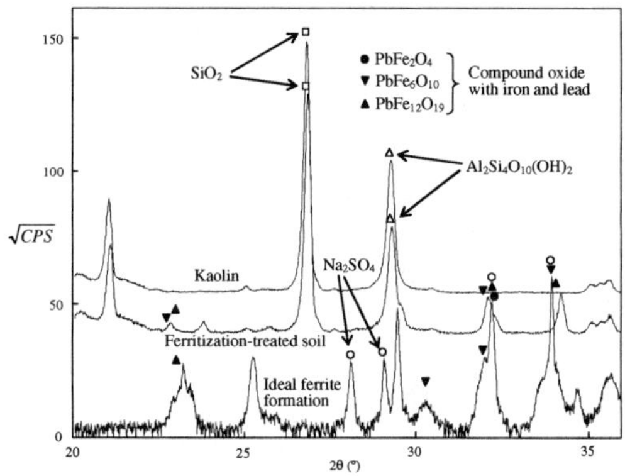

FIG. 4. Results of X-ray diffraction of each sample.

c) Immobilization by Ferritization Treatment on Pb-Contaminated Water and Soil

The immobilization property of the ferritization treatment by the oxidation method on Pb-Contaminated water is shown in Fig.5. Since lead is an amphoteric element, the amount of release of Pb released in contaminated water without treatment becomes very large in both conditions of strongly acidic and strongly alkaline conditions. However, the solubility of Pb following the ferrite treatment decreases remarkably. In particular, the concentration of Pb is less than that established by environmental quality standard when pH lies in the range of 6~11. It was confirmed that the ferrite method is very effective for the immobilization of Pb-contaminated water.

Figure 6 shows the immobilization effect of the ferritization treatment by the oxidation method on Pb-Contaminated soil. The concentration of Pb in the ferritization-treated soil is very low in comparison with that of the contaminated soil without treatment. In particular, the amount of Pb released is controlled in the presence of a strong alkali. It was also confirmed that a similar immobilization effect has been obtained for the ferritization treatment by the coprecipitation method. Thus, the ferrite method can be applied to the Pb-contaminated soil by adding sufficient ferrous ions.

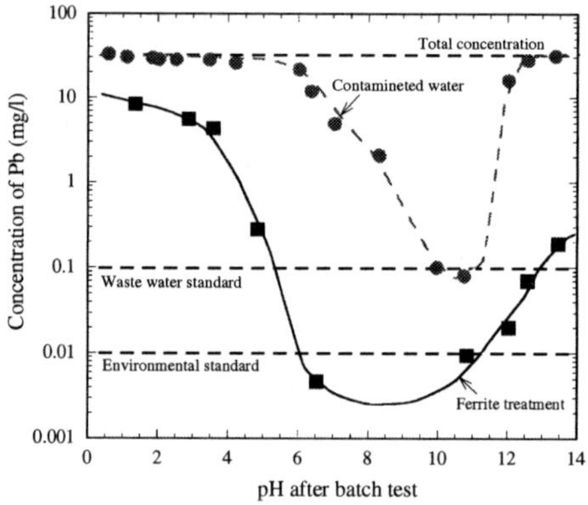

FIG. 5. Immobilization property of the ferritization treatment by oxidation method on Pb-Contaminated water.

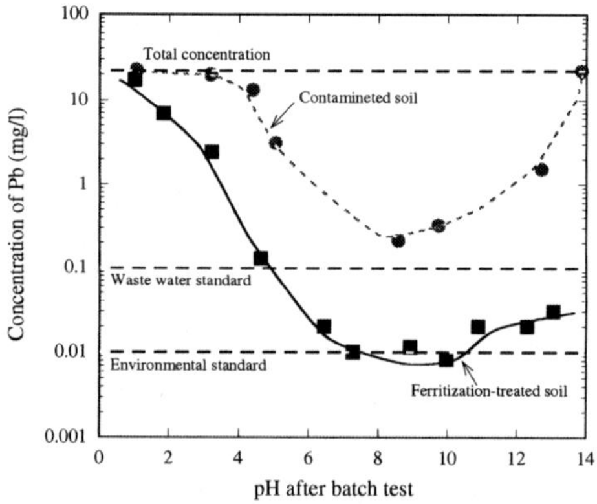

FIG. 6. Immobilization property of ferritization treatment by oxidation method on Pb-Contaminated soil.

APPLICAION OF SUPERCONDUCTING MAGNETIC SEPARAION

a) Concept of Magnetic Separation

A superconducting magnet enables a high gradient magnetic separation. The concept of magnetic separation systems is that ferromagnetic particles are retained by a magnetic filter when a high gradient magnetic field is applied. This magnetic separation system has been applied to environmental fields such as in the removal of heavy metal ions or purification of waste water.

b) Test Method

It is easier to produce the ferrite formation by the coprecipitation method using ions of Fe^{2+} and Fe^{3+} than by the oxidation method with the processes of air bubbling and heating. The ferritization treatment by the coprecipitation method was used for the application of magnetic separation. It was expected that a large amount of clay particles would be obtained as a by-product of ferrite formation. Therefore, a certain amount of magnetite is added into the contaminated soil as adsorptive particulates. The mixing conditions of Pb and Fe are listed in Table 1 and the process of ferritization treatment using adsorptive particulates is shown in Fig.7. The Pb-contaminated soil was prepared in the same manner as that mentioned above and the Pb content was adjusted to 1000 mg/kg-dry soil. The total mixing ratio of Pb and Fe in mol is 80 for all samples of Case-1~Case-3.

The schematic diagram of the superconducting magnet apparatus is shown in Fig.8. The sample in the slurry is passed through several magnetic filters. The inflow speed of the sample is approximately 40 ml/min and the magnetic field intensity is 1 T.

Table 1. Mixing conditions of Pb and Fe

	Pb content (mg/kg-dry soil)	Mixing ratio of Pb and Fe in mol		
		Pb	Fe in magnetite particulates	Fe in iron sulfate on ferritization treatment
Case-1	1000	1	0	80
Case-2	1000	1	40	40
Case-3	1000	1	60	20

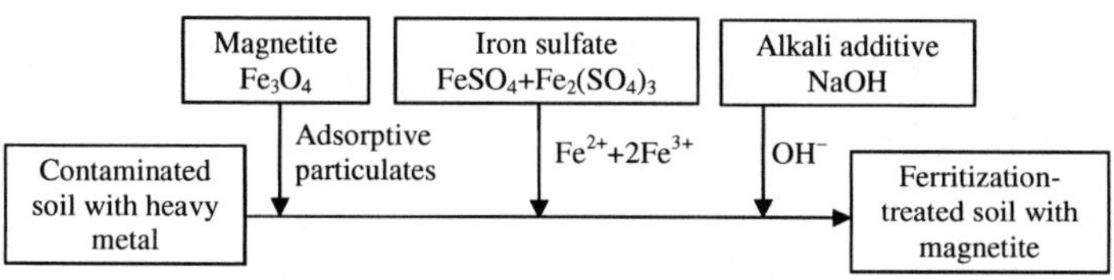

FIG. 7. Process of ferritization treatment by the coprecipitation method and by using adsorptive particulates.

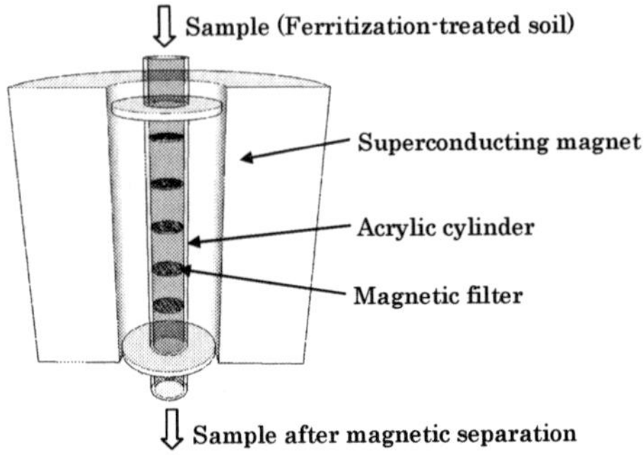

FIG. 8. Schematic diagram of superconducting magnet apparatus.

c) Test Results and Discussions

Pictures of the samples before and after magnetic separation are shown in Fig.9. The sample of Case-1 without magnetite is still dark after the magnetic separation. The samples of Case-2 and Case-3 with magnetite appear similar to Kaolin clay and are whitish. This indicates that the magnetite plays the part of an adsorptive material to realize ferrite formation.

In order to confirm the effect of magnetic separation, a test for lead content was performed on the samples after they passed through the magnetic filters. The content was measured by extraction in hydrochloric acid (1 mol/L). Although this is not a strict method, the accuracy of the measurement has been confirmed by the preliminary test. Figure 10 shows the test results of the Pb content the samples after magnetic separation. The Pb-content in Case-1 without magnetite decreases to approximately 300 mg/kg and the rate of removal for lead is 70%. On the other hand, the Pb-content in Case-2 and Case-3 with magnetite becomes lesser than 100 mg/kg and thus satisfies the environmental quality standard of content in Japan.

It was observed that a large amount of clay particles were retained by the magnetic filters together along with the magnetite. In further studies, the effective test conditions for the removal of Pb should be clarified.

Before After (Case-1) After (Case-2) After (Case-3)

FIG. 9. Comparison of samples before and after magnetic separation.

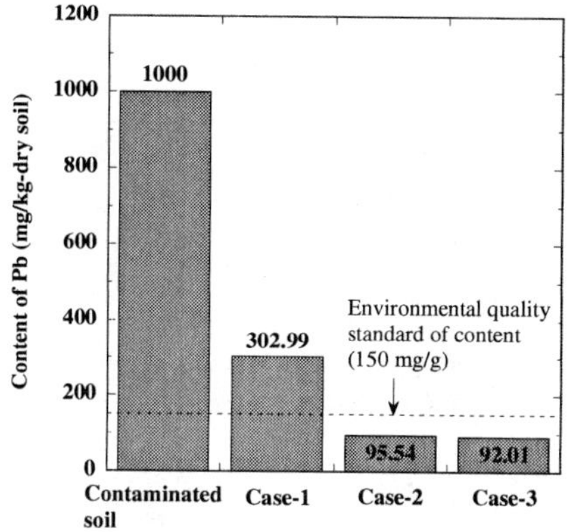

FIG. 10. Test results of the Pb-Content in the samples after magnetic separation.

CONCLUSIONS

The main conclusions drawn from this study are as follows:

1) The ferritization-treated soil exhibits significant insolubility over a wide range of pH. It was also confirmed that the ferritization-treated soil is ferromagnetic.
2) The magnetic separation for the removal of Pb was applied to the ferritization-treated soil. When magnetite was used as an adsorptive material, more than 90% of the lead could be removed from the contaminated soil.

This is a fundamental study on the ferrite-induced immobilization of Pb-contaminated soil and application of magnetic separation. For a practical application of this method, its efficiency and the cost of treatment must be considered.

ACKNOWLEDGEMENTS

This research was supported by the Steel Industry Foundation for the Advancement of Environmental Protection Technology, Japan.

REFERENCES

Okamoto, S. (1976). "Utility of magnetic scavengers for waste water treatment", *Ceramics*, Vol.11, No.3, pp.234-241 (in Japanese).

Omine, K. Ochiai, H. and Yasufuku, N. (2006). "A fundamental study on remediation of contaminated soil with heavy metal using electrokinetic and magnetic properties", *Fourth International Conference on Soft Soil Engineering*, pp.673-680.

Takada, T. (1977). "Removal of heavy metal ions from waste water by ferritization", *Journal of Environmental Pollution Control*, Vol.13, No.1, pp.37-41 (in Japanese).

The Assessment and Remediation of Chromite Ore Processing Residue at Former Disposal Sites, Glasgow, Scotland: Current Status (2007)

J.G. Farmer[1], S. Hillier[2], D.G. Lumsdon[2], M.C. Graham[1], and R.J.F. Bewley[3]

[1]School of GeoSciences, University of Edinburgh, Edinburgh EH9 3JN, UK; J.G.Farmer@ed.ac.uk
[2]The Macaulay Institute, Craigiebuckler, Aberdeen AB15 8QH, UK
[3]URS Corporation, St. James's Building, Oxford Street, Manchester M1 6EJ, UK

ABSTRACT: Analytical, experimental and modelling work has identified the mineral phases and processes responsible for the retention and release of hexavalent chromium (Cr(VI)) from high-lime chromite ore processing residue (COPR) under prevailing field conditions in Glasgow, Scotland. Key processes are precipitation/dissolution of Cr-containing minerals, inter-layer anion-exchange, sorption and co-precipitation of Cr(VI), and organic complexation in pore waters. A mineral implicated in Cr(VI) dynamics is a hydrocalumite-type phase, which has anion-exchange properties, exchanging chromate (CrO_4^{2-}) for sulfate (SO_4^{2-}) in laboratory experiments. Such phases, however, are often transitory to the more stable hydrogarnet, which has now been identified as a major COPR phase capable of hosting Cr(VI). Both the nature of mineral phase retention and the high pH of the COPR militate against direct remediative treatment of the source material, for example by the application of generic methods such as ferrous sulfate to reduce Cr(VI) to Cr(III). Nevertheless, calcium polysulfide (CaS_x) has recently been successfully applied to the remediation of high-lime COPR and Cr(VI)-contaminated groundwater in a series of laboratory experiments, leading to ongoing field trials.

INTRODUCTION

Previously disposed high-lime COPR (~ 2.5 x 10^6 t, 1830-1968) from the former J.&.J. White's chemical works in S.E. Glasgow continues to release toxic and carcinogenic Cr(VI)) to groundwaters, which are highly contaminated with $Cr(VI)O_4^{2-}$ (Farmer et al. 1999). Chromium contamination arising from COPR extends across an area of 500 ha, including much of the 73 ha Shawfield 'supersite', at the core of the Clyde Gateway Initiative, one of Scotland's largest urban regeneration projects. Hitherto, horizontal capping of sites has reduced airborne contamination, minimising the carcinogenic risk to the lungs from the most serious exposure route of dust inhalation. However, while an engineered containment system might protect human health, it would not prevent Cr(VI) leaching to groundwater. Geochemical investigation of COPR from these sites (Farmer et al. 2002, Geelhoed et al. 2002, 2003, Hillier et al. 2003), funded by the UK Natural

Environment Research Council Urban Regeneration and the Environment Programme (1998-2002), has illustrated why certain remediation techniques originally assessed in exploratory field trials (Bewley et al., 2000) were of limited effectiveness and has informed subsequent remediation strategies (Farmer et al. 2006). One of these, funded by the Environmental Development and Educational Network, has involved the application of CaS_x to remediation of COPR in laboratory trials (Graham et al. 2006).

MINERALOGY OF COPR

Using a combination of quantitative mineralogical analysis (Rietveld method) of X-ray powder diffraction (XRPD) data and phase chemical compositions from electron microscopy and microprobe (EM) analysis, Thomas et al. (2001) outlined initial results for COPR from three sites in Glasgow. Chromium was identified in at least four distinct phases, namely chromite, brownmillerite, hydrocalumite and hydrogarnet, the latter two the most important with respect to Cr(VI). Hiller et al. (2003) provided more details (Table 1) and rationalised the crystalline phases in COPR into three categories: residual unreacted chromite ore, high-temperature phases produced during the high-lime roasting process, mainly brownmillerite and periclase, and low-temperature phases, including brucite, calcite, aragonite, ettringite, hydrocalumite and hydrogarnet. This latter group is assumed to have formed by processes such as hydration, occurring while the COPR has been exposed to ambient conditions and to leaching at the various disposal sites. Analogous reactions occur during the hydration of cement clinkers, the cement literature greatly aiding understanding of the mineralogical/phase composition of COPR.

The combination of quantitative mineralogical data and the assumed or measured composition of each mineral phase also allows a complete chemical speciation of COPR to be calculated, including oxidation states (Hillier et al. 2003). This is a powerful approach that provides a phase-specific insight to complement speciation determined independently by bulk chemical or spectroscopic methods, as subsequently successfully applied to COPR from New Jersey (Crysochoou and Dermatas 2007). For the Glasgow COPR, Hillier et al. (2003) demonstrated that unreacted chromite and brownmillerite host virtually all of the Cr (III) present in COPR, typically ~30000 mg kg^{-1}. In the cement literature, brownmillerite is a well known high-temperature clinker phase, often termed ferrite, and its hydration reactions have been studied in detail. Hydrocalumite (HYDROus mineral containing CAlcium and ALUMinum) is amongst the phases that may form by hydration of brownmillerite and is a member of the so-called layered double hydroxide (LDH) group of phases, or 'anionic clays'. Hydrocalumite-type phases occur with variable states of hydration and anion compositions and in the cement literature hydrocalumite belongs to the group of phases termed AFm. These phases consist of layers of composition $[Ca_2(Al,Fe)(OH)_6]^+ \cdot 2\, H_2O$, the positive charge from which is balanced by interlayer anions (Fig. 1). Water is also present in the interlayer, and hydration state is easily changed. In Glasgow COPR, Hillier et al. (2003) showed that hydrocalumite is an important host for approximately half of the Cr(VI) content, amounting to ~4000 mg kg^{-1} Cr(VI), present in an exchangeable anionic form.

Table 1. Summary of Mineralogical Data for Glasgow COPR (Hillier et al. 2003)

Crystalline phases and average abundance (wt. %)	Indicative formulae and important substitutions [ ]	Notation for some cement analogues	Known chromium substitutions
Chromite (6%)	$(Fe,Mg)Cr_2O_4$		
Brownmillerite (14%)	$Ca_2(Fe,Al,Cr)_2O_5$	$C_2(A,F)$ (ferrite) Solid solution	Cr^{3+} for Fe, Al
Periclase (7%)	MgO		
Brucite (7%)	$Mg(OH)_2$		
Calcite (5%)	$CaCO_3$		
Aragonite (1%)	$CaCO_3$		
Hydrocalumite (5%)	$Ca_2(Al,Fe)(OH)_6(OH).3H_2O$ [Other anions for OH^-]	$C_4(A,F)H_{13}$ (AFm)	Cr^{3+} for Al, Fe CrO_4^{2-} as the anion
Hydrogarnet (23%)	$Ca_3(Al,Fe)_2(H_4O_4)_3$ [SiO_4^{4-} for $H_4O_4^{4-}$]	$C_3(A,F)(H,S)_6$ Solid solution	Cr^{3+} for Al, Fe Cr^{6+} as CrO_4^{2-} for $H_4O_4^{4-}$
Ettringite (1%)	$Ca_6Al_2(SO_4)_3(OH)_{12}.26H_2O$ [Other oxyanions for SO_4^{2-}]	(AFt)	CrO_4^{2-} for SO_4^{2-}

Previously, hydrocalumite in which CrO_4^{2-} is the interlayer anion has been described for Cr-contaminated concrete (Palmer 2000) and AFm phases with CrO_4^{2-} as the interlayer anion have also been synthesised in the laboratory, in a variety of hydration states (e.g. Perkins and Palmer 2001). Subsequently, in relation to potential remediation strategies employing SO_4^{2-}, Geelhoed et al. (2003) provided direct evidence of anion exchange of CrO_4^{2-} from the hydrocalumite phase and further highlighted the importance of anion exchange as a process liable to lead to Cr(VI) release from COPR.

More recently, the importance of hydrocalumite-type or AFm phases has been recognised in COPR from the USA by Chrysochoou and Dermatas (2007), who use the term calcium aluminium chromates (CACs) for phases of this general type. With time, such phases may react to give the more stable calcium aluminate known as hydrogarnet (Taylor 1997), although it may also form by direct hydration of the ferrite phase. In the Glasgow COPR, hydrogarnet is the most abundant crystalline phase present, accounting for an average of ~23 wt.% (Hillier et al. 2003, Graham et al. 2006). Hillier et al. (2003) speculated that minor amounts of Cr identified in hydrogarnet from COPR were most likely present as Cr(VI). The capacity of hydrogarnet to host Cr(VI) was subsequently confirmed experimentally by application of spectroscopic and diffraction methods to laboratory-synthesised hydrogarnets doped with CrO_4^{2-} (Hillier et al. 2007). The maximum amount of Cr(VI) that can be introduced synthetically corresponds to a replacement of about one out of every 24 hydroxyl tetrahedra in the unit cell by a CrO_4^{2-} tetrahedron (Fig. 1), agreeing closely with the amount of Cr measured directly by microprobe analysis in hydrogarnets from COPR. Assuming an average Cr(VI) content in hydrogarnet of 0.17 atoms per formula unit (Hillier et al. 2003) and typical hydrogarnet contents of 20-30%, then it can be calculated that hydrogarnet is likely to host between 4500 and 6500 mg kg^{-1} of the Cr(VI) present in the Glasgow COPR.

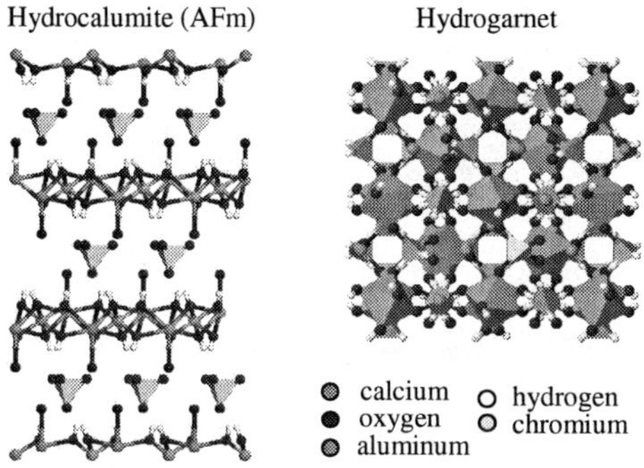

FIG. 1. Schematic representation of Cr-hydrocalumite (AFm) and chromate-substituted hydrogarnet.

Ettringite is the only other phase thus far identified that can also host Cr(VI), and indeed the capacity for CrO_4^{2-} to substitute for SO_4^{2-} in ettringite is well documented (Perkins and Palmer 2000). In the cement literature, ettringite belongs to a group of phases termed AFt. Although in the analyses of Glasgow COPR, ettringite had up until 2006 been identified at only minor (<3%) to trace concentrations, more recent work on samples of COPR from a site at Shawfield Road (National Grid ref. NS605630) indicates that ettringite and the similar AFt phase thaumasite can be considerably more abundant, as much as 12% in combination. Indeed, preliminary observations on three field-moist samples from this site indicate several other phase differences. These include much greater calcite contents than previously observed, ~ 40% in one case, together with common aragonite (5%), much lower brownmillerite contents (<7%), little evidence (<2%) for hydrocalumite-type phases, the presence of significant amounts of gypsum (5-10%), and higher contents of quartz along with cristobalite in one sample. These mineralogical differences indicate that the COPR samples from this site are likely to be chemically quite different and would be expected to be richer in Ca and Si, and poorer in Mg and Al, compared with those examined previously (Hillier et al. 2003, Graham et al. 2006). Such compositional differences also tally with the observation that thaumasite appears more abundant than ettringite in these samples, though it is not always easy to differentiate between these phases by XRPD and some degree of solid solution exists between them. To our knowledge, Cr(VI)-substituted thaumasite has not been described previously, although the structural analogies with ettringite suggest that CrO_4^{2-} substitution in thaumasite is very likely. Thaumasite may therefore be a host for Cr(VI) at the Shawfield locality.

In addition to the crystalline phases, the Glasgow COPR has an X-ray amorphous content of ~30% (Hillier et al. 2003), although its nature is not yet well understood. Both glassy and C-S-H type phase(s) similar to those formed by hydration of Portland cement may be present. Mass balance accounting suggests that the amorphous phase(s) is not an important reservoir for Cr(VI) in COPR (Hillier et al. 2003), but it should be noted that there is evidence that Cr(VI) may be able to substitute in C-S-H in cement systems. Independent application of the Rietveld method to the quantification of COPR from sites in New Jersey indicates similar amorphous contents of ~30% and highlights the importance of further characterization of this component of COPR. Recent mineralogical studies of New Jersey COPR (Dermatas et al. 2006, 2007, Chrysochoou and Dermatas 2007) have also documented the same basic suite of mineral phases as found in Glasgow COPR. Interestingly, residual chromite is not consistently reported in these investigations, whereas Megooda et al. (2000) emphasised its presence.

No doubt the variability noted in the proportions of the various phases between sites and samples reflects a variety of factors, but, aside from admixture with other materials and variation in process raw materials, foremost amongst these is the progress of hydration reactions or, in other terms, the extent of weathering. Potentially, this gives us a framework to organise our observations and thereby to understand how COPR has evolved and how it may behave in response to external factors.

GEOCHEMICAL MODELLING AND THE SOLUBILITY OF Cr(VI) IN COPR

Previous geochemical modelling studies (Geelhoed et al. 2002) showed that a reasonable description of soluble Cr(VI), Ca, Al, Si and Mg in the presence of COPR for the pH range 9.5-12 could be obtained from a model in which thermodynamic equilibrium between an assemblage of discrete mineral phases (including those in Table 1) was assumed. Modelling Cr(VI) solubility as a function of pH in the range 12.5 to ~9.5, calculations indicated that the mineral phases controlling the solubility of Cr(VI) at pH>11 are Cr(VI)-substituted hydrogarnet and Cr(VI)-hydrocalumite. In the pH range 11-8.5, the model description of the Cr(VI) concentration in solution was strongly improved by the incorporation of Cr(VI)-ettringite, which precipitates as a secondary mineral when hydrocalumite dissolves. The proposed model for leaching of COPR at high pH included Cr(VI)-bearing hydrogarnet, Cr(VI)-hydrocalumite, Cr(VI)-ettringite, brucite, calcite, $Ca_2Al_2(OH)_{10}{\cdot}3\,H_2O$, CaH_2SiO_4 and gehlenite hydrate and accurately predicts the concentrations of Cr(VI), Ca, Al, Si and Mg in solution in the pH range 10-12 as well as the pH buffering behaviour. Below pH 8, a decrease in the Cr(VI) concentration in solution was observed, attributable to sorption of Cr(VI) onto freshly precipitated Al and Fe hydroxide surfaces. The solubility constant for the Cr(VI)-substituted hydrogarnet was inferred from the Ion Activity Product, derived from the soluble amounts of Cr and other elements found using batch equilibration experiments (Geelhoed et al. 2002), as there were no relevant thermodynamic solubility data.

Published work on modelling the action of remediation treatments (e.g. CaS_x) on COPR is virtually non-existent. By combining the COPR solubility model (Geelhoed et al. 2002) with redox reactions for species of Cr and sulfur (S), however, it is possible to calculate how Cr(VI) in COPR will behave in the presence of polysulfide (S_x) species, for example. To allow this calculation the following redox reactions are needed:

$Cr^{3+} + 4H_2O \leftrightarrow CrO_4^{2-} + 8H^+ + 3e^-$	log K = 76.99	(1)
$Cr(OH)_{3[amorphous]} + H_2O \leftrightarrow CrO_4^{2-} + 5H^+ + 3e^-$	log K = 68.13	(2)
$HS^- \leftrightarrow S_{[solid]} + H^+ + 2e^-$	log K = 2.11	(3)
$H^+ + HS^- \leftrightarrow H_2S$	log K = 6.99	(4)
$HS^- \leftrightarrow S^{2-} + H^+$	log K = -12.92	(5)
$4H_2O + HS^- \leftrightarrow SO_4^{2-} + 9H^+ + 8e^-$	log K = -33.66	(6)

The S_x species (see below) can be defined using HS^- as the chemical component within a chemical equilibrium scheme, e.g. ORCHESTRA (Meeussen, 2003), but the mass balance equation for total S must include the S atoms contained in each S_x species.

$HS^- \leftrightarrow S_2^{2-} + H^+$	log K = -14.53	(7)
$HS^- \leftrightarrow S_3^{2-} + H^+$	log K = -13.28	(8)
$HS^- \leftrightarrow S_4^{2-} + H^+$	log K = -9.83	(9)
$HS^- \leftrightarrow S_5^{2-} + H^+$	log K = -9.59	(10)
$HS^- \leftrightarrow S_6^{2-} + H^+$	log K = -9.88	(11)

The model calculations suggested that Cr(VI) concentrations would be reduced by S_x treatments of COPR at the 1:1 and 5:1 stoichiometries (S_x:Cr(VI)) of the laboratory trials and confirmed that $Cr(III)(OH)_3$ and elemental S would be formed (Graham et al. 2006).

CALCIUM POLYSULFIDE REMEDIATION OF Cr(VI) FROM COPR

Graham et al. (2006) described the first application of CaS_x to the remediation of high-lime COPR and contaminated groundwater in a series of laboratory experiments demonstrating the effectiveness of the treatment in quantitatively and rapidly reducing Cr(VI) to Cr(III) over the pH range (8-12.5) typically found at the Glasgow sites. From the batch equilibration experiments, the ~5:1 ratio of S_x:COPR Cr(VI) presented a stoichiometric excess of the active S_x species and Cr(VI) was not detected either in solution or in the residual solid phase. From a column leaching experiment, only ~0.3% of Cr(VI) was leached from the treated column and only ~1% of original Cr(VI) remained as Cr(VI) in the residual solid phase. Wazne et al. (2007), in subsequent laboratory and pilot-scale (~450 kg) treatability studies on New Jersey COPR, found that a CaS_x dosage of twice the stoichiometric requirement proved effective in meeting the NJDEP total Cr(VI) and the USEPA TCLP regulatory standards. In the pilot-scale study, however, which lasted 15 months, X-ray absorption near edge structure (XANES) analysis indicated that only 4678 of 7588 mg Cr(VI) kg^{-1} was reduced after 10 months.

For the Glasgow COPR material, changes in mineralogical composition as a result of the reduction process were surprisingly small, with only a small decrease in concentration of the main Cr(VI)-containing phases, hydrocalumite and hydrogarnet (Graham et al. 2006). No new Cr(III) phases (e.g. $Cr(OH)_3$, $Fe_{(1-x)}Cr_{(x)}(OH)_3$) were detected by XRPD, consistent with the formation of amorphous rather than crystalline material. Elemental S, however, was observed in ~75% of treated samples, consistent with the model prediction and with the typical mechanism proposed:

$$2CrO_4^{2-} + 3CaS_5 + 10H^+ \leftrightarrow 2Cr(OH)_{3(s)} + 3Ca^{2+} + 15S_{(s)} + 2H_2O \quad (12)$$

Under oxidising conditions this $S_{(s)}$ may be converted to SO_4^{2-} but, in the Glasgow-based laboratory trials (Graham et al. 2006), gypsum ($CaSO_4$) was identified only in certain of the treated solids at lower pH (~8-9) from near-surface COPR-contaminated soil samples, and there was no marked increase in the amount of gypsum formed with increasing extent of treatment. In trials on New Jersey material, Wazne et al. (2007) also observed the formation of gypsum but again only at lower pH values (9.5), which also triggered the formation of ettringite and a concomitant decrease in the amount of brownmillerite. The possible formation of ettringite is a potential problem of all S-based treatment methods because of geotechnical concern over ettringite-induced heave with its implications for the longer term use of the remediated sites (Dermatas et al. 2006, Moon et al., 2007). In the Glasgow trials, ettringite (<1%) was detected only in the products from the high pH (12) treatment of the near-surface layer.

CONCLUSION

On the basis that CaS_x has provided favourable results using Glasgow-derived COPR at bench scale, a field trial involving treatment of ~50 m^3 of COPR has been designed and is currently undergoing technical review. If the findings of this approach prove to be acceptable, such treatment could be the major strategy for remediation of Cr(VI) within the Glasgow conurbation and facilitate successful redevelopment of the Clyde Gateway.

REFERENCES

Bewley, R.J.F., Jeffries, R., and Bradley, K. (2000). "Chromium contamination: field and laboratory remediation trials." Project Report 39, CIRIA, London, 110 p.

Chrysochoou, M. and Dermatas, D. (2006). "Evaluation of ettringite and hydrocalumite formation for heavy metal immobilization: Literature review and experimental study." *J. Hazard. Mater.* 136: 20-33.

Chrysochoou, M. and Dermatas, D. (2007). "Application of the Rietveld method to assess chromium(VI) speciation in chromite ore processing residue." *J. Hazard. Mater.* 141: 370-377.

Dermatas, D., Chrysochoou, M., Moon, et al. (2006). "Ettringite-induced heave in chromite ore processing residue (COPR) upon ferrous sulfate treatment." *Environ. Sci. Technol.* 40: 5786-5792.

Dermatas, D., Chrysochoou, M., Pardali, S., et al. (2007). "Influence of X-ray diffraction sample preparation on quantitative mineralogy: Implications for chromate waste treatment." *J. Environ. Qual.* 36: 487-497.

Farmer, J.G., Graham, M.C., Thomas, R.P., et al. (1999). "Assessment and modelling of the environmental chemistry and potential for remediative treatment of chromium-contaminated land." *Environ. Geochem. Health* 21: 331-337.

Farmer, J.G., Thomas, R.P., Graham, M.C., et al. (2002). "Chromium speciation and fractionation in ground and surface waters in the vicinity of chromite ore processing residue disposal sites." *J. Environ. Monit.* 4: 235-243.

Farmer, J.G., Paterson, E., Bewley, R.J.F., et al. (2006) "The implications of integrated assessment and modelling studies for the future remediation of chromite ore processing residue disposal sites." *Sci. Total Environ.*.360: 90-97.

Geelhoed, J.S., Meeussen,J.C.L., Hillier, S., et al. (2002). "Identification and geochemical modeling of processes controlling leaching of Cr(VI) and other major elements from chromite ore processing residue." *Geochim. Cosmochim. Acta* 66: 3927-3942.

Geelhoed, J.S., Meeussen, J.C.L., Roe, M.J., et al. (2003). "Chromium remediation or release? Effect of Fe(II) sulfate addition on chromium(VI) leaching from columns of chromite ore processing residue." *Environ. Sci. Technol.* 37: 3206-3213.

Graham, M.C., Farmer, J.G., Anderson, P., et al. (2006). "Calcium polysulfide remediation of hexavalent chromium contamination from chromite ore processing residue." *Sci. Total Environ.* 364: 32-44.

Hillier, S., Roe, M.J., Geelhoed, J.S., et al. (2003). "Role of quantitative mineralogical analysis in the investigation of sites contaminated by chromite ore processing residue." *Sci. Total Environ.* 308: 195-210.

Hillier, S., Lumsdon, D.G., Brydson, R., et al. (2007). "Hydrogarnet: A host phase for Cr(VI) in chromite ore processing residue (COPR) and other high pH wastes." *Environ. Sci. Technol.* 41: 1921-1927.

Meeussen, J.C.L. (2003). "ORCHESTRA, a new object-oriented framework for implementing chemical equilibrium models." *Environ. Sci. Technol.* 37: 1175-1182.

Megooda, J.N., Kamolpornwijit, W., and Charleston, G. (2000). "Construction use of vitrified chromium-contaminated soils." *Pract. Period. Haz. Toxic Radioact. Waste Mgmt.* 4: 89-98.

Moon, D.H., Wazne, M., Dermatas, D., et al. (2007). "Long-term treatment issues with chromite ore processing residue (COPR): Cr^{6+} reduction and heave." *J. Hazard. Mater.* 143: 629-635.

Palmer, C.D. (2000). "Precipitates in a Cr(VI)-contaminated concrete." *Environ. Sci. Technol.* 34: 4185-4192.

Perkins, R.B. and Palmer, C.D. (2000). "Solubility of $Ca_6[Al(OH)_6](CrO_4)_3{\cdot}26H_2O$, the chromate analog of ettringite, 5-75°C." *Appl. Geochem.* 15: 1203–1218.

Perkins, R.B. and Palmer, C.D. (2001). "Solubility of chromate hydrocalumite ($3CaO{\bullet}Al_2O_3{\bullet}CaCrO_4{\bullet}nH_2O$), 5-75°C." *Cem. Concr. Res.* 31: 983-992.

Taylor, H.F.W. (1997). *Cement Chemistry*, 2nd Edn., Thomas Telford Publishing, London.

Thomas, R.P, Hillier, S.J., Roe, M.J., et al. (2001). "Analytical characterisation of solid- and solution-phase chromium species at COPR-contaminated sites." *Environ. Geochem. Health* 23: 195-199.

Wazne, M., Jagupilla, S.C., Moon, D.H., et al. (2007). "Assessment of calcium polysulfide for the remediation of hexavalent chromium in chromite ore processing residue (COPR)." *J. Hazard. Mater.* 143: 620-628.

Reductive treatment of Chromite Ore Processing Residue (COPR): lessons from a field study

Maria Chrysochoou[1], Dimitris Dermatas, M. ASCE[2], Deok Hyun Moon[3] and Christos Christodoulatos[4]

[1]Assistant Professor, Department of Civil and Environmental Engineering, University of Connecticut, Storrs, CT, 06929-2037.
[2] Senior Associate, [3]Research Professor, and [4] Professor, respectively, Department of Civil and Environmental Engineering, Stevens Institute of Technology, Castle Point on Hudson, Hoboken, NJ 07030.

ABSTRACT: A bench- and pilot-scale treatability study conducted on Chromite Ore Processing Residue (COPR) from a deposition site in New Jersey showed that regulatory compliance to 20 (current) or 240 (up to 2006) mg/kg Cr(VI) was not possible using several common reductants (zero valent iron, pyrite, ferrous sulfate) due to the slow release of Cr(VI) from the binding minerals and the competing oxidation of iron and ferrous by oxygen. Initial success as evidenced by alkaline digestion was an artifact of the method, which facilitated Cr(VI) reduction during the test itself. X-ray Absorption Near-Edge Structure (XANES) analyses showed that the actual Cr(VI) concentrations in the solid were significantly higher (>1,000 mg/kg) in both ferrous sulfate and calcium polysulfide treated samples. Calcium polysulfide decreased Cr(VI) to non-detectable values (<2 mg/kg) for long monitoring times (up to 1½ years), as measured by alkaline digestion. The apparent success of calcium polysulfide is attributed to its slow oxidation, so that sulfide is available during analysis to reduce Cr(VI) that slowly diffuses out of the COPR matrix. The comparison of the in situ sulfide oxidation and Cr(VI) release rates can facilitate the design of a successful reductive treatment.

INTRODUCTION

Chromite Ore Processing Residue (COPR) is a granular waste generated by the high lime (CaO) process once used to extract Cr from chromite ore. Due to its granular, sand-like texture, it was widely used as construction fill in several states in the U.S. and elsewhere until approximately the 1960's. However, COPR turned out to be a hazardous waste according to more recent environmental laws due to its high concentrations of carcinogenic hexavalent chromium (Cr(VI)). Reported concentrations in New Jersey sites were up to 46,000 mg/kg total Cr, 30% of which were Cr(VI) (Lioy et al. 1992). As a measure for comparison, the current regulatory limit for Cr(VI) for residential soil use in the state of New Jersey (NJ) is 20 mg/kg (NJDEP 2007). One of the largest COPR deposition sites in NJ was SA7, a 34-acre lot located on Route 440 in Jersey City that was previously a commercially developed site. Dermatas et al. (2006) provide an overview of the SA7 history and COPR properties. Briefly, environmental litigation launched an extensive investigation by the site owner, aiming to understand COPR properties and find remediation methods alternative to excavation and disposal. This paper reports the major findings of a pilot study undertaken at SA7 for Cr(VI) treatment directly in the solid phase.

The most common treatment method for Cr(VI) is based on its reduction to non-toxic trivalent Cr(III). A variety of agents can be used to this end, such as biological and organic agents, sulfides, ferrous iron, and zero-valent iron (ZVI). Various authors have employed different combinations of agents, with varying success. The main challenges associated with COPR are a) its high pH and alkalinity, and, b) matrix-bound Cr(VI) that requires dissolution prior to treatment.

In 2003, when the SA7 investigation was initiated, there was no known treatment that could successfully reduce the total Cr(VI) content below 240 mg/kg (the NJDEP residential soil cleanup criterion at the time). Higgins et al. (1998) employed biologically-mediated treatment for Cr(VI); however, their results pertained only to pH below 9.7, while bacterial activity was inhibited at higher pH values. Natural COPR pH exceeds 12, so that biological treatment would require a large decrease in pH and thus large amounts of acid. Furthermore, a research group working on COPR sites in Glasgow, Scotland, had reported that commonly used reductive agents had only limited success (Bewley et al. 1998). COPR mineralogy seemed to have a strong influence on Cr(VI) reactions. Hillier et al. (2003) investigated Cr(VI) speciation in Glasgow COPR and found that is was bound in cementitious minerals, primarily hydrocalumite ($Ca_4Al_2(OH)_{12}(CrO_4)\cdot 6H_2O$) and Cr(VI)–substituted hydrogarnet ($Ca_3Al_2(OH)_{12}$). Geelhoed et al. (2003) subsequently reported that ferrous sulfate injected in COPR columns facilitated Cr(VI) release form those minerals through anion exchange of chromate with sulfate, while ferrous precipitated quickly as ferric at the top of the column due to the high COPR pH; consequently, injection of ferrous sulfate resulted in Cr(VI) release, rather than treatment.

Because of the strong influence of COPR geochemistry on treatment success, as well as the lack of comprehensive information about the various reductants, an extensive investigation was conducted on SA7 COPR, including geochemical and mineralogical characterization, as well as bench- and pilot-scale treatability studies. Individual results of these tests have been presented elsewhere (Wazne et al. 2005; Dermatas et al. 2006; Chrysochoou 2006; Chrysochoou and Dermatas 2007; Moon et al. 2007; Wazne et al. 2007). This paper focuses on the lessons learned from these investigations.

GEOCHEMISTRY OF SA7 COPR

Two zones of COPR with different geochemical properties were identified at SA7. The lower zone consisted of a clayey material that was dominated by Ca-bearing minerals (portlandite $Ca(OH)_2$ and calcite $CaCO_3$). This type of COPR is believed to be an artifact of the COPR depositional process (slurried through a pipe) that accumulated fine particles in the lower zone. It has never been encountered at another site thus far and will therefore be excluded from this discussion. The bulk material that was found in the upper SA7 zone, was a grey-black (GB) granular material that resembles mineralogically Glasgow COPR, as well as COPR encountered in other sites in New Jersey and Maryland. Another type of COPR, a hard brown material of high Cr(VI) content, was also sporadically encountered at SA7. However, reductive treatment of this material has not been attempted thus far.

SA7 GB COPR consists primarily of brownmillerite (Ca_2FeAlO_5), at an average concentration of 30 % wt. and periclase (MgO) (2% wt.), both of which formed during the roasting process, with smaller amounts of hydration and carbonation products. Hydration products include hydrogarnet, hydrotalcite ($Al_2Mg_4(OH)_{12}(CO_3)(H_2O)_3$), Calcium Aluminum Chromium Oxide Hydrates or CACs ($Ca_4Al_2O_6(CrO_4)\cdot nH_2O$ n=9,12,14) and ettringite ($Ca_6Al_2(SO_4)_3(OH)_{12}\cdot 26H_2O$). Cr(VI), with an average concentration of 5,000 mg/kg, could be bound in the CACs, or in hydrotalcite, hydrogarnet and ettringite. Quantitative XRPD coupled with mass balance calculations showed that CACs retained approximately 50% of the total Cr(VI) in their structure (Chrysochoou and Dermatas 2007). SEM/EDX analyses showed that ettringite was an insignificant sink of Cr(VI), forming preferentially the sulfate species (Chrysochoou and Dermatas, 2006). SEM/EDX analyses also showed that the hydrotalcite phase was a mixed cation and anion phase that was part of a cementing matrix and thus not easy to dissolve (Chrysochoou 2006). No conclusive evidence were obtained for the existence of Cr(VI)-bearing hydrogarnet in SA7 GB COPR. However, based on mass balance calculations, and on the experimental observations of Hillier et al. (2007), its existence cannot be excluded.

An essential prerequisite for technically feasible Cr(VI) reduction is its release from the aforementioned minerals to the liquid phase, where redox reactions occur much faster compared to the solid phase. Cr(VI) release can be facilitated by anion exchange with sulfate, as observed by Geelhoed et al. (2003), or by shifting the pH outside the stability domain of the binding minerals. Since hydrogarnets, CACs, and hydrotalcites are all cementitious compounds, they dissolve upon pH decrease. Geochemical modeling and XRPD analyses showed that CACs and hydrogarnet dissolve at pH 11 (Chrysochoou and Dermatas 2006; Chrysochoou 2006). However, Cr(VI) is first bound in Cr(VI)-ettringite, which dissolves at lower pH values (10.3 observed in SA7 COPR, 9.1 according to Geelhoed et al. (2002)). Acid Neutralization Capacity (ANC) analyses also showed that hydrotalcites were more pH resistant phases, persisting at pH 8 or lower.

COPR possesses high amounts of alkalinity, as a result of the high-lime process that generates it. ANC analyses of SA7 GB COPR showed that 1.2 eq/kg H^+ (or 6% wt. H_2SO_4) were necessary to decrease the pH from 12.5 to 11; this value increased to ~6 eq/kg (30% wt. H_2SO_4) to reduce pH to 8 and facilitate (theoretically) complete dissolution of Cr(VI). Thus, a substantial amount of acid is required to this end, with all the engineering difficulties and health and safety issues associated with its large-scale application.

A further complication to the dissolution of Cr(VI) bearing phases is the fact that they are not present as individual particles, readily accessible to chemicals in solution, but rather form a cementing matrix around brownmillerite nodules. Among the three groups, CACs are formed on the rim of those nodules as large crystals (up to 50 μm diameter), as evidenced by SEM/EDX analyses, and are thus more readily accessible to solution. Hydrogarnets and hydrotalcites were, on the contrary, as small crystals (5 μm) in the interior of the COPR cement matrix, and may thus by more challenging to dissolve. Particle size reduction may facilitate Cr(VI) release, even though it presents an engineering challenge to substantially reduce the particle size of a relatively hard matrix (brownmillerite hardness is 5.5 on the Moh's scale).

TREATABILITY STUDIES OF SA7 COPR

Several were screened in bench scale treatability studies at the beginning of the SA7 COPR investigation in 2003; elemental iron (Fe), ferrous sulfate ($FeSO_4{\cdot}7H2O$), pyrite (FeS_2), calcium polysulfide (CaSx), sodium bisulfide (NaSH), and sodium dithionite ($Na_2S_2O_4$). The results of the regulatory test, alkaline digestion (AD), showed that several of the commonly used reductants (ZVI, pyrite, sodium dithionite) were not successful in reducing Cr(VI) (Wazne et al. 2005). The main reason was the competing oxidation of the active ingredients (elemental iron, ferrous iron) by oxygen, which proceeded quickly at high pH. Conversely, decreasing the pH to 9 yielded more satisfactory results for most reductants. Ferrous sulfate and calcium polysulfide showed good performance at both low and high pH at the bench scale level and were chosen as the most promising reductants to perform a pilot scale treatability study.

The pilot study involved the screening of three types of mixing equipment: in situ Horizontal Rotary Mixers (HRM), in situ Vertical Augers (VA) and ex situ pugmills. Three stoichiometries of ferrous sulfate were tested, amounting to 3X, 5X and 8X the necessary amount to reduce the average Cr(VI) content of GB COPR (4,000 mg/kg). Those were combined with variable amounts (0, 10, 20 and 30%) sulfuric acid (H_2SO_4). Particle size reduction prior to pugmilling was also attempted, and the target water content was varied between 30 and 60%. All in all, five HRM, three VA and 23 pugmill combinations were tested for ferrous sulfate. Additionally, two pugmill mixes were tested for calcium polysulfide (2X and 2.25X stoichiometries at natural pH). All mixes were monitored for pH, water content, total Cr(VI) and TCLP Cr at various intervals for over 600 days following treatment. Selected samples were also tested for Cr(VI) using X-ray Absorption Near Edge Spectroscopy (XANES) at 240 days curing. Details of the experimental methods and selected results can be found in Dermatas et al. (2006) and Wazne et al. (2007).

The Cr(VI) results obtained from the treatment plots immediately following mixing were satisfactory (<240 mg/kg) for practically all combinations. However, follow up analyses at 30, 60, 120 and 600 days yielded consistent upward trends for Cr(VI) in all ferrous sulfate plots, eventually exceeding 240 mg/kg, the NJ residential regulatory limit at the time of the pilot test. To obtain an independent measure of the total Cr(VI) in the treatment plots, XANES analyses were conducted after 240 days of curing for selected mixes (see Dermatas et al. 2006). These were consistently higher than the concentrations obtained by AD at the same curing time.

The following questions then arise: a) what caused the apparent increase in Cr(VI) concentrations following treatment; and, b) what caused the discrepancy between AD and XANES?

If Cr(VI) reduction were complete following mixing, as indicated by AD, then re-oxidation of Cr(III) to Cr(VI) would be necessary to explain the progressive Cr(VI) rebound. However, this reaction is very slow, if not absent, under most conditions encountered in natural systems. Cr(III) precipitates as $Cr(OH)_3$ or $(Fe,Cr)(OH)_3$, both of which are highly insoluble; thus very little is available in solution (He et al. 2004). Furthermore, oxygen is an ineffective oxidant for Cr(III) (Eary and Rai 1987), and only manganese oxides have the ability to oxidize Cr(III) directly in the solid phase via a diffusion mechanism (Manceau and Charlet 1992). Thus, it is considered that Cr(VI) increase following treatment was not caused by Cr(III) reoxidation, but rather an artifact produced by the analytical method. This brings forward the second question of the discrepancy between AD and XANES. Both are explained by the following hypothesis: AD itself promotes Cr(VI) reduction in the presence of sufficient amounts of reductants. Thus, when ferrous was available in ample amounts immediately following mixing, complete reduction of the released Cr(VI) took place during AD, yielding satisfactory results. Ferrous was progressively oxidized by oxygen, so that decreasing amounts were available during testing that occurred at later times, yielding higher and higher Cr(VI) concentrations.

This scenario also highlights the fact that Cr(VI) was only slowly released from the COPR matrix over 600 days of curing; otherwise, treatment would have been successful at the initial stages of mixing. According to its geochemistry, Cr(VI) should have been released at pH lower than 10. The pH of ferrous sulfate treated COPR dropped initially to lower levels than those required for the stability of Cr(VI) compounds. COPR pH varied between 5 and 8 in all ferrous sulfate mixes. However, equilibrium with the solid phase was apparently not immediately established, as pH also increased over 600 days of treatment, as a result of the progressive dissolution of COPR alkaline minerals, primarily the dissolution of brownmillerite (Dermatas et al. 2006). Thus, complete Cr(VI) dissolution is not facilitated by the mere decrease in pH, exactly due to the slow dissolution of the cementitious matrix. Even the addition of 30% sulfuric acid to 5X ferrous sulfate could not facilitate fast and thorough dissolution of Cr(VI)-bearing minerals, yielding eventually Cr(VI) concentrations above 240 mg/kg.

The attempted particle size reduction that would theoretically facilitate the faster attainment of equilibrium was not successful. Particle size analyses prior to and following pulverization showed that it did break up coarse (gravel sized) particles, but it was unsuccessful in reducing the size of smaller particles. This phenomenon is attributed to the cementation of the COPR matrix, as well as to the high hardness of brownmillerite, the primary constituent of grey-black COPR.

The comparison between the different equipment used to mix ferrous sulfate with COPR (see Dermatas et al. 2006) showed that different rates of Cr(VI) increase were observed in the HRM, VA and pugmill samples. The slowest rate was observed in the VA and the fastest in the HRM, which are attributed to the different physical conditions prevailing in the mixes. The HRM facilitates only shallow mixing, where the material is exposed to the atmosphere and the moisture conditions may quickly change by exposure to rain and evaporation. The HRM treated COPR changed from a green sludgy material to fine grained brown sand within 3 weeks of treatment. Conversely, the VA plots contained a thick sludge of very high water content (up to

90%) after 600 days of curing. The physical conditions are believed to affect the rate of ferrous oxidation. Exposure to the atmospheric conditions and to wet and dry cycles favor quick ferrous oxidation by oxygen, leading to fast ferrous depletion and treatment failure. Conversely, the prevalence of saturated conditions facilitate ferrous oxidation only by diffusion of oxygen into the liquid phase, a slow process that prolongs the presence of ferrous and inhibits Cr(VI) increase over time.

The results of the calcium polysulfide (CaSx) mixes were also consistent with these observations. Cr(VI) reduction was found complete by AD, yielding non-detectable Cr(VI) (<2 mg/kg) up to 120 days (Wazne et al. 2007) and 3 mg/kg at 600 days for both the 2X and the 2.25X stoichiometries. Thus, CaSx is the only treatment that conforms with the current NJDEP regulatory standard of 20 mg/kg for Cr(VI), as measured by AD. These observations agree with the results of Graham et al. (2006), who found that CaSx was highly effective for treatment of Glasgow COPR. However, XANES analyses for total Cr(VI) showed that CaSx performance was again an artifact of AD and yielded a Cr(VI) concentration of 2910 mg/kg (Wazne et al. 2007) that was even higher than the 1848 mg/kg obtained for the ferrous sulfate HRM 8X mix (Dermatas et al. 2006). The reason for the inconsistencies between the XANES and AD results for the two reductants are again to be sought in Cr(VI) speciation.

CaSx treatment occured at natural pH and no pH decrease was caused through its addition to the matrix, contrary to ferrous sulfate, which acidified COPR upon mixing. This means that no mechanism for Cr(VI) dissolution was provided upon CaSx addition, other than equilibration between the solid phase and the solution upon depletion of dissolved Cr(VI) by sulfide. This mechanism is, however, slower compared to pH-induced dissolution of Cr(VI)-bearing minerals. Furthermore, oxygen-mediated oxidation of sulfide is slow at the strongly alkaline COPR conditions that do not favor or even allow microbial activity. Thus, the staying power of sulfide ions in solution is much higher compared to ferrous iron, so that it is available over long periods of time to reduce Cr(VI) during AD. In the case of SA7 COPR, enough sulfide was available to reduce residual Cr(VI) after 600 days of curing. Apparently, treatment will continue to be successful (in terms of AD) as long as sulfide is available. If sulfide is still available, when all Cr(VI) diffuses out of the solid and is reduced, then treatment will be permanently successful, no matter which method is used to assess it. Thus, the rates of Cr(VI) dissolution, sulfide oxidation, and the quantities of both are necessary knowledge to design a Cr(VI) treatment that meets the regulatory criteria in the long term.

CONCLUSIONS

Reductive treatment of SA7 COPR to meet the current regulatory criteria for Cr(VI) in the state of New Jersey (20 mg/kg) was not possible using a variety of reductants (zero-valent iron, pyrite, sodium dithionite, ferrous sulfate). The slow dissolution of Cr(VI)-binding minerals and the competing oxidation of ferrous iron by oxygen were the primary reasons for this failure. Increased addition of acid to ferrous sulfate did result in improved treatment performance, however, not nearly adequate to satisfy the 20 mg/kg requirement. The use of different mixing equipment (horizontal rotary mixers, vertical augers, pugmills) was found to affect treatment performance due to the creation of different physical conditions.

A caveat to reliably assess treatment performance is that Cr(VI) reduction takes place during the regulatory test, alkaline digestion. When the reductant is depleted, reduction can no longer take place during the test and the treatment fails the regulatory criteria. The rate of oxidation of the reductant and the rate of Cr(VI) dissolution from the matrix determine the time point of treatment failure. XANES analyses can be used to determine the actual Cr(VI) concentration in the solid.

The best performance in terms of regulatory compliance was obtained for calcium polysulfide, which yielded undetectable or very low (3 mg/kg) Cr(VI) over 1½ years of monitoring. Apparently, sulfide has a high staying power in the COPR matrix and is available to reduce Cr(VI) that diffuses very slowly under the alkaline conditions of the treatment. A more reliable assessment of the rate of sulfide oxidation and Cr(VI) release under in situ conditions will allow the design of a reductive treatment for COPR that can meet the regulatory standards and ultimately lead to complete reduction of Cr(VI).

ACKNOWLEDGMENTS

We thank Honeywell International Inc. for the financial support of the COPR investigation and Geosyntec Consultants for the execution of the pilot scale study and for providing the analytical data.

REFERENCES

Bewley, R.I.E., Watson, S., Jeffries, R. and Sansom M. (1998). "Field trials for evaluation of remedial technologies for chromium-contaminated soils in Glasgow." *Land Reclamation: Achieving Sustainable Benefits*, Balkema, Toterdam, 457-463.

Chrysochoou, M. (2006). "Investigation of environmental and geotechnical implications associated with deposition of Chromite Ore Processing Residue (COPR)." Ph.D. thesis, Stevens Institute of Technology, Hoboken, NJ.

Chrysochoou, M. and Dermatas, D. (2006). "Evaluation of Ettringite and Hydrocalumite Formation for Heavy Metal Immobilization: Literature Review and Experimental Stud." *J. Haz. Mat.* Vol.136: 20-33.

Chrysochoou M. and Dermatas D. (2007). "Application of the Rietveld method to assess Cr(VI) speciation in Chromite Ore Processing Residue." *J. Haz. Mat.* Vol. 141: 370-377.

Dermatas, D., Chrysochoou, M., Moon, D.H., Grubb, D.G, Wazne, M. and Christodoulatos, C. (2006). "Ettringite-Induced Heave in Chromite Ore Processing Residue (COPR) upon Ferrous Sulfate Treatment." *Environ. Sci. Tech.* Vol.40: 5786-5792.

Eary, L.E. and Rai, D. (1987). "Kinetics of Cr(III) oxidation by manganese dioxide." *Environ. Sci. Tech.* Vol. 21: 1187-1193.

Geelhoed, J.S., Meussen, J.C.L., Hillier, S., Lumsdon, D.G., Thomas, R.P., Farmer, J.G., and Paterson, E. (2002). "Identification and geochemical modeling of processes controlling leaching of Cr(VI) and other major elements from chromite ore processing residue." *Geochim. Cosmochim. Acta* Vol. 66(22): 3927-3942.

Geelhoed, J. S., Meussen, J. C. L., Roe, M. J., Hillier, S., Thomas, R. P., Farmer, J. G. and Paterson E. (2003). "Chromium remediation or release? Effect of iron(II) sulfate addition on chromium(VI) leaching from columns of chromite ore processing residue." *Environ. Sci. Tech.* Vol. 37: 3206-3213.

Graham, M.C., Farmer, J.G., Anderson, P., Paterson, E., Hillier, S., Lumsdon, D.G. and Bewley, R.J.F. (2006). "Calcium polysulfide remediation of hexavalent chromium contamination from chromite ore processing residue." *Sci. Tot. Environ.* Vol.364: 32-44.

He, Y.T., Chen, C. and Traina, S.J. (2004). "Inhibited Cr(VI) reduction by aqueous Fe(II) under hyperalkaline conditions" *Environ. Sci. Tech.* Vol.38: 5535-5539.

Higgins, T.E., Halloran, A.R., Dobbins, M.E. and Pittignano, A.J. (1998) "In situ reduction of hexavalent chromium in alkaline soils enriched with Chromite Ore Processing Residue" *J. Air and Waste Manage. Assoc.* Vol.48: 1100-1106.

Hillier, S., Roe, M.J., Geelhoed, J.S., Fraser, A.R., Farmer, J.G. and Paterson E. (2003) "Role of quantitative mineralogical analysis in the investigation of sites contaminated by chromite ore processing residues." *Sci. Tot. Environ.* Vol. 308: 195-210.

Hillier, S., Lumsdon, D.G., Brydson, R. and Paterson, E. (2007). "Hydrogarnet: A host phase for Cr(VI) in Chromite Ore Processing Residue (COPR) and other high pH wastes." *Environ. Sci. Tech.* Vol. 41: 1921-1927.

Manceau, A and Charlet, L. (1992). "X-ray absorption spectroscopic study of the sorption of Cr(III) at the oxide-water interface I. Molecular mechanism of Cr(III) oxidation on Mn oxides." *J. Colloid Interface Sci.* Vol.148: 425-442.

Moon, D.H., Wazne, M., Dermatas, D., Christodoulatos, C., Sanchez, A.M., Grubb, D.G., Chrysochoou, M., and Kim ,M.G. (2007)."Long-term treatment issues with chromite ore processing residue (COPR): Cr^{6+} reduction and heave." *J. Haz. Mat.* 143(3): 629-635.

Lioy, P.J., Freeman, N.C.G., Wainman, T., Stern, A.H., Boesch, R., Howell, T., and Shupack, S.I. (1992). "Microenvironmental analysis of residential exposure to chromium-laden wastes in and around New Jersey homes." *Risk Analysis* Vol.12: 287-299.

NJDEP (New Jersey Department of Environmental Protection). (2007). "Chromium Moratorium", Memorandum, February 8.

Wazne, M., Papazoglou, P., Papastavrou, M., Meng, X., Dermatas, D. and Christodoulatos, C. (2005). "Remediation of Chromite Ore Processing Residue by Chemical Reductants" *Conference Proceedings In situ and on site bioremediation Symposium* Battelle, Columbus, Ohio.

Wazne, M. Jagupilla, S.C., Moon, D.H., Jagupilla, S.C., Christodoulatos, C. and Kim, M.G. (2007). "Assessment of calcium polysulfide for the remediation of hexavalent chromium in chromite ore processing residue (COPR)." *J. Haz. Mat.* doi:10.1016/j.jhazmat.2007.01.012.

Weng, C.H., Huang, C.P., Allen, H.E., Leavens, P.B. and Sanders, P.F. (1996). "Chemical interactions between Cr(VI) and hydrous concrete." Environ. Sci. Techn, 30:371-376.

Advancements in the Management of Dredged Material in the State of New Jersey

Douglas A. Gaffney[1], P.E., M. ASCE, Edward S. Gorleski[2] A.M. ASCE and Genevieve Boehm Clifton[3].

[1]Regional Director, Ocean and Coastal Consultants, Inc., 20 East Clementon Road, Suite 201N, Gibbsboro, NJ 08026; dgaffney@ocean-coastal.com
[2]Engineer, Ocean and Coastal Consultants, Inc., 20 East Clementon Road, Suite 201N, Gibbsboro, NJ 08026; egorleski@ocean-coastal.com
[3]Program Manager, New Jersey Department of Transportation, 1035 Parkway Ave., Trenton, NJ 08625-0600; Genevieve.boehm-clifton@dot.state.nj.us

ABSTRACT: Demand for dredging of New Jersey's waterways, coupled with the lack of capacity in Confined Disposal Facilities (CDFs), are the catalysts for recent advances in the management of dredged material (DM) in the State. The New Jersey Department of Transportation's Office of Maritime Resources (NJDOT/OMR), through the I-BOAT NJ program, is attempting to enhance the maritime industry in the State for the benefit of the general boating public. Several grants have specifically targeted management techniques, processes and end uses to foster beneficial use of DM. A recently completed project in Cape May, New Jersey, separated 1,000 cubic yards of clean sand from the existing DM. This material was subsequently utilized as fill material for a State highway project. Other projects under way include enhancement of habitat and reduction of road-side mortality of Diamondback Terrapins through the beneficial use of DM, and mapping the volume and character of DM found in the state's coastal CDF's. As part of these studies, the State is examining the hydraulic and strength properties of various dredged materials, alone and as amended. A follow-on study will compare the properties of various DM products to existing classes of aggregates to establish the utility of DM as a viable alternative to virgin geo-materials.

1. INTRODUCTION

An urgent need for dredging of New Jersey's waterways, coupled with the lack of capacity in local Confined Disposal Facilities (CDFs), are the catalysts for recent advances in the management of Dredged Material (DM) in the State. The New Jersey Department of Transportation's Office of Maritime Resources (NJDOT/OMR), through the I-BOAT NJ program, is administering projects to enhance boating along New Jersey's 130 miles of coastline, 1,792 miles of shoreline, and coastal waterways and bays. The I-BOAT NJ program provides partnership funding to selected applicants who promote, improve and enhance the maritime industry in the State of New Jersey for the benefit of the general boating public.

Dredging, and the subsequent management of the resultant DM, are two very separate issues. There are many reasons why this historically was, and remains, disjointed. In the past, DM was placed wherever was convenient or the most cost-effective. This was typically a ringed dike in the wetlands now referred to as a CDF. A lack of dedicated funding for waterway maintenance and an uncertainty regarding the responsibility has lead to the deferment of dredging. No coordinated system or model to follow has traditionally existed for the management of DM, and a common familiarity and expertise in dredging, waterways management and the attendant upland social, economic and environmental implications, simply did not exist within the resource management community.

DM "management", a system that includes regulatory, administrative, public, planning, environmental and operational components, often ends with "disposal." In the past, material was disposed on an *ad hoc* basis with no tracking or future management plans. In part, this was due to economics, less-strict environmental regulations, and the fact that there was abundant space to place the material without neighboring concerns.

1.1 Regional Management

Today, NJDOT/OMR is fostering multi-agency cooperation through renewable capacity approaches (PHL, Baltimore), improved regulatory oversight, environmental sensitivity and research. Much time has been spent achieving dedicated State Channel Dredging Program dollars, working toward the development of a prioritized system of state channel dredging, creating a multi-agency Dredged Material Management System (DMMS), educating state leaders that the DM management process should be managed as a public priority, and developing sensible regional management designed with the support of current science and technology.

Working cooperatively with NJDEP, NJDOT/OMR is prioritizing and acquiring previously utilized CDFs for regionalized DM management. NJDOT/OMR endeavors to implement a regional system of material management that reflects a highest and best use opportunity for the local environment, citizenry, safe navigation and the overall marine transportation system. Regional management will allow for local DM to be managed efficiently, and create an economic and transportation benefit to the area. The concept of a "sediment-shed" can be used to describe the regional management of sediments and DM, and has many parallels to watershed management. Regional management at a government level provides a lower cost alternative (and in some cases, the only alternative) for marinas, municipalities, counties, homeowners associations and others that have dredging needs.

Through the I-BOAT NJ Program, the volume and character of DM found in the state's coastal CDF's (see Figure 1) and channels are being mapped. Figure 1 depicts a typical CDF and its location in the marsh, typically inaccessible to land-based equipment. Three dimensional mapping, coupled with DM characterization, allows for accurate volume calculations and capacity estimates. The future volumetric dredging needs of the state's waterways (see Figure 2) are also being determined. Figure 2 shows the relationship between the types of sediments to be dredged, their location and the water depth. This type of information will dramatically improve the ability of the NJDOT to manage and foster beneficial use of DM. The study program consists of three separate but closely integrated projects that will establish a regional scale framework and baseline for maintenance dredging related activities. These projects include: (1) compiling an up-to-date GIS database of known DM disposal facilities and other large legacy sites, which minimally would include characterization of DM holdings, estimated quantity, use history and ownership, (2) characterization of the bay waters through sediment sampling and distribution mapping of the *in situ* bay floor resources, which will highlight areas likely in need of dredging and characterize the materials to be excavated, and (3) spatial data analysis and modeling to determine sites suitable for prospective dredged material processing centers and to identify and prioritize areas in need of maintenance dredging, both of which will be based upon environmental, socio-economic, distance, available placement options and beneficial use potential.

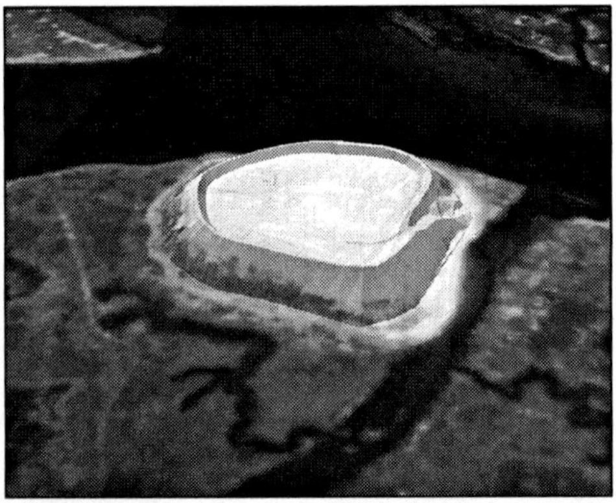

FIG. 1. Three dimensional view of a CDF developed through GIS allowing volume calculations. Source: The Richard Stockton College of New Jersey, Coastal Research Center.

The study program is complementary to, and can bridge between, local-scale, site-specific efforts (i.e. flounder spawning studies, disposal facility studies and material management technology development studies) and regional-scale reconnaissance efforts providing a framework across which data and findings can be extrapolated. One of the long-term interests of regional management is the reclassification of dissimilar and geographically dispersed DM into a commodity. This requires the

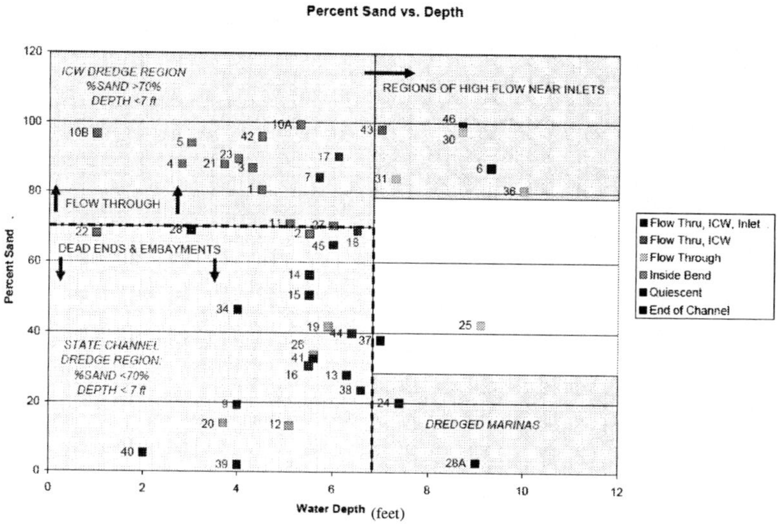

FIG. 2. Relationship of sediment samples to water depth and percent sand indicating shoaling regions.

development of a regional-scale DM excavation, transportation, processing and distribution system that could provide a predictable flow of materials with a consistent character and with pre-approved applications. In this manner, DM management may become a single, well-planned process that produces a product (the commodity), which is attractive to potential end-users such as aggregate and construction companies. For this to be feasible, DM management must be viewed at the regional scale rather than the local scale, where municipalities with navigable waterways lack the funding to properly manage this commodity.

In 2006, NJDOT/OMR spearheaded a Memorandum of Agreement (MOA) with NJDEP which capitalizes on the Bureau of Coastal Engineering's contracting and coordination expertise. The MOA established a financial process for regional management sites to be independently and sustainably managed. Non-state entities wishing to utilize space in a regional management site will pay a user fee which in turn would be placed in a dedicated NJDOT account. Funds collected in the account will be applied toward future site management and maintenance.

One of the benefits of the regional management concept is that it may allow multiple legacy disposal sites to be taken off the table in favor of one CDF that can be utilized sustainably. Important qualities for regional site selection include: Ownership and cost, historic dredged material placement and use, and the nature of transportation access. As most CDFs exist along coastal waterways, choosing a site that provides the overall highest and best choice between preferred geography and least environmental impact is important.

1.2 Coordinated Planning Efforts

NJDOT/OMR dedicates over $3 million annually toward the State Channel Dredging Program and is working on two major projects which will enhance dredged material management capabilities: A multi-agency DMMS and a Dynamic Waterway Linear Segmentation mapping project of the state's waterways. The I BOAT NJ grant program is currently funding over $2 million in DM management technologies, strategies and systems planning. All of these projects aim to lend additional information to, and support for, a coordinated dredging and material management program in New Jersey.

2.0 DREDGED MATERIAL SEPARATION

Among the many grants awarded through the I-BOAT NJ program, several have specifically targeted better management techniques, and the beneficial use of DM. A demonstration project was recently completed at a CDF at the U. S. Coast Guard Training Center (TRACEN) in Cape May, New Jersey. The purpose of the demonstration was to separate clean sand from the existing DM, beneficially use that material, and design a mobile system to cost-effectively excavate material from other coastal New Jersey CDFs. To support the beneficial use, the Cape May County Municipal Utilities Authority supplied recycled mixed-color glass cullet.

A wet separation unit was designed for this demonstration project (see Figure 3). The specifications for wet separation included: (1) separating the sand fraction from the silt and clay (fine) fraction of the DM to a degree sufficient for meeting several NJDOT Standard Soil Aggregate Gradations, (2) dewatering the fine fraction following separation from the sand and leaving the fines within the CDF, and (3) minimizing process water use and retaining the water in the CDF. In this project, sand was defined as material greater than 0.074 mm (No. 200 sieve). Material smaller than this comprised the silt and clay (fine) fractions of the DM. Fines are typically difficult to dewater and have few uses due to poor hydraulic and geotechnical properties.

The separation process proceeded as follows. Raw DM was selectively excavated and brought to the separation unit. Material was mechanically loaded onto the conveyor through the first screen to a second vibrating screen which had smaller openings. This screen removed the bulk of the vegetation and clay balls. The material fell through the vibrating screen into the hopper at the base of the sand screw. Through a washing and agitation process, the fines were removed from the sand particles which were transported up the sand screw. The water flow rate into the hopper and over the weir, combined with the rotational speed of the screw, defined the "cut point" or sand size of the final product. The dirty water had a coagulant added to it and was pumped to a geotextile tube for dewatering. The fines were retained within the tube. The closed-loop process water system recycles the effluent water from the dewatering process and reuses it in the separation process, thus eliminating discharges into adjacent water bodies.

A test plan for process water and sediments was established in accordance with the Acceptable Use Determination issued to the USCG in July 2006. DM and water samples were collected from various points along the separation process and various times during the demonstration project. Only the separated portion of the fines contained analytes in concentrations above regulatory standards. Additionally, several analytes above NJ Ground Water Quality Standards were detected in decant water from the process. These analytes were also detected in, and may be a result of, site water from the harbor used during the separation process. Testing results confirmed that the amended material met NJDOT I-13 material specifications.

FIG. 3. The DM separation plant in operation.

Approximately 1,066 cubic yards of clean sand were produced during the 32-day field demonstration. A total quantity of 1,000 cubic yards of sand was mixed with 1,600 cubic yards of crushed glass (CG), producing a total of 2,000 cubic yards of I-13 material. The glass is crushed and screened to a size of 0.5 inches (12.5 mm) or less, but is predominantly less than 0.375 inches (9.5 mm). In December 2006, the sand/ glass blend material was used as structural fill in NJDOT's Route 52 causeway reconstruction project between Somers Point and Ocean City, New Jersey. To be accepted for this use, the material had to meet gradation requirements, and was tested for Angle of Internal Friction (39°), resistivity and pH.

A cost analysis including mobilization, setup and support, material processing, material mixing, site restoration and demobilization was conducted based on the demonstration project. Costs for separating approximately 1,066 cubic yards during the demonstration project were $90.92 per cubic yard (cy). However, as the scale of the project increases, the cost per cubic yard will dramatically decrease. Based on the costs of the demonstration project, and using an implementation cost model similar to a mining operation, the cost for separating 500,000 cubic yards of DM is an estimated $17.67 /cy.

3.0 BENEFICIAL USES OF DREDGED MATERIAL

The NJDOT/OMR is investigating a wide variety of beneficial uses of DM. In addition to fill for road construction projects, an innovative approach is to enhance nesting habitat and reduce road-side mortality of Diamondback Terrapins. Terrapins face several obstacles to maintaining a healthy and stable population. These include loss of nesting habitats through shoreline erosion and road kills of females during nesting migrations. Along the Atlantic coast of New Jersey, the search for nesting sites on highway embankments has resulted in large numbers of road kills. The intent of this project is to enhance terrapin habitat and reduce mortality through beneficial use of existing DM, and thereby provide capacity for new DM placement. Modification of existing DM disposal sites along New Jersey's Intracoastal Waterway can create new nesting habitat for terrapins and other species. This study will also investigate the use of DM in concrete blocks that would be designed to replace silt fence in keeping terrapins out of the road. Uses of DM in the large-scale production of construction materials such as blocks offer a predictable and consistent use of future dredged sediments.

3.1 PROPERTIES OF DREDGED MATERIALS AND AMENDMENTS

As part of these studies, the hydraulic and strength properties of various DM, alone and as amended, are being investigated. For instance, the CG-sand mix produced by the separation process resulted in a construction material having a Constant Head Permeability (AASHTO T215) of 11.3 ft/day and a California Bearing Ratio (CBR) of between 51 and 69. Another amendment is Recycled Concrete Aggregate (RCA) which is added to the clean sand from the separation process. RCA tends to have very low permeability and the addition of sand improves water flow through the material. Testing of the RCA/separated sand mix showed increased permeability with an acceptable decrease in the CBR (see Figure 4). A coarser sand would increase permeability even more. A follow-on study will compare the properties of various DM products to existing classes of aggregates to establish the utility of DM as a viable alternative to virgin geo-materials.

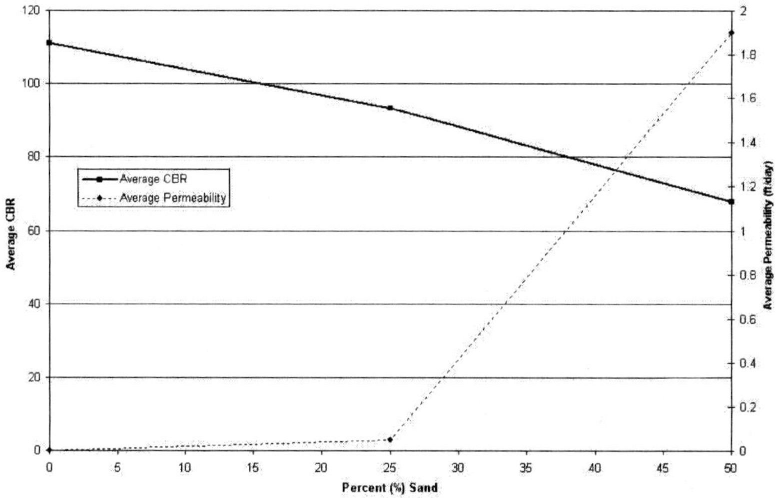

FIG. 4. Relationship of CBR and permeability to the percent sand added to RCA.

4.0 CONCLUSIONS

Regional DM management, increased capacity in existing CDFs and new strategies for the beneficial use of DM, are the best opportunities available to keep New Jersey's waterways navigable. The approach can be thought of as a "sediment-shed" and is best managed on a regional scale. New Jersey is taking a pro-active approach to developing the DM management tools; from sediment in the channels, through dredging, dewatering and placement, to processing and beneficial use. The projects described in this paper demonstrate new opportunities and methods to harvest and beneficially use this resource. Benefits accrue to the environment, the boating economy of the state, and will eventually result in a growing list of uses to replace virgin geo-materials.

The successful completion of the dredge separation demonstration project suggests that large scale CDF mining to separate sandy material from fines is a viable method for reclaiming CDF capacity. Additionally, amending the sandy fraction with other recycled materials (CG or RCA) can provide a resource for various industries including construction and aggregate production. The cost of DM separation must be viewed in light of the societal benefits of recycling DM and renewing DM placement capacity. The NJDOT/OMR is in the early stages of quantifying monetary benefits related to boating and dredging, and substituting DM for virgin geo-materials from quarries.

POPULATING A "DREDGED MATERIAL" FAMILY OF COMPACTION CURVES

Alexandra Srebro[1], Dennis G. Grubb, M. ASCE[2] and Allen W. Cadden, M. ASCE[3]

[1]Graduate Student Researcher, GeoEngineering Program, 440 Davis Hall, University of California, Berkeley, 94720.
[2]Senior Associate, Schnabel Engineering North LLC, 510 East Gay St., West Chester, PA 19380; PH: (610) 696-6066; FAX (610) 696-7771; dgrubb@schnabel-eng.com.
[3]Principal, Schnabel Engineering North LLC, 510 East Gay St., West Chester, PA 19380.

ABSTRACT: This paper presents the results of compaction testing for 100 samples of dredged material (DM) for the enhancement and refinement of a one-point Proctor curve for DM. A revised DM family of compaction curves is presented and discussed.

INTRODUCTION

There is an emerging crisis in materials management for the construction of large fills in an around major coastal metropolitan areas, namely from Norfolk to Boston. The Virginia Port Authority (VPA) currently estimates that it requires approximately 20 to 25 million (MM) cubic yards (CY) of fill to build a commercial cargo facility immediately adjacent to the US Army Corps of Engineers (USACE) Craney Island confined disposal facility (CDF) in Portsmouth, VA. The deepening of the Delaware River (Philadelphia to the sea) to 45 ft is currently hampered by insufficient storage space for the approximately 26 MM CY of fresh dredged material (DM) that will be generated during the project by the USACE. Further a field, in New Orleans, the crisis in DM issues has taken the analogous form of new dike and levee construction and improvements, which is currently estimated to be in excess of 170 MM CY of material that simply isn't locally available. In San Francisco, the airport runway extensions are on hold for this and other reasons.

Meanwhile, in these same areas, millions of cubic yards of DM are routinely generated and its beneficial use is overlooked for many reasons. The impending choice with DM management always seems to be: 1) build new CDFs which take up valuable land and are expensive to permit and operate; or, 2) develop renewable capacity approaches to sustain existing CDFs, which requires the excavation and use of DM in conventional construction. Unfortunately, the second is often more difficult than the first, though USACE-Philadelphia District appears to have crossed this threshold (USACE, 2006).

While the coarse fraction (sands and gravel) of DM can be easily separated from the fines and the construction industry has an interest in mining these materials, the DM fines (OH, MH, CH, OL, ML, and CL soils) continue to accumulate in CDFs and will ultimately result in long term storage capacity issues. Also, due to contamination, DM fines often require some form of pozzolanic-based stabilization/solidification (S/S) for beneficial use purposes including fill; the typical dosage being on the order of 10% Portland Cement (PC).

However, non- to moderately-contaminated DM is caught in a Catch-22. The main materials management strategy historically utilized has been the hydraulic placement of DM in CDFs, and other forms of upland disposal have focused on pozzolanic treatment. Other disposal approaches (ocean disposal dumping, side casting, and beach replenishment) which constitute the bulk of the USACE operations, also require no compaction data for implementation. So, on the one hand, the standard approach for the beneficial use of DM as fill is PC-based S/S, which is expensive and prices DM out of the conventional fill market. Conversely, there is essentially no Standard Proctor compaction data on pure DM for use on its own. As such, when presented with the use of DM as fill, the already uncomfortable construction industry has no idea how to accommodate the variability of DM in fill construction because there is no database, local knowledge or pre-established methods to deal with compaction control where DM is concerned.

Ultimately, the successful use of DM as construction fill will be linked to the reliability of quality control methods associated with DM placement and compaction, either pure and variable in nature, or when blended with other recycled materials such as crushed curbside collected glass (CG). On conventional fill projects, if the incoming fill material changes, the field inspector can rapidly check or adjust the compaction criteria by performing a one-point Proctor in the field (i.e., AASHTO T272). By this method (Joslin, 1959), the compacted moist unit weight and water content of the incoming soil material are determined via the Standard Proctor method. These data are then compared to a family of compaction curves, such as those initially developed by the Ohio Department of Transportation (ODOT) based on approximately 10,000 Standard Proctor tests, from which the maximum dry density (MDD) of the new fill material can be estimated. The process takes less than 30 minutes and fill construction proceeds uninterrupted by delays associated with developing full compaction curves in the laboratory every time a source or fill material changes.

Along these lines, Cadden et al. (2007) compared 56 sets of DM compaction data to the ODOT family of curves and compaction data organized by USCS soil type (Scavuzzo 2000). They found that by reducing the MDD of the ODOT curves by a factor of 0.91 to account for the lower specific gravity of DM (~2.48 vs. ~2.67) and extending the line of optimums downward and to the right (on a compaction plot), a family of curves was produced which adequately represented the behavior of a wide variety of DM. However, given the relative scarcity of DM compaction data (only 56 data sets), there was concern that the work of Cadden et al. (2007) was too limited for adoption by industry. Accordingly, under the auspices of a National Science Foundation Small Grant for Exploratory Research (NSF-SGER), this study sought to refine the DM family of curves (Cadden et al. 2007) by adding 100 DM compaction data sets that specifically targeted DM fines.

DATA COLLECTION

One hundred samples of DM were collected from Cell A of the USACE Fort Mifflin CDF in Philadelphia, PA. The surface crust (0 to 2 ft below the ground surface, bgs) of Cell A was avoided so that excessive roots or other plant matter would not affect results. Relatively undisturbed DM samples from various depths (~2 to 10 feet bgs) were collected at or near their natural moisture content. The DM sampling locations were recorded by a GPS unit (Trimble Pro XRS), as shown in Figure 1.

Two 5-gallon sealable buckets were used to store each fresh DM sample, which was then shipped to our soils laboratory for testing. The DM from both buckets was homogenized prior to testing by the Unified Soil Classification System (USCS; ASTM D2487) and Standard Proctor compaction (ASTM D698) procedures. Strict adherence to the USCS definition of organic soils was used, using the wet procedure (ASTM D4318, Method A).

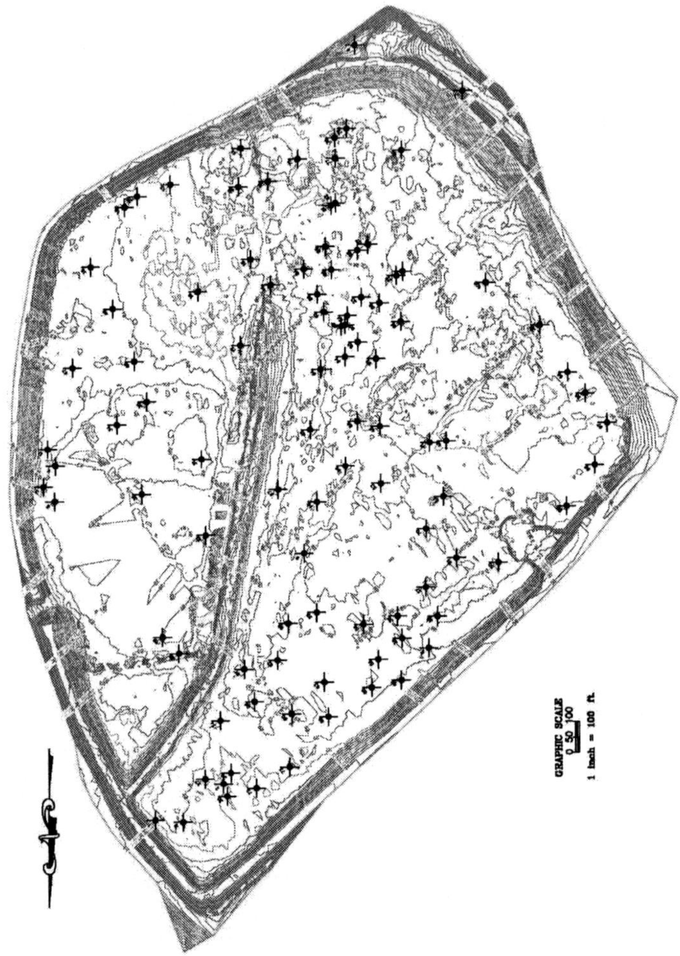

FIG. 1. Dredged material sampling locations, Cell A, USACE-Fort Mifflin Confined Disposal Facility.

RESULTS AND DISCUSSION

Figure 1 shows a survey of Cell A (~100 acres) from the center of the perimeter dike system inward, including the central baffle dike. Figure 1 illustrates that the sampling program did reasonable job of covering the entire Cell A.

Figure 2 shows the compaction curves for all 100 DM samples based on a minimum of 5 points, plotted using the spline function (with tension factor=2) in Grapher, Version 6.3.28 (Golden Software, Denver, CO). Shown for comparison in Figure 2 are zero air voids curves based on specific gravities of 2.67 (consistent with ODOT data) and 2.48 (Cadden et al. 2007). While the actual optimum points (MDD, w_{opt}) were determined from hand-fitted curves, the spline function fitted data (which forces the curves through all data points) captures the main essence of each curve.

Of the 100 DM samples, a total (n) of 4, 20, 2, 71 and 3 DM samples classified as SM, MH, ML, OH and OL soils, respectively. The granular (SM), low plasticity (ML, OL) and high plasticity (MH, OH) soils are respectively grouped as dot-dash, dash and solid lines in Figure 2. Many soils classified as MH and ML despite having loss on ignition (LOI) values between 5 and 20% by ASTM D2974. This is in contrast to the Navy designation of soils as "organic," based on a LOI value exceeding 5% (NAVFAC, 1987).

One main observation to be made based on the data shown in Figure 2 is that the curves on the dry side of w_{opt} become increasingly flat with decreasing MDD. That the curves appear to have the same general shape on the dry side of optimum (especially MH samples) adds confidence from a practitioner's perspective that a significantly larger data set for DM samples may not be necessary to support the use of One-Point Proctor curves based on Figure 2. Moreover, incorporated into these compaction curves is the influence of organic matter at LOIs between 5 and 20%.

Figure 3 presents the ODOT line of optimums (converted to dry density), the lines of optimums determined by Scavuzzo (2000) for six soil types (CH, CL, SM, SC, MH, ML) from the Western US states, and compaction data for 64 DM samples compiled by Bartos (1977) and Cadden et al. (2007) for several types of DM (SM, OH, OL, MH, ML soils). Co-plotted on Figure 3 are the new 100 sets of optimums from this study (from hand-fitted compaction curves, as described above), along with a revised "extended" line of optimums. The clustering of the DM data around the modified ODOT line of optimums is encouraging that Gs~2.48 can be taken to be representative of DM from the mid-Atlantic states. The revision of the extended line of optimums presented by Cadden et al. (2007) was minimal--the leftmost point of the extension remained stationary while the rightmost point was lowered by approximately 0.5 kN/m^3 (~3.1 lb/ft^3).

Table 1 presents a summary of the statistics for the compaction data for the 100 DM samples. The standard deviations on the MDD and w_{opt} are comparable with the ranges identified by Scavuzzo (2000) for fine grained soils. Thus, the revised "extended" line of optimums reflects the similar levels of variability than conventional soils. Nevertheless, the user should consider that for given MDD, w_{opt} can vary by up to 20% so tight controls on water content may not be realistic. Likewise, for the same $w_{opt,}$ the MDD may vary by up to 2.8 kN/m^3 (28 lb/ft^3) suggesting (in both cases) that compaction compliance strategies based on a minimum dry density or maximum allowable moisture content may be preferable for field quality assurance activities depending on the critical nature of the fill (structural vs. general).

Figure 4 presents the generalized DM family of curves based on Figures 2 and 3, where data was incrementally grouped every 5 lb/ft^3 of MDD, in keeping with the method adopted by Joslin (1959). In this way, hand-drawn lines overlain on Figure 2 were digitized to create Figure 4. The line of optimums in Figure 4 was taken from Figure 3. Of note in Figure 4 is that the data sets generated in this study allowed the portions of the curves on the dry side of optimum to be extended to significantly lower water contents than shown by either Joslin

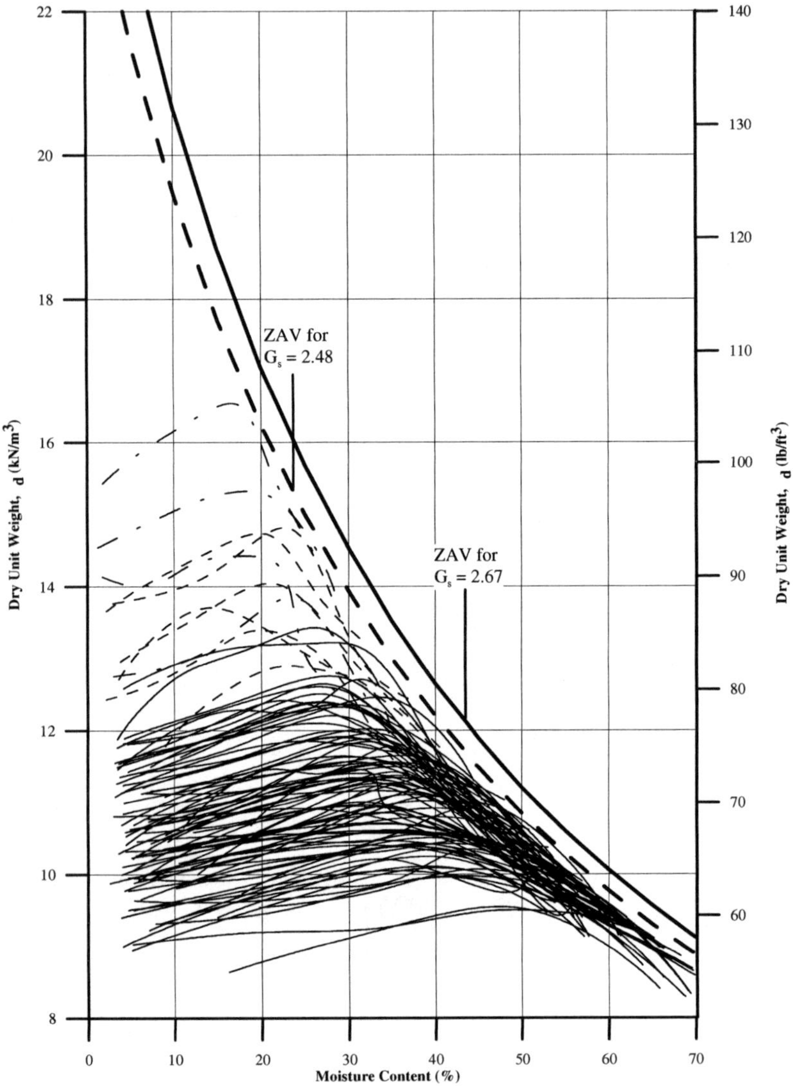

FIG. 2. Compaction curves for all 100 DM samples utilizing a spline fit.

(1959) or Scavuzzo (2000) for lightweight soils such as DM. The "flatness" of the curves also suggests that drying back wet DM may not necessarily yield significantly more dense DM. Figure 4 therefore becomes an important tool for working with compacted DM fills. Given the general consistency of the shape of the curves shown in Figure 2, Figure 4

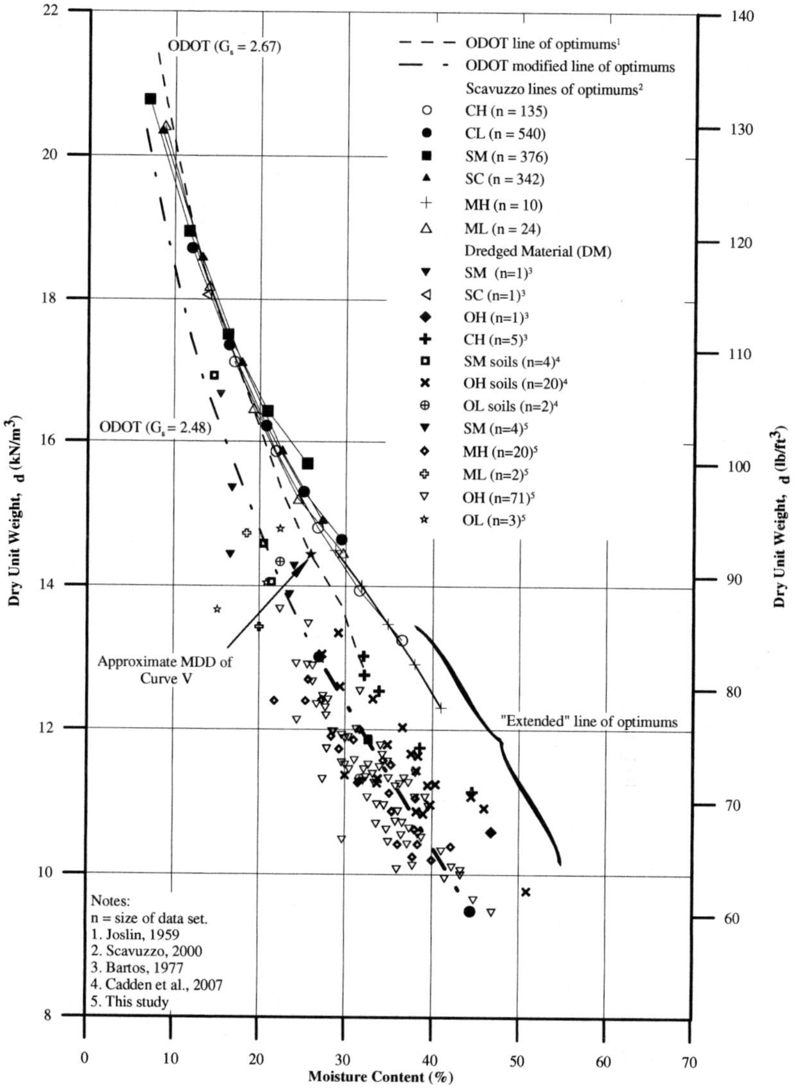

FIG. 3. DM optimum compaction indices with extended line of optimums versus Scavuzzo and ODOT lines of optimums.

illustrates, as with typical soils, that DM variability can be handled in a rational fashion.

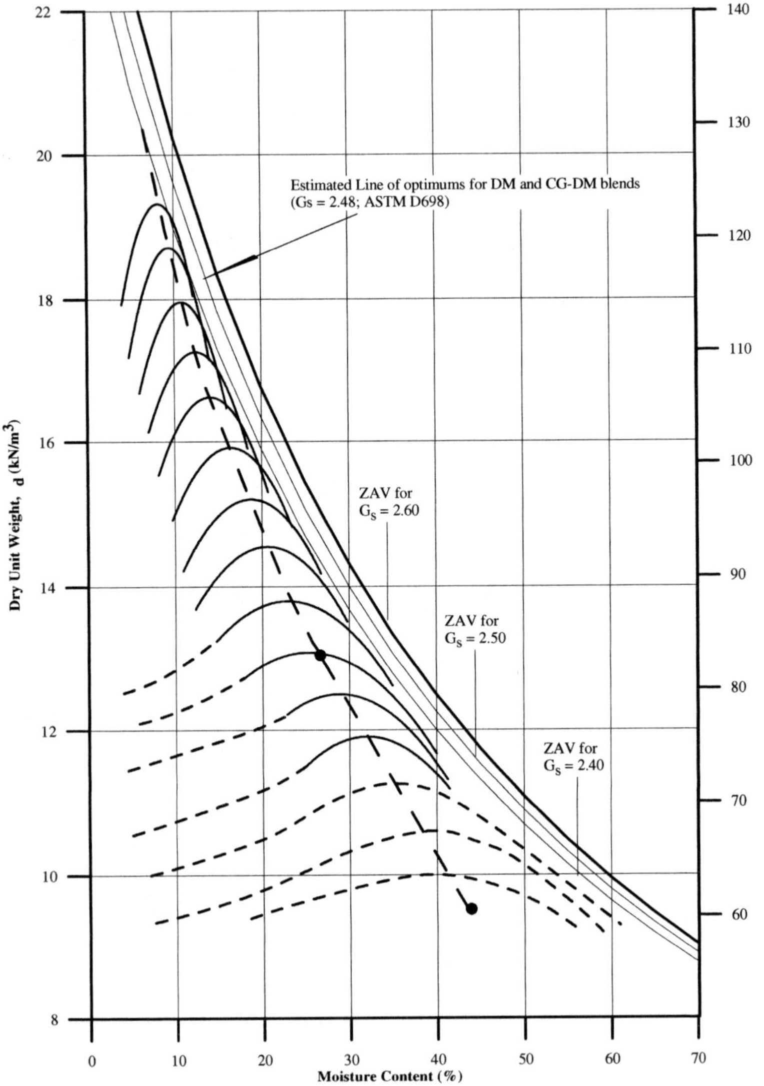

FIG. 4. Revised DM family of compaction curves.

CONCLUDING REMARKS

The DM family of compaction curves originally presented in Cadden et al. (2007) was refined based on new data set of 100 DM samples. While the line of optimums for DM was shifted slightly, the new data allowed for the extension of the family of curves to lower

TABLE 1: Summary of DM compaction data

USCS Soil type	Parameter	MDD lb/ft^3	W_{opt} %	USCS Soil type	Parameter	MDD lb/ft^3	W_{opt} %
MH (n=20)	average	72.3	33.1	OH (n=71)	average	72.2	33.5
	stddev.	5.2	5.5		stddev.	5.6	5.3
	max. value	80.8	42.2		max. value	87.1	46.9
	min. value	64.9	21.8		min. value	60.3	22.4
ML (n=2)	average	89.7	19.3	OL (n=3)	average	90.2	19.4
	stddev.	5.9	1.1		stddev.	3.7	3.8
	max. value	93.8	20.0		max. value	94.2	22.4
	min. value	85.5	18.5		min. value	87.0	15.1
SM (n=4)	average	96	18.1				
	stddev.	7.7	3.7				
	max. value	106	23.5				
	min. value	88.4	15.4				

maximum dry densities and higher water contents than previously indicated. The degree of material variability was comparable to prior research on traditional soils.

ACKNOWLEDGMENTS

A NSF SGER (CMMI-0714825) supported this study. The USACE-Philadelphia District provided the DM samples. We are grateful to Scott Rashke, Nicholas Malasavage, Hollis Graham, Trevor Reilly, Steve Penzone, Luke Geake, Mark Natale, Julie Delaney, Stephanie Olson, Kortney Brown, Andrew Horniger and Phil Shull for their assistance in this work. Any opinions, findings, and conclusions or recommendations expressed in this paper are those of the author(s) and do not necessarily reflect the views of the project sponsors.

REFERENCES

AASHTO (2004). "Family of curves--One point method," *AASHTO T272-04*, American Association of State, Highway and Transportation Officials, Washington, DC.

Bartos Jr., M.J. (1977). *Classification and engineering properties of dredge material*, USACE Technical Report D-77-18, Waterways Experimental Station, Vicksburg, MI, September, p. 175.

Cadden, A.W., Grubb, D.G., Miller, D.D., and Mibroda, J. (2007). "A proposed 'Dredged Material family of compaction curves' for field compaction control purposes," *Geo-Denver 2007: New Peaks in Geotechnics*, GSP 172, pp. 16.

Joslin, J.G. (1959). "Ohio's Typical Moisture-Density Curves", *Symposium on Application of Soil Testing in Highway Design and Construction,* STP 239, American Society for Testing and Materials, Philadelphia PA, 111-118.

Naval Facilities Engineering Command (NAVFAC) (1987). *Soil Mechanics: Design Manual 7.01*, Revalidated by Change 1 September 1986, p. 388.

Scavuzzo, R. (2000). "Construction quality control testing of compacted fills: Optimum moisture density values," *Constructing and Controlling Compaction and Earth Fills*, STP 1384, D.W. Shanklin, K.R. Rademacher and J.R. Talbot, (eds.), American Society for Testing and Materials, Conshohoken, PA, 126-136.

USACE, (2006). United States Army Corps of Engineers, Philadelphia District Contract No. W912BU-06-C-0024, awarded to Fort Mifflin Reclamation Associates, Inc., 36 Hog Island Rd., Philadelphia, PA.

Weston. (2002). *Dredged material characterization report, Fort Mifflin confined disposal facility – Area A, Philadelphia, Pennsylvania*, Prepared for: Sunoco, Inc., 10 Penn Center, 1801 Market Street, Philadelphia, PA by: Roy F. Weston, Inc., 1400 Weston Way, West Chester, PA, February 1.

DREDGED MATERIAL STABILIZATION: THE ROLE OF MELLOWING ON CURED PROPERTIES

Dennis G. Grubb, M. ASCE[1], Maria Chrysochoou[2] and Charles J. Smith[3]

[1]Senior Associate, Schnabel Engineering North LLC, 510 East Gay St., West Chester, PA 19380; PH: (610) 696-6066; FAX (610) 696-7771; dgrubb@schnabel-eng.com
[2]Assistant Professor, Department of Civil and Environmental Engineering, University of Connecticut, Storrs, CT, 06929-2037.
[3]Associate, Schnabel Engineering South, LLC, 1901 S. Main Street, Suite 11, Blacksburg, VA, 24060.

ABSTRACT: This study presents the results of a treatability study for DM stabilization using fifteen combinations of stabilizing agents (lime, Cement Kiln Dust (CKD), fly ash). Two paste studies were conducted, both employing a mellowing period of up to 5 days under open and closed conditions, respectively. To illustrate the impacts of the different mellowing conditions on the characteristics of the stabilized DM, select physicochemical, geotechnical and mineralogical differences of the two sample sets are presented. Mellowing under closed conditions yielded non-compactable material for 10 of 15 blends, as the moisture content was too high. Conversely, open mellowing yielded compactable material for all blends as early as 2 days. The mineralogical analyses showed that cement reactions proceeded more slowly under closed conditions. Compaction curves generated for two blends under open and closed conditions yielded contradictory results in terms of blend performance, the higher densities being obtained under open conditions. Consequently, since open mellowing more closely resembles field conditions, it is recommended as a procedure prior to curing in future S/S paste studies.

INTRODUCTION

This study was motivated by the possibility of using stabilized dredged material (DM) blends as potential fill material as part of the Virginia Port Authority (VPA) Craney Island Marine Terminal (CIMT) expansion project (2007 to 2030). VPA currently estimates that it requires approximately 20 to 25 million cubic yards of fill to build a commercial cargo facility immediately adjacent to the US. Army Corps of Engineers (USACE) Craney Island confined disposal facility (CDF) in Portsmouth, VA. The immediate proximity of the CDF and enormous supply of DM fines lead to a situation where VPA is compelled to consider using stabilized DM blends for large scale construction of the backlands fill (cargo container staging and lay down areas behind wharf structures).

Large scale applications of DM stabilization using pozzolanic materials such as lime (L), Portland Cement (PC), Cement Kiln Dust (CKD) and Fly ash (FA) usually involve the execution of bench scale treatability studies that seek to optimize the amount of added reagents, by monitoring properties such as the moisture content (MC), the Unconfined Compressive Strength (UCS) and the environmental (leaching) performance of different mix designs. While the cement reactions are ongoing from the time of blending, we distinguish between *mellowing* and *curing*, where curing begins when the stabilized DM is mined from a production stockpile, re-worked and placed/compacted in its final configuration (e.g., when no further sample disturbance can break cementation bonds).

The main issue to consider for the mellowing period is its representativeness to the intended operating conditions in the field for the specified application. The two main factors are its duration (usually 24 to 72 hours) and the exposure conditions (open vs. closed). The first factor relates to the progress of the cement reactions before the stabilized DM is workable and compactable, which will be influenced by the physical constraints of the project such as available stockpile area, and the rates of generation and fill construction. While some water will be incorporated into the cement reaction products, the second factor relates to the ability of water to evaporate (under open conditions), either being driven off by the exothermic cement reaction or by wind action. Open conditions also allow for the uptake of atmospheric carbon dioxide (CO_2) under the strong alkaline cement conditions (pH~11-13) that may promote the formation of calcite ($CaCO_3$) and reduce Ca availability for cement reactions, i.e., formation of the calcium silicate hydrates (CSH) and calcium aluminate hydrates (CAH) that are responsible for strength development.

The moisture content of fresh DM from a dredging scow or barge is on the order of 100 to 200%. Blending of amendments to DM for stabilization/solidification (S/S) purposes is usually referenced to 100% DM (wet basis) plus the individual additives (dry). The sample nomenclature used here refers to 100% DM (wet) plus the lime/CKD/FA contents added on a dry basis. For example, DM-10/0/0 refers to DM (MC=130%) plus 10% lime (dry). In this regard, the blend percentages do not sum to 100%.

An important drying agent for high water content DM which also provides for high quantities of reactive calcium is quicklime (CaO), which is usually applied between 5 and 10% and more often at 8%. The use of quicklime requires that other sources of reactive Si and Al have to be present for cement reactions to take place, either from the DM, of from another additive (in this case, CKD or FA). PC is less effective than lime from a drying perspective, and the required dosage for similar performance may approach up to 2x (10 to 20%; 8% to 12% common in NY/NJ harbor (PBQD, 1999; Grubb et al. 2006). However, PC contains substantial quantities of reactive silicates and aluminates to be highly self cementing. In each case, the goal is to minimize the lime/PC cost by replacing them with CKD and FA. The main strategy in using CKD is to provide sufficient lime (CaO) and some reactive silicates and aluminates to enable pozzolanic reactions to occur at a high pH (>11), but at a significantly reduced cost (60%) compared to L/PC. The drawback with CKD is that it may contain sulfur in concentrations that may contribute to ettringite ($Ca_6Al_2(SO4)_3(OH)_{12} \cdot 26H_2O$) formation, an expansive mineral associated with swell and failure in lime-stabilized soils and cements (Dermatas, 1995, Taylor et al., 2001).

With FA, doses have been traditionally limited to 10 to 30% to avoid bulking and minimize DM disposal costs. The limited use of FA, however, does not necessarily apply to a fill project where bulking may provide benefits. Significantly more FA can be added, which may or may not be sufficient on its own to significantly improve the performance of the DM through pozzolanic reactions, depending on its geochemistry (class F or class C). Class C FA contains reactive Ca that can raise the pH and is self-cementing, while class F FA does not contain alkalinity, but only provides reactive silica for cement reactions. High quantities of dry FA (50 to 75%) may be also desirable to make initially very wet DM workable.

MATERIALS & METHODS

Fresh DM classifying as a CH soil with a natural MC of approximately 130% was acquired from the Craney Island CDF. The DM was obtained from below the surface crust (6 inches), and had the consistency of toothpaste (MC~130%; e_{insitu}~3.35), readily pouring from the excavator bucket into 55-gallon drums (Figure 1). Decant water collected from the CDF cell drainage structure was used in all phases of experimentation to maintain a water chemistry consistent with the DM. The FA (MC ~10-20%) used for this study was acquired from the Dominion-Virginia Power Yorktown station from the conditioning silos prior to offsite transport. Lime and CKD were provided by LaFarge North America (Whitehall, PA). Key geo-environmental parameters of the DM, L, CKD and FA materials are presented in Table 1. It should be noted that the bulk chemistry provided by x-ray fluorescence (XRF) in Table 1 for the reagents do not necessarily reflect their actual geochemistry; for example, the Ca content in lime represents CaO, while Ca in CKD is mostly bound in unreactive calcite ($CaCO_3$) and dolomite ($(Ca,Mg)CO_3$), with only small amounts of free CaO, as evidenced by XRPD analyses (unpublished data). Si in FA was primarily amorphous, while Si in CKD was mainly bound in reactive alite ($3CaO \cdot SiO_4$) and larnite ($2CaO \cdot SiO_4$).

The DM had a plasticity index of 37, liquid limit of 62, approximately 92% fines and a specific gravity on the order of 2.75. The maximum dry densities by standard and modified Proctor compaction were 95.5 and 107.7 lbs/ft^3 (15 and 16.92 kN/m^3) with optimum moisture contents of 22.1% and 17.5%, respectively (vs. $MC_{initial}$~130%).

FIGURE 1: Photograph of DM sampling event. Removal of thin surface crust (left); flowable DM obtained from below crust (right).

The treatability study was initiated with paste studies to determine the workability and compactability of the stabilized DM blends for up to 5 days of mellowing by tracking the moisture content (ASTM D2216) under closed and open conditions. However, as a practical matter, industry generally limits stockpiling to 72 hours maximum. DM at a MC of 130% was used for blending purposes. The sequence for adding admixtures to the wet DM on a % dry basis is as follows. DM and FA were first mixed and homogenized to reduce the DM MC. The pozzolan (L and/or CKD) were then added and blending continued for five minutes either by hand or electric hand mixer depending on the volume and stiffness of the mix. Each blend was then spread across the bottom of a plastic container to a depth of roughly 2.0 cm (~0.75 in). A lid was placed on the container for the *closed* mellowing paste study; drying assistance (i.e., fans or drying lights) was not used during *open* mellowing. Sub-samples from each blend were removed every 24 hours for MC testing.

Table 1: Bulk Chemistry and Select Environmental Parameters for lime (L), cement kiln dust (CKD), fly ash (FA) and dredged material (DM)

Chemistry (%)	Lime	CKD	FA	DM
Silicon Dioxide (SiO_2)	1.09	15.16	43.5	
Aluminum Oxide (Al_2O_3)	0.33	3.96	25.16	
Iron Oxide (Fe_2O_3)	0.09	2.21	11.53	
Calcium Oxide (CaO)	97.73	46.99	1.85	
Magnesium Oxide (MgO)	1.12	2.8	1.04	
Sodium Oxide (Na_2O)	<.01	0.35	0.43	
Potassium Oxide (K_2O)	0.08	2.08	2.26	
Sulfur Trioxide (SO_3)	0.06	6.42	0.21	
Miscellaneous				
pH	12.14	13.03	8.4	6.9-7.4
Moisture Content (%)	0.13	0.15	24.83	130
Loss On Ignition (LOI, %)	2.14	19.17	10.76	3.3

A mix design (ASTM D558) was undertaken for DM with varying initial moisture contents (80% to 150%) to generate a set of compaction curves for the 3-day mellowing period to accommodate the natural MC variability in the field. DM from the shipping drums at MC ~130% was either dried back or wetted with Craney Island decant water to achieve the targeted initial MC of the DM. While the DM water content varied for the mix design, blending on a wet basis with the prescribed blending ratios of the pozzolans was maintained. Batch mixes were prepared and mellowed for 3 days. Afterwards, the DM-L/CKD/FA blends were removed from their mellowing trays and were pushed through a 0.95-cm (3/8-in) sieve to simulate the re-working process of DM prior to compaction.

The mineralogy of the 31-day old (3 days mellowing plus 28 days curing) specimens from the closed and open mellowing paste studies was evaluated by X-ray powder diffraction (XRPD) with Rietveld quantitative analysis (RQA). The 28-d period was chosen to be consistent with UCS testing of compacted blends (not presented here). 20 g were extracted from each blend and air-dried for 24-h. Each air-dried sample was pulverized to pass a 0.4 mm sieve, homogenized again and a 2-g subsample was pulverized to pass a U.S. 400 standard sieve (38 μm). The resulting powder was mixed with α-corundum (Al_2O_3) (Sawyer, Lot. No. C04-AO-41) on a 80/20 basis as an internal standard to quantify the amorphous phase. Step-scanned XRPD data were collected using a Rigaku DXR 3000 computer-automated diffractometer with Bragg-Brentano geometry. The diffractometry was conducted at 40-kV and 40-mA using diffracted beam graphite-monochromator with Cu radiation. The data were collected in the range of two-theta values between 5° to 65° with a step size of 0.02° and a count time of 3 seconds per step. XRPD patterns were analyzed by the Jade software, version 7.5 (MDI, 2005) and reference to the patterns of the International Centre for Diffraction Data database (ICDD, 2002) and the Inorganic Crystal Structure Database (ICSD, 2007).

RESULTS & DISCUSSION

Tables 2 and 3 present the results of the closed and open mellowing of the stabilized DM blends. The testing program was initiated with a closed mellowing approach since this imparts a higher degree of control on experimentation than does open mellowing. However, a limit was put on the maximum of duration for mellowing of 3-days, to reflect actual field operations, which is also characterized by (more) open conditions. The criteria for *compactibility* was based on plasticity and the ability to physically compact the material. ASTM D558-04 uses compaction equipment analogous to that used in the Standard Proctor, where a 2-inch diameter rammer is used to compact the material inside a 4-inch diameter mold, e.g., an aspect of ratio of 2 between the mold and rammer diameter.

Soils at MCs above their plastic limit (PL) will tend to deform in the mold around the rammer without or limiting densification. Hence while our MC ranges for testing were comparable to Dermatas et al. (1999) for wet DM, their compaction results were based on an aspect ratio of 1.27, which introduces additional confinement that may allow wetter stabilized DM blends to be artificially compacted in the laboratory. This is a significant departure between the two studies and while this study assumed a greater aspect ratio for procedural purposes, it is not known how closely either compaction approach mimics the ability of stabilized DM to shift and deform during lift placement and proof-rolling in the field.

There are important differences in MC reduction between the blends and mellowing conditions. Under closed conditions, the major MC reduction mechanism is the incorporation of water into cement reaction products and the uptake and dilution of water by the additives (especially CKD, FA). In general, the small change (<10%) in the water contents from the time of blending through 5 days reflects the inability of water to evaporate under closed conditions. The threshold water content for all blends under closed mellowing appears to be in the 55 to 60% range, which is comparable to the MCs of harvestable DM crusts in CDFs. The use of the industry standard (5 and 10% quicklime) proved insufficient to produce a compactable material under closed conditions. The heat of lime hydration quickly dissipated upon blending of the relatively small amounts of DM used in the paste study (and any paste study, for that matter). Upon sealing, water could no longer evaporate, and the rate of cement reactions was apparently too slow to produce improved properties after 5 days of mellowing. This issue will be discussed in greater detail below (mineralogical results).

The use of CKD alone also proved inadequate to reduce MC for compaction. Note that the higher water content of 10% CKD compared to the 10% lime is consistent with the lower free lime content of

Table 2: Paste Study - Closed (Sealed) Mellowing

Mix ID	MC (ASTM D2216) per Day					Compactable at 3 days?
	1	2	3	4	5	
DM-5/0/0	112.4	112.7	111.4	108.9	107.8	No
DM-10/0/0	94.6	94.0	91.7	91.3	88.8	No
DM-0/10/0	107.1	107.3	106.0	104.9	102.0	No
DM-0/15/0	97.5	96.1	94.5	93.2	91.9	No
DM-0/20/0	88.7	86.8	85.7	83.9	82.6	No
DM-10/0/25	71.4	71.4	72.2	71.2	70.8	No
DM-10/0/75	53.8	53.8	52.8	52.0	52.0	Maybe
DM-0/10/25	78.4	77.7	78.4	77.2	70.6	No
DM-0/10/75	57.0	57.3	53.8	54.6	53.8	Maybe
DM-0/20/25	69.0	67.7	67.2	66.8	67.4	No
DM-0/20/75	53.0	51.7	52.2	51.5	49.9	Likely
DM-2.5/7.5/25	76.2	76.9	77.3	75.4	76.2	No
DM-2.5/7.5/75	56.6	56.0	55.9	54.2	55.2	Likely
DM-2.5/17.5/25	66.5	65.7	66.5	67.1	63.4	No
DM-2.5/17.5/75	49.6	51.2	50.4	48.3	49.8	Likely

Table 3: Paste Study - Open (Unsealed) Mellowing

Mix ID	MC (ASTM D2216) per Day					Compactable at what day?
	0	1	2	3	4	
DM-5/0/0	110.3	90.6	71.7	51.5	35.8	3
DM-10/0/0	92.6	75.3	59.3	41.9	28.2	3
DM-0/10/0	106.0	85.6	66.9	53.7	43.7	3
DM-0/15/0	96.0	75.3	59.9	48.2	35.9	3
DM-0/20/0	87.8	65.6	54.6	40.7	31.7	3
DM-10/0/25	72.6	53.8	41.9	28.8	22.4	Maybe 2
DM-10/0/75	49.5	33.5	25.3	19.0	11.1	2
DM-0/10/25	79.2	60.2	44.3	29.3	18.3	Maybe 2
DM-0/10/75	56.5	42.7	32.3	18.9	9.7	Maybe 2
DM-0/20/25	65.1	52.6	39.9	27.6	21.9	Maybe 2
DM-0/20/75	51.1	39.5	31.4	20.1	10.8	Maybe 2
DM-2.5/7.5/25	78.0	57.7	42.3	27.6	17.7	Maybe 2
DM-2.5/7.5/75	55.9	40.2	31.8	21.4	11.5	Maybe 2
DM-2.5/17.5/25	67.7	51.2	39.0	25.0	12.8	Maybe 2
DM-2.5/17.5/75	50.7	36.1	26.7	15.2	8.3	Maybe 2

CKD. The addition of large amounts of FA (75%) appeared necessary, whether the Ca source was lime or CKD. The individual and combined use of lime and CKD with 75% FA yielded the best results. The 25% FA blends were always approximately 15 to 20% wetter than the 75% FA blends, and thus did not necessarily yield compactable material. However, the MC results for the pure (L, CKD) and hybrid (L/CKD) blends at the 10% or 20% total pozzolan dose for both FA levels (25 and 75) are comparable, suggesting very limited differences in reaction rates.

The situation was entirely different for open mellowing. Table 3 shows that major changes in MC are recorded for each blend under open conditions, so that all stabilized DM blends were compactable after 2 or 3 days of mellowing, stressing the role of evaporation in making DM workable and compactable. The trends between the different blends remained consistent, with lime providing better drying effect compared to CKD when applied alone and the 75% FA blends yielding lower MC due to dilution. Again, the MCs for blends of the same overall dosing rates (total pozzolans, FA) were comparable within 2 to 5%.

The differences in performance between the open and closed mellowing conditions are also illustrated in the respective mix designs. Figure 2 shows the moisture-density relationships for open and closed conditions for two stabilized DM blends, 10% lime (DM-10/0/0) and 2.5% lime, 7.5% CKD and 25% FA (DM-2.5/7.5/25). In each case, the DM had to be dried back from its natural MC by at least 33 MC points prior to use. Both the DM-10/0/0 and DM-2.5/7.2/25 blends under closed conditions only yielded a partial compaction curve on the wet side of optimum, with the 10L providing significantly higher dry densities for the two initial DM MCs tested (80%, 97.5%). This strongly suggests, on the basis of both MC and dry density, that 10L will outperform the DM-2.5/7.5/25 blend in terms of UCS strength.

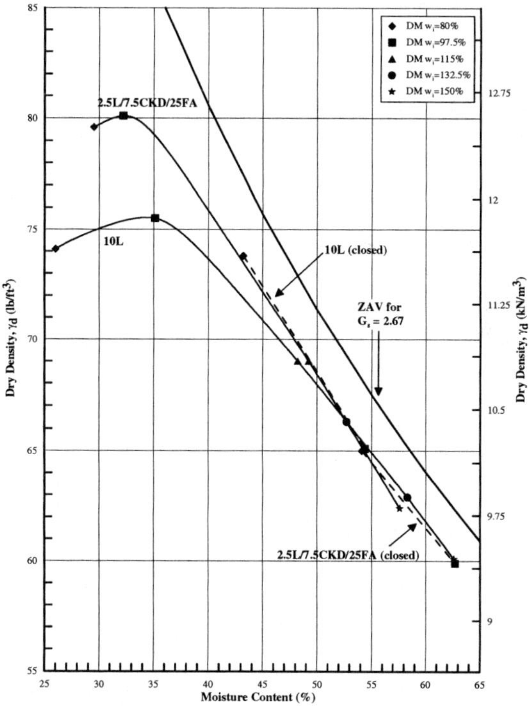

FIG. 2: Moisture-Density Relationships (ASTM D558) for Open vs. Closed mellowed samples

However, when comparing DM-10/0/0 and DM-2.5/7.5/25 under open conditions from the mix design, while both are capable of stabilizing DM up to an initial MC of 150%, the hybrid blends produced greater dry densities below a 3-day mellowed MC of approximately 53%. In this case, a higher capability for increased strength is indicated for the hybrid blends. Consequently, the results of the compaction curves for open and closed mellowing are somewhat contradictory in terms of evaluating density and strength for different blends. This is an important conclusion for S/S treatability studies. Even though closed conditions provide more control over the experimental design, they may be too far from reality to yield meaningful conclusions.

The differences between open and closed mellowing conditions were also investigated by mineralogical analyses. Table 4 presents select XRPD data for the DM-10/0/0 and DM-2.5/7.5/25 blends for the 31-day pastes from open and closed mellowing experiments. The mineralogy of the DM-10/0/0 indicates that approximately 54% to 69% of the initial CaO which hydrated to portlandite remained unreacted, presumably due to the absence of soluble silica and alumina in the DM. The highest portlandite content occurred in the closed mellowing sample indicating that it was an environment with limited reactivity as evidenced by the lowest calcite (no CO_2 uptake) and hydrocalumite contents, and absence lack of ettringite. The presence of unreacted portlandite would also suggest limited strength development from the formation of cement reaction products (even though the density was higher in FIG. 2 under closed conditions).

Table 4: Summary of XRPD data pastes mellowed under open and closed conditions

Mineral	Chem. formula	Source	DM-10/0/0		DM-2.5/7.5/25	
			Open[1]	Closed[1]	Open	Closed
Quartz	SiO_2	DM	20.5	22.6	17.3	19.7
Albite	$(Na,Ca)Al(Si,Al)_3O_8$	DM	3.2	4.8	4.7	3.2
Orthoclase		DM	2.5	3.4	3.2	3.1
Muscovite	$(K,Na)(Al,Mg,Fe)_2(Si_{3.1}Al_{0.9})O_{10}(OH)_2$	DM	6.8	7.3	8.1	6.8
Chlorite	$(Mg,Al)_6(Si,Al)_4O_{10}(OH)_8$	DM	4.1	5.0	5.2	5.4
Kaolinite	$Al_2Si_2O_5(OH)_4$	DM	2.7	3.7	--	--
Pyrite	FeS_2	DM	0.9	1.2	0.7	1.1
Halite	NaCl	DM	--	0.7	--	0.7
Calcite	$CaCO_3$	DM, CKD, new	5.1	4.4	6.7	8.8
Dolomite, Ankerite	$(Ca,Mg,Fe)CO_3$	DM, CKD	0.6	0.5	0.7	2.0
Portlandite	$Ca(OH)_2$	L	5.4	6.9	--	--
Mullite	$Al_6Si_2O_{13}$	FA	--	--	3.4	1.5
Hydrocalumite	$Ca_4Al_2Cl_2(OH)_{12}\cdot 6H_2O$	FA, new	1.5	1.6	--	0.9
Ettringite	$Ca_4Al_2(SO_4)_3(OH)_{12}\cdot 26H_2O$	New	2.1	--	3.8	1.5
Amorphous		All, new	44.5	37.9	46.2	45.4

Conversely, the absence of portlandite in the DM-2.5/7.5/25 blends reflects a lower initial lime content and lime-based calcium reactions running to completion. This, coupled with the higher initial sulfur content (in CKD), results in greater quantities of ettringite formation. While the lime content (5 to 15%) of typical CKD is obviously lower than pure lime, its lime and reactive silicate contents contribute to higher quantities of cement reaction products. The ettringite content is higher in the open vs. the closed mellowed blend, indicating that open conditions may accelerate reactions.

CONCLUSIONS

Treatability studies for stabilization of dredged material involve the execution of paste studies that monitor parameters such as water content, density, and strength, to evaluate and anticipate material behavior in the field. To understand the extent of the differences between the mellowing conditions, a DM paste study was undertaken using 15 different blends of lime, Cement Kiln Dust and Class F Fly Ash, added to DM having an initial moisture content (MC) of 130%.

Under closed conditions, the MC of 10 of the blends was too high to produce a compactable material after 5 days of mellowing. Given that the stabilized DM is usually not stockpiled for more than 24 to 72 hours in the field; theoretically, none of these blends were considered satisfactory for large-scale implementation. Only a mixture of lime and CKD, or 20% CKD alone, in conjunction with 75% FA yielded marginally compactable material after 3 days of closed mellowing. Conversely, all 15 stabilized DM blends yielded compactable material after 3 days of open mellowing.

Compaction curves obtained under closed conditions only yielded partial compaction curves for the initial DM MC range of interest (80 to 150%), limiting potential flexibility for field operations. Also, mineralogical analyses showed that cementation reactions proceed more slowly under closed conditions, with more unreacted portlandite in lime (L) blends, and less ettringite in hybrid (L/CKD) blends. Based on these results, the conditions for paste studies should be chosen to reflect the desired end use to ensure the greatest representativeness and applicability of the results.

ACKNOWLEDGMENTS

Dominion-Virginia Power provided the Class F fly ash to support this research. Messrs. David Bristow and Ron Birckhead (Dominion) are thanked for their support and involvement. Messrs. Sam McGee and Carlos M. Quinones of USACE Craney Island facilitated the DM sample collection. The lime and CKD were provided by Mr. Jeff Fair of LaFarge North America (Coopersburg, PA). Dr. Chrysochoou completed much of the mineralogical work while at the Stevens Institute of Technology (Hoboken, NJ). Messrs. David Cinsavich and Matt Farley of Schnabel assisted with the geotechnical testing program. Ms. Alexandra Srebro of Schnabel (now UC Berkeley) prepared most of the graphics for this paper. Any opinions, findings, and conclusions or recommendations expressed in this material are those of the author(s) and do not necessarily reflect the views of the project sponsors.

REFERENCES

Dermatas, D. (1995). "Ettringite-induced heaving in soils: State-of-the-art," *Appl. Mech Rev., 38(10),* 659-673.

Dermatas, D., Dutko, P. and Moon, D.H. (1999). "Use of Dredged Materials as Fill in Transportation-Related Projects, Center for Environmental Engineering, Stevens Institute of Technology, Hoboken, NJ, December, pp. 71 (with appendices).

Grubb, D.G., Davis, A., Sands, S.C., Carnivale III, M., Wartman, J., and Gallagher, P.M. (2006). "Field evaluation of crushed glass-dredged material blends," *Journal of Geotechnical and Geoenvironmental Engineering*, 132(5), 577-590.

MDI (Material's Data Inc.), (2005). Jade, Version 7.5, California, U.S.A

ICDD (International Centre for Diffraction Data). (2002). Powder Diffraction File, PDF-2 Database Release.

ICSD (Inorganic Crystal Structure Database). (2007). Fachinformationszentrum Karlsruhe, Germany.

PBQD (Parsons Brinckerhoff, Quade and Douglas), (1999). *Guidance document for processing and beneficial use of dredged material as fill*, Prepared for: The Port Authority of New York and New Jersey, May, pp. 40 (with appendices).

Taylor, H.F.W., Famy, C., and Scrivener, K.L. (2001). "Review Delayed Ettringite Formation." *Cem. Conc. Res.*, 31, 683-693.

Beneficial Reuse of Contaminated Dredge Spoils: Capping of a Harborside Railyard Brownfields Site

Jay A. Borkland[1], Karen Hartel[2], Paul Craffey[3], John Simpson[4]

[1]Waterways Program Manager, Apex Companies, LLC, 115 Broad Street Suite 200, Boston, MA 02110; jborkland@apexcos.com
[2]Principal Engineer, Apex Companies, LLC, 115 Broad Street Suite 200, Boston, MA 02110; khartel@apexcos.com
[3]Project Manager, MA DEP, One Winter Street, Boston, MA 02108
[4]Consultant, Waterways Information Services, 6 Piccadilly Road, South Dennis, MA 02660

ABSTRACT: New Bedford Harbor (Massachusetts) represents a highly contaminated harbor; PCB laden DMs range from 2 to 10,000 ppm. USEPA cleanup of the harbor is limited to PCB's over 50 ppm, leaving residual contamination over much of the active Port. Maintenance dredging has not occurred for over 50 years, forcing many piers to cease operations due to DM buildup. In order to kick-start needed maintenance dredging, the City of New Bedford embarked on an ambitious plan to dredge the slip adjacent to the City's most active pier – the New Bedford State Pier. Economic disposal of the contaminated dredged DM became the most challenging aspect of the project. After evaluating many options, the City settled upon a plan to *beneficially reuse* the material at a railyard brownfields site adjacent to the Harbor. Engineering for the project involved designing an efficient method for transferring the material to the railyard site and a scheme for amending the DM to make the material suitable for permanent placement as landscape berms at the railyard. The project was successfully completed, and the reuse of contaminated dredge material in this case also allowed for the rehabilitation and reuse of not one, but two, distressed portions of the New Bedford waterfront.

1. INTRODUCTION

The Port of New Bedford, Massachusetts represents the largest commercial fishing Port by dollar volume of catch in the United States. Despite its national prominence as a center for fishing and maritime commerce, maintenance of Port infrastructure has been suspended for over 50 years due to the presence of significant levels of PCB's in DMs within the Harbor. The entire harbor was designated a superfund site as a result of widespread distribution of contaminants, predominantly PCBs and metals, in the fine-grained DM. Remediation of the "New Bedford Harbor Superfund Site," is currently projected to take over 40 years. The USEPA cleanup action levels range from 10 to 50 ppm, which means that even after cleanup is complete, more than 2,000,000 CY of DMs containing PCB's will be left behind, impacting most (if not all) future navigational dredging projects required for maintenance of the commercial Port. As there are no licensed facilities in Massachusetts which can accept the (PCB contaminated) material for upland disposal, options for disposal are limited.

Faced with the pending loss of existing commercial Port business due to inadequate draft as a result of non-maintained channels and slips, the City of New Bedford lobbied the State and Federal Authorities to grant the Harbor special regulatory status in order to offset some of the difficulties faced by parties attempting to conduct maintenance dredging in the Port. The City also worked with Coastal Zone Management (CZM) in order to obtain approval for Confined Aquatic Disposal (CAD) facilities to be built within the Harbor for the on-site disposal of contaminated DMs. Recognizing that these initiatives would take some time to achieve the full approval of all regulatory agencies, the City took the bold move of pushing ahead a *Quick Start* project: the maintenance dredging of the State Pier within the central wharf district and the Designated Port Area (DPA) of the Harbor. The goal of the dredging was to return the slip at the State Pier to its authorized depth of -30 feet mean lower low water. The City raised $5 million dollars in funding for the project through the blending of City money and State and Federal grant money, and made the project a centerpiece of its Port development plans. Over a ten month period, the City permitted, designed, and brought the project through construction to its successful conclusion.

Figure 1: Aerial view of the State Pier project area. State Pier dredge area lies south of State Pier and the Railyard disposal site lies north of Fish Island.

2. DISPOSAL ISSUES

Because of the nature of the contamination in the DM (PCBs at greater than 2 ppm), there were very limited options for disposal. Offshore, in-water, and landfill disposal within Massachusetts was not an option, leaving only out-of-state landfilling at approved hazardous waste landfills, and beneficial re-use at brownfields sites. An investigation of off-site disposal alternatives revealed that the most viable options represented trucking or truck-to-rail combinations to landfills in either Michigan,

Utah, or upstate New York. The cost for these options was in the $250 per cubic yard (CY) range. With a required dredge volume of 70,000 CY in order to make the State Pier slip operational, the off-site disposal option was simply not economically feasible. That left the brownfield beneficial reuse option. Engineers estimated of the likely costs to dispose of the dredged material (DM) at a local brownfield site were in the $75/CY range for disposal (see Cost Estimate Summary, Table 1). While still significantly higher than a typical open ocean clean dredge project disposal cost (usually about $25/CY), the brownfield option was significantly less than other upland options, and could fit within the City's budget.

Table 1: Project Cost Estimate Summary

Disposal Option	Est. Cost per CY	CY	Projected Total Cost
Out-of-State (MI)[1]	$ 250	70,000	$17,500,000
Local Landfill[2]	$ 125	70,000	$ 8,750,000
Rail Yard Reuse[3]	$ 75	70,000	$ 5,250,000

Notes:
1. Estimated cost to dewater and ship contaminated DM from New Bedford to disposal site in Michigan by rail. Trucked from New Bedford to Boston, then rail from Boston to Michigan. Estimate generated using USEPA data from the 1998 ROD for the Superfund Site.
2. Estimated cost to dewater and ship contaminated DM to *Crappo Hill Landfill* in New Bedford, approximately 5-miles from the harbor and inland. Cost data for T&D solicited from the industry.
3. Estimated cost to amend DM and reuse on the rail yard site adjacent to the Harbor. Calculated costs based upon project bid data.

A canvassing of brownfields sites in the New Bedford area revealed that several sites in various stages of remediation or redevelopment were located within a 10-mile radius of the Harbor. Most of these were either currently inactive (and were expected to stay that way for some time), or were too far along to benefit from receiving material from the dredge project. Two possible sites emerged. A former landfill that was in the process of final closure needed material to cap an area that had been contaminated with PCBs from industrial waste that had been disposed of in the landfill in the past. The landfill, known as Crappo Hill, needed approximately 50,000 to 100,000 CY of low permeability capping and contour material. The other possible brownfield reuse site was a railyard site that was located adjacent to the Harbor. This site had the obvious advantage of proximity to the Harbor, but also had a significant amount of land. Also contaminated with PCBs from former railyard operations, the site also needed approximately 20,000 to 50,000 CY of low permeability capping material. Additionally, the railyard site was in redevelopment to a functioning railyard and terminal.

A review of these two most promising potential reuse sites indicated that there were several other logistical hurdles to overcome: 1) much of the Crappo landfill area that required capping was wetland, and the potential impact the capping on the wetland resources of the area introduced political and regulatory obstacles; and 2) the sufficient acreage to dewater the DM (the DM would have had to be dewatered elsewhere, necessitating the use of expensive active dewatering processes). The railyard site had the benefit of both proximity to the dredge site and abundant open land for passive dewatering. State approval was rapidly granted (in less than two months) when it became apparent that there were significant advantages to both projects. Permits were submitted to the relevant regulatory agencies while at the same time the design was advanced to completion and plans and specifications were developed for bidding.

3. ENGINEERING AND DESIGN

With a beneficial reuse site selected for the disposal alternative, the DM removal and final placement were designed. Samples of the DM were collected and tested for composition, which determined that all of the material to be dredged was fine grained organic silts and clays (see Table 2 below). After discussion with numerous dredging contractors, DM processing experts and regulators, the project team decided on a process that involved the dredging of the contaminated DM from the Harbor floor using an excavator equipped with an environmental bucket (see Figure 2), and placement of the DM into scows for transport. The scows took the DM approximately ½ mile north in the Harbor to a position just off the shore adjacent to the railyard site. From that position, only a few hundred yards separated the sediment barge from the railyard reuse site, and the slurried DM was transferred to shore via a pipeline.

Figure 2: Environmental bucket dredging in New Bedford Harbor for the State Pier Dredge Project.

Once off-loaded, the sediment would be amended with a mixture of lime and cement (4% by dry weight) and "farmed" out in a sectioned off portion of the site to dewater passively. Once the DM had dewatered, it was graded into berms and covered with topsoil and seeded with wild grasses.

4. MATERIAL COMPOSITION

Reuse of the DM at the railyard site was predicated upon the ability of the engineers to design a process that would render the material suitable for placement in berms that were expected to be approximately 30 feet high. A pilot program was thus established to demonstrate the workability of the final stabilized DM. The sample was obtained by box core (of the upper two to three feet of sediment), which was considered representative of the maintenance DM. (see Table 2 below). The sediment sample was placed in a 5-gallon container for transport to the laboratory. The free liquid was decanted in the laboratory prior to actual testing. Water content, grain size and Atterberg Limits were measured on two samples, obtained from the top (Sample 1) and bottom (Sample 2) of the sediment contained in the bucket. The organic content of one sample of DM was measured.

Table 2: Dredge DM Bulk Sample[1] Analysis - Geotechnical Parameters

Analysis	Sample 1[2]	Sample 2[3]	Comment
Organic Content:	17.4%		ASTM Standard 2974
Soil Moisture:	67.7%	67.4%	ASTM Standard 2974
Soil Moisture: (Dry Mass)	209.3%	206.3%	ASTM Standard 2974
Gravel %:	0.5%	0.1%	ASTM D422
Sand %:	11.4%	8.9%	ASTM D422
Silt %:	70.4%	62.4%	ASTM D422 Hydrometer
Clay %:	17.7%	28.5%	ASTM D422 Hydrometer

Notes:

1. Sample represents bulk 5-gallon sample collected from upper 3-feet of DM composited from 5 randomly chosen locations throughout the dredge footprint.
2. Sample 1 = Sub-sample collected from upper ½ of the 5-gallon bulk sample container.
3. Sample 2 = Sub-sample collected from lower ½ of the 5-gallon bulk sample container.

In addition, Atterberg Limits were determined for DM samples mixed with 4% and 8% of lime and cured for approximately one week. The treated samples were then subjected to water content and Atterberg testing to determine an optimal mixture that would result in a useable product. The tests indicated that a minimum admixture of 4% lime was required to achieve a plastic limit of greater than 95 (see Table 3 below).

Table 3: Dredge DM Admixture Analysis - Atterberg Limits

Analysis	Composite Sample[1]	4% Lime	8% Lime A
Liquid Limit:	125.1	135.8	140.9
Plastic Limit:	87.2	96.0	104.9
Plasticity Index	37.9	39.8	36.0

Notes:

1. Composite Sample = Composite sample collected from 5-gallon bulk sample container. No admixture.
2. Lime Admixture = Dry Weight.

In order to build as much flexability into the project as possible (for those bidding the work), while at the same time reducing the risk to the City (of having an unworkable material), a performance specification was developed that involved a minimum addition of 4% lime/cement, coupled with a requirement that the material pass the paint-filter test. The bidding process resulted in four competitive bids. The project was awarded to a local contractor (Cashman Construction) on the same day that the final permits for the project were received by the City.

5. CONSTRUCTION

For the project, the contractor mobilized a dredge barge with excavator dredge equipment (equipped with the "environmental bucket") and several scows for transport of the material to the shore transport plant. The DM transfer plant consisted of a large screened hopper that fed into a series of large concrete pumps. The DM was pumped to shore via a floating pipeline. Once on shore, the pipeline was buried at a shallow depth to run underneath the road adjacent to the Harbor. The pipeline discharged onto a conveyor that ran under dual lime and cement silos for amendment (see Figure 3). The amended material was then extruded onto a paved area of the railyard site, where it was "farmed" into thin (approximately two to three foot) lifts for initial curing (see Figure 4). The material was tested by paint filter and once it met the performance specification for stackability, the material was moved to a vacant portion of the site for stacking into mounds for further drying and setting (see Figure 5).

Figure 3: Lime silo at the railyard amendment site.

Figure 4: "Farming" amended material out for "curing" at the railyard beneficial reuse site.

Figure 5: Amended DM spread out at the railyard beneficial reuse site. As the material set up to the desired moisture content, it was stacked into berms and moved to its final location.

6. SUMMARY AND CONCLUSIONS

Once the material had dried out sufficiently (approximately six months after the completion of the dredging), crews moved the material into place in sculpted berms at the railyard in the designated location called for by the cap design (see Figure 6).

Figure 6: Final orientation of the vegetated DM berms at the railyard site.

Today the railyard is active and the DM berms stand as testament to the fact that cooperation between local, state, and federal officials can benefit multiple parties with greats result for all. The 600-foot 1,000 passenger cruise ship the *Regal Empress* docked at the State Pier in New Bedford approximately 10 months after the project kickoff meeting for the dredge project (see Figure 7). The *Quick Start* State Pier Dredge Project became known as Phase I of the New Bedford Harbor Maintenance Dredge Program, which is currently in Phase III with nearly 14 separate dredge projects completed or nearing completion to date.

Figure 7: The Cruise Ship Regal Empress enters New Bedford Harbor after the successful completion of the State Pier Dredge Project.

7.0 REFERENCES

Sturgis, T. C. and Gunnison, D. (1988). "New Bedford Harbor Superfund Project, Acushnet River Estuary Engineering Feasibility Study of Dredging and Dredged Material Disposal Alternatives. *Department of the Army Waterways Experiment Station*, Vicksburg, MS.

USEPA Record of Decision for the New Bedford Harbor Superfund Site (1998), US Environmental Protection Agency.

Navigating the Regulatory Environment: Beneficial Use of Dredged Sand in New Bedford Harbor

Karen Hartel[1], P.E. and Jay Borkland[2], P.G.

[1]Principal Engineer, Apex Companies, LLC, 115 Broad Street Suite 200, Boston, MA 02110; khartel@apexcos.com
[2]Program Manager, Apex Companies, LLC, 115 Broad Street Suite 200, Boston, MA 02110; jborkland@apexcos.com

ABSTRACT: Navigational dredging in New Bedford Harbor has become prohibitively expensive due to issues associated with dredging and off-site disposal of contaminated sediment. As a result, The City of New Bedford has designed and constructed a series of confined aquatic disposal (CAD) cells within the harbor for disposal of contaminated dredged sediment. This paper focuses on the reuse of clean sand excavated from a CAD cell within the harbor as cap material for an area of contaminated sediment in the outer harbor. Design of the United States Environmental Protection Agency (USEPA) Superfund cap was performed through a partnership with the Army Corps of Engineers (USACE), USEPA and Massachusetts Department of Environmental Protection (MADEP). Construction of the cap involved placement of sand using split hull barges in shallow water conditions (less than 10 ft), while maintaining minimum and maximum cap thickness between 2 and 6 ft.

1. INTRODUCTION

Navigational dredging of New Bedford port facilities was suspended in 1970, when the entire harbor was designated a Superfund site due to widespread contamination in the fine-grained sediment, predominantly PCBs and metals. Ongoing remediation of the New Bedford Harbor Superfund site is currently projected to be completed in 2045. USEPA established cleanup levels ranging from 10 ppm to 50 ppm for PCBs, with the higher cleanup level in effect over much of the harbor. Completion of the superfund cleanup will leave behind more than 2,000,000 cy of hazardous material which will impact future navigational dredging projects. Dredging of PCB-contaminated fine-grained sediment at the levels encountered in New Bedford Harbor post-cleanup is prohibitively expensive due to absence of local disposal options.

Faced with the loss of use of its port facilities, the City of New Bedford partnered with the USEPA, MADEP, the USACE and other federal, state and local regulatory agencies to perform necessary maintenance dredging projects since 2002.

The City has constructed on site CAD cells to allow for disposal of contaminated fine-grained sediments resulting from maintenance dredging within the designated port area.

2. REGULATORY SETTING

Prior to designation of the harbor as a Superfund Site in 1970, maintenance dredging of the federal navigational channel was performed by the USACE. From 1970 to 2001, dredging within the port was limited to that performed by the USEPA as part of the Superfund remediation. Much of the navigational channel and commercial port area contains sediment contaminated with between 2 and 50 ppm PCBs, and hence falls outside the purview of the superfund removal action. The City of New Bedford initiated maintenance dredging of the commercial port area through formation of a Portfields committee, involving participation of the USEPA, MADEP, and other relevant agencies having jurisdiction.

The Portfields Committee adopted a provision of the Superfund Record of Decision (ROD) called the "State Enhanced Remedy" (or SER) to facilitate design and oversight of City dredge projects. The provision was created specifically to address future maintenance or improvement projects in the Harbor which would involve dredging and disposal of sediment with contamination at levels that precluded in-state disposal. The SER provision was designed to allow the Port to emulate the Superfund process in the way it would deal with contaminated sediments that were outside the Superfund cleanup areas.

Through the SER process, the City proposed developing a series of CAD cells within the harbor to be used for disposal of 1,800,000 cy of contaminated maintenance dredge material projected to be generated over a 20 yr period. The SER provision, administered by the USEPA, USACE and MADEP, allowed the initial CAD cell project to advance under the regulatory authority of the Superfund program, thus eliminating the requirement to obtain permits from each of the oversight agencies. Design and construction of the project was performed in accordance with the intent of the permitting requirements, however, the long time frames (anticipated at greater than a year) associated with review and issuance of the required permits was avoided. Regulatory approval of project plans and specifications by the City was accomplished in a matter of several months, and as such was considerably expedited through the SER process.

3. CAD CELL DESIGN

The design process resulted in an initial CAD cell sited in the inner harbor, as shown on Figure 1. The CAD cell location was selected to provide the following: minimum thickness of contaminated sediment at the site, existing bathymetry sufficient for draft requirements of dredges and barges, sufficient depth to bedrock to allow optimizing cell depth (and volume), and adequate distance from active shipping traffic. To accommodate the first phase of navigational dredging, the CAD cell was designed for a disposal volume of approximately 90,000 cy. Generalized stratigraphy within the CAD cell area consisted of a surficial deposit of very soft to soft organic silt, underlain by medium dense to dense fine sand, in turn underlain by dense glacial till over parts of the site, and bedrock. Thickness of the contaminated organic silt layer averaged approximately 3.5 ft within the footprint of the CAD cell. The underlying fine sands were shown to be uncontaminated.

FIG. 1 Site Plan

Excavation of the CAD cell generated approximately 20,000 cy of contaminated organic silt and 80,000 cy of clean fine sand. Dredging of the organic silt was accomplished with a clamshell dredge fitted with an environmental bucket, and the sediment was disposed of in an adjacent existing borrow pit. Grain size analyses indicated that the uncontaminated material to be excavated from the CAD cell consisted of fine sand, little medium sand, trace coarse sand and silt. Permit applications were filed by the design team early in the design phase to allow for offshore disposal at the Cape Cod Disposal Site (CCDS), located approximately 45 nautical miles from New Bedford harbor.

4. OPERABLE UNIT #3 CAP

However, during the initial phase of construction of the CAD cell, the USEPA identified an alternative option for reuse of clean sand generated from the CAD cell excavation. A 26.5 acre area in New Bedford's outer harbor contained sediment contaminated at levels above the Superfund cleanup criteria. Although PCB levels were elevated, remediation of the area designated Operable Unit #3 (OU#3, see Figure 1), was a low priority, and was targeted for remediation late in the overall Superfund cleanup horizon. The City modified the CAD cell contract documents to accommodate transport and placement of the clean sand within the OU#3 area, rather than disposal at the CCDS, with the provision that the modification did not result in an increase to the negotiated unit rates for dredging and disposal. A capping steering committee was formed to facilitate a rapid design for the cap itself.

4.1 Site Conditions

Bathymetry collected within the OU#3 cap area indicated the mudline elevation varied from El. –4 MLLW at the west side of the area, to El. –13 MLLW at the east side. Subsurface investigations indicated that the cap site was underlain by a surficial deposit of soft non plastic silt with sand, varying in thickness from 0.5 to 4.5 ft across the site. The sand content within the silt layer ranged from 17% to 44%. Denser interbedded sand with varying quantities of silt and gravel were present immediately below the silt deposit. Laboratory analytical testing was performed to provide PCB concentration data to supplement the USEPA data. PCB concentrations were interpolated between sample points to produce a PCB concentration contour plan over the footprint of the OU#3 area. The interpreted 10 ppm PCB concentration contour became the proposed cap limits, based on the Superfund cleanup criteria (10 ppm) for the area, see Figure 2.

4.2 Cap Thickness

Required cap thickness was determined based on consideration of the following design components: chemical isolation, bioturbation, cap material consolidation and erosion. Thickness for chemical isolation (14 inches) and bioturbation (8 inches) had previously been evaluated by the USACE in conjunction with the engineering feasibility study conducted for the New Bedford Harbor superfund project (Sturgis and Gunnison 1988). An estimate of cap consolidation was performed assuming hydraulic placement of the cap material would result in a loose initial condition of the sand, which would densify under the cap self weight, resulting in a conservative estimate of cap consolidation of less than ½ inch. Erosion potential of the cap material was evaluated by comparing measured current velocities within the cap area to available threshold reference values, for the given grain size distribution of the cap material. Due to the location of the site in a relatively protected cove area, erosion due to the low measured current velocities was predicted to be minimal. Erosion due to storm events was considered a cap maintenance issue. Summation of the various components yielded a design cap thickness of 2 ft.

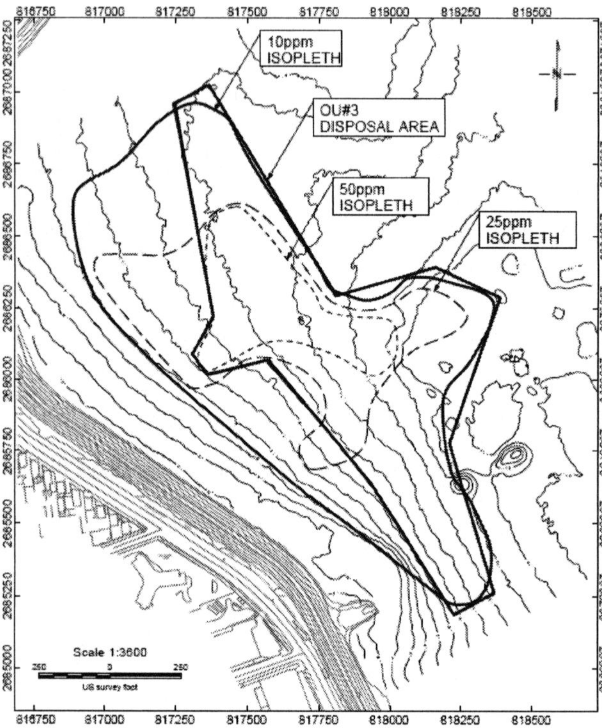

FIG. 2 OU#3 Cap Site Plan

4.3 Cap Placement Constraints

A combination of existing conditions within the OU#3 area and limitations resulting from the contract modification negotiations provided challenges relative to cap design and placement. The area encompassed by the 10 ppm PCB concentration contour was approximately 26.5 acres; a 2 ft thick cap would require approximately 89,000 cy of material, versus the anticipated 80,000 cy of available capping material. Given the relatively shallow water in the cap area, and the finite volume of cap material available, the optimum placement method in terms of providing a uniform cap thickness was hydraulic. However, the contractor determined that hydraulic placement would result in additional operations costs above the transport and disposal for the CCDS alternative.

Evaluation of the draft requirements of the contractor's split hull barges relative to the shallow water depths in the cap area showed that portions of the OU#3 area could only be capped at high tide, and areas shallower than El -9 MLLW (the west edge of the footprint) could not be capped given the existing barge fleet. In addition, discharge of the cap material from the split hull barges would result in a non-uniform thickness of cap material above mudline. The steering committee determined that the overriding design parameter was to provide a *minimum* cap thickness of 2 ft, recognizing that the non-uniform bottom distribution and limited supply of cap material would result in a smaller cap footprint.

Detailed analysis of barge dimensions, capacities and associated draft allowed prediction of the bottom distribution of the cap material immediately post discharge. Based on the predicted distributions, a preliminary barge discharge plan was created for the cap footprint, with estimated maximum lane widths to maintain the minimum required cap thickness, through overlapping of discharges. This information was used (along with draft restrictions) to prioritize areas to be capped within the 10 ppm PCB concentration contour. The resulting design limits of the cap are indicated on Figure 2. The cap limits were designed to completely encompass the 50 ppm PCB concentration contour, as well as most of the 25 ppm PCB concentration contour. Full coverage of the 50 ppm concentration area required capping a small portion of the footprint with existing mudline shallower than El. –9 MLLW. Capping of an adjacent area containing sediment with PCB concentrations between 25 and 50 ppm was not attempted due to inadequate water depths.

Specifications for the capping operation contained performance criteria negotiated with the contractor relative to minimum and maximum cap thickness (2 and 6 ft respectively), and area of coverage. The performance criteria also outlined modifications to the dumping sequence to be implemented in the event that performance criteria were exceeded. These modifications included adjusting lane widths and scow loading as necessary to optimize coverage based on post-discharge surveys.

5. CAP CONSTRUCTION

Construction of the cap commenced in June 2005. The capping sequence was designed to progress from north to south, given that the priority areas (concentrations above 25 ppm PCBs) covered the north end of the cap footprint. The cap area was divided into east-west oriented discharge lanes, allowing for optimum barge orientation relative to maximization of available draft. Control of barge position during discharge was a critical concern in order to avoid grounding in shallow water. For this reason, the contractor elected to position a dredge barge along the south edge of the discharge lane, to act as a fixed (spudded) platform. Loaded barges were tied off to the fixed platform during discharge. Bathymetric data was collected immediately following the initial discharges, and evaluated to determine the as-built distribution of cap material. The contractor utilized a fleet consisting of three split hull barges, each having different geometries, capacities and associated drafts. Optimum distribution of discharged cap material given the shallow water conditions at the site resulted from a combination of a large barge footprint and shallow overall barge depth. The lack of draft within the cap footprint meant that there was potential for grounding the barges during discharge events, in the open hull position.

Lane widths and load amounts were adjusted by trial and error after review of the post-discharge bathymetry, to meet the performance criteria relative to minimum and maximum cap thickness. In actual practice, optimum lane widths varied for each of the three barges. Better coverage could be achieved utilizing a single barge for all discharges within a given lane, although this was not always possible without schedule impact to the contractors dredging operation, given the barge cycle times and tidal cycles. The contractor attempted to address areas between discharge lanes where minimum cap thickness was not achieved. Lightly loaded barges were floated at high tide over the target area, and discharged. This technique required using tugs to position the barge during discharge, resulting in reduced accuracy of barge position as compared to utilization of the fixed platform, and hence, increased risk of barge grounding. The technique was successful over a limited area, generally only along the east edge of the cap footprint where water depth was greatest.

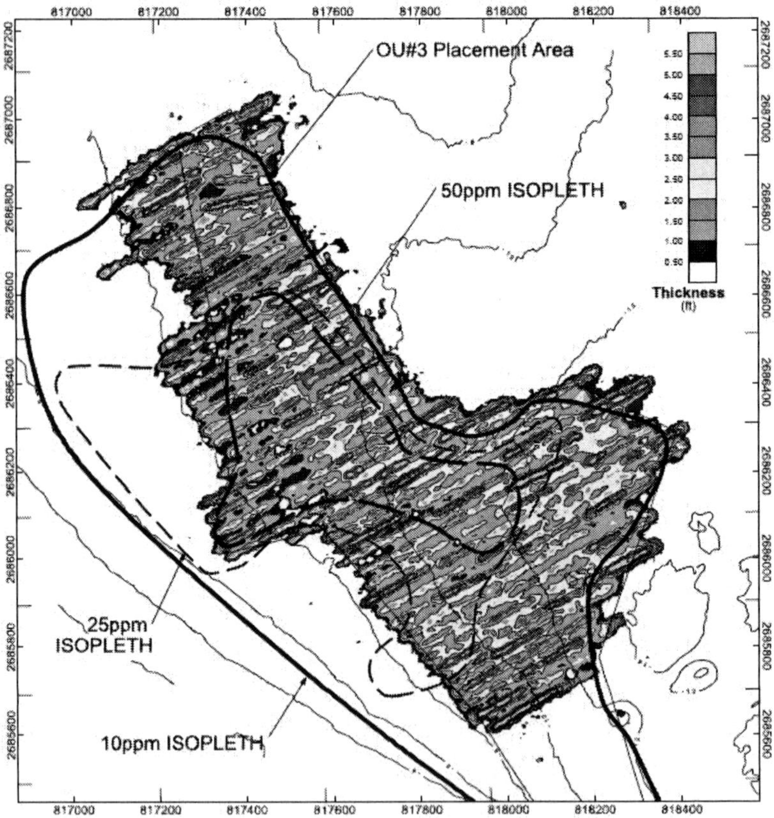

FIG 3. Cap Thickness July 2005

Dredging and capping was completed by the end of July, 2005. A total of 84,193 cy of cap material was placed, resulting in a final cap footprint of 18.9 acres. Post-cap bathymetry data indicated that the average thickness of cap material over the footprint was 2.2 ft, and minimum and maximum thicknesses were 0 and 8 ft, respectively. The design minimum cap thickness of 2 ft was achieved over 65% of the cap footprint, while 95% of the footprint had a minimum cap thickness greater than one ft. The distribution of cap thickness is indicated on Figure 3.

Additional bathymetric data was collected 6 months after cap completion, which indicated redistribution of the cap material has occurred, possibly as a result of wave action. In general, the data indicate that the cap material is tending to level off slightly. The maximum cap thickness observed in January 2006 was 7.4 ft, down from 8 ft immediately post-cap placement, and the area of cap with a minimum thickness of 2 ft had increased from 65% to 70%.

6. CONCLUSIONS

The successful completion of the projects described herein is a direct result of the collaboration between multiple entities advancing the common goal of improving the port of New Bedford. Representation of the City of New Bedford, the USEPA, the MADEP, fisheries, etc. on the portfields committee resulted in adoption of the SER, which allowed the projects to advance in a cost effective, timely manner. Dredging of the CAD cell, transport to the OU#3 area and placement of the cap was accomplished at a cost of $25/cy. Formation of the Capping Steering Committee facilitated rapid design and construction of the Superfund pilot cap in the OU#3 area. The OU#3 project duration, from initiation of cap design through completion of construction, was less than four months. Neither of these projects would have been accomplished within the relatively short time frame without cooperation among all of the referenced entities. These projects can be considered as a model for maintenance and development of other ports and harbors facing similar challenges.

Completion of the pilot capping study at the OU#3 area in the outer harbor demonstrated that placement of cap material by discharge from split hull barges can be a viable, cost-effective methodology, even in shallow water. Establishment of performance criteria relative to cap thickness prior to bidding would have allowed bidders to optimize the dredge plant relative to the performance requirements, and potentially result in increased uniformity of coverage and cap thickness.

7. REFERENCES

Sturgis, T. C. and Gunnison, D. (1988). "New Bedford Harbor Superfund Project, Acushnet River Estuary Engineering Feasibility Study of Dredging and Dredged Material Disposal Alternatives, Report 6 Laboratory Testing for Subaqueous Capping." *Department of the Army Waterways Experiment Station*, Vicksburg, MS.

Feasibility of Treating Contaminated Dredged Sediments Using Ultrasound with Acoustic and Flow Fields

Jay N. Meegoda, M. ASCE

Professor, Department of Civil & Env. Eng., New Jersey Institute of Technology, University Heights, Newark, NJ 07102; meegoda@njit.edu

ABSTRACT: The US requires long-term alternatives to dispose more than 230 million cubic meters of dredged material per year. The dredged sediments in a slurry form need to be de-watered, remediated and disposed. The contaminants (heavy metals and organics) in the dredged sediments are usually attached to the clay and silt fractions. A comprehensive treatment technology coupling ultrasound with acoustic and flow fields was proposed to remove contaminants and water from dredged residues. A complete treatment technology with three different processes was proposed for commercial implementation. In the first process coarse sediments are decontaminated using ultrasound, and fines and the bulk dredged suspension are removed using a sub-atmospheric pressure. The second process treats the fines and bulk solution from the first process using ultrasound and fine sediments are separated from bulk fluid containing contaminants with water using acoustic and flow fields. The treated sediments can be ocean dumped as clean material or used in construction. The extracted organic and inorganic contaminants can be treated using advanced oxidation or concentrated using membrane technology and disposed in hazardous waste landfills. This technology can be easily added to a dredging vessel to minimize material handling. A conceptual arrangement of onboard treatment of sediments inside a dredging vessel is presented. The result of this research indicated that the proposed technology would provide viable alternatives for remediation of contaminated dredge sediments.

Keywords: Dredged sediments, contaminated, ultrasound, acoustic field, flow field, reuse

INTRODUCTION

To ensure safe, navigable waters, rivers and waterways must be dredged. Dredging in the US requires long-term alternatives as more than 230 million cubic meters of dredged material per year need to be disposed. In many locations containment facilities are almost full. Sites for additional capacity are scarce and expensive to construct. Dredged material research has been on going for a number of years. However, the disposal techniques have concentrated on sea disposal, artificial islands or disposal into lagoons for either contaminant or for long term land reclamation. The chemical status of the material determined the end use. The process has been a fairly passive one. The United States Army Corps of Engineers in their Dredged Material Reach Program have postulated a series of options for the use of dredged material including transfer for use at other sites

It is estimated that dredging of NY/NJ ports will generate 7 million tons of residues per year. The dredged sediments in a slurry form need to be de-watered, remediated and disposed. The contaminants (heavy metals and organics) are usually attached to the clay and silt fractions. The best remediation should have a process for separation of particles based on size followed by treatment of wastewater. The ultrasound technology may be applied to decontaminate dredged sediments.

This manuscript describes a novel technology using ultrasound with acoustic and flow fields to treat the contaminated dredged sediments while being dredged.

PROPSED TECHNOLOGY

Ultrasound: Ultrasound energy can be used to decontaminate soil water slurry such as dredged sediments. Ultrasound is widely used in industry for a variety of applications such as in medical imaging, cleaning metal parts, removing oxide films, oil, grease and other contaminants from solid surfaces and plastic welding, and even to removing contaminants from soils. In the area of Sono-chemistry, high frequency ultrasound is used to destroy complex organic compounds. Ultrasound refers to sound waves having frequencies in the range 16 kHz to 500 MHz, inaudible to the human ear, and can be transmitted through any medium with elastic properties, including water, gas-saturated water and aqueous particle suspensions. It is transmitted in the form of mechanical energy.

The ultrasonic cleaning is an application similar to the proposed technology and it works by providing shear forces to remove the material adhered to surface. This shear force is developed by cavitation. Ultrasound causes high-energy acoustic cavitation: the formation, growth and implosive collapse of bubbles in a liquid. That is the formation of microscopic vapor bubbles in the low-pressure (rarefied) part of the ultrasonic wave. These bubbles collapse in the compression part of the wave creating minute, but high-energy movements of the solvent that result in localized high shear forces. During cavitational collapse, intense heating of bubbles occur. These localized hot spots have temperatures of roughly 5000°C, pressures of 500 atmospheres, and a lifetime of a few microseconds (Suslick, 1990). Shock waves from cavitation in liquid-solid slurries produce high-velocity inter-particle collisions, the impact of which is sufficient to melt most metals. Applications to chemical reactions exist in both liquids (homogeneous) and in liquid-solid (heterogeneous) systems. Of special synthetic use is the ability of ultrasound to create clean, highly reactive surfaces on metals. Ultrasound has also found important uses for initiation or enhancement of catalytic reactions, in both homogeneous and heterogeneous systems. These "cavities" or areas of low pressure provide a sink of low concentration or partial pressure on contaminants causing adsorbed materials to desorb. Also, cavitation is the most important factor in determining dispersion and de-aggregation of soil in ultrasonic system (Watson, 1971).

Combined use of acoustic and flow fields: In an acoustic field, forces exerted on particles in the medium are proportional to the local velocities of the fluid and when averaged it becomes zero. In an acoustic standing wave field, the averaged forces acting on a particle are not zero and play an important part as they arise from second order effects (Gor'kov, 1962). The dominant force is the primary axial acoustic radiation force (F_{PARF}). Immediately after the application of an acoustic field, particles experience a time-averaged primary axial acoustic force, F_{ac} (F_{PARF}), which is generated by the interaction between particles and the wave field. The primary axial radiation force drives dispersed particles toward the velocity antinodes of the resonance field, so that the average distance between the particles will considerably decrease. The magnitude of the force depends on the compressibility and density difference between the particle and the medium. The axial primary radiation force has sinusoidal variation and is proportional to the acoustic energy density. This force acts in the direction parallel to the propagation of waves. The response of any solid suspended in a fluid to a resonant acoustic field depends on the acoustic contrast factor, G which is a function of the compressibilities of the fluid and particle respectively, and the densities of fluid and particle respectively. If G is positive, the particles move towards the pressure nodes and if G is negative, the particles move towards the pressure anti-nodes. Hence the above force can be used to change the particle trajectory of moving particle suspension and concentrate the particles to a node or anti node based on G value.

Proposed Technology: In this research ultrasound energy was used to clean the course sediments and to separate fines in dredged sediments using sub-atmospheric pressures. Then in a separate process ultrasound is applied again to the decontaminate fines and the fines are removed from contaminants and water using acoustic and flow fields. The membrane technology can be used to remove water and concentrate the contaminants extracted from the sediments. The complete technology is shown in Figure 1.

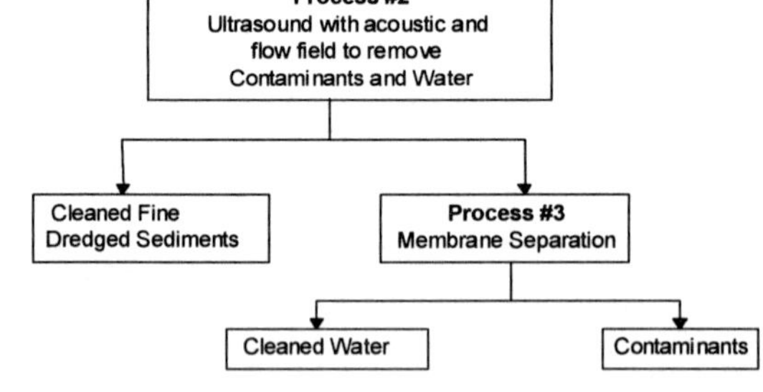

Figure 1 A Flow Chart of the Proposed Technology

Process #1: Meegoda and Perera (2001) and Meegoda and Veerawat (2002) designed an experimental apparatus to separate fines from the bulk-dredged suspension and remove the fines using sub-atmospheric pressure while treating the coarse particles. Figure 2 shows the experimental apparatus specially designed to implement process #1. Laboratory model was built as a cylindrical container separated into two parts with a US#200 mesh sieve, which was used to separate the fines from coarse fraction. The outlet of the model was connected to a vacuum through a sedimentation tank, which was used to collect fines and water. The ultrasound probe was dipped into the sample slurry. The total volume of sediment slurry was kept constant (500mL) during the experiments. To find the initial contaminant concentration, the prepared sediment sample was sieved using a US#200 mesh sieve and the retained fraction, which was considered as the coarse fraction was analyzed to determine the contaminant concentration. The slurry was made with the measured amount of sediments knowing the solids to water ratio and keeping the total volume constant at 500mL. The slurry was placed inside the container as shown in Figure 2, and ultrasound energy with the required power setting and desired vacuum pressure were applied for given experimental condition selected from the experimental design. Water was added as needed to maintain the constant volume at 500mL. At the end of application of ultrasound, coarse fraction retained was oven dried, measured, to find the separating efficiency, and was analyzed to determine the final contaminant concentration. After oven drying the separated fines were used as the feed material to evaluate porcess#2.

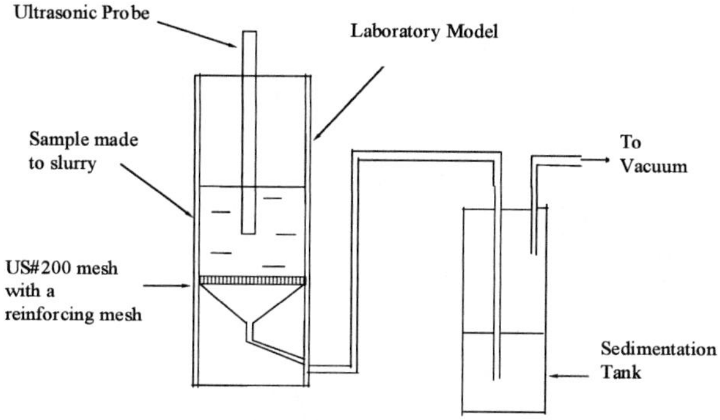

Figure 2 Laboratory Setup to Evaluate Process #1

Process #2: Meegoda and Perera (2001) and Meegoda and Veerawat (2002) designed an experimental apparatus to decontaminate fines and to separate the contaminated water using a sub atmospheric pressure. Figure 3 gives the laboratory set up for process #2. A laboratory batch test module in the shape of a box was made of five porous stones as the sample container, where the ultrasound probe was dipped, and has an outer box to apply vacuum pressure to the porous box. The specially selected porous stones were used to separate the water from fines. The outlet from the outer box was connected to a retention tank to collect wastewater, which was in turn connected to the vacuum system. Collected fine sediments from process #1 were used as the feed material to evaluate process #2. These sediments were tested for the contaminant concentration after process #1. This concentration was considered as the initial concentration for the process #2. The total volume of sediment slurry was maintained constant (1000mL) during each experiment. The slurry was made with the measured amount of sediment, each time according to the solids to water ratio, with 1000mL as the total volume. The slurry was placed in the batch device and ultrasound energy with the required power setting and the maximum vacuum pressure was applied for a required time period selected from the experimental design. Water was added as needed to maintain the constant volume. At the end of the experiment the retained slurry was oven dried and the final contaminant concentration was determined. The analysis of contaminants in the separated fine fractions was performed using the procedure used for process #1.

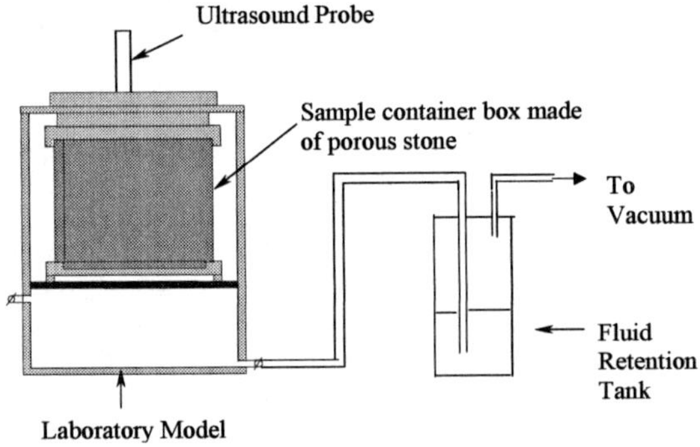

Figure 3 Laboratory Setup to Evaluate Decontamination in Process #2

The technology based on ultrasound consisted of two coupled processes to separate and to treat both coarse (Process #1) and fine (Process #2) fractions of the sediments. Selected variables for evaluation of Process#1 were power, soil-water ratio, vacuum pressure and dwell time, and for evaluation of process #2 was power, soil-water ratio and dwell. Experimental data showed that the process #1 could effectively separate and treat the coarse fraction of the sediment with 100% separation efficiency and 92% removal efficiency for inorganic contaminants and 100% separation efficiency and 99% removal efficiency for organic contaminants. The process #2 was not that effective in treating the silt fraction of the dredged sediments with 100% separation efficiency and 82% removal efficiency for inorganic contaminants and 100% separation efficiency and a maximum removal efficiency of 56% for organic contaminants. However, there was difficulty in decontaminating or separating ultra fine particles such as clays.

Therefore, Process #2 was modified to decontaminate fines and remove cleaned fines from contaminants and water using acoustic and flow fields. Nazhat et al., 2003 and 2005 developed a flow chamber with two transducers at opposite ends to generate an acoustic standing wave. The acoustic chamber was made of clear plexi-glass. Two lead zirconate titanate transducers from American Piezo Ceramics Inc. (transducer #APC 880), with a fundamental resonance frequency of 320 kHz, were fixed into the inside walls of the chamber facing each other and flush with the inside walls as shown in Figure 4. Vertical walls of the chamber were placed parallel to each other with a narrow gap. The top end of the chamber was connected to a vacuum by a short rubber tube, while the bottom was connected to a glass tube to pump in the solution of water and suspended particles.

The back electrode of the load transducer was connected to an Anritsu Synthesizer/level signal generator MG 443B through an ENI 1040L power amplifier to power the transducer. The other transducer (acting as a receiver) was connected to a Tektronix T912 Oscilloscope. Different electrical units were interconnected using Bayonet Neil Concelman (BNC) coaxial cables.

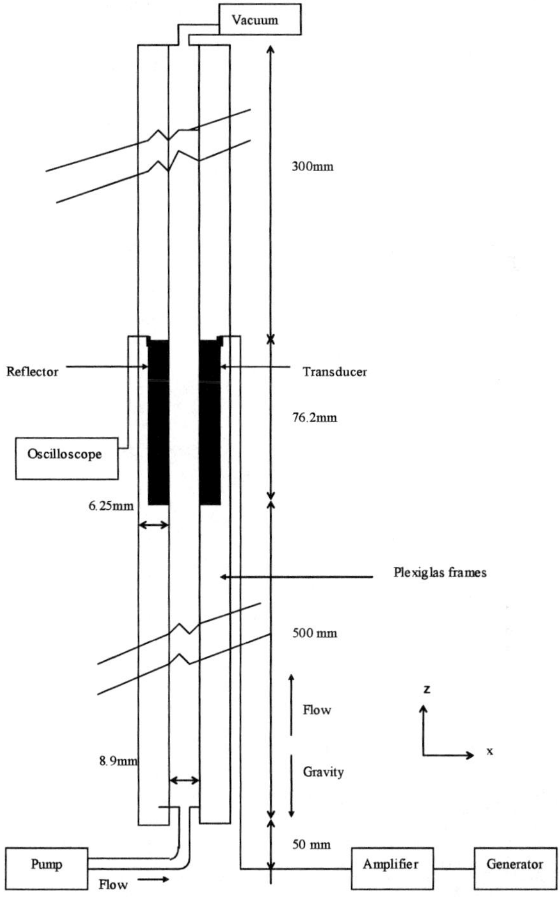

Figure 3 Laboratory Setup to Evaluate Particle Separation in Process #2

The feasibility of using the proposed technology for silica fume (SiO_2) and silicon oxide (SiC) in deionized water was studied. For input power levels between 3 W and 5 W, the experimental data matched the mathematically predicted trajectories (Nazhat et al., 2003 and 2005). Hence based on preliminary results it was concluded that acoustic field can be used either to separate or fractionate fine particles from the bulk fluids.

DEVELOPMENT OF A PROTOTYPE

Conceptual diagram of continuous flow treatment module incorporating process #1 is shown in Figure 4. The ultrasound transducers will have six different power supplies and each power supply will be made with a phase difference of 60°. The purpose of this phase shift is to minimize the vibrations on supports while delivering the total power to the dredged sediments. An inner perforated tube with #200 mesh will be fitted to extract fines (Figure 4). The fluid loss due to extraction of fines will be supplemented by adding treated/clean water from process #3.

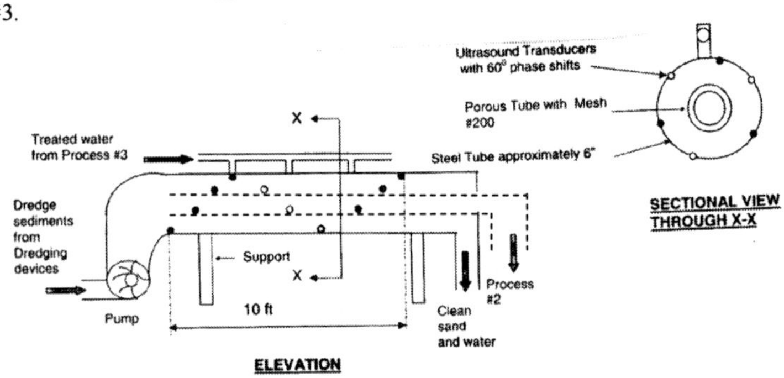

Figure 4 Conceptual Design of Scaled up Process #1

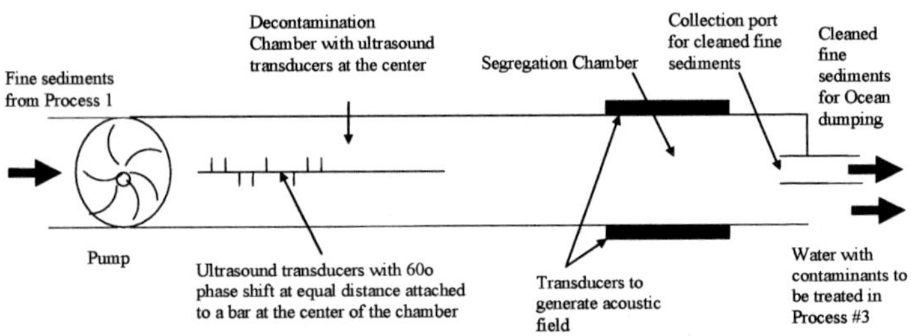

Figure 5 Conceptual Design of Scaled up Process #2

Conceptual diagram of continuous flow treatment module incorporating process #2 is shown in Figure 5. The treated sediments from process #1 will be saved and used in process #2. In order to maximize the extraction area per unit volume the contaminated fine sediments from process #1 will be pumped into the process #2. The ultrasound transducers will be mounted onto a rod at the center of this tube. The ultrasound transducers will be connected to different power supplies with 60° phase shifts as in process #1. The fluid loss due to extraction contaminated water will be supplemented by adding treated/clean water from process #3.

It is proposed that the technology described can be scaled up for continuous treatment of dredged sediments using decontamination module shown in Figures 4 and 5. It is anticipated that several such modules can be retrofitted into dredging vessels to treat all the sediments dredged. The exact number of modules will be determined based on the average volume of sediments dredged per day. The sediment holding tank in the dredging vessel/ship will be divided into clean and dirty compartments. The decontamination modules will be connected between the clean and dirty tanks. Those modules will operate 24 hours or will be off during the hours of dredging. The decontamination modules will treat all sediments within 24hours and treated/clean sediments can be disposed before beginning the dredging operation on the following day. The decontaminated sediments can be either dumped in the ocean as clean material or used in construction. Based on the laboratory data, the extracted organic and inorganic waste will be concentrated using membrane technology to approximately 8% the original volume using process #3 and can be disposed in hazardous waste landfills. Initial cost estimate based on the laboratory data shows that the treatment cost including the amortized capital expenditure to be around $35 per ton.

SUMMARY AND CONCLUSIONS

A comprehensive treatment technology coupling ultrasound and acoustic and flow fields was proposed to decontaminate contaminated dredged sediments and remove contaminants and water from dredged residues. A complete treatment technology with three different processes was proposed for commercial implementation. The process #1- coarse sediments are decontaminated using ultrasound, and fines and the bulk dredged suspension are removed using a sub-atmospheric pressure. The process #2- the fines and bulk solution from process #1 are treated using ultrasound and fine sediments are separated from bulk fluid containing contaminants with water are removed using acoustic and flow fields. Two new processes #1 and #2 were evaluated in laboratory scale batch tests. The laboratory scale tests showed the feasibility of the proposed technology. A scaled up conceptual designs were developed for commercial implementation.

ACKNOWLEDGMENTS

This research was sponsored by a grant from the National Science Foundation (Grant # CMS-9700318). Author would also like to acknowledge the NSF grant ECS-9214571 to NJIT to create a state-of-the-art Geo-environmental laboratory.

REFERENCES

Aboobaker, N., Meegoda J. N. and Blackmore, D., (2003), "Fractionation and Segregation of Suspended Particles using Acoustic and Flow Fields" ASCE Journal of Environmental Engineering, Vol. 129, No. 5, pp. 427-434.

Aboobaker, N., Blackmore, D., and Meegoda, J. N., (2005), "Mathematical modeling of the movement of suspended particles subjected to acoustic and flow fields" Applied Mathematical Modeling, Volume 29, Issue 5, pp. 515-532. 6.

Gor'kov, L. P., "On the Forces Acting on a Small Particle in an Acoustic Field in an Ideal Fluid," Soviet Phys.-Doklady, Vol. 6, pp. 773-775, 1962.

Meegoda, J. N., and Perera, R., (2001), "Ultrasound to Decontaminate Metals in Dredged Sediments" Journal of Hazardous Materials Vol. #85, pp. 73-89, paper # 2656.

Meegoda, J. N., and Veerawat, K., (2002), "Ultrasound to Decontaminate Organic Compounds in Dredged Sediments" Soil & Sediment Contamination: An International Journal, Vol. 11 #1.

Suslick, K. S., 1990. "Sonochemistry" *Science*, 257:1439- 1445

Watson, J. R., 1971, "Ultrasonic Vibration as a Method of Soil Dispersion," Soils and Fertilizers, Vol. #34, pp. 127-134.

Index Properties and Compaction Characteristics of Kerosene Contaminated Clayey Soil

C. Vipulanandan, Ph.D., P.E.[1], and P. Elesvwarapu, M.S.[2]

[1]Chairman, Professor and Director of Center For Innovative Grouting Materials And Technology, Dept. of Civil and Environmental Engineering, University of Houston, Houston, TX 77204-4003
[2]Former Student, Dept. of Civil and Envir. Engrg., Univ. of Houston, Houston, TX, 77204-4003.

ABSTRACT: In order to excavate and transport kerosene contaminated soils from construction sites, it is critical to understand the changes in geotechnical proportions of the soil when it is contaminated with kerosene. In this study, soils consisting of 80% sand and 20% clay with and without organic matter (5%) were synthetically contaminated up to 7.5% of kerosene (based on dry soil weight). Effect of organic matter and kerosene content on the index properties and compaction characteristics (Harvard miniature) of the clay soils was quantified. Addition of kerosene increased the liquid limit and reduced the maximum dry density of the soil. The liquid limit of the soil was increased by 75 % and the maximum density was reduced by 6 % due to the addition of 7.5% of kerosene.

INTRODUCTION

The contamination at the construction sites comprises of both hazardous and non-hazardous waste. The presence of contamination does not automatically require expensive treatment strategies. But if the contamination is due to hazardous substances and/or petroleum products which will subsequently pose threat to environment as well as human life, the site should be remediated before/during/after construction (Vipulanandan 1995). When soil and groundwater contamination are found at a construction site, completing the project on time and within the budget is always a challenge. Early detection before construction begins is the best defense against added costs and delays. The physical risk involved with handling and disposing of contaminated materials is substantial and often requires special training and equipment.

Contamination will affect the geotechnical properties of soils (Meegoda et al. 1998). In order to implement the contamination management plan, it is important to understand the behavior of soil. Kerosene has been identified as one of the potential contaminants in the transportation projects by the Oklahoma Department of Transportation. Kerosene was the predominant petroleum contaminant in 5 of the 83 sites reported in the NCHRP Synthesis 226 (1996). Kerosene is a major ingredient in jet fuel and is also widely used in oil refineries. Only limited information on the geotechnical properties of contaminated soils is available in the literature. Most of the studies on contaminated soil in the literature involved leachability and treatment studies. Hence the behavior of kerosene-contaminated soil (80% sand, 20% kaolinite, and up to 5% organic matter) was investigated in this study.

OBJECTIVES

The overall objective of this research was to investigate the effect of kerosene contamination on the index properties and compaction characteristics of artificial clay soils.

MATERIALS AND METHODS

Commercially ASTM C-33 sand, kaolinite clay and a garden mulch (representing organic matter) were used to prepare the artificial soil for this study. For the sand, coefficient of uniformity (C_u) and coefficient of curvature (C_c) are 3.15 and 1.15 respectively for the sand. The C_u and C_c values indicate that the ASTM C-33 sand is poorly graded and classified as SP. Based on the hydrometer test, the coefficient of uniformity (C_u) and coefficient of curvature (C_c) are 2.82 and 0.57 respectively for the kaolinite. The C_u and C_c values indicate that the kaolinite is poorly graded. Test parameters considered in the study are soil composition and kerosene content.

Kerosene: The carbon content in kerosene varies from C_{11} to C_{15}. The typical boiling point is 185°C with an average specific gravity of 0.8. The flash point is 52°C and is sparingly soluble in water (< 0.1%).

Organic Matter: The organic matter used was garden mulch and was available in large-sized particles. Hence it was grounded to the size that is compatible for mixing with the sand and clay and used in the testing. Particle size distribution, unit weight (loose), unit weight (compacted), pH, moisture content, and total carbon content were measured for the organic mulch used.

Grain Size Analysis

The grain size analysis was performed on a 500 gm sample of organic matter that was grounded using a hand-operated grinder. The coefficient of uniformity (C_u) and coefficient of curvature (C_c) are 5.68 and 1.2 respectively for the organic matter. The C_u and C_c values indicate that the organic matter was well graded.

Moisture Content

The moisture content in the organic matter was measured over a period of 24 hours at two temperatures and the percent change in weight based on dry weight of soil (weight of solids). The percentage weight changes observed at 60 °C and 110 °C are 15% and 35% respectively. The loss in moisture was rapid in the initial 5 hours and remained constant after 11 hours for both the samples maintained at 60°C and 110°C. The variation of moisture content with time for kaolinite was monitored for 24 hours and the change in weight or moisture loss remained unchanged after 10 hours.

pH

The pH of organic matter and kaolinite were tested in the laboratory using a pH meter. A solid-to-liquid ratio of 1:1 was used. The average value of pH for organic matter was found to be 6.2 from three recorded values. The average value of pH of kaolinite was found to be 5.4 from three values.

Total Carbon Content

The total carbon content in the organic matter was determined by heating the sample at 450°C for three hours in the furnace. The weight loss is reported as the total carbon content. The total carbon content was calculated as the average of three values and was determined to be 75.8% with a range of 73.2% to 77.5%.

Unit Weight

The unit weight of organic matter was determined by compacting the organic matter using Harvard Miniature Compaction apparatus. The maximum dry unit weight was 4.7 kN/m^3 at a moisture content of 113%

The experimental program included testing for physical and engineering properties of artificially contaminated soils to study the effect of kerosene on the geotechnical properties of the soil.

The soil compositions used in this study are summarized in the Table 1.

Table 1. Composition of soils (by Weight)

Soil Designation	Sand (%)	Kaolinite (%)	Organic Matter (%)	Kerosene (%)
SK	80	20	0	0
SKK (2.5%)	80	20	0	2.5
SKK (5%)	80	20	0	5
SKK (7.5%)	80	20	0	7.5
SKO (2%)	78	20	2	0
SKO (5%)	75	20	5	0
SKOK (2/2.5%)	78	20	2	2.5

Kerosene Contamination

The kerosene was added to the clayey soil before or after adding the water.

ANALYSIS AND DISCUSSION

Liquid Limit and Plastic Limit (ASTM D 4138)

The mechanical properties of a fine grained soil, mostly organic clay, are altered by changing the water content. The plasticity index for cohesionless soils is zero since they do not exhibit plastic state. The results of liquid limit and plastic limit tests on different soil mixes with and without kerosene are summarized in Table 2.

Table 2. Summary of liquid limit and plastic limit for soils

Soil Designation	Liquid Limit (%)	Plastic Limit (%)	Plasticity Index (PI)	USCS Classification
SK	17.6	8.5	8.7	CL-ML
SKK (2.5%)	25.1	11.1	12.0	CL
SKK (5%)	27.5	14.6	12.9	CL
SKK (7.5%)	30.7	16.9	13.9	CL
SKO (2%)	48.6	24.9	23.7	CH
SKO (5%)	69.7	40.2	29.5	OH
Kaolinite	52.0	32.0	20.0	CH

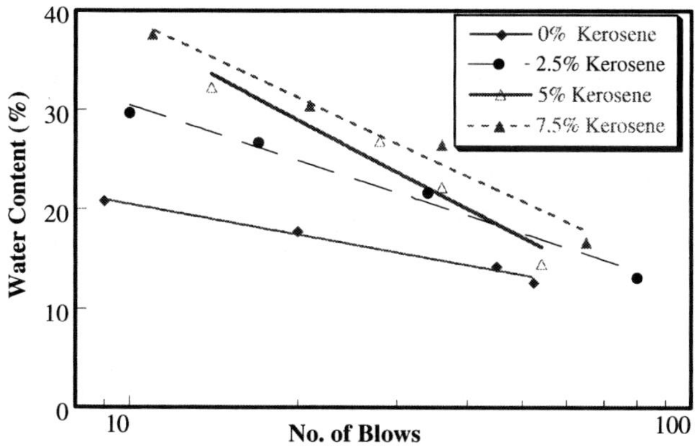

Figure 1. Liquid Limit for Soils with Various Kerosene Percentage

The liquid limit and plastic limit of kaolinite were 52 and 32 respectively. Head (1992) reported a liquid limit of 40-60 and a plastic limit of 10-25 for kaolinite.

Effect of Kerosene

Figure 1. shows the variation of liquid limit with kerosene content in the soil. The liquid limit and plastic limit values increased as the kerosene content in the soil varied up to 7.5%. As the kerosene content in the soil increased from 0 to 7.5%, the liquid limit increased by 75%, from 17.6% to 30.7%. The plasticity index was increased by 60% with 7.5% increase in the kerosene content in the soil and the changes are shown in Fig. 2 on a plasticity chart.

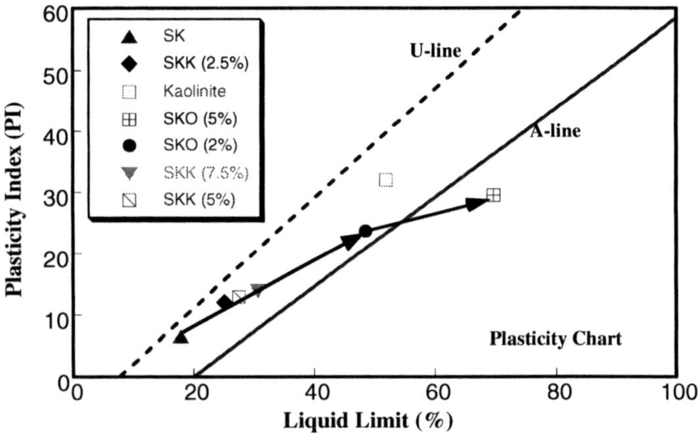

Figure 2. Effect of organic matter and kerosene on the index properties of the soil

Compaction Characteristics

For the kaolinite clay, the maximum unit weight of 14.4 kN/m^3 was obtained at an optimum liquid content of 31%.

Organic Matter

The maximum dry unit weight of 4.7 kN/m^3 was obtained at an optimum liquid content of 114%. Das (2005) reported a dry unit weight of 6 to 8 kN/m^3 at a moisture content of 90 to120% for soft organic clays.

Effect of Contaminant

Figure 3 shows the compaction relationships for soil mixes with varying kerosene content. As the kerosene percentage in the soil was increased, the maximum dry unit weight reduced and optimum liquid content increased.

The compaction test for 80% sand, 20% kaolinite, with 2.5% kerosene was performed by adding kerosene before (BW) and after adding water (AW) to the soil.

The results indicated that if the soil was first contaminated with kerosene, it had a maximum dry unit weight of 21.1 kN/m^3 at an optimum liquid content of 10.2%. The soil contaminated with kerosene after adding water had a maximum dry unit weight of 21.1 kN/m^3 at the optimum liquid content of 9.9%. The optimum liquid content was reduced by 3% while there was no change in the maximum dry unit weight. Thus, it can be concluded that the order of addition of 2.5% kerosene to the soil has negligible effect on the soil density and optimum liquid content.

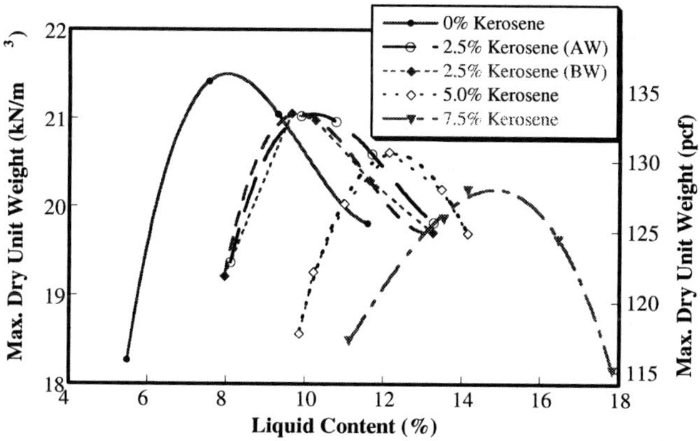

Figure 3. Moisture - unit weight relationships for soils with varying kerosene content

Hence the order of adding kerosene to the soil was not important and all the tests were performed by adding kerosene before adding water to the soil.

Figure 4 shows the variation of optimum liquid content with kerosene content in the soil. The optimum liquid content varied from 7% to 15% (doubled) when the kerosene content was varied from 0 to 7.5 % in the soil. Figure 5 shows the variation of maximum dry unit weight with kerosene content in the soil. The maximum dry unit weight reduced from 21.5 kN/m^3 to 20.2 kN/m^3, a 6.4% reduction in maximum dry unit weight as the kerosene content in the soil varied from 0 to 7.5%.

CONCLUSIONS

Effect of kerosene contamination on the index properties and the compaction characteristics of a clayey soil, with and without organic matter, was investigated. Based on the experimental study, the following conclusions are advanced:

1. Liquid limit increased with increased kerosene content. As the kerosene content was increased to 7.5%, the liquid limit was increased from 17.6% to 30.7% indicating a 75% increase in the liquid limit values.

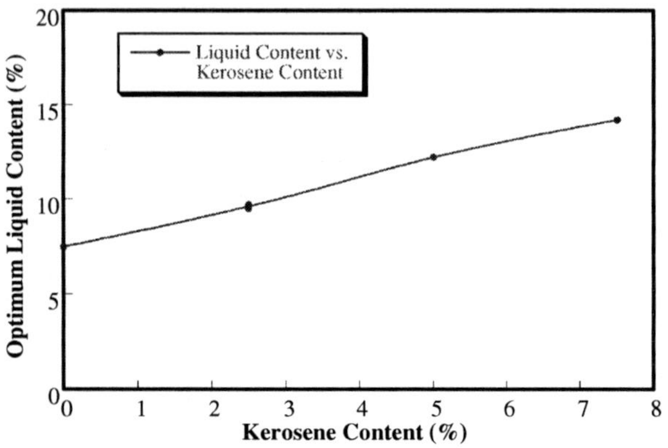

Figure 4. Variation of liquid content with kerosene content

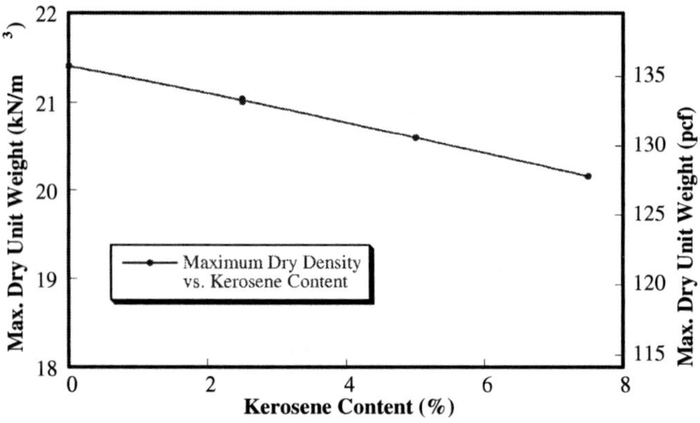

Figure 5. Variation of maximum dry density with kerosene content

2. The maximum dry unit weight for the clayey soil decreased as the kerosene was increased. As the kerosene content in the soil increased from 0% to 7.5%, the maximum dry density was reduced from 21.5 kN/m^3 to 20.2 kN/m^3, corresponding to 6% reduction in the value.
3 The optimum water content of the soils increased as the kerosene content was increased. The optimum liquid content increased from 8.0% to 14.8% as the kerosene content varied from 0% to 7.5%. This corresponds to 85% increase in the optimum liquid content of the soil.

4. Liquid limit increased with increased organic matter content in the soil. As the organic matter content was increased to 5%, the liquid limit was increased from 17.6% to 69.7% indicating a 300% increase in the liquid limit.
5. The maximum dry unit weight for the clayey soil decreased as the organic matter content was increased. As the organic matter content in the soil increased from 0% to 5%, the maximum dry density was reduced from 21.5 kN/m^3 to 19.13 kN/m^3, corresponding to 11% reduction in the value.
6. The optimum water content of the soils increased as the organic matter content was increased. The optimum liquid content increased from 8.0% to 11.7% as the organic matter content varied from 0% to 5%. This corresponds to 46% increase in the optimum liquid content of the soil.

ACKNOWLEDGEMENT

The study was supported by the Center for Innovative Grouting Materials and Technology (CIGMAT) at the University of Houston.

REFERENCES

Das, M. B. (2005) Principals of Geotechnical Engineering, Thomson Brooks/Cole, Pacific grove, CA.

Head, H. K. (1992). Manual for Soil laboratory Testing, Second Edition, Vol. 1, John Wiley & Sons, New York.

Holtz, R.D., and Kovacs, W.D. (1981), An introduction to geotechnical engineering: Englewood Cliffs, N.J., Prentice-Hall, 733 p.

Meegoda, N. J., Chen, B., Gunasekera, S. D. and Pederson, P. (1998) "Compaction Characteristics of Contaminated Soils – Reuse as a Road Base Material" Recycled Materials in Geotechnical Applications, GSP 79, ASCE, pp. 195-209.

NCHRP Synthesis 226 (1996), Remediation of Petroleum Contaminated soils: A Synthesis of Highway practice, National academy Pres, Washington D. C.

Vipulanandan, C. (1995), "Effect of Clays and Cement on the Solidification/Stabilization of Phenol contaminated Soils," Waste Management, Vol. 15, No. 5 pp. 309-406.

Immobilizing Arsenic in Contaminated Soil Using Humic Mineral Concentrates

Alexander I.Shulgin[1] and D. Joseph Hagerty[2], F. ASCE, PE

1. Research Professor, Chemical Engineering Dept., U. Louisville, Louisville, KY 40292 a.shulgin@insightbb.com
2. Professor, Civil & Environmental Engineering Dept., U. Louisville, Louisville, KY 40292 hagerty@louisville.edu

ABSTRACT: Immobilization of heavy metals, especially arsenic, may involve both cationic and anionic forms. Arsenic may occur as both cations and anions. Humic mineral concentrates (HMC), including chemically active functional groups, can form complexes to immobilize both cationic and anionic forms.

The efficiency of HMC in binding both cations and anions of heavy metals was demonstrated in lab-scale tests. Then, field tests were carried out on a former industrial disposal site destined for construction of commercial/residential structures. Field test results showed efficiencies of immobilization for zinc of 60%, cadmium of 66%, nickel of 72%, lead of 73%, and strontium of 87%. Arsenic immobilization efficiency was 85%.

The source of initial arsenic cations was AsCl3 and the source of the anion form was NaAsO2. Both were triple valent ions. Arsenic content in soil samples ranged from 2 mg/kg up to 200 mg/kg. Tests showed that the average efficiency of initial cationic arsenic immobilization was 90%, for an initial arsenic concentration in soil of 2 mg/kg and 82% for an initial arsenic concentration of 20 mg/kg. Average efficiency of immobilization of anionic arsenic was 85% for an initial arsenic soil concentration of 2 mg/kg and 76% for an initial arsenic concentration of 20 mg/kg.

INTRODUCTION

Increased concentrations of arsenic in soils are caused by emissions from industrial activities (metallurgical, power, chemical, etc.), by leaching from landfills or burial sites for a variety of industrial wastes (including processing residues and mineral enrichment wastes), and also by application of mineral fertilizers and pesticides (Zirin and Sadovnikova 1985). Under stable conditions, arsenic is most often found as anion forms in water and soil, but the composition of compounds including cationic forms of arsenic in contaminated soil is not clear.

Migration of arsenic and other heavy metals in groundwater and into soils is closely connected to alkalinity-acidity and redox conditions, but also is affected significantly by the presence of organic substances in soil. In general, the major sink where polluting substances collect is the sequence of uppermost soil horizons enhanced by humus. These layers constitute a biogeochemical barrier against the migration of chemical elements and substances. Organic compounds in humus bind, in order of priority, Pb, Zn, Cd, Ba, Cr, Cu, Hg and other heavy metals; particularly effective are humic acids (Warshal et al., 1993).

In contrast to the atmosphere in constant movement, and the dynamic aquatic environment, soil is immobile, and therefore is most vulnerable to pollution. On the one hand, soils lie on pathways for migrating ecotoxicants, and on the other hand, soils work as traps or pollutant accumulators. Resultant concentrations of contaminants may be harmful and even dangerous for people.

The ecological danger inherent in polluted soils is determined by the exposure of species in a given situation (potentially affected population, severity of exposure, etc.), but soil can serve as a secondary source of pollution of air by microscopic particles of arsenic or arsenic contaminated soil, and of water when arsenic is leached from the soil into receiving waters. Epidemiological risks may be increased by suppression of native soil microorganisms through contact with chemical pollutants, while pathogens carried to a site may survive under the same conditions (Goncharuk and Sidorenko 1986).

HUMIC ACIDS AND ARSENIC – A FEASIBILITY STUDY

Humic acids present as carboxyl, phenolic, carbonyl and other functional groups interact with heavy metals and arsenic in a variety of ways, including: ionic exchange with metal cations; formation of complex heteropolar salts with metal cations; and formation of complexes with anions, including anionic forms of metals. Prior research (Shulgin 2003) demonstrated that sulfate and nitrate anions play a significant role in formation of complexes between metals and humic acids. As a result of such interactions, the concentrations of available metals and the aforementioned anions are reduced significantly.

Arsenic may exist under natural environmental conditions in forms with valence of 5+, 3+, 3-, and 2+, and in the elemental form (Zirin and Sadovnikova 1985). While arsenic in soil tends to be bound and not excessively soluble or mobile, it may be released into the environment under certain conditions. Although not well understood, this mechanism appears to occur under reduced, or anoxic conditions, which cause arsenic to be released into water and to rapidly migrate through groundwater (Jackson 2007). It is well known that in solution in soil porewater, arsenic is most often found as arsenite or arsenate. But the time required for arsenic cations newly released into contaminated soil to be converted into anions through hydrolysis is unknown. The investigation of the interactions of arsenic in both forms with humic acids has great scientific and practical interest.

Examples of practical applications of such interaction include the successful development of an effective technology for remediation of soils contaminated with arsenic, and detoxification of arsenical waste using humic materials (Shulgin et al. 1998; Shulgin et al. 2004). This technology is based on application of humic acid in the form of a Humic Mineral Concentrate (HMC) (Patent 2000).

Laboratory tests were performed for preliminary assessment of the interaction of HMC with both cationic and anionic arsenic forms in samples of podsolic, sandy soils with organic content of about 0.4 percent from an industrial site.

In the tests, the soil samples were contaminated by anionic arsenic in $NaAsO_2$ and cationic arsenic in $AsCl_3$. No attempt was made to determine the changes in form of the arsenic during these tests. Standard solutions of these compounds were introduced into soil samples in doses of 2 mg / kg and 20 mg / kg. Anionic arsenic was added also at a concentration of 200 mg / kg. HMC was introduced into the polluted soil samples one week after the samples had been contaminated with arsenic. HMC dosages were 0.5 % and 2 % (dry weight to soil sample dry weight). Determination of the arsenic content in the soil samples was done by nuclear absorption spectrophotometry using a Perkin Elmer model 603 spectrophotometer after the arsenic was leached from the soil samples using an ammonium acetate solution with pH of 4.8. Each determination was repeated three times, with an average accuracy of 24.2%. This procedure allows extraction of the mobile (water soluble) forms of the arsenic.

Using the same technique, the arsenic level in untreated soil, in a "control sample" was determined to be 0.6 mg/kg. Results for mobile forms of arsenic found in these tests are shown in Table 1.

TABLE 1. Concentration of As Mobile Forms in Soil Samples Before Treatment

Sample	**Initial As level (mg/kg)**	**Arsenic chemical formula**	**Extractable As content (mg/kg)**
Control	—	—	0.6
1	2	$NaAsO_2$	1.93
2	2	$NaAsO_2$	1.99
3	2	$AsCl_3$	1.85
4	2	$AsCl_3$	1.80
5	20	$NaAsO_2$	18.98
6	20	$NaAsO_2$	19.23
7	20	$AsCl_3$	17.90
8	20	$AsCl_3$	17.65
9	200	$NaAsO_2$	190.65
10	200	$NaAsO_2$	190.23

Table 1 shows that arsenic anion forms were adsorbed by soil less strongly than were arsenic cation forms. Most of the added arsenic in both forms was mobile and available for extraction. The results of similar extraction tests on mobile forms of arsenic after treatment with HMC are presented in Table 2.

The second set of tests showed that the average efficiency of cationic arsenic immobilization was 90% for an initial arsenic concentration in soil of 2 mg/kg and 82% for an initial arsenic concentration of 20 mg/kg. Average efficiency of immobilization of anionic arsenic was 85% for an initial arsenic soil concentration of 2 mg/kg, and 76% for an initial arsenic soil concentration of 20 mg/kg.

TABLE 2. Immobilization of Mobile As Forms by Humic Mineral Concentrate

Test	Arsenic initial level, (mg/kg)	Arsenic chemical formula	Potentially mobile As content in the soil (mg/kg)		Efficiency of fixation (%)
			Before treatment	After treatment	
1	2	$NaAsO_2$	1.93	0.29	85.0
2	2	$NaAsO_2$	1.99	0.27	86.4
3	2	$AsCl_3$	1.85	0.19	89.7
4	2	$AsCl_3$	1.8	0.16	91.1
5	20	$NaAsO_2$	18.98	4.55	76.0
6	20	$NaAsO_2$	19.23	4.26	77.8
7	20	$AsCl_3$	17.9	3.12	82.6
8	20	$AsCl_3$	17.65	3.01	82.9
9	200	$NaAsO_2$	190.65	42.55	77.7
10	200	$NaAsO_2$	190.23	41.36	78.3

Average efficiency of immobilization of anionic arsenic was virtually identical to cationic As immobilization for an initial arsenic soil concentration of 200 mg/kg. HMC worked well in these trials, binding cationic arsenic better than anionic arsenic.

One of the main purposes of arsenic immobilization in soil is a reduction or elimination of its toxicity. Immobilized arsenic may be less available for uptake by the root systems of some plants and even unavailable to other plants. To investigate this aspect of the remediation, tests on oats were done to study the impact of arsenic on seed germination and seedling development and to determine arsenic content in the plants. HMC dosage was 0.5 % and 2% for each of two groups of soil samples. Results are presented in Table 3.

TABLE 3. Impact of As on Seed Germination Modified by HMC Treatment

Sample	Arsenic initial level (mg/kg)	Arsenic chemical formula	HMC dose (%)	Seed germination (%)
Control	0.6	—	—	91.0
1	2	$NaAsO_2$	0.50	95.0
2	2	$NaAsO_2$	2.00	96.0
3	2	$AsCl_3$	0.50	98.0
4	2	$AsCl_3$	2.00	97.0
5	20	$NaAsO_2$	0.50	91.0
6	20	$NaAsO_2$	2.00	90.0
7	20	$AsCl_3$	0.50	97.0
8	20	$AsCl_3$	2.00	96.0
9	200	$NaAsO_2$	0.50	86.0
10	200	$NaAsO_2$	2.00	85.0

Under low arsenic levels in the soil (2-20 mg/kg), HMC treated soil samples showed more seed germination than did even the uncontaminated soil (control). A trend was noted, that anionic arsenic is more toxic than cationic arsenic for seeds. For example, for an arsenic initial level of 20 mg/kg, the difference in seed germination between samples containing anionic As and those containing cationic As was 6%. In contrast, no difference was noted in impact between HMC doses of 0.5 % and 2%.

Table 4 contains data on arsenic content in dried oat plants grown on the arsenic-contaminated soil samples treated by HMC. The tests showed that arsenic content in the plants is 1-2% for an initial extractable arsenic concentration of 2-20 mg/kg and 4.8 to 4.98% for an initial concentration of 200 mg/kg. No plant distress or death was detected during the tests.

Table 4. Arsenic Content in Oat Plants

Test	Initial Arsenic level (mg/kg)	Arsenic chemical formula	Extractable As, after treatment (mg/kg)	HMC dose (%)	As content (mg/g)[1]	As in dry plants (%)[2]
1	2	$NaAsO_2$	0.29	0.50	5.8×10^{-3}	2.00
2	2	$NaAsO_2$	0.27	2.00	5.88×10^{-3}	2.17
3	2	$AsCl_3$	0.19	0.50	3.55×10^{-3}	1.87
4	2	$AsCl_3$	0.16	2.00	3.42×10^{-3}	2.14
5	20	$NaAsO_2$	4.55	0.50	1.24×10^{-1}	2.72
6	20	$NaAsO_2$	4.26	2.00	1.21×10^{-1}	2.84
7	20	$AsCl_3$	3.12	0.50	8.42×10^{-2}	2.70
8	20	$AsCl_3$	3.01	2.00	8.28×10^{-2}	2.75
9	200	$NaAsO_2$	42.55	0.50	2.12	4.98
10	200	$NaAsO_2$	41.36	2.00	1.98	4.80

1. Measured on basis of dry plant weight.
2. Measured as percent of extractable arsenic in the soil.

FIELD STUDIES

All the tests listed above showed that there is no significant difference between anionic and cationic arsenic in terms of plant uptake; for practical purposes, the initial As form does not matter. On the basis of these tests, an area of 2 hectares of soil contaminated by heavy metals and arsenic was treated with HMC. Mobile forms of heavy metals and arsenic were isolated from the soil before and after HMC treatment, using an ammonium acetate solution with pH of 4.8. Pollutant concentrations were determined using the aforementioned atomic-absorption method. Results are presented in Table No. 5, on the next page.

The results obtained show that significant fractions of the mobile forms of heavy metals and arsenic were immobilized. Humic Mineral Concentrate showed especially high efficiency in binding strontium (reduction of 7.9 times, or to 12.7 percent of initial concentration), chromium (4.4 times, or to 22.7 percent of initial concentration), lead (3.7 times, or to 26.7 percent of initial concentration), nickel (3.6 times, or to 28 percent of initial concentration) and arsenic (6.8 times, or to

14.7 percent of initial concentration).

CONCLUSIONS

It is possible to draw conclusions based on the tests carried out.

1. Humic acids in the form of Humic Mineral Concentrates work well for immobilization of both anionic and [initially] cationic arsenic. Mechanisms of both ion exchange and formation of complexes operate.

2. The efficiency of arsenic immobilization in the lab scale tests was from 76% up to 91% depending on the initial soil contamination level. The efficiency

Table 5. Heavy Metal/As Immobilization with Humic Mineral Concentrate

Element	Avg. Conc. Before treatment (mg/kg)	Avg. Conc. After treatment (mg/kg)	Efficiency, %
Zn	25.0	10.1	60
Mn	7.7	6.24	19
Sr	45.6	5.78	87
Ni	9.6	2.7	72
Cr	23.7	5.38	77
Cu	4.3	2.46	43
Pb	34.8	9.32	73
As	12	1.76	85
V	8.4	6.18	26
Sb	0.7	0.27	62
Sn	12.2	5.28	57
Cd	0.99	0.34	66
Co	9.3	3.66	61

of arsenic immobilization in the field scale was 85%. Thus, both the lab scale tests and the practical field scale work showed the same result.

3. Through arsenic immobilization, it is possible to reduce the toxic impact of arsenic on seeds and plants. Seed germination was increased and arsenic content in the plants was reduced by treatment with HMC.

4. Typical cationic forms of heavy metals (except chromium, which may exist in both cationic and anionic forms) are bound by HMC differently, but immobilization efficiency looks acceptable for practical application of HMC for contaminated soil detoxification and remediation.

REFERENCES

Goncharuk, E.I. and Sidorenko, G.I. (1986) *Hygienic Normalization of Chemical Substances in Soil.* Medicine [Publisher],Moscow, 320 pp..

Jackson, J.D. (2007) "Removing Arsenic Sustainable." *Civil Engineering.* 77 (4) April, 48.

Shulgin, A. I. (2003). "A study of heavy metal interaction with humic acids in a form of humic mineral concentrate." *Humic Substances in the Biosphere.* Proceedings of the II International Conference in Moscow, Science.168-70.

Shulgin, A. I., Shapovalov, A.A.. and Putsikin, Y.G. (1998). "Activated humic acids and problems of chemistry of the environment." *Humic Substance Downunder: Understanding and Managing Organic Matter in Soils, Sediments and Waters.* Proceedings, 9th International Meeting of the International Humic Substances Society, Adelaide, Australia, 212.

Shulgin, A. I., Nunes, R.A., Brocchi, E.A. and Faller, M.C. (2004). "Humic substance applied to detoxify solid waste from INGA Brazilian Zinc Company." *Humic Substances and the Soil and Water Environment.* Proceedings, XII International Meeting of International Humic Substances Society, July 25-30, São Pedro, Brazil. Editors: L. Martin-Neto, D.M.B.P. Milori and W.T.L. da Silva, 474-6.

Warshal G.M., Weluhanova T.K., and Koscheeva I.A. (1993) "Geochemical role of Humic Acids in Element Migration." *Humic Substances in theBiosphere.* Proceedings of the First International Conference in Moscow, Science [Publisher]. 97-115.

Zirin I.G.and Sadovnikova L.K. (1985).*The Chemistry of Heavy Metals, Arsenic and Molybdenum in Soil.* Moscow State University. Moscow, 137.

A Method (and system) for Obtaining Humic-Mineral Concentrates, Patent Ru, No. 2175651, 2000.

Volume Change Behavior of Calcitic Soil influenced with Sulfuric Acid

P.V. Sivapullaiah[1], B. Guru Prasad[2], and Allam. M. M[3].

[1, 3] Professor, [2] Research Scholar, Department of Civil Engineering, Indian Institute of Science, Bangalore-560 012, India ; e-mail:siva@civil.iisc.ernet.in, gurubpch@civil.iisc.ernet.in

ABSTRACT: The paper brings out the role of calcium carbonate ($CaCO_3$) on the volume change behaviour of natural black cotton soil with 1N sulfuric acid (H_2SO_4) as pore fluid. Natural black cotton soil contained predominantly montmorillonite [$Ca_{0.2}(Al,Mg)_2Si_4\,O_{10}\,(OH)_2\,.4H_2O$] along with other minerals such as amesite [$(Mg\,Fe)_2\,Al\,(Si\,Al)_2\,O_5\,(OH)_4$], kalsilite [$KAlSiO_4$] and quartz [$SiO_2$]. The calcitic soil, reacted with H_2SO_4 during consolidation testing, showed the presence of the new mineral yavapaiite [$K\,Fe(SO_4)_2$]. Consequently, the carbonate soil treated with 1N H_2SO_4 led to higher swell at seating load and more compression upon loading than the soil with no carbonate. The swelling increased with increase in the amount of carbonate present in the soil.

INTRODUCTION

The presence of calcium carbonate ($CaCO_3$) is common in soils of arid and semiarid climates. The amount of calcite in soils can vary widely. Many industrial structures are founded on soils containing calcite because of its wide spread uses. There have been attempts to use sulfuric acid (H_2SO_4) to stabilize soil (Budnikov et al 1946, Mace et al., 1999). It is thought that an acid like sulfuric acid might have been injected into limestone to construct Adam's bridge across India and Srilanka (Schuiling, 2004). However, the influence and causes of sulfuric acid solution on volume change behavior of calcareous clay is not known. Hence, this paper attempts to understand the impact of dilute sulfuric acid (of concentration of about 1N) on volume change behaviour. Because of the complexity associated in obtaining natural soil from the site with uniform calcite content, natural soil mixed with known amounts is used. To understand the mineralogical changes in soils due to acid treatment, X-ray diffraction (XRD), scanning electron microscopy (SEM) and energy dispersive analysis of X-ray (EDAX) were conducted.

EXPERIMENTAL PROGRAM

Soil and Chemicals Used

The natural soil sample was collected from the Belgaum area of the Karnataka, India, from an open excavation, from a depth of 1 m below the ground surface. The collected soil sample was oven dried (105 ± 5° C) and powdered in such a way that it passes through IS 425 μm. The geotechnical properties of soil are presented in Table 1. Commercial pure calcite and sulfuric acid were obtained from standard manufacturers. Calcite was oven dried (105 ± 5° C) for 48 h to prepare anhydrite calcite and powdered before mixing with natural black cotton soil with 0%, 5% and 10% by dry weight of soil. The tests were performed on duplicate specimens to obtain average results and to check reproducibility. If the test results had significant variation, additional tests were performed before averaging.

Table 1. Geotechnical characteristics of natural black cotton soil

Property	Value
Atterberg Limits (ASTM D2487):	
Liquid limit, LL (%)	82
Plastic limit, PL (%)	35
Plasticity index, PI (%)	47
Activity, A	0.77
Standard Proctor (ASTM D698)	
Maximum dry unit weight (kN/m^3)	12.9
Optimum water content (%)	37
Specific gravity, G_S (ASTM D854)	2.79
Grain-size distribution (ASTM D422)	
Clay (%)	61
Silt (%)	36
Sand (%)	3
Soil classification (ASTM D2487)	CH
Primary mineral (JCPDS 1999)	Montmorillonite
Cation exchange capacity, CEC (meq/100g) (ASTM 9081)	30
Organic Content (%) (ASTM D2974)	<5
Initial pH (ASTM D4972)	6.35
Free swell (cc/g)	2.2

Volume Change Behaviour

Swelling and consolidation behaviour of samples of black cotton soil with calcium carbonate were evaluated using a conventional oedometer test. The dried soil samples were manually mixed at optimum water content and placed in watertight polyethylene bags for about 7 days before compaction. Then the moisture conditioned soil sample was statically compacted to a thickness of 14 mm to the maximum dry density in a 60-mm diameter and 20-mm high fixed consolidation ring. The consolidation cell with the compacted soil specimen was positioned in the consolidation equipment, and the samples were inundated with dilute sulfuric acid at a seating pressure of 6.25 kPa. After completion of swelling cycle, loading cycle was undertaken. A load increment ratio of one was adopted and the load increment duration was kept as 24 h or until primary consolidation was complete. The test was performed up to a pressure of 800 kPa. Then samples were removed and dried in oven.

Mineralogical Composition

X-Ray Diffraction, Scanning Electron Microscopy and EDAX Analysis

The XRD analysis was carried out using a Philips diffractometer (Pw 3710), with generated radiations at 40 kV and 40 mA. The scanning speed was 0.02° 2**θ**/s and the angle scanned was 3°–70°. Randomly oriented samples were prepared by manual grinding oven-dried material in a glass slide using silica gel as adhesive. To identify mineral types, qualitative assessment was made by comparing the diffraction pattern of each sample with the standard patterns prepared by the Joint Committee of Powder Diffraction Data Service (JCPDS, 1999). SEM attached with EDAX, SIRION (XL40, series) was used to determine the morphology and elemental composition of the soil samples. The oven-dried (105 ± 5°C) soil samples placed onto an aluminum stub covered with double sided carbon tape and the surface of the sample were coated with fine gold film by vacuum sputtering, to prevent any charging effect over the sample.

RESULSTS AND DISCUSSION

Mineralogical Changes in Black Cotton (BC) Soil Due to Acid Treatment

X-ray diffraction (XRD) patterns of natural BC soil showed pronounced (Fig. 1a) peaks at 17.6, 3.58, 1.69 and 1.52Å, which are due to montmorillonite mineral. However, the other minerals like amesite (peaks at 7.06, 2.64, 4.63 and 1.54 Å), kalsilite (peaks at 3.12, 3.98, and 2.17 Å), and quartz (peaks at 3.37, 1.83 and 2.36 Å) were also found in the soil. Amesite is similar to serpentine with replacement of part of the magnesium and part of silica by aluminum and iron. Greater mineralogical complexity was induced in the soil by kalsilite due to unfavorable Si-O-Al angle of 180° along the hexagonal c-axis (Christopher and Michael, 1989).

The XRD studies of BC soil treated with 1N sulfuric acid is presented in Fig. 1b. New peaks (d-spacing of 8.03 and 3.33 Å) due to iron magnesium aluminum silicate hydrate, and larnite-calcium silicate-Ca_2SiO_4- with d-spacing of 2.79, 2.74, 2.28 and 1.40 Å were found in the sample treated with 1N sulfuric acid (Fig. 1b). Since tetrahedral layer is stable, the mineralogical changes may be due to alterations in octahedral layer in dilute 1N sulfuric acid. Tyagi et al, (2006) have noted that only the octahedral layer is affected at low acid concentration. The mineralogical alterations can be attributed to:

a) Attack of aluminosilicates at the Si-O-Al bonds with the escape of aluminum from the lattice and its subsequent combination with the SO_4^{3-} anion.
b) The liberated negative charges are compensated by hydrogen ions with the formation of hydroxyl groups Si-OH. This was calculated by subtracting the exchangeable base cations from the total cation exchange capacity.
c) Apart from the presence of hydroxyl groups, the addition of water molecules due to hydrogen bonding.

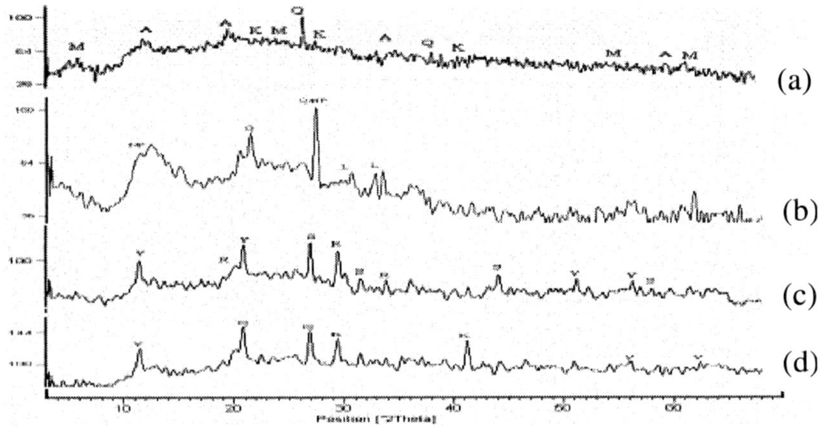

Fig. 1: Xrd patterns of (a) natural black cotton soil; (b) BC inundated with 1N H_2SO_4;(c) BC+5%$CaCO_3$ inundated with 1NH_2SO_4;(d) BC+10%$CaCO_3$ inundated with 1NH_2SO_4. (Q-Quartz; M-Montmorillonite; A-Amesite; K-Kalsilite; NP-New Peak; L-Larnite; Y-Yavapaiite; R-Rankinite; S-Sillimanite)

Mineralogical Changes in Calcium Carbonate Samples after Acid Treatment

X-ray diffraction patterns of BC mixed with 5% $CaCO_3$ inundated with 1N H_2SO_4 after completion of consolidation are presented in Fig. 1c. The new minerals yavapaiite - K $Fe(SO_4)_2$-potassium iron sulfate(d-spacing of 7.83 Å) and krausite-K $Fe(SO_4)_2$ H_2O- potassium iron sulfate hydrate (3.09 and 2.23 Å) were formed (Fig. 1c). Yavapaiite, a sulphate bearing mineral, might have formed in the soil treated with sulfuric acid by interaction of potassium leached from kalsilite mineral, and iron form amesite mineral, both present in original soil. Latest studies conducted by Melissa et al (2007) shown the presence of yavapaiite on Mars. Foshag (1931) identified the occurrence of mineral krausite-potassium-iron sulfate, at sulfur hole after the end of 1st world war in San Bernardino county, California.

The X-ray diffraction patterns of BC mixed with 10% $CaCO_3$ inundated with 1N H_2SO_4 after completion of consolidation are presented in Fig. 1d. Apart from yavapaiite, other minerals such as rankinite and sillimanite were formed in the soil containing 10% $CaCO_3$. Rankinite-calcium silicate-$Ca_3Si_2O_7$ (d-spacings of 2.72, 3.18, and 4.48Å) was formed because of presence of high calcium content and sillimanite-Al_2SiO_5-aluminum silicate (d-spacings of 3.38, 2.22, 2.9, 2.1, and 1.60 Å) was reformed from the other minerals. Sillimanite is one of three alumino-silicate polymorphs. Watson (1948) described fibrolitic sillimanite needles arranged at angles of 60° or 90° to one another in feldspar, and also as parallel groups of needles in vein quartz and feldspar, the orientation of the groups being controlled partly by the orientation of the enclosing grains and partly by the local attitude of the wall of the vein. The presence of these minerals indicates that the soil containing Mg, K and Fe and carbonate can undergo reactions to produce this mineral.

Morphological Changes in Natural BC Soil due to Acid Treatment

From Fig. 2a, untreated BC soil clearly exhibits comparatively close type of microstructure. Montmorillonite clay minerals, under electron microscopy, appear as aggregates of irregular or hexagonal flakes or, less commonly, of thin laths (Grim 1968). The aggregation of particles in BC soil might be due to cementing nature of iron compounds. The occurrence of peaks on the EDAX spectrums describes the presence of silica (Si), aluminum (Al), magnesium (Mg), ferrous (Fe), calcium (Ca) and potassium (K). The presence of Fe peak is attributed to the occurrence of iron oxide in the natural soil sample.
Fig. 2b shows partial disintegration based on a relatively open type of microstructure compared with natural BC soil. The dispersed nature of the fabric has been attributed to the cation exchange processes, which results in replacing of interlayer and inter-grain with proton (H^+). The exchangeable cations, such as calcium, magnesium and potassium are exchanged by hydrogen ions. The changes in microstructure are also due to mineralogical changes, particularly those with amesite and kalsilite, in presence of H_2SO_4. Further, from the spectrums of soil sample inundated with 1N H_2SO_4 after completion of consolidation, aluminum-to-silica ratios were evaluated for each sample using the weight percentage results and are presented in Table 2. Based on Table 2, the decrease in aluminum-to-silica ratios (0.54 to 0.35) with inundation of sulfuric acid results from the decrease in aluminum due to a transfer of Al^{3+} cations and/or an increase in silica, which leads to the formation of new minerals as described earlier.

Morphological Changes in Carbonatic BC soil due to Acid Treatment

The SEM micrographs of BC mixed with 5% and 10% $CaCO_3$ remoulded with water and inundated with 1N H_2SO_4 after completion of consolidation are shown in Figs 2c

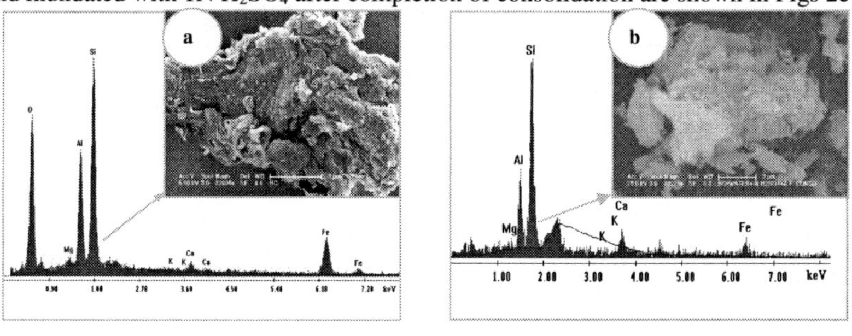

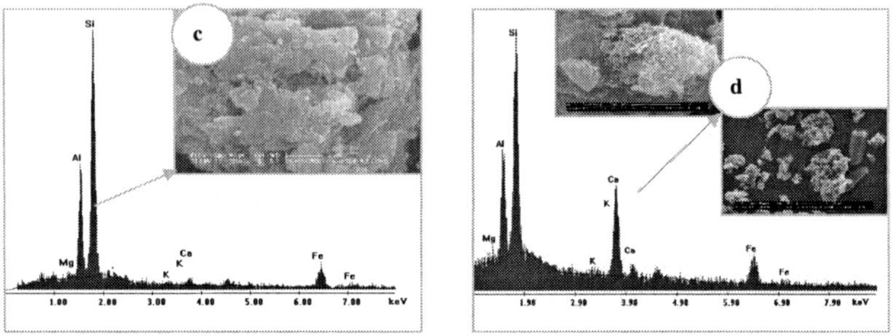

Fig. 2: SEM images attached with EDAX spectrums of (a) natural black cotton soil; (b) BC inundated with 1N H_2SO_4;(c) BC+5% $CaCO_3$ inundated with 1NH_2SO_4;(d) BC+10% $CaCO_3$ inundated with 1NH_2SO_4

Table 2. Elemental composition of black cotton soil with additives

Consoli -dation pose	Pore fluid	Weight (%)						
		Mg	Al	Si	K	Ca	Fe	Al/Si
Before	0% $CaCO_3$ + Water	0.30	13.89	25.73	0.60	1.58	16.87	0.54
After	0% $CaCO_3$ + 1N H_2SO_4	2.14	19.42	55.57	0.24	5.49	17.38	0.35
	5% $CaCO_3$ + 1N H_2SO_4	1.72	20.62	52.45	1.20	2.62	21.39	0.39
	10% $CaCO_3$ + 1N H_2SO_4	1.55	17.12	41.01	1.38	21.29	17.65	0.42

and 2d, respectively.The surface texture resembles a cauliflower with existence of aggregates of spherical or semi-spherical in shape. Based on the EDAX spectrum (fig. 2c), the addition of 1N H_2SO_4 the Al/Si ratio decreased from 0.54 to 0.39 (Table 2) proving further conformation of mineralogical changes. As can be seen from Fig. 2d, and from Table 2, that the aluminum-to-silica ratio (0.39 to 0.42) increased with further addition of $CaCO_3$ from 5 to 10%. In general, at low pH, aluminum is soluble and could be removed from the clay lattice, resulting in a decrease in the Al/Si ratio, and vice versa. However, the changes in the ratios of Al to Si lead to the changes in the d-spacing of the clay structure, which results in the formation of new minerals.

Volume Change Behaviour

Oedometer Swelling

Swelling was defined as ratio of increase in height of the specimen to the original thickness of the specimen expressed as a percentage. The swelling in the specimens compacted with water and inundated with 1N H_2SO_4 solution under nominal surcharge pressure of 6.25kPa was observed for about 90 d.

Oedometer Swelling With 1N Sulfuric Acid Solution as Inundating Fluid

The variation of swelling with time for natural BC soil (1), BC mixed with 5%$CaCO_3$ (3) and BC mixed with 10% $CaCO_3$ (2) specimens compacted with water and inundated with 1N H_2SO_4 solution at seating pressure of 6.25 kPa is presented in Fig.3(a). Swelling started almost immediately in all the samples. Natural BC showed

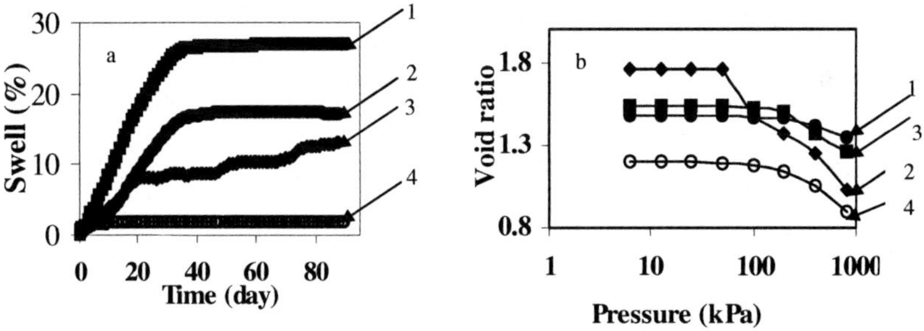

Fig. 3: BC soil remoulded with water and inundated with 1N H_2SO_4 shows (a) Variation in percent swelling with time; (b) Variation in void ratio-pressure relationship.

an equilibrium swelling of about 12% in 1N H_2SO_4 solution and is obtained within 7 d. This is very high compared to the amount of swell in natural soil with water as inundating fluid (2%). This might be due to replacement of exchangeable ions of the clay with hydrogen ions. About 7 meq/100g of exchangeable ions (CEC = 23 meq/100gm) are replaced by hydrogen ions. Also dissolution of minor minerals acting as binding agents restricting the swelling of montmorillonite could have contributed to an increase the swelling of the expansive minerals. This observation is supported by the SEM and X-ray diffraction studies. The equilibrium swelling obtained was about 17% for the sample containing 5% $CaCO_3$ after about 40 d. This shows that the swelling in the soil increased with increasing $CaCO_3$ content in the presence of 1N H_2SO_4. The equilibrium swelling in soil containing 10% $CaCO_3$ was about 27% after 35 d. In the soil containing carbonate, new minerals, as discussed in X-ray diffraction studies, were formed. The original minerals such as amesite and kalsilite were transformed to new minerals such as yavapaiite, krausite, rankinite, and sillimanite. Higher swelling in the presence of 10% carbonate indicates that this swelling is sensitive to the amount of carbonate present in soil.

Consolidation Behaviour of Calcareous Soil Contaminated With 1N H_2SO_4

The void ratio-pressure relationships of the BC soil (1), BC with 5% (3) and 10%

$CaCO_3$ (2) by 1N H_2SO_4 solution are compared in Fig. 3(b). The natural BC soil relatively high compression compared to soil containing carbonate. This might be due to high swelling that has taken place in the soil with the presence of carbonate.

CONCLUSIONS

The following conclusions are drawn from the study

i. Natural BC soil containing montmorillonite exhibited less swelling due to the presence of minerals such as amesite and kalsilite, which acted as binding agents.
ii. The increase in swelling in the BC soil with the acid solution is attributed to predominantly due to cation exchange; the higher is the concentration of carbonte, the higher is the swelling.
iii. The minerals present in BC soil such as amesite and kalsilite are transformed to yavapaiite, krausite, rankinite, and sillimanite in the soil containing carbonate and sulfuric acid.
iv. The swelling in BC soil containing carbonate in very dilute acid is controlled more by ion exchange than by mineralogical change.

REFERENCES

Budnikov, P. P., Nekrich, M. I., and Kvitnitskii, A. B. (1946). "Effect of phosphoric acid on clays." *Zh. Prild. Khim.*, 15(5):10–15.

Christopher, C., and Michael, C. (1989). "Thermally induced changes in kalsilite ($KAlSiO_4$)." *American Mineralogist*, 74(7-8): 797–811.

Foshag, W.F. (1931). "Krausite, a new sulfate from California." *American Mineralogist*, 16(1): 352–341.

Grim, R.E. (1968). "*Clay mineralogy*." 2nd Ed., New York, McGraw-Hill.

Mace, J.E., Amrhein, C., and Oster, J.D. (1999). "Comparison of gypsum and sulfuric acid for sodic soil reclamation." *Arid Soil Research Rehabilitation*, 13(2): 171–188.

Melissa D. Lane., Janice L. Bishop., M. Darby Dyar., Mario Parente., Penelope L. King., and Brendt C. Hyde. (2007). "Identifying the phosphate and ferric sulfate minerals in the paso robles soils (gusev crater, mars) using an integrated spectral approach." *Lunar and Planetary Science*, XXXVIII. 2176.pdf.

PCPDFWIN (1999). "Joint committee for powder diffraction studies (JCPDS)." *The powder diffraction file*, International centre for diffraction data (ICDD), Newtown Square, PA,USA.

Schuiling, R. D., (2004). "Palk strait: repairing Adam's bridge with gypsum?." *Current Science*, 86(10): 1351–1352.

Tyagi, B., and Chudasama Jasra, R.V. (2006). "Determination of structural modification in acid activated montmorillonite clay by FT-IR spectroscopy." *Spectrochimica Acta Part A*, 64 (3): 273–278.

Watson Janet. (1948). "Late sillimanite in the migmatites of Kildonan, Sutherland." *Geological magazine*, 85(3): 149–162.

Effect of Acid Effluent on the Characterization and Physio-Chemical Behaviour of Clayey and Sandy Soil

Murugaiyan[1], V., Saravanane[2], R. and Sundararajan[3], T.

[1]Geotechnology Laboratory, Department of Civil Engineering, Pondicherry Engineering College, Puducherry – 605014, India, email: vpmpl@yahoo.com
[2]Environmental Engineering Laboratory, Department of Civil Engineering, Pondicherry Engineering College, Puducherry – 605014, India,
email: saravananae@gmail.com
[3]Hydraulics and Machinery laboratory, Department of Civil Engineering, Pondicherry Engineering College, Puducherry – 605014, India

ABSTRACT: In the recent years, acute land pollution due to indiscriminate discharge of industrial effluents especially organic acid based effluents has become one of the major concerns due to erosion in the prioritization of sustainable development under Indian Scenario. The aim of this work is to study the effect of chemical changes on soil properties due to amino acid effluent contamination on soil and its impact on soil erosion behavior. The changes in the soil properties of effluent contaminated clay and silty soil were experimentally observed to be due to the interaction of organic and inorganic constituents of effluent on index properties of soil. The surface reactivity of soil mass is experimentally reported to be varying proportionately with respect to clay (%) in the soil. Also higher surface reactivity of clay increases the Solid Retention Time (SRT) with increase in rate of chemical transformation of organic and inorganic solids within soil. Hence soil S_1 (high clay content 54.5%) showed increased SRT and soil S_3 (low clay content, high silt content 94.9%) showed decreased SRT and long transformation stage.

INTRODUCTION

Groundwater pollution is mainly due to waste disposal sources like landfills, mining waste, sludge management via land spreading, etc., non-disposal sources like abandoned wells, agricultural chemical practices, petroleum exploration, etc., and depletion of ground water like increased salinity and salt water encroachment. The degree of the contamination threat to ground water supplies from land fills and waste water ponds depends on several factors: toxicity and the volume of the contaminants generated at each site, the nature of the geological medium underlying the site and the hydrological condition dominant in the area. Hence the changes in physical, chemical and hydraulic properties of soil subjected to leachate transport should be considered to reduce the rate of ground water pollution. Previous research investigations on organic acid contaminated land especially amino acid (high molecular weight) contaminated soil are notably sparse. Hence the present study aims to experiment on amino acid based effluent contaminated soil. The scope and objective of the present study are as follows: (i)to study the effect of chemical changes on soil properties and extent of erosion due to amino acid effluent contamination. (ii)to study and evaluate the extent of chemical and biological transformation that takes place on the soil mass and hence determine the contaminated soil properties.

Soil and Pollution

Sources of Soil Pollution

Soil Pollution is a complex process and it is generally characterized by pH of solution of pore fluid, temperature of the ground soil, and ionic characteristics of the solution. The major sources which cause ground / soil pollution can be summarized as: (1) acid rain, (2) animal waste, and (3) disposal of oil-field brines etc. (Cooper and Istok, 1988). The chemical characteristics of the pollutants depend on the type of the industries and their products. The chemical constituents present in the pollutant waste will have a direct influence on the behaviour of the soils. The parameters of pollutant which may influence the soil pollutants interaction are (1) Characteristics of pollutants (2) Temperature (3) pH (4) retention of pollutants in soil (5) Viscosity of the solution.

Effect of acid soil properties

Borglin, et al. (1996) investigated the desorption of organic chemicals from sediments. Khan et al. (1999) reported on the mobility of aminoacid in soils amended with pesticides in agricultural field. Increased use of nitrogenous fertilizers, sanitary waste disposal, changes in land use pattern and geological deposits were responsible for high nitrate levels. Foreman and Deniel (1986) investigated the permeation of compacted clay with organic chemicals. Effect of leaching properties of chemicals on clay was also reported. Lo et al. (1997), investigated the modification of clays for waste contaminant and pollutant attenuation. The authors also reported on the solidification / stabilization of phenolic waste on organic-clay complex. The solidification /stabilization (S/S) process is applicable to soil contaminated by metals and other inorganic. However, this process was not currently viewed by the U.S. Environmental Protection Agency (EPA) as applicable for remediation of soils contaminated solely with volatile organic compounds, because these organic compounds had been shown to have a detrimental effect on the macro structural properties of ordinary Portland cement (OPC) and could be released during the mixing and curing process.

MATERIALS AND METHODS

Scheme of experimentation

The fabricated laboratory set - up was used for the two stages of experimental study, namely (i) Batch mode operation, (ii) Column mode operation. The batch mode was operated to analyze and report chemical stabilization of soils (S_1, S_2, S_3) with respect to amino acid effluent. The column mode operation was aimed at analyzing and reporting soil pollutant interaction [with respect to hydraulic retention time (HRT)] 8 hr and 16 hr as applicable to field condition as shown in Table 1.

Table 1. Flow rate for continuous mode of operation

Sl.n	Column mode	Soil - S_1	Soil - S_2	Soil - S_3
1	8 hr HRT (ml/min)	1.85	2.1	3.65
2	16 hr HRT (ml/min)	0.925	1.05	1.825

Amino Acid Effluent – Variation of Concentration and Detention Time

The summary of results based on the concentration of amino acid effluent with respect to Hydraulic Retention Time (HRT) and Solid Retention Time (SRT) could be further refined in this section to highlight the relevance of physico-chemical behaviour of soil mass (under batch and column operation). The chemical behaviour of soil mass detailed to explain the effect of concentration of organic and inorganic constituents of amino acid effluent.

Effect of inorganic solids :

The effect of inorganic solids on soils (S_1, S_2 and S_3) is detailed with respect to the variation of chloride and sulphate content in the aqueous phase. These variations are presented in Figs. 1-3 respectively. The following salient points have been deduced from the Figures.

(i) The chloride concentration for soil S_1 was found to attain peak variation (13,800 to 16,160 mg/L) during batch mode, operation (1 to 70 days). The soil S_1 attained steady state on 19th day. During the 16 hr HRT the variation of chloride concentration in soil mass was relatively high (with respect to variation during batch mode) measuring in the range of 1000 to 1134 mg/L.

(ii) The sulphate concentration in soil S_1 was varying was varying from 1000 to 398 mg/L during batch mode showed high permeation of chloride solids during 8 hr HRT. This variation was found to be in the range of 5500 to 6000 mg/L. The sulphate variation in S_1 attained a steady Solid Retention Time (SRT) during the days of 16hr HRT.

The variation of chloride interaction in soil S_1 (54.4% clay) was mainly due to the inorganic permeability of clay content, possessing less Solid Retention Time (SRT) during batch mode operation. The variation during 8 hr and 16 hr HRT was less and it signifies that the soil attain higher SRT stability towards the end of 16 hr HRT operations. The retention of soil S_1, for Sulphate contents was characterized by high rate of permeation and low SRT during batch mode operation and attained a minimum and stabilized variation (indicating maximum SRT shown in Fig.) during 16 hr HRT.

(iii) The soil S_2 (35.17% clay and 64.8% silt) and soil S_3 (4% clay and 96% silt) showed comparable variations in chloride and sulphate concentrations during Batch & column mode of operations. The chloride & sulphate content for soil S_2 was found to attain maximum variation during batch mode operation (44000 mg/L to 5284) and relatively attained less variation (due to progressive increase in Solid Retention Time (SRT) developed within soil mass) during 8 hr HRT and found ultimately attained peak SRT during 137-141 days (16 hr HRT period). The chloride & sulphate concentration in soil S_3 was experimentally observed to signify their variations in relation to Solid Retention Time (SRT) and rate of permutation and transformation.

The chloride retention was less (evident from outlet concentration of 26000 mg/L to 5000 mg/L) during batch mode and progressive increase in Solid Retention Time (SRT) was observed during 8 hr HRT operation and subsequent steady state was attained at the end of 8 hr HRT. The sulphate variation was relatively independent of SRT and the variation during batch a mode was presumed to be due to high rate permutation through pores and capillary retention. The retention of sulphate content developed in the soil mass was less due to the progressive decrease in permeation rate observed from the beginning of 8 hr HRT (82nd day) to end of 16 hr HRT (141 day).

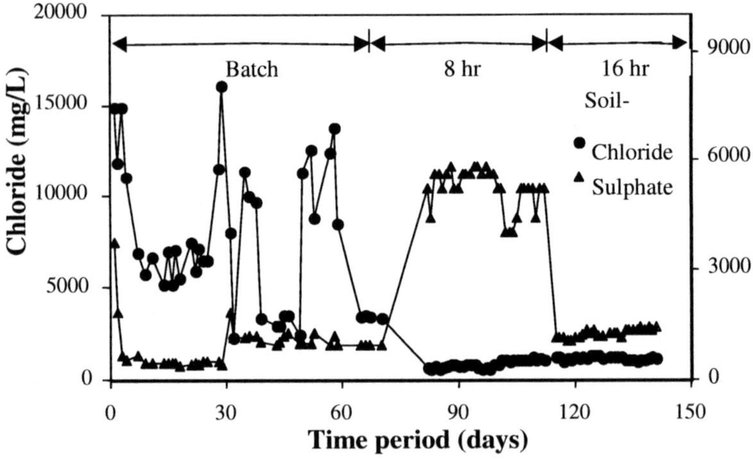

FIG. 1. Variation of Chloride and sulphate for Soil-S_1

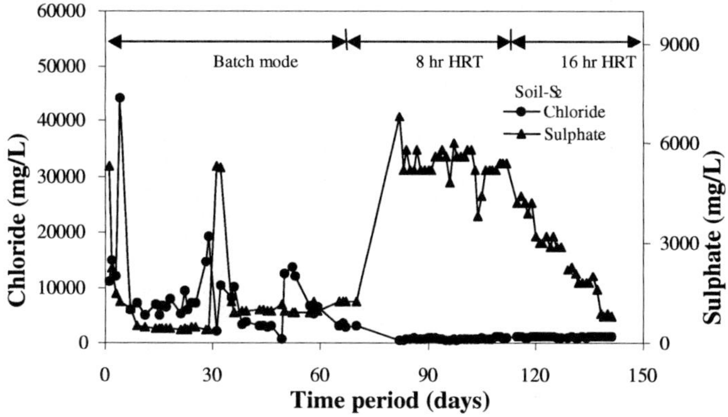

FIG. 2. Variation of chloride and sulphate for soil – S_2

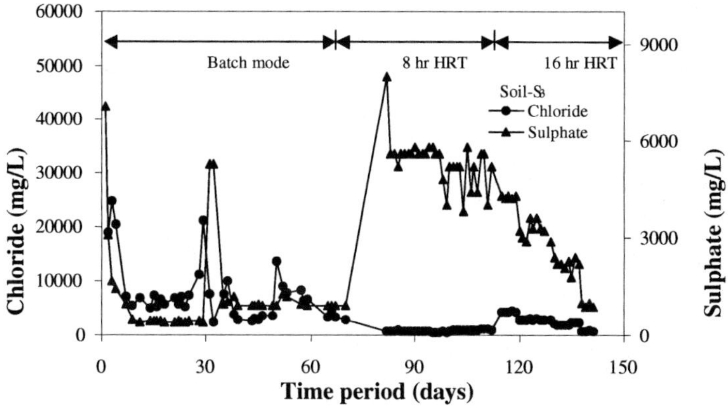

FIG. 3. Variation of chloride and sulphate for soil – S_3

Effect of organic solids

The effect of organic solids on soils S1, S_2 and S_3 were observed in relation to the variation of concentration of Total solid, volatile solids and COD in soil mass. These variations are presented in Figs. 4 and 5. The salient points are discussed below. The Total Solids and Volatile Solids for soil S_1 one found to attain 10,000 to 24,000 mg/L during batch mode operation, signifying the fact that the Solid Retention Time (SRT) was gradually increased from 1st day to 70th day. The variation in concentration of Total Volatile Solids during 8 hr and 16 hr HRT was less due to a relatively low SRT developed and maintained uniformly over the period of 8 hr and 16 hr HRT. The COD variation in S_1 was observed to allow more permeation of organic solids during batch mode, and undergo a state of uniform retention organic solids during 8 hr and 16 hr HRT . The uniform concentration in Figs, 4 and 5, reflects the fact of steady and long biological transformation during the period of 8 hr & 16 hr HRT. The Total Solids and Total Volatile Solids variations for soil S_2 were high during batch mode operation (upto 70th day) and started declining towards the end of 16 hr HRT period. It is seen that the retention of total solids was significant (as observed from the Figures), the outlet concentration, 2100 mg/L to 300 mg/L during 8 hr & 16 hr HRT as compared to Total Volatile Solids retention (2400 – 600 mg/L) which was found to attain 1000 mg/L concentration during the period of 16 hr HRT (115th day to 141st day). The COD variation during batch mode was highly permeable in nature, allowing more solids to pass through. The COD concentration contributed by organic solids was relatively less permeative and attain a steady uniform SRT during the period of 16 hr HRT. These variations were due to the cumulative effect of clay (35.17%) and silt (64.8%) which would possibly contribute to the retention of Total Solids and Total Volatile Solids and retention of and transformation of organic solids (COD). The TS was found to vary from 19800 to 2000 mg/L during batch mode, and 2000 to 300 mg/L during column mode. This shows a clear indication of Total Solids undergoing uniform and high retention (high SRT) as experimentally observed from the effluent collected at the outlet of soil column.

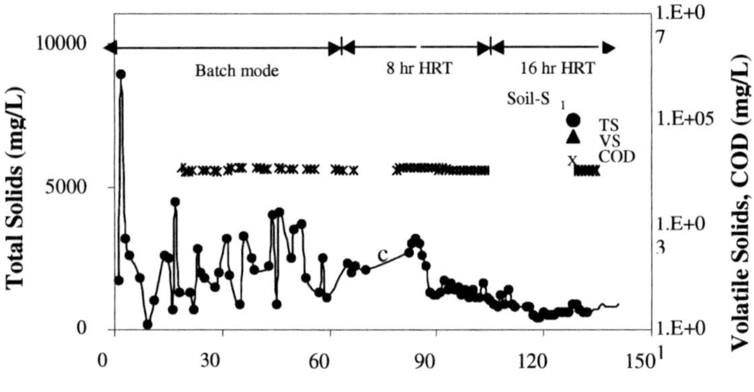

FIG. 4. Variation of Total Solids, Total Volatile Solids and COD for soil-S_1

The Total Volatile Solids variations as observed from outlet were relatively more comparable to TS depicting the fact of non-uniform SRT developed within soil mass during batch and continuous mode operation. This is due to the fact that the concentration of Volatile Solids in aqueous phase, is varying with respect to SRT and transformation to gaseous phase during 8 hr HRT & 16 hr HRT.

Effect of Amino Acid Effluent on Soil Properties

The effect of chemical and biological composition of amino acid effluent on the index properties of three locally available soils (S_1 : % of silt and clay : 93% (54.4% clay and 45.6% silt; S_2 : % of silt and clay : 96.5% (35.17% clay and 64.8% silt); S_3 : % of silt and clay : 98.9% (4% clay and 96% silt) having different % of silt and clay. Hence the results have been interpreted with the view of analyzing the effect of chemical composition of effluent on soil properties, viz. liquid limit, plastic limit and shrinkage limit with related indices (plasticity index, flow index, toughness index, liquidity index and consistency indices. The volume (%) of contaminated soil mass was also interpreted in terms of shrinkage ratio, volumetric shrinkage (%), Linear shrinkage (%) and shrinkage limit from specific gravity (%).

The contaminated retention of organic and inorganic solids in soil mass contribute for the changes in liquid limit, plastic limit and shrinkage limit of soil mass. Maximum value of liquid limit (99.5%) plastic limit (76.5%) (for soil S_3) and shrinkage limit 28.32% (for soils S_3) was experimentally determined due to higher % of silt and clay (S_2 : 96.5); S_3 : 98.9%) which would accumulate and undergo transformation for a longer period as compared to soils S_2 and S_3.

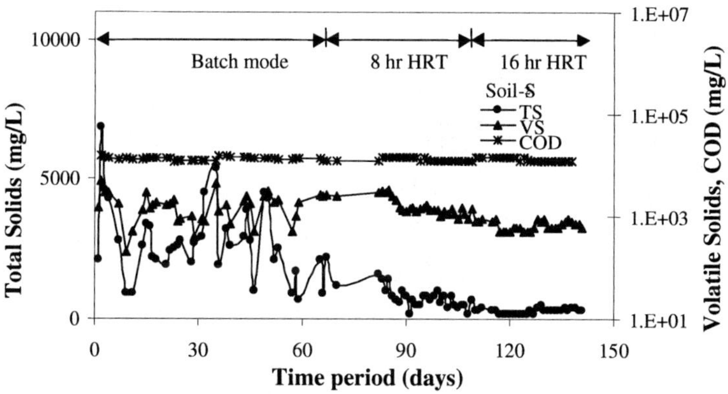

FIG. 5. Variation of Total solids, Volatile solids and COD for Soil S_2

CONCLUSIONS

Based on the results and interpretation the following conclusions were made:

i. The changes in the soil properties of effluent contaminated clay and silty soil were experimentally observed to be due to the interaction of organic and inorganic constituents of effluent on index properties of soil.

ii. The surface reactivity of soil mass is experimentally reported to be varying proportionately with respect to clay (%) in the soil. Also higher surface reactivity of clay increases the Solid Retention Time (SRT) with increase in rate of chemical transformation of organic and inorganic solids within soil. Hence soil S_1 (high clay content 54.5%) showed increased SRT and soil S_3 (low clay content, high silt and clay content 94.9%) showed decreased SRT and long transformation stage.

iii. The Solid Retention Time (SRT) was found to be less for the interaction of inorganic solids (chlorides and sulphates) with soils S_1 (% of silt & clay 93%), S_2 (% of silt and clay 96.5%) and S_3 (% of silt and clay 98.9%) and SRT was found to be longer for the interaction of organic solids with soils S_1, S_2 & S_3.

iv. Maximum value of liquid limit (99.5%) plastic limit (76.5%) (for soil S_3) and shrinkage limit 28.32% (for soils S_3) was experimentally determined due to higher % of silt and clay (S_2 : 96.5); S_3 : 98.9%) which would accumulate and undergo transformation for a longer period as compared to soils S_2 and S_3.

REFERENCES

Cooper, R.M., and Istok, J.D. (1988) "Geostatistics applied to Groundwater Contamination. I: Methodology", *Journal. of Env. Eng.*, 114(2): 270-286.

Borglin, S., Wilke, A., Jepson, R., and Lick, W. (1996), "Parameters affecting the desorption of hydrophobic organic chemicals from suspended sediments". *Envir. Toxicology and Chem.*, Vol. (15): 2254- 2262.

Foreman, D.E. and Daniel, D.E. (1986). "Permeation of compacted clay with organic chemicals", *J. Geotech. Eng.*, Vol. 122(7): 669-689.

Khan, S.V., Shafiullah and Khan, J.A. (1999). "Mobility of amino acids through soils amended with some pesticides", *Indian J. Env. Health*, Vol. 41(2): 83-89.

Lo, I.M.C., Mak, K.M. and Lee, C.H. (1997), "Modified Clays for Waste Contaminant and Pollutant Attenuation", J. Env. Eng., Vol.123(1): 25-32.

Characteristics of phosphatic clay for immobilizing heavy metals

Anna Hwang[1], and Jeehyeong Khim[2]

[1]Ph.D. Student, Dept. of Civil, Environmental and Architectural Eng., Korea Univ., Anam-Dong, Seongbuk-Gu, Seoul, 136-713, Republic of Korea; harvy0627@korea.ac.kr
[2]Professor, Dept. of Civil, Environmental and Architectural Eng., Korea Univ., Anam-Dong, Seongbuk-Gu, Seoul, 136-713, Republic of Korea; hyeong@korea.ac.kr

ABSTRACT: In this study, phosphatic clay was used as a phosphate-containing material. The phosphorus was fractionated by CRM BCR-684 protocol, and inorganic phosphorus, mostly apatite phosphorus, was determined to be the major form. The elemental composition of the phosphatic clay was identified using X-ray fluorescence (XRF) spectrometry and the main compositions were CaO and P_2O_5. The specific surface area, pore volume and average pore diameter were also measured. The leaching test results showed that the phosphate in the phosphatic clay was easily dissolved to aqueous phase, enabling the phosphate clay to be used as a source material for the immobilization process of heavy metals.

INTRODUCTION

Heavy metals such as lead and cadmium are major contaminants in soil and groundwater. Though various industrial activities, vehicles, paints and atmospheric deposition have been identified as the sources of heavy metals, while mining, including processing and smelting, activities are the major source of heavy metal problems of soil and groundwater.

In order to solve heavy metal pollution in the environment, practical solutions are critical. Some available treatment technologies for the removal of heavy metal ions from aqueous solutions are chemical precipitation, ion exchange, bioremediation and sorption/adsorption. Chemical precipitation is one of the most important mechanisms for the fixation of heavy metals.

Phosphate materials such as phosphoric acid (Xu and Schwarts 1994; Yang et al. 2001) phosphate rock (Ma et al. 1995) and bones have been used for the remediation of metals. Specifically, the primary mechanism of metal ion removal by phosphate clay is shown below:

$$Ca_{10}(PO_4)_6F_2 + 12H^+ \rightarrow 10Ca^{2+} + 6H_2PO_4^- + 2F^- \text{ (Dissolution)}$$

$$10M^{2+} + 6H_2PO^{4-} + 2F^- \rightarrow M_{10}(PO_4)_6F_2 + 12H^+ \text{ (Precipitation)}$$

As some heavy metals form a new mineral having extremely low solubility with phosphate, significant research efforts have been focused on evaluating the effectiveness of phosphate (Ma et al., 1995; Hettiarachchi et al., 2000; Yang et al., 2001). The release rate of phosphate ions from phosphate mineral varies significantly according to environmental conditions such as temperature and pH, mineral properties such as chemical composition, phosphate type and available surface.

This study evaluated the release rate of phosphate from phosphatic clay and the effect of pH on the release rate.

EXPERIMENTAL PROCEDURE

The phosphatic clay was obtained from Vietnam mining and a portion of size 0.15~0.075 mm was used in this research.

The elemental composition of the phosphatic clay was identified using an X-ray fluorescence (XRF) spectrometer (Philips PW2404) and the Barrett–Joyner–Halenda (BJH ASAP 2020, Micromeritics, USA) desorption method was used to measure the specific surface area, pore volume and average pore diameter.

The total phosphorus contents and fractions in the materials were determined by the extraction procedure based on the Williams procedure of CRM BCR-684 (Pardo et al. 2003), which was one of the most widely used methods for the fractionation of phosphorus in soils and sediments. This protocol afforded the following five types of phosphorus: total phosphorus(TP), inorganic phosphorus(IP), apatite phosphorus(AP), non-apatite inorganic phosphorus(NAIP) and organic phosphorus(OP).

Tests of dissolution were carried out in a batch reactor of 100 ml. After adding 1 g of the materials to 100 ml of de-ionized water at 25 ℃, the solution was continuously stirred in a rotary shaker at 200rpm. The phosphate concentrations were measured after 5 and 20 min, and 1, 2, 4, 8, 16, 24 and 48 hours. The initial solution pH was adjusted with 0.1 M HCl and 0.1 M NaOH.

MECHANISM

The primary mechanism of metal ion removal by phosphate clay is governed by its dissolution in acidic condition followed by precipitation. Experimental evidence (Saxena et al. 2006)suggests that the removal is governed by a dissolution–precipitation mechanism, as shown in Fig. 1.

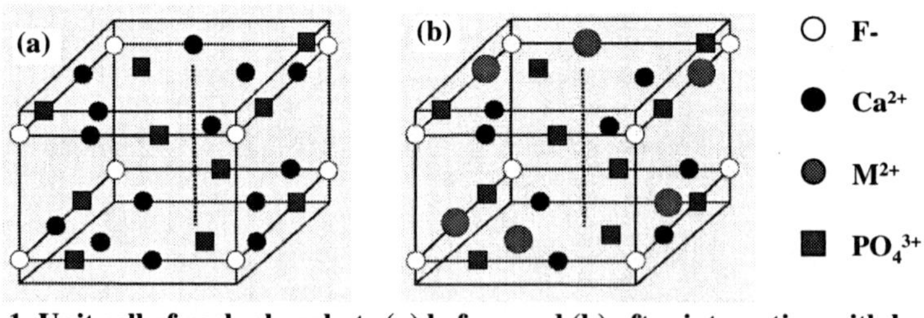

FIG. 1. Unit cell of rock phosphate (a) before and (b) after interaction with heavy metal ions (Saxena et al. 2006).

MATERIAL CHARACTERISTICS

The elemental composition as determined using XRF spectrometer is shown in

Table 1. The main components in the phosphatic clay were CaO and P_2O_5, followed by SiO_2 and Al_2O_3. The high phosphorus content in the material confirmed its potential as a phosphate source in the process of heavy metal immobilization.

Table 1. Elemental analysis using XRF spectrometry

Element	Al_2O_3	CaO	Fe_2O_3	K_2O	MgO	MnO	Na_2O	P_2O_5	SiO_2	L.O.I
w/w(%)	2.94	41.52	1.37	0.31	0.25	0.43	0.04	32.80	16.00	2.23

*L.O.I : Loss of Ignition

Table 2 presents the phosphorus types. As the content of each phosphorus type was analyzed independently, the relationships between the types were not exact but they did show relatively good agreement. The sums of IP and OP were 98.68 % of TP while those of AP and NAIP were 98.68 % of IP in the phosphatic clay.

Table 2. Phosphorus types in the phosphatic clay (mg/kg)

TP(OP+IP)	OP	IP(AP+NAIP)	AP	NAIP
1808.3	173.33	1611.1	1543.4	46.3

The result of the physical properties of the phosphatic clay is given in the Table 3. The specific surface area was considered as the representative physical property of the phosphate-containing materials, which could help researchers explain many physical phenomena such as the mass transfer phenomenon. Ma et al. (1995) and Chen et al. (1997) used phosphate rocks with specific surface areas of 4.22 ~ 18.7 m^2/g and 27 m^2/g, respectively, in heavy metal immobilization. Phosphatic clay with a specific surface area of 33.65 ± 1.19 m^2/g was used for heavy metal removal by Singh et al. (2001, 2006). The porosity of the phosphatic clay was very low: about the same as that of granite (Fetter, 1994).

Table 3. Physical properties of the phosphatic clay

Specific surface area (m^2/g)	Pore volume (cm^3/g)	Average pore diameter (cm)
7.56	1.30×10^{-2}	7.66×10^{-7}

MASS TRANSFER

The first-order mass transfer equation model has been used for the mass transfer of heterogeneous systems in various situations: nuclide leaching (Kim et al., 2002), mercury emission from stabilized waste (Hamilton and Bowers, 1997) phosphate release from sediments (Kim et al., 2004) and dissolution and precipitation of minerals (Dunn and Yen, 1999).

The nonlinear curve fitting procedure was used to estimate the mass transfer coefficient and the saturated concentration; the values are given in Table 4. From these results, the mass transfer coefficient was decreased with increasing pH because the solubility was lower and thus the driving force of dissolution was weaker at higher pH.

The amount of leached phosphate varied at different pH due to the difference of the solubility of apatite-related minerals such as calcium phosphate in the materials at different pH (Vernon and David, 1980).

The mass transfer rate coefficients for the phosphate release from the lake sediments were 3.28×10^{-6} cm/s ~ 2.04×10^{-6} cm/s (Kim et al., 2004) and 5.0×10^{-5} cm/s ~ 5.0×10^{-4} cm/s (Lijklema, 1986), both of which were greater than the values obtained in the present study. However, the values seemed reasonable considering the difference of solution pH, solid matrix and phosphorus types between the three experiments.

The solubility product constants of calcium phosphates vary significantly according to the compounds: 9.50×10^{-7} for dicalcium phosphate anhydrate but 9.21×10^{-58} for hydroxyapatite at 25 °C (Montastruc et al., 2003). The results of the saturated concentration suggested the presence in the phosphatic clay of various calcium phosphates such as brushite ($Ca(HPO_4)\cdot2H_2O$), tricalcium phosphate ($Ca_3(PO_4)_2$), hydroxyapatite ($Ca_5(PO_4)_3(OH)$) and fluorapatite ($Ca_5(PO_4)_3(F)$).

Table 4. Fitted mass transfer data according to the mass transfer model

pH	K_L(cm/s)	K_{La} (s^{-1})	C_s(mg/L)
4	3.84×10^{-7}	2.75×10^{-4}	0.47
7	1.24×10^{-7}	8.89×10^{-5}	0.36
10	4.66×10^{-8}	3.34×10^{-5}	0.35

CONCLUSIONS

Phosphorus was fractionated by CRM BCR-684 protocol and apatite phosphorus was found to be the major form. The XRF analysis results identified a P_2O_5 content in the phosphatic clay of about 35% and a phosphorus level in the aqueous phase after 24 hours of about 0.01% at pH 7. The phosphatic clay was easily dissolved to aqueous phase, thereby demonstrating its potential use as a phosphate source for heavy metal immobilization processes.

REFERENCES

Chen X., Wright J.V., Concha J.L., and Peurrung, L.M. (1997). "Effects of pH on heavy metal sorption on mineral apatite." *Environ. Sci. Technol.*, 31(3), 624-631.

Dunn, K., and Yen, T.F. (1999). "Dissolution of barium sulfate scale deposits by chelating agents." *Environ. Sci. Technol.*, 33(16), 2821-2824.

Fetter C.W., Applied hydrogeology, 3rd ed., Prentice-Hall, New Jersey, 1994.

Hamdi, M, and Ide-Ektessabi, A. (2007). "Dissolution behavior of simultaneous vapor deposited calcium phosphate coatings in vitro." *Mat. Sci. Eng.*, C27(4), 670-674.

Hamilton, W.P., and Bowers, A. R. (1997). "Determination of acute Hg emissions from solidified/stabilized cement waste forms." *Waste Manag.*, 17(1), 25-32.

Hettiarachchi, G.M., Pierzynski, G.M., and Ransom, M.D. (2000). "In situ stabilization of soil lead using phosphorous and manganese oxide." *Environ. Sci. Technol.*, 34(21), 4614-4619.

Kim, J.Y., Kim, C.K. and Chung, C.J. (2002). "Modeling of nuclide release from low-level radioactive paraffin waste: a comparison of simulated and real waste." *J. Haz. Mat.,* B94(2), 161-178.

Kim, L.H., Choi, E.S., Gil, K.I., Stenstrom, M.K. (2004). "Phosphorus release rates from sediments and pollutant characteristics in Han river, Seoul, Korea." *Sci. Total. Environ.,* 321(1-3), 115-125.

Lijklema, L. (1986). "Phosphorus accumulation in sediments and internal loading." *Hydrobiological Bulletin,* 20(1-2), 213–224.

Ma, Q.Y., Logan, T.J., and Traina, S.J. (1995). "Lead immobilization from aqueous solutions and contaminated soils using phosphate rocks." *Environ. Sci. Technol.,* 29(4), 1118-1126.

Montastruc, L., Azzaro-Pantel, C., Biscans, B., Cabassud, M., and Domenech, S. (2003). "A thermochemical approach for calcium phosphate precipitation modeling in a pellet reactor" *Chem. Eng. J.*, 94 (1), 41–50.

Pardo, P., López-Sánchez, J.F., and Rauret, G. (2003). "Relationships between phosphorus fractionation and major components in sediments using the SMT harmonised extraction procedure." *Anal. Bioanal.,* 376(2), 248–254.

Saxena, S., and D'Souza, S.F. (2006). "Heavy metal pollution abatement using rock phosphate mineral." *Environmental International*, 32, 199-202.

Singh, S.P., Ma, L.Q. and Harris, W.G. (2001). "Heavy Metal Interactions with Phosphatic Clay: Sorption and Desorption Behavior." *J. Environ. Qual.*, 30(6), 1961-1968.

Singh, S.P., Ma, L.Q., and Hendry, M.J. (2006). "Characterization of aqueous lead removal by phosphatic clay: equilibrium and kinetic studies." *J. Haz. Mat.*, B136(3), 654-662.

Vernon, L.S., and David, J., (1980). "Water chemistry." John Wiley & Sons, New York.

Xu, Y., and Schwarts, F.W. (1994). "Lead immobilization by hydroxyapatite in aqueous solutions." *J. Contam. Hydrol.,* 15(3), 187-206.

Yang, J., Mosby, D.E., Career, S.W., and Blanchar, R.W. (2001). "Lead immobilization using phosphoric acid in a smelter-contaminated urban soil." *Environ. Sci. Technol.,* 35(17), 3553-355.

Evaluation of Ground Movement Due to COPR Expansion

Carlos A. Lazarte[1], M.ASCE, Ph.D., P.E. and
Rudolph Bonaparte[2], M.ASCE, Ph.D., P.E.

[1]Senior Engineer, Geosyntec Consultants, 10015 Old Columbia Road, Suite A-200, Columbia, MD 21046, clazarte@geosyntec.com
[2]Principal, Geosyntec Consultants, 2002 Summit Blvd NE, Suite 885, Atlanta, Georgia 30319; rbonaparte@geosyntec.com

ABSTRACT: At Area 1501 of the Dundalk Marine Terminal (DMT) in Baltimore, Maryland, weathering processes caused chromite ore processing residue (COPR) to indurate, expand, and damage infrastructure. An evaluation is presented of ground movement caused by COPR expansion around a 96-inch-diameter reinforced concrete stormdrain pipe. The movement was assessed before and after a stress relief trench (SRT) was constructed next to the pipe to reduce lateral earth pressures and absorb COPR lateral expansion. It was found that the magnitude of ground movement due to COPR expansion, as measured in inclinometers, increases near "free boundaries" and adjacent to the SRT. COPR movements were observed to be "block-like" and the movement of these blocks caused shear deformations in a 3- to 5- ft thick soil zone beneath the COPR. Based on the monitoring results, three stages of COPR movement are hypothesized to occur after SRT installation: (i) an initial stage resulting from lateral stress relief due to SRT installation; (ii) an intermediate stage, with lower rates of movement following stress relief; and (iii) a long-term stage, with rates of movement associated with steady COPR expansion. These observations are essential for optimizing SRT layouts to protect structures against COPR expansion and for developing numerical models to predict COPR/SRT interaction and performance.

INTRODUCTION

COPR, a slag-like by-product of the sodium dichromate production process, was deposited as fill material to reclaim land at DMT (Fig. 1), Port of Baltimore, Maryland in the late 1960's and early 1970's. DMT was originally created by filling a tract of land with dredge spoils from the adjacent Patapsco River. Subsequently, the terminal was expanded by placing COPR fill over an approximately 120-acre area at the south end of the site, including Area 1501 (Figs. 1 and 2), the DMT parcel of interest in this paper. After COPR placement, Area 1501 was covered with clean fill and paved to elevation 21 feet above mean sea level (ft-msl).

After COPR placement, exposure to water and air produced varying degrees of COPR weathering, mineral change, induration (i.e., hardening due to cementation), and expansion (Kremser et al., 2005; Dermatas et al., 2005). At DMT, COPR expansion has locally deformed and cracked asphalt and concrete pavements, caused collapse of small pipelines, and cracked larger pipes. In Area 1501, COPR lateral expansion damaged the south pipe of a twin-pipe stormdrain system (Fig. 2), consisting of reinforced concrete, 96-inch-(in.) diameter pipes installed in the late 1980's.

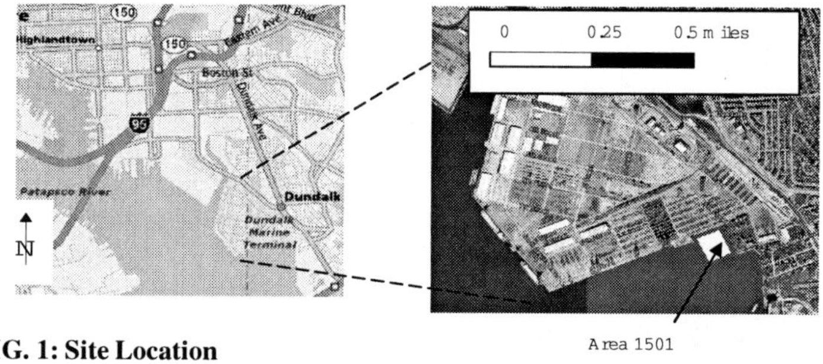

FIG. 1: Site Location

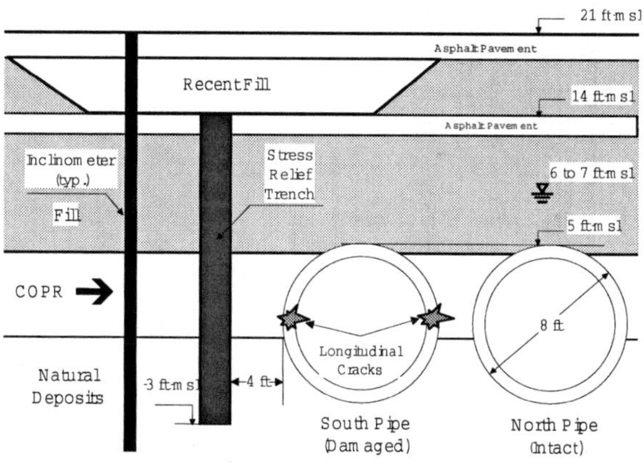

FIG. 2: Schematic Cross Section of Pipes and SRT

In 2001, the south pipe, which had been placed within a COPR layer, was observed to be deforming inwards at the springline. In contrast, the north pipe was barely deformed. Starting in June 2002, the Maryland Port Authority (MPA) commissioned the monitoring of pipe compression at the south pipe springline (i.e., diameter shortening). Monitoring occurred over 16 months along a 180-ft long pipe section. During this period, compression of the south pipe increased by 1.7 in. (Fig. 3a) and longitudinal cracking was observed. At the start of monitoring, the apparent compression rate was approximately 2 in./year (Fig. 3b). After approximately four months, pipe compression continued, but at a lower rate, approximately 1.2 in./year, possibly due to stress relaxation in the COPR layer following pipe cracking. In contrast, the diameter compression of the north pipe was less than 0.1 in. in the same period and this pipe remained fairly intact thereafter. It is believed that the south pipe deformed and cracked, not the north pipe, because the lateral extent of COPR on the south side is larger than on the north. In addition, once the south pipe cracked, it became less stiff, localizing subsequent movements.

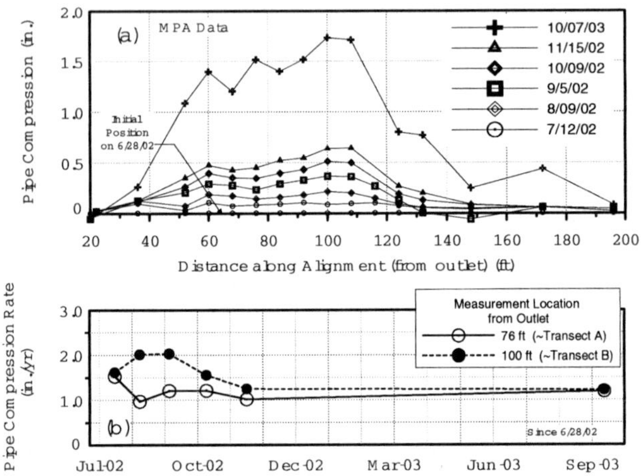

FIG. 3: (a) Pipe Compression and (b) Pipe Compression Rate

DESCRIPTION OF STRESS RELIEF TRENCH

Based on monitoring results, the MPA retained Mueser Rutledge Consulting Engineers to evaluate methods to prevent further crushing of the pipe. The selected technique was a stress relief trench (SRT), which essentially is an excavated trench with compressible backfill. An SRT will deform at much lower lateral stresses than a concrete pipe and thereby can absorb COPR lateral expansion before the onset of pipe distress. In mid-2004, contractors constructed an SRT parallel to, and 4 ft offset from, the damaged pipe in mid-2004 (Fig. 2). The SRT is approximately 185 ft long, 17 ft deep, and 2 ft wide, and extends beyond the deformed pipe section and deeper than the COPR stratum. The SRT was backfilled with slurry consisting of bentonite, COPR spoils, and styrofoam pellets. A few days after SRT installation, the pipe compression rate decreased to essentially zero, while COPR/soil movements continued.

GEOTECHNICAL MONITORING PROGRAM

After SRT installation, a geotechnical monitoring program using inclinometers was developed to measure COPR response to the SRT. The program was implemented approximately three months after SRT installation. The specific goals of the monitoring program were to: (i) measure the magnitude of COPR lateral movement (δ) in the vicinity of the SRT; (ii) observe spatial variation of ground movement; and (iii) assess the rate of COPR lateral movement with time. Six inclinometers were installed along Transects A and B (Fig. 4), perpendicular to the pipe alignment within the zone of maximum pipe compression, approximately 75 and 100 ft from the outlet structure, respectively. Instruments for measuring horizontal earth pressures and settlement were also installed, but results from these instruments are not discussed in this paper.

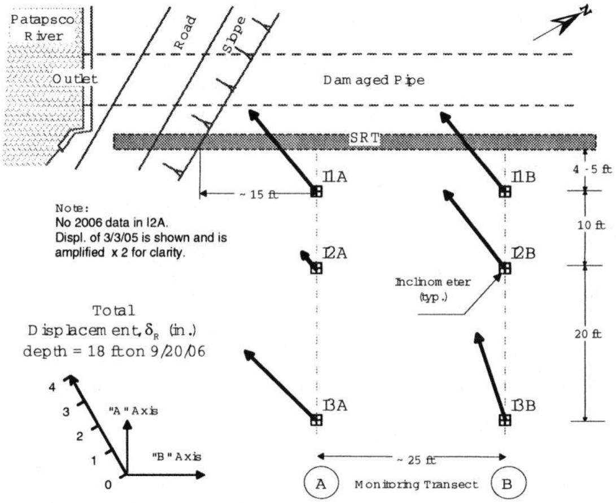

FIG. 4: Schematic Plan View of Inclinometers

Horizontal movement of the ground perpendicular and parallel to the SRT (δ_A, along axis "A" and δ_B, along axis "B," respectively, Fig. 4) was monitored in inclinometers installed at increasing distances (*d*) from the SRT. Inclinometers consisted of 3.34-in. diameter plastic casing and were installed through recent fill, COPR (about 4 to 7 ft thick), and natural soils (Fig. 2). Terminal depths of inclinometers were approximately 32 to 35 ft-bgs, where the ground was mostly unaffected by COPR movement.

Attempts were made to protect the inclinometer casings from COPR expansion and thus extend their longevity. To this end, the inclinometer casings were installed inside larger polyvinyl chloride (PVC) pipes extending to the bottom of the borehole and the annuli were backfilled with a grout that had a consistency softer than conventional grout to provide protection against lateral crushing and a shear interface against vertical heave, but at the same time, stiff enough to prevent PVC casing deflection into this grout. The PVC pipe/borehole annuli were backfilled with conventional stiff grout.

RESULTS

Displacements. All six inclinometers were monitored monthly between December 2004 and March 2005, and five inclinometers were monitored in August and September 2006. The sixth inclinometer was not monitored in 2006 due to inaccessibility. Fig. 5 shows profiles of inclinometer deflections perpendicular to the SRT alignment (i.e., component δ_A) along Transect B, where pipe compression was largest, as measured between December 2004 and September 2006. Measurements began approximately three months after SRT installation; therefore, excavation-related displacements were not measured by the inclinometers.

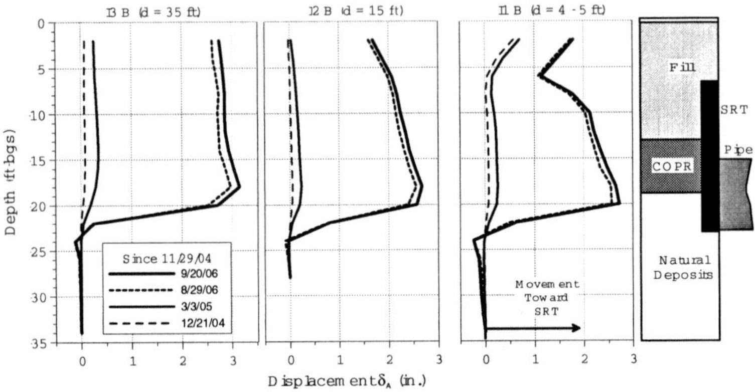

FIG. 5: Displacement Profiles – Perpendicular to SRT - Transect B

At the end of the monitoring period: (i) the largest displacements δ_A, perpendicular to the SRT, occurred in all three inclinometers at about 18 to 20 ft-bgs, more or less at the bottom of the COPR layer; (ii) the maximum δ_A was about 3.3 in.; (iii) a 3- to 5-ft thick zone of shear deformation occurred in the natural deposits immediately below the COPR layer; (iv) deformations in the natural deposits were negligible below the highly sheared zone; (v) soil movement above COPR was similar to that of COPR for the most part, suggesting that the COPR layer and upper soils moved mostly as a block, although the COPR layer tended to move slightly ahead of the soil above; and (vi) displacement patterns were consistent with time. The profile shapes for Transect A (not shown) are, for the most part, comparable to that of Transect B. The magnitude of δ_A for Transect A is about 70 to 85 percent of δ_A for Transect B. Additionally, the block-type of interaction between the COPR layer and upper soils is not as strong along Transect A.

Fig. 4 includes a plan view of the resultants of inclinometer displacement [calculated as $\delta_R = (\delta_A^2 + \delta_B^2)^{1/2}$] at 18 ft-bgs, within the COPR layer, as computed in all inclinometers at the end of the monitoring period. As no monitoring could be made for I2A in 2006, the results for March 05 are presented. In general, all resultants of inclinometer displacement, even that for I2A, have consistent orientations. The magnitude of δ_R is largest at inclinometers (I1A and I3A) closest to the slope at the edge of Area 1501 (e.g., "free boundary," where confinement is reduced), intermediate at inclinometers adjacent to the SRT, and smallest at the inclinometers farthest from the SRT and the slope (Table 1). The observation that inclinometer displacements near the SRT are comparable to those associated with a "free boundary" is an indication that the ground movement due to COPR expansion displacement is effectively being accommodated by the SRT.

Table 1. Magnitude of Displacement Resultants at the End of Monitoring

Inclinometer	Transect A	Transect B
	Displacement Resultants, δ_R (in.)	
I1	3.7	3.5
I2	Not measured	3.4
I3	4.3	3.2

Apparent Strain. Apparent horizontal strains (ε) between inclinometers were estimated in directions parallel, oblique, and perpendicular to the SRT as $\varepsilon = \Delta\delta/\Delta d \times 100$ ($\Delta\delta$= relative inclinometer displacement and Δd = inclinometer spacing) and are presented on Fig. 6.

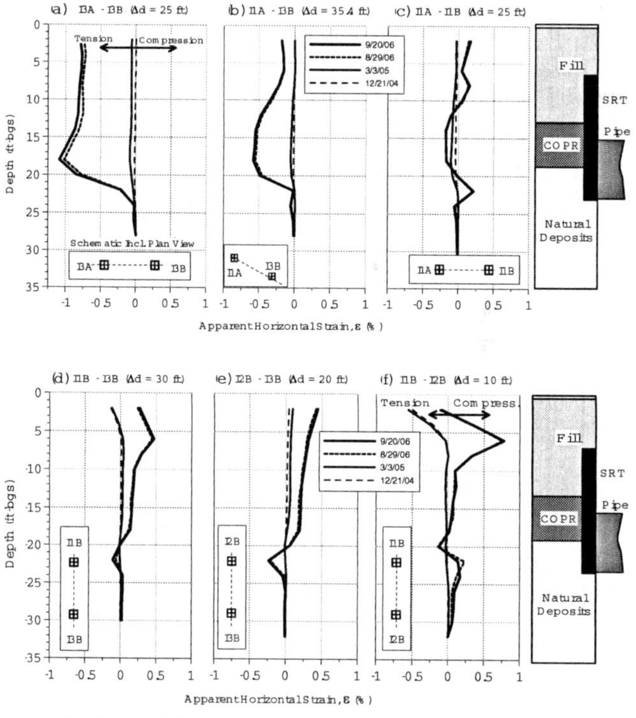

FIG. 6: Apparent Horizontal Strain between Inclinometers

Strains in a direction parallel to the SRT exhibit the highest tension in the COPR stratum between I3A-I3B [i.e., farthest from the SRT, Fig. 6(a)]. Maximum tensile strains are about -0.75 to -0.90 percent in this layer along I3A-I3B. These high tension strains are another manifestation of the "free boundary" effect near the slope. Tensile strains calculated in an oblique direction (approximately 45 degrees) with respect to the SRT [i.e., between I1A-I3B, Fig. 6(b)] are smaller. Close to the SRT, in the direction I1A-I1B, the tensile strains are even smaller [Fig. 6(c)]. Strains in an oblique direction and near the SRT may represent a transition between a field of tensile strains (as in I3A-I3B) and one with moderate compressive strains perpendicular to the pipe, as described below.

Profiles of strain perpendicular to the SRT [Fig. 6(d-f)] show that in this direction the COPR stratum is mostly in compression as a result of its expansion toward the SRT and the reaction of the SRT. Farthest from the SRT (and the slope), compressive strains at the COPR stratum are maximum [i.e., ε is about 0.25 percent, Fig. 6(e)]. At the SRT, compressive strains in the COPR stratum decrease to about half this value [Fig. 6(f)], which is an indication that the relative high compression away from the SRT (most likely a pre-SRT condition) tends to be relieved by the SRT. Within the layers of recent fill placed near the SRT (see Fig. 2), higher compression occurs likely because the recent fill located immediately above the SRT is significantly stiffer than the SRT backfill and is not as effective in absorbing displacements induced by underlying COPR expansion.

Displacement Rates. Calculated deflection rates within the COPR layer at 18 ft-bgs are presented for the A-axis displacement [Fig. 7(a)] and displacement resultants [Fig. 7(b)]. Deflection rates of most inclinometers tended to decrease through March 2005. Between March 2005 and September 2006, the displacement rate increased from an average of about 1.0 to 2 in./year for Axis A. For the displacement resultant, the rate increased from approximately 1.3, at the beginning of the inclinometer monitoring, to about 3 in./year at the end of this period.

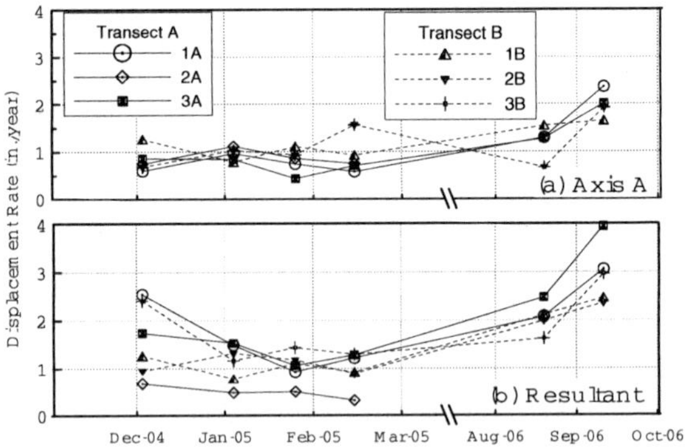

FIG. 7: Displacement Rates at Depth of 18 ft-bgs – Axis A and Resultant

Three stages of COPR movement are hypothesized as occurring after installation of an SRT: (i) an initial stage resulting from COPR lateral stress relief toward the SRT; (ii) an intermediate stage, in which the rate of movement drops and remains relatively low, possibly because the COPR expansion rate is temporarily reduced after the initial stress relief; and (iii) a long-term stage, with a rate of movement associated with steady COPR expansion. It is believed that the free boundary effect accounts for the higher rates of movement at inclinometers closer to the Area 1501 slope [e.g., I3A and I1A in Fig. 7(b)]. Displacement rates along the A axis in the long-term (1.6 to 2.4 in. per year) are slightly higher than pipe compression rates (1.3 to 2.0 in. per year) observed before SRT installation. This suggests that the SRT is effectively absorbing COPR lateral expansion.

CONCLUSIONS

The interaction of a newly-installed SRT, expanding COPR, and a "free boundary" condition at a nearby slope created complex distributions of displacement and deformation at this site. Horizontal displacements in all inclinometers are of the same order of magnitude suggesting that COPR expansion controls ground movement over the entire study area. The SRT appears to be effective in absorbing these ground movements and relieving high compressive stresses previously absorbed by a nearby concrete stormwater pipe. Inclinometers showed a significant displacement component parallel to the SRT, likely due to the decreased lateral confinement near a slope at the edge of Area 1501. Along displacement resultants, the predominant deformation mode is tension, again believed to be associated with the "free boundary" slope condition. Displacement rates perpendicular to the SRT are slightly higher than pipe compression rates observed before SRT installation. This observation is consistent with an SRT that is effectively absorbing COPR lateral expansion toward the pipe. Based on the results of this instrumentation program, three stages of COPR ground movement are hypothesized after SRT installation: initial, intermediate, and long-term. The monitoring results presented herein provide an opportunity to: (i) assess the performance of SRTs as a mechanism to absorb COPR lateral expansion and protect infrastructure; (ii) develop efficient SRT layouts and details; (iii) quantify free boundaries effects near expanding COPR layers; and (iv) obtain data to support and calibrate numerical modeling of SRT/COPR structural interaction.

REFERENCES

Dermatas, D., Chrysochoou, M., Moon, D.H., Christodoulatos, C., Lazarte, C.A., Pendleton, C.H., Bonaparte, R., Briggs, R., Myers, M., French, C., Morris, J., and Kaouris, M. (2005). "Mineralogical Characterization of COPR at Dundalk Marine Terminal Area 1800 to Elucidate Heave Mechanism." *In Situ and On-Site Bioremediation, Eighth International Symposium*, Baltimore, MD, June 6-9.

Kremser, D.T., Sass, B.M., Clark, G.I., Bhargava, M., and French, C. (2005). "Environmental Forensics Investigation of Buried Chromite Ore Process Residue." *In Situ and On-Site Bioremediation, Eighth International Symposium*, Baltimore, MD, June 6-9.

Subject Index

Page number refers to first page of paper

Author Index

Page number refers to first page of paper